U.S. Customary and Metric Comparisons	To Change From	To	Multiply By
Length			
1 meter =	meters	inches	39.37
39.37 inches	inches	meters	0.0254
1 meter =	meters	feet	3.2808
3.2808 feet	feet	meters	0.3048
1 meter =	meters	yards	1.0936
1.0936 yards	yards	meters	0.9144
1 centimeter =	centimeters	inches	0.3937
0.3937 inch	inches	centimeters	2.54
1 millimeter =	millimeters	inches	0.03937
0.03937 inch	inches	millimeters	25.4
1 kilometer =	kilometers	miles	0.6214
0.6214 mile	miles	kilometers	1.6093
Weight or Mass			
1 gram =	grams	ounces	0.0353
0.0353 ounce	ounces	grams	28.3286
1 kilogram =	kilograms	pounds	2.2046
2.2046 pounds	pounds	kilograms	0.4536
Liquid Capacity			
1 liter =	liters	quarts	1.0567
1.0567 quarts	quarts	liters	0.9463
Capacity or Volume			
1 cubic inch =	cubic inches	cubic centimeters	16.387
16.387 cubic centimeters	cubic centimeters	cubic inches	0.0610
1 cubic inch =	cubic inches	liters	0.01639
0.01639 liters	liters	cubic inches	61.0128
1 cubic foot =	cubic feet	cubic meters	0.0283
0.0283 cubic meter	cubic meters	cubic feet	35.3357
1 teaspoon =	teaspoons	milliliters	4.93
4.93 milliliters	milliliters	teaspoons	0.2028
1 tablespoon =	tablespoons	milliliters	14.97
14.97 milliliters	milliliters	tablespoons	0.0668
1 fluid ounce =	fluid ounces	milliliters	29.57
29.57 milliliters	milliliters	fluid ounces	0.0338
1 cup = 0.24 liters	cups	liters	0.24
	liters	cups	4.1667
1 pint = 0.47 liters	pints	liters	0.47
	liters	pints	2.1277
1 gallon =	gallons	cubic meters	0.00379
0.00379 cubic meters	cubic meters	gallons	263.85

Special Algebra Patterns for Factoring

$a^2 + 2ab + b^2 = (a + b)^2$

$a^2 - b^2 = (a + b)(a - b)$

$a^3 + b^3 = (a + b)(a^2 - ab + b^2)$

$a^3 - b^3 = (a - b)(a^2 + ab + b^2)$

Symbols

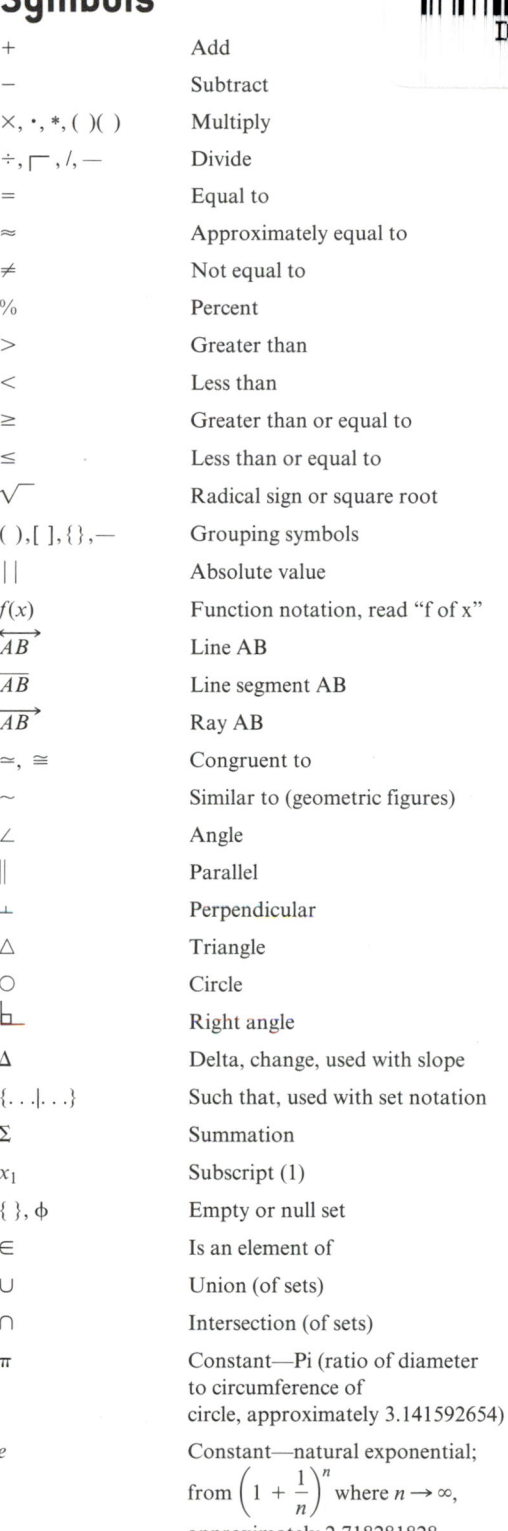

+	Add
−	Subtract
×, ·, *, ( )( )	Multiply
÷, ⌐, /, —	Divide
=	Equal to
≈	Approximately equal to
≠	Not equal to
%	Percent
>	Greater than
<	Less than
≥	Greater than or equal to
≤	Less than or equal to
$\sqrt{\ }$	Radical sign or square root
( ), [ ], { }, —	Grouping symbols
| |	Absolute value
$f(x)$	Function notation, read "f of x"
$\overleftrightarrow{AB}$	Line AB
$\overline{AB}$	Line segment AB
$\overrightarrow{AB}$	Ray AB
≃, ≅	Congruent to
~	Similar to (geometric figures)
∠	Angle
∥	Parallel
⊥	Perpendicular
△	Triangle
○	Circle
∟	Right angle
Δ	Delta, change, used with slope
{. . .|. . .}	Such that, used with set notation
Σ	Summation
x_1	Subscript (1)
{ }, φ	Empty or null set
∈	Is an element of
∪	Union (of sets)
∩	Intersection (of sets)
π	Constant—Pi (ratio of diameter to circumference of circle, approximately 3.141592654)
e	Constant—natural exponential; from $\left(1 + \dfrac{1}{n}\right)^n$ where $n \to \infty$, approximately 2.718281828
i	The square root of −1; $\sqrt{-1}$
∞	Infinity
∴	Therefore
∃	There exists
∀	For every

TENTH EDITION

COLLEGE MATHEMATICS
FOR TRADES AND TECHNOLOGIES

Cheryl Cleaves
Southwest Tennessee Community College

Margie Hobbs
Southwest Tennessee Community College

330 Hudson Street, New York, NY 10013

Director, Portfolio Manager: Michael Hirsch
Courseware Portfolio Manager: Matthew Summers
Courseware Portfolio Management Assistant: Shannon Bushee
Content Producer: Tamela Ambush
Managing Producer: Karen Wernholm
Producer: Jonathan Wooding
Manager, Content Development: Eric Gregg
Manager, Courseware QA: Mary Durnwald
Associate Content Producer: Sneh Singh
Product Marketing Manager: Alicia Frankel
Field Marketing Manager: Jennifer Crum and Lauren Schur
Marketing Assistant: Brooke Imbornone
Senior Author Support/Technology Specialist: Joe Vetere
Manager, Rights and Permissions: Gina Cheselka
Manufacturing Buyer: Carol Melville, LSC Communications
Senior Designer: Barbara Atkinson
Composition: iEnergizer Aptara®, Ltd.
Cover Image: Bloom Icon/Shutterstock

Library of Congress Cataloging-in-Publication Data
Names: Cleaves, Cheryl, 1944– author. | Hobbs, Margie, 1943– author.
Title: College mathematics for trades and technologies / Cheryl Cleaves,
    Southwest Tennessee Community College, Margie Hobbs, Southwest Tennessee
    Community College.
Other titles: Basic mathematics for trades and technologies
Description: 10th edition. | Boston : Pearson, [2019] | Previous edition:
    College mathematics / Cheryl Cleaves, Margie Hobbs (Boston : Pearson,
    2014) | Includes index.
Identifiers: LCCN 2017028244 | ISBN 9780134690339 (casebound) | ISBN
    0134690338 (casebound)
Subjects: LCSH: Mathematics—Textbooks.
Classification: LCC QA39.3 .C535 2019 | DDC 512/.1—dc23
LC record available at https://lccn.loc.gov/2017028244

1 17

ISBN 13: 978-0-134-69033-9
ISBN 10:   0-134-69033-8

Contents

Preface

In *College Mathematics for Trades and Technologies,* Tenth Edition, we have preserved all the features in previous editions that have made this one of the most appropriate texts on the market for a comprehensive study of mathematics in general education and in career programs. We continue to use real-life situations as a context for applied problems.

What's New

The tenth edition retains the writing style and pedagogy that has made the first nine editions successful, while incorporating new features and new material to keep abreast of the curriculum and technology changes that are occurring globally.

New Features:

▶ Stop and Check exercises reinforce new concepts and appear in the margins opposite the examples. The answers are located directly beneath to allow students to check their understanding, and worked-out solutions can be found online in MyLab Math.

▶ Examples are coded to specific exercises in the Section Exercises to facilitate the mastery of the concepts.

▶ Definitions are featured in the margin as they are introduced. These definitions are also included in the Glossary/Index.

▶ Learning Catalytics is an interactive student response tool that uses students' own devices to engage them in the learning process. Learning Catalytics is accessible through MyLab Math, where instructors can generate class discussion, promote peer-to-peer learning, and use real-time data to adjust instructional strategy. As an introduction to this exciting new tool we have provided premade Learning Catalytics questions that cover pre-requisite skills at the start of each section to check students' preparedness for the new section. Each question is tagged for searching **CleavesCol#,** where **#** is the chapter number. For example, search **CleavesCol8** for all questions from Chapter 8. Learn more about Learning Catalytics in the Instructor Resources tab in MyLab Math.

New Material:

▶ Trigonometric functions of cosecant, secant, and cotangent are introduced.

Four online appendices are available in MyLab Math, now with accompanying assignable exercises:

▶ Appendix A: Measuring Instruments (calipers, micrometers, uniform and non-uniform scales, and gauges)

▶ Appendix B: Number Systems (decimal, binary, octal, and hexadecimal)

▶ Appendix C: Matrices and Determinants (including solving systems of equations using Cramer's rule)

▶ Appendix D Conic Sections (parabolas, ellipses, circles, and hyperbolas)

Career Coding:

▶ Examples, section exercises, and chapter review exercises have been coded using 15 different career categories. This will strengthen students' ability to make connections between mathematical concepts and career applications. An index of applications by career code can be found at the end of the book.

Code	Category	Description
AG/H	Agriculture/Horticulture/Landscaping	Agriculture, Horticulture, Landscaping
AUTO	Auto/Diesel Technology	Automobile and Diesel Mechanics
AVIA	Aviation Technology	Aviation, Geographical Information Systems
BUS	Business/Accounting/Real Estate	Business Administration, Accounting, Real Estate
CAD/ARC	CAD/Drafting/Architecture/Surveying	CAD, Drafting, Architecture, Graphic Communication, Surveying
COMP	Computer Technologies	Computer Tech, Information Systems, Information Technology, Network Technology
CON	Construction Trades	Construction, Carpentry, Electrical, Plumbing, HVAC, Pipe Fitting
ELEC	Electronics Technology	Electronics Technology, Computer Electronics
HELPP	Helping Professionals/Education/Criminal Justice/Fire Fighting	Criminal Justice, Fire Science, Counseling, Education
HLTH/N	Allied Health/Nursing/EMS	Healthcare, Nursing, Nutrition, EMS
HOSP	Hospitality/Culinary/Food Technology	Hotel and Restaurant Management, Culinary Arts, Food Technology
INDTEC	Industrial Technology/Manufacturing/Machine Technology/Engineering Technology	Manufacturing, Industrial, Machine, Engineering Technologies
INDTR	Industrial Trades/Welding/Machine Tool/Industrial Maintenance	Welding, Machine Tools, Industrial Maintenance
PFIN	Personal Finance	Home Loans, Credit Cards, Making Purchases, Best Buys
TELE	Telecommunications	Telecommunication Technology

Our goal is to present a systematic framework for successful learning in mathematics that will strengthen students' *mathematical sense* and give students a greater appreciation for the power of mathematics in everyday life and in the workplace. The new material in this edition has been added to broaden the usefulness of the text. Many of the explanations have been enhanced with carefully constructed visualizations. Exercises have been updated and new ones added.

Commitment to Improving Mathematics Education

The author continues to be active in the standards and other initiatives of the American Mathematical Association of Two-Year Colleges (AMATYC). We enthusiastically promote the standards and guidelines encouraged by AMATYC, NCTM, MAA, and the SCANS document.

Calculator Usage

Calculator tips appropriate for both scientific and graphing calculators are periodically included. These generic tips guide students to use critical thinking to determine how their calculator operates without referring to a user's manual. In addition, specific instructions for the TI-84 calculator are given.

We continue to emphasize the calculator as a tool that *facilitates* learning and understanding. Assessment strategies are included throughout the text and supplementary materials to enable students to test their understanding of a concept independently of their calculator.

To the Student

The mathematics you learn from this book will help you advance on your career path. We have given much thought to the best way to teach mathematics and have done extensive research on how students learn. We have provided a wide variety of features and resources so that you can customize your study to your needs and circumstances. The following features are key to helping you learn the mathematics in this text.

Table of Contents.   The table of contents is your "roadmap" to this text. Study it carefully to determine how the topics are arranged. This will aid you in relating topics to each other.

Glossary/Index.   An extensive glossary/index is an important part of every mathematics book. Use the index to cross-reference topics and to locate other topics that relate to the topic you are studying.

In Great Company.   Each chapter opens with an interesting article that gives a situation that resulted from errors made in mathematics. These articles are intended to motivate you to learn from your own errors.

In Great
Company

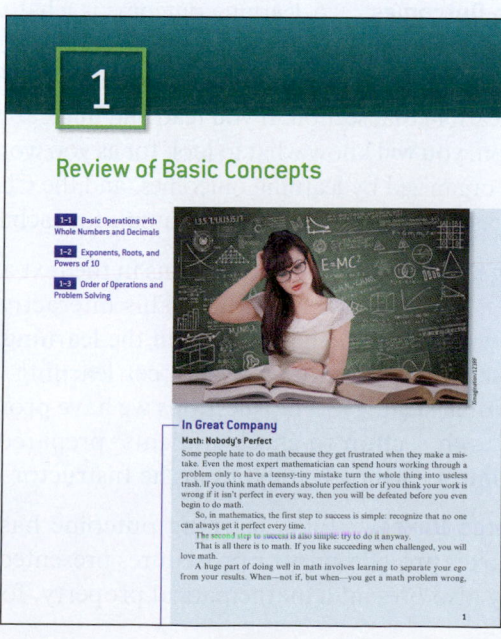

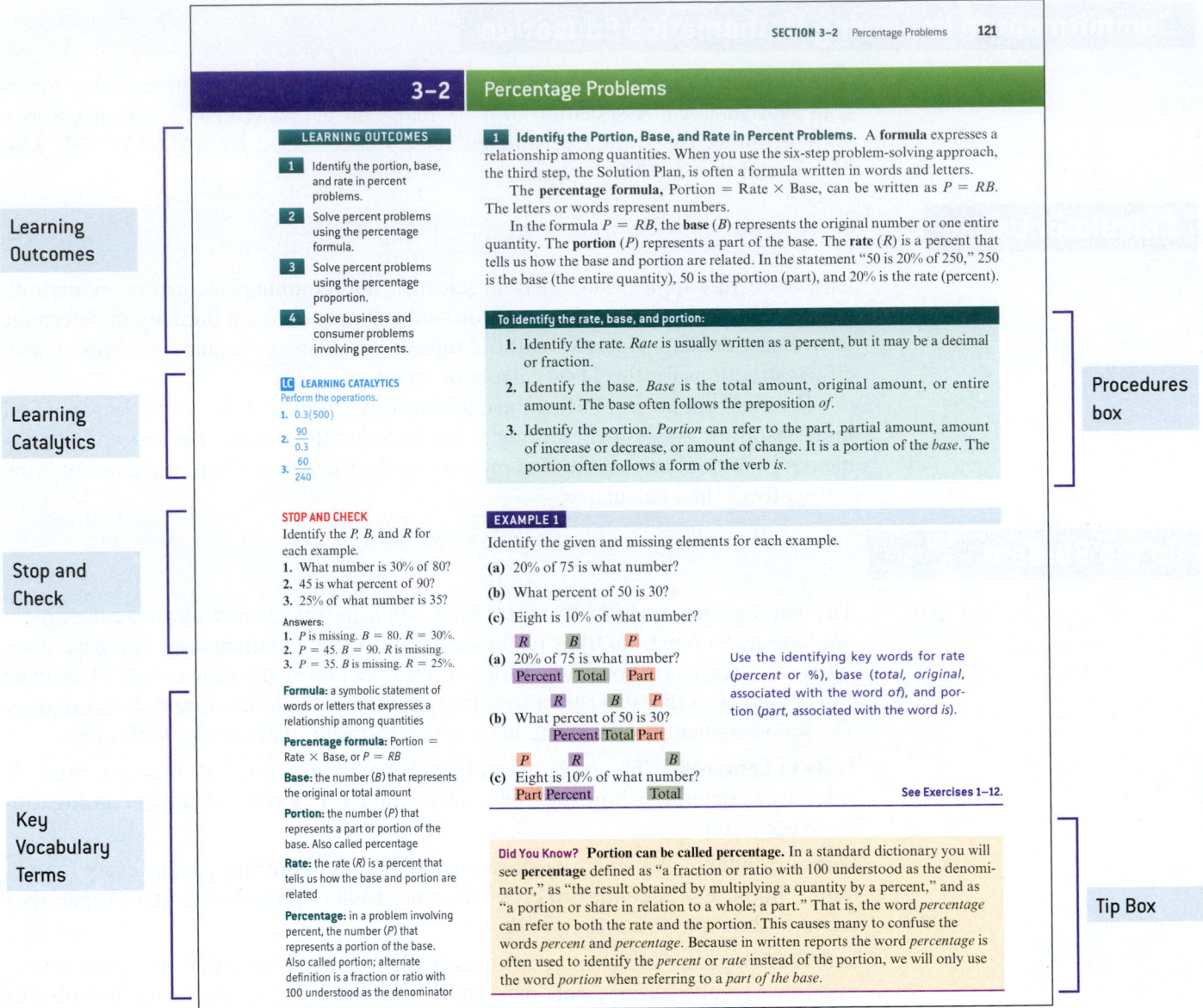

Learning Outcomes.   A learning outcome is what you should be able to do when you master a concept. These outcome statements can guide you through your study plan. Each section begins with a statement of learning outcomes that shows you what you should look for and learn in that section. If you read and think about these outcomes before you begin the section, you will know what to look for as you work through the section. Section Exercises are organized by learning outcomes, and the Chapter Review of Key Concepts give procedures to review and a worked example for each learning outcome.

Learning Catalytics.   These questions in the text are featured in Learning Catalytics available through MyLab Math. This interactive student response tool uses students' own devices to engage them in the learning process. Instructors can generate class discussion, promote peer-to-peer learning, and use real-time data to adjust instructional strategy. The questions we have provided test pre-requisite skills at the start of each section to check students' preparedness for the new section. Learn more about Learning Catalytics in the Instructor Resources tab in MyLab Math.

Procedures Boxes.   Each learning outcome has one or more procedures boxes. These boxes provide rules or procedures presented as numbered steps. A procedures box may also present a mathematical property, formula, or fact.

Tip Boxes.   These boxes give helpful hints for doing mathematics, and they draw your attention to important observations and connections that you may have missed in an example.

Stop and Check Exercises.   These extra exercises located in the margin give students immediate practice with answers as new concepts are introduced, so they can master every outcome.

Key Terms.   These important vocabulary words are highlighted in bold in the text and called out in the margin with their definitions.

Career coding in examples

Six-Step Problem Solving Example with Explanatory Comments

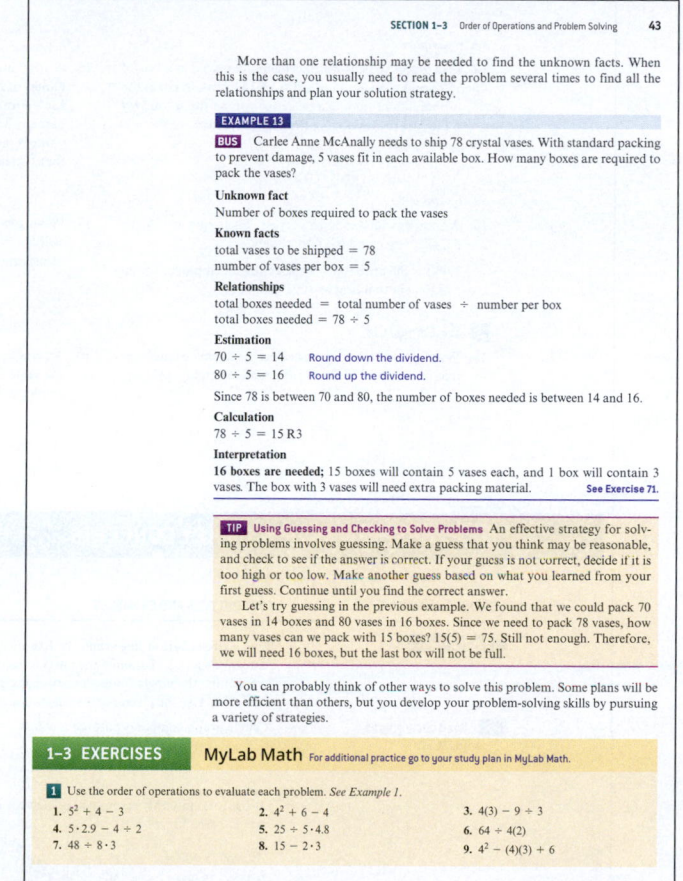

Career-Coded Examples and Exercises.   Applied problems focus on a wide variety of careers available as a course of study at your community college, technology center, or university. These careers are grouped into 14 categories, and the examples and exercises are coded to these categories as appropriate. There is also a category for personal finance which is appropriate for all career choices. An index of applications is provided in the back of the book for your convenience.

Six-Step Approach to Problem Solving.   Successful problem solvers use a systematic, logical approach. We use a six-step approach to problem solving. This approach gives you a system for solving a variety of math problems. You will learn how to organize the information given and how to develop a logical plan for solving the problem. You are asked to analyze and compare and to estimate as you solve problems. Estimation helps you decide whether your answer is reasonable. You will learn to interpret the results of your calculations within the problem's context, a skill you will use on the job.

Use of Color in the Text.   As you read the text and work through the examples, notice the items shaded with color. These will help you follow the logic of working through the example. Color also highlights important items and boxed features such as the Tips, Learning Outcomes, rules, procedures, and formulas.

Using Your Calculator.   Calculators are useful in all levels of mathematics. Some tips introduce easy-to-follow calculator strategies. The tips show you how to analyze the procedure and set up a problem for a calculator solution; a sample series of keystrokes is often included. In addition, the tips help you determine how your type of calculator operates for various mathematical processes.

Section Exercises with Reference to Examples

Chapter Review of Key Concepts

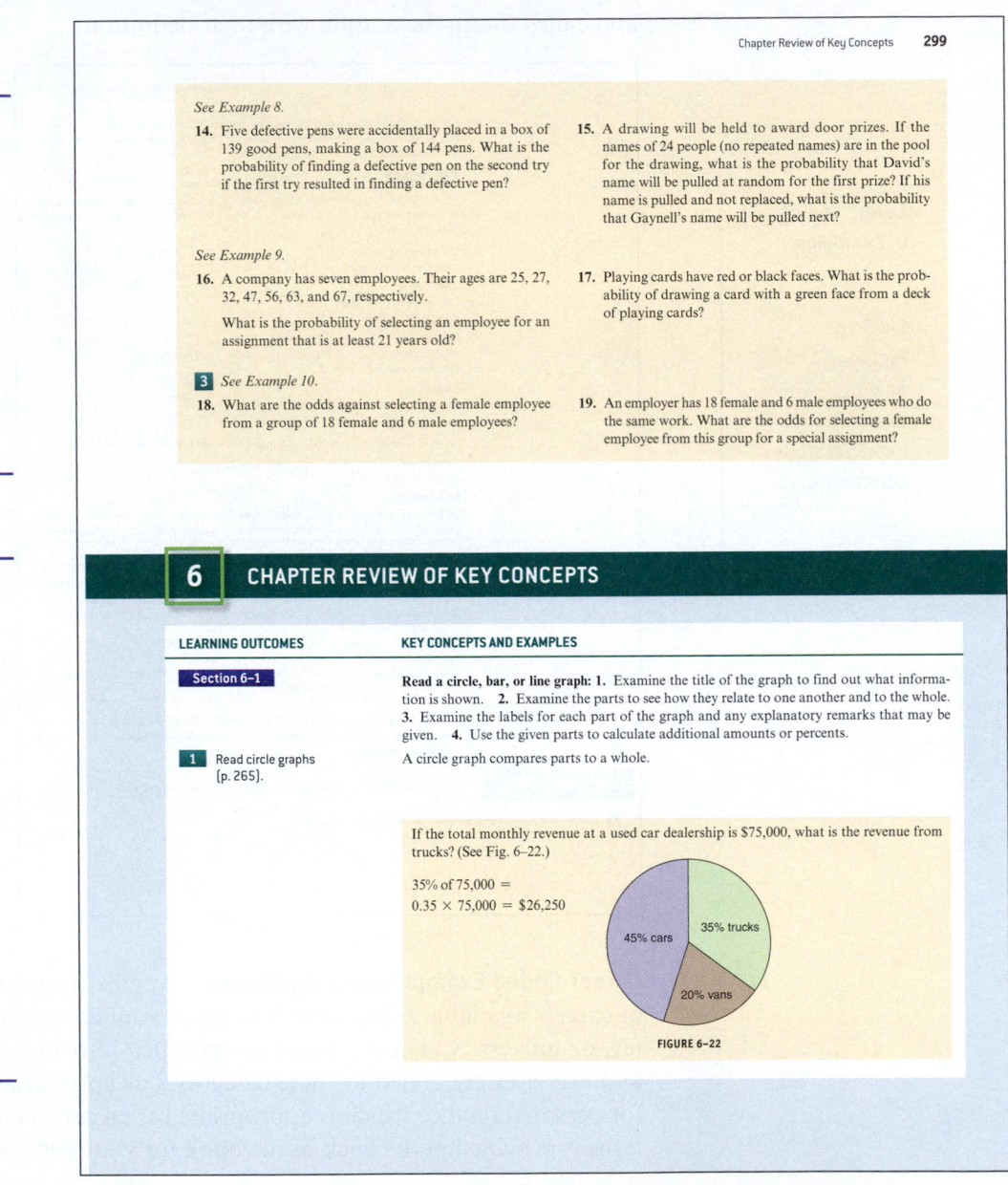

Section Exercises.   These practice sets are keyed to the learning outcomes and appear at the end of each section. Use these exercises to check your understanding of the section. **The answers to every exercise are at the end of the text,** so you can get immediate feedback on whether you understand the concepts. Each example in the section has at least one exercise that is referenced to that example.

Chapter Review of Key Concepts.   Each chapter includes a summary in the form of a two-column chart. The first column lists the learning outcomes of the chapter. The second column gives procedures and examples for each outcome. Page references are included to facilitate your preview or review of the chapter.

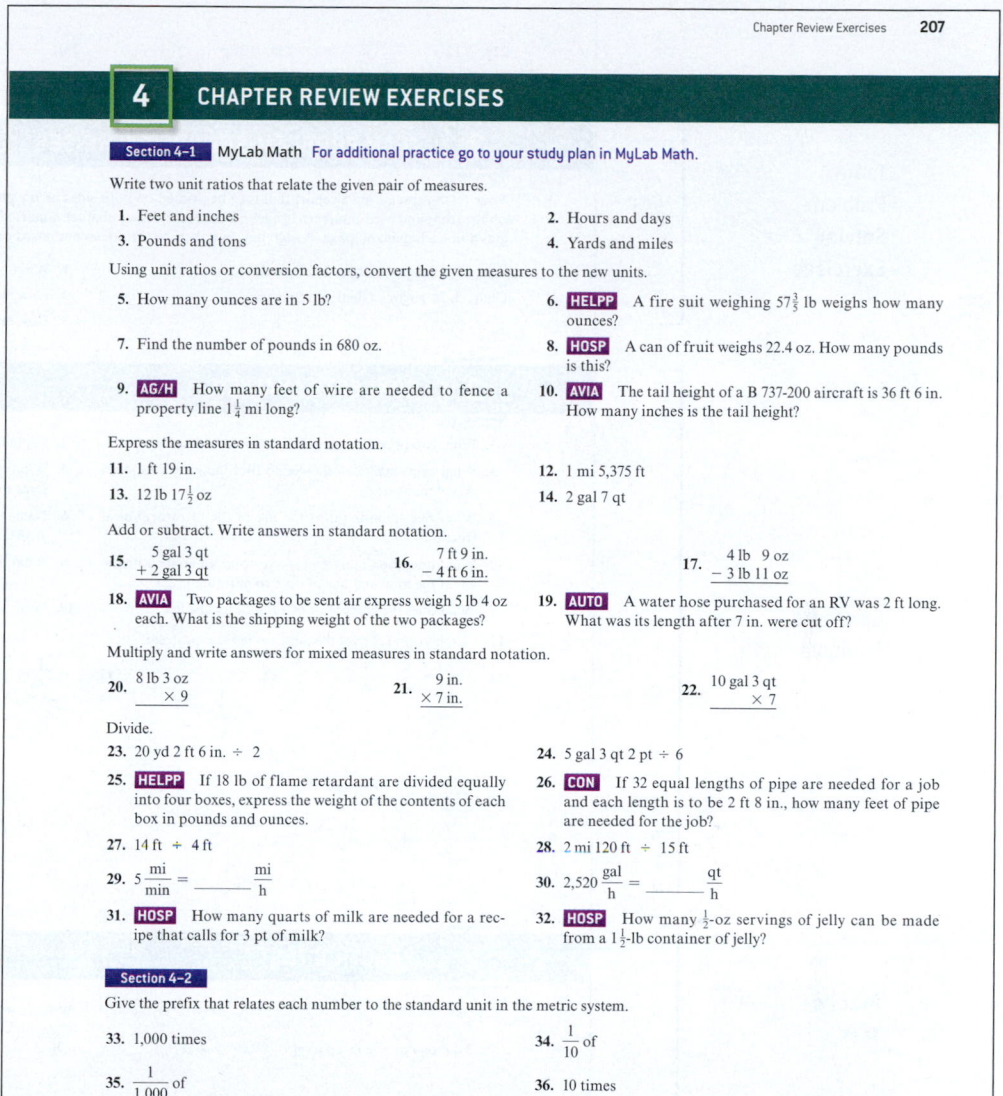

Chapter Review Exercises **207**

4 CHAPTER REVIEW EXERCISES

Section 4–1 MyLab Math For additional practice go to your study plan in MyLab Math.

Write two unit ratios that relate the given pair of measures.

1. Feet and inches
2. Hours and days
3. Pounds and tons
4. Yards and miles

Using unit ratios or conversion factors, convert the given measures to the new units.

5. How many ounces are in 5 lb?
6. **HELPP** A fire suit weighing $57\frac{3}{5}$ lb weighs how many ounces?
7. Find the number of pounds in 680 oz.
8. **HOSP** A can of fruit weighs 22.4 oz. How many pounds is this?
9. **AG/H** How many feet of wire are needed to fence a property line $1\frac{1}{4}$ mi long?
10. **AVIA** The tail height of a B 737-200 aircraft is 36 ft 6 in. How many inches is the tail height?

Express the measures in standard notation.

11. 1 ft 19 in.
12. 1 mi 5,375 ft
13. 12 lb $17\frac{1}{2}$ oz
14. 2 gal 7 qt

Add or subtract. Write answers in standard notation.

15. 5 gal 3 qt
    + 2 gal 3 qt
16. 7 ft 9 in.
    − 4 ft 6 in.
17. 4 lb 9 oz
    − 3 lb 11 oz
18. **AVIA** Two packages to be sent air express weigh 5 lb 4 oz each. What is the shipping weight of the two packages?
19. **AUTO** A water hose purchased for an RV was 2 ft long. What was its length after 7 in. were cut off?

Multiply and write answers for mixed measures in standard notation.

20. 8 lb 3 oz
    × 9
21. 9 in.
    × 7 in.
22. 10 gal 3 qt
    × 7

Divide.

23. 20 yd 2 ft 6 in. ÷ 2
24. 5 gal 3 qt 2 pt ÷ 6
25. **HELPP** If 18 lb of flame retardant are divided equally into four boxes, express the weight of the contents of each box in pounds and ounces.
26. **CON** If 32 equal lengths of pipe are needed for a job and each length is to be 2 ft 8 in., how many feet of pipe are needed for the job?
27. 14 ft ÷ 4 ft
28. 2 mi 120 ft ÷ 15 ft
29. $5\frac{mi}{min} = \underline{\hspace{1cm}} \frac{mi}{h}$
30. $2{,}520\frac{gal}{h} = \underline{\hspace{1cm}} \frac{qt}{h}$
31. **HOSP** How many quarts of milk are needed for a recipe that calls for 3 pt of milk?
32. **HOSP** How many $\frac{1}{2}$-oz servings of jelly can be made from a $1\frac{1}{2}$-lb container of jelly?

Section 4–2

Give the prefix that relates each number to the standard unit in the metric system.

33. 1,000 times
34. $\frac{1}{10}$ of
35. $\frac{1}{1{,}000}$ of
36. 10 times

Career
Coding in
Chapter
Review
Exercises

Chapter Review Exercises. An extensive set of exercises appears at the end of each chapter so you can review all the learning outcomes presented in the chapter. These exercises, organized by section, may be assigned as homework, or you may want to work them on your own for additional practice. **Answers to the odd-numbered exercises are given at the end of the text,** and worked-out solutions appear in a separate Student Solutions Manual available for purchase. Your instructor has the solutions to the even-numbered exercises in the Instructor's Resource Manual.

Team Problem-Solving Exercises. Employers value an employee's ability to interact with others in a team environment. These exercises will allow you to develop and refine your team-interaction skills.

Concepts Analysis. Too often we focus on the *how to* and overlook the *why* of mathematical concepts. The Concepts Analysis questions further your understanding of a concept and help you see the connections between concepts. Some concepts questions present incorrect solutions to exercises to give you practice in analyzing and correcting errors. Error analysis also reinforces your understanding of concepts. As an added bonus, these exercises strengthen your writing skills. Suggested responses (answers) are found in the Instructor's Resource Manual.

Team Problem-Solving Exercises

Concepts Analysis

Practice Test

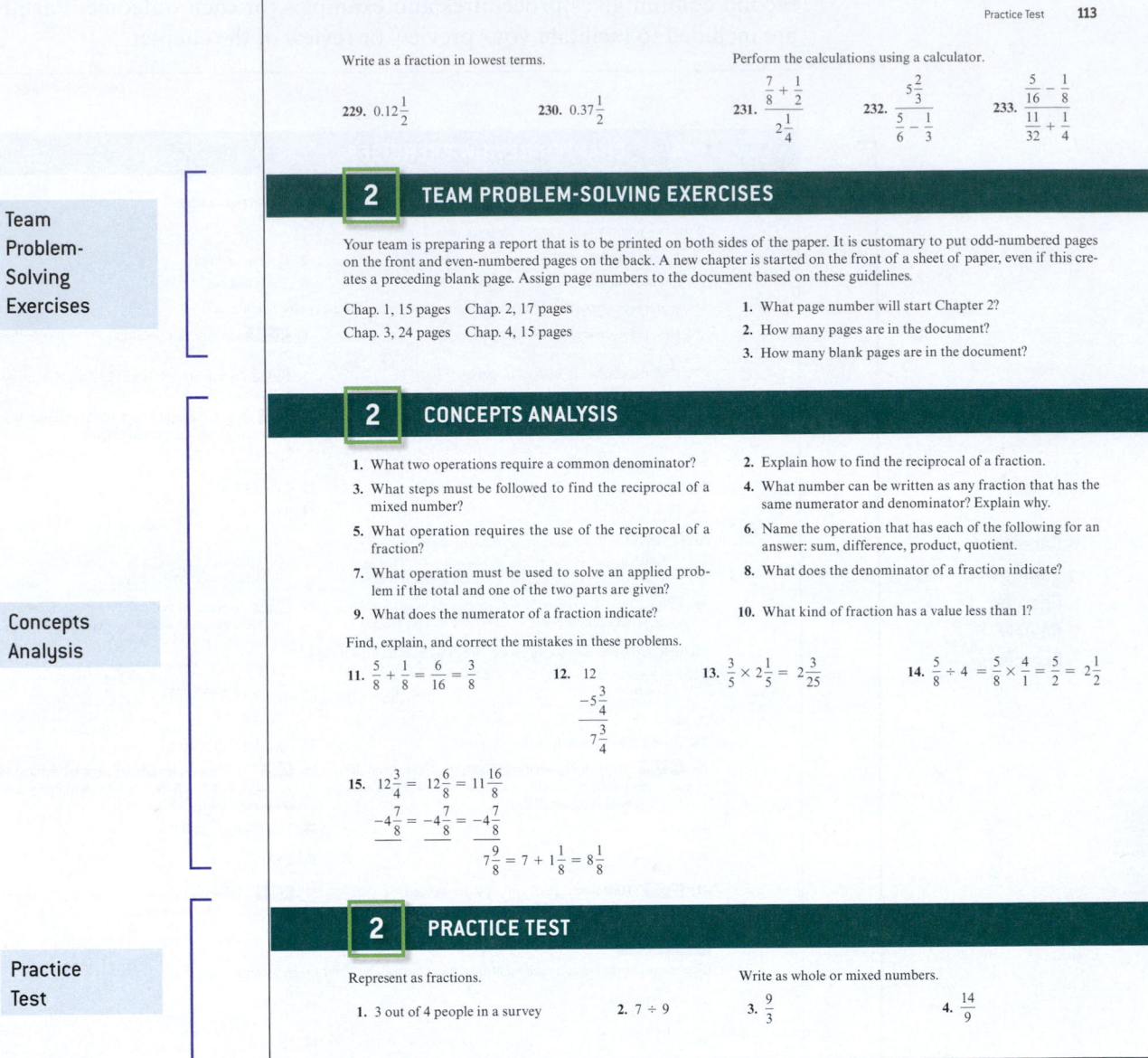

Practice Test. The practice test at the end of each chapter lets you check your understanding of the chapter learning outcomes. You should be able to work each problem without referring to any examples in your text or your notes. Take this test before you take the class test to check and verify your understanding of the chapter material. **Answers to the odd-numbered exercises appear at the end of the text,** and their solutions appear in a separate Student Solutions Manual. Your instructor has the solutions to the even-numbered exercises in the Instructor's Resource Manual.

Cumulative Practice Tests. Practice tests for a group of chapters are included after Chapters 3, 6, 10, 15, 18, and 20. These tests will help you prepare for mid-course or end-of-course exams. Periodically reviewing previously learned material will help you retain the concepts for a longer period of time. **Answers to the odd-numbered exercises appear at the end of the text,** and their solutions appear in a separate Student Solutions Manual. Your instructor has the solutions to the even-numbered exercises in the Instructor's Resource Manual.

Cheryl Cleaves

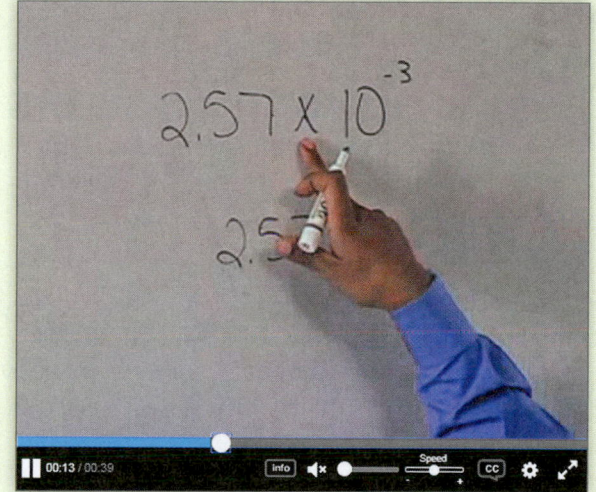
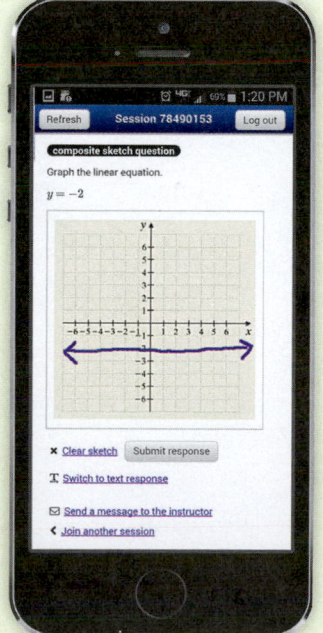

Resources for Success

Motivate and Support Students with Updated Resources

NEW! A trade application question library provides a wide range of exercises available to assign to work for any instructor's class dynamics. Available in the MyLab Math Assignment Manager, this library of exercises now allows instructors to pull in additional application questions from particular trades or industries.

Expanded MyLab exercise coverage now includes coverage of all online appendices, giving students the chance to practice and reinforce skills from all content areas.

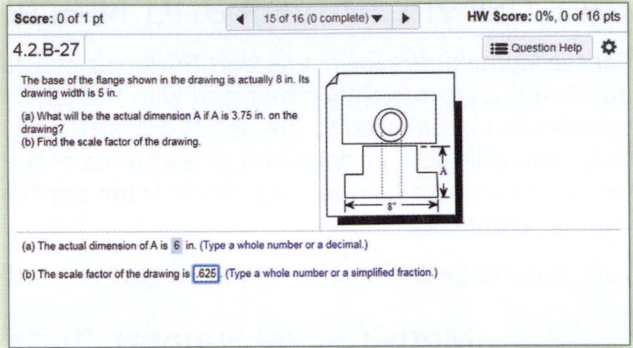

Resources for Instructors

The following resources can be downloaded from the Instructor's Resource Center on www.pearson.com, or in the MyLab Math course.

Instructor's Resource Manual

This resource includes teaching tips and activity ideas for each chapter, which can be used for individual or group work, as well as solutions to even-numbered exercises for Chapter Reviews, Practice Tests, Cumulative Practice Tests, and solutions to all Concept Analysis exercises.

UPDATED! TestGen

TestGen® (www.pearsoned.com/testgen) enables instructors to build, edit, print, and administer tests using a computerized bank of questions developed to cover all the objectives of the text. Updated for the 10th edition, TestGen is now algorithmically based, allowing instructors to create multiple but equivalent versions of the same question or test with the click of a button. Instructors can also modify test bank questions or add new questions.

Resources for Students

Student Solutions Manual

This manual contains completely worked-out solutions for all the odd-numbered exercises in the Chapter Reviews, Practice Tests, and Cumulative Practice Test in the text.

pearson.com/mylab/math

Acknowledgments

A project such as this does not come together without help from many people. My first avenue for input is through students and faculty who use the text. Their comments and suggestions have been invaluable.

I wish to express thanks to all the people who helped make this edition a reality, especially team leaders Matt Summers and Tamela Ambush, and project manager Amy Kopperude at iEnergizer Aptara®, Ltd.

The teaching of mathematics over time produces a wealth of knowledge about instructional strategies and specific content. I am grateful for the many valuable suggestions received in these areas. I wish to thank the following individuals:

Stan Adamski, Owens Community College, Ohio (OH)
Milton Clark, Florence Darlington Technical College (SC)
Virginia Dewey, York Technical College, South Carolina (SC)
Terry B. Gaalswyk, Western Iowa Technical Community College (IA)
John Gillis, Portsmouth Naval Shipyard Apprentice Program, under the auspices of New Hampshire Community Technical College (Portsmouth, NJ), and York County Community College (Wells, ME)
Brent Hamilton, North Iowa Area Community College (IA)
Ryan Harper, Spartanburg Community College (SC)
Crystal Heshmat, Hudson Valley Community College (NY)
Mara Jorgensen, Western Iowa Technical Community College (IA)
Stephanie Krehl, Arkansas State University Mid-South (AR)
Edwin G. Landauer, Clackamas Community College (OR)
Michelle McClain, Delgado Community College (LA)
Lisa J. Music, Big Sandy Community and Technical College (KY)
Nicole Muth, Lakeshore Technical College (WI)
Karen Newson, Triangle Tech (PA)
Justin Ostrander, Manhattan Area Technical College, Kansas (KS)
Scott Randby, University of Akron (OH)
Henry Regis, Valencia Community College, Florida (FL)
Behnaz Rouhani, Athens Area Technical Institute (GA)
Susan Rowland, Middle Georgia Technical College (GA)
Kathy Ryan, Front Range Community College (CO)
David C. Shellabarger, Lane Community College (OR)
Joseph Sukta, Moraine Valley Community College (IL)
Jimmie A. Van Alphen, Ozarks Technical Community College (MO)

I'd like to thank Deana Richmond for accuracy checking the manuscript and always providing constructive suggestions, and Shirley Riddle for preparing the glossary, index, and list of career applications.

Finally, I want to dedicate this book to Margie Hobbs. Margie and I began working together at State Technical Institute at Memphis in 1975 and published our first textbook in 1979. Through the years we were more than just best friends, we were family. Margie had a unique set of qualities that truly set her apart: her attention to detail, her passion and persistence, and her willingness to see the job through made the people around her better—made me better. Her commitment to her family, to the profession of teaching, to her students, to her colleagues, and to this project was unquestioned. She was a dedicated professional to the end, but somehow she managed to always put the needs of others first—which is a truly remarkable thing. Her legacy will live on not only in this textbook, but in our hearts. Thank you for everything you did for us, Margie, and for making me part of your family.

Cheryl Cleaves

Dr. Margie Johnson Hobbs
June 13, 1943–March 16, 2016

1

Review of Basic Concepts

Ximagination/123RF

In Great Company

Math: Nobody's Perfect

Some people hate to do math because they get frustrated when they make a mistake. Even the most expert mathematician can spend hours working through a problem only to have a teensy-tiny mistake turn the whole thing into useless trash. If you think math demands absolute perfection or if you think your work is wrong if it isn't perfect in every way, then you will be defeated before you even begin to do math.

So, in mathematics, the first step to success is simple: recognize that no one can always get it perfect every time.

The second step to success is also simple: try to do it anyway.

That is all there is to math. If you like succeeding when challenged, you will love math.

A huge part of doing well in math involves learning to separate your ego from your results. When—not if, but when—you get a math problem wrong, that wrong answer does not mean you are stupid. All it means, is you got that one wrong. Everyone gets math wrong, some of the time. Look at all the *great company* you have!

Steve Kellmeyer, Math Instructor
Art Institute of Dallas

Scientists Can't Add (November 9, 1993)

In 1987, U.S. physicists convinced Congress that the United States was falling behind in high-energy particle physics. We must leap ahead of the competition! We must build a machine capable of carrying out such high-energy work by the next century! The physicists, the mathematicians, the Nobel laureates had it all worked out. They wanted to build what was called a Superconducting Super Collider (SSC). And they could build it, they said, for the low, low price of just $4.4 billion. Chicken feed, really.

The U.S. government bought into the idea, and Congress approved the funds. In 1991, construction began in a little town just south of Dallas, Texas, the town of Waxahachie. The workers drilled 200 feet down and then began boring a circular tunnel through the bedrock. But even before construction began, prices had begun to rise. Worse, international collaboration, which was supposed to bring in an additional $2 billion in funding, never actually happened.

By 1993, cost estimates for the project rose from the original $4.4 billion to over $12 billion. Prices had already tripled, but only 14 miles of the 54 miles (25%) of underground tunnels had been dug. Equipment still hadn't been bought. Worse, the International Space Station was competing for the same funds the SSC was supposed to get. It was too much. Congress threw in the towel on October 21, 1993, and officially canceled the SSC.

So, what did the United States get for its $2 billion investment? A big hole in the ground in Texas, topped by a lot of empty buildings, the whole thing surrounded by rusting cyclone fences.

The mathematicians and the Department of Energy officials who oversaw the project apparently couldn't add very well. They kept miscalculating costs and failed to foresee basic difficulties. How can the American public have faith in mathematical models put forward by mathematicians who cannot add up a few dollars to arrive at realistic costs?

Remember this the next time you make a basic addition or subtraction mistake. You aren't the first to do it. You won't be the last. And it isn't nearly as humiliating for you to make such a mistake as it was for all the engineers, mathematicians, physicists, and government employees involved with the SSC. After all, they made essentially the same error, and their mistakes were splashed across the front pages of newspapers across the nation.

Decimal-number system: the system of numbers that uses 10 individual symbols called digits (0, 1, 2, 3, 4, 5, 6, 7, 8, 9) and place values of powers of 10

Digit: one of the symbols 0, 1, 2, 3, 4, 5, 6, 7, 8, 9

Place value: the value of a digit based on its position in a number

---

1–1   Basic Operations with Whole Numbers and Decimals

LEARNING OUTCOMES

1. Compare whole numbers.
2. Write fractions with power-of-10 denominators as decimal numbers.
3. Compare decimal numbers.

Our system of numbers, the **decimal-number system**, uses 10 symbols called **digits**: 0, 1, 2, 3, 4, 5, 6, 7, 8, 9. A number can be represented with one or more digits. When a number contains two or more digits, each digit must be in the correct place for the number to have the value we intend it to have. Each place in the system has a specific **place value**.

The numbering system is made up of many different types of numbers. The first two types of numbers that we review are natural numbers and whole numbers. The

4 Round a whole number or a decimal number to a specified place value.

5 Add and subtract whole numbers and decimals.

6 Multiply and divide whole numbers and decimals.

LC **LEARNING CATALYTICS**

Perform the indicated operations.
1. $1 - 0.03$
2. $0.2(0.03)$
3. $1.5 \div 0.3$

Natural numbers: the set of counting numbers beginning with 1, 2, 3, and continuing indefinitely; also called counting numbers

Ellipsis (...): a notation to show that the pattern established before the ellipsis continues

Counting numbers: same as natural numbers

Whole numbers: the set of natural numbers and the number 0; numbers made up of one or more digits

Periods: groups of three place values—ones, tens, hundreds—for periods such as units, thousands, millions, billions, and trillions

Units: period formed by up to three digits on the right end of a whole number

Thousands: period formed by up to three digits to the left of the units period

Millions: period formed by up to three digits to the left of the thousands period

Billions: period formed by up to three digits to the left of the millions period

Trillions: period formed by up to three digits to the left of the billions period

Number line: visual representation of the relationship of numbers by size

Infinity (∞): a concept that numbers continue without end

Inequality: a mathematical statement showing two numbers that are not equal. Inequality symbols are $<$ and $>$

Less than ($<$): a symbol placed between two numbers showing the left number is smaller than the right number

Greater than ($>$): a symbol placed between two numbers showing the left number is larger than the right number

natural numbers begin with the number 1 and continue indefinitely (1, 2, 3, 4, 5, . . . , 101, 102, 103, . . .). Three periods that follow a list of numbers are called an **ellipsis** and mean that the pattern established before the ellipsis continues. The natural numbers are also called **counting numbers.** The set of **whole numbers** includes all the natural numbers and the number 0. Other types of numbers will be introduced as appropriate.

The whole-number place values are arranged in **periods,** or groups of three (Fig. 1–1) reading from right to left. The first period of three is called **units,** the second period of three is called **thousands,** the third period is called **millions,** the fourth period is called **billions,** and the fifth period is called **trillions.** Commas are used to separate these periods. The commas make larger numbers easier to read because we can locate specific place values and interpret numbers more easily. Each group of three digits has a hundreds place, a tens place, and a ones place.

Trillions Period			Billions Period			Millions Period			Thousands Period			Units Period		
Hundred trillions (100,000,000,000,000s)	Ten trillions (10,000,000,000,000s)	Trillions (1,000,000,000,000s)	Hundred billions (100,000,000,000s)	Ten billions (10,000,000,000s)	Billions (1,000,000,000s)	Hundred millions (100,000,000s)	Ten millions (10,000,000s)	Millions (1,000,000s)	Hundred thousands (100,000s)	Ten thousands (10,000s)	Thousands (1,000s)	Hundreds (100s)	Tens (10s)	Ones (1s)

FIGURE 1–1 Whole-number place values and periods.

Did You Know? In four-digit numbers, the comma separating the units period from the thousands period is optional. Thus, 4,575 and 4575 are both acceptable.

1 **Compare Whole Numbers.** Whole numbers can be arranged on a **number line** to show a visual representation of the relationship of numbers by size. The most common arrangement is to begin with zero and place numbers on the line from left to right as they get larger.

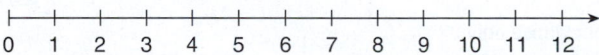

All numbers have a place on the number line and the numbers continue indefinitely without end. A term that is often used to describe this concept is **infinity** and the symbol is ∞.

Whole numbers can be compared by size by determining which of the two numbers is larger or smaller. If two numbers are positioned on a number line, the smaller number is positioned to the left of the larger number. The order relationship can be written in a mathematical statement called an **inequality.** An inequality shows that two numbers are not equal; that is, one is larger than the other. Symbols for showing inequalities are the **less than** symbol $<$ and the **greater than** symbol $>$.

$5 < 7$　Five is less than seven.

$7 > 5$　Seven is greater than five.

Cardinal number: a number that shows *how many*

Ordinal number: a number that shows *order* or position

Fraction: a number that is a part of a whole number

Fraction notation: writing the numerator and denominator of a fraction separated by a horizontal bar or slash

Denominator: the denominator of a fraction is the number of parts one unit has been divided into. It is the bottom number of a fraction or the divisor of the indicated division

STOP AND CHECK

1. Write two inequalities comparing the numbers 203 and 230.

Answer:

1. $203 < 230$ and $230 > 203$

Numerator: the numerator of a fraction is the number of the parts being considered. It is the top number of a fraction or the dividend of the indicated division

Decimal fraction: a fraction whose denominator is always 10 or some power of 10; a fractional notation that uses the decimal point and the place values to its right to represent a fraction whose denominator is 10 or some power of 10, such as 100, 1,000, and so on. A decimal fraction is also referred to as a decimal, a decimal number, or a number using decimal notation

Power of ten: a number whose only nonzero digit is 1: 10, 100, and 1,000 are examples of powers of 10

Decimal number: an alternate name for decimal

Decimal: a fractional notation based on place values for fractions with a denominator of 10 or a power of 10

Decimal notation: a notation for writing a fraction as an equivalent decimal

Decimal point: the symbol (period) placed between the ones place and the tenths place to identify the place value of each digit

---

Did You Know? The inequality symbols $<$ and $>$ always point to the smaller number.

To compare whole numbers:

1. Mentally position the numbers on a number line.

2. Select the number that is farther to the left to be the smaller number.

3. Write an inequality using the *less than* symbol.   smaller number $<$ larger number

   or

   Write an inequality using the *greater than* symbol.   larger number $>$ smaller number

---

EXAMPLE 1

Write two inequalities comparing the numbers 12 and 19:

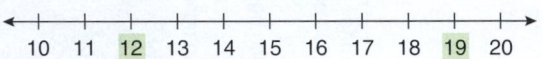

Mentally position the numbers on the line.

12 is the smaller number.   12 is to the *left* of 19.

$12 < 19$ or $19 > 12$ Use appropriate inequality symbol.   **See Exercises 1–10.**

Numbers are used to show *how many* and to show *order*. **Cardinal numbers** show *how many* and **ordinal numbers** show *order* or position (such as first, second, third, fourth, etc.). For example, in the statement "three students are doing a presentation," three is a cardinal number (showing how many). In the statement "Margaret is the third tallest student in the class," third is an ordinal number (showing order).

2 Write Fractions with Power-of-10 Denominators as Decimal Numbers. A **fraction** is a number that is a part of a whole number. A notation for writing numbers that are parts of a whole number is called **fraction notation.** In fraction notation, we write one number over another number.

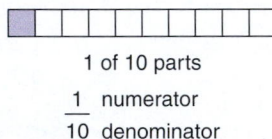

1 of 10 parts

$\dfrac{1}{10}$ numerator
      denominator

The bottom number, the **denominator,** represents the number of parts that a whole unit contains. The top number, the **numerator,** represents the number of parts being considered.

A special type of fraction is called a **decimal fraction.** Other types of fractions are covered in Chapter 2.

A decimal fraction is a fraction whose denominator is 10 or some power of 10. A **power of 10** is a whole number whose only nonzero digit is 1 (10, 100, 1,000, 10,000, and so on). Often the terms decimal fraction, **decimal number,** and **decimal** are used interchangeably. In fraction notation, 3 out of 10 parts is written as $\frac{3}{10}$. In **decimal notation,** the denominator 10 is not written but is implied by position on the place-value chart (Fig. 1–2). A **decimal point** (.) separates whole-number amounts on the left and fractional parts on the right. The fraction $\frac{3}{10}$ can be written in decimal notation as 0.3.

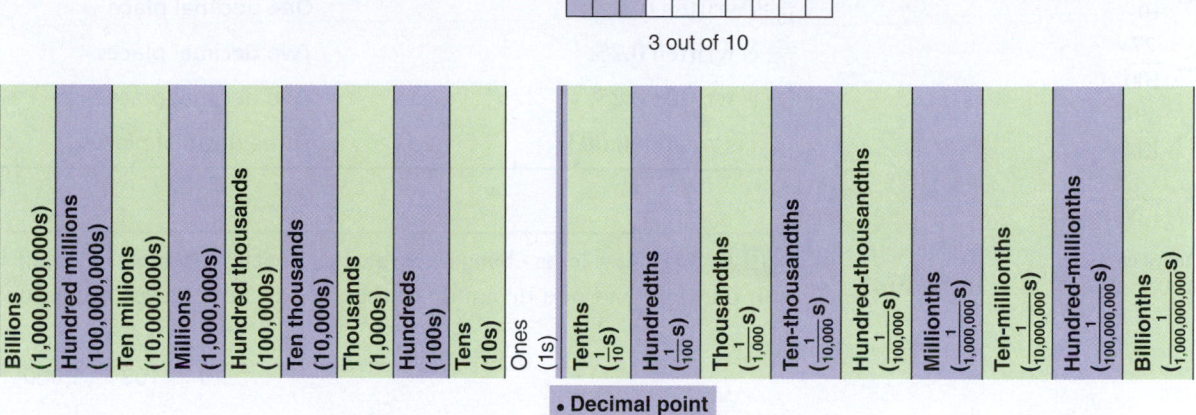

3 out of 10

FIGURE 1–2 Place-value chart for whole numbers and decimals.

To extend the place-value chart to include parts of whole amounts, we place a decimal point (.) after the ones place. The place on the right of the ones place is called the *tenths* place. A decimal point is placed between the ones place and the tenths place to distinguish between whole amounts and fractional amounts.

A number that has both a whole-number part and a fraction or decimal part is called a **mixed number.** 3.6 is a mixed number.

Mixed number: a number that has both a whole-number part and a fraction or decimal part

---

TIP   **Informal Use of the Word Point** Informally, the decimal point is sometimes read as "point." 3.6 is read "three and six tenths" or "three point six."

0.0162 is read as "one hundred sixty-two ten-thousandths," or "point zero one six two," or "zero point zero one six two." This informal process is often used in verbal communication to ensure that numbers are not miscommunicated.

Unwritten Decimal Points

When we write whole numbers, we usually omit the decimal point; the decimal point is understood to be at the right end of the whole number. Therefore, any whole number, such as 32, can be written without a decimal (32) or with a decimal (32.).

Fractions like $\frac{1}{10}$ and $\frac{75}{100}$ have denominators that are powers of 10. Any fraction whose denominator is 10, 100, 1,000, 10,000, and so on, can be written as a decimal number without performing any calculations.

To write a fraction that has a denominator of 10, 100, 1,000, 10,000, and so on, as a decimal:	
1. Use the denominator to find the number of decimal places.	Write $\frac{17}{1,000}$ as a decimal.
$\begin{aligned} 10 &\rightarrow 1 \text{ place} \\ 100 &\rightarrow 2 \text{ places} \\ 1,000 &\rightarrow 3 \text{ places} \\ 10,000 &\rightarrow 4 \text{ places} \end{aligned}$	0.___ Three decimal places are needed.
2. Place the numerator so that the last digit is in the farthest place on the right.	0._17
3. Fill in any blank spaces with zeros.	0.017

STOP AND CHECK
Write as decimal numbers.

1. $\dfrac{9}{10}$

2. $\dfrac{27}{100}$

3. $\dfrac{307}{1,000}$

4. $\dfrac{43}{1,000}$

Answers:
1. 0.9 **2.** 0.27 **3.** 0.307 **4.** 0.043

EXAMPLE 2

Write $\frac{3}{10}, \frac{25}{100}, \frac{425}{100}$, and $\frac{3}{1,000}$ as decimal numbers.

$\frac{3}{10}$ is written **0.3.**   One decimal place

$\frac{25}{100}$ is written **0.25.**   Two decimal places

$\frac{425}{100}$ is written **4.25.**   Two decimal places

$\frac{3}{1,000}$ is written **0.003.**   Three decimal places   See Exercises 11–16.

TIP **Do Ending Zeros Change the Value of a Decimal Number?** When we attach zeros on the *right* end of a decimal number, we do not change the value of the number.

$$0.5 = 0.50 = 0.500 \qquad \frac{5}{10} = \frac{50}{100} = \frac{500}{1,000}$$

See equivalent fractions on page 69.

---

3 **Compare Decimal Numbers.** As with whole numbers, we often need to compare decimals by size. To make valid comparisons, we must compare like amounts. Whole numbers compare with whole numbers, tenths compare with tenths, thousandths compare with thousandths, and so on.

To compare decimal numbers:

1. Compare whole-number parts.

2. If the whole-number parts are equal, compare digits place by place, starting at the tenths place and moving to the right.

3. Stop when two digits in the same place are different.

4. The digit that is larger determines the larger decimal number.

STOP AND CHECK

1. Which is larger, 3.51 or 3.508?

2. Arrange in order from smallest to largest: 24.6, 24, and 24.55

Answers:
1. 3.51 **2.** 24, 24.55, 24.6

EXAMPLE 3

(a) Which is larger, 32.47 or 32.48?

32.47   Look at the whole-number parts. They are the same.

32.48   Look at the tenths place for each number. Both numbers have a 4 in the tenths place.

Look at the hundredths place. They are different and 8 is larger than 7.

32.48 is the larger number.

(b) Arrange the numbers in order from the smallest to the largest. Use appropriate inequality symbols. 4.1, 4.05, and 4

All three numbers have a whole-number part of 4.

4 has no decimal part so it is the smallest.

4.1 has a one in the tenths place and 4.05 has a zero in the tenths place. 4.05 is smaller.

The numbers in order are 4, 4.05, and 4.1.

$4 < 4.05 < 4.1$ See Exercises 17–29.

> **TIP** **Common Denominators in Decimals** The denominator of a decimal fraction is determined by the number of decimal places in the number. Decimal fractions have a common denominator if they have the same number of digits to the right of the decimal point. See common denominators of fractions on page 78.

Approximate number: another name for a rounded amount

Round or rounding: to express a number as an approximation

4 **Round a Whole Number or a Decimal Number to a Specified Place Value.** In many situations, an approximate number is used in place of an exact number. An **approximate number** is a rounded amount.

 Rounding a number means finding the closest approximate number to a given number. For example, if 37 is rounded to the nearest ten, is 37 closer to 30 or 40? Locate 37 on the number line.

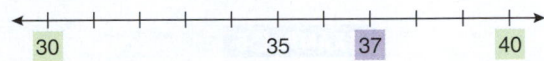

Exact number: a number that is not rounded

37 is closer to 40 than 30. Thus, 40 is the approximation to the nearest ten for 37. Another way to say this is that 37 rounded to the nearest ten is 40. In this situation, 37 is the exact number and 40 is the approximate number. An **exact number** is a number that is not rounded.

 When rounding a number to a certain place value, we must make sure that we are as accurate as the specific situation requires. Generally, the size of the number and its use dictate the decimal place to which it should be rounded.

To round a whole or decimal number to a given place value:

1. Locate the digit that occupies the rounding place. Then examine the digit to the immediate right.

2. If the digit to the right of the rounding place is 0, 1, 2, 3, or 4, do not change the digit in the rounding place. If the digit to the right of the rounding place is 5, 6, 7, 8, or 9, add 1 to the digit in the rounding place.

3. Replace all digits to the *right* of the digit in the rounding place with zeros if they are to the left of the decimal point. Drop digits that are to the right of the digit in the rounding place *and also* to the right of the decimal point.

STOP AND CHECK

1. In making a presentation about the population of Wisconsin, what is a reasonable approximate number for a population of 5,686,986?

Answer:

1. 6,000,000 and 5,700,000 are both reasonable approximations.

EXAMPLE 4

AG/H Oregon has a land area of 96,187 square miles. What would be a reasonable approximate number for this land area?

96,187 rounds to the following approximate numbers:

96,190 to the nearest ten

96,200 to the nearest hundred

96,000 to the nearest thousand

100,000 to the nearest ten thousand

Deciding to which place to round a number is a judgment depending on what use you will make of the rounded or approximate number.

Both 96,000 and 100,000 are reasonable approximations. See Exercises 30–31.

STOP AND CHECK

1. Round 46.897 to the nearest tenth.

Answer:
1. 46.9

EXAMPLE 5

Round 46.897 to the hundredths place.

46.897   9 is in the hundredths place.

46.897   The next digit to the right is 7, so add 1 to 9. (9 + 1 = 10 and 89 + 1 = 90.)

Drop the 7 in the thousandths place because it is to the right of the rounding place and also to the right of the decimal.

46.90   **See Exercises 32–35.**

> **TIP**   **Nine Plus One Still Equals Ten**   When the digit in the rounding place is 9 and must be rounded up, it becomes 10. The 0 replaces the 9 and the 1 is regrouped to the next place to the left.

STOP AND CHECK

1. Round $87.49 to the nearest dollar.
2. Round $3.394 to the nearest cent.

Answers:
1. $87   2. $3.39

EXAMPLE 6

(a) Round $293.48 to the nearest dollar.

$293.48   When we round to the nearest dollar, we are rounding to the *ones* place.

$293

(b) Round $71.8986 to the nearest cent.

$71.8986   One cent is 1 hundredth of a dollar; to round to the nearest cent is to round to the *hundredths* place.

$71.90   **See Exercises 36–37.**

Addition: a basic operation for combining two or more numbers

Addends: in addition, the numbers being added

Sum or total: the answer of an addition problem

Commutative property of addition: values being added may be added in any order and the sum remains the same

Associative property of addition: values being added may be grouped in any manner and the sum remains the same

Binary operation: an operation that involves working with two numbers at a time

5   Add and Subtract Whole Numbers and Decimals.   Addition is a basic operation for combining two or more numbers. In an addition problem, the numbers being added are called **addends** and the answer is called the **sum or total.**

Two useful properties of addition are the **commutative** and **associative properties**. By the *commutative* property, we mean that numbers can be added in any *order*. We can add 7 and 6 in any order and still get the same answer:

$$7 + 6 = 13 \qquad 6 + 7 = 13$$

By the *associative* property, we mean that we can *group* numbers together any way we want when we add and still get the same answer. To add $7 + 4 + 6$, we can group the $4 + 6$ to get 10; then we add the 10 to the 7:

$$7 + (4 + 6) = 7 + 10 = 17$$

Or we can group the $7 + 4$ to get 11; then we add the 6 to the 11:

$$(7 + 4) + 6 = 11 + 6 = 17$$

Addition is a **binary operation;** that is, the rules of addition apply to adding *two* numbers at a time. The associative property of addition shows how addition is extended to more than two numbers.

> **TIP**   **Using Symbols to Write Properties and Definitions**   Many properties and definitions can be written symbolically. Symbolic representation allows a quick recall of the rule or definition.
>
> Some properties are restricted to certain types of numbers and others are appropriate for practically all types of numbers. When writing properties symbolically, we will also note any restrictions and refer to the most inclusive type of numbers that are appropriate. Many properties apply to all *real numbers*.

Real numbers: initial informal definition is the set of numbers including whole numbers, fractions, and decimals

Real numbers are introduced later, but for now, whole numbers, fractions, and decimals are included in the set of real numbers.

Commutative Property of Addition: Two numbers may be added in any order and the sum remains the same.

$$a + b = b + a \quad \text{where } a \text{ and } b \text{ are any real numbers.}$$

Associative Property of Addition: Three numbers may be added using different groupings and the sum remains the same.

$$a + (b + c) = (a + b) + c \quad \text{where } a, b, \text{ and } c \text{ are any real numbers.}$$

The associative and commutative properties of addition also allows other possible groupings and extends to more than three numbers.

$$7 + 4 + 6 = 13 + 4 = 17$$
$$3 + 5 + 7 + 9 = 8 + 16 = 24$$

Adding zero to any number results in the same number. This property is called the **zero property of addition** and zero is called the **additive identity.**

Zero property of addition: the sum of zero and any number is the number itself

Additive identity: zero is the additive identity because $a + 0 = a$ for all values of a

Zero property of addition:

Adding zero to any number results in the same number.

$n + 0 = n$ or $0 + n = n$, where n is any real number
$5 + 0 = 5$ or $0 + 5 = 5$

When adding whole numbers of two or more digits, the same place values must be aligned under one another so that all the ones are in the far right column, all the tens in the next column, and so on.

To add whole numbers of two or more digits:

1. Arrange the numbers in columns so that the ones place values are in the same column.

2. Add the ones column, then the tens column, then the hundreds column, and so on, until all the columns have been added. **Regroup** whenever the sum of a column is more than one digit. This is also called **carrying.**

Regroup: in addition, the process of adding the tens digit to the next digit to the left when the sum of a column of digits is 10 or more

Carrying: also called regrouping

STOP AND CHECK

1. Arrange in columns and add: 2,380, 515, 30, and 9.

Answer:
1. 2,934

EXAMPLE 7

Shipping fees are often charged by the total weight of the shipment. Find the total weight of this order: nails, 250 pounds (lb); tacks, 75 lb; brackets, 12 lb; and screws, 8 lb. Arrange in columns and add.

```
 11
 250    The sum of the digits in the right column is 15. Record the 5 in the ones column
  75    and regroup the 1 to the tens column.
  12    Add the tens column. 1 + 5 + 7 + 1 = 14. Regroup the 1.
+  8
 345
```

The total weight is 345 lb.   See Exercises 42–44.

When adding, decimal numbers are aligned so that all decimal points fall in the same vertical line. Aligning decimal points has the same effect as using *like* denominators when adding (or subtracting) fractions. Like denominators are discussed in Chapter 2.

To add decimal numbers:

1. Arrange the numbers so that the decimal points are in one vertical line.

2. Add each column, regrouping as necessary.

3. Align the decimal point for the sum in the same vertical line.

STOP AND CHECK

1. Add 23.82 + 31 + 0.6.

Answer:

1. 55.42

EXAMPLE 8

Add 42.3 + 17 + 0.36.

$$
\begin{array}{r}
42.3 \\
17 \\
0.36 \\
\hline
\end{array}
$$
Note that the decimal point in 17 is understood to be at the right end.

$$
\begin{array}{r}
42.30 \\
17.00 \\
0.36 \\
\hline
59.66
\end{array}
$$
If we prefer, we may write each number so that all have the same number of decimal places by attaching zeros on the right.

See Exercises 45–47.

Estimating is an increasingly important skill to develop and it relates in part to your *number sense.* Number sense must be developed. We develop and strengthen our number sense by estimating and doing mental calculations. In making calculations, it is important to **estimate** and **check** your work. To estimate a calculation, round the values in a calculation before performing the calculation to get a sense of the value of the result. To check a calculation, rework the calculation to verify the initial result.

In estimating or rounding, we often refer to a nonzero digit. A **nonzero digit** is a 1, 2, 3, 4, 5, 6, 7, 8, or 9. That is, it is a digit that is not zero.

Estimate: to round the values in a calculation before performing the calculation to get a sense of the value of the result

Check: to rework a problem to verify the initial result

Nonzero digit: a digit that is not zero (1, 2, 3, 4, 5, 6, 7, 8, or 9)

To estimate the answer to an addition problem:

1. Round each addend to a specific place value or to the place of the first nonzero digit.

2. Add the rounded addends.

To check an addition problem:

1. Add the numbers a second time and compare with the first sum.

2. Use a different order or grouping if convenient.

STOP AND CHECK

1. Round to the place of the first digit and add.

24,285 + 5,875 + 312

Answer:

1. 26,300

EXAMPLE 9

Elston Home Renovators spent the following amounts on a job: $16,466.15, $23,963.10, and $5,855.20. Estimate the total amount by rounding to thousands. Then find the exact amount and check your answer.

Thousands Place	Estimate	Exact	Check
$16,466.15	$16,000	$16,466.15	$16,466.15
23,963.10	24,000	23,963.10	23,963.10
5,855.20	6,000	5,855.20	5,855.20
	$46,000	$46,284.45	$46,284.45

The estimate and exact answer are close. The exact answer is reasonable.

See Exercises 54–60.

TIP **Decimal Point and Zeros on the Calculator** The decimal key $\boxed{\cdot}$ is most often located near the number keys on a calculator. This key is pressed when the decimal point appears in the number being entered.

▶ Does the zero to the left of the decimal have to be entered before the decimal in a number like 0.2? Try it both ways: Add $3 + 0.2$ on the calculator.

Options: $3\boxed{+}0\boxed{\cdot}2\boxed{=} \Rightarrow 3.2$ entering the zero The symbol $\Rightarrow$ is

$3\boxed{+}\boxed{\cdot}2\boxed{=} \Rightarrow 3.2$ not entering the zero used to indicate the calculator result.

▶ Do zeros that follow a decimal or that are on the right end of a decimal number have to be entered? Try it both ways: Add $3.00 + 1.50$ on the calculator.

Options: $3\boxed{\cdot}00\boxed{+}1\boxed{\cdot}50\boxed{=} \Rightarrow 4.5$ entering ending zeros

$3\boxed{+}1\boxed{\cdot}5\boxed{=} \Rightarrow 4.5$ not entering ending zeros

Ending zeros to the right of the decimal are usually dropped in a calculator display unless the calculator is set to display a specific number of decimal places.

Subtraction: the inverse operation of addition

Inverse operations: pairs of operations that "do" and "undo" an operation; addition/subtraction and multiplication/division are pairs of inverse operations

Difference: the answer to a subtraction problem

Remainder: the answer to a subtraction problem

Minuend: the initial quantity in a subtraction problem

Subtrahend: the quantity being subtracted in a subtraction problem

STOP AND CHECK

1. Subtract $9 - 7 - 1$.

Answer:
1. 1

Not equal ($\neq$): symbol used to show two quantities are not equal

Subtraction is the **inverse operation** of addition. In addition, we add numbers to get their total (such as $5 + 4 = 9$), but to solve the subtraction problem $9 - 5 = ?$ we ask, "What number must be added to 5 to give us 9?" The answer is 4 because 4 added to 5 gives a total of 9. When we subtract two numbers, the answer is called the **difference** or **remainder**. The initial quantity is the **minuend**. The amount being subtracted from the initial quantity is the **subtrahend**.

When subtracting two or more numbers, if no grouping symbols are included, perform subtractions from left to right.

EXAMPLE 10

Subtract $8 - 3 - 1$.

$8 - 3 - 1 = 5 - 1 = 4$ Subtract from left to right. **See Exercises 65–66.**

Did You Know? Subtraction is *not* commutative. $8 - 3 = 5$, but $3 - 8$ does not equal 5; that is, $3 - 8 \neq 5$. The symbol $\neq$ is read **is not equal to.**

Subtraction is *not* associative. Does $9 - (5 - 1)$ equal $(9 - 5) - 1$?

$$9 - (5 - 1) = 9 - 4 = 5, \text{ but } (9 - 5) - 1 = 4 - 1 = 3$$

$5 \neq 3$ (Read "Five does not equal 3".)

Subtracting zero from a number results in the same number:

$$n - 0 = n, \quad \text{where } n \text{ is any real number} \quad 7 - 0 = 7$$

> **TIP**   **Subtraction and Zeros**   Subtracting a number from zero is not the same as subtracting zero from a number; that is, $7 - 0 = 7$, but $0 - 7$ does not equal 7. *Negative numbers are introduced later.*

To subtract whole numbers that have two or more digits:

1. Arrange the numbers in columns, with the minuend at the top and the subtrahend at the bottom.

2. Make sure the ones digits are in a vertical line on the right.

3. Subtract the ones column first, then the tens column, the hundreds column, and so on.

4. To subtract a larger digit from a smaller digit in a column, **regroup** by subtracting 1 from the digit in the next column to the left. This is the equivalent to *one* group of 10; thus, add 10 to the digit in the given column, then continue subtracting. The concept of *regrouping* is also referred to as **borrowing.**

Regroup: in subtraction, the process of subtracting 1 from the digit in the next column to the left and adding 10 to the digit in the column being subtracted when the digit in the subtrahend is larger than the digit in the minuend

Borrowing: also called regrouping

STOP AND CHECK

1. Subtract $4,385 - 2,189$.

Answer:

1. 2,196

EXAMPLE 11

Subtract $9,327 - 3,514$.

Arrange in columns.

$$
\begin{array}{r}
\overset{8\,13}{9,327} \\
-3,514 \\
\hline
5,813
\end{array}
$$

Arrange in columns. Subtract each column from the right.
In the hundreds place, 5 is more than 3. Regroup by subtracting 1 group of 10 from 9. $9 - 1 = 8$, $10 + 3 = 13$.   **See Exercises 67–69.**

> **TIP**   **Words and Phrases that Imply Subtraction**   These phrases indicate subtraction in applied problems:
>
> how many are left   how many more
> how much less   how much larger
> how much smaller
>
> Also, some applied problems require more than one operation. Many problems that require more than one operation involve parts and a total. If we know the total and all the parts but one, we can add all the known parts and subtract the result from the total to find the missing part.

EXAMPLE 12

The Froehlichs left Memphis and drove 356 miles (mi) on the first day of their vacation. They drove 426 mi on the second day. If they are traveling to Albuquerque, which is 1,050 mi from Memphis, how many more miles do they have to drive?

The phrase *how many more* indicates subtraction.

$$1{,}050 \text{ mi} = \text{total mi}$$
$$356 \text{ mi} + 426 \text{ mi} + \text{mi left to drive} = 1{,}050 \text{ mi}$$

To find the miles left to drive, add $356 \text{ mi} + 426 \text{ mi}$ and subtract the result from 1,050 mi.

$$356 \text{ mi} + 426 \text{ mi} = \boxed{782 \text{ mi}} \quad 1{,}050 \text{ mi} - \boxed{782 \text{ mi}} = \mathbf{268 \text{ mi}}$$

See Exercises 70–71.

Rick Grainger/Shutterstock

To subtract decimals:

1. Arrange the numbers so that the decimal points align vertically.

2. Subtract each column beginning at the right, regrouping as necessary.

3. Interpret blank places as zeros.

4. Place the decimal in the difference in the same vertical line.

STOP AND CHECK

1. Subtract 3.7 from 9.

Answer:

1. 5.3

EXAMPLE 13

Subtract 7.18 from 15.

Take care to align the decimals properly.

$$\begin{array}{r} 15. \\ -7.18 \\ \hline \end{array}$$ Because 15 is a whole number, its decimal point is placed after the 5.

$$\begin{array}{r} 15.00 \\ -7.18 \\ \hline 7.82 \end{array}$$ To subtract, we put zeros in the tenths and hundredths places of 15 and then regroup.

See Exercises 72–75.

Tolerance: the amount the part can vary from a specification and still be acceptable

Plus or minus (±): a symbol used to show that the value following the symbol is both added to and subtracted from the value preceding the symbol for two distinct results

Limit dimensions: the largest and smallest possible size of an object with a plus or minus tolerance value on a blueprint specification

When a worker machines an object using a blueprint as a guide, a certain amount of variation from the blueprint specification is allowed for the machining process. This variation is called the **tolerance.** Thus, if a blueprint calls for a part to be 9.47 inches (in.) with a tolerance of ± 0.05 in. (read **plus or minus** five hundredths), this means that the actual part can be 0.05 in. *more* or 0.05 in. *less* than the specification. To find the *largest* acceptable measure of the object, we add 9.47 in. + 0.05 in. = 9.52 in. To find the *smallest* acceptable size of the object, we subtract 9.47 in. − 0.05 in. = 9.42 in. The dimensions 9.52 in. and 9.42 in. are called the **limit dimensions** of the object. That is, 9.52 in. is the largest acceptable measure, and 9.42 in. is the smallest acceptable measure.

Did You Know? Mathematical symbols can indicate that two different operations are to be performed. The symbol ± (read "plus or minus") indicates that the number following should first be added, and then as a separate calculation the number is to be subtracted.

EXAMPLE 14

CON   Find the limit dimensions of an object with a blueprint specification of 8.097 in. and a tolerance of ± 0.005 in. (This is often written 8.097 in. ± 0.005 in.)

8.097 in. = the blueprint specification for the dimension of an object.
± 0.005 in. = tolerance of object's dimension.

Largest dimension for object = blueprint specification + tolerance.
Smallest dimension for object = blueprint specification − tolerance.

Estimation

The tolerance of the part is very small—just five thousandths—so the limit dimensions for the part should be only a few thousandths of an inch smaller or larger than the blueprint specification.

8.097 in. − 0.005 in. = 8.092 in.
8.097 in. + 0.005 in. = 8.102 in.

The smallest acceptable dimension for the object is 8.092 in. and the largest acceptable dimension is 8.102 in.

See Exercises 76–77.

Estimating a subtraction problem is similar to estimating an addition problem. *The numbers in the problem are rounded before the subtraction is performed.*

To check subtraction, we can use the inverse relationship between addition and subtraction. If $9 - 5 = 4$, then $4 + 5$ should equal 9.

> **To estimate the difference for a subtraction problem:**
>
> 1. Round each number to the desired place value or to a number with one nonzero digit.
> 2. Subtract the rounded numbers.

> **To check a subtraction problem:**
>
> 1. Add the subtrahend (number being subtracted) and difference.
> 2. Compare the result of Step 1 with the minuend. If the two numbers are equal, the subtraction is correct.

STOP AND CHECK

1. Estimate by rounding to the first digit on the left, then find the exact difference, and check.

   $4,284.17 - 2,847$

Answer:
1. Estimate: 1,000; Exact: 1,437.17

EXAMPLE 15

Estimate by rounding to hundreds, then find the exact difference, and check.

$$427.45 - 125$$

	Estimate	Exact	Check
427.45	400	427.45	125.00
−125.00	−100	−125.00	+302.45
	300	302.45	427.45

See Exercises 78–79.

When using a calculator, it is advisable to estimate first and then use a calculator.

STOP AND CHECK

1. Estimate by rounding to the nearest thousand and then find the difference using a calculator.

   $81,396.49 - $56,388.11

Answer:
1. Estimate: $25,000; Exact: $25,008.38

EXAMPLE 16

Find the difference between \$53,943.76 and \$34,256.45 using a calculator.
  One option:

Estimate: $50,000 - 30,000 = 20,000$ Round first.

53943 $\boxed{\cdot}$ 76 $\boxed{-}$ 34256 $\boxed{\cdot}$ 45 $\boxed{=}$ $\boxed{\text{ENTER}}$ or $\boxed{\text{EXE}}$ may replace $\boxed{=}$

Calculator display: 19687.31

The difference is \$19,687.31.

Check: 34256 $\boxed{\cdot}$ 45 $\boxed{+}$ 19687 $\boxed{\cdot}$ 31 $\boxed{=}$ 53943.76   See Exercises 80–81.

A real-world situation can require more than one operation.

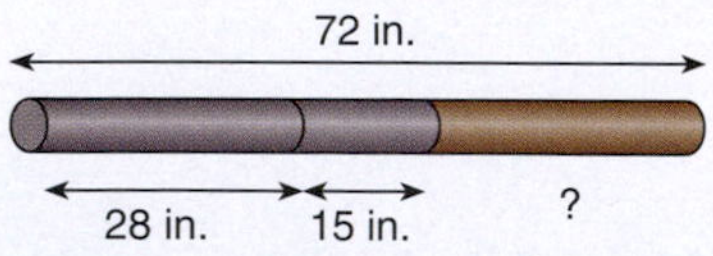

FIGURE 1–3 Lengths cut from pipe.

EXAMPLE 17

INDTEC   Two cuts are made from a 72-in. pipe (see Fig. 1–3). The two lengths cut from the pipe are 28 in. and 15 in. How much of the pipe is left after these cuts are made?

$$72 - 28 - 15 =$$ Subtract from left to right.
$$44 - 15 =$$ $$72 - 28 = 44$$
$$29$$ $$44 - 15 = 29$$

There are 29 in. of pipe left.   See Exercise 86.

Often there is more than one way to solve a problem. In the previous example, we could have found the sum of the cuts first and then subtracted.

28 in. + 15 in. = 43 in.   Sum of two cuts.
72 in. − 43 in. = 29 in.   Remaining length of pipe.

Multiplication: repeated addition

Multiplicand: the first number in a multiplication problem

Multiplier: the number to multiply by

Factor: amounts involved in multiplication, either the multiplicand or the multiplier

Product: the result of multiplication

6 Multiply and Divide Whole Numbers and Decimals. **Multiplication** of whole numbers is repeated addition. If we have three $10 bills, we have $10 + $10 + $10, or $30. Using multiplication, we see that this is the same as 3 times $10, or $30.

When we multiply two numbers, the first number is called the **multiplicand,** and the number we multiply by is called the **multiplier.** Either number is referred to as a **factor.** The answer or result of multiplication is called the **product.**

$$
\begin{array}{ccccc}
2 & \times & 3 & = & 6 \\
\text{multiplicand} & & \text{multiplier} & & \text{product} \\
\text{or factor} & & \text{or factor} & &
\end{array}
$$

> **TIP** **Various Notations for Multiplication** Besides the familiar × or "times" sign, parentheses ( ), a raised dot (·), and an asterisk (∗) are also used to show multiplication. Parentheses are most often used as notation for multiplication.
>
> $$2(3) = 6, \quad (2)(3) = 6, \quad 2 \cdot 3 = 6, \quad 2 * 3 = 6$$

Multiplication is commutative and associative, just like addition. The **commutative property of multiplication** permits two numbers to be multiplied in any order. In symbols, $a(b) = b(a)$ for all real numbers.

$$4(5) = 20, \quad 5(4) = 20$$

Commutative property of multiplication: values being multiplied can be multiplied in any order

Associative property of multiplication: values being multiplied may be grouped in any manner

When more than two numbers are multiplied, the numbers must be grouped, and the **associative property of multiplication** permits the numbers to be grouped in any way. In symbols, $a(b \cdot c) = (a \cdot b)c$ for all real numbers.

$$
\begin{array}{ccc}
2(3 \cdot 5) & \text{or} & (2 \cdot 3)5 \\
2(15) & & (6)5 \\
30 & & 30
\end{array}
$$

The product of a number and zero is zero:

Zero property of multiplication: the product of zero and any number is 0

$$n(0) = 0, 0(n) = 0, \text{ where } n \text{ is any real number} \qquad 4(0) = 0, \qquad 0(4) = 0$$

This is called the **zero property of multiplication.**

STOP AND CHECK

1. Multiply 4(7)(5).
2. Multiply 3(0)(12).

Answer:
1. 140   2. 0

EXAMPLE 18

(a) Multiply 3(2)(9).

$$
\begin{array}{ll}
3(2)(9) & \text{Group any two factors.} \\
(3 \cdot 2)(9) & \text{Multiply grouped factors.} \\
6(9) & \text{Multiply the factors: 6 and 9.}
\end{array}
$$

54

(b) Multiply 2(5)(0)(7).

    (2)(5)(0)(7)   Use the zero property of multiplication.

      0

See Exercises 87–90.

To multiply whole-number factors of two or more digits:

1. Arrange the factors one under the other.

2. Multiply each digit in the multiplicand by each digit in the multiplier. The product of the multiplicand and each digit in the multiplier gives a **partial product.**

   (a) To start, multiply the ones digit in the multiplier by the multiplicand from right to left.

   (b) Align each partial product with its right-most digit directly under the multiplier digit.

3. Add the partial products.

Partial product: the product of the multiplicand and one digit of the multiplier

STOP AND CHECK

1. Multiply 502(307).

Answer:

1. 154,114

EXAMPLE 19

Multiply 204(103).

$$
\begin{array}{r}
204 \\
\times\ \ 103 \\
\hline
612 \\
0\ 00 \\
20\ 4 \\
\hline
\mathbf{21{,}012}
\end{array}
$$

Multiply: $3 \times 204 = 612$. Align 612 under 3 in the multiplier.

Multiply: $0 \times 204 = 000$. Align 000 under 0 in the multiplier.

Multiply: $1 \times 204 = 204$. Align 204 under 1 in the multiplier.

Add the partial products as they are aligned.

Using a calculator: 204 ☒ 103 ▣
Calculator display: 21012

See Exercises 91–92.

Since decimals are fractions, multiplication of decimals causes us to rethink or expand our basic number sense of multiplication. The product of any number and a decimal number less than 1 is less than the original number.

To multiply decimal numbers:

1. Align the numbers as if they were whole numbers and multiply.

2. Count the total number of digits to the right of the decimal in each factor.

3. Place the decimal in the product so that the number of decimal places is the sum of the number of decimal places in the factors.

STOP AND CHECK

Multiply.

1. 12.7(0.3)

2. 0.204(0.15)

Answer:

1. 3.81   2. 0.0306

EXAMPLE 20

(a) Multiply 1.36(0.2).

$$
\begin{array}{r}
1.36 \\
\times\ \ 0.2 \\
\hline
\mathbf{0.272}
\end{array}
$$

In multiplication, the decimals do not have to be aligned vertically.

The product has three places to the right of the decimal. Place a zero in the ones place so that the decimal point will not be overlooked.

(b) Multiply 0.309(0.17).

▫ 309 ☒ ▫ 17 ▣

Calculator display: **0.05253**

See Exercises 97–104.

EXAMPLE 21

AG/H The outside diameter of a circular flower bed is 7.82 meters (m) (see Fig. 1–4). If the brick walk surrounding the bed is 1.56 m thick, find the inside diameter of the flower bed.

The **diameter** of a circle is the distance across the center of the circle.

7.82 m = outside diameter of the flower bed

1.56 m = width of brick walk surrounding the flower bed

Outside diameter − two widths (one on each end of the inside diameter) of the walk = inside diameter

Estimation

The outside diameter is nearly 8 m, and the total of the two widths to be subtracted is about 3 m, so the inside diameter should be about 5 m.

$7.82 - (2 \cdot 1.56) = 7.82 - 3.12 = 4.70$

The inside diameter of the flower bed is 4.7 m.   See Exercise 105.

Diameter: the distance across the center of a circle

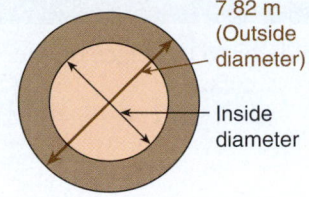

7.82 m (Outside diameter)

Inside diameter

FIGURE 1–4

Distributive property of multiplication: the property stating that multiplying a grouping containing sums or differences of terms by a factor is equivalent to multiplying each term of the grouping by the factor

The **distributive property of multiplication** means that multiplying a grouping containing sums or differences of terms by a factor is equivalent to multiplying each term of the grouping by the factor.

Distributive property of multiplication:

1. Add or subtract the terms within the grouping.

2. Multiply the result of Step 1 by the factor outside the grouping.

or

1. Multiply each term inside the grouping by the factor outside the grouping.

2. Add or subtract the products from Step 1.

   Symbolically, $a(b + c) = ab + ac$ or $a(b - c) = ab - ac$ for all real numbers.

TIP **Additional Notations for Multiplication**

▶ Parentheses show multiplication when the distributive property is used: $a(b + c)$ means $a \times (b + c)$.

▶ The letters represent numbers.

▶ Letters written together with no operation sign between them imply multiplication: ab means $a \times b$; ac means $a \times c$.

The distributive property is found in many formulas. One example is the formula for the perimeter of a rectangle. The *formula* for finding the *perimeter of a rectangle* is

$$P = 2(l + w) \quad \text{or} \quad P = 2l + 2w$$

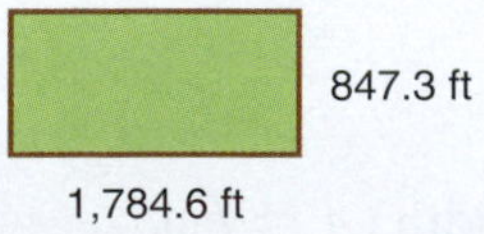

847.3 ft

1,784.6 ft

FIGURE 1–5 Perimeter of rectangle.

EXAMPLE 22

AG/H Find the number of feet of fencing needed to enclose a rectangular pasture that is 1,784.6 feet (ft) long and 847.3 ft wide. See Fig. 1–5.

$$P = 2(l + w) \qquad \text{or} \qquad P = 2l + 2w$$
$$P = 2(1{,}784.6 + 847.3) \qquad\qquad P = 2(1{,}784.6) + 2(847.3)$$
$$P = 2(2{,}631.9) \qquad\qquad\qquad P = 3{,}569.2 + 1{,}694.6$$
$$P = 5{,}263.8 \qquad\qquad\qquad\qquad P = 5{,}263.8$$

The amount of fencing needed is 5,263.8 ft.   **See Exercise 106.**

To estimate the product for a multiplication problem:

1. Round both factors to a chosen or specified place value or to the place of the first digit from the left that is not a zero.

2. Multiply the rounded numbers.

To check a multiplication problem:

1. Multiply the numbers a second time and check the product.

2. Interchange the factors if convenient.

EXAMPLE 23

AG/H   Find the approximate (estimate) and exact costs of 48 flower bulbs if each bulb costs $2.15.

48 = total number of flower bulbs

$2.15 = cost of each bulb.

Total cost of flower bulbs = number of bulbs × cost of each bulb

Estimation

Estimate to the place of the first digit that is not a zero. If 50 bulbs were purchased at $2 each, the total cost would be $100. So the exact cost should be close to $100.

Exact Cost

48($2.15) = $103.20   Number of bulbs × Cost of each bulb = Exact cost

The total cost of 48 flower bulbs is $103.20, which is approximately $100, as estimated.   **See Exercise 107.**

Multiplication is used in many formulas. The formula for finding the *area of a rectangle* is $A = lw$, where l is the length of the rectangle and w is the width. The area is expressed as a square measure.

EXAMPLE 24

AG/H   Maintenance Consultants needs to apply fertilizer to a customer's lawn. The lawn is 223.4 ft long and 132.8 ft wide. The fertilizer costs $0.004 per square foot to apply. What is the cost of applying the fertilizer? Use the formula $A = lw$. The area will be expressed in square feet (ft^2) (see Fig. 1–6).

132.8 ft

223.4 ft

FIGURE 1–6 Area of a rectangle.

Lawn is in the shape of a rectangle.
Length of the lawn = 223.4 ft.
Width of the lawn = 132.8 ft.
Fertilizer costs $0.004 per square foot to apply.
Area of a rectangle is the product of the length times the width ($A = lw$).
Total cost = total number of square feet × cost per square foot.

$$A = lw$$
$$A = 223.4(132.8)$$
$$A = 29{,}667.52 \text{ ft}^2 \qquad \text{Area of lawn}$$

$$\text{cost} = 29{,}667.52(\$0.004)$$
$$\text{cost} = \$118.67 \qquad \text{Rounded to the nearest cent}$$

The total cost of applying fertilizer to the lawn is \$118.67.   **See Exercise 108.**

When either or both factors of a multiplication problem end in zeros, a shortcut process such as the one in the following example can be used.

STOP AND CHECK

1. Multiply 3,200(60).

Answer:

1. 192,000

EXAMPLE 25

Multiply 2,600(70).

1. $\begin{array}{r} 2600 \\ \times\ 70 \\ \hline \end{array}$ Separate the ending zeros from the other digits.

2. $\begin{array}{r} 2600 \\ \times\ 70 \\ \hline 182 \end{array}$ Multiply the other digits as if the ending zeros were not there (26 × 7 = 182).

3. $\begin{array}{r} 2600 \\ \times\ 70 \\ \hline \mathbf{182{,}000} \end{array}$ Attach the ending zeros to the basic product. Note that the number of zeros affixed to the basic product is the same as the sum of the number of zeros at the end of each factor.   **See Exercises 109–111.**

TIP   **The Mind Is Often Quicker than the Fingers** Don't use your calculator as a crutch. It is a tool! When multiplying 2,500 times 30, you can multiply 25 times 3 mentally. 25(3) = 75. Then, attach three zeros to that product.

$$2{,}500(30) = 75{,}000$$

This skill does not come automatically. Like playing a musical instrument or mastering a sport, you don't develop skill by watching. You have to practice!

Division: the inverse operation of multiplication

Dividend: the number being divided

Divisor: the number to divide by

Quotient: the result of a division problem

Division is the **inverse operation** of multiplication. Since 4(7) = 28, then 28 divided by 7 is 4 and 28 divided by 4 is 7. The number being divided is called the **dividend.** The number to divide by is the **divisor.** The result is the **quotient.**

Division is *not* commutative. 12 ÷ 6 = 2, but 6 ÷ 12 does not equal 2; that is, 6 ÷ 12 ≠ 2.

Division is *not* associative. (12 ÷ 6) ÷ 2 = 2 ÷ 2 = 1. But 12 ÷ (6 ÷ 2) = 12 ÷ 3 = 4. That is, (12 ÷ 6) ÷ 2 ≠ 12 ÷ (6 ÷ 2).

TIP   **To Write Division Symbolically**

1. To use the *divided by* symbol (÷), write the dividend first.

$$28 \div 7 = 4 \longleftarrow \text{quotient}$$

dividend ——↑   ↑—— divisor

2. To use the *long-division symbol* ($\overline{}$), write the dividend under the bar.

$$\begin{array}{r} 4 \longleftarrow \text{quotient} \\ 7\,\overline{)\,28} \end{array}$$

divisor ——↑   ↑—— dividend

3. To use the *division bar* or *slash symbol,* write the dividend on top or first. The bar may also be referred to as the *fraction line* or *fraction bar*.

$$\frac{28}{7} = 4 \longleftarrow \text{quotient} \quad \text{or} \quad 28/7 = 4 \longleftarrow \text{quotient}$$

dividend — quotient — divisor

Remainder: the final difference after all the digits of the dividend have been used in long division

Partial dividend: the first group of digits of the dividend that is equal to or larger than the divisor

Partial quotient: the quotient of a partial dividend and the divisor

Long division, like long multiplication, involves using one-digit multiplication facts repeatedly to find the quotient.

When the quotient is not a whole number, the quotient may have a *whole-number part* and a **remainder.** When a dividend has more digits than a divisor, parts of the dividend are called **partial dividends,** and the quotient of a partial dividend and the divisor is called a **partial quotient.**

To divide whole numbers:

1. Beginning with its leftmost digit, identify the first group of digits of the dividend that is larger than or equal to the divisor. This group of digits is the first *partial dividend.*

2. For each partial dividend in turn, beginning with the first:

   (a) Divide the partial dividend by the divisor. Write the partial quotient above the rightmost digit of the partial dividend.

   (b) Multiply the partial quotient by the divisor. Write the product below the partial dividend, aligning places.

   (c) Subtract the product from the partial dividend. Write the difference below the product, aligning places. The difference must be less than the divisor.

   (d) Next to the ones place of the difference, write the next digit of the dividend. This is the new partial dividend.

3. When all the digits of the dividend have been used, write the final difference in Step 2c as the remainder (unless the remainder is 0). The whole-number part of the quotient is the number written above the dividend.

STOP AND CHECK

1. Divide and show remainder in the quotient.
   497 by 23

Answer:
1. 21 R14

EXAMPLE 26

Divide 881 by 35 and show remainder in the quotient:

$$35\overline{)881}$$ The first partial dividend is 88.

$$\begin{array}{r} 2 \\ 35\overline{)881} \\ \underline{70} \\ 18 \end{array}$$ The partial quotient for $88 \div 35$ is 2. Multiply $2(35) = 70$. Then subtract $88 - 70 = 18$. The difference 18 is less than the divisor 35.

$$\begin{array}{r} 2 \\ 35\overline{)881} \\ \underline{70} \\ 181 \end{array}$$ 1 from the dividend is written next to 18 to form the next partial dividend.

$$\begin{array}{r} 25 \\ 35\overline{\smash{)}881} \\ 70 \\ \hline 181 \\ 175 \\ \hline 6 \end{array}$$

The partial quotient for $181 \div 35$ is 5. The product of $5(35)$ is 175. The difference of $181 - 175$ is 6. The remainder is 6.

881 divided by 35 = **25 R6.**

See Exercises 112–113.

Special properties with division, zero and one.

Zero divided by a nonzero number is zero:

$$0 \div n = 0, \qquad n\overline{\smash{)}0}, \quad \frac{0}{n} = 0 \qquad \text{where } n \text{ is a nonzero real number}$$

$$0 \div 5 = 0, \qquad 5\overline{\smash{)}0}, \quad \frac{0}{5} = 0$$

A number divided by zero is **undefined** or **indeterminate:**

$$n \div 0 \text{ is undefined,} \qquad 0\overline{\smash{)}\overset{\text{undefined}}{n}}, \qquad \frac{n}{0} \text{ is undefined}$$

$$12 \div 0 \text{ is undefined}$$

$$0 \div 0 \text{ is indeterminate}$$

Dividing any nonzero number by itself yields 1:

$$n \div n = 1, \qquad \text{if } n \text{ is not equal to zero;} \qquad 12 \div 12 = 1$$

Dividing any number by 1 yields the same number:

$$n \div 1 = n, \qquad 5 \div 1 = 5$$

Undefined: a number divided by zero

Indeterminate: zero divided by zero

Since division is *not* associative, if there are no grouping symbols, we divide from left to right.

STOP AND CHECK

1. Divide $36 \div 6 \div 2$.

Answer:

1. 3

EXAMPLE 27

Divide $16 \div 4 \div 2$.

$16 \div 4 \div 2 = 4 \div 2 = 2$ Divide from left to right.

See Exercises 114–115.

To divide decimal numbers written in long-division form:

1. Move the decimal in the divisor so that it is on the right side of all digits. (By moving the decimal, you are multiplying by 10, 100, 1,000, and so on.)

2. Move the decimal in the dividend to the right as many places as the decimal was moved in the divisor. Attach zeros if necessary. (This is multiplying the dividend by the same number as was used in Step 1.)

3. Write the decimal point in the answer directly above the new position of the decimal point in the dividend. (Do this *before* dividing.)

4. Divide as you would whole numbers.

STOP AND CHECK
1. Divide $5.4 \div 9$.
Answer:
1. 0.6

EXAMPLE 28

Divide $4.8 \div 6$.

$$6\overline{)4.8}\;\overset{.}{}$$ Insert a decimal point above the decimal point in the dividend.

When the divisor is a whole number, the decimal is understood to be to the right of 6 and is not moved. The decimal is placed in the quotient directly above the decimal in the dividend.

$$6\overline{)4.8}\quad \textbf{0.8}$$ Divide.

See Exercises 116–117.

STOP AND CHECK
1. Divide $6.12 \div 1.8$.
Answer:
1. 3.4

EXAMPLE 29

Divide $3.12 \div 1.2$.

$$1.2\overline{)3.12}$$ Move the decimal in the divisor and dividend.

$$12\overline{)31.2}\;\overset{.}{}$$ Write the decimal in the quotient, then divide.

$$
\begin{array}{r}
\textbf{2.6} \\
12\overline{)31.2} \\
\underline{24} \\
7\,2 \\
\underline{7\,2}
\end{array}
$$

See Exercises 121–123.

> **To round a quotient to a place value:**
>
> 1. Divide to one place past the desired rounding place.
>
> 2. Attach zeros to the dividend after the decimal if necessary to carry out the division.
>
> 3. Round the quotient to the place specified.

STOP AND CHECK
1. Divide and round to the nearest tenth. $4.1\overline{)18.42}$
Answer:
1. 4.5

EXAMPLE 30

Divide and round the quotient to the nearest tenth.

$$3.2\overline{)15.27}$$

$$
\begin{array}{r}
4.77 \\
3.2\overline{)15.2\,70} \\
\underline{12\,8} \\
2\,4\,7 \\
\underline{2\,2\,4} \\
2\,30 \\
\underline{2\,24} \\
6
\end{array}
$$

Since we are rounding to the nearest tenth, divide to the hundredths.

4.77 rounds to 4.8.   See Exercises 128–129.

Estimating division is similar to estimating other operations. The numbers are rounded *before* the calculation is made.

Because division and multiplication are inverse operations, division is checked by multiplication.

> **To estimate division of whole numbers:**
>
> 1. Round the divisor and dividend to one nonzero digit.
> 2. Find the first digit of the quotient.
> 3. Attach a zero in the quotient for each remaining digit in the dividend.

> **To check division:**
>
> 1. Multiply the divisor by the quotient.
> 2. Add any remainder to the product in Step 1.
> 3. The result of Step 2 should equal the dividend.

EXAMPLE 31

Estimate, find the exact answer, and check $913 \div 22$.

Estimate:
$$\begin{array}{r} 40 \\ 20\overline{)900} \end{array}$$
Round 913 to 900 and 22 to 20
20 divides into 90 four whole times. Attach a zero after 4.

Exact:
$$\begin{array}{r} 41 \text{ R}11 \\ 22\overline{)913} \\ \underline{88} \\ 33 \\ \underline{22} \\ 11 \end{array}$$
or
$$\begin{array}{r} 41.5 \\ 22\overline{)913.0} \\ \underline{88} \\ 33 \\ \underline{22} \\ 11\,0 \\ \underline{11\,0} \end{array}$$

Check:
$$\begin{array}{r} 41 \\ \times\,22 \\ \hline 82 \\ \underline{82} \\ 902 \end{array} \qquad \begin{array}{r} 902 \\ +\,11 \\ \hline 913 \end{array}$$
or
$$\begin{array}{r} 41.5 \\ 2\,2 \\ \hline 83\,0 \\ \underline{830} \\ 913.0 \end{array}$$

The answer checks.

See Exercises 130–131.

We use averages to make comparisons, such as when we compare the average mileage different cars get per gallon of gasoline. There are several types of averages.

In most courses students take, their numerical grade is determined by a process called **numerical averaging.** This average is also called the **arithmetic average,** or **mean.** If the grades are 92, 87, 76, 88, 95, and 96, we can find the average grade by adding the grades and dividing by the number of grades. Because we have six grades, we divide the sum by 6.

$$\frac{92 + 87 + 76 + 88 + 95 + 96}{6} = \frac{534}{6} = 89$$

Numerical average: the sum of a list of values divided by the number of values. Also called mean

Arithmetic average: the sum of the quantities in the data set divided by the number of quantities. Also called mean

Mean: the sum of the quantities in the data set divided by the number of quantities. Also called arithmetic average

> **To find the average of a group of numbers or like measures:**
>
> 1. Add the numbers or like measures.
> 2. Divide the sum by the number of addends.

EXAMPLE 32

AUTO A car involved in an energy efficiency study had the following miles per gallon (mpg or $\frac{mi}{gal}$) listings for five tanks of gasoline: 21.7, 22.4, 26.9, 23.7, and 22.6 $\frac{mi}{gal}$. Find the average miles per gallon for the five tanks of gasoline. Round to the nearest tenth.

Estimate: The low value is 21.7 and the high value is 26.9. The average will be between the two values.

Exact: $\dfrac{21.7 + 22.4 + 26.9 + 23.7 + 22.6}{5} = \dfrac{117.3}{5} = 23.46$

$= 23.5 \dfrac{mi}{gal}$ Rounded

The approximate average is 23.5 $\frac{mi}{gal}$.

See Exercises 134–139.

1–1 EXERCISES

MyLab Math For additional practice go to your study plan in MyLab Math.

1 Write two inequalities to compare each pair of numbers. *See Example 1.*

1. 6 and 8
2. 42 and 32
3. 196 and 148
4. 2,802 and 2,517
5. 7,809 and 8,902
6. 44,000 and 42,999

7. **BUS** A house that sold for $183,500 four years ago has just sold for $198,900. Write two inequalities to compare the housing prices.

8. **AG/H** Touliatas Nursery sold 786 flats of annual bedding plants and 583 flats of perennial bedding plants. Write two inequalities to compare the number of plants of each type.

9. **HOSP** The Orlando Renaissance Resort sold 758 rooms for a horticulture convention and 893 rooms for a motorcycle trade show. Write two inequalities to compare the number of rooms sold for each of the events.

10. **COMP** The first day a spam filter was installed at Phoenix College, 5,982 spam e-mails were blocked. On the second day 2,807 spam e-mails were blocked. Write two inequalities to compare the number of blocked spam e-mails.

2 Write the fractions as decimal numbers. *See Example 2.*

11. $\dfrac{5}{10}$
12. $\dfrac{23}{100}$
13. $\dfrac{7}{100}$
14. $\dfrac{683}{100}$
15. $\dfrac{79}{1,000}$
16. $\dfrac{468}{1,000}$

3 Compare the number pairs and identify the larger number. *See Example 3.*

17. 3.72, 3.68
18. 7.08, 7.06
19. 0.23, 0.3

20. **INDTEC** Two micrometer readings are recorded as 0.837 in. and 0.81 in. Which is larger? Write your answer as an inequality.

21. **INDTEC** A micrometer reading for a part is 3.85 in. The specifications call for a dimension of 3.8 in. Which is larger, the micrometer reading or the specification?

22. **INDTR** A washer has an inside diameter of 0.33 in. Will it fit a bolt that has a diameter of 0.325 in.? In other words, is the inside diameter of the washer larger than the diameter of the bolt?

23. **INDTEC** Aluminum sheeting can be purchased in thicknesses of 0.04 in. or 0.035 in. Which sheeting is thicker?

24. **CON** If No. 14 copper wire has a diameter of 0.064 in., and No. 10 wire has a diameter of 0.09 in., which has the larger diameter?

25. **HLTH/N** A nurse recorded the weights of two patients as 64.8 kilograms and 72.3 kilograms. Which weight is greater?

Arrange in order from smallest to largest. See Example 3.

26. 1.9, 1.87, 1.92

27. 72.1, 72.07, 73

28. INDTEC   A 100-watt bulb that burns continuously for two minutes uses 0.003 kilowatt-hour of electricity, and an 800-watt toaster uses 0.026 kilowatt-hour for a piece of toast. Which uses the greater number of kilowatt-hours?

29. To change centimeters to inches, multiply by 0.394, and to change kilometers to miles, multiply by 0.621. Which factor is larger?

4 Give a reasonable approximate number for the following. *See Example 4.*

30. A national debt of 19,954,864,714,083

31. U.S. population of 324,351,823

Round to the place value indicated. See Example 5.

32. Nearest hundred: 468

33. Nearest ten thousand: 429,207

34. Nearest billion: 82,629,426,021

35. Nearest ten million: 297,384,726

See Example 6.

36. Nearest dollar: $549.87

37. Nearest cent: $3,295.8627

38. CON   A micrometer measure is listed as 0.7835 in. Round this measure to the nearest thousandth.

39. ELEC   To the nearest tenth, what is the current of a 2.836-amp (A) motor?

40. HOSP   If round steak costs $2.78 per pound, what is the cost per pound to the nearest dollar?

41. AUTO   The average response times (in seconds) for drivers braking when they first see a road hazard were measured as follows: driver A, 0.0275; driver B, 0.0264; driver C, 0.0234; driver D, 0.0284; and driver E, 0.0379. Round each response time to hundredths to identify the three drivers whose response times were most similar.

5 Write in columns and add. *See Example 7.*

42. 4,582 + 86,724 + 482 + 5,826

43. 6,017 + 893 + 15 + 82

44. 17 + 5,804 + 23,907 + 405

See Example 8.

45. 4.2 + 3.6 + 7.9

46. 12.8 + 13.52 + 7.86

47. 83.37 + 42 + 1.6 + 3

48. CON   A 0.103-in.-thick pipe has an inside diameter of 2.871 in. Find the outside diameter of the pipe. (*Hint:* The thickness of the pipe is on both sides of the inside diameter.)

49. ELEC   The total current in amps in a parallel circuit is found by adding the individual currents. If a circuit has individual currents of 3.98 A, 2.805 A, and 8.718 A, find the total current.

50. PFIN   A part-time hourly worker earned $25.97 on Monday, $7.48 on Tuesday, $5.88 on Wednesday, $65.45 on Thursday, and $76.47 on Friday. Find the total week's wages.

51. CON   A four-sided residential lot that measures 100.8 ft, 87.3 ft, 104.7 ft, and 98.6 ft is to be fenced. How many feet of fencing are required?

52. HLTH/N   A patient's normal body temperature registered at 98.2°F and his temperature rose 2.7°F. What was his increased body temperature?

53. PFIN   Your investment portfolio totals $25,915.53 at the beginning of the year and it increases by $2,418.48 over the one-year period. What is the value of your portfolio at the end of the year?

Estimate the sum for Exercises 54–55 by rounding to thousands and Exercises 56–58 by rounding to tens. Then, add to find the exact sum. See Example 9.

54.	**55.**	**56.**	**57.**	**58.**
24,003	52,843	0.935	34.07	24.381
5,874	17,497	12.4	15.962	1.1
319,467	13,052	152.07	5.81	17.92
+ 52,855	+ 821	+ 18	+ 0.523	+ 38

59. **CON** Palmer Associates provided the following prices for items needed to build a sidewalk: concrete, $2,583.45; wire, $43.25; frame material, $18.90; labor, $798. Estimate the cost by rounding each amount to the nearest ten. Find the exact total.

60. **PFIN** Antonio's expenses for one semester are food, $1,500; lodging, $1,285; books, $288; supplies, $130; transportation, $162. Estimate his expenses by rounding each amount to the nearest hundred. Calculate the exact amount.

61. **CON** A hardware store filled an order for nails: 25 lb, $2\frac{1}{2}$-in. common; 16 lb, 4-in. common; 12 lb, 2-in. siding; 24 lb, $2\frac{1}{2}$-in. floor brads; 48 lb, 2-in. roofing; and 34 lb, $2\frac{1}{2}$-in. finish. Find the total weight.

62. **AG/H** If four containers have a capacity of 12 gal, 27 gal, 55 gal, and 21 gal, can 100 gal of fuel be stored in these containers? (Find the total capacity of the containers first.)

63. **BUS** A printer has three printing jobs that require the following numbers of sheets of paper: 185, 83, and 211. Will one ream of paper (500 sheets) be enough to finish the three jobs?

64. **CON** How many feet of fencing are needed to enclose the area shown in Fig. 1–7?

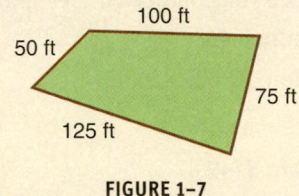

FIGURE 1–7

Subtract. *See Example 10.*

65. $15 - 8 - 3$

66. $38 - 18 - 6$

Subtract. *See Example 11.*

67. $3,672 - 2,652$

68. $946 - 831$

69. $53,867 - 831$

See Example 12.

70. **CON** If a mason orders 75 bags of cement for a job and uses only 53, how many bags are left?

71. **BUS** An inventory sheet shows that 468 outlet boxes were in stock on March 1. Sales during March were 127. How many outlet boxes were left at the end of the month?

Subtract. *See Example 13.*

72. Subtract 24.38 from 316.2.

73. Subtract 13.5 from 21.

74. Subtract 67.2 from 378.

75. Find the difference between 42 and 37.6.

See Example 14.

76. **INDTR** According to a blueprint, the length of an object is 12.09 in. If the tolerance is ±0.01 in., what are the limit dimensions of the object?

77. **CAD/ARC** Find the limit dimensions of an object whose blueprint dimension is 4.195 in. ± 0.006 in.

See Example 15.

78. Estimate by rounding to hundreds, then find the exact difference and check. $583.924 - 385$.

79. Estimate by rounding to tens, then find the exact difference and check. $152.83 - 127.34$.

See Example 16.

80. Use a calculator to find the difference between $47,956.32 and $29,458.77.

81. Use a calculator to find the difference between $185,274.48 and $92,863.95.

82. **HTH/N** Nurse Jennings recorded a temperature of 103.6°F for his patient. If the patient's normal body temperature was 98.6°F, how many degrees of fever did the patient have?

83. **INDTR** One box of rivets weighs 52.6 lb and another box weighs 37.5 lb. How much more does the first box of rivets weigh?

84. **CON** Two lengths of copper tubing measure 63.6 cm and 3.77 cm. What is the difference in their lengths?

85. **CON** A bricklayer laid 1,283 bricks on one day. A second bricklayer laid 1,097 bricks. How many more bricks did the first bricklayer lay?

See Example 17.

86. CON   Find the missing dimension in Fig. 1–8.

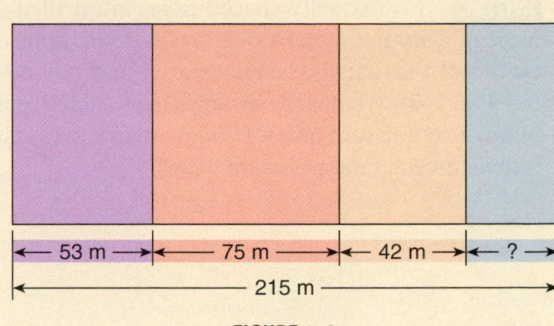

FIGURE 1–8

6 Multiply the following. *See Example 18.*

87. 3(7)(9)(2)   **88.** 5(3)(7)(12)   **89.** 8(4)(0)(5)   **90.** 21(15)(0)

See Example 19.

91.   709
     × 306

92.   503
     × 204

93. BUS   Helene Wright counted 8 unopened boxes of washers. Each box contained 512 washers. What total number of washers will be shown on the inventory sheet?

94. AUTO   Bill Weppner repaired 6 automobiles a week over a period of 14 weeks. How many automobiles did Bill repair during this time period?

95. HELPP   Each officer in the Public Safety office wrote, on average, 25 tickets a week. If there are 7 officers, how many tickets were written over a 4-week period?

96. BUS   Amanda Wimberly is planning to sell candy bars in her Smart Shop. She receives 12 boxes, and each box contains 24 candy bars. If Amanda sells the bars for $1 each, how much money will she get for all the bars?

See Example 20.

97.   37.7
     × 1.5

98.   9.27
     × 0.35

99.   0.215
     × 0.27

100.   0.271
     × 0.32

101. 73.806(2.305)   **102.** 1.9067 · 0.2013   **103.** 8.2037 · 0.602   **104.** 42(0.73)

105. CON   A plastic pipe has an inside diameter of 4.75 in. Find the outside diameter if the pipe wall is 0.25 in. thick. *See Example 21.*

106. How many feet of fencing are needed to enclose a rectangular backyard that is 187 ft by 125 ft? *See Example 22.*

107. BUS   A retailer purchases 15 cases of potato chips at $8.67 per case. If the chips are sold at $12.95 per case, how much profit does the retailer make? Estimate then find the exact amount of profit. *See Example 23.*

108. BUS   Premier Yard Service treats a lawn that is 215.8 feet long and 196.7 feet wide and will apply weed killer that costs $0.007 per square foot. Find the cost of applying the weed killer. *See Example 24.*

See Example 25.

109.   90,000
     × 7,000

110.   254,000
     × 9,000

111.   46,000
     × 5,200

Perform the long-division problems. *See Example 26.*

112. 5⟌215   **113.** 37⟌1,739

Divide. *See Example 27.*

114. 18 ÷ 3 ÷ 2   **115.** 36 ÷ 4 ÷ 3

See Example 28.

116. 5.12 ÷ 4

117. 182.52 ÷ 12

118. CON   In a building where 46 outlets are installed, 1,472 ft of cable are used. Estimate then find the approximate number of feet and find the exact number of feet of cable used per outlet.

119. INDTR   Twelve water tanks are constructed in a welding shop at a total contract price of $14,940. What is the price per tank?

120. **INDTEC**   Five equally spaced holes are drilled in a piece of $\frac{1}{4}$-in. flat metal stock that is 28 in. long. The centers of the first and last holes are 2 in. from the end (see Fig. 1–9). What is the distance between the centers of any two adjacent holes? (*Caution:* How many equal center-to-center distances are there?)

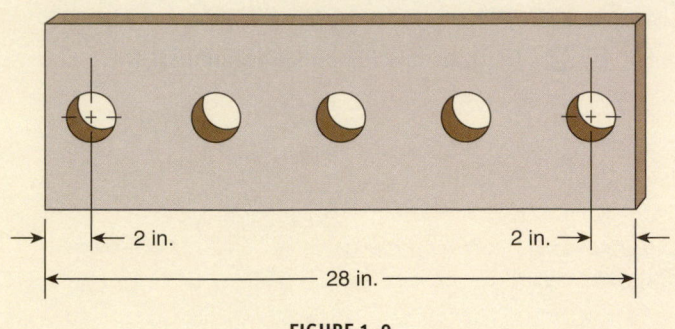

FIGURE 1–9

Divide. *See Example 29.*

121. $6.3\overline{)68.67}$

122. $0.23\overline{)0.0437}$

123. $8 \div 0.32$

124. **CON**   A 7.3-ft-long pipe weighs 43.8 lb. What is the weight of 1 ft of pipe?

125. **CON**   A room requires 770.5 ft² of wallpaper, including waste. How many whole single rolls are needed for the job if a roll covers 33.5 ft² of surface?

126. **CON**   The feed per revolution of a drill is 0.012 in. A hole 7.2 in. deep will require how many revolutions of the drill?

127. **INDTR**   A piece of channel iron 5.6 ft long is cut into eight pieces. Assuming that there is no waste, what is the length of each piece?

Divide and round the quotient to the place indicated. *See Example 30.*

128. Nearest hundredth: $25\overline{)3.897}$

129. Nearest whole number: $4.1\overline{)34.86}$

Estimate, find the exact quotient, and check. *See Example 31.*

130. $834 \div 12$

131. $589 \div 21$

132. **INDTR**   If 12 lathes cost $6,895, find the cost of each lathe to the nearest dollar.

133. **ELEC**   If 12 electrolytic capacitors cost $23.75, find the cost of one capacitor to the nearest cent.

Find the average, then round to the same place value used in the exercises. *See Example 32.*

134. Test scores: 86, 73, 95, 85

135. **AG/H**   Weight of crates of oranges: 515 lb, 468 lb, 435 lb, 396 lb

136. **BUS**   Monthly income: $873.46, $598.21, $293.85, $546.83, $695.83, $429.86, $955.34, $846.95, $1,025.73, $1,152.89, $957.64, $807.25

137. **AG/H**   Average rainfall: 1.25 in., 0.54 in., 0.78 in., 2.35 in., 4.15 in., 1.09 in.

138. **CON**   Amperes of current: 3.0 A, 2.5 A, 3.5 A, 4.0 A, 4.5 A

139. **AG/H**   Crates of strawberries: 347, 623, 491, 387, 519

1–2   Exponents, Roots, and Powers of 10

LEARNING OUTCOMES

1 Simplify expressions that contain exponents.

2 Square numbers and find the square roots of numbers.

3 Use powers of 10 to multiply and divide.

1 Simplify Expressions that Contain Exponents. The product of repeated factors can be written in shorter form using natural-number exponents. Exponents that are not natural numbers will have a different interpretation.

In the example, $4(4)(4) = 4^3$, the 4 is called the **base** and is the repeated factor. The 3 is the **exponent** and indicates the number of times the factor is used in multiplication. The expression 4^3 is written in **exponential notation** and has a value of 64. The number 64 is written in **standard notation.** Both the exponential notation and the standard notation are called the **power.** The expression 4^3 is read four **cubed** or

Base: the value used in repeated multiplication when raising to a natural-number power

Exponent: a natural-number exponent indicates how many times a repeated factor (base) is used

STOP AND CHECK

Identify the base and exponent, and write in standard notation.
1. 8^2 2. 2.1^3

Answers:
1. Base, 8; exponent, 2; 64
2. Base, 2.1; exponent, 3; 9.261

Exponential notation: a value written with a base and an exponent

Standard notation: numbers written with digits in the appropriate place-value positions; a power written as an ordinary number

Power: another term for the exponent or the result of raising a value to a power

Cubed: a number raised to the third power

Squared: the product that results from multiplying any number times itself; a number raised to the second power

Square or perfect square: the result of using a natural number as a factor 2 times

STOP AND CHECK

Find the square.
1. 9   2. 5.3

Answers:
1. 81   2. 28.09

Principal square root or square root: the number that was used as a factor twice to equal a perfect square

Radical sign [$\sqrt{\ }$]: an operational symbol indicating that a square root is to be taken of the number under the bar portion of the radical sign

Radicand: the number under the radical sign

Radical expression: an expression including the radical sign and radicand that indicates a root is to be taken such as square root

four to the third power or four raised to the third power. In expressions where 2 is an exponent, such as 4^2, the expression is usually read as four **squared;** however, it may also be read as four to the second power.

To change from natural-number exponential notation to standard notation:

1. Use the base as a factor as many times as indicated by the exponent.

2. Perform the multiplication.

EXAMPLE 1

Identify the base and exponent of the expressions, and write in standard notation.

(a) 5^3 **(b)** 1.5^2

(a) 5^3 **5 is the base; 3 is the exponent.**
$5^3 = 5(5)(5) = 125$ Standard notation

(b) 1.5^2 **1.5 is the base; 2 is the exponent.**
$1.5^2 = 1.5(1.5) = 2.25$ Standard notation   **See Exercises 1–14.**

Any number with an exponent of 1 is the number itself:

$a^1 = a,$ for any base a where a is a real number $8^1 = 8$ $2.3^1 = 2.3$

---

2 **Square Numbers and Find the Square Roots of Numbers.** The result of using a natural number as a factor 2 times is a **square** number or a **perfect square.** In the expression $7^2 = 49$, the 49 is a perfect square. This terminology evolved from a common formula for finding the area of a square. $A = s^2$, where A represents the area and s represents the length of one side. When we square 3, we write $3^2 = 3(3) = 9$. We say 9 is the square of 3.

EXAMPLE 2

Find the square.

(a) 2   **(b)** 7   **(c)** 3.2

(a) 2; $2^2 = 2 \times 2 = 4$

(b) 7; $7^2 = 7 \times 7 = 49$

(c) 3.2; $3.2^2 = 3.2 \times 3.2 = 10.24$ **See Exercises 17–20.**

The inverse operation of squaring is taking the square root of a number. The **principal square root** of a perfect square is the number that was used as a factor twice to equal that perfect square. The principal square root of 9 is 3 because 3^2 or $3(3) = 9$.

The **radical sign** $\sqrt{\ }$ indicates that the square root is to be taken of the number under the bar. This bar serves as a grouping symbol just like parentheses. The number under the bar is called the **radicand.** The entire expression is called a **radical expression.**

radical sign ⟶ ⌐ bar
$\sqrt{25} = 5$ ⟵ principal square root
radicand ⟶

> **To find the square root of a perfect square by estimation:**
>
> 1. Select a trial estimate of the square root.
>
> 2. Square the estimate.
>
> 3. If the square of the estimate is less than the original number, adjust the estimate to a larger number. If the square of the estimate is more than the original number, adjust the estimate to a smaller number.
>
> 4. Square the adjusted estimate from Step 3.
>
> 5. Continue the adjusting process until the square of the trial estimate is the original number.

EXAMPLE 3

Find **(a)** $\sqrt{256}$, and **(b)** $\sqrt{12}$ to the nearest tenth.

(a) Select 15 as the estimated square root: 15^2 or $15(15) = 225$. The number 225 is less than 256, so the square root of 256 is larger than 15. We adjust the estimate to 17: 17^2 or $17(17) = 289$. The number 289 is more than 256, so the square root of 256 must be smaller than 17. Now adjust the estimate to 16.

Because $16^2 = 256$, 16 is the square root of 256.

(b) 12 is between 9 and 16 so the square root of 12 is between 3 and 4. Try 3.2. $3.2^2 = 10.24$. $3.4^2 = 11.56$. $3.5^2 = 12.25$.

12 is closer to 12.25 than 11.56 so $\sqrt{12}$ to the nearest tenth is 3.5.

See Exercises 21–28.

The squares of the numbers 1 through 10 are 1, 4, 9, 16, 25, 36, 49, 64, 81, and 100. These are the only natural numbers from 1 through 100 that are perfect squares.

3 Use Powers of 10 to Multiply and Divide. A *nonzero digit* is any digit except zero. That is, 1, 2, 3, 4, 5, 6, 7, 8, and 9 are nonzero digits. *Powers of 10* are numbers whose only nonzero digit is 1. Thus, 10, 100, 1,000, and so on are powers of 10 because each value can be written in exponential form with a base of 10.

Compare the number of zeros in standard notation with the exponent in exponential notation.

One million	$1{,}000{,}000 = 10^6$	6 zeros
One hundred thousand	$100{,}000 = 10^5$	5 zeros
Ten thousand	$10{,}000 = 10^4$	4 zeros
One thousand	$1{,}000 = 10^3$	3 zeros
One hundred	$100 = 10^2$	2 zeros
Ten	$10 = 10^1$	1 zero
One	$1 = 10^0$	0 zeros

The exponents in powers of 10 indicate the number of zeros used in standard notation. Zero exponents will be discussed in Chapter 4.

EXAMPLE 4

Express as powers of 10.

(a) 10,000,000   **(b)** 100,000,000   **(c)** 100,000,000,000

STOP AND CHECK
Express as powers of 10.
1. 10,000,000,000
2. 1,000,000,000
Express in standard notation.
3. 10^{14}
4. 10^{16}

Answers:
1. 10^{10} 2. 10^9
3. 100,000,000,000,000
4. 10,000,000,000,000,000

(a) 10,000,000 $= 10^7$
(b) 100,000,000 $= 10^8$
(c) 100,000,000,000 $= 10^{11}$

Express in standard notation.
(d) 10^5 (e) 10^{13}

(d) $10^5 =$ **100,000**

(e) $10^{13} =$ **10,000,000,000,000**   See Exercises 29–36.

In some applications, powers of 10 are used to simplify multiplication and division problems. Compare the examples to find the pattern for multiplying a number by a power of 10.

$5(100) = 500$ $5(10^2) = 500$ Decimal point moved two places to the right and two zeros attached.

$27(1,000) = 27,000$ $27(10^3) = 27,000$ Decimal point moved three places to the right and three zeros attached.

To multiply a number by a power of 10 written in standard notation:

1. Move the decimal point to the *right* as many places as the number of zeros in 10, 100, 1,000, and so on.

2. Attach zeros to the right if necessary.

STOP AND CHECK
Multiply.
1. 315(1,000)
2. 14.23(10,000)

Answers:
1. 315,000 2. 142,300

EXAMPLE 5

Multiply (a) 237(100)   and   (b) 36.2 (1,000).

(a) $237(100) = 237.00(100) =$ **23,700**   Attach two zeros to the *right* of the 7 in 237, and insert the appropriate comma.

(a) $36.2(1,000) = 36.200(1,000) =$ **36,200**   Move the decimal point three places to the *right*. Two zeros need to be attached.   See Exercises 37–40.

Examine these divisions.

$32 \div 10 = 3.2$ Decimal point moved one place to the *left*.
$78.9 \div 100 = 0.789$ Decimal point moved two places to the *left*.
$52,900 \div 1,000 = 52.9$ Decimal point moved three places to the *left*.

To divide a number by a power of 10 written in standard notation:

1. Move the decimal point to the *left* as many places as the number of zeros in 10, 100, and 1,000, and so on.

2. Attach zeros to the left if necessary.

3. You may drop zeros to the right of the decimal point if they follow the last nonzero digit of the quotient.

Compare this rule with the rule for multiplying decimal numbers by 10, 100, 1,000, and so on. For multiplication, the decimal shifts to the right; for division, the decimal shifts to the left.

EXAMPLE 6

If 100 lb of floor cleaner costs $63, what is the cost of 1 lb?

Divide $63 by 100.

$$63 \div 100 = 063. \div 100 = \$0.63$$

Decimal is after 3 in 63. Move decimal two places to the left.

The cost of 1 lb is $0.63.

See Exercises 41–44.

Some scientific and graphing calculators have specific power keys, such as keys for squares and cubes. They also have a "general power" key that can be used for all powers. Even though the "general power" key can be used to square and cube numbers, the "square" and "cube" keys require fewer keystrokes.

> **TIP** **Labels on Calculator Keys Are Not Universal** To show calculator steps, we often use a box to identify a common label for a function. The exact label will vary with the specific calculator model. Some calculators also provide many functions above the keys or on menus instead of on specific keys. Even though we use a key notation in this text to show a calculator function, this function may actually appear above a key or on a menu. Check your calculator manual for exact location and labeling of functions.
>
> **Common Labels for Power Keys:** $\boxed{x^2}$ $\boxed{x^3}$ $\boxed{\wedge}$ $\boxed{x^y}$
>
> **Common Labels for Root Keys:** $\boxed{\sqrt{}}$ $\boxed{\sqrt{x}}$ $\boxed{\sqrt[x]{}}$ $\boxed{x^{1/y}}$

EXAMPLE 7

Use a calculator to find

(a) 35^2,   **(b)** 2.3^5,   **(c)** Evaluate $\sqrt{529}$,   **(d)** Evaluate $\sqrt{10.89}$, and

(e) Evaluate $\sqrt{5}$ to the nearest hundredth.

(a) 35 $\boxed{x^2}$ $\boxed{=}$ On some calculators, pressing the equal key may not be required.
1,225

(b) 2 $\boxed{\cdot}$ 3 $\boxed{\wedge}$ 5 $\boxed{=}$ $\boxed{\wedge}$ and $\boxed{x^y}$ are the most common labels for the general power key.
64.36343

(c) $\boxed{\sqrt{}}$ 529 $\boxed{=}$ **(d)** $\boxed{\sqrt{}}$ 10 $\boxed{\cdot}$ 89 $\boxed{=}$ **(e)** $\boxed{\sqrt{}}$ 5 $\boxed{=}$ $\Rightarrow$ 2.236067977
23   **3.3**   **2.24**   Rounded

See Exercises 47–58.

> **Did You Know?** Some calculators require that the radicand be entered *before* pressing the square root key. Other calculators will open a parenthesis when the square root key is pressed. Parentheses are automatically closed when the equal or enter key is pressed.
>
> **Get to know your calculator! Test your calculator with an example you can do mentally.**

1–2 EXERCISES MyLab Math For additional practice go to your study plan in MyLab Math.

1 Identify the base and exponent of the expressions. *See Example 1.*

1. 4^3 **2.** 9^4 **3.** 2.7^9 **4.** 15^2

Write the exponential expressions in standard notation. *See Example 1.*

5. 10^3 **6.** 2^4 **7.** 3.4^2 **8.** 15^1 **9.** 8^1

10. 9^2 **11.** 8^2 **12.** 18^2 **13.** 1.4^2 **14.** 13^2

15. Explain what an exponent of 5 means. **16.** Explain what an exponent of 3 means.

2 Square the numbers. *See Example 2.*

17. 8 **18.** 12 **19.** 4.3 **20.** 8.9

Perform the operations. Use estimation where appropriate. *See Example 3.*

21. $\sqrt{25}$ **22.** $\sqrt{49}$ **23.** $\sqrt{81}$ **24.** $\sqrt{36}$

25. $\sqrt{196}$ **26.** $\sqrt{441}$ **27.** $\sqrt{529}$ **28.** $\sqrt{1,024}$

Express as powers of 10. *See Example 4.*

29. 1,000,000 **30.** 1,000,000,000,000 **31.** 1,000,000,000 **32.** 10,000,000,000

Express in standard notation. *See Example 4.*

33. 10^{14} **34.** 10^9 **35.** 10^{12} **36.** 10^4

3 Multiply the whole numbers using powers of 10. *See Example 5.*

37. $10(10^2)$ **38.** $12(10^5)$ **39.** $2(10^4)$ **40.** $102(100)$

Divide the whole numbers using powers of 10. *See Example 6.*

41. $250 \div 10$ **42.** $210 \div 10$ **43.** $\dfrac{300}{10^2}$ **44.** $2,500 \div 10$

45. What does the exponent of a power of 10 mean when you are multiplying? **46.** What does the exponent of a power of 10 mean when you are dividing?

Use a calculator to evaluate. *See Example 7.*

47. 15^2 **48.** 7^3 **49.** 5^7 **50.** 12^4

51. $\sqrt{324}$ **52.** $\sqrt{784}$ **53.** $\sqrt{1,089}$ **54.** $\sqrt{196}$

Round to the nearest hundredth. *See Example 7.*

55. $\sqrt{39}$ **56.** $\sqrt{72}$ **57.** $\sqrt{210}$ **58.** $\sqrt{1,095}$

1–3 Order of Operations and Problem Solving

LEARNING OUTCOMES

1 Apply the order of operations to a series of operations.

2 Evaluate a formula.

3 Solve applied problems using problem-solving strategies.

1 **Apply the Order of Operations to a Series of Operations.** Whenever several mathematical operations are performed, the proper **order of operations** must be followed.

To apply the order of operations:

1. **Parentheses (grouping symbols):** Perform operations within parentheses (or other grouping symbols), beginning with the innermost set of parentheses; or apply the distributive property.

LC LEARNING CATALYTICS
Perform the operations.
1. $3 + 4(2)$
2. $\dfrac{5(8)}{2}$

Order of operations: the order in which a series of mathematical operations are performed; the correct order is grouping, exponents and roots, multiplication and division, addition and subtraction

Grouping symbols: parentheses, brackets, braces, and a bar

Brackets: a pair of symbols, [ ], that are used for grouping

Braces: a pair of symbols, { }, that are used for grouping

Bar: a horizontal line that is used as a grouping symbol

2. **Exponents and roots:** Evaluate exponential operations and find square roots in order from left to right.

3. **Multiply and divide** in order from left to right.

4. **Add and subtract** in order from left to right.

To summarize, use the following key words:

<u>P</u>arentheses (grouping), <u>E</u>xponents (and roots), <u>M</u>ultiplication and <u>D</u>ivision, <u>A</u>ddition and <u>S</u>ubtraction

Other grouping symbols are **brackets** [ ], **braces** { }, and a **bar.** The bar can combine with other symbols like the radical sign and be used as a grouping symbol, $\sqrt{4 + 5} = \sqrt{9} = 3$.

> **TIP** **A Memory Aid for the Order of Operations** To remember the order of operations, use the sentence, "<u>P</u>lease <u>E</u>xcuse <u>M</u>y <u>D</u>ear <u>A</u>unt <u>S</u>ally."
>
> ▶ <u>P</u>arentheses (grouping)   ▶ <u>E</u>xponents (roots)
>
> ▶ <u>M</u>ultiply/<u>D</u>ivide   ▶ <u>A</u>dd/<u>S</u>ubtract
>
> Be sure that you do not separate ***My Dear*** into two steps. Multiplications and divisions are done as one step as they appear from left to right. Also, ***Aunt Sally*** does not separate into two steps. Additions and subtractions are done as one step as they appear from left to right.

Parentheses may show both a grouping and multiplication. When a number is multiplied by a sum or a difference, the distributive property can also be applied.

STOP AND CHECK
Evaluate.
1. $15 - 3(3)$

2. $3(6) + \dfrac{4 + 11}{3}$

Answers:
1. 6   2. 23

EXAMPLE 1

(a) Evaluate $3(2 + 3)$ using the distributive property and using the order of operations.

Using the distributive property:

$$3(2 + 3) = 3(2) + 3(3) \qquad \text{Distribute.}$$
$$= 6 + 9 \qquad\qquad\quad \text{Add.}$$
$$= \mathbf{15}$$

Using the order of operations:

$$3(2 + 3) = 3(5) \qquad \text{Do operation in parentheses first.}$$
$$= \mathbf{15} \qquad\quad\ \text{Multiply.}$$

(b) Evaluate $4^2 - 5(2) \div (4 + 6)$ by performing the operations in the correct order.

$4^2 - 5(2) \div \boxed{(4 + 6)}$	Do operation within parentheses first: $4 + 6 = 10$.	**P**
$4^2 - 5(2) \div \boxed{10}$	Evaluate exponential notation: $4^2 = 16$.	**E**
$16 - \boxed{5(2)} \div 10$	Multiply: $5(2) = 10$.	**M** D
$16 - \boxed{10 \div 10}$	Divide: $10 \div 10 = 1$.	M **D**
$16 - 1$	Subtract last.	A **S**
$16 - 1 = \mathbf{15}$		See Exercises 1–27.

> **TIP** **Parentheses Indicate Multiplication, the Distributive Property, or a Grouping** Parentheses can indicate multiplication or an operation that should be done first. If the parentheses contain an operation, they indicate a grouping. Otherwise, they indicate multiplication. The expression $5(2)$ indicates multiplication, while $(4 + 6)$ indicates a grouping. The expression $3(2 + 3)$ is an example of the distributive property.
>
> Many calculators have parentheses keys $(\boxed{(}\ \boxed{)})$. These keys are used for all types of grouping symbols.

> **Did You Know?** A division bar serves as a grouping symbol in the same way parentheses do.
>
> Part b in Example 1 could be written as $4^2 - \dfrac{5(2)}{4 + 6}$. The steps for evaluating will be similar.
>
> $$4^2 - \frac{5(2)}{4 + 6} = 16 - \frac{10}{10} = 16 - 1 = 15$$

STOP AND CHECK

1. $2 \cdot \sqrt{9} + 4 - [11 - (2 \cdot 5)]$
2. $2.8^2 - 3\sqrt{7 - 2(3)}$

Answers:
1. 9   2. 4.84

Did You Know? In Example 2b, $\sqrt{21 - 5}(2)$ can also be written as $2\sqrt{21 - 5}$ since multiplication is commutative. Writing the factor of 2 in front of the radical sign would make you less likely to mistakenly think that the factor of 2 was under the radical grouping.

Formula: a procedure written as a symbolic statement that indicates a relationship among numbers

EXAMPLE 2

Evaluate **(a)** $5 \cdot \sqrt{16} - 5 + [15 - (3 \cdot 2)]$ and **(b)** $3.2^2 + \sqrt{21 - 5}(2)$.

(a)

$5 \cdot \sqrt{16} - 5 + [15 - (3 \cdot 2)]$	Work innermost grouping: $3 \cdot 2$.	P
$5 \cdot \sqrt{16} - 5 + [15 - 6]$	Work remaining grouping: $15 - 6$.	P
$5 \cdot \sqrt{16} - 5 + 9$	Find square root: $\sqrt{16} = 4$.	E
$5 \cdot 4 - 5 + 9$	Multiply: $5 \cdot 4$.	M D
$20 - 5 + 9$	Add and subtract from left to right.	A S
$15 + 9 = \mathbf{24}$		

Continuous calculator sequence (keystrokes will vary depending on the calculator model used):

$5\ \boxed{\sqrt{}}\ 16\ \boxed{)}\ \boxed{-}\ 5\ \boxed{+}\ \boxed{(}\ \boxed{(}\ 15\ \boxed{-}\ \boxed{(}\ 3\ \boxed{\times}\ 2\ \boxed{)}\ \boxed{)}\ \boxed{)}\ \boxed{=}$

(b)

$3.2^2 + \sqrt{21 - 5}(2)$	Do operation within grouping first; the bar of the radical symbol is a grouping symbol: $21 - 5 = 16$.
$3.2^2 + \sqrt{16}(2)$	Evaluate exponent and square root from left to right: $3.2^2 = 10.24$; $\sqrt{16} = 4$.
$10.24 + 4(2)$	Then multiply: $4(2) = 8$.
$10.24 + 8$	Add last.
$10.24 + 8 = \mathbf{18.24}$	

Continuous calculator sequence (keystrokes will vary depending on the calculator model used):

$3.2\ \boxed{x^2}\ \boxed{+}\ \boxed{\sqrt{}}\ 21\ \boxed{-}\ 5\ \boxed{)}\ \boxed{\times}\ 2\ \boxed{=}$

See Exercises 28–39.

2 **Evaluate a Formula.** **Formulas** are procedures that have been used so frequently to solve certain types of problems that they have become the accepted means of solving these problems. Most formulas are expressed with one or more letter terms rather than words, and these procedures are written as symbolic equations, such as $P = 2(l + w)$, the formula for the perimeter of a rectangle. Electronic spreadsheets and calculator and computer programs are developed through formulas.

Evaluate: to evaluate a formula is to substitute known numerical values for some variables and perform the indicated operations to find the value of the remaining variable in question

The most common use of formulas is for finding missing values. If we know values for all but one letter of a formula, we can find the missing value. To **evaluate** a formula is to substitute known values for the appropriate letters of the formula and perform the indicated operations to find the missing value.

To evaluate a formula:
1. Write the formula.
2. Rewrite the formula substituting known values for letters of the formula.
3. Perform the indicated operations, applying the order of operations.
4. Interpret the solution within the context of the formula.

Polygon: a plane or flat, closed figure described by straight-line segments and angles

Some of the first formulas that we use are the formulas for basic geometric shapes called polygons. A **polygon** is a plane or flat, closed figure described by straight-line segments and angles. Polygons have different numbers of sides and different properties. Some common polygons are the parallelogram, rectangle, and square (Fig. 1–10).

Parallelogram   Rectangle   Square

FIGURE 1–10

Base: the horizontal side of any polygon or a side that would be horizontal if the polygon's orientation were modified

Adjacent side: the side of any polygon that has an end point in common with the base

Parallelogram: a four-sided polygon whose opposite sides are parallel

Parallel: lines that never meet or intersect. They are always the same distance apart

Rectangle: a parallelogram with angles that are all right angles (square corners)

Right angle: a right angle (90°) represents one-fourth of a circle or one-fourth of a complete rotation, square corners

Square: a parallelogram with all sides of equal length and with all right angles; a rectangle with all sides of equal length

Perimeter: the total length of the sides of a plane figure

The **base** of any polygon is the horizontal side or a side that would be horizontal if the polygon's orientation were modified.

An **adjacent side** of any polygon is the side that has an end point in common with the base.

A **parallelogram** is a four-sided polygon with opposite sides that are **parallel** (never meet or intersect).

A **rectangle** is a parallelogram with angles that are all **right angles** (square corners).

A **square** is a parallelogram with all sides of equal length and with all right angles. A square can also be described as a rectangle with all sides of equal length.

The **perimeter** is the total length of the sides of a plane figure. As we saw earlier, some common applications for perimeter are finding the amount of trim molding for a room, determining the amount of fencing for a yard, finding the amount of edging for a flower bed, and so on.

A general procedure for finding the perimeter of any shape is to add the lengths of the sides. However, shortcuts based on the properties of the shape are given as formulas.

Perimeter			
Parallelogram	$P = 2b + 2s$ or $P = 2(b + s)$	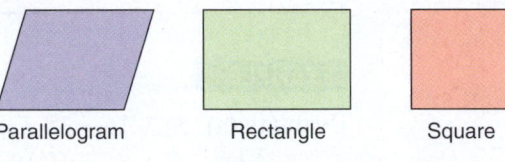	b is the base s is an adjacent side
Rectangle	$P = 2l + 2w$ or $P = 2(l + w)$		l is length w is width
Square	$P = 4s$		s is length of a side

The perimeter of a parallelogram, like the perimeter of a rectangle, is the sum of its four sides. The sides of the parallelogram are called *base* and *adjacent side* (instead of length and width). Notice the locations of the base and adjacent side in Fig. 1–11.

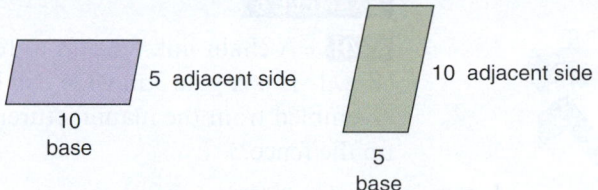

5 adjacent side 10 adjacent side

10 base 5 base

FIGURE 1–11

EXAMPLE 3

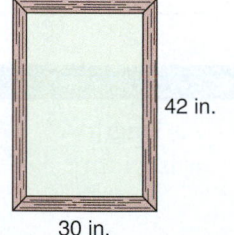

8 in.

16 in.

FIGURE 1–12

Find the perimeter of a parallelogram with a base of 16 in. and an adjacent side of 8 in. (Fig. 1–12).

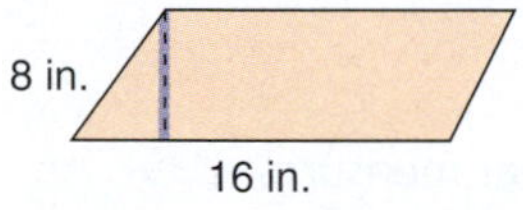

Visualize the parallelogram.

$P_{\text{parallelogram}} = 2(b + s)$	Select the perimeter of parallelogram formula.
$P_{\text{parallelogram}} = 2(16 \text{ in.} + 8 \text{ in.})$	Substitute: $b = 16, s = 8$.
$P_{\text{parallelogram}} = 2(24 \text{ in.})$	Evaluate.
$\mathbf{P_{\text{parallelogram}} = 48 \text{ in.}}$	**See Exercises 40–41.**

EXAMPLE 4

BUS A shop that makes custom picture frames has an order for a frame that has outside measurements of 42 in. by 30 in. (Fig. 1–13). How many inches of picture frame molding are needed for the job?

$P_{\text{rectangle}} = 2(l + w)$	The figure is a rectangle. Select the appropriate formula.
$P_{\text{rectangle}} = 2(42 \text{ in.} + 30 \text{ in.})$	Substitute and evaluate the formula.
$P_{\text{rectangle}} = 2(72 \text{ in.})$	
$P_{\text{rectangle}} = 144 \text{ in.}$	

When analyzing the dimensions, inches are added to inches, so the result is written in inches. Then the inches are multiplied by a number and the final result is inches.

The frame requires 144 in. of molding. **See Exercises 42–45.**

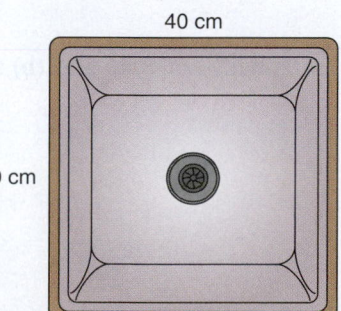

42 in.

30 in.

FIGURE 1–13

TIP **Using Subscripts** Subscripted words, letters, or numbers are a handy way to provide additional information. Because we will examine the formulas for the perimeter of several different shapes, we sometimes use subscripts to distinguish among them. For instance, the perimeter of a square, rectangle, or parallelogram can be indicated as P_{square}, $P_{\text{rectangle}}$, or $P_{\text{parallelogram}}$, respectively.

40 cm

40 cm

EXAMPLE 5

CON How much aluminum edge molding is needed to surround a stainless steel kitchen sink that measures 40 cm on each side (Fig. 1–14)?

$P_{\text{square}} = 4s$	The figure is a square. Select the appropriate formula.
$P_{\text{square}} = 4(40 \text{ cm})$	Substitute and evaluate the formula.
$P_{\text{square}} = 160 \text{ cm}$	

The sink requires 160 cm of molding. **See Exercises 46–47.**

FIGURE 1–14

In some applications, we need to decrease the total perimeter to account for doorways and other openings in the perimeter.

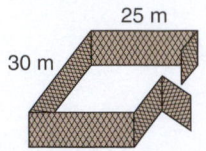

FIGURE 1–15

Area: the amount of surface of a plane figure

EXAMPLE 6

AG/H   A chain-link fence is to be installed around a yard measuring 25 m by 30 m (Fig. 1–15). A gate 3 m wide will be installed over the driveway. The gate comes pre-assembled from the manufacturer. How much fencing does the installer need to put up the fence?

$P_{\text{rectangle}} = 2(l + w) - 3$	Select the perimeter of a rectangle formula. Subtract the length of the gate.
$P_{\text{rectangle}} = 2(30 \text{ m} + 25 \text{ m}) - 3 \text{ m}$	Substitute and perform operations.
$P_{\text{rectangle}} = 2(55 \text{ m}) - 3 \text{ m}$	Multiply.
$P_{\text{rectangle}} = 110 \text{ m} - 3 \text{ m}$	Subtract.
$P_{\text{rectangle}} = 107 \text{ m}$	Perimeter is a linear measure.

The job requires 107 m of fencing.   **See Exercises 55–56.**

area $= 15 \text{ m}^2$

3 m wide

5 m long

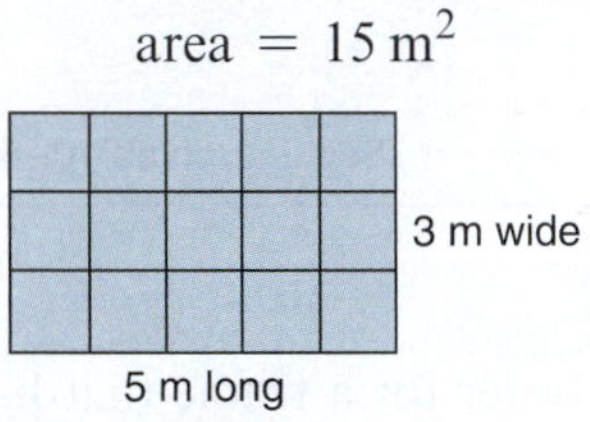

FIGURE 1–16

The **area** of a polygon is the amount of surface of a plane figure. Area is expressed in square units. For example, if a rectangle is 3 m wide and 5 m long, there are 15 m² in the area (Fig. 1–16).

> **TIP**   **Writing Square Units**   The exponent 2 following a unit of measure indicates *square measure* or area. Thus, 15 m² = 15 square meters, 23 ft² = 23 square feet, 120 cm² = 120 square centimeters, and so on. This is a shortcut way to express a measure that has *already* been "squared."

Did You Know? There are different conventions for abbreviations for measures. The most common for measures is to not follow the abbreviation with a period unless the abbreviation forms a word in the English language. For example, the abbreviation for inch is in. and the abbreviation for feet is ft with no period. An exception for the abbreviation of inch is if the measure is square inch (in²) or cubic inch (in³).

Area

Rectangle	$A = lw$	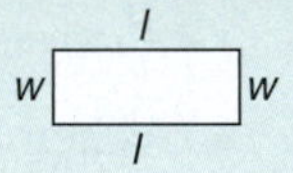	l is length w is width
Square	$A = s^2$	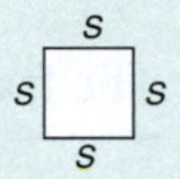	s is length of a side
Parallelogram	$A = bh$	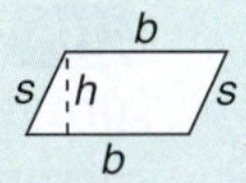	b is base h is height

EXAMPLE 7

Find the area of **(a)** a rectangle with a length of 14 in. and a width of 8 in. and **(b)** a square with each side measuring 6 ft.

(a) $A = lw$ when $l = 14$ in. and $w = 8$ in.
     $A = 14 \text{ in.}(8 \text{ in.})$
     $A = 112 \text{ in}^2$ in.(in.) = in²

(b) $A = s^2$ when $s = 6$ ft
     $A = (6 \text{ ft})^2$
     $A = 36 \text{ ft}^2$ ft(ft) = ft²   **See Exercises 58–61.**

Since a rectangle is a parallelogram, the areas of both types of polygons are related.

The **height** of a parallelogram is the perpendicular distance between two parallel sides. Height is also called **altitude.**

If triangle ABC ($\triangle ABC$) in Fig. 1–17 were transposed to the right side of the parallelogram ($\triangle A'B'C'$), we would have a rectangle. Triangles are presented in detail in Chapters 17 and 18.

The base of the original parallelogram is the same as the length of the newly formed rectangle. Similarly, the height of the parallelogram is the same as the width of the rectangle. Thus, the area of the rectangle and parallelogram is 10 in.(4 in.) or 40 in^2.

Height: the perpendicular distance from the base to the highest point of the side of the polygon above the base; also called altitude

Altitude: also called height

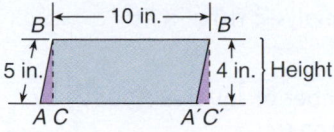

FIGURE 1–17

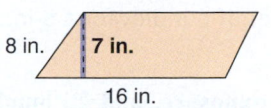

FIGURE 1–18

EXAMPLE 8

Find the area of a parallelogram with a base of 16 in., an adjacent side of 8 in., and a height of 7 in. (Fig. 1–18).

Visualize the parallelogram.

$A_{\text{parallelogram}} = bh$	Substitute.
$A_{\text{parallelogram}} = 16 \text{ in. } (7 \text{ in.})$	Multiply.
$A_{\text{parallelogram}} = \mathbf{112 \text{ in}^2}$	

See Exercises 62–63.

When do you need the area of a polygon? Some common applications for area are finding the amount of carpeting needed to cover a floor, the amount of paint needed to paint a surface, the amount of fertilizer needed to treat a lawn, and so on.

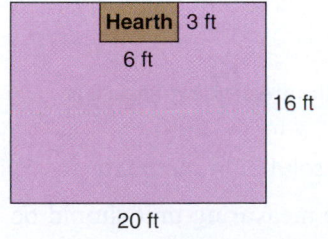

FIGURE 1–19

EXAMPLE 9

CON A carpet installer is carpeting a room measuring 16 ft by 20 ft. Projecting out from one wall is a fireplace whose hearth measures 3 ft by 6 ft (Fig. 1–19). How many square feet of carpet does the installer require for the job? How much is wasted?

First, compute the area of the room without considering the area of the hearth. This is the amount of carpet needed. Second, compute the area of the fireplace hearth. This is the amount wasted.

Let A_{room} = the area of the room and A_{hearth} = the area of the hearth.

$A_{\text{room}} = lw$	Select the appropriate formula and substitute.
$A_{\text{room}} = 20 \text{ ft}(16 \text{ ft})$	Multiply.
$A_{\text{room}} = 320 \text{ ft}^2$	Area is a square measure (ft $\times$ ft = ft^2).
$A_{\text{hearth}} = lw$	Select the appropriate formula and substitute.
$A_{\text{hearth}} = 6 \text{ ft}(3 \text{ ft})$	Multiply.
$A_{\text{hearth}} = 18 \text{ ft}^2$	Area is a square measure (ft $\times$ ft = ft^2).

This job requires 320 ft^2 of carpet. Of this, 18 ft^2 is waste from the hearth.

See Exercise 64.

The preceding example brings up some interesting questions. Is carpet purchased in square feet? If carpet can be purchased only in 12-ft or 15-ft widths, will there be additional waste? Is it more practical to purchase extra carpet that will be wasted or to spend extra labor costs to install the carpet with several seams? You may want to investigate common practices in the industry.

Christina Richards/Shutterstock

EXAMPLE 10

CON   Asphalt roofing shingles are sold in bundles. The number of bundles needed to cover a square (100 ft^2) depends on the overlap when the shingles are installed. If the overlap allows 4 in. of each shingle to be exposed, then four bundles are needed per square. With a 5-in. exposure, 3.2 bundles are needed per square. Figure the number of bundles for a 4-in. exposure and for a 5-in. exposure for a roof measuring 30 ft $\times$ 20 ft.

$$A_{\text{roof}} = 30 \text{ ft}(20 \text{ ft})$$ Dimension analysis: $\text{ft}(\text{ft}) = \text{ft}^2$
$$A_{\text{roof}} = 600 \text{ ft}^2$$
$$600 \text{ ft}^2 \div 100 \text{ ft}^2 = 6 \text{ squares}$$ Find the number of squares (100 ft^2) by dividing by 100 ft^2.

$$6(4) = 24 \text{ bundles}$$ Find the number of bundles for a 4-in. exposure.

$$6(3.2) = 19.2 \text{ or } 20 \text{ bundles}$$ Find the number of bundles for a 5-in. exposure.

Therefore, 24 bundles of shingles are required for a 4-in. exposure, and 20 bundles (from 19.2 bundles) are required for a 5-in. exposure.   **See Exercise 65.**

TIP   **Dimension Analysis**   When we evaluate formulas, the measuring units are sometimes omitted from the written steps. However, it is very important to use the correct unit in your calculations so that the unit in the solution is correct. Compare examples involving perimeter and area.

Example 5 (molding for sink)

$$P_{\text{square}} = 4s$$
$$P_{\text{square}} = 4(40 \text{ cm})$$ The measuring unit is centimeters; the measure is multiplied by a number.

$$P_{\text{square}} = 160 \text{ cm}$$ The measuring unit in the solution is centimeters.

Perimeter is always a linear measure, which means the measuring unit should be to the first power.

Example 10 (roof)

$$A_{\text{roof}} = lw$$
$$A_{\text{roof}} = (30 \text{ ft})(20 \text{ ft})$$ The measuring unit is feet and a measure is multiplied by a measure.

$$A_{\text{roof}} = 600 \text{ ft}^2$$ The measuring unit in the solution is square feet.

Area is always a square measure, which means that measures of area are to the second power or have an exponent of 2.

Volume: the amount of space a three-dimensional geometric figure occupies, measured in terms of three dimensions (length, width, and height). Measures of volume will always be cubic measures

The *volume* of an object, such as a container, is used to estimate how many containers can be loaded into a given-size storage area or shipped in a container of certain dimensions.

The **volume** of a three-dimensional geometric figure is the amount of space it occupies, measured in terms of three dimensions (length, width, and height).

If we have a rectangular box measuring 1 ft long, 1 ft wide, and 1 ft high (Fig. 1–20), it will be a cube representing 1 cubic foot (ft^3). Its volume is calculated

Cubic measure: a measure with an exponent of 3

Cubed: another term for cubic measure

Prism: a box

Right rectangular prism: a prism whose faces and bases are all rectangles or squares and all corners are square corners. Commonly called a box

Cube: a right rectangular prism with the length, wide, and height equal

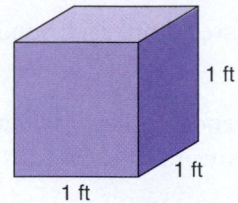

FIGURE 1–20

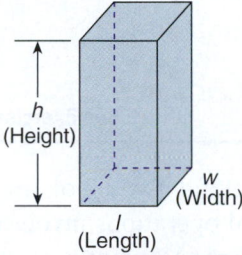

Right rectangular prism

FIGURE 1–21

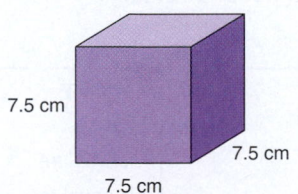

FIGURE 1–22

by multiplying length × width × height, or 1 ft × 1 ft × 1 ft = 1 ft³. We indicate a **cubic measure** with an exponent 3 after the unit of measure, meaning that the measure is **cubed.**

From this concept of 1 ft³ or 1 cubic foot comes the formula for the volume of a rectangular box as length × width × height or $V = lwh$. The mathematical term for a box is a **prism.** Specifically, if the sides and bases all make square corners and the faces and bases are rectangles or squares, we refer to the prism as a **right rectangular prism** (Fig. 1–21). If the length, width, and height of a right rectangular prism are equal, the prism is a **cube.** Prisms will be discussed in more detail in Chapter 18.

Formula for the volume of a right rectangular prism or a cube:

$V_{\text{prism}} = lwh$, where l is the length, w is the width, and h is the height.

$V_{\text{cube}} = s^3$, where s is the length of one side.

EXAMPLE 11

Find the volume of the cube in Fig. 1–22 if each side is 7.5 cm. Round to the nearest tenth.

$V_{\text{cube}} = s^3$	Select the appropriate formula and substitute 7.5 cm for s.
$V_{\text{cube}} = (7.5 \text{ cm})^3$	Evaluate.
$V_{\text{cube}} = 421.875 \text{ cm}^3$	Round.

The volume of the cube is 421.9 cm³.   **See Exercises 66–69.**

---

3 **Solve Applied Problems Using Problem-Solving Strategies.** Problem solving is an important skill in the workplace and in everyday life. In developing good problem-solving skills, it is helpful to use a systematic problem-solving plan. A plan gives you a framework for approaching problems.

You will encounter many different plans for solving problems. They help you organize the problem details so that you can find an appropriate procedure for solving the problem. Our plan, like others, is very structured. As you develop confidence and skill in solving problems, you will probably use less structured and more intuitive approaches.

Six-Step Problem-Solving Plan

1. **Unknown Facts.** What facts are missing from the problem? What are you trying to find?
2. **Known Facts.** What relevant facts are known or given? What facts must you bring to the problem from your own background?
3. **Relationships.** How are the known facts and the unknown facts related? What formulas or definitions can you use to establish a model for solving the problem?
4. **Estimation.** What are some of the characteristics of a reasonable solution? For instance, should the answer be more than a certain amount or less than a certain amount?
5. **Calculations.** Perform the operations identified in the relationships.
6. **Interpretation.** What do the results of the calculations represent within the context of the problem? Is the answer reasonable? Have all unknown facts been found? Do the unknown facts make sense in the context of the problem?

EXAMPLE 12

BUS   The 7th Inning buys baseball cards from eight different vendors. In November, the company purchased 8,832 boxes of cards. If an equal number of boxes was purchased from each vendor, how many boxes of cards were supplied by each vendor?

Unknown fact

Number of boxes of cards supplied by each vendor

Known facts

8,832 = total number of boxes purchased
8 = total number of vendors
An equal number of boxes was purchased from each vendor.

Relationships

total boxes purchased ÷ number of vendors = boxes purchased from each vendor

Estimation

If 8,000 boxes were purchased in equal amounts from eight vendors, then 1,000 were purchased from each vendor. Since more than 8,000 boxes were purchased, then more than 1,000 were purchased from each vendor.

Calculations

$$\frac{1,104}{8\,\overline{)\,8,832}}$$

Interpretation

Each vendor supplied 1,104 boxes of cards.   **See Exercise 70.**

In identifying the relationships of a problem, it is often helpful if you look for key words or phrases that give you clues about the mathematical operations involved in the relationships (Table 1–1). Key words give clues as to whether one quantity is added to, subtracted from, or multiplied or divided by another quantity. For example, if a problem tells you that Carol's salary in 2018 exceeds her 2017 salary by $2,500, you know that you should add $2,500 to her 2017 salary to find her 2018 salary.

Table 1–1   Key Words and What They Generally Imply in Word Problems

Addition	Subtraction	Multiplication	Division	Equality
The sum of	Less than	Times	Divide(s)	Equals
Plus/total	Decreased by	Multiplied by	Divided by	Is/was/are
Increased by	Subtracted from	Of	Divided into	Is equal to
More/more than	Difference between	The product of	Half of (divided by 2)	The result is
Added to	Diminished by	Twice (2 times)	Third of (divided by 3)	What is left
Exceeds	Take away	Double (2 times)	Per	What remains
Expands	Reduced by	Triple (3 times)	How big is each part?	The same as
Greater than	Less/minus	Half of ($\frac{1}{2}$ times)	How many parts can be made from	Gives/giving
Gain/profit	Loss	Third of ($\frac{1}{3}$ times)		Makes
Longer	Lower			Leaves
Older	Shrinks			
Heavier	Smaller than			
Wider	Younger			
Taller	Slower			
Larger than	Shorter			

More than one relationship may be needed to find the unknown facts. When this is the case, you usually need to read the problem several times to find all the relationships and plan your solution strategy.

EXAMPLE 13

BUS   Carlee Anne McAnally needs to ship 78 crystal vases. With standard packing to prevent damage, 5 vases fit in each available box. How many boxes are required to pack the vases?

Unknown fact

Number of boxes required to pack the vases

Known facts

total vases to be shipped $= 78$
number of vases per box $= 5$

Relationships

total boxes needed $=$ total number of vases $\div$ number per box
total boxes needed $= 78 \div 5$

Estimation

$70 \div 5 = 14$ Round down the dividend.
$80 \div 5 = 16$ Round up the dividend.

Since 78 is between 70 and 80, the number of boxes needed is between 14 and 16.

Calculation

$78 \div 5 = 15\,\text{R}3$

Interpretation

16 boxes are needed; 15 boxes will contain 5 vases each, and 1 box will contain 3 vases. The box with 3 vases will need extra packing material.   **See Exercise 71.**

TIP   **Using Guessing and Checking to Solve Problems**   An effective strategy for solving problems involves guessing. Make a guess that you think may be reasonable, and check to see if the answer is correct. If your guess is not correct, decide if it is too high or too low. Make another guess based on what you learned from your first guess. Continue until you find the correct answer.

Let's try guessing in the previous example. We found that we could pack 70 vases in 14 boxes and 80 vases in 16 boxes. Since we need to pack 78 vases, how many vases can we pack with 15 boxes? $15(5) = 75$. Still not enough. Therefore, we will need 16 boxes, but the last box will not be full.

You can probably think of other ways to solve this problem. Some plans will be more efficient than others, but you develop your problem-solving skills by pursuing a variety of strategies.

1-3 EXERCISES

MyLab Math   For additional practice go to your study plan in MyLab Math.

1 Use the order of operations to evaluate each problem. *See Example 1.*

1. $5^2 + 4 - 3$

2. $4^2 + 6 - 4$

3. $4(3) - 9 \div 3$

4. $5 \cdot 2.9 - 4 \div 2$

5. $25 \div 5 \cdot 4.8$

6. $64 \div 4(2)$

7. $48 \div 8 \cdot 3$

8. $15 - 2 \cdot 3$

9. $4^2 - (4)(3) + 6$

10. $17 - 4 \cdot 2$

11. $6 \times \sqrt{36} - 2 \times 3$

12. $3 \cdot \sqrt{81} - (3)(4)$

13. $4^2 \cdot 3^2 + (4 + 2)(2)$

14. $2^2 \cdot 5^2 + (2 + 1)(3)$

15. $54 - 3^3 - \dfrac{8}{2}$

16. $156 - 2^3 - \dfrac{9}{3}$

17. $3 - 2 + 3 \cdot 3 - \sqrt{9}$

18. $2 - 1 + (4)(4) - \sqrt{4}$

19. $2^4 \times (7 - 2) \cdot 2$

20. $3^4 \times (9 - 3) \cdot 3$

21. $124 - 8 \cdot 7 + 12$

22. $5 \times 12 \div 6$

23. $72 \div 9 \cdot 3$

24. $32 - 2.05 \cdot 4^2 \div 2$

25. $5.2^2 - 3 \cdot 2^2 \div 6$

26. $4 + 15 \div 3 - \dfrac{0}{7}$

27. $3 + 8 \div 4 - \dfrac{0}{5}$

Evaluate. *See Example 2.*

28. $4^3 + 14 - 8$

29. $2^3 + 12 - 7$

30. $2 \cdot \sqrt{16} + \left(8 - \sqrt{25}\right)$

31. $4 \cdot \sqrt{49} + \left(9 - \sqrt{64}\right)$

32. $3(2^2 + 1) - 30 \div 3$

33. $4(3^2 + 2) - 60 \div 12$

34. $25 - 5^2 \div (7 - 2)$

35. $36 - 6^2 \div (8 - 2)$

36. $2(3.1^2 + 2) - \sqrt{7.29}$

37. $6 \times 9^2 - \dfrac{12}{4}$

38. $7 \times 8^2 - \dfrac{10}{2}$

39. $3^4 - 2 \times 4.6 \div \left(\dfrac{10}{2}\right)$

2 *See Example 3.*

40. Find the perimeter of the parallelogram in Fig. 1–23.

41. Find the perimeter of the parallelogram in Fig. 1–24.

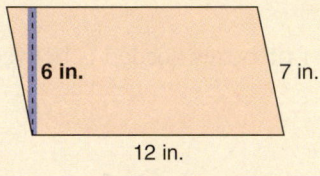

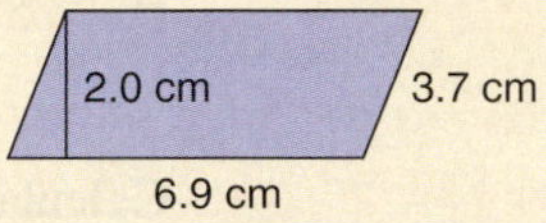

6 in.   7 in.

12 in.

FIGURE 1–23

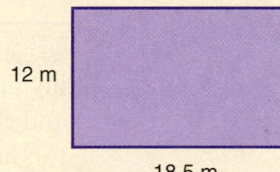

2.0 cm   3.7 cm

6.9 cm

FIGURE 1–24

See Example 4.

Find the perimeter of the rectangular crop fields in Figs. 1–25 and 1–26.

42.

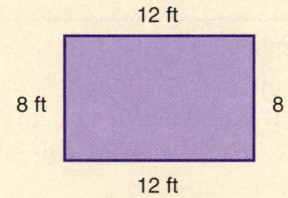

12 ft

8 ft   8 ft

12 ft

FIGURE 1–25

43.

12 m

18.5 m

FIGURE 1–26

Find the perimeter of each rectangular garden illustrated in Figs. 1–27 and 1–28. Check your work.

44.

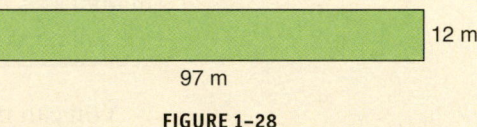

8.7 ft

14.2 ft

FIGURE 1–27

45.

12 m

97 m

FIGURE 1–28

See Example 5.

46. Find the perimeter of Fig. 1–29.

47. Find the perimeter of Fig. 1–30.

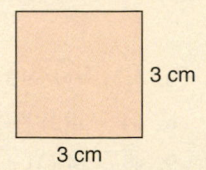

3 cm

3 cm

FIGURE 1–29

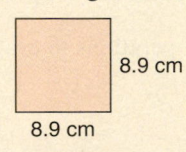

8.9 cm

8.9 cm

FIGURE 1–30

Solve the problems.

48. CON An illuminated sign in the main entrance of a hospital is a parallelogram with a base of 48 in. and an adjacent side of 30 in. How many feet of aluminum molding are needed to frame the sign?

49. AUTO A customized van has a window cut in each side in the shape of a parallelogram with a base of 20 in. and an adjacent side of 11 in. How many inches of trim are needed to surround the two windows?

50. CON A contemporary building has a window in the shape of a parallelogram with a base of 50 in. and an adjacent side of 30 in. How many inches of trim are needed to surround the window?

51. CON A table for a reading lab has a top in the shape of a parallelogram with a base of 36 in. and an adjacent side of 18 in. How many inches of edge trim are needed to surround the tabletop?

52. CON A rectangular parking lot is 340 ft by 125 ft. Find the perimeter of the parking lot.

53. CON A room is 15 ft by 12 ft. How many feet of chair rail are needed for the room? Disregard openings.

54. CON The swimming pool in Fig. 1–31 measures 32 ft by 18 ft. How much fencing is needed, including material for a gate, if the fence is to be built 7 ft from each side of the pool? *See Example 5.*

55. CON How many feet of quarter-round molding are needed to finish around the baseboard after sheet vinyl flooring is installed if the room is 16 ft by 18 ft and there are three 3-ft-wide doorways? *See Example 6.*

56. CON The square parking lot of a doctor's office is to have curbs built on all four sides. If the lot is 150 ft on each side, how many feet of curb are needed? Allow 10 ft for a driveway into the parking lot.

57. CON A border of 4-in. × 4-in. wall tiles surrounds the floor of a shower stall that is 48 in. × 48 in. How many tiles are needed for this border? Disregard spaces for grout (connecting material between the tiles).

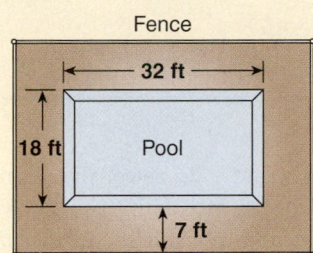

FIGURE 1–31

See Example 7.

58. Find the area of the rectangle in Fig. 1–32.

59. Find the area of the rectangle in Fig. 1–33.

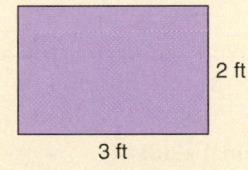

FIGURE 1–32

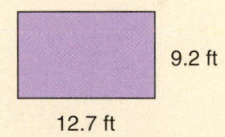

9.2 ft

12.7 ft

FIGURE 1–33

60. Find the area of the square in Fig. 1–29 in Exercise 46.

61. Find the area of the square in Fig. 1–30 in Exercise 47.

See Example 8.

62. Find the area of the parallelogram in Fig. 1–23 in Exercise 40.

63. Find the area of the parallelogram in Fig. 1–24 in Exercise 41.

64. CAD-ARC An architect is designing a rectangular building with a protected entrance as illustrated in Fig. 1–34. If the entrance is not considered part of the square footage of the building, what is the square footage of the building? *See Example 9.*

65. CAD-ARC If the roof for the building in Fig. 1–34 is to include the protected entrance and the slant of the roof will add 125 ft^2 to the area of the roof, how many bundles of asphalt roofing will be required if 3.2 bundles are needed for each 100 ft^2? *See Example 10.*

66. Find the volume of a cube that is 4.7 ft on each side. *See Example 11.*

67. Find the volume of a cube that is 12.8 cm on each side. *See Example 11.*

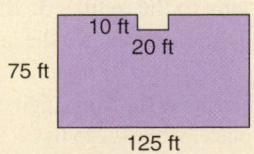

FIGURE 1–34

68. Find the volume of a prism if the length is 18 cm, the width is 17 cm, and the height is 21 cm.

69. HELPP   Fire retardant is stored in a bin that is a prism with a length of 2.9 m, a width of 1.6 m, and a height of 3.6 m. Find the volume of the bin.

3

70. BUS   The On-The-Square Card and Gift Shop buys cards from six vendors. In November, the company purchased 6,480 boxes of cards. If the shop purchased an equal number of boxes from each vendor, how many boxes of cards did each vendor supply? *See Example 12.*

71. BUS   If you have 348 packages of Halloween candy to rebox for shipment to a discount store and you can pack 12 packages in each box, how many boxes will you need? *See Example 13.*

1   CHAPTER REVIEW OF KEY CONCEPTS

LEARNING OUTCOMES	**KEY CONCEPTS AND EXAMPLES**

Section 1–1

1 Compare whole numbers (pp. 3–4).

1. Mentally position the numbers on a number line.   **2.** Select the number that is farther to the left to be the smaller number.   **3.** Write an inequality using the *less than symbol*. smaller number < larger number or Write an inequality using the *greater than symbol*. larger number > smaller number

> Write two inequalities comparing 5 and 9.
>
> $5 < 9$ or $9 > 5$

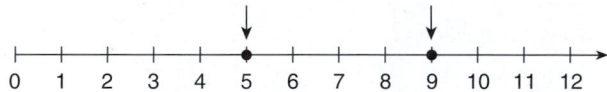

2 Write fractions with power-of-10 denominators as decimal numbers (pp. 4–6).

1. Use the denominator to find the number of decimal places.

$$10 \rightarrow 1 \text{ place}$$
$$100 \rightarrow 2 \text{ places}$$
$$1{,}000 \rightarrow 3 \text{ places}$$
$$10{,}000 \rightarrow 4 \text{ places}$$

2. Place the numerator so that the last digit is in the farthest place on the right.   **3.** Fill in any blank spaces with zero.

> Express $\dfrac{53}{1{,}000}$ as a decimal: 0.053.

3 Compare decimal numbers (pp. 6–7).

1. Compare whole-number parts.   **2.** If the whole-number parts are equal, compare digits place by place, starting at the tenths place and moving to the right.   **3.** Stop when two digits in the same place are different.   **4.** The digit that is larger determines the larger decimal number.

> Which decimal is larger, 0.23 or 0.225? Both numbers have the same digit, 2, in the tenths place. 0.23 is larger because it has a 3 in the hundredths place, while 0.225 has a 2 in that place.

LEARNING OUTCOMES	KEY CONCEPTS AND EXAMPLES

4 Round a whole number or a decimal number to a specified place value (pp. 7–8).

1. Locate the digit that occupies the rounding place. Then examine the digit to the immediate right.   **2.** If the digit to the right of the rounding place is 0, 1, 2, 3, or 4, do not change the digit in the rounding place. If the digit to the right of the rounding place is 5, 6, 7, 8, or 9, add 1 to the digit in the rounding place.   **3.** Replace all digits to the *right* of the rounding place with zeros if they are to the left of the decimal point. Drop the digits that are to the right of the digit in the rounding place *and also* to the right of the decimal point.

Round 3,624 to the tens place: 2 is in the tens place, 4 is the digit to the right; 4 is less than 5, so round down. The rounded value is 3,620.

Round 5.847 to the nearest tenth: 8 is in the tenths place; 4 is less than 5, so leave 8 as is and drop 4 and 7. The rounded value is 5.8.

5 Add and subtract whole numbers and decimals (pp. 8–15).

Addition is a binary operation that is commutative and associative.

Zero property of addition (additive identity): $0 + n = n + 0 = n$, where n is any real number.

Add whole numbers: 1. Arrange the numbers in columns so that the ones place values are in the same column.   **2.** Add the ones column, then the tens column, then the hundreds column, and so on, until all the columns have been added. **Regroup** whenever the sum of a column is more than one digit. This is also called **carrying.**

$5 + 7 = 7 + 5 = 12$	Commutative property of addition
$(3 + 2) + 6 = 3 + (2 + 6) = 11$	Associative property of addition
$0 + 3 = 3 + 0 = 3$	Zero property of addition

4,824	Addend
+ 745	Addend
5,569	Sum

Add decimal numbers: 1. Arrange the numbers so that the decimal points are in one vertical line.   **2.** Add each column, regrouping as necessary.   **3.** Align the decimal for the sum in the same vertical line.

Add: $43.35 + 3.7 + 0.462$

43.35	Addend
3.7	Addend
+ 0.462	Addend
47.512	Sum

Estimate addition: 1. Round each addend to a specified place value or to the place of the first nonzero digit.   **2.** Add the rounded addends.

Check addition: 1. Add the numbers a second time and compare with the first sum.   **2.** Use a different order or grouping if convenient.

Estimate the sum by rounding to the nearest hundred: $483 + 723$; $500 + 700 = 1,200$. Exact sum: 1,206.

Subtraction is a binary operation that is *not* commutative or associative. Addition and subtraction are inverse operations.

Zero property of subtraction: $n - 0 = n$, where n is any real number.

LEARNING OUTCOMES	KEY CONCEPTS AND EXAMPLES

Subtract whole numbers: 1. Arrange the numbers in columns, with the minuend at the top and the subtrahend at the bottom. **2.** Make sure the ones digits are in a vertical line on the right. **3.** Subtract the ones column first, then the tens column, the hundreds column, and so on. **4.** To subtract a larger digit from a smaller digit in a column, **regroup** by subtracting 1 from the digit in the next column to the left. This is the equivalent to *one* group of 10; thus, add 10 to the digit in the given column, then continue subtracting. The concept of *regrouping* is also referred to as **borrowing.**

If $5 + 4 = 9$, then $9 - 5 = 4$ or $9 - 4 = 5$.	Inverse operations
$7 - 0 = 7$	Zero property of subtraction
$\quad 4,227$	Regroup in tens column: $12 - 4 = 8$
$\underline{-\quad 745}$	Regroup in hundreds column:
$\quad 3,482$	$11 - 7 = 4$

Subtract decimal numbers: 1. Arrange the numbers so that the decimal points align vertically. **2.** Subtract each column beginning at the right, regrouping as necessary. **3.** Interpret blank places as zeros. **4.** Place the decimal in the difference in the same vertical line.

Subtract: $53.824 - 4.0423$	
$\quad 53.824$	Minuend
$\underline{-\quad 4.0423}$	Subtrahend
$\quad 49.7817$	Difference

Estimate subtraction: 1. Round each number to the desired place value or to a number with one nonzero digit. **2.** Subtract the rounded numbers.

Check subtraction: 1. Add the subtrahend (number being subtracted) and difference. **2.** Compare the result of Step 1 with the minuend. If the two numbers are equal, the subtraction is correct.

Estimate the difference by rounding to the nearest hundred: $783 - 423$
$800 - 400 = 400$. Exact difference: $783 - 423 = 360$.
Check: $360 + 423 = 783$.

6 Multiply and divide whole numbers and decimals (pp. 15–24).

Multiplication is a binary operation that is commutative and associative.

Zero property of multiplication: $n(0) = 0(n) = 0$, where n is any real number.

Multiply whole number factors of two or more digits: 1. Arrange the factors one under the other. **2.** Multiply each digit in the multiplicand by each digit in the multiplier. The product of the multiplicand and each digit in the multiplier gives a **partial product.** (a) To start, multiply the ones digit in the multiplier by the multiplicand from right to left. (b) Align each partial product with its first digit directly under its multiplier digit. **3.** Add the partial products.

$5(7) = 7(5) = 35$	Commutative property of multiplication
$(3 \cdot 2) \cdot 6 = 3 \cdot (2 \cdot 6) = 36$	Associative property of multiplication
$5(0) = 0(5) = 0$	Zero property of multiplication
$\quad\quad 259$	Multiplicand
$\underline{\times \quad\quad 23}$	Multiplier
$\quad\quad 777$	Partial product
$\underline{\quad 5\ 18}$	Partial product
$\quad 5,957$	Product

LEARNING OUTCOMES	KEY CONCEPTS AND EXAMPLES

Multiply decimal numbers: 1. Align the numbers as if they were whole numbers and multiply. **2.** Count the total number of digits to the right of the decimal in each factor. **3.** Place the decimal in the product so that the number of decimal places is the sum of the number of decimal places in the factors.

Multiply: 3.25(0.53)

$$
\begin{array}{rl}
3.25 & \text{Two decimal places} \\
\times\ 0.53 & \text{Two decimal places} \\
\hline
9\ 75 & \\
1\ 62\ 5 & \\
\hline
1.72\ 25 & \text{Four decimal places}
\end{array}
$$

Apply the distributive property: 1. Add or subtract the numbers within the grouping. **2.** Multiply the result of Step 1 by the factor outside the grouping.

or

1. Multiply each number inside the grouping by the factor outside the grouping. **2.** Add or subtract the products from Step 1.

Symbolically,

$a(b + c) = ab + ac$ or $a(b - c) = ab - ac$ for all real numbers.

$$
\begin{array}{rl}
3(5 + 6) & = 3(5) + 3(6) \\
3(11)\ \ \ & = 15 + 18 \\
33 & = 33
\end{array}
$$

Estimate multiplication: 1. Round both factors to a chosen or specified place value or to the place of the first digit from the left that is not a zero. **2.** Multiply the rounded numbers.

Check multiplication: 1. Multiply the numbers a second time and check the product. **2.** Interchange the factors if convenient.

Estimate the product by rounding to one nonzero digit: 483(72); 500(70) = 35,000. Exact product: 34,776.

Use symbols to indicate division:

The division a divided by b can be written as

$$a \div b \qquad b\overline{)a} \qquad \frac{a}{b} \qquad a/b$$

The divisor, b, cannot be zero.

Division is *not* commutative and is *not* associative.

Write 12 divided by 4 in four ways.

$$12 \div 4 \qquad 4\overline{)12} \qquad \frac{12}{4} \qquad 12/4$$

Divide whole numbers: 1. Beginning with its leftmost digit, identify the first group of digits of the dividend that is larger than or equal to the divisor. This group of digits is the first *partial dividend*. **2.** For each partial dividend in turn, beginning with the first: (a) Divide the partial dividend by the divisor. Write the partial quotient above the rightmost digit of the partial dividend. (b) Multiply the partial quotient by the divisor. Write the product below the partial dividend, aligning places. (c) Subtract the product from the partial dividend. Write the difference below the product, aligning places. The difference must be less than the divisor. (d) Next to the ones place of the difference, write the next digit of the dividend. This is the new partial product. **3.** When all the digits of the dividend have been used, write the final difference in Step 2c as the remainder (unless the remainder is 0). The whole-number part of the quotient is the number written above the dividend.

$$
\begin{array}{r}
20 \text{ R } 15 \\
23\overline{)475} \\
\underline{46} \\
15 \\
\underline{0} \\
15
\end{array}
$$

Divide decimal numbers: 1. Move the decimal in the divisor so that it is on the right side of all digits. (By moving the decimal, you are multiplying by 10, 100, 1,000, and so on.) **2.** Move the decimal in the dividend to the right as many places as the decimal was moved in the divisor. Attach zeros if necessary. (This is multiplying the dividend by the same number as was used in Step 1.) **3.** Write the decimal point in the answer directly above the new position of the decimal in the dividend. (Do this *before* dividing.) **4.** Divide as you would in whole numbers.

Round a quotient to a place value: 1. Divide to one place past the desired rounding place. **2.** Attach zeros to the dividend after the decimal if necessary to carry out the division. **3.** Round the quotient to the place specified.

Divide:

$$
\begin{array}{r}
3.8 \\
2.1\overline{)7.9\,8} \\
\underline{6\,3} \\
1\,6\,8 \\
\underline{1\,6\,8}
\end{array}
$$

Divide and round to tenths:

$$
\begin{array}{r}
1.70 \;\approx\; 1.7 \\
15\overline{)25.60} \\
\underline{15} \\
10\,6 \\
\underline{10\,5} \\
10
\end{array}
$$

Estimate division: 1. Round the divisor and dividend to one nonzero digit. **2.** Find the first digit of the quotient. **3.** Attach a zero in the quotient for each remaining digit in the dividend.

Check division: 1. Multiply the divisor by the quotient. **2.** Add any remainder to the product in Step 1. **3.** The result of Step 2 should equal the dividend.

Estimate $2,934 \div 42$.

$$
\begin{array}{r}
70 \\
40\overline{)3,000}
\end{array}
$$

Exact quotient $= 69 \text{ R}36$

Check:

$69(42) = 2,898$

$2,898 + 36 = 2,934$

$$
\begin{array}{r}
69 \\
42\overline{)2,934} \\
\underline{2\,52} \\
414 \\
\underline{378} \\
36
\end{array}
$$

LEARNING OUTCOMES	KEY CONCEPTS AND EXAMPLES

Average of a group of numbers or like measures: 1. Add the numbers or like measures. **2.** Divide the sum by the number of addends.

> Find the average of 74, 65, and 85 to the nearest whole number.
>
> $74 + 65 + 85 = 224$ — Find the sum of the values.
>
> $224 \div 3 = 74.6$ or 75 (rounded) — Divide by the number of values.

Section 1-2

1 Simplify expressions that contain exponents (pp. 28–29).

Change from natural-number exponential notation to standard notation: 1. Use the base as a factor as many times as indicated by the exponent. **2.** Perform the multiplication.

$a^1 = a$, for any base a

> $5^3 = 5(5)(5) = 125$ — Five is used as a factor 3 times.
>
> $7^1 = 7$ — Any number raised to the first power is the number.

2 Square numbers and find the square roots of numbers (pp. 29–30).

Squaring and finding square roots are inverse operations.

Find the square root of a perfect square by estimation: 1. Select a trial estimate of the square root. **2.** Square the estimate. **3.** If the square of the estimate is less than the original number, adjust the estimate to a larger number. If the square of the estimate is more than the original number, adjust the estimate to a smaller number. **4.** Square the adjusted estimate from Step 3. **5.** Continue the adjusting process until the square of the trial estimate is the original number.

> $7^2 = 49$
>
> $\sqrt{49} = 7$

3 Use powers of 10 to multiply and divide (pp. 30–32).

Multiply a decimal number by a power of 10 written in standard notation: 1. Move the decimal point in the number to the *right* as many places as the number of zeros in 10, 100, 1,000, and so on. **2.** Attach zeros to the right if necessary.

> **Multiply:**
>
> $18(100) = 1,800$
>
> $23.52(1,000) = 23,520$

Divide a decimal number by a power of 10 written in standard notation: 1. Move the decimal point to the *left* as many places as the number of zeros in 10, 100, 1,000, and so on. **2.** Attach zeros to the left if necessary. **3.** You may drop zeros to the right of the decimal point if they follow the last nonzero digit of the quotient.

> **Divide:**
>
> $3.52 \div 10 = 0.352$
>
> $400 \div 10^2 = 4$
>
> $23,000 \div 10^3 = 23$

| LEARNING OUTCOMES | KEY CONCEPTS AND EXAMPLES |

Section 1–3

1 Apply the order of operations to a series of operations (pp. 33–35).

Apply the order of operations: 1. Parentheses (grouping symbols): Perform operations within parentheses (or other grouping symbols), beginning with the innermost set of parentheses; or apply the distributive property. **2.** Exponents and roots: Evaluate exponential operations and find square roots in order from left to right. **3.** Multiply and divide from left to right. **4.** Add and subtract from left to right.

Simplify:

$3^2 + 5(6 - 4) \div 2$	Work inside parentheses.
$3^2 + 5(2) \div 2$	Raise to power.
$9 + 5(2) \div 2$	Multiply.
$9 + 10 \div 2$	Divide.
$9 + 5$	Add.
14	

2 Evaluate a formula (pp. 35–41).

1. Write the formula. **2.** Rewrite the formula substituting known values for letters of the formula. **3.** Perform the indicated operations, applying the order of operations. **4.** Interpret the solution within the context of the formula.

Evaluate the formula $P = 2(l + w)$ for $l = 17$ cm and $w = 11$ cm.

$P = 2(l + w)$	Substitute known values.
$P = 2(17 \text{ cm} + 11 \text{ cm})$	Add numbers in grouping.
$P = 2(28 \text{ cm})$	Multiply.
$P = 56 \text{ cm}$	

3 Solve applied problems using problem-solving strategies (pp. 41–43).

Apply the Six-Step Problem-Solving Plan:

A shipment of textbooks to a college bookstore is sent in two boxes. One box weighs 9 lb, and the second box weighs 14 lb more than the first box. What is the total weight of the boxes?

Unknown facts
Weight of second box
Total weight of the two boxes together

Known facts
First box weighs 9 lb.
Second box weighs 14 lb more than first box.

Relationships
weight of first box $= 9$
weight of second box $= 9 + 14$
total weight $= 9 + 9 + 14$

Estimation
Since the second box weighs more than twice as much as the first, the two boxes together weigh approximately $10 + 20$ or 30 lb.

Calculation
$9 + 9 + 14 = 32$

Interpretation
The two boxes weigh 32 lb together.

1 CHAPTER REVIEW EXERCISES

Section 1–1 MyLab Math For additional practice go to your study plan in MyLab Math.

1. Write two inequalities comparing 142 and 187.

Write as decimal numbers.

2. (a) $\dfrac{75}{1,000}$ (b) $\dfrac{21}{10}$ (c) $\dfrac{652}{100,000}$

3. (a) $\dfrac{3}{10}$ (b) $\dfrac{15}{100}$ (c) $\dfrac{4}{100}$

4. Which of these decimal numbers is smaller: 0.83 or 0.825?

5. Which of these decimal numbers is larger: 4.783 or 4.79?

6. Two measurements of an object are recorded. If the measures are 4.831 in. and 4.820 in., which is larger?

7. Write these decimal numbers in order of size from smallest to largest: 0.021, 0.0216, 0.02.

8. Two parts are machined from the same stock. They measure 1.023 in. and 1.03 in. after machining. Which part has been machined more; that is, which part is now smaller?

9. The decimal equivalent of $\frac{7}{8}$ is 0.875. The decimal equivalent of $\frac{6}{7}$ is approximately 0.857. Which fraction is larger?

10. The population of Canada in 2017 was 36,469,872. What would be a reasonable approximation for reporting this population?

11. Round to the indicated place.
    (a) Nearest hundred: 468
    (b) Nearest ten thousand: 49,238
    (c) Nearest tenth: 41.378
    (d) Nearest hundredth: 6.8957
    (e) Nearest ten-thousandth: 23.46097

12. Round to the indicated place.
    (a) Nearest ten: 98
    (b) Nearest ten: 94
    (c) Nearest thousand: 25,786
    (d) Nearest hundredth: 0.0736
    (e) Nearest whole number: 7.93
    (f) Nearest whole number: 1.876

13. Round $83.48 to the nearest dollar.

14. Round $0.096 to the nearest cent.

15. Add.
    (a) 8 + 5 + 3 + 6 + 2 + 4
    (b) 7 + 4 + 3 + 2 + 5 + 4

16. Add.
    (a) 6.2 + 32.7 + 46.82 + 0.29 + 4.237
    (b) 86.3 + 9.2 + 70.02 + 3 + 2.7

17. An air conditioner uses 10.4 kW (kilowatts), a stove uses 15.3 kW, a washer uses 2.9 kW, and a dryer uses 6.3 kW. What is the total number of kilowatts used?

18. A do-it-yourself project requires $57.32 for concrete, $74.26 for fence posts, and $174.85 for fence boards. Estimate the cost by rounding to numbers with one nonzero digit, then find the exact cost.

19. Subtract: 28 − 13 − 5.

20. Subtract: 31 − 18 − 9.

21. Subtract.
    (a) 21.34 − 16.73 (b) 15.934 − 12.807
    (c) 284.73 − 79.831 (d) 13,342 − 1,202

22. Estimate by rounding to hundreds, then find the exact answer.
    (a) $12,346.87 − $4,468.63 (b) 3,495 − 3,090
    (c) 6,767 − 478 (d) 293.86 − 148

23. For a moving sale, a family sold a sofa for $75 and a table for $25. If a newspaper ad for the sale cost $12.75, estimate and then find the exact amount the family cleared on the two items sold?

24. Subtract: 8 − 3.78.

25. Subtract: 143 − 78.2.

Use Fig. 1–35 for Exercises 26–29.

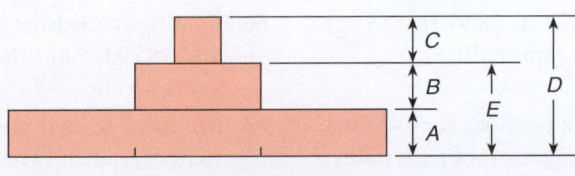

FIGURE 1–35

26. What is the dimension of A if $E = 4.86$ in., and $B = 1.972$ in.?

27. Find the length of A if $D = 4.237$ in., $B = 1.861$ in., and $C = 1.946$ in.

28. What dimension should be listed for C if D measures 3.7 in. and E measures 1.6 in.?

29. Give the limit dimensions of D if $D = 8.935$ in. with a ± 0.005-in. tolerance.

30. A blueprint calls for the length of a part to be 8.296 in. with a tolerance of ± 0.005 in. What are the limit dimensions of the part?

Multiply.

31. $2(6)(7)$

32. $6 \cdot 3 \cdot 2 \cdot 4$

33. $305(45)$

34. $236 \cdot 244$

35. $56,002 \cdot 7,040$

36. A college bookstore sold 327 American history textbooks for $39 each. How much did the bookstore receive for the books?

37. A mail-order supplier sells computer keyboards for $67 each. How much would a business pay for 21 keyboards?

38. An automotive tire dealer ran a special on heavy-duty, deluxe whitewall truck tires. If the dealer sold 105 tires for $112 each, how much did the dealer take in on the sale?

39. If a wholesaler ordered 144 computers for $305 each, how much did the dealer pay for the order?

40. A luxury car dealer pays a sound system installer $33.25 per hour. How much is the sound system installer paid for 37 hours of work? Estimate by rounding to a number with one nonzero digit, then find the exact answer.

41. A worker is offered a job that pays $365 per week. If the worker takes the job for 36 weeks, how much will the worker earn? Estimate by rounding to tens, then find the exact answer.

42. Find the area of a field 234.6 ft by 123.2 ft. Estimate the area by rounding to one nonzero digit, then find the exact answer. Check your answer. Express the area in square feet ($A = lw$).

43. A parcel of land measures 1,940.7 ft by 620.4 ft. Estimate the area by rounding to hundreds, then find the exact area. Express the area in square feet ($A = lw$).

44. A piecework employee averages 178.6 pieces per day. If the employee earns $0.28 per item, how much is earned in 5 days?

45. If a steel tape expands 0.00014 in. for each inch when heated, how much will a tape 864 in. long expand?

46. If the outside diameter of a pipe (Fig. 1–36) is 1.25 in. and the thickness of the pipe is 0.125 in., what is the inside diameter of the pipe?

47. If the inside diameter of the pipe in Fig. 1–36 is 3.5 cm and the outside diameter is 3.7 cm, what is the thickness of the pipe?

48. A rectangle has a length of 25.3 cm and a width of 15.8 cm. Find the perimeter.

49. Estimate the cost of 218 printed photos if the cost of each photo is $0.23. Then find the exact cost.

FIGURE 1–36

Multiply.

50. $2,400(70)$

51. $580(600)$

52. $14,000(250)$

53. $53,500(450)$

Divide.

54. $325 \div 25$

55. $30,126 \div 15$

56. $10,160 \div 20$

57. $54 \div 9 \div 2$

58. $60 \div 20 \div 3$

59. $364.8 \div 6$

60. $29.25 \div 0.36$

61. Divide 8,367 by 12.6 and round to the nearest tenth.

62. Estimate the quotient of 924.17 divided by 8. Then find the exact answer rounded to the nearest hundredth.

63. A group of 27 volunteers is seeking contributions to send a first-grade class to the circus. There are 632 envelopes for the collection to be divided equally among the 27 volunteers. How many will each receive? How many will be left over?

64. A school marching band is in a formation of 7 rows, each with the same number of students. If the band has 56 members, how many are in each row?

65. Find the average measure for 42.34 ft, 38.97 ft, 51.95 ft, and 61.88 ft. Round to the nearest hundredth.

66. Five light fixtures cost $74.98, $23.72, $51.27, $125.36, and $85.93. Find the average cost of the fixtures to the nearest cent.

67. Judy Ackerman purchases 5 pairs of shoes for $43, $68, $72, $59, and $21. What is the average cost of each pair of shoes to the nearest dollar?

68. Makisha Brown records the high temperatures for the week of August 13 to be 78°F, 72°F, 86°F, 88°F, 90°F, 85°F, and 82°F. What is the average daily temperature?

Section 1–2

69. Give the base and exponent of each expression, then write in standard notation.
    (a) 7^3 (b) 2.3^4 (c) 8^4

70. Give the base and exponent of each expression, then write in standard notation.
    (a) 5^6 (b) 1.2^2 (c) 10^6

71. Evaluate:
    (a) 1^2 (b) 125^2 (c) 5.6^2 (d) 21^2

72. Find the square root.
    (a) $\sqrt{2500}$ (b) $\sqrt{1.44}$ (c) $\sqrt{289}$ (d) $\sqrt{81}$

73. Express as powers of 10.
    (a) 10   (b) 1,000
    (c) 10,000   (d) 100,000

74. Multiply by using powers of 10.
    (a) 3×100 (b) $75 \times 10,000$
    (c) $2.2 \times 1,000$ (d) 5×100
    (e) 40.6×10

75. Divide by using powers of 10.
    (a) $700 \div 100$
    (b) $40.56 \div 1,000$
    (c) $60.5 \div 100$
    (d) $23,079 \div 10,000$

76. Use a calculator to write in standard notation.
    (a) $3,162.3^2$ (b) 21.53^3

77. Use a calculator to write in standard notation to the nearest hundredth.
    (a) $\sqrt{38}$ (b) $\sqrt{14.5}$

Section 1–3

Evaluate.

78. $2 + 3 \cdot 3 \div 3$

79. $4^2 \cdot (12 - 7) - 8 + 3$

80. $18 \div 6 - 3$

81. $5 + 21 \div 3 \cdot 7$

82. $82 + 4 \div 2 \times 5$

83. $21 + 7 \cdot 2 - 5 \cdot 4$

84. $15 - 6 \cdot 2 + 3$

85. $18 - 5 \cdot 2 + 7$

86. $24 \div 4 - 18 \div 6$

87. $5 - 2 \cdot 2 + 12$

88. $26 + 8 \div 2 - 3 \cdot 3$

89. $3.1 \cdot 4 \cdot \sqrt{16} - 6^2$

90. $\sqrt{12.25} \cdot (4 - 2) + 8$

91. $5.2^3 - \sqrt{81} \cdot (2 + 1)$

92. $2^4 \div 2 - \sqrt{10 - 1}$

93. $4 + 5 - 2 \cdot 3$

94. $4 + \dfrac{8.6}{2}(2)$

95. $12 \div 4(6)$

96. $8^2 - (3 - 1.5)(5.2)$

97. $27 \div 3(14 - 8) - 2 + 3^2$

98. $5.13 \div (6.2 - 4.3) + 8.6$

Find the perimeter of Figs. 1–37 through 1–42.

99.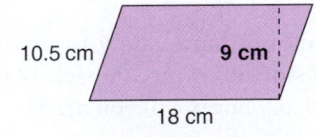
10.5 cm **9 cm** 18 cm

FIGURE 1–37

100.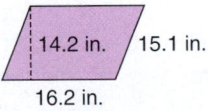
14.2 in. 15.1 in. 16.2 in.

FIGURE 1–38

101.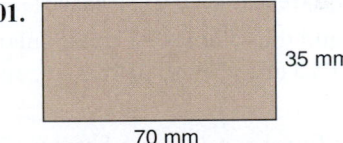
35 mm 70 mm

FIGURE 1–39

102.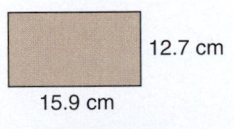
12.7 cm 15.9 cm

FIGURE 1–40

103.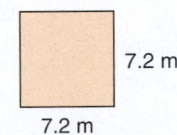
7.2 m 7.2 m

FIGURE 1–41

104.
2.5 ft 2.5 ft

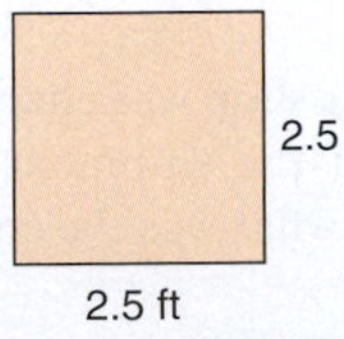

FIGURE 1–42

105. The Tennessee Highway Department has signs in the form of a parallelogram. One set of parallel sides each measures 15 ft and one set of parallel sides each measures 18 ft. Find the perimeter of the sign.

106. Antique tiles were often made in the form of a square that is 6 in. on each side. What is the perimeter of a tile?

107. A rectangular tablecloth measures 84 in. by 60 in. What length of lace is required to trim the edges of the cloth?

108. Find the number of feet of roll fencing needed to fence a square storage area measuring 15.5 ft on a side. A preassembled gate 4 ft wide will be installed.

Find the area of Figs. 1–43 through 1–46.

109.

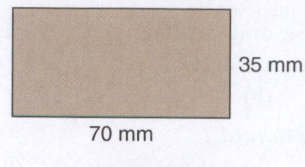

35 mm

70 mm

FIGURE 1–43

110.

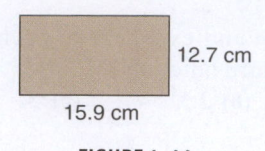

12.7 cm

15.9 cm

FIGURE 1–44

111.

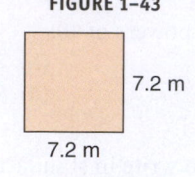

7.2 m

7.2 m

FIGURE 1–45

112.

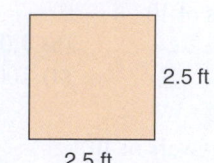

2.5 ft

2.5 ft

FIGURE 1–46

113. An office 18 ft by 16.5 ft is to be carpeted. How many square yards of carpeting are needed? Square yards $(\text{yd}^2) = \text{ft}^2 \div 9$.

114. A hall wall with no windows or doors measures 25 ft long by 8 ft high. Find the number of square feet to be covered if paneling is installed on the wall.

115. A roof measuring 16 ft by 20 ft is to be covered with asphalt roofing cement. How much would the project cost if the asphalt roofing cement spreads at the rate of 150 square feet per gallon and costs $4.75 per gallon? The cement is purchased by the gallon only.

116. Vincent Ores, a contractor, is to brick the storefront of a landscape service that has a doorway measuring 7 ft by 6 ft. How many bricks are needed if the storefront is 20 ft by 12 ft and the bricks cover at the rate of 6 per square foot using $\frac{1}{2}$-in. mortar joints?

Find the area of Figs. 1–47 and 1–48.

117.

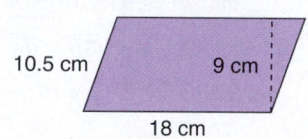

10.5 cm   9 cm

18 cm

FIGURE 1–47

118.

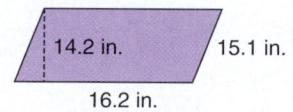

14.2 in.   15.1 in.

16.2 in.

FIGURE 1–48

119. A parking lot for a new hospital, in the shape of a parallelogram, measures 275 ft by 150 ft and has a height (distance between the 275-ft sides) of 120 ft. How many square feet need to be paved?

120. Find the volume of a cube that is 14 in. on each side.

121. Find the volume of a rectangular prism that has a length of 5.3 cm, a width of 12.2 cm, and a height of 8 cm.

122. If you have 584 packages of hard candy to rebox for shipment to a discount store and you can pack 12 packages in each box, how many boxes will you need?

123. If European Internet Service (EIS) has an annual payroll of $7,460,174,000 for its 194,582 employees, what is the average salary of an EIS employee?

124. Cottage House Antique Mall has 88 booths that rent for $1.40 per square foot. If 48 of the booths have 100 ft^2 and 40 booths have 110 ft^2, what is the total rental from the booths?

125. A repair shop was charged $1,017.61 for a shipment of 18 tires. If the shipment included a shipping charge of $48.85, what was the cost of each tire?

1   TEAM PROBLEM-SOLVING EXERCISES

1. Your team is to design a rectangular playground that has 2,160 yd^2 of space.

  (a) Examine different options for the length and width of the playground if one piece of equipment requires at least 15 yd of length. Consider only whole-number options.

  (b) Give three practical options for the dimensions of the playground and explain why you selected each of these options.

  (c) Of all possible rectangular designs, which design requires the least amount of fencing? Consider only whole-number options.

2. An important practice in minimizing calculation errors when using a calculator is to anticipate some characteristics of your answer.
 (a) Find the squares of five different decimal numbers when each has one decimal place.
 (b) Find the squares of five different decimal numbers when each has two decimal places.
 (c) Find the squares of five different decimal numbers when each has three decimal places.
 (d) Examine the patterns established with the squares in parts a, b, and c. Write a statement describing each of these patterns.

1   CONCEPTS ANALYSIS

1. Addition and subtraction are inverse operations. Write the following addition problem as a subtraction problem, and find the value of the number represented by the letter n: $1.2 + n = 1.7$.

3. Squaring and finding square roots are inverse operations. Write the following square root as a squaring problem, and find the value of the number represented by the letter n: $\sqrt{n} = 6$.

5. Give an example that shows division is not commutative.

Find and explain the mistake, then rework each problem correctly.

7. $2.5 + 4.9$
$$\begin{array}{r} 2.5 \\ + 4.9 \\ \hline 6.14 \end{array}$$

9. $\sqrt{9} = 81$

11. Without making any calculations, do you think 0.004 is a perfect square? Why or why not?

13. Can you find at least one exception to the generalization that a perfect square decimal has an even number of decimal places? Illustrate your answer.

2. Multiplication and division are inverse operations. Write the following multiplication problem as a division problem, and find the value of the number represented by the letter n: $5 \times n = 4.5$.

4. Give an example that shows subtraction is not associative.

6. Give the steps in the order of operations.

8. $2 + 5(4) =$
$7(4) = 28$

10. How are the procedures for adding whole numbers and adding decimals related?

12. Without making any calculations, do you think 0.008 is a perfect cube? Why or why not?

1   PRACTICE TEST

1. Which number is smaller, 5.09 or 5.1?

3. Round 48.3284 to the nearest tenth.

Perform the indicated operations.

5. $37 + 158 + 764 + 48$

7. $\$13,207(702)$

9. $3^2 + 5^3$

11. $3 \cdot 6^2 - 4 \div 2$

13. If a man has a bill for $165 and his paycheck is $475, estimate how much of his paycheck is left after paying the bill by rounding to tens. Find the exact amount left.

15. A softball coach paid $126 for 9 pizzas for a party after a successful season. Estimate the cost of each pizza using numbers with one nonzero digit. Find the exact cost per pizza.

2. Round 4.018 to the nearest hundredth.

4. Round $4.834 to the nearest cent.

6. $\$61,532 - \$47,245$

8. $\$25,600 \div 12$

10. 46×10^3

12. $5^3 - (3 + 2) \times \sqrt{9}$

14. A corporation buys 45 DVD players for $335 each. Use numbers with one nonzero digit to estimate the total cost. Find the exact cost.

16. A mathematics professor promises to give 2 extra points for each set of exercises a student works. If one student works 17 sets of exercises, how many points should be given?

17. Find the product: $42.73 \times 1,000$.

18. Divide: $25\overline{)27.75}$.

19. Round answer to the nearest tenth: $7.2\overline{)83.41}$.

20. Divide: $52.38 \div 10,000$.

21. Find the average of these test scores: 82, 95, 76, 84, 72, and 91. Round to the nearest whole number.

22. Estimate the sum by rounding to the nearest whole number: $3.85 + 7.46$.

23. Estimate the difference by rounding each number to the nearest tenth: $0.87 - 0.328$.

24. The blueprint specification for a machined part calls for its thickness to be 1.485 in. with a tolerance of ± 0.010 in. Find the limit dimensions of the part.

25. Heating oil costs $1.75 per gallon. What is the cost of 10,000 gal?

26. A construction job requires 16 pieces of steel, each 7.96 ft long. What length of steel is needed?

Find the perimeter and area of Figs. 1–49 through 1–51.

27.

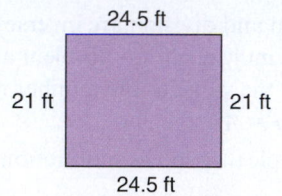

24.5 ft
21 ft   21 ft
24.5 ft

FIGURE 1–49

28.

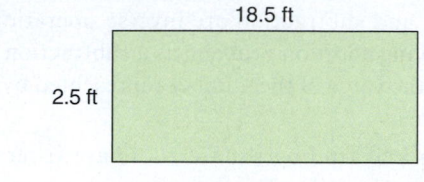

18.5 ft
2.5 ft

FIGURE 1–50

29.

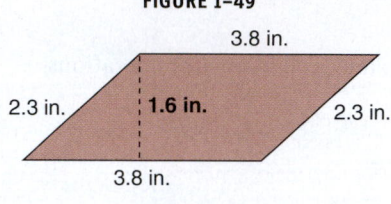

3.8 in.
2.3 in.   **1.6 in.**   2.3 in.
3.8 in.

FIGURE 1–51

30. Evaluate: $\sqrt{121}$

2

Review of Fractions

Yuri Arcurs/DigitalVision/Getty Images

In Great Company

When Fractions Kill (April 2010)

In Mansfield, central England, 23-year-old Michael Lee Bedford decided to go to a party. It was a high-energy party, and Michael wanted to keep up with the rest of the partygoers. So, he decided a little caffeine would do the trick. A friend had recently purchased a bottle of caffeine powder on the Internet for about $5, and Michael asked for a sample. He mixed two spoonfuls of the powder into his energy drink and drank the concoction.

Fifteen minutes later, he began sweating heavily. Then he vomited blood. His friends got him to a hospital as fast as they could, but it was too late. He died that evening at King's Mill Hospital in Nottinghamshire, central England. Six months later, a court of inquest ruled that Michael had died of accidental self-poisoning.

What happened?

It turned out that both Michael and his friend had failed to read the instructions that came with the caffeine powder. The powder was not just caffeine, but caffeine concentrate. The instructions recommended that the user take no more than one-sixteenth of a teaspoon. By taking two full spoons of the powder, Michael had ingested 70 times more caffeine than is usually found in a single high-energy drink.

Although caffeine is unregulated, it is a psychoactive drug that acts as a stimulant to the central nervous system and the metabolic system. More than 250 mg of caffeine in a day can lead to nervousness, irritability, headaches, and heart palpitations. In extremely high doses, caffeine can even lead to mania and psychosis.

However, caffeine is also very useful for treatment of certain conditions. Asthmatics, for instance, commonly take a drug called theophylline to open the lung's bronchial passage. Theophylline is simply a high dose of caffeine. Because asthmatics take it frequently, their body builds up a tolerance to higher doses. If an asthmatic has an attack but does not have his inhaler, the asthmatic attack can sometimes be helped by drinking a high-caffeine energy drink or a couple of strong cups of coffee.

The kind of calculation that Michael and his friend needed to do is common. Recipes, for instance, often require halving or doubling of ingredients in order to prepare the right amount of food for the number of people expected. Many people find fractions difficult because each fraction problem requires twice as many calculations as a "normal" math problem—both the top and bottom of fractions often need adjustment. It is important to work slowly, step-by-step, and double-check the math involved.

2–1   Multiples and Factors

LEARNING OUTCOMES

1. Find multiples of a natural number.

2. Find all factor pairs of a natural number.

3. Determine the prime factorization of composite numbers.

4. Find the least common multiple and greatest common factor of two or more numbers.

LC LEARNING CATALYTICS

1. Even numbers are multiples of what number?
2. Which number is prime, 3 or 4?

Common fraction: the division of one integer by a nonzero integer

In Chapter 1, we examined a special type of fraction called a *decimal fraction*. A *fraction* is a number that can be expressed as the quotient of two whole numbers.

1   Find Multiples of a Natural Number. Symbolically, a **common fraction** is written as $\frac{a}{b}$ or a/b, where a and b are whole numbers and b cannot equal zero ($b \neq 0$). The symbol $\neq$ is read "is not equal to." If one unit is divided into four parts, we can write the fraction $\frac{4}{4}$ to represent this single unit. Fig. 2–1 illustrates a unit divided into four equal parts. The fraction $\frac{4}{4}$ is an example of a *common fraction*. The bottom number, the *denominator*, indicates the number of *equal parts* that make up one whole unit. The top number, the *numerator*, tells *how many of these parts* are being considered. Figs. 2–2 and 2–3 illustrate the fractions $\frac{1}{4}$ and $\frac{3}{7}$.

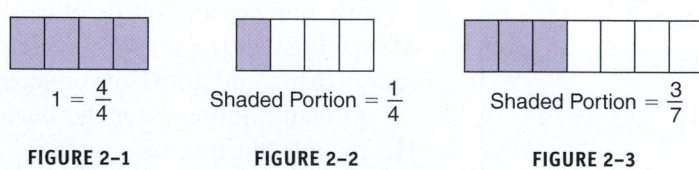

$1 = \frac{4}{4}$ Shaded Portion $= \frac{1}{4}$ Shaded Portion $= \frac{3}{7}$

FIGURE 2–1   **FIGURE 2–2**   **FIGURE 2–3**

A fraction shows the relationship between two numbers. If 3 of 500 manufactured parts are defective, then the fraction $\frac{3}{500}$ shows the fraction of the parts that are defective. Fractions also show division. 12 divided by 3 can be written as $\frac{12}{3}$. Similarly, 5 divided by 9 can be written as $\frac{5}{9}$.

Proper fraction: a common fraction whose value is less than one unit; that is, the numerator is less than the denominator

Improper fraction: a fraction that represents one or more than one unit; that is, the numerator is equal to or larger than the denominator

Slash: a type of division symbol; separates the numerator and denominator. If used with a grouping in either a numerator or denominator, grouping symbols are required

Fraction line or fraction bar: a horizontal line that separates the numerator and denominator of a fraction. It also serves as a division symbol and a grouping symbol

EXAMPLE 1

Write as a fraction. **(a)** 5 out of 8 **(b)** $3 \div 10$

(a) To show the relationship of 5 out of 8, the numerator is 5 and the denominator is 8. $\frac{5}{8}$

(b) $3 \div 10$ is written as a fraction with a numerator of 3 and a denominator of 10. $\frac{3}{10}$

See Exercises 1–2.

The unit is the *standard* amount when writing fractions. Thus, $\frac{4}{4}$ and $\frac{3}{3}$ each represent one unit. Fractions that represent less than one unit (less than 1), for example $\frac{1}{4}$ or $\frac{3}{7}$, are called **proper fractions.** Fractions that represent one or more units, for example $\frac{7}{5}$, are called **improper fractions.** Figs. 2–4 and 2–5 illustrate the improper fractions $\frac{7}{5}$ and $\frac{10}{5}$.

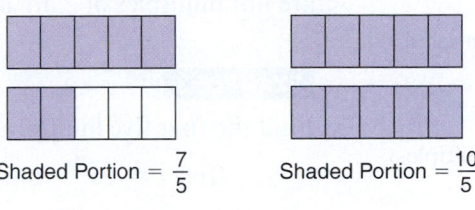

Shaded Portion = $\frac{7}{5}$ Shaded Portion = $\frac{10}{5}$

FIGURE 2–4   **FIGURE 2–5**

Fractions and division are related. In the fraction $\frac{10}{5}$ (two units), $10 \div 5 = 2$. This relationship to division is important to our understanding of fractions.

The numerator and denominator of a fraction are most often separated by a horizontal bar, although sometimes a slash is used. This horizontal bar or **slash** is the **fraction line** or **fraction bar,** and it also serves as a division symbol.

When using fraction terminology or notation to describe division, *the numerator is divided by the denominator in all cases.*

$$\text{numerator} \div \text{denominator} \qquad \frac{\text{numerator}}{\text{denominator}} \qquad \text{numerator/denominator}$$

An improper fraction can be written as a whole number when the numerator is evenly divided by the denominator $\left(\frac{10}{5} = 2\right)$. When the fraction is more than one unit and the numerator cannot be evenly divided by the denominator, the improper fraction can be written as a combination of a whole number and a fractional part, such as $\frac{7}{5} = 1\frac{2}{5}$.

A *mixed number* consists of both a whole number and a fraction. The whole number and fraction are added together. For example, $1\frac{2}{5}$ means 1 whole unit and $\frac{2}{5}$ of another unit, or $1\frac{2}{5} = 1 + \frac{2}{5}$.

Fractions indicate division. Multiplication and division are inverse operations. Let's look at some relationships involving multiplication and division.

If we count by 3s, such as 3, 6, 9, 12, 15, 18, we obtain natural numbers that are *multiples* of 3. Each is the product of 3 and a natural number; that is,

$$3 = 3(1), \qquad 6 = 3(2), \qquad 9 = 3(3)$$
$$12 = 3(4), \qquad 15 = 3(5), \qquad 18 = 3(6)$$

Multiple: the product of a given number and any natural number

A **multiple** of a number is the product of that number and a natural number.

To find a multiple of a natural number:

1. Multiply the given number by any natural number.

EXAMPLE 2

Show that 2, 4, 6, 8, and 10 are multiples of 2 by writing each as the product of 2 and a natural number.

$$2 = 2(1), \quad 4 = 2(2), \quad 6 = 2(3), \quad 8 = 2(4), \quad 10 = 2(5)$$

See Exercises 2–8.

Natural numbers that are multiples of 2 are **even numbers.** Natural numbers that are not multiples of 2 are **odd numbers.**

Even numbers: natural numbers that are multiples of 2

Odd numbers: natural numbers that are not multiples of 2

EXAMPLE 3

Find the first five multiples of 16.

$$16(1) = 16, \quad 16(2) = 32, \quad 16(3) = 48, \quad 16(4) = 64, \quad 16(5) = 80$$

16, 32, 48, 64, and 80 are multiples of 16.

See Exercises 9–14.

We say that a number is **divisible** by another number if the quotient has no remainder or if the dividend is a multiple of the divisor.

It is desirable to be able to determine divisibility **by inspection.** This means that we can examine the number being divided (dividend) and decide if it is divisible by a divisor without actually having to perform the division.

Is 35 divisible by 7? 35 is divisible by 7 if $35 \div 7$ has no remainder or if 35 is a multiple of 7.

Divisible: one number is divisible by a second number if the quotient has no remainder or if the dividend is a multiple of the divisor

By inspection: the process of mentally determining the answer to a problem

$$35 \div 7 = 5 \quad \text{or} \quad 35 = 7(5)$$

Yes, 35 is divisible by 7.

Rules or tests can help us decide by inspection if certain numbers are divisible by other numbers. These mental processes help us develop our number sense.

Tests for divisibility:

A number is divisible by

1. 2 if the last digit is an even number (0, 2, 4, 6, or 8).
2. 3 if the sum of its digits is divisible by 3.
3. 4 if the last two digits form a number divisible by 4.
4. 5 if the last digit is 0 or 5.
5. 6 if the number is divisible by *both* 2 *and* 3.
6. 7 if the division has no remainder.
7. 8 if the last three digits form a number divisible by 8.
8. 9 if the sum of its digits is divisible by 9.
9. 10 if the last digit is 0.

EXAMPLE 4

Use the tests for divisibility to identify which number in each pair is divisible by the given divisor.

Numbers	Divisor	Answer
874 or 873	2	**874**; the last digit is an even number (4).
275 or 270	2	**270**; the last digit is an even number (0).
427 or 423	3	**423**; the sum of the digits is divisible by 3: $4 + 2 + 3 = 9$.
5,912 or 5,913	4	**5,912**; the last two digits form a number divisible by 4: $12 \div 4 = 3; 9 \div 3 = 3$.
80 or 82	5	**80**; the last digit is 0.
56 or 65	5	**65**; the last digit is 5.
804 or 802	6	**804**; the last digit is even and the sum of the digits is divisible by 3: $8 + 0 + 4 = 12; 12 \div 3 = 4$.
58 or 56	7	**56**; it divides by 7 with no remainder.
3,160 or 3,162	8	**3,160**; the last three digits form a number divisible by 8: $160 \div 8 = 20$.
477 or 475	9	**477**; the sum of the digits is divisible by 9: $4 + 7 + 7 = 18. 18 \div 9 = 2$.
182 or 180	10	**180**; the last digit is 0.

See Exercises 15–23.

2   Find All Factor Pairs of a Natural Number. Any natural number can be expressed as the product of two natural numbers. These two natural numbers are called a **factor pair** of the number.

Factor pair: a factor pair of a natural number consists of two natural numbers whose product equals the given natural number

Every natural number greater than 1 has at least one factor pair, the number itself and 1.

$$1 \text{ and } 3 \text{ form a factor pair for } 3: 1(3) = 3.$$
$$1 \text{ and } 5 \text{ form a factor pair for } 5: 1(5) = 5.$$

Many natural numbers have more than one factor pair. For example, list all factor pairs of 12. Start with the pair 1×12. *Every number has a factor pair of the number and 1.* Next, examine each number from 2 to 11.

2	12 is divisible by 2.	$12 \div 2 = 6$	2(6) is a factor pair of 12.
3	12 is divisible by 3.	$12 \div 3 = 4$	3(4) is a factor pair of 12.
4	12 is divisible by 4.	$12 \div 4 = 3$	4(3) is a factor pair of 12.
5	12 is not divisible by 5.		
6	12 is divisible by 6.	$12 \div 6 = 2$	6(2) is a factor pair of 12.
7	12 is not divisible by 7.		
8	12 is not divisible by 8.		
9	12 is not divisible by 9.		
10	12 is not divisible by 10.		
11	12 is not divisible by 11.		

Are the factor pairs 2×6 and 6×2 different pairs? No. Both pairs have the same numbers, and since multiplication is commutative, they count as one pair. Similarly, 3×4 and 4×3 are the same factor pair. Is it necessary to examine every number less than 12? No. Once we get the first repeat, 4 × 3, we can assume we have found all the factor pairs of the number.

The factor pairs of 12 are 1(12), 2(6), and 3(4).

> **To find all factor pairs of a natural number:**
>
> 1. Write the factor pair of 1 and the given natural number.
>
> 2. Check to see if the given number is divisible by 2. If so, write the factor pair of 2 and the quotient of the given number and 2.
>
> 3. Check the next natural number for divisibility; if the given number is divisible by this number, write the factor pair.
>
> 4. Continue Step 3 until you reach a number that already has been found as a quotient in a previous factor pair.

STOP AND CHECK
1. List all the factor pairs of 28.

Answer:
1. 1 and 28, 2 and 14, and 4 and 7

EXAMPLE 5

List all the factor pairs of 18.

1(18)	
2(9)	18 is divisible by 2: 18 ÷ 2 = 9.
3(6)	18 is divisible by 3: 18 ÷ 3 = 6.
4 or 5	18 is not divisible by 4 or 5.
6	18 is divisible by 6; 6 was found in the factor pair 3 × 6, so we stop.

The factor pairs of 18 are 1 and 18, 2 and 9, and 3 and 6. **See Exercises 24–33.**

Once we have listed all factor pairs of a number, we can list all factors of a number. From the factor pairs, we list every different factor that appears in any factor pair. The factors of 12 are 1, 2, 3, 4, 6, and 12. The factors of 18 are 1, 2, 3, 6, 9, and 18.

STOP AND CHECK
1. List all the factors of 28.

Answer:
1. 1, 2, 4, 7, 14, and 28

EXAMPLE 6

List all the factors of 48 by first listing all the factor pairs of 48. Then write each distinct factor in order from smallest to largest.

1(48)	
2(24)	48 is divisible by 2: 48 ÷ 2 = 24.
3(16)	48 is divisible by 3: 48 ÷ 3 = 16.
4(12)	48 is divisible by 4: 48 ÷ 4 = 12.
5	48 is not divisible by 5.
6(8)	48 is divisible by 6: 48 ÷ 6 = 8.
7	48 is not divisible by 7.
8	48 is divisible by 8; 8 was found in the factor pair 6 × 8, so we stop.

Factor pairs: 1 and 48, 2 and 24, 3 and 16, 4 and 12, 6 and 8.
Factors: 1, 2, 3, 4, 6, 8, 12, 16, 24, 48. **See Exercises 33–43.**

3 **Determine the Prime Factorization of Composite Numbers.** When all the factors of a natural number are listed, some numbers have no factor other than the number itself and 1. These numbers form a special set of numbers called prime numbers. A **prime number** is a whole number *greater than 1* that has factors of only the number itself and 1. Note that *1 is not a prime number.*

Prime number: a whole number greater than 1 that has only one factor pair, the number itself and 1

STOP AND CHECK

Identify each number as prime or not prime.
1. 14
2. 17

Answers:
1. Not prime 2. Prime

EXAMPLE 7

Identify the prime numbers by examining the factors of the numbers.

(a) 8 (b) 1 (c) 3 (d) 9 (e) 7

(a) **8 is *not* a prime number** because its factors are 1, 2, 4, and 8.

(b) **1 is *not* a prime number** because a prime must be greater than 1.

(c) **3 is a prime number** because it has only factors 1 and 3.

(d) **9 is *not* a prime number** because its factors are 1, 3, and 9.

(e) **7 is a prime number** because it has only factors 1 and 7. **See Exercises 44–55.**

Composite number: a whole number greater than 1 that is not a prime number

A **composite number** is a whole number greater than 1 that is not a prime number. In the preceding example, 8 and 9 are composite numbers. A composite number has at least one factor other than the number itself and 1.

STOP AND CHECK

Identify each number as composite or not composite.
1. 26
2. 33

Answers:
1. Composite 2. Composite

EXAMPLE 8

Identify the composite numbers by examining the factors of the numbers.

(a) 4 (b) 10 (c) 13 (d) 12 (e) 5

(a) **4 is a composite number** because its factors are 1, 2, and 4.

(b) **10 is a composite number** because its factors are 1, 2, 5, and 10.

(c) **13 is *not* a composite number** because its only factors are 1 and 13. It is a prime number.

(d) **12 is a composite number** because its factors are 1, 2, 3, 4, 6, and 12.

(e) **5 is *not* a composite number** because its only factors are 1 and 5. It is a prime number. **See Exercises 44–55.**

Sieve of Eratosthenes: process for finding prime numbers

TIP **Prime Numbers Less than 50** We can find all the prime numbers that are 50 or less using an ancient technique developed by the mathematician Eratosthenes. This technique is called **sieve of Eratosthenes.**

Step 1. List the numbers from 1 through 50.

Step 2. Eliminate numbers that are not prime, using the systematic process:
(a) 1 is not prime. Eliminate 1.
(b) 2 is prime. Eliminate all multiples of 2.
(c) 3 is prime. Eliminate all multiples of 3.
(d) 4 has already been eliminated.
(e) 5 is prime. Eliminate all multiples of 5.
(f) 6 has already been eliminated.
(g) 7 is prime. Eliminate all multiples of 7.

Step 3. Circle remaining numbers as prime numbers.

1	②	③	4	⑤	6	⑦	8	9	10
⑪	12	⑬	14	15	16	⑰	18	⑲	20
21	22	㉓	24	25	26	27	28	㉙	30
㉛	32	33	34	35	36	㊲	38	39	40
㊶	42	㊸	44	45	46	㊼	48	49	50

All numbers not eliminated are prime. Why? The numbers 8, 9, and 10 have already been eliminated as multiples of 2, 3, and 5, respectively. Multiples of 11 that are less than 50 have already been eliminated: $11 \times 2 = 22$, $11 \times 3 = 33$, $11 \times 4 = 44$. The product $11 \times 5 = 55$ is greater than 50. Similarly, all other composite numbers have been eliminated.

Prime factorization: the process of writing a composite number as the product of *only* prime numbers

Prime factor: a factor that is divisible only by 1 and the number itself

A composite number can be expressed as a product of prime numbers. **Prime factorization** refers to the process of writing a composite number as the product of *only* prime numbers. These factors are called **prime factors.**

To find the prime factors of a composite number:

1. Test each prime number to see if the composite number is divisible by the prime number.

2. Make a factor pair using the first prime number that passes the test in Step 1.

3. Carry forward the prime factors and test the remaining factors by repeating Steps 1 and 2.

EXAMPLE 9

Find the prime factorization of **(a)** 30, **(b)** 16, and **(c)** 210.

(a) $30 = 2(15)$
first prime ⤴

30 is divisible by 2. Factor 30 into a factor pair using its smallest prime factor, 2.

$30 = 2(3)(5)$
last primes ⤴⤴

Carry the prime factor 2 forward. Factor the composite number 15 using its smallest prime factor, 3. Because 5 is also prime, the factoring is complete.

The prime factorization of 30 is 2(3)(5).

(b) $16 = 2(8)$
first prime ⤴

Factor 16 into two factors using its smallest prime factor.

$16 = 2(2)(4)$
second prime ⤴

Factor 8 into two factors using its smallest prime factor.

$16 = 2(2)(2)(2)$
last two primes ⤴⤴

Factor 4 into two factors using its smallest prime factor.

The prime factorization of 16 is 2(2)(2)(2). We can write this expression in exponential notation as 2^4.

(c) $210 = 2(105)$
first prime ⤴

Factor 210 into two factors using its smallest prime factor.

$210 = 2(3)(35)$
second prime ⤴

Factor 105 into two factors using its smallest prime factor.

$210 = 2(3)(5)(7)$
last two primes ⤴⤴

Factor 35 into two factors using its smallest prime factor.

The prime factorization of 210 is 2(3)(5)(7).   **See Exercises 56–85.**

Least common multiple (LCM): the least common multiple (LCM) of two or more natural numbers is the smallest number that is a multiple of each number. The LCM is divisible by each number

4 **Find the Least Common Multiple and Greatest Common Factor of Two or More Numbers.** The **least common multiple (LCM)** of two or more natural numbers is the smallest number that is a multiple of each number. The LCM is divisible by each number.

To find the least common multiple of 3 and 5, examine the multiples of each number.

multiples of 3:   3, 6, 9, 12, 15, 18, 21, 24, 27, 30, 33, 36, 39, . . .
multiples of 5:   5, 10, 15, 20, 25, 30, 35, 40, . . .

The common multiples in these lists that are less than 40 are 15 and 30; 15 is the *least common multiple* of 3 and 5.

Prime factorization can also be used to find the least common multiple of two or more numbers.

> **TIP** The least common multiple (LCM) will be greater than or equal to the largest number in the group.

> **To find the least common multiple (LCM) of two or more natural numbers using the prime factorization of the numbers:**
>
> 1. List the prime factorization of each number writing repeat factors in exponential notation.
>
> 2. List the prime factorization of the least common multiple by including the prime factors appearing in *each* number. If a prime factor appears in more than one number, use the factor with the *largest* exponent.
>
> 3. Write the resulting expression in standard notation.

STOP AND CHECK

Find the least common multiple of
1. 8 and 26
2. 9 and 26

Answers:
1. 104   2. 234

EXAMPLE 10

Find the least common multiple of 12 and 40 by prime factorization.

$$12 = 2(2)(3) \qquad = 2^2(3) \qquad \text{Prime factorization of 12.}$$
$$40 = 2(2)(2)(5) \quad = 2^3(5) \qquad \text{Prime factorization of 40.}$$
$$\text{LCM} = 2^3(3)(5) \qquad\qquad \text{Prime factorization of LCM.}$$
$$\textbf{LCM} = \textbf{120} \qquad\qquad\quad \text{LCM in standard notation.} \qquad \textbf{See Exercises 86–109.}$$

Greatest common factor: the greatest common factor (GCF) of two or more numbers is the largest factor common to each number. Each number is divisible by the GCF

The **greatest common factor (GCF)** of two or more numbers is the largest factor common to each number. Each number is divisible by the GCF.

Let's take the numbers 30 and 42. The prime factors are

$$30 = 2(3)(5) \quad 42 = 2(3)(7)$$

The *common* prime factors of both 30 and 42 are 2 and 3, which represent the composite factor 6. The *greatest* common factor is the product of the common prime factors, $2(3) = 6$.

> **TIP** The greatest common factor (GCF) will be less than or equal to the smallest number in the group.

> **To find the greatest common factor (GCF) of two or more natural numbers:**
>
> 1. List the prime factorization of each number writing repeat factors in exponential notation.
>
> 2. List the prime factorization of the greatest common factor by including each prime factor appearing in *every* number. If a prime factor appears more than one time in any number (that is, the exponent is greater than 1), use the factor with the *smallest* exponent. If there are no common prime factors, the GCF is 1.
>
> 3. Write the resulting expression in standard notation.

STOP AND CHECK

Find the greatest common factor of

1. 48, 18, and 30
2. 15, 23, and 8

Answers:
1. 6   2. 1

EXAMPLE 11

Find the greatest common factor of **(a)** 15, 30, and 45 and **(b)** 10, 12, and 13.

(a)

	$15 = 3(5)$	$= 3(5)$	Prime factorization of 15.
	$30 = 2(3)(5)$	$= 2(3)(5)$	Prime factorization of 30.
	$45 = 3(3)(5)$	$= 3^2(5)$	Prime factorization of 45.
	GCF $= 3(5)$		Common prime factors.
	GCF $= 15$		GCF in standard notation.

(b)

	$10 = 2(5)$	$= 2(5)$	Prime factorization of 10.
	$12 = 2(2)(3)$	$= 2^2(3)$	Prime factorization of 12.
	$13 = 13$	$= 13$	Prime factorization of 13.
	GCF $= 1$		No common prime factors. **See Exercises 110–124.**

> **TIP** **LCM Versus GCF** The least common multiple (LCM) and greatest common factor (GCF) are easily confused. The LCM is the *smallest* of the common multiples of the original numbers. The GCF is the *largest* of the common factors of the original numbers.

2–1 EXERCISES

MyLab Math For additional practice go to your study plan in MyLab Math.

1 *See Example 1.*

1. **HLTH/N** On a given day at the Good Health Hospital, 9 out of 248 emergency room (ER) patients were admitted to the hospital. What fraction of the patients were admitted?

2. Write $7 \div 15$ as a fraction.

Show that each number is a multiple of the first number by writing each as the product of the first number and a natural number. *See Example 2.*

3. 5, 10, 15, 20, 25, 30
4. 6, 12, 18, 24, 30, 36
5. 8, 16, 24, 32, 40, 48
6. 9, 18, 27, 36, 45, 54
7. 10, 20, 30, 40, 50, 60
8. 30, 60, 90, 120, 150, 180

Find the first five multiples of the given number. *See Example 3.*

9. 6
10. 12
11. 13
12. 3
13. 50
14. 4

Use the tests for divisibility to determine which number is divisible by the given number. Explain. *See Example 4.*

15. 2,434 by 6
16. 230 by 5
17. 2,434 by 4
18. 1,221 by 3
19. 756 by 7
20. 920 by 8
21. 621 by 3
22. 426 by 6
23. 1,232 by 2

2 List all the factor pairs of each number. *See Example 5.*

24. 4
25. 8
26. 12
27. 15
28. 16
29. 20
30. 24
31. 30
32. 36
33. 38

List all the factors of each number. *See Example 6.*

34. 40
35. 46
36. 52
37. 64
38. 72
39. 81
40. 85
41. 92
42. 96
43. 98

3 Identify the prime numbers and the composite numbers by examining the factors of the numbers. *See Examples 7 and 8.*

44. 2	**45.** 6	**46.** 9	**47.** 11	**48.** 14	**49.** 15
50. 16	**51.** 51	**52.** 52	**53.** 53	**54.** 66	**55.** 67

Find the prime factorization. *See Example 9.*

56. 12	**57.** 18	**58.** 20	**59.** 24	**60.** 25	**61.** 27
62. 29	**63.** 30	**64.** 35	**65.** 40	**66.** 47	**67.** 49
68. 50	**69.** 52	**70.** 65	**71.** 75	**72.** 100	**73.** 105
74. 108	**75.** 115	**76.** 121	**77.** 144	**78.** 156	**79.** 157

Write the prime factorization using exponential notation. *See Example 9b.*

80. 72	**81.** 112	**82.** 124	**83.** 164	**84.** 568	**85.** 900

4 Find the least common multiple. *See Example 10.*

86. 2 and 3	**87.** 5 and 6	**88.** 7 and 8
89. 3 and 4	**90.** 18 and 24	**91.** 10 and 12
92. 12 and 24	**93.** 9 and 18	**94.** 4, 8, and 12
95. 20, 25, and 35	**96.** 3, 9, and 27	**97.** 2, 8, and 16
98. 6, 15, and 18	**99.** 20, 24, and 30	**100.** 12, 18, and 20
101. 6, 10, and 15	**102.** 8, 12, and 32	**103.** 8, 12, and 18
104. 10, 15, and 20	**105.** 30, 50, and 60	**106.** 6, 11, and 33
107. 8, 13, and 39	**108.** 2, 7, and 14	**109.** 12, 16, and 18

Find the greatest common factor. *See Example 11.*

110. 2 and 3	**111.** 5 and 6	**112.** 7 and 8
113. 18 and 24	**114.** 10 and 15	**115.** 12 and 24
116. 9 and 18	**117.** 40 and 55	**118.** 2, 8, and 16
119. 18, 30, and 36	**120.** 20, 25, and 35	**121.** 6, 15, and 18
122. 12, 18, and 20	**123.** 6, 10, and 12	**124.** 8, 12, and 18

2-2   Equivalent Fractions and Decimals

LEARNING OUTCOMES

1 Write equivalent fractions with different denominators.

2 Write improper fractions as whole numbers or mixed numbers.

3 Write whole numbers or mixed numbers as improper fractions.

4 Write decimals as fractions and fractions as decimals.

5 Compare fractions, mixed numbers, and decimals.

1 **Write Equivalent Fractions with Different Denominators.** There are many different ways to express the same value in fractional form. For example, the whole number 1 can be written as $\frac{1}{1}, \frac{4}{4}, \frac{7}{7}, \frac{15}{15}$, and so on.

Fractions are equivalent if they represent the same value. Compare the illustrations in Fig. 2–6, where line a is one whole unit divided into only 1 part $\left(\frac{1}{1}\right)$. Line b is one unit divided into 2 parts $\left(\frac{2}{2}\right)$. Lines c and d are divided into 4 parts $\left(\frac{4}{4}\right)$, line e is divided into 8 parts $\left(\frac{8}{8}\right)$, and line f is divided into 16 parts $\left(\frac{16}{16}\right)$.

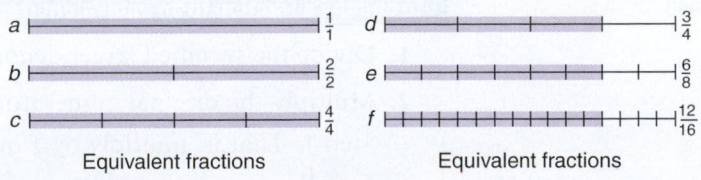

Equivalent fractions   Equivalent fractions

FIGURE 2–6

LC **LEARNING CATALYTICS**
1. Reduce $\frac{10}{15}$ to lowest terms.
2. Change $2\frac{3}{4}$ to an improper fraction.

Equivalent fractions: fractions that represent the same value

Family of fractions: a set of fractions consisting of a fraction in lowest terms and equivalent fractions found by multiplying both the numerator and denominator of the fraction in lowest terms by the same natural number

Lowest terms: a fraction that has no natural number that divides evenly into both the numerator and denominator except the number 1

STOP AND CHECK

1. Find five fractions that are equivalent to $\frac{2}{3}$.

Answer:

1. $\dfrac{4}{6}, \dfrac{6}{9}, \dfrac{8}{12}, \dfrac{10}{15}, \dfrac{12}{18}$

Again, look at lines d, e, and f. Line d is divided into 4 parts and 3 of them are shaded. Line e is divided into 8 parts and 6 of them are shaded. Line f is divided into 16 parts and 12 of them are shaded.

Look at the shaded portions of lines d, e, and f. They are the same length, even though they are divided into a different number of parts. Thus, $\frac{3}{4}$, $\frac{6}{8}$, and $\frac{12}{16}$ are **equivalent fractions** since they represent the same shaded part of one whole unit. That is,

$$\frac{3}{4} = \frac{6}{8} = \frac{12}{16}$$

Equivalent fractions are in the same **family of fractions.** The first member of the "family" is the fraction in **lowest terms;** that is, no natural number divides evenly into *both* the numerator and the denominator except the number 1. Other "family members" are found by multiplying both the numerator and the denominator by the same natural number.

EXAMPLE 1

Find five fractions that are equivalent to $\frac{1}{2}$.

$\dfrac{1(2)}{2(2)}$ or $\dfrac{1}{2}\left(\dfrac{2}{2}\right) = \dfrac{2}{4}$ $\qquad$ $\dfrac{1(5)}{2(5)}$ or $\dfrac{1}{2}\left(\dfrac{5}{5}\right) = \dfrac{5}{10}$

$\dfrac{1(3)}{2(3)}$ or $\dfrac{1}{2}\left(\dfrac{3}{3}\right) = \dfrac{3}{6}$ $\qquad$ $\dfrac{1(6)}{2(6)}$ or $\dfrac{1}{2}\left(\dfrac{6}{6}\right) = \dfrac{6}{12}$

$\dfrac{1(4)}{2(4)}$ or $\dfrac{1}{2}\left(\dfrac{4}{4}\right) = \dfrac{4}{8}$

Other equivalent fractions can be found. $\qquad\qquad\qquad\qquad$ **See Exercises 1–2.**

TIP **Multiplication, Division, and 1** In the preceding example, $\frac{1}{2}$ is multiplied by a fraction whose value is 1, and 1 times any number does not change the value of that number. Written symbolically,

$$\frac{n}{n} = 1 \text{ when } n \text{ is a real number and } n \neq 0 \quad \text{and}$$

$$1(n) = n \text{ where } n \text{ is any real number}$$

Fractions in the same family can be generated by multiplying the fraction by 1 in the form of $\frac{2}{2}$, $\frac{3}{3}$, $\frac{4}{4}$, and so on. Even though we only give natural number illustrations, the principle applies to any real numbers.

Fundamental principle of fractions: if the numerator and denominator of a fraction are multiplied by the same nonzero number, the value of the fraction remains unchanged

The concept presented in the preceding example and tip is referred to as the **fundamental principle of fractions.** If the numerator and denominator of a fraction are multiplied by the same nonzero number, the value of the fraction remains unchanged.

To change a fraction to an equivalent fraction with a specified larger denominator:

1. Divide the specified larger denominator by the original denominator.

2. Multiply the original numerator and denominator by the quotient found in Step 1. That is, multiply by 1 in the form of $\dfrac{n}{n}$ when n is a real number and $n \neq 0$.

EXAMPLE 2

Change $\frac{5}{8}$ to an equivalent fraction whose denominator is 32.

$$\frac{5}{8} = \frac{?}{32}$$ $32 \div 8 = 4$. Apply the fundamental principle of fractions.

$$\frac{5}{8}\left(\frac{4}{4}\right) = \frac{20}{32}$$

See Exercises 3–10.

Because each fraction has an unlimited number of equivalent fractions, we usually work with fractions in *lowest terms*. When we find an equivalent fraction with smaller numbers and there are no common factors in the numerator and denominator, we have **reduced to lowest terms**.

Reduce to lowest terms: to find an equivalent fraction that has smaller numbers and has no common factors in the numerator and denominator

To change a fraction to an equivalent fraction with a smaller denominator or to reduce a fraction to lowest terms:

1. Find a common factor greater than 1 for the numerator and denominator.

2. Divide both the numerator and the denominator by this common factor.

3. Continue until the fraction is in lowest terms or has the desired smaller denominator.

Note: To reduce to lowest terms in the fewest steps, find the greatest common factor (GCF) in Step 1.

EXAMPLE 3

Reduce $\frac{8}{10}$ to lowest terms.

Prime factors of 8: 2 (2)(2) or 2^3
Prime factors of 10: 2 (5)
The greatest common factor (GCF) is 2.

$$\frac{8 \div 2}{10 \div 2} \qquad \text{or} \qquad \frac{8}{10} \div \frac{2}{2} = \frac{4}{5}$$

See Exercises 11–23.

TIP **Reducing and the Properties of 1** To reduce the fraction $\frac{8}{10}$, we divide by the whole number 1 in the form of $\frac{2}{2}$. That is, $\frac{8}{10} \div \frac{2}{2} = \frac{4}{5}$. A nonzero number divided by itself is 1, and to divide a number by 1 does not change the value of the number. Symbolically,

$$\frac{n}{n} = 1 \text{ when } n \text{ is a real number and } n \neq 0 \text{ and } n \div 1 = n \text{ or } \frac{n}{1} = n \text{ when } n \text{ is}$$

a real number.

EXAMPLE 4

Reduce $\frac{18}{24}$ to lowest terms.

Prime factors of 18: 2(3) (3)   or $2 \cdot 3^2$
Prime factors of 24: 2(2) (2)(3)   or $2^3 \cdot 3$
The GCF is 2(3) or 6.

$$\frac{18 \div 6}{24 \div 6} \qquad \text{or} \qquad \frac{18}{24} \div \frac{6}{6} = \frac{3}{4}$$

See Exercises 11–23.

Ratio: a comparison between two quantities by division; a fraction

Proportion: two equal ratios. Two equivalent fractions form a proportion

Cross product: the product of the numerator of one fraction and the denominator of another; a proportion has two cross products

Cross products of a proportion: the cross products of a proportion are equal

Another name for fraction is *ratio*. Like a fraction, a **ratio** is a comparison between two quantities by division. In expressing the ratio of one quantity to another, the first quantity is the numerator or dividend and the second quantity is the denominator or divisor. If two ratios are equal, they form a **proportion.** That is, *equivalent fractions form a proportion*. In Example 4, $\frac{18}{24} = \frac{3}{4}$ is a proportion.

An important property of proportions is that *the cross products of a proportion are equal*. The **cross products of a proportion** are products of the numerator of one fraction and the denominator of the other. In the proportion $\frac{18}{24} = \frac{3}{4}$, the cross products are 18(4) and 24(3). Each of these products equals 72.

Property of Proportions

The cross products in a proportion are equal. Symbolically, if $\frac{a}{b} = \frac{c}{d}$ and b and d are not equal to zero, then $a \cdot d = b \cdot c$. Also, if $a \cdot d = b \cdot c$, then $\frac{a}{b} = \frac{c}{d}$.

STOP AND CHECK

1. Is $\frac{8}{18} = \frac{4}{9}$?

2. Is $\frac{12}{18} = \frac{3}{6}$?

Answers:

1. Yes **2.** No

EXAMPLE 5

Use the property of proportions to determine if the following statements are true or false.

(a) $\frac{5}{8} = \frac{15}{24}$ The cross products are 5(24) = 120 and 8(15) = 120. The statement is **true.**

(b) $\frac{6}{9} = \frac{10}{15}$ The cross products are 6(15) = 90 and 9(10) = 90. The statement is **true.**

(c) $\frac{9}{12} = \frac{12}{18}$ The cross products are 9(18) = 162 and 12(12) = 144. The statement is **false.**   See Exercises 24–25.

U.S. customary rule: a measuring device for measuring length in inches and parts of an inch

The **U.S. customary rule** is divided into inches and parts of an inch and is used to measure length. Each inch is subdivided into fractional parts, usually 8, 16, 32, or 64.

The rule in Fig. 2–7 shows each inch divided into 16 equal parts, so each part is $\frac{1}{16}$ in.; that is, the first mark from the left edge represents $\frac{1}{16}$ in. The left end of the rule represents zero (0).

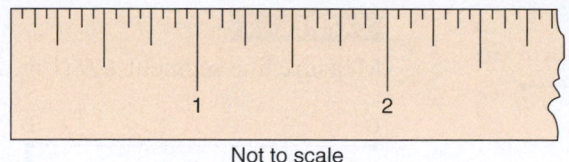

Not to scale

FIGURE 2-7 U.S. customary rule.

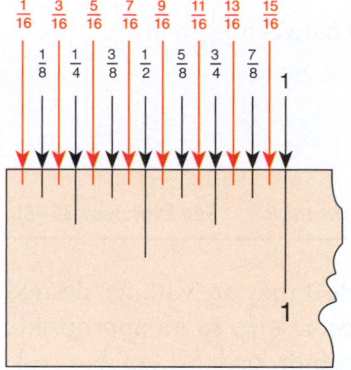

FIGURE 2-8 1 in.

Fig. 2–8 labels division marks for the parts of 1 in. The second mark from the left edge represents $\frac{2}{16}$ or $\frac{1}{8}$ in. This mark is slightly longer than the first mark.

The fourth mark from the left is labeled $\frac{1}{4}$; that is, $\frac{4}{16} = \frac{1}{4}$. In each case, fractions are always reduced to lowest terms. Notice that the $\frac{1}{4}$ mark is slightly longer than the $\frac{1}{8}$ mark.

The division marks are different lengths to make the rule easier to read. The shortest marks represent fractions that, in lowest terms, are sixteenths $\left(\frac{1}{16}, \frac{3}{16}, \frac{5}{16}, \frac{7}{16}, \frac{9}{16}, \frac{11}{16}, \frac{13}{16}, \frac{15}{16}\right)$. The fractions that reduce to eighths are slightly longer than the sixteenths marks $\left(\frac{1}{8}, \frac{3}{8}, \frac{5}{8}, \frac{7}{8}\right)$. The marks representing fractions that reduce to fourths are slightly longer than the eighths $\left(\frac{1}{4}, \frac{3}{4}\right)$. The mark for one-half $\left(\frac{1}{2}\right)$ is longer than the fourths, and the inch mark is the longest.

Did You Know? Some rules do not start the zero mark at the left end of the rule. It is inset slightly (see Fig. 2–9). Be sure to locate the zero mark before you begin measuring.

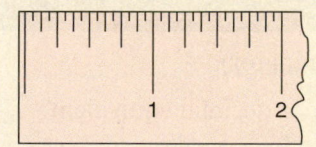

FIGURE 2-9 Zero mark for some rules.

STOP AND CHECK

1. Measure line segment *EF*.

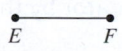

E F

Answer:

1. $\frac{5}{8}$ in.

Measure a line segment (or object) using a U.S. customary rule.

1. Align the left end of the line segment (or object) with the left end of the rule or the zero mark.

2. Determine the last whole unit that the segment passes.

3. Examine the fraction mark and determine the value of the fraction mark where the line segment ends on the right.

4. Write the measure as a mixed number by combining the results from Steps 2 and 3.

EXAMPLE 6

Measure line segment *AB* (Fig. 2–10).

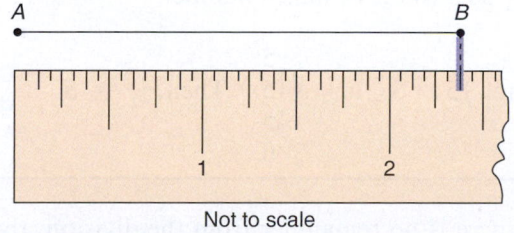

A B

Not to scale

FIGURE 2-10 Line segment AB.

Align point *A* with zero. Line segment *AB* goes past the 2-in. mark but not up to the 3-in. mark. Therefore, the measure of *AB* will be a mixed number between 2 and 3. Point *B* is $\frac{3}{8}$ in. past 2.

AB **is** $2\frac{3}{8}$ **in.**   **See Exercises 26–33.**

TIP **Judge to the Closest Mark** A line segment may not always align exactly with a division mark. Use eye judgment to decide which mark is closer to the end of the line segment.

EXAMPLE 7

Measure line segment *CD* (Fig. 2–11).

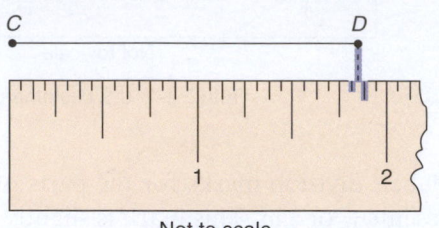

C   D

Not to scale

FIGURE 2–11

Point *D* aligns between $1\frac{13}{16}$ and $1\frac{7}{8}$.

Measurements are always approximations; using our best eye judgment, point *D* seems to be halfway between $1\frac{13}{16}$ and $1\frac{7}{8}$.

We say *CD* is $1\frac{13}{16}$ **in.** or $1\frac{7}{8}$ **in.** *to the nearest sixteenth of an inch.*   **See Exercises 34–35.**

In practice, measurements are considered acceptable if they are within a desired *tolerance.* In the preceding example, the smallest division is $\frac{1}{16}$, so an appropriate tolerance would be plus or minus one-half of one-sixteenth, or $\pm\frac{1}{32}$ ($\frac{1}{2}$ of $\frac{1}{16} = \frac{1}{32}$). That is, the acceptable measure can be $\frac{1}{32}$ more than or $\frac{1}{32}$ less than the ideal measure.

If the ideal measure is halfway between $1\frac{13}{16}$ and $1\frac{7}{8}$ (or equal to $1\frac{27}{32}$) and the tolerance is $\frac{1}{32}$ in., the interval of acceptable values is from $1\frac{27}{32} - \frac{1}{32}$ to $1\frac{27}{32} + \frac{1}{32}$. The acceptable interval is from $1\frac{26}{32}$ to $1\frac{28}{32}$, or $1\frac{13}{16}$ to $1\frac{7}{8}$.

2 Write Improper Fractions as Whole Numbers or Mixed Numbers. Earlier we noted that an improper fraction is a fraction whose value is equal to or greater than one unit, such as $\frac{6}{3}$, $\frac{15}{7}$, or $\frac{5}{5}$. These fractions can be changed to equivalent whole numbers or mixed numbers.

To write an improper fraction as a whole or mixed number:

1. Perform the division indicated (numerator ÷ denominator).

2. Express any remainder of the division as a fraction or decimal equivalent.

EXAMPLE 8

Write $\frac{15}{3}$ as a whole or mixed number.

$\frac{15}{3}$ means $15 \div 3$ or $3\overline{)15}$. Then, $\frac{15}{3} =$ **5.**   Divide the numerator by the denominator.

See Exercises 36–50.

If there is no remainder from the division, the improper fraction converts to a whole number.

TIP Writing a Whole or Mixed Number Is Different from Reducing Do not confuse converting an improper fraction to a whole or mixed number with reducing a fraction to lowest terms. An improper fraction is in lowest terms if its numerator and denominator have no common factor other than 1. Therefore, the improper fraction $\frac{10}{7}$ is written in lowest terms. The improper fraction $\frac{10}{4}$ is not in lowest terms. It will reduce to $\frac{5}{2}$, which is in lowest terms.

When writing an improper fraction as a whole or mixed number, we must make sure that the fraction is in lowest terms. We can reduce to lowest terms either before dividing or after dividing.

STOP AND CHECK
Write as a mixed number.

1. $\dfrac{62}{5}$

2. $\dfrac{38}{6}$

Answers:

1. $12\dfrac{2}{5}$ **2.** $6\dfrac{1}{3}$

EXAMPLE 9

Write $\dfrac{28}{8}$ as a mixed number.

$\dfrac{28}{8} = \dfrac{7}{2}$ $2\overline{\smash{)}7} = 3\dfrac{1}{2}$ Fraction is reduced before dividing.
$\phantom{2\overline{)7}}\dfrac{6}{1}$

or

$\dfrac{28}{8}$ $8\overline{\smash{)}28} = 3\dfrac{4}{8} = 3\dfrac{1}{2}$ Fraction is reduced after dividing.
$\dfrac{24}{4}$

See Exercises 36–50.

3 **Write Whole Numbers or Mixed Numbers as Improper Fractions.** Some computation processes require whole numbers or mixed numbers to be expressed as equivalent improper fractions.

To write a whole number as an improper fraction:

1. Write the whole number as the numerator.

2. Write 1 as the denominator.

STOP AND CHECK
1. Change 4 to thirds.
2. Change 12 to halves.

Answers:

1. $\dfrac{12}{3}$ **2.** $\dfrac{24}{2}$

EXAMPLE 10

Change 8 to fifths.

$8 = \dfrac{8}{1}$ Write 8 as a fraction with a denominator of 1.

$\dfrac{8(5)}{1(5)} = \dfrac{40}{5}$ Multiply by 1 in the form $\dfrac{5}{5}$.

See Exercises 51–55.

To write a mixed number as an improper fraction:

1. Multiply the denominator of the fractional part by the whole number.

2. Add the numerator of the fractional part to the product; this sum becomes the numerator of the improper fraction.

3. The denominator of the improper fraction is the same as the denominator of the fractional part of the mixed number.

STOP AND CHECK
Write as an improper fraction.

1. $5\dfrac{4}{5}$

2. $7\dfrac{8}{9}$

Answers:

1. $\dfrac{29}{5}$ **2.** $\dfrac{71}{9}$

EXAMPLE 11

Change $6\dfrac{2}{3}$ to an improper fraction.

$6\dfrac{2}{3} = \dfrac{(3 \cdot 6) + 2}{3} = \dfrac{20}{3}$ Multiply the denominator times the whole number and add the numerator.

See Exercises 56–70.

TIP Sometimes it is helpful to visualize a process. Look at the process for changing $3\frac{1}{5}$ to an improper fraction represented in words, visually, and symbolically.

In words,
five times three
plus one written
over five.

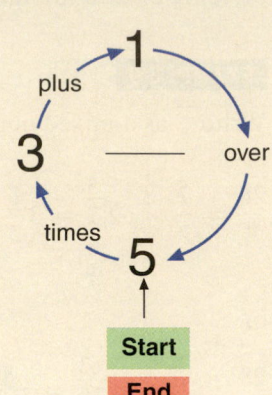

In symbols,

$$3\frac{1}{5} = \frac{5 \times 3 + 1}{5} = \frac{16}{5}$$

4 **Write Decimals as Fractions and Fractions as Decimals.** When a decimal is written as a fraction, the number of digits that follow the decimal point determines the denominator of the fraction.

To write a decimal number as a fraction or mixed number in lowest terms:

1. Write the digits without the decimal point and leading zeros as the numerator.

2. Write the denominator as a power of 10 with as many zeros as there are places after the decimal point.

3. Reduce and, if the fraction is improper, convert to a mixed number. *See Outcome 2 of this section.*

STOP AND CHECK
Write as a fraction or mixed number.
1. 0.21
2. 0.06
3. 6.033
Answers:
1. $\frac{21}{100}$ 2. $\frac{3}{50}$ 3. $6\frac{33}{1,000}$

EXAMPLE 12

Write as a fraction or mixed number **(a)** 0.4, **(b)** 0.075, and **(c)** 3.17.

(a) $0.4 = \frac{4}{10} = \frac{2}{5}$ Tenths indicates a denominator of 10. Reduce to lowest terms.

(b) $0.075 = \frac{75}{1,000} = \frac{3}{40}$ Thousandths indicates a denominator of 1,000. Reduce to lowest terms.

(c) $3.17 = 3\frac{17}{100}$ Hundredths indicates a denominator of 100.

See Exercises 71–88.

Writing decimal numbers as fractions was introduced in greater detail in Chapter 1, Section 1, Outcome 2.

With increased calculator and computer use, we find it convenient to work with decimals.

The bar separating the numerator and denominator of a fraction indicates division: $\frac{2}{5}$ also means $2 \div 5$ or $5\overline{)2}$.

If we show the decimal after the 2 and attach a zero in the tenths place, we can divide.

$$\begin{array}{r} 0.4 \\ 5\overline{)2.0} \end{array}$$

Therefore, $\frac{2}{5} = 0.4$.

To convert a fraction to a decimal number:
1. Place a decimal point after the numerator.
2. Divide the numerator by the denominator using long division.
3. Attach zeros after the decimal point in the dividend as needed for division.

STOP AND CHECK

1. Change $\frac{5}{16}$ to a decimal.

2. Change $\frac{5}{9}$ to a decimal to the nearest thousandth.

3. Change $\frac{3}{17}$ to a decimal to the nearest thousandth.

Answers:
1. 0.3125 2. 0.556 3. 0.176

EXAMPLE 13

Change (a) $\frac{7}{8}$ to a decimal. Change (b) $\frac{1}{3}$ and (c) $\frac{4}{11}$ to a decimal to the nearest thousandth using a calculator.

(a) $7 \div 8$ or

$$\begin{array}{r} 0.875 \\ 8\overline{)7.000} \\ 6\,4 \\ \hline 60 \\ 56 \\ \hline 40 \\ 40 \\ \hline \end{array}$$

Attach zeros and divide until the division terminates; that is, it has no remainder.

(b) $1 \boxed{\div} 3 \approx 0.3333333333 \approx 0.333$ rounded to the nearest thousandth

(c) $4 \boxed{\div} 11 \approx 0.3636363636 \approx 0.364$ rounded to the nearest thousandth

See Exercises 89–106.

When fractions are changed into decimals that do not terminate, the decimals are called **repeating decimals.** A repeating pattern will be established. Repeating decimals can be written with a line over the digits that repeat. Another notation that is used is an **ellipsis** (three periods) to show that the established pattern continues.

Repeating decimal: when the division indicated by a fraction does not terminate. A repeating pattern will be established

Ellipsis (…): three periods used to show that the established pattern continues

Approximate decimal equivalent: rounded decimal equivalent

Exact decimal equivalent: decimal equivalent that terminates or a repeating decimal equivalent that uses appropriate repeating decimal notation

$$\frac{1}{3} = 0.\overline{3} \quad \text{or} \quad 0.33\ldots \qquad \frac{4}{11} = 0.\overline{36} \quad \text{or} \quad 0.3636\ldots$$

The decimal equivalent can be rounded to any desirable place. The rounded decimal equivalent is an **approximate decimal equivalent,** and the repeating decimal notation is an **exact decimal equivalent.**

Did You Know? Approximate numbers are often represented using the $\approx$ symbol. $\frac{1}{3} \approx 0.333$ is read "one-third is approximately equal to 0.333." $\frac{1}{3} = 0.\overline{3}$ and $\frac{1}{3} = 0.333\ldots$ show exact equivalents; whereas $\frac{1}{3} \approx 0.333$ shows an approximate equivalent.

To write a mixed number as a decimal number:
1. The whole-number part remains the same.
2. Write only the fraction part as a decimal by dividing the numerator by the denominator.

STOP AND CHECK

1. Write $3\frac{7}{10}$ as a decimal equivalent.

2. Write $25\frac{3}{4}$ as a decimal equivalent.

Answers:
1. 3.7   2. 25.75

EXAMPLE 14

Change $3\frac{2}{5}$ to a decimal equivalent.

$2 \div 5$ or $\begin{array}{r} 0.4 \\ 5\overline{)2.0} \end{array}$ Write the fraction part as an equivalent decimal.

Then, $3\frac{2}{5} = 3.4.$

See Exercises 107–110.

Like fractions: fractions that have the same denominator

Common denominators: when comparing two or more fractions, a denominator that each denominator will divide into evenly

Least common denominator (LCD): the smallest common denominator for two or more fractions; the least common multiple of the denominators

5 **Compare Fractions, Mixed Numbers, and Decimals.** **Like fractions** are fractions that have the same denominator. When fractions are not like fractions, they can be changed to equivalent fractions with like denominators. We call these like denominators **common denominators.**

The **least common denominator (LCD)** for two or more fractions is the *least common multiple* (LCM) of the denominators.

The least common denominator often can be found by inspection. By inspection, we mean examine each denominator and intuitively (or mentally) determine the LCD. For fractions with larger denominators, you may need to use the procedure for finding the LCM discussed in Section 2–1, Outcome 4.

TIP   **LCM and LCD. Are They the Same? Are They Different?** To find the LCM or LCD, you use the same procedure. The only difference in these two concepts is that the LCD specifically uses the denominators of fractions and the LCM is the least common multiple of numbers in general and may be used for other purposes.

STOP AND CHECK

1. Find the least common denominator for $\frac{3}{4}$, $\frac{11}{18}$, and $\frac{9}{10}$.

Answer:
1. 180

EXAMPLE 15

Find the least common denominator for the fractions $\frac{5}{12}$, $\frac{4}{15}$, $\frac{3}{8}$.

12 =	**15** =	**8** =	Find the prime factorization of the
2(6) =	3(5) =	2(4) =	denominators.
2(2)(3) =		2(2)(2) =	
$2^2(3)$		2^3	

$$\text{LCM or LCD} = 2^3(3)(5)$$
$$= 8(3)(5)$$
$$= \mathbf{120}$$

See Exercises 115–119.

TIP   **Alternative Procedure for Finding the LCM or LCD** We can also find the LCM or LCD of several fractions by dividing duplicated factors and then multiplying.

1. Arrange the denominators horizontally.

2. Divide by any prime factor that divides evenly into *at least two* denominators.

3. Continue Step 2 until there is no common factor for at least two of the denominators.

4. The LCM or LCD is the product of the primes and remaining factors.

5. Look at the denominators from the preceding example.

2	8	12	15	Divide by the prime factor 2.
2	4	6	15	Divide by the prime factor 2.
3	2	3	15	Divide by the prime factor 3.
	2	1	5	

Primes: 2, 2, 3
Remaining factors: 2, 1, 5

$$\text{LCM} = 2 \cdot 2 \cdot 3 \cdot 2 \cdot 1 \cdot 5$$
$$= 120$$

To compare fractions, the denominators must be the same. To compare $\frac{3}{7}$ and $\frac{5}{7}$, which have the same denominators, we compare the numerators, and $\frac{3}{7}$ is smaller than $\frac{5}{7}$.

To compare fractions:

1. Find the least common denominator (LCD).

2. Change each fraction to an equivalent fraction with the least common denominator (LCD) as its denominator.

3. Compare the numerators. The larger numerator indicates the larger fraction.

To compare mixed numbers:

1. Compare the whole-number parts.

2. If the whole-number parts are equal, compare the fraction parts.

3. Compare the numerators after changing fractions to equivalent fractions with the LCD.

STOP AND CHECK

1. Which fraction is larger, $\frac{2}{3}$ or $\frac{3}{4}$?

Answer:

1. $\frac{3}{4}$

EXAMPLE 16

Which fraction is larger, $\frac{1}{4}$ or $\frac{5}{16}$?

The LCD for 4 and 16 is 16 since 16 is evenly divisible by 4.

Change $\frac{1}{4}$ to an equivalent fraction with a denominator of 16.

$$\frac{1}{4} = \frac{1 \cdot 4}{4 \cdot 4} = \frac{4}{16}$$

$\frac{5}{16}$ is larger than $\frac{1}{4}$ since $\frac{5}{16}$ is larger than $\frac{4}{16}$.   **See Exercises 120–129.**

Did You Know? The property of proportions can be used when comparing two fractions. Look at Example 16 again.

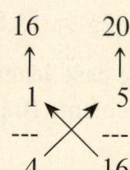

Find the cross products.

Since $16 < 20, \frac{1}{4} < \frac{5}{16}$.
 For the process to work, notice that the cross products are written above the numerators.

EXAMPLE 17

CON   Is it possible to have a pipe with an outside diameter of $\frac{5}{8}$ in. and an inside diameter of $\frac{21}{32}$ in. (see Fig. 2–12)?

To answer this question, we need to compare the two fractions $\frac{5}{8}$ and $\frac{21}{32}$.

$$\frac{5}{8} = \frac{5 \cdot 4}{8 \cdot 4} = \frac{20}{32}$$ Change $\frac{5}{8}$ to 32nds.

Is $\frac{20}{32}$, which is equivalent to $\frac{5}{8}$, larger than $\frac{21}{32}$? No, so **$\frac{5}{8}$ in. cannot be the outside diameter of a pipe with an inside diameter of $\frac{21}{32}$ in.**   **See Exercises 130–139.**

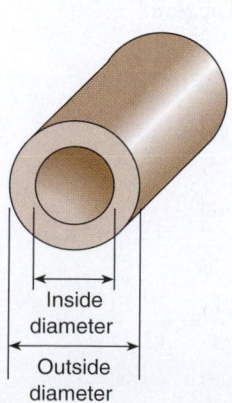

Inside diameter

Outside diameter

FIGURE 2–12

Dmitry Naumov/123RF

STOP AND CHECK

1. Which is smaller, $7\frac{3}{16}$ or 7.2?

Answer:

1. $7\frac{3}{16}$

EXAMPLE 18

CON   Two drill bits have diameters of $\frac{3}{8}$ in. and $\frac{5}{16}$ in., respectively. Which drill bit makes the larger hole?

$$\frac{3}{8} = \frac{6}{16}; \qquad \frac{5}{16} = \frac{5}{16}$$ The least common denominator is 16. Change $\frac{3}{8}$ to 16ths.

Compare the numerators: $\frac{6}{16}$ is larger than $\frac{5}{16}$ because 6 is larger than 5, so $\frac{3}{8}$, which is equivalent to $\frac{6}{16}$, is larger than $\frac{5}{16}$.

The drill bit with a $\frac{3}{8}$-in. diameter will drill the larger hole.   See Exercises 130–139.

TIP   Comparing Fractions Using Decimal Equivalents We can compare fractions by changing them to decimal equivalents. Review comparing decimals in Chapter 1, Section 1, Outcome 3.

To compare a fraction and a decimal, we change one number so that both numbers are either in fraction form or decimal form. Unless otherwise specified, you may use either form.

EXAMPLE 19

COMP   The specifications of an electronic tablet state the length is 15.8 in. and the width is 10 in. A carrying case is $15\frac{7}{8}$ in. by $10\frac{1}{8}$ on the inside. Will the tablet fit inside the case?

$10\frac{1}{8}$ in. is longer than 10 in.

Change $15\frac{7}{8}$ to a decimal.   Change the fractional part of the mixed number to an equivalent decimal. $7 \div 8 = 0.875$

$$15\frac{7}{8} = 15.875$$ Compare decimals.

$15.875 > 15.8$

The case length measurement, 15.875 in., is greater than 15.8 in. and the width measurement, $10\frac{1}{8}$ in. is greater than 10 in., so the tablet will fit into the case.
    See Exercises 140–144.

2-2 EXERCISES   MyLab Math For additional practice go to your study plan in MyLab Math.

1 *See Example 1.*

1. Find five fractions that are equivalent to $\frac{4}{5}$.

2. Find five fractions that are equivalent to $\frac{7}{10}$.

3. Find a fraction that is equivalent to $\frac{3}{4}$ and has a denominator of 24. *See Example 2.*

Find the equivalent fractions with the indicated denominators. *See Example 2.*

4. $\frac{3}{8} = \frac{?}{16}$

5. $\frac{4}{7} = \frac{?}{21}$

6. $\frac{9}{11} = \frac{?}{44}$

7. $\frac{1}{3} = \frac{?}{15}$

8. $\frac{5}{6} = \frac{?}{24}$

9. $\frac{7}{8} = \frac{?}{24}$

10. $\frac{2}{5} = \frac{?}{30}$

Reduce the fractions to lowest terms. *See Examples 3 and 4.*

11. $\dfrac{4}{8}$　　　**12.** $\dfrac{6}{10}$　　　**13.** $\dfrac{12}{16}$　　　**14.** $\dfrac{10}{32}$　　　**15.** $\dfrac{16}{32}$

16. $\dfrac{28}{32}$　　　**17.** $\dfrac{20}{64}$　　　**18.** $\dfrac{2}{10}$　　　**19.** $\dfrac{8}{40}$　　　**20.** $\dfrac{12}{50}$

21. $\dfrac{10}{16}$　　　**22.** $\dfrac{4}{16}$　　　**23.** $\dfrac{24}{32}$

Use the property of proportions to determine if the statements are true or false. *See Example 5.*

24. $\dfrac{3}{8} = \dfrac{24}{32}$　　　　　　**25.** $\dfrac{11}{16} = \dfrac{22}{32}$

Measure line segments 26–35 in Fig. 2–13 to the nearest sixteenth of an inch (tolerance $= \pm\frac{1}{32}$ in.). *See Examples 6 and 7.*

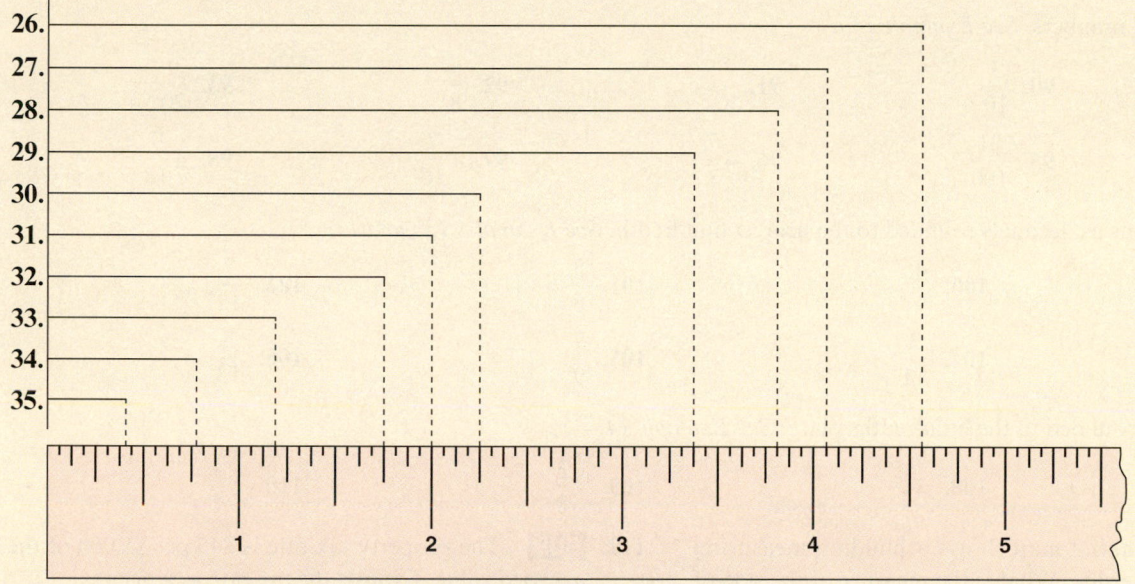

FIGURE 2-13

2 Write the improper fractions as whole or mixed numbers. *See Examples 8 and 9.*

36. $\dfrac{12}{5}$　　　**37.** $\dfrac{10}{7}$　　　**38.** $\dfrac{12}{12}$　　　**39.** $\dfrac{32}{7}$　　　**40.** $\dfrac{24}{6}$

41. $\dfrac{15}{7}$　　　**42.** $\dfrac{23}{9}$　　　**43.** $\dfrac{47}{5}$　　　**44.** $\dfrac{86}{9}$　　　**45.** $\dfrac{38}{21}$

46. $\dfrac{57}{15}$　　　**47.** $\dfrac{64}{4}$　　　**48.** $\dfrac{72}{10}$　　　**49.** $\dfrac{19}{2}$　　　**50.** $\dfrac{36}{4}$

3 Change the whole numbers to an equivalent fraction with the given denominator. *See Example 10.*

51. $5 = \dfrac{?}{3}$　　　**52.** $9 = \dfrac{?}{2}$　　　**53.** $7 = \dfrac{?}{8}$　　　**54.** $8 = \dfrac{?}{4}$　　　**55.** $3 = \dfrac{?}{16}$

Write the mixed numbers as improper fractions. *See Example 11.*

56. $2\dfrac{1}{3}$　　　**57.** $3\dfrac{1}{8}$　　　**58.** $1\dfrac{7}{8}$　　　**59.** $6\dfrac{5}{12}$　　　**60.** $9\dfrac{5}{8}$

61. $3\dfrac{7}{8}$　　　**62.** $7\dfrac{5}{12}$　　　**63.** $6\dfrac{7}{16}$　　　**64.** $8\dfrac{1}{32}$　　　**65.** $1\dfrac{5}{64}$

66. $7\frac{3}{10}$ **67.** $8\frac{2}{3}$ **68.** $33\frac{1}{3}$ **69.** $66\frac{2}{3}$ **70.** $12\frac{1}{2}$

4 Change the decimals to their fraction or mixed-number equivalents, and reduce answers to lowest terms. *See Example 12.*

71. 0.5   **72.** 0.1   **73.** 0.2   **74.** 0.7

75. 0.25   **76.** 0.025   **77.** 3.9   **78.** 4.8

79. 0.378   **80.** 0.875   **81.** 0.375   **82.** 0.625

83. A measure of 0.75 in. represents what fractional part of an inch?

84. What fraction represents 0.1875 ft?

85. **INDTEC**   The length of a screw is 2.375 in. Represent this length as a mixed number.

86. An instrument weighs 0.83 lb. Write this as a fraction of a pound.

87. **INDTEC**   Some sheet metal is 0.3125 in. thick. What is the thickness expressed as a fraction?

88. **COMP**   A predrilled PC board is 3.125 in. long. Write this length as a mixed number.

Change to decimal numbers. *See Example 13a.*

89. $\frac{3}{5}$ **90.** $\frac{3}{10}$ **91.** $\frac{7}{8}$ **92.** $\frac{3}{8}$ **93.** $\frac{9}{20}$

94. $\frac{49}{50}$ **95.** $\frac{21}{100}$ **96.** $3\frac{7}{8}$ **97.** $1\frac{7}{16}$ **98.** $4\frac{9}{16}$

Write these fractions as decimals rounded to the nearest hundredth. *See Examples 13b and 13c.*

99. $\frac{1}{6}$ **100.** $\frac{4}{9}$ **101.** $\frac{5}{6}$ **102.** $\frac{7}{12}$

103. $\frac{2}{3}$ **104.** $\frac{3}{11}$ **105.** $\frac{7}{9}$ **106.** $\frac{5}{13}$

Write as decimals rounded to the hundredths place. *See Example 14.*

107. $1\frac{3}{7}$ **108.** $3\frac{5}{11}$ **109.** $2\frac{3}{8}$ **110.** $5\frac{4}{7}$

111. **TELE**   An aerial map shows a building measuring $2\frac{3}{64}$ in. on one side. What is the measure of the side of the building in decimal numbers?

112. **BUS**   The property tax rate is $45 per $1,000 of the assessed value. Express the tax rate as a decimal.

113. **CON**   A plan allows a gap of $\frac{1}{8}$ in. between vinyl flooring and the wall for expansion. What is the gap measure in decimal notation?

114. **AVIA**   A B-767-200 aircraft has a wingspan of 47.6 m. What is the wingspan written as a mixed number?

5 Find the least common denominator. *See Example 15.*

115. $\frac{5}{8}, \frac{4}{9}$ **116.** $\frac{3}{10}, \frac{4}{15}$ **117.** $\frac{9}{10}, \frac{4}{25}$ **118.** $\frac{7}{12}, \frac{9}{16}, \frac{5}{8}$ **119.** $\frac{2}{3}, \frac{5}{12}, \frac{7}{8}$

Show which fraction is larger. *See Example 16.*

120. $\frac{2}{3}, \frac{3}{5}$ **121.** $\frac{5}{12}, \frac{7}{16}$ **122.** $\frac{8}{9}, \frac{7}{8}$ **123.** $\frac{5}{8}, \frac{11}{16}$ **124.** $\frac{15}{32}, \frac{29}{64}$

125. $\frac{7}{12}, \frac{9}{16}$ **126.** $\frac{3}{8}, \frac{4}{5}$ **127.** $\frac{7}{11}, \frac{9}{10}$ **128.** $\frac{4}{15}, \frac{3}{16}$ **129.** $\frac{1}{2}, \frac{7}{16}$

See Examples 17 and 18.

130. **INDTEC**   Will a pipe with a $\frac{5}{16}$-in. outside diameter fit inside a pipe with a $\frac{3}{8}$-in. inside diameter? *See Example 17.*

131. **INDTEC**   Is the thickness of a $\frac{3}{16}$-in. sheet of metal greater than the length of a $\frac{15}{64}$-in. sheet metal screw?

132. **CON** A range top is $29\frac{1}{8}$ in. long by $19\frac{1}{2}$ in. wide. Is it smaller than an existing opening $29\frac{3}{16}$ in. long by $19\frac{9}{16}$ in. wide?

133. **INDTEC** A hollow-wall fastener has a grip range up to $\frac{7}{16}$ in. Is it long enough to fasten a thin sheet metal strip to a plywood wall if the combined thickness of the wall is $\frac{3}{8}$ in.?

134. **INDTEC** Charles Bryant has a wrench marked $\frac{25}{32}$, but it is too large. Would a $\frac{7}{8}$-in. wrench be smaller?

135. **INDTEC** A wrench is marked $\frac{5}{8}$ at one end and $\frac{19}{32}$ at the other. Which end is larger? *See Example 18.*

136. **INDTEC** A plastic anchor for a No. 6 × 1-in. screw requires that at least a $\frac{3}{16}$-in. diameter hole be drilled. Will a $\frac{1}{4}$-in. drill bit be large enough?

137. **INDTEC** For a do-it-yourself project, Brenda Jinkins needs to cut a piece of sheet metal slightly longer than the $10\frac{21}{32}$ in. required in the plans and to trim it down to size. Brenda cuts the piece $10\frac{3}{4}$ in. Is it too short?

138. **INDTEC** Is a $\frac{3}{8}$-in. wrench too large or too small for a $\frac{1}{2}$-in. bolt?

139. **INDTEC** The plastic anchor in Exercise 136 requires a minimum hole depth of $\frac{7}{8}$ in. Is a $\frac{3}{4}$-in. hole deep enough?

Show which common fraction or decimal is smaller. *See Example 19.*

140. $\frac{3}{8}$, 0.37

141. $\frac{4}{5}$, 0.82

142. $\frac{5}{12}$, 0.42

143. $\frac{1}{2}$, 0.65

144. $\frac{3}{4}$, 0.34

2–3 Adding and Subtracting Fractions and Mixed Numbers

LEARNING OUTCOMES

1 Add fractions and mixed numbers.

2 Subtract fractions and mixed numbers.

LC LEARNING CATALYTICS

1. Change $\frac{3}{4}$ and $\frac{5}{6}$ to equivalent fractions with a common denominator.

2. Change $\frac{19}{12}$ to a mixed number.

1 Add Fractions and Mixed Numbers. Adding fractions requires that all fractions being added have the same denominator. Trying to add unlike fractions is like trying to add unlike objects or measures. Before adding unlike fractions, we change the fractions to equivalent fractions with a common denominator.

To add fractions:

1. If the denominators are not the same, find the least common denominator.

2. Change each fraction not already expressed in terms of the common denominator to an equivalent fraction with the common denominator.

3. Add the numerators only.

4. The common denominator is the denominator of the sum.

5. Reduce the sum to lowest terms and change improper fractions to whole or mixed numbers.

STOP AND CHECK

1. Find the sum of $\frac{3}{4} + \frac{5}{6}$.

2. Find the sum of $\frac{3}{8} + \frac{5}{16} + \frac{3}{4}$.

Answers:

1. $1\frac{7}{12}$ 2. $1\frac{7}{16}$

EXAMPLE 1

Find the sum of **(a)** $\dfrac{3}{8} + \dfrac{1}{8}$ and   **(b)** $\dfrac{5}{32} + \dfrac{3}{16} + \dfrac{7}{8}$.

(a) Because the denominators are the same, start with Step 3 of the addition procedure.

$$\frac{3}{8} + \frac{1}{8} = \frac{4}{8} = \frac{1}{2} \qquad \text{Add numerators and reduce.}$$

(b) The least common denominator may be found by inspection. Both 8 and 16 divide evenly into 32, so we use 32 as the common denominator.

$$\frac{5}{32} = \frac{5}{32}, \quad \frac{3}{16} = \frac{6}{32}, \quad \frac{7}{8} = \frac{28}{32}$$

Change each fraction to an equivalent fraction whose denominator is 32.

$$\frac{5}{32} + \frac{6}{32} + \frac{28}{32} = \frac{39}{32}$$

Add the numerators.

$$\frac{39}{32} = 1\frac{7}{32}$$

Change to a mixed number.

See Exercises 1–12.

EXAMPLE 2

CON A plumber uses a $\frac{9}{16}$-in.-diameter copper pipe wrapped with $\frac{5}{8}$-in. insulation (Fig. 2–14). What size hole must he bore in the stud (wall support) to install the insulated pipe?

To find the total diameter of the pipe and insulation, add $\frac{9}{16} + \frac{5}{8} + \frac{5}{8}$. The thickness of the insulation is added twice because it counts in the total diameter of the pipe and insulation two times.

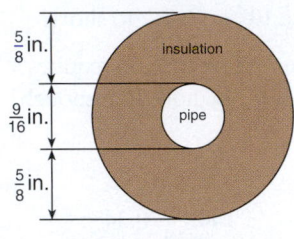

$\frac{5}{8}$ in.
insulation
$\frac{9}{16}$ in. pipe
$\frac{5}{8}$ in.

FIGURE 2–14

$$\frac{9}{16} = \frac{9}{16}$$

The LCD is 16. Change each fraction to 16ths.

$$\frac{5}{8} = \frac{10}{16}$$

$$+\frac{5}{8} = \frac{10}{16}$$

Add numerators. Keep the common denominator.

$$\frac{29}{16} = 1\frac{13}{16}$$

Change to a mixed number.

The total diameter is $1\frac{13}{16}$ in., and the diameter of the hole must be at least this large.

See Exercise 16.

To add mixed numbers:
1. Add the whole-number parts.
2. Add the fractional parts and reduce to lowest terms.
3. Change improper fractions to whole or mixed numbers.
4. Add whole-number parts.

STOP AND CHECK

1. Add $8\frac{3}{4} + 9\frac{4}{5} + 3\frac{2}{3}$.

Answer:

1. $22\frac{13}{60}$

EXAMPLE 3

Add $5\frac{2}{3} + 7\frac{3}{8} + 4\frac{1}{2}$.

$$5\frac{2}{3} = 5\frac{16}{24}$$

The LCD is 24. Change fractions to equivalent fractions.

$$7\frac{3}{8} = 7\frac{9}{24}$$

Add whole numbers.

$$+4\frac{1}{2} = 4\frac{12}{24}$$

Add fractional parts.

$$16\frac{37}{24}$$

$\frac{37}{24} = 1\frac{13}{24}$. Change improper fraction to a mixed number.

$$16 + 1\frac{13}{24} = 17\frac{13}{24}$$

Add whole-number parts.

See Exercises 18–29.

2 Subtract Fractions and Mixed Numbers. The steps for subtracting fractions and mixed numbers are very similar to the steps for adding fractions and mixed numbers.

> **To subtract fractions:**
>
> 1. If the denominators are not the same, find the least common denominator.
> 2. Change each fraction not expressed in terms of the common denominator to an equivalent fraction having the common denominator.
> 3. Subtract the numerators.
> 4. The common denominator will be the denominator of the difference.
> 5. Reduce the difference to lowest terms.

STOP AND CHECK

1. Subtract $\frac{4}{5} - \frac{7}{10}$.

Answer:

1. $\frac{1}{10}$

EXAMPLE 4

Subtract $\frac{3}{8} - \frac{7}{32}$.

$$\frac{3}{8} = \frac{12}{32} \qquad \text{Change } \frac{3}{8} \text{ to an equivalent fraction with a denominator of 32.}$$

$$-\frac{7}{32} = \frac{7}{32} \qquad \text{Subtract numerators and keep the common denominator.}$$

$$\frac{5}{32}$$

See Exercises 36–39.

> **To subtract mixed numbers:**
>
> 1. If the fractional parts of the mixed numbers do not have the same denominator, change them to equivalent fractions with a common denominator.
> 2. When the fraction in the minuend is larger than the fraction in the subtrahend, go to Step 6.
> 3. When the fraction in the subtrahend is larger than the fraction in the minuend, regroup (borrow) by taking one whole number from the whole-number part of the minuend. This makes the whole number 1 less.
> 4. Change the whole number that was borrowed to an improper fraction with the common denominator. For example, $1 = \frac{3}{3}, 1 = \frac{8}{8}, 1 = \frac{n}{n}$, where n is the common denominator.
> 5. Add the borrowed fraction $\left(\frac{n}{n}\right)$ to the fraction already in the minuend.
> 6. Subtract the fractional parts and the whole-number parts.
> 7. Reduce the difference to lowest terms.

STOP AND CHECK

1. Subtract $23\frac{2}{3} - 8\frac{1}{6}$.
2. Subtract $14\frac{7}{8}$ from $19\frac{3}{4}$.

Answers:

1. $15\frac{1}{2}$ 2. $4\frac{7}{8}$

EXAMPLE 5

(a) Subtract $15\frac{7}{8} - 4\frac{1}{2}$. **(b)** Subtract $15\frac{3}{4}$ from $18\frac{1}{2}$. **(c)** Subtract 27 from $45\frac{1}{3}$.

(a)
$$15\frac{7}{8} = 15\frac{7}{8} \qquad \text{LCD is 8.}$$

$$-4\frac{1}{2} = 4\frac{4}{8} \qquad \begin{array}{l}\text{Subtract like fractions.}\\ \text{Subtract whole numbers.}\end{array}$$

$$11\frac{3}{8}$$

(b) $18\frac{1}{2} = 18\frac{2}{4} = 17\frac{4}{4} + \frac{2}{4} = 17\frac{6}{4}$ LCD = 4. Change $\frac{1}{2}$ to $\frac{2}{4}$.

$-15\frac{3}{4} = 15\frac{3}{4} = 15\frac{3}{4}\phantom{+ \frac{2}{4}} = 15\frac{3}{4}$

Regroup. $18 - 1 = 17, 1 = \frac{4}{4}, \frac{4}{4} + \frac{2}{4} = \frac{6}{4}$

Subtract fractions.

$2\frac{3}{4}$

Subtract whole numbers.

(c) $45\frac{1}{3} = 45\frac{1}{3}$

$-27\phantom{\frac{1}{3}} = 27\frac{0}{3}$ Write 27 as a mixed number. Subtract.

$18\frac{1}{3}$

See Exercises 40–43.

EXAMPLE 6

HOSP $127\frac{1}{2}$ lb of sugar is used from an inventory of $433\frac{3}{8}$ lb. How many pounds of sugar remain in inventory?

$433\frac{3}{8} = 433\frac{3}{8} = 432\frac{11}{8}$ $433 - 1 = 432, 1 = \frac{8}{8}, \frac{8}{8} + \frac{3}{8} = \frac{11}{8}$

$-127\frac{1}{2} = 127\frac{4}{8} = 127\frac{4}{8}$ Subtract fractions. Subtract whole numbers.

$\phantom{-127\frac{1}{2} = 127\frac{4}{8} = }305\frac{7}{8}$

$305\frac{7}{8}$ lb of sugar remain in inventory.   See Exercises 44–45.

EXAMPLE 7

TELE   How many feet of coaxial cable are left on a 100-ft roll if $27\frac{1}{4}$ ft are used from the roll?

$100 = 100\frac{0}{4} = 99\frac{4}{4}$ Write 100 as a mixed number. Regroup.

$-27\frac{1}{4} = 27\frac{1}{4} = 27\frac{1}{4}$ Subtract.

$\phantom{-27\frac{1}{4} = 27\frac{1}{4} = }72\frac{3}{4}$

$72\frac{3}{4}$ ft of cable are left on the roll.   See Exercises 46–47.

Oksana Tkachuk/123RF

There are times when you will need to both add and subtract fractions and mixed numbers in completing a series of applications.

STOP AND CHECK

1. Perform the indicated operations. $3\frac{5}{8} + 7\frac{1}{3} - 6\frac{1}{2}$.

Answer:

1. $4\frac{11}{24}$

EXAMPLE 8

Perform the indicated operations. $5\frac{2}{3} + 1\frac{1}{2} - 3\frac{2}{5}$.

$5\frac{2}{3} + 1\frac{1}{2} - 3\frac{2}{5} =$ The LCD is $(3)(2)(5) = 30$.

$$5\frac{2\cdot10}{3\cdot10} + 1\frac{1\cdot15}{2\cdot15} - 3\frac{2\cdot6}{5\cdot6} =$$ Change each fraction to a fraction with the LCD.

$$5\frac{20}{30} + 1\frac{15}{30} - 3\frac{12}{30} =$$ Add the first two mixed numbers.

$$6\frac{35}{30} - 3\frac{12}{30} =$$ Do not change improper fraction to mixed number. Subtract.

$$3\frac{23}{30}$$

See Exercises 48–51.

EXAMPLE 9

INDTR Three lengths measuring $5\frac{1}{4}$ in., $7\frac{3}{8}$ in., and $6\frac{1}{2}$ in. are cut from a 64-in. bar of angle iron. If $\frac{3}{16}$ in. is wasted on each cut, how many inches of angle iron remain?

Visualize the problem by making a sketch (see Fig. 2–15). Then find the total amount of angle iron used. This includes the three lengths and the waste for three cuts.

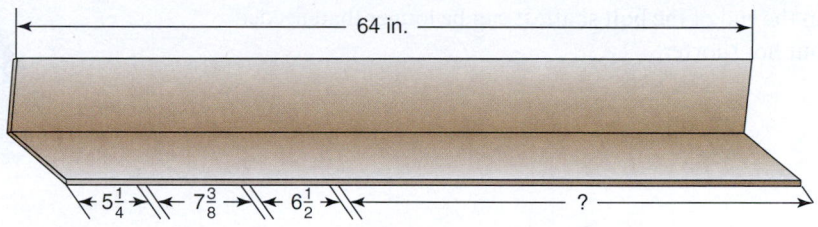

FIGURE 2–15

Total removed and wasted = 3 lengths + 3 cuts

Three Lengths	**Three Cuts**	
$5\frac{1}{4} + 7\frac{3}{8} + 6\frac{1}{2} +$	$\frac{3}{16} + \frac{3}{16} + \frac{3}{16} =$	Change fractions to equivalent fractions with LCD of 16 and add $5 + 7 + 6 = 18$ $4 + 6 + 8 + 3 + 3 + 3 = 27.$

$$5\frac{4}{16} + 7\frac{6}{16} + 6\frac{8}{16} + \frac{3}{16} + \frac{3}{16} + \frac{3}{16} = 18\frac{27}{16}$$ Regroup mixed number.

$$18\frac{27}{16} = 18 + \frac{27}{16} = 18 + 1\frac{11}{16} = 19\frac{11}{16}$$ Total amount removed and wasted. Regroup and subtract.

amount of angle iron remaining = $\dfrac{\text{beginning}}{\text{length}}$ − $\dfrac{\text{total iron removed}}{\text{and wasted}}$

$$64 - 19\frac{11}{16} =$$ Total remaining.

$$63\frac{16}{16} - 19\frac{11}{16} = 44\frac{5}{16}$$

$44\frac{5}{16}$ in. of angle iron remain.

See Exercises 52–54.

2-3 EXERCISES

MyLab Math For additional practice go to your study plan in MyLab Math.

1 Add; reduce answers to lowest terms and write any improper fractions as whole or mixed numbers. *See Example 1.*

1. $\dfrac{5}{16} + \dfrac{1}{16}$ **2.** $\dfrac{1}{2} + \dfrac{1}{8} + \dfrac{3}{4}$ **3.** $\dfrac{1}{8} + \dfrac{1}{2}$ **4.** $\dfrac{3}{8} + \dfrac{5}{32} + \dfrac{1}{4}$ **5.** $\dfrac{5}{16} + \dfrac{1}{4}$ **6.** $\dfrac{15}{16} + \dfrac{1}{2}$

7. $\dfrac{3}{32} + \dfrac{5}{64}$ **8.** $\dfrac{7}{8} + \dfrac{3}{5}$ **9.** $\dfrac{3}{4} + \dfrac{8}{9}$ **10.** $\dfrac{7}{8} + \dfrac{5}{24}$ **11.** $\dfrac{3}{5} + \dfrac{4}{5}$ **12.** $\dfrac{5}{7} + \dfrac{4}{21}$

13. **CON** What is the thickness of a countertop made of $\frac{7}{8}$-in. plywood and $\frac{1}{16}$-in. Formica?

14. **INDTR** Three pieces of steel are joined together. What is the total thickness if the pieces are $\frac{1}{2}$ in., $\frac{7}{16}$ in., and $\frac{29}{32}$ in.?

15. **BUS** Three books are placed side by side. They are $\frac{5}{16}$ in., $\frac{7}{8}$ in., and $\frac{3}{4}$ in. wide. What is the total width of the books if they are polywrapped in one package?

16. **AUTO** Find the outside diameter of a hose (Fig. 2–16) whose wall is $\frac{1}{2}$ in. thick if the inside diameter is $\frac{7}{8}$ in. *See Example 2.*

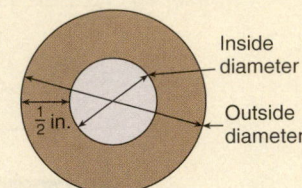

FIGURE 2–16

17. **INDTEC** What is the minimum bolt length needed to fasten two pieces of metal each $\frac{7}{16}$ in. thick if a $\frac{1}{8}$-in. lock washer and a $\frac{1}{4}$-in. nut are used (Fig. 2-17)? The length of a bolt is measured from the bottom of the bolt head to the end of the bolt shaft. It can be longer than needed but not shorter.

Kaban-Sila/Shutterstock

FIGURE 2–17

Add; reduce answers to lowest terms and write any improper fractions as whole or mixed numbers. *See Example 3.*

18. $2\dfrac{3}{5} + 4\dfrac{1}{5}$ **19.** $1\dfrac{5}{8} + 2\dfrac{1}{2}$ **20.** $3\dfrac{3}{4} + 7\dfrac{3}{16} + 5\dfrac{7}{8}$ **21.** $\dfrac{1}{6} + \dfrac{7}{9} + \dfrac{2}{3}$

22. $2\dfrac{1}{4} + 2\dfrac{9}{16}$ **23.** $1\dfrac{5}{16} + 4\dfrac{7}{32}$ **24.** $3\dfrac{1}{4} + 1\dfrac{7}{16}$ **25.** $4\dfrac{1}{2} + 9$

26. $518\dfrac{7}{12} + 483\dfrac{5}{18}$ **27.** $291\dfrac{8}{15} + 78\dfrac{31}{40}$ **28.** $309\dfrac{11}{18} + 805\dfrac{13}{30}$ **29.** $78\dfrac{17}{20} + 46\dfrac{9}{32}$

30. **CON** The studs (interior supports) of an outside wall are $5\frac{3}{4}$ in. thick. The inside wallboard is $\frac{7}{8}$ in. thick, and the outside covering is $2\frac{3}{16}$ in. thick. What is the total thickness of the wall?

31. **CAD/ARC** A blueprint calls for a piece of bar stock $3\frac{7}{8}$ in. long. If a tolerance of $\pm \frac{1}{16}$ in. is allowed, what is the longest acceptable measurement for the bar stock?

32. **HOSP** If $4\frac{3}{8}$ gal of water are used to dilute $7\frac{1}{4}$ gal of juice, how many gallons are in the mixture?

33. **CON** How much bar stock is needed to make bars of the following lengths: $10\frac{1}{4}$ in., $8\frac{7}{16}$ in., $5\frac{15}{32}$ in.? Disregard waste.

34. **HLTH/N** Three pieces of bandage material each measuring $7\frac{5}{8}$ in. are needed to complete a job. How much bandage material is needed?

35. **HOSP** If $5\frac{1}{8}$ cups of water are mixed with $\frac{3}{4}$ cup of Kool Aid, how many cups are in the mixture?

2 Subtract; reduce when necessary. *See Example 4.*

36. $\dfrac{9}{16} - \dfrac{3}{8}$ **37.** $\dfrac{7}{16} - \dfrac{1}{4}$ **38.** $\dfrac{5}{8} - \dfrac{1}{2}$ **39.** $\dfrac{5}{32} - \dfrac{1}{64}$

See Example 5.

40. $23\dfrac{3}{16} - 5\dfrac{7}{16}$ **41.** $9\dfrac{1}{4} - 4\dfrac{5}{16}$ **42.** $9\dfrac{1}{32} - 3\dfrac{3}{8}$ **43.** $14\dfrac{1}{7} - 12\dfrac{3}{7}$

See Example 6.

44. **INDTR**   A length of bar stock $16\frac{3}{8}$ in. long is cut so that a piece only $7\frac{9}{16}$ in. long remains. What is the length of the cutoff piece? Disregard waste.

45. **INDTR**   A casting is machined so that $22\frac{1}{5}$ lb of metal remain. If the casting weighed $25\frac{3}{10}$ lb, how many pounds were removed by machine?

See Example 7.

46. **AG/H**   A flower bed includes $7\frac{7}{8}$ in. of base fill. If the bed is to be 18 in. thick, how thick must the topsoil be?

47. **AVIA**   An airport cargo handler is loading containers on a plane. The linear space available is 50 ft and the containers being loaded will take up $15\frac{1}{4}$ ft of linear space. How much linear space will still be available for additional cargo?

Perform the indicated operations. See Example 8.

48. $2\frac{5}{8} + 7\frac{1}{4} - 6\frac{1}{12}$

49. $21\frac{5}{12} - 7\frac{5}{16} + 4\frac{1}{4}$

50. $183\frac{2}{5} + 555\frac{7}{25} - 388\frac{8}{35}$ 51. $843 - 115\frac{7}{8} + 32\frac{3}{14}$

See Example 9.

52. **AVIA**   A Delta Airlines freight container is 60.4 in. wide. A freight forwarder has three boxes that have widths of $12\frac{3}{16}$ in., $32\frac{3}{8}$ in., and $15\frac{3}{4}$ in. Will these three boxes fit in the container?

53. **AUTO**   A bolt $2\frac{5}{8}$ in. long fastens two pieces of metal 1 in. and $1\frac{7}{32}$ in. thick. If a $\frac{3}{32}$-in.-thick lock washer and a $\frac{1}{8}$-in.-thick flat washer are used, what thickness is the nut if it is even with the end of the bolt? The measure of a bolt length does not include the bolt head.

54. **CAD/ARC**   Find the missing length in Fig. 2–18.

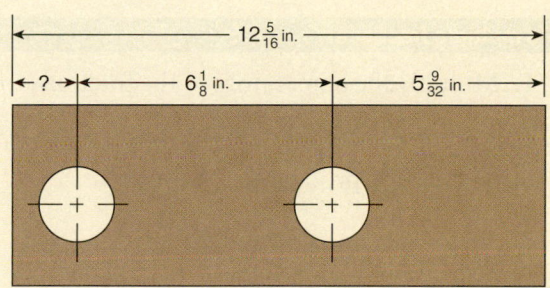

FIGURE 2–18

2-4   # Multiplying and Dividing Fractions and Mixed Numbers

LEARNING OUTCOMES

1   Multiply fractions and mixed numbers.

2   Raise a fraction to a power.

3   Divide fractions and mixed numbers.

4   Perform calculations involving fractions with a calculator.

When multiplying a fraction by a fraction, we are finding *a part of a part*.

1   Multiply Fractions and Mixed Numbers. In multiplication, $\frac{1}{2} \times \frac{1}{2}$ is $\frac{1}{2}$ of $\frac{1}{2}$ (Fig. 2–19). The word "of" is the clue that we must multiply to find the part we are looking for.

$$\frac{1}{2} \times \frac{1}{2} = \frac{1}{4}$$

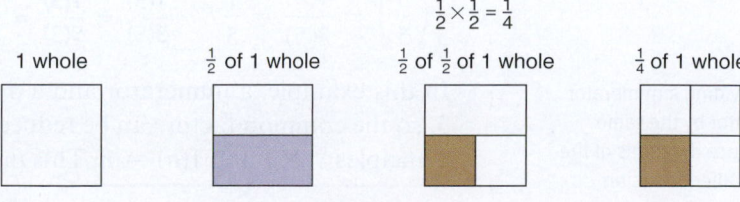

1 whole $\frac{1}{2}$ of 1 whole $\frac{1}{2}$ of $\frac{1}{2}$ of 1 whole $\frac{1}{4}$ of 1 whole

FIGURE 2–19

LC **LEARNING CATALYTICS**

1. What is $\frac{1}{2}$ of 4?
2. What is $\frac{1}{2}$ of $\frac{1}{4}$?

Adding or subtracting fractions and mixed numbers requires a common denominator. In multiplying fractions, we do *not* change fractions to equivalent fractions with a common denominator. Look at two more examples of taking a part of a part (Figs. 2–20 and 2–21).

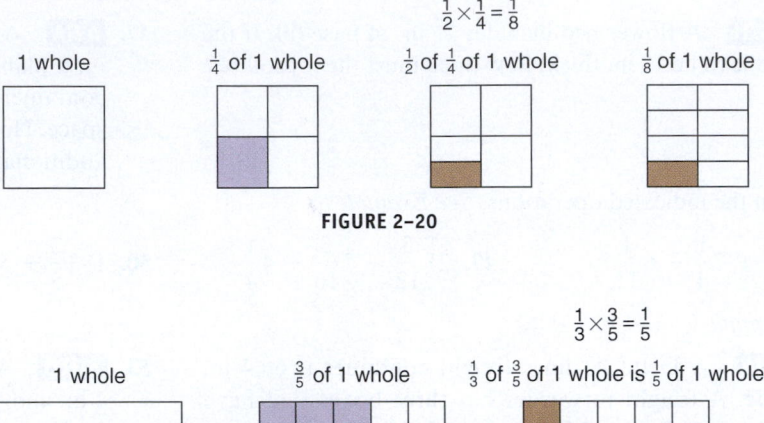

$$\frac{1}{2} \times \frac{1}{4} = \frac{1}{8}$$

1 whole $\frac{1}{4}$ of 1 whole $\frac{1}{2}$ of $\frac{1}{4}$ of 1 whole $\frac{1}{8}$ of 1 whole

FIGURE 2–20

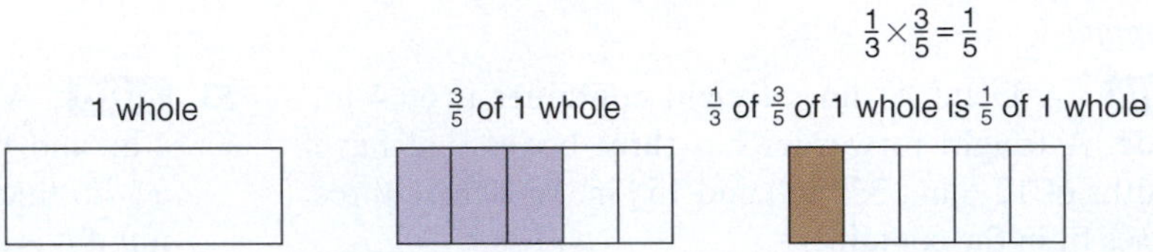

$$\frac{1}{3} \times \frac{3}{5} = \frac{1}{5}$$

1 whole $\frac{3}{5}$ of 1 whole $\frac{1}{3}$ of $\frac{3}{5}$ of 1 whole is $\frac{1}{5}$ of 1 whole

FIGURE 2–21

> **To multiply fractions:**
>
> 1. Multiply the numerators of the fractions to get the numerator of the product.
> 2. Multiply the denominators to get the denominator of the product.
> 3. Reduce the product to lowest terms.

STOP AND CHECK

1. Find $\frac{1}{4}$ of $\frac{2}{3}$.

Answer:

1. $\frac{1}{6}$

EXAMPLE 1

Find **(a)** $\frac{1}{2}$ of $\frac{1}{4}$ and   **(b)** $\frac{1}{3}$ of $\frac{3}{5}$.

(a) $\dfrac{1}{2}\left(\dfrac{1}{4}\right) = \dfrac{1}{8}$ Multiply numerators.
Multiply denominators.

(b) $\dfrac{1}{3}\left(\dfrac{3}{5}\right) = \dfrac{3}{15} = \dfrac{1}{5}$ Multiply numerators.
Multiply denominators. Reduce.   **See Exercise 1–4.**

> **TIP**   **Reduce, or Cancel, Before Multiplying** In the preceding example, reducing is possible. When multiplying fractions, we can reduce common factors before multiplying them.
>
> $$\frac{1}{3}\left(\frac{3}{5}\right) = \frac{1(\overset{1}{\cancel{3}})}{\underset{1}{\cancel{3}}(5)} = \frac{1}{5} \qquad \frac{1(3)}{3(5)} = \frac{1(3)}{5(3)} = \frac{1}{5}\left(\frac{3}{3}\right) = \frac{1}{5}(1) = \frac{1}{5}$$
>
> In this example, a numerator and a denominator both have a common factor of 3, so the common factor can be reduced before multiplying. Reducing applies the principles $\frac{n}{n} = 1$ and $1(n) = n$. This process is also referred to as **canceling.**

Canceling: dividing a numerator and denominator by the same amount to reduce the terms of the fraction; also called reducing

EXAMPLE 2

Multiply **(a)** $\dfrac{2}{3}\left(\dfrac{5}{9}\right)\left(\dfrac{1}{6}\right)$ and   **(b)** $\dfrac{2}{5}\left(\dfrac{10}{21}\right)\left(\dfrac{6}{12}\right)$.

(a) $\dfrac{\overset{1}{\cancel{2}}}{3}\left(\dfrac{5}{9}\right)\left(\dfrac{1}{\underset{3}{\cancel{6}}}\right) = \dfrac{5}{81}$ 2 is a common factor of both a numerator and a denominator. Reduce before multiplying.

(b) $\dfrac{\overset{1}{\cancel{2}}}{\underset{1}{\cancel{5}}}\left(\dfrac{\overset{2}{\cancel{10}}}{21}\right)\left(\dfrac{\overset{1}{\cancel{6}}}{\underset{2}{\cancel{12}}}\right) = \dfrac{2}{21}$ 5 and 10 are diagonal to each other.
6 is above 12.
2 and 2 are separated by another fraction.

Other patterns of reducing could also have been used.   **See Exercises 5–6.**

TIP   **Reduce from Any Numerator to Any Denominator** Common factors that are reduced can be diagonal to each other, one above the other, or separated by another fraction, but one factor *must* be in the numerator and the other in the denominator.

$$\dfrac{2}{\underset{1}{\cancel{3}}}\left(\dfrac{\overset{1}{\cancel{3}}}{5}\right) = \dfrac{2}{5} \qquad \dfrac{\overset{3}{\cancel{6}}}{\underset{4}{\cancel{8}}}\left(\dfrac{3}{5}\right) = \dfrac{9}{20} \qquad \dfrac{\overset{1}{\cancel{2}}}{7}\left(\dfrac{1}{3}\right)\left(\dfrac{5}{\underset{4}{\cancel{8}}}\right) = \dfrac{5}{84}$$

When multiplying mixed numbers or combinations of whole numbers, fractions, and mixed numbers, we first change each mixed number or whole number to an improper fraction. Then we proceed as in multiplying fractions.

To multiply mixed numbers, fractions, and whole numbers:

1. Change each mixed number or whole number to an improper fraction.
2. Reduce as much as possible.
3. Multiply remaining numerators.
4. Multiply remaining denominators.
5. Write the answer as a whole or mixed number as appropriate.

EXAMPLE 3

Multiply **(a)** $3\left(4\dfrac{1}{3}\right)$ and **(b)** $2\dfrac{1}{2}\left(5\dfrac{1}{3}\right)$.

(a) $3\left(4\dfrac{1}{3}\right) =$

$\dfrac{3}{1}\left(\dfrac{13}{3}\right) =$ Change whole number and mixed number to improper fractions.

$\dfrac{\overset{1}{\cancel{3}}}{1}\left(\dfrac{13}{\underset{1}{\cancel{3}}}\right) =$ Reduce.

$\dfrac{1}{1}\left(\dfrac{13}{1}\right) =$ Multiply numerators. Multiply denominators.

$\dfrac{13}{1} =$

13 Change the product to a whole number.

(b) $2\dfrac{1}{2}\left(5\dfrac{1}{3}\right) =$

$\dfrac{5}{2}\left(\dfrac{16}{3}\right) =$ Change each mixed number to an improper fraction.

$\dfrac{5}{\overset{}{\underset{1}{2}}}\left(\dfrac{\overset{8}{16}}{3}\right) =$ Reduce.

$\dfrac{5}{1}\left(\dfrac{8}{3}\right) =$ Multiply numerators. Multiply denominators.

$\dfrac{40}{3} =$ Change the product to a mixed number.

$13\dfrac{1}{3}$

See Exercises 7–12.

EXAMPLE 4

AG/H   Bedding plants are to be planted $3\dfrac{5}{8}$ in. apart and $3\dfrac{5}{8}$ in. from the end of the planter. What length planter is needed for 9 plants (Fig. 2–22)?

$3\dfrac{5}{8}(10) = \dfrac{29}{\overset{}{\underset{4}{8}}}\left(\dfrac{\overset{5}{10}}{1}\right) = \dfrac{145}{4} = 36\dfrac{1}{4}$ 10 equal spaces of $3\dfrac{5}{8}$ in. are required.

$3\dfrac{5}{8}$ $3\dfrac{5}{8}$ $3\dfrac{5}{8}$

FIGURE 2–22

A $36\dfrac{1}{4}$ in. planter will be needed.   See Exercises 13–14.

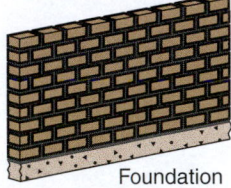

Foundation

FIGURE 2–23

EXAMPLE 5

CON   Bricks that are $2\dfrac{1}{4}$ in. thick form a brick wall with $\dfrac{3}{8}$-in. mortar joints (Fig. 2–23). What is the height of the wall above the foundation after nine courses?

Use the Six-Step Problem-Solving Plan.

Unknown facts

Height of the wall after nine courses of brick have been laid.

Known facts

$2\dfrac{1}{4}$ in.   thickness of each brick

$\dfrac{3}{8}$ in.   thickness of each mortar joint

9   number of courses (or rows) of brick and mortar joints

Relationships

Height of wall = thickness of each brick × number of rows of brick + thickness of each mortar joint × number of mortar joints

Estimation

Each brick is a little more than 2 in. thick, so the wall should be at least 2×9 or 18 in. high. Since the mortar joint is not quite $\frac{1}{2}$ in. and the fractional portion of the brick's thickness is less than $\frac{1}{2}$ in., the combined thickness of the brick and mortar joint must be less than 3 in. So the total height of the wall should be less than 3×9 or 27 in. Thus, we estimate the wall height to be between 18 and 27 in.

Calculation

$$\left(9 \cdot 2\frac{1}{4}\right) + \left(9 \cdot \frac{3}{8}\right)$$ Write as improper fractions.

$$\left(\frac{9}{1} \cdot \frac{9}{4}\right) + \left(\frac{9}{1} \cdot \frac{3}{8}\right)$$ Multiply.

$$\frac{81}{4} + \frac{27}{8}$$ Change to mixed numbers.

$$20\frac{1}{4} + 3\frac{3}{8}$$ Write fractions with a common denominator.

$$20\frac{2}{8} + 3\frac{3}{8} = 23\frac{5}{8}$$ Add.

Interpretation

The wall will be $23\frac{5}{8}$ in. high.

See Exercise 18.

---

2   Raise a Fraction to a Power.

To raise a fraction or quotient to a power:

1. Raise the numerator to the power.

2. Raise the denominator to the power. The denominator cannot be zero.

Symbolically,

$$\left(\frac{a}{b}\right)^n = \frac{a^n}{b^n} \qquad b \neq 0 \text{ and } a, b, \text{ and } n \text{ are real numbers}$$

STOP AND CHECK

Raise to the indicated powers.

1. $\left(\dfrac{4}{5}\right)^2$ **2.** $\left(\dfrac{2}{3}\right)^3$

Answers:

1. $\dfrac{16}{25}$ **2.** $\dfrac{8}{27}$

EXAMPLE 6

Raise the fractions to the indicated powers.

(a) $\left(\dfrac{2}{3}\right)^2$ **(b)** $\left(\dfrac{1}{2}\right)^3$

(a) $\left(\dfrac{2}{3}\right)^2 = \dfrac{2^2}{3^2} = \dfrac{4}{9}$ Raise the numerator to the power.
Raise the denominator to the power.

(b) $\left(\dfrac{1}{2}\right)^3 = \dfrac{1^3}{2^3} = \dfrac{1}{8}$ $(1)(1)(1) = 1; (2)(2)(2) = 8$

See Exercises 19–23.

---

3   Divide Fractions and Mixed Numbers. If we compare $12 \div 3$ and $\frac{1}{3} \times 12$, we find that both answers are 4. That is, $12 \div 3 = \frac{1}{3}(12)$ or $12(\frac{1}{3})$. Not only is there a relationship between multiplication and division, there is also a relationship between numbers like 3 and $\frac{1}{3}$. Pairs of numbers like $\frac{1}{3}$ and 3 are called *reciprocals*.

Reciprocals: two numbers are reciprocals if their product is 1. Symbolically, n and $\frac{1}{n}$ are reciprocals

Multiplicative inverse: the reciprocal of a number

Multiplicative identity: the number 1 is the multiplicative identity because $a \cdot 1 = 1 \cdot a$ for all values of a. A number times its multiplicative inverse (reciprocal) is the multiplicative identity (1)

Two numbers are **reciprocals** if their product is 1. Thus, $\frac{1}{3}$ and 3 are reciprocals because $\frac{1}{3}(3) = 1$, and $\frac{2}{3}$ and $\frac{3}{2}$ are reciprocals because $\frac{2}{3}\left(\frac{3}{2}\right) = 1$. The **multiplicative inverse** of a number is its reciprocal. A number times its multiplicative inverse is 1, the **multiplicative identity.**

To find the reciprocal of a number:

1. Write the number in fractional form.

2. Interchange the numerator and the denominator so that the numerator is the denominator and the denominator is the numerator.

Inverting: interchanging the numerator and the denominator of a fraction

> **TIP** **Reciprocals and Inverting** Interchanging the numerator and the denominator of a fraction is commonly called **inverting** the fraction.

STOP AND CHECK
Find the reciprocal.

1. $\dfrac{7}{12}$ **2.** 5

3. $3\dfrac{2}{5}$ **4.** 0.4

Answers:

1. $\dfrac{12}{7}$ or $1\dfrac{5}{7}$ **2.** $\dfrac{1}{5}$

3. $\dfrac{5}{17}$ **4.** $\dfrac{5}{2}$ or 2.5

EXAMPLE 7

Find the reciprocal of $\frac{4}{7}, \frac{1}{5}, 3, 2\frac{1}{2}, 0.8, 1$, and 0.

The reciprocal of $\frac{4}{7}$ is $\frac{7}{4}$ or $1\frac{3}{4}$.	Interchange the numerator and the denominator.
The reciprocal of $\frac{1}{5}$ is $\frac{5}{1}$ or **5.**	Write $\frac{5}{1}$ as a whole number.
The reciprocal of 3 is $\frac{1}{3}$.	Write 3 as an improper fraction. $3 = \frac{3}{1}$.
The reciprocal of $2\frac{1}{2}$ is $\frac{2}{5}$.	Write $2\frac{1}{2}$ as an improper fraction. $2\frac{1}{2} = \frac{5}{2}$.
The reciprocal of 0.8 is $\frac{5}{4}$ or **1.25.**	Write 0.8 as a common fraction. $0.8 = \frac{8}{10} = \frac{4}{5}$.
The reciprocal of 1 is **1.**	Write 1 as an improper fraction. $1 = \frac{1}{1}$.
0 has no reciprocal.	$0 = \frac{0}{1}$ and $\frac{1}{0}$ is undefined. **See Exercises 24–28.**

Let's review the terminology of division.

$$\begin{array}{ccccc} 15 & \div & 3 & = & 5 \\ \text{dividend} & & \text{divisor} & & \text{quotient} \end{array}$$

To identify the divisor, remember that the symbol $\div$ is always read "divided by."

To divide fractions:

1. Change the division to an equivalent multiplication by replacing the divisor with its reciprocal and replacing the division notation with multiplication notation.

2. Perform the resulting multiplication.

STOP AND CHECK

1. Find $\frac{3}{7} \div \frac{3}{5}$.

Answer:

1. $\dfrac{5}{7}$

EXAMPLE 8

Find $\dfrac{5}{8} \div \dfrac{2}{3}$.

$$\frac{5}{8} \div \frac{2}{3} = \frac{5}{8} \cdot \frac{3}{2} = \frac{15}{16}$$ Change division to an equivalent multiplication.

See Exercises 29–33.

> **TIP** **Put Rules into Your Own Words** Some common phrases for stating the division of fractions rule are:
>
> ▶ Invert the divisor and multiply.
>
> ▶ Invert the second number and multiply.
>
> ▶ Invert the number after the division sign and multiply.
>
> A rule in your own words is often easier for you to remember. Be sure the words guide you to an appropriate process.

EXAMPLE 9

CON An auger bit advances $\frac{1}{16}$ in. for each turn (see Fig. 2–24). How many turns are needed to drill a hole $\frac{5}{8}$ in. deep? ($\frac{5}{8}$ in. can be divided into how many $\frac{1}{16}$-in. parts?)

$$\frac{5}{8} \div \frac{1}{16} = \frac{5}{8} \cdot \frac{\overset{2}{\cancel{16}}}{1} = 10 \qquad \text{Multiply by the reciprocal of the divisor.}$$

Ten turns are needed.   See Exercises 34–35.

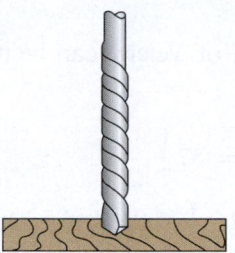

FIGURE 2-24

To divide mixed numbers and whole numbers, we first write the mixed numbers or whole numbers as improper fractions and then follow the rule for dividing fractions.

> **To divide mixed numbers, fractions, and whole numbers:**
>
> 1. Change each mixed number or whole number to an improper fraction.
>
> 2. Convert to an equivalent multiplication problem using the reciprocal of the divisor.
>
> 3. Multiply according to the rules for multiplying fractions.

STOP AND CHECK

1. Find $1\frac{2}{3} \div 2\frac{1}{4}$.

2. Find $5\frac{2}{5} \div 2$.

Answers:

1. $\frac{20}{27}$ 2. $2\frac{7}{10}$

EXAMPLE 10

Find **(a)** $2\frac{1}{2} \div 3\frac{1}{3}$ and **(b)** $5\frac{3}{8} \div 3$.

(a) $2\frac{1}{2} \div 3\frac{1}{3} =$ Change each mixed number to an improper fraction.

$\quad\;\; \frac{5}{2} \div \frac{10}{3} =$ Change division to equivalent multiplication.

$\quad\;\; \frac{\overset{1}{\cancel{5}}}{2} \cdot \frac{3}{\underset{2}{\cancel{10}}} =$ Reduce and multiply.

$\quad\;\; \frac{3}{4}$

(b) $5\frac{3}{8} \div 3 =$ Change mixed number and whole number to improper fractions.

$\quad\;\; \frac{43}{8} \div \frac{3}{1} =$ Change division to equivalent multiplication.

$\quad\;\; \frac{43}{8} \cdot \frac{1}{3} =$ Multiply.

$\quad\;\; \frac{43}{24} = 1\frac{19}{24}$

See Exercises 36–40.

EXAMPLE 11

BUS   A developer subdivides $5\frac{1}{4}$ acres into lots; each lot is $\frac{7}{10}$ of an acre. How many lots are made?

$$5\frac{1}{4} \div \frac{7}{10} = \frac{21}{4} \div \frac{7}{10} = \frac{\overset{3}{\cancel{21}}}{\underset{2}{\cancel{4}}} \cdot \frac{\overset{5}{\cancel{10}}}{\underset{1}{\cancel{7}}} = \frac{15}{2} = 7\frac{1}{2}$$

Seven lots are made and each is $\frac{7}{10}$ of an acre. The $\frac{1}{2}$ lot is left over or combined with one of the other lots.   See Exercises 41–42.

EXAMPLE 12

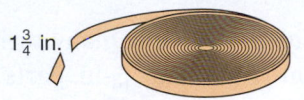

$1\frac{3}{4}$ in.

FIGURE 2–25

HLTH/N   How many Velcro fasteners, each requiring $1\frac{3}{4}$ in. of Velcro, can be made from a roll containing 100 in. of Velcro? See Fig. 2–25.

$$100 \div 1\frac{3}{4} = \frac{100}{1} \div \frac{7}{4} = \frac{100}{1} \cdot \frac{4}{7} = \frac{400}{7} = 57\frac{1}{7}$$

Fifty-seven fasteners can be cut to the desired length. The extra $\frac{1}{7}$ of the desired length is considered waste.   See Exercises 43–45.

EXAMPLE 13

FIGURE 2–26

INDTEC   A piece of trophy column stock that is $21\frac{1}{2}$ in. long is cut into four equal trophy columns (Fig. 2–26). If $\frac{1}{16}$ in. is wasted on each cut, find the length of each piece:

Use the Six-Step Problem-Solving Strategy.

Unknown facts

Length of cuts to be made.

Known facts

4   Note the number of pieces needed.

3   Note the number of cuts to be made ($4 - 1 = 3$).

$\frac{1}{16}$ in. Note the waste for each cut.

$21\frac{1}{2}$ in. Note the length of trophy column stock.

Relationships

Length of each piece = (total length − 3 cuts × amount wasted for each cut) ÷ 4 pieces of stock needed.

Estimation

If the stock is 20 in. long and cut into 4 pieces and if waste is disregarded, each piece will be 5 in.

Calculation

Three cuts are to be made and each cut wastes $\frac{1}{16}$ in. Find the amount wasted.

$\dfrac{1}{16} \cdot 3 = \dfrac{3}{16}$ Total waste.

$21\dfrac{1}{2} - \dfrac{3}{16}$ Subtract to find the amount of stock that will be left to be divided equally into four trophy columns.

$21\dfrac{1}{2} = 21\dfrac{8}{16}$ Align vertically and use common denominators.

$-\dfrac{3}{16} = \dfrac{3}{16}$

$\overline{\qquad 21\dfrac{5}{16}}$ in.   Amount of stock left to be divided.

Interpretation

Find the length of each trophy column.

$$21\frac{5}{16} \div 4 = \frac{341}{16} \div \frac{4}{1} = \frac{341}{16}\left(\frac{1}{4}\right) = \frac{341}{64} = 5\frac{21}{64}$$

Each trophy column is $5\frac{21}{64}$ in. long.   **See Exercises 46–47.**

Complex fraction: a fraction in which either the numerator or the denominator or both contain a fraction or a mixed number

A **complex fraction** is a fraction in which either the numerator or the denominator or both contain a fraction or a mixed number. Fractions indicate division as we read from top to bottom. The large fraction line is read as "divided by." For example,

$$\frac{2\frac{1}{2}}{7} \text{ is read } \text{ "}2\frac{1}{2}\text{ divided by 7"}$$

$$\frac{4}{1\frac{1}{2}} \text{ is read } \text{ "4 divided by }1\frac{1}{2}\text{"}$$

If we think of a fraction as division, then a complex fraction is another way of writing division.

> **To simplify a complex fraction:**
> 1. Rewrite the fraction with the division symbol $\div$.
> 2. Perform the indicated division.

STOP AND CHECK

1. Simplify $\dfrac{5\frac{3}{4}}{1\frac{1}{3}}$.

Answer:

1. $4\frac{5}{16}$

EXAMPLE 14

Simplify $\dfrac{6\frac{3}{8}}{4\frac{1}{2}}$.

$$\frac{6\frac{3}{8}}{4\frac{1}{2}} = \qquad \text{Rewrite using the } \div \text{ symbol.}$$

$$6\frac{3}{8} \div 4\frac{1}{2} = \qquad \text{Perform the indicated division. Write mixed numbers as improper fractions.}$$

$$\frac{51}{8} \div \frac{9}{2} = \qquad \text{Change division to multiplication.}$$

$$\frac{\overset{17}{\cancel{51}}}{\underset{4}{\cancel{8}}} \cdot \frac{\overset{1}{\cancel{2}}}{\underset{3}{\cancel{9}}} = \qquad \text{Reduce and multiply.}$$

$$\frac{17}{12} = \mathbf{1\frac{5}{12}} \qquad \text{Write as a mixed number.}$$

See Exercises 48–49.

We sometimes use complex fractions to change mixed decimals to fractions.

1. Write $0.16\frac{2}{3}$ as a fraction.

Answer:

1. $\frac{1}{6}$

EXAMPLE 15

Write $0.33\frac{1}{3}$ as a fraction.

$$0.33\frac{1}{3} = \frac{33\frac{1}{3}}{100} = 33\frac{1}{3} \div 100$$

Count places for digits only; that is, do not count the fraction $\frac{1}{3}$ as a place.

$$33\frac{1}{3} \div 100 = \frac{100}{3} \div \frac{100}{1} = \frac{\overset{1}{\cancel{100}}}{3} \cdot \frac{1}{\underset{1}{\cancel{100}}} = \frac{1}{3}$$

Change division to multiplication. Reduce and multiply.

See Exercise 50.

TIP **What Place Does $\frac{1}{3}$ Hold if the Decimal Is $0.33\frac{1}{3}$?** When changing decimals to fractions, we divide by the place value of the last digit. Is $\frac{1}{3}$ in $0.33\frac{1}{3}$ in the hundredths or thousandths place? Hundredths. The fraction attaches to the last digit.

$0.12\frac{1}{2}$ is read "twelve and one-half hundredths"

$0.008\frac{1}{3}$ is read "eight and one-third thousandths"

4 **Perform Calculations Involving Fractions with a Calculator.** Some calculators have a special key for entering fractions. The most common label for a fraction key is $\boxed{a\frac{b}{c}}$. Other calculators have a menu choice that allows the result of a calculation to be displayed in fraction form. Investigate various options and limitations of your calculator or refer to the user's manual if necessary.

To use a calculator fraction key to perform calculations:

To enter a fraction using the $\boxed{a\frac{b}{c}}$ key:

1. Enter the numerator.

2. Press the fraction key.

3. Enter the denominator.

To enter a fraction using division and the fraction display menu choice:

1. Use parentheses.

2. Enter the numerator divided by the denominator. Press $\boxed{\text{ENTER}}$.

3. Display the result in fraction form ($\boxed{\text{MATH}}$, $\boxed{\triangleright\text{Frac}}$, $\boxed{\text{ENTER}}$).

To enter a mixed number using the $\boxed{a\frac{b}{c}}$ key:

1. Enter the whole-number portion of the mixed number.

2. Press the fraction key.

3. Enter the numerator.

4. Press the fraction key.

5. Enter the denominator.

To enter a mixed number using division and the fraction display menu choice:

1. Use parentheses.

2. Enter the whole-number portion of the mixed number followed by the addition operation.

3. Enter the fraction portion of the mixed number as numerator divided by the denominator. Press ENTER.

4. Display the result in fraction form (MATH, ▷Frac, ENTER). The display for a value greater than 1 will be an improper fraction in lowest terms.

STOP AND CHECK

Perform the calculations using a calculator.

1. $\dfrac{5}{8}\left(\dfrac{3}{10}\right)$

2. $\dfrac{\dfrac{5}{16} - \dfrac{1}{4}}{4\dfrac{1}{2}}$

Answers:

1. $\dfrac{3}{16}$ 2. $\dfrac{1}{72}$

EXAMPLE 16

Perform the calculations using a calculator.

(a) $\dfrac{3}{4}\left(\dfrac{5}{6}\right)$ (b) $\dfrac{\dfrac{7}{8} + \dfrac{1}{4}}{3\dfrac{1}{2}}$

(a) $\dfrac{3}{4}\left(\dfrac{5}{6}\right)$

Using the fraction key:

$$3\;\boxed{a\dfrac{b}{c}}\;4\;\boxed{\times}\;5\;\boxed{a\dfrac{b}{c}}\;6\;\boxed{=}\;\Rightarrow\;\dfrac{5}{8}$$

Using the fraction menu choice:

$$\boxed{(}\;3\;\boxed{\div}\;4\;\boxed{)}\;\boxed{(}\;5\;\boxed{\div}\;6\;\boxed{)}\;\boxed{\text{ENTER}}\;\boxed{\text{MATH}}\;\boxed{\text{▷Frac}}\;\boxed{\text{ENTER}}\;\Rightarrow\;\dfrac{5}{8}$$

(b) $\dfrac{\dfrac{7}{8} + \dfrac{1}{4}}{3\dfrac{1}{2}}$

Using the fraction key:

$$\boxed{(}\;7\;\boxed{a\dfrac{b}{c}}\;8\;\boxed{+}\;1\;\boxed{a\dfrac{b}{c}}\;4\;\boxed{)}\;\boxed{\div}\;\boxed{(}\;3\;\boxed{a\dfrac{b}{c}}\;1\;\boxed{a\dfrac{b}{c}}\;2\;\boxed{)}\;\boxed{=}\;\Rightarrow\;\dfrac{9}{28}$$

Using the fraction menu choice:

$$\boxed{(}\;7\;\boxed{\div}\;8\;\boxed{+}\;1\;\boxed{\div}\;4\;\boxed{)}\;\boxed{\div}\;\boxed{(}\;3\;\boxed{+}\;1\;\boxed{\div}\;2\;\boxed{)}\;\boxed{\text{ENTER}}\;\boxed{\text{MATH}}\;\boxed{\text{▷Frac}}\;\boxed{\text{ENTER}}\;\Rightarrow\;\dfrac{9}{28}$$

See Exercises 51–60.

2–4 EXERCISES **MyLab Math** For additional practice go to your study plan in MyLab Math.

1 Multiply and reduce answers to lowest terms. Write improper fractions as whole or mixed numbers. *See Examples 1, 2, and 3.*

See Example 1.

1. $\dfrac{3}{4}\left(\dfrac{1}{8}\right)$

2. $\dfrac{1}{2}\left(\dfrac{7}{16}\right)$

3. $\dfrac{5}{8}\left(\dfrac{7}{10}\right)$

4. $\dfrac{2}{3}\left(\dfrac{7}{8}\right)$

See Example 2.

5. $\frac{1}{2}\left(\frac{3}{4}\right)\left(\frac{8}{9}\right)$

6. $\frac{3}{8}\left(\frac{5}{6}\right)\left(\frac{1}{2}\right)$

See Example 3.

7. $7\left(3\frac{1}{8}\right)$

8. $\frac{3}{5}(125)$

9. $2\frac{3}{4} \cdot 1\frac{1}{2}$

10. $9\frac{1}{2} \cdot 3\frac{4}{5}$

11. $\frac{1}{5} \cdot 7\frac{5}{8}$

12. $\frac{2}{3} \cdot 3\frac{1}{4}$

See Example 4.

13. **HELLP** A firefighter wants to make a horizontal coat rack by placing coat hooks on a board. Six coat hooks are to be placed $4\frac{1}{4}$ in. from each end of the board and each hook should be $4\frac{1}{4}$ in. apart. What length of board is needed?

14. **HOSP** Bellmen want to make a key rack for organizing the keys of valet parking customers. The plaque is $17\frac{7}{8}$ in. wide and each row on the key rack will need to hold 10 keys. If each key holder is to be the same length from each side of the plaque and the same distance apart, how far apart should each holder be placed?

15. **AUTO** A fuel tank that holds 75 liters (L) of fuel is $\frac{1}{4}$ full. How many liters of fuel are in the tank?

16. **CON** If steps are 12 risers high and each riser is $7\frac{1}{2}$ in. high, what is the total rise of the steps? *See Example 4.*

17. **INDTEC** An alloy, which is a substance composed of two or more metals, is $\frac{11}{16}$ copper, $\frac{7}{32}$ tin, and $\frac{3}{32}$ zinc. How many kilograms of each metal are needed to make 384 kg of alloy?

18. **AG/H** A flower bed border is being built of 8-in. by 8-in. by 16-in. concrete blocks. If nine blocks are laid end to end with $\frac{1}{2}$ in. mortar joints, how long is the border? *See Example 5.*

2 Raise the fractions to the indicated power. *See Example 6.*

19. $\left(\frac{2}{5}\right)^2$

20. $\left(\frac{1}{4}\right)^2$

21. $\left(\frac{3}{5}\right)^3$

22. $\left(\frac{1}{3}\right)^3$

23. $\left(\frac{7}{9}\right)^2$

3 Give the reciprocal. *See Example 7.*

24. $\frac{5}{8}$

25. $2\frac{1}{5}$

26. 8

27. 0.9

28. 1.8

Divide and reduce answers to lowest terms. *See Example 8.*

29. $\frac{1}{2} \div \frac{7}{12}$

30. $\frac{4}{5} \div \frac{8}{9}$

31. $\frac{11}{32} \div \frac{3}{8}$

32. $\frac{3}{4} \div \frac{3}{8}$

33. $\frac{7}{8} \div \frac{3}{16}$

See Example 9.

34. **TELE** A cable installer needs to drill holes that are $\frac{3}{8}$ in. deep for inserting brackets on a wall. If the drill bit advances $\frac{3}{32}$ in. with each rotation of the bit, how many rotations are needed to drill the hole?

35. **INDTR** A 4-in. earth auger will dig a hole $2\frac{1}{2}$-in. deep with each complete turn. How many turns will need to be made to dig a hole 15-in. deep?

Convert improper fractions to whole or mixed numbers. *See Example 10.*

36. $10 \div \frac{3}{4}$

37. $3\frac{1}{8} \div \frac{1}{4}$

38. $2\frac{1}{2} \div 4$

39. $1\frac{1}{7} \div \frac{2}{7}$

40. $3\frac{3}{4} \div 1\frac{1}{2}$

See Example 11.

41. **CON** A truck will hold 21 yd³ (cubic yards) of gravel. If an earth mover has a shovel capacity of $1\frac{3}{4}$ yd³, how many shovelfuls are needed to fill the truck?

42. **CAD/ARC** If $\frac{1}{8}$ in. represents 1 ft on a drawing, find the dimensions of a room that measures $2\frac{1}{2}$ in. by $1\frac{7}{8}$ in. on the drawing. (How many $\frac{1}{8}$s are there in $2\frac{1}{2}$; how many $\frac{1}{8}$s are there in $1\frac{7}{8}$?)

See Example 12.

43. **CON** How many $17\frac{5}{8}$-in. strips of quarter-round molding can be cut from a piece $132\frac{3}{4}$ in. long? Disregard waste.

44. **INDTEC** A segment of I-beam is $10\frac{1}{2}$ ft long. Into how many whole $2\frac{1}{4}$-ft pieces can it be divided? Disregard waste.

45. INDTEC How many $9\frac{1}{4}$-in. drinking straws can be cut from a $216\frac{1}{2}$-in. length of stock? How much stock is left over?

See Example 13.

46. CON Three shelves of equal length are cut from a 72-in. board. If $\frac{1}{8}$ in. is wasted on each cut, what is the maximum length of each shelf? (Two cuts are made to divide the entire board into three equal lengths.)

47. AG/H A gardener has a board that is 8 ft long and he wants to cut it into 5 shelves for pots. If each cut wastes $\frac{1}{8}$ in., what is the length of each shelf?

Simplify. *See Example 14.*

See Example 15.

48. $\dfrac{5\frac{3}{5}}{1\frac{3}{4}}$

49. $\dfrac{10}{1\frac{1}{5}}$

50. Write $0.83\frac{1}{3}$ as a fraction.

4 Perform the calculations using a calculator. *See Example 16.*

51. $5\frac{1}{2} + 8\frac{7}{15} + 3\frac{5}{24}$
52. $124\frac{8}{35} - 42\frac{6}{49}$
53. $4\frac{5}{12}\left(7\frac{7}{15}\right)$
54. $1\frac{7}{9}\left(3\frac{5}{16}\right)$
55. $5\frac{5}{6} \div 1\frac{1}{14}$

56. $8\frac{1}{3} \div 2\frac{7}{9}$
57. $\dfrac{\frac{7}{10} + \frac{3}{5}}{1\frac{11}{15}}$
58. $\dfrac{6\frac{1}{4}}{\frac{5}{8} + \frac{7}{8}}$
59. $\dfrac{\frac{5}{12} + \frac{3}{8}}{\frac{4}{19} + \frac{1}{38}}$
60. $\dfrac{\frac{7}{8} - \frac{3}{4}}{\frac{1}{12} + \frac{1}{9}}$

2 CHAPTER REVIEW OF KEY CONCEPTS

LEARNING OUTCOMES	KEY CONCEPTS AND EXAMPLES
Section 2–1	The *numerator* (top number) of a fraction is the number of parts of the whole amount we are considering. The *denominator* (bottom number) of a fraction is the number of parts a whole amount has been divided into. *Proper fractions* are less than 1. *Improper fractions* are equal to or larger than 1. A *mixed number* consists of a whole number and a proper fraction written together and indicates addition of the whole number and fraction. A *decimal fraction* is a fraction whose denominator is 10 or a power of 10 and it can be written in decimal notation with the place value representing the denominator of the fraction.
1 Find multiples of a natural number (pp. 60–63).	To find a multiple of a natural number: **1.** multiply the given number by any natural number.

Find the first five multiples of 7:

$$7 \times 1 = 7, \quad 7 \times 2 = 14, \quad 7 \times 3 = 21, \quad 7 \times 4 = 28, \quad 7 \times 5 = 35$$

The first five multiples of 7 are 7, 14, 21, 28, and 35.

LEARNING OUTCOMES	KEY CONCEPTS AND EXAMPLES

2 Find all factor pairs of a natural number (pp. 63–64).

1. Write the factor pair of 1 and the given natural number.   **2.** Check to see if the given number is divisible by 2. If so, write the factor pair of 2 and the quotient of the given natural number and 2.   **3.** Check the next natural number for divisibility; if the given number is divisible by this number, write the factor pair.   **4.** Continue Step 3 until you reach a number that already has been found as a quotient in a previous factor pair.

> Find all the factors of 24: 1(24), 2(12), 3(8), 4(6); 5 is not a factor, 6(4) is a repeat. Factors are 1, 2, 3, 4, 6, 8, 12, 24.

3 Determine the prime factorization of composite numbers (pp. 64–66).

1. Test each prime number to see if the composite number is divisible by the prime number.   **2.** Make a factor pair using the first prime number that passes the test in Step 1.   **3.** Carry forward the prime factors and test the remaining factors by repeating Steps 1 and 2.

> Find the prime factorization of 28:
> $$28 = 2(14)$$
> $$= 2(2)(7) \text{ or } 2^2(7)$$

4 Find the least common multiple and greatest common factor of two or more numbers (pp. 67–68).

Find the least common multiple (LCM) of two or more natural numbers using the prime factorization of the numbers: 1. List the prime factorization of each number writing repeat factors in exponential notation.   **2.** List the prime factorization of the least common multiple by including the prime factors appearing in *each* number. If a prime factor appears in more than one number, use the factor with the *largest* exponent.   **3.** Write the resulting expression in standard notation.

> Find the LCM of 12, 15, and 30:
> $$12 = 2^2(3) \qquad 15 = 3(5) \qquad 30 = 2(3)(5)$$
> $$\text{LCM} = 2^2(3)(5) \text{ or } 60$$

Find the greatest common factor (GCF) of two or more natural numbers: 1. List the prime factorization of each number using exponential notation when appropriate.   **2.** List the prime factorization of the greatest common factor by including each prime factor appearing in *every* number. If a prime factor appears more than one time in any number (that is the exponent is greater than 1), use the factor with the *smallest* exponent. If there are no common prime factors, the GCF is 1.   **3.** Write the resulting expression in standard notation.

> Find the GCF of 12, 15, 30:
> $$12 = 2^2(3) \qquad 15 = 3(5) \qquad 30 = 2(3)(5)$$
> $$\text{GCF} = 3$$

Section 2–2

1 Write equivalent fractions with different denominators (pp. 69–74).

Change a fraction to an equivalent fraction with a specified larger denominator: 1. Divide the specified larger denominator by the original denominator.   **2.** Multiply the original numerator and denominator by the quotient found in Step 1. That is, multiply by 1 in the form of $\frac{n}{n}$ when n is a real number and $n \neq 0$.

> Change $\frac{5}{9}$ to an equivalent fraction whose denominator is 36.
> $$\frac{5}{9}\left(\frac{4}{4}\right) = \frac{20}{36} \qquad 36 \div 9 = 4$$

LEARNING OUTCOMES   **KEY CONCEPTS AND EXAMPLES**

Change a fraction to an equivalent fraction with a smaller denominator or reduce a fraction to lowest terms: 1. Find a common factor greater than 1 for the numerator and denominator. **2.** Divide both the numerator and the denominator by this common factor.   **3.** Continue until the fraction is in lowest terms or has the desired smaller denominator. *Note:* To find the lowest terms in the fewest steps, find the greatest common factor (GCF) in Step 1.

Reduce $\frac{12}{16}$ to lowest terms.

$$\frac{12}{16} = \frac{12}{16} \div \frac{4}{4} = \frac{3}{4} \qquad \text{GCF is 4.}$$

Use the property of proportions: The cross products in a proportion are equal. Symbolically, if $\frac{a}{b} = \frac{c}{d}$ and b and d are not equal to zero, then $a \cdot d = b \cdot c$. Also, if $a \cdot d = b \cdot c$, then $\frac{a}{b} = \frac{c}{d}$.

Is $\frac{10}{14} = \frac{15}{21}$? The cross products are $(10)(21) = 210$ and $(14)(15) = 210$.

  Yes, the fractions are equal.

Measure a line segment (or object) using a U.S. customary rule: 1. Align the left end of the line segment (or object) with the left end of the rule or the zero mark.   **2.** Determine the last whole unit that the segment passes.   **3.** Examine the fraction mark and determine the value of the fraction mark where the line segment ends on the right.   **4.** Write the measure as a mixed number by combining the results from Steps 2 and 3.

  If the line segment ends between two marks on the rule, use eye judgment to estimate closeness of the object to a mark on the rule.

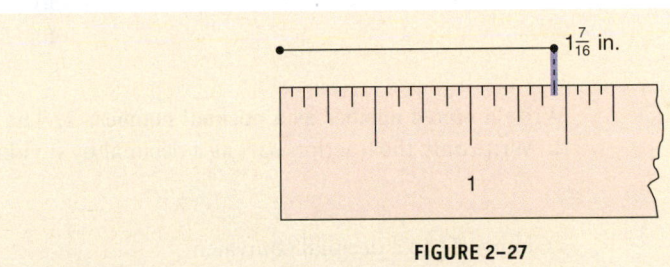

$1\frac{7}{16}$ in.

FIGURE 2–27

2 Write improper fractions as whole numbers or mixed numbers (pp. 74–75).

Write an improper fraction as a whole or mixed number: 1. Perform the division indicated (numerator ÷ denominator).   **2.** Express any remainder of the division as a fraction or decimal equivalent.

Write $\frac{18}{6}$ and $\frac{15}{4}$ as whole or mixed numbers.

$$\frac{18}{6} = 3 \qquad \frac{15}{4} = 3\frac{3}{4}$$

3 Write whole numbers or mixed numbers as improper fractions (pp. 75–76).

Write a whole number as an improper fraction: 1. Write the whole number as the numerator. **2.** Write 1 as the denominator.

Write a mixed number as an improper fraction: 1. Multiply the denominator of the fractional part by the whole number.   **2.** Add the numerator of the fractional part to the product; this sum becomes the numerator of the improper fraction.   **3.** The denominator of the improper fraction is the same as the denominator of the fractional part of the mixed number.

LEARNING OUTCOMES	KEY CONCEPTS AND EXAMPLES

Change 7 and $4\frac{7}{8}$ to improper fractions.

$$7 = \frac{7}{1} \qquad 4\frac{7}{8} = \frac{(8 \cdot 4) + 7}{8} = \frac{39}{8}$$

4 Write decimals as fractions and fractions as decimals (pp. 76–77).

Write a decimal number as a fraction or mixed number in lowest terms: 1. Write the digits without the decimal point and leading zeros as the numerator.   **2.** Write the denominator as a power of 10 with as many zeros as there are places after the decimal point.   **3.** Reduce and, if the fraction is improper, convert to a mixed number.

Write 0.23 as a fraction:

$$\frac{23}{100}$$ Two decimal digits so the denominator is 100.

Convert a fraction to a decimal number: 1. Place a decimal point after the numerator. **2.** Divide the numerator by the denominator using long division.   **3.** Attach zeros after the decimal point in the dividend as needed for division.

Change $\frac{5}{8}$ to a decimal:

$$\begin{array}{r} 0.625 \\ 8\overline{)5.000} \\ \underline{4\ 8} \\ 20 \\ \underline{16} \\ 40 \\ \underline{40} \end{array}$$ Divide by the denominator.

Write a mixed number as a decimal number: 1. The whole-number part remains the same. **2.** Write only the fraction part as a decimal by dividing the numerator by the denominator.

Change $5\frac{3}{4}$ to a decimal equivalent.

$3 \div 4$ or $\begin{array}{r} 0.75 \\ 4\overline{)3.00} \end{array}$ Write the fraction part as an equivalent decimal.

Then, $5\frac{3}{4} = \mathbf{5.75}$

5 Compare fractions, mixed numbers, and decimals (pp. 78–80).

Find the least common denominator (LCD): Use the process for finding the least common multiple. See Learning Outcome 2–1, 4.

Find the least common denominator for $\frac{7}{18}$ and $\frac{5}{24}$.

$$18 = 2 \cdot 3 \cdot 3 = 2 \cdot 3^2$$
$$24 = 2 \cdot 2 \cdot 2 \cdot 3 = 2^3 \cdot 3$$
$$\text{LCM} = \text{LCD} = 2^3 \cdot 3^2 = 2 \cdot 2 \cdot 2 \cdot 3 \cdot 3 = 72$$

The smallest number that can be divided evenly by both 18 and 24 is 72.

Compare fractions: 1. Find the least common denominator (LCD).   **2.** Change each fraction to an equivalent fraction with the least common denominator (LCD) as its denominator. **3.** Compare the numerators. The larger numerator indicates the larger fraction.

LEARNING OUTCOMES   **KEY CONCEPTS AND EXAMPLES**

Compare mixed numbers: 1. Compare the whole-number parts.   **2.** If the whole-number parts are equal, compare the fraction parts.   **3.** Compare the numerators after changing fractions to equivalent fractions with the LCD.

Which fraction is smaller, $\frac{2}{5}$ or $\frac{5}{12}$?

$$\frac{2}{5} = \frac{24}{60}$$ LCD = 60

$$\frac{5}{12} = \frac{25}{60}$$ Since $\frac{24}{60}$ is smaller than $\frac{25}{60}$, $\frac{2}{5}$ is smaller than $\frac{5}{12}$.

Section 2–3

1 Add fractions and mixed numbers (pp. 83–84).

Compare fractions and decimals: Change fractions to decimal equivalents. The process for comparing decimals is found in Learning Outcome 1–1, 3.

Add fractions: 1. If the denominators are not the same, find the least common denominator.   **2.** Change each fraction not already expressed in terms of the least common denominator to an equivalent fraction with the common denominator.   **3.** Add the numerators only.   **4.** The common denominator is the denominator of the sum.   **5.** Reduce the sum to lowest terms and change improper fractions to whole or mixed numbers.

Add:

$$\frac{1}{7} + \frac{3}{7} + \frac{2}{7} = \frac{6}{7} \qquad \frac{5}{8} + \frac{3}{4} = \frac{5}{8} + \frac{6}{8} = \frac{11}{8} = 1\frac{3}{8}$$

Add mixed numbers: 1. Add the whole-number parts.   **2.** Add the fractional parts and reduce to lowest terms.   **3.** Change improper fractions to whole or mixed numbers.   **4.** Add whole-number parts.

Add $4\frac{3}{4} + 5\frac{2}{8} + 1\frac{1}{2}$.

$$4\frac{3}{4} = 4\frac{6}{8}$$ Change each fraction to an equivalent fraction with the LCD.

$$5\frac{2}{8} = 5\frac{2}{8}$$

$$1\frac{1}{2} = 1\frac{4}{8}$$ Add fractions.
Add whole numbers.

$$10\frac{12}{8}$$ Simplify. $\frac{12}{8} = 1\frac{4}{8}$

$$10 + 1\frac{4}{8} = 11\frac{4}{8} = 11\frac{1}{2}$$

---

2 Subtract fractions and mixed numbers (pp. 85–87).

Subtract fractions: 1. If the denominators are not the same, find the least common denominator.   **2.** Change each fraction not expressed in terms of the common denominator to an equivalent fraction having the common denominator.   **3.** Subtract the numerators.   **4.** The common denominator will be the denominator of the difference.   **5.** Reduce the difference to lowest terms.

LEARNING OUTCOMES	KEY CONCEPTS AND EXAMPLES

Subtract $\frac{5}{8} - \frac{7}{16}$.

$$\frac{5}{8} = \frac{10}{16}$$

Convert each fraction to an equivalent fraction with a common denominator.

$$-\frac{7}{16} = \frac{7}{16}$$

Subtract the numerators.

$$\frac{3}{16}$$

Subtract mixed numbers: 1. If the fractional parts of the mixed numbers do not have the same denominator, change them to equivalent fractions with a common denominator. **2.** When the fraction in the minuend is larger than the fraction in the subtrahend, go to Step 6. **3.** When the fraction in the subtrahend is larger than the fraction in the minuend, regroup (borrow) by taking one whole number from the whole-number part of the minuend. This makes the whole number 1 less. **4.** Change the whole number that was borrowed to an improper fraction with the common denominator. For example, $1 = \frac{3}{3}$, $1 = \frac{8}{8}$, and $1 = \frac{n}{n}$, where n is the common denominator. **5.** Add the borrowed fraction ($\frac{n}{n}$) to the fraction already in the minuend. **6.** Subtract the fractional parts and the whole-number parts. **7.** Reduce the difference to lowest terms.

Subtract $5\frac{3}{8} - 3\frac{9}{16}$.

$$5\frac{3}{8} = 5\frac{6}{16} = 4\frac{22}{16}$$

Convert fraction to 16ths. Regroup.

$$-3\frac{9}{16} = 3\frac{9}{16} = 3\frac{9}{16}$$

Subtract fractions. Subtract whole numbers.

$$1\frac{13}{16}$$

Section 2–4

1 Multiply fractions and mixed numbers (pp. 89–93).

Multiply fractions: 1. Multiply the numerators of the fractions to get the numerator of the product. **2.** Multiply the denominators to get the denominator of the product. **3.** Reduce the product to lowest terms.

Multiply $\frac{4}{5}\left(\frac{7}{10}\right)\left(\frac{15}{35}\right)$.

$$\frac{\overset{2}{\cancel{4}}}{\underset{1}{\cancel{5}}}\left(\frac{\overset{1}{\cancel{7}}}{\underset{5}{\cancel{10}}}\right)\left(\frac{\overset{3}{\cancel{15}}}{\underset{5}{\cancel{35}}}\right) = \frac{6}{25}$$

Reduce and multiply.

Multiply mixed numbers, fractions, and whole numbers: 1. Change each mixed number or whole number to an improper fraction. **2.** Reduce as much as possible. **3.** Multiply remaining numerators. **4.** Multiply remaining denominators. **5.** Write the answer as a whole or mixed number as appropriate.

Multiply $4\left(3\frac{1}{5}\right)\left(\frac{2}{7}\right)$.

$$4\left(3\frac{1}{5}\right)\left(\frac{2}{7}\right) = \frac{4}{1}\left(\frac{16}{5}\right)\frac{2}{7} = \frac{128}{35} = 3\frac{23}{35}$$

LEARNING OUTCOMES	KEY CONCEPTS AND EXAMPLES

2 Raise a fraction to a power (p. 93).

Raise a fraction or quotient to a power: 1. Raise the numerator to the power. **2.** Raise the denominator to the power. The denominator cannot be zero.

Symbolically,

$$\left(\frac{a}{n}\right)^n = \frac{a^n}{b^n} \qquad b \neq 0 \text{ and } a, b, \text{ and } n \text{ are real numbers.}$$

Raise $\left(\frac{3}{5}\right)^3$ to the indicated power.

$$\left(\frac{3}{5}\right)^3 = \frac{3^3}{5^3} = \frac{3 \cdot 3 \cdot 3}{5 \cdot 5 \cdot 5} = \frac{27}{125}$$

3 Divide fractions and mixed numbers (pp. 93–98).

Find the reciprocal of a number: 1. Write the number in fractional form. **2.** Interchange the numerator and the denominator so that the numerator is the denominator and the denominator is the numerator.

Find the reciprocal of $\frac{3}{5}$, 6, $2\frac{3}{4}$, and 0.2.

The reciprocal of $\frac{3}{5}$ is $\frac{5}{3}$; the reciprocal of 6 is $\frac{1}{6}$; the reciprocal of $2\frac{3}{4}$ or $\frac{11}{4}$ is $\frac{4}{11}$.

The reciprocal of 0.2 or $\frac{2}{10}$ is $\frac{10}{2}$ or 5.

Divide fractions: 1. Change the division to an equivalent multiplication by replacing the divisor with its reciprocal and replacing the division notation with multiplication notation. **2.** Perform the resulting multiplication.

Divide $\frac{4}{5} \div \frac{8}{9}$.

$$\frac{4}{5} \div \frac{8}{9} = \frac{\overset{1}{4}}{5}\left(\frac{9}{\underset{2}{8}}\right) = \frac{9}{10} \qquad \text{Change to an equivalent multiplication and multiply.}$$

Divide mixed numbers, fractions, and whole numbers: 1. Change each mixed number or whole number to an improper fraction. **2.** Convert to an equivalent multiplication problem using the reciprocal of the divisor. **3.** Multiply according to the rules for multiplying fractions.

Divide $4\frac{2}{3} \div 1\frac{1}{6}$.

$$4\frac{2}{3} \div 1\frac{1}{6} = \frac{14}{3} \div \frac{7}{6} = \frac{\overset{2}{14}}{\underset{1}{3}} \times \frac{\overset{2}{6}}{\underset{1}{7}} = \frac{4}{1} = 4$$

Simplify a complex fraction: 1. Rewrite the fraction with the divided by symbol ÷. **2.** Perform the indicated division.

Simplify $\dfrac{2\frac{1}{2}}{1\frac{2}{3}}$.

$$\frac{2\frac{1}{2}}{1\frac{2}{3}} = 2\frac{1}{2} \div 1\frac{2}{3} = \frac{5}{2} \div \frac{5}{3} = \frac{5}{2} \cdot \frac{3}{5} = \frac{3}{2} = 1\frac{1}{2} \qquad \text{Convert to division, then multiplication.}$$

LEARNING OUTCOMES	KEY CONCEPTS AND EXAMPLES

4 Perform calculations involving fractions with a calculator (pp. 98–99).

The most common label for a fraction key is $\boxed{a\frac{b}{c}}$. Other calculators have a menu choice ($\boxed{\text{MATH}}$, $\boxed{\triangleright\text{Frac}}$) that allows the result of a calculation to be displayed in fraction form. Investigate options and limitations of your calculator. Additional details are given on pp. 98–99.

Add $\frac{3}{4} + \frac{5}{6}$.

Using the fraction key:

$3 \boxed{a\frac{b}{c}} 4 \boxed{+} 5 \boxed{a\frac{b}{c}} 6 \boxed{=} \Rightarrow 1\frac{7}{12}$

Using the fraction menu choice:

$\boxed{(} 3 \boxed{\div} 4 \boxed{)} \boxed{+} \boxed{(} 5 \boxed{\div} 6 \boxed{)} \boxed{\text{ENTER}} \boxed{\text{MATH}} \boxed{\triangleright\text{Frac}} \boxed{\text{ENTER}} \Rightarrow \frac{19}{12}$

$\frac{19}{12} = 1\frac{7}{12}$

2 | CHAPTER REVIEW EXERCISES

Section 2–1   MyLab Math   For additional practice go to your study plan in MyLab Math.

1. A report states that 7 out of every 10 high school graduates pursue some type of postsecondary education. What fraction of the graduates pursue some type of postsecondary education?

2. Write $9 \div 26$ as a fraction.

Show that each number is a multiple of the first number by writing each as the product of the first number and a natural number.

3. 11, 22, 33, 44, 55

4. 12, 24, 36, 48, 60

Write the first five multiples of each number.

5. 21   6. 14   7. 7   8. 11   9. 8   10. 15

Is the number divisible by the given number? Explain.

11. 153 by 3   12. 8,234 by 4   13. 8,726 by 6   14. 5,986 by 5
15. 240 by 10   16. 5,845 by 5   17. 63,539 by 9   18. 52,428 by 8

List all factor pairs for each number, then write the factors in order from smallest to largest.

19. 48   20. 50   21. 51   22. 63   23. 74

Identify each number as *prime or composite*. Explain.

24. 17   25. 18   26. 20   27. 21   28. 29

Find the prime factorization of each number. Write in factored form and then in exponential notation.

29. 42   30. 48   31. 98   32. 120

Find the least common multiple.

33. 18 and 40   34. 12 and 18   35. 12, 18, and 30   36. 6, 10, and 12

Find the greatest common factor.

37. 10 and 12   **38.** 12 and 18   **39.** 12, 18, and 30   **40.** 4, 9, and 16

Section 2–2

41. Find five fractions that are equivalent to $\dfrac{2}{3}$.

42. Find five fractions that are equivalent to $\dfrac{3}{5}$.

Find the equivalent fractions using the indicated denominators.

43. $\dfrac{5}{12} = \dfrac{?}{60}$ **44.** $\dfrac{4}{5} = \dfrac{?}{40}$ **45.** $\dfrac{2}{3} = \dfrac{?}{15}$ **46.** $\dfrac{4}{9} = \dfrac{?}{18}$ **47.** $\dfrac{3}{4} = \dfrac{?}{32}$

48. $\dfrac{1}{6} = \dfrac{?}{30}$ **49.** $\dfrac{1}{5} = \dfrac{?}{55}$ **50.** $\dfrac{7}{8} = \dfrac{?}{64}$ **51.** $\dfrac{4}{5} = \dfrac{?}{20}$ **52.** $\dfrac{1}{6} = \dfrac{?}{15}$

Reduce to lowest terms.

53. $\dfrac{6}{12}$ **54.** $\dfrac{8}{12}$ **55.** $\dfrac{4}{32}$ **56.** $\dfrac{26}{64}$ **57.** $\dfrac{2}{8}$

58. $\dfrac{8}{32}$ **59.** $\dfrac{34}{64}$ **60.** $\dfrac{16}{64}$ **61.** $\dfrac{12}{32}$ **62.** $\dfrac{45}{90}$

Use the property of proportions to determine if the statements are true or false.

63. $\dfrac{15}{42} = \dfrac{20}{56}$ **64.** $\dfrac{8}{12} = \dfrac{16}{20}$

Measure line segments 65–74 in Fig. 2–28 (tolerance $= 6\frac{1}{32}$ in.).

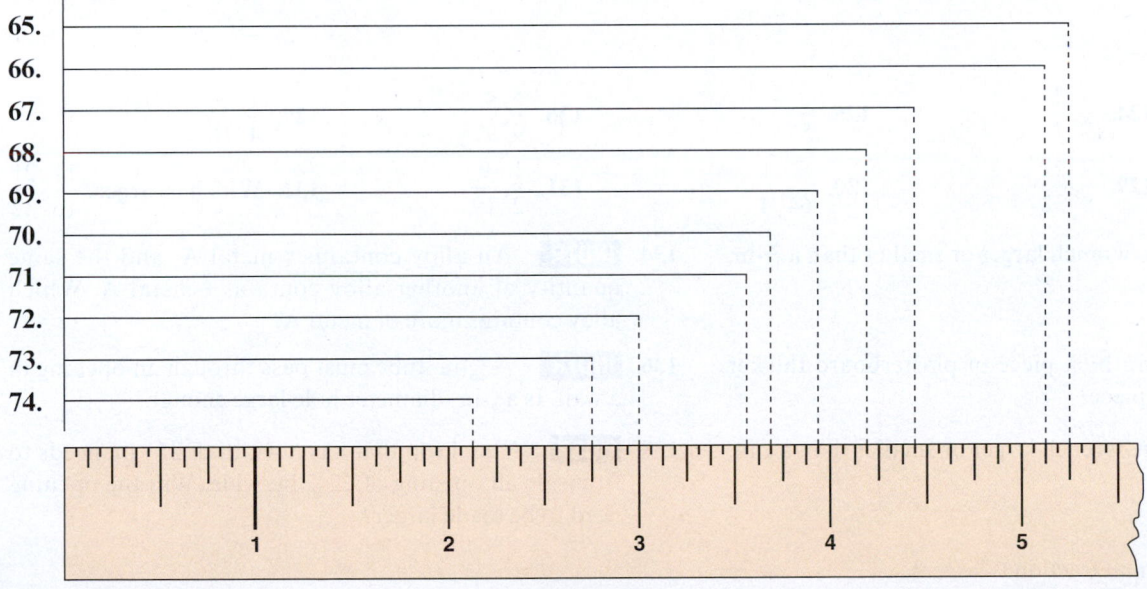

FIGURE 2–28

Use a U.S. customary rule to measure each line segment.

75. $A \bullet \!\!-\!\!-\!\!-\!\!-\!\!-\!\! \bullet B$ **76.** $C \bullet \!\!-\!\!-\!\!-\!\!-\!\!-\!\!-\!\!-\!\! \bullet D$

77. $E \bullet \!\!-\!\!-\!\!-\!\!-\!\!-\!\! \bullet F$ **78.** $G \bullet \!\!-\!\!-\!\!-\!\!-\!\!-\!\!-\!\!-\!\! \bullet H$

Write the improper fractions as whole or mixed numbers.

79. $\dfrac{18}{5}$ **80.** $\dfrac{27}{6}$ **81.** $\dfrac{39}{8}$ **82.** $\dfrac{21}{15}$ **83.** $\dfrac{43}{8}$

84. $\dfrac{22}{7}$ **85.** $\dfrac{175}{2}$ **86.** $\dfrac{135}{3}$ **87.** $\dfrac{18}{12}$ **88.** $\dfrac{35}{7}$

Write the whole or mixed numbers as improper fractions.

89. 8 **90.** $10\dfrac{1}{2}$ **91.** $7\dfrac{1}{8}$ **92.** $5\dfrac{7}{12}$ **93.** $9\dfrac{3}{16}$ **94.** $7\dfrac{8}{17}$

95. $4\dfrac{3}{5}$ **96.** $9\dfrac{1}{9}$ **97.** 12 **98.** $16\dfrac{2}{3}$ **99.** $5\dfrac{1}{3}$ **100.** $6\dfrac{1}{2}$

Change the whole number to an equivalent fraction using the indicated denominator.

101. $2 = \dfrac{?}{10}$ **102.** $6 = \dfrac{?}{4}$ **103.** $11 = \dfrac{?}{3}$ **104.** $7 = \dfrac{?}{5}$

Change the decimals to fractions in lowest terms.

105. 0.7 **106.** 0.83 **107.** 0.95 **108.** 0.25

109. 0.872 **110.** 0.081 **111.** 0.02 **112.** 0.005

Change the fractions to decimals. If necessary, round to the nearest thousandth.

113. $\dfrac{1}{5}$ **114.** $\dfrac{1}{10}$ **115.** $\dfrac{5}{8}$ **116.** $\dfrac{3}{7}$ **117.** $\dfrac{9}{11}$

Find the least common denominator.

118. $\dfrac{7}{8}, \dfrac{2}{3}$ **119.** $\dfrac{3}{4}, \dfrac{1}{16}$ **120.** $\dfrac{1}{12}, \dfrac{3}{4}$ **121.** $\dfrac{5}{12}, \dfrac{3}{10}, \dfrac{13}{15}$ **122.** $\dfrac{1}{12}, \dfrac{3}{8}, \dfrac{15}{16}$

Which fraction is smaller?

123. $\dfrac{5}{8}, \dfrac{3}{8}$ **124.** $\dfrac{3}{7}, \dfrac{2}{7}$ **125.** $\dfrac{3}{8}, \dfrac{4}{8}$ **126.** $\dfrac{5}{9}, \dfrac{4}{9}$ **127.** $\dfrac{1}{4}, \dfrac{3}{16}$

128. $\dfrac{5}{8}, \dfrac{11}{16}$ **129.** $\dfrac{7}{8}, \dfrac{27}{32}$ **130.** $\dfrac{7}{64}, \dfrac{1}{4}$ **131.** $\dfrac{1}{2}, \dfrac{9}{19}$ **132.** Which is larger? $\dfrac{11}{16}, \dfrac{21}{32}$

133. **INDTR**   Is a $\frac{5}{8}$-in. wrench larger or smaller than a $\frac{9}{16}$-in. wrench?

134. **INDTR**   An alloy contains $\frac{2}{3}$ metal A, and the same quantity of another alloy contains $\frac{3}{5}$ metal A. Which alloy contains more of metal A?

135. **INDTR**   Is a $\frac{3}{8}$-in.-thick piece of plasterboard thicker than a $\frac{1}{2}$-in.-thick piece?

136. **INDTR**   A $\frac{9}{16}$-in. tube must pass through an opening in a wall. Is a $\frac{3}{4}$-in.-diameter hole large enough?

137. **INDTR**   Is a $\frac{19}{32}$-in. wrench larger or smaller than a $\frac{7}{8}$-in. bolt head?

138. **HOSP**   A cooktop that has a width of $22\frac{1}{4}$ in. needs to fit inside an opening of $22\frac{5}{16}$ in. wide. Will the opening need to be made larger?

Which common or decimal fraction is larger?

139. $0.38, \dfrac{2}{5}$ **140.** $0.127, \dfrac{1}{8}$ **141.** $0.26, \dfrac{3}{4}$ **142.** $0.08335, \dfrac{1}{6}$ **143.** $0.272, \dfrac{3}{11}$

Section 2–3

Add; reduce sums to lowest terms and convert improper fractions to mixed numbers or whole numbers.

144. $\dfrac{1}{8} + \dfrac{5}{16}$ **145.** $\dfrac{3}{16} + \dfrac{9}{64}$ **146.** $\dfrac{3}{14} + \dfrac{5}{7}$ **147.** $\dfrac{3}{5} + \dfrac{5}{6}$

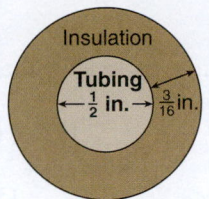

FIGURE 2-29

148. **CON** A hollow-wall fastener has a grip range up to $\frac{3}{4}$ in. Is it long enough to fasten three sheets of metal $\frac{5}{16}$ in. thick, $\frac{3}{8}$ in. thick, and $\frac{1}{16}$ in. thick?

149. **CON** Fig. 2–29 shows $\frac{1}{2}$-in. copper tubing wrapped in insulation. What is the distance across the tubing and insulation?

150. **CON** In Exercise 149, what would be the overall distance across the tubing and insulation if $\frac{3}{8}$-in.-ID (inside diameter) tubing were used?

151. $2\frac{5}{8} + 4 + 3\frac{3}{4}$

152. $3\frac{7}{8} + 7 + 5\frac{1}{2}$

153. $2\frac{7}{16} + 6\frac{5}{32}$

154. $9\frac{7}{8} + 5\frac{3}{4}$

155. $3\frac{7}{8} + 5\frac{3}{16} + 1\frac{7}{32}$

156. $2\frac{1}{4} + 3\frac{7}{8}$

157. **HELPP** A forest fire advanced $7\frac{5}{8}$ mi in one day and $10\frac{7}{16}$ mi the next. How far did the fire advance?

158. **CON** Find the total thickness of a wall if the outside covering is $3\frac{7}{8}$ in. thick, the studs (interior supports) are $3\frac{7}{8}$ in., and the inside covering is $\frac{5}{16}$-in. paneling.

159. **INDTEC** If $7\frac{5}{16}$ in. of a piece of square bar stock is turned (machined) so that it is cylindrical and $5\frac{9}{32}$ in. remains square, what is the total length of the original bar stock?

160. **AVIA** Two carts of airline food weigh $27\frac{1}{2}$ lb and $20\frac{3}{4}$ lb. What is the total weight of the two carts?

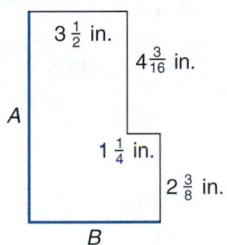

161. **INDTEC** Three metal rods measuring $3\frac{1}{8}$ in., $5\frac{3}{32}$ in., and $7\frac{9}{16}$ in. are welded together end to end. How long is the welded rod?

162. **CON** In Fig. 2–30, what is the length of side A and side B?

FIGURE 2-30

Subtract; reduce to lowest terms when necessary.

163. $\frac{5}{9} - \frac{2}{9}$

164. $\frac{11}{32} - \frac{5}{64}$

165. $3\frac{5}{8} - 2$

166. $7 - 4\frac{3}{8}$

167. $8\frac{7}{8} - 2\frac{29}{32}$

168. $7 - 2\frac{9}{16}$

169. $12\frac{11}{16} - 5$

170. $48\frac{5}{12} - 12\frac{11}{15}$

171. $122\frac{1}{2} - 87\frac{3}{4}$

172. **INDTR** Pins of $2\frac{3}{8}$ in. and $3\frac{7}{16}$ in. are cut from a drill rod 12 in. long. If $\frac{1}{16}$ in. of waste is allowed for each cut, how many inches of drill rod are left?

173. **CON** A bolt 2 in. long fastens a piece of $\frac{7}{8}$-in.-thick wood to a piece of metal. If a $\frac{3}{16}$-in.-thick lock washer, a $\frac{1}{16}$-in. washer, and a $\frac{7}{16}$-in.-thick nut are used, what is the thickness of the metal if the nut is even with the end of the bolt after tightening?

174. **TELE** Four lengths measuring $6\frac{1}{4}$ in., $9\frac{3}{16}$ in., $7\frac{1}{8}$ in., and $5\frac{9}{32}$ in. are cut from 48 in. of cable. How much cable remains? Disregard waste.

175. **INDTR** A piece of tapered stock has a diameter of $2\frac{5}{16}$ in. at one end and a diameter of $\frac{55}{64}$ in. at the other end. What is the difference in the diameters?

Section 2–4

Multiply and reduce answers to lowest terms. Convert improper fractions to whole or mixed numbers.

176. $\dfrac{3}{5}\left(\dfrac{10}{21}\right)$

177. $\dfrac{1}{3}\left(\dfrac{7}{8}\right)$

178. $\dfrac{2}{5}\left(\dfrac{7}{10}\right)$

179. $\dfrac{7}{9}\left(\dfrac{3}{8}\right)$

180. $\dfrac{2}{3}\left(\dfrac{5}{8}\right)\left(\dfrac{3}{16}\right)$

181. $\dfrac{15}{16}\left(\dfrac{4}{5}\right)\left(\dfrac{2}{3}\right)$

182. $5\left(\dfrac{3}{4}\right)$

183. $\dfrac{7}{16}(18)$

184. $\dfrac{3}{16}(184)$

185. $1\dfrac{1}{2}\cdot\dfrac{4}{5}$

186. $3\dfrac{1}{3}\cdot 4\dfrac{1}{2}$

187. $1\dfrac{3}{4}\cdot 1\dfrac{1}{7}$

188. **CON** In a concrete mixture, $\dfrac{4}{7}$ of the total volume is sand. How much sand is needed for 135 cubic yards (yd³) of concrete?

189. **CON** Concrete blocks are 8 in. high. If a $\dfrac{3}{8}$-in. mortar joint is used, how high will a wall of 12 courses of concrete blocks be? (*Hint:* There are 12 rows of mortar joints.)

190. **INDTEC** An adjusting screw will move $\dfrac{3}{64}$ in. for each full turn. How far will it move in four full turns?

191. **HOSP** A chef is making a dessert that is $\dfrac{3}{4}$ the original recipe. How much flour should be used if the original recipe calls for $3\dfrac{2}{3}$ cups of flour?

192. **INDTEC** If an alloy is $\dfrac{3}{5}$ copper and $\dfrac{2}{5}$ zinc, how many pounds of each metal are in a casting weighing $112\dfrac{1}{2}$ lb?

193. **CON** A water pipe has an outside diameter of $18\dfrac{3}{4}$ cm. What is the width of eight pipes that have the same diameter?

Raise the fractions to the indicated power.

194. $\left(\dfrac{1}{5}\right)^2$

195. $\left(\dfrac{3}{4}\right)^2$

196. $\left(\dfrac{5}{6}\right)^2$

197. $\left(\dfrac{4}{9}\right)^2$

198. $\left(\dfrac{1}{7}\right)^3$

199. $\left(\dfrac{1}{2}\right)^3$

200. $\left(\dfrac{7}{8}\right)^2$

201. $\left(\dfrac{9}{10}\right)^2$

Give the reciprocal.

202. $\dfrac{7}{8}$

203. 4

204. $2\dfrac{3}{5}$

205. 0.7

206. 1.8

Divide and reduce answers to lowest terms. Convert improper fractions to whole or mixed numbers.

207. $\dfrac{7}{8}\div\dfrac{3}{4}$

208. $\dfrac{4}{9}\div\dfrac{5}{16}$

209. $\dfrac{7}{8}\div\dfrac{3}{32}$

210. $8\div\dfrac{2}{3}$

211. $18\div\dfrac{3}{4}$

212. $35\div\dfrac{5}{16}$

213. $5\dfrac{1}{10}\div 2\dfrac{11}{20}$

214. $27\dfrac{2}{3}\div\dfrac{2}{3}$

215. $7\dfrac{1}{5}\div 12$

216. **CAD/ARC** On a house plan, $\dfrac{1}{4}$ in. represents 1 ft. Find the dimensions of a porch that measures $4\dfrac{1}{8}$ in. by $6\dfrac{1}{2}$ in. on the plan. (How many $\dfrac{1}{4}$s are there in $4\dfrac{1}{8}$; how many $\dfrac{1}{4}$s are there in $6\dfrac{1}{2}$?)

217. **CON** A pipe that is 12 in. long is cut into four equal parts. If $\dfrac{3}{16}$ in. is wasted per cut, what is the maximum length of each pipe? (It takes three cuts to divide the entire length into four equal parts.)

218. **CON** A stack of $\dfrac{5}{8}$-in. plywood is $21\dfrac{7}{8}$ in. high. How many sheets of plywood are in the stack?

219. **CON** A rod $1\dfrac{1}{8}$ yd long is cut into six equal pieces. What is the length of each piece? Disregard waste.

220. **AG/H** If $7\dfrac{1}{2}$ gal of liquid are distributed equally among five containers, what is the number of gallons per container?

221. **BUS** Fabric that is $22\dfrac{1}{2}$ yd long is cut into lengths of $\dfrac{5}{8}$ yd. How many equal lengths can be made?

Perform the operations and reduce to lowest terms. Convert improper fractions to whole or mixed numbers.

222. $\dfrac{\frac{5}{8}}{2\frac{1}{8}}$

223. $\dfrac{\frac{1}{3}}{\frac{1}{6}}$

224. $\dfrac{\frac{4}{4}}{\frac{4}{5}}$

225. $\dfrac{\frac{8}{1}}{1\frac{1}{2}}$

226. $\dfrac{3\frac{1}{4}}{5}$

227. $\dfrac{2\frac{1}{5}}{8\frac{4}{5}}$

228. $\dfrac{16\frac{2}{3}}{3\frac{1}{3}}$

Write as a fraction in lowest terms.

229. $0.12\frac{1}{2}$

230. $0.37\frac{1}{2}$

Perform the calculations using a calculator.

231. $\dfrac{\frac{7}{8}+\frac{1}{2}}{2\frac{1}{4}}$

232. $\dfrac{5\frac{2}{3}}{\frac{5}{6}-\frac{1}{3}}$

233. $\dfrac{\frac{5}{16}-\frac{1}{8}}{\frac{11}{32}+\frac{1}{4}}$

2   TEAM PROBLEM-SOLVING EXERCISES

Your team is preparing a report that is to be printed on both sides of the paper. It is customary to put odd-numbered pages on the front and even-numbered pages on the back. A new chapter is started on the front of a sheet of paper, even if this creates a preceding blank page. Assign page numbers to the document based on these guidelines.

Chap. 1, 15 pages   Chap. 2, 17 pages
Chap. 3, 24 pages   Chap. 4, 15 pages

1. What page number will start Chapter 2?
2. How many pages are in the document?
3. How many blank pages are in the document?

2   CONCEPTS ANALYSIS

1. What two operations require a common denominator?

2. Explain how to find the reciprocal of a fraction.

3. What steps must be followed to find the reciprocal of a mixed number?

4. What number can be written as any fraction that has the same numerator and denominator? Explain why.

5. What operation requires the use of the reciprocal of a fraction?

6. Name the operation that has each of the following for an answer: sum, difference, product, quotient.

7. What operation must be used to solve an applied problem if the total and one of the two parts are given?

8. What does the denominator of a fraction indicate?

9. What does the numerator of a fraction indicate?

10. What kind of fraction has a value less than 1?

Find, explain, and correct the mistakes in these problems.

11. $\dfrac{5}{8}+\dfrac{1}{8}=\dfrac{6}{16}=\dfrac{3}{8}$

12. $\begin{array}{r}12\phantom{\frac{3}{4}}\\-5\frac{3}{4}\\\hline 7\frac{3}{4}\end{array}$

13. $\dfrac{3}{5}\times 2\dfrac{1}{5}=2\dfrac{3}{25}$

14. $\dfrac{5}{8}\div 4=\dfrac{5}{8}\times\dfrac{4}{1}=\dfrac{5}{2}=2\dfrac{1}{2}$

15. $\begin{array}{r}12\frac{3}{4}=12\frac{6}{8}=11\frac{16}{8}\\-4\frac{7}{8}=-4\frac{7}{8}=-4\frac{7}{8}\\\hline 7\frac{9}{8}=7+1\frac{1}{8}=8\frac{1}{8}\end{array}$

2   PRACTICE TEST

Represent as fractions.

1. 3 out of 4 people in a survey

2. $7\div 9$

Write as whole or mixed numbers.

3. $\dfrac{9}{3}$

4. $\dfrac{14}{9}$

Write as improper fractions.

5. $4\frac{6}{7}$

6. $3\frac{1}{10}$

Write the least common multiple (LCM).

10. 48 and 64

11. 36, 45, and 54

14. Reduce $\frac{18}{24}$ to lowest terms.

15. Find the measure of the line segment AB in Fig. 2–31.

16. Write $\frac{7}{20}$ as a decimal.

17. Write $\frac{21}{2}$ as a decimal.

18. Write $5\frac{3}{8}$ as an improper fraction.

19. Which fraction is smaller, $\frac{4}{5}$ or $\frac{7}{10}$?

Write the prime factors in exponential notation.

7. 96   **8.** 132   **9.** 62

Write the greatest common factor (GCF).

12. 15 and 35   **13.** 12, 18, and 36

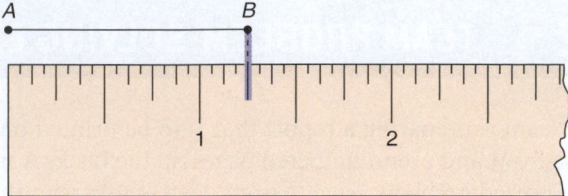

FIGURE 2–31

Perform the indicated operations.

20. $\frac{7}{12} + \frac{5}{6}$

21. $3\frac{4}{15} + 4\frac{3}{10}$

22. $\frac{5}{6}\left(\frac{3}{10}\right)$

23. $2\frac{2}{9}\left(1\frac{3}{4}\right)$

24. $\frac{7}{8} - \frac{1}{4}$

25. $6\frac{1}{4} - 2\frac{3}{4}$

26. $\frac{5}{12} \div \frac{5}{6}$

27. $7\frac{1}{2} \div \frac{5}{9}$

28. $5\frac{2}{3} \div 1\frac{1}{9}$

Solve the problems.

29. Two of the seven security employees at the local community college received safety awards from the governor. Represent the part of the total number of security employees who received an award as a fraction.

30. A candy-store owner mixes $1\frac{1}{2}$ lb of caramels, $\frac{3}{4}$ lb of chocolates, and $\frac{1}{2}$ lb of candy corn. What is the total weight of the mixed candy?

31. A homemaker has $5\frac{1}{2}$ cups of sugar on hand to make a batch of cookies requiring $1\frac{2}{3}$ cups of sugar. How much sugar is left?

32. If $6\frac{1}{4}$ ft of wire is needed to make one electrical extension cord, how many extension cords can be made from $68\frac{3}{4}$ ft of wire?

33. A costume maker figures one costume requires $2\frac{2}{3}$ yd of red satin. How many yards of red satin are needed to make three costumes?

34. Will a $\frac{5}{8}$-in.-wide drill bit make a hole wide enough to allow a $\frac{1}{2}$-in. (outside diameter) copper tube to pass through?

3

Percents

JLBvdWolf/Alamy Stock Photo

In Great Company

When Rounding Errors Kill

In 1991, during the first Gulf War, Iraqi forces frequently fired conventional ballistic missiles at targets in Israel and Saudi Arabia. The United States responded by deploying a 1970s-era antimissile system, dubbed "The Patriot."

The Patriot system's radar identifies an incoming missile based on flight characteristics. A ballistic missile doesn't fly in a straight line, like a plane does. Instead, it flies in a very predictable parabola, like a thrown rock. The system identifies an object as a missile only if it flies through the air like a rock.

The Patriot system bounces radar waves off every flying object, but missiles travel very quickly. It takes time for a radar wave to travel to an object, hit it, and then travel back to the radar system. The Patriot system takes that into account by "guessing" where the object would be at the next radar "ping." It calculates where the object should be if it were really a missile and not something else, like a plane, and then looks in that "window" of space to see if it is there.

To do this, the Patriot radar system needs a very accurate clock. By February 11, 1991, Israeli soldiers had noticed that the Patriot system clock became increasingly inaccurate the longer the system ran. Specifically, it was off by 20% after 8 h of use. Upon investigation, the designers discovered that the system clock was rounding off 0.000000095ths of a second for each second that the system was being operated. If the system were run for 20 h, the "window" calculation would be off by about 50%. At that point, the system would begin to fail.

No one thought the system was being run for more than 8 h at a time, but five days later, on February 16, the designers issued a software fix anyway. Because there were no secure data connections, the fix had to be flown from the United States to the Patriot sites in Saudi Arabia.

By February 25, the Patriot system protecting Dharhan airbase had been running continuously for about 100 hours. By the time a Scud missile was launched toward its army barracks, the error had magnified to 0.34ths of a second. This was five times too large for the Patriot battery to detect the missile. Because of the round-off error, the Patriot system ignored the incoming Scud missile. Twenty-eight American soldiers died; more than 100 were wounded. The software fix arrived by plane the next day. See https://fas.org/nuke/guide/iraq/missile/scud_info/scud_info_s07.htm for illustrations.

3–1   Percent and Number Equivalents

LEARNING OUTCOMES

1 Write any number as a percent equivalent.

2 Write any percent as a numerical equivalent.

LEARNING CATALYTICS

1. Mentally multiply 0.3 times 100.
2. Mentally divide 8 by 100.

Percent: a fractional part of 100 expressed with a percent sign (%), or as a fraction or a decimal

Mixed percent: a percent with a mixed number or mixed decimal

With fractions and decimals, we compare only like quantities, that is, fractions with common denominators and decimals with the same number of decimal places. We can standardize our representation of quantities so that they can be more easily compared. We standardize by expressing quantities in relation to a standard unit of 100. This relationship, called a **percent**, is used to solve many different types of business problems.

The word *percent* means *hundredths* or *out of 100* or *per 100* or *over 100* (in a fraction). That is, 44 percent means 44 hundredths, or 44 out of 100, or 44 per 100, or 44 over 100. We can write 44 hundredths as 0.44 or $\frac{44}{100}$.

The symbol for *percent* is %. You can write 44 percent using the percent symbol: 44%; using fractional notation: $\frac{44}{100}$; or using decimal notation: 0.44.

$$44\% = 44 \text{ percent} = 44 \text{ hundredths} = \frac{44}{100} = 0.44$$

Percents can contain whole numbers, decimals, fractions, mixed numbers, or mixed decimals. Percents with mixed numbers and mixed decimals are often referred to as **mixed percents.** Examples are $33\frac{1}{3}\%$, $0.05\frac{3}{4}\%$, and $0.23\frac{1}{3}\%$.

1 **Write Any Number as a Percent Equivalent.** The businessperson must be able to write whole numbers, decimals, or fractions as percents, and to write percents as whole numbers, decimals, or fractions. First we examine writing whole numbers, decimals, and fractions as percents.

Hundredths and percent have the same meaning: per hundred. Just as 100 cents is the same as 1 dollar, 100 percent is the same as 1 whole quantity.

$$100\% = 1$$

This fact is used to write percent equivalents of numbers and to write numerical equivalents of percents. It is also used to calculate markups, markdowns, discounts, and numerous other business applications.

When we multiply a number by 1, the product has the same value as the original number. $N \times 1 = N$. We have used this concept to change a fraction to an equivalent fraction with a higher denominator. For example,

$$1 = \frac{2}{2} \quad \text{and} \quad \frac{1}{2}\left(\frac{2}{2}\right) = \frac{2}{4}$$

We can also use the fact that $N \times 1 = N$ to change numbers to equivalent percents.

$$1 = 100\% \qquad \frac{1}{2} = \frac{1}{2}(100\%) = \frac{1}{\overset{}{\underset{1}{2}}}\left(\frac{\overset{50}{\cancel{100}}\%}{1}\right) = 50\%$$

$$0.5 = 0.5(100\%) = 0\underset{\smile}{50}.\% = 50\%$$

In each case when we multiply by 1 in some form, the value of the product is equivalent to the value of the original number even though the product *looks different*.

To write a number as its percent equivalent:	
1. Multiply the number by 1 in the form of 100%.	Write 0.3 as a percent.
2. The product has a % symbol.	$0.3 = 0.3(100\%) = 0\underset{\smile}{30}.\% = 30\%$

STOP AND CHECK

Write as a percent.
1. 0.71   2. 3.6
3. 0.009   4. 4

Answers:
1. 71%   2. 360%
3. 0.9%   4. 400%

EXAMPLE 1

Write the decimal or whole number as a percent.

(a) 0.27   **(b)** 0.875   **(c)** 1.73   **(d)** 0.004   **(e)** 2

(a) $0.27 = 0.27(100\%) = 0\underset{\smile}{27}.\% = 27\%$ Multiply 0.27 by 100% (move the
0.27 as a percent is 27%.   decimal point two places to the right).

(b) $0.875 = 0.875(100\%) = 0\underset{\smile}{87}.5\% = 87.5\%$ Multiply 0.875 by 100% (move the
0.875 as a percent is 87.5%.   decimal point two places to the right).

(c) $1.73 = 1.73(100\%) = 1\underset{\smile}{73}.\% = 173\%$ Multiply 1.73 by 100% (move the
1.73 as a percent is 173%.   decimal point two places to the right).

(d) $0.004 = 0.004(100\%) = 0\underset{\smile}{00}.4\% = 0.4\%$ Multiply 0.004 by 100% (move the
0.004 as a percent is 0.4%.   decimal point two places to the right).

(e) $2 = 2(100\%) = 2\underset{\smile}{00}.\% = 200\%$ Multiply 2 by 100% (move the decimal
2 as a percent is 200%.   point two places to the right).

See Exercises 1–12.

TIP **Multiplying by 1 in the Form of 100%** To write a number as its percent equivalent, identify the number as a fraction, whole number, or decimal. If the number is a whole number or decimal, multiply by 100% by using the shortcut rule for multiplying by 100. If the number is a fraction, multiply it by 1 in the form of $\frac{100\%}{1}$. In each case, the percent equivalent will be expressed with a percent symbol.

As you can see, the procedure is the same regardless of the number of decimal places in the number and regardless of whether the number is greater than, equal to, or less than 1.

STOP AND CHECK

Write as a percent.

1. $\dfrac{42}{100}$ **2.** $\dfrac{2}{5}$

3. $6\dfrac{3}{10}$ **4.** $\dfrac{5}{3}$

Answers:
1. 42%   **2.** 40%

3. 630%   **4.** $166\dfrac{2}{3}\%$

EXAMPLE 2

Write the fraction or mixed number as a percent.

(a) $\dfrac{67}{100}$ **(b)** $\dfrac{1}{4}$ **(c)** $3\dfrac{1}{2}$ **(d)** $\dfrac{7}{4}$ **(e)** $\dfrac{2}{3}$

(a) $\dfrac{67}{100} = \dfrac{67}{\overset{}{\underset{1}{100}}}\left(\dfrac{\overset{1}{100}\%}{1}\right) = \mathbf{67\%}$ Reduce and multiply.

(b) $\dfrac{1}{4} = \dfrac{1}{\overset{}{\underset{1}{4}}}\left(\dfrac{\overset{25}{100}\%}{1}\right) = \mathbf{25\%}$ Reduce and multiply.

(c) $3\dfrac{1}{2} = 3\dfrac{1}{2}\left(\dfrac{100\%}{1}\right) = \dfrac{7}{\overset{}{\underset{1}{2}}}\left(\dfrac{\overset{50}{100}\%}{1}\right) = \mathbf{350\%}$ Change to an improper fraction, reduce, and multiply.

(d) $\dfrac{7}{4} = \dfrac{7}{\overset{}{\underset{1}{4}}}\left(\dfrac{\overset{25}{100}\%}{1}\right) = \mathbf{175\%}$ Reduce and multiply.

(e) $\dfrac{2}{3} = \dfrac{2}{3}\left(\dfrac{100\%}{1}\right) = \dfrac{200\%}{3} = \mathbf{66\dfrac{2}{3}\%}$ Multiply.

See Exercises 13–24.

TIP **What Happens to the % (Percent) Sign?** In multiplying fractions we reduce or cancel common factors from a numerator to a denominator. Percent signs and other types of labels also cancel.

$$\dfrac{\%}{\%} = 1$$

2 **Write Any Percent as a Numerical Equivalent.** When a number is divided by 1, the quotient has the same value as the original number. $N \div 1 = N$ or $\dfrac{N}{1} = N$. We have used this concept to reduce fractions. For example,

$$1 = \dfrac{2}{2} \qquad \dfrac{2}{4} \div \dfrac{2}{2} = \dfrac{1}{2}$$

We can also use the fact that $N \div 1 = N$ or $\dfrac{N}{1} = N$ to change percents to numerical equivalents.

$$50\% \div 100\% = \dfrac{50\%}{100\%} = \dfrac{50}{100} = \dfrac{1}{2}$$

$$50\% \div 100\% = 50 \div 100 = 0.50 = 0.5$$

To write a percent as a number:

1. Divide the number by 1 in the form of 100% or multiply by $\dfrac{1}{100\%}$.

2. The quotient does not have a % symbol.

STOP AND CHECK

Write as a decimal.

1. 45%   **2.** 4.3%
3. 315%   **4.** 0.5%
5. $5\dfrac{7}{10}\%$ **6.** $5\dfrac{1}{9}\%$

Answers:
1. 0.45   **2.** 0.043   **3.** 3.15
4. 0.005   **5.** 0.057
6. 0.051 to the nearest thousandth

EXAMPLE 3

Write the percent as a decimal.

(a) 37%   **(b)** 26.5%   **(c)** 127%   **(d)** 7%   **(e)** 0.9%   **(f)** $2\dfrac{19}{20}\%$ **(g)** $167\dfrac{1}{3}\%$

(a) $37\% = 37\% \div 100\% = 0.37 = \mathbf{0.37}$ Divide by 100 mentally.

(b) $26.5\% = 26.5\% \div 100\% = 0.265 = \mathbf{0.265}$ Divide by 100 mentally.

(c) $127\% = 127\% \div 100\% = 1.27 = \mathbf{1.27}$ Divide by 100 mentally.

(d) $7\% = 7\% \div 100\% = 0.07 = \mathbf{0.07}$ Divide by 100 mentally.

(e) $0.9\% = 0.9\% \div 100\% = 0.009 = \mathbf{0.009}$ Divide by 100 mentally.

(f) $2\dfrac{19}{20}\% = 2.95\% \div 100\% = 0.\underset{\smile}{0}295 = \mathbf{0.0295}$

Write the mixed number in front of the percent symbol as a decimal before dividing by 100%.

(g) $167\dfrac{1}{3}\% = 167.3\overline{3}\% \div 100\%$

$\qquad = 1.\underset{\smile}{6}7\overline{3} \div \mathbf{1.67\overline{3}}$ **or 1.673 (rounded)**

Write the mixed number in front of the percent symbol as a repeating decimal before dividing by 100.

See Exercises 25–49.

STOP AND CHECK

Write as a fraction, mixed number, or whole number.

1. 45% **2.** $\dfrac{1}{5}\%$

3. 325% **4.** $16\dfrac{2}{3}\%$

5. 87.5% **6.** 700%

Answers:

1. $\dfrac{9}{20}$ **2.** $\dfrac{1}{500}$ **3.** $3\dfrac{1}{4}$

4. $\dfrac{1}{6}$ **5.** $\dfrac{7}{8}$ **6.** 7

EXAMPLE 4

Write the percent as a fraction or mixed number.

(a) 65% **(b)** $\frac{1}{4}\%$ **(c)** 250% **(d)** $83\frac{1}{3}\%$ **(e)** 12.5% **(f)** 600%

(a) $65\% = 65\% \div 100\% = \dfrac{\overset{13}{\cancel{65\%}}}{1}\left(\dfrac{1}{\underset{20}{\cancel{100\%}}}\right) = \dfrac{13}{20}$

Convert division to multiplication.

(b) $\dfrac{1}{4}\% = \dfrac{1}{4}\% \div 100\% = \dfrac{1\%}{4}\left(\dfrac{1}{100\%}\right) = \dfrac{1}{400}$

(c) $250\% = 250\% \div 100\% = \dfrac{\overset{5}{\cancel{250\%}}}{1}\left(\dfrac{1}{\underset{2}{\cancel{100\%}}}\right) = \dfrac{5}{2} = 2\dfrac{1}{2}$

(d) $83\dfrac{1}{3}\% = 83\dfrac{1}{3}\% \div 100\% = \dfrac{\overset{5}{\cancel{250\%}}}{3}\left(\dfrac{1}{\underset{2}{\cancel{100\%}}}\right) = \dfrac{5}{6}$

Convert to improper fraction.

(e) $12.5\% = 12\dfrac{1}{2}\% \div 100\% = \dfrac{\overset{1}{\cancel{25\%}}}{2}\left(\dfrac{1}{\underset{4}{\cancel{100\%}}}\right) = \dfrac{1}{8}$

Convert decimal to mixed number.

(f) $600\% = 600\% \div 100\% = \dfrac{\overset{6}{\cancel{600\%}}}{1}\left(\dfrac{1}{\underset{1}{\cancel{100\%}}}\right) = 6$

See Exercises 50–71.

3-1 EXERCISES

MyLab Math For additional practice go to your study plan in MyLab Math.

1 *See Example 1.*

Change the numbers to their percent equivalents.

1. 0.2 **2.** 0.14 **3.** 0.007 **4.** 0.0125 **5.** 5

6. 8 **7.** 3.05 **8.** 7.2 **9.** 15.1 **10.** 36.25

11. From a recent U.S. Census Bureau report, the portion of the U.S. population under 18 years old was 0.273. What percent of the population was under 18 years old?

12. **PFIN** A recent consumer expenditures survey showed that the portion of total expenditures for telephone services that was spent on cellular phone service for persons under 25 was 0.752. What percent of this age group's total expenditures on telephone service was spent for cellular phone services?

See Example 2.

Change the numbers to their percent equivalents. Round to the nearest tenth of a percent if appropriate.

13. $\dfrac{5}{8}$　　　　**14.** $\dfrac{7}{9}$　　　　**15.** $\dfrac{7}{1000}$　　　　**16.** $\dfrac{1}{350}$

17. $1\dfrac{1}{3}$　　　　**18.** $3\dfrac{1}{2}$　　　　**19.** $4\dfrac{3}{10}$　　　　**20.** $2\dfrac{1}{5}$

21. **ELEC** If $\dfrac{2}{5}$ of the electricians in a city are self-employed, what percent are self-employed?

22. **CON** If $\dfrac{7}{10}$ of the bricklayers in a city are male, what percent are male?

23. The report, Global Video Game Market Forecast, projects that online game revenues will account for approximately $\dfrac{3}{5}$ of total software revenue in the next three years. What percent of the total software revenue will online game revenues be within three years?

24. According to recent data from the U.S. Census Bureau, approximately $\dfrac{9}{10}$ of the U.S. population was less than 65 years of age. What percent of the population was less than 65?

2 *See Example 3.* Change to decimal equivalents. Round to the nearest ten-thousandth if appropriate.

25. 36%　　**26.** 45%　　**27.** 20%　　**28.** 75%　　**29.** $6\dfrac{1}{4}$%　　**30.** 62.5%

31. $66\dfrac{2}{3}$%　　**32.** 0.6%　　**33.** $\dfrac{1}{5}$%　　**34.** 0.05%　　**35.** $8\dfrac{1}{3}$%　　**36.** 18.75%

Change to equivalent whole numbers.

37. 800%　　　　　　　　　　　　　　**38.** 400%

Change to mixed-decimal equivalents.

39. 250%　　**40.** 425%　　**41.** 176%　　**42.** 380%　　**43.** $137\dfrac{1}{2}$%　　**44.** 387.5%

45. 115.3%　　**46.** $212\dfrac{1}{2}$%　　**47.** $106\dfrac{1}{4}$%

48. **PFIN** A recent consumer expenditures survey showed that 54.8% of all expenditures in the United States for annual telephone services was allocated to cellular phone service. Write the percent as a decimal.

49. Recent statistics showed that 25.7% of California's population was under 18 years old and 0.4% of the state's population was Native Hawaiian or Other Pacific Islander. Express these two percents as decimals.

See Example 4.

Change to fractional equivalents.

50. 45%　　**51.** 36%　　**52.** 75%　　**53.** 20%　　**54.** 62.5%　　**55.** $6\dfrac{1}{4}$%

56. 0.6%　　**57.** $66\dfrac{2}{3}$%　　**58.** 0.05%　　**59.** $\dfrac{1}{5}$%　　**60.** 18.75%　　**61.** $8\dfrac{1}{3}$%

Change to mixed-number equivalents.

62. 425%　　　　　　**63.** 250%　　　　　　**64.** 380%　　　　　　**65.** 176%

66. 387.5%　　　　　**67.** $137\dfrac{1}{2}$%　　　　　**68.** $166\dfrac{2}{3}$%　　　　　**69.** $316\dfrac{2}{3}$%

70. Statistics from FedStats, a governmental website, showed that approximately 30% of firms in California were owned by women and 15% were owned by Hispanics. Express each percent as a fraction.

71. Statistics from FedStats, a governmental website, showed that approximately 0.5% of Florida's population was American Indian or Alaskan Native. Express the percent as a fraction.

3-2 Percentage Problems

LEARNING OUTCOMES

1. Identify the portion, base, and rate in percent problems.

2. Solve percent problems using the percentage formula.

3. Solve percent problems using the percentage proportion.

4. Solve business and consumer problems involving percents.

LC LEARNING CATALYTICS

Perform the operations.

1. $0.3(500)$

2. $\dfrac{90}{0.3}$

3. $\dfrac{60}{240}$

STOP AND CHECK

Identify the *P, B,* and *R* for each example.

1. What number is 30% of 80?
2. 45 is what percent of 90?
3. 25% of what number is 35?

Answers:

1. *P* is missing. *B* = 80. *R* = 30%.
2. *P* = 45. *B* = 90. *R* is missing.
3. *P* = 35. *B* is missing. *R* = 25%.

Formula: a symbolic statement of words or letters that expresses a relationship among quantities

Percentage formula: Portion = Rate × Base, or *P = RB*

Base: the number (*B*) that represents the original or total amount

Portion: the number (*P*) that represents a part or portion of the base. Also called percentage

Rate: the rate (*R*) is a percent that tells us how the base and portion are related

Percentage: in a problem involving percent, the number (*P*) that represents a portion of the base. Also called portion; alternate definition is a fraction or ratio with 100 understood as the denominator

1 Identify the Portion, Base, and Rate in Percent Problems. A **formula** expresses a relationship among quantities. When you use the six-step problem-solving approach, the third step, the Solution Plan, is often a formula written in words and letters.

The **percentage formula,** Portion = Rate × Base, can be written as $P = RB$. The letters or words represent numbers.

In the formula $P = RB$, the **base** (*B*) represents the original number or one entire quantity. The **portion** (*P*) represents a part of the base. The **rate** (*R*) is a percent that tells us how the base and portion are related. In the statement "50 is 20% of 250," 250 is the base (the entire quantity), 50 is the portion (part), and 20% is the rate (percent).

To identify the rate, base, and portion:

1. Identify the rate. *Rate* is usually written as a percent, but it may be a decimal or fraction.

2. Identify the base. *Base* is the total amount, original amount, or entire amount. The base often follows the preposition *of*.

3. Identify the portion. *Portion* can refer to the part, partial amount, amount of increase or decrease, or amount of change. It is a portion of the *base*. The portion often follows a form of the verb *is*.

EXAMPLE 1

Identify the given and missing elements for each example.

(a) 20% of 75 is what number?

(b) What percent of 50 is 30?

(c) Eight is 10% of what number?

(a) $\quad$ *R* $\qquad$ *B* $\qquad$ *P*
$\quad$ 20% of 75 is what number?
$\quad$ Percent $\;$ Total $\quad$ Part

(b) $\qquad$ *R* $\qquad$ *B* $\quad$ *P*
$\quad$ What percent of 50 is 30?
$\qquad$ Percent Total Part

(c) $\;$ *P* $\qquad$ *R* $\qquad\qquad$ *B*
$\quad$ Eight is 10% of what number?
$\quad$ Part Percent $\qquad$ Total

Use the identifying key words for rate (*percent* or *%*), base (*total, original,* associated with the word *of*), and portion (*part,* associated with the word *is*).

See Exercises 1–12.

Did You Know? **Portion can be called percentage.** In a standard dictionary you will see **percentage** defined as "a fraction or ratio with 100 understood as the denominator," as "the result obtained by multiplying a quantity by a percent," and as "a portion or share in relation to a whole; a part." That is, the word *percentage* can refer to both the rate and the portion. This causes many to confuse the words *percent* and *percentage*. Because in written reports the word *percentage* is often used to identify the *percent* or *rate* instead of the portion, we will only use the word *portion* when referring to a *part of the base.*

2 **Solve Percent Problems Using the Percentage Formula.** The percentage formula, Portion = Rate × Base, can be written as $P = RB$. When the numbers are put in place of the letters, the formula guides you through the calculations.

The three forms of the percentage formula are

Portion = Rate × Base	$P = RB$	For finding the portion.
Base = $\dfrac{\text{Portion}}{\text{Rate}}$	$B = \dfrac{P}{R}$	For finding the base.
Rate = $\dfrac{\text{Portion}}{\text{Base}}$	$R = \dfrac{P}{B}$	For finding the rate.

Percentage formula form for portion: Portion = Rate × Base, or $P = RB$

Percentage formula form for base:

Base = Portion/Rate or $B = \dfrac{P}{R}$

Percentage formula form for rate:

Rate = Portion/Base, or $R = \dfrac{P}{B}$

Circles can help us visualize these formulas. The shaded part of each circle in Fig. 3–1 represents the missing amount. The unshaded parts represent the known amounts. If the unshaded parts are *side by side, multiply* their corresponding numbers to find the unknown number. If the unshaded parts are *one on top of the other, divide* the corresponding numbers, divide top by bottom, to find the unknown number.

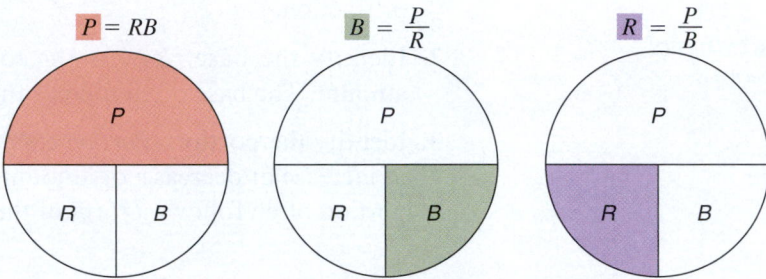

FIGURE 3–1 Forms of the Percentage Formula.

To use the percentage formula to solve percentage problems:

1. Identify and classify the two known values and the one unknown value.

2. Choose the appropriate percentage formula for finding the unknown value.

3. Substitute the known values into the formula. For the rate, use the decimal or fractional equivalent of the percent.

4. Perform the calculation indicated by the formula.

5. Interpret the result. If finding the rate, convert decimal or fractional equivalent of the rate to a percent.

STOP AND CHECK

Solve the problems.
1. What number is 30% of 80?
2. 45 is what percent of 90?
3. 25% of what number is 35?

Answers:
1. 24   2. 50%   3. 140

EXAMPLE 2

Solve the problems.

(a) 20% of 400 is what number?

(b) 20% of what number is 80?

(c) 80 is what percent of 400?

(a) 20% = Rate	Identify known values and unknown value.
400 = Base	
Portion is unknown	
$P = RB$	Choose the appropriate formula.
$P = 0.2(400)$	Substitute values using the decimal equivalent of 20%.
$P = 80$	Perform calculation.
20% of 400 is 80.	Interpret result.

(b) 20% = Rate   Identify known values and unknown value.
80 = Portion
Base is unknown

$B = \dfrac{P}{R}$ Choose the appropriate formula.

$B = \dfrac{80}{0.2}$ Substitute values. Perform calculation.

$B = 400$
20% of 400 is 80. Interpret result.

(c) 80 = Portion   Identify known values and unknown value.
400 = Base
Rate is unknown

$R = \dfrac{P}{B}$ Choose the appropriate formula.

$R = \dfrac{80}{400}$ Substitute values. Perform calculation.

$R = 0.2$ or 20%
80 is 20% of 400. Interpret result. 0.2 = 20%.   **See Exercises 13–22.**

Very few percentage problems that you encounter in business tell you the values of P, R, and B directly. Percentage problems are usually written in words that must be interpreted before you can tell which form of the percentage formula you should use.

Shebeko/Shutterstock

EXAMPLE 3

HOSP During a special one-day sale, 600 customers bought the on-sale pizza. Of these customers, 20% used coupons. The manager will run the sale again the next day if more than 100 coupons were used. Should she run the sale again?

Unknown facts

Quantity of coupon-using customers
Should the manager run the sale again?

Known facts

Total customers: 600
Coupon-using customers as a percent of total customers: 20%

Relationship

The quantity of coupon-using customers is a *portion* of the *base* of total customers, at a *rate* of 20% (Fig. 3–2).

$$P = RB$$

Quantity of coupon-using customers = RB

Calculation

$P = RB$ P is unknown; R = 20%; B = 600
$P = 20\%(600)$ Substitute known values. Change % to decimal equivalent.
$P = 0.2(600)$ Multiply.
$P = 120$

Interpretation
The quantity of coupon-using customers is 120.
Because 120 is more than 100, the manager should run the sale again.

             See Exercises 23–24.

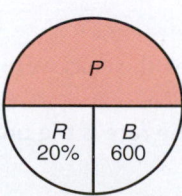

FIGURE 3–2

EXAMPLE 4

BUS   If $66\frac{2}{3}\%$ of the 900 employees in a company choose the Preferred Provider insurance plan, how many people from that company are enrolled in the plan?

First, identify the terms. The rate is the percent, and the base is the total number of employees. The portion is the quantity of employees enrolled in the plan.

$P = RB$ The portion is the unknown value (Fig. 3–3).

$P = 66\frac{2}{3}\% (900)$ The rate is $66\frac{2}{3}\%$; the base is 900. Write $66\frac{2}{3}\%$ as a fraction.

$P = \frac{2}{\cancel{3}_{1}}\left(\frac{\overset{300}{\cancel{900}}}{1}\right) = 600$ Multiply.

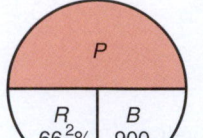

FIGURE 3–3

The Preferred Provider plan has 600 people enrolled.   See Exercises 25–26.

TIP   **Noncontinuous Calculator Sequence Versus Continuous Calculator Sequence**   We can write the fractional equivalent of the percent as a rounded decimal and divide using a calculator.

$$\boxed{\text{AC}}\; 2 \;\boxed{\div}\; 3 \;\boxed{=}\; \Rightarrow 0.6667$$
$$\boxed{\text{AC}}\; 900 \;\boxed{\times}\; .6667 \;\boxed{=}\; \Rightarrow 600.03$$

As one continuous sequence, enter

$$\boxed{\text{AC}}\; 2 \;\boxed{\div}\; 3 \;\boxed{\times}\; 900 \;\boxed{=}\; \Rightarrow 600$$

Note slight discrepancies from rounding when using two separate calculations. The answer obtained by using a continuous sequence of steps is more accurate.

EXAMPLE 5

PFIN   Stan sets aside 15% of his weekly income for rent. If he sets aside $150 each week, what is his weekly income?

Identify the terms: The rate is the number written as a percent, 15%. The portion is given, $150; it is a portion of his weekly income, the unknown base.

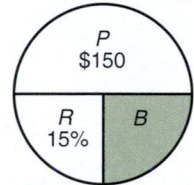

FIGURE 3–4

$B = \dfrac{P}{R}$ The rate is 15% and the portion is $150 (Fig. 3–4). The base is the weekly income to be found.

$B = \dfrac{\$150}{15\%}$ Convert 15% to a decimal equivalent.

$B = \dfrac{150}{0.15}$ Divide.

$B = \$1,000$

Stan's weekly income is $1,000.   See Exercises 27–30.

EXAMPLE 6

If 20 cars were sold from a lot that had 50 cars, what percent of the cars were sold?

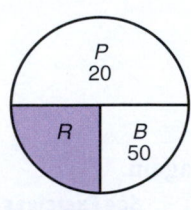

FIGURE 3–5

$R = \dfrac{P}{B}$ The portion is 20; the base is 50 (Fig. 3–5). The rate is the unknown to find.

$R = \dfrac{20}{50}$ Divide.

$R = 0.4$ Convert to % equivalent.

$R = 0.4(100\%)$

$R = 40\%$

Of the cars on the lot, 40% were sold.   See Exercises 31–34.

Many students mistakenly think that the portion can never be larger than the base. The portion (percentage) is smaller than the base only when the rate is less than 100%. The portion is larger than the base when the rate is greater than 100%.

STOP AND CHECK

Solve the problem.

1. 75 is what percent of 25?

Answer:

1. 300%

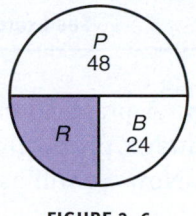

FIGURE 3–6

EXAMPLE 7

48 is what percent of 24?

$R = \dfrac{P}{B}$ The rate is unknown. The portion is 48. The base is 24 (Fig. 3–6).

$R = \dfrac{48}{24}$ Divide.

$R = 2$ Rate written as a whole number.

$R = 2(100\%)$

$R = \mathbf{200\%}$ Rate written as a percent.   See Exercises 35–36.

It is advisable to anticipate the approximate size of an answer before making any calculations. This approximation helps you discover mistakes, especially when using a calculator.

> **TIP** **Estimating a Percent of a Number** Estimate a percent by comparing it to a percent that you can calculate mentally.
>
> **To find**
>
> | | | |
> |---|---|---|
> | 1% of a number | *Multiply by 0.01* | Move decimal two places to the left. |
> | 2% of a number | $2 \cdot 1\%$ of a number | |
> | 3% of a number | $3 \cdot 1\%$ of a number | |
> | 10% of a number | *Multiply by 0.1* | Move decimal one place to the left. |
> | 5% of a number | $\frac{1}{2}$ of 10% of a number | |
> | 20% of a number | $2 \cdot 10\%$ of a number | |
> | 30% of a number | $3 \cdot 10\%$ of a number | |
> | 50% of a number | $\frac{1}{2}$ *of a number* | Divide by 2. |
> | 25% of a number | $\frac{1}{4}$ *of a number* | Divide by 4. |
> | 75% of a number | $3 \cdot 25\%$ of a number | |
> | $33\frac{1}{3}\%$ of a number | $\frac{1}{3}$ *of a number* | Divide by 3. |
> | $66\frac{2}{3}\%$ of a number | $2 \cdot 33\frac{1}{3}\%$ of a number | |

EXAMPLE 8

Mentally find 10% of $51.00.

10% of $51.00 = **$5.10.**   Mentally, move the decimal point in the number one place to the left.   See Exercises 37–38.

Asia Images Group/Shutterstock

EXAMPLE 9

PFIN   Find a 15% tip on a restaurant bill of $48.18.

10% of $48.00 = $4.80   Find 10% of the rounded total bill.
                                          $48.18 rounded to $48.00.

5% of $48.00 = $\dfrac{1}{2}$ of $4.80 = $2.40   Find 5% of the rounded total bill.

15% = 10% + 5%
$4.80 + $2.40 =   Add the amounts for 10% and 5%.
$7.20   15% tip.   **See Exercise 39.**

EXAMPLE 10

PFIN   A taxi fare is $24.00. Find the amount of a 20% tip.

10% of $24.00 = $2.40   Find 10% of the fare.
2 · $2.40 = $4.80   Double the 10% amount.

A 20% tip is $4.80.   **See Exercise 40.**

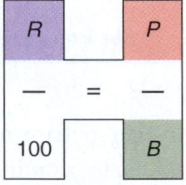

FIGURE 3–7

Percentage proportion: a formula in the form of a proportion for solving percent problems.
$R/100 = P/B$

3   **Solve Percent Problems Using the Percentage Proportion.** A practical method for solving percent problems involves solving the percentage proportion. In Chapter 2 we looked at proportions in comparing fractions. Now we will use proportions to find missing values.

The percentage formula $P = RB$ can be used to form the **percentage proportion.** The relationships among the rate, portion, base, and the standard unit of 100 are represented by two fractions that are equal to each other (Fig. 3–7).

> **Percentage proportion formula:**
>
> $\dfrac{R}{100} = \dfrac{P}{B}$ where $\dfrac{R}{100}$ = fractional form of rate or percent, P = portion or part,
>
> B = base or total

In a percent problem, if we know any two of the elements, we can find the third element using the property of proportions and cross multiplication. This process is called *solving the proportion.*

> **To use the percentage proportion to solve percentage problems:**
>
> 1. Cross multiply to find the cross products.
>
> 2. Divide the product of the two known factors by the factor with the letter.
>
> *Note:* An algebraic explanation for solving a proportion is presented in Chapter 8.

STOP AND CHECK

1. What is 35% of 60?

Answer:

1. 21

EXAMPLE 11

What is 20% of 75?

The rate is 20%, the base is 75, and the portion is missing.

$\dfrac{R}{100} = \dfrac{P}{B}$ Set up the proportion and substitute the known elements (Fig. 3–8).

$\dfrac{20}{100} = \dfrac{P}{75}$ Cross multiply.

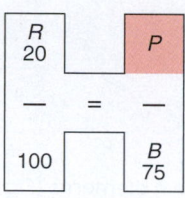

FIGURE 3–8

$20(75) = 100 \cdot P$

$1,500 = 100 \cdot P$ Divide to find P.

$\dfrac{1,500}{100} = P$

$15 = P$

The portion is 15.   See Exercises 41–50.

TIP **Multiply, Then Divide** Solving percentage proportions always involves one multiplication and one division. One cross product will involve two numbers—this is the multiplication. The other cross product will involve one number and one letter—this is the division. We divide the cross product with two numbers by the one number in the other cross product.

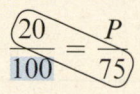

Cross product with two numbers is $20 \cdot 75$.
Cross product with one number is $100 \cdot P$. Divide by 100.

$20 \cdot 75 \div 100 = P$

$1,500 \div 100 = P$

$15 = P$

To solve proportions with a calculator, we follow a continuous series of steps.

$20 \boxtimes 75 \boxdiv 100 \boxed{=} \Rightarrow 15$

TIP **Advantages of Using the Percentage Proportion** Using percentage proportions gives us several advantages:

▶ Only one formula is needed to solve problems for portion, rate, and base.

▶ A standardized approach to percentage problems works in all cases.

▶ The standard 100 takes care of having to convert decimal numbers or fractions to percents or to convert percents to decimal or fraction equivalents.

▶ All problems involve one multiplication and one division step.

▶ Multiplication by 100 or division by 100 can be done mentally.

STOP AND CHECK

1. 70% of what number is 210?

Answer:

1. 300

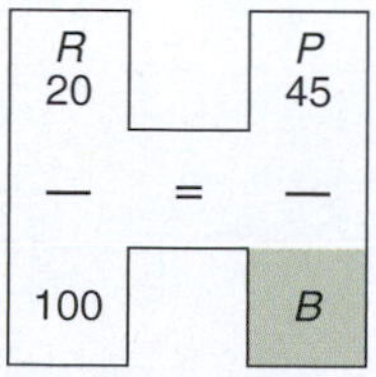

EXAMPLE 12

20% of what number is 45?

This time we know the rate, 20%, and the portion, 45. We are looking for the base, as indicated by the key word *of* (Fig. 3–9). The base is the original or whole amount. The base is larger than the portion when the rate is less than 100%.

FIGURE 3–9

Option 1

$\dfrac{20}{100} = \dfrac{45}{B}$

$20 \cdot B = 100 \cdot 45$

$20 \cdot B = 4,500$

$B = \dfrac{4,500}{20}$

$B = 225$

Option 2

$\dfrac{1}{5} = \dfrac{45}{B}$

$B = 5 \cdot 45$

$B = 225$

Option 3

$\dfrac{0.2}{1} = \dfrac{45}{B}$

$0.2 \cdot B = 45$

$B = \dfrac{45}{0.2}$

$B = 225$

Option 4

$B = \dfrac{P}{R}$

$B = \dfrac{45}{0.2}$

$B = 225$

20% of 225 is 45.   See Exercises 51–60.

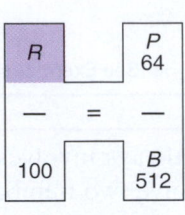

FIGURE 3–10

64 is what percent of 512?

The rate is missing, the base is 512, and the portion is 64.

$$\frac{R}{100} = \frac{P}{B}$$ Set up the proportion and substitute the known elements (Fig. 3–10).

$$\frac{R}{100} = \frac{64}{512}$$ Cross multiply.

$$R \cdot 512 = 100(64)$$

$$R \cdot 512 = 6{,}400$$ Divide to find R.

$$R = \frac{6{,}400}{512}$$

$$R = 12.5\%$$

The rate is 12.5%.   See Exercises 61–70.

When solving applied problems, our first task is to identify the two given parts and determine which part is missing.

INDTR   If a type of solder contains 55% tin, how many pounds of tin are needed to make 10 lb of solder?

Solder: a mixture of metals

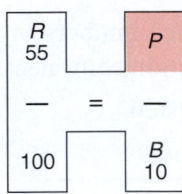

FIGURE 3–11

First, let's be sure we understand the word *solder*. **Solder** is a mixture of metals. In this problem, the *total amount* or the *base* is the 10 lb of solder, and it is made of tin and other metals.

Known facts

55% of 10 lb of solder is tin.
Rate: percent of tin = 55%
Base: amount of solder = 10 lb

Unknown fact

Portion or number of pounds of tin = P

Relationships

Percentage proportion (Fig. 3–11)

$$\frac{R}{100} = \frac{P}{B}$$ Substitute known values.

$$\frac{55}{100} = \frac{P}{10\text{ lb}}$$

Estimation

To estimate, 55% is more than $\frac{1}{2}$. $\frac{1}{2}$ of 10 = 5. So the amount of tin should be more than 5 lb.

Calculations

$$\frac{55}{100} = \frac{P}{10}$$ Solve for *P*. Cross multiply.

$$55 \cdot 10 = 100 \cdot P$$

$$550 = 100 \cdot P$$ Divide by 100.

$$\frac{550}{100} = P$$

$$5\frac{1}{2} = P$$

Interpretation

Thus, $5\frac{1}{2}$ lb of tin are needed to make 10 lb of solder.

To check the reasonableness of the answer, $5\frac{1}{2}$ lb is a little more than 5 lb.

See Exercises 71–79.

EXAMPLE 15

AUTO   If a 150-horsepower (hp) engine delivers only 105 hp to the driving wheels of a car, what is the efficiency of the engine?

Efficiency: the *percent* the output is of the total amount the engine is capable of delivering

Efficiency means the *percent* the output (105 hp) is of the total amount (150 hp) the engine is capable of delivering. Thus, the base amount is 150 hp, the portion delivered is 105 hp, and the percent of 150 represented by 105 is the rate or efficiency.

Known facts

Base: Total amount of horsepower = 150 hp
Portion: Amount of horsepower the engine delivers = 105 hp

Unknown facts

Efficiency or percent of the total horsepower

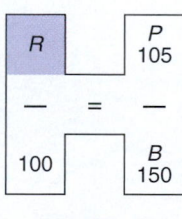

FIGURE 3–12

Relationships

Percentage proportion

$$\frac{R}{100} = \frac{P}{B}$$ Substitute known values (Fig. 3–12).

$$\frac{R}{100} = \frac{105}{150}$$

Estimation

Because the engine is not operating at full capacity (150 hp), we expect the efficiency to be less than 100%. Since 105 is more than $\frac{1}{2}$ of 150, the percent or rate will be more than 50%. A more precise estimate is

$$\frac{100}{150} = \frac{2}{3} = 66\frac{2}{3}\%$$ Round 105 to 100.

Since $105 > 100$ > is read "is greater than."

$$\text{Rate} > 66\frac{2}{3}\%$$

Calculations

$$\frac{R}{100} = \frac{105}{150}$$ Cross multiply.

$$R \cdot 150 = 100 \cdot 105$$

$$R \cdot 150 = 10{,}500$$ Divide.

$$R = \frac{10{,}500}{150}$$

$$R = 70$$

Interpretation

The engine is 70% efficient.   **See Exercises 80–85.**

> **TIP** **Must I Write All the Steps in the Six-Step Problem-Solving Plan for Every Problem?** As you develop confidence in your problem-solving ability you will begin to instinctively think through the steps without writing them down. We will begin to omit some of the steps in our examples, but will continue to draw special attention to the two *often neglected* steps: Estimation and Interpretation.

EXAMPLE 16

ELEC The effective value of current or voltage in an AC circuit is 71.3% of the maximum voltage. If a voltmeter shows a voltage of 110 volts (V) as the effective value in a circuit, what is the maximum voltage?

71.3% of the maximum voltage is 110 V. The maximum voltage is the *base*, and the amount of voltage shown on the voltmeter (110 V) is the *portion*.

Estimation

The rate is less than 100%. Therefore, the base is larger than the portion. $B > 110$.

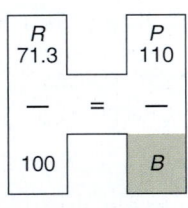

$$\begin{array}{cc} \dfrac{R}{71.3} & \dfrac{P}{110} \\[4pt] \underline{} = \underline{} \\[4pt] 100 & B \end{array}$$

FIGURE 3–13

$$\frac{71.3}{100} = \frac{110}{B}$$ Substitute known values (Fig. 3–13).

$$71.3 \cdot B = 100 \cdot 110$$ Cross multiply.

$$71.3 \cdot B = 11{,}000$$ Divide.

$$B = \frac{11{,}000}{71.3}$$

$$B = 154.2776999$$

or

$$B = 154 \text{ V}$$ To the nearest volt

Interpretation

The maximum voltage is 154 V.   **See Exercises 86–92.**

4   Solve Business and Consumer Problems Involving Percents. No matter what career you pursue, you will encounter business and consumer applications of percents. Spreadsheets are commonly used to solve these applications, and spreadsheets use formulas. Generally, using the appropriate version of the percentage formula is the most efficient method for working with spreadsheets. If you are solving an application using a calculator, either the percentage formula or percentage proportion can be used.

EXAMPLE 17

BUS A 5% sales tax is levied on an order of building supplies costing $127.32. What is the amount of sales tax to be paid? What is the total bill?

To find the amount of sales tax to be paid, we need to find 5% of $127.32. The amount of sales tax is the portion, and the cost of the supplies is the base.

Estimation

Tax: 10% of $127.32 = $12.73. $\frac{1}{2}$ of 10% = 5%. $\frac{1}{2}$ of $12.73 > $6.
Therefore, 5% would be greater than $6.
Total bill would be greater than $127 + $6 or $133.

$P = R \cdot B$	*P* is the amount of sales tax. 5% is the percent or sales tax rate. *B* is the initial cost. Substitute known values.
$P = 5\% \cdot \$127.32$	Change 5% to a decimal equivalent.
$P = 0.05(\$127.32)$	Multiply.
$P = \$6.366$	Sales tax

Money amounts are usually rounded to the nearest cent. **Thus, the sales tax is $6.37.**

To find the total amount to be paid, we add the cost of the supplies and the sales tax.

$\$127.32 + \$6.37 = \mathbf{\$133.69}$ Total bill = Cost of supplies + Sales tax.

Interpretation

The amount of sales tax to be paid is $6.37 and the total bill is $133.69. This amount is very close to our estimation of $133 for the total bill we pay.   See Exercises 93–94.

EXAMPLE 18

PFIN If the rate of social security tax is 6.2% of the first $118,500 of earnings in a given year, how much social security tax is withheld on a weekly paycheck of $425 if the total earnings for the year are less than $118,500?

The amount of pay before any deductions are made is called the *gross pay*, and it is the base. The amount of social security tax is the portion. The rate is 6.2%.

Estimation

10% of $425 = $42.50	
5% of $425 = $\frac{1}{2}$ of $42.50 = $21.25	6.2% is between 10% and 5%.
6.2% of $425 < $42.50	< is read "is less than."
6.2% of $425 > $21.25	> is read "is greater than."
$P = R \cdot B$	*P* is the amount of social security deduction.
$P = 6.2\% \cdot \$425$	*R* is 6.2%, *B* = $425. Substitute known values.
$P = 0.062(\$425)$	Change 6.2% to its decimal equivalent.
$P = \$26.35$	Multiply.

Interpretation

$26.35 is withheld for social security tax.   See Exercises 95–98.

Commission: the salary of a salesperson that is based on the amount of sales made

Salaries of salespeople are sometimes based on the amount of sales made. We call this selling on **commission.** Such a salary is usually a certain percent of the sales on which the commission is to be paid.

EXAMPLE 19

PFIN An automotive parts salesperson earns a salary of $325 per week and 8% commission on all sales over $2,500 per week. The sales during one week were $4,875. What is the salesperson's total pay for that week?

Estimation

To estimate, round the sales to $5,000.

$5,000 − $2,500 = $2,500 eligible for commission.	Round 8% to 10%.
10% of $2,500 = $250	Estimated commission
$250 + $325 = $575	Estimated total pay

Commission rate and sales are both rounded up so the estimate is more than the actual amount; therefore, the commission is less than $250 and the total salary is less than $575.

First, we need to determine the amount of sales on which the salesperson will earn a commission.

$4,875 − $2,500 = $2,375   Sales that earn a commission

The salesperson receives an 8% commission on $2,375 (base).

$P = R \cdot B$	P is the amount of the commission. $2,375 is the base on which the commission is earned. The rate is 8%.
$P = 8\%(\$2,375)$	Change 8% to its decimal equivalent.
$P = 0.08(\$2,375)$	Multiply.
$P = \$190$	Commission

The salesperson will receive a $325 base salary and $190 in commission.

$325 + $190 = $515

Interpretation

The total pay for that week is $515.   **See Exercises 99–104.**

Interest: the amount charged for borrowing or loaning money, or the amount of money earned when money is saved or invested

Amount of interest: the portion in problems dealing with simple interest

Principal: the total amount invested or borrowed or the base in problems dealing with interest

Percent of interest: the rate in problems dealing with interest

Rate of interest: a percent per time period

Per annum: per year

Interest is the amount charged for borrowing or lending money, or the amount of money earned when money is saved or invested. The **amount of interest** is the *portion,* the total amount invested or borrowed is the **principal** or *base,* and the **percent of interest** is the *rate.*

The **rate of interest** is expressed as a percent per time period. For example, the rate of interest on a loan may be 8% per year, or **per annum.** The rate of interest or finance charge on a charge account may be $1\frac{1}{2}\%$ per month.

EXAMPLE 20

PFIN A credit-card company charges a finance charge (interest) of $1\frac{1}{2}\%$ per month on the average daily balance of the account. If the average daily balance on an account is $157.48, what is the finance charge for the month?

Estimation

1% of $157.48 = $1.57

Since $1\frac{1}{2}\% > 1\%$, the finance charge $> \$1.57$

$P = R \cdot B$	P is the amount of finance charge for 1 month (Fig. 3–14).
$P = 1.5\%(\$157.48)$	$1\frac{1}{2}$ is the percent of finance charge for 1 month. $1\frac{1}{2} = 1.5$.

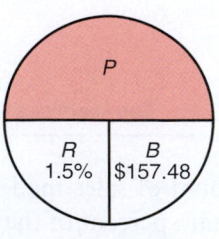

FIGURE 3–14

Annual percentage rate (APR):
the rate of interest paid on loans

Annual percentage yield (APY):
the rate of interest earned on investments

Simple interest formula:
Interest = Principal × Rate × Time, or $I = PRT$

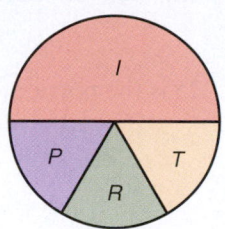

FIGURE 3-15

$P = 0.015(\$157.48)$ Multiply.
$P = \$2.3622$
$P = \$2.36$ To the nearest cent

Interpretation

The finance charge is $2.36.   See Exercises 105–106.

There are many different ways to figure interest. Most banks or loan institutions use compound interest; some institutions use the exact time of a loan; others use an approximate time of a loan, such as 30 days per month or 360 days per year. However, the truth-in-lending law requires that all businesses equate their interest rate to an annual simple interest rate known as the **annual percentage rate (APR)** or **annual percentage yield (APY).** This allows consumers to compare rates of various institutions and to understand exactly what rate they are earning or are being charged.

Simple interest for one time period can be found using the formula $I = PRT$, which is similar to the percentage formula (Fig. 3–15). The unit of time for T should match the time period for the rate.

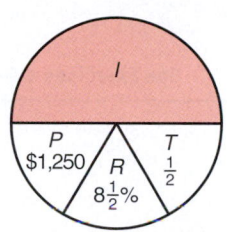

FIGURE 3-16

EXAMPLE 21

PFIN $1,250 is invested for 6 months at an interest rate of $8\frac{1}{2}\%$ per year. Find the simple interest earned. Use the simple interest formula $I = PRT$.

Estimation

10% of $1,250 = $125   Estimated interest for one year

6 months = $\frac{1}{2}$ year; $\frac{1}{2}$ of $125 = $62.50   Estimated interest for six months

Since $8\frac{1}{2}\%$ is less than 10%, the interest < $62.50

Use the simple interest formula $I = PRT$ (Fig. 3–16).

$I = PRT$ P is the principal or base of $1,250
$I = (\$1,250)(8\frac{1}{2}\%)(\frac{1}{2})$ $8\frac{1}{2}\%$ is the rate R for 1 year. The time T is 6 months, or $\frac{1}{2}$ year
$I = (\$1,250)(0.085)(0.5)$ Change $8\frac{1}{2}\%$ and $\frac{1}{2}$ to decimal equivalents.
$I = \$53.125$ Multiply.
$I = \$53.13$ To the nearest cent

Interpretation

The interest earned is $53.13.   See Exercises 107–110.

TIP   **Identify the Rate, Base, and Percentage in Applied Problems** Applied problems that involve percents are solved using the percentage formula or percentage proportion. For each type of application it is still important to identify the rate, base, and portion.

	Rate	Base	Portion
Sales tax	Sales-tax rate	Purchase price Marked price	Amount of sales tax
Discount	Discount rate	Original price	Amount of discount
Commission	Commission rate	Commissable sales	Amount of commission
Interest	Interest rate per year	Total investment Total loan Principal	Amount of interest per year

EXAMPLE 22

PFIN   In applying for a home loan, two factors that a lender examines are the prospective buyer's gross monthly income (income before deductions) and the total monthly debt payments. If the maximum debt to income percent allowed is 43%, what is the necessary gross monthly income to the nearest dollar for a buyer whose total monthly debt payments are $1,225?

Estimation

50% of what amount is $1,200? Since 50% represents one half, then the amount would be about twice $1,200, or $2,400.

Using the percentage formula $B = \dfrac{P}{R}$, 43% is the rate and $1,225 is the portion. The base is missing.

$B = \dfrac{\$1,225}{43\%}$ Change 43% to 0.43.

$B = \dfrac{\$1,225}{0.43}$ Divide.

$B = \$2,848.837209$

$B = \$2,849$ to the nearest dollar

Interpretation

The necessary gross monthly income is $2,849.   See Exercises 119–120.

3-2 EXERCISES   MyLab Math   For additional practice go to your study plan in MyLab Math.

1   Identify the given and missing elements as R (rate), B (base), and P (portion). Do not solve. *See Example 1.*

1. 40% of 18 is what number?

2. **HLTH/N**   A mail-order pharmacy offers Allegra® at 66% of the retail price of $35.99 for a month's supply. What is the mail-order price?

3. **HLTH/N**   Premarin® tablets have a discounted price of $12.21, which is 27% of the retail price. What is the retail price?

4. **HLTH/N**   A nurse spent 83% of a 12-h shift caring for patients. How many hours were devoted to patient care?

5. What percent of 10 is 2?

6. 2 is 20% of what number?

7. Three books is what percent of four books?

8. What percent of 25 students is 5 female students?

9. 6 of 15 motorists is what percent?

10. How many nurses is 20% of the 15 nurses on duty?

11. 35% of how many pieces of sod is 70 pieces?

12. What percent of a total bill of $45 is $3.15?

2   Solve using the percentage formula. *See Example 2.*

13. 30% of 18 is what number?

14. 42% of 600 is what number?

15. 1.8% of 100 is what number?

16. What is $33\frac{1}{3}\%$ of 78?

17. What percent of 18 is 9?

18. What percent of 15.6 is 52?

19. What percent of 88 is 77?

20. 150 is what percent of 100? *See Example 7.*

21. 125% of what number is 80?

22. 0.01% of what number is 7?

Applications
See Example 3.

23. PFIN At the Evans Formal Wear department store, all suits are reduced 20% from the retail price. If Charles Stewart purchased a suit that originally retailed for $258.30, how much did he save?

24. PFIN Joe Passarelli earns $8.67 per hour working for Dracken International. If Joe earns a merit raise of 12%, how much is his raise?

See Example 4.

25. An ice cream truck began its daily route with 95 gallons of ice cream. The truck driver sold $87\frac{1}{2}\%$ of the ice cream. How many whole gallons of ice cream were sold?

26. PFIN Ethan Thomas earns $132,500 annually. He received a $33\frac{1}{3}\%$ bonus on sales and stock management. How much was his annual bonus to the nearest dollar?

See Example 5.

27. PFIN Nancy Botano expects to receive $18,000 in bonuses for managing the media center at the Olympic Games. If the bonus is 24% of her annual salary, what is her annual salary?

28. Stacy Bauer sold 80% of the tie-dyed T-shirts she took to the Green Valley Music Festival. If she sold 44 shirts, how many shirts did she take?

29. PFIN A stockholder sold her shares and made a profit of $1,466. If this is a profit of 23%, how much were the shares worth when she originally purchased them?

30. The Drammelonnie Department Store sold 30% of its shirts in stock. If the department store sold 267 shirts, how many shirts did the store have in stock?

See Example 6.

31. Ali gave correct answers to 23 of the 25 questions on the driving test. What percent of the questions did he get correct?

32. A soccer stadium in Manchester, England, has a capacity of 78,753 seats. If 67,388 seats were filled, what percent of the stadium seats were vacant? Round to the nearest hundredth of a percent.

33. PFIN Holly Hobbs purchased a magazine at the Atlanta airport for $2.99. The tax on the purchase was $0.18. What is the tax rate at the Atlanta airport? Round to the nearest percent.

34. A receipt from Walmart in Memphis showed $4.69 tax on a subtotal of $53.63. What is the tax rate? Round to the nearest hundredth percent.

See Example 7.

35. 162 is what percent of 54?

36. 416.25 is what percent of 185?

See Example 8.

37. Mentally find 10% of $28,000.

38. Mentally find 1% of 34,200.

39. PFIN Find a 15% tip on a restaurant bill of $31.15. *See Example 9.*

40. PFIN Find a 20% tip for a taxi fare of $43.00. *See Example 10.*

3 Solve using the percentage proportion.
See Example 11.

41. 20% of 375 is what number?

42. 75% of 84 is what number?

43. $66\frac{2}{3}\%$ of 309 is what number?

44. 34.5% of 336 is what number?

45. $\frac{3}{4}\%$ of 90 is what number?

46. 0.2% of 470 is what number?

47. What number is 134% of 115?

48. Find 275% of 84.

49. 400% of 231 is what number?

50. $37\frac{1}{2}\%$ of 920 is what number?

See Example 12.

51. 50% of what number is 36?

52. 60% of what number is 30?

53. $12\frac{1}{2}\%$ of what number is 43?

54. 15.87 is 34.5% of what number?

55. $\frac{2}{3}\%$ of what number is $2\frac{2}{3}$?

56. 0.3% of what number is 0.825?

57. 150% of what number is $112\frac{1}{2}$?

58. 43% of what number is 107.5?

59. 92 is 500% of what number?

60. $133\frac{1}{3}\%$ of what number is 348?

See Example 13.

61. What percent of 348 is 87?

62. What percent of 350 is 105?

63. 72 is what percent of 216?

64. 28 is what percent of 85 (to the nearest tenth percent)?

65. 37.8 is what percent of 240?

66. What percent of 175 is 28?

67. 32 is what percent of 4,000?

68. What percent is 2 out of 300?

69. What percent of 125 is 625?

70. 173.55 is what percent of 156?

See Example 14.

71. Find the percentage if the base is 75 and the rate is 5%.

72. Find the percentage if the base is 25 and the rate is 2.5%.

73. If the rate is $12\frac{1}{2}\%$ and the base is 75, find the percentage.

74. **AG/H** If wrought iron contains 0.07% carbon, how much carbon is in a 30-lb bar of wrought iron?

75. **INDTEC** Cast iron contains 4.25% carbon. How much carbon is contained in a 25-lb bar of cast iron?

76. **AG/H** A landscape contractor figures it costs $\frac{1}{2}\%$ of the total cost of a job to make a bid. What would be the cost of making a bid on a $115,000 job?

77. **INDTR** In a welding shop, 104,000 welds are made. If 97% of them are acceptable, how many are acceptable?

78. $\frac{1}{10}\%$ of 400 is what number?

79. $\frac{1}{8}\%$ of 320 is what number?

See Example 15.

80. Find the rate when the base is 80 and the percentage is 30.

81. What is the rate if the base is 10.5 and the percentage is 7?

82. **AG/H** 3,645 rolls of landscape fabric are manufactured during one day. After being inspected, 121 of these rolls are rejected as imperfect. What percent of the rolls is rejected? (Round to the nearest whole percent.)

83. Find the rate when the percentage is 15 and the base is 75.

84. **CON** A contractor makes a profit of $12,350 on a $115,750 job. What is the percent of profit? (Round to the nearest whole percent.)

85. **ELEC** The voltage of a generator is 120 V. If 6 V are lost in a supply line, what is the rate of voltage loss?

See Example 16.

86. Find the base when the rate is 15% and the percentage is 52.5.

87. If the percentage is 4.75 and the rate is $33\frac{1}{3}\%$, find the base.

88. If the percentage is 11 and the rate is 5%, find the base.

89. **AUTO** An engine operating at 82% efficiency transmits 164 hp. What is the engine's maximum capacity (base) in horsepower?

90. What is the base if the percentage is 35 and the rate is 17.5%?

91. **INDTEC** A certain ore yields an average of 67% iron. How much ore is needed to obtain 804 lb of iron?

92. **HLTH/N** 385 defective alcohol swabs were produced during a day. If 4% of the alcohol swabs produced were defective, how many alcohol swabs were produced in all?

4

See Example 17.

93. **BUS** Find the sales tax and the total bill on an order of office supplies costing $75.83 if the tax rate is 6%. Round to the nearest cent.

94. **AG/H** Materials to landscape a property total $785.84. What is the total bill if the sales tax rate is $5\frac{1}{2}$%? Round to the nearest cent.

See Example 18.

95. **PFIN/BUS** If the rate of social security tax is 6.2%, find the tax on gross earnings of $375.80. Round to the nearest cent.

96. **PFIN/BUS** An employee's gross earnings for a pay period are $895.65. The net pay for this salary is $675.23. What percent of the gross pay are the total deductions? Round to the nearest whole percent.

97. **PFIN/BUS** An employee has a net salary of $576.89 and a gross salary of $745.60. What percent of the gross salary is the total of the deductions? Round to the nearest whole percent.

98. **PFIN/BUS** Madison Duke earns $2,892 in one pay period. Medicare tax is 1.45% of the earnings. How much is deducted for Medicare tax?

See Example 19.

99. **PFIN/BUS** A real estate salesperson earns 4% commission. What is the commission on a property that sold for $295,800?

100. **INDTEC** A manufacturer gives a 2% discount to customers paying cash. If a parts store pays cash for an order totaling $875.84, what amount is saved? Calculate to the nearest cent.

101. **HLTH/N** Find the cash price for an order of hospital supplies totaling $3,985.57 if a 3% discount is offered for cash orders. Calculate to the nearest cent.

102. **PFIN** What commission is earned by a salesperson who sells $18,890 in merchandise if a 5% commission is paid on all sales?

103. **TELE** A telecommunications representative is paid a salary of $140 per week and 7% commission on all sales over $3,200 per week. The sales for a recent week were $7,412. What is the representative's salary for that week?

104. **PFIN/COMP** A computer store manager is paid a salary of $2,153 monthly plus a bonus of 1% of the net earnings of the business. Find the total salary for a month when the net earnings of the business are $105,275.

See Example 20.

105. **ELEC** An electrician purchases $650 worth of electrical materials. A finance charge of $1\frac{1}{2}$% per month is added to the bill. What is the finance charge for 1 month?

106. **PFIN** If $10,000 is invested for 3 months at 6% per year, how much interest is earned?

See Example 21.

107. **PFIN** Find the interest on a loan of $5,840 at 10% per year for 2 years, 6 months.

108. **PFIN** Find the interest on a loan of $2,450 at 7% per year for 1 year.

109. **PFIN/HLTH/N** A nurse paid $10.24 in monthly interest on a credit-card account that had an average daily balance of $584.87. Find the monthly rate of interest. Round to the nearest hundredth of a percent.

110. **PFIN** Find the annual interest rate if a deposit of $5,000 earns $125 in 1 year.

111. **HLTH/N** The adult human skeleton consists of 206 bones. The face has 14 bones. What percent of human bones are found in the face? Round to the nearest tenth of a percent.

112. **HLTH/N** Each lower limb of the adult human skeleton has 31 bones. What percent of the 206 bones are found in both lower limbs? Round to the nearest tenth of a percent.

113. **HLTH/N** Approximately 60% of an adult man's body is water. A male that weighs 185 lb has approximately how many pounds of water?

114. **HLTH/N** The average American consumes about $3\frac{1}{2}$ lb of sodium each year. If salt is 40% sodium, how many pounds of salt are consumed, on average, each year by an individual?

115. **CON** Home heat loss through poor-fitting doors and windows can account for 15% of a homeowner's energy cost. A January energy bill is $219.42. Find the cost of unnecessary heat loss.

116. **INDTEC** Alumina makes up 45% of high-grade aluminum ore. If 48,000 lb of high-grade aluminum ore are mined, how many pounds of alumina can be obtained?

117. **INDTEC** How many pounds of pure sulfur are contained in 3,400 lb of coal that has 5% pure sulfur content?

118. **INDTEC** ACME Products estimates a product costs $42.80 to manufacture. How much did it cost to make the product if a 12% cost overrun occurred?

See Example 22.

119. **PFIN** Find the monthly income a prospective buyer must have to qualify for a home that has a monthly cost (including taxes and insurance) of $1,200 if the mortgage company requires the monthly cost to be no more than 28% of the buyer's gross monthly income.

120. **PFIN** Chloe Duke earns $4,872 monthly and wants to purchase a home, but the mortgage firm requires the monthly payment to be no more than 28% of her gross monthly income. What is the maximum monthly mortgage payment, including taxes and insurance, she will be allowed?

3-3 Increase and Decrease

LEARNING OUTCOMES

1 Find the amount of increase or decrease in percentage problems.

2 Find the new amount directly in increase or decrease problems.

3 Find the rate or base in increase or decrease problems.

LC LEARNING CATALYTICS

1. If an amount goes from 352 to 321, did the amount increase or decrease?
2. If an amount goes from 112 to 205, did the amount increase or decrease?

New amount: when working with the amount of increase or decrease in percent problems, the original amount plus or minus the amount of change (increase or decrease)

STOP AND CHECK

1. If a salary increased from $48,250 to $53,825, find the amount of increase.

Answer:
1. $5,575

In many applications an original amount is increased or decreased to give a new amount. The **new amount** is the original amount plus or minus the amount of change (increase or decrease). Some examples of increases are the sales tax on a purchase, the raise in a salary, and the markup on a wholesale price. Some examples of decreases are the deductions on your paycheck and the markdown or the discount on an item for sale.

1 Find the amount of increase or decrease in percentage problems. The amount of increase or decrease is the amount that an original number changes. Subtraction is used to find the amount of change when the beginning and ending (or new) amounts are known.

> **To find the amount of increase or decrease from the beginning and ending amounts:**
>
> 1. To find the amount of increase (when the new amount is larger than the beginning amount):
>
>   **Amount of increase = new amount − beginning amount**
>
> 2. To find the amount of decrease (when the new amount is smaller than the beginning amount):
>
>   **Amount of decrease = beginning amount − new amount**

EXAMPLE 1

PFIN/BUS David Spear's salary increased from $58,240 to $63,190. What is the amount of increase?

$$\text{Beginning amount} = \$58,240$$
$$\text{New amount} = \$63,190$$
$$\text{Increase} = \text{new amount} - \text{beginning amount}$$
$$= \$63,190 - \$58,240$$
$$= \$4,950$$

David's salary increase was $4,950. **See Exercise 1.**

STOP AND CHECK

1. A food processor that originally costs $59.99 was reduced to $40. What is the amount of markdown?

Answer:
1. $19.99

EXAMPLE 2

BUS A coat was marked down from $98 to $79. What is the amount of markdown?

Beginning amount = $98

New amount = $79

Decrease = beginning amount − new amount

= $98 − $79

= $19

The coat was marked down $19.   See Exercise 2.

Percent of change: the amount of change (increase or decrease) expressed as a percent of the original or beginning amount

A change in a value is often expressed as a **percent of change.** The amount of change (increase or decrease) is a percent of the original or beginning amount.

> **To find the amount of change (increase or decrease) from a percent of change:**
>
> 1. Identify the original or beginning amount and the percent or rate of change.
>
> 2. Multiply the decimal or fractional equivalent of the rate of change times the original or beginning amount.
>
> **Amount of change = percent of change × original amount**

STOP AND CHECK

1. An employee making $21.35 an hour receives a 2% wage increase. What is the new wage per hour?

Answer:
1. $21.78

EXAMPLE 3

PFIN/BUS Technical assistants at Best Buy are to receive a 4% increase in wages per hour. If they were making $19.25 an hour, what is the amount of increase per hour (to the nearest cent)? Also, what is the new wage per hour?

The original wage per hour is the base, and we want to find the amount of increase (portion).

Estimation

10% of $19.25 is $1.92. Therefore, 5% of $19.25 would be $\frac{1}{2}$($1.92). The 4% increase will be less than $1. The new wage per hour will be approximately $20.

$P = RB$	*P* represents amount of increase, 4% is the rate of increase, and *B* is the original wage per hour.
$P = 4\%(\$19.25)$	4% = 0.04
$P = 0.04(19.25)$	Multiply.
$P = 0.77$	$0.77 to the nearest cent is the amount of increase.

Interpretation
The technical assistants will receive a $0.77 per-hour increase in wages.

$19.25 + $0.77 = $20.02   New amount = original amount + amount of increase.

Their new hourly wage will be $20.02.   See Exercises 3–6.

STOP AND CHECK

1. If a product shrinks in volume 0.6% during the cooling process, how much remains to the nearest hundredth of a cubic centimeter when a product of 38 cm³ is cooled?

Answer:
1. 37.77 cm³

EXAMPLE 4

INDTEC Molten iron shrinks 1.2% while cooling. What is the cooled length of a piece of iron if it is cast in a 24-cm mold?

First, we find the amount of shrinkage (amount of decrease, portion). The original amount, 24 cm, is the base.

Estimation

The cooled piece will be less than 24 cm.

$P = RB$ P is the amount of decrease, 1.2% is the rate of
                decrease, and the base is the original amount, or 24 cm.

$P = 1.2\%(24)$ 1.2% = 0.012

$P = 0.012(24)$ Multiply.

$P = 0.288$ Amount of shrinkage (decrease)

Interpretation

The amount of shrinkage is 0.288 cm, so the length of the cooled piece (new amount) is

$24 - 0.288 = 23.712$ cm   New amount = original amount − amount of decrease.

See Exercises 7–8.

2   Find the New Amount Directly in Increase or Decrease Problems. Often in increase or decrease problems we are more interested in the new amount than the amount of change. We can find the new amount directly by adding or subtracting percents first. The original or beginning amount is always considered to be our *base* and is represented by 100%.

To find the new amount directly in a percent problem:

1. Find the rate of the new amount.

   For increase: 100% + rate of increase

   For decrease: 100% − rate of decrease

2. Find the new amount.

   $$P = RB$$

   New amount = rate of new amount × original amount

EXAMPLE 5

PFIN   Medical assistants are to receive a 9% increase in wages per hour. If they were making $15.25 an hour, what is the *new wage per hour* to the nearest cent?

Rate of new amount = 100% + rate of increase

                   = 100% + 9%

                   = 109%

New amount = rate of new amount × original amount

           = 109%($15.25)   Change % to its decimal equivalent.

           = 1.09($15.25)   Multiply.

           = $16.6225   New amount

           = $16.62   Nearest cent

The new hourly wage is $16.62.   See Exercises 9–14.

EXAMPLE 6

A pair of jeans that originally cost $49.99 now is advertised as 70% off. What is the sale price of the jeans?

Rate of new amount = 100% − rate of decrease

                   = 100% − 70%

                   = 30%

Monkey Business Images/Shutterstock

STOP AND CHECK

Find the new amount directly
1. A product costing $45 is marked up 25%.
2. A product selling for $89 is marked down 40%. Write the answer as dollars and cents.

Answers:
1. $56.25   2. $53.40

New amount = rate of new amount × original amount

$$= 30\%(\$49.99) \quad \text{Change \% to its decimal equivalent.}$$
$$= 0.3(\$49.99) \quad \text{Multiply.}$$
$$= \$14.997 \quad \text{New amount}$$
$$= \$15.00 \quad \text{Nearest cent}$$

The sale price of the jeans is $15.　　　　　　　　　　　　See Exercises 15–18.

EXAMPLE 7

INDTR A 3% error is acceptable for a machine part to be usable. If the part is intended to be 57 cm long, what is the range of measures that is acceptable for this part?

The machine part can be ±3% from the ideal length of 57 cm. The range of acceptable measures is found by calculating the smallest acceptable measure and the largest acceptable measure. Recall that the symbol ± is read *plus or minus*. It means that the measure can be more (+) or less (−) than the designated amount. In this case, the part can be 3% longer than or shorter than 57 cm and still be usable. The smallest acceptable value is 97% of the ideal length (100% − 3%).

$$P = RB \qquad P \text{ is the smallest acceptable amount because 97\% is the } \textit{smallest}$$
　　　　　　　acceptable percent. The base is 57 cm.
$$P = 97\%(57) \qquad 97\% = 0.97$$
$$P = 0.97(57) \qquad \text{Multiply.}$$
$$P = 55.29$$

Interpretation

The smallest acceptable measure is 55.29 cm.

The largest acceptable value is 103% of the ideal length (100% + 3%).

$$P = RB \qquad P \text{ is the largest acceptable amount because 103\% is the } \textit{largest}$$
　　　　　　　acceptable percent. The base is 57 cm.
$$P = 103\%(57) \qquad 103\% = 1.03$$
$$P = 1.03(57) \qquad \text{Multiply.}$$
$$P = 58.71$$

Interpretation

The largest acceptable value is 58.71 cm.
The range of acceptable measures is from 55.29 cm to 58.71 cm.　　See Exercises 19–20.

It is helpful in developing our number sense with percents to think of percents in pairs. 100% is one whole quantity. A percent that is less than 100% represents part of a quantity. For example, 30% represents part of one quantity. Then, 70% is the rest of the quantity if 30% is removed. This percent, 70%, is the complement of 30%. The **complement of a percent** is the difference between 100% and a given percent.

Complement of a percent: the difference between 100% and a given percent

---

To find the complement of a percent:

Subtract the percent from 100%.

EXAMPLE 8

Find the complement of **(a)** 25%, **(b)** 80%, and **(c)** 36%.

(a) $100\% - 25\% =$ **75%**   Subtract from 100%.

(b) $100\% - 80\% =$ **20%**   Subtract from 100%.

(c) $100\% - 36\% =$ **64%**   Subtract from 100%.   See Exercises 21–26.

To estimate the amount you pay for a sale item:

1. Round the original price and the percent to numbers you can work with mentally.

2. Find the complement of the rounded percent.

3. Relate the complement to 10% by dividing it by 10%.

4. Find 10% of the rounded original price.

5. Multiply the results from Steps 3 and 4.

EXAMPLE 9

BUS   Estimate the amount you pay on a $49.99 item advertised at 70% off.

$49.99 rounds to $50.

$100\% - 70\% = 30\%$ Complement of 70% (percent you pay).

$30\% \div 10\% = 3$ Relate to 10%.

10% of $\$50 = \$5$ Move decimal one place to the left.

$3 \cdot \$5 =$ **$15**   Three times 10% of $50.   See Exercises 27–28.

3 **Find the Rate or Base in Increase or Decrease Problems.** Many kinds of increase or decrease problems involve finding either the rate or the base.

The rate is the *percent of change* or the **percent of increase or decrease.** The base is still the *original amount.*

Percent of increase or decrease: same as the percent of change

---

To find the rate or base in increase or decrease problems:

1. Identify or find the amount of change (increase or decrease).

2. To find the rate of increase or decrease, use the percentage formula $R = \dfrac{P}{B}$.

$$R = \frac{\text{amount of change}}{\text{original amount}}$$

3. To find the base or original amount, use the percentage formula $B = \dfrac{P}{R}$.

$$B = \frac{\text{amount of change}}{\text{rate of change}}$$

EXAMPLE 10

AUTO   A worn brake lining is measured to be $\frac{3}{32}$ in. thick. If the original thickness was $\frac{1}{4}$ in., what is the percent of wear?

First, the amount of wear is $\frac{1}{4} - \frac{3}{32}$.

$$\frac{8}{32} - \frac{3}{32} = \frac{5}{32}$$ Find the common denominator. Subtract.

The amount of wear (decrease) is the portion. The base is the original amount.

$$R = \frac{P}{B} \cdot 100\%$$ R is the percent of wear.
P is the amount of wear, or $\frac{5}{32}$ in. B is $\frac{1}{4}$ in.

$$R = \frac{\frac{5}{32}}{\frac{1}{4}} \cdot 100\%$$ Divide.

$$R = \frac{5}{32} \div \frac{1}{4} \cdot 100\%$$ Change to an equivalent multiplication.

$$R = \frac{5}{\overset{}{\underset{8}{32}}} \cdot \frac{\overset{1}{4}}{1} \cdot 100\%$$ Multiply.

$$R = \frac{5}{8} \cdot 100\%$$ Change to an equivalent percent.

$$R = \frac{5}{\underset{2}{8}} \cdot \frac{\overset{25}{100\%}}{1}$$ Reduce and multiply.

$$R = 62\frac{1}{2}\%$$

Interpretation

The percent of wear is $62\frac{1}{2}\%$. See Exercises 29–35.

In some cases you may not have enough information to determine the amount of increase or decrease with the previous procedure. Then we must match the rate with the information we are given.

Silver John/123RF

EXAMPLE 11

BUS   At Best Buy the price of a DVD player dropped by 20% to $179. What was the original price to the nearest dollar?

Reduced price = new amount = $179

Rate of decrease = 20%

Rate of reduced price = 100% − rate of decrease

Use the formula to find base, $B = \dfrac{P}{R}$

$$\text{Original price} = \frac{\text{reduced price}}{\text{rate of reduced price}}$$

Rate of reduced price = 100% − 20% = 80%

$$\text{Original price} = \frac{\$179}{80\%}$$ Convert % to decimal equivalent.

$$= \frac{179}{0.8}$$ Divide.

$$= \$223.75$$ Round to nearest dollar.

$$= \$224$$

The original price of the DVD player was $224. See Exercises 36–40.

> **TIP** **Be Sure to Use the Correct Rate** When using the percentage formula, the description for the rate must match the description of the portion.
>
> | | **Example 11 Above**
DVD Problem | **Example 6**
Jeans Problem |
> |---|---|---|
> | Form of percentage formula | $B = \dfrac{P}{R}$ | $P = RB$ |
> | Description of rate
Description of portion | Rate of *reduced price*
Reduced price | Rate of *new amount*
New amount |

3–3 EXERCISES

MyLab Math For additional practice go to your study plan in MyLab Math.

1 Solve. *See Example 1.*

1. Kamryn Pastner's salary increased from $28,500 to $32,100. What is the amount of increase?

See Example 2.

2. A pair of shoes was marked down from $45.99 to $38. What is the amount of markdown?

See Example 3.

3. Find the amount of increase if 432 is increased by 25%.

4. If 78 is increased by 40%, what is the new amount?

5. Jobs as athletic trainers totaled 18,200 in 2010. According to the Bureau of Labor Statistics, jobs in this occupation are expected to grow by 30.2% by 2020. How many new jobs will there be in 2020? Find the total number of jobs expected in this occupation in 2020.

6. In 2010, the number of jobs as personal financial advisors was 206,800. According to the Bureau of Labor Statistics, jobs in this fast-growing occupation are expected to grow by 32.1% by 2020. Find the number of new jobs expected and the total number of jobs expected to be available in 2020.

See Example 4.

7. Find the amount of decrease if 68 is decreased by 15%.

8. If 135 is decreased by 75%, what is the new amount?

2 *See Example 5.*

9. **CON** If 17% extra flooring is needed to allow for waste when boards are laid diagonally, how much flooring should be ordered to cover 2,045 board feet of floor? Answer to the nearest whole board foot.

10. **CON** A contractor needs 1,650 board feet of 1-in. × 8-in. common boards to subfloor a house. If she needs 10% extra flooring to allow for waste when the boards are laid square, how much flooring should she order?

11. **CON** When making an estimate on a job, a contractor wants to make a 10% profit. If all the estimated costs are $15,275, what is the total bid of cost and profit for the job?

12. **CON** A construction company requires 25,400 bricks for a job. If it allows 2% more bricks for breakage, how many bricks must it order?

13. **PFIN/BUS** Ciara Walker was earning $49,860 and received a 7% raise. Find her new annual earnings directly.

14. **BUS** LaTreas Walker received a 3% salary increase on her weekly earnings of $1,982. What are her new weekly earnings?

See Example 6.

15. **INDTR** Steel rods shrink 10% when cooled from furnace temperature to room temperature. If a tie rod is 30 in. long at furnace temperature, how long is the cooled tie rod?

16. **CON** Rock must be removed from a highway right-of-way. If 976 cubic yards (yd^3) of unblasted rock are to be removed, how many cubic yards is this after blasting? Blasting causes a 40% swell in volume.

17. **PFIN/BUS** David Dawson earned $4,290 but paid 6.2% in social security taxes and 1.45% of his earnings in Medicare taxes. What were his net earnings after paying social security and Medicare taxes?

18. **BUS** Megan Anders purchased a swimsuit that was priced at $84 before a 30% discount. What was the sale price?

See Example 7.

19. AMATYC ordered 1,500 copies of the conference program. The printer has a policy that the delivered count on print jobs will be within 2% of the ordered number. What is the least acceptable number of programs to be delivered?

20. In anticipation of some paper jams or other problems, the printer set up the presses to run 3% more than the 1,500 AMATYC conference programs that were ordered. What number of programs were set to run?

Solve. *See Example 8.*

21. Find the complement of 40%.

22. Find the complement of 18%.

23. Find the complement of 86.3%.

24. Find the complement of $33\frac{1}{3}$%.

25. Find the complement of 0.09%.

26. Find the complement of 1%.

See Example 9.

27. **PFIN** Estimate the amount you pay on a handbag marked $38, if the handbag is advertised as 30% off.

28. **PFIN** Estimate the amount you pay on a camera that costs $52 if the sales-tax rate on the camera is 9.25%.

3 Solve. *See Example 10.*

29. **CON** The cost of a pound of nails increased from $2.36 to $2.53. What is the percent of increase to the nearest whole-number percent?

30. **AG/H** A landscape contractor estimated that materials, shrubs, saplings, and labor for a job would cost $5,385. An estimate 1 year later for the same job was $5,816, due to inflation. Find the percent of increase due to inflation to the nearest whole number.

31. **AG/H** A chicken farmer bought 2,575 baby chicks. Of this number, 2,060 lived to maturity. What percent loss was experienced by the chicken farmer?

32. **ELEC** An electrician recorded costs of $1,297 for a job. If he received $1,232 for the job, what was the percent of money lost on the job? Round to the nearest tenth of a percent.

33. **CON** A floor that would normally require 2,580 board feet is to be laid diagonally. What percent waste allowance was allowed for flooring laid diagonally, if 3,019 board feet were ordered? Round to the nearest whole percent.

34. **CON** The cost of No. 1 pine studs increased from $3.85 each to $4.62 each. Find the percent of increase.

35. **HLTH/N** A month's supply of Prevacid® retails for $147.09 and is discounted to $129.06. Find the percent savings for the discounted drug to the nearest whole percent.

See Example 11.

36. **AG/H** A contractor ordered 10.5 yd³ of sand for a job that included a 5% allowance for waste. How much sand was needed for the job?

37. **AUTO** An engine that has a 4% loss of power has an output of 336 hp. What is the input (base) horsepower of the engine?

38. **BUS** A shop manager records a 14% loss on rivets for waste. If the shop ordered 28.5 lb of rivets to compensate for loss due to waste, how many pounds of rivets were needed?

39. **INDTR** Steel bars shrink 10% when cooled from furnace temperature to room temperature. If a cooled steel bar is 36 in. long, how long was it when it was formed?

40. **INDTEC** A 141-hp output is required for an engine. If there is a 6% loss of power, what amount of input horsepower (or base) is needed?

3 | CHAPTER REVIEW OF KEY CONCEPTS

LEARNING OUTCOMES	KEY CONCEPTS AND EXAMPLES

Section 3–1

1 Write any number as a percent equivalent (pp. 116–118).

1. Multiply the number by 1 in the form of 100%   **2.** The product has a % symbol. For a shortcut, move the decimal two places to the right.

Change $\frac{1}{2}$, 1.2, and 7 to percents.

$$\frac{1}{2} \cdot 100\% = \frac{100\%}{2} = 50\% \qquad 1.2 \cdot 100\% = 120\% \qquad 7 \cdot 100\% = 700\%$$

2 Write any percent as a numerical equivalent (pp. 118–120).

1. Divide the number by 1 in the form of 100% or multiply by $\frac{1}{100}$%.   **2.** The quotient does not have a % symbol. For a shortcut, move the decimal two places to the left.

Change 7% to a fraction.

$$7\% \div 100\% = \frac{7}{100}$$

Change $1\frac{1}{4}$% to a fraction.

$$1\frac{1}{4}\% \div 100\% = \frac{5\%}{4} \div 100\% = \frac{5\%}{4} \cdot \frac{1}{100\%} = \frac{5}{400} = \frac{1}{80}$$

Convert 3.5% to a decimal.

$$3.5\% \div 100\% = 0.035 = 0.035$$

Convert 245% to a mixed number.

$$245\% \div 100\% = \frac{245\%}{100\%} = 2\frac{45}{100} = 2\frac{9}{20}$$

Convert 124.5% to a decimal.

$$124.5\% \div 100\% = 1.245 = 1.245$$

Convert $100\frac{1}{4}$% to a decimal.

$$100\frac{1}{4}\% = 100.25\%$$

$$100.25\% \div 100\% = 1.0025$$

Section 3–2

1 Identify the portion, base, and rate in percent problems (p. 121).

1. Identify the rate. *Rate* is usually written as a percent, but it may be a decimal or fraction. **2.** Identify the base. *Base* is the total amount, original amount, or entire amount. The base often follows the preposition *of*.   **3.** Identify the portion. *Portion* can refer to the part, partial amount, amount of increase or decrease, or amount of change. It is a portion of the *base*. The portion often follows a form of the verb *is*.

$$\qquad\quad R \qquad\; B \;\; P$$
(a) What percent of 10 is 5?

$$\qquad\quad R \qquad\qquad\quad B \qquad\;\; P$$
(b) 25% of what number is 3?

$$\qquad\;\; P \;\; R \qquad\qquad\quad B$$
(c) 3 is 20% of what number?

LEARNING OUTCOMES	KEY CONCEPTS AND EXAMPLES

2 Solve percent problems using the percentage formula (pp. 122–126).

Forms of the Percentage Formula:

$$P = RB \qquad B = \frac{P}{R} \qquad R = \frac{P}{B}$$

Use the percentage formula to solve percentage problems:

1. Identify and classify the two known values and the one unknown value.   **2.** Choose the appropriate percentage formula for finding the unknown value.   **3.** Substitute the known values into the formula. For the rate, use the decimal or fractional equivalent of the percent. **4.** Perform the calculations indicated by the formula.   **5.** Interpret the result. If finding the rate, convert the decimal or fractional equivalent of the rate to a percent.

What amount is 5% of $200? *Of* identifies $200 as the base.
The percent or rate is 5%. The missing element is the portion.

$P = RB$	Select the appropriate percentage formula and substitute given amounts.
$P = 5\%(\$200)$	Change 5% to a fraction or decimal equivalent by dividing by 100%.
$P = 0.05(200)$	Multiply.
$P = \$10$	Portion.

What percent of 6 is 2?
Is suggests 2 is the portion. *Of* identifies 6 as the base. The rate is missing.

$R = \dfrac{P}{B}(100\%)$	Select the appropriate percentage formula and substitute given amounts.
$R = \dfrac{2}{6}(100\%)$	Reduce fraction or change to decimal equivalent by dividing.
$R = \dfrac{1}{3}(100\%)$	Change to a percent by multiplying by 100%.
$R = 33\dfrac{1}{3}\%$	Rate $\left(\dfrac{1}{3} \times \dfrac{100\%}{1} = \dfrac{100\%}{3} = 33\dfrac{1}{3}\%\right)$.

12 is 24% of what number?
12 is the part or portion and is suggested by *is*. The rate is 24%. *Of* identifies "what number" as the missing base.

$B = \dfrac{P}{R}$	Select the appropriate percentage formula and substitute given amounts.
$B = \dfrac{12}{24\%}$	Change 24% to a decimal equivalent by dividing by 100%.
$B = \dfrac{12}{0.24}$	Divide.
$B = 50$	Base.

3 Solve percent problems using the percentage proportion (pp. 126–130).

Percentage Proportion:

$$\frac{R}{100} = \frac{P}{B}$$

Use the percentage proportion to solve percentage problems:

1. Cross multiply to find the cross products.   **2.** Divide the product of the two known factors by the factor with the letter.

LEARNING OUTCOMES	KEY CONCEPTS AND EXAMPLES

What amount is 15% of $40? *Of* identifies $40 as the base.
The percent or rate is 15%. The missing element is the portion.

$$\frac{R}{100} = \frac{P}{B} \qquad \text{Identify the rate, base, and portion and substitute given amounts into the percentage proportion.}$$

$$\frac{15}{100} = \frac{P}{\$40} \qquad \text{Multiply cross products.}$$

$$15(\$40) = P(100)$$

$$\$600 = P(100) \qquad \text{Divide by 100.}$$

$$\frac{\$600}{100} = P$$

$$\$6 = P \qquad \text{Portion.}$$

4 Solve business and consumer problems involving percents (pp. 130–134).

Business and consumer applications are often solved using the percentage formula and an electronic spreadsheet or calculator.

If the sales tax rate is 7%, find the tax and total bill on a purchase of $30.

$P = RB$	Estimation:
$P = 7\%(\$30)$	10% of $30 = $3
$P = 0.07(\$30)$	Sales tax < $3
$P = \$2.1$	Total bill < $30 + $3 or < $33.
$P = \$2.10$	Interpretation:
$\$30 + \$2.10 = \$32.10$	Sales tax is $2.10 and total cost is $32.10.

Section 3-3

1 Find the amount of increase or decrease in percentage problems (pp. 138–140).

Find the amount of increase or decrease from the beginning and ending amounts:
1. To find the amount of increase (when the new amount is larger than the beginning amount):

Amount of increase = new amount − beginning amount

2. To find the amount of decrease (when the new amount is smaller than the beginning amount):

Amount decrease = beginning amount − new amount

The price of gasoline increased from $2.85 per gallon to $4.10 per gallon during a two-year period. What was the amount of increase?

Amount of increase = new amount − beginning amount

$$= \$4.10 - \$2.85$$

$$= \$1.25$$

Find the amount of change (increase or decrease) from a percent of change: 1. Identify the original or beginning amount and the percent or rate of change. **2.** Multiply the decimal or fractional equivalent of the rate of change times the original or beginning amount.

Amount of change = percent of change × original amount

LEARNING OUTCOMES	KEY CONCEPTS AND EXAMPLES

Julio made \$15.25 an hour but took a 20% pay cut. What was the new hourly pay?

Estimation:

$P = 20\%(\$15.25)$ 10% of \$15 is \$1.50.

$P = 0.2(15.25)$ 20% of \$15 is \$3.00. \$15 − \$3 = \$12

$P = 3.05$ **Hourly pay cut (decrease)**

$\$15.25 - \$3.05 = \$12.20$ **Original amount − decrease = new amount**

2 Find the new amount directly in increase or decrease problems (pp. 140–142).

Find the new amount directly in a percent problem:

1. Find the rate of the new amount. For increase: 100% + rate of increase. For decrease: 100% − rate of decrease.   **2.** Find the new amount. $P = RB$ or

New amount = rate of new amount × original amount

A project requires 5 lb of galvanized nails. If 15% of the nails will be wasted, how many pounds must be purchased?

$P = 115\%(5)$ **$B = 5$ lb, $R = 100\% + 15\% = 115\%$**
P is missing.

$P = 1.15(5)$ Estimation:
10% of 5 = 0.5.

More than 0.5 or $\frac{1}{2}$ lb of nails must be added for waste.

Order > 5.5 lb.
Interpretation:

$P = 5.75$ lb   5.75 lb must be ordered.

3 Find the rate or base in increase or decrease problems (pp. 142–144).

1. Identify or find the amount of change (increase or decrease).   **2.** To find the rate of increase or decrease, use the percentage formula $R = \dfrac{P}{B}$.

$$R = \frac{\text{amount of change}}{\text{original amount}}$$

3. To find the base or original amount, use the percentage formula $B = \dfrac{P}{R}$.

$$R = \frac{\text{amount of change}}{\text{rate of change}}$$

A 5-in. power edger blade now measures $4\frac{3}{4}$ in. What is the percent of wear?

$R = \left(\dfrac{\frac{1}{4}}{5}\right)(100\%)$ $4\frac{4}{4} - 4\frac{3}{4} = \frac{1}{4}$. The portion is $\frac{1}{4}$. The base or original amount is 5 in. R is missing.

$R = \left(\dfrac{1}{4} \div 5\right)(100\%)$

LEARNING OUTCOMES	KEY CONCEPTS AND EXAMPLES

Estimation:

10% of 5 in. $= 0.5$ or $\frac{1}{2}$ in.

Wear $= \frac{1}{4}$ in. which is $< \frac{1}{2}$ in.

Percent of wear $< 10\%$.

$R = \frac{1}{20}(100\%)$

$R = 5\%$

Interpretation:
The percent of wear is 5%.

A PC has 20% of its hard drive storage capacity filled. If the PC now has 5.12 terabytes (T) of storage capacity available, what was the original storage capacity?

Current storage capacity available is $100\% - 20\% = 80\%$ of the original storage capacity. 5.12 T is the portion. The original storage capacity is the base.

$B = \frac{5.12}{80\%}$

$B = \frac{5.12}{0.8}$

$B = 6.4$ T

---

3   CHAPTER REVIEW EXERCISES

Section 3–1   **MyLab Math**   For additional practice go to your study plan in MyLab Math.

Change to percent equivalents.

1. 0.7   **2.** 0.35   **3.** 0.83   **4.** 18

5. 125   **6.** 17.3   **7.** $\frac{72}{100}$ **8.** $\frac{9}{100}$

9. $\frac{23}{100}$ **10.** $\frac{87}{100}$ **11.** $\frac{5}{6}$ **12.** $4\frac{1}{3}$

13. $3\frac{1}{5}$ **14.** $2\frac{1}{2}$

Change to decimal equivalents.

15. 227.2%   **16.** 73.8%   **17.** 340%   **18.** 92%

19. 83%   **20.** 52.7%   **21.** $62\frac{1}{2}\%$ **22.** $37\frac{1}{2}\%$

Change to fraction equivalents.

23. 72%   **24.** 40%   **25.** $12\frac{1}{2}\%$ **26.** $16\frac{2}{3}\%$ **27.** $\frac{2}{3}\%$

28. $\frac{3}{5}\%$ **29.** 275%   **30.** 124%   **31.** $112\frac{1}{2}\%$ **32.** $183\frac{1}{3}\%$

Section 3–2

Identify the given and missing elements as R (rate), B (base), and P (portion). Do not solve.

33. 5% of 180 is what number?

34. $15 is what percent of $120?

35. 45% of how many dollars is $36?

36. What percent of 10 syringes is 2 syringes?

37. 6 is what percent of 25 sacks of grass seed?

38. How many landscape contractors is 15% of 40 landscape contractors?

39. 18% of 150 pieces of sod is how many?

40. What percent of a total of 28 students is 8 students?

Solve using the percentage formula.

41. 5% of 480 is what number?

42. $62\frac{1}{2}$% of 120 is what number?

43. $\frac{1}{4}$% of 175 is what number?

44. $233\frac{1}{3}$% of 576 is what number?

45. 39 is what percent of 65?

46. What percent of 118 is 42.48?

47. What percent of 65 is 162.5?

48. 80% of what number is 116?

49. A hotel chain offers a 10% discount for guests who are senior citizens. Estimate the discount for a room bill of $159 before taxes.

50. Estimate the amount of a tip to include if you want to leave a 15% tip on a restaurant bill of $28.

51. Estimate the amount of a tip to include if you want to leave a 20% tip on a spa bill of $35.

52. Estimate the amount of a tip to include if you want to leave a 20% tip on a taxi bill of $8.

Solve using the percentage proportion.

53. 83% of 163 is what number?

54. 4.75% of 348.2 is what number?

55. 24% of what number is 19.92?

56. 7.56 is $6\frac{3}{4}$% of what number?

57. 260% of what number is 395.2?

58. 3 is 0.375% of what number?

59. $10\frac{1}{3}$% of what number is 8.68?

60. 38.25 is what percent of 250?

61. What percent of 26 is 130?

62. **INDTEC** Specifications for bronze call for 80% copper. How much copper is needed to make 300 lb of bronze?

63. **INDTEC** Zinc makes up 84 lb of an alloy that weighs 224 lb. What percent of the alloy is zinc?

64. **CON** When a subfloor using common 1-in. × 8-in. boards is laid diagonally, 17% is allowed for waste. How many board feet will be wasted out of 1,250 board feet?

65. **INDTEC** Of the 2,374 pieces produced by a particular machine, 27 were defective. What percent (to the nearest hundredth of a percent) were defective?

66. **ELEC** The voltage loss in a line is 2.5 V. If this is 2% of the generator voltage, what is the generator voltage?

67. **PFIN** If a family spends 28% of its income on food, how much of a $950 paycheck goes for food?

68. If a freshman class of 1,125 college students is made up of 8% international students, how many international students are in the class?

69. A city prosecuted 1,475 individuals with traffic citations. If 36,875 individuals received traffic citations, what percent was prosecuted?

70. **INDTEC** A survey studied 600 people for their views on nuclear power plants near their towns. Of these, 75 people said that they approved of nuclear power plants near their towns. What percent approved?

71. In one college, 67 students made the dean's list. If this was 33.5% of the student body, what was the total number of students in the college?

72. **BUS** It is estimated that only 19% of the licensed big game hunters on a state wildlife management area are successful. If 95 big game hunters were successful, what was the total number of big game hunters in a particular hunting season?

73. **BUS** Denise Knighton's gross pay is $3,296. She paid 6.2% in social security taxes. How much social security tax did she pay?

Solve.

74. **CON**   Discontinued paneling is sale priced at 25% off the regular price. If the regular price is $12.50 per sheet, what is the sale price per sheet rounded to the nearest cent?

75. **BUS**   A parts distributor is paid a weekly salary of $250 plus an 8% commission on all sales over $2,500 per week. What is the distributor's salary for a given week if the sales for that week totaled $4,873?

76. **BUS**   An employee's gross earnings for a month is $1,750. If the net pay is $1,237, what percent of the gross pay is the total of the deductions? Round to the nearest whole percent.

77. **CON**   An order of lumber totals $348.25. If a 5% sales tax is added to the bill, what is the total bill? Round to the nearest cent.

78. **CON**   A builder purchases a concrete mixer for $785. The builder does not know the sales tax rate, but the total bill is $828.18. Find the sales tax rate. Round to the nearest tenth of a percent.

79. **BUS**   An employer must match employees' social security contributions. If the weekly payroll is $27,542 and if the social security rate is 6.2%, what are the employer's contributions? (All employees' year-to-date salaries are under the maximum salary subject to the social security tax.)

80. **CON**   A contractor gets a 2% discount on a monthly sand and gravel bill of $1,655.75 if the bill is paid within 10 days of the statement date. How much is saved by paying the bill within the 10-day period?

81. **BUS**   A businessperson is charged a $4.96 monthly finance charge on a bill of $283.15. What is the monthly interest rate on the account? Round to the nearest hundredth of a percent.

82. **BUS**   Find the interest on a loan of $3,200 if the annual interest rate is 6% and the loan is for 9 months.

83. **BUS**   A property owner sells a house for $127,500 and pays off all outstanding mortgages. If she has $52,475 cash left from this transaction and the money is invested at 5% interest per year for 18 months, how much interest does she earn?

84. **BUS**   The sales tax in one city is 8.75% of the purchase price. How much is the sales tax on a purchase of $78.56?

85. **BUS**   A small town charges 3.25% of the purchase price for sales tax. What is the sales tax on a purchase of $27.45?

86. **BUS**   Interest for 1 year on a loan of $2,400 is $192. What is the interest rate?

87. **BUS**   A $2,000 certificate of deposit earns $170 interest in 1 year. What is the interest rate?

88. **BUS**   If a loan for $500 at 8.75% interest per annum (year) is paid in 3 months, how much is the interest?

89. **BUS**   What is the interest on a business loan for $6,500 at 9.75% interest per annum (year) paid in 7 months?

90. **BUS**   A salesperson earns a commission of 15% of the total monthly sales. If the salesperson earns $2,145, how much were the total sales?

91. **BUS**   A salesclerk in a store is paid a salary plus 3% commission. If the salesclerk earns $10.65 commission for a weekend, how much were the sales?

Section 3–3

92. Pam Cox's salary increased from $72,350 to $76,125. How much did her salary increase?

93. Jacob Palmer improved his GPA from 3.09 to 3.23. What was the increase in his GPA?

94. A refrigerator priced at $2,149 was reduced to $1,799. What was the amount of the reduction?

95. A casting of a steel beam originally is 3.12 m in length and decreases to 3.05 m when it is cooled. What is the amount of decrease?

96. **INDTR**   A steel beam expands 0.01% of its length when exposed to the sun. If a beam measures 49.995 ft after being exposed to the sun, what is its cooled length?

97. **CON**   A contractor ordered 800 board feet of lumber for a job that required 750 board feet. What percent, to the nearest whole number, of the required lumber was ordered for waste?

98. **BUS**   A brick mason received a 12% increase in wages, amounting to $35.40. Find the amount of wages received before the increase. Find the amount of wages received after the increase.

99. **INDTR**   A lathe costing $600 was sold for $516. What was the percent of decrease in the price of the lathe?

100. **AG/H**   A mixture of uncut loam and clay soil will have a 20% earth swell when it is excavated. If 300 yd^3 of uncut earth are to be removed, how many cubic yards will have to be hauled away if the earth swell is taken into account?

101. **INDTR**   A wet casting weighing 145 kg has a 2% weight loss in the drying process. How much does the dried casting weigh?

102. **INDTR**   According to specifications, a machined part may vary from its specified measure by ±0.4% and still be usable. If the specified measure of the part is 75 in. long, what is the range of measures acceptable for the part?

103. **CAD/ARC**   A blueprint specification for a part lists its overall length as 62.5 cm. If a tolerance of ±0.8% is allowed, find the limit dimensions for the part's length.

104. Find the complement of 23%.

105. Find the complement of 12.9%.

106. Find the complement of 7%.

107. Find the complement of 21.5%.

108. Find the complement of $18\frac{1}{2}\%$.

109. Find the complement of $32\frac{2}{5}\%$.

110. Find the complement of 2%.

111. Find the complement of 92%.

112. A washing machine that sells for $389 is on sale for 30% off. Estimate the sale price using the complement of the sale percent.

113. Estimate the discounted price using the complement of the discount rate if a shower enclosure that sells for $3,983 is discounted 20%.

114. **BUS**   A book that did sell for $18.50 now sells for 20% more. How much does the book now sell for?

115. **COMP**   A computer disk that once sold for $2.25 now sells for 25% less. How much does the computer disk sell for?

116. **BUS**   A paperback dictionary originally sold for $4. It now sells for $1 more. What is the percent of the increase?

117. **COMP**   A laptop computer originally priced at $2,400 now sells for $300 more. What is the percent of the increase?

118. **AG/H**   When earth is dug, it usually increases in volume or expands by 20%. How much earth will a contractor have to haul away if 150 ft³ are dug?

119. **INDTR**   A shipping carton is rated to hold 50 lb. A larger carton that holds 15% more weight will be used. How much weight will the larger carton hold?

120. **INDTR**   A 20-in. bar of iron measures 20.025 in. when it is heated. What is the percent of increase?

121. **BUS**   An 18-in. pearl necklace is exchanged for a 24-in. one. What is the percent of increase in the length of the necklace?

122. **INDTR**   A casting weighed 130 ounces (oz) when first made. After it dried, it weighed 127.4 oz. What was the percent of weight loss caused by drying?

123. **HELPP**   A dieter went from 168 lb to 160 lb in 1 week. To the nearest tenth of a percent, what was the percent of weight loss?

124. **BUS**   Workers took a 10% pay cut to help their company stay open during economic hard times. What is the reduced annual salary of a worker who originally earned $35,000?

125. **AUTO**   A motorist trades an older car with 350 hp for a new car with 17.4% less horsepower. What is the horsepower of the new car to the nearest whole number?

126. The Bureau of Labor Statistics notes there were 8,000,000 jobs requiring an associate degree in 2010 and that number is expected to increase to 9,444,000 by 2020. What is the percent of increase?

127. The Bureau of Labor Statistics predicts that the veterinary technologist and technician occupation that had 80,200 jobs in 2010 is expected to have 121,900 jobs by 2020. What is the percent of increase to the nearest whole percent?

128. According to the Bureau of Labor Statistics, the profession of home health and personal care aides is projected to have the fastest rate of growth from 2010 to 2020. The industry had 1,878,700 jobs in 2010 and expects to have 3,191,900 jobs in 2020. What is the percent of increase to the nearest whole percent?

129. **AUTO**   A motorist with an annual income of $18,250 each year spends $3,420 on automobile financing, $652 on gasoline, $625 on insurance, and $150 on maintenance and repair. To the nearest tenth, what percent of the motorist's annual income is used for automotive transportation?

130. **PFIN**   A homeowner with an annual family income of $35,500 spends in a year $6,900 for a home mortgage, $950 for property taxes, $380 for homeowner's insurance, $2,400 for utilities, and $200 for maintenance and repair. To the nearest tenth, what percent of the homeowner's annual income is spent for housing?

131. There are 25 women in a class of 35 students. Find the percent of men in the class to the nearest tenth of a percent.

132. **INDTR**   A 78-lb alloy of tin and silver contains 69.3 lb of tin. Find the percent of silver in the alloy to the nearest tenth of a percent.

3 | TEAM PROBLEM-SOLVING EXERCISES

Your team wants to analyze the percent of income that is spent on housing and transportation. Assume you have a family annual income of $55,000.

1. In a year $6,900 is spent for a home mortgage, $950 for property taxes, $380 for homeowner's insurance, $2,400 for utilities, and $200 for maintenance and repair. What percent, to the nearest tenth percent of the family's annual income, is spent for housing?

2. In a year $3,420 is spent on automobile financing, $1,152 on gasoline, $625 on insurance, and $150 on maintenance and repair. What percent, to the nearest tenth percent of the family's annual income, is spent for transportation?

3 | CONCEPTS ANALYSIS

1. Under what conditions are two fractions proportional?

2. Solving a proportion with one missing term requires two computations. In the proportion, $\frac{R}{100} = \frac{65}{26}$, what two computations do you need to perform to find the value of R?

3. Give some clues for determining if a value in a percent problem represents a rate.

4. Give some clues for determining if a value in a percent problem represents a portion.

5. Give some clues for determining if a value in a percent problem represents a base.

6. Is the amount of sales tax required on a purchase in your state determined by a percent? What is the sales tax rate in your state?

7. If the total bill, including sales tax, on a purchase is 107% of the original amount, what is the sales tax rate? Explain.

8. If a dress is marked 25% off the original price, what percent of the original price does the buyer pay? Explain.

9. If a quantity increases 50%, is the new amount twice the original amount? Explain.

10. If the cost of an item decreases 50%, is the new amount half the original amount? Explain.

Find and explain any mistakes in the following, then rework the incorrect problems.

11. $$\frac{R}{100} = \frac{4}{5}$$
$$\frac{R}{25} = \frac{1}{5}$$
$$5 \cdot R = 25 \cdot 1$$
$$5 \cdot R = 25$$
$$R = 25 \div 5$$
$$R = 5\%$$

12. $3\% = 0.3$

13. What percent of 25 is 75?
$$\frac{R}{100} = \frac{25}{75}$$
$$75 \cdot R = 100 \cdot 25$$
$$75 \cdot R = 2,500$$
$$R = 2,500 \div 75$$
$$R = 33\frac{1}{3}\%$$

14. 26 is 0.5% of what number?
$$\frac{0.5}{100} = \frac{26}{B}$$
$$0.5 \cdot 26 = 100 \cdot B$$
$$13 = 100 \cdot B$$
$$\frac{13}{100} = B$$
$$0.13 = B$$

15. If the cost of a $15 shirt increases 10%, what is the new cost of the shirt?
$$\frac{10}{100} = \frac{P}{15}$$
$$100 \cdot P = 10 \cdot 15$$
$$100 \cdot P = 150$$
$$P = \frac{150}{100}$$
$$P = 1.5$$

The shirt increased $1.50 in price.

3 | PRACTICE TEST

1. Write $\frac{4}{5}$ as a percent.
2. Write 85% as a fraction.
3. Write 0.3% as a decimal.

Identify the given and missing elements as *R* (rate), *B* (base), and *P* (percent).

4. $10 is what percent of $35?
5. 40% of 10 X-ray technicians is how many?
6. What percent of 24 syringes is 8?
7. 9 is what percent of 27 dogwood trees?
8. How many books is 30% of 40 books?
9. 12% of 50 grass plugs is how many?

Solve.

10. 10% of 150 is what number?
11. What number is $6\frac{1}{4}$% of 144?
12. What percent of 275 is 33?
13. 45.75 is 15% of what number?
14. 55 is what percent of 11?
15. 250% of what number is 287.5?
16. What percent of 360 is 1.2?
17. 245% of what number is 164.4? Round to the nearest hundredth.
18. 5.4% of 57 is what number?
19. Find the complement of 88%.
20. Find the complement of 13%.
21. $15,000 is invested at 14% per year for 3 months. How much interest is earned on the investment?
22. A casting measuring 48 cm when poured shrinks to 47.4 cm when cooled. What is the percent of decrease?
23. An electronic parts salesperson earned $175 in commission. If the commission is 7% of sales, how much did the salesperson sell?
24. Electronic parts increased 15% in cost during a certain period, amounting to an increase of $65.15 on one order. How much would the order have cost before the increase? Round to the nearest cent.
25. In 2005, an area vocational school had an enrollment of 325 men and 123 women. In 2006, there were 149 women. What was the percent of increase of women students? Round to the nearest hundredth percent.
26. The payroll for a hobby shop for 1 week is $1,500. If federal income and social security taxes average 28%, how much is withheld from the $1,500?
27. During one period, a bakery rejected 372 items as unfit for sale. In the following period, the bakery rejected only 323 items, a decrease in unfit bakery items. What was the percent of decrease? Round to the nearest hundredth percent.
28. Materials to landscape a new home cost $643.75. What is the amount of tax if the rate is 6%? Round to the nearest cent.
29. A casting weighed 36.6 kg. After milling, it weighs 34.7 kg. Find the percent of weight loss to the nearest whole percent.
30. After soil was excavated for a project, it swelled 15%. If 275 yd^3 were excavated, how many cubic yards of soil were there after excavation?
31. The total bill for machinist supplies was $873.92 before a discount of 12%. How much was the discount? Round to the nearest cent.
32. A business paid a $5.58 finance charge on a monthly balance of $318.76. What was the monthly rate of interest? Round to the nearest hundredth.

1–3 | CUMULATIVE PRACTICE TEST

1. Evaluate: $12 + 5(4) \div 2 - 8$.
2. Evaluate: 6.7^2.
3. Find the perimeter of a rectangle that has a width of 24 cm and length of 40 cm.
4. Find the area of the rectangle in Exercise 3.
5. Round 42.8196 to the nearest hundredth.
6. Evaluate: $5^3 - \sqrt{49}(3 - 1)$.

7. Write in factored form and using exponential notation the prime factorization of 420.

8. Evaluate: $37 - (2 + 8 \cdot 2 - 10)$.

9. $42.8 + 23.06 + 15.9$

10. $196.7 - 84.92$

11. $52.06(8.723)$

Evaluate and simplify.

12. $5\frac{3}{8} + 4 + 12\frac{1}{8}$

13. $3\frac{7}{8} + 9\frac{3}{4} + 5\frac{1}{2}$

14. $15\frac{3}{5} - 9\frac{3}{8}$

15. $4\frac{1}{2}\left(\frac{4}{9}\right)$

16. $\frac{3}{5} \div \frac{7}{10}$

17. $1\frac{3}{4} \div 1\frac{5}{12}$

18. 12^3

19. 3.5^2

20. 123^2

21. $\sqrt{1.96}$

22. $\sqrt{361}$

23. $\sqrt{3600}$

24. How many pints are in 268 qt?

25. A shipping container has a width of 125 in. How many feet is this?

26. An LD-8 airplane shipping container has a maximum net weight of 5,100 lb. How many tons can it hold?

27. The length of an MD-88 aircraft is 147 ft 11 in. Express the length in inches.

28. 28% of 15 is what number?

29. What percent of 24 is 20? Round to the nearest percent.

30. 30% of what number is 12?

31. A salary of $42,000 is increased to $45,000. What is the percent of increase? Round to the nearest hundredth percent.

32. Find the complement of 62%.

33. Find the complement of 14%.

34. To calculate the percent of protein in pet food on a *dry matter basis* (DM basis), divide the percent of protein shown on the Guaranteed Analysis label by the complement of the percent of moisture. PMI Nutrition® dry dog food has 21.0% protein and 14% moisture on its guaranteed analysis label. What is the DM basis percent of protein?

35. Alpo Prime Cuts® canned dog food has 8.0% protein and 82.0% moisture on its guaranteed analysis label. Using the procedure described in Exercise 34, what is the DM basis percent of protein in this dog food?

Measurement

Detlev Van Ravenswaay/Science Source

In Great Company

"The Metric Mixup" (1998)

On December 11, 1998, after years of research and design, NASA launched the Mars Climate Orbiter (MCO) into space. Designed to study Martian climate, atmosphere, and surface features, it was also intended to act as the communications relay for the upcoming Mars Polar Lander. But, as with any Mars mission, the successful launch was only the first part of the job. The satellite traveled through the empty reaches of the solar system for 286 days until it finally reached Mars, where it promptly slid into the upper atmosphere and disintegrated. Years of work, months of waiting, and nearly $328 million burned up in the Martian atmosphere.

What happened?

The investigation discovered that when NASA engineers wrote the guidance software that controlled the Orbiter thrusters, they assumed that all values for the thrusters were going to be calculated using metric units, in this case, newtons. So, the guidance software was designed to fire the thrusters using those units. Unfortunately, the ground crew engineers had not been kept up-to-date on that expectation. Once the MCO actually entered orbit, the ground crew engineers assumed that data was supposed to be calculated and entered in imperial units (pounds of force).

Because the ground crew didn't enter numbers that had been properly converted, the thruster engines fired, but the thruster fire was not long enough. The MCO descended too far into the atmosphere, the atmosphere caused the MCO to slow down even more, which dropped it even farther into the atmosphere, and, well . . . so long, MCO.

How big was the difference? The units differ by a factor of 4.45. So that would be something like the difference between a 100-second burn and a 445-second burn. It wasn't small.

This error was so humiliating that it even got its own name: "the Metric Mixup." After that failure, NASA created a special check on all software and engineering manuals to make sure it would not happen again.

If you ever wondered why math and science teachers get so picky about having the right units attached to calculations, it is because many of them know this story. When it comes to doing math, having the right numbers is pleasant and praiseworthy. But if you don't also have the right units to go with the numbers, then your years of sweat might produce nothing but a $300 million debt and a smoking hole in the ground.

Sad.

4–1 The U.S. Customary System of Measurement

LEARNING OUTCOMES

1. Convert one unit of measure to another using unit ratios.

2. Convert one unit of measure to another using conversion factors.

3. Add and subtract U.S. customary measures.

4. Multiply and divide U.S. customary measures.

5. Change one U.S. customary rate measure to another.

LC LEARNING CATALYTICS

1. Multiply $60\left(\frac{5,280}{1}\right)\left(\frac{1}{60}\right)\left(\frac{1}{60}\right)$.

U.S. customary system of measurement: the system of measurement commonly used in the United States. Formerly known as the English or British system

Mass: the quantity of material that makes up an object

Weight: a measure of the Earth's gravitational pull on an object

The **U.S. customary system of measurement** evolved from the English or British system of measurement, and many of the original units in this system are now obsolete. In our discussion of the U.S. customary system of measurement, we will include only selected measuring units.

1 Convert One Unit of Measure to Another Using Unit Ratios.

Length

Four basic units in the U.S. customary system are commonly used to measure length. They are the inch, the foot, the yard, and the mile. Table 4–1 gives the relationships among these measurements of length.

Table 4–1 U.S. Customary Units of Length or Distance

12 inches (in.)[a] = 1 foot (ft)[b]	36 inches (in.) = 1 yard (yd)
3 feet (ft) = 1 yard (yd)	5,280 feet (ft) = 1 mile (mi)

[a]The symbol ″ means inches (8″ = 8 in.) or seconds (60″ = 60 seconds).
[b]The symbol ′ means feet (3′ = 3 ft) or minutes (60′ = 60 minutes).

Weight or Mass

The terms *weight* and *mass* are commonly used interchangeably. In technical, engineering, and scientific applications, the **mass** of an object is the quantity of material that makes up the object. The **weight** is a measure of the Earth's gravitational pull on the object. Three commonly used measuring units for weight or mass in the U.S. customary system are the ounce, pound, and ton. Table 4–2 gives the relationships among these measures of weight or mass.

Table 4-2   U.S. Customary Units of Weight or Mass

16 ounces (oz) = 1 pound (lb)
2,000 pounds (lb) = 1 ton (T)

Capacity or Volume

The U.S. customary system includes units for both liquid and dry capacity measures; however, the dry capacity measures are seldom used. It is more common to express dry measures in terms of weight than in terms of capacity.

Common U.S. customary units of measure for capacity or volume are the ounce, cup, pint, quart, and gallon. Table 4–3 gives the relationships among the liquid measures for capacity or volume. In the U.S. customary system the term *ounce* represents both weight and liquid capacity. The measures are different and have no common relationship. Also, the abbreviation for ton and tablespoon is T. The context of the problem will suggest whether the unit for weight or capacity is meant.

Table 4-3   U.S. Customary Units of Liquid Capacity or Volume

3 teaspoons (t) = 1 tablespoon (T)	2 tablespoons (T) = 1 ounce (oz)
8 ounces (oz) = 1 cup (c)	4 cups (c) = 1 quart (qt)
2 cups (c) = 1 pint (pt)	4 quarts (qt) = 1 gallon (gal)
2 pints (pt) = 1 quart (qt)	

> **TIP**   It is important to understand that for a unit ratio to be equal to 1, the numerator and denominator **must** have measures that are equal in value.

Unit ratio: a ratio that has a value of 1; a fraction with one unit of measure in the numerator and a different, but equivalent, unit of measure in the denominator

Per: a common way for expressing the dimensions of a ratio in words, as in inches per foot (in./ft)

Using the relationship between two units of measure, we can form a ratio in two different ways that *has a value of 1*. We call this type of ratio a **unit ratio.**

A *ratio* is a fraction. A unit ratio, then, is a fraction with one unit of measure in the numerator and a different, but equivalent, unit of measure in the denominator. Some examples of unit ratios are

$$\frac{12 \text{ in.}}{1 \text{ ft}}, \quad \frac{1 \text{ ft}}{12 \text{ in.}}, \quad \frac{3 \text{ ft}}{1 \text{ yd}}, \quad \frac{1 \text{ mi}}{5,280 \text{ ft}}$$

In each unit ratio, the value of the numerator equals the value of the denominator. A ratio with the numerator and denominator equal has a value of 1. When we convert from one unit of measure to another, we use a unit ratio that contains the original unit and the new unit. A common way for expressing the dimensions of a ratio in words uses the word **per,** as in inches per foot (in./ft).

> **To write two equivalent measures as a unit ratio:**
>
> 1. Write one measure in the numerator.
>
> 2. Write the equivalent measure in the denominator. The ratio has a value of 1.

STOP AND CHECK

Write two unit ratios that relate the pair of measures.

1. Cups and quarts
2. Tons and pounds

Answers:

1. $\dfrac{1 \text{ qt}}{4 \text{ c}}, \dfrac{4 \text{ c}}{1 \text{ qt}}$ 2. $\dfrac{2,000 \text{ lb}}{1 \text{ T}}, \dfrac{1 \text{ T}}{2,000 \text{ lb}}$

EXAMPLE 1

Write two unit ratios that relate the pair of measures.

(a) Ounces and pounds   **(b)** Cups and pints

(a) The relationship between ounces and pounds is 1 lb contains 16 oz. The unit ratios involving ounces and pounds are

$$\frac{1 \text{ lb}}{16 \text{ oz}} \quad \text{and} \quad \frac{16 \text{ oz}}{1 \text{ lb}}$$

(b) The relationship between cups and pints is 1 pint contains 2 cups. The unit ratios involving cups and pints are

$$\frac{1 \text{ pt}}{2 \text{ c}} \quad \text{and} \quad \frac{2 \text{ c}}{1 \text{ pt}}$$

See Exercises 1–4.

Unit ratios help us convert from one unit of measure to another.

> **To change from one U.S. customary unit of measure to another using unit ratios:**
>
> 1. Write the original measure in the numerator of a fraction with 1 in the denominator.
>
> 2. Multiply this fraction by a unit ratio with the original unit of measure in the denominator and the new unit of measure in the numerator.
>
> 3. Reduce like units of measure and all numbers wherever possible.

STOP AND CHECK

1. Find the number of feet in 3 miles.
2. Find the number of teaspoons in 5 tablespoons.

Answers:
1. 15,840 ft   2. 15 t

EXAMPLE 2

Find **(a)** the number of inches in 5 ft and **(b)** how many pints are in 4.5 quarts.

(a) Multiply 5 ft by a unit ratio that contains both inches and feet.

Because 5 ft is a whole number, we write it with 1 as the denominator.

$$\frac{5 \text{ ft}}{1}\left(\right)$$ Place the original unit with the 5 in the *numerator* of the first fraction.

$$\frac{5 \text{ ft}}{1}\left(\frac{}{\text{ft}}\right)$$ We are changing *from* feet, so we place ft in the *denominator* of the unit ratio, which is shown in parentheses. This allows us to reduce the units later.

$$\frac{5 \text{ ft}}{1}\left(\frac{\text{in.}}{\text{ft}}\right)$$ To change *to* inches, place inches in the *numerator* of the unit ratio.

$$\frac{5 \text{ ft}}{1}\left(\frac{12 \text{ in.}}{1 \text{ ft}}\right) = 60 \text{ in.}$$ Place in the unit ratio the numerical values that make these two units of measure equivalent (1 ft = 12 in.). Complete the calculation, reducing wherever possible.

$$5 \text{ ft} = \mathbf{60 \text{ in.}}$$

(b) $$\frac{4.5 \text{ qt}}{1}\left(\frac{2 \text{ pt}}{1 \text{ qt}}\right) = 9 \text{ pt}$$ From quart (denominator) to pint (numerator).

$$4.5 \text{ qt} = \mathbf{9 \text{ pt}}$$

See Exercises 5–12.

Dimensions: measurements

Dimension analysis: a systematic examination of the appropriate measuring units of a solution

When working with units of measure, it is very important to include the measuring unit in our analysis. Measurements are also referred to as **dimensions,** and the systematic examination of the appropriate measuring units of a solution is referred to as **dimension analysis.**

Sometimes it is necessary to convert a U.S. customary unit to a unit that is *not* the next larger or smaller unit of measure.

> **TIP** **Changing to Any Larger or Smaller Unit** To change from a U.S. customary unit to one other than the next larger or smaller unit, proceed as before, but multiply the original amount by as many unit ratios as needed to attain the new U.S. customary unit.
>
> For instance, to change from yards to inches: yards → feet → inches.
> To change from gallons to ounces: gallons → quarts → pints → cups → ounces.

EXAMPLE 3

How many inches are in $2\frac{1}{3}$ yd?

$$2\frac{1}{3}\text{ yd} = \frac{7}{3}\text{ yd}$$

Write $2\frac{1}{3}$ as an improper fraction.

$$\frac{7\text{ yd}}{3}\left(\frac{\text{ft}}{\text{yd}}\right)\left(\frac{\text{in.}}{\text{ft}}\right)$$

Multiply the improper fraction by two unit ratios. To change from yards to inches, first change from yards to feet, and then from feet to inches. Place the original unit in the *numerator* of the improper fraction, $\frac{7}{3}$.

$$\frac{7\text{ yd}}{\overset{}{\underset{1}{3}}}\left(\frac{\overset{1}{3}\text{ ft}}{1\text{ yd}}\right)\left(\frac{12\text{ in.}}{1\text{ ft}}\right) = 84\text{ in.}$$

Insert the appropriate numerical values for each unit ratio (3 ft = 1 yd; 12 in. = 1 ft), and multiply, reducing wherever possible.

$$2\frac{1}{3}\text{ yd} = \textbf{84 in.}$$

Alternative method

If we use the relationship for inches and yards, 36 in. = 1 yd, we need only one unit ratio for the calculation. That is, we convert $2\frac{1}{3}$ yd $\left(\frac{7}{3}\text{ yd}\right)$ to inches as follows:

$$\frac{7\text{ yd}}{3}\left(\frac{36\text{ in.}}{1\text{ yd}}\right)$$

Set up the unit ratio using 36 in. = 1 yd.

$$\frac{7\text{ yd}}{\underset{1}{3}}\left(\frac{\overset{12}{36}\text{ in.}}{1\text{ yd}}\right) = 84\text{ in.}$$

Reduce units and numbers; then multiply.

$$2\frac{1}{3}\text{ yd} = \textbf{84 in.}$$

See Exercises 13–14.

TIP   **Focus on One Thing at a Time**   Sometimes the steps in a multistep problem can be overwhelming. It is often helpful to focus on one aspect of the problem at a time. In the example changing yards to inches, focus first on just the units of measure or dimensions.

$$\frac{\text{yd}}{1}\left(\frac{\text{ft}}{\text{yd}}\right)\left(\frac{\text{in.}}{\text{ft}}\right)$$

Reduce as appropriate. Yards reduce to 1. Feet reduce to 1. The only measuring unit left is inches. Therefore, the result will be in inches.

Next, focus on the numbers.

$$\frac{7\text{ yd}}{3}\left(\frac{3\text{ ft}}{1\text{ yd}}\right)\left(\frac{12\text{ in.}}{1\text{ ft}}\right) \rightarrow \frac{7}{3}\left(\frac{3}{1}\right)\left(\frac{12}{1}\right) = 84$$

Putting the numbers and units together, we have 84 in.

---

2   **Convert One Unit of Measure to Another Using Conversion Factors.**   Unit ratios can be used to develop conversion factors. A **conversion factor** is a standard value used to convert one measure to another. With conversion factors, you *always* multiply to change from one measuring unit to another.

Conversion factor: a standard value used to convert one measure to another

To develop a conversion factor for converting from one measure to another:

1. Write a unit ratio that changes the given unit to the new unit.

2. Change the fraction (or ratio) to its decimal equivalent by dividing the numerator by the denominator.

EXAMPLE 4

Develop two conversion factors relating pounds and ounces.

Pounds to Ounces

$$\frac{\text{pounds}}{1}\left(\frac{\text{ounces}}{\text{pounds}}\right)$$

$$\frac{\text{pounds}}{1}\left(\frac{16\ \text{ounces}}{1\ \text{pound}}\right)$$

$$\frac{16}{1} = 16$$

pounds $\times$ 16 = ounces

Ounces to Pounds

$$\frac{\text{ounces}}{1}\left(\frac{\text{pounds}}{\text{ounces}}\right)$$

$$\frac{\text{ounces}}{1}\left(\frac{1\ \text{pound}}{16\ \text{ounces}}\right)$$

$$\frac{1}{16}\ \text{or}\ 1 \div 16 = 0.0625$$

ounces $\times$ 0.0625 = pounds   **See Exercises 15–16.**

To change from one U.S. customary unit of measure to another using conversion factors:

1. Select the appropriate conversion factor.

2. Multiply the original measure by the conversion factor.

Table 4–4 U.S. Customary Conversion Factors*

	TO CHANGE		
	From	**To**	**Multiply By**
Length or Distance			
12 inches (in.) = 1 foot (ft)	feet	inches	12
	inches	feet	0.0833333
3 feet (ft) = 1 yard (yd)	yards	feet	3
	feet	yards	0.3333333
36 inches (in.) = 1 yard (yd)	yards	inches	36
	inches	yards	0.0277778
5,280 feet (ft) = 1 mile (mi)	miles	feet	5,280
	feet	miles	0.0001894
Weight or Mass			
16 ounces (oz) = 1 pound (lb)	pounds	ounces	16
	ounces	pounds	0.0625
2,000 pounds (lb) = 1 ton (T)	tons	pounds	2,000
	pounds	tons	0.0005
Liquid Capacity or Volume			
8 ounces (oz) = 1 cup (c)	cups	ounces	8
	ounces	cups	0.125
2 cups (c) = 1 pint (pt)	pints	cups	2
	cups	pints	0.5
2 pints (pt) = 1 quart (qt)	quarts	pints	2
	pints	quarts	0.5
4 quarts (qt) = 1 gallon (gal)	gallons	quarts	4
	quarts	gallons	0.25

*Additional conversion factors are found on the inside covers of the text.

STOP AND CHECK

1. Use a conversion factor from Table 4-4 to convert 62 ounces to cups.

Answer:
1. 7.75 c

EXAMPLE 5

Use a conversion factor from Table 4-4 to convert 56 ounces to pounds.

ounces $\times$ 0.0625 = pounds   Conversion factor for ounces to pounds.

56(0.0625) = 3.5 pounds   Multiply.

56 ounces is 3.5 pounds.   **See Exercises 17–22.**

TIP   **Estimation and Dimension Analysis** When estimating unit conversions, first see if the new unit is larger or smaller than the original unit.

Larger to Smaller

Each larger unit can be divided into smaller units. Larger-to-smaller conversions mean *more* smaller units. *More* implies multiplication by a number greater than one.

    To convert a U.S. customary unit to a desired *smaller* unit: *Multiply* by a conversion factor that is *greater than one*.

2 yd = _____ ft   The smaller unit is feet: 3 ft = 1 yd. Use the conversion factor 3 ft per yd or 3 ft/yd.

2 yd(3 ft/yd) = 6 ft   Multiply number of yards by 3 ft/yd.

2 yd = 6 ft   Dimension analysis: $\dfrac{2 \text{ yd}}{1}\left(\dfrac{3 \text{ ft}}{1 \text{ yd}}\right) = 6 \text{ ft}$

The key word clues are:

larger to smaller unit → obtain more units → multiply by number greater than one

Smaller to Larger

Several small units combine to make one large unit. Thus, smaller-to-larger conversions mean *fewer* large units. *Fewer* implies multiplication by a number less than one.

    To convert a U.S. customary unit to a *larger* unit: *Multiply* by a conversion factor that is *less than one*.

12 ft = _____ yd   The larger unit is yards: 1 yd = 3 ft. Use the conversion factor 0.3333333 yd per ft or yd/ft.

12 ft $\times$ 0.3333333 yd/ft = 4 yd   Multiply by 0.3333333 yd/ft to get yards.

    Dimension analysis: $\dfrac{12 \text{ ft}}{1}\left(\dfrac{1 \text{ yd}}{3 \text{ ft}}\right) = 4 \text{ yd}$

12 ft = 4 yd   Rounded

Using key word clues:

smaller to larger unit → obtain fewer units → multiply by number less than one

Estimation can catch errors in setting up the problem, but it will not likely catch calculation errors.

Mixed measures: measures expressed by using two or more units of measure

Standard notation of mixed measures: when the number associated with each unit of measure is smaller than the number required to convert to the next larger unit

    Measures that use two or more units are called **mixed measures**. The **standard notation of mixed measures** is when the number associated with each unit of measure is smaller than the number required to convert to the next larger unit.

> **To express mixed measures in standard notation:**
>
> 1. Start with the smallest unit of measure and determine if there are enough units to make one or more of the next larger unit.
>
> 2. Regroup to make as many of the larger units as possible.
>
> 3. Combine like units.
>
> 4. Repeat the process with each given measuring unit.

STOP AND CHECK

1. Express 1 ft 15 in. in standard notation.
2. Express 2 gal 5 qt 3 pt in standard notation.

Answers:

1. 2 ft 3 in.   **2.** 3 gal 2 qt 1 pt

EXAMPLE 6

Express **(a)** 8 lb 20 oz, **(b)** 1 gal 5 qt, and **(c)** 2 yd 4 ft 16 in. in standard notation.

(a) 8 lb 20 oz   20 oz = 1 lb 4 oz
8 lb 20 oz = 8 lb + 1 lb 4 oz = **9 lb 4 oz**   Standard notation.

(b) 1 gal 5 qt   5 qt = 1 gal 1 qt
1 gal 5 qt = 1 gal + 1 gal 1 qt = **2 gal 1 qt**   Standard notation.

(c) 2 yd 4 ft 16 in. = 2 yd 4 ft + 1 ft 4 in.   Regroup inches.
                                      16 in. = 1 ft 4 in.
= 2 yd 5 ft 4 in. = 2 yd + 1 yd 2 ft 4 in.   Regroup feet.
                                      5 ft = 1 yd 2 ft
= **3 yd 2 ft 4 in.**   Standard notation.

See Exercises 23–30.

> **TIP** **Standard Conventions** We can say that 3 ft 15 in. is 4 ft 3 in. in standard notation. Why not change 4 ft 3 in. to 1 yd 1 ft 3 in.? There may be situations when 1 yd 1 ft 3 in. is the desirable form; however, in general, we keep the same units of measure in standard notation as used in the original measure.

3 Add and Subtract U.S. Customary Measures. We can add U.S. customary measures *only* when their units are the same. Measures with the same units are **like measures.** Measures with different units are **unlike measures.**

Like measures: measures with the same measuring units

Unlike measures: measures with different measuring units

> **To add unlike U.S. customary measures:**
>
> 1. Convert each measure to a measure with a common U.S. customary unit.
>
> 2. Add. The unit of measure of the sum is the common unit.

STOP AND CHECK

1. Add 2 lb + 5 oz.

Answer:

1. 37 oz

EXAMPLE 7

Add 3 ft + 2 in.

Because 2 in. is a fraction of a foot, we can avoid working with fractions by converting the larger unit (feet) to the smaller unit (inches).

$$3 \text{ ft} = \frac{3 \text{ ft}}{1}\left(\frac{12 \text{ in.}}{1 \text{ ft}}\right) = 36 \text{ in.}$$ Convert ft to in., and add like measures.

36 in. + 2 in. = **38 in.**   **See Exercises 31–32.**

To add mixed U.S. customary measures:

1. Align like measures in columns.

2. Add common units of measure.

3. Express the sum in standard notation.

STOP AND CHECK

Add and write the answer in standard form.
1. 2 ft 5 in. and 4 ft 9 in.
2. 2 gal 1 qt + 1 gal 2 qt

Answers:
1. 7 ft 2 in.   2. 3 gal 3 qt

EXAMPLE 8

Add and write the answer in standard form: 6 lb 7 oz and 3 lb 13 oz.

```
   6 lb  7 oz
 + 3 lb 13 oz
   9 lb 20 oz
```
Write in standard notation, 20 oz = 1 lb 4 oz.

Thus, 9 lb 20 oz = 10 lb 4 oz.   See Exercises 33–42.

To subtract unlike U.S. customary units:

1. Convert each measure to a measure with a common U.S. customary unit.

2. Subtract. The unit of measure of the difference is the common unit.

STOP AND CHECK

1. Subtract 18 oz from 2 lb.

Answer:
1. 14 oz

EXAMPLE 9

Subtract 15 in. from 2 ft.

Changing 15 in. to feet gives us a mixed number, so it is more convenient to convert 2 ft to inches.

2 ft = 24 in.   Convert ft to in. and subtract.

24 in. − 15 in. = **9 in.**   See Exercises 43–46.

To subtract mixed U.S. customary measures:

1. Align like measures in columns.

2. Subtract common units of measure beginning with the smallest unit of measure.

STOP AND CHECK

1. Subtract 3 T 500 lb from 5 T 1,000 lb.

Answer:
1. 2 T 500 lb

EXAMPLE 10

Subtract 5 ft 3 in. from 7 ft 4 in.

```
  7 ft 4 in.
 −5 ft 3 in.
  2 ft 1 in.
```
Align like measures in a vertical line, and then subtract.

See Exercises 47–48.

When we subtract mixed measures, we use our knowledge of regrouping to subtract a larger unit from a smaller one.

To subtract a larger U.S. customary unit from a smaller unit in mixed measures:

1. Align the common measures in vertical columns.

2. Regroup by subtracting one unit from the next larger unit of measure in the minuend, convert to the equivalent smaller unit, and add it to the smaller unit.

3. Subtract common measures beginning with the smallest unit of measure.

EXAMPLE 11

Subtract 3 lb 12 oz from 7 lb 8 oz.

$$7 \text{ lb } 8 \text{ oz}$$
$$-3 \text{ lb } 12 \text{ oz}$$

We always begin subtraction with the smallest unit, which should be on the *right*. 12 oz cannot be subtracted from 8 oz.

7 lb 8 oz = 6 lb 16 oz + 8 oz =	6 lb 24 oz	Rewrite 7 lb as 6 lb 16 oz. Add
3 lb 12 oz =	−3 lb 12 oz	16 oz to 8 oz to get 24 oz.
	3 lb 12 oz	**See Exercises 49–56.**

4 **Multiply and Divide U.S. Customary Measures.** Suppose a tank of weed killer contains 21 gal 3 qt. What is the total amount of weed killer in eight tanks? To find the total amount of weed killer, we multiply 21 gal 3 qt by 8.

> **To multiply a U.S. customary measure by a number:**
>
> 1. Multiply the numbers associated with each unit of measure by the given number.
>
> 2. Write the resulting measure in standard notation.

EXAMPLE 12

AG/H   A tank holds 21 gal 3 qt of weed killer. How much do eight containers hold?

$$21 \text{ gal } 3 \text{ qt}$$ Multiply each unit of measure by 8.
$$\times \quad 8$$
$$\overline{168 \text{ gal } 24 \text{ qt}}$$ Write in standard notation, 24 qt = 6 gal.

168 gal 24 qt = 168 gal + 6 gal = 174 gal in standard notation

The eight containers hold 174 gal.   **See Exercises 57–62.**

> **To multiply a length measure by a like length measure:**
>
> 1. Multiply the numbers associated with each like unit of measure.
>
> 2. The unit of measure of the product is a square measure or an area.

EXAMPLE 13

A desktop is 2 ft × 3 ft (Fig. 4–1). What is the number of square feet in the surface?

2 ft (3 ft) = **6 ft^2**   Multiply numbers. Product is ft^2.   **See Exercises 63–70.**

For a more detailed discussion of area see Chapter 1, Section 3, Outcome 2. We are frequently required to divide measures by a number.

> **To divide a U.S. customary measure by a number that divides evenly into each measure:**
>
> 1. Divide the numbers associated with each unit of measure by the given number.
>
> 2. Write the resulting measure in standard notation.

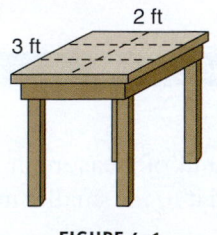

FIGURE 4–1

Sevenke/Shutterstock

EXAMPLE 14

HOSP   How much milk is needed for a half-recipe if the original recipe calls for 2 gal 2 qt?

$$2\overline{)\text{2 gal 2 qt}}\qquad\text{Divide each measure by 2.}$$

$$\begin{array}{c}\text{1 gal 1 qt}\\ \hline 2\overline{)\text{2 gal 2 qt}}\end{array}$$

The half-recipe requires 1 gal 1 qt of milk.   **See Exercise 71.**

If a given number does not divide evenly into each measure, there will be a remainder.

> **To divide a U.S. customary measure by a number that does not divide evenly into each measure:**
>
> 1. Set up the problem and proceed as in long division.
>
> 2. When a remainder occurs after subtraction, convert the remainder to the same unit used in the next smaller measure, and add it to the quantity in the next smaller measure.
>
> 3. Divide the given number into this next smaller unit.
>
> 4. If a remainder occurs when the smallest unit is divided, express the remainder as a fractional part of the smallest unit.

STOP AND CHECK

1. Divide 6 ft 9 in. by 3.
2. Divide 5 yd 2 ft 5 in. by 2.

Answers:

1. 2 ft 3 in.   2. 2 yd 2 ft $8\frac{1}{2}$ in.

EXAMPLE 15

Divide 5 gal 3 qt 1 pt by 3.

$$\begin{array}{llll} & \text{1 gal} & \text{3 qt} & 1\frac{2}{3}\text{ pt}\\ 3\overline{)} & \text{5 gal} & \text{3 qt} & \text{1 pt}\\ & \underline{\text{3 gal}}\\ & \text{2 gal} = & \underline{\text{8 qt}}\\ & & \text{11 qt}\\ & & \underline{\text{9 qt}}\\ & & \text{2 qt} = & \underline{\text{4 pt}}\\ & & & \text{5 pt}\\ & & & \underline{\text{3 pt}}\\ & & & \text{2 pt} \end{array}$$

5 gal ÷ 3 = 1 gal, remainder 2 gal
2 gal = 8 qt
3 qt + 8 qt = 11 qt
11 qt ÷ 3 = 3 qt, remainder 2 qt
2 qt = 4 pt
1 pt + 4 pt = 5 pt
5 pt ÷ 3 = 1 pt, remainder 2 pt

Write the final remainder, **2**, as a fraction $\frac{2}{3}$ pint, and add to 1 pint to get $1\frac{2}{3}$ pints.

Thus, 1 gal 3 qt $1\frac{2}{3}$ pt is the solution.   **See Exercises 72–79.**

When dividing a measure by a measure, we express both measures in the same unit, just as in adding and subtracting measures. We generally convert to the *smallest* unit used in the example.

> **To divide a U.S. customary measure by a U.S. customary measure:**
>
> 1. Convert both measures to the same unit if they are different.
>
> 2. Write the division as a fraction, including the common unit in the numerator and the denominator.
>
> 3. Reduce the units and divide the numbers to simplify the fractions.

STOP AND CHECK
1. Divide 3 gal 1 qt by 2 qt.
Answer:
1. $6\frac{1}{2}$

EXAMPLE 16

INDTR If tubing is manufactured in lengths of 8 ft 4 in. and a part uses tubing that is 10 in. long, how many parts can be cut from the length of tubing if we do not account for waste? Divide 8 ft 4 in. by 10 in.

$$8 \text{ ft} = \frac{8 \text{ ft}}{1}\left(\frac{12 \text{ in.}}{1 \text{ ft}}\right) = 96 \text{ in.} \qquad \text{Convert 8 ft to inches. Then add to 4 in.}$$

$$8 \text{ ft 4 in.} = 96 \text{ in.} + 4 \text{ in.} = 100 \text{ in.} \qquad \text{Write the mixed measure as a measure with one measuring unit.}$$

$$100 \text{ in.} \div 10 \text{ in.} = 10 \qquad \text{Divide the total length by the length of each part.}$$

If we write this division in fraction form, we can see more easily that the common units reduce. In other words, the answer will be a number (not a measure) telling *how many equal parts* can be cut from the tubing.

$$\frac{100 \text{ in.}}{10 \text{ in.}} = 10$$

Therefore, 10 parts of equal length can be cut from the tubing. **See Exercises 80–86.**

Rate measure: a ratio of two different kinds of measures. It is often referred to as a rate

5 **Change One U.S. Customary Rate Measure to Another.** A **rate measure** is a ratio of two different kinds of measures. A rate measure is often referred to as a **rate.** Some examples of rates are 55 miles per hour and 20 cents per mile. In each of these rates, the word *per* means *divided by.*

The rate 55 miles per hour means 55 miles ÷ 1 hour or $\frac{55 \text{ mi}}{1 \text{ h}}$. The rate 20 cents per mile means 20 cents ÷ 1 mile or $\frac{20 \text{ cents}}{1 \text{ mi}}$.

Many rate measures involve measures of time. The units we use to measure time are universally accepted. The basic units of time are the year, month, week, day, hour, minute, and second. Table 4–5 gives the relationships among the units for time. These units are often used when working with rates.

Table 4–5   Units of Time

1 year (yr) = 12 months (mo)	1 minute (min) = 60 seconds (s)[c]
1 year (yr) = 365 days (da)	1 millisecond (ms) = $\frac{1}{1,000}$ s
1 week (wk) = 7 days (da)	
1 day (da) = 24 hours (h)[a]	
1 hour (h) = 60 minutes (min)[b]	1 nanosecond (ns) = $\frac{1}{1,000,000,000}$ s

[a]The abbreviation for hour can also be hr.
[b]The symbol ′ means feet (3′ = 3 ft) or minutes (60′ = 60 minutes).
[c]The symbol ″ means inches (8″ = 8 in.) or seconds (60″ = 60 seconds). The abbreviation for second can also be sec.

To change one U.S. customary rate measure to another:

1. Compare the units of both numerators and both denominators to determine which units will change.

2. Multiply the original rate measure by one or more unit ratios containing the new units so that the original units to be changed will reduce.

STOP AND CHECK

1. Change $150\,\dfrac{\text{ft}}{\text{min}}$ to $\dfrac{\text{ft}}{\text{sec}}$.

Answer:

1. $2\dfrac{1}{2}\dfrac{\text{ft}}{\text{sec}}$ or $2.5\dfrac{\text{ft}}{\text{sec}}$

EXAMPLE 17

Change $8\,\dfrac{\text{pt}}{\text{min}}$ to $\dfrac{\text{qt}}{\text{min}}$.

Estimation

Pints to quarts is *smaller* to *larger,* so there will be fewer quarts.

Examine the rates:
    numerators—pints change to quarts
    denominators—no change

$$\dfrac{\cancel{\text{pt}}}{\text{min}}\left(\dfrac{\text{qt}}{\cancel{\text{pt}}}\right) = \dfrac{\text{qt}}{\text{min}}$$
Develop a unit ratio with pints in the denominator.

$$\dfrac{\overset{4}{\cancel{8}}\,\cancel{\text{pt}}}{\text{min}}\left(\dfrac{1\,\text{qt}}{\underset{1}{\cancel{2}}\,\cancel{\text{pt}}}\right) = \dfrac{4\,\text{qt}}{\text{min}}$$
Insert numbers in the unit ratio and reduce. Multiply.

Interpretation

Thus, $8\,\dfrac{\textbf{pt}}{\textbf{min}}$ equals $4\,\dfrac{\textbf{qt}}{\textbf{min}}$.

See Exercises 87–89.

A separate unit ratio is used for each unit in the rate that changes. For example, when both the numerator and the denominator of a rate measure change, we multiply by at least two unit ratios to make the conversion.

STOP AND CHECK

1. Change $120\,\frac{\text{gal}}{\text{min}}$ to $\frac{\text{qt}}{\text{sec}}$.

Answer:

1. $8\,\dfrac{\text{qt}}{\text{sec}}$

EXAMPLE 18

AUTO   Change 60 miles per hour to feet per second.

$$60 \text{ miles per hour} = 60\,\dfrac{\text{mi}}{\text{h}}$$
Write rate as a fraction.

$$\text{feet per second} = \dfrac{\text{ft}}{\text{s}}$$

Numerators—miles change to feet
Denominators—hours change to seconds
Examine the changes in the measures.

$$\underset{\substack{(\text{miles} \\ \text{to} \\ \text{feet})}}{} \underset{\substack{(\text{hours} \\ \text{to} \\ \text{minutes})}}{} \underset{\substack{(\text{minutes} \\ \text{to} \\ \text{seconds})}}{}$$

$$\cancel{60}\,\dfrac{\cancel{\text{mi}}}{\cancel{\text{h}}}\left(\dfrac{5{,}280\,\text{ft}}{1\,\cancel{\text{mi}}}\right)\left(\dfrac{1\,\cancel{\text{h}}}{\cancel{60}\,\cancel{\text{min}}}\right)\left(\dfrac{1\,\cancel{\text{min}}}{60\,\text{s}}\right)$$
Develop appropriate unit ratios and reduce.

$$\dfrac{5{,}280\,\text{ft}}{60\,\text{s}} = 88\,\dfrac{\text{ft}}{\text{s}}$$
Divide.

At 60 mi/h you are traveling 88 ft/s.

See Exercises 90–92.

4–1 EXERCISES   **MyLab Math** For additional practice go to your study plan in MyLab Math.

1 Write two unit ratios that relate the *given* pair of measures. *See Example 1.*

   1. Pints and quarts
   2. Feet and miles
   3. Inches and feet
   4. Feet and yards

Use unit ratios to convert the units. *See Example 2.*

5. 4 ft = _____ in.

6. 7 yd = _____ ft

7. $2\frac{1}{2}$ mi = _____ yd

8. Find the number of yards in 28 ft.

9. Find the number of pounds in $36\frac{4}{5}$ oz.

10. How many ounces are in 45.8 lb?

11. How many quarts are in 5 gal?

12. How many pints are in $6\frac{1}{2}$ qt?

See Example 3.

13. How many cups are in 3 gal?

14. 72 ounces will make how many quarts?

2 Develop two conversion factors for the pairs of units. *See Example 4.*

15. Feet and yards

16. Quarts and gallons

Use conversion factors to convert the units of measure. *See Example 5.*

17. 460 oz is equivalent to how many pints?

18. How many pints are in 46 qt?

19. How many pounds are in 580 oz?

20. If a cabinet is $12\frac{3}{4}$ ft long, how many inches is this?

21. How many feet are in 1.5 mi?

22. How many quarts are in 7 gal?

Express the measures in standard notation. *See Example 6.*

23. 2 ft 20 in.

24. 1 mi 6,375 ft

25. 2 lb $19\frac{1}{2}$ oz

26. 1 gal 5 qt

27. 2 ft 10 in.

28. 5 lb 25 oz

29. 3 gal 5 qt 48 oz

30. 6 qt 20 oz

3 Add and write the answer in standard notation.

See Example 7.

31. 5 oz + 2 lb

32. 4 ft + 7 in.

35. 5 qt 1 pt
    + 2 qt $1\frac{1}{2}$ pt

36. 8 gal 3 qt
    + 5 gal 2 qt

See Example 8.

33. 8 lb 2 oz
    + 7 lb 9 oz

34. 5 ft 45 in.
    + 7 ft 30 in.

37. 7 yd 2 ft
    + 1 yd 2 ft

38. 4 yd 2 ft 7 in.
    + 2 yd 1 ft 10 in.

Solve the problems. When necessary, express the answers in standard notation.

39. CON   A plumber has a 2-ft length of copper tubing and a 7-in. length of copper tubing. What is the total in inches?

40. COMP   How many feet are two computer cables together if one is 6 ft and the other is 60 in.?

41. HOSP   A mixture for hamburgers contains 2 lb 8 oz of ground round steak and 3 lb 7 oz of regular ground beef. How much does the hamburger mixture weigh?

42. HLTH/N   According to hospital maternity records, one infant twin weighed 6 lb 1 oz and the other weighed 5 lb 15 oz at birth. What was their total weight?

Subtract and write the answer in standard form. *See Example 9.*

43. 2 ft − 18 in.

44. 2 qt − 3 pt

45. 3 yd − 7 ft

46. 6 lb − 18 oz

See Example 10.

47. 3 lb 12 oz
    − 2 lb 6 oz

48. 12 lb 14 oz
    − 5 lb 12 oz

See Example 11.

49. 2 ft 30 in.
    − 1 ft 40 in.

50. 5 gal 3 qt 1 pt
    − 1 gal 3 qt $1\frac{1}{2}$ pt

51. AUTO   A mechanic has a length of hose 5 ft long. What is its length after 10 in. is cut off?

52. CON   A cabinetmaker cut 9 in. from a 3-ft shelf in a medical lab. How long was the shelf after it was cut?

53. COMP   If a computer sorts a list of names in 1 min 30 s and a faster computer does the same job in 45 s, how much time is saved by using the faster computer?

54. AUTO   A car with a 2.0-L engine accelerates a certain distance in 1 min 27 s. A car with a 5.0-L engine accelerates the same distance in 48 s. How much faster does the car with the 5.0-L engine accelerate?

55. COMP A package containing a laser printer weighs 74 lb 3 oz. The container and packing material weigh 4 lb 12 oz. How much does the laser printer weigh?

56. AG/H To weigh a baby pig, Stacey held the pig while standing on a scale. If the scale showed 132 lb 6 oz and Stacey weighs 115 lb 8 oz, how much does the pig weigh?

4 Multiply and write the answers for mixed measures in standard notation. *See Example 12.*

57. 12 mi
  $\times\ 5$

58. 18 gal
  $\times\ 6$

59. 7 lb 3 oz
  $\times\ \ \ 8$

60. 7 ft 3 in.
  $\times\ \ \ 8$

61. HOSP Tuna is packed in 1-lb 8-oz cans. If a case contains 24 cans, how much does a case weigh?

62. AUTO A car used 1 qt 1 pt of oil each month for 5 months. Find the total amount of oil used.

Multiply. *See Example 13.*

63. 5 in. × 7 in.

64. 12 ft × 9 ft

65. 15 yd × 12 yd

66. 4 mi × 27 mi

67. CON A room is to be covered with square linoleum tiles that are 1 ft by 1 ft. If the room is 18 ft by 21 ft, how many tiles (square feet) are needed?

68. AG/H A horticulturist stores a stock solution of fertilizer in two tanks, each with a capacity of 23 gal 9 oz. How much liquid fertilizer is needed to fill both tanks?

69. ELEC Latonya has three containers, each containing 1 qt 3 pt of photographic solution. How much total photographic solution is in all three containers?

70. AG/H A package of grass seed weighs 1 lb 4 oz. How much would five packages weigh?

Divide. *See Example 14.*

See Example 15.

71. 20 yd 2 ft 6 in. ÷ 2

72. 3 days 6 h ÷ 2

73. 4 yd 1 ft 9 in. ÷ 3

74. HELPP Sixty feet of crime-scene tape are required to complete eight jobs. If each job requires an equal amount of tape, find the amount of tape required for one job. (Express your answer in feet and inches.)

75. AG/H A vat holding 10 gal 2 qt of defoliant is emptied equally into three tanks. How many gallons and quarts are in each tank?

76. CON How many pieces of $\frac{1}{2}$-in. OD (outside diameter) plastic pipe 8 in. long can be cut from a piece 72 in. long?

77. ELEC A roll of No. 14 electrical cable 150 ft long is divided into 30 equal sections. How long is each section?

78. AG/H A greenhouse attendant has a container with 6 gal 2 qt 10 oz of potassium nitrate solution that will be divided equally into two smaller containers. How much solution will be stored in each smaller container?

79. HOSP For a catering order, Mr. Sonnier prepared 96 lb 12 oz of boiled crawfish. He brought the crawfish to the site in four containers containing equal amounts. How much did the crawfish in each container weigh?

Divide. *See Example 16.*

80. 36 ft ÷ 12 ft

81. 6 lb 12 oz ÷ 6 oz

82. 2 ft 6 in. ÷ 10 in.

83. INDTR How many 6-in. pieces can be cut from 48 in. of pipe?

84. HOSP How many 2-lb boxes can be filled from 18 lb of candy?

85. HOSP How many 15-oz cans are in a case if the case weighs 22 lb 8 oz?

86. BUS How many $8 tickets can be purchased for $72?

5 Change to the indicated rate measure. *See Example 17.*

87. $\dfrac{45\ \text{lb}}{\text{h}} = \underline{\hphantom{xx}}\dfrac{\text{lb}}{\text{min}}$

88. $\dfrac{3\ \text{mi}}{\text{h}} = \underline{\hphantom{xx}}\dfrac{\text{ft}}{\text{h}}$

89. $\dfrac{144\ \text{lb}}{\text{min}} = \underline{\hphantom{xx}}\dfrac{\text{oz}}{\text{min}}$

See Example 18.

90. $\dfrac{30\ \text{gal}}{\text{min}} = \underline{\hphantom{xx}}\dfrac{\text{qt}}{\text{s}}$

91. INDTEC A pump that can pump $45\frac{\text{gal}}{\text{h}}$ can pump how many quarts per minute?

92. INDTEC A pump can dispose of sludge at the rate of $3,200\frac{\text{lb}}{\text{h}}$. How many pounds can be disposed of per minute? Round to the nearest tenth.

4–2   Introduction to the Metric System

LEARNING OUTCOMES

1 Identify uses of metric measures of length, mass, weight, and capacity.

2 Convert from one metric unit of measure to another.

3 Make calculations with metric measures.

LC LEARNING CATALYTICS

1. Mentally multiply 51.3(100).
2. Mentally multiply 3.78(0.001)

Metric system: an international system of measurement that uses standard units and power-of-10 prefixes to indicate other units of measure

International System of Units (SI): an international system of measurement that uses standard units and power-of-10 prefixes to indicate other units of measure; also called the metric system

Standard unit: the unit that assumes the position of the ones place on a place-value chart and is the word to which the prefixes attach

Meter: the standard unit for measuring length or distance in the metric system (also spelled metre)

Gram: the standard unit for measuring mass or weight in the metric system

Liter: the standard metric unit of capacity or volume (also spelled litre)

Base unit: the unit of measure that is used most often in practice; often is the standard unit

Deci-: the metric prefix used for a unit that is 1/10 of the standard unit

Centi-: the metric prefix used for a unit that is 1/100 of the standard unit

Milli-: the prefix used for a metric unit that is 1/1000 of the standard unit

Deka-: the metric prefix used for a unit that is 10 times the standard unit

Hecto-: the prefix used for a metric unit that is 100 times the standard unit

Kilo-: the metric prefix used for a unit that is 1,000 times the standard unit

The **metric system** is an international system of measurement that uses standard units and power-of-10 prefixes to indicate other units of measure.

In the metric system, or the **International System of Units (SI),** a *standard unit* represents each type of measurement. A **standard unit** is the unit that assumes the position of the ones place on the place-value chart and is the word to which the prefixes attach. The **meter** is used for length or distance, the **gram** is used for mass or weight, and the **liter** is used for capacity or volume. A prefix is affixed to the standard unit to indicate a measure greater than the standard unit or less than the standard unit.

In most cases the standard unit is also the base unit for the type of measure. The **base unit** is the unit that is used most often in practice.

An instance when the base unit is not the standard unit is the kilogram for the base unit for mass or weight.

1 Identify Uses of Metric Measures of Length, Mass, Weight, and Capacity. The most common prefixes for units *smaller* than the standard unit are

$$\text{deci-}\frac{1}{10}\text{ of}\qquad \text{centi-}\frac{1}{100}\text{ of}\qquad \text{milli-}\frac{1}{1,000}\text{ of}$$

The most common prefixes for units *larger* than the standard unit are

deka- 10 times   **hecto-** 100 times   **kilo-** 1,000 times

We can compare the decimal place-value chart with the prefixes (Fig. 4–2). The standard unit (whether meter, gram, or liter) corresponds to the *ones* place. All the places to the left are multiples of the standard unit. That is, the value of *deka-* (some sources use *deca-*) is 10 times the standard unit; the value of *hecto-* is 100 times the standard unit; the value of *kilo-* is 1,000 times the standard unit; and so on. All the places to the right of the standard unit are subdivisions of the standard unit. That is, the value of *deci-* is $\frac{1}{10}$ of the standard unit; the value of *centi-* is $\frac{1}{100}$ of the standard unit; the value of *milli-* is $\frac{1}{1,000}$ of the standard unit; and so on.

Kilo- Thousands (1,000)	Hecto- Hundreds (100)	Deka- Tens (10)	STANDARD UNIT Units or ones (1)	Deci- Tenths ($\frac{1}{10}$)	Centi- Hundredths ($\frac{1}{100}$)	Milli- Thousandths ($\frac{1}{1,000}$)

• Decimal point

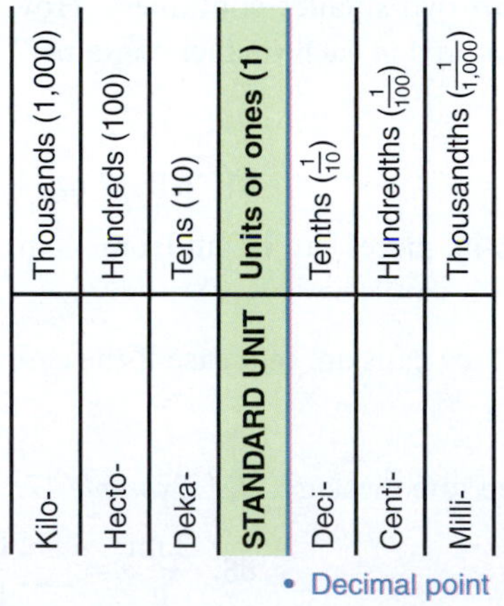

FIGURE 4–2 Place-Value Metric-Value Comparison.

EXAMPLE 1

Give the value of the metric unit using the standard unit (gram, liter, or meter).

(a) Kilogram (kg) = 1,000 times 1 gram or **1,000 g**

(b) Deciliter (dL) = $\frac{1}{10}$ of 1 liter or **0.1 L**

(c) Hectometer (hm) = 100 times 1 meter or **100 m**

(d) Dekaliter (dkL) = 10 times 1 liter or **10 L**

(e) Milliliter (mL) = $\frac{1}{1,000}$ of 1 liter or **0.001 L**

(f) Centigram (cg) = $\frac{1}{100}$ of 1 gram or **0.01 g**

See Exercise 1.

Other metric prefixes are used for very large and very small amounts. For measurements smaller than one-thousandth of a unit or larger than one thousand times a unit, prefixes that align with periods on a place-value chart are commonly used.

Powers of 10 for decimal places are written with negative exponents. The relationship between the exponent and the value of the number is discussed more completely in Chapter 5.

Metric Prefixes

Prefix	Relationship to Standard Unit*
atto-(a)	quintillionth part ($\times$ 0.000000000000000001 or 10^{-18})
femto-(f)	quadrillionth part ($\times$ 0.000000000000001 or 10^{-15})
pico-(p)	trillionth part ($\times$ 0.000000000001 or 10^{-12})
nano-(n)	billionth of ($\times$ 0.000000001 or 10^{-9})
micro-(µ)	millionth of ($\times$ 0.000001 or 10^{-6})
milli-(m)	thousandth of ($\times$ 0.001 or 10^{-3})
centi-(c)	hundredth of ($\times$ 0.01 or 10^{-2})
deci-(d)	tenth of ($\times$ 0.1 or 10^{-1})
deka-/deca-(dk)	ten times ($\times$ 10 or 10^{1})
hecto-(h)	hundred times ($\times$ 100 or 10^{2})
kilo-(k)	thousand times ($\times$ 1,000 or 10^{3})
mega-(M)	million times ($\times$ 1,000,000 or 10^{6})
giga-(G)	billion times ($\times$ 1,000,000,000 or 10^{9})
tera-(T)	trillion times ($\times$ 1,000,000,000,000 or 10^{12})
peta-(P)	quadrillion times ($\times$ 1,000,000,000,000,000 or 10^{15})
exa-(E)	quintillion times ($\times$ 1,000,000,000,000,000,000 or 10^{18})

*See Chapter 5, Section 5 for powers of 10.

All metric units of length, weight, and volume are expressed either as a standard unit or as a standard unit with a prefix. We examine some common metric units to develop an intuitive sense of their size.

Length

Meter: The *meter* is the standard unit for measuring length. Both *meter* and *metre* are acceptable spellings for this unit of measure. A meter is about 39.37 in. It is 3.37 in. longer than a yard (Fig. 4–3). We use the meter to measure lengths and distances like room dimensions, land dimensions, lengths of poles, heights of mountains, and heights of buildings. The abbreviation for meter is m.

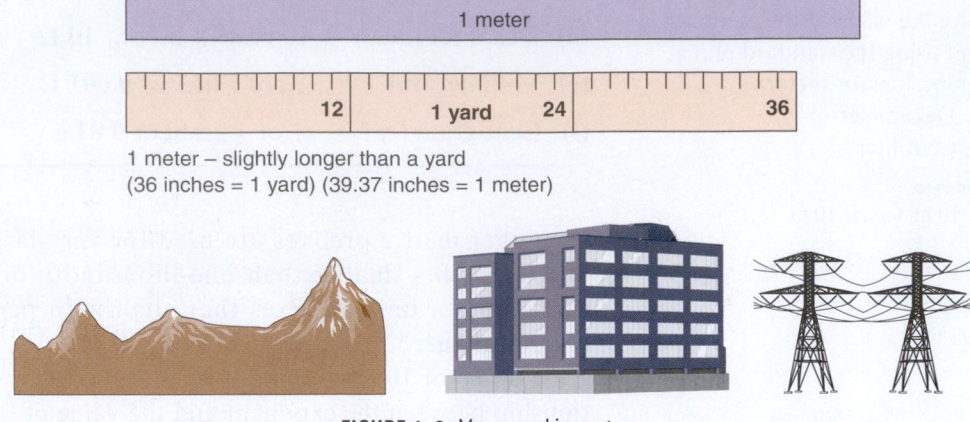

FIGURE 4–3 Measured in meters.

Kilometer: the unit used most often for measuring longer distances in the metric system. It is 1,000 meters

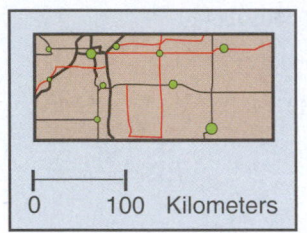

FIGURE 4–4

Centimeter: the most common metric unit used to measure objects less than 1 meter long. It is $\frac{1}{100}$ of a meter

Kilometer: A **kilometer** is 1,000 m and is used for longer distances. The abbreviation for kilometer is km. The prefix *kilo* means 1,000. We measure the distance from one city to another, one country to another, or one landmark to another in kilometers (Fig. 4–4). Driving at a speed of 55 mi (or 90 km) per hour, we would travel 1 km in about 40 seconds (s). An average walking speed is 1 km (approximately 5 city blocks) in about 10 min (Fig. 4–5).

FIGURE 4–5 Measured in kilometers.

Centimeter: To measure objects less than 1 m long, we commonly use the **centimeter** (cm). The prefix *centi* means "$\frac{1}{100}$ of," and a centimeter is one-hundredth of a meter. A centimeter is about the width of a thumbtack head, somewhat less than $\frac{1}{2}$ in. (Fig. 4–6). We use centimeters to measure medium-sized objects such as tires, clothing, textbooks, and television screens.

1 centimeter—about the width of a thumbtack or large paper clip

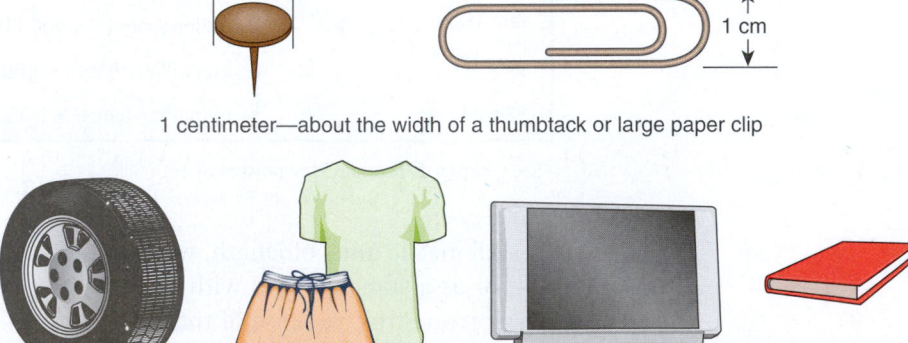

FIGURE 4–6 Measured in centimeters.

Millimeter: the most common metric unit used to measure objects less than 1 cm long. It is $\frac{1}{1,000}$ of a meter

Millimeter: Many objects are too small to be measured in centimeters, so we use a **millimeter** (mm), which is "$\frac{1}{1,000}$ of" a meter. It is about the thickness of a plastic credit card or a dime (Fig. 4–7). Bolts and nuts, insects, and similar items are measured in millimeters.

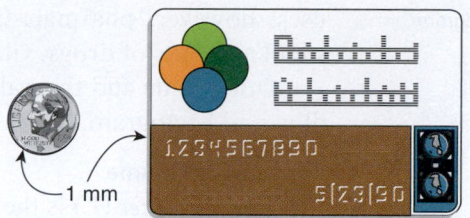

1 millimeter—about the thickness
of a dime or a credit card

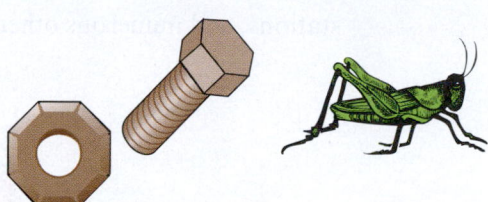

FIGURE 4-7 Measured in millimeters.

Other units of length and their abbreviations are decimeter, dm; dekameter, dkm; and hectometer, hm.

Weight or Mass

Mass and weight are often used interchangeably, but in technical or scientific terms they are different. The weight of an object is a measure of the earth's gravitational pull on the object. As an illustration, an object may have a specific weight and mass on Earth. As an object moves away from Earth, out into space, the mass remains constant (the same), whereas the weight decreases. When the object is in a "weightless" state, it floats freely in space.

Gram: A metric unit for measuring mass in the metric system is the *gram*. A gram is the mass of 1 cubic centimeter (cm^3) of water at its maximum density. The metric unit for measuring weight is the Newton (N); however, in common usage the gram is used in comparing metric and U.S. customary units of weight. A cubic centimeter is a cube whose edges are each 1 centimeter long. It is a little smaller than a sugar cube. The abbreviation for gram is g. We use grams to measure small or light objects such as paper clips, cubes of sugar, coins, and bars of soap (Fig. 4–8).

Kilogram: the most often used unit for measuring mass and weight in the metric system. It is 1,000 grams

Milligram: a metric unit of measure used to measure the weight of very small objects; 1/1000 of a gram

1 gram—about the weight of two paper clips

FIGURE 4-8 Measured in grams.

Kilogram: A **kilogram** (kg) is 1,000 grams. Since a cube 10 cm on each edge will be 1,000 cm^3, the weight of water required to fill this cube is 1,000 grams or 1 kilogram. A kilogram is approximately 2.2 lb. The kilogram is used to measure the weight of people, books, meat, grain, automobiles, and so on (Fig. 4–9). The kilogram is probably the most commonly used metric measure of mass. It is sometimes referred to as the base metric unit for mass.

Milligram: The gram is used to measure small objects, and the **milligram** ($\frac{1}{1,000}$ of a gram) is used to measure *very* small objects. Milligrams are too small for ordinary

Hong Vo/Shutterstock

FIGURE 4-9

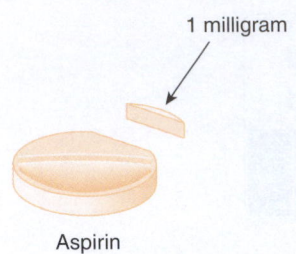

1 milligram

Aspirin

FIGURE 4–10

uses; however, pharmacists and manufacturers use milligrams (mg) to measure small amounts of drugs, vitamins, and medications (Fig. 4–10).

Other units and their abbreviations are decigram, dg; centigram, cg; dekagram, dkg; and hectogram, hg.

Capacity or Volume

Liter: A *liter* (L) is the volume of a cube 10 cm on each edge. It is the standard metric unit of capacity. Like the meter, it may be spelled *liter* or *litre;* we use the spelling *liter*. A cube 10 cm on each edge filled with water weighs approximately 1 kg, so 1 L of water weighs about 1 kg. One liter is just a little larger than a liquid quart. Soft drinks are sold in both 1-L and 2-L bottles, gasoline is sold by the liter at some service stations, and numerous other products are sold in liter containers (Fig. 4–11).

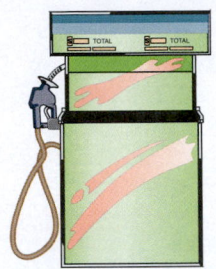

1 liter—the volume of a large, individual sized soft drink

FIGURE 4–11

Milliliter: a metric unit of measure used to measure small capacities or volume; 1/1,000 of a liter

Milliliter: A liter is 1,000 cm^3, so $\frac{1}{1,000}$ of a liter, or a **milliliter,** has the same volume as a cubic centimeter. Most liquid medicine is labeled and sold in milliliters (mL) or cubic centimeters (cc or cm^3). Medicines, perfumes, and other very small quantities are measured in milliliters (Fig. 4–12).

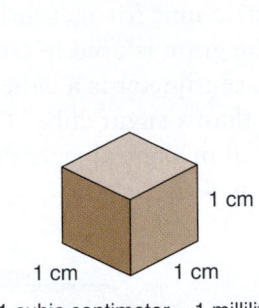
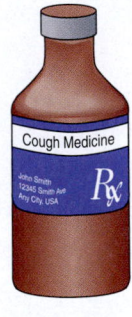

Cough Medicine

1 cm

1 cm   1 cm

1 cubic centimeter = 1 milliliter

FIGURE 4–12 Measured in mililiters.

STOP AND CHECK
Choose the most reasonable metric measure.
1. Length of a pencil
  a. kilometer   b. meter
  c. centimeter   d. millimeter
2. Weight of a candy bar
  a. kilogram   b. gram
  c. centigram   d. milligram
Answers:
1. c. centigram   2. b. gram

Other units and their abbreviations are deciliter, dL; centiliter, cL; dekaliter, dkL; hectoliter, hL; and kiloliter, kL.

Other standard units in the metric system exist, but those discussed here are the ones most commonly used.

EXAMPLE 2

Choose the most reasonable metric measure.

1. Distance from Jackson, Mississippi, to New Orleans, Louisiana

  (a) 322 m   **(b)** 322 km   **(c)** 322 cm   **(d)** 322 mm

2. Weight of an adult woman

(a) 56 g **(b)** 56 mg **(c)** 56 kg **(d)** 56 dkg

3. Bottle of eyedrops

(a) 30 dL **(b)** 30 dkL **(c)** 30 L **(d)** 30 mL

4. Weight of an aspirin

(a) 352 mg **(b)** 352 dg **(c)** 352 g **(d)** 352 kg

1. (b) 322 km (about 200 mi)
2. (c) 56 kg (about 120 lb)
3. (d) 30 mL (about 1 oz)
4. (a) 352 mg (one regular-strength aspirin)

See Exercises 2–16.

2 Convert from One Metric Unit of Measure to Another. To understand how we change one metric unit to another, let's arrange the units into a place-value chart like the one used for decimals (Fig. 4–13). The units are arranged from left to right and from largest to smallest.

Kilo-	Hecto-	Deka-	STANDARD UNIT	Deci-	Centi-	Milli-
1,000 × standard units	100 × standard units	10 × standard units	**meter, gram, liter**	$\frac{1}{10}$ × standard unit	$\frac{1}{100}$ × standard unit	$\frac{1}{1,000}$ × standard unit

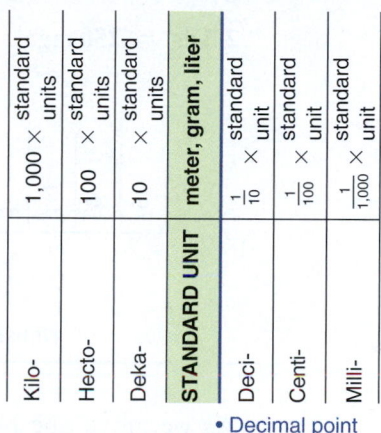

• Decimal point

FIGURE 4–13 Metric-value chart.

1 meter

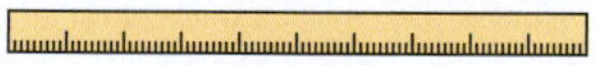

10 decimeters

FIGURE 4–14

As we move from any place on the chart one place to the *right*, the metric unit changes to the next smaller unit. The larger unit is broken down into 10 smaller units, so we are *multiplying* the larger unit by 10 when we move one place to the right (Fig. 4–14).

To change from one metric unit to *any smaller* metric unit:

1. Mentally position the measure on the metric-value chart so that the decimal point immediately follows the original measuring unit.
2. Move the decimal point to the right so that it *follows* the new measuring unit. (Attach zeros if necessary.)

STOP AND CHECK

Find the missing measures.

1. 22 hg = _______ g

2. 3.15 cL = _______ mL

Answers:
1. 2,200 g **2.** 31.5 mL

EXAMPLE 3

Find the missing measures. **(a)** 43 dkm = _______ cm? **(b)** 2.5 dg = _______ mg? (Note the decimal location.)

(a) Place 43 dkm on the metric-value chart so that the last digit is in the dekameters place; that is, the understood decimal that follows the 3 will be *after* the dekameters place (see Fig. 4–15). To change to centimeters, move the decimal point so that it follows the centimeters place. Fill in the empty places with zeros (see Fig. 4–16).

Note the shortcut to multiplication by powers of 10: Move the decimal point one place to the right for each time 10 is used as a factor.

43 dkm = 43,000 cm.

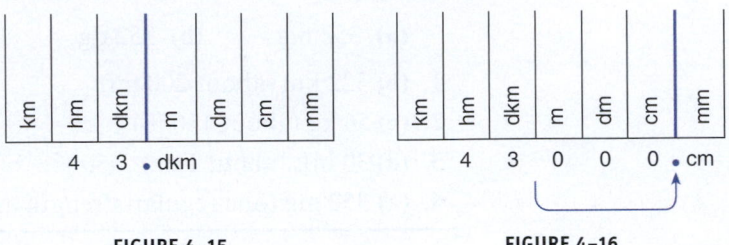

FIGURE 4–15   FIGURE 4–16

(b) Place the number 2.5 on the chart so that the decimal point follows the decigrams place (see Fig. 4–17). (Note the decimal after the decigrams place.) To change to milligrams, shift the decimal two places to the right so that it follows the milligrams place (see Fig. 4–18). (Note the decimal after the milligrams place.)

2.5 dg = 250 mg.

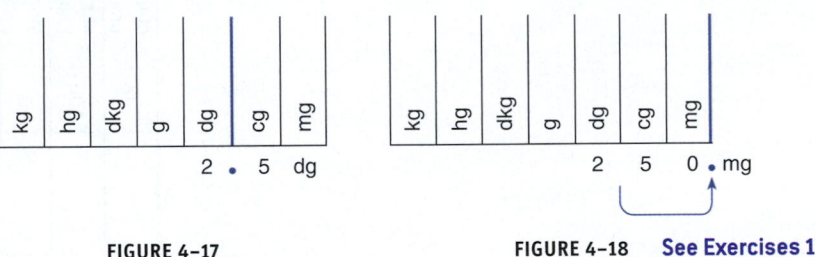

FIGURE 4–17   FIGURE 4–18   **See Exercises 17–42.**

As we move one place to the *left* on the metric-value chart, the metric unit changes to the next larger unit. The smaller units are combined into one larger unit 10 times larger than each smaller unit, so we are *dividing* the smaller unit by 10 when we move one place to the left.

To change from one metric unit to *any larger* metric unit:

1. Mentally position the measure on the metric-value chart so that the decimal point immediately follows the original unit.

2. Move the decimal point to the left so that it *follows* the new unit. (Attach zeros if necessary.)

STOP AND CHECK

Find the indicated measures.
1. 425 mg = _____ dkg
2. 8,223 m = _____ km

Answers:
1. 0.0425 dkg   2. 8.223 km

EXAMPLE 4

Find the indicated measures. **(a)** 3,495 L = _____ kL?
(b) 2.78 cm = _____ dkm?

(a) Place the number on the chart so the digit 5 is in the liters place; that is, the decimal point follows the liters place (see Fig. 4–19). (Note the understood decimal point after the liters place.) To change to kiloliters, move the decimal *three* places to the left so the decimal follows the kiloliters place (see Fig. 4–20).

Thus, 3,495 L = 3.495 kL.

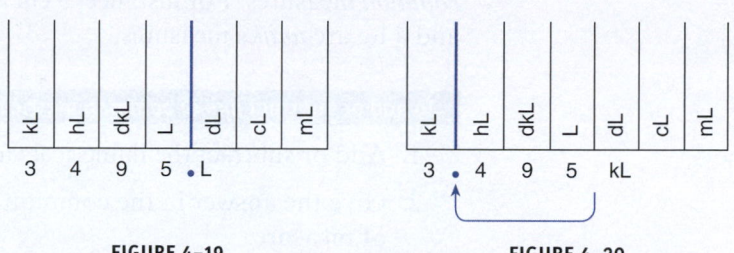

FIGURE 4–19   **FIGURE 4–20**

(b) Place the number on the chart so that the decimal follows the centimeters place (see Fig. 4–21). (Note the decimal position after the centimeters place.) Move the decimal three places to the left so that it follows the dekameters place (see Fig. 4–22). (Note the decimal position after the dekameters place.)

Therefore, 2.78 cm = 0.00278 dkm.

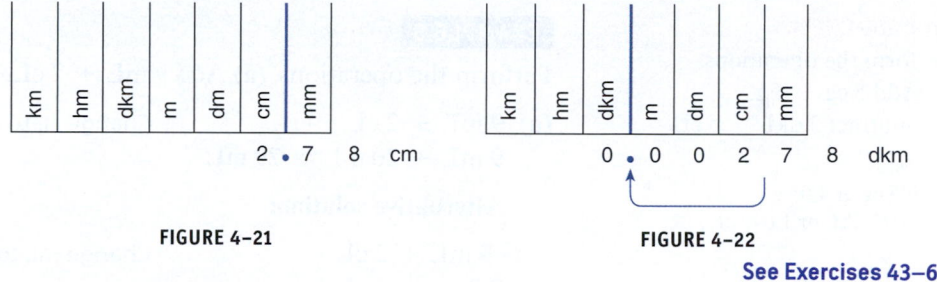

FIGURE 4–21   **FIGURE 4–22**

See Exercises 43–64.

TIP   **How Far and Which Way?** To determine the movement of the decimal point when changing from one metric unit to another, answer these questions:

1. *How far* is it from the original unit to the new unit (how many places)?

2. *Which way* is the movement on the chart (left or right)?

   Change 28.392 cm to m.

How far is it from cm to m (Fig. 4–23)? **Two places**
Which way? **Left**

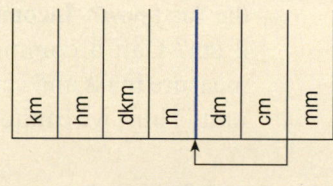

FIGURE 4–23

Move the decimal in the original measure *two places* to the *left*.

$$28.392 \text{ cm} = 0.28392 \text{ m}$$

Conversion factors can also be used to convert from one metric measure to another. Some of the most commonly used conversion factors are given on the inside front cover of this text.

3 **Make Calculations with Metric Measures.** We can add or subtract only *like* or *common* measures. For instance, 5 cm and 3 cm are *like* measures. In contrast, 17 kg and 4 hg are *unlike* measures.

> **To add or subtract *like* or *common* metric measures:**
>
> 1. Add or subtract the numerical values. $7 \text{ cm} + 4 \text{ cm} = 11 \text{ cm}$
>
> 2. Give the answer in the common unit $15 \text{ dkg} - 3 \text{ dkg} = 12 \text{ dkg}$
>   of measure.

> **To add or subtract *unlike* metric measures:**
>
> 1. Change the measures to a common unit of measure.
>
> 2. Add or subtract the numerical values.
>
> 3. Give the answer the common unit of measure.

STOP AND CHECK
Perform the operations.
1. Add 5 cg + 4 g.
2. Subtract 2 dkL − 5 cL.

Answers:
1. 405 cg or 4.05 g
2. 1.095 dkL or 1,095 cL

EXAMPLE 5

Perform the operations. **(a)** Add 9 mL + 2 cL. **(b)** Subtract: 14 km − 34 hm.

(a) 9 mL + 2 cL   Change cL to mL; that is, 2 cL = 20 mL.
    9 mL + 20 mL = **29 mL**

Alternative solution:

  9 mL + 2 cL   Change mL to cL. 9 mL = 0.9 cL.
  0.9 cL + 2 cL =

    0.9 cL   Note alignment of decimals.
    2   cL   An understood decimal follows the addend 2.
    2.9 cL   2.9 cL = 29 mL

The sum is 29 mL or 2.9 cL.

(b) 14.0 km   Change hm to km; that is, 34 hm = 3.4 km.
    −3.4 km   Caution: Notice alignment of decimals.
    10.6 km

The difference is 10.6 km, or 106 hm.   See Exercises 65–76.

Did You Know? **Incompatible measures cannot be combined!** Can 5 g be added to 2 cm? Can a common unit be found for centimeters and grams? No, grams measure mass and centimeters measure length (Fig. 4–24), so there is no common unit. Thus, we cannot add 5 g + 2 cm.

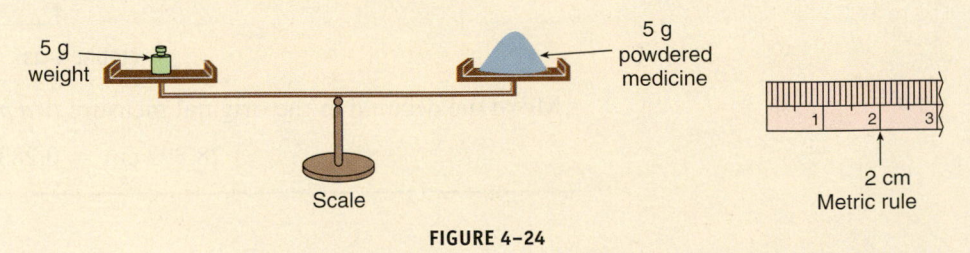

FIGURE 4-24

Metric measures can be multiplied by a number. If we want to find the total length of three pieces of landscape timber that are each 2.7 m long, we multiply 2.7 m by 3.

To multiply a metric measure by a number:	
1. Multiply the numerical values.	2.7 m(3)
2. Give the answer the same unit as the original measure.	= 8.1 m

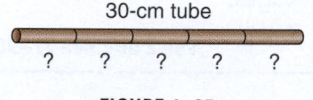

30-cm tube

? ? ? ? ?

FIGURE 4-25

Suppose we have a tube that is 30 cm long and we need it cut into five equal parts (Fig. 4–25). How long is each part? To solve this problem, we divide 30 cm by 5.

$$\frac{30 \text{ cm}}{5} = 6 \text{ cm}$$

To divide a metric measure by a number:	
1. Divide the numerical values.	30 cm ÷ 5
2. Give the quotient the same unit as the original measure.	= 6 cm

We can divide all measures by a number because we are dividing a quantity into parts.

In some situations, we need to divide a measure by a measure. Suppose we have to administer a 250-mg dose of ascorbic acid in 100-mg tablets. How many tablets are needed per dose?

To divide a metric measure by a like measure:	
1. If unlike measures are used, first change them to like measures.	$\dfrac{250 \cancel{\text{ mg}}}{100 \cancel{\text{ mg}}}$
2. Divide the numerical values.	
3. Give the quotient as a number that tells how many parts there are.	= 2.5 tablets

The answer, 2.5, is labeled tablets rather than mg because we are looking for *how many* tablets are needed.

STOP AND CHECK

1. What is the total weight of 6 smart phones weighing 218 g each and a shipping carton of 153 g?
2. A 3-kg package of trail mix is to be repackaged into 40-g packages. How many packages will there be?

Answers:
1. 1,461 g or 1.461 kg
2. 75 packages of trail mix

EXAMPLE 6

Solve.

(a) Four micrometers weighing 752 g each will fit into a shipping carton. If the shipping carton and filler weigh 217 g, what is the total weight of the shipment?

(b) A 5-m-long board is to be cut into four equal parts. How long is each part?

(c) How many 25-g packages of seed can be made from 2 kg of seed?

(a) Total weight = carton and filler + 4 micrometers
= 217 g + (4 · 752 g)   Multiply.
= 217 g + 3,008 g   Add.
= 3,225 g, or 3.225 kg

The total weight is 3,225 g, or 3.225 kg.

(b) $\dfrac{5\,\text{m}}{4} = 1.25\,\text{m}$ Divide length in meters by number of parts. Result is expressed in meters.

Each part is 1.25 m long.

(c) $\dfrac{2\,\text{kg}}{25\,\text{g}} = \dfrac{2{,}000\,\cancel{\text{g}}}{25\,\cancel{\text{g}}}$ Convert kg to g; then divide. Reduce units.

$= 80$ Result expresses the number of packages.

80 packages of seed can be made.   **See Exercises 77–90.**

4–2 EXERCISES   MyLab Math   For additional practice go to your study plan in MyLab Math.

1 Give the value of the metric units in standard units. *See Example 1.*

1. (a) kilometer (km) **(b)** dekaliter (dkL) **(c)** decigram (dg)
   (d) millimeter (mm) **(e)** hectogram (hg) **(f)** centiliter (cL)

Choose the most reasonable metric measure. *See Example 2.*

2. Height of a 6-year-old pediatric patient
   (a) 1.5 km   **(b)** 1.5 m
   (c) 1.5 cm   **(d)** 1.5 mm

3. Diameter of a dime
   (a) 1.5 m   **(b)** 1.5 cm
   (c) 1.5 mm   **(d)** 1.5 km

4. Distance from Los Angeles to San Francisco
   (a) 800 m   **(b)** 800 mm
   (c) 800 km   **(d)** 800 cm

5. Length of a pencil
   (a) 20 km   **(b)** 20 cm
   (c) 20 m   **(d)** 20 mm

6. Overnight accumulation of snowfall
   (a) 8 cm   **(b)** 8 m
   (c) 8 km   **(d)** 8 dkm

7. Width of a DVD case
   (a) 12 m   **(b)** 12 km
   (c) 12 cm   **(d)** 12 mm

8. Weight of the average male adult
   (a) 70 mg   **(b)** 70 kg
   (c) 70 g   **(d)** 7 dg

9. Weight of a can of tuna
   (a) 184 mg   **(b)** 184 kg
   (c) 184 g   **(d)** 184 cg

10. Weight of a teaspoonful of sugar
   (a) 2 g   **(b)** 2 kg
   (c) 2 mg   **(d)** 2 hg

11. Weight of a vitamin C tablet
   (a) 250 g   **(b)** 250 mg
   (c) 250 kg   **(d)** 250 dkg

12. Weight of a dinner plate
   (a) 350 kg   **(b)** 350 mg
   (c) 350 g   **(d)** 350 cg

13. Weight of a bag of dog food
   (a) 25 g   **(b)** 25 kg
   (c) 25 mg   **(d)** 25 dg

14. Volume of a tank of pesticide
   (a) 60 L   **(b)** 60 mL

15. Dose of cough syrup
   (a) 100 L   **(b)** 100 mL

16. Glass of juice
   (a) 0.12 mL   **(b)** 0.12 L

2 Change to the measure indicated. When using the metric-value chart, place the decimal immediately *after* the measuring unit. *See Example 3.*

17. 4 m = _____ dm

18. 7 g = _____ dg

19. 58 km = _____ hm

20. 8 hL = _____ dkL

21. 0.25 km = _____ hm

22. 21 dkL = _____ L

23. 8.5 cm = _____ mm

24. 14.2 dg = _____ cg

25. How many milliliters are there in 15.3 cL?

26. How many meters are there in 46 dkm?

27. How many dekagrams are there in 7.5 hg?

28. How many millimeters are there in 16 cm?

29. 4 L = _____ cL

30. 8 m = _____ mm

31. 58 km = _____ m

32. 8 hg = _____ g

33. 0.25 km = _____ m

34. 21 dkg = _____ dg

35. 10.25 hm = _____ cm

36. 8.33 L = _____ mL

37. 2 km = _____ mm

38. 0.7 g = _____ cg

39. Change 2.36 hL to liters.

40. Change 0.467 dkm to centimeters.

41. Change 3.8 kg to decigrams.

42. Change 13 dkm to centimeters.

See Example 4.

43. 28 m = _____ dkm

44. 238 hL = _____ kL

45. 101 mg = _____ cg

46. 60 hm = _____ km

47. 29 dkL = _____ hL

48. 192.5 g = _____ dkg

49. 17 cm = _____ dm

50. How many decimeters are in 4,389 cm?

51. How many grams are in 47 dg?

52. How many deciliters are in 2.25 cL?

53. 2,743 mm = _____ m

54. 385 g = _____ kg

55. 15 dkm = _____ km

56. 8 cL = _____ L

57. 296,484 m = _____ hm

58. 29.83 dg = _____ dkg

59. 0.3 cm = _____ dkm

60. 40 dL = _____ kL

61. 2,857 mg = _____ kg

62. 15,285 m = _____ km

63. Change 297 cm to hectometers.

64. Change 0.03 mL to liters.

3 Add or subtract as indicated. *See Example 5.*

65. 3 m + 8 m

66. 7 hL + 5 hL

67. 15 cg − 9 cg

68. 2 dm + 4 cm

69. 5 cL + 9 mL

70. 4 m + 2 L

71. 14 kL − 39 hL

72. 1 g − 45 cg

73. 3 cg − 5 mL

74. 7 km + 2 m

75. **HLTH/N** A patient absorbs 175 mL of fluid through an IV. If the IV bag has 825 mL left, how much fluid was in the bag to begin with?

76. **HLTH/N** 653 dkL of orange juice concentrate is removed from a vat containing 8 kL of the concentrate. How much concentrate remains in the vat?

Multiply. *See Example 6.*

77. 43 m(12)

78. 3.4 m(12)

79. 50.32 dm(3)

80. **CAD/ARC** A plot of ground is divided into seven plots, each with road frontage of 138.5 m. What is the total road frontage of the plot of ground?

81. **AVIA** Earth consists of a series of relatively thin plates that are in constant motion. These plates move at different velocities. The Australian plate moves about 60 mm per year. How many millimeters will the plate move in 78 years?

Divide. *See Example 6.*

82. 48 g ÷ 3

83. 39 m ÷ 3

84. $\dfrac{54 \text{ cL}}{6}$

85. **INDTEC** A block of silver weighing 978 g is cut into six equal pieces. How much does each piece weigh?

86. **INDTEC** Two pieces of steel each 12 m long are cut into a total of 60 equal pieces. How long is each piece?

87. 2.5 cg ÷ 0.5 cg

88. 3 m ÷ 10 cm

89. **HLTH/N** How many 250-mL prescriptions can be made from a container of 4 L of decongestant?

90. **AG/H** How many 500-g containers are needed to hold 40 kg of grass seed?

91. Add 4.6 cL + 5.28 dL of photographic developer.

92. Add 3 m + 2 dkm of fabric for draperies.

93. Subtract 19.8 km − 32.3 hm of paved highway.

94. Subtract 13 kL − 39 hL of stored liquid.

95. Multiply 0.25 cL of cologne by 5.

96. Multiply a 35-mm film size by 2.

97. **INDTEC** A length of satin fabric 30 dm long is cut into 4 equal pieces. How long is each piece?

98. **HLTH/N** An IV bag holding 250 mL of an antibiotic is calibrated (marked off) into five equal sections. How many milliliters are represented by each section?

99. **INDTEC** How many containers of jelly can be made from 8,500 L of jelly if each container holds 4 dL of jelly?

100. **HELPP** How many 2-kg vials of fire retardant (HC1) can be obtained from 38 kg of fire retardant?

4–3   Time, Temperature, and Other Measures

LEARNING OUTCOMES

1   Convert from one unit of time to another.

2   Make calculations with measures of time.

3   Convert between Fahrenheit temperatures and Celsius temperatures.

4   Examine other useful measures.

LC LEARNING CATALYTICS

1. If the following units of measures are used in a dimension analysis, what is the final unit of measure?

$$\text{qt}\left(\frac{\text{pt}}{\text{qt}}\right)\left(\frac{\text{c}}{\text{pt}}\right)\left(\frac{\text{oz}}{\text{c}}\right)$$

2. How many seconds are in an hour?

STOP AND CHECK

1. Convert 3 days to hours.

Answer:

1. 72 h

As we examined rate measures in Chapter 4, we introduced the units of time. We convert from one unit of time to another just as we convert other compatible units.

1 Convert from One Unit of Time to Another. The relationships among various units of time are given in Chapter 4, Section 1, Outcome 5. We can use unit ratios or conversion factors to convert from one unit of time to another.

EXAMPLE 1

Convert 3 h to minutes.

Estimation

Hours to minutes → larger to smaller → more minutes

Use a unit ratio to make the conversion:

$$3\text{ h}\left(\frac{\text{min}}{\text{h}}\right) \qquad \text{Use an appropriate unit ratio.}$$

$$3\text{ h}\left(\frac{60\text{ min}}{1\text{ h}}\right) = 3(60\text{ min}) = 180\text{ min}$$

Interpretation

There are 180 min in 3 h.

Or use a conversion factor to make the conversion:

1 h = 60 min   Conversion factor = 60.

3 h(60 min per hour) = 180 min   Multiply hours by 60 min.   **See Exercises 1–8.**

2 Make Calculations with Measures of Time. Suppose we have a 1-hour meeting and five items to be covered on our agenda. If we are to devote the same amount of time to each item, how much time should we give to each item?

$$1\text{ h} \div 5 = \frac{1}{5}\text{h}$$

How long is $\frac{1}{5}$ h? Is this the usual way to express an amount of time? No. When less than 1 h is involved, we often change to minutes.

$$\frac{1}{5}\text{ h}\left(\frac{\overset{12}{60}\text{ min}}{1\text{ h}}\right) = 12\text{ min}$$

EXAMPLE 2

BUS   A technician can assemble one precut picture frame in 7 min. How long will it take to assemble 25 frames?

Estimation

If the frame could be assembled in 6 min, then the technician could assemble 10 per hour (6 min × 10 = 60 min = 1 h). Thus, it would take 2 h to assemble 20 frames and $2\frac{1}{2}$ h to assemble 25 frames. Since it actually takes 7 min per frame, it will take more than $2\frac{1}{2}$ h.

$$\frac{7 \text{ min}}{\text{frame}}(25 \text{ frames}) = 175 \text{ min}$$

$$175 \text{ min}\left(\frac{1 \text{ h}}{60 \text{ min}}\right) = 2.91\overline{6} \text{ h} \qquad \text{Convert min to h using a unit ratio.}$$

If we want to know how many hours and minutes, 2 h = 120 min:

175 min − 120 min = 55 min

Alternative method for converting a part of an hour to minutes:

$$0.91\overline{6} \text{ h}\left(\frac{60 \text{ min}}{1 \text{ h}}\right) = 55 \text{ min} \qquad \text{To calculate minutes from the decimal part of an hour, use the decimal portion of } 2.91\overline{6} \text{ h.}$$

Interpretation

It will take 2 h 55 min to assemble 25 frames. **See Exercises 9–18.**

---

3 **Convert Between Fahrenheit Temperatures and Celsius Temperatures.** The **Kelvin** scale is one scale used to measure temperature in the metric system of measurement. Units on the scale are abbreviated with a capital *K* (without the symbol ° because these units are called *kelvins*) and are measured from absolute zero, the temperature at which *all* heat is said to be removed from matter. Another metric temperature scale is the **Celsius** scale (abbreviated °C), which has as its zero the freezing point of water. The Kelvin and Celsius scales are related such that absolute zero on the Kelvin scale is the same as −273°C on the Celsius scale. Each unit of change on the Kelvin scale is equal to 1 degree of change on the Celsius scale; that is, the size of a kelvin and a Celsius degree is the same on both scales.

The U.S. customary system temperature scale that starts at absolute zero is called the **Rankine** scale. It is related to the more familiar **Fahrenheit** scale, which places the freezing point of water at 32°. One degree of change on the Rankine scale equals 1 degree of change on the Fahrenheit scale. Absolute zero (the zero for the Rankine scale) corresponds to 460 degrees *below* zero (−460°) on the Fahrenheit scale. Water freezes at 492 degrees and water boils at 672 degrees on the Rankine scale.

The Celsius and Fahrenheit scales are the most common temperature scales used for reporting air and body temperatures. The formulas for converting temperatures using these two scales are more complicated than the previous ones because 1 degree of change on the Celsius scale does *not* equal 1 degree of change on the Fahrenheit scale (Fig. 4–26).

> **To convert degrees Fahrenheit to degrees Celsius:**
>
> Use the formula: $°C = \frac{5}{9}(°F − 32)$, where °C = degrees Celsius and °F = degrees Fahrenheit.

Using the formula we can verify that 212°F (the boiling point of water) is equivalent to 100°C.

$$°C = \frac{5}{9}(°F − 32) \qquad \text{Substitute 212 for °F in the formula.}$$

$$°C = \frac{5}{9}(212 − 32) \qquad \text{Work within grouping; subtract } 212 − 32.$$

$$°C = \frac{5}{9}(180) \qquad \text{Multiply. } \frac{5}{\underset{1}{9}}\left(\frac{\overset{20}{180}}{1}\right) = 100.$$

$$°C = 100$$

212°F = 100°C.

Sirtravelalot/Shutterstock

Kelvin: a scale to measure temperature in the metric system in which units are measured from absolute zero

Celsius: a scale to measure temperature in the metric system in which units are measured from the freezing point of water

Rankine: a scale to measure temperature in the U.S. customary system in which units are measured from absolute zero

Fahrenheit: a scale to measure temperature in the U.S. customary system in which units are measured from the freezing point of water at 32° and 460° above absolute zero on the Rankine scale

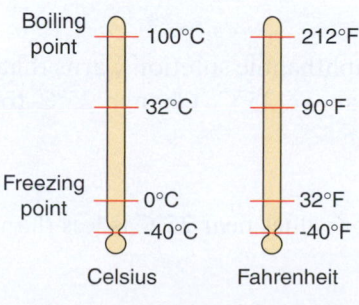

FIGURE 4–26

STOP AND CHECK

1. Change 86°F to degrees Celsius.

Answer:

1. 30°C

EXAMPLE 3

HLTH/N   According to Dr. Wimberly, an antifungal powder containing tolnaftate may be stored at a room temperature of 77°F. Change 77°F to degrees Celsius.

Estimation

The Celsius temperature is a smaller value than the Fahrenheit temperature for values near 77°F. From Fig. 4–26 we see that 90°F is approximately 32°C. Then 77°F will be less than 32°C.

$$°C = \frac{5}{9}(°F - 32)$$ Substitute 77 for °F.

$$°C = \frac{5}{9}(77 - 32)$$ Work grouping. $77 - 32 = 45$

$$°C = \frac{5}{9}(45)$$ Multiply. $\frac{5}{\overset{}{9}}\left(\frac{\overset{5}{45}}{1}\right)$

$$°C = 25$$

Interpretation

77°F = 25°C.   Based on the estimation, 25°C is reasonable.

See Exercises 19–30.

To convert degrees Celsius to degrees Fahrenheit:

Use the formula: $°F = \frac{9}{5}°C + 32$, where °F = degrees Fahrenheit and °C = degrees Celsius.

The formula can also be written as $°F = 1.8°C + 32$.

Now, use this formula to show that 100°C is equivalent to 212°F.

$$°F = \frac{9}{5}°C + 32$$ Substitute 100 for °C.

$$°F = \frac{9}{5}(100) + 32$$ Multiply first. $\frac{9}{\overset{}{5}}\left(\frac{\overset{20}{100}}{1}\right) = 180.$

$$°F = 180 + 32$$ Add.

$$°F = 212$$

100°C = 212°F.

STOP AND CHECK

1. Change 20°C to degrees Fahrenheit.

Answer:

1. 68°F

EXAMPLE 4

HLTH/N   The label on a dropper bottle of ofloxacin ophthalmic solution warns that the medicine must not be stored at a temperature above 25°C. Change 25°C to degrees Fahrenheit.

Estimation

In Fig. 4–26 we see that the Fahrenheit temperature for values near 25°C is less than 90°F.

$$°F = \frac{9}{5}°C + 32$$ Substitute 25 for °C.

$$°F = \frac{9}{5}(25) + 32$$ Multiply. $\frac{9}{\overset{}{5}}\left(\frac{\overset{5}{25}}{1}\right)$

$$°F = 45 + 32 \qquad \text{Add.}$$
$$°F = 77$$

Interpretation

$25°C = 77°F.$ Based on the estimate, 77°F is reasonable.

See Exercises 31–42.

See Exercises 31–42.

Base unit of measure: a unit of measure for a type of measure that has been adopted as the fundamental unit of a physical quantity.

4 **Examine Other Useful Measures.** All SI metric measures use the same system of prefixes for multiples and submultiples. There prefixes attach to a stem word that is called the standard unit; however, the **base unit** of a type of measure is the unit of measure that has been adopted as the fundamental unit of a physical quantity. In most cases, the base unit and the standard unit are the same. The exception is the kilogram for measuring mass. Table 4–6 summarizes the SI base units and some other commonly used metric units.

Table 4–6 SI Metric Units

Base Unit and Abbreviation	Standard Unit When Different from the Base Unit	Type of Measure
meter (m)		length
kilogram (kg)	gram (g)	mass
liter (L)		volume or capacity
second (s)		time
kelvin (K)		temperature
ampere (A)		electric current
candela (cd)		light intensity
mole (mol)		molecular substance
newton (N)		force
joule (J)		energy
watt (W)		power
square meter (m²)		area
meter per second (m/s)		speed
cubic meter (m³)		volume
volt (V)		voltage
ohm (Ω)		resistance
hertz (Hz)		frequency
farad (F)		capacitance
henry (H)		inductance
coulomb (C)		charge

Metric prefixes for very large and small units are given in Section 2, page 173.

STOP AND CHECK

1. 3 megatons is how many tons?

Answer:

1. 3,000,000 T or 3 million T

EXAMPLE 5

12 nanoseconds (ns) is what part of a second?

1 ns = 1 billionth of a second, or 0.000000001 s

12 ns = 12(0.000000001) = **0.000000012 s**

Using unit ratios:

$$12 \text{ ns}\left(\frac{0.000000001 \text{ s}}{1 \text{ ns}}\right) = \mathbf{0.000000012 \text{ s}}$$

See Exercises 43–44.

EXAMPLE 6

Give the meaning of the following units.

(a) 115 μA   (b) 3.2 MW

(a) μA means microamperes, microamps, or millionths of an amp.

115μA = **115 microamps, or 0.000115 A**

(b) MW means megawatts, or 1 million watts.

3.2 MW = **3.2 megawatts or 3,200,000 W**   See Exercises 45–48.

4-3 EXERCISES   MyLab Math For additional practice go to your study plan in MyLab Math.

1 Use ratios or conversion factors to convert the measures of time. Round to hundredths. *See Example 1.*

1. How many days are in 36 h?

2. Find the number of minutes in 580 s.

3. How many minutes are in 2.5 h?

4. A physical test took 5.3 min to perform. How many seconds is this?

5. If you studied for 3.5 h, how many minutes did you study?

6. A process takes 182 s to perform. How many minutes is this?

7. Convert 7.2 h to minutes.

8. Convert 88 s to minutes.

2 Make the calculations and write the answer as a rate. *See Example 2.*

9. **INDTEC**   A conveyor belt can move 72 lb of rice per minute. How many pounds can be moved on the conveyor belt in 1 h?

10. **AVIA**   Aircraft fuel moves through a pipeline at 40 gal per minute. How many gallons per hour can move through the pipeline?

11. **BUS**   A dog kennel uses 30 lb of dog food per day. How many pounds are used in a year?

12. **BUS**   A toll station can accommodate on average 28 vehicles per minute. How many vehicles can be accommodated in an hour?

13. **AUTO**   A car traveling at the rate of 60 mph (miles per hour) is traveling how many miles per second? Round to the nearest ten-thousandth.

14. **INDTR**   A pump that can pump 125 $\frac{\text{gal}}{\text{h}}$ can pump how many gallons per minute? Round to the nearest ten-thousandth.

15. **INDTR**   A pump can dispose of sludge at the rate of 3,600 $\frac{\text{lb}}{\text{h}}$. How many pounds can be disposed of per minute?

16. **HELPP**   If water flows through a fire hose at the rate of 96 $\frac{\text{gal}}{\text{min}}$, how many gallons will flow per second?

17. **AG/H**   Scientists estimate approximately 137 species of life forms become extinct every day. How many to the nearest thousand become extinct each year?

18. **AG/H**   Scientists estimate forests are being destroyed on Earth's surface at a rate of 149 acres per minute. Use unit ratios to find the number of acres to the nearest thousand being destroyed each day.

3 Change the Fahrenheit temperatures to Celsius. *See Example 3.*

19. 95°F   **20.** 32°F   **21.** 113°F   **22.** 41°F   **23.** 59°F

24. 50°F   **25.** 149°F   **26.** 122°F   **27.** 176°F   **28.** 248°F

29. **HLTH/N**   Normal skin temperature in cool weather is 90° to 93°F. What are the Celsius temperatures to the nearest tenth?

30. **HLTH/N**   Persons experiencing mild hypothermia have a core body temperature of 95° to 98.6°F. Express as Celsius temperatures.

Change the Celsius temperatures to Fahrenheit. *See Example 4.*

31. 70°C **32.** 15°C **33.** 45°C **34.** 50°C **35.** 20°C

36. 215°C **37.** 310°C **38.** 410°C **39.** 185°C **40.** 0°C

41. HLTH/N A patient with a core body temperature of 32.2°C is experiencing severe hypothermia. Express as a Fahrenheit temperature to the nearest whole degree.

42. HLTH/N A person who is wet, improperly dressed, and intoxicated with alcohol can become hypothermic when the air temperature is approximately 20°C. Express as a Fahrenheit temperature.

4 Use the table of metric prefixes on p. 173 and Table 4–6 on p. 187. *See Example 5.*

43. 45 ms is what part of a second?

44. 805 ns is what part of a second?

See Example 6.

45. TELE One hertz is a frequency of one cycle per second. Frequencies of radio and television waves are often measured in kilohertz or megahertz. How many hertz are in 500 MHz?

46. ELEC The henry is a large unit. Inductances in circuits are often measured in millihenrys (mH) or microhenrys (μH). How many henrys are in 420 mH?

47. ELEC The watt (W) is the unit used for measuring electrical power. An electrical device that used 1,400 kW has used how many watts?

48. HLTH/N The average human visual system requires about 50 ms to take in an image. How many different images can be processed per second before they blur?

4–4 Metric–U.S. Customary Comparisons

LEARNING OUTCOME

1 Convert between U.S. customary measures and metric measures.

LC LEARNING CATALYTICS
1. Which is larger, l m or 1 yd?
2. Which is larger, l lb or 1 kg?

Converting a measure from the U.S. customary system to the metric system, and vice versa, is often necessary because both systems are used in the United States. Most industries use one system exclusively, making conversions from one system to another uncommon. However, an industry that uses the U.S. customary system in the United States often needs to convert to the metric system to market its products in other countries.

1 **Convert Between U.S. Customary Measures and Metric Measures.** To convert measures from the U.S. customary system to the metric system, and vice versa, we need only one conversion relationship for each type of measure (length, weight, and capacity). The following equivalency relationships have been rounded to the nearest ten thousandth of a unit:

For length: 1 m = 1.0936 yd
For weight: 1 kg = 2.2046 lb
For capacity: 1 L = 1.0567 liquid qt (liquid measure)

Additional conversion relationships, such as centimeters to inches and kilometers to miles, can be derived from these relationships using unit ratios.

The most common procedure for converting between the U.S. customary and metric systems is to use conversion factors. Conversion factors always involve multiplication, whereas unit ratios may involve multiplication or division. Many conversion factors are given in Table 4–7. Calculators sometimes have conversion factors programmed internally for greater accuracy.

Table 4–7   U.S. Customary and Metric Conversion Factors

	From	To	Multiply By
Length			
1 meter = 39.37 inches	meters	inches	39.37
	inches	meters	0.0254
1 meter = 3.2808 feet	meters	feet	3.2808
	feet	meters	0.3048
1 meter = 1.0936 yards	meters	yards	1.0936
	yards	meters	0.9144
1 centimeter = 0.3937 inch	centimeters	inches	0.3937
	inches	centimeters	2.54
1 millimeter = 0.03937 inch	millimeters	inches	0.03937
	inches	millimeters	25.4
1 kilometer = 0.6214 mile	kilometers	miles	0.6214
	miles	kilometers	1.6093
Weight			
1 gram = 0.0353 ounce	grams	ounces	0.0353
	ounces	grams	28.3286
1 kilogram = 2.2046 pounds	kilograms	pounds	2.2046
	pounds	kilograms	0.4536
Liquid Capacity			
1 liter = 1.0567 quarts	liters	quarts	1.0567
	quarts	liters	0.9463

To convert between U.S. customary measures and metric measures:

1. Select the appropriate conversion factor for the measures involved.

2. Multiply the original measure by the conversion factor and use the appropriate new measuring unit to label the product.

STOP AND CHECK

1. How many inches are in 20 cm?

Answer:
1. 7.874 in.

EXAMPLE 1

Change 50 ft to meters.

feet to meters: Conversion factor is 0.3048.

$50(0.3048) = \mathbf{15.24\ m}$ Multiply.   See Exercises 1–31.

STOP AND CHECK

1. How many liters are in 5 gallons?

Answer:
1. 18.926 L

EXAMPLE 2

How many full cups are in a 2-L soft-drink bottle?

Either conversion factors or unit ratios or a combination of the two can be used.

$$\frac{2\ L}{1}\left(\frac{1.0567\ qt}{1\ L}\right)\left(\frac{2\ pt}{1\ qt}\right)\left(\frac{2\ c}{1\ pt}\right)$$ Set up unit ratios and multiply.

$= 2(1.0567)(2)(2)\ c = 8.4536\ c$ 8 full cups and a portion of a cup left.

There are 8 full cups of soft drink in a 2-L bottle.   See Exercises 32–39.

Body mass index (BMI): a standard unit for measuring a person's degree of obesity or emaciation

Body mass index (BMI) is the standard unit for measuring a person's degree of obesity or emaciation. BMI is body weight in kilograms (kg) divided by height in meters squared, or $BMI = \dfrac{w}{h^2}$.

According to federal guidelines in the United States, a BMI greater than 25 means that you are overweight. Surveys show that 59% of men and 49% of women have BMIs greater than 25. Extreme obesity is defined as a BMI greater than 40.

To calculate body mass index (BMI):

1. Multiply the weight in pounds by 0.4536 to convert to kilograms.

2. Convert the height to inches.

3. Multiply the inches by 0.0254 to get meters.

4. Square the number found in Step 3.

5. Divide the weight in kilograms (Step 1) by the result from Step 4, and round to the nearest whole number. The result is your BMI.

STOP AND CHECK

1. What is the BMI of an adult male who is 6 ft 2 in. tall and weighs 238 lb?

Answer:

1. 31

EXAMPLE 3

HLTH/N Alexa May is 5 feet 6 inches and weighs 138 pounds. Find her body mass index.

$138(0.4536) = 62.5968$ Change pounds to kilograms.

$5 \text{ ft } 6 \text{ in.} = 5(12) + 6$ Change feet to inches.

$\qquad\qquad = 60 + 6$

$\qquad\qquad = 66 \text{ in.}$

$66(0.0254) = \boxed{1.6764}$ Change inches to meters.

$BMI = \dfrac{w}{h^2}$ $w = 62.5968, h = 1.6764$

$BMI = \dfrac{62.5968}{(1.6764)^2}$

$BMI = \dfrac{62.5968}{2.81031696}$

$\mathbf{BMI = 22.27392885}$

$\mathbf{BMI = 22}$ Rounded to nearest whole number

See Exercises 40–41.

4–4 EXERCISES

MyLab Math For additional practice go to your study plan in MyLab Math.

1 Use unit ratios or conversion factors to change to the units indicated. Do not round. *See Example 1.*

1. 9 m to inches

2. 120 m to yards

3. 42 km to miles

4. 6 L to quarts

5. 10 qt to liters

6. 27 kg to pounds

7. 50 lb to kilograms

8. 7 in. to centimeters

9. 18 ft to meters

10. 39 mL to ounces (1 qt = 32 oz)

11. $5\frac{3}{4}$ gal to liters (1 gal = 4 qt)

12. 7.3 m to feet

13. 12.7 cm to inches

14. 12 mm to inches

15. 235 km to miles

16. 500 km to miles
17. 28 in. to millimeters
18. 36 in. to centimeters
19. 48 in. to meters
20. 5,280 ft to meters
21. 100 yd to meters
22. 200 mi to kilometers
23. 50 L to quarts
24. 100 qt to liters
25. 24 g to ounces
26. 20 kg to pounds
27. 150 lb to kilograms
28. 45 lb to kilograms
29. 2 oz to grams
30. 0.5 oz to grams
31. 3 kg to pounds

See Example 2.

32. **ELEC** A spool of wire contains 100 ft of wire. How many meters of wire are on the spool?

33. **INTDR** A 60-lb sheet of metal weighs how many kilograms?

34. **CAD/ARC** Two cities 150 mi apart are how many kilometers apart?

35. **AG/H** A field that is 30 m wide is how many yards wide?

36. **HOSP** A tourist in Europe traveled 200 km, 60 km, and 120 km by car. How many total miles was this?

37. **HLTH/N** A patient in therapy jogged 5 km, 4 km, and 3 km. How many miles did the patient jog?

38. **HOSP** A container holds 12 qt. How many liters will the container hold?

39. **ELEC** A spool of electrical wire contains 100 m of wire. How many feet of wire are on the spool?

See Example 3.

40. **HLTH/N** Gary Druckemiller is 6 ft 7 in. and weighs 192 lb. What is his body mass index?

41. **HLTH/N** Jo Ella Stearns weighs 121 lb and is 5 ft 8 in. What is her body mass index?

4–5   Accuracy, Precision, and Error

LEARNING OUTCOMES

1 Determine the significant digits of a number.

2 Find the precision and the greatest possible error of a measurement.

3 Determine the relative error and the percent error of a measurement.

4 Determine an appropriate approximation of measurement calculations.

5 Read a metric rule.

6 Find the distance and the midpoint of a line segment.

LC LEARNING CATALYTICS

1. What is $\frac{1}{2}$ of $\frac{1}{32}$?

2. Add $2.6 + 3 + 0.05$.

Accuracy: the closeness of the measured value to the true or accepted value

1 **Determine the Significant Digits of a Number.** Approximate numbers that represent measured values may have varying degrees of accuracy. The **accuracy** of a measurement refers to how close the measured value is to the true or accepted value. **Precision** is the degree to which the measuring process gives consistent results. (See Fig. 4–27.)

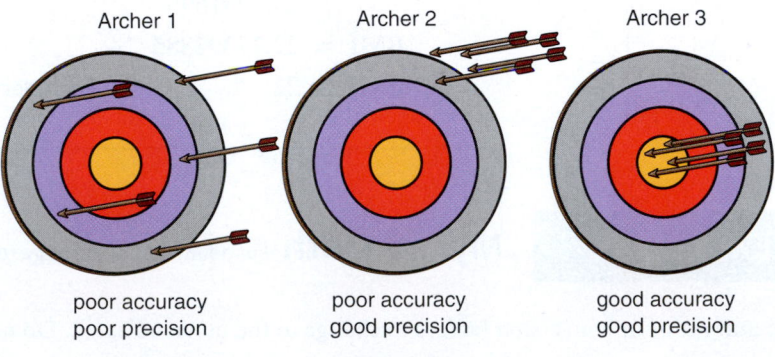

Archer 1   Archer 2   Archer 3

poor accuracy   poor accuracy   good accuracy
poor precision   good precision   good precision

FIGURE 4–27

One indicator of the degree of precision of a measurement is the number of **significant digits** in a measure. The numbers 2,500, 250, 25, 2.5, 0.25, and 0.025 all have two significant digits. When a number contains no zeros, all the digits are significant. However, zeros are special digits. Sometimes zeros are significant digits, and sometimes they are not significant digits.

Precision: the degree to which the measuring process gives consistent results

Significant digits: the significant digits of a whole number or integer are the digits beginning with the first nonzero digit on the left and ending with the last nonzero digit on the right. The significant digits of a decimal number are the digits beginning with the first nonzero digit on the left and ending with the last digit on the right of the decimal point

To determine the number of significant digits of a number:

For whole numbers:	*Determine the number of significant digits in the following:*
1. Start with the leftmost nonzero digit.	200 has 1 significant digit.
2. Count each digit (including zeros) through the rightmost nonzero digit.	28 has 2 significant digits. 1,320 has 3 significant digits. 2,005 has 4 significant digits.

For decimals and mixed decimals:

1. Start with the leftmost nonzero digit.
2. Count each digit (including zeros) through the last digit.

0.07 has 1 significant digit.
2.0 has 2 significant digits.
1.20 has 3 significant digits.
0.01250 has 4 significant digits.

EXAMPLE 1

Indicate the number of significant digits in (a) 5,205 (b) 12,000 (c) 0.003 (d) 0.0105 (e) 3.02

(a) 5,205 has 4 significant digits. All digits are significant.

(b) 12,000 has 2 significant digits. Only the 1 and 2 are significant.

(c) 0.003 has 1 significant digit. Only the 3 is a significant digit.

(d) 0.0105 has 3 significant digits. The digits 105 are significant.

(e) 3.02 has 3 significant digits. All digits are significant.   **See Exercises 1–10.**

2 **Find the Precision and the Greatest Possible Error of a Measurement.** The concept of *precision* is associated with approximate numbers. Exact numbers have infinite precision. The precision of a measure is based on the place value or significant digits of a measure. The average of the measurements of 2.3 cm, 4.25 cm, and 5.125 cm can be no more precise than the least precise of the measurements (tenths).

For the calculations $\dfrac{2.3 + 4.25 + 5.125}{3}$, we would round the results to the nearest tenth. The number 3 is an exact number (exactly three measures) and has an infinite precision, so tenths is still the least precise.

$$\frac{2.3 + 4.25 + 5.125}{3} = \frac{11.675}{3} = 3.891666667 = 3.9 \text{ (rounded to tenths)}$$

The **greatest possible error** of a measurement is half of the precision of the measurement.

Greatest possible error: the greatest possible error of a measurement is half of the precision of the measurement

To find the precision and the greatest possible error of a measurement:

	Find the greatest possible error of a measurement of $1\frac{3}{4}$ ft.
1. Determine the precision of the measurement, which is the smallest subdivision.	Precision $= \frac{1}{4}$ ft
2. The greatest possible error is one-half of the precision.	$\frac{1}{2}\left(\frac{1}{4}\right) = \frac{1}{8}$ ft

STOP AND CHECK

Find the precision and the greatest possible error of the measurements.

1. $4\frac{5}{16}$ in.

2. 2.32 kg

Answers:

1. Precision $= \frac{1}{16}$ in.; greatest possible error $= \frac{1}{32}$ in.

2. Precision $= 0.01$ kg; greatest possible error $= 0.005$ kg

EXAMPLE 2

Find the precision and the greatest possible error of the measurements.

(a) $2\frac{5}{8}$ in.   **(b)** 3.5 cm.

(a) $2\frac{5}{8}$ in.

Precision $= \dfrac{1}{8}$ **in**.

$\dfrac{1}{2}\left(\dfrac{1}{8}\right) = \dfrac{1}{16}$ **in.**   Greatest possible error is one-half the precision.

(b) 3.5 cm

Precision $= \mathbf{0.1\ cm}$

$0.5(0.1) = \mathbf{0.05\ cm}$ Greatest possible error is one-half the precision.

See Exercises 11–26.

3 **Determine the Relative Error and the Percent Error of a Measurement.** Because factories and other industrial concerns are extremely interested in issues related to quality control, most regularly measure parts to ensure that they fall within preestablished tolerance limits for quality. Every measurement taken has some amount of error. There are three ways of reporting the error: as an absolute error, a relative error, and a percent of error. Reporting the error as an absolute error does not give much information, so relative error and percent error are frequently calculated.

► The absolute value of the difference between the observed measurement and the true value is the **absolute error.** That is, the value is positive no matter which value is larger.

► The quotient of the absolute error and the true value is the **relative error.**

► The relative error converted to a percent is the **percent error.**

The concept of absolute value is discussed in detail in Section 5–1 on page 215.

Absolute error: the absolute value of the difference between the observed measurement and the true value of the measurement

Relative error: the quotient of the absolute error and the true value of the measurement

Percent error: the relative error converted to a percent

> **To find the absolute error, the relative error, and the percent error of a measurement:**
>
> **Absolute Error:**
> 1. Find the absolute value of the difference between the observed measurement and the true value.
>
> **Relative Error:**
> 1. Find the quotient of the absolute error and the true value.
>
> **Percent Error:**
> 1. Change the relative error to a percent by multiplying by 1 in the form of 100%.

STOP AND CHECK

Find the absolute error, relative error, and percent error.

1. An observed value of 3.4 m and a true value of 3.45 m

Answer:

1. absolute error $= 0.05$; relative error $= 0.0144927536$; percent error $= 1.45\%$ to the nearest hundredth of a percent

EXAMPLE 3

IND/TR   The blueprint for a part calls for it to be 32.112 mm. A measurement of an actual part is recorded as 32.155 mm. Find the absolute error, relative error, and percent error.

$$\text{absolute error} = |\text{observed value} - \text{true value}|$$
$$\text{absolute error} = |32.155 - 32.112|$$
$$\mathbf{absolute\ error\ =\ 0.043\ mm}$$

$$\text{relative error} = \frac{\text{absolute error}}{\text{true value}}$$

$$\text{relative error} = \frac{0.043}{32.112}$$

relative error = 0.0013390633

$$\text{percent error} = \text{relative error} \times 100\%$$

$$\text{percent error} = 0.0013390633(100\%)$$

percent error = 0.134%   Rounded   See Exercises 27–32.

---

4   Determine an Appropriate Approximation of Measurement Calculations. It is often necessary to round the results of calculations. In some instances specific industry standards are appropriate. In other cases, specific directions will be given. However, in many instances, you will determine the appropriate rounding place. When adding or subtracting measurements, we examine the precision of the measurements being added or subtracted.

> **To round the sum or difference of measurements of different precisions:**
>
> 1. If necessary, change all measurements to a common unit of measure.
>
> 2. Add or subtract the common units of measure.
>
> 3. Round the result to have the same precision (place value) as the *least precise* measurement.

STOP AND CHECK
1. Add 34.6 g, 31 cg, 3.9 g, and 15.32 cg. Round to the appropriate precision.

Answer:
1. 39.0 g

EXAMPLE 4

Add the measurements 12.5 m, 38 cm, 2.9 m, 43.25 cm.

12.5 m	Change all measurements to meters.
38 cm = 38(0.01 m) = 0.38 m	
2.9 m	
43.25 cm = 43.25(0.01 m) = 0.4325 m	Add the like measurements.

$$
\begin{array}{r}
12.5 \\
0.38 \\
2.9 \\
+\ 0.4325 \\
\hline
16.2125
\end{array}
$$

The least precise measurement is tenths.

16.2125 m rounds to 16.2 m.   See Exercises 33–36.

When multiplying or dividing measurements, the convention for rounding is to round so that the product or quotient has the same number of significant digits as the measurement with the fewest number of significant digits.

> **To round the product or quotient of measurements of different accuracy:**
>
> 1. Multiply or divide the measurements.
>
> 2. Round the product or quotient to have the same number of significant digits as the measurement with the fewest number of significant digits.

STOP AND CHECK
1. Multiply 20 cm, 315 cm, and 104 cm. Round to the appropriate number of significant digits.

Answer
1. 700,000 cm^3

EXAMPLE 5

Multiply the measurements: 150 m(105 m)(50 m).

150 m(105 m)(50 m) = 787,500 m^3 m(m)(m) = m^3

50 m is the measurement with the fewest significant digits, one.

787,500 m³ rounds to 800,000 m³.   **See Exercises 37–40.**

Metric rule: a measuring device for measuring length in centimeters and millimeters

5   Read a Metric Rule. Many standard rulers have both a U.S. customary scale and a metric scale. The metric rule is usually calibrated in centimeters or millimeters. The **metric rule** illustrated in Fig. 4–28 shows centimeters (cm) as the major divisions, represented by the longest lines. Each centimeter is divided into 10 mm, which are the shortest lines. A line slightly longer than the millimeter line divides each centimeter into two equal parts of 5 mm each.

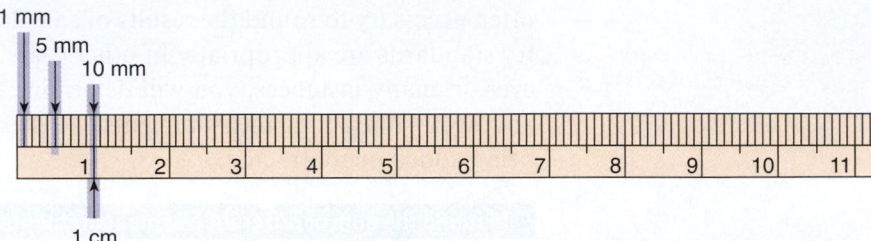

FIGURE 4–28 The Metric Rule.

The metric rule is read like the U.S. customary rule with the exception that only two measures are indicated, centimeters and millimeters. Other measures are calculated in relation to millimeters or centimeters.

Since 1 mm is $\frac{1}{10}$ cm, we can write metric measures in decimals. For example, a measure of 3 cm and 4 mm is written as either 3.4 cm or 34 mm.

> **To read a metric rule:**
>
> 1. Align the rule along the object. Count the number of millimeters and/or centimeters (10 mm = 1 cm) to determine the approximate length.
>
> 2. Use eye judgment to estimate closeness of the object to a mark on the rule.

STOP AND CHECK

1. Measure the length of the line segment *EF*.

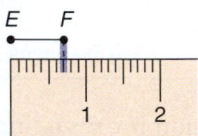

Answer:
1. 7 mm or 0.7 cm to the nearest millimeter

EXAMPLE 6

Find the length of line segment **(a)** *AB* (Fig. 4–29) and **(b)** line segment *CD* (Fig. 4–30) to the nearest millimeter.

(a)

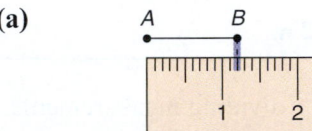

FIGURE 4–29

The line segment *AB* extends two marks past 1 cm.

Line segment *AB* is 12 mm or 1.2 cm to the nearest millimeter.

(b)

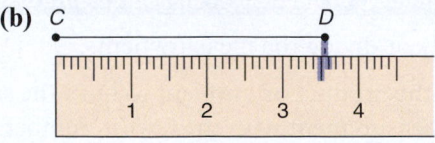

FIGURE 4–30

The end of line segment *CD* falls approximately halfway between the 35-mm mark and the 36-mm mark, by eye judgment, measuring about 35.5 mm. Both 35 and 36 mm would be acceptable measures of the line segment *CD*.

35 mm, 35.5 mm, and 36 mm are acceptable approximations for line segment *CD*.

See Exercises 41–54.

6 **Find the Distance and the Midpoint of a Line Segment.** Two points determine a *line* that is infinite in length. However, the two points also determine a **line segment** that has a certain length. The two points are called the **end points of a line segment.**

The distance between two points on a line is the difference between the coordinates of the points on the line. A **coordinate** is the number that identifies a specific location on a number line or measuring device.

Line segment or segment: a line segment, or segment, consists of all points on the line between and including two points that are called end points

End points of a line segment: the two points that determine a line segment

Coordinate: the number that identifies a specific location on a number line or measuring device

> **To find the distance between two points on a line:**
>
> 1. Determine the coordinate for each point.
>
> 2. Subtract the value of the coordinate of the leftmost point from the value of the coordinate of the rightmost point.
>
> $$\text{Distance} = P_2 - P_1$$
>
> where P_1 and P_2 are points on the number line or measuring device and P_1 is the leftmost point.

STOP AND CHECK

1. Find the distance from $2\frac{5}{8}$ to $5\frac{1}{4}$.

Answer:

1. $2\frac{5}{8}$ units

EXAMPLE 7

To find the distance from $2\frac{1}{2}$ to $4\frac{1}{4}$:

Visualize the points on a number line (Fig. 4–31).

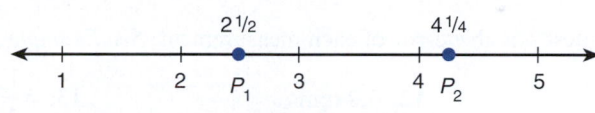

FIGURE 4–31

$4\frac{1}{4} - 2\frac{1}{2} =$ Subtract the coordinate of the leftmost point (P_1) from the coordinate of the rightmost point (P_2).

$4\frac{1}{4} - 2\frac{2}{4} =$ Regroup. $4\frac{1}{4} = 3 + \frac{4}{4} + \frac{1}{4} = 3\frac{5}{4}$.

$3\frac{5}{4} - 2\frac{2}{4} =$ Subtract.

$1\frac{3}{4}$ **units**   Distance from $2\frac{1}{2}$ to $4\frac{1}{4}$.

See Exercises 55–60.

On a number line or measuring device, the coordinate of the **midpoint of a line segment** between two points is the average of the coordinates of the points.

Midpoint of a line segment: the coordinate of the point that is halfway beween the end points. It is the average of the coordinates of the end points

> **To find the midpoint of a line segment between two points on a line:**
>
> 1. Determine the coordinate for each end point.
>
> 2. Average the coordinates of the end points (P_1 and P_2).
>
> $$\text{Midpoint} = \frac{P_1 + P_2}{2}$$
>
> where P_1 and P_2 are the end points of the line segment on the number line or measuring device.

EXAMPLE 8

Find the midpoint between two points on a metric rule at 2.8 cm and 5.6 cm.

$$\text{Midpoint} = \frac{P_1 + P_2}{2} \qquad \text{Let } P_1 = 2.8 \text{ and } P_2 = 5.6. \text{ Average the two points.}$$

$$\text{Midpoint} = \frac{2.8 + 5.6}{2} \qquad \text{Add.}$$

$$\text{Midpoint} = \frac{8.4}{2} \qquad \text{Divide.}$$

$$\text{Midpoint} = 4.2 \text{ cm}$$

The midpoint of the points 2.8 and 5.6 is 4.2 cm.   See Exercises 61–66.

4-5 EXERCISES

MyLab Math For additional practice go to your study plan in MyLab Math.

1 Indicate the number of significant digits in each number. *See Example 1.*

1. 583,000	**2.** 702,500	**3.** 0.0057	**4.** 82.07	**5.** 7.200
6. 860	**7.** 5,080	**8.** 2.0091	**9.** 572,080	**10.** 1,010

2 Find the greatest possible error of each measurement. *See Example 2.*

11. $7\frac{3}{16}$ in.	**12.** 7.2 mm	**13.** $5\frac{1}{4}$ in.	**14.** 19 oz
15. 8 L	**16.** $5\frac{1}{8}$ in.	**17.** $12\frac{1}{32}$ in.	**18.** $1\frac{5}{8}$ oz
19. 6.2 mi	**20.** 7.9 cm	**21.** 8.17 dg	**22.** 5.3 cg
23. 6.7 cL	**24.** 1.4 km	**25.** 12.7 km	**26.** 125.3 km

3 *See Example 3.*

27. A measurement is observed to be 52.02 cm, against a true measure of 52 cm. Find the absolute error, relative error, and percent error.

28. **INDTR** A blueprint for a machine part specifies a measure of 52.2 mm. An actual part has a measurement of 52.4 mm. Find the absolute error, relative error, and percent error.

29. **CAD/ARC** The blueprint for a part specifies a measurement of 48.7 cm. An actual part has a measurement of 50.2 cm. Find the absolute error, relative error, and percent error.

30. **AUTO** The blueprint for a racing steering wheel specifies a measurement of 14.25 in. An actual wheel has a measurement of 14.54 in. Find the absolute error, relative error, and percent error.

31. **INDTEC** A blueprint for the gasoline tank of an industrial lawn mower specifies that it holds 50.05 L of gasoline. Upon close measurement, an actual tank is found to hold 50.25 L of gasoline. Find the absolute error, relative error, and percent error.

32. **AUTO** An actual gear is measured to be 15.4 in. in diameter. The blueprint specifies that the diameter of the part be 15 in. Find the absolute error, relative error, and percent error.

4 Add. Give the sum using the appropriate precision. *See Example 4.*

33. 12.84 m, 13.5 m, 182.3 m, 6.732 m

34. 0.93 g, 1.8 g, 42.7 g, 18.932 g

35. 14.3 cm, 0.7 m, 4.9 m, 23.45 cm

36. 12.78 g, 0.23 kg, 4.752 kg, 0.3466 kg

Give the product or quotient using the appropriate number of significant digits. *See Example 5.*

37. 202 m(45 m)(100 m)

38. 125 m(403 m)(340 m)

39. 1,500 g ÷ 35 g

40. 1,968 km ÷ 30 km

5 Measure line segments 41–50 in Fig. 4–32 to the nearest millimeter. Express answers in millimeters or centimeters. *See Example 6.*

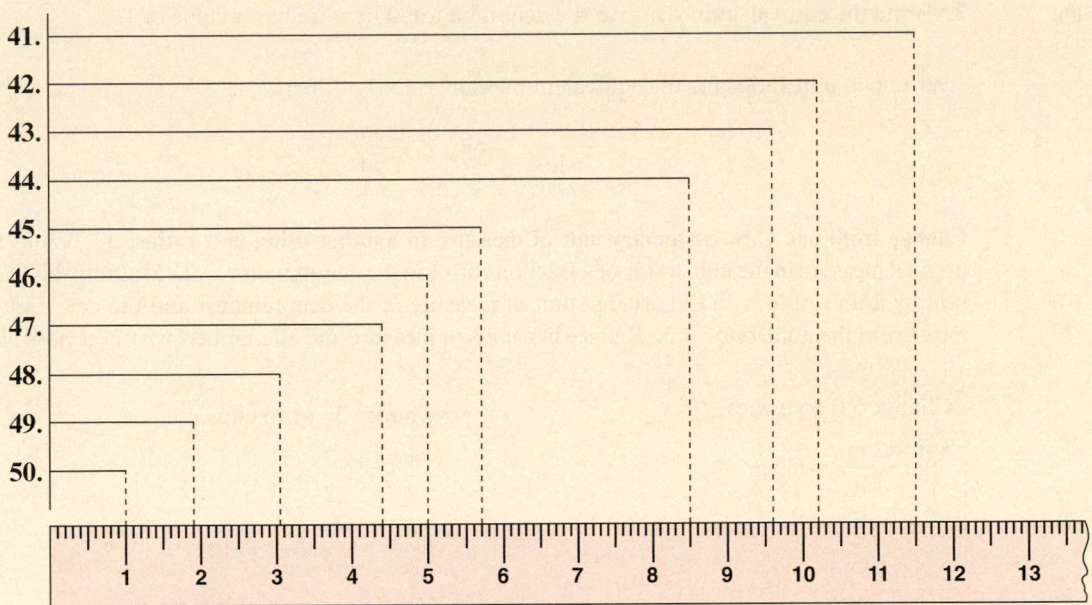

FIGURE 4–32

Use a metric rule to measure each line segment to the nearest tenth centimeter.

51. A ———————————— B

52. C ——————————— D

53. E ————— F

54. G —————————— H

6 Find the distance between each pair of points. *See Example 7.*

55. 5 in. and 8 in.

56. $2\frac{1}{2}$ in. and $5\frac{1}{4}$ in.

57. 8.2 cm and 15.7 cm

58. 4.9 cm and 18.5 cm

59. $12\frac{3}{8}$ in. and $23\frac{5}{8}$ in.

60. $14\frac{7}{8}$ in. and $25\frac{3}{4}$ in.

Find the midpoint between each pair of points. *See Example 8.*

61. 1.8 cm and 7 cm

62. 3.5 cm and 6.3 cm

63. 5.5 cm and 10.3 cm

64. 3 cm and 5.2 cm

65. $7\frac{1}{2}$ in. and $15\frac{1}{4}$ in.

66. 2 in. and $9\frac{3}{4}$ in.

4 | CHAPTER REVIEW OF KEY CONCEPTS

LEARNING OUTCOMES

KEY CONCEPTS AND EXAMPLES

Section 4-1

1 Convert one unit of measure to another using unit ratios (pp. 158–161).

Write two equivalent measures as a unit ratio: 1. Write one measure in the numerator. **2.** Write the equivalent measure in the denominator. The ratio has a value of 1.

> Write two unit ratios for the equivalent measures: 1 yd = 36 in.
>
> $$\frac{1 \text{ yd}}{36 \text{ in.}} \quad \text{or} \quad \frac{36 \text{ in.}}{1 \text{ yd}}$$

Change from one U.S. customary unit of measure to another using unit ratios: 1. Write the original measure in the numerator of a fraction with 1 in the denominator. **2.** Multiply this fraction by a unit ratio with the original unit of measure in the denominator and the new unit of measure in the numerator. **3.** Reduce like units of measure and all numbers wherever possible.

> Change 5 ft to inches.
>
> $$\frac{5 \text{ ft}}{1} \cdot \frac{12 \text{ in.}}{1 \text{ ft}} = 60 \text{ in.}$$
>
> Change 285 ft to yards.
>
> $$\frac{285 \text{ ft}}{1} \cdot \frac{1 \text{ yd}}{3 \text{ ft}} = 95 \text{ yd}$$
>
> Change $3\frac{1}{2}$ qt to cups.
>
> $$3\frac{1}{2} \text{ qt} = \frac{7}{2} \text{ qt}$$
>
> $$\frac{7 \text{ qt}}{2} \cdot \frac{2 \text{ pt}}{1 \text{ qt}} \cdot \frac{2 \text{ c}}{1 \text{ pt}} = 14 \text{ c}$$
>
> More than one unit ratio may be needed.

2 Convert one unit of measure to another using conversion factors (pp. 161–164).

When we use conversion factors to convert units, we always multiply. Two equivalent measures have two conversion factors.

Develop a conversion factor for converting from one measure to another: 1. Write a unit ratio that changes the given unit to the new unit. **2.** Change the fraction (or ratio) to its decimal equivalent by dividing the numerator by the denominator.

> Develop two conversion factors relating pints and quarts. 1 qt = 2 pt
>
> **Pints to Quarts**
>
> $$\frac{\text{pt}}{1}\left(\frac{\text{qt}}{\text{pt}}\right)$$ Write appropriate labels.
>
> $$\frac{\text{pt}}{1}\left(\frac{1 \text{ qt}}{2 \text{ pt}}\right)$$ Insert numbers to form a unit ratio.
>
> $$\frac{1}{2} = 0.5$$ Convert fraction to whole number, decimal, or mixed decimal equivalent.
>
> $$\text{pt} \times 0.5 = \text{qt}$$
>
> **Quarts to Pints**
>
> $$\frac{\text{qt}}{1}\left(\frac{\text{pt}}{\text{qt}}\right)$$
>
> $$\frac{\text{qt}}{1}\left(\frac{2 \text{ pt}}{1 \text{ qt}}\right)$$
>
> $$\frac{2}{1} = 2$$
>
> $$\text{qt} \times 2 = \text{pt}$$

Change from one U.S. customary unit of measure to another using conversion factors: 1. Select the appropriate conversion factor. **2.** Multiply the original measure by the conversion factor.

> Change 9 pt to quarts.
>
> $$9 \text{ pt } (0.5) = 4.5 \text{ qt} \qquad \text{Conversion factor is 0.5.}$$

LEARNING OUTCOMES	KEY CONCEPTS AND EXAMPLES

Standard notation means that each measure is converted to the next larger unit of measure when possible.

Express mixed measures in standard notation: 1. Start with the smallest unit of measure and determine if there are enough units to make one or more of the next larger unit. **2.** Regroup to make as many of the larger units as possible. **3.** Combine like units. **4.** Repeat the process with each given measuring unit.

> 2 ft 16 in. = 2 ft + 1 ft 4 in. = 3 ft 4 in.   16 in. = 1 ft 4 in.

3 Add and subtract U.S. customary measures (pp. 164–166).

Add or subtract unlike U.S. customary measures: 1. Convert each measure to a measure with a common U.S. customary unit. **2.** Add or subtract. The unit of measure of the sum or difference is the common unit. **3.** Express the sum or difference in standard notation.

> Add 5 lb and 7 oz.   **Convert 5 lb to oz.**
>
> $\dfrac{5 \text{ lb}}{1} \times \dfrac{16 \text{ oz}}{1 \text{ lb}} = 80 \text{ oz}$ **Multiply by unit ratio $\dfrac{16 \text{ oz}}{1 \text{ lb}}$.**
>
> 80 oz + 7 oz = 87 oz   **Combine like measures.**

Add or subtract mixed U.S. customary measures: 1. Align like measures in columns. **2.** Add or subtract common units of measure, carrying or regrouping as necessary.

> Subtract 2 ft 8 in. from 7 ft.
>
> $\begin{array}{rl} 7 \text{ ft} = & 6 \text{ ft } 12 \text{ in.} \\ - & 2 \text{ ft } 8 \text{ in.} \\ \hline & 4 \text{ ft } 4 \text{ in.} \end{array}$ **Rewrite 7 ft as 6 ft 12 in.**
> **Subtract.**

4 Multiply and divide U.S. customary measures (pp. 166–168).

Multiply a U.S. customary measure by a number: 1. Multiply the numbers associated with each unit of measure by the given number. **2.** Write the resulting measure in standard notation.

> Multiply 2 gal 3 qt by 5.
>
> $\begin{array}{r} 2 \text{ gal } 3 \text{ qt} \\ \times \phantom{2 \text{ gal }} 5 \\ \hline 10 \text{ gal } 15 \text{ qt} \\ 13 \text{ gal } 3 \text{ qt} \end{array}$ 15 qt = 3 gal 3 qt

Multiply a length measure by a like length measure: 1. Multiply the numbers associated with each like unit of measure. **2.** The unit of measure of the product is a square measure or an area.

> Multiply 5 yd by 12 yd.
>
> $5 \text{ yd } (12 \text{ yd}) = 60 \text{ yd}^2$

Divide a measure by a number: 1. Set up the problem and proceed as in long division. **2.** When a remainder occurs after subtraction, convert the remainder to the same unit used

LEARNING OUTCOMES	KEY CONCEPTS AND EXAMPLES

in the next smaller measure, and add it to the quantity in the next smaller number.   **3.** Divide the given number into this next smaller unit.   **4.** If a remainder occurs when the smallest unit is divided, express the remainder as a fractional part of the smallest unit.

Divide 7 lb 5 oz by 5.

$$
\begin{array}{r}
1\ \text{lb} \qquad 7\frac{2}{5}\text{oz} \\
5\overline{)7\ \text{lb} \qquad 5\ \text{oz}} \\
\underline{5\ \text{lb}} \\
2\ \text{lb} \ = \ 32\ \text{oz} \\
\overline{37\ \text{oz}} \\
\underline{35\ \text{oz}} \\
2
\end{array}
$$

Divide a U.S. customary measure by a U.S. customary measure: 1. Convert both measures to the same unit if they are different.   **2.** Write the division as a fraction, including the common unit in the numerator and the denominator.   **3.** Reduce the units and divide the numbers to simplify the fraction.

How many 5-oz glasses of juice can be poured from 1 gallon of juice?

$$\frac{1\ \text{gal}}{1}\left(\frac{4\ \text{qt}}{1\ \text{gal}}\right)\left(\frac{2\ \text{pt}}{1\ \text{qt}}\right)\left(\frac{2\ c}{1\ \text{pt}}\right)\left(\frac{8\ \text{oz}}{1\ c}\right) = 128\ \text{oz} \qquad \text{Change 1 gal to ounces.}$$

$$\frac{128\ \text{oz}}{5\ \text{oz}} = 25\frac{3}{5}\ \text{glasses}$$

5 Change one U.S. customary rate measure to another (pp. 168–169).

1. Compare the units of both numerators and both denominators to determine which units will change.   **2.** Multiply each unit that changes by a unit ratio (or ratios) containing the new unit so that the unit to be changed will reduce.

Change $10\ \dfrac{\text{mi}}{\text{hr}}$ to $\dfrac{\text{ft}}{\text{hr}}$.

$$\frac{10\ \text{mi}}{1\ \text{hr}}\ \text{to}\ \frac{\text{ft}}{\text{hr}} \qquad \text{Miles change to feet. Hours stay the same.}$$

$$\frac{10\ \text{mi}}{1\ \text{hr}}\left(\frac{5{,}280\ \text{ft}}{1\ \text{mi}}\right) = 52{,}800\ \frac{\text{ft}}{\text{hr}}$$

Section 4–2

1 Identify uses of metric measures of length, mass, weight, and capacity (pp. 172–177).

Associate small, medium, and large objects with appropriate measuring units of comparable size. Power-of-10 prefixes are used, such as *kilo-* for 1,000.

Perfume, mL; soda, L; travel, km; racetrack, m; vitamin, mg; potatoes, kg; eyedrops, mL

2 Convert from one metric unit of measure to another (pp. 177–179).

1. Mentally position the measure on the metric-value chart so that the decimal point immediately follows the original unit.   **2.** Move the decimal point in the original measure to the left or right as many places as necessary so that it *follows* the new unit. (Attach zeros if necessary.)

Change 5.04 cL to liters. Move the decimal two places to the left.

$$5.04\ \text{cL} = 0.0504\ \text{L}$$

LEARNING OUTCOMES	KEY CONCEPTS AND EXAMPLES

3 Make calculations with metric measures (pp. 180–182).

Add or subtract metric measures: 1. Change the measures to a common unit of measure if necessary. **2.** Add or subtract the numerical values. **3.** Give the answer in the common unit of measure.

Multiply or divide a metric measure by a number: Multiply or divide the numerical values and give the answer the same unit as the original measure.

Divide a metric measure by a like measure: 1. If unlike measures are used, first change them to like measures. **2.** Divide the numerical values. **3.** Give the quotient as a number that tells how many parts there are.

$$7 \text{ km} + 34 \text{ m} = \qquad\qquad 7 \text{ mg}(3) = 21 \text{ mg}$$
$$7{,}000 \text{ m} + 34 \text{ m} = 7{,}034 \text{ m} \qquad 5 \text{ m} \div 25 \text{ cm} = 500 \text{ cm} \div 25 \text{ cm}$$

or

$$7 \text{ km} + 0.034 \text{ km} = 7.034 \text{ km} \qquad \frac{500 \cancel{\text{ cm}}}{25 \cancel{\text{ cm}}} = 20$$

Section 4-3

1 Convert from one unit of time to another (p. 184).

Use unit ratios or conversion factors to convert one unit of time to another.

How many days are in 3,420 hours?
$$1 \text{ day} = 24 \text{ h}$$
$$3{,}420 \text{ h}\left(\frac{1 \text{ day}}{24 \text{ h}}\right) = 142.5 \text{ days}$$

2 Make calculations with measures of time (pp. 184–185).

Measures of time that are the same unit of measure can be added or subtracted by adding or subtracting the quantities and keeping the same unit of measure. A time measure can be multiplied or divided by a number by multiplying or dividing the measure by the number and keeping the same measuring unit.

A printer prints 1 page in 30 s. How long in hours does it take to print 45 pages?

$1 \text{ min} = 60 \text{ s}; 1 \text{ h} = 60 \text{ min}$

$30 \text{ s} (45) = 1{,}350 \text{ s}$ Seconds to complete job.

$1{,}350 \text{ s}\left(\dfrac{1 \text{ min}}{60 \text{ s}}\right) = 22.5 \text{ min}$ Minutes to complete job.

$22.5 \text{ min}\left(\dfrac{1 \text{ h}}{60 \text{ min}}\right) = 0.375 \text{ h}$ Portion of hour to complete job.

The printer can print 45 pages in 22.5 min or 0.375 h.

3 Convert between Fahrenheit temperatures and Celsius temperatures (pp. 185–187).

Convert degrees Fahrenheit to degrees Celsius: Use the formula: $°C = \dfrac{5}{9}(°F - 32)$.

Change 50°F to °C.

$°C = \dfrac{5}{9}(°F - 32)$ Substitute 50 for °F.

$°C = \dfrac{5}{9}(50 - 32)$ Perform operation in parentheses.

$°C = \dfrac{5}{9}(18)$ Multiply.

$°C = 10$

LEARNING OUTCOMES	KEY CONCEPTS AND EXAMPLES

Convert degrees Celsius to degrees Fahrenheit: Use the formula: $°F = \frac{9}{5}°C + 32$.

Change 28°C to °F.

$°F = \frac{9}{5}°C + 32$ Substitute 28 for °C.

$°F = \frac{9}{5}(28) + 32$ Multiply.

$°F = \frac{252}{5} + 32$ Divide.

$°F = 50.4 + 32$ Add.

$°F = 82.4$

4   Examine other useful measures (pp. 187–188).

All SI metric measures use the same system of prefixes for multiples and submultiples. Table 4–6 on page 187 summarizes the SI base units and some other commonly used metric units.

What is the meaning of 3 μA?

$$1 \; \mu A = 1 \text{ microampere}$$
$$1 \; \mu A = 1 \text{ millionth of an ampere, or } 0.000001 \text{ A, or } 10^{-6} \text{ A}$$
$$3 \; \mu A = 3(0.000001) = \mathbf{0.000003 \text{ A, or } 3 \times 10^{-6} \text{ A}}$$

Using a unit ratio:

$$3 \; \mu A \left(\frac{0.000001 \text{ A}}{1 \; \mu A} \right) = 0.000003 \text{ A}$$

Section 4–4

1   Convert between U.S. customary measures and metric measures (pp. 189–191).

1. Select the appropriate conversion factor for the measures involved.   **2.** Multiply the original measure by the conversion factor and use the appropriate new measuring unit to label the product.

How many feet are in 14 m?

$$\frac{14 \; \cancel{m}}{1} \left(\frac{3.2808 \text{ ft}}{1 \; \cancel{m}} \right) = 45.9312 \text{ ft} \qquad \text{Use the conversion factor 1 m} = 3.2808 \text{ ft.}$$

Calculate body mass index (BMI): 1. Multiply the weight in pounds by 0.4536 to convert to kilograms.   **2.** Convert the height to inches.   **3.** Multiply the inches by 0.0254 to get meters.   **4.** Square the number found in Step 3.   **5.** Divide the weight in kilograms (Step 1) by the result from Step 4, and round to the nearest whole number. The result is your BMI.

Matthew Bennett is 6 ft 1 in. and weighs 175 pounds. Find his body mass index.

$175(0.4536) = 79.38$

$6 \text{ ft 1 in.} = 72 \text{ in.} + 1 \text{ in.} = 73 \text{ in.}$ $73(0.0254) = 1.8542$ $(1.8542)^2 = 3.43805764$

$$\frac{79.38}{3.43805764} = 23.08861814$$

Mathew's BMI is 23.

LEARNING OUTCOMES	KEY CONCEPTS AND EXAMPLES

Section 4–5

1 Determine the significant digits of a number (pp. 192–193).

For whole numbers: 1. Start with the leftmost nonzero digit.   **2.** Count each digit (including zeros) through the rightmost nonzero digit.

> 258 has three significant digits.
> 1,050 has three significant digits.

For decimal numbers: 1. Start with the leftmost nonzero digit.   **2.** Count each digit (including zeros) through the last digit.

> 0.023 has two significant digits.
> 1.50 has three significant digits.

2 Find the precision and the greatest possible error of a measurement (pp. 193–194).

1. Determine the precision of the measurement, which is the smallest subdivision.   **2.** The greatest possible error is one-half of the precision.

> The precision of $2\frac{5}{8}$ in. is $\frac{1}{8}$ in.
>
> $\frac{1}{2}\left(\frac{1}{8}\right) = \frac{1}{16}$ in.
>
> The greatest possible error is $\frac{1}{16}$ in.

3 Determine the relative error and the percent error of a measurement (pp. 194–195).

To find the absolute error, find the absolute value of the difference between the observed measurement and the true value.

To find the relative error, find the quotient of the absolute error and the true value.

To find the percent error, convert the relative error to a percent by multiplying by 1 in the form of 100%.

> The actual measure of a widget is 15.963 cm and the blueprint specifies 15.94 cm for the widget.
>
> absolute error $= |15.963 - 15.94|$
> $\qquad\qquad\quad = 0.023$ cm
>
> relative error $= \dfrac{0.023}{15.94} = 0.0014429109$
>
> percent error $= 0.0014429109(100\%) = 0.1443\%$ (rounded)

4 Determine an appropriate approximation of measurement calculations (pp. 195–196).

Round the sum or difference of measurements of different precisions: 1.   If necessary, change all measurements to a common unit of measure.   **2.** Add or subtract the common units of measure. **3.** Round the result to have the same precision (place value) as the *least precise* measurement.

> Add the measurements: 34.5 m, 15 cm, 4.6 mm.
>
> Change all measurements to meters.
> $\qquad$ 34.5 m $=$ 34.5 m
> $\qquad\quad$ 15 cm $=$ 15(0.01 m) $=$ 0.15 m
> $\qquad$ 4.6 mm $=$ 4.6(0.001 m) $=$ 0.0046 m

LEARNING OUTCOMES	KEY CONCEPTS AND EXAMPLES

Add the like measurements.

$$
\begin{array}{r}
34.5 \\
0.154 \\
\underline{0.0046} \\
34.6546
\end{array}
$$

The least precise measurement is tenths.

34.6546 m rounds to 34.7 m.

Round the product or quotient of measurements of different accuracy: 1. Multiply or divide the measurements. **2.** Round the product or quotient to have the same number of significant digits as the measurement with the fewest significant digits.

Multiply the measurements: 12.3 cm(5 cm)(0.25 cm).

$$12.3 \text{ cm}(5 \text{ cm})(0.25 \text{ cm}) = 15.375 \text{ cm}^3 \qquad \text{cm(cm)(cm)} = \text{cm}^3$$

5 cm is the measurement with the fewest significant digits, one.

15.375 cm³ rounds to 20 cm³.

---

5 Read a metric rule
(pp. 197–198).

1. Align the rule along the object. Count the number of millimeters and/or centimeters (10 mm = 1 cm) to determine the approximate length. **2.** Use eye judgment to estimate closeness of the object to a mark on the rule.

Read the measure in Fig. 4–33.

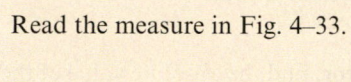 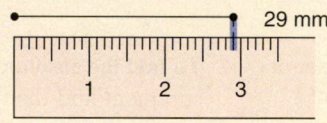   29 mm

FIGURE 4–33

---

6 Find the distance and the midpoint of a line segment
(p. 197).

Find the distance between two points on a line: 1. Determine the coordinate for each point. **2.** Subtract the value of the coordinate of the leftmost point from the value of the coordinate of the rightmost point.

Distance = $P_2 - P_1$, where P_1 and P_2 are points on the number line or measuring device and P_1 is the leftmost point.

Find the distance between the points 3.5 cm and 6.3 cm on a metric rule.

$$
\begin{aligned}
\text{Distance} &= 6.3 - 3.5 \qquad \text{Let } P_1 = 3.5 \text{ cm and } P_2 = 6.3 \text{ cm.} \\
&= 2.8 \text{ cm}
\end{aligned}
$$

Find the midpoint of a line segment between two points on a line: 1. Determine the coordinate for each end point. **2.** Average the coordinates of the end points (P_1 and P_2).

Midpoint = $\dfrac{P_1 + P_2}{2}$, where P_1 and P_2 are end points of the line segment on the number line or measuring device.

Find the midpoint between the points 3.5 cm and 6.3 cm on a metric rule.

$$
\begin{aligned}
\text{Midpoint} &= \frac{3.5 + 6.3}{2} \qquad \text{Let } P_1 = 3.5 \text{ cm and } P_2 = 6.3 \text{ cm.} \\
&= \frac{9.8}{2} \\
&= 4.9 \text{ cm}
\end{aligned}
$$

4   CHAPTER REVIEW EXERCISES

Section 4–1   MyLab Math   **For additional practice go to your study plan in MyLab Math.**

Write two unit ratios that relate the given pair of measures.

1. Feet and inches

2. Hours and days

3. Pounds and tons

4. Yards and miles

Using unit ratios or conversion factors, convert the given measures to the new units.

5. How many ounces are in 5 lb?

6. **HELPP**   A fire suit weighing $57\frac{3}{5}$ lb weighs how many ounces?

7. Find the number of pounds in 680 oz.

8. **HOSP**   A can of fruit weighs 22.4 oz. How many pounds is this?

9. **AG/H**   How many feet of wire are needed to fence a property line $1\frac{1}{4}$ mi long?

10. **AVIA**   The tail height of a B 737-200 aircraft is 36 ft 6 in. How many inches is the tail height?

Express the measures in standard notation.

11. 1 ft 19 in.

12. 1 mi 5,375 ft

13. 12 lb $17\frac{1}{2}$ oz

14. 2 gal 7 qt

Add or subtract. Write answers in standard notation.

15. $\begin{array}{r} 5 \text{ gal } 3 \text{ qt} \\ + \ 2 \text{ gal } 3 \text{ qt} \\ \hline \end{array}$

16. $\begin{array}{r} 7 \text{ ft } 9 \text{ in.} \\ - \ 4 \text{ ft } 6 \text{ in.} \\ \hline \end{array}$

17. $\begin{array}{r} 4 \text{ lb } \ \ 9 \text{ oz} \\ - \ 3 \text{ lb } 11 \text{ oz} \\ \hline \end{array}$

18. **AVIA**   Two packages to be sent air express weigh 5 lb 4 oz each. What is the shipping weight of the two packages?

19. **AUTO**   A water hose purchased for an RV was 2 ft long. What was its length after 7 in. were cut off?

Multiply and write answers for mixed measures in standard notation.

20. $\begin{array}{r} 8 \text{ lb } 3 \text{ oz} \\ \times \ 9 \\ \hline \end{array}$

21. $\begin{array}{r} 9 \text{ in.} \\ \times \ 7 \text{ in.} \\ \hline \end{array}$

22. $\begin{array}{r} 10 \text{ gal } 3 \text{ qt} \\ \times \ 7 \\ \hline \end{array}$

Divide.

23. 20 yd 2 ft 6 in. ÷ 2

24. 5 gal 3 qt 2 pt ÷ 6

25. **HELPP**   If 18 lb of flame retardant are divided equally into four boxes, express the weight of the contents of each box in pounds and ounces.

26. **CON**   If 32 equal lengths of pipe are needed for a job and each length is to be 2 ft 8 in., how many feet of pipe are needed for the job?

27. 14 ft ÷ 4 ft

28. 2 mi 120 ft ÷ 15 ft

29. $5\dfrac{\text{mi}}{\text{min}} = \underline{\hspace{1cm}} \dfrac{\text{mi}}{\text{h}}$

30. $2{,}520\dfrac{\text{gal}}{\text{h}} = \underline{\hspace{1cm}} \dfrac{\text{qt}}{\text{h}}$

31. **HOSP**   How many quarts of milk are needed for a recipe that calls for 3 pt of milk?

32. **HOSP**   How many $\frac{1}{2}$-oz servings of jelly can be made from a $1\frac{1}{2}$-lb container of jelly?

Section 4–2

Give the prefix that relates each number to the standard unit in the metric system.

33. 1,000 times

34. $\dfrac{1}{10}$ of

35. $\dfrac{1}{1{,}000}$ of

36. 10 times

37. $\dfrac{1}{100}$ of

38. 100 times

Give the value of the prefixes based on a standard measuring unit.

39. dekameter (dkm)

40. hectogram (hg)

41. milligram (mg)

42. centigram (cg)

43. kiloliter (kL)

44. deciliter (dL)

Choose the most reasonable answer.

45. Height of the Washington Monument
   (a) 200 m   **(b)** 200 cm
   (c) 200 mm   **(d)** 200 km

46. Height of Mt. Rushmore
   (a) 1.6 km   **(b)** 1.6 m
   (c) 1.6 cm   **(d)** 1.6 mm

47. Weight of an egg
   (a) 50 g   **(b)** 50 kg
   (c) 50 mg

48. Weight of an aspirin tablet
   (a) 50 kg   **(b)** 50 mg
   (c) 50 g

49. Weight of a man's shoe
   (a) 0.25 g   **(b)** 0.25 mg
   (c) 0.25 kg

50. Carton of milk
   (a) 4 L   **(b)** 4 mL

51. Bottle of medicine
   (a) 50 L   **(b)** 50 mL

Change to the unit indicated.

52. 0.4 dkm = _____ hm

53. 67.1 m = _____ dkm

54. 4 m = _____ dm

55. 2.3 m = _____ mm

56. 5 cm = _____ mm

57. 0.123 hm = _____ mm

58. How many millimeters are in 0.432 km?

59. 23 dkm = _____ mm

60. 42.7 cm = _____ dkm

61. 41,327 dkm = _____ km

62. A board is 1.82 m long. How many centimeters long is the board?

63. 394.5 g = _____ hg

64. 2.7 hg = _____ dg

65. 3,000,974 cg = _____ kg

Perform the operations indicated.

66. 25 mm − 14 mm

67. 12 g + 5 m

68. 17 mg − 8 mL

69. 8 g − 52 cg

70. 43 dkg(7)

71. 6.83 cg(9)

72. $\dfrac{18\ \text{cm}}{9}$

73. 7.5 kg ÷ 0.5 kg

74. $\dfrac{8\ \text{hL}}{20\ \text{L}}$

75. 34 hL ÷ 4

76. 2.4 m ÷ 5 cm

77. **BUS**   Fabric must be purchased to make seven garments, each requiring 2.7 m of fabric. How much fabric must be purchased?

78. **BUS**   Candy weighing 526 g is mixed with candy weighing 342 g. What is the weight of the mixture?

79. **HOSP**   A recipe calls for 5 mL of vanilla flavoring and 24 cL of milk. How much liquid is this?

80. **BUS**   20 boxes, each weighing 42 kg, are to be moved. How much weight must be moved?

81. **CON**   A metal rod 42 m long is cut into seven equal pieces. How long is each piece?

82. **INDTEC**   32 kg of a chemical are distributed equally among 16 chemistry students. How many kilograms of chemical does each student receive?

83. **HOSP**   A serving of punch is 25 cL. How many servings can be obtained from 25 L of punch?

84. **HOSP**   How many containers of jelly can be made from 8,548 L of jelly if each container holds 4 dL of jelly?

Section 4-3

Use unit ratios or conversion factors to convert the measures of time.

85. How many days are in 72 h?

86. How many minutes are in 2.4 h?

87. Convert 158 min to hours.

88. **HLTH/N**   A doctor ordered a surgery patient not to drive for 3 weeks. How many days is this?

89. **HLTH/N** A heart surgery patient was held in intensive care for 96 h. How many days is this?

90. **HLTH/N** A bone marrow patient remained hospitalized for 72 days. How many weeks is this?

91. **BUS** If you have worked for your employer for 39 months, how many years have you worked?

92. There are 96 days remaining in a year. How many weeks remain?

Make the temperature conversions.

93. $15°C =$ _____ °F

94. $86°F =$ _____ °C

95. $95°C =$ _____ °F

96. The freezing point of benzene is 5°C. What is this temperature on the Fahrenheit scale?

97. **HLTH/N** The label on a container of the acid-reducer famotidine states that storage above 40°C should be avoided. What is 40°C on the Fahrenheit scale?

98. **AUTO** Summer road surface temperatures of 122°F give what reading on the Celsius scale?

99. **HOSP** Candy that should reach a cooking temperature of 365°F should reach what temperature using the Celsius scale?

Make the appropriate conversions.

100. **AVIA** A jet plane that travels at 250 m/s travels how many kilometers per second?

101. **HLTH/N** A hospital patient has a temperature reading of 37.9°C. What is the temperature in degrees Fahrenheit to the nearest tenth?

102. A picosecond is what part of a second?

103. What is the meaning of a terabyte?

Section 4–4

Make the conversions using unit ratios or conversion factors. Round to thousandths.

104. $7 m =$ ____ inches

105. $215 m =$ ____ yards

106. $69 km =$ ____ miles

107. $15 L =$ ____ quarts

108. $12 qt =$ ____ liters

109. $32 kg =$ ____ pounds

110. $10 lb =$ ____ kilograms

111. $9 in. =$ ____ centimeters

112. $21 ft =$ ____ meters

113. $14.8 dkL =$ ____ quarts

114. $3\frac{1}{2} gal =$ ____ liters

115. How many meters long is 200 ft of pipe?

116. **CON** Concrete weighing 90 lb weighs how many kilograms?

117. Two cities 175 mi apart are how many kilometers apart?

118. **CON** A room 10 m wide is how many feet wide?

119. **HLTH/N** Wanda Williams is 5 ft 9 in. and weighs 142 lb. What is her body mass index?

120. **HLTH/N** Ravi Mehra weighs 168 lb and is 5 ft 7 in. What is his body mass index?

Section 4–5

Determine the number of significant digits of each number.

121. 304,243

122. 2,401,000

123. 4.010

124. 0.023

Find the greatest possible error of each measurement.

125. $2\frac{1}{2}$ in.

126. $7\frac{3}{8}$ in.

127. $3\frac{5}{16}$ ft

128. $7\frac{9}{32}$ in.

129. 5.8 cm

130. 12.2 cm

131. 15.3 cm

132. 7.5 oz

133. **BUS** A gas pump registers 4.95 gal when the measuring standard of 5.00 gal is used. What is the percent error?

134. **HLTH/N** A pharmacy balance reads 2.03 g when a standard 2.00-g weight is used in calibration. What is the percent error?

135. How many significant digits are there in 0.5010?

136. How many significant digits are there in 203.07?

137. What is the greatest possible error of the measurement 4.7 cg?

138. What is the greatest possible error of the measurement $2\frac{5}{32}$ in.?

139. Add and write the sum using appropriate precision:

$$4.2\,\text{m} + 508\,\text{cm} + 31.72\,\text{m} + 5.46\,\text{m}$$

140. Multiply and write the product using the appropriate number of significant digits.

$$132\,\text{m}(80\,\text{m})(27.5\,\text{m})$$

Measure line segments 141–150 in Fig. 4–34 to the nearest millimeter.

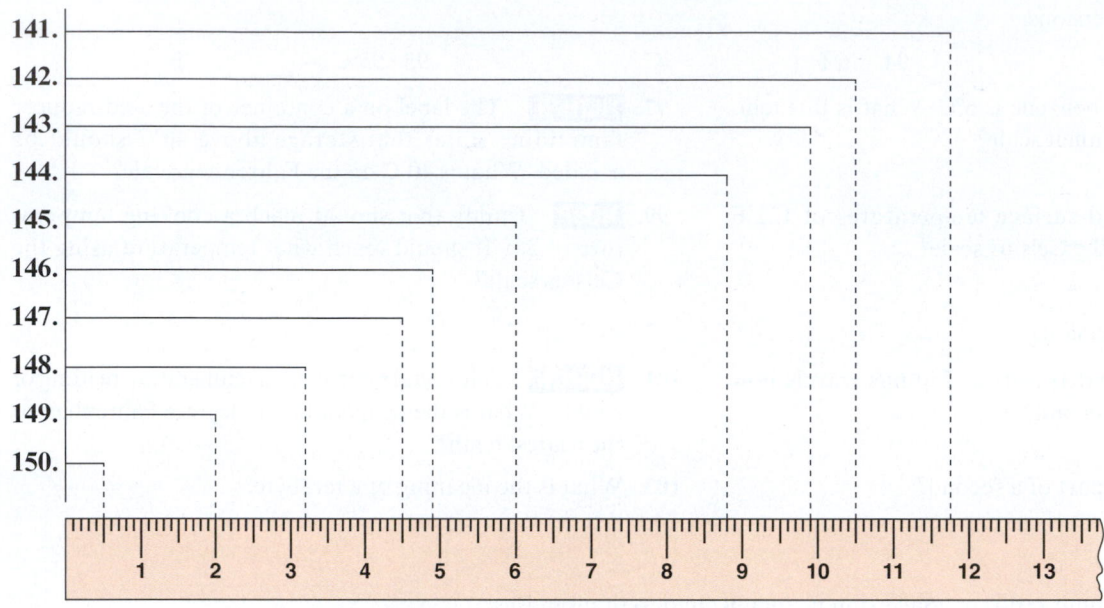

FIGURE 4–34

Find the distance between each pair of points.

151. $7\frac{5}{8}$ in. and $12\frac{7}{8}$ in.

152. $9\frac{5}{16}$ in. and $11\frac{3}{8}$ in.

153. $3\frac{5}{32}$ in. and 8 in.

154. 9 in. and $16\frac{1}{4}$ in.

155. 17.8 cm and 20.5 cm

156. 31.9 cm and 37 cm

Find the midpoint between each pair of points.

157. $15\frac{3}{8}$ in. and $25\frac{1}{8}$ in.

158. $12\frac{1}{2}$ in. and $15\frac{3}{4}$ in.

159. 2.3 cm and 8.9 cm

160. 1.2 cm and 7 cm

161. 5 cm and 9.8 cm

162. 8 cm and 8.3 cm

4   TEAM PROBLEM-SOLVING EXERCISES

A *lumen* is the most common measurement of light output, or luminous flux. The lumen rating of a light source is a measure of the total light output of the light source. Light sources are labeled with an output rating in lumens. For example, an R30, 65-W incandescent indoor flood lamp may have a rating of 750 lumens. A *watt* is the unit used to measure electrical power. It defines the rate of energy consumption by an electrical device when it is in operation. The energy cost of operating an electrical device is calculated as its wattage time in hours of use.

The measure of electrical energy from which electricity billing is determined is the kilowatt-hour (kWh). For example, a 100-W bulb operated for 1,000 h would consume 100 kWh. (100 W × 1,000 h = 100,000 Wh, or 100 kWh)

The types of light bulbs that are commonly available are incandescent light bulbs, compact fluorescent light bulbs (CFLs), and light-emitting diodes (LEDs).

1. A conference room has five R30, 65-W incandescent indoor flood lamps. How many lumens are produced if all 5 lamps are lighted?

2. How many kilowatt-hours are consumed if the 5 lamps are lighted for 36 h?

3. At the electrical billing rate of $0.12 per kWh, how much will it cost to operate the five lamps for 1,000 h?

4. If the incandescent bulbs in Exercise 1 are replaced with LED bulbs that produce the same number of lumens using 13 watts, how much will it cost to operate the five lamps for 1,000 h?

5. The life expectancy for the LED is fifteen times the life expectancy for the incandescent, the cost of each incandescent bulb is $3.33, and the cost of each LED bulb is $19.88. Based on cost alone, which is the better deal?

6. What factors other than cost might be considered in the decision to change from incandescent to LED bulbs?

4   CONCEPTS ANALYSIS

1. Give four items that are measured in kilograms.

2. Give four items that are measured in meters.

3. If you were building a house, what units of metric linear measure would you be likely to use?

4. Explain how metric units for length, weight, and capacity are similar.

5. Which measure is longer, a yard or a meter?

6. Which metric measure would you use to dispense a liquid medicine?

7. If your medicine bottle reads 5 mg, is the medicine more likely to be in liquid or capsule form?

8. The kilometer is usually associated with what U.S. customary unit?

Identify the mistake, explain why it is wrong, and correct the mistake.

9. 3.252 dkL = 325.2 kL

10. 418 yd = __ m?

$$418 \, \cancel{yd}\left(\frac{0.9144 \text{ m}}{1 \, \cancel{yd}}\right) = 362 \text{ yd}$$

11. 24 qt = __ L?

$$24 \, \cancel{qt}\left(\frac{1 \text{ L}}{0.9463 \, \cancel{qt}}\right) = 25 \text{ L}$$

12. Compare and contrast the U.S. customary and the metric systems of measurement.

13. Why do you think the metric system has not fully replaced the U.S. customary system in the United States?

4   PRACTICE TEST

Change to the U.S. customary unit indicated.

1. 276 in. = _____ ft

2. 5 gal = _____ cups

Perform the indicated operations. Write result in standard notation.

3. 5 yd 1 ft 7 in. – 2 yd 2 ft 9 in.

4. 7(6 lb 12 oz)

Change to the metric unit indicated.

5. 298 m = _____ km

6. 8 dm = _____ mm

7. 5.2 dL of liquid are poured from a container holding 10 L. How many liters of liquid remain in the container?

8. A bar of soap weighs 175 g. How much do 15 bars of soap weigh (in kilograms)?

9. 75 mi = _____ km

10. 25 kg = _____ lb

11. 4 L = _____ pt

12. How many liters of weed killer are contained in a 55-gal drum?

13. Change 48°F to degrees Celsius to the nearest whole degree.

14. Change 70°C to degrees Fahrenheit.

15. Find the number of minutes in 42 h.

16. How many hours are in 5,700 min?

17. A GIS Specialist worked 47 h a week for 5 consecutive weeks. How many hours did she work?

18. A machine was in use for 20 h a day during the 30 days in June. How many hours was the machine producing?

19. The number 840 has how many significant digits?

20. A tool measures 3.5 cm. What is the greatest possible error?

21. A hip replacement part measures 24.75 cm and the blueprint specifies the part is 24.72 cm. What is the relative error and percent error?

22. Find the sum of 32.5 cg, 4.76 cg, 503.2 cg, and 16.3 cg and use an appropriate approximation to report the sum.

Measure line segments 23–24 in Fig. 4–35 to the nearest millimeter.

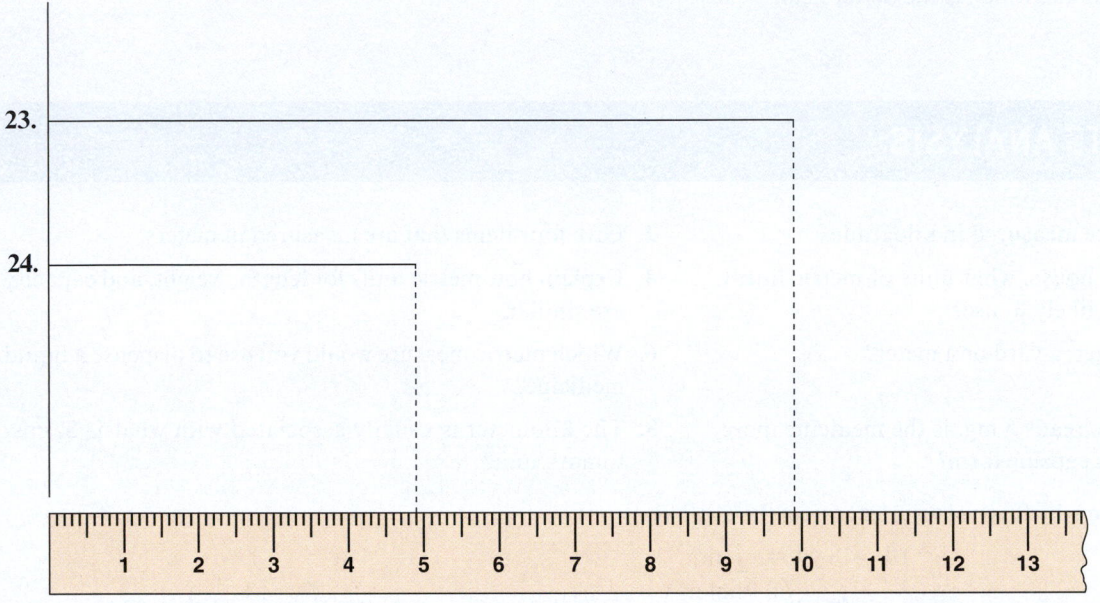

FIGURE 4–35

25. A 0.243-caliber bullet travels 2,450 $\frac{ft}{s}$ for the first 200 yd. Using conversion factors, change 2,450 $\frac{ft}{s}$ to $\frac{m}{s}$ to the nearest whole number.

26. Lyla Jackson weighs 172 lb and is 5 ft 10 in. Find her body mass index.

27. Find the midpoint of the two measures taken on the metric rule: 3.8 cm and 5.9 cm.

28. Find the distance between the points on a U.S. customary rule: $17\frac{15}{32}$ in. and $23\frac{5}{16}$ in.

5

Signed Numbers and Powers of 10

Michaeljung/Shutterstock

In Great Company

Deadly Word Problems (January 2008)

Some people hate hospitals, hate even the smell of hospitals. Nowadays, people don't go there unless they are sick or they work there—and no one likes to do either one. It gets worse. Have you ever considered that half the people who graduate with a medical degree were in the lower half of their class? Or that one quarter of them were in the lowest quarter of their class? Or that one in ten medical professionals were in the lowest one-tenth of their class? And you don't know which one is treating you.

This is important because clinical error and negligence are responsible for injuries that lead to about 1 in 25 hospital admissions. Most of these are caused by drug dosage problems—either the wrong drug or the wrong amount of the right drug.

Now, these facts are not something anyone likes to think about, so the medical journal *Annals of Internal Medicine* decided to think about them. In January 2008, the *Annals of Internal Medicine* published a study in which researchers asked doctors to calculate the correct dose of epinephrine for a hypothetical 5-year-old with an allergic reaction.

Epinephrine is simply another name for adrenalin, the stuff your body naturally produces when you are excited, scared, or exercising hard. It increases heart rate, constricts blood vessels, and dilates air passages. As a result, it is often used to assist those who are having trouble breathing. In the test for this study, the physicians were told the patient was 5 years old and suffering from a peanut allergy. Such an allergic reaction requires quick treatment to prevent serious injury or death.

The correct dose was 0.12 mg. Half the 28 volunteer doctors were given a bottle labeled "1 milligram in a 1 milliliter solution," while the others got bottles labeled "1 milliliter of a 1:1000 solution." This is just two different ways of saying the same thing. The good news: 11 of the first 14 doctors got the math right, and they figured it out in roughly 30 s, on average. The bad news: only 2 in the second group did, and it took them an average of over 2 min to calculate what turned out to be the wrong answer. In fact, one doctor in the second group administered what would have been eight times the correct amount.

Converting from milligrams per milliliter can lead to errors by factors of 10. That's more than enough to kill the patient.

5-1   Adding Signed Numbers

LC LEARNING CATALYTICS

Perform the operations.
1. $8 - 3 + 7 - 4$
2. $12 - 3 - 5 - 1$

Integers: the set of whole numbers plus the opposite of each natural number

Opposite: a number that is the same number of units from zero but in the opposite direction of the original number

1   Compare Signed Numbers. Many physical phenomena have values that are less than zero. Integers are needed to express these values. When the opposite of each natural number is included, the set of whole numbers can be expanded to form the set of **integers.** Fig. 5–1 shows how the set of whole numbers is extended to include all integers. A number line for integers continues indefinitely in *both* directions. Numbers get smaller as we proceed to the left and larger as we proceed to the right.

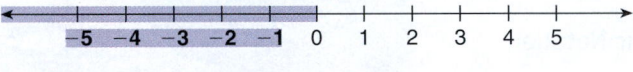

FIGURE 5–1 Integers.

The **opposite** of a number is the same number of units from zero but in the opposite direction. Fig. 5–2 shows a number line with numbers and their opposites.

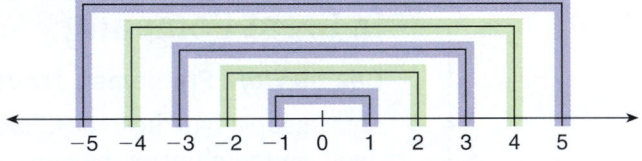

FIGURE 5–2 Opposite numbers.

The relationship among the three sets of numbers—natural numbers, whole numbers, and integers—is shown in Fig. 5–3.

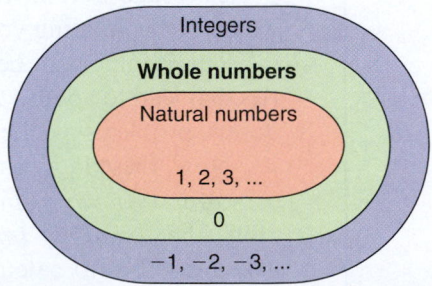

FIGURE 5–3 Relating natural numbers, whole numbers, and integers.

> **TIP** **Are Positive Signs Necessary?** As you can see from the number line in Fig. 5–1, it is not necessary to include the positive sign for positive values. You may sometimes want to include the sign to draw attention to it or to emphasize its direction. Negative signs, however, must always be included.

Signed numbers: all types of numbers that are either positive or negative or 0

Integers are just one type of signed numbers. Fractions, mixed numbers, and decimals also have values less than zero. **Signed numbers** are all types of numbers that are either positive or negative or 0. We begin our study of signed numbers with integers and expand to other types of signed numbers later in the chapter.

If a number is positioned to the left of another number on a number line, it represents a smaller value. Larger values are positioned to the right of a number on the number line. As with whole numbers, fractions, and decimals, we use the inequality symbols $<$ and $>$ to compare signed numbers.

> **To compare signed numbers:**
>
> 1. Mentally position the two numbers on a number line.
>
> 2. The leftmost number is the smaller number. If the smaller number is written first, use the $<$ symbol between the numbers. If the larger number is written first, use the $>$ symbol between the numbers.

STOP AND CHECK

Use the symbol $<$ or $>$ to show the relationship between each pair of numbers.
1. -2 ____ -5
2. -4 ____ 0
3. -3 ____ 1

Answers:
1. $>$ 2. $<$ 3. $<$

EXAMPLE 1

Use the symbol $<$ or $>$ to show the relationship between each pair of numbers.

(a) 7 ____ 9 **(b)** 0 ____ -1 **(c)** -4 ____ -2 **(d)** -2 ____ 0

(a) $7 < 9$ 7 is smaller—to the left of 9 on the number line.

(b) $0 > -1$ 0 is larger—to the right of -1 on the number line.

(c) $-4 < -2$ -4 is smaller—to the left of -2 on the number line.

(d) $-2 < 0$ -2 is smaller—to the left of 0 on the number line.   **See Exercises 1–8.**

Absolute value (| |): the number of units or distance the number is from zero

The distance between each pair of consecutive integers is the same for all integers located on the number line. This concept of *distance between numbers* on the number line is related to the concept of *absolute value*. The **absolute value** of a number is often described as the number of units or distance the number is from zero. The symbol for absolute value is $|\ |$; $|3|$ is read "the absolute value of 3."

> **TIP** **What Does Nonnegative Mean?** Are positive numbers also nonnegative? Yes, every positive number is nonnegative.
>
> Are any nonnegative numbers not positive? Yes, zero is nonnegative *and* nonpositive. Therefore, when we say **nonnegative,** we mean positive or zero. Similarly, **nonpositive** means negative or zero.

Nonnegative: positive or zero, $n \geq 0$

Nonpositive: negative or zero, $n \leq 0$

Distance is a physical property, and it cannot have a negative value. The absolute value of a number is always a nonnegative value.

TIP   **Write Facts Symbolically**   Positive numbers are represented symbolically as $a > 0$, where a is any real number greater than zero.

Negative numbers are represented by $a < 0$, where a is any real number less than zero.

The symbols $\geq$ which means "is greater than or equal to" and $\leq$ which means "is less than or equal to" are used to represent nonnegative and nonpositive numbers, respectively.

Practice reading symbolic statements in words.

$a > 0$	positive numbers	$a \geq 0$	nonnegative numbers (positives and zero)
$a < 0$	negative numbers	$a \leq 0$	nonpositive numbers (negatives and zero)

To find *the absolute value* of a number:

Apply the appropriate relationship.

$|a| = a$, for $a > 0$ The absolute value of a positive number is equal to itself.

$|a| = -a$, for $a < 0$ The absolute value of a negative number is equal to its opposite.

$|0| = 0$ The absolute value of zero is zero.

STOP AND CHECK
Evaluate.
1. $|-2|$
2. $|0|$
3. $|5|$

Answers:
1. 2   2. 0   3. 5

EXAMPLE 2

Give the absolute value of the following quantities: **(a)** $|9|$ **(b)** $|-4|$

(a) $|9| = 9$ 9 is positive. Its absolute value is the number itself.

(b) $|-4| = 4$ -4 is negative. Its absolute value is the opposite of -4.

See Exercises 9–12.

Did You Know?   **There are two components of a signed number** Integers and other signed numbers tell us *two things*. For the integer -8, the negative sign tells us the number is to the left of zero and is referred to as the *directional sign* of the number. The 8 tells us how many units the number is from zero; it is the *absolute value* of the number.

Every number except zero has an opposite. The opposite of a number is also called the **additive inverse** of the number. A number and its additive inverse (opposite) have the same absolute value but different directional signs.

Additive inverse: the opposite of a number is the additive inverse of the number because $a + (-a) = 0$ for all values of a

To find the additive inverse:

1. Find the value that has the same absolute value and opposite sign.

STOP AND CHECK
Find the opposite of each number and show your answer on the number line.
1. 3   2. -4

Answers:
1. The opposite is -3.

2. The opposite is 4.

EXAMPLE 3

Find the opposite of each number and show your answer on the number line.

(a) -8 **(b)** 6   **(c)** 0

(a) The opposite of -8 is 8.

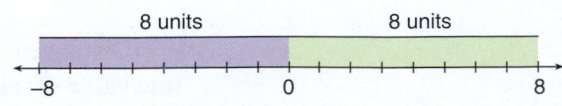

(b) The opposite of 6 is −6.

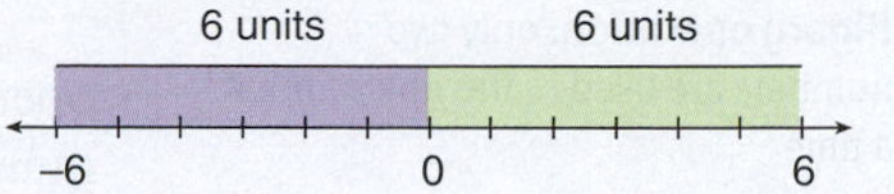

(c) Zero is its own opposite.

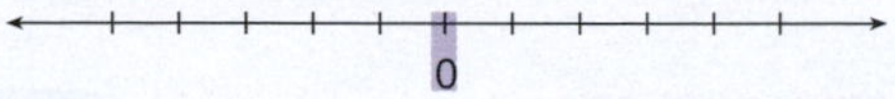

See Exercises 13–16.

---

2 **Add Signed Numbers with Like Signs.** Since our earliest experiences with addition we have added positive numbers. In the addition problem $3 + 2 = 5$, all three numbers are positive. Numbers are understood to be positive when no sign is given. To emphasize that numbers are positive, we can write the problem as $+3 + +2 = +5$. We can also write the problem using parentheses to separate the two plus signs, $+3 + (+2) = +5$. The plus symbol serves two purposes in mathematical expressions. It is used as the directional sign for a positive number, and it is used to show addition. In the expression $+3 + (+2) = +5$, the plus sign in front of the parentheses shows the operation of addition. The plus sign inside the parentheses identifies the number as positive.

Adding integers builds on our knowledge of addition of positive numbers.

To add 3 and 2 on the number line, we begin at zero and move three units to the right (positive direction). This move takes us to $+3$. From the $+3$ position, we move two units to the right. This second move takes us to the $+5$ position (Fig. 5–4). In other words, $3 + 2 = 5$.

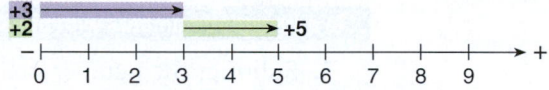

FIGURE 5–4 Adding using a number line.

To add two negative integers on the number line, $(-3) + (-2)$, we begin at zero and move three units to the left (negative direction). This takes us to -3 on the number line. Then, from -3 we move two units to the left. This second move takes us to -5 (Fig. 5–5). Thus, $-3 + (-2) = -5$.

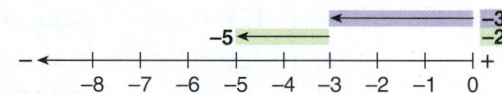

FIGURE 5–5 Adding two negative numbers using a number line.

When adding the two positive numbers or the two negative numbers, the sum could have been obtained using the following rule.

To add signed numbers with like signs:

1. Add the absolute values of the numbers.

2. Give the sum the common or like sign.

EXAMPLE 4

Add −12 and −5.

$$-12 + (-5) = -17 \quad \text{Add absolute values. Keep common negative sign.}$$

See Exercises 17–26.

Binary operation: only two numbers are used in the operation at a time

Addition is a **binary operation.** This means that only two numbers are used in the operation. If more than two numbers are added, two are added and then the next number is added to the sum of the first two. Rules developed for addition apply to two numbers at a time.

STOP AND CHECK

1. Add $-4 + (-2) + (-5)$.

Answer:

1. -11

EXAMPLE 5

Add $-3 + (-4) + (-1)$.

$-3 + (-4) + (-1) =$ Add the first two numbers.

$\quad\quad -7 + (-1) = -8$ Add the sum to remaining number.

See Exercises 27–31.

3 **Add Signed Numbers with Unlike Signs.** When adding numbers with unlike signs, we move in the direction indicated by the sign of each number, positive to the right and negative to the left. To add $-4 + 3$, we first move four units to the left of zero (see Fig. 5–6). This takes us to -4. Then, from the -4 position, we move three units to the right. This takes us to -1. Thus, $-4 + 3 = -1$. This process is generalized by a rule using absolute values.

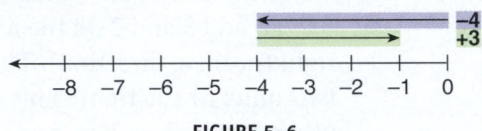

FIGURE 5–6

To add signed numbers with unlike signs:
1. Subtract the number with the smaller absolute value from the number with the larger absolute value.
2. Give the sum the sign of the number with the larger absolute value.

STOP AND CHECK

Add.

1. $9 + (-3)$
2. $3 + (-4)$
3. $-2 + 8$

Answers:

1. 6   2. -1 3. 6

EXAMPLE 6

Add $7 + (-12)$.

$7 + (-12) =$ Signs are unlike.

$7 + (-12) = -5$ Subtract absolute values: $12 - 7 = 5$. Keep the sign of the -12 (negative), the number with the larger absolute value.

See Exercises 32–39.

TIP **To Add, You . . . Subtract?** Phrases we used in arithmetic for adding integers, such as "find the sum" or "add," can be confusing. In addition of integers, sometimes we add absolute values and sometimes we subtract absolute values. We often use the word "**combine**" to imply the addition of numbers with either like or unlike signs.

Combine: to add numbers with either like or unlike signs

STOP AND CHECK

Combine.

1. $-6 + 9$
2. $-3 + 1$

Answers:

1. 3   2. -2

EXAMPLE 7

Combine $-5 + 7$.

$-5 + 7 =$ Signs are unlike.

$-5 + 7 = 2$ Subtract absolute values: $7 - 5 = 2$. Keep the sign of the 7 (positive), the number with the larger absolute value.   **See Exercises 40–44.**

When an addition involves several positive numbers and several negative numbers, it is convenient to group all the positive numbers and all the negative numbers, and to add each group separately. We then add the sums of each group. We can do this because of the *commutative* and *associative* properties of addition of real numbers. In the next example, we combine the commutative and associative properties of addition to add several signed numbers.

STOP AND CHECK
Add.
1. $3 + (-2) + 5 + (-12)$
2. $-2 + 5 + (-3) + (-7)$

Answers:
1. -6 **2.** -7

EXAMPLE 8

Add $8 + (-9) + 13 + (-15)$.

$8 + (-9) + 13 + (-15) =$	Signs are unlike. Rearrange and group numbers with like signs.
$[8 + 13] + [-9 + (-15)] =$	The brackets, [ ], show the grouping and separate the operational plus sign from the directional sign of -9.
	Add groups of numbers with like signs.
$21 + (-24) = -3$	Combine resulting numbers using the rule for adding numbers with unlike signs.

See Exercises 45–62.

The properties of addition that involve the number zero apply to integers as well as to whole numbers, fractions, and decimals. When zero is added to any real number, the number is not changed. We say that zero is the *additive identity*. The sum of a number and its opposite is zero.

> **TIP** **Write Definitions Using Symbols** Zero is the *additive identity* because $a + 0 = a$ for all real values of a.
>
> The opposite of a number is the *additive inverse* of the number because $a + (-a) = 0$ for all real values of a.

STOP AND CHECK
Add.
1. $0 + (-8)$
2. $-3 + 3$
3. $3 + (-1) + (-2)$

Answers:
1. -8 **2.** 0 **3.** 0

EXAMPLE 9

Add: **(a)** $-2 + 0$ **(b)** $5 + (-5)$ **(c)** $0 + (-3) + 4$ **(d)** $5 + (-2) + (-5)$

(a) $-2 + 0 = -2$	Additive identity.
(b) $5 + (-5) = 0$	Additive inverse.
(c) $0 + (-3) + 4 =$	Additive identity.
$-3 + 4 = 1$	Add numbers with unlike signs.
(d) $5 + (-2) + (-5) =$	Apply the commutative and associative properties.
$5 + (-5) + (-2) =$	Additive inverse.
$0 + (-2) = -2$	Additive identity.

See Exercises 63–68.

5–1 EXERCISES

MyLab Math For additional practice go to your study plan in MyLab Math.

1 Use the symbol $>$ or $<$ to show the relationship between each pair of numbers. *See Example 1.*

1. 5 ___ 8   **2.** -3 ___ -2 **3.** -7 ___ -9 **4.** 0 ___ 5

5. -8 ___ 0   **6.** 12 ___ 7   **7.** -3 ___ 2   **8.** 5 ___ -7

Write the absolute value. *See Example 2.*

9. $|7|$ **10.** $|-17|$ **11.** $|8|$ **12.** $|-7|$

Write the opposite of each number. *See Example 3.*

13. 42   **14.** -17 **15.** -78 **16.** 57

2 Add. *See Example 4.*

17. $8 + 15$ **18.** $-9 + (-7)$ **19.** $-5 + (-8)$ **20.** $(-12) + (-17)$

21. $-24 + (-31)$ **22.** $-78 + (-46)$ **23.** $12 + 87$ **24.** $-21 + (-38)$

25. $-32 + (-16)$ **26.** $(-58) + (-103)$

See Example 5.

27. $7 + 10 + 12$ **28.** $-5 + (-8) + (-7)$ **29.** $-5 + (-12) + (-36)$

30. $52 + (+92) + (+88)$ **31.** $-107 + (-502) + (-396)$

3 Add. *See Examples 6.*

32. $5 + (-3)$ **33.** $8 + (-4)$ **34.** $9 + (-3)$

35. $3 + (-9)$ **36.** $-4 + (+2)$ **37.** $4 + (-6)$

38. $21 + (-14)$ **39.** $32 + (-72)$

See Example 7.

40. $-3 + 5$ **41.** $-8 + 4$ **42.** $-18 + 8$ **43.** $-4 + 6$ **44.** $-3 + 7$

See Example 8.

45. $17 + (-4) + 3 + (-1)$ **46.** $-3 + 2 + (-7)$ **47.** $47 + (-82) + 2$

48. $14 + (-6) + 1$ **49.** $-7 + (-3) + (-1)$ **50.** $4 + 2 + (-3) + 10$

51. $2 + (-1) + 8$ **52.** $-2 + 1 + (-8) + 12$ **53.** $6 + (-3) + 5 + (-8)$

54. $12 + (-5) + 8 + (-18)$ **55.** $4 + (-7) + 6 + (-15)$ **56.** $18 + (-7) + (-9) + (-16)$

57. $8 + (-4) + (-13) + 22$ **58.** $18 + (-3) + 5 + (-32)$ **59.** $12 + (-15) + 73 + (-46)$

60. $(-83) + (-21) + 42 + (-36)$ **61.** $(-5) + (-18) + 32 + (-7)$ **62.** $15 + (-32) + 46 + (-19)$

See Example 9.

63. $15 + (-15)$ **64.** $-7 + 7$ **65.** $92 + (-92)$

66. $(-396) + (396)$ **67.** $-3 + 0$ **68.** $0 + (-7)$

69. **PFIN/BUS**   You open a bank account by depositing $242. You then write checks for $21, $32, and $123. What is your new account balance?

70. **PFIN/BUS**   A stock priced at $42 has the following changes in one week: -3, $+8$, -6, -7, $+2$. What is the value of the stock at the end of the week?

71. **AVIA**   Liquid hydrogen, fuel in the space shuttle, must be kept below $-253°C$. A tank of liquid hydrogen is stored at $-320°C$, but the temperature is raised by $25°C$. What is the new temperature of the hydrogen?

72. The temperature at the South Pole is recorded as $-38°F$ at midnight. The temperature rises $15°F$ by noon. Determine the temperature reading at noon.

5-2	Subtracting Signed Numbers

LEARNING OUTCOMES

1 Subtract signed numbers.

2 Combine addition and subtraction.

LC LEARNING CATALYTICS

1. $15 + 6 - 12$

2. $9 - 2 + 12$

1 Subtract Signed Numbers. Addition and subtraction are related. Subtracting a number is the *same as* adding the opposite of the number.

$6 - 9$	Show directional signs of the numbers.
$+6 - (+9)$	Subtract $+9$.
$+6 + (-9)$	Add the opposite of $+9$ or add -9.

Let's write this relationship as a rule.

To subtract signed numbers:

1. Change the subtraction sign to addition.

2. Change the sign of the second number (subtrahend) to its opposite.

3. Apply the appropriate rule for adding signed numbers.

STOP AND CHECK

Subtract.

1. $5 - 12$

2. $-5 - 12$

3. $-5 - (-12)$

Answers:

1. -7 **2.** -19 **3.** 7

EXAMPLE 1

Subtract: **(a)** $2 - 6$ **(b)** $-9 - 5$ **(c)** $12 - (-4)$ **(d)** $-7 - (-9)$

(a)

$2 - 6 =$	Subtract positive 6 from positive 2. Write sign of subtrahend.
$2 - (+6) =$	Change subtraction to addition.
$2 + (-6) =$	Add the opposite of $+6$, which is -6, to 2.
$2 + (-6) = \mathbf{-4}$	Apply the rule for adding numbers with unlike signs.

(b)

$-9 - 5 =$	Subtract positive 5 from negative 9. Write sign of subtrahend.
$-9 - (+5) =$	Change subtraction to addition.
$-9 + (-5) =$	Add the opposite of $+5$, which is -5, to -9.
$-9 + (-5) = \mathbf{-14}$	Apply the rule for adding numbers with like signs.

(c)

$12 - (-4) =$	Subtract negative 4 from positive 12.
$12 + (+4) =$	Add the opposite of -4, which is $+4$, to 12.
$12 + 4 = \mathbf{16}$	Apply the rule for adding numbers with like signs.

(d)

$-7 - (-9) =$	Subtract negative 9 from negative 7.
$-7 + (+9) =$	Add the opposite of -9, which is $+9$, to -7.
$-7 + 9 = \mathbf{2}$	Apply the rule for adding numbers with unlike signs.

See Exercises 1–24.

TIP **Writing a Subtraction as an Equivalent Addition** When writing a subtraction as an equivalent addition, you make *two* changes.

1. Change the operation from subtraction to addition.

2. Change the subtrahend to its opposite.

$2 - (+6)$	$-9 - (+5)$	$12 - (-4)$	$-7 - (-9)$
$2 + (-6)$	$-9 + (-5)$	$12 + (+4)$	$-7 + (+9)$
↑ ↑	↑ ↑	↑ ↑	↑ ↑
1. 2.	1. 2.	1. 2.	1. 2.

Subtractions involving zero and opposites are interpreted first as equivalent addition problems. Since zero has no opposite, +0 and −0 are still 0.

STOP AND CHECK
Evaluate.
1. $12 - 0$
2. $0 - 12$
3. $-15 - (-15)$
4. $-8 - 8$

Answers:
1. 12 **2.** −12 **3.** 0 **4.** −16

EXAMPLE 2

Evaluate: **(a)** $8 - 0$ **(b)** $0 - 15$ **(c)** $-32 - (-32)$ **(d)** $-15 - 15$

(a) $8 - 0 =$ | Subtract zero from positive 8.
$8 + (0) =$ | Add zero to positive 8.
$8 + (0) = \mathbf{8}$ | Zero added to any number is that number (additive identity).

(b) $0 - 15 =$ | Subtract positive 15 from zero. Write sign of subtrahend.
$0 - (+15) =$ | Change subtraction to addition.
$0 + (-15) =$ | Add the opposite of $+15$, which is -15, to zero.
$0 + (-15) = \mathbf{-15}$ | Any number added to zero (additive identity) is that number.

(c) $-32 - (-32) =$ | Subtract negative 32 from negative 32.
$-32 + (+32) =$ | Add the opposite of -32, which is $+32$, to -32.
$-32 + 32 = \mathbf{0}$ | A number added to its opposite (additive inverse) is zero.

(d) $-15 - 15$ | Subtract positive 15 from negative 15.
$-15 - (+15) =$ | Change subtraction to addition.
$-15 + (-15) =$ | Add the opposite of $+15$, which is -15, to -15.
$-15 + (-15) = \mathbf{-30}$ | Apply the rule for adding numbers with like signs.

See Exercises 25–36.

Did You Know? Subtracting an opposite from a number is the same as adding the number to itself.

$$3 - (-3) = 3 + 3 = 6.$$

2 **Combine Addition and Subtraction.** When we express the sum and difference of signed numbers with all the appropriate operational and directional signs, we have an expression with both an operational sign and a directional sign between every two numbers. We have four different possibilities when two signs are written together. We can simplify these four possibilities.

TIP **Omitting Signs** Writing mathematical expressions that include every operational and directional sign is cumbersome. In general practice, we omit as many signs as possible. When two signs are written between two numbers, we can write a simplified expression with only one sign.

	Double Signs		Single Sign
Like Signs	Plus, Plus: $+3 + (+5)$	Add $+3$ and $+5$.	$3 + 5$
	Minus, Minus: $+3 - (-5)$	Change to addition.	
		$+3 + (+5)$	$3 + 5$
Unlike Signs	Plus, Minus: $+3 + (-5)$	Add $+3$ and -5.	$3 - 5$
	Minus, Plus: $+3 - (+5)$	Change to addition.	
		$+3 + (-5)$	$3 - 5$

To generalize, *two like signs between signed numbers,* whether both plus or both minus, translate to adding a positive number. Use just *one plus sign:*

$$+ + \rightarrow + \qquad - - \rightarrow + \qquad \text{like signs} \rightarrow +$$

Two unlike signs between signed numbers, either a plus/minus or a minus/plus, translate to adding a negative number. Use just *one minus sign:*

$$+ - \rightarrow - \qquad - + \rightarrow - \qquad \text{unlike signs} \rightarrow -$$

To add and subtract more than two signed numbers:

1. Rewrite the problem so that all integers are separated by only one sign. Or, change all subtractions to equivalent additions.

2. Add the series of signed numbers from left to right.

STOP AND CHECK
Evaluate.
1. $-2 - (-3) - 5$
2. $8 + 3 - (-4)$

Answers:
1. -4 2. 15

EXAMPLE 3

Evaluate: **(a)** $3 - (-5) - 6$ **(b)** $-8 + 10 - (-7)$

(a) $3 - (-5) - 6 =$	Rewrite with only one sign between signed numbers.
$3 + 5 - 6 =$	Add numbers with like signs.
$8 - 6 = 2$	Apply the rule for adding numbers with unlike signs.
(b) $-8 + 10 - (-7) =$	Rewrite with only one sign between signed numbers.
$-8 + 10 + 7 =$	Add numbers with like signs.
$-8 + 17 = 9$	Apply the rule for adding numbers with unlike signs.

See Exercises 37–60.

The key to solving applied problems with signed numbers is to identify which numbers are positive and which are negative.

Positive key words: profits, gains, money in the bank, temperatures above zero, receipts, income, winnings, and so on.

Negative key words: losses, deficits, checks that cleared the bank, temperatures below zero, drops, declines, payments, and so on.

EXAMPLE 4

 **BUS**   A landscaping business makes a profit of $345 one week, has a loss of $34 the next week, and makes a profit of $235 the third week. What is the total, or net, profit?

The total profit is the sum of the weekly profits and losses.

$\$345 - \$34 + \$235$	Interpret profits as positive and losses as negative.
profit loss profit	
$345 + 235 - 34$	Rearrange and add positives.
$580 - 34 = 546$	Apply rule for adding numbers with unlike signs.

The total or net profit for the three weeks is $546.   Interpret answer.   **See Exercise 61.**

EXAMPLE 5

In a recent year, 98°F was the highest temperature in Boston, and −2°F was the lowest temperature. What was the temperature range for the city that year? (The range is the difference between the highest and lowest values.)

$$98 - (-2) =$$ Subtract −2 from 98.

$$98 + 2 = 100$$ Rewrite with only one sign between integers, and apply the appropriate rule for adding integers.

The temperature range for Boston that year was 100°F.   Interpret answer.   **See Exercises 62–63.**

5-2 EXERCISES   MyLab Math For additional practice go to your study plan in MyLab Math.

1 Subtract. *See Example 1.*

1. −3 − 9	**2.** 8 − 2	**3.** 9 − 15
4. (−3) − (−7)	**5.** −11 − 14	**6.** (−6) − (−3)
7. 5 − (−3)	**8.** 8 − 11	**9.** −8 − 1
10. 11 − (−2)	**11.** (−8) − (−7)	**12.** (−15) − (−7)
13. (20) − (−42)	**14.** (−38) − (−27)	**15.** (−42) − (+16)
16. (−21) − (+36)	**17.** (−18) − (+15) − (−18)	**18.** (−12) − (+21) − (+72)
19. 14 − (−21) − 17	**20.** (−14) − (−21) − 24	**21.** −142 − (+46) − (−21)
22. 217 − (−38) − (+172)	**23.** 802 + (196) − (−415)	**24.** 72 − (−23) − (+198)

See Example 2.

25. 15 − 0	**26.** 0 − 8	**27.** −12 − 0
28. 0 − (−8)	**29.** 0 − (−7)	**30.** 10 − 0
31. 28 − (−28)	**32.** −46 − 46	**33.** 7 − (−7)
34. −18 − 18	**35.** 5 − 5	**36.** 21 − 21

2 Evaluate. *See Example 3.*

37. 5 + 8 − 9	**38.** 6 − 4 + 5	**39.** 9 + 2 − 5
40. −1 + 1 − 4	**41.** 5 + 3 − 7	**42.** 7 + 3 − (−4)
43. −8 − 2 − (−7)	**44.** −3 + 4 − 7 − 3	**45.** 2 − 4 − 5 − 6 + 8
46. 8 − 3 + 2 − 1 + 7	**47.** −5 − 3 + 8 − 2 + 4	**48.** 6 − (−3) + 5 − 6 − 9
49. −8 + 2 − 7 + 14	**50.** 3 − 5 + 8 − 11 − 15	**51.** 5 − 2 + 9 − 6 − 12 + 32
52. 18 + 12 − 16 − 32 − 81	**53.** −28 + 32 − 17 − 32	**54.** 78 − 64 + 32 − 78
55. 52 − 96 − 102 + 86	**56.** 42 − 17 + 86 − 191	**57.** 77 − 96 − 102 + 86
58. 149 − 23 + 6 − 82	**59.** 54 − 87 + 33 − 21	**60.** 228 − 21 + 33 − 54

61. **BUS** Computing Solutions records a profit of $28,296 one quarter (three months), a loss of $1,896 for the second quarter, a profit of $52,597 for the third quarter, and a profit of $36,057 for the fourth quarter. What is its net profit for the year? *See Example 4.*

62. New Boston, Texas, registered −8°F as its lowest temperature one year and 99°F as its highest temperature for the same year. What was the temperature range for New Boston? *See Example 5.*

63. The temperatures for Bowling Green, Kentucky, ranged from 102°F to −5°F. What was the temperature range for the city? *See Example 5.*

64. Explain the difference between subtracting zero from a number and subtracting a number from zero.

Use the table showing the U.S. trade balance with Aruba for Exercises 65–68.

65. **BUS** Express the difference in trade balances in May 2012 and April 2012 with a signed number.

66. **BUS** Express the difference in trade balances in July 2012 and June 2012 with a signed number.

67. **BUS** Find the difference in trade balances for June 2012 and May 2012.

68. **BUS** Use your knowledge of operations with signed numbers to find the overall trade balance (sum of the four trade balances) from April 2012 through July 2012.

U.S. Trade with Aruba (in millions of U.S. dollars)

Month	Trade Balance
April 2012	−55.4
May 2012	91.5
June 2012	−5.3
July 2012	58.3

Source: U.S. Census Bureau, Foreign Trade Division, Data Dissemination Branch, Washington, DC 20233.

5-3 Multiplying and Dividing Signed Numbers

LEARNING OUTCOMES

1. Multiply signed numbers.
2. Evaluate powers of signed numbers.
3. Divide signed numbers.

LC LEARNING CATALYTICS
Evaluate.
1. $3(4)(5)$
2. $\dfrac{5+9}{2}$
3. 2^3

Multiplying absolute values of signed numbers is exactly the same as multiplying whole numbers, fractions, and decimals. However, the rules for multiplying signed numbers must also include the assignment of the proper sign to the product.

1 Multiply Signed Numbers. Just as with whole numbers, fractions, and decimals, multiplication of signed numbers is a binary operation that is commutative and associative; that is, two factors at a time can be multiplied in any order. More than two factors can be grouped in any manner. These examples show how we assign signs to the product of two signed numbers:

Like Signs

$4(6) = 24, \qquad -3(-7) = 21$

Unlike Signs

$-10(2) = -20, \qquad 8(-2) = -16$

To multiply two signed numbers:

1. Multiply the absolute values of the numbers.

2. If the factors have like signs, the sign of the product is positive: $(+)(+) = +; (-)(-) = +$

3. If the factors have unlike signs, the sign of the product is negative: $(+)(-) = -; (-)(+) = -$

STOP AND CHECK

Multiply.
1. $-6(-9)$
2. $4 \cdot 7$
3. $30(-5)$
4. $-8 * 3$

Answers:
1. 54 2. 28 3. −150 4. −24

EXAMPLE 1

Multiply: **(a)** $-12(-2)$ **(b)** $10 \cdot 3$ **(c)** $25(-3)$ **(d)** $-5 * 7$

(a) $-12(-2) = \mathbf{24}$ Like signs give a positive product.

(b) $10 \cdot 3 = \mathbf{30}$ Like signs give a positive product.

(c) $25(-3) = \mathbf{-75}$ Unlike signs give a negative product.

(d) $-5 * 7 = \mathbf{-35}$ Unlike signs give a negative product.   **See Exercises 1–12.**

When the number 1 is multiplied by any real number, the result is the number. The number 1 is the *multiplicative identity* because $a \cdot 1 = 1 \cdot a = a$ for all real values of a.

STOP AND CHECK
Multiply.
1. $3(-2)(5)$
2. $-5(6)(-4)$
3. $-2(-1)(-5)(-3)$

Answers:
1. -30 2. 120 3. 30

EXAMPLE 2

Multiply: **(a)** $4(-2)(6)$ **(b)** $-3(4)(-5)$ **(c)** $-2(-8)(-3)$
(d) $-2(-3)(-4)(-1)$

(a) $4(-2)(6) =$	Multiply the first two factors and apply the rule for factors with unlike signs.
$-8(6) = -48$	Multiply and apply the rule for factors with unlike signs.
(b) $-3(4)(-5) =$	Multiply the first two factors and apply the rule for factors with unlike signs.
$-12(-5) = 60$	Multiply and apply the rule for factors with like signs.
(c) $-2(-8)(-3) =$	Multiply the first two factors and apply the rule for factors with like signs.
$16(-3) = -48$	Multiply and apply the rule for factors with unlike signs.
(d) $-2(-3)(-4)(-1) =$	Multiply the first two and the last two factors.
$6(4) = 24$	Multiply and apply the rule for factors with like signs.

See Exercises 13–18.

Did You Know? Multiplying by -1 is the same as finding the opposite. Also, when multiplying more than two factors, the multiplications can be ordered or grouped in a variety of ways. If there is a combination of multiplications and divisions, the left-to-right order is appropriate.

If we examine the multiplications in the preceding example more closely, we see that the number of negative factors affects the sign of the answer.

	Number of Negative Factors	**Sign of Product**
$4(-2)(6)$	1	$-$
$-3(4)(-5)$	2	$+$
$-2(-8)(-3)$	3	$-$
$-2(-3)(-4)(-1)$	4	$+$

To determine the sign of the product when multiplying three or more factors:

1. The sign of the product is *positive* if the number of negative factors is *even*.

2. The sign of the product is *negative* if the number of negative factors is *odd*.

STOP AND CHECK
Multiply.
1. $5(-4)(-3)(-1)$
2. $-3(7)(-2)(4)$

Answers:
1. -60 2. 168

EXAMPLE 3

Multiply: **(a)** $-2(6)(-1)(-3)$ **(b)** $2(-5)(1)(-3)$

(a) $-2(6)(-1)(-3) = -36$	Multiply absolute values. The odd number of negative factors makes the product negative.
(b) $2(-5)(1)(-3) = 30$	Multiply absolute values. The even number of negative factors makes the product positive.

See Exercises 19–25.

The *zero property of multiplication* extends to integers. The product of zero and any real number is zero: $a \cdot 0 = 0$. If we have two or more factors and one factor is zero, we can immediately write the product as zero without having to work through the steps.

EXAMPLE 4

Multiply: **(a)** $3(-21)(2)(0)$ **(b)** $-9(-2)(8)(-1)$ **(c)** $5(-2) + (-8)(0)$

(a) $3(-21)(2)(0) = \mathbf{0}$ Zero is a factor.

(b) $-9(-2)(8)(-1) = \mathbf{-144}$ Zero is not a factor.

(c) $5(-2) + (-8)(0) = -10 + 0 = \mathbf{-10}$ Zero is only a factor of the second product of the expression.

See Exercises 26–41.

Applied problems often require multiplication of integers.

EXAMPLE 5

BUS   In a 3-week period, a technology stock declined approximately 2 points each week. How many points did the stock decline in the 3 weeks?

Because there are equal declines each week, we multiply the amount of weekly decline times the number of weeks: $-2(3) = -6$. Thus, **the stock declined (negative) a total of 6 points over the 3-week period.**   See Exercises 42–43.

2   **Evaluate Powers of Signed Numbers.** Raising a number to a natural-number power is an extension of multiplication, so determining the sign of the result is similar to multiplying several integers. Observe the pattern for determining the sign of the result:

Positive Base   **Negative Base**

$(+4)^2 = (+4)(+4) = +16$ $(-4)^2 = (-4)(-4) = +16$

$(+4)^3 = (+4)(+4)(+4) = +64$ $(-4)^3 = (-4)(-4)(-4) = -64$

> **To raise signed numbers to a natural-number power, use the following patterns:**
>
> **1.** A positive number raised to any natural-number power is positive.
>
> **2.** Zero raised to any natural-number power is zero.
>
> **3.** To raise a negative to a power requires parentheses. A negative number raised to an even natural-number power is positive.
>
> **4.** A negative number raised to an odd natural-number power is negative.

EXAMPLE 6

Evaluate the powers: **(a)** 4^3 **(b)** 0^8 **(c)** $(-2)^4$ **(d)** -2^4 **(e)** $(-3)^5$ **(f)** -3^5

(a) $4^3 = 4(4)(4) = \mathbf{64}$ Base is positive.

(b) $0^8 = \mathbf{0}$ Base is zero.

(c) $(-2)^4 = (-2)(-2)(-2)(-2) = \mathbf{16}$ Base is negative, exponent is even.

(d) $-2^4 = -(2)(2)(2)(2) = -16$ Opposite of positive base, exponent is even.

(e) $(-3)^5 = (-3)(-3)(-3)(-3)(-3) = \mathbf{-243}$ Base is negative, exponent is odd.

(f) $-3^5 = -(3)(3)(3)(3)(3) = -243$ Opposite of positive base, exponent is odd.   See Exercises 46–65.

TIP   **Negative Base Versus an Opposite**   Look again at Example 6, parts c–f. $(-2)^4$ is not the same expression as -2^4. *To raise a negative number to a power requires parentheses.*

$(-2)^4$ is a *negative base* raised to a power. -2^4 is the *opposite* of 2^4.

Sometimes the values of two expressions are equal, but the interpretation is different. The expressions $(-3)^5$ and -3^5 give the same result because the exponent is an odd number.

$$(-3)^5 = (-3)(-3)(-3)(-3)(-3) = -243 \qquad -3^5 = -(3)(3)(3)(3)(3) = -243$$

EXAMPLE 7

ELEC   Security systems sometimes use a four-digit code to activate the system. How many different codes can be made with four digits? Identify codes that may be impractical.

The number system has 10 digits: 0, 1, 2, 3, 4, 5, 6, 7, 8, and 9, so we can fill each one of the four slots of the code in 10 different ways. We allow the same digit to be used repeatedly. (This is called **sampling with replacement.**) To find the total number of different codes, we multiply $10 \cdot 10 \cdot 10 \cdot 10$. This product can be written as 10^4, or 10,000. **There are 10,000 ways to make a four-digit security code. Some codes, such as 0000, may be impractical.**   **See Exercise 66.**

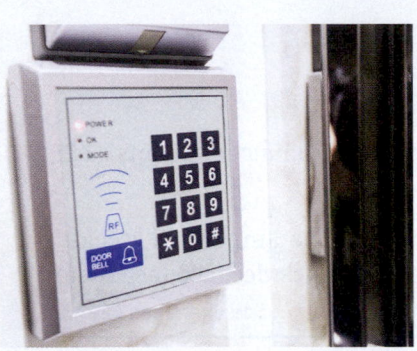

Zhudifeng/123RF

Sampling with replacement: in determining a sample, repeats are allowed

3 **Divide Signed Numbers.**   The rules for determining the sign when dividing signed numbers are similar to the rules for multiplying signed numbers.

To divide two signed numbers:

1. Divide the absolute values of the dividend and the divisor.

2. If the dividend and the divisor have like signs, the sign of the quotient is positive: $(+) \div (+) = +; (-) \div (-) = +$.

3. If the dividend and the divisor have unlike signs, the sign of the quotient is negative: $(+) \div (-) = -; (-) \div (+) = -$.

STOP AND CHECK

Divide.

1. $\dfrac{12}{-2}$ **2.** $\dfrac{-15}{3}$ **3.** $\dfrac{-2}{-1}$

Answers:
1. -6 **2.** -5 **3.** 2

EXAMPLE 8

Divide: **(a)** $\dfrac{-8}{-2}$ **(b)** $\dfrac{6}{3}$ **(c)** $\dfrac{10}{-2}$ **(d)** $\dfrac{-9}{1}$

(a) $\dfrac{-8}{-2} = 4$ Like signs give a positive quotient.

(b) $\dfrac{6}{3} = 2$ Like signs give a positive quotient.

(c) $\dfrac{10}{-2} = -5$ Unlike signs give a negative quotient.

(d) $\dfrac{-9}{1} = -9$ Unlike signs give a negative quotient.

                                                                    See Exercises 67–78.

Division with zero works the same for signed numbers as it does for whole numbers. Any real number divided by 0 is not defined because there is not a real number that can be multiplied by 0 to give the original number. We say $\frac{0}{0}$ is indeterminate because *any* real number will multiply by 0 to give 0.

To evaluate division with zero:

Zero divided by any nonzero number is zero.

$$\frac{0}{a} = 0; \qquad a \neq 0 \text{ and } a \text{ is any real number} \qquad \frac{0}{-5} = 0$$

Division by zero is either undefined or indeterminate.

$$\frac{a}{0} \text{ is undefined; } a \neq 0 \text{ and } a \text{ is any real number}$$

$$\frac{-5}{0} \text{ is undefined.} \qquad \frac{0}{0} \text{ is indeterminate.}$$

STOP AND CHECK

Divide.

1. $\dfrac{0}{-3}$ **2.** $\dfrac{-8}{0}$

Answers:

1. 0   **2.** Undefined

EXAMPLE 9

Evaluate: **(a)** $\dfrac{-3}{0}$ **(b)** $\dfrac{0}{0}$ **(c)** $\dfrac{0}{-5}$

(a) $\dfrac{-3}{0}$ is undefined.

(b) $\dfrac{0}{0}$ is indeterminate.

(c) $\dfrac{0}{-5} = 0$

See Exercises 79–88.

5–3 EXERCISES

MyLab Math For additional practice go to your study plan in MyLab Math.

1 Multiply. *See Example 1.*

1. $5 \cdot 8$ **2.** $-4(-3)$ **3.** $7 * 5$ **4.** $-3(-7)$ **5.** $-8(-3)$ **6.** $-2(-3)$

7. $5(-3)$ **8.** $-2(5)$ **9.** $-4 * 8$ **10.** $-3 \cdot 4$ **11.** $-7 * 8$ **12.** $6(-4)$

See Example 2.

13. $8(-3)(-2)(7)$ **14.** $5(-2)(3)(2)$ **15.** $6(1)(-3)(-2)$

16. $4(0)(-12)(3)$ **17.** $15(-2)(-3)$ **18.** $5(2)(-3)(0)$

See Example 3.

19. $-3(2)(-7)(-1)$ **20.** $9(-1)(3)(-2)$ **21.** $-7(-5)(-6)$

22. $-3(-9)(-12)(-7)$ **23.** $7(-3)(-10)(12)(-8)$ **24.** $7(8)(-5)(-3)$

25. $-3(-8)(-2)(5)$

See Example 4.

26. $-8(0)$ **27.** $5(0)$ **28.** $0(-12)$ **29.** $-15(0)$

30. $18(0)$ **31.** $0 \cdot 3$ **32.** $0(-15)$ **33.** $0(-17)$

34. $-28 \cdot 0$ **35.** $46 \cdot 0$ **36.** $5(1)(-3)(2) + 5(0)$ **37.** $-8(1)(3)(-7) + 7(0)$

38. $5(0) + 2(-3)(-7)$ **39.** $4(-2) + 3(0)(-5)$ **40.** $-3(-6)(-8)(0)$ **41.** $2(-3)-5(2) + 7(0)(-4)$

See Example 5.

42. **PFIN**   Madison Duke had four checks returned for insufficient funds, and her bank charged her a $28 service charge for each check. Use multiplication of signed numbers to show how these transactions affected her checking account balance.

43. **PFIN**   Chloe Duke made seven withdrawals of $40 each from her checking account. Use signed numbers to show how these transactions affected her checking account balance.

44. Review the definitions for additive inverse and additive identity and write a similar definition for multiplicative inverse.

45. Illustrate the commutative property of multiplication using one positive and one negative integer.

2 Evaluate. *See Example 6.*

46. $(-3)^2$

47. $(-2)^3$

48. $(-5)^2$

49. 0^{10}

50. -2^3

51. $(-8)^3$

52. $(5)^4$

53. 3^4

54. 7^4

55. $(-42)^2$

56. $(5)^1$

57. $(-28)^1$

58. 12^1

59. $(-8)^1$

60. -10^1

61. -8^1

62. $(-3)^3$

63. -11^2

64. $(-11)^2$

65. -5^3

See Example 7.

66. In Tennessee, the pattern for license plates is three numbers followed by three letters. How many different plates can be made using this pattern? Disregard any impractical or inappropriate combinations.

3 Divide. *See Example 8.*

67. $\dfrac{-25}{-5}$

68. $-15 \div 5$

69. $\dfrac{8}{-2}$

70. $\dfrac{-24}{6}$

71. $-28 \div 7$

72. $\dfrac{-32}{-4}$

73. $\dfrac{-50}{-10}$

74. $\dfrac{36}{-6}$

75. $\dfrac{-48}{-6}$

76. $\dfrac{-185}{5}$

77. **PFIN**   You have decreased your house loan balance by $1,800 over the past 6 months. Show this figure as a signed number, and find the average monthly decrease.

78. A temperature of 10°C fell to −8°C in 3 h. Find the average hourly change expressed as a signed number.

See Example 9.

79. $\dfrac{12}{0}$

80. $\dfrac{-15}{0}$

81. $\dfrac{0}{+3}$

82. $\dfrac{0}{-12}$

83. $\dfrac{0}{-8}$

84. $\dfrac{-20}{0}$

85. $\dfrac{-100}{0}$

86. $\dfrac{1}{0}$

87. $\dfrac{0}{8}$

88. $\dfrac{-350}{0}$

89. Which two operations have the same rules for handling signs when working with signed numbers?

90. A nor'easter storm blew into Green Bay, Wisconsin, and the temperature changed from 38°F to −22°F between 1 P.M. and 7 P.M. Represent the average hourly change in temperature with a signed number.

91. **BUS**   A business started the year with a net worth of −$4,852. During the first six months of the year, the business recovered and increased its net worth to a value of $15,983. Find the average monthly increase in net worth for the 6-month period.

92. What value or values must be in the numerator and denominator of a division in order for the result to be zero?

5–4 Signed Rational Numbers

Rational number: any type of real number that can be written as the quotient of two integers; natural numbers, whole numbers, integers, fractions, decimals, and signed numbers are included

Signed fraction: a fraction that is positive or negative

Fractions and decimals can also have negative values. This extends our types of numbers to include rational numbers. A **rational number** is any type of number that can be written as the quotient of two integers (Fig. 5–7).

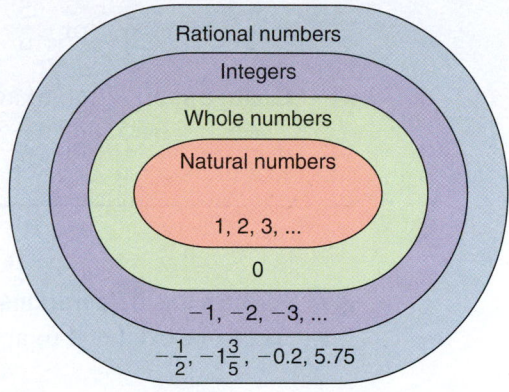

FIGURE 5–7 Rational numbers.

1 Change a Signed Fraction to an Equivalent Signed Fraction. A fraction has three signs: the sign of the fraction, the sign of the numerator, and the sign of the denominator. One way to express the fraction $\frac{2}{3}$ as a **signed fraction** is $+\frac{+2}{+3}$. When a signed fraction has negative signs, we sometimes change the signed fraction to an equivalent signed fraction.

To find an equivalent signed fraction:
1. Identify the three signs of the fraction (sign of the fraction, sign of the numerator, and sign of the denominator).
2. Change any two of the three signs to the opposite sign.

STOP AND CHECK

1. Change $-\frac{3}{-4}$ to three equivalent fractions.

Answer:

1. $\frac{-3}{-4}, \frac{3}{4},$ and $-\frac{-3}{4}$

EXAMPLE 1

Change $-\frac{-2}{-3}$ to three equivalent signed fractions.

$$= \frac{-2}{-3} = +\frac{+2}{-3} = \frac{2}{-3} \qquad \text{Change the signs of the fraction and the numerator.}$$

$$= \frac{-2}{-3} = +\frac{-2}{+3} = \frac{-2}{3} \qquad \text{Change the signs of the fraction and the denominator.}$$

$$-\frac{-2}{-3} = -\frac{+2}{+3} = -\frac{2}{3} \qquad \text{Change the signs of the numerator and the denominator.}$$

See Exercises 1–5.

Changing the signs of a fraction is a manipulation tool that simplifies our work when we perform basic operations with signed fractions.

> **TIP**   **Why Would We Want to Change the Signs of a Fraction?** When changing any two signs of a fraction, we can accomplish these desirable outcomes.
>
> ▶ Avoid dealing with negatives.
>
> $$-\frac{-3}{+4} = +\frac{+3}{+4} \quad \text{or} \quad \frac{3}{4}$$
>
> $$+\frac{-6}{-7} = +\frac{+6}{+7} \quad \text{or} \quad \frac{6}{7}$$
>
> $$-\frac{+5}{-9} = +\frac{+5}{+9} \quad \text{or} \quad \frac{5}{9}$$
>
> ▶ Change subtraction to addition.
>
> $$-\frac{+5}{+6} = +\frac{-5}{+6} \quad \text{or} \quad \frac{-5}{6}$$
>
> ▶ Avoid negative denominators.
>
> $$-\frac{+3}{-4} = +\frac{+3}{+4} \quad \text{or} \quad \frac{3}{4}$$
>
> $$+\frac{+2}{-5} = +\frac{-2}{+5} \quad \text{or} \quad \frac{-2}{5}$$
>
> ▶ Deal with fewer negatives.
>
> $$-\frac{-5}{-8} = +\frac{-5}{+8} \quad \text{or} \quad \frac{-5}{8}$$

2   **Perform Basic Operations with Rational Numbers.** The rules for operating with integers can be extended to apply to all rational numbers.

> **To perform basic operations with rational numbers:**
>
> 1. Determine the indicated operations.
> 2. Apply the appropriate rules for signed numbers and for rational numbers.

STOP AND CHECK

Perform the indicated operations.

1. $-\frac{2}{3} + \frac{-5}{6}$ **2.** $\frac{4}{5} - \frac{-3}{10}$

3. $\frac{-2}{7}\left(\frac{5}{-9}\right)$ **4.** $\left(-\frac{5}{6}\right)^2$

Answers:

1. $-\frac{3}{2}$ or $-1\frac{1}{2}$ **2.** $\frac{11}{10}$ or $1\frac{1}{10}$

3. $\frac{10}{63}$ **4.** $\frac{25}{36}$

EXAMPLE 2

Perform the indicated operations. **(a)** Add $\frac{-3}{4} + \frac{5}{-8}$ **(b)** Subtract $\frac{-3}{7} - \frac{-5}{7}$

(c) Multiply $\frac{-4}{5}\left(\frac{3}{-7}\right)$ **(d)** Simplify $\left(\frac{-2}{3}\right)^3$

(a) $\dfrac{-3}{4} + \dfrac{5}{-8} = \dfrac{-3}{4} + \dfrac{-5}{8} =$ Change the signs of the numerator and denominator in the second fraction so that both denominators are positive.

Change to equivalent fractions with a common denominator.

$$\dfrac{-6}{8} + \dfrac{-5}{8} =$$ Add numerators, applying the rule for adding numbers with like signs.

$$\dfrac{-11}{8} = -1\dfrac{3}{8}$$ Change to a mixed number. The sign of the mixed number is determined by the rule for dividing numbers with unlike signs.

(b) $\dfrac{-3}{7} - \dfrac{-5}{7} = \dfrac{-3}{7} + \dfrac{5}{7} = \dfrac{2}{7}$ Change subtraction to addition by changing the signs of the second fraction and the numerator.

Apply the rule for adding numbers with unlike signs.

(c) $\dfrac{-4}{5}\left(\dfrac{3}{-7}\right) = \dfrac{-4}{5} \cdot \dfrac{3}{-7} =$ Multiply the numerators and denominators.

$$\dfrac{-12}{-35} = \dfrac{12}{35}$$ Apply rule for dividing numbers with like signs to simplify signs.

(d) $\left(\dfrac{-2}{3}\right)^3 =$ Cube the numerator and cube the denominator.

$$\dfrac{(-2)^3}{3^3} =$$ $(-2)(-2)(-2) = -8;\ (3)(3)(3) = 27$

$$\dfrac{-8}{27} \quad \text{or} \quad -\dfrac{8}{27}$$ Manipulate signs if desired.

See Exercises 6–32.

STOP AND CHECK

Perform the indicated operations.

1. Add -7.09 and 3.82.
2. Subtract -2.5 from -6.1.
3. Multiply -4.12 and 2.3.
4. Divide 3.5 by -0.7.

Answers:

1. -3.27 2. -3.6
3. -9.476 4. -5

EXAMPLE 3

Perform the indicated operations. **(a)** Add -5.32 and -3.24 **(b)** Subtract -3.7 from 8.5 **(c)** Multiply 3.91 and -7.1 **(d)** Divide (-1.2) by (-0.4)

(a)
$$\begin{array}{r} -5.32 \\ -3.24 \\ \hline -8.56 \end{array}$$
Align decimals and use the rule for adding numbers with like signs.

(b) $8.5 - (-3.7) = 8.5 + 3.7 = \mathbf{12.2}$ Change subtraction to an equivalent addition and add numbers with like signs.

(c)
$$\begin{array}{r} 3.91 \\ \times\ -7.1 \\ \hline 391 \\ 27\ 37 \\ \hline -27.761 \end{array}$$
Apply rule for multiplying numbers with unlike signs.

(d)
$$-0.4\overline{)-1.2}\ \ ^{+3.}$$
Shift decimal points and divide. Use rule for dividing numbers with like signs.

See Exercises 33–44.

---

3 **Apply the Order of Operations with Signed Numbers.** Evaluating operations with signed numbers follows the same order of operations as evaluating with whole numbers.

Perform operations in the following order as they appear from left to right:	
1. Parentheses used as groupings and other grouping symbols.	**P**
2. Exponents (powers and roots).	**E**
3. Multiplication and division.	**MD**
4. Addition and subtraction.	**AS**

STOP AND CHECK

1. Evaluate $3(7 - 12)$.

Answer:

1. -15

EXAMPLE 4

Evaluate $8(4 - 6)$.

Work within the grouping symbols first:

$8(4 - 6) =$ Add or subtract inside parentheses. $4 - 6 = -2$. **P**

$8(-2) = -16$ Multiply 8 by -2. **MD**

Or use the distributive principle:

$8(4 - 6) =$ Distribute.

$8(4) + 8(-6) =$ Multiply each term in the parentheses by the factor 8.

$32 + -48 = -16$ Add 32 and -48. **See Exercises 45–46.**

EXAMPLE 5

Evaluate $-12 \div 3 - (2)(-5)$.

$-12 \div 3 - (2)(-5) =$	Divide and multiply first.
$-4 - (-10) =$	Change to a single sign between the numbers.
$-4 + 10 = \mathbf{6}$	Add integers with unlike signs.

See Exercises 47–56.

Did You Know? **Parentheses are sometimes optional.** The problem in Example 5 can also be written without parentheses around the 2. The problem is then expressed as

$$-12 \div 3 - 2(-5)$$

We divide and multiply first, but notice how we perform the multiplication.

$-12 \div 3 - 2(-5) =$	Consider the minus sign between 3 and 2 as the sign of 2. $-12 \div 3 = -4$ and $-2(-5) = 10$.
$-4 + 10 =$	Add the results of the division and the multiplication.
$-4 + 10 = \mathbf{6}$	Add integers with unlike signs.

Divisions are often written in fraction form. Example 5 could be written as $\dfrac{-12}{3} - (2)(-5)$. Then, performing the division and the multiplication you have $-4 + 10$. Finally, performing the addition you have 6.

The fraction bar as a symbol for division is also a grouping symbol.

EXAMPLE 6

Evaluate **(a)** $\dfrac{3-5}{2} - \dfrac{9}{2+1} + 4(5)$ and   **(b)** $10 - 3(-2)$.

(a)

$\dfrac{3-5}{2} - \dfrac{9}{2+1} + 4(5) =$	Perform operations grouped by the fraction bar.
$\dfrac{-2}{2} - \dfrac{9}{3} + 4(5) =$	Divide and multiply.
$-1 - 3 + 20 =$	Add.
$-4 + 20 = \mathbf{16}$	Add.

(b)

$10 - 3(-2) =$	Think of the multiplication as -3 times -2.
$10 + 6 =$	Interpret as addition.
$10 + 6 = \mathbf{16}$	Add integers with like signs.

See Exercises 57–72.

Only after parentheses and all multiplications and/or divisions are completed do we perform the final additions and/or subtractions from left to right.

> **TIP** **The Order of Operations Is Important** Note what happens if we proceed *out of order* in the preceding example!
>
> **Incorrectly Worked**
>
>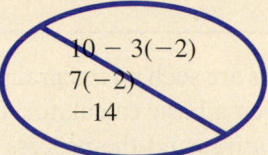
>
> *Incorrectly* subtract first instead of last.
> Multiply last instead of second.
> Incorrect solution.
>
> We get an incorrect answer. *The order of operations must be followed to arrive at a correct solution.*

EXAMPLE 7

STOP AND CHECK

Evaluate.

1. $4 - (-4)^2 + 2(7 - 8)$

2. $\dfrac{1 + |5 - 9| - 5^2}{2(3 + 4)}$

3. $\dfrac{2}{3} - 5\left(\dfrac{2}{5} - \dfrac{7}{10}\right)$

4. $-2.1 + 1.1(0.3)^2$

Answers:

1. -14 2. $-1\dfrac{3}{7}$

3. $2\dfrac{1}{6}$ 4. -2.001

Evaluate **(a)** $5 + (-2)^3 - 3(4 + 1)$ **(b)** $\dfrac{5 + |3 - 7| - 3^3}{3(3 - 1)}$

(c) $-\dfrac{1}{2} + 4\left(\dfrac{3}{8} - \dfrac{5}{8}\right)$ **(d)** $3.2 - 1.7(0.2)^3$

(a) $5 + (-2)^3 - 3(4 + 1) =$ Perform operation in parentheses.

$\qquad 5 + (-2)^3 - 3(5) =$ Raise to a power.

$\qquad 5 + (-8) - 3(5) =$ Multiply.

$\qquad 5 + (-8) - 15 =$ Change to one sign only between numbers.

$\qquad 5 - 8 - 15 =$ Perform first calculation on the left.

$\qquad\qquad -3 - 15 =$ Perform remaining calculation.

$\qquad\qquad\qquad -18$

(b) $\dfrac{5 + |3 - 7| - 3^3}{3(3 - 1)} =$ Perform operations in absolute value term and in parentheses.

$\qquad \dfrac{5 + 4 - 3^3}{3(2)} =$ Raise to power in numerator. Multiply in denominator.

$\qquad \dfrac{5 + 4 - 27}{6} =$ Add and subtract in numerator.

$\qquad \dfrac{-18}{6} =$ Divide.

$\qquad -3$

(c) $-\dfrac{1}{2} + 4\left(\dfrac{3}{8} - \dfrac{5}{8}\right) =$ Perform operation inside grouping. $\dfrac{3}{8} - \dfrac{5}{8} = \dfrac{-2}{8} = -\dfrac{1}{4}$.

$\qquad -\dfrac{1}{2} + 4\left(-\dfrac{1}{4}\right) =$ Multiply. $4\left(-\dfrac{1}{4}\right) = -1$

$\qquad -\dfrac{1}{2} - 1 =$ Change -1 to an equivalent fraction with a denominator of 2.

$\qquad -\dfrac{1}{2} - \dfrac{2}{2} =$ Add like fractions with like signs.

$\qquad -\dfrac{3}{2}$ or $-1\dfrac{1}{2}$ Change to signed mixed number if desired.

(d) $\quad 3.2 - 1.7(0.2)^3 =$ $\qquad$ Raise 0.2 to the third power. $(0.2)(0.2)(0.2) = 0.008$

$\qquad 3.2 - 1.7(0.008) =$ $\qquad$ Multiply. $-1.7(0.008) = -0.0136$

$\qquad\qquad 3.2 - 0.0136 =$ $\qquad$ Subtract (add opposite).

$\qquad\qquad\qquad$ **3.1864** $\qquad\qquad\qquad\qquad\qquad$ See Exercises 73–78.

STOP AND CHECK

Use a calculator to evaluate.
1. $17 - 23$
2. $-5(4)$
3. $\dfrac{-18}{3} + 7$

Answers:
1. -6 $\quad$ 2. -20 $\quad$ 3. 1

Since negative numbers are such an integral part of our everyday lives, practically all types of calculators, even a basic calculator, can deal with negative values. However, the notation for negatives and the process for entering negatives varies widely from calculator to calculator. Some of the most common options are discussed.

Some calculators use the subtraction key for both subtracting and entering negative numbers. Others have a special key for entering a **negative sign** that is labeled as a negative sign enclosed in parentheses $\boxed{(-)}$. A common option with basic calculators is the **sign-change key** $\boxed{+/-}$. This key is a **toggle key** and changes the sign of the number in the display to the opposite sign.

EXAMPLE 8

Use a calculator to evaluate.

(a) $2 - 7$ $\qquad$ **(b)** $(-7)(2)$ $\qquad$ **(c)** $\dfrac{-4}{2} + 14$

The most common options:

(a) $2 - 7$ $\qquad$ $2 \boxed{-} 7 \boxed{=} \Rightarrow \mathbf{-5}$ $\qquad\qquad$ $\boxed{=}$ may be labeled $\boxed{\text{ENTER}}$ or $\boxed{\text{EXE}}$.

(b) $(-7)(2)$ $\qquad$ $\boxed{(-)} 7 \boxed{\times} 2 \boxed{=} \Rightarrow \mathbf{-14}$

$\qquad\qquad\qquad$ or

$\qquad\qquad\qquad$ $\boxed{(-)} 7 \boxed{(} 2 \boxed{)} \boxed{=} \Rightarrow \mathbf{-14}$ $\qquad$ Some calculators interpret parentheses with no operational symbol preceding the parentheses as multiplication.

$\qquad\qquad\qquad$ or

$\qquad\qquad\qquad$ $7 \boxed{+/-} \boxed{\times} 2 \boxed{=} \Rightarrow \mathbf{-14}$ $\qquad$ The sign-change key is entered after the absolute value of the number.

(c) $\dfrac{-4}{2} + 14$ $\qquad$ $\boxed{(-)} 4 \boxed{\div} 2 \boxed{=} \boxed{+} 14 \boxed{=} \Rightarrow \mathbf{12}$ $\qquad$ The first $\boxed{=}$ may not be required if your calculator applies the order of operations. A slash $\boxed{/}$ may be used for division on some calculators.

$\qquad\qquad\qquad\qquad\qquad\qquad\qquad\qquad\qquad\qquad\qquad$ See Exercises 79–81.

You may impose the appropriate order of operations by using the $\boxed{=}$ key when you perform a series of calculations. This is necessary with a basic calculator when parentheses keys are not available.

When a bar is used as both a grouping and a division symbol on a calculator, we must instruct the calculator to work the grouping first by enclosing the grouping in parentheses.

STOP AND CHECK

1. Evaluate using a calculator:
$\dfrac{27 - 3}{2(6)} + 5.$

Answer:
1. 7

EXAMPLE 9

Evaluate using a calculator: $\dfrac{5 + 1}{3} + 2$

$\dfrac{5 + 1}{3} + 2$ $\qquad$ $\boxed{(} 5 \boxed{+} 1 \boxed{)} \boxed{\div} 3 \boxed{+} 2 \boxed{=} \Rightarrow \mathbf{4}$ $\qquad$ Using parentheses for grouping.

$\qquad\qquad\qquad$ or

$\qquad\qquad$ $5 \boxed{+} 1 \boxed{=} \boxed{\div} 3 \boxed{+} 2 \boxed{=} \Rightarrow \mathbf{4}$ $\qquad$ Using equal for grouping.

$\qquad\qquad\qquad\qquad\qquad\qquad\qquad\qquad\qquad$ See Exercises 82–83.

5-4 EXERCISES   MyLab Math For additional practice go to your study plan in MyLab Math.

1 Change each fraction to three equivalent signed fractions. *See Example 1.*

1. $+\dfrac{+5}{+8}$ **2.** $-\dfrac{3}{4}$ **3.** $\dfrac{-2}{-5}$ **4.** $-\dfrac{-7}{-8}$ **5.** $\dfrac{7}{8}$

2 Perform the operations. *See Example 2.*

6. $\dfrac{-7}{8} + \dfrac{5}{8}$ **7.** $\dfrac{-4}{5} + \left(\dfrac{-3}{10}\right)$ **8.** $\dfrac{3}{8} + \left(-\dfrac{7}{16}\right)$ **9.** $\left(-\dfrac{5}{16}\right) + \dfrac{1}{2}$

10. $\dfrac{1}{3} + \left(-\dfrac{5}{9}\right)$ **11.** $\left(-3\dfrac{1}{4}\right) + \left(-7\dfrac{3}{8}\right)$ **12.** $1\dfrac{7}{8} + \left(-4\dfrac{5}{12}\right)$ **13.** $-7\dfrac{1}{2} + \left(-1\dfrac{3}{4}\right) + \left(5\dfrac{3}{8}\right)$

14. $\dfrac{1}{2} - \left(\dfrac{-3}{5}\right)$ **15.** $\left(-\dfrac{3}{4}\right) + \left(-\dfrac{5}{8}\right)$ **16.** $-\dfrac{3}{8} + \left(+\dfrac{5}{16}\right)$ **17.** $1\dfrac{7}{8} - \left(-\dfrac{5}{6}\right)$

18. $3\dfrac{3}{4} - \left(+4\dfrac{1}{4}\right)$ **19.** $\dfrac{-3}{5}\left(\dfrac{10}{-11}\right)$ **20.** $\left(-\dfrac{4}{9}\right)\left(-\dfrac{5}{6}\right)$ **21.** $\left(-7\dfrac{1}{3}\right)\left(-4\dfrac{1}{11}\right)$

22. $\dfrac{1}{4} \div (-8)$ **23.** $-5\left(\dfrac{-2}{-5}\right)$ **24.** $\left(-\dfrac{2}{7}\right)\left(-\dfrac{14}{15}\right)\left(-\dfrac{3}{8}\right)$ **25.** $\left(-\dfrac{3}{5}\right)\left(\dfrac{-1}{-2}\right)\left(\dfrac{10}{21}\right)$

26. $-\dfrac{5}{8} \div \dfrac{4}{5}$ **27.** $(-1)^5$ **28.** -9^2 **29.** $(-9)^2$

30. $-(-4)^3$ **31.** $-(-2)^4$ **32.** $-(-2)^5$

Round to hundredths if necessary. *See Example 3.*

33. $5.823 - 32.12$ **34.** $-8.32 + 7.21$ **35.** $-84.23 - 7.21$ **36.** $34.6(-3.2)$

37. $-7.2(8.2)$ **38.** $-83.1(-4.1)$ **39.** $83.2 \div (-3)$ **40.** $-0.826 \div (-2)$

41. $-3.2 + 7.8$ **42.** $4.23 - 4.2$ **43.** $4.6 \div (-2)$ **44.** $-3.8 + (-1.7)$

3 Evaluate. *See Example 4.*

45. $3(5 - 8)$ **46.** $(7 + 8) - (2 - 7)$

See Example 5.

47. $3(7) + 9 \div 3$ **48.** $5 - 3(2)$ **49.** $-12 + 2(3) - 7$ **50.** $12 \div 4(7)$

51. $15 + 7(-4)$ **52.** $-8 - 7(-3)$ **53.** $9 - 3(-2)$ **54.** $-6(-2) - 4$

55. $-6(-4)$ **56.** $7(1 - 4) \div 3 + 2$

See Example 6.

57. $\dfrac{2 - 8}{3} + 5(-2) \div 2$ **58.** $\dfrac{5 - 9}{4} + 4(-3) \div 6$ **59.** $4(8) - 7(3^2) + 18 \div 6$

60. $5(3 - 4) - 7(2 - 5) \div (-3)$ **61.** $142(3 - 21) + 48(27)$ **62.** $24 - 3(2 + 7) \div 3 + 12$

63. $\dfrac{12 - 18}{3} + \dfrac{15 - 81}{3}$ **64.** $14 - \dfrac{3 + 7}{5} \div 2 + 5(3)$ **65.** $\dfrac{2(3 - 12)}{4(6)} - 8$

66. $2(4 - 3) - 7 + 4(2 - 8)$ **67.** $7 - 21 + 138 - 256$ **68.** $\dfrac{4}{5} + 3\left(-\dfrac{2}{3} + \dfrac{1}{3}\right)$

69. $\dfrac{-5}{8} + 7\left(\dfrac{1}{8} - \dfrac{7}{8}\right)$ **70.** $\dfrac{1}{2} - 3\left(2\dfrac{3}{4} - \dfrac{1}{4}\right)$ **71.** $-\dfrac{3}{5} + \dfrac{1}{2}(2 - 8)$

72. $5.2 + 3.8(-4.1)$

See Example 7.

73. $-\dfrac{3}{5} + \left(\dfrac{2}{7}\right)^2$

74. $\left(-\dfrac{2}{3}\right)^2 - \dfrac{1}{2}\left(\dfrac{3}{8}\right)$

75. $1.3 - (2.1)^2(1.2)$

76. $(0.3)^2 + 5.7(-2.1)$

77. $\dfrac{4 - |1 - 7| - 2^2}{2(3)}$

78. $\dfrac{|7 - 12| + 5(3)}{-|4 - 9|}$

See Example 8.

79. $5(8 - 7 + 3) - 2$

80. $7(8 + 2) - 4(7 - 5 - 10)$

81. $7(2) - 4(15 - 8 - 5(3))$

See Example 9.

82. $5 - \dfrac{9 - 3}{2}$

83. $\dfrac{128}{4(15 - 4 + 3 - 10)}$

5–5   Powers of 10

LEARNING OUTCOMES

1   Multiply and divide by powers of 10.

2   Raise a power of 10 to a power.

LC   **LEARNING CATALYTICS**

1. Multiply mentally 100(10,000).

2. Divide mentally $\dfrac{100,000}{100}$.

3. Divide mentally $\dfrac{10}{1,000}$.

Exponent of zero: a number with an exponent of zero is equal to 1; 10^0 is the power-of-10 equivalent of the ones place on the place-value chart

Negative integer exponent: a notation for writing reciprocals

1   **Multiply and Divide by Powers of 10.** We learned that our number system is based on the number 10; that is, each place value is a *power of 10*. To show how each place value is related to the base 10, we need to define two special exponents. A number with an **exponent of zero** is defined to have a value of 1. On the place-value chart, the ones place is shown to be 10^0. The decimal places have the power of 10 in the denominator of the fraction representing the place value. That is, the place value is the *reciprocal* of a power of 10 with a positive exponent. To illustrate these decimal places as powers of 10, we use **negative integer exponents.** Look at the place-value chart in Fig. 5–8 to see how our number system relates to powers of 10.

> **TIP**   **Zero and Negative Integer Exponent Reciprocals**   Any nonzero number raised to the zero power is equal to 1.
>
> For any real number n, $n^0 = 1$ if $n \neq 0$.
>
> An expression with a *negative integer exponent* can be written as an equivalent expression with a positive exponent.
>
> For any real number n, $n^{-1} = \dfrac{1}{n}$ $\dfrac{1}{n^{-1}} = n$ if $n \neq 0$.
>
> A nonzero number times its reciprocal equals 1.
>
> For any real number n, $n \cdot \dfrac{1}{n} = 1$ $n^1 \cdot n^{-1} = n^0 = 1$ if $n \neq 0$.

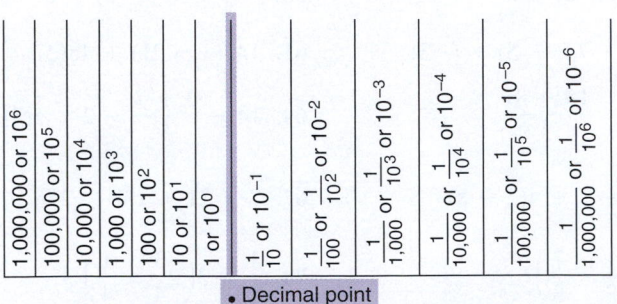

FIGURE 5–8   Base-10 place-value chart.

Whole-Number Part		Fractional or Decimal Part	
Millions	$1{,}000{,}000 = 10^6$	$\dfrac{1}{10} = 0.1 = 10^{-1}$	Tenths
Hundred thousands	$100{,}000 = 10^5$	$\dfrac{1}{100} = 0.01 = 10^{-2}$	Hundredths
Ten thousands	$10{,}000 = 10^4$	$\dfrac{1}{1{,}000} = 0.001 = 10^{-3}$	Thousandths
Thousands	$1{,}000 = 10^3$	$\dfrac{1}{10{,}000} = 0.0001 = 10^{-4}$	Ten-thousandths
Hundreds	$100 = 10^2$	$\dfrac{1}{100{,}000} = 0.00001 = 10^{-5}$	Hundred-thousandths
Tens	$10 = 10^1$	$\dfrac{1}{1{,}000{,}000} = 0.000001 = 10^{-6}$	Millionths
Ones	$1 = 10^0$		

> **TIP**   **What Does the Exponent in a Power of 10 Tell Us?**
>
> ▶ When the exponent of 10 is a *positive integer,* the exponent is the same as the *number of zeros* in the equivalent whole-number place value in standard notation.
>
> ▶ When the exponent of 10 is a *negative integer,* the exponent is the same as the *number of zeros* in the denominator of the equivalent fraction represented by the place value.
>
> ▶ When the exponent of 10 is a *negative integer,* the exponent is the same as the *number of decimal digits* in the decimal equivalent.

To multiply a number by a power of 10:

1. If the exponent is positive, shift the decimal point to the *right* the number of places indicated by the *positive* exponent. Attach zeros as necessary.

2. If the exponent is negative, shift the decimal point to the *left* the number of places indicated by the *negative* exponent. Insert zeros as necessary.

STOP AND CHECK

Perform the multiplications using powers of 10.

1. $252(100)$
2. $0.7(10)$
3. $5.17(0.01)$
4. $0.14(10^3)$
5. $0.38(10^{-2})$

Answers:
1. $25{,}200$ 2. 7 3. 0.0517
4. 140 5. 0.0038

EXAMPLE 1

Perform the multiplications using powers of 10.

(a) $275(10)$ **(b)** $0.18(100)$ **(c)** $2.4(1{,}000)$ **(d)** $43(0.1)$ **(e)** $5.72(10^5)$
(f) $2.3(10^{-3})$

(a) $275(10) = 275(10^1) = \mathbf{2{,}750}$ The exponent is $+1$, so move the decimal point one place to the right. Attach one zero.

(b) $0.18(100) = 0.18(10^2) = \mathbf{18}$ The exponent is $+2$, so move the decimal point two places to the right.

(c) $2.4(1{,}000) = 2.4(10^3) = \mathbf{2{,}400}$ The exponent is $+3$, so move the decimal point three places to the right. Attach two zeros.

(d) $43(0.1) = 43(10^{-1}) = \mathbf{4.3}$ The exponent is -1, so move the decimal point one place to the left.

(e) $5.72(10^5) = \mathbf{572,000}$

The exponent is $+5$, so move the decimal five places to the right.

(f) $2.3(10^{-3}) = \mathbf{0.0023}$

The exponent is -3, so move the decimal three places to the left. **See Exercises 1–12.**

To divide a number by a power of 10:

1. Change the division to an equivalent multiplication by multiplying by the reciprocal of the divisor.

2. Use the rule for multiplying a number by a power of 10.

STOP AND CHECK

Perform the divisions using powers of 10.

1. $12.3 \div 100$

2. $\dfrac{3.7}{1,000}$

3. $0.4 \div 10^3$

4. $\dfrac{14.8}{10^4}$

Answers:
1. 0.123 **2.** 0.0037 **3.** 0.0004
4. 0.00148

EXAMPLE 2

Perform the divisions using powers of 10.

(a) $3.14 \div 10$ **(b)** $0.48 \div 100$ **(c)** $20.1 \div 1,000$ **(d)** $0.1 \div 10^2$

(e) $\dfrac{2.5}{10^3}$

(a) $3.14 \div 10 = 3.14\left(\frac{1}{10}\right)$ Change the division to an equivalent multiplication.

$\qquad\qquad\quad = 3.14(10^{-1})$ Express the fraction $\frac{1}{10}$ as a power of 10.

$\qquad\qquad\quad = \mathbf{0.314}$ Because the exponent of 10 is -1, move the decimal point one place to the *left*.

(b) $0.48 \div 100 = 0.48\left(\frac{1}{100}\right) = 0.48(10^{-2}) = \mathbf{0.0048}$

Move the decimal point two places to the left. Insert two zeros.

(c) $20.1 \div 1,000 = 20.1\left(\frac{1}{1,000}\right) = 20.1(10^{-3}) = \mathbf{0.0201}$

Move the decimal point three places to the left. Insert one zero.

(d) $0.1 \div 10^2 = 0.1 \div \frac{1}{100} = 0.1(10^{-2}) = \mathbf{0.001}$

Move the decimal point two places to the left. Insert two zeros.

(e) $\dfrac{2.5}{10^3} = \dfrac{2.5}{1,000} = 2.5(10^{-3}) = \mathbf{0.0025}$

Move the decimal point three places to the left. Insert two zeros.

See Exercises 13–22.

Did You Know? Reciprocal power-of-ten factors have exponents that are opposites.

The reciprocal of 10^{-3} is 10^3.
The reciprocal of 10^5 is 10^{-5}.

To multiply or divide a power of 10 by another power of 10, we can use special applications of the *laws of exponents*. These laws are introduced in Chapter 11, but we will look at these special cases now.

To multiply powers of 10 written in exponential notation:

1. The product will be a power of 10.

2. Add the exponents for the power-of-10 factors.

Symbolically, $10^a \cdot 10^b = 10^{a+b}$, where a and b are integers.

> **To divide powers of 10 written in exponential notation:**
>
> 1. The quotient will be a power of 10.
>
> 2. Subtract the exponents for the power-of-10 factors (numerator minus denominator).
>
> 3. Powers of 10 with negative exponents can be written as reciprocal expressions with positive exponents, if desired.
>
> Symbolically, $\dfrac{10^a}{10^b} = 10^{a-b}$, where a and b are integers; $10^{-a} = \dfrac{1}{10^a}$.

STOP AND CHECK

Multiply or divide using the laws of exponents.

1. $10^6(10^1)$ 2. $10^5(10^0)$

3. $10^{-10}(10^2)$ 4. $\dfrac{10^2}{10^3}$

5. $\dfrac{10^5}{10^5}$ 6. $\dfrac{10^{-3}}{10^{-1}}$

Answers:

1. 10^7 2. 10^5 3. 10^{-8} or $\dfrac{1}{10^8}$

4. 10^{-1} or $\dfrac{1}{10}$ 5. 10^0 or 1

6. 10^{-2} or $\dfrac{1}{10^2}$

EXAMPLE 3

Multiply or divide using the laws of exponents.

(a) $10^5(10^2)$ **(b)** $10^{-1}(10^2)$ **(c)** $10^0(10^3)$ **(d)** $\dfrac{10}{10^3}$

(e) $\dfrac{10^5}{10^4}$ **(f)** $\dfrac{10^2}{10^2}$ **(g)** $\dfrac{10^{-2}}{10^3}$

(a) $10^5(10^2) = 10^{5+2} = \mathbf{10^7}$ — Add exponents.

(b) $10^{-1}(10^2) = 10^{(-1+2)} = \mathbf{10^1}$ **or 10** — Add exponents.

(c) $10^0(10^3) = 10^{0+3} = \mathbf{10^3}$ — Add exponents.

(d) $\dfrac{10}{10^3} = 10^{1-3} = \mathbf{10^{-2}}$ **or** $\dfrac{1}{\mathbf{10^2}}$ — Subtract exponents.

(e) $\dfrac{10^5}{10^4} = 10^{5-4} = \mathbf{10^1}$ **or 10** — Subtract exponents.

(f) $\dfrac{10^2}{10^2} = 10^{2-2} = \mathbf{10^0}$ **or 1** — Subtract exponents.

(g) $\dfrac{10^{-2}}{10^3} = 10^{-2-3} = \mathbf{10^{-5}}$ **or** $\dfrac{1}{\mathbf{10^5}}$ — Subtract exponents.

See Exercises 23–40.

2 Raise a Power of 10 to a Power. Recall that raising a number to a natural-number power is an extension of multiplication. Powers of 10 can be raised to a power by applying this same concept and thus establishing another law of exponents.

> **To raise a power of 10 to a power:**
>
> 1. The result will be a power of 10.
>
> 2. The exponent will be the product of the original exponent and the exponent of the power.
>
> Symbolically, $(10^a)^b = 10^{ab}$.

STOP AND CHECK

Raise the powers of 10 to the indicated power.

1. $(10^5)^2$ 2. $(10^{-2})^{-1}$

3. $(10^4)^{-2}$

Answers:

1. 10^{10} 2. 10^2 3. 10^{-8}

EXAMPLE 4

Raise the powers of 10 to the indicated power.

(a) $(10^3)^2$ **(b)** $(10^{-3})^4$ **(c)** $(10^{-2})^{-3}$

(a) $(10^3)^2 = 10^{3\cdot2} = \mathbf{10^6}$ — Multiply exponents.

(b) $(10^{-3})^4 = 10^{-3(4)} = \mathbf{10^{-12}}$ — Multiply exponents.

(c) $(10^{-2})^{-3} = 10^{-2(-3)} = \mathbf{10^6}$ — Multiply exponents.   **See Exercises 41–45.**

5–5 EXERCISES   MyLab Math For additional practice go to your study plan in MyLab Math.

1 Perform the operations. *See Example 1.*

1. $453(100)$
2. $0.27(100)$
3. $5.82(1{,}000)$
4. $8.97(0.1)$
5. $523(0.01)$
6. $8.06(0.001)$
7. $0.37(10^2)$
8. $1.82(10^3)$
9. $5.6(10^{-1})$
10. $142(10^{-2})$
11. $78(10^4)$
12. $62(10^0)$

See Example 2.

13. $5.73 \div 10$
14. $0.293 \div 100$
15. $45.7 \div 1{,}000$
16. $85.79 \div \dfrac{1}{10}$
17. $43.7 \div \dfrac{1}{100}$
18. $8.37 \div \dfrac{1}{1{,}000}$
19. $4.6 \div 10^4$
20. $6.1 \div 10$
21. $7.2 \div 10^1$
22. $42 \div 10^0$

See Example 3.

23. $10^7(10^5)$
24. $10^{-3}(10^4)$
25. $10^0(10^2)$
26. $\dfrac{10^7}{10^3}$
27. $\dfrac{10^5}{10^5}$
28. $\dfrac{10^3}{10^5}$
29. $\dfrac{10^{-5}}{10^7}$
30. $\dfrac{10}{10^7}$
31. $10^4(10^6)$
32. $10^{-3}(10^{-4})$
33. $10^0(10^{-3})$
34. $10^{-3}(10^4)$
35. $10(10^2)$
36. $\dfrac{10^4}{10^2}$
37. $\dfrac{10}{10^4}$
38. $\dfrac{10^4}{10^4}$
39. $\dfrac{10^{-2}}{10^3}$
40. $\dfrac{10^0}{10^1}$

2 Raise the powers of 10 to the indicated exponents. *See Example 4.*

41. $(10^2)^3$
42. $(10^3)^4$
43. $(10^{-2})^2$
44. $(10^{-5})^{-2}$
45. $(10^4)^{-2}$

5–6   Scientific Notation

LEARNING OUTCOMES

1. Change a number from scientific notation to ordinary notation.
2. Change a number from ordinary notation to scientific notation.
3. Multiply and divide numbers in scientific notation.
4. Raise a number in scientific notation to a power.
5. Change among engineering, scientific, and ordinary notations.

LC LEARNING CATALYTICS

Identify the values that are greater than or equal to 1 and less than 10.
1. 0.8   2. 3.7   3. 9.8   4. 11.3

1 Change a Number from Scientific Notation to Ordinary Notation. A number is expressed in **scientific notation** if it is the product of *two factors* that meet the following conditions. The absolute value of the first factor is a number greater than or equal to 1 but less than 10. That is, the first factor has only one digit to the left of the decimal point and it is not a zero. The second factor is a power of 10. The expression, 3.5×10^3 is written in scientific notation.

A number that is written strictly according to place value is an **ordinary number.** 3,500 is written in ordinary notation.

TIP Characteristics of Scientific Notation

▶ Numbers between 0 and 1 (proper fractions and their decimal equivalents) and between −1 and 0 require negative exponents when written in scientific notation.

▶ The first factor in scientific notation always has an absolute value that is greater than or equal to one but less than 10. There is only one digit to the left of the decimal and that digit cannot be a zero.

Scientific notation: a number is expressed in scientific notation if it is the product of two factors that meet the following conditions. The absolute value of the first factor is a number greater than or equal to 1 but less than 10. The second factor is a power of 10

Ordinary number: a number that is written according to its place value

STOP AND CHECK

Which of the terms is expressed in scientific notation?

1. 14×10^3
2. 9×10^{-2}
3. -3×10^4
4. 0.5×10^{-1}

Answers:
1. Not scientific notation
2. Scientific notation
3. Scientific notation
4. Not scientific notation

▶ The use of the times sign ($\times$) for multiplication is the most common representation for scientific notation.

▶ The second factor is a power of 10.

EXAMPLE 1

Which of the terms is expressed in scientific notation?

(a) 4.7×10^2 (b) 0.2×10^{-1} (c) -3.4×5^2 (d) 2.7×10^0

(e) 34×10^4 (f) 8×10^{-6} (g) $-2.8 \div 10^3$

(a) 4.7 is more than 1 and less than 10. 10^2 is a power of 10. Multiplication is indicated. **Thus, the term is in scientific notation.**

(b) Even though 10^{-1} is a power of 10, **this term is not in scientific notation** because the first factor (0.2) is less than 1.

(c) The absolute value of -3.4 is more than 1 and less than 10, but 5^2 is not a power of 10. **Thus, this term is not in scientific notation.**

(d) 2.7 is more than 1 and less than 10. 10^0 is a power of 10. **Thus, this term is in scientific notation.**

(e) 34 is greater than 10. Even though 10^4 is a power of 10, **this term is not in scientific notation** because the first factor (34) is 10 or more.

(f) 8 is more than 1 and less than 10. 10^{-6} is a power of 10. **Thus, this term is in scientific notation.**

(g) The absolute value of -2.8 is more than 1 and less than 10, but division rather than multiplication is the indicated operation. **This term is not in scientific notation.**

See Exercises 1–2.

To change a number written in scientific notation to an ordinary number:

1. Perform the indicated multiplication by moving the decimal point in the first factor the appropriate number of places as indicated by the power-of-10 factor. Affix or insert zeros as necessary.

2. Omit the power-of-10 factor.

Did You Know? When we multiply by a power of 10, the exponent of 10 tells us how many places and in which direction to move the decimal.

STOP AND CHECK

Change to ordinary numbers.

1. 7.2×10^1
2. 5.4×10^{-3}
3. -6.4×10^0

Answers:
1. 72   2. 0.0054   3. -6.4

EXAMPLE 2

Change to ordinary numbers.

(a) 3.6×10^4 (b) 2.8×10^{-2} (c) 1.1×10^0 (d) -6.9×10^{-5}

(e) -9.7×10^6

(a) $3.6 \times 10^4 = 36000. = \mathbf{36{,}000}$ Move the decimal point four places to the right.

(b) $2.8 \times 10^{-2} = 0.028 = \mathbf{0.028}$ Move the decimal point two places to the left.

(c) $1.1 \times 10^0 = 1.1$ Move the decimal point zero places.

(d) $-6.9 \times 10^{-5} = -0.000069 = -0.000069$ Move the decimal point five places to the left.

(e) $-9.7 \times 10^6 = -9700000. = -9,700,000$ Move the decimal point six places to the right.   **See Exercises 3–16.**

---

2   Change a Number from Ordinary Notation to Scientific Notation. To express an ordinary number in scientific notation, we reverse the procedures we used before. Shifting the decimal point changes the value of a number. The power-of-10 factor is used to offset or balance this change so the original value of the number is maintained.

> **To change a number written in ordinary notation to scientific notation:**
>
> 1. Insert a caret ($\wedge$) in the proper place to indicate where the decimal point should be positioned in the ordinary number so that the absolute value of the number is valued at 1 or between 1 and 10.
>
> 2. Determine the number of places and in which direction the decimal point shifts from the **new** position (caret) to the **old** position (decimal point). This number indicates the exponent of the power of 10.

> **TIP   A Balancing Act: Why Count from the New to the Old?** Moving the decimal in the ordinary number changes the value of the number unless you balance the effect of the move in the power-of-10 factor. When a decimal is moved in the first factor, the value is changed. To offset this change, an opposite change must be made in the power-of-10 factor. Counting from the new to the old position indicates the proper number of places and the direction (positive or negative) for balancing with the power-of-10 factor.
>
> Remember the word **NO.** Count from **N**ew to **O**ld.
>
> $$3,800 = 3.8 \times 10^3 \qquad 3_\wedge 800. \times 10^3 \, \text{N} \rightarrow \text{O} = +3.$$
> $$0.0045 = 4.5 \times 10^{-3} \qquad 0.004_\wedge 5 \times 10^{-3} \, \text{N} \rightarrow \text{O} = -3.$$

STOP AND CHECK

Express in scientific notation.
1. 64
2. 0.04
3. 8,200
4. 0.00472

Answers:
1. 6.4×10^1 2. 4×10^{-2}
3. 8.2×10^3 4. 4.72×10^{-3}

EXAMPLE 3

Express in scientific notation.

(a) 285   (b) 0.007   (c) 9.1   (d) 85,000   (e) 0.00074

(a) $285 \rightarrow 2_\wedge 85 = \mathbf{2.85 \times 10^2}$
The unwritten decimal is after the 5. Place the caret between 2 and 8 so the number 2.85 is between 1 and 10. Count *from* the caret *to* the decimal to determine the exponent of 10. A move two places to the right represents the exponent +2.

(b) $0.007 \rightarrow 0.007_\wedge = \mathbf{7 \times 10^{-3}}$
7 is between 1 and 10. Count *from* the caret *to* the decimal. A move three places to the left represents the exponent −3.

(c) $9.1 = \mathbf{9.1 \times 10^0}$
9.1 is already between 1 and 10, so the decimal does not move; that is, the decimal moves zero places.

(d) $85{,}000 \rightarrow 8_\wedge 5000 = \mathbf{8.5 \times 10^4}$
From the caret *to* the decimal is four places to the right.

(e) $0.00074 \rightarrow 0.0007_\wedge 4 = \mathbf{7.4 \times 10^{-4}}$
From the caret *to* the decimal is four places to the left.   **See Exercises 17–28.**

> **TIP** **Between 1 and 10: One Digit to the Left of the Decimal and That Digit Cannot Be Zero** The expression "between 1 and 10" means any number that is greater than or equal to 1 but less than 10. The first factor in scientific notation must be equal to 1 *or* between 1 and 10. That means **there will be one and only one digit to the left of the decimal point and that digit cannot be zero.**

Occasionally a number is written in power-of-10 notation but not in scientific notation because the first factor is not equal to 1 or is not between 1 and 10. Write the first factor in scientific notation and multiply the power-of-10 factors.

EXAMPLE 4

Express in scientific notation.

(a) 37×10^5 **(b)** 0.03×10^3

(a) $3_\wedge 7 \times 10^5 = 3.7 \times 10^1 \times 10^5$ Write the first factor in scientific notation and
$\qquad\qquad\quad = \mathbf{3.7 \times 10^6}$ multiply the power-of-10 factors.

(b) $0.03_\wedge \times 10^3 = 3 \times 10^{-2} \times 10^3$ Write the first factor in scientific notation and
$\qquad\qquad\quad = \mathbf{3 \times 10^1}$ multiply the power-of-10 factors.

See Exercises 29–36.

---

3 **Multiply and Divide Numbers in Scientific Notation.** We can multiply or divide numbers expressed in scientific notation without first having to convert them to ordinary numbers.

> **To multiply numbers in scientific notation:**
>
> 1. Multiply the first factors using the rules of signed numbers.
>
> 2. Multiply the power-of-10 factors using the laws of exponents (add exponents).
>
> 3. Examine the first factor of the product (Step 1) to see if its absolute value is equal to 1 or is between 1 and 10.
>
>   **(a)** If so, write the results of Steps 1 and 2.
>
>   **(b)** If not, write the first factor in scientific notation and multiply the power-of-10 factors.

EXAMPLE 5

Multiply and express the product in scientific notation.

(a) $(4 \times 10^2)(2 \times 10^3)$ **(b)** $(3.7 \times 10^3)(2.5 \times 10^{-1})$

(c) $(8.4 \times 10^{-2})(5.2 \times 10^{-3})$

(a) $(4 \times 10^2)(2 \times 10^3) = \mathbf{8 \times 10^5}$
The product is in scientific notation.

(b) $(3.7 \times 10^3)(2.5 \times 10^{-1}) = \textbf{9.25} \times \textbf{10}^2$
The product is in scientific notation.

(c) $(8.4 \times 10^{-2})(5.2 \times 10^{-3}) = 43.68 \times 10^{-5}$
43.68×10^{-5} is not in scientific notation.

$43.68 \to 4_\wedge 3.68 \qquad$ or $\qquad 4.368 \times 10^1 \qquad$ Write first factor in scientific notation.
$4.368 \times 10^1 \times 10^{-5} = 4.368 \times 10^{1-5} \qquad$ Multiply the powers-of-10 factors.
$= \textbf{4.368} \times \textbf{10}^{-4} \qquad$ **See Exercises 37–40.**

Division involving numbers in scientific notation is similar to multiplication.

> **To divide numbers in scientific notation:**
>
> 1. Divide the first factors using the rules of signed numbers.
>
> 2. Divide the power-of-10 factors using the laws of exponents (subtract exponents).
>
> 3. Examine the first factor of the quotient (Step 1) to see if its absolute value is equal to 1 or between 1 and 10.
>
>   **(a)** If so, write the results of Steps 1 and 2.
>
>   **(b)** If not, write the first factor in scientific notation and multiply the power-of-10 factors.

STOP AND CHECK
Divide and express the quotient in scientific notation.
1. $\dfrac{7 \times 10^4}{5 \times 10^1}$ **2.** $\dfrac{2.1 \times 10^{-4}}{7 \times 10^{-2}}$

3. $\dfrac{7.5 \times 10^2}{15}$

Answers:
1. 1.4×10^3 **2.** 3×10^{-3}
3. 5×10^1

EXAMPLE 6

Divide and express the quotient in scientific notation.

(a) $\dfrac{3 \times 10^5}{2 \times 10^2}$ **(b)** $\dfrac{1.44 \times 10^{-3}}{6 \times 10^{-5}}$ **(c)** $\dfrac{9.6 \times 10^{29}}{3.2 \times 10^{111}}$ **(d)** $\dfrac{1.25 \times 10^3}{5}$

(a) $\dfrac{3 \times 10^5}{2 \times 10^2} = \dfrac{3}{2} \times 10^{5-2} = \textbf{1.5} \times \textbf{10}^3$

The first factor is usually written in decimal notation.

(b) $\dfrac{1.44 \times 10^{-3}}{6 \times 10^{-5}} = \dfrac{1.44}{6} \times 10^{-3-(-5)} = 0.24 \times 10^{-3+5} = 0.24 \times 10^2$

0.24 is less than 1, so adjustments are necessary.

$0.24 \to 0.2_\wedge 4 = 2.4 \times 10^{-1} \qquad$ Write the first factor in scientific notation.
Then $2.4 \times 10^{-1} \times 10^2 = \textbf{2.4} \times \textbf{10}^1$. $\qquad$ Multiply the powers-of-10 factors.

(c) $\dfrac{9.6 \times 10^{29}}{3.2 \times 10^{111}} = \dfrac{9.6}{3.2} \times 10^{29-111} = \textbf{3} \times \textbf{10}^{-82}$

(d) $\dfrac{1.25 \times 10^3}{5} \qquad\qquad\qquad$ 5 is the same as 5×10^0.

$\dfrac{1.25 \times 10^3}{5 \times 10^0} = \dfrac{1.25}{5} \times 10^{3-0} = 0.25 \times 10^3 \qquad$ 0.25 is less than 1.

$0.25 \to 0.2_\wedge 5 = 2.5 \times 10^{-1} \qquad$ Write the first factor in scientific notation and multiply the power-of-10 factors.

Then, $2.5 \times 10^{-1} \times 10^3 = 2.5 \times 10^{-1+3} = \textbf{2.5} \times \textbf{10}^2$. $\qquad$ **See Exercises 41–45.**

TIP **Mental Adjustment of Exponents** Once we understand the concept of balancing the effect of moving a decimal by adjusting the exponent of the power-of-10 factor, we can perform this adjustment mentally.

$$43.68 \times 10^{-5} = 4_\wedge 3.68 \times 10^{-5} \qquad \text{N} \rightarrow \text{O} = +1$$
$$= 4.368 \times 10^{-5+1} \qquad \text{Adjust mentally.}$$
$$= 4.368 \times 10^{-4}$$
$$0.25 \times 10^{3} = 0.2_\wedge 5 \times 10^{3} \qquad \text{N} \rightarrow \text{O} = -1$$
$$= 2.5 \times 10^{3-1} \qquad \text{Adjust mentally.}$$
$$= 2.5 \times 10^{2}$$

TIP **Scientific Notation and the Calculator**

Power-of-10 Key

The power-of-10 key, labeled $\boxed{10^x}$, $\boxed{\text{EXP}}$, or $\boxed{\text{EE}}$ on most calculators, is a shortcut key for entering the following keys:

$$\boxed{\times}\ 10\ \boxed{x^y}$$

The shortcut key is used only for power-of-10 factors, and only the *exponent* of 10 is entered. If you enter $\times 10$, then $\boxed{\text{EXP}}$, your answer will have one extra factor of 10.

Look at $\dfrac{(3 \times 10^5)}{(2 \times 10^2)}$ on the calculator. *Steps will vary on various calculators.*

$\boxed{(}\ 3\ \boxed{10^x}\ 5\ \boxed{)}\ \boxed{\div}\ \boxed{(}\ 2\ \boxed{10^x}\ 2\ \boxed{)}\ \boxed{=}\ \Rightarrow 1{,}500$ or $3\ \boxed{\text{EXP}}\ 5\ \boxed{\div}\ 2\ \boxed{\text{EXP}}\ 2\ \boxed{=}\ \Rightarrow 1{,}500$

This result may be expressed in scientific notation if desired: $1{,}500 = 1.5 \times 10^3$

The internal program of a calculator has predetermined how the output of a calculation is displayed. For example, even if you would like an answer displayed in scientific notation, it may fall within the guidelines for display as an ordinary number. You must make the conversion to scientific notation yourself. The reverse may also be true.

Experiment using examples for which you already know the answer to determine the limitations and requirements of your calculator. Some calculators require the parentheses and some do not.

EXAMPLE 7

AVIA A star is 4.2 light-years from Earth. If 1 light-year is 5.87×10^{12} mi, how many miles from Earth is the star? Express your answer in scientific notation.

To solve this problem, multiply the number of light years times the distance of 1 light year.

Estimation

By rounding, we estimate that the star will be approximately 4 times 6×10^{12} mi away.

$$4(6 \times 10^{12}) = 24 \times 10^{12} = 2.4 \times 10^{13}$$
$$4.2(5.87 \times 10^{12}) = \qquad\qquad 4.2 = 4.2 \neq 10^0$$
$$24.654 \times 10^{12} = \qquad\qquad \text{Perform scientific notation adjustment.}$$
$$2.4654 \times 10^{12+1} = \qquad\qquad \text{N} \rightarrow \text{O} = +1$$

Titonz/123RF

$$2.4654 \times 10^{13}$$

or $\quad 2.5 \times 10^{13}$ Round the first factor to tenths.

Interpretation

The star is 2.5×10^{13} mi from Earth. **See Exercise 46.**

EXAMPLE 8

ELEC An angstrom (Å) is 1×10^{-7} mm. What is the length in millimeters of 14.82 Å?

Solve the problem by multiplying the number of angstrom units times the length of 1 Å.

Estimation

14.82 angstrom units are more than 10 times longer than 1×10^{-7} mm, or more than 1×10^{-6}.

$$14.82(1 \times 10^{-7}) =$$
$$14.82 \times 10^{-7} = \quad \text{\color{blue}Adjust the first factor and exponent.}$$
$$1.482 \times 10^{-7+1} = \quad \text{\color{blue}N} \rightarrow \text{\color{blue}O} = \text{\color{blue}+1}$$
$$1.482 \times 10^{-6} \text{ mm}$$

Interpretation

The length of 14.82 Å is 1.482×10^{-6} mm. **See Exercise 46.**

EXAMPLE 9

ELEC One coulomb (C) is approximately 6.28×10^{18} electrons. How many coulombs do 2.512×10^{21} electrons represent?

Divide the total number of electrons by the number of electrons in one coulomb.

$$\frac{2.512 \times 10^{21}}{6.28 \times 10^{18}}$$

Estimation

By rounding the first factors we have

$$\frac{3 \times 10^{21}}{6 \times 10^{18}} = 0.5 \times 10^{21-18} = 0.5 \times 10^{3} = 500 \text{ C}$$

$$\frac{2.512 \times 10^{21}}{6.28 \times 10^{18}} = \quad \text{\color{blue}Divide coefficients, subtract exponents.}$$
$$0.4 \times 10^{3} = \quad \text{\color{blue}Adjust the first factor.}$$
$$4 \times 10^{3-1} = \quad \text{\color{blue}N} \rightarrow \text{\color{blue}O} = \text{\color{blue}-1}$$
$$4 \times 10^{2} \text{ or } 400 \text{ C} \quad \text{\color{blue}Write as an ordinary number.}$$

Interpretation

2.512×10^{21} electrons represent 400 C. **See Exercise 47.**

---

4 **Raise a Number in Scientific Notation to a Power.** Two more of the laws of exponents are applied when raising a number in scientific notation to a power. These laws address raising more than one factor to a power and raising a power to a power. Again, we will examine these laws in Chapter 11; but for now we will examine the laws as they apply to scientific notation.

> **To raise a number in scientific notation to a power:**
>
> 1. Raise the first factor to the power.
>
> 2. Raise the power of 10 to the power by multiplying exponents.
>
> 3. Adjust the first factor and power of 10 so that the first factor is greater than or equal to 1 or less than 10.
>
> Symbolically, $(n \times 10^a)^b = n^b \times 10^{ab}$, n is any real number, and $n \neq 0$. Adjust so that the standard notation for $n^b \geq 1$ and $n^b < 10$ if appropriate.

EXAMPLE 10

Raise the following to the indicated powers: **(a)** $(2 \times 10^3)^2$ **(b)** $(4.2 \times 10^{-4})^3$ **(c)** $(7.3 \times 10^{-2})^{-5}$

(a) $(2 \times 10^3)^2 = 2^2 \times 10^{3(2)}$ Square each factor.

$\qquad = 4 \times 10^6$

(b) $(4.2 \times 10^{-4})^3 = 4.2^3 \times 10^{-4(3)}$ Cube each factor. Evaluate.

$\qquad = 74.088 \times 10^{-12}$ Adjust to scientific notation.

$\qquad = 7.4088 \times 10^{-12+1}$

$\qquad = \mathbf{7.4088 \times 10^{-11}}$

(c) $(7.3 \times 10^{-2})^{-5} = 7.3^{-5} \times 10^{-2(-5)}$ Raise each factor to the -5 power. Evaluate.

$\qquad = 0.00004823760083 \times 10^{10}$ Adjust to scientific notation.

$\qquad = 0.00004\,823760083 \times 10^{10}$

$\qquad = 4.823760083 \times 10^{-5} \times 10^{10}$

$\qquad = 4.823760083 \times 10^5$ **See Exercises 48–64.**

5 Change Among Engineering, Scientific, and Ordinary Notations. In Chapter 4 we examined some metric prefixes for very large and very small units of measure. In this chapter we examined power-of-10 exponents and scientific notation. We now examine power-of-10 exponents that are multiples of 3.

Metric Prefixes

Prefix	Abbreviation	Meaning	Power of 10
Yotta-	Y	one septillion times	$\times 10^{24}$
Zetta-	X	one sextillion times	$\times 10^{21}$
Exa-	E	one quintillion times	$\times 10^{18}$
Peta-	P	one quadrillion times	$\times 10^{15}$
Tera-	T	one trillion times	$\times 10^{12}$
Giga-	G	one billion times	$\times 10^9$
Mega-	M	one million times	$\times 10^6$
kilo	k	one thousand times	$\times 10^3$
	Standard Unit		$\times 10^0$
milli-	m	one thousandth of	$\times 10^{-3}$
micro-	μ	one millionth of	$\times 10^{-6}$
nano-	n	one billionth of	$\times 10^{-9}$
pico-	p	one trillionth of	$\times 10^{-12}$
femto-	f	one quadrillionth of	$\times 10^{-15}$
atto-	a	one quintillionth of	$\times 10^{-18}$
zepto-	z	one sextillionth of	$\times 10^{-21}$
yocto-	y	one septillionth of	$\times 10^{-24}$

Engineering notation: a number is expressed in engineering notation if it is the product of two factors. The absolute value of the first factor is a number greater than or equal to 1 and less than 1,000. The second factor is a power of 10 that is a multiple of 3

These prefixes correlate with a modification of scientific notation called **engineering notation**. Engineering notation, like scientific notation, has a first factor and a power-of-10 factor. The absolute value of the first factor is greater than or equal to 1 and less than 1,000. The power-of-10 factor is a multiple of three.

To change an ordinary number to engineering notation:

1. Indicate with a caret ($\wedge$) where the decimal point should be positioned.

   (a) If the number is greater than or equal to 1 and less than 1,000, the decimal point will not shift and the power-of-10 factor will be 10^0.

   (b) If the number is greater than or equal to 1,000, insert commas as appropriate to separate the place-value periods, and place the caret (the *new* position of the decimal point) at the leftmost comma.

   (c) If there are no nonzero digits to the left of the decimal, count from the decimal to the right in groups of three places until you have at least one, but no more than three, significant digits to the left of the caret (*new* position of the decimal point).

2. Determine the exponent of the power-of-10 factor and its sign by counting the number of places from the *new* position of the decimal point to the *old* position. The resulting exponent will be a multiple of 3.

STOP AND CHECK

Change to engineering notation.
1. 5,240
2. 86
3. 0.5
4. 0.00000075

Answers:
1. 5.24×10^3 2. 86×10^0
3. 500×10^{-3} 4. 750×10^{-9}

EXAMPLE 11

Change to engineering notation:

(a) 2,400,000 (b) 2,400 (c) 24
(d) 0.24 (e) 0.024 (f) 0.0000024

(a) 2,400,000 $2_\wedge 400,000$ Place the caret at the leftmost comma.
2.4×10^6 N→O = +6

(b) 2,400 $2_\wedge 400$ Place the caret at the leftmost comma.
2.4×10^3 N→O = +3

(c) 24 $24_\wedge$ Place the caret at the decimal point. The power-of-10 exponent can be only 0 or a multiple of 3.
24×10^0 N→O = 0

(d) 0.24 $0.240_\wedge$ Place the caret after the third decimal place value. There can be no more than 3 significant digits on the left of the caret.
240×10^{-3} N→O = −3

(e) 0.024 $0.024_\wedge$ Place the caret after the third decimal value. There can be no more than 3 significant digits on the left of the caret.
24×10^{-3} N→O = −3

(f) 0.0000024 $0.000002_\wedge 4$ Place the caret after the sixth decimal value. There can be no more than 3 significant digits on the left of the caret.
2.4×10^{-6} N→O = −6 **See Exercises 65–80.**

Measures that are expressed in metric units can be written in engineering notation by showing the power-of-10 factor with the standard unit or by translating the power-of-10 factor to the appropriate metric prefix. When using metric prefixes, a prefix replaces the power-of-10 factor. If the measure has a metric unit other than a standard unit and a first factor that is *less than* 1 or *greater than* 1,000, the metric unit will need to be adjusted. The table of metric prefixes with powers of 10 is on p. 249 and the table of the types of SI measuring units is on p. 187.

STOP AND CHECK

Write the engineering notation using metric prefixes.

1. $12,396,205\ \Omega$
2. 0.000078 J
3. $3,225$ kW
4. $8,925$ GHz

Answers:
1. $12.396205\ \mu\Omega$
2. 78 mJ
3. 3.225 MW
4. 8.925 THz

EXAMPLE 12

Write the engineering notation using metric prefixes.

(a) $4,382,000\ \Omega$ **(b)** $0.00001\ \text{Å}$ **(c)** $5,870\ \mu\text{W}$ **(d)** $3,500$ MHz

(a) $4,382,000\ \Omega = 4.382 \times 10^6\ \Omega$ $10^6 = $ M   See p. 249.
$\qquad\qquad\qquad = 4.382\ \text{M}\Omega$

(b) $0.00001\ \text{Å} = 10 \times 10^{-6}\ \text{Å}$ $10^{-6} = \mu$
$\qquad\qquad = 10\ \mu\text{Å}$

(c) $5,870\ \mu\text{W} = 5.87 \times 10^3\ \mu\text{W}$ $\mu = 10^{-6}$
$\qquad\qquad = 5.87 \times 10^3 \times 10^{-6}\ \text{W}$ Multiply power-of-10 factors.
$\qquad\qquad = 5.87 \times 10^{-3}\ \text{W}$ $10^{-3} = $ m
$\qquad\qquad = 5.87\ \text{mW}$

(d) $3,500\ \text{MHz} = 3.5 \times 10^3\ \text{MHz}$ M $= 10^6$
$\qquad\qquad\quad = 3.5 \times 10^3 \times 10^6\ \text{Hz}$ Multiply power-of-10 factors.
$\qquad\qquad\quad = 3.5 \times 10^9\ \text{Hz}$ $10^9 = $ G
$\qquad\qquad\quad = 3.5\ \text{GHz}$

See Exercises 81–100.

5–6 EXERCISES   MyLab Math For additional practice go to your study plan in MyLab Math.

1 *See Example 1.*

1. Is -5.1×10^5 in scientific notation? Explain why or why not.

2. Is 35×10^{-3} in scientific notation? Explain why or why not.

Write as ordinary numbers. *See Example 2.*

3. 4.3×10^2 **4.** 6.5×10^{-3} **5.** 2.2×10^0 **6.** 8.3×10^4

7. 5.8×10^{-3} **8.** 8×10^4 **9.** 6.732×10^0 **10.** 5.89×10^{-3}

11. 7.83×10 **12.** 1.59×10^3 **13.** 3.97×10^5 **14.** 4.723×10^{-4}

15. 9.91×10^{-6} **16.** 1.03×10^6

2 Express in scientific notation. *See Example 3.*

17. 392 **18.** 0.02 **19.** 7.03 **20.** $42,000$

21. 0.081 **22.** 0.0021 **23.** 23.92 **24.** 0.101

25. 1.002 **26.** 721 **27.** 42×10^4 **28.** 32.6×10^3

See Example 4.

29. 0.213×10^2 **30.** 0.0062×10^{-3} **31.** $56,000 \times 10^{-3}$ **32.** 0.197×10^{-5}

33. 745×10^{-1} **34.** 18×10^3 **35.** 0.701×10^2 **36.** $72,500 \times 10^{-5}$

3 Perform the indicated operations. Express answers in scientific notation. *See Example 5.*

37. $(6.7 \times 10^4)(3.2 \times 10^2)$

38. $(1.6 \times 10^{-1})(3.5 \times 10^4)$

39. $(5.0 \times 10^{-3})(4.72 \times 10^0)$

40. $(8.6 \times 10^{-3})(5.5 \times 10^{-1})$

See Example 6.

41. $\dfrac{3.15 \times 10^5}{4.5 \times 10^2}$

42. $\dfrac{4.68 \times 10^3}{7.2 \times 10^7}$

43. $\dfrac{4.55 \times 10^{-1}}{6.5 \times 10^{-4}}$

44. $\dfrac{7.84 \times 10^{-2}}{9.8 \times 10^0}$

45. $\dfrac{1.96 \times 10^{-3}}{8.0 \times 10^{-5}}$

46. AVIA   A star is 5.5 light years from Earth. If one light-year is 5.87×10^{12} miles, how many miles from Earth is the star? *See Examples 7 and 8.*

47. ELEC   An angstrom (Å) is 1×10^{-7} mm. How many angstroms are in 4.2×10^{-5} mm? *See Example 9.*

4 Perform the indicated operation and express the result in scientific notation. Round appropriately. *See Example 10.*

48. $(8.3 \times 10^2)^3$

49. $(7 \times 10^{-2})^4$

50. $(1.93 \times 10^{-4})^2$

51. $(7.2 \times 10^2)^2(3.1 \times 10^1)^2$

52. $\dfrac{(8.2 \times 10^{-3})^2}{(5.73 \times 10^4)^3}$

53. $63,000 \times 7,000$

54. $2,500(61,000)(300)$

55. $\dfrac{42,000}{60,000}$

56. $(52,000)^2$

57. $(0.00013)^{-2}$

58. $(17,000)^2$

59. $(83,000)^2(5,200)^2$

60. $\dfrac{(0.0071)^3}{(0.02)^4}$

61. $\left(\dfrac{210 \times 300}{510}\right)^2$

62. $\left(\dfrac{5,000 \times 820}{4,000}\right)^3$

63. $\left(\dfrac{16.7 \times 5,200}{0.16 \times 820}\right)^3$

64. $\left(\dfrac{0.0061 \times 2,300}{0.23 \times 46,000}\right)^3$

5 Write each number in engineering notation. *See Example 11.*

65. 3,800,000

66. 5,600

67. 78

68. 52,000

69. 80,000,000

70. 5,830,000

71. 1,736,500,000

72. 41,980,000

73. 0.78

74. 0.33

75. 0.0000011

76. 0.000000008

77. 0.0009832

78. 0.0000719

79. 0.0120307

80. 0.000675

Change each unit to engineering notation using metric prefixes. *See Example 12.*

81. 428,000 Ω

82. 5,700,000 V

83. 3,520,000,000 W

84. 79,000,000 Hz

85. 0.000081 s

86. 0.0973 Å

87. 0.00000000541 s

88. 0.89 s

89. 0.00058 MΩ

90. 0.00077 μs

91. 2,980,000 ps

92. 7,810,000 mW

93. 42,300 kV

94. 1,572,000 kW

95. 5,096,000 Ω

96. 0.000000008 Å

97. 182,000 Hz

98. 1,600 Ω

99. 5,200,000 V

100. 97,000,000 W

5 | CHAPTER REVIEW OF KEY CONCEPTS

| LEARNING OUTCOMES | KEY CONCEPTS AND EXAMPLES |

Section 5–1

1 Compare signed numbers (pp. 214–217).

Positive numbers are to the right of zero and negative numbers are to the left of zero on the number line.

> Arrange from smallest to largest: $5, -3, 0, 8, -5$
>
> $$-5, -3, 0, 5, 8$$

Compare signed numbers: 1. Mentally position the two numbers on a number line. **2.** The leftmost number is the smaller number. If the smaller number is written first, use the $<$ symbol between the numbers. If the larger number is written first, use the $>$ symbol between the numbers.

  The "less than" symbol is $<$.
  The "greater than" symbol is $>$.

> Use $<$ or $>$ to make a true statement: $5 \_ -3$
>
> $$5 > -3 \quad \text{or} \quad -3 < 5$$

The absolute value of a number is its *distance* from zero without regard to direction.

> Evaluate the following absolute values: $|-3|, |5|$
>
> $$|-3| = 3, \quad |5| = 5$$

Opposites are numbers that have the same absolute value but opposite signs.

> Give the opposite of the following: $8, -4, -2, +4$
>
> $$-8, \quad 4, \quad 2, -4$$

2 Add signed numbers with like signs (pp. 217–218).

1. Add the absolute values of the numbers. **2.** Give the sum the common or like sign.

> Add: $13 + 7 = 20 \quad -35 + (-13) = -48$

3 Add signed numbers with unlike signs (pp. 218–219).

1. Subtract the smaller absolute value from the larger absolute value. **2.** Give the sum the sign of the number with the larger absolute value.

> Add: $-12 + 7$
>
> Subtract absolute values: $12 - 7 = 5$. Give the 5 the negative sign because -12 has the larger absolute value.
>
> $$-12 + 7 = -5$$

LEARNING OUTCOMES	KEY CONCEPTS AND EXAMPLES

When zero is added to a signed number, the result is unchanged: $0 + a = a + 0 = a$.

> Add: $-42 + 0 = -42$ $0 + 38 = 38$

Section 5-2

1 Subtract signed numbers (pp. 221–222).

1. Change the subtraction sign to addition.   **2.** Change the sign of the second number (subtrahend), to its opposite.   **3.** Apply the appropriate rule for adding signed numbers.

> Subtract: $-32 - (-28)$
>
> $$-32 - (-28) = -32 + (+28)$$
> $$= -4$$

Subtract with zero: 1. Change the subtraction to addition.   **2.** Use the appropriate rule for addition.

> $-8 - 0 = -8 + (0) = -8$ $0 - (-4) = 0 + (+4) = 4$

Subtracting an opposite from a number is the same as adding the number to itself.

> $8 - (-8) = 8 + 8 = 16$ $-12 - 12 = -24$

2 Combine addition and subtraction (pp. 222–224).

Add and subtract more than two signed numbers: 1. Rewrite the problem so that all integers are separated by only one sign. Or, change all subtractions to equivalent additions.   **2.** Add the series of signed numbers from left to right.

> Simplify: $5 - 3 + 2 - (-4)$
>
> $$5 + (-3) + 2 + 4 =$$
> $$2 + 2 + 4 = 4 + 4 = 8$$

Section 5-3

1 Multiply signed numbers (pp. 225–227).

Multiply two signed numbers with like signs: 1. Multiply the absolute values.   **2.** Make the sign of the product positive.

> $-6(-7) = +42$ $8 \cdot 6 = 48$

Multiply two signed numbers with unlike signs: 1. Multiply the absolute values of the numbers.   **2.** The sign of the product is negative.

> $-7(2) = -14$ $8(-3) = -24$

Determine the sign of the product when multiplying three or more factors: 1. The sign of the product is *positive* if the number of negative factors is even.   **2.** The sign of the product is *negative* if the number of negative factors is odd.

> $5(-2)(3) = -10(3) = -30$

LEARNING OUTCOMES	KEY CONCEPTS AND EXAMPLES

Any number (including signed numbers) multiplied by zero results in zero: $a \times 0 = 0 \times a = 0$

$$0 \times 3 = 0 \qquad -5 \times 0 = 0 \qquad 0(-7) = 0$$

2 Evaluate powers of signed numbers (pp. 227–228).

Raise signed numbers to a natural-number power: 1. A positive number raised to any natural-number power is positive. **2.** Zero raised to any natural-number power is zero. **3.** To raise a negative number to a power requires parentheses. A negative number raised to an even natural-number power is positive. **4.** A negative number raised to an odd natural-number power is negative.

$$(3)^3 = 27 \qquad (-2)^4 = 16 \qquad -2^4 = -16$$
$$0^5 = 0 \qquad (-2)^3 = -8 \qquad -2^3 = -8$$

3 Divide signed numbers (pp. 228–229).

Divide two signed numbers: 1. Divide the absolute values of the dividend and the divisor. **2.** If the dividend and divisor have like signs, the sign of the quotient is positive. **3.** If the dividend and divisor have unlike signs, the sign of the quotient is negative.

$$-12 \div (-4) = 3 \qquad 15 \div (-3) = -5$$

Evaluate division with zero: Zero divided by any nonzero number is zero.

$$\frac{0}{a} = 0; \qquad a \neq 0 \text{ and } a \text{ is any real number} \qquad \frac{0}{-5} = 0$$

Division by zero is either undefined or indeterminate.

$$\frac{a}{0} \text{ is undefined; } a \neq 0 \text{ and } a \text{ is any real number}$$

$$\frac{-5}{0} \text{ is undefined.} \quad \frac{0}{0} \text{ is indeterminate.}$$

$$0 \div 12 = 0 \qquad -7 \div 0 \text{ is undefined.}$$
$$0 \div 0 \text{ is indeterminate.}$$

Section 5–4

1 Change a signed fraction to an equivalent signed fraction (pp. 231–232).

1. Identify the three signs of the fraction (sign of the fraction, sign of the numerator, and sign of the denominator). **2.** Change any two of the three signs to the opposite sign.

Write three equivalent signed fractions for $-\frac{7}{8}$.

$$-\frac{7}{8} = -\frac{+7}{+8} \quad \text{Equivalent fractions are } +\frac{+7}{-8} \text{ or } -\frac{-7}{-8} \text{ or } +\frac{-7}{+8}$$

2 Perform basic operations with rational numbers (pp. 232–233).

1. Determine the indicated operations. **2.** Apply the appropriate rules for signed numbers and for rational numbers.

LEARNING OUTCOMES

KEY CONCEPTS AND EXAMPLES

Add $\dfrac{-5}{8} + \dfrac{7}{8}$

$$\dfrac{-5}{8} + \dfrac{7}{8} = \dfrac{2}{8} = \dfrac{1}{4}$$ Add numerators with unlike signs. Reduce.

Add $4.37 + (-2.91)$

$$4.37 - 2.91 = 1.46$$ Add decimals with unlike signs.

3 Apply the order of operations with signed numbers (pp. 233–237).

Perform operations in the following order as they appear from left to right: 1. Parentheses. **2.** Exponents (powers and roots). **3.** Multiplication and division. **4.** Addition and subtraction.

Evaluate: $5 - 4(7 + 2)^2 - 15 \div 5$

$5 - 4(7 + 2)^2 - 15 \div 5 =$	Perform operation inside parentheses.
$5 - 4(9)^2 - 15 \div 5 =$	Evaluate exponent.
$5 - 4(81) - 15 \div 5 =$	Multiply.
$5 - 324 - 15 \div 5 =$	Divide.
$5 - 324 - 3 =$	Add signed numbers.
$-319 - 3 = -322$	

Evaluate $-\dfrac{1}{4} - \dfrac{2}{3}\left(-\dfrac{1}{2}\right)$

$-\dfrac{1}{4} - \dfrac{2}{3}\left(-\dfrac{1}{2}\right) =$	Multiply using rules for multiplying fractions with like signs.
$-\dfrac{1}{4} + \dfrac{1}{3} =$	Change to equivalent fractions with the LCD.
$-\dfrac{3}{12} + \dfrac{4}{12} = \dfrac{1}{12}$	Use the rule for adding numbers with unlike signs.

Evaluate $(-0.3)^2 - 2.8$

$(-0.3)^2 - 2.8 =$	Square signed decimal.
$0.09 - 2.8 = -2.71$	Add decimals with unlike signs.

Section 5–5

1 Multiply and divide by powers of 10 (pp. 238–241).

Multiply a number by a power of 10: 1. If the exponent is positive, shift the decimal point to the *right* the number of places indicated by the *positive* exponent. Attach zeros as necessary. **2.** If the exponent is negative, shift the decimal point to the *left* the number of places indicated by the *negative* exponent. Insert zeros as necessary.

$$481(1,000) = 481(10)^3 = 481,000 \qquad 6.9(0.0001) = 6.9(10)^{-4} = 0.00069$$

Divide a number by a power of 10: 1. Change the division to an equivalent multiplication by multiplying by the reciprocal of the divisor. **2.** Use the rule for multiplying a number by a power of 10.

$$4,386 \div 100 = 4,386 \div 10^2 = 4,386(10)^{-2} = 43.86$$
$$4.5 \div 0.001 = 4.5 \div 10^{-3} = 4.5 \div 10^3 = 4,500$$

LEARNING OUTCOMES	KEY CONCEPTS AND EXAMPLES

Multiply powers of 10 written in exponential notation: 1. The product will be a power of 10. **2.** Add the exponents for the power-of-10 factors.

> Multiply. $10^6(10^7) = 10^{6+7} = 10^{13}$

Divide powers of 10 written in exponential notation: 1. The quotient will be a power of 10. **2.** Subtract the exponents for the power-of-10 factors (numerator minus denominator). **3.** Powers of 10 with negative exponents can be written as reciprocal expressions with positive exponents, if desired.

Symbolically, $\frac{10^a}{10^b} = 10^{a-b}$, where a and b are integers; $10^{-a} = \frac{1}{10^a}$.

> Divide. $10^3 \div 10^5 = 10^{3-5} = 10^{-2}$

2 Raise a power of 10 to a power (p. 241).

1. The result will be a power of 10. **2.** The exponent will be the product of the original exponent and the exponent of the power.

Symbolically, $(10^a)^b = 10^{ab}$.

> Raise the powers of 10 to the indicated exponents.
>
> $(10^4)^2 = 10^{4 \cdot 2} = 10^8$ $\qquad$ $(10^{-3})^2 = 10^{-3 \cdot 2} = 10^{-6}$

Section 5–6

1 Change a number from scientific notation to ordinary notation (pp. 242–244).

Change a number written in scientific notation to an ordinary number: 1. Perform the indicated multiplication by moving the decimal point in the first factor the appropriate number of places as indicated by the power-of-10 factor. Affix or insert zeros as necessary. **2.** Omit the power-of-10 factor.

> Write 3.27×10^{-4} as an ordinary number.
> $00003.27 = 0.000327.$ $\qquad$ **Shift the decimal 4 places to the *left*.**

2 Change a number from ordinary notation to scientific notation (pp. 244–245).

Change a number written in ordinary notation to scientific notation: 1. Insert a caret in the proper place to indicate where the decimal point should be positioned so that the absolute value of the number is valued at 1 or between 1 and 10. **2.** Determine the number of places and in which direction the decimal point shifts from the *new* position (caret) to the *old* position (decimal point). This number indicates the exponent of the power of 10.

> Write 54,000 in scientific notation.
> $5_\wedge 4000. = 5.4 \times 10^4$ $\qquad$ **From New to Old is 4 places to the right so the exponent is +4.**

3 Multiply and divide numbers in scientific notation (pp. 245–248).

Multiply numbers in scientific notation: 1. Multiply the first factors using the rules of signed numbers. **2.** Multiply the power-of-10 factors using the laws of exponents (add exponents). **3.** Examine the first factor of the product (Step 1) to see if its absolute value is equal to 1 or is between 1 and 10. **(a)** If so, write the results of Steps 1 and 2. **(b)** If not, write the first factor in scientific notation and multiply the power-of-10 factors.

LEARNING OUTCOMES	KEY CONCEPTS AND EXAMPLES

Multiply.

$$(4.5 \times 10^{89})(7.5 \times 10^{36}) =$$ Multiply first factors then power-of-10 factors.
$$33.75 \times 10^{125} =$$ Write first factor in scientific notation.
$$3_{\wedge}375 \times 10^{1} \times 10^{125} =$$ Multiply powers of 10.
$$3.375 \times 10^{126}$$

Divide numbers in scientific notation: 1. Divide the first factors using the rules of signed numbers. **2.** Divide the power-of-10 factors using the laws of exponents (subtract exponents). **3.** Examine the first factor of the quotient (Step 1) to see if its absolute value is equal to 1 or between 1 and 10. **(a)** If so, write the results of Steps 1 and 2. **(b)** If not, write the first factor in scientific notation and multiply the power-of-10 factors.

Divide.

$$(3 \times 10^{-3}) \div (4 \times 10^{2}) =$$ Divide first factors, then power-of-10 factors.
$$0.75 \times 10^{-5} =$$ Write first factor in scientific notation.
$$07_{\wedge}5 \times 10^{-1} \times 10^{-5} =$$ Multiply powers of 10.
$$7.5 \times 10^{-6}$$

4 Raise a number in scientific notation to a power (pp. 248–249).

1. Raise the first factor to the power. **2.** Raise the power of 10 to the power by multiplying exponents. **3.** Adjust the first factor and power of 10 so that the first factor is greater than or equal to 1 or less than 10.

Symbolically, $(n \times 10^{a})^{b} = n^{b} \times 10^{ab}$. Adjust so that the standard notation for $n^{b} \geq 1$ and $n^{b} < 10$ if appropriate.

Raise the following to the indicated powers:

(a) $(3 \times 10^{4})^{2}$ **(b)** $(2.3 \times 10^{-3})^{3}$

(a) $(3 \times 10^{4})^{2} = 3^{2} \times 10^{4 \cdot 2} = 9 \times 10^{8}$ Square each factor.

(b) $(2.3 \times 10^{-3})^{3} = 2.3^{3} \times 10^{-3 \cdot 3}$ Cube each factor.
$$= 12.167 \times 10^{-9}$$
$$= 1.2167 \times 10^{-9+1}$$ Adjust the first factor and the power-of-10 factor.
$$= 1.2167 \times 10^{-8}$$

5 Change among engineering, scientific, and ordinary notations (pp. 249–251).

Change an ordinary number to engineering notation: 1. Indicate with a caret where the decimal point should be positioned. (a) If the number is greater than or equal to 1 and less than 1,000, the decimal point will not shift and the power-of-10 factor will be 10^{0}. (b) If the number is greater than or equal to 1,000, insert commas as appropriate to separate the place-value periods and place the caret (the *new* position of the decimal point) at the leftmost comma. (c) If there are no nonzero digits to the left of the decimal point, count from the decimal point to the right in groups of three places until you have at least one, but no more than three, significant digits to the left of the caret (*new* position of the decimal point). **2.** Determine the exponent of the power-of-10 factor and its sign by counting the number of places from the *new* position of the decimal point to the *old* position. The resulting exponent will be a multiple of 3.

Write 82,000,000 Hz in engineering notation.

$$82,000,000 \text{ Hz} = 82 \times 10^{6} \text{ Hz or } 82 \text{ MHz}$$

Write 0.00017 μs in engineering notation.

$$0.00017\mu s = 170 \times 10^{-6} \mu s \text{ or } 170 \text{ ps}$$

Section 5–1   MyLab Math   For additional practice go to your study plan in MyLab Math.

Use the symbol $<$ or $>$ to show the relationship between each pair of numbers.

1. -2 _____ -6

2. -5 _____ 1

Give the value of each number:

3. $|5|$ **4.** $|-8|$

5. $|+7|$ **6.** $|-52|$

Give the opposite of each number:

7. -12 **8.** 8 **9.** -2

10. -13 **11.** 87

Add.

12. $-3 + (-8)$ **13.** $(-15) + (-8) + (-3)$

14. $7 + (-11)$ **15.** $-25 + 0 + 12 + 7$

16. **BUS** A publicly traded company has a profit of $256,872 for one year and a loss of $38,956 for the following year. What is the net profit over the two-year period?

17. A football team gained and lost the following yardage during a series of plays beginning with first down: $+4, -5, +9$. What is the net yardage for the three plays?

18. **AVIA** Because of stormy weather, a pilot flying at 35,000 ft descends 8,000 ft. What is his new altitude?

19. **PFIN** Agnes opens a checking account by depositing $500. She then writes checks for $42, $18, and $21. What is her balance after depositing another $150?

Section 5–2

Evaluate.

20. $8 - 5$ **21.** $-9 - 4$

22. $-3 - (-3)$ **23.** $-7 - 7$

24. $-7 - (-2)$ **25.** $11 - (-3)$

26. $12 + 3 + (-8) - 5$ **27.** $-6 + 3 - 5 - 7$

28. Temperatures in northern Canada ranged as high as 37°F one summer. That same year the lowest temperature was -28°F. What was the range of temperatures for the year?

29. What is the difference (or range) in temperatures of 43°F above zero and 27°F below zero?

30. What is the difference in temperatures of 47°F below zero and 28°F below zero?

31. **HLTH/N** Two successive recordings for a surgery patient's temperature were 103.2°F and 97.8°F. Express the temperature change with a signed number.

32. **CON** A 10-ft-long fence post is placed in a hole that is 3 ft deep. How much of the post is above the ground?

33. If the temperature changes from -22°C to 14°C, what is the change?

Use the table showing the number of travelers to the nearest thousand to the United States for Exercises 34–37.

34. Express the difference in the number of travelers in 2000 from Japan and from the United Kingdom as a signed number.

35. Express the difference in the number of travelers in 2014 from Japan and from the United Kingdom as a signed number.

36. Which country had the highest amount of change in travelers from 2000 and 2014?

37. Which country had the highest percent of decrease in travelers from 2000 to 2014?

Country	2000	2014	Percent Change from 2000 to 2014
Canada	14,594,000	22,975,000	57.4
Mexico	10,322,000	17,334,000	67.9
Japan	5,061,000	3,579,000	−29.3
United Kingdom	4,703,000	3,973,000	−15.5

Source: U.S. Department of Commerce, International Trade Administration.

Section 5–3

Evaluate.

38. $-3(-7)$ **39.** $7(-2)$

40. $-7(3)$ **41.** $2(3)(-7)(0)$

42. $5(-2)(-1)(-3)$ **43.** $4(3)(-2)(7)$

44. $(-3)^2$ **45.** $(7)^3$

46. $(-4)^3$ **47.** -4^2 **48.** -2^3 **49.** 5^2

50. **BUS** A stock dropped $2.00 for each of seven straight weeks. What was the total change in the stock value?

51. On one winter day, the temperature dropped 2° each hour for 5 h. What was the total drop in temperature?

52. If the temperature in Exercise 51 was 8° originally, what was the temperature at the end of the 5-h period?

53. **BUS** An article states, "XYZ stock has dropped 4 points each week for the past 5 weeks." Was the stock price higher or lower 5 weeks ago than it is today? How much higher or lower? Use negative numbers to express drops in prices. Use a signed number to express the amount the stock price changed.

Divide.

54. $-8 \div (-4)$ **55.** $12 \div 3$ **56.** $\dfrac{14}{-7}$ **57.** $\dfrac{-20}{-5}$ **58.** $\dfrac{16}{-4}$

59. $\dfrac{-51}{-3}$ **60.** $\dfrac{0}{-8}$ **61.** $\dfrac{-7}{0}$ **62.** $\dfrac{-51}{3}$ **63.** $\dfrac{51}{-17}$

64. **BUS** A company records the following gains and losses in net profit for a six-month period: $22,973; $-$12,357; $-$2,791; $32,872; $18,930; and $2,093. Find the net gain or loss for the six-month period.

Section 5–4

Write three equivalent signed fractions for each fraction.

65. $-\dfrac{2}{3}$ **66.** $\dfrac{-5}{-6}$

Perform the indicated operations.

67. $\dfrac{-4}{5} \div -\dfrac{7}{15}$ **68.** $3.23 + (-4.61)$ **69.** $-12.4 \div 0.2$ **70.** $\dfrac{-11}{12} - \left(\dfrac{-7}{8}\right)$

71. $-\dfrac{7}{8} + \left(-\dfrac{5}{12}\right)$ **72.** $1\dfrac{3}{5} \div \left(-7\dfrac{5}{8}\right)$ **73.** $-2\dfrac{5}{8}\left(4\dfrac{1}{2}\right)$ **74.** $-2 - \dfrac{5 - 19}{7}$

Use the order of operations to evaluate the following. Verify the results with a calculator.

75. $7(3 + 5)$ **76.** $-2(3 - 1)$ **77.** $\dfrac{15 - 7}{8}$ **78.** $-20 \div 4 - 3(-2)$

79. $4 + (-3)^4 - 2(5 + 1)$ **80.** $(-3)^3 + 1 - 8$ **81.** $0.2 - 3.1(-7.6)$ **82.** $-0.7 - (-7.2 + 5)$

83. $4 - (-3)^3 + 5(-2 + 5)$ **84.** $5^2 + (-2)^4 - 3(7 + 1)$ **85.** $\dfrac{5 - |7 - 12| - 2^3}{2(3 + 2)}$ **86.** $\dfrac{6(-2 + 7) + (-3^2)}{4 + |3 - 9|}$

87. $2.7 - 3.1(0.7)^2$ **88.** $\dfrac{3}{5} - 2\left(\dfrac{1}{5} - \dfrac{3}{10}\right)$ **89.** $-\dfrac{5}{6} + \left(-\dfrac{2}{3}\right)^2$ **90.** $\dfrac{3}{4} - \left(-\dfrac{1}{2}\right)^2$

Section 5–5

Perform the indicated operations. Express as ordinary numbers.

91. $10^5 \cdot 10^7$ **92.** $10^{-2} \cdot 10^8$ **93.** $10^7 \cdot 10^{-10}$

94. $4.2(10^5)$ **95.** $8.73 \div 10^{-3}$ **96.** $5.6 \div 10^{-2}$

97. Is -7.5×10^{-2} in scientific notation?   **98.** Is 12.6×10^4 in scientific notation?

Write as ordinary numbers.

99. 3.75×10^5 **100.** 4.23×10^4 **101.** 3.87×10^{-5} **102.** 7.37×10^{-9}

Write in scientific notation.

103. 52,000   **104.** 4,500   **105.** 0.00017   **106.** 3,800,000   **107.** 0.000000008

108. 0.7×10^{-2} **109.** 47.3×10^{3} **110.** 287×10^{-2} **111.** 0.0017×10^{3}

Perform the indicated operation and write the result in scientific notation.

112. $(4.2 \times 10^{5})(3.9 \times 10^{-2})$

113. $\dfrac{1.25 \times 10^{3}}{3.7 \times 10^{-8}}$

Raise to the indicated power and express in scientific notation. Round appropriately.

114. $(5.2 \times 10^{3})^{4}$ **115.** $(8.3 \times 10^{-2})^{3}$ **116.** $(4.5 \times 10^{-2})^{-2}$

117. The United States population is approximately 325 million. Write the number in scientific notation.

118. One coulomb (C) is approximately 6.28×10^{18} electrons. How many coulombs do 4.87×10^{15} electrons represent?

Write each number in engineering notation.

119. 0.00000092   **120.** 0.004   **121.** 8,400,000   **122.** 5,900   **123.** 41

124. 31,000   **125.** 17,000,000   **126.** 129,000,000   **127.** 3,084,000,000   **128.** 0.0982

129. 0.0000018   **130.** 0.000989   **131.** 0.007   **132.** 0.12   **133.** 0.00000000035

134. 5,200,000,000   **135.** 0.00049 s   **136.** 0.00052 Å   **137.** 0.588 Å   **138.** 10.53 s

Change each unit to engineering notation using metric prefixes.

139. 246.7 V   **140.** 5,082 W   **141.** 42,000 mW   **142.** 7,800 kΩ

143. 5,729 μW   **144.** 25,000 ps   **145.** 4,800 GHz   **146.** 4,000 ns

5   TEAM PROBLEM-SOLVING EXERCISES

1. Your team is examining the patterns that could be used for automobile license plates. Determine how many license plates can be made using the following patterns. Assume that no plate will be discarded from these patterns.
  (a) 3 digits followed by 3 letters and both digits and letters can be repeated
  (b) the same option from part (a) with either the group of letters or numbers coming first
  (c) 6 letters and all letters can be repeated

2. The national debt at one point was approximately $7.5 trillion. Assuming that a dollar bill is approximately 0.2 mm thick, answer the following questions.
  (a) Suppose the national debt is represented with dollar bills stacked on top of each other. How many kilometers high will the stack reach?
  (b) The distance from Earth to the moon is 380,000 km. How many times will the stack of bills represented by the national debt reach from Earth to the moon?
  (c) If there are approximately 300 million people in the United States, how much would it take per person to pay off the national debt?

5   CONCEPTS ANALYSIS

1. What two operations for integers use similar rules for handling the signs? Explain the rules for these operations.

2. Explain what is meant by "the absolute value of a number." Give an example.

3. What operation with 0 is not defined?

4. Describe the process of adding two integers that have different signs.

5. Write a statement using the symbol for "is greater than."

6. Describe the correct order for operations with integers in words.

7. Explain how to find the sign of a power if the base is a negative integer. Give an example for an even exponent and for an odd exponent.

8. Give an example of multiplying two negative integers, and give the product.

9. Draw a number line that shows positive and negative integers and zero, and place the following integers on the number line: $-3, 8, -2, 0, 3, 5$.

10. Find and correct the mistakes in the following problem.

$$(-8)^2 - 3(2)$$
$$16 - 3(2)$$
$$13(2)$$
$$26$$

5 | PRACTICE TEST

Use the symbols $>$ and $<$ to write the following as *true* statements.

1. -8 is less than 0

2. 2 is less than 3

3. -5 is more than -10

Answer the questions.

4. What is the value of $|-12|$?

5. What is the opposite of 8?

Perform the operations.

6. $-8 - 2$

7. $-3 + 7$

8. $\dfrac{8}{-2}$

9. $2(6)(-4)$

10. $-\dfrac{2}{5} + \dfrac{1}{10}$

11. $-1.3 - 2.4 + 5.8$

12. $-7 + (-3)$

13. $(-8)(3)(0)(-1)$

14. $-3 + 5 + 0 + 2 + (-5)$

15. $\dfrac{-7}{0}$

16. $4(-13)$

17. $\dfrac{4}{-2}$

18. $2 + 10(2) + 6(7)$

19. $\dfrac{2(3 - 9)}{2^2} + 7$

20. $5(2 - 3) + \dfrac{16}{4}$

21. $\dfrac{5 - |3 - 5|}{2(-3)^2}$

22. $\dfrac{10^{-5}}{10^3}$

23. $(10^3)^2$

24. Which expression is in scientific notation, -4.7×10^0 or 12×10^5?

Write as ordinary numbers.

25. 42×10^3

26. 0.83×10^2

27. 5.9×10^{-2}

Write in scientific notation.

28. 5.2301

29. 0.021

30. 52.3×10^2

31. 783×10^{-5}

Perform the indicated operations. Express the answers in scientific notation.

32. $(5.9 \times 10^5)(3.1 \times 10^4)$

33. $\dfrac{5.25 \times 10^4}{1.5 \times 10^2}$

34. A star is 3.4 light-years from Earth. If 1 light-year is 5.87×10^{12} mi, how many miles from Earth is the star?

35. The total resistance (in ohms) of a dc series circuit equals the total voltage divided by the total amperage. If the total voltage is 3×10^3 V and the total amperage is 2×10^{-3} A, find the total resistance (in ohms) expressed as an ordinary number.

36. In Grand Rapids, Minnesota, the temperature ranged from 42°F at 12 noon to -14°F at 7 P.M. Represent the change in temperature with a signed number.

37. Temperatures around the world may range from 135°F in Seville, Spain, to -40°F in Fairbanks, Alaska. What is the range (difference) of temperatures?

Change to engineering notation.

38. 0.000001

39. 0.0047

Change to engineering notation using metric prefixes.

40. 0.0000829 kW

41. 0.000000023 fs

Statistics

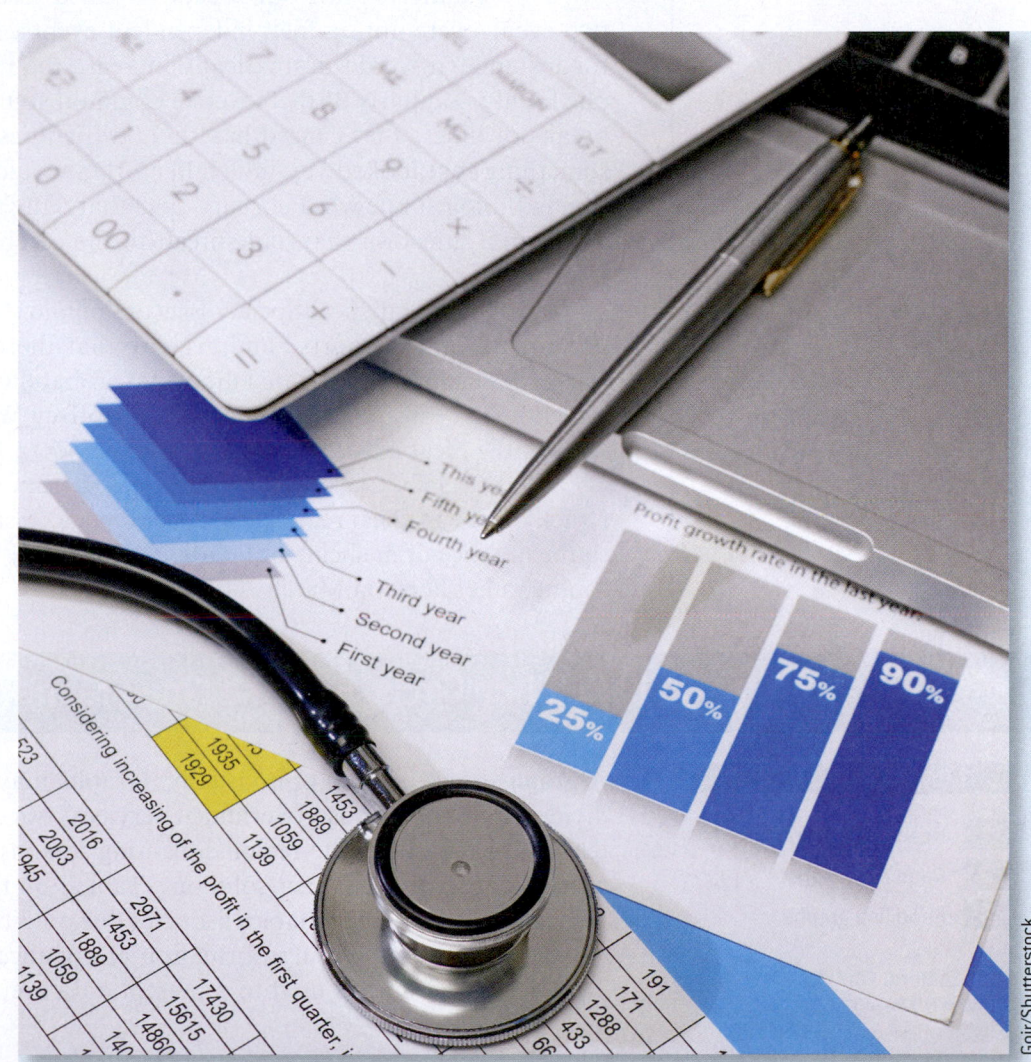

Goir/Shutterstock

In Great Company

Statistics Destroys Lives (November 9, 1999)

On November 9, 1999, at Chester Crown Court, Sally Clark, a lawyer from the town of Cheshire, England, was convicted, by 10–2 majority, of smothering her two baby boys. Clark's first son died suddenly within a few weeks of his birth in 1996. In 1998, when her second son died in similar circumstances, she was arrested and tried for the murder of both sons.

Convicted, she was sent to jail, where she began to serve a life sentence for the two murders. There was only one problem. The children hadn't been murdered. Sally Clark was innocent.

At her trial, Professor Sir Roy Meadows, a pediatrician who served as an expert witness on children's health conditions, testified that the chance of two children from an affluent family suffering sudden infant death syndrome (SIDS) was 1 in 73 million. Sudden infant death syndrome occurs in about 1 out of 8,500 births. So, Professor Meadows treated the two cases as independent statistical events, something like the chances of getting two heads in a row in two coin flips. He calculated his probability by squaring 1 in 8,500 in order to determine how often crib death should happen in similar circumstances.

Unfortunately, Professor Meadows failed to consider the possibility that the two events were not, in fact, independent events. If there were a crib death gene, for instance, it would dramatically increase the likelihood of two crib deaths in one family. Similarly, if there were a common source of infection or frequent outbreaks of illness that caused breathing difficulties in children, the probability of something that looked very much like SIDS would greatly increase. The prosecution pathologist knew in February 1998 that Sally's second son was suffering from a bacterial infection that had spread to the cerebrospinal fluid. He didn't point this out.

The Royal Statistical Society issued a public statement concerning the "misuse of statistics in the courts" and arguing that there was "no statistical basis" for Meadows' claim. Experts rated the actual probability of SIDS at somewhere between 1 in 100 and 1 in 8,500. The journalist Geoffrey Wansell called Clark's experience "one of the great miscarriages of justice in modern British legal history."

In 2003, Sally Clark was finally released from prison. Professor Meadows was struck off the medical registrar in 2005, but successfully petitioned to be readmitted to the practice of medicine in February 2006. Sally Clark died of acute alcohol poisoning in her home in March 2007.

6-1 Reading Circle, Bar, and Line Graphs

LEARNING OUTCOMES

1. Read circle graphs.
2. Read bar graphs.
3. Read line graphs.

LC LEARNING CATALYTICS
1. What is 32% of 136?
2. 92 is what percent of 368?

Graph: a visual representation of information

A **graph** shows information visually. Graphs may show how our tax dollars are divided among various government services, trace the fluctuations in a patient's temperature, or illustrate regional planting seasons. Other graphs may show equations, inequalities, and their solutions. Tables, on the other hand, usually list data. For example, income tax tables list taxes due on different incomes.

Graphs give us useful information at a glance if we interpret them properly. Three common graphs used to represent data are the circle graph, the bar graph, and the line graph.

To read a circle, bar, or line graph:

1. Examine the title of the graph to find out what information is shown.
2. Examine the parts to see how they relate to one another and to the whole.
3. Examine the labels for each part of the graph and any explanatory remarks that may be given.
4. Use the given parts to calculate additional amounts or percents.

Circle graph: a graph that uses a circle to show pictorially how a total amount is divided into parts

1 Read Circle Graphs. A **circle graph** uses a circle to show pictorially how a whole quantity is divided into parts.

The complete circle represents one whole quantity. The circle is divided into parts so that the sum of all the parts equals the whole quantity. These parts can be expressed as fractions, decimals, or percents. Fig. 6–1 is a circle graph.

When we "read" a graph, we examine the information on the graph.

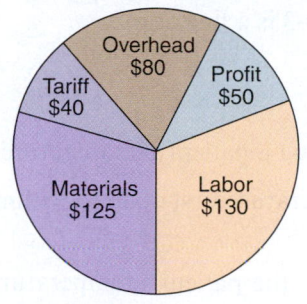

FIGURE 6–1 Distribution of wholesale price for a $425 color television.

EXAMPLE 1

BUS   Use Fig. 6–1 to answer these questions.

(a) What percent of the wholesale price is the cost of labor?

(b) What percent of the wholesale price is the cost of materials?

(c) What would the wholesale price be if no tariff (tax) were paid on imported parts?

(a) $R = \dfrac{P}{B}(100\%)$ Use the percentage formula to find R.

$R = \dfrac{\$130}{\$425}(100\%)$ R is the percent of the wholesale price ($425) that is attributed to labor cost. The labor cost is $130.

$R = 0.3058823529(100)$ Round to tenths.

$\boldsymbol{R = 30.6\% \text{ (labor)}}$

(b) $R = \dfrac{P}{B}(100\%)$ Use the percentage formula to find R.

$R = \dfrac{\$125}{\$425}(100\%)$ R is the percent of the wholesale price ($425) that is attributed to materials cost. The materials cost is $125.

$R = 0.2941176471(100\%)$ Round to tenths.

$\boldsymbol{R = 29.4\% \text{ (materials)}}$

(c) **Price − tariff = $425 − $40 = $385 (cost without tariff)**   **See Exercises 1–12.**

Bar graph: a graph that uses two or more bars to show pictorially how two or more amounts compare to each other rather than to a total

Axis of a bar graph: one axis, also called reference line, is a scale to identify the amounts associated with the length of the bars. The other axis or reference line gives a description of the bars

Reference line of a bar graph: also called axis of a bar graph

2 Read Bar Graphs. Different types of graphs allow us to access different types of information. A **bar graph** uses two or more bars to compare two or more amounts.

The bar lengths represent the amounts being compared. Bars can be drawn either horizontally or vertically.

The **axis,** or **reference line of a bar graph** that runs along the length of the bars is a scale of the amounts associated with the lengths of the bars; in Fig. 6–2 this line is horizontal. The other axis or reference line (vertical in this case) labels the bars or gives a description of the bars. Fig. 6–2 is a horizontal bar graph.

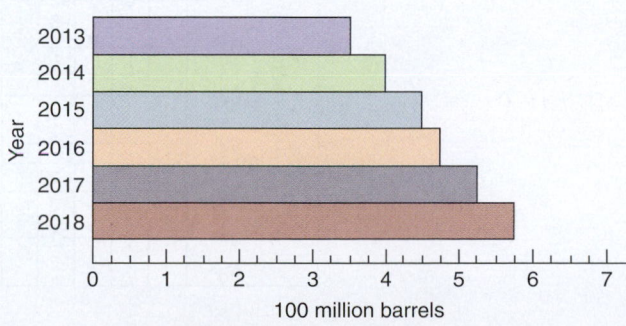

FIGURE 6–2 Company oil production.

STOP AND CHECK

Use Fig. 6–2 to answer these questions.
1. What is the percent of increase from 2014 to 2018?
2. What is the difference between the oil production in 2017 and in 2018?

Answers:
1. 43.75%
2. 50,000,000 barrels

Line graph: a graph that uses one or more lines to show changes in data

STOP AND CHECK

Use Fig. 6–3 to answer these questions.
1. When was the first increase of temperature?
2. What was the temperature change from 12 P.M. to 4 P.M. on 4-10-2019?

Answers:
1. 12 P.M. on 4-10-2019   2. 1.0°

EXAMPLE 2

BUS   Use Fig. 6–2 to answer these questions.

(a) How many 100 million barrels of oil did the company produce in 2014?

(b) Judging from the graph, should company oil production in 2019 be more or less than in 2017?

(c) How many more 100 million barrels of oil are indicated for the company in 2018 than in 2014?

(a) Four hundred million barrels in 2014.

(b) More, because the trend has been toward greater production.

(c) 2018 production − 2014 production = 5.75 − 4.00 = 1.75 hundred million barrels.
See Exercises 13–24.

3 **Read Line Graphs.** Line graphs are encountered in industrial reports, handbooks, and the like. A **line graph** uses one or more lines to show changes in data.

The horizontal axis on a line graph usually represents periods of time or specific times. The vertical axis commonly represents numerical amounts. Line graphs show trends in data and high and low values at a glance. Fig. 6–3 is a line graph.

EXAMPLE 3

HLTH/N   Use Fig. 6–3 to answer these questions regarding a patient's temperature.

(a) On what date and time of day did the patient's temperature first drop to within 0.2° of normal (98.6°F)?

(b) On what post-op (post-operative) period of time did the patient's temperature remain within 0.2° of normal?

(c) What was the highest temperature recorded for the patient?

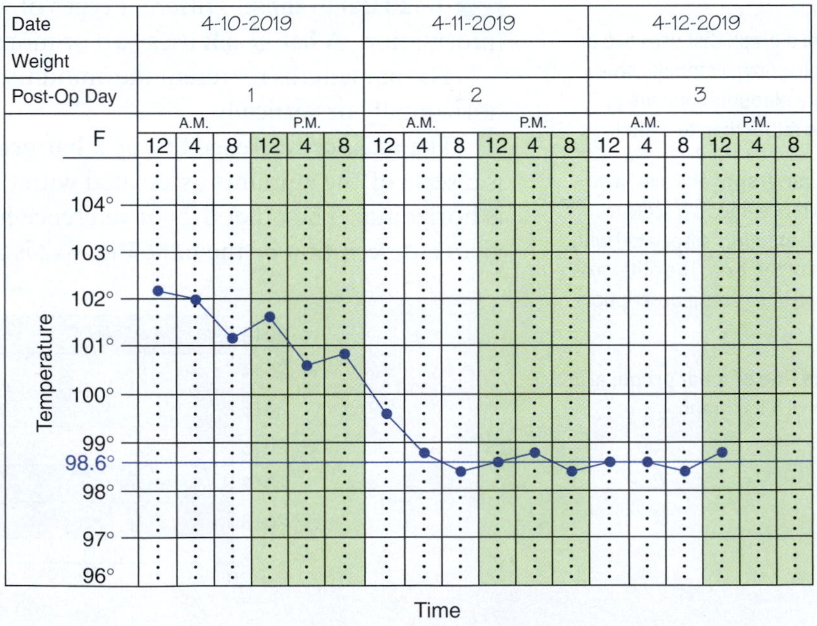

Date	4-10-2019	4-11-2019	4-12-2019
Weight			
Post-Op Day	1	2	3

FIGURE 6-3 Patient's graphic temperature chart.

(a) **4-11-2019 at 4 a.m.** (Each "dot" is 0.2°, so the temperature was 98.8°F.)

(b) From 4 a.m., 4-11-2019 through 12 p.m., 4-12-2019

(c) **102.2 degrees** (recorded at 12 a.m. on 4-10-2019)   **See Exercises 25–42.**

6–1 EXERCISES   **MyLab Math** For additional practice go to your study plan in MyLab Math.

1 Use Fig. 6–4 to answer Exercises 1–3. *See Example 1.*

1. What percent of the gross salary goes into retirement?

2. What percent of the take-home pay is federal income tax? Round to tenths.

3. What percent of the gross pay is the take-home pay? Round to tenths.

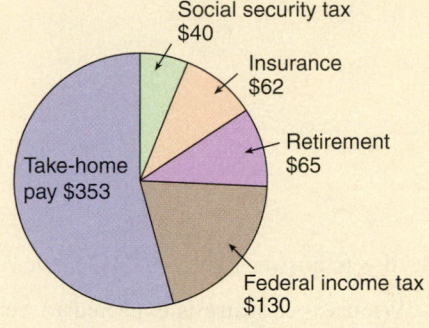

FIGURE 6–4 Distribution of weekly salary of $650.

Use Fig. 6–5 to answer Exercises 4–6. Round to tenths. *See Example 1.*

4. What percent of the day is spent working?

5. What percent of the day is spent sleeping?

6. The amount of time spent studying is what percent of the time spent in class?

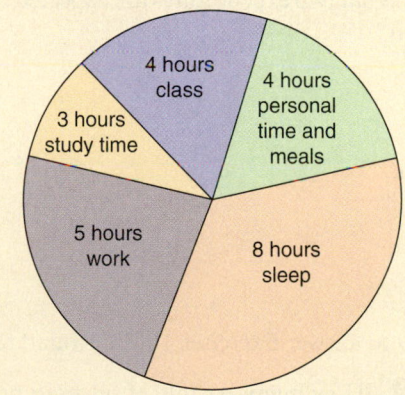

FIGURE 6–5 Distribution of a student's typical day.

Use Fig. 6–6 to answer Exercises 7–10. Round to two significant digits. *See Example 1.*

7. **AG/H** How many bushels of wheat are used for livestock feed?

8. **AG/H** How many bushels of wheat are exported annually?

9. **AG/H** If 1 bushel of wheat yields 42 lb of flour after milling, how many pounds of flour are consumed in the United States each year?

10. **AG/H** If U.S. wheat farmers harvested 48,653,000 acres of wheat, how many whole bushels per acre were averaged?

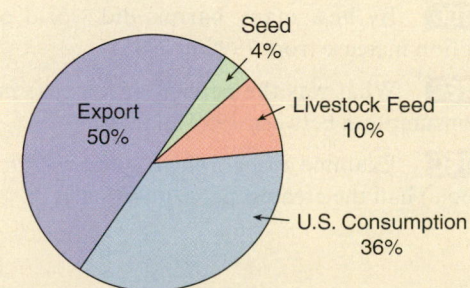

FIGURE 6–6 Distribution by percent of annual wheat production of 2.3 billion bushels.

Use Fig. 6–7 to answer Exercises 11–12. *See Example 1.*

11. **HLTH/N**   What type of hospital discharge accounted for the greatest percentage of discharges?

12. **HLTH/N**   Hospital discharges attributed to long-term care and home-health care accounted for what percentage of discharges?

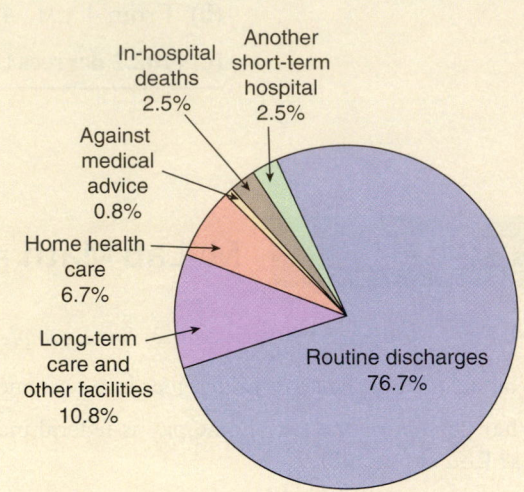

FIGURE 6-7   Hospital discharges in a recent year.

2   Use Fig. 6–8 to answer Exercises 13–15. *See Example 2.*

13. **BUS**   What expenditure is expected to be the same next year as this year?

14. **BUS**   What two expenditures are expected to increase next year?

15. **BUS**   What two expenditures are expected to decrease next year?

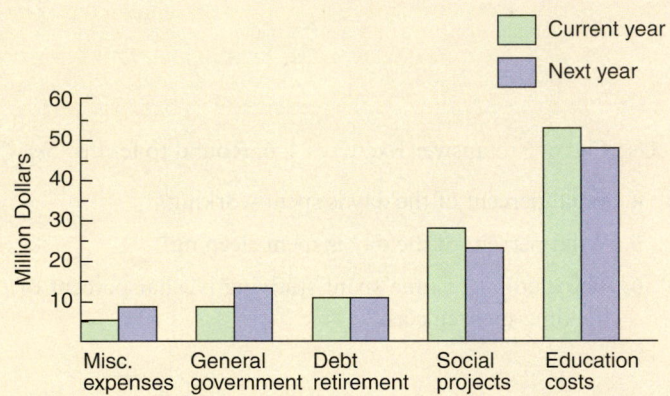

FIGURE 6-8   Distribution of local tax dollars.

Use Fig. 6–9 to answer Exercises 16–20. Round to one significant digit. *See Example 2.*

16. **INDTEC**   How many barrels of oil were produced in 2010?

17. **INDTEC**   Find the percent increase in world oil consumption from 2010 to 2015.

18. **INDTEC**   By how many barrels did world oil consumption increase from 1990 to 2015?

19. **INDTEC**   What was the percent of increase in world oil consumption between 1990 and 2005?

20. **INDTEC**   Examine the graph to guess which 5-year period(s) had the greatest percentage increase.

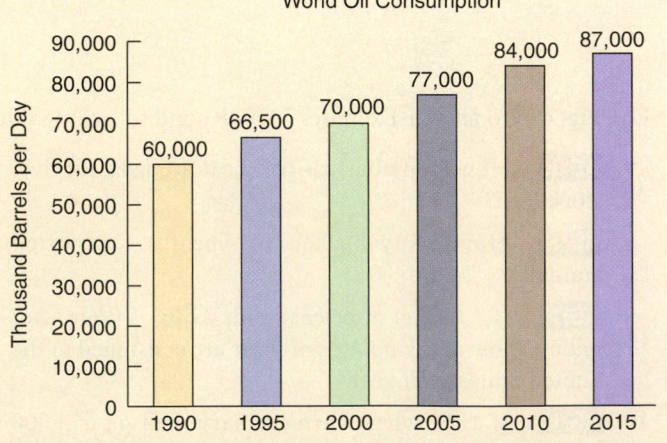

FIGURE 6-9   World oil consumption. *Source: Energy Information Administration, Department of Energy, U.S. Government, 2016.*

Use Figure 6–10 to answer Exercises 21–24. *See Example 2.*

21. What year had the largest daily number of barrels of oil produced by the United States?

22. In 2015, what was the number of barrels of oil produced daily by the world, not including the United States?

23. By what percent did the daily world oil production increase from 1995 to 2015?

24. What percent of the world's oil was produced by the United States in 2015?

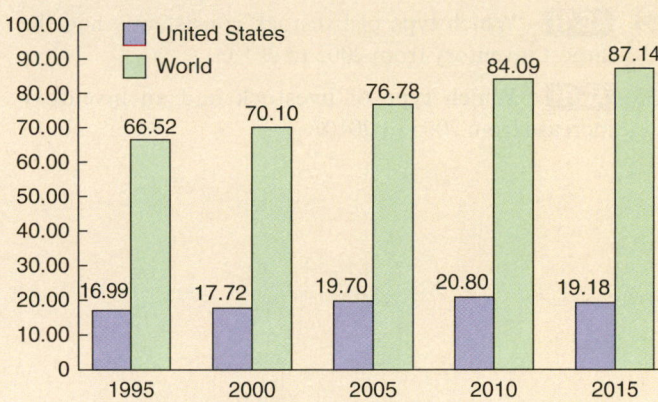

FIGURE 6–10   Daily crude oil production (million barrels).
Source: Energy Information Administration, Department of Energy, U.S. Government, 2016.

---

3 Use Fig. 6–11 to answer Exercises 25–28. *See Example 3.*

25. **ELEC**   How many amperes of current are produced by 50 V when the resistance is 10Ω?

26. **ELEC**   How many volts are needed to produce 2 A of current when the resistance is 25Ω?

27. **ELEC**   Approximately how many volts are required to produce a current of 3.5 A when the resistance is 10 ohms?

28. **ELEC**   Find the resistance when 100 V is needed to produce a current of 4 A.

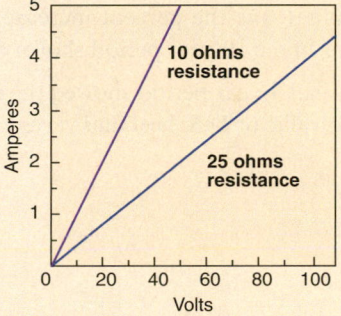

FIGURE 6–11   Amperage produced by voltage across two resistances.

Use Fig. 6–12 to answer Exercises 29–33.

29. **AUTO**   What year was the motor gasoline supply 8,796,000 barrels of oil?

30. **AUTO**   Which petroleum product consistently has the greatest supply?

31. **AUTO**   Which 5-year period had the greatest increase in the supply of motor gasoline?

32. **AVIA**   What 5-year period saw the greatest percentage decrease in jet fuel?

33. In 2011, the supply of jet fuel was what percent of the motor gasoline supply? Round to tenths.

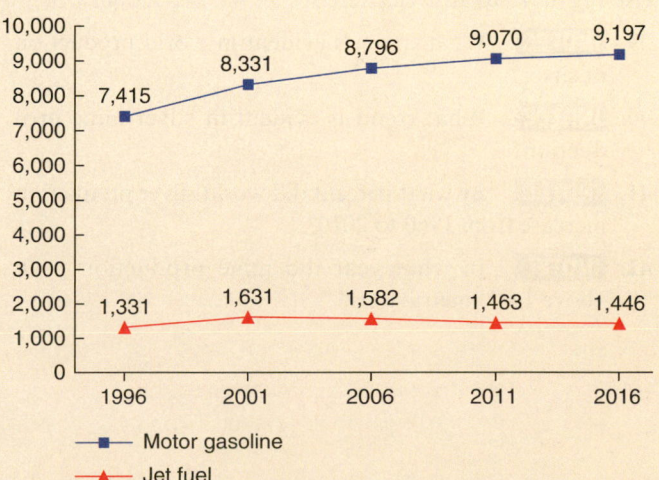

FIGURE 6–12   Petroleum products supplied by type in the United States (thousand barrels). *Source: Energy Information Administration, Department of Energy, U.S. Government, 2016.*

Use Fig. 6–13 to answer Exercises 34–35.

34. **AG/H**   Which type of livestock consistently had the largest inventory from 2005 to 2010?

35. **AG/H**   Which type of livestock had an inventory increase from 2005 to 2010?

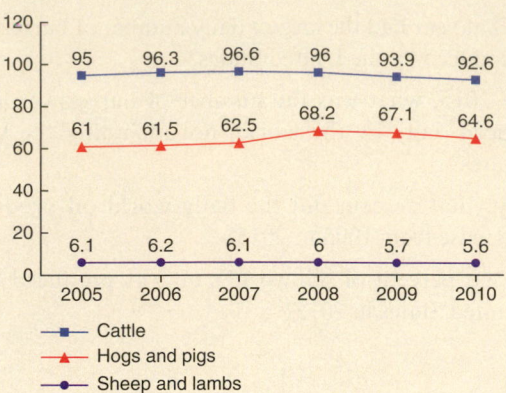

FIGURE 6–13   U.S. livestock inventories, 2005–2010.

Source: U.S. Census Bureau Statistical Abstract of the United States, 2012.

Use Fig. 6–14 to answer Exercises 36–38.

36. **AG/H**   How many metric tons of U.S. beef and variety meat exports were exported in 2010?

37. **AG/H**   What was the percent increase in the export metric tons in the 10-year period shown on the graph?

38. **AG/H**   What 4-year period showed the most dramatic increase in value of U.S. beef and variety meat exports?

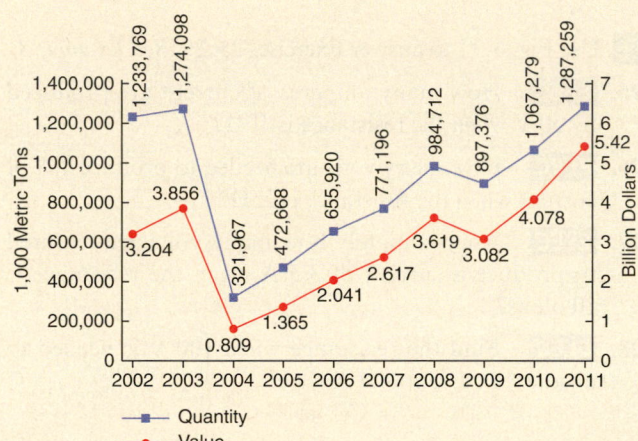

FIGURE 6–14   U.S. beef and variety meat exports.

Source: U.S. Meat Export Federation, 2012.

Use Fig. 6–15 to answer Exercises 39–42. *See Example 3.*

39. **INDTEC**   What trend is evident in world production of silver?

40. **INDTEC**   What trend is evident in silver mine production?

41. **INDTEC**   By what percent did world silver production increase from 1960 to 2010?

42. **INDTEC**   In what year did mine production peak above 1,000 metric tons?

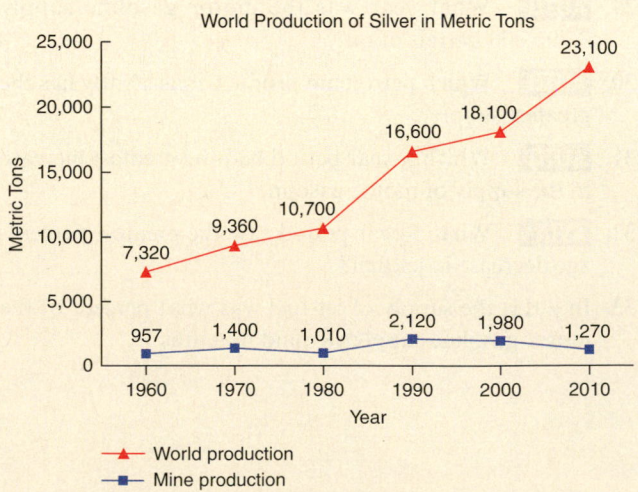

FIGURE 6–15   World production of silver in metric tons.

Source: U.S. Biological Survey, 2011.

6-2 Measures of Central Tendency

LC LEARNING CATALYTICS

1. Find the average of 29, 19, 28, and 20.
2. Find the value halfway between 78 and 92.

Data: facts or information from which conclusions may be drawn

Statistical measurement or statistic: a standardized, meaningful measure of a set of data that reveals a certain feature or characteristic of the data

Descriptive statistics: statistics that give a concise summary of data

Average: an approximate number that is a central value of a set of data

Arithmetic mean, arithmetic average, or statistical mean: the sum of the quantities in a data set divided by the number of quantities

Sigma (Σ): Greek capital letter that is a summation symbol; indicates the addition of a set of values

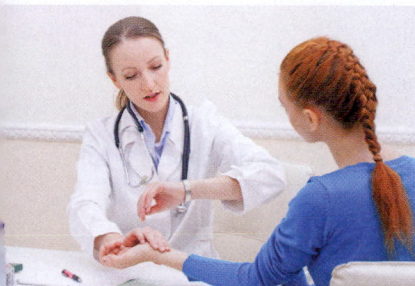

Lurii Sokolov/123RF

STOP AND CHECK

1. Find the mean of 126, 317, 285, 168, and 225 to the nearest tenth.

Answer:

1. 224.2

1 Find the Arithmetic Mean. In this age of information explosion we have massive amounts of data available to us. **Data** are facts or information from which conclusions may be drawn. To use these data effectively in the decision-making process, we need to examine the data and summarize key trends and characteristics. This summary is generally in the form of statistical measurements. A **statistical measurement** or **statistic** is a standardized, meaningful measure of a set of data that reveals a certain feature or characteristic of the data. Statistics that give a concise summary of data are called **descriptive statistics.**

An **average** is an approximate number that is a central value of a set of data. The most common average is the arithmetic mean. The **arithmetic mean,** also called **arithmetic average** or **statistical mean,** is the sum of the quantities in the data set divided by the number of quantities.

Symbols can be used to express the procedures for statistical measures. In symbols, the formula for the mean for a set of data is

$$\overline{x} = \frac{\Sigma x_i}{n}$$

The formula is read as "the mean $\overline{x}$ (read "x-bar") equals the sum Σ of all values of x_i (read "x sub i") divided by the number of values, n. The Greek capital letter **sigma,** Σ, is a summation symbol and indicates the addition of a set of values. The notation x_i identifies each data value with a subscript: x_1 is the first value, x_2 is the second value, and x_n is the nth value.

To find the arithmetic mean:

1. Add the quantities.
   Σx_i

2. Divide the sum by the number of quantities.

Find the mean of 22, 31, and 37.
$$x_1 = 22, x_2 = 31, x_3 = 37$$
$$\Sigma x_i = 22 + 31 + 37$$
$$\Sigma x_i = 90$$
$$\overline{x} = \frac{\Sigma x_i}{n} = \frac{90}{3}$$
$$\overline{x} = 30$$

Symbolically, $\overline{x} = \frac{\Sigma x_i}{n}$, where $\overline{x}$ = arithmetic mean; Σ means sum; x_i represents each x value; n = number of values.

EXAMPLE 1

HLTH/N Find the mean of each set of quantities.

(a) Pulse rates: 68, 84, 76, 72, 80

There are five pulse rates, so we find their sum and divide by 5:

$$\overline{x} = \frac{\Sigma x_i}{n} = \frac{68 + 84 + 76 + 72 + 80}{5} = \frac{380}{5} = 76$$

The mean pulse rate is 76.

(b) Pounds: 21, 33, 12.5, 35.2 (to the nearest pound)

There are four weights, so we find their sum and divide by 4.

$$\overline{x} = \frac{\Sigma x_i}{n} = \frac{21 + 33 + 12.5 + 35.2}{4} = \frac{101.7}{4} = 25.425$$

The mean weight is 25 lb (to the nearest pound).   **See Exercises 1–14.**

TIP **Using a Calculator to Make Lists and Find Statistical Measures** A graphing calculator has many features to facilitate finding statistical measures. The first step is to build a list containing the data to be measured.

To build a list using a TI-83 or a TI-84 calculator:
Clear any previous list that you no longer care to keep.

[STAT] [4:ClrList] [LIST] [1:L₁] [ENTER].

Enter your new list as List 1.

[STAT] [1:Edit...] Enter each data amount, followed by [ENTER]. After the last data amount is entered, press [QUIT].

Remember if a feature is labeled above a key, an access key [2ND] is pressed first. On the TI-83 or TI-84, the [QUIT] and numbered lists (L₁, L₂, L₃, . . . , L₆) features are labeled above a key.

Using the data from Part (a) of the previous example:

[STAT] [1:Edit...] 68 [ENTER] 84 [ENTER] 76 [ENTER] 72 [ENTER] 80 [ENTER] [QUIT].

To find the mean:

[LIST] [MATH] [3:mean] [LIST] [1:L₁] [ENTER] [ENTER] ⇒ 76

To access the [MATH] menu from the [LIST], use the right-arrow keys.

STOP AND CHECK

1. If your car traveled 382.4 miles on 18.3 gallons of gasoline, how many miles per gallon to the nearest tenth did the car average?

Answer:

1. $20.9 \frac{\text{mi}}{\text{gal}}$

EXAMPLE 2

AUTO An automobile used 41 gal of regular gasoline on a trip of 876 mi. What was the average miles per gallon $\left(\frac{\text{mi}}{\text{gal}}\right)$ to the nearest gallon?

The number of miles traveled on the 41 gal of gasoline is 876 and represents the total miles. Divide the total miles by 41 gal.

$$\frac{876 \text{ mi}}{41 \text{ gal}} = 21.36585366$$

The car averaged 21 $\frac{\text{mi}}{\text{gal}}$ to the nearest gallon.   **See Exercises 15–20.**

EXAMPLE 3

AG/H Table 6–1 lists the final grades of a horticulture student. Find the student's QPA (quality point average) to the nearest hundredth based on a 4-point system.

To find the QPA for a term, the quality points for the letter grade of each course are multiplied by the credit hours of each course to obtain the total quality points for each course. This total then is divided by the total credit hours earned. The GPA is commonly rounded to the nearest hundredth. The quality points awarded are A = 4, B = 3, C = 2, D = 1, and F = 0.

Table 6-1 Quality Points for Final Grades

Subject	Grade	Credit Hours		Quality Points per Hour		Total Points
Algebra 101	B	3	×	3	=	9
Spray chemicals 102	A	3	×	4	=	12
Landscape 301	A	3	×	4	=	12
English 101	C	4	×	2	=	8
		13 hr				41 points

$$\frac{41}{13} = 3.153846154 \quad \text{or} \quad 3.15 \qquad \text{To nearest hundredth.}$$

Thus, the student's QPA for the term is 3.15. See Exercise 21.

EXAMPLE 4

A community college student has the following grades in Physics 101: 73, 84, 80, 62, and 70. What grade is needed on the last test for the student to get a C on the final grade or a 75 average?

One way to find the needed grade is to assume that each grade is 75 for the 6 tests:

$$\frac{75 + 75 + 75 + 75 + 75 + 75}{6} = \frac{450}{6} = 75 \qquad \text{The sum of the six grades is 450.}$$

Then a total of 450 points are needed on six tests to have an average of 75.

$73 + 84 + 80 + 62 + 70 = 369$ Add the five test grades.

$450 - 369 = 81$ Find the difference between the sum of the five grades and 450.

The student must earn a score of 81 or higher on the last test to have an average of 75 or higher.

$$\text{Check:} \quad \frac{73 + 84 + 80 + 62 + 70 + 81}{6} = \frac{450}{6} = 75$$

See Exercise 22.

2 Find the Median and the Mode. Besides the arithmetic mean, we also use the *median* and the *mode* to describe groups of data. These are called measures of *central tendency*. **Measures of central tendency** are descriptive statistics that determine the center of a data set from a different perspective.

The **median** is the middle value when the data values are arranged in order of size. The median is frequently used for data relating to annual incomes, housing price ranges, etc.

Measures of central tendency: descriptive statistics that show typical or central characteristics of a set of data. Examples are the mean, the median, and the mode of a data set

Median: the middle value of a data set when the given values are arranged in order of size

To find the median:

1. Arrange the values in order of size, either smaller to larger or larger to smaller.

2. If the number of values is odd, the median is the middle value.

3. If the number of values is even, the median is the average of the two middle values.

STOP AND CHECK

1. Find the median of 45, 23, 78, 43, 65, 29, and 55.

Answer:

1. 45

EXAMPLE 5

HLTN/N   A TPR chart shows a patient's temperature, pulse rate, and respiration rate. The following pulse rates were recorded on a TPR chart: 68, 88, 76, 64, 72. What is the patient's median pulse rate?

In descending order of size: 88
76
72   ← median or middle value in an odd number of values
68
64

The median pulse rate is 72.   See Exercises 23–25.

STOP AND CHECK

1. Find the median of 135, 120, 170, and 155.

Answer:

1. 145

EXAMPLE 6

The following temperatures were recorded: 56°F, 48°F, 66°F, and 62°F. What is the median temperature?

The number of temperatures is even, so we find the average of the two middle values.

In ascending order of size   48

$$\left.\begin{array}{c}56\\62\end{array}\right\} \frac{56+62}{2}=\frac{118}{2}=59$$

66

The median temperature is 59°.   See Exercises 26–30.

TIP **Find the Median Using a Graphing Calculator** A median feature is included on many graphing calculators.

Examine the calculator steps to find the median for Example 6 above.

Build a list:

STAT 4:ClrList LIST 1:L$_1$ ENTER.   Clear previous list in L$_1$.

Enter your new list as List 1. (Calculator does not require numbers to be arranged in increasing or decreasing order.)

STAT 1:Edit... 56 ENTER 48 ENTER 66 ENTER 62 ENTER QUIT.

Find the median:

LIST MATH 4:median(LIST 1:L$_1$ ENTER ⇒ 59

Mode: the value or values that occur most frequently in a data set

The **mode** is the value or values that occurs most frequently in the data set.

To find the mode:

1. Identify the value or values that occur with the greatest frequency as the mode.

2. If no value occurs more than another value, there is *no mode* for the data set.

3. If more than one value occurs with the greatest frequency, the modes of the data set are the values that have the greatest frequency.

EXAMPLE 7

BUS The hourly pay rates at a local fast-food restaurant are as follows: cooks, $8.50; servers, $7.15; bussers, $7.15; dishwashers, $7.25; managers, $10.50. Find the mode.

Identify the value or values that appear most often.

The hourly pay rate of $7.15 occurs more than any other rate. It is the mode.

See Exercises 33–42.

EXAMPLE 8

The daily work shifts at a mall clothing store are 4, 6, and 8 hours. Find the mode.

No shift occurs more frequently than another, so there is no mode.

See Exercises 33–42.

3 **Make and Interpret a Frequency Distribution.** The instructor records the following grades for a class of 25 students:

76 91 71 83 97 87 77 88 93 77 93 81 63
79 74 77 76 97 87 89 68 90 84 88 91

It is difficult to make sense of all these numbers as they appear here. But the instructor can arrange the scores into several smaller groups, called **class intervals.** The word *class* means a special category.

These scores can be grouped into class intervals of 5, such as 60–64, 65–69, 70–74, 75–79, 80–84, 85–89, 90–94, and 95–99. Each class interval has an odd number of scores. The *middle score* of each interval is a **class midpoint.**

The instructor can now **tally** the number of scores that fall into each class interval to get a **class frequency,** the number of scores in each class interval.

A compilation of class intervals, midpoints, tallies, and class frequencies is called a **grouped frequency distribution.**

Class intervals: the groupings of a large amount of data into smaller groups of data

Class midpoint: the middle value of each class interval

Tally: a process for determining the number of values in each class interval

Class frequency: the number of values in each class interval

Grouped frequency distribution: a compilation of class intervals, midpoints, tallies, and class frequencies

EXAMPLE 9

Examine the grouped frequency distribution in Table 6–2, and answer the following questions.

Table 6–2   Frequency Distribution of 25 Scores

Class Interval	Midpoint	Tally	Class Frequency
60–64	62	/	1
65–69	67	/	1
70–74	72	//	2
75–79	77	LHT /	6
80–84	82	///	3
85–89	87	LHT	5
90–94	92	LHT	5
95–99	97	//	2

(a) How many students scored 70 or above?

$2 + 6 + 3 + 5 + 5 + 2 = 23$ Add the frequencies for class intervals with scores 70 or higher.

23 students scored 70 or above.

(b) How many students made As (90 or higher)?

$5 + 2 = 7$ Add the frequencies for class intervals 90–94
                                                 and 95–99.

7 students made As (90 or higher).

(c) What percent of the total grades were As (90s)?

$$\frac{7 \text{ As}}{25 \text{ total}} = \frac{7}{25} = 0.28 = \textbf{28\% As}$$ The portion or part is 7 and the base
                                                                     or total is 25.

(d) Were the students prepared for the test?

The relatively high number of 90s (7) compared to the relatively low number of
60s (2) suggests that **in general, most students were prepared for the test.**

(e) What is the ratio of As (90s) to Fs (60s)?

$$\frac{7 \text{ As}}{2 \text{ Fs}} = \frac{7}{2}$$

The ratio is $\frac{7}{2}$.

See Exercises 43–52.

EXAMPLE 10

Students in a history class reported their credit-hour loads as shown. Make a grouped
frequency distribution of their credit hours. Credit hours carried: 3, 12, 15, 3, 6, 6,
12, 9, 12, 9, 6, 3, 12, 18, 6, 9.

To establish a class interval with an easy-to-find midpoint, use an odd number of
points in the interval. Here, an interval of 5 is used; that is, 0–4 contains five possibilities: 0, 1, 2, 3, and 4. The middle number is the midpoint, 2. Make a tally mark
for each time the credit hours of a student falls in the interval. Then count the tally
marks to get the class frequency (Table 6–3).

Table 6–3   Frequency Distribution of Credit-Hour Loads

Class Interval	Midpoint	Tally	Class Frequency
0–4	2	///	3
5–9	7	LH1 //	7
10–14	12	////	4
15–19	17	//	2

See Exercises 53–68.

---

4 | **Find the Mean of Grouped Data.** When data are grouped, it may be desirable to
find the mean of the grouped data. To do this we extend our frequency distribution.

To find the mean of grouped data:

1. Make a frequency distribution.

2. Find the product (xf) of the midpoint (x) of the interval and the frequency
   (f) of the interval for each interval.

3. Find the sum of the frequencies (Σf).

4. Find the sum of the products (Σxf).

5. Divide the sum of the products by the sum of the frequencies.

Symbolically, $\text{mean}_{\text{grouped}} = \dfrac{\Sigma xf}{\Sigma f}.$

STOP AND CHECK

1. Find the grouped mean to the nearest whole number of the data in the frequency distribution in Table 6–2.

Answer:

1. 83

EXAMPLE 11

Find the grouped mean to the nearest whole number of the data in the frequency distribution in Table 6–4.

Table 6–4   Frequency Distribution of Credit-Hour Loads

Class Interval	Midpoint x	Frequency f	Product xf
0–4	2	3	6
5–9	7	7	49
10–14	12	4	48
15–19	17	2	34
Total		16	137

$\Sigma f = 16$ Add the frequencies.

$\Sigma xf = 137$ Add the products.

$\text{mean}_{\text{grouped}} = \dfrac{\Sigma xf}{\Sigma f}$ Substitute.

$\text{mean}_{\text{grouped}} = \dfrac{137}{16}$ Divide.

$\text{mean}_{\text{grouped}} = 8.5625$ Round.

$\text{mean}_{\text{grouped}} = 9$

See Exercises 69–72.

> **TIP   Is the Mean of Grouped Data Exact?** No. The mean of grouped data is based on the assumption that all the data in an interval have a mean that is exactly equal to the midpoint of the interval. Because this is usually not the case, the mean of grouped data is a reasonable approximation.

6-2 EXERCISES

MyLab Math For additional practice go to your study plan in MyLab Math.

1 Find the mean of the given values. Round to hundredths if necessary. *See Example 1.*

1. 12, 14, 16, 18, 20
2. 13, 15, 17, 19, 21
3. 68, 54, 73, 69
4. 85, 68, 77, 65
5. 37.6, 29.8
6. 65.3, 67.9
7. 32°F, 41°F, 54°F
8. 10°C, 13°C, 15°C
9. $27, $32, $65, $29, $21
10. $32, $43, $22, $63, $36
11. 11 in., 17 in., 16 in.
12. 9 in., 7 in., 8 in.
13. Respiration rates: 16, 24, 20
14. Pulse rates: 68, 84, 76

See Example 2.

15. A baseball player batted 276 home runs over a 16-year period. What was the average number of home runs per year to the nearest tenth?

16. Noel Womack scored 87, 96, 86, 92, and 93 in English 101. What must he score on the last test to earn an average score of 92?

17. **AUTO**   An automobile used 32 gal of regular gasoline on a 786-mi trip. What was the average miles per gallon to the nearest tenth?

18. **AUTO**   A pickup truck used 25 gal of regular gasoline on a 256-mi trip. What was the average miles per gallon to the nearest tenth?

19. **CON**   U.S. cement export data for the last 10 years in thousand tons:

    746, 625, 633, 759, 803,
    791, 743, 694, 738, 834

    Find the mean number of tons exported for the period.

20. **CON**   World cement production data for the last 20 years in million tons:

    916.6; 941.1; 959.4; 1,008; 1,053; 1,118; 1,042; 1,043; 1,185; 1,123; 1,291; 1,370; 1,445; 1,493; 1,547; 1,540; 1,600; 1,650; 1,730; 1,800

    Find the mean number of tons of cement produced in the world. Report two significant digits.

See Example 3.

21. Based on a 5-point system (A = 4, B = 3, C = 2, D = 1, F = 0), determine the quality point average (QPA) to the nearest hundredth for the following semester grades:

Subject	Grade	Credit Hours
Calculus	B	4
Spanish I	A	3
English I	B	3
Fluid Dynamics	A	3
Writing Lab	B	1

See Example 4.

22. Sharon Pennington is trying to have an A average (94 or above) on the four exam scores in Nursing I. If her first three scores are 88, 96, and 92, what must she score on the fourth exam?

2 Find the median for each data set. *See Example 5.*

23. 32, 56, 21, 44, 87

24. 78, 23, 56, 43, 38

25. $22, $35, $45, $30, $29

See Example 6.

26. 21, 33, 18, 32, 19, 44

27. 12, 21, 14, 18, 15, 16

28. $66, $54, $76, $55, $69

29. **BUS** The following hourly pay rates are used at fast-food restaurants: cooks, $8.15; servers, $8.25; bussers, $8.15; dishwashers, $8.25; managers, $11.25. Find the median pay rate.

30. **BUS** The following hourly pay rates are used at a locally owned store: clerks, $8.45; bookkeepers, $9.25; operators, $8.15; assistant managers, $10.95. Find the median pay rate.

31. **CON** Find the median number of tons of concrete exported for the 10-year period in Exercise 19. Report three significant digits.

32. **CON** Find the median number of tons of concrete produced in the world for the data in Exercise 20.

Find the mode for each data set. *See Examples 7 and 8.*

33. 2, 4, 6, 2, 8, 2

34. 5, 12, 5, 5, 20

35. 21, 32, 67, 34, 23, 22

36. 32, 45, 41, 23, 56, 77

37. $56, $67, $32, $78, $67, $20, $67, $56

38. $32, $87, $67, $32, $32, $87, $77, $22

39. **BUS** These weekend work shifts are in effect at a mall clothing store: 4 hours in A.M., 6 hours in P.M., 4 hours in P.M. Find the mode for the number of hours.

40. **BUS** These special prices are in effect at a fast-food restaurant: $1.75, hamburgers; $1.97, hot ham sandwiches; $2.38, chicken fillet sandwiches; $1.97, roast beef sandwiches. Find the mode.

41. **CON** Find the mode for the data in Exercise 19.

42. **CON** Find the mode for the data in Exercise 20.

3 Use Table 6–5 to answer these questions. The frequency distribution shows the ages of 25 college students in a landscaping class. *See Example 9.*

43. How many students are 22 or younger?

44. How many students are older than 34?

45. What is the ratio of the number of students 38–40 to the number of students 17–19?

46. What is the ratio of the smallest class frequency to the largest class frequency?

47. What percent of the total class are students aged 17–19?

48. What percent of the total class are students aged 20–22?

49. What two age groups make up the smallest number of students in the class?

Table 6–5 Frequency Distribution of 25 Ages

Class Interval	Midpoint	Tally	Class Frequency
38–40	39	/	1
35–37	36	/	1
32–34	33	//	2
29–31	30	///	3
26–28	27	//	2
23–25	24	LH1 /	6
20–22	21	LH1 //	7
17–19	18	///	3

50. What two age groups make up the largest number of students in the class?

51. How many students are over age 28?

52. How many students are under age 26?

See Example 10.

Use the given hourly pay rates (rounded to the nearest whole dollar) for 33 support employees in a private college to complete a frequency distribution using the format shown in Table 6–6.

Table 6–6   Pay Rates of 33 Support Employees

	Class Interval	Midpoint	Tally	Class Frequency
53.	$14–16			
54.	$11–13			
55.	$8–10			
56.	$5–7			

$6	$6	$10	$7	$6	$6	$6
$6	$7	$7	$8	$8	$6	$6
$11	$10	$7	$11	$8	$16	$6
$6	$9	$6	$7	$9	$6	$6
$12	$13	$7	$15	$5		

Use the given 40 test scores of two physics classes to complete a frequency distribution using the format in Table 6–7.

Table 6–7   Test Scores of 40 Physics Students

	Class Interval	Midpoint	Tally	Class Frequency
57.	91–95			
58.	86–90			
59.	81–85			
60.	76–80			
61.	71–75			
62.	66–70			
63.	61–65			
64.	56–60			

57	91	76	89	82	59	72	88
76	84	67	59	77	66	56	76
77	84	85	79	69	88	75	58
85	65	67	66	93	83	69	81
80	64	78	76	72	90	79	90

See Example 10.

65. Students recorded the number of hours they studied each week as: 3, 15, 18, 0, 2, 9, 12, 16, 7, 8, 5, 10, 14, 9, 7, 3, 4, 14, 17, 16, 6, 4, 8, 11, 14, 13, 10, 15, 16, 5.

Create a grouped frequency distribution with four classes beginning with the class 0–4. Show the class interval, midpoint, tally, and class frequency for each class interval.

66. The number of animals at the city animal shelter varies daily. Use the data to make a frequency distribution with five classes beginning with the class 11–20. Show the class interval, midpoint, tally, and class frequency for each class interval: 22, 31, 32, 27, 29, 16, 12, 18, 21, 30, 46, 52, 43, 51, 42, 26, 42, 17, 19, 25.

67. Use the data in Exercise 19 to create a grouped frequency distribution with three classes beginning with the class 601–700 thousand tons. Show the class interval, midpoint, tally, and class frequency for each class interval.

68. Use the data in Exercise 20 to create a grouped frequency distribution with five classes beginning with the class 751–1,000 million tons. Show the class interval, midpoint, tally, and class frequency for each class interval.

4 *See Example 11.*

69. Find the grouped mean for the data in Table 6–5. Round to the nearest whole number.

70. Find the grouped mean for the data in Table 6–6. Round to the nearest dollar.

71. Find the grouped mean for the data in Table 6–7. Round to the nearest whole number.

72. Find the grouped mean for the data in Exercise 65. Round to the nearest whole number.

6-3 | Measures of Dispersion

LC LEARNING CATALYTICS
Evaluate.
1. $(3.7)^2$
2. $\sqrt{584}$ to the nearest tenth.

STOP AND CHECK

1. Find the range for the data given in Example 5 on page 274.
2. Find the range for the data given in Example 6 on page 274.

Answers:
1. 24 **2.** 18

Measures of variation or dispersion: measures that show the dispersion or spread of a set of data. Examples are the range and the standard deviation

Spread: another term for the variation or dispersion of a set of data

Range: the difference between the highest value and the lowest value in a set of data

Midrange: the mean of the lowest and highest values of a data set and is a measure of central tendency

Outliers: data items that are unusually large or unusually small when compared to the other data items

1 Find the Range. The mean, the median, and the mode are *measures of central tendency.* Other statistical measures are **measures of variation or dispersion.** The variation or dispersion of a set of data may also be referred to as the **spread.** One of these measures of dispersion is the **range.** The range is the difference between the highest value and the lowest value in a set of data.

> **To find the range:**
>
> 1. Find the highest and lowest values.
> 2. Find the difference between the highest and lowest values.

EXAMPLE 1

BUS   Find the range for the data described in Example 7 for fast-food restaurant hourly pay rates on page 275.

The high value is $10.50. The low value is $7.15.

$$\text{range} = \$10.50 - \$7.15 = \mathbf{\$3.35} \qquad \textbf{See Exercises 1–13.}$$

> **TIP   Use More Than One Statistical Measure**   A common mistake when making conclusions or inferences from statistical measures is to examine only one statistic, such as the range. To obtain a complete picture of the data requires looking at more than one statistic.

In some sets of data the midrange might be a statistic of interest. The **midrange** is the mean of the lowest and highest values of the data set and is a measure of central tendency. The midrange is highly sensitive to outliers since it is determined by only two data points (highest and lowest) in the set and it is only useful when the data set has no outliers. An **outlier** is a data item that is unusually large or unusually small when compared to the other data items.

> **To find the midrange of a data set:**
>
> 1. Find the highest and lowest values.
> 2. Find the mean of the highest and lowest values.
>
> $$\text{midrange} = \frac{\text{highest value} + \text{lowest value}}{2}$$

STOP AND CHECK

1. Find the midrange for the data given in Example 5 on page 274.
2. Find the midrange for the data given in Example 6 on page 274.

Answers:
1. 76 **2.** 57

EXAMPLE 2

Find the midrange for the data described in Example 7 for fast-food restaurant hourly pay rates on page 275.

The highest value is $10.50. The lowest value is $7.15.

$$\text{midrange} = \frac{\$10.50 + \$7.15}{2} = \frac{\$17.65}{2} = \mathbf{\$8.825}$$

The midrange hourly pay rate is $8.83 rounded to the nearest cent.

See Exercises 14–17.

Relative frequency: the relative frequency of a specific category or class in a frequency distribution is the fraction of times that category occurs

2 **Find Measures of Relative Position.** The more ways that you look at data, the more insights you find. Another way of examining data is to look at the relative position of data. First, we will determine the relative frequency of data and the cumulative relative frequency. The **relative frequency** of a specific category or class in a frequency distribution is the fraction of times that a category occurs. *The sum of the relative frequencies for a distribution is 1.*

To make a relative frequency distribution:

1. Make a frequency distribution.

2. Find the relative frequency of each class or category by dividing the frequency by the total number in the data set. Express as a fraction in lowest terms or a decimal equivalent.

$$\text{relative frequency of a class} = \frac{\text{frequency of a class}}{\text{total number in data set}}$$

3. To check the distribution, the sum of all the relative frequencies is 1.

Cumulative frequency distribution: a frequency distribution that also shows the accumulated frequency for each class as you continue to add classes. The cumulative frequency of the last class is equal to the total items in the data set

A **cumulative frequency distribution** is made by adding the frequency of a given class to the accumulated frequency of the previous classes. The cumulative frequency of the last class is the total number in the data set.

To make a cumulative frequency distribution:

1. Make a frequency distribution.

2. The cumulative frequency of the first (lowest) class is the same as the frequency of the class.

3. The cumulative frequency of each subsequent class is the frequency of the class plus the sum of the frequencies of all previous classes.

4. The cumulative frequency of the last (highest) class is equal to the total number of items in the data set.

Cumulative relative frequency: the accumulation of the relative frequencies. The cumulative relative frequency of the last class in a distribution is 1

Similarly, a **cumulative relative frequency** is the accumulation of the relative frequencies. The cumulative relative frequency of the last class in a distribution is 1.

Make a cumulative relative frequency distribution:

1. Make a relative frequency distribution.

2. The cumulative relative frequency of the first class is the same as the relative frequency.

3. The cumulative relative frequency of each subsequent class is the relative frequency of the class plus the sum of the relative frequencies of all previous classes.

4. The cumulative relative frequency of the last class is 1.

EXAMPLE 3

Make a relative frequency distribution, a cumulative frequency distribution, and a cumulative relative frequency distribution for the 25 scores given in Example 9 on page 275.

Complete the table.

Class Interval	Class Frequency	Relative Frequency	Cumulative Frequency	Cumulative Relative Frequency
60–64	1	$\frac{1}{25} = 0.04$	1	0.04
65–69	1	$\frac{1}{25} = 0.04$	$1 + 1 = 2$	$0.04 + 0.04 = 0.08$
70–74	2	$\frac{2}{25} = 0.08$	$2 + 2 = 4$	$0.08 + 0.08 = 0.16$
75–79	6	$\frac{6}{25} = 0.24$	$4 + 6 = 10$	$0.16 + 0.24 = 0.4$
80–84	3	$\frac{3}{25} = 0.12$	$10 + 3 = 13$	$0.4 + 0.12 = 0.52$
85–89	5	$\frac{5}{25} = 0.2$	$13 + 5 = 18$	$0.52 + 0.2 = 0.72$
90–94	5	$\frac{5}{25} = 0.2$	$18 + 5 = 23$	$0.72 + 0.2 = 0.92$
95–99	2	$\frac{2}{25} = 0.08$	$23 + 2 = 25$	$0.92 + 0.08 = 1$

See Exercises 18–19.

When data sets are compared, we often define benchmarks in the data that we will use for making the comparisons. This process is used to report scores from norm-referenced tests. The most common of these benchmarks are *quartiles* and *percentiles*.

Quartiles and percentiles are based on data that are ranked (arranged in order of size) in ascending order (from smallest to largest). First, we will look at quartiles.

Quartiles are boundaries that divide the data set into four parts. Three boundaries will divide the data into four parts. There is more than one method for finding these division boundaries, but we will use the method for finding the median of a set of data.

Quartiles: boundaries that divide the data set into four parts

Lower quartile: lowest quarter of data bounded by the median of the lower half of the data

Upper quartile: highest quarter of data bounded by the median of the upper half of the data

To divide a set of data into quartiles:

1. Find the median of the data set. This divides the data into two halves (lower half and upper half).

2. Find the median of each half of the data.

3. The median of the lower half of data is the boundary for the **lower quartile**. The median for the upper half of data is the boundary for the **upper quartile**.

STOP AND CHECK

1. Find the quartiles for the data set of 16 credit-hour loads found in Example 10 on page 276.

Answer:

1. Lower quartile: 6; median: 9; upper quartile: 12

EXAMPLE 4

Find the quartiles for the set of 25 student grades found on page 275.

Arrange the scores in ascending order.

63, 68, 71, 74, 76, 76, 77, 77, 77, 79, 81, 83, 84, 87, 87, 88, 88, 89, 90, 91, 91, 93, 93, 97, 97

Find the median.

$\frac{25}{2} = 12.5$ The middle value is the 13th value or 84. This is also the boundary of the second quartile.

Find the median of the lower half of data. That is, find the median of the values below 84.

$\frac{12}{2} = 6$ The median of the lower half of data is between the 6th and 7th values.

$\frac{(76 + 77)}{2} = 76.5$ This is the boundary of the lower quartile.

Find the median of the upper half of data that is the median of the values above 84. There are also 12 scores in the upper half of data. $\frac{12}{2} = 6$. To find the median of the upper half of data, you will start with the score that is 6 places above the median.

$13 + 6 = 19.$ The median of the upper half of data is between the 19th and 20th values.

$\frac{(90 + 91)}{2} = 90.5$ This is the boundary of the upper quartile.

63, 68, 71, 74, 76, 76, 77, 77, 77, 79, 81, 83, 84, 87, 87, 88, 88, 89, 90, 91, 91, 93, 93, 97, 97

<table>
<tr><td>↑</td><td>↑</td><td>↑</td></tr>
<tr><td>Lower quartile
Below 76.5</td><td>Median
84</td><td>Upper quartile.
Above 90.5</td></tr>
</table>

See Exercises 20–21.

TIP **Ordinal Numbers** When data items of a set of data are arranged in ascending order, the position of a data item is identified by a value called an ordinal number. An *ordinal number* is a number that identifies an ordered position. Examples of ordinal numbers are first, second, third, and so on.

Inner quartiles: the second and third quartiles of a set of data

Inner-quartile range: the difference between the upper-quartile boundary and the lower-quartile boundary

Let's look more closely at the second and third quartiles (two quartiles in the middle). These two quartiles are referred to as the **inner quartiles.** When we looked at the midrange, we said that this statistic was not very useful because it was greatly affected by extreme values or outliers. If we just look at the inner quartiles, we eliminate all the outliers. The **inner-quartile range** is the difference between the upper-quartile boundary and the lower-quartile boundary. The inner-quartile range can be a useful statistic in examining a set of data. Half of the data will be within the inner-quartile range.

To find the inner-quartile range:

1. Arrange the data in ascending order.

2. Find the boundary for the lower quartile.

3. Find the boundary for the upper quartile.

4. Find the difference between the upper-quartile boundary and the lower-quartile boundary.

   inner-quartile range = upper-quartile boundary − lower-quartile boundary

STOP AND CHECK

1. Find the inner-quartile range for the data set of 16 credit-hour loads found in Example 10 on page 276.

Answer:
1. 6

Percentile: a boundary that determines a data item (score) below which a certain percent of the data will fall

Percentile rank: the percent of data items that is less than or equal to the given data item

---

EXAMPLE 5

In the set of 25 scores given on page 275, find the inner-quartile range.

The upper-quartile boundary is 90.5.   From Example 4
The lower-quartile boundary is 76.5.   From Example 4

The inner-quartile range is $90.5 - 76.5 = 14$.   See Exercises 22–23.

Percentiles are boundaries that determine a data item (score) below which a certain percent of the data will fall. To relate percentiles and quartiles, the first quartile boundary is also the 25th percentile. The second quartile boundary is the 50th percentile and the third quartile boundary is the 75th percentile. As with quartiles, computer software uses various techniques for calculating a percentile. Again, we will use a process that can be done by hand with a small data set. This process requires that data be arranged in ascending order and each data item be assigned an ordinal number to show its position. We first identify the position of a given percentile by finding the percentile rank. The **percentile rank** of a data item is the percent of data items that is less than or equal to the given data item.

To find a specified percentile:

1. Arrange the data in ascending order.

2. Assign each data item an ordinal number that corresponds to each data point. The lowest data item will correspond to the first (1st) ordinal number.

3. Find the ordinal number (n) that corresponds to the given percentile rank by using the formula:

$$\text{corresponding ordinal number} = \frac{\text{number of ordered values}}{100}(\text{percentile rank})$$

or
$$n = \frac{N}{100}(p)$$

If the result is not a whole number, round up to the next whole number.

4. Find the data item that corresponds to the ordinal number found in Step 3.

Note: The percentile rank p in the formula in Step 3 is the actual percentile rank, not a percent converted to a decimal equivalent.

EXAMPLE 6

Using the set of 25 student grades found on page 275, find the following percentiles:

(a) 25%   (b) 80%   (c) 90%

(a) $n = \dfrac{25}{100}(25) = 6.25$ The next whole number is 7.

The 7th score is 77.

25% of the scores are at or below 77.

(b) $n = \dfrac{25}{100}(80) = 20$

The 20th score is 91.

80% of the scores are at or below 91.

STOP AND CHECK

Using the data set of 16 credit-hour loads found in Example 10 on page 276, find the following percentiles:

1. 40%
2. 90%

Answers:
1. 40% of the credit hours are at or below 6.
2. 90% of the credit hours are at or below 12.

(c) $n = \dfrac{25}{100}(90) = 22.5$ The next whole number is 23.

The 23rd score is 93.

90% of the scores are at or below 93.

See Exercises 24–25.

> **TIP** **Working with an Electronic Spreadsheet** When data are entered into an electronic spreadsheet, the data can be arranged in ascending order by sorting the data on the data column. If another column is used to count the data items (Item 1, Item 2, etc.), this column will give the rank of the sorted data. Then the 7th, 20th, or 23rd value can be found at a glance.

3 **Find the Standard Deviation.** Although the range gives us some information about dispersion, it does not tell us whether the highest or lowest values are typical values or extreme outliers. We can get a clearer picture of the data set by examining how much each value *differs* or *deviates* from the mean.

Deviation from the mean: the difference between a specific value and the mean

The **deviation from the mean** of a data value is the difference between a specific value and the mean.

To find the deviations from the mean:

Data set: 38, 43, 45, 44.

1. Find the mean of a set of data.

$$\overline{x} = \frac{\text{sum of data values}}{\text{number of values}} = \frac{\Sigma x_i}{n}$$

$$\frac{38 + 43 + 45 + 44}{4} = \frac{170}{4} = 42.5$$

2. Find the amount that each data value deviates or is different from the mean.

deviation from the mean =
data value − mean = $x_i - \overline{x}$

$38 - 42.5 = -4.5$ (below the mean)
$43 - 42.5 = 0.5$ (above the mean)
$45 - 42.5 = 2.5$ (above the mean)
$44 - 42.5 = 1.5$ (above the mean)

For values smaller than the mean, the difference is represented by a *negative* number indicating the value is *below* or less than the mean. For values larger than the mean, the difference is represented by a positive number indicating the value is *above* or greater than the mean. *The absolute value of the sum of the deviations below the mean should equal the sum of the deviations above the mean.* In the example in the box, only one value is below the mean, and its deviation is −4.5. Three values are above the mean, and the sum of these deviations is 0.5 + 2.5 + 1.5 = 4.5. We say that *the sum of all deviations from the mean is zero.* This is true for all sets of data.

We have not gained any statistical insight or new information by analyzing the sum of the deviations from the mean or even by analyzing the average of the deviations.

$$\text{average deviation} = \frac{\text{sum of deviations}}{\text{number of values}} = \frac{0}{n} = 0$$

1. Find the deviations from the mean for the set of data 114, 142, 145, and 152.

Answer:
1. | Value | Deviation |
|---|---|
| 114 | -24.25 |
| 142 | 3.75 |
| 145 | 6.75 |
| 152 | 13.75 |

EXAMPLE 7

Find the deviations from the mean for the set of data 45, 63, 87, and 91.

$$\overline{x} = \frac{\Sigma x_i}{n} = \frac{45 + 63 + 87 + 91}{4} = \frac{286}{4} = 71.5 \qquad \text{Mean.}$$

To find the deviation from the mean, subtract the mean, $\overline{x}$, from each value of x. We arrange these values in a table.

Values x_i	Deviations $x_i - \overline{x}$
45	$45 - 71.5 = $ **-26.5**
63	$63 - 71.5 = $ **-8.5**
87	$87 - 71.5 = $ **15.5**
91	$91 - 71.5 = $ **19.5**
Sum 286	0

See Exercises 26–34.

As we might expect, the sum of the deviations in the example equals zero because the sum of the negative deviations $(-26.5 + -8.5 = -35)$ equals the sum of the positive deviations $(15.5 + 19.5 = 35)$. To compensate for this situation, mathematicians employ a statistical measure called the **standard deviation**, which uses the square of each deviation from the mean. The square of a negative value is always positive. The sum of the squared deviations is divided by 1 less than the number of values, and the result is called the sample **variance:**

Standard deviation: a statistical measure of dispersion that shows how data vary about the mean; the square root of the variance

Variance: the sum of the squared deviations divided by 1 less than the number of values in the data set

$$\text{sample variance} = v = \frac{\Sigma(x_i - \overline{x})^2}{n - 1}$$

The square root of the variance is the standard deviation. Various formulas exist for finding the standard deviation of a set of values, but we examine only one formula. Several calculations are necessary and are best organized in a table.

$$\text{sample standard deviation} = s = \sqrt{\frac{\Sigma(x_i - \overline{x})^2}{n - 1}}$$

To find the standard deviation of a set of sample data:

1. Find the mean, $\overline{x}$.

2. Find the deviation of each value from the mean: $(x_i - \overline{x})$

3. Square each deviation: $(x_i - \overline{x})^2$

4. Find the sum of the squared deviations: $\Sigma(x_i - \overline{x})^2$

5. Divide the sum of the squared deviations by *one less than* the number of values in the data set. The quotient is called the *sample variance:*

$$v = \frac{\Sigma(x_i - \overline{x})^2}{n - 1}.$$

6. Find the sample standard deviation by taking the square root of the sample variance:

$$s = \sqrt{\frac{\Sigma(x_i - \overline{x})^2}{n - 1}}.$$

STOP AND CHECK

STOP AND CHECK

1. Find the sample standard deviation to the nearest tenth for the set of data 114, 142, 145, and 152.

Answer:

1. 16.7

EXAMPLE 8

Find the sample standard deviation to the nearest tenth for the values 45, 63, 87, and 91.

From Example 7 on page 286, the mean is 71.5 and the number of values is 4.

$$\bar{x} = 71.5, n = 4$$

Values x_i	Deviations from the mean $x_i - \bar{x}$	Squares of the deviations from the mean $(x_i - \bar{x})^2$
45	$45 - 71.5 = -26.5$	$(-26.5)^2 = 702.25$
63	$63 - 71.5 = -8.5$	$(-8.5)^2 = 72.25$
87	$87 - 71.5 = 15.5$	$(15.5)^2 = 240.25$
91	$91 - 71.5 = 19.5$	$(19.5)^2 = 380.25$
Sum of values 286	Sum of deviations 0	Sum of squared deviations 1,395

$$v = \frac{\Sigma(x_i - \bar{x})^2}{n - 1} \qquad \frac{\text{Sum of squared deviations}}{n - 1}.$$

$$v = \frac{1,395}{3} = 465$$

$$s = \sqrt{v} \qquad \text{Square root of variance.}$$

$$s = \sqrt{465} = 21.56385865 \; or \; \textbf{21.6 sample standard deviation (rounded).}$$

See Exercises 26–34.

TIP **Find the Standard Deviation Using a Graphing Calculator** A standard deviation feature is included on many graphing calculators.

Examine the calculator steps used to find the median in the tip box on page 274.

Build a list:

STAT 4:ClrList LIST 1:L₁ ENTER   Clear previous list in L_1.

Enter your new list as List 1. Use the data from Example 8 above.

STAT 1:Edit... 45 ENTER 63 ENTER 87 ENTER 91 ENTER QUIT

Find the standard deviation:

LIST MATH 7:stdDev(LIST 1:L₁ ENTER ⇒ 21.56385865.

A small standard deviation indicates that the mean is a typical value in the data set. A large standard deviation indicates that the mean is not typical, and other statistical measures should be examined to better understand the characteristics of the data set.

Let's examine the various statistics for the data set on a number line (Fig. 6–16).

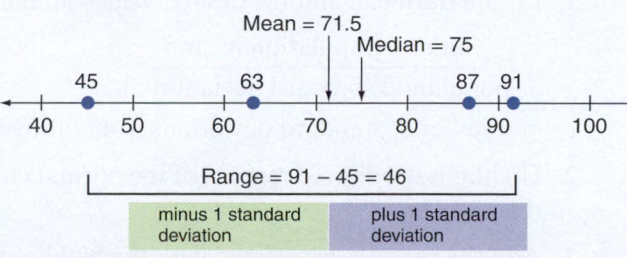

$$\text{Median} = \frac{63 + 87}{2}$$

$$= \frac{150}{2} = 75.$$

Mean − 1 standard deviation
= 71.5 − 21.6 = 49.9.

Mean + 1 standard deviation
= 71.5 + 21.6 = 93.1.

FIGURE 6–16

Normal distribution: an arrangement of a data set in which most of the values cluster around the mean and the rest taper off symmetrically toward the two ends. The graph representing a normal distribution is a bell-shaped curve

We can confirm visually that the dispersion of the data is broad and the mean is not a typical value in the data set.

Another interpretation of the standard deviation is in its relationship to the normal distribution. A **normal distribution** is an arrangement of a data set in which most of the values cluster around the mean and the rest taper off symmetrically toward the two ends. The graph representing a normal distribution is a bell-shaped curve. If we assume that data are normally distributed, we make speculations about where data are located in relation to the population mean of the data set. Procedures for determining whether data are normally distributed are presented in advanced studies of statistics.

The graph of a normal distribution is a bell-shaped curve, as in Fig. 6–17. The curve is *symmetrical*; that is, if folded at the highest point of the curve, the two halves would match. The mean of the data set is at the highest point or fold line. Then, half the data (50%) are to the left or *below* the mean and half the data (50%) are to the right or *above* the mean. Other characteristics of the normal distribution are:

68.26% of the data are within 1 standard deviation of the mean.
95.44% of the data are within 2 standard deviations of the mean.
99.74% of the data are within 3 standard deviations of the mean.

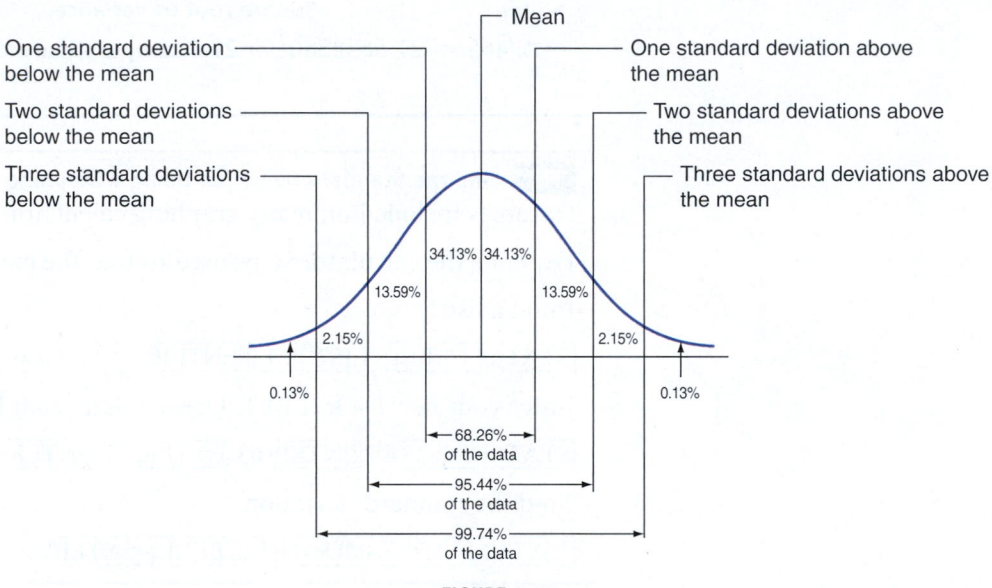

FIGURE 6–17

To solve applied problems involving the mean, standard deviation, and a normal distribution of a population:

1. Locate the mean and the desired values on the normal curve.

$$\frac{\text{value} - \text{population mean}}{\text{population standard deviation}}$$

= number of standard deviations from (above or below) the mean

2. Highlight the desired regions of the normal curve based on the conditions of the problem.

3. Add the percents associated with the highlighted regions in Step 2.

1. A brand of automobile tires has a population mean life of 50,000 miles with a standard deviation of 5,000 miles. In an inventory of 250 tires, what percent do you expect to last no more than 45,000 miles?

Answer:

1. 15.87% of the tires should last 45,000 miles or less.

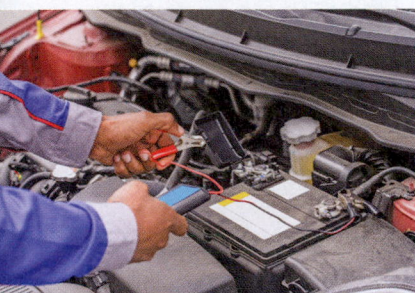
Andrey Popov/Shutterstock

1. Using the facts in the previous Stop and Check problem, how many of the 250 tires do you expect to last longer than 45,000 miles?

Answer:

1. 210 tires are expected to last 45,000 miles or more.

EXAMPLE 9

AUTO   An Auto Zone Duralast Gold automobile battery has a population mean life of 46 months with a population standard deviation of 4 months. In an order of 100 batteries, what percent do you expect to last less than 54 months?

$$\frac{\text{value} - \text{population mean}}{\text{population standard deviation}}$$

= number of standard deviations *from* (above or below) the mean

$$\frac{54 - 46}{4} = \frac{8}{4} = 2 \quad \text{Standard deviations above the mean.}$$

Highlight all regions that are below the mean and two standard deviations above the mean (Fig. 6–18).

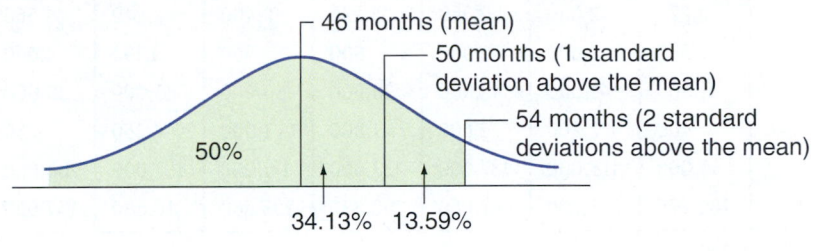

FIGURE 6–18

$50\% + 34.13\% + 13.59\% = 97.72\%$ Add the percents associated with the highlighted regions.

97.72% of the batteries should last less than 54 months.   See Exercises 35–37.

EXAMPLE 10

AUTO   Using the facts from Example 9, how many batteries do you expect to last at least 54 months? Round to the nearest battery.

The phrase "at least 54 months" is the same as the phrase "54 months or more." Therefore, locate regions on the normal curve that are more than 2 standard deviations above the mean (Fig. 6–19).

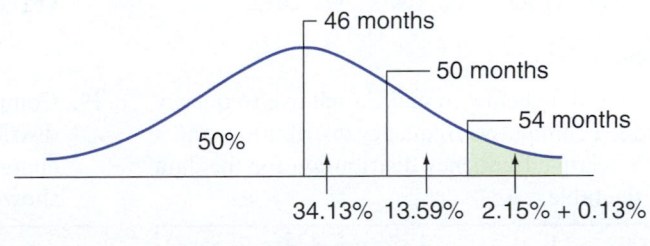

FIGURE 6–19

$2.15\% + 0.13\% = 2.28\%$ Add the percents associated with the shaded region.

Alternative method for finding the percent:
$50\% + 34.13\% + 13.59\% = 97.72\%$ Add the percents associated with the unshaded regions.

$100\% - 97.72\% = 2.28\%$ Complement of 97.72%

$2.28\% (100) \text{ batteries} = 0.0228(100) = 2.28 \text{ batteries}$ Percent of 100 batteries

2 batteries (rounded) should last at least 54 months.   See Exercises 38–40.

6-3 EXERCISES   MyLab Math   For additional practice go to your study plan in MyLab Math.

1 Find the range for each data set. *See Example 1.*

1. 22, 36, 41, 41, 17

2. 28, 33, 36, 13, 28

3. 10, 23, 12, 17, 13, 16

4. 23, 23, 18, 32, 29, 14

5. $25, $15, $25, $40, $19

6. $36, $44, $26, $52, $19

7. 23°F, 37°F, 29°F, 54°F, 46°F, 71°F, 67°F

Table 6–8   Production in Tons of Sweet Cherries by State

State	2002	2003	2004	2005	2006	2007	2008	2009	2010	2011	2012*
CA	55,500	65,600	73,000	52,700	42,100	85,000	86,000	78,000	97,000	75,000	85,000
ID	1,700	2,900	3,100	1,700	3,800	1,500	1,900	6,000	1,900	2,800	4,000
MI	2,700	13,000	24,700	27,000	21,500	27,300	26,500	28,700	15,100	18,600	3,300
MT	2,220	2,060	2,360	1,230	2,400	2,440	1,560	23,900	2,470	2,015	
NY	350	600	900	800	960	1,190	1,050	1,240	1,000	700	250
OR	31,000	41,000	43,000	28,600	50,000	35,000	30,000	67,000	38,150	43,200	53,000
UT	400	2,200	1,600	1,800	1,800	1,250	50	1,540	1,100	800	1,600
WA	86,000	118,000	137,000	137,000	171,000	157,000	100,000	245,000	156,000	200,000	235,000
Total U.S.	180,225	245,700	282,060	250,830	293,560	310,680	247,060	429,870	312,720	343,115	382,150

*Forecast for 2012
Source: National Agricultural Statistics Service (NASS), Agricultural Statistics Board, U.S. Department of Agriculture (USDA), 2012.

Use Table 6–8 to answer Exercises 8–17. *See Example 1.*

8. **AG/H**   Find the range for cherry production in states in 2003.

9. **AG/H**   Find the range for cherry production in states in 2004.

10. **AG/H**   Find the range for cherry production in states in 2011.

11. **AG/H**   Find the range for cherry production for Washington State across the years 2002–2012.

12. **AG/H**   Find the range for cherry production for the state of California across the years 2002–2012.

13. **AG/H**   Find the range for cherry production for New York State across the years 2002–2012.

See Example 2.

14. Find the midrange for cherry production in states in 2002.

15. Find the midrange for cherry production in states in 2011.

16. Find the midrange for cherry production in the state of Washington (WA) across the years 2002–2012.

17. Find the midrange for cherry production in the state of Oregon (OR) across the years 2002–2012.

2 *See Example 3.*

18. Complete the table below to make a relative frequency distribution, a cumulative frequency distribution, and a cumulative relative frequency distribution for the data shown in the table.

Class	Class Frequency	Relative Frequency	Cumulative Frequency	Cumulative Relative Frequency
30–39	4			
40–49	3			
50–59	2			
60–69	8			
70–79	5			
80–89	6			
90–99	4			

19. Complete the table below to make a relative frequency distribution, a cumulative frequency distribution, and a cumulative relative frequency distribution for the data shown in the table.

Class	Class Frequency	Relative Frequency	Cumulative Frequency	Cumulative Relative Frequency
100–119	45			
120–139	54			
140–159	72			
160–179	63			
180–199	81			
200–219	45			

See Example 4.

20. Find the quartiles for the set of AlC measures (an AlC test measures average blood sugar levels over time by taking a sample of hemoglobin AlC cells) for a 56-year-old person.

6.2 5.9 7.1 6.1 5.8 4.9 6.3 6.4 6.7 5.8 5.9 4.5

21. Find the quartiles for the set of total blood serum cholesterol levels recorded by a lab.

205 194 188 220 245 218 255 231 214 199
187 201 238 229 231 267 188 191 203 205

See Example 5.

22. Find the inner-quartile range for the data in Exercise 20.

23. Find the inner-quartile range for the data in Exercise 21.

See Example 6.

24. Find the following percentiles for the scores in Exercise 20:
   (a) 25%   **(b)** 95%

25. Find the following percentiles for scores in Exercise 21:
   (a) 40%   **(b)** 98%

3 Find the standard deviation for each data set. Round to the nearest hundredth. *See Examples 7 and 8.*

26. 12, 14, 16, 18, 20

27. 68, 54, 73, 69

28. 32°F, 41°F, 54°F

29. $27, $32, $65, $29, $21

30. Respiration rates: 16, 24, 20

31. Pulse rates: 68, 84, 76

Use Table 6–8 to answer Exercises 32–34.

32. **AG/H** Find the standard deviation for cherry production for Washington State for the years reported. Round to the nearest ton.

33. **AG/H** Find the standard deviation for cherry production for California for the years reported. Round to the nearest ton.

34. **AG/H** Find the standard deviation for cherry production for New York State for the years reported. Round to the nearest ton.

See Example 9.

35. **HLTH/N** The population mean length of a hospital stay for surgery is 5.8 days and the standard deviation is 1.9 days. What percentage of patients are hospitalized for 3.9 days or less?

36. **HLTH/N** Research has documented that the mean brain weight of people with Alzheimer's disease is 1,076.8 g and the standard deviation is 105.8. What percent of patients have a brain weight greater than 1,288.4 g?

37. **HLTH/N** Pediatricians work an average of 50 h per week. The standard deviation is 16 hours. What percentage of pediatricians work less than 18 h per week?

See Example 10.

38. **HLTH/N** In a sample of 80 pediatricians, how many work less than 18 h per week if the mean is 50 h per week and the standard deviation is 16 h?

39. **HLTH/N** In a sample of 200 surgery patients, how many are expected to stay 9.7 or more days if the mean length of stay is 5.8 days and the standard deviation is 3.9 days?

40. **HLTH/N** The mean brain weight of people with Alzheimer's disease is 1,076.8 g and the standard deviation is 105.8. In a sample of 500 Alzheimer's patients, how many will have a brain weight less than 865.2 g?

6–4 Counting Techniques and Simple Probabilities

LC LEARNING CATALYSTS

1. Evaluate 2^6.
2. Evaluate $8(7)(6)(5)$.

STOP AND CHECK

1. List and count the ways the numbers 1, 2, and 3 can be arranged without repetition.

Answer:

1. 123, 132, 213, 231, 312, 321; 6 ways

Set: a well-defined group of objects or elements

Element: another name for the member of a set of data

Counting: determining all the possible ways the elements in a set can be arranged

Without replacement or without repetition: an arrangement of some or all the elements of a set that has no element appearing more than one time

Tree diagram: a method of counting the ways the elements in a set can be arranged; it allows each new set of possibilities to branch out from a previous possibility

STOP AND CHECK

1. Use a tree diagram to determine the number of ways the numbers 1, 2, 3, and 4 can be arranged without repetition.

Answer:

1. 24 ways

---

1 Count the Number of Ways Objects in a Set Can Be Arranged. A **set** is a well-defined group of objects or **elements.** The numbers 2, 4, 6, 8, and 10 can be a set of even numbers between 1 and 12. Women, men, and children can be a set of people. *A*, *B*, and *C* can be a set of the first three capital letters in the alphabet.

Counting, in this section, means determining all the possible ways the elements in a set can be arranged. One way to count is to *list* all possible arrangements and then count the number of arrangements. In a single arrangement, all elements are included and no element appears more than one time. We refer to this process as arranging **without replacement or without repetition.**

EXAMPLE 1

List and count the ways the elements in the set *A*, *B*, and *C* can be arranged without repetition.

A is first,	*B* is first,	*C* is first,
2 choices	2 choices	2 choices
ABC	*BAC*	*CAB*
ACB	*BCA*	*CBA*

Therefore, *A*, *B*, and *C* can be arranged in six ways. Each of these arrangements can also be called a set. **See Exercises 1–2.**

If more than three elements are in the set, the procedure becomes more challenging. It may be helpful to use a **tree diagram,** which allows each new set of possibilities to branch out from a previous possibility.

To make a tree diagram of possible arrangements of items in a set:

1. List the choices for putting an item in the first slot.

2. From each choice, list the remaining choices for the second slot (first branch).

3. From the end of each branch, make a branch to the remaining choices for the third slot.

4. Continue the process until there are no remaining choices.

5. To list an arrangement, start with the first slot and follow a branch to its end and record the choices.

6. To count the total number of possible arrangements, count the number of branch ends in the last slot.

EXAMPLE 2

Make a tree diagram and count the number of ways the elements in the set containing letters *W*, *X*, *Y*, and *Z* can be arranged without repetition.

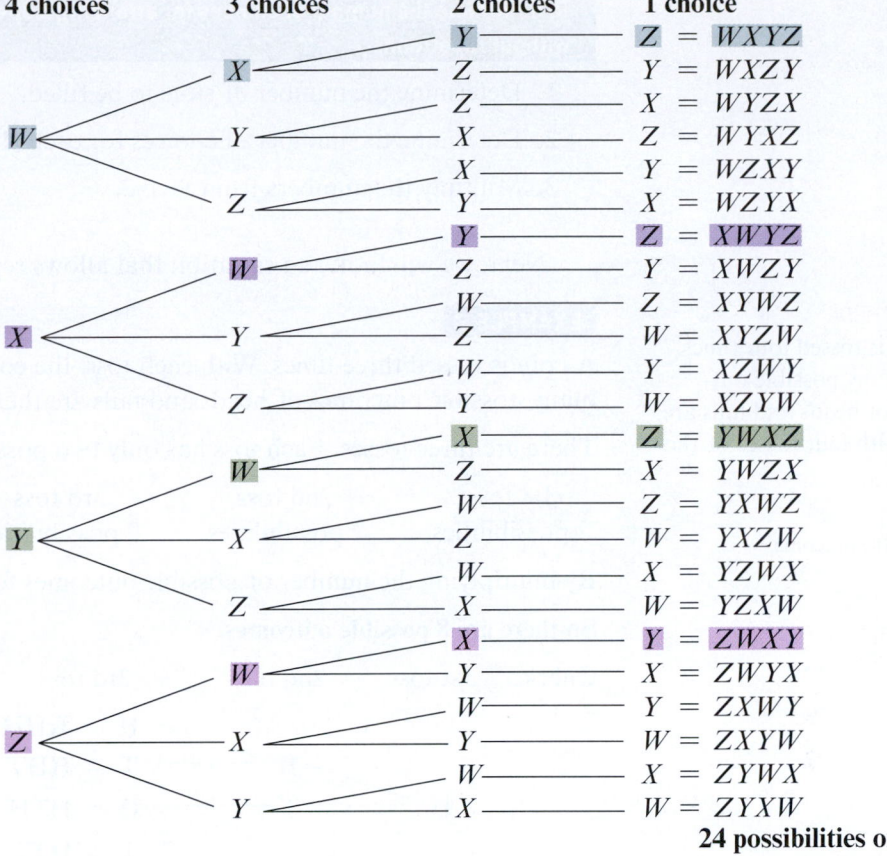

4 choices	3 choices	2 choices	1 choice	
	X	Y	Z =	WXYZ
		Z	Y =	WXZY
W	Y	Z	X =	WYZX
		X	Z =	WYXZ
	Z	X	Y =	WZXY
		Y	X =	WZYX
	W	Y	Z =	XWYZ
		Z	Y =	XWZY
X	Y	W	Z =	XYWZ
		Z	W =	XYZW
	Z	W	Y =	XZWY
		Y	W =	XZYW
	W	X	Z =	YWXZ
		Z	X =	YWZX
Y	X	W	Z =	YXWZ
		Z	W =	YXZW
	Z	W	X =	YZWX
		X	W =	YZXW
	W	X	Y =	ZWXY
		Y	X =	ZWYX
Z	X	Y	W =	ZXWY
		W	Y =	ZXYW
	Y	Y	X =	ZYWX
		X	W =	ZYXW

24 possibilities or sets

See Exercises 1–2.

As evident in Example 2, the greater the number of elements in a set, the greater the complexity and the time required to list all possible arrangements. We can use logic and common sense to obtain a count of the possible arrangements of a set of elements.

There are four possibilities for the first letter: *W, X, Y,* or *Z.* For each of the four possible first letters, we have three choices for second letters. Now, we have 4 × 3 or 12 possibilities. For each of the 12 possibilities, two choices remain for the third letter: 12 × 2 = 24. Then, for each of the 24 three-letter combinations, only one choice is left: 24 × 1 = 24. So, we have a total of 24 possibilities.

Another way to visualize this is to think of drawing letters from a bag or bowl (Fig. 6–20).

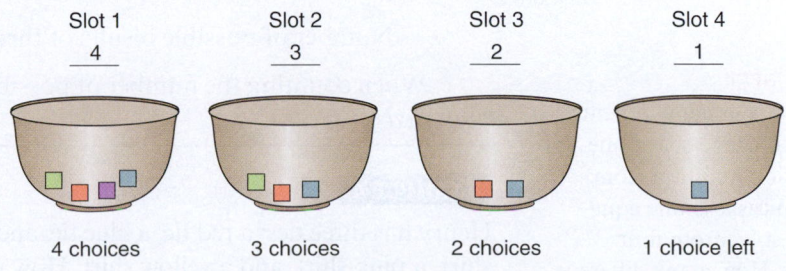

Slot 1	Slot 2	Slot 3	Slot 4
4	3	2	1

4 choices   3 choices   2 choices   1 choice left

FIGURE 6-20

By multiplying the number of possible choices for each position, we can determine the total number of possibilities without listing them: 4 · 3 · 2 · 1 = 24.

To determine the number of choices for arranging a specified number of items without repetition:

1. Determine the number of slots to be filled.

2. Determine the number of choices for each slot.

3. Multiply the numbers from Step 2.

Next, we will look at a situation that allows repetition.

EXAMPLE 3

A coin is tossed three times. With each toss, the coin falls heads up or tails up. How many possible outcomes of heads and tails are there with three tosses of the coin?

There are three tosses. Each toss has only two possibilities, heads or tails; that is,

1st toss	2nd toss	3rd toss
2 possibilities	2 possibilities	2 possibilities

By multiplying the number of possible outcomes for each toss, we get $2 \cdot 2 \cdot 2 = 8$.

So there are 8 possible outcomes.

Check:

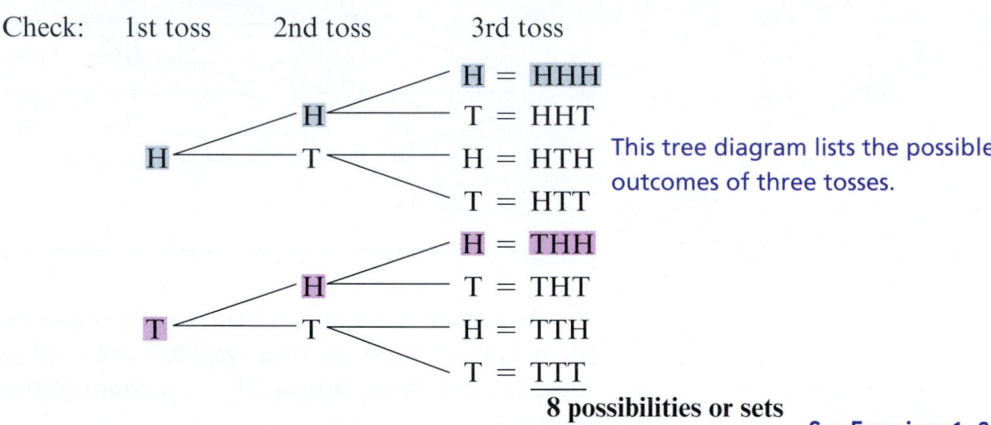

This tree diagram lists the possible outcomes of three tosses.

8 possibilities or sets

See Exercises 1–2.

TIP To Repeat or Not Repeat In some situations, as in arranging W, X, Y, and Z, once a choice or selection is made, that choice cannot be repeated. In these situations the number of choices decreases with each selection.

Number of possible arrangements of W, X, Y, $Z = 4 \cdot 3 \cdot 2 \cdot 1 = 24$

In other situations, as in tossing a coin, every coin toss can result in a head or a tail. The result of any coin toss can repeat the result of a previous toss.

Number of possible results of three coin tosses $= 2 \cdot 2 \cdot 2 = 2^3 = 8$

When counting the number of possible outcomes, *first determine if repeats are allowed or not.*

EXAMPLE 4

Henry has three ties: a red tie, a blue tie, and a green tie. He also has three shirts: a white shirt, a pink shirt, and a yellow shirt. How many sets of shirts and ties are possible?

If we start with the shirts, there are three possibilities (white, pink, and yellow). For each shirt, there are three possible ties (red, blue, and green). **So we have $3 \cdot 3 = 9$ possible outcomes.**

See Exercises 3–5.

STOP AND CHECK

1. In a state lottery, one option is to pick four single-digit numbers. If repeats are allowed, how many possible choices are there for picking four numbers?

Answer:
1. 10,000 ways

EXAMPLE 5

Given the digits 1, 2, 3, 4, 5, and 6, how many three-digit numbers can be made without repeating a digit?

The numbers are to contain three digits, so there are three positions to fill. We have six digits to work with. The first digit can be one of six. For each of these six digits, there are five possible second digits. For each of the five second digits, there are four possible third digits.

Positions:	1st	2nd	3rd
Possibilities:	6	5	4

By multiplying the possibilities for each position, we get $6 \cdot 5 \cdot 4 =$ **120 possible outcomes or ways to make a three-digit number.**

See Exercises 6–7.

Probability: the chance of an event occurring if an activity is repeated over and over; expressed as a ratio or a percent

2 Determine the Probability of an Event Occurring if an Activity Is Repeated Over and Over. Probability means the chance of an event occurring if an activity is repeated over and over. The probability of an event occurring is expressed as a ratio or a percent.

> **TIP** **Theoretical Probability Versus Empirical Probability** There are many variations in the definition of the word probability and the distinction between theoretical and empirical probability. While both concepts are often referred to as probability, the most common distinction is:
>
> **Theoretical probability** is the likelihood of an event occurring as determined through reasoning and calculation.
>
> **Empirical probability** is the likelihood of an event occurring as determined through experimentation.

Theoretical probability: the likelihood of an event occurring as determined through reasoning and calculation

Empirical probability: the likelihood of an event occurring as determined through experimentation

Weather forecasting is an example of empirical probability. Numerous studies and observations are used to develop expected weather outcomes for various circumstances. Weather forecasters use percents when they forecast a 60% chance of rain or a 20% chance of snow. This text will use ratios like $\frac{3}{5}$ for 3 chances out of 5 or $\frac{2}{3}$ for 2 chances out of 3. The decimal equivalent of a ratio can also be used to express a probability: $\frac{3}{5} = 0.6$.

When the expected outcomes of events are equally likely to occur and there is a specified number of outcomes, theoretical probabilities can be determined. For example, when a coin is tossed, two outcomes are possible, heads or tails. But only one side will be up. The probability of tossing heads is 1 out of 2, $\frac{1}{2}$, 0.5, or 50%.

To express the probability of an event occurring successfully:

1. Determine the total number of elements in the set (total possible outcomes).

2. Determine the number of elements that are defined as successful.

3. Make a ratio with the number of choices that are defined as successful divided by the total number of elements in the set.

$$\frac{\text{number of successful possibilities}}{\text{number of total possibilities}}$$

4. Express the ratio in lowest terms. The ratio can be expressed as a decimal or percent if desired.

Invalid probability: the probability of an event that is defined to be less than 0 or greater than 1

FIGURE 6–21

Olga Bonchuk/123RF

The probability of an event occurring ranges from 0 for an impossible event to 1 for a certain event.

The probability of an event that is defined to be less than 0 or greater than 1 is an **invalid probability.**

EXAMPLE 6

When a die is rolled, what is the probability that a 3 will appear on the top face (Fig. 6–21)?

A die has six sides where dots represent 1–6. Each side is an element in the set of six sides, so the set has a total of six elements.

Only one element is a 3.

The probability of rolling a 3 is $\frac{1}{6}$.   See Exercises 8–9.

EXAMPLE 7

BUS   A holiday gift shopper wrapped eight men's ties in separate boxes. There were two solid-color ties and six striped ties. If the gift boxes were given at random to eight men, what is the probability of a man receiving a solid-color tie?

eight ties   Total possibilities
two solid-color ties   Successful possibilities

$$\text{probability} = \frac{\text{successful possibilities}}{\text{total possibilities}} = \frac{2}{8} = \frac{1}{4}$$

The probability of receiving a solid-color tie is $\frac{1}{4}$ or 0.25 or 25%.   See Exercises 10–13.

EXAMPLE 8

HLTH/N   A practical nurse has a box of 144 syringes in individual sterile packets. Three of the syringes have torn packets. What is the probability that the first syringe picked will have a torn packet? If the packet for the first syringe selected is torn, what is the probability of picking a second syringe with a torn packet?

On the first pick, the probability of picking one syringe with a torn packet is $\frac{3}{144}$, which reduces to $\frac{1}{48}$. We assume that the first pick was a syringe with a torn packet, so this leaves a total of 143 syringes and now only 2 have torn packets. On the second pick, the probability of picking a syringe with a torn packet is $\frac{2}{143}$.   See Exercises 14–15.

EXAMPLE 9

Determine the probabilities of the following events:

(a) A standard pair of six-sided dice is rolled. What is the probability that the sum of the two amounts showing on the tops of the dice is 14?

(b) A standard pair of six-sided dice is rolled. What is the probability that the sum of the two amounts showing on the tops of the dice is at least 2?

(a) **The probability of a sum of 14 from the rolling of a pair of dice is 0.** The highest number on a single die is 6. The maximum sum of a pair of dice would be 12.

(b) **The probability of a sum of at least 2 from the rolling a pair of dice is 1.** The lowest number on a single die is 1. The sum of the dice would be at least 2.

See Exercises 16–17.

STOP AND CHECK

A standard pair of six-sided dice is rolled.

1. What is the probability that the numbers 2 and 3 are showing on the tops of the dice?

2. What is the probability that the sum of the two amounts showing on the tops of the dice is 1?

3. What is the probability that the sum of the two amounts showing on the tops of the dice is 12?

Answers:

1. $\frac{1}{18}$ 2. 0   3. $\frac{1}{36}$

3 Determine the Odds of an Event Occurring. In some circumstances it is desirable to compare the probability that an event occurs to the probability that an event does not occur. This establishes a ratio that is referred to as the odds for the occurrence of an event. Similarly, you can establish the odds against the occurrence of an event.

Odds for: the ratio of the number of successful possible outcomes to the number of unsuccessful possible outcomes

Odds against: the ratio of the number of unsuccessful possible outcomes to the number of successful possible outcomes

The **odds for** the occurrence of an event is the ratio of the number of successful possible outcomes to the number of unsuccessful possible outcomes. The **odds against** the occurrence of an event is the ratio of the number of unsuccessful possible outcomes to the number of successful possible outcomes.

> **To find the odds for or the odds against an event occurring:**
>
> 1. Determine the number of possible successful outcomes.
>
> 2. Determine the number of possible unsuccessful outcomes by subtracting the number of successful outcomes from the total possible outcomes.
>
> Unsuccessful outcomes = total possible outcomes − possible successful outcomes
>
> 3. Use the appropriate formula to find the odds for or the odds against an event occurring. Express the result as a ratio of the numerator to the denominator, a fraction, decimal equivalent, or percent.
>
> $$\text{Odds for an event} = \frac{\text{possible successful outcomes}}{\text{possible unsuccessful outcomes}}$$
>
> $$\text{Odds against an event} = \frac{\text{possible unsuccessful outcomes}}{\text{possible successful outcomes}}$$

STOP AND CHECK
A bag contains 18 black pairs of socks and 12 gray pairs of socks.
1. What is the probability of selecting a black pair of socks?
2. What are the odds for selecting a black pair of socks?
3. What are the odds against selecting a black pair of socks?

Answers:
1. $\frac{2}{5}$ or 0.4 or 40% 2. $\frac{3}{2}$ or 1.5
3. $\frac{2}{3}$

EXAMPLE 10

Twenty wrapped gifts contain 12 blue shirts and 8 green shirts.

(a) What is the probability of selecting a blue shirt?

(b) What are the odds for selecting a blue shirt?

(c) What are the odds against selecting a blue shirt?

The event is defined as selecting a blue shirt. There are 12 possible successful selections. There are 8 possible unsuccessful selections. The total possible selections are $12 + 8 = 20$.

(a) $\text{Probability} = \dfrac{\text{possible successful outcomes}}{\text{total possible outcomes}}$

$\text{Probability} = \dfrac{12}{20}$

$\textbf{Probability} = \dfrac{3}{5} \textbf{ or 0.6 or 60\%}$

(b) $\text{Odds for selecting a blue shirt} = \dfrac{\text{possible successful outcomes}}{\text{possible unsuccessful outcomes}}$

$\text{Odds for selecting a blue shirt} = \dfrac{12}{8}$

$\textbf{Odds for selecting a blue shirt} = \dfrac{3}{2} \textbf{ or 3 to 2}$

(c) Odds against selecting a blue shirt = $\dfrac{\text{possible unsuccessful outcomes}}{\text{possible successful outcomes}}$

Odds against selecting a blue shirt = $\dfrac{8}{12}$

Odds against selecting a blue shirt $= \dfrac{2}{3}$ or 2 to 3

See Exercises 18–19.

6-4 EXERCISES   MyLab Math   For additional practice go to your study plan in MyLab Math.

1 Complete the exercises using counting techniques. *See Examples 1–3.*

1. List and count all the possible arrangements for Keaton, Brienne, and Renee to be seated in three adjacent seats at a basketball game.

2. List and count all the possible outcomes for arranging books A, B, C, and D on a shelf.

See Example 4.

3. **BUS** Count all the possible outcomes for Jim Riddle to arrange a T-shirt, a sport shirt, a dress shirt, and a sweater on a shelf for display.

4. **HLTH/N** How many outcomes are possible for arranging containers of cotton balls, gauze pads, swabs, tongue depressors, and adhesive tape in a row on a shelf in a doctor's examining room?

5. **AG/H** A landscaping process involves five steps. The steps can be arranged in any order. The landscaping company efficiency officer wants to determine the most efficient order for the five steps. How many arrangements of steps are possible?

See Example 5.

6. **AG/H** A farmer has three plots of land and intends to plant corn, beans, tomatoes, rice, or cotton. How many ways can he arrange his plantings without repeating any crop?

7. Given the letters *A*, *B*, *C*, *D*, and *E*, and accepting any two letters as a word, how many two-letter words can be made from these letters if repeated letters are not allowed?

2 Use simple probability to complete the exercises.

See Example 6.

8. When a single die is rolled, what is the probability that a 1 will appear?

9. When a die is rolled, what is the probability that an even number will appear on the top face?

See Example 7.

10. A jar holds 10 lock washers and 15 flat washers. What is the probability of drawing a lock washer at random?

11. A TV quiz program puts all questions in a box. If the box contains five hard questions, five average questions, and five easy questions, what is the probability of being asked an easy question?

12. A box of greeting cards contains 20 friendship cards, 10 get-well cards, and 10 congratulations cards. What is the probability of picking a get-well card at random?

13. Mimi tosses 21 pennies, 16 nickels, and 11 dimes into a container. What is the probability of reaching in and picking a dime?

See Example 8.

14. Five defective pens were accidentally placed in a box of 139 good pens, making a box of 144 pens. What is the probability of finding a defective pen on the second try if the first try resulted in finding a defective pen?

15. A drawing will be held to award door prizes. If the names of 24 people (no repeated names) are in the pool for the drawing, what is the probability that David's name will be pulled at random for the first prize? If his name is pulled and not replaced, what is the probability that Gaynell's name will be pulled next?

See Example 9.

16. A company has seven employees. Their ages are 25, 27, 32, 47, 56, 63, and 67, respectively.

What is the probability of selecting an employee for an assignment that is at least 21 years old?

17. Playing cards have red or black faces. What is the probability of drawing a card with a green face from a deck of playing cards?

3 *See Example 10.*

18. What are the odds against selecting a female employee from a group of 18 female and 6 male employees?

19. An employer has 18 female and 6 male employees who do the same work. What are the odds for selecting a female employee from this group for a special assignment?

---

6   CHAPTER REVIEW OF KEY CONCEPTS

LEARNING OUTCOMES	KEY CONCEPTS AND EXAMPLES

Section 6–1

Read a circle, bar, or line graph: 1. Examine the title of the graph to find out what information is shown.   **2.** Examine the parts to see how they relate to one another and to the whole.   **3.** Examine the labels for each part of the graph and any explanatory remarks that may be given.   **4.** Use the given parts to calculate additional amounts or percents.

1 Read circle graphs (p. 265).

A circle graph compares parts to a whole.

If the total monthly revenue at a used car dealership is $75,000, what is the revenue from trucks? (See Fig. 6–22.)

35% of 75,000 =
$0.35 \times 75,000 = \$26,250$

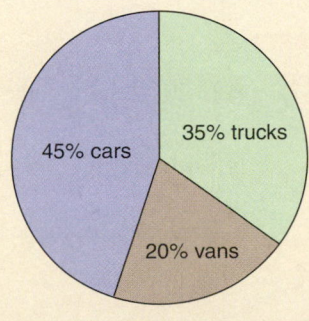

FIGURE 6–22

LEARNING OUTCOMES	KEY CONCEPTS AND EXAMPLES

2 Read bar graphs (pp. 265–266).

Bar graphs are used to compare values to each other.

What is the ratio of men's salaries to women's salaries in the pants department? (See Fig. 6–23.)

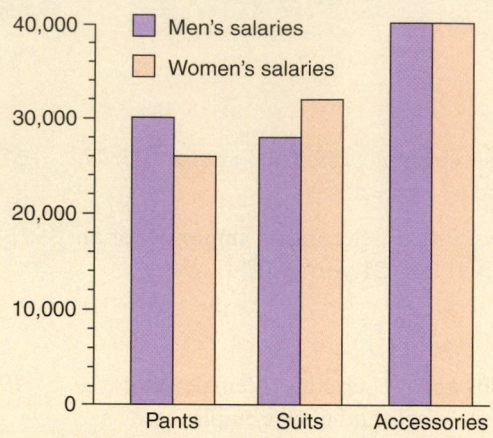

FIGURE 6–23

Men's salary in the pants department = $30,000

Women's salary in the pants department = $26,000

Ratio of men's salaries to women's salaries = $\frac{30}{26} = \frac{15}{13}$

3 Read line graphs (pp. 266–267).

A line graph shows how an item changes with time.

Use the line graph of the average prices for ebooks to find the year when ebooks averaged $67 per book. (See Fig. 6–24.)

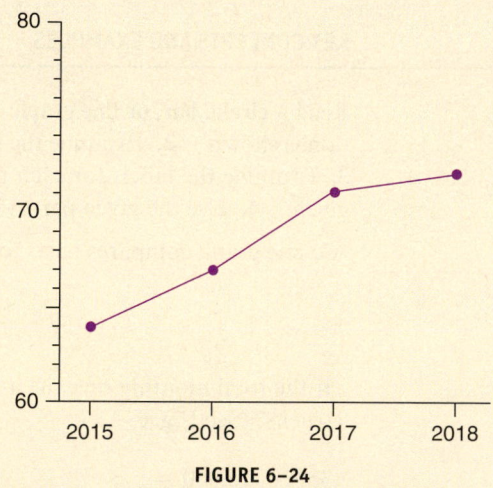

FIGURE 6–24

In 2016, the average price of ebooks was $67.

Section 6-2

1 Find the arithmetic mean (pp. 271–273).

1. Add the quantities.   **2.** Divide the sum by the number of quantities.

Find the arithmetic mean of the test scores: 76, 86, 93, 87, 68, 76, 88

$$76 + 86 + 93 + 87 + 68 + 76 + 88 = 574$$

$$574 \div 7 = 82$$

The mean is 82.

LEARNING OUTCOMES	KEY CONCEPTS AND EXAMPLES

2 Find the median and the mode (pp. 273–275).

1. Arrange the values in order of size, either smaller to larger or larger to smaller.   **2.** If the number of values is odd, the median is the middle value.   **3.** If the number of values is even, the median is the average of the two middle values.

The mode is the value that occurs most frequently in a data set. A set of values may have no mode or more than one mode.

Find the median and mode of the set of test scores:

68, 76, 76, 86, 87, 88, 93   Already arranged in order.
   ↑ ↑

The median is 86.   Middle value.

The mode is 76.   Most frequent value.

3 Make and interpret a frequency distribution (pp. 275–276).

Make a frequency distribution: Determine the appropriate interval for classifying the data. Tally the data.

Make a frequency distribution with the following data, indicating leave days for State College employees. (See Table 6–9.)

2	2	4	4	4	5	5	6	6	8
8	8	9	12	12	12	14	15	20	20

Table 6–9   Annual Leave Days of 20 State College Employees

Class Interval	Midpoint	Tally	Class Frequency
16–20	18	//	2
11–15	13	LH1	5
6–10	8	LH1 /	6
1–5	3	LH1 //	7

4 Find the mean of grouped data (pp. 276–277).

1. Make a frequency distribution.   **2.** Find the product (xf) of the midpoint (x) of the interval and the frequency (f) of the interval for each interval.   **3.** Find the sum of the frequencies (Σf).   **4.** Find the sum of the products (Σxf).   **5.** Divide the sum of the products by the sum of the frequencies.

Symbolically, $\text{mean}_{\text{grouped}} = \dfrac{\Sigma xf}{\Sigma f}$.

Find the mean of the grouped data in the frequency distribution in Table 6–10.

Table 6–10   Annual Leave Days of 20 State College Employees

Class Interval	Midpoint x	Frequency f	Product xf
16–20	18	2	36
11–15	13	5	65
6–10	8	6	48
1–5	3	7	21
Total		20	170

$\Sigma f = 20$ Add the frequencies.

$\Sigma xf = 170$ Add the products.

$\text{mean}_{\text{grouped}} = \dfrac{\Sigma xf}{\Sigma f}$ Substitute.

$\text{mean}_{\text{grouped}} = \dfrac{170}{20}$ Divide.

$\text{mean}_{\text{grouped}} = 8.5$

LEARNING OUTCOMES	KEY CONCEPTS AND EXAMPLES

Section 6-3

1 Find the range (p. 280).

1. Find the highest and lowest values.   **2.** Find the difference between the highest and lowest values.

> Find the range of the set of test scores:
>
> $$68, 76, 76, 86, 87, 88, 93$$
>
> range $= 93 - 68 = 25$

Find the midrange of a data set: 1. Find the highest and lowest values.   **2.** Find the mean of the highest and lowest values.

$$\text{midrange} = \frac{\text{highest value} + \text{lowest value}}{2}$$

> Find the midrange of the set of test scores: 68, 76, 76, 86, 87, 88, 93
>
> Highest value $= 93$ Found in previous example.
> Lowest value $= 68$ Found in previous example.
> $$\text{Midrange} = \frac{93 - 68}{2} = \frac{25}{2} = 12.5$$

2 Find measures of relative position (pp. 281–285).

Make a relative frequency distribution: 1. Make a frequency distribution.   **2.** Find the relative frequency of each class or category by dividing the frequency by the total number in the data set. Express as a fraction in lowest terms or a decimal equivalent.

$$\text{relative frequency of a class} = \frac{\text{frequency of a class}}{\text{total number in data set}}$$

3. To check the distribution, the sum of all the relative frequencies is 1.

Make a cumulative frequency distribution: 1. Make a frequency distribution.   **2.** The cumulative frequency of the first (lowest) class is the same as the frequency of the class.   **3.** The cumulative frequency of each subsequent class is the frequency of the class plus the sum of the frequencies of all previous classes.   **4.** The cumulative frequency of the last (highest) class is equal to the total number of items in the data set.

Make a cumulative relative frequency distribution: 1. Make a relative frequency distribution. **2.** The cumulative relative frequency of the first class is the same as the relative frequency. **3.** The cumulative relative frequency of each subsequent class is the relative frequency of the class plus the sum of the relative frequencies of all previous classes.   **4.** The cumulative relative frequency of the last class is 1.

> Make a relative frequency distribution, a cumulative frequency distribution, and a cumulative relative frequency distribution for the data set with class frequencies of annual leave days of 20 State College employees (p. 301).
>
> | Class Interval | Class Frequency | Relative Frequency | Cumulative Frequency | Cumulative Relative Frequency |
> |---|---|---|---|---|
> | 16–20 | 2 | $\frac{2}{20} = \frac{1}{10} = 0.1$ | $18 + 2 = 20$ | $0.9 + 0.1 = 1$ |
> | 11–15 | 5 | $\frac{5}{20} = \frac{1}{4} = 0.25$ | $13 + 5 = 18$ | $0.65 + 0.25 = 0.9$ |
> | 6–10 | 6 | $\frac{6}{20} = \frac{3}{10} = 0.3$ | $7 + 6 = 13$ | $0.35 + 0.3 = 0.65$ |
> | 1–5 | 7 | $\frac{7}{20} = 0.35$ | 7 | 0.35 |

LEARNING OUTCOMES	KEY CONCEPTS AND EXAMPLES

Divide a set of data into quartiles: 1. Find the median of the data set. This divides the data into two halves (lower half and upper half).   **2.** Find the median of each half of the data. **3.** The median of the lower half of data is the boundary for the lower quartile. The median for the upper half of data is the boundary for the upper quartile.

Find the quartiles for the annual leave days of the 20 State College employees in the previous example.

Arrange the scores in ascending order.

2, 2, 4, 4, 4, 5, 5, 6, 6, 8, 8, 8, 9, 12, 12, 12, 14, 15, 20, 20

Find the median. $\frac{20}{2} = 10$ The middle value is between the 10th and 11th items.

Both of these values are 8, so the median is 8. This is also the upper boundary of the second quartile.

Find the median of the values below the median. There are 10 scores in the lower half of data. $\frac{10}{2} = 5$ The median of the lower half of data is between the 5th and 6th values. $\frac{(4 + 5)}{2} = 4.5$ This is the upper boundary of the lower quartile.

Find the median of the values above the median. There are also 10 scores in the upper half of data. $\frac{10}{2} = 5$ To find the median of the upper half, you will start with the score that is 5 places above the median. $10 + 5 = 15$ The median of the upper half of data is between the 15th and 16th values. Both of these values are 12, so the lower boundary of the upper quartile is 12.

2, 2, 4, 4, 4, 5, 5, 6, 6, 8, 8, 8, 9, 12, 12, 12, 14, 15, 20, 20

$\uparrow$ $\uparrow$ $\uparrow$

Lower quartile   Median   Upper quartile.

Below 4.5   8   Above 12

Find the inner-quartile range: 1. Arrange the data in ascending order.   **2.** Find the boundary for the lower quartile.   **3.** Find the boundary for the upper quartile.   **4.** Find the difference between the upper-quartile boundary and the lower-quartile boundary.

inner-quartile range = upper-quartile boundary − lower-quartile boundary

In the set of 20 scores in the previous example, find the inner-quartile range.

The upper-quartile boundary is 12.   From the previous example

The lower-quartile boundary is 4.5   From the previous example

The inner-quartile range is 12 − 4.5 = 7.5.

Find a specified percentile: 1. Arrange the data in ascending order.   **2.** Assign each data item an ordinal number that corresponds to each data point. The lowest data item will correspond to the first (1st) ordinal number.   **3.** Find the ordinal number (n) that corresponds to the given percentile rank by using the formula:

corresponding ordinal number $= \dfrac{\text{number of ordered values}}{100} \text{(percentile rank)}$ or $n = \dfrac{N}{100}(p)$

If the result is not a whole number, round up to the next whole number.   **4.** Find the data item that corresponds to the ordinal number found in Step 3.

LEARNING OUTCOMES	KEY CONCEPTS AND EXAMPLES

Using the set of data in the previous example, find the following percentiles:

(a) 25%   **(b)** 80%

(a) $n = \dfrac{20}{100}(25) = 5$

The fifth item is 4.

25% of the employees used 4 or fewer days of sick leave.

(b) $n = \dfrac{20}{100}(80) = 16$

The 16th score is 12.

80% of the employees used 12 or fewer days of sick leave.

3 Find the standard deviation (pp. 285–289).

Find the deviations from the mean: 1. Find the mean of a set of data.

$$\overline{x} = \frac{\text{sum of data values}}{\text{number of values}} = \frac{\Sigma x_i}{n}$$

2. Find the amount that each data value deviates or is different from the mean.

$$\text{deviation from the mean} = \text{data value} - \text{mean} = x_i - \overline{x}$$

Find the standard deviation of a set of sample data: 1. Find the mean, $\overline{x}$.   **2.** Find the deviation of each value from the mean: $(x_i - \overline{x})$.   **3.** Square each deviation: $(x_i - \overline{x})^2$.   **4.** Find the sum of the squared deviations: $\Sigma(x_i - \overline{x})^2$.   **5.** Divide the sum of the squared deviations by *one less* than the number of values in the data set. This quotient is called the *sample variance*,

$v = \dfrac{\Sigma(x_i - \overline{x})^2}{n - 1}$.   **6.** Find the sample standard deviation by taking the square root of the sample variance.

$$s = \sqrt{\frac{\Sigma(x_i - \overline{x})^2}{n - 1}}$$

Find the standard deviation of the test scores: 68, 76, 76, 86, 87, 88, 93

$$\overline{x} = \frac{68 + 76 + 76 + 86 + 87 + 88 + 93}{7} = \frac{574}{7} = 82$$

$x_i - \overline{x}$	$(x_i - \overline{x})^2$
$68 - 82 = -14$	196
$76 - 82 = -6$	36
$76 - 82 = -6$	36
$86 - 82 = 4$	16
$87 - 82 = 5$	25
$88 - 82 = 6$	36
$93 - 82 = 11$	121

$\Sigma(x_i - \overline{x})^2 = 466$

$$v = \frac{\Sigma(x_i - \overline{x})^2}{n - 1} = \frac{466}{6} = 77.66666667$$

$$s = \sqrt{\Sigma\frac{(x_i - \overline{x})^2}{n - 1}} = \sqrt{77.66666667} = 8.812869378 = 8.8 \qquad \text{Round.}$$

Solve applied problems involving the mean, standard deviation, and a normal distribution of a population: 1. Locate the mean and the desired values on the normal curve.

$$\frac{\text{value} - \text{population mean}}{\text{population standard deviation}} = \text{number of standard deviations from (above or below) the mean}$$

LEARNING OUTCOMES	KEY CONCEPTS AND EXAMPLES

2. Highlight the desired regions of the normal curve based on the conditions of the problem.   **3.** Add the percents associated with the highlighted regions in Step 2.

Section 6-4

1 Count the number of ways objects in a set can be arranged (pp. 292–295).

Multiply the number of choices for each position in the arrangement.

> Renee Smith's closet has two new blazers (navy and red) and four new skirts (gray, black, tan, and brown). How many outfits can she make from the new clothes?
>
> $$1 \text{ blazer} + 1 \text{ skirt} = 1 \text{ outfit}$$
> $$\begin{pmatrix} \text{blazer} \\ \text{choices} \end{pmatrix} \cdot \begin{pmatrix} \text{skirt} \\ \text{choices} \end{pmatrix} = \begin{pmatrix} \text{possible} \\ \text{outfits} \end{pmatrix}$$
> $$2 \quad \cdot \quad 4 \quad = \quad 8$$

2 Determine the probability of an event occurring if an activity is repeated over and over (pp. 295–296).

The probability of an event occurring is the ratio of the number of possible successful outcomes to the number of possible outcomes.

> A box in a doctor's office contains thirty $\frac{1}{2}$-in. adhesive strips and seventy $\frac{3}{4}$-in. adhesive strips. What is the probability of picking a $\frac{3}{4}$-in. adhesive strip at random?
>
> $$\frac{70}{100} = \frac{7}{10} \qquad \text{Total items from which to select} = 30 + 70 = 100$$
>
> The probability of picking a $\frac{3}{4}$-in. adhesive strip is $\frac{7}{10}$ or 0.7 or 70%.

3 Determine the odds of an event occurring (pp. 297–298).

Find the odds for or the odds against an event occurring: 1. Determine the number of possible successful outcomes.   **2.** Determine the number of possible unsuccessful outcomes by subtracting the number of successful outcomes from the total possible outcomes. Unsuccessful outcomes = total possible outcomes − possible successful outcomes   **3.** Use the appropriate formula to find the odds for or the odds against an event occurring. Express the result as a ratio of the numerator to the denominator, a fraction, decimal equivalent, or percent.

$$\text{Odds for an event} = \frac{\text{possible successful outcomes}}{\text{possible unsuccessful outcomes}}$$

$$\text{Odds against an event} = \frac{\text{possible unsuccessful outcomes}}{\text{possible successful outcomes}}$$

> Fifty tickets are available to the school play. Thirty are available for the matinee performance and 20 are available for the evening performance. The tickets are placed in a basket and selected at random.
>
> **(a)** What is the probability of selecting an evening ticket?
>
> **(b)** What are the odds for selecting an evening ticket?
>
> **(c)** What are the odds against selecting an evening ticket?
>
> The successful event is defined as selecting an evening ticket. There are 20 possible successful selections. There are 30 possible unsuccessful selections. The total possible selections are 20 + 30 = 50.
>
> **(a)** $\text{Probability} = \dfrac{\text{possible successful outcomes}}{\text{total possible outcomes}}$
>
> $\qquad \text{Probability} = \dfrac{20}{50}$
>
> $\qquad \textbf{Probability} = \dfrac{\textbf{2}}{\textbf{5}} \textbf{ or 0.4 or 40\%}$

LEARNING OUTCOMES	KEY CONCEPTS AND EXAMPLES

(b) Odds for selecting an evening ticket $= \dfrac{\text{possible successful outcomes}}{\text{possible unsuccessful outcomes}}$

Odds for selecting an evening ticket $= \dfrac{20}{30}$

Odds for selecting an evening ticket $= \dfrac{2}{3}$ or 2 to 3

(c) Odds against selecting an evening ticket $= \dfrac{\text{possible unsuccessful outcomes}}{\text{possible successful outcomes}}$

Odds against selecting an evening ticket $= \dfrac{30}{20}$

Odds against selecting an evening ticket $= \dfrac{3}{2}$ or 3 to 2

6 CHAPTER REVIEW EXERCISES

Section 6–1 MyLab Math For additional practice go to your study plan in MyLab Math.

Use Fig. 6–25 to answer Exercises 1–3.

1. **CON** What percent of the total cost is the cost of the lot? Round to the nearest tenth.

2. **CON** What percent of the total cost is the cost of the house? Round to the nearest tenth.

3. **CON** The cost of the lot and landscaping represents what percent of the total cost? Round to the nearest tenth.

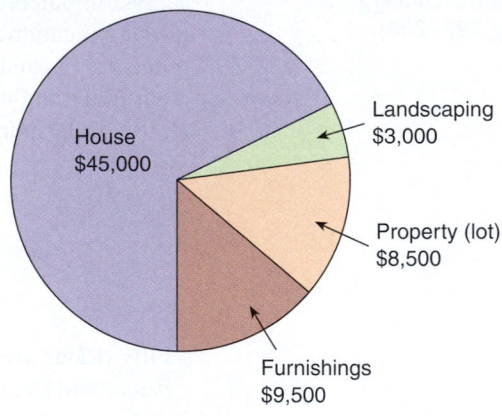

FIGURE 6–25 Distribution of costs for a $66,000 home.

Use Fig. 6–26 to answer Exercises 4–7.

4. **AG/H** The cranberry crop 2012 forecast for Oregon was 39.4% above the 2010 actual production. What was the actual production in 2010? Round to the nearest thousand barrels.

5. **AG/H** What was the total U.S. production in barrels of cranberries forecast for 2012?

6. **AG/H** Which state produces the greatest percentage of cranberries in the United States?

7. **AG/H** Massachusetts produces what percentage of U.S. cranberries? Round to tenths.

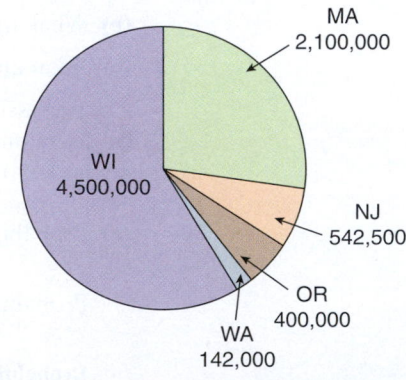

FIGURE 6–26 U.S. cranberry 2012 forecasted production by state in barrels. *Source: National Agricultural Statistics Service (NASS), Agricultural Statistics Board, U.S. Department of Agriculture (USDA), 2012.*

Use Fig. 6–27 to answer Exercises 8–11.

8. In what year(s) did men use about five sick days?

9. In what year(s) did women use more sick days than men?

10. What was the greatest number of sick days for men?

11. In what year(s) did men use more sick days than women?

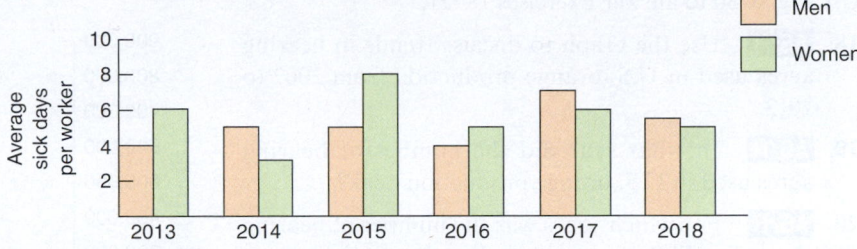

FIGURE 6–27 Comparison of sick days for men and women.

Use Fig. 6–28 to answer Exercises 12–15.

12. **AG/H** In what year was utilized production of U.S. oranges the greatest?

13. **AG/H** What was the percent increase in orange production from 2003 to 2004? Round to the nearest whole percent.

14. **AG/H** What was the percent of decrease in orange production from 2006 to 2007? Round to the nearest whole percent.

15. **AG/H** In what year were 11,545,000 tons of oranges produced in the United States?

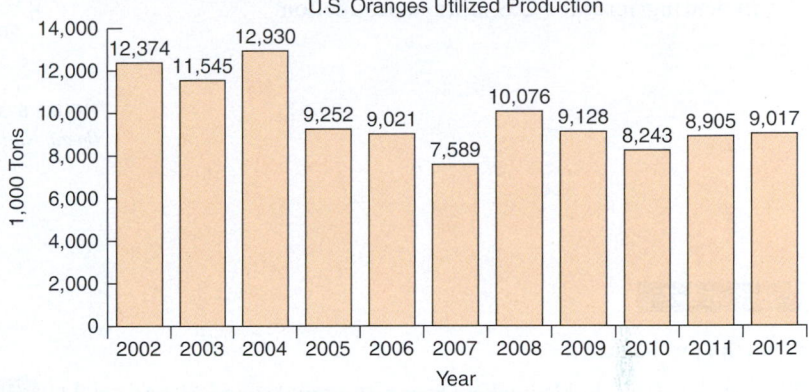

FIGURE 6–28 U.S. oranges: Bearing acres. *Source: Agricultural Statistics Board, National Agricultural Statistics Service (NASS), U.S. Department of Agriculture (USDA), 2012.*

Use Fig. 6–29 to answer Exercises 16–17.

16. **HLTH/N** What was the patient's highest pulse rate? The highest respiration rate?

17. **HLTH/N** On what date and at what time were the patient's pulse rate and respiration rate at their highest level?

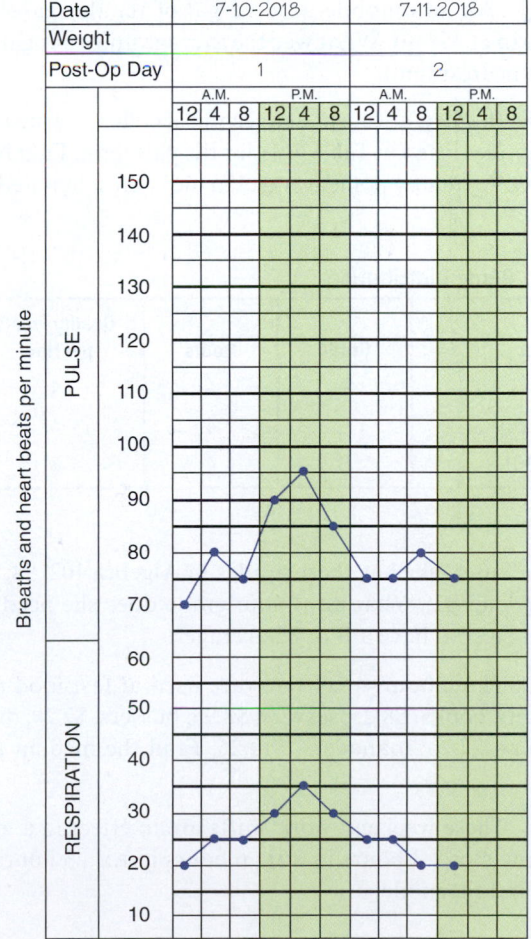

FIGURE 6–29 Graphic respiration/pulse chart.

Use Fig. 6–30 to answer Exercises 18–21.

18. **AG/H**   Use the graph to discuss trends in bearing acres used in U.S. orange production from 2002 to 2012.

19. **AG/H**   In what year did the number of bearing acres used in U.S. orange production peak?

20. **AG/H**   For which years was the number of bearing acres for U.S. orange production greater than 800,000 acres?

21. **AG/H**   Identify the year that had a slight increase in bearing acres for U.S. orange production.

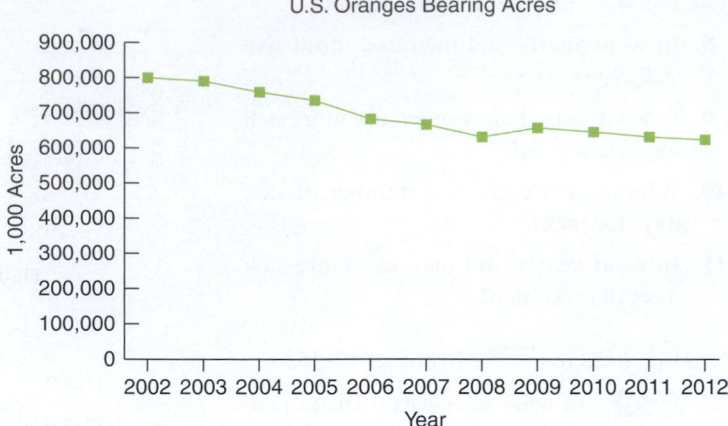

FIGURE 6–30 U.S. oranges: Bearing acres. *Source: Agricultural Statistics Board, National Agricultural Statistics Service (NASS), U.S. Department of Agriculture (USDA), 2012.*

Section 6–2

Solve the problems.

22. Jim Smith made 176 baskets over a 16-game period. What was the average number of baskets per game to the nearest tenth?

23. **BUS**   Lee Vance sold 63 new cars over a 5-month period. What was the average number of cars sold per month to the nearest tenth?

24. **AUTO**   An automobile used 22 gal of regular gasoline on a trip of 358 mi. What was the average miles per gallon to the nearest tenth?

25. **AUTO**   A delivery truck used 21 gal of regular gasoline on a trip of 289 mi. What was the average miles per gallon to the nearest tenth?

26. Natalie Bradley, a student at a technical college, earned the final grades listed in Table 6–11 for the past term. Find Natalie's QPA (quality point average) to the nearest hundredth.

27. Ezell Allen earned the final grades listed in Table 6–12 for the past term. Find Ezell's QPA (quality point average) to the nearest hundredth.

Table 6–11   Grade Distribution

Subject	Grade	Hours	Quality Points per Hour
Electronics 101	A	4	4
Circuits 201	A	4	4
Algebra 101	B	4	3

Table 6–12   Grade Distribution

Subject	Grade	Hours	Quality Points per Hour
English 201	A	3	4
History 202	B	3	3
Philosophy 101	A	3	4

28. Janice Van Dyke has these grades in Algebra 102: 98, 82, 87, 72, and 82. What minimum grade does she need on the last test to have a B or 85 average?

29. Sarah Smith has the following grades in American History: 99, 93, 91, 88, and 86. What grade does she need on the last test to get an A or 90 average?

30. **BUS**   These hourly pay rates are used at fast-food restaurants: cooks, $8.25; servers, $8.95; bussers, $7.20; dishwashers, $7.20; managers, $10.25. Find the median pay rate.

31. **BUS**   These hourly pay rates are used at a locally owned store: office assistants, $9.85; bookkeepers, $10.20; cashiers, $9.45; assistant managers, $11.90. Find the median pay rate.

32. **BUS**   These weekend work shifts are in effect in a mall clothing store: 3 hours in A.M., 6 hours in P.M., 3 hours in P.M. Find the mode.

33. **BUS**   These special prices are in effect at a fast-food restaurant: $1.85, hamburgers; $1.98, hot ham sandwiches; $2.28, chicken sandwiches; $1.85, roast beef sandwiches. Find the mode.

A regional horticultural association is composed of 54 members of several clubs. Use the format shown in Table 6–13 and make a frequency distribution of the members' ages: 17, 18, 20, 21, 21, 24, 24, 29, 29, 29, 31, 31, 33, 33, 34, 35, 35, 38, 38, 38, 39, 41, 42, 43, 43, 43, 43, 45, 45, 47, 47, 48, 48, 48, 49, 50, 51, 51, 52, 56, 56, 58, 58, 60, 60, 62, 64, 64, 65, 66, 68, 70, 71, 71.

Table 6–13 Ages of Club Members of Regional Horticultural Association

	Class Interval	Midpoint	Tally	Class Frequency
34.	66–75			
35.	56–65			
36.	46–55			
37.	36–45			
38.	26–35			
39.	16–25			

Answer Exercises 40–47 based on the information in Table 6–13.

40. In what age group is the least number of members?

41. In what age group is the greatest number of members?

42. How many members are under age 36?

43. How many members are over age 55?

44. What is the ratio of the number of members aged 66–75 to members aged 16–25?

45. What is the ratio of the number of members aged 66–75 to the number of members aged 46–55?

46. What is the ratio of the number of members aged 46–55 to the members aged 16–25?

47. What percent (to the nearest tenth of a percent) of the total number of members are the members aged 46–55?

48. Use seven class intervals beginning with 66–70 to make a frequency distribution of these scores on an English language test: 68, 70, 74, 77, 78, 82, 82, 84, 86, 86, 86, 88, 89, 90, 90, 93.

49. Use three class intervals beginning with 20–24 as the first interval to make a frequency distribution of these miles per gallon reported in one week by customers to an automotive rental company for six-cylinder cars: 22, 22, 23, 23, 23, 24, 24, 25, 26, 26, 27, 29, 29, 30, 30, 31, 31.

50. Find the grouped mean for the data in Table 6–13. Round to tenths.

51. Find the grouped mean for the data in Exercise 49. Round to tenths.

Section 6-3

52. Find the range of the test scores: 67, 87, 76, 89, 70, 69, 82. Round to hundredths.

53. Find the range for Marcus Johnson's test scores of 92, 83, 39, 98, 88, 90.

54. Find the midrange for the test scores in Exercise 52.

55. Find the midrange for the test scores in Exercise 53.

56. Make a relative frequency distribution, a cumulative frequency distribution, and a cumulative relative frequency distribution for the data in Exercise 48.

57. Use the data from Table 6–13 on this page to make a relative frequency distribution, a cumulative frequency distribution, and a cumulative relative frequency distribution.

58. Find the quartiles for the set of 16 scores in Exercise 48.

59. Find the quartiles for the set of 17 scores in Exercise 49.

60. Find the inner-quartile range for the scores given in Exercise 48.

61. Find the inner-quartile range for the scores given in Exercise 49.

62. Use the set of scores in Exercise 48 to find the 35th percentile.

63. Use the set of scores in Exercise 49 to find the 80th percentile.

64. Find the standard deviation of the test scores in Exercise 52. Round to hundredths.

65. Find the standard deviation of the test scores in Exercise 53. Round to hundredths.

66. **HLTH/N** The heights of 25-year-old men are normally distributed with a mean of 1.72 m and a standard deviation of 0.27 m. What percent of men have heights between 1.45 m and 2.26 m?

67. **HLTH/N** In a sample of 1,000 25-year-old men, how many will be taller than 1.45 m? The population mean is 1.72 m and standard deviation is 0.27 m.

68. **HLTH/N** Juice sold in 12-oz containers that have amounts that are normally distributed. The contents have a mean of 11.9 oz and a standard deviation of 0.2 oz. What percent of containers hold 11.7 oz to 12.1 oz?

69. **HLTH/N** Refer to Exercise 68 to determine how many containers in a batch of 3,000 containers of juice are filled to 12.1 oz or more.

70. For her first professional job, Martha Deskin, a recent college graduate, purchased four pairs of shoes, three business suits, and five blouses. How many outfits are possible?

71. HLTH/N There are three magazines on horses and ten magazines on fashion in Dr. Nelson Company's waiting room. If Shirley Riddle sends her toddler to get two magazines, what is the probability that he will randomly pick a fashion magazine on the first pick? If he is successful, what is the probability of his picking a fashion magazine on the second pick?

72. Two coins are tossed three times. Count the number of possible outcomes of heads and tails with three tosses of the two coins.

73. If four coins are tossed, what is the number of possible outcomes of heads and tails? *Hint:* HHHT and HTHH are different outcomes.

74. Harper has five blouses and three skirts. How many outfits of one blouse and one skirt are possible from these selections?

75. A digital lock allows the owner to select four-digit codes. How many codes are available if no digit can be repeated? How many codes are available if digits can be repeated?

76. A lottery jackpot can be won by selecting the six winning numbers. The first five numbers represent the five white balls from a total of 50 white balls numbered from 1 through 50. The sixth number represents the one red ball from a total of 20 red balls numbered from 1 through 20. How many different number selections are possible?

77. A single die is rolled. What is the probability that an odd number will appear on the top?

78. Cy Pipkin is puzzled over a true–false question on a test and does not know the answer. What is the probability that he will pick the right answer by chance?

79. A box is filled with 32 quarters, 28 dimes, and 156 pennies. What is probability of drawing a nickel from the box?

80. What is the probability of drawing a dime from the box in Exercise 79?

81. What are the odds for drawing a dime from the box in Exercise 79?

82. What are the odds against drawing a penny from the box in Exercise 79?

6 | TEAM PROBLEM-SOLVING EXERCISES

1. Statistical reports are an important part of a statistical study. Conduct a study on a topic of interest to the team and prepare a statistical report that includes the following components:
 (a) Question or questions that you are attempting to answer
 (b) Description of the process for collecting data
 (c) Summary of the data with appropriate statistics
 (d) Conclusion

2. Most statistical reports include graphical representations of the data collected. Computer software such as Excel™ allows you to prepare graphs or charts from the data entered into a spreadsheet.
 (a) Prepare spreadsheets with appropriate formulas to make the necessary calculations for the data in Exercise 1.
 (b) Prepare charts to illustrate the data collected in Exercise 1.

6 | CONCEPTS ANALYSIS

1. What type of information does a circle graph show?

2. Describe a situation where it would be appropriate to organize the data in a circle graph.

3. What type of information does a bar graph show?

4. Describe a situation where it would be appropriate to organize the data in a bar graph.

5. What type of information does a line graph show?

6. Describe a situation where it would be appropriate to organize the data in a line graph.

7. Explain why the grouped mean may not be the same as the mean for the data set.

8. Explain the differences among the three types of averages (measures of central tendency): the mean, the median, and the mode.

9. Use an example to show how using a tree diagram for counting gives the same result as multiplying the number of choices for each position in the set.

10. If there are five red marbles and seven blue marbles in a bag, is the probability of drawing a red marble $\frac{5}{7}$? Explain your answer.

6 | PRACTICE TEST

1. A _____ graph shows how different data values relate to each other.

2. A _____ graph shows how a whole quantity is related to its parts.

3. A _____ graph shows how an item or items change over time.

Answer Questions 4 and 5 from the line graph in Fig. 6–31.

4. How many degrees warmer was it indoors at midnight than outdoors?

5. What was the change in outdoor temperature between 11:00 A.M. and noon?

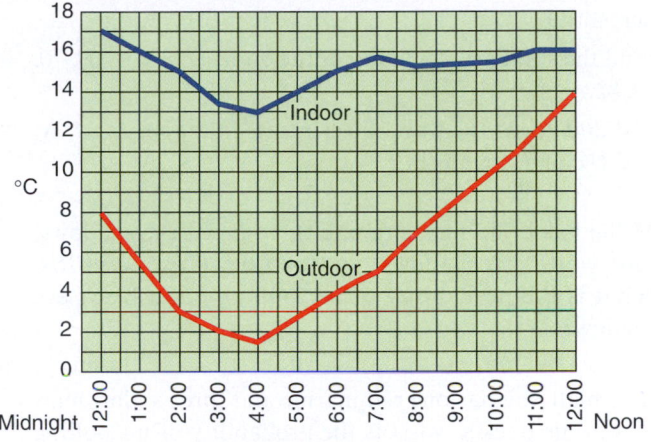

FIGURE 6–31 Indoor-outdoor temperature.

Answer the questions about the manufacturing costs for the electronic game as shown in the circle graph in Fig. 6–32.

6. What percent of the total cost is materials?

7. What would the profit be if there were no tariff on imported parts?

8. How much are overhead and materials together?

9. What percent of the total cost is the profit?

10. What is the ratio of labor to total cost?

11. What is the ratio of the tariff to total cost?

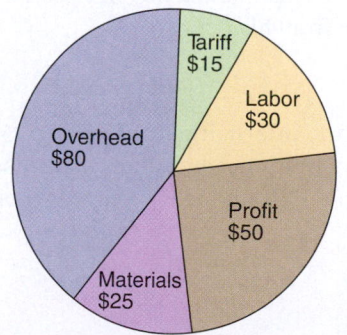

FIGURE 6–32 Manufacturing costs of a $200 electronic game.

Use the bar graph in Fig. 6–33 to answer Questions 12–15 about the academic-year starting salaries of women and men college professors in various academic departments of a college.

12. In what department(s) do men make more than women?

13. In what departments do women make more than men?

14. What percent of women's salaries are men's salaries in the English Department (to the nearest whole percent)?

15. What percent of men's salaries are women's salaries in the Electronics Department (to the nearest whole percent)?

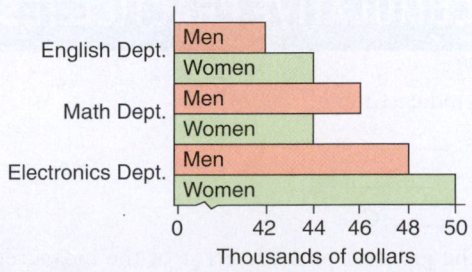

FIGURE 6–33 Salaries of women and men college professors.

Use the frequency distribution shown in Table 6–14 to answer Questions 16–18. The distribution shows the number of correct answers on a 30-question test in a science class.

16. How many students scored more than 25 correct?

17. What is the ratio of students scoring 6–10 correct to those scoring 21–25 correct?

18. Find the grouped mean for the data in Table 6–14.

19. Find the range, mean, median, and mode for these test scores: 81, 78, 69, 75, 81, 93, 68. Round to tenths.

20. Find the midrange for the data in Exercise 19.

21. Use the data in Table 6–14 to make a relative frequency distribution, a cumulative frequency distribution, and a cumulative relative frequency distribution.

Table 6–14   Frequency Distribution of Correct Answers

Class Interval	Midpoint	Tally	Class Frequency
26–30	28	///	3
21–25	23	////	4
16–20	18	LHH //	7
11–15	13	///	3
6–10	8	//	2
1–5	3	/	1

22. Find the quartiles for the set of test scores: 72, 86, 99, 82, 78, 62, 79, 92, 67, 83, 88, 75, 76, 80, 88, 90, 91, 96, 46, 87, 58.

23. Find the inner-quartile range for the data set in Exercise 22.

24. Use the set of scores in Exercise 22 to find the 95th percentile.

25. Use the set of scores in Exercise 22 to find the 80th percentile.

26. Find the standard deviation of the data: 77, 87, 77, 89, 70, 69, 82.

27. Find the standard deviation of the data in Exercise 19. Round to hundredths.

28. List and count the ways the elements in the set L, M, N, and O can be arranged.

29. Rayford has two ties: a red tie and a blue tie. He also has three shirts: a white shirt, a green shirt, and a yellow shirt. How many combinations of shirts and ties are possible?

30. Millie has in her makeup bag three green eye shadows, four white eye shadows, and two black eye shadows. What is the probability of her pulling out a black eye shadow?

31. An envelope contains the names of two men and three women to be interviewed for a promotion at Washington's Landscape Service. The interviewer draws names to determine the order of the interviews. What is the probability of drawing a woman's name first?

32. If a small boy has one red marble and three yellow marbles in his pocket, what is the probability of his pulling out the red one on the first try?

33. What are the odds for the boy in Exercise 32 pulling a yellow marble from his pocket?

34. What are the odds against the boy in Exercise 32 pulling a yellow marble from his pocket?

35. **HLTH/N**   The average hospital cost of an angioplasty/stint surgery for heart attack is $57,417 with a standard deviation of $3,584. What percent of hospital stays cost between $53,833 and $57,417 if the data are normally distributed?

36. **HLTH/N**   Use the information in Exercise 35 to find in 562 angioplasty/stint procedures how many cost between $57,417 and $61,001. Round to the nearest whole number.

4–6   CUMULATIVE PRACTICE TEST

Change to the indicated unit.

1. 14.3 dm = _____ m

2. 12 m = _____ cm

3. 30,002 dg = _____ kg

4. 0.159 dkm = _____ dm

5. 86°F = _____ °C

6. 45°C = _____ °F

7. What is the greatest possible error of the measurement $3\frac{1}{4}$ cm?

8. How many significant digits are in 40.240?

9. How many significant digits are in 0.00356?

10. A scale calibrated to 10 grams actually reads 10.05 grams. What is the percent error?

11. Add and write the sum using appropriate precision: 5.7 cm + 3.05 cm + 21.46 cm

Perform the indicated operations and simplify.

12. $5 + (-2) + (-8)$

13. $-3 - 8 - 5 - (-2)$

14. $-8(2)(-4)$

15. $-24 \div 4$

16. $(-2)^3 - 5 + (-6) - 2^2$

17. $-5\frac{1}{2} + 4\frac{1}{4}$

18. Write 58,000 in scientific notation.

19. Write 5.3×10^{-2} as an ordinary number.

20. Write 4.7×10^4 as an ordinary number.

21. Write 0.0035 in scientific notation.

22. Write 42,300,000 in engineering notation.

23. Write 0.0001457 in engineering notation.

Simplify and write the result in scientific notation.

24. $(5.1 \times 10^2)(4.1 \times 10^3)$

25. $\dfrac{4.6 \times 10^4}{2.3 \times 10^{-2}}$

26. $(4 \times 10^2)^3$

Use Figure 6–34 to answer Questions 27–29.

27. How many cwts (100 lbs) of long grain rice are produced by Arkansas?

28. Which state produces the most medium grain rice?

29. Which two states produce only long grain rice?

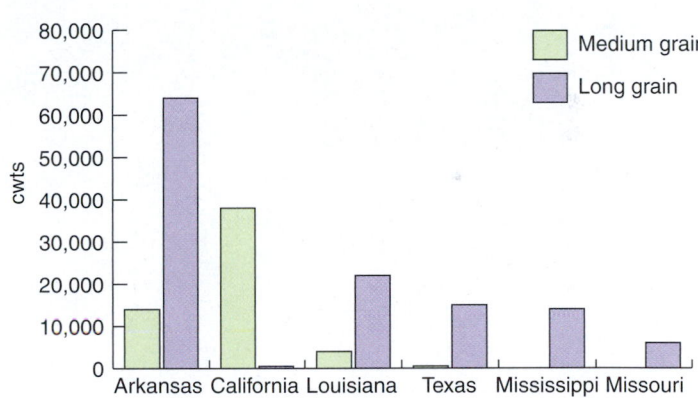

FIGURE 6–34 Top Six Rice Producing States in United States.
Source: http://www.sagevfoods.com/MainPages/Rice101/Production.htm

Use Figure 6–35 to answer Questions 30–32.

30. If 560 million metric tons of rice were grown worldwide, how many metric tons were produced by India?

31. Which country produces more rice than any other?

32. How many metric tons of the world rice production are produced by the USA?

World Rice Production

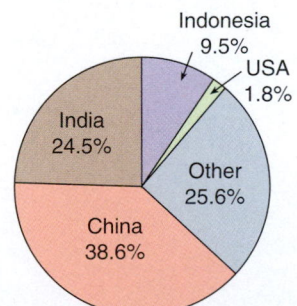

FIGURE 6–35 *Source: http://www.sagevfoods.com/MainPages/Rice101/Production.htm*

The data represent the number of miles driven each day for a cross country moving van. 583, 417, 456, 498, 523, 456.

33. Find the mean, median, and mode to the nearest whole number for the data set.

34. Find the range and standard deviation for the data set.

35. A container has 30 assorted coins. Seventeen are quarters, 8 are nickels, 3 are pennies, and 2 are dimes. What is the probability that one coin selected from the container is a nickel if each coin is equally likely to be selected?

36. Find the quartiles for the glucose readings for a patient:

89, 107, 103, 121, 95, 88, 82, 124, 114, 96, 86, 94

37. Find the inner-quartile range for the data in Exercise 36.

39. Forty of the 62 troops in Company B are being deployed. If each has an equally likely chance of being deployed, what are the odds for Lorinda Brown, who is a member of Company B, being deployed?

40. Complete the data table to make a relative frequency distribution, a cumulative frequency distribution, and a cumulative relative frequency distribution.

38. Find the 80th percentile for the scores in Exercise 36.

Class Interval	Class Frequency	Relative Frequency	Cumulative Frequency	Cumulative Relative Frequency
1–20	8			
21–40	10			
41–60	5			
61–80	35			
81–100	22			

Linear Equations and Inequalities

Alexfan32/Shutterstock

In Great Company

Copying Answers Is a Bad Idea (1996)

June 4, 1996, was a beautiful spring day in Kourou, French Guiana, on the northeast coast of South America. It was a beautiful day to launch a rocket. Which was a good thing, because the European Space Agency (ESA) intended to launch its newest unmanned rocket, the *Ariane 5*.

But the *Ariane 5* wasn't just going through a test launch. No, after $7 billion in development and testing costs, the *Ariane 5*, the successor to the ESA's wildly successful *Ariane 4* rocket, was going on its first live mission. *Ariane 5* carried four ESA spacecraft designed to study the Earth's magnetic fields over the course of an entire solar cycle. This mission and these spacecraft, called the Cluster II, were first proposed in 1982; they were approved in 1986, and they took ten years to build. Each of the four spacecraft cost $100 million. The rocket was itself worth about $100 million.

And 37 s after launch, the entire thing—ten years' worth of work, roughly half a billion dollars' worth of rocket and satellite—was blown apart by the missile's automatic self-destruct system. Imagine that. Ten years of your life's work. Poof. Gone.

What happened?

Well, the *Ariane 4* software had been so darned good, ESA ported huge chunks of the code over to *Ariane 5* without looking at it very closely. Sadly, nobody thought about the fact that the *Ariane 5* was designed for entirely different launch loads and flight stresses. That's the point of building a new rocket, right? It is meant to be bigger and better than the old one. Right.

So, after liftoff, the guidance computer ran into sideways rocket velocities much higher than previously encountered. The software was only designed to handle a 16-bit number—it was getting a 64-bit number. The number was too big, and an overflow error resulted. The guidance system cursed softly and shut itself down.

The good news: Before the guidance system shut itself down, it passed control to the redundant backup unit. The bad news: The backup unit had an identical design. It also failed immediately because it was running the same software.

Adding insult to injury, in the postmortem of the disaster, an internal memo was discovered. One of the software engineers had recommended that the code be rewritten to be a little more robust. Big sigh.

That, ladies and gentlemen, is how you burn through half a billion dollars in under 60 seconds.

7–1   Variable Notation

LEARNING OUTCOMES

1 Identify equations, terms, factors, constants, variables, and coefficients.

2 Write verbal interpretations of symbolic statements.

3 Translate verbal statements into symbolic statements using variables.

4 Simplify variable expressions.

LC LEARNING CATALYTICS
Apply the distributive property.
1. $3(5 + 7 - 3)$
2. $-4(2 - 5 + 1)$

Equation: a symbolic statement that two expressions or quantities are equal in value

STOP AND CHECK
Are the statements true or false equations?
1. $-5(-3) = -15$
2. $15 - 3 = 2(6)$

Answers:
1. False 2. True

1 Identify Equations, Terms, Factors, Constants, Variables, and Coefficients.
Before we work with equations to solve problems, we need to understand the basic concepts and terminology of equations. In mathematics, we use the symbol "=" to show that quantities are "equal to" each other. We write the statement "5 is equal to 2 plus 3" as $5 = 2 + 3$. This symbolic statement is called an *equation*.

An **equation** is a symbolic statement that two expressions or quantities are equal in value. An equation may be true or false.

To verify that an equation is true:

1. Find the value of the expression on each side of the equal sign.

2. Compare the values for each side.

   (a) If the values for each side are equal, the equation is true.

   (b) If the values for each side are not equal, the equation is not true.

EXAMPLE 1

Verify that the statements are true equations.

(a) $3(8) = 24$ (b) $12 - 3 = 2 + 7$

(a) $3(8) = 24$ 3 times 8 is 24.
      $24 = 24$ The equation is true.

(b) $12 - 3 = 2 + 7$ 12 minus 3 is 9. 2 plus 7 is 9.
         $9 = 9$ The equation is true.   See Exercises 1–6.

Solve an equation: to find the missing number that makes the equation *true*

Variable: a letter that represents an unknown or missing number in an equation

Some types of equations contain a *missing* or *unknown* number. We **solve an equation** by finding the missing number that makes the equation *true*. We use letters such as *x*, *a*, or *z* to represent the missing number in the equation. These letters are called **variables**. In each equation, the letter has a certain but unknown value that depends on the other numbers and relationships in the equation.

EXAMPLE 2

Find the value of the variable that makes the equation true.

(a) $x = 3 + 8$ (b) $\dfrac{18}{3} = n$ (c) $y = 3(4) - 5$

(a) $x = 3 + 8$ 8 added to 3 is 11.
$\quad x = 11$

(b) $\dfrac{18}{3} = n$ 18 divided by 3 is 6.
$\quad 6 = n$

(c) $y = 3(4) - 5$ Multiply first. 3 times 4 is 12.
$\quad y = 12 - 5$ Add 12 and -5. 12 plus $-5 = 7$.
$\quad y = 7$

See Exercises 7–12.

Because both sides of an equation are equivalent, either side can come first in the equation. The equation, $x = 3 + 8$ means the same as $3 + 8 = x$. Also $n = 6$ is equivalent to $6 = n$. The unknown, or variable, may appear on the left or the right side of an equation. This is an application of the **symmetric property of equality**.

Symmetric property of equality: a property of equality that states if the sides of an equation are interchanged, equality is maintained

Symmetric property of equality:

If the sides of an equation are interchanged, equality is maintained.

Symbolically, if $a = b$, then $b = a$, where *a* and *b* are any real numbers.

Terms: algebraic expressions that are single quantities or products or quotients of quantities.

In algebra we carefully distinguish between terms and factors. *Factors* are numbers or variables that are multiplied. **Terms** are algebraic expressions that are single quantities or products or quotients of quantities.

A term can be a single letter; a single number; the product of numbers and letters; the product of numbers, letters, and/or groupings; or the quotient of numbers, letters, and/or groupings. In a division, the fraction line implies that the numerator and/or the denominator is grouped.

TIP **Multiplication Notation Conventions** When a term is the product of letters alone (such as *ab* or *xyz*) or the product of a number and one or more letters (such as $3x$ or $2ab$), parentheses or other symbols of multiplication are usually omitted. Thus, *ab* means *a* times *b*, and $3x$ means 3 times *x*.

To identify terms:

1. Separate an expression into terms by identifying addition or subtraction signs that are not within a grouping.

2. The sign of each term is the sign that precedes the term.

$$3a + b \qquad 3(a + b) \qquad -\dfrac{3}{a + b}$$

two terms one term one term

TIP **Signs that separate expressions into terms serve a dual role.** They are both operational signs (addition or subtraction) and directional signs (positive or negative). The directional sign goes with the term that follows.

Constant or number term: a term that contains only numbers

Variable term or letter term: a term that contains only one letter, several letters used as factors, or a combination of letters and numbers used as factors

Coefficient: each factor of a term containing more than one factor is the coefficient of the remaining factor(s). The numerical factor, also called the numerical coefficient, is often implied when the term *coefficient* is used

Numerical coefficient: the numerical factor of a term

EXAMPLE 3

Identify the terms in each expression by highlighting each term.

(a) $3x + 5$ (b) $2ab + 4a - b + 2(a + b)$

(c) $\dfrac{2a + 1}{3}$ (d) $\dfrac{2a}{3} + 1$

(a) $3x + 5$

Terms are separated by "+" and "−" signs that are not within a grouping. The expression has two terms.

(b) $2ab + 4a - b + 2(a + b)$

The plus sign within the grouping does not separate terms. The expression has four terms.

(c) $\dfrac{2a + 1}{3}$

The fraction bar is a grouping symbol. The numerator is a grouping, $2a + 1$. The expression has one term.

(d) $\dfrac{2a}{3} + 1$

The expression has two terms.

See Exercises 13–21.

A term that contains only numbers is a **constant** or **number term.** A term that contains only one letter, several letters used as factors, or a combination of letters and numbers used as factors is a **variable term** or **letter term.**

When a term has more than one factor, each factor is the **coefficient** of the remaining factor or factors. The numerical factor of a term is the **numerical coefficient.** Unless otherwise specified, we will use *coefficient* to mean the *numerical coefficient.* Generally, the numerical factor is written *in front* of the variable.

To identify the numerical coefficient of a term:

1. Since a term contains only factors, the coefficient is the numerical factor or the product of all numerical factors.

2. If the term is a fraction (an indicated division),

   (a) Write the numerical factor in the denominator as an equivalent multiplication.

   (b) Write the product of all numerical factors.

   (c) The numerical coefficient is the product from Step 2b.

EXAMPLE 4

Identify the numerical coefficient in each term.

(a) $2x$ (b) $-3ab$ (c) $4(x + 3)$ (d) $-\dfrac{n}{3}$ (e) $\dfrac{2b}{5}$

(a) The coefficient of $2x$ is **2.**

(b) The coefficient of $-3ab$ is **−3.**

(c) The coefficient of $4(x + 3)$ is **4.**

(d) $-\dfrac{n}{3}$ is the same as $-\dfrac{1}{3}n$, so the coefficient is $-\dfrac{1}{3}$.

(e) $\dfrac{2b}{5} = \dfrac{1}{5}(2b) = \dfrac{1}{5}(2)b = \dfrac{2}{5}b$. The coefficient is $\dfrac{2}{5}$.

See Exercises 22–30.

Did You Know? **Coefficients of 1 or −1 are not always written.** When a variable term has no written numerical coefficient, the coefficient is understood to be 1, so $x = 1x$. Similarly, the numerical coefficient of $-x$ is −1, so $-x = -1x$.

2 **Write Verbal Interpretations of Symbolic Statements.** In symbolic statements we commonly use letters to represent missing values. That is, we write $5 + x = 8$ or $5 + y = 8$ or $5 + a = 8$. The choice of letters is not usually important. The position of the letter and the other conditions of the statement are important. We can phrase a mathematical statement several ways: $5 + x = 8$ can be stated

"What number added to 5 gives 8?"

or

"5 plus a number has a result of 8. What is the missing number?"

To write verbal interpretations of symbolic statements:

1. Locate the variable or variable terms.

2. Examine the operations that link the factors and terms.

3. Write a verbal statement or statements that describe all the conditions of the symbolic statement.

STOP AND CHECK

State the equations in words.
1. $3x = 24$
2. $4x - 7 = 13$
3. $4(x - 7) = 12$

Answers: Answers may vary.
1. When a number is multiplied by 3, the result is 24.
2. If 7 is subtracted from 4 times a number, the result is 13.
3. If 7 is subtracted from a number and the difference is multiplied by 4, the result is 12.

EXAMPLE 5

State the equations in words.

(a) $x - 7 = 4$ **(b)** $\dfrac{x}{5} = 3$ **(c)** $2x + 3 = 15$ **(d)** $2(x + 3) = 14$

Each equation can be stated several ways. One choice is given for each equation.

(a) **When 7 is subtracted from a number, the result is 4.**

(b) **A number divided by 5 is 3.**

(c) **15 is the result when 3 is added to 2 times a number.**

(d) **If the sum of a number and 3 is doubled, the result is 14.** **See Exercises 31–34.**

3 **Translate Verbal Statements into Symbolic Statements Using Variables.** Symbolic representations of written statements or real-life situations allow us to determine the value of missing amounts more systematically.

To translate verbal statements into symbolic statements using variables:

1. Assign a letter to represent the missing number.

2. Identify key words or phrases that imply or suggest specific operations.

3. Translate words into symbols.

EXAMPLE 6

Translate the statements into symbols.

(a) The sum of 12, 23, and a third number is 52.

The third number is missing.

Let x represent the third number.

$$12 + 23 + x = 52$$ Translate the entire statement. *Sum* indicates addition. *Is* translates to *equals*.

(b) When a number is subtracted from 45, the result is 17.

The number being subtracted (the subtrahend or second number) is missing.

Let x represent the missing number.

$$45 - x = 17$$ 45 is the minuend or first number, and *the result* translates to *equals*.   **See Exercises 35–40.**

Key words can be examined to identify operations.

Addition:	the sum of, plus, increased by, more than, added to, exceeds, longer, total, heavier, older, wider, taller, gain, greater than, more, expands
Subtraction:	less than, decreased by, subtracted from, the difference between, diminished by, take away, reduced by, less, minus, shrinks, younger, lower, shorter, narrower, slower, loss
Multiplication:	times, multiply, of, the product of, multiplied by
Division:	divide, divided by, divided into, how big is each part, how many parts can be made from

Some words that imply multiplication or division may indicate a specific number in the multiplication or division. Examples include twice (2 times), double (2 times), triple (3 times), and half of (1/2 times or divided by 2).

EXAMPLE 7

Translate the statement into symbols.

How many shelves that are each 3 feet long can be made from a board that is 12 feet long?

The missing number is the number of shelves that can be made from one 12-ft board. Let x represent the number of shelves. The number of shelves, x, times the length of each shelf, 3, equals 12.

$$3x = 12$$ **See Exercises 41–44.**

EXAMPLE 8

Explain the difference between the two statements.

4 times the difference between a number and 8 is 24.

The difference between 4 times a number and 8 is 24.

4 times the difference between a number and 8 is 24. This statement is written symbolically as $4(x - 8) = 24$. It shows that the difference is taken first and the result is multiplied by 4.

The difference between 4 times a number and 8 is 24. This statement is written symbolically as $4x - 8 = 24$. A number is multiplied by 4 and then 8 is subtracted from the result.   **See Exercises 45–46.**

Like terms: terms are like terms if they are numbers or if they are variable terms with exactly the same letter factors (including exponents)

4 **Simplify Variable Expressions.** Terms are **like terms** if they are numbers or if they are variable terms with exactly the same letter factors.

The terms $4y$ and $2y$ are like terms because both contain the same letter (y). Similarly, 3 and 1 are like terms because both are numbers. We *cannot* combine 3 and $4y$ because they are unlike terms (constant and variable term), and we *cannot* combine $4y$ and $2x$ because they are also unlike terms (different variables).

To simplify variable expressions:

1. Change subtraction to addition if appropriate.

2. Add the constants using the appropriate rule for adding signed numbers. The sum is a signed number.

3. Add the numerical coefficients of the like variables using the appropriate rule for adding signed numbers. The sum has the same variable or variables as the like terms being added.

EXAMPLE 9

Simplify the expressions by combining like terms.

(a) $5a + 2a - a$ **(b)** $3x + 5y + 8 - 2x + y - 12$

(a) $5a + 2a - a = \mathbf{6a}$ All terms are like terms. Add coefficients of a: $5 + 2 - 1 = 6$.

(b) $3x + 5y + 8 - 2x + y - 12$ Add like terms.
$\quad = \mathbf{x + 6y - 4}$ $3x - 2x = x$
                              $5y + y = 6y$
                              $8 - 12 = -4$ **See Exercises 47–52.**

The *distributive property* can be extended to include multiplying numbers and variables. When an expression has an instance of the distributive principle, first remove the grouping symbol by multiplying by the factor or factors in front of the grouping.

EXAMPLE 10

Simplify by applying the distributive property.

(a) $3(5x - 2)$ **(b)** $-7(x + y - 3z)$ **(c)** $-(-3x + 4)$

(a) $3(5x - 2) = 3(5x) - 3(2)$ Apply the distributive property
$\quad\quad\quad\quad = \mathbf{15x - 6}$ by multiplying $(5x - 2)$ by 3.

(b) $-7(x + y - 3z) = -7(x) - 7(y) - 7(-3z)$ Apply the distributive property
$\quad\quad\quad\quad\quad\quad = \mathbf{-7x - 7y + 21z}$ by multiplying each term of the grouping by -7.

(c) $-(-3x + 4) = -1(-3x) - 1(+4)$ Apply the distributive property
$\quad\quad\quad\quad = \mathbf{3x - 4}$ by multiplying each term of the grouping by -1.

                                          See Exercises 53–60.

7–1 EXERCISES   MyLab Math For additional practice go to your study plan in MyLab Math.

1 Verify that the statements are true. *See Example 1.*

1. $5(-3) = -15$

2. $17 - 6 = 11$

3. $12 - 5(2) = -7 + 9$

4. $8 - 3(5) = 2(-1 - 3) + 1$

5. $7(3 - 8) = 4(-8) - 3$

6. $4(-7) + 3(-2) = -5(8 - 2) - 4$

Find the value of the variable that makes the equation true. *See Example 2.*

7. $n = 4 + 7$

8. $m = 8 - 2$

9. $5 - 9 = y$

10. $\dfrac{12}{2} = x$

11. $p = 2(5) - 1$

12. $b = 6 - 3(5)$

Identify the terms in each expression by drawing a box around each term. *See Example 3.*

13. $7 + c$

14. $4a - 7$

15. $3x - 2(x + 3)$

16. $\dfrac{a}{3}$

17. $7xy + 3x - 4 + 2(x + y)$

18. $14x + 3$

19. $\dfrac{7}{(a + 5)}$

20. $\dfrac{4x}{7} + 5$

21. Write an algebraic expression that contains three terms.

Identify the numerical coefficient of each term. *See Example 4.*

22. $5x$

23. $-4xy$

24. $\dfrac{n}{5}$

25. $\dfrac{2a}{7}$

26. $6(x + y)$

27. $\dfrac{-4}{5x}$

28. $7(3x - 2y)$

29. $-(x - 3)$

30. Write an expression that has one term and a numerical coefficient of -15.

2 State the equations in words. *See Example 5.*

31. $x + 4 = 7$

32. $x - 5 = 2$

33. $3x = 15$

34. $3x + 1 = 7$

3 Write the statements in symbols. *See Example 6.*

35. 5 more than a number is 12.

36. A number divided by 6 is 9.

37. 4 times the difference of a number and 3 is 12.

38. 3 less than 4 times a number is 12.

39. The sum of 12, 7, and a third number is 17.

40. 7 more than twice a number is 21.

See Example 7.

41. **INTDR**   If the temperature rises 15°, it will be 48°. What is the temperature?

42. **HLTH/N**   If 15 mL of water are added to a medicine, there are 45 mL in all. What is the volume of the medicine?

43. **INDTEC**   How many 5-ft pieces of I-beam can be cut from a piece that is 45 ft long?

44. **CON**   A piece of oak flooring that is 18 ft long has an unacceptable flaw that requires 3 ft to be trimmed from one end. How many 5-ft boards can be cut from the remaining length of flooring?

Write the following statements in symbols. *See Example 8.*

45. Three times the sum of a number and 12 is 45.

46. The sum of three times a number and 12 is 45.

4 Simplify by combining like terms. *See Example 9.*

47. $3a - 7a + a$

48. $-8x + y - 3y$

49. $5x - 3y + 2x + y$

50. $-4a + b + 9 + a - b - 3$

51. $2x + 8 - x + 4 - x - 2$

52. $3a + 5b + 8c + 1 + b$

Simplify by applying the distributive property and combining like terms if appropriate. *See Example 10.*

53. $5(2x - 4)$ **54.** $4 + 3(2x + 3)$ **55.** $3 - (4x - 2)$

56. $5 - 2(6a + 1)$ **57.** $-5(x - y + 2z)$ **58.** $5 - (2a - 3b + 7)$

59. $8(x + 2y) - 3(7x - 3y + 5)$ **60.** $-3(2x + 5y - 6) - (5y - 2)$

7-2   Solving Linear Equations

LEARNING OUTCOMES

1 Solve linear equations using the addition axiom.

2 Solve linear equations using the multiplication axiom.

3 Solve linear equations with like terms on the same side of the equation.

4 Solve linear equations with like terms on opposite sides of the equation.

5 Solve linear equations that contain parentheses.

LC **LEARNING CATALYTICS**

Evaluate.

1. $\dfrac{5}{8} - \dfrac{2}{3}$ **2.** $\dfrac{-3.8}{0.19}$

Linear equation in one variable: an equation in which the same letter is used in all variable terms and the exponent of the variable is 1

Solution or root: the value of the variable that makes the equation true

Basic principle of equality: to preserve equality, if we perform an operation on one side of an equation, we must perform the *same* operation on the other side

Isolate: to perform operations so that the variable with a coefficient of 1 is by itself on one side of the equation

Addition axiom: the same quantity can be added to both sides of an equation without changing the equality of the two sides. Casually, we describe applying the addition axiom as sorting or transposing terms

A **linear equation in one variable** is an equation in which the same letter is used in all variable terms and the exponent of the variable is 1.

1   **Solve Linear Equations Using the Addition Axiom.** An equation is *solved* when the letter or variable is alone on one side of the equal sign. That is, the coefficient of the variable is +1 and no other terms are on the same side as the variable. The number on the side opposite the variable is called the **solution** or **root** of the equation. The solution is the number that makes the equation true.

The underlying principle for solving equations is the **basic principle of equality**.

> **Basic principle of equality:**
>
> To preserve equality, if an operation is performed on one side of an equation, the *same* operation must be performed on the other side.

We can illustrate this principle by visualizing a balanced scale. If we have a scale with 1 oz on one side and 1 oz on the other side, the scale is balanced. If we increase or decrease the weight on one side, we need to do the same on the other side or one side would be heavier than the other and the equality of both sides would be lost (see Fig. 7–1). If $\frac{1}{2}$ oz is taken from both sides, the scale is still balanced.

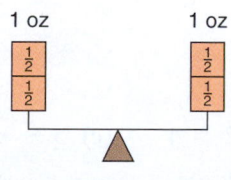
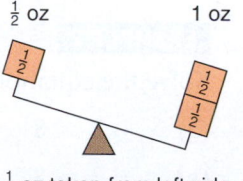
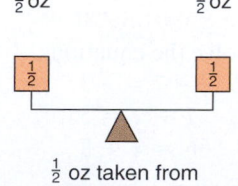

$\frac{1}{2}$ oz taken from left side $\frac{1}{2}$ oz taken from both sides

FIGURE 7–1

The purpose in solving an equation is to **isolate** the variable on one side of the equation. That is, the variable is by itself on one side of the equation. Terms can be moved from one side of an equation to the other using the addition axiom. The **addition axiom** states that the same quantity can be added to both sides of an equation without changing the equality of the two sides. Symbolically, if $a = b$, then $a + c = b + c$ for all real numbers a, b, and c.

Did You Know? Since subtraction is the same as adding the opposite, the addition axiom applies to both addition and subtraction.

To solve a linear equation using the addition axiom:

1. Locate the variable in the equation.

2. Identify the constant that is associated with the variable by addition (or subtraction).

3. Add the opposite of the constant to both sides of the equation.

STOP AND CHECK

1. Solve the equation
   $x + 7 = 3$.

Answer:
1. $x = -4$

EXAMPLE 1

Solve the equation $x + 4 = 8$.

$$x + 4 = 8$$ 4 is added to x. Add -4, the opposite of 4, to both sides (addition axiom).

$$x + 4 - 4 = 8 - 4$$ $4 - 4 = 0, 8 - 4 = 4$

$$x + 0 = 4$$ $x + 0 = x$

$$x = 4$$

See Exercises 1–8.

TIP **Adding Opposites and Zero** A number plus its opposite equals zero.

$$n + (-n) = 0, \text{ where } n \text{ is any real number}$$

Zero plus a number or variable leaves the number or variable unchanged.

$$n + 0 = n, \text{ where } n \text{ is any real number}$$

These two properties and the addition axiom allow us to *mentally* move a term to the other side of the equation as its opposite.

$x + 3 = 8$	becomes	$x = 8 - 3$	Mentally add -3 to both sides.
$x - 2 = 4$	becomes	$x = 4 + 2$	Mentally add $+2$ to both sides.
$5 + x = 2$	becomes	$x = 2 - 5$	Mentally add -5 to both sides.
$10 = x - 1$	becomes	$10 + 1 = x$	Mentally add $+1$ to both sides.

STOP AND CHECK

Solve the equations.
1. $x - 7 = -3$
2. $-2 + x = 5$
3. $-6 = x + 8$

Answers:
1. $x = 4$
2. $x = 7$
3. $x = -14$

EXAMPLE 2

Solve the equations.

(a) $x + 3 = 8$ **(b)** $x - 2 = 4$ **(c)** $5 + x = 2$ **(d)** $10 = x - 1$

(a) $x + 3 = 8$ To isolate x, mentally add -3 to both sides.

$$x = 8 - 3$$ Combine like terms.

$$x = 5$$

(b) $x - 2 = 4$ To isolate x, mentally add $+2$ to both sides.

$$x = 4 + 2$$ Combine like terms.

$$x = 6$$

(c) $5 + x = 2$ To isolate x, mentally add -5 to both sides.

$$x = 2 - 5$$ Combine like terms.

$$x = -3$$

(d)

$$10 = x - 1 \qquad \text{To isolate } x, \text{ mentally add } +1 \text{ to both sides.}$$

$$10 + 1 = x \qquad \text{Combine like terms.}$$

$$\mathbf{11 = x} \qquad \text{Apply the symmetric property of equality.}$$

or

$$x = 11 \qquad \qquad \text{See Exercises 9–16.}$$

> **TIP Use Simple Cases to Learn and Understand a Process or Procedure** Many equations can be solved mentally, or by using basic arithmetic and our knowledge of the relationships among the operations of addition, subtraction, multiplication, and division. In the example of $x + 3 = 8$, we can determine mentally that the missing number is 5. Our goal is to learn a procedure for solving simple equations, so that we can apply the same procedure to more complex equations.

Multiplication axiom: both sides of an equation may be multiplied by the same *nonzero* quantity without changing the equality of the two sides. This axiom also applies to dividing both sides of an equation by the same *nonzero* quantity

2 Solve Linear Equations Using the Multiplication Axiom. The basic principle of equality can be extended to multiplication and division. The **multiplication axiom** states that both sides of an equation may be multiplied or divided by the same *nonzero* quantity without changing the equality of the two sides. Symbolically, if $a = b$ and $c \neq 0$, then $ac = bc$. Similarly, if $a = b$ and $c \neq 0$, then $\dfrac{a}{c} = \dfrac{b}{c}$, for all real numbers a, b, and c.

To solve a linear equation using the multiplication axiom:

1. Multiply both sides of the equation by the reciprocal of the coefficient of the variable term; or

2. Divide both sides of the equation by the coefficient of the variable term.

STOP AND CHECK

Solve the equations.
1. $5x = -65$
2. $12x = 108$

Answers:
1. $x = -13$ 2. $x = 9$

EXAMPLE 3

Solve the equation $4x = 48$.

Multiplication and division are inverse operations, so dividing by a number is the same as multiplying by its multiplicative inverse—its reciprocal.

$$4x = 48 \qquad \text{Multiply by the}$$
$$\frac{1}{4}(4x) = \frac{1}{4}(48) \qquad \text{reciprocal of the coefficient of the}$$
$$x = 12 \qquad \text{variable term.}$$

$$4x = 48 \qquad \text{Divide by the}$$
$$\frac{4x}{4} = \frac{48}{4} \qquad \text{coefficient of the variable term.}$$
$$x = 12 \qquad \text{See Exercises 17–25.}$$

STOP AND CHECK

Solve the equations.
1. $28 = 0.4x$
2. $-14.4 = 1.2x$

Answers:
1. $x = 70$ 2. $x = -12$

EXAMPLE 4

Solve the equation $45 = 0.5x$.

$$\frac{45}{0.5} = \frac{0.5x}{0.5} \qquad \text{Divide both sides by the coefficient of the variable term.}$$

$$\mathbf{90 = x} \qquad \text{Apply the symmetric property of equality.}$$

or

$$x = 90 \qquad \qquad \text{See Exercises 26–28.}$$

> **TIP** **On Which Side Should the Variable Be?** Multiplication and division are effective regardless of which side of the equation contains the variable term. Once the equation is solved, however, many people prefer to put the variable on the left ($x = 90$ instead of $90 = x$). Either way is correct.

STOP AND CHECK
Solve the equations.

1. $\frac{1}{5}n = 3$

2. $\frac{2}{3}n = 6$

Answers:
1. $n = 15$ 2. $n = 9$

EXAMPLE 5

Solve the equation $\frac{1}{4}n = 7$.

Using multiplication: $\frac{1}{4}n = 7$ — Multiply both sides by the reciprocal of the coefficient of the variable term.

$$\left(\frac{4}{1}\right)\frac{1}{4}n = \left(\frac{4}{1}\right)7$$ — The reciprocal of $\frac{1}{4}$ is $\frac{4}{1}$.

$$n = 28$$

Using division: $\frac{1}{4}n = 7$ — Divide both sides by the coefficient of the variable term.

$$\frac{\frac{1}{4}n}{\frac{1}{4}} = \frac{7}{\frac{1}{4}}$$ $7 \div \frac{1}{4} = \frac{7}{1}\cdot\frac{4}{1} = 28$

$$n = 28$$

See Exercises 29–32.

> **Did You Know?** **You have a choice.** When the equation contains only whole numbers or decimals as coefficients or constant terms, it is generally more convenient to divide by the coefficient of the variable term than to multiply by the reciprocal of the coefficient of that variable.
>
> When the equation contains a fraction, it is usually preferable to multiply by the reciprocal of the coefficient of the variable.

An equation is *solved* when the coefficient of the variable term is $+1$. If the coefficient is -1, we apply the multiplication axiom to complete the solution.

STOP AND CHECK
Solve the equations.
1. $-x = 6$
2. $-15 = -x$
3. $17 = -x$

Answers:
1. $x = -6$ 2. $x = 15$
3. $x = -17$

EXAMPLE 6

Solve the equation $-n = 25$.

The coefficient of n is -1, so the equation is not solved. The coefficient of n must be $+1$ for the equation to be solved.

$$\frac{-1n}{-1} = \frac{25}{-1}$$ — Divide by the coefficient of the variable term.

$$n = -25$$

See Exercises 33–36.

When an equation has been solved, the *solution*, or *root*, can be checked to verify it is the solution.

> **To verify or check the solution of an equation:**
> 1. Substitute the solution in place of the variable in the original equation.
> 2. Perform all indicated operations on each side of the equation.
> 3. If the solution is correct, the value of the left side of the equation should equal the value of the right side.

STOP AND CHECK

1. In Example 4, the solution was found to be 90. Check this root.

Answer:

1. $45 = 0.5(90)$
   $45 = 45$

EXAMPLE 7

In Example 3, the solution was found to be 12. Check this root.

$4(x) = 48$	Substitute 12 for x.
$4(12) = 48$	Perform the multiplication.
$\mathbf{48 = 48}$	The root is verified if the final equation is true. **See Exercises 37–38.**

3 **Solve Linear Equations with Like Terms on the Same Side of the Equation.** We can combine only *like* terms and only if they are on the *same side* of the equation. That means constant terms can be added only to other constant terms. Variable terms can be added only to other like variable terms. By *combine* we mean add like terms on the same side of the equation according to the appropriate signed number rule.

> **To solve linear equations with like terms on the same side of the equation:**
>
> 1. Combine (add or subtract) like terms that are on the same side of the equal sign.
>
> 2. *Multiply* both sides of the equation by the reciprocal of the coefficient of the variable or *divide* both sides of the equation by the coefficient of the variable.

STOP AND CHECK

Solve the equations.

1. $9x - 4x = 70 + 15$
2. $6x - 10x = -24 + 4$

Answers:

1. $x = 17$ 2. $x = 5$

EXAMPLE 8

Solve the equation $4x - x = 7 + 8$.

$4x - x = 7 + 8$	Combine like terms on each side of the equal sign.
$3x = 15$	Divide both sides of the equation by 3, the coefficient of x.
$\dfrac{3x}{3} = \dfrac{15}{3}$	
$x = 5$	**See Exercises 39–47.**

4 **Solve Linear Equations with Like Terms on Opposite Sides of the Equation.** Like terms do not always appear on the same side of the equal sign. In such cases, the like terms must be manipulated so that they appear on the same side of the equal sign *before* they can be combined. The basic strategy for solving an equation is to *isolate* the variable terms on one side of the equation.

> **To solve linear equations with like terms on opposite sides of the equation:**
>
> 1. Identify the like terms and determine which term or terms should be moved to create an equation with like terms on each side.
>
> 2. Add to both sides of the equation the opposite of each term that is to be moved (addition axiom).
>
> 3. Combine like terms on each side of the equation.
>
> 4. Solve the resulting equation by using the multiplication axiom.

EXAMPLE 9

Solve the equations **(a)** $2x + 4 = 8$ and **(b)** $9 - 4x = 8x$.

(a) $2x + 4 = 8$ 4 and 8 are like terms on opposite sides of the equal sign. Mentally add -4, the opposite of 4, to both sides (addition axiom).

$$2x = 8 - 4$$ Combine like terms. $8 - 4 = 4$

$$2x = 4$$ Divide both sides by the coefficient of the variable term, 2 (multiplication axiom).

$$\frac{2x}{2} = \frac{4}{2}$$

$$x = 2$$

(b) Manipulate the terms in this equation so that both variable terms are isolated on one side of the equation and the number term is on the other side.

$$9 - 4x = 8x$$ $-4x$ and $8x$ are like terms on opposite sides of the equal sign. Mentally add $4x$, the opposite of $-4x$, to both sides (addition axiom).

$$9 = 8x + 4x$$ $-4x + 4x = 0$, $8x + 4x = 12x$.

$$9 = 12x$$ Divide both sides by the coefficient of x, which is 12 (multiplication axiom).

$$\frac{9}{12} = \frac{12x}{12}$$

$$\frac{9}{12} = x$$ Reduce to lowest terms.

$$\frac{3}{4} = x \quad \text{or} \quad x = \frac{3}{4}$$ Apply the symmetric property of equality if desired.

See Exercises 48–55.

Did You Know? **Sorting is not a mathematical term.** Manipulating terms by applying the addition axiom can be *described* as **sorting the terms** so that like terms are on the same side of the equation.

Sorting the terms: applying the addition axiom to move terms in an equation so that like terms are on the same side of the equation

There are circumstances in solving equations when the variable is eliminated. This presents two possible interpretations for the solution.

EXAMPLE 10

Solve the equations **(a)** $4x - 3 = 3x - 3 + x$ and **(b)** $3x - 7 = x + 2x - 11$.

(a) $4x - 3 = 3x - 3 + x$ Combine like terms. $3x + x = 4x$

 $4x - 3 = 4x - 3$ Sort terms (addition axiom). Add $-4x$ to both sides. Add $+3$ to both sides.

 $4x - 4x = -3 + 3$ Combine like terms. $4x - 4x = 0$; $-3 + 3 = 0$

 $0 = 0$ Since $0 = 0$ is a *true* statement, the original equation is true for any real number value of x.

The solution is all real numbers.

Equations like $4x - 3 = 3x - 3 + x$ that have many solutions are called *identities*.

(b) $3x - 7 = x + 2x - 11$ Combine like terms. $x + 2x = 3x$
$3x - 7 = 3x - 11$ Sort terms (addition axiom). Add $-3x$ to both sides.
Add $+7$ to both sides.

$3x - 3x = -11 + 7$ Combine like terms. $3x - 3x = 0$. $-11 + 7 = -4$.
$0 = -4$ Zero and -4 are not equal. Therefore, there is no solution of the equation.

No solution   **See Exercises 56–57.**

When the variable is eliminated in solving an equation, we examine the resulting statement. A true statement indicates the solution set is the *set of all real numbers*. We call this type of equation an **identity**. A false statement indicates the equation has *no solution*.

Identity: an equation for which the solution is all real numbers

---

TIP **Align Equal Signs! One Equal Sign per Line!** Organize the steps of the solution of an equation to reduce errors.

▶ Arrange steps under each other and align the equal signs in a vertical line. This helps you determine if a term has moved from one side to the other. As a term is eliminated from one side, move it to the other side as its opposite.

▶ Have only one equal sign per line.

Each line should be one complete equation or statement.

Good Form	Poor Form
$x = \dfrac{2}{4}$	$x = \dfrac{2}{4} = \dfrac{1}{2}$
$x = \dfrac{1}{2}$	

An equation can be solved two ways: (1) by isolating the variable terms on the left and (2) by isolating the variable terms on the right. When we choose the side for the variable terms, the constants must be on the opposite side.

STOP AND CHECK

Solve the equation two ways.
1. $4x - 3 = x + 12$

Answer:
1. $x = 5; 5 = x$

EXAMPLE 11

Solve the equation $9x + 2 = 6x - 10$ two ways.

Variable Terms Isolated on Left		**Variable Terms Isolated on Right**
$9x + 2 = 6x - 10$	Sort.	$9x + 2 = 6x - 10$
$9x - 6x = -10 - 2$	Combine like terms.	$2 + 10 = 6x - 9x$
$3x = -12$	Divide.	$12 = -3x$
$\dfrac{3x}{3} = \dfrac{-12}{3}$		$\dfrac{12}{-3} = \dfrac{-3x}{-3}$
$x = -4$		$-4 = x$

See Exercises 59–60.

STOP AND CHECK

Solve the equation.
1. $3x - 2 - 8x = 7 - 2x + 3$

Answer:
1. $x = -4$

EXAMPLE 12

Solve the equation $2x - 3 + 5x = 8 - 6x$.

$$2x - 3 + 5x = 8 - 6x \qquad \text{Combine the like terms } 2x \text{ and } 5x \text{ on the left side.}$$

$$7x - 3 = 8 - 6x \qquad \text{Sort to collect like terms on the same side.}$$

$$7x + 6x = 8 + 3 \qquad \text{Combine the like terms on each side of the equation.}$$

$$13x = 11 \qquad \text{Divide by the coefficient of } x.$$

$$\frac{13x}{13} = \frac{11}{13}$$

$$x = \frac{11}{13} \qquad \qquad \qquad \qquad \text{See Exercises 61–62.}$$

---

5 **Solve Linear Equations That Contain Parentheses.** When an equation contains an addition or subtraction in parentheses and that quantity in parentheses is multiplied by another factor, we have an example of the *distributive property*. (See Chapter 1, Section 1, Outcome 6.)

> **To solve linear equations that contain parentheses:**
>
> **1.** Apply the distributive property to *remove parentheses.*
>
> **2.** *Combine like terms* on each side of the equation.
>
> **3.** *Sort terms* to collect the variable terms on one side and constants on the other (addition axiom).
>
> **4.** *Combine like terms* on each side of the equation.
>
> **5.** *Multiply* by the reciprocal of the coefficient of the variable term or *divide* by the coefficient of the variable term (multiplication axiom).

STOP AND CHECK

Solve the equation.
1. $32 = 42 - 2(x + 3)$

Answer:
1. $x = 2$

EXAMPLE 13

Solve the equation $28 = 7x - 3(x - 4)$.

$$28 = 7x - 3(x - 4) \qquad \text{Each term in the parentheses is multiplied by } -3 \text{ using the distributive property. } -3(x) = -3x. \; -3(-4) = 12.$$

$$28 = 7x - 3x + 12 \qquad \text{Combine like terms on the right.}$$

$$28 = 4x + 12 \qquad \text{Sort terms.}$$

$$28 - 12 = 4x \qquad \text{Combine like terms on the left.}$$

$$16 = 4x \qquad \text{Divide.}$$

$$\frac{16}{4} = \frac{4x}{4}$$

$$x = 4 \qquad \qquad \qquad \qquad \text{See Exercises 63–84.}$$

As equations get more involved, the importance of checking becomes more apparent. To check the root 4 in Example 13, we substitute 4 for x in the equation.

$$28 = 7x - 3(x - 4) \qquad \text{Substitute 4 for } x.$$

$$28 = 7(4) - 3(4 - 4) \qquad \text{Simplify the grouping } 4 - 4 = 0.$$

$$28 = 7(4) - 3(0) \qquad \text{Multiply.}$$

$$28 = 28 - 0 \qquad \text{Subtract.}$$

$$\mathbf{28 = 28} \qquad \text{Solution checks.}$$

STOP AND CHECK
Solve the equation.
1. $11 - (x - 4) = 4x$

Answer:
1. $x = 3$

EXAMPLE 14

Solve the equation $6 - (x + 3) = 2x$ for x.

Since $(x + 3)$ is in parentheses, it is a grouping and we handle it as one term. The sign of the term is negative and the numerical coefficient is understood to be -1.

The first step in solving this equation is to apply the distributive property by multiplying $(x + 3)$ by its understood coefficient, -1.

$6 - (x + 3) = 2x$	-1 is the understood coefficient of $(x + 3)$.
$6 - 1(x + 3) = 2x$	Distribute.
$6 - x - 3 = 2x$	Combine. $6 - 3 = 3$
$3 - x = 2x$	Sort.
$3 = 2x + x$	Combine. $2x + x = 3x$
$3 = 3x$	Divide.
$\dfrac{3}{3} = \dfrac{3x}{3}$	
$x = 1$	**See Exercises 85–86.**

STOP AND CHECK
Solve the equations.
1. $\dfrac{3x}{5} = 21$

2. $\dfrac{1}{3}x = \dfrac{1}{2} + \dfrac{1}{3} + \dfrac{1}{18}$

3. $5.4x = 32.4$

Answers:

1. $x = 35$ **2.** $x = \dfrac{8}{3}$ **3.** $x = 6$

EXAMPLE 15

Solve the equations **(a)** $\dfrac{5x}{8} = 30$ **(b)** $\dfrac{1}{2}z = \dfrac{1}{4} + \dfrac{1}{12} + \dfrac{1}{3}$, and **(c)** $2.8x = 1.4$.

(a)

$$\frac{5x}{8} = 30$$

$$\frac{8}{5}\left(\frac{5x}{8}\right) = \frac{8}{5}(30) \qquad \text{Divide by } \frac{5}{8}, \text{ which is the same as multiplying by } \frac{8}{5}.$$

$$x = 48$$

(b)

$$\frac{1}{2}z = \frac{1}{4} + \frac{1}{12} + \frac{1}{3}$$

$$\frac{1}{2}z = \frac{3}{12} + \frac{1}{12} + \frac{4}{12} \qquad \text{Change fractions on the right to equivalent fractions with a common denominator.}$$

$$\frac{1}{2}z = \frac{8}{12}$$

$$\frac{1}{2}z = \frac{2}{3} \qquad \text{Reduce to lowest terms.}$$

$$2\left(\frac{1}{2}z\right) = 2\left(\frac{2}{3}\right) \qquad \text{Divide by } \frac{1}{2}, \text{ which is the same as multiplying by 2.}$$

$$z = \frac{4}{3}$$

(c) $2.8x = 1.4$

$$\frac{2.8x}{2.8} = \frac{1.4}{2.8} \qquad \text{Divide by 2.8.}$$

$$x = 0.5 \qquad\qquad\qquad\qquad\qquad \textbf{See Exercises 87–100.}$$

The usefulness of solving linear equations is in solving applied problems. We will continue to use the Six-Step Problem-Solving Plan as our guide for solving applied problems.

EXAMPLE 16

AG/H A horticulturist marks off a 32-m-wide rectangular nursery plot with 158 m of fencing. Because the plants must be properly spaced, she needs to know the length of the plot. Find the length in meters.

Unknown facts

The length of the nursery plot in meters.

Known facts

The nursery plot is rectangular. The width is 32 m. The fencing gives us the perimeter of 158 m.

Relationships

Using the formula for the perimeter of a rectangle, we know that the perimeter is twice the sum of the length and width, or $P = 2(l + w)$.

Estimation

Two widths ($2 \cdot 32$ m) will use 64 m of fencing. There will be less than 100 m of fencing for the two lengths. The length will be less than 50 m.

Calculations

$P = 2(l + w)$	Substitute in the formula.
$158 = 2(l + 32)$	Distribute.
$158 = 2(l) + 2(32)$	Multiply.
$158 = 2l + 64$	Sort.
$158 - 64 = 2l$	Combine.
$94 = 2l$	Divide.
$\dfrac{94}{2} = \dfrac{2l}{2}$	
$47 = l$	

Interpretation

Thus, the length of the nursery plot is 47 m.

Check:		
	$P = 2(l + w)$	Substitute.
	$158 = 2(47 + 32)$	Combine within grouping.
	$158 = 2(79)$	Multiply.
	$158 = 158$	**See Exercises 101–110.**

EXAMPLE 17

INDTEC A machine part weighs 2.7 kg and is to be shipped in a carton weighing x kg. If the total weight of three packaged machine parts is 9.3 kg, how much does each carton weigh? (See Fig. 7–2.)

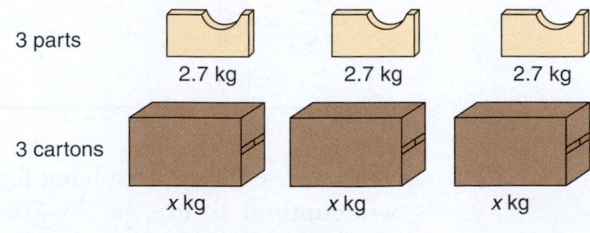

FIGURE 7–2

Unknown facts

The weight of each empty carton.

Known facts

There are three parts and three cartons. Each part weighs 2.7 kg. The three packaged parts (in cartons) weigh a total of 9.3 kg.

Relationships

Let x equal the weight of one empty carton. Each packaged part weighs the sum of the part (2.7 kg) plus the weight of a carton (x kg). The sum of the three packaged parts equals 9.3 kg; that is, $3(2.7 + x) = 9.3$.

Estimation

Three packaged parts weigh approximately 9 kg, or 3 kg per packaged carton. Since one part weighs 2.7 kg, then the carton weighs about 0.3 kg.

Calculations

$3(2.7 + x) = 9.3$	Distribute.
$3(2.7) + 3(x) = 9.3$	Multiply.
$8.1 + 3x = 9.3$	Sort.
$3x = 9.3 - 8.1$	Combine.
$3x = 1.2$	Divide.
$\dfrac{3x}{3} = \dfrac{1.2}{3}$	
$x = 0.4$	

Interpretation

Each carton weighs 0.4 kg.

Check:		
	$3(2.7 + \boxed{x}) = 9.3$	Substitute.
	$3(2.7 + \boxed{0.4}) = 9.3$	Add within grouping.
	$3(3.1) = 9.3$	
	$9.3 = 9.3$	**See Exercises 101–110.**

TIP **Additional Problem-Solving Strategies**

▶ Read the problem carefully. Read it several times and read it phrase by phrase.

▶ Understand all the words in the problem.

▶ Analyze the problem:

What are you asked to find?

What facts are given?

What facts are implied?

▶ Visualize the problem.

▶ State the conditions or relationships of the problem "symbolically."

▶ Examine the options.

▶ Develop a *plan* for solving the problem.

▶ Write your *plan* symbolically; that is, write an equation.

▶ Anticipate the characteristics of a reasonable solution.

▶ Solve the equation.

▶ Verify your answer with the conditions of the problem.

There are many different problem-solving plans. Additional strategies can be considered.

7–2 EXERCISES   MyLab Math   For additional practice go to your study plan in MyLab Math.

1 Solve the equations. Use the addition axiom. *See Example 1.*

1. $x - 3 = 5$ **2.** $x - 7 = 8$ **3.** $x - 9 = -12$ **4.** $x - 11 = -19$

5. $x + 5 = 9$ **6.** $x + 8 = -3$ **7.** $x + 11 = -15$ **8.** $x + 14 = 17$

See Example 2.

9. $12 = x + 7$ **10.** $-15 = x - 8$ **11.** $14 = x - 2$ **12.** $18 = x + 9$

13. $7 = 5 - x$ **14.** $5 = 9 - x$ **15.** $17 = 11 + x$ **16.** $21 - x = 12$

2 Solve the equations. Use the multiplication axiom. *See Example 3.*

17. $7x = 56$ **18.** $2x = 18$ **19.** $15 = 3x$

20. $-28 = -3b$ **21.** $-3a = 27$ **22.** $-18 = 9b$

23. $-7c = -49$ **24.** $-72 = 8x$ **25.** $-6n = 15$

See Example 4.

26. $36 = 1.2n$ **27.** $0.5x = 32$ **28.** $36 = 0.8y$

See Example 5.

29. $\dfrac{3}{5}x = 2$ **30.** $-\dfrac{3}{4}y = 5$ **31.** $-\dfrac{1}{2}x = -6$ **32.** $21 = \dfrac{3}{8}n$

See Example 6.

33. $-x = 7$ **34.** $-2 = -y$ **35.** $\dfrac{1}{3}x = 5$ **36.** $4 = \dfrac{1}{2}y$

See Example 7.

37. Show the steps to check Exercise 31.   **38.** Show the steps to check Exercise 32.

3 Solve. *See Example 8.*

39. $3x + 7x = 60$ **40.** $42 = 8m - 2m$ **41.** $5a - 6a = 3$

42. $3m - 9m = 3$ **43.** $y + 3y = 32$ **44.** $0 = 2x - x$

45. $8y - 6y = -14$ **46.** $5y - y = -16$ **47.** $-6x - 12x = 36$

4 Solve. *See Example 9.*

48. $-36 = 9x + 18$ **49.** $b + 6 = 5$ **50.** $1 = x - 7$ **51.** $5t - 18 = 12$

52. $10 - 2x = 4$ **53.** $2y + 7 = 17$ **54.** $3x + 7 = x$ **55.** $3a - 8 = 7a$

See Example 10.

56. $4x = 5x + 8$ **57.** $4x + 7 = 8 + 4x$ **58.** $10x + 18 = 8x + 18 + 2x$

See Example 11.

59. $2t + 6 = t + 13$ **60.** $12 + 5x = 6 - x$

See Example 12.

61. $4y - 8 = 2y + 14$ **62.** $8 - 7y = y + 24$

5 Solve. *See Example 13.*

63. $4(x - 5) = 8$ **64.** $3(y - 7) = -15$ **65.** $6(x + 3) = -12$

66. $5(x + 7) = 25$ **67.** $-3(x + 8) = -39$ **68.** $-2(x - 7) = 26$

69. $3(2x - 1) = 3$ **70.** $4(x - 1) = 28$ **71.** $-4(-3x + 1) = 8$

72. $-9(2x - 3) = 27$ **73.** $7(x - 3) = 7x + 21$ **74.** $5x + 15 = 5(x + 3)$

75. $8(x - 5) = 8x + 12$ **76.** $6x + 7 = 3(2x + 4)$ **77.** $2(3x - 5) = 7x - 12$

78. $5(3x - 7) + 2 = 7x + 7$ **79.** $2x - 3 + 6x = 4(3x - 2) + 3$ **80.** $7(x - 1) = 2(x + 3) + 4(x - 7)$

81. $3(x - 4) - 2(3x - 1) = 17$ **82.** $6 - 2(3x - 1) = 14$ **83.** $12 - 5(2x - 3) = -5$

84. $8 - 3(x - 2) = x - 6$

See Example 14.

85. $6x - 4 + 2x = -(x - 5)$ **86.** $8x - (2x - 9) = -3$

See Example 15.

87. $\dfrac{2x}{3} = 18$ **88.** $P = \dfrac{1}{2} + \dfrac{1}{3}$ **89.** $x + \dfrac{1}{7}x = 16$

90. $\dfrac{2}{5} - x = \dfrac{1}{2}x + \dfrac{4}{5}$ **91.** $\dfrac{3}{7}m - \dfrac{1}{2} = \dfrac{2}{3}$ **92.** $\dfrac{1}{4}s = \dfrac{1}{4} + \dfrac{1}{10} + \dfrac{1}{20}$

93. $m = 2 + \dfrac{1}{4}m$ **94.** $\dfrac{7}{9} + 3 = \dfrac{1}{2}T$ **95.** $2.3x = 4.6$

96. $4.5x - 5.1 = 3.9$ **97.** $0.8R = 0.6$ (round to nearest tenth)

98. $0.33x + 0.25x = 3.5$ (round to nearest hundredth) **99.** $0.04x = 0.08 - x$ (round to nearest hundredth)

100. $0.47 = R + 0.4R$ (round to nearest hundredth)

Write the statements as equations and solve. *See Examples 16 and 17.*

101. The sum of x and 4 equals 12. Find x.

102. 4 less than 2 times a number is 6. Find the number.

103. **INDTEC** Three parts totaling 27 lb are packaged for shipping. Two parts weigh the same. The third part weighs 3 lb less than each of the two equal parts. Find the weight of each part.

104. **HLTH/N** How many gallons of water must be added to 24 gal of pure disinfectant to make 60 gal of diluted disinfectant?

105. **INDTEC** A wet casting weighing 4.03 kg weighs 3.97 kg after drying. Write and solve an algebraic equation to find the weight loss due to drying.

106. **CON** A plumber needs 3 times as much perforated pipe as solid pipe to lay a drain line 400 ft long. How much of each type of pipe is needed?

107. **ELEC** An engineering student purchased a new circuits text, a used Spanish text, and a graphing calculator. She remembered that the calculator cost twice as much as the Spanish book and that the circuits text cost $70.00. The total before tax was $235.00. What was the cost of the calculator and the Spanish text?

108. **CON** Mary Jefferson purchased a home on a square lot 150 ft on each side. She wants to enclose the entire lot with a cedar fence. One estimate for the job was for $14.00 per linear foot. How much will the fence cost?

109. **AG/H** Jake Drewrey, owner of Jake's Landscape Service, knows that one of his fertilizer tanks holds twice as many gallons of liquid fertilizer as a second tank. The two tanks together hold 325 gal. If both tanks are filled to capacity, how many gallons of fertilizer does each tank hold?

110. **AG/H** Natalie Bradley hired a stone mason to build a rectangular flower bed at one end of her patio. She needs enough mulch to cover 60 ft^2, the area of the flower bed. If the flower bed has a length of 10 ft, how wide is the bed?

7–3   Solving Linear Equations with Fractions and Decimals by Clearing the Denominators

1   Solve Fractional Equations by Clearing the Denominators. Thus far, we have used mostly integers in equations. However, real-world situations require us to deal with various fraction and decimal quantities when we solve equations.

The techniques we used in Section 7–2 will also help us solve equations with fractions and decimals. Other techniques minimize the calculations with fractions and decimals and allow more steps to be performed mentally. One of these techniques involves *clearing* the equation of all denominators in the first step. The resulting equation, which contains no fractions, is then solved using the procedures we learned earlier.

This process, called **clearing the fractions**, is another application of the multiplication axiom.

To clear an equation of a single fraction:

Apply the multiplication axiom by multiplying the entire equation by the fraction's denominator.

EXAMPLE 1

Solve **(a)** $\frac{4d}{3} = 12$ and **(b)** $2x + \frac{3}{4} = 1$ by clearing the fractions.

(a)
$$\frac{4d}{3} = 12$$ Multiply both sides by the denominator 3. Reduce where possible.

$$(\overset{1}{\cancel{3}})\frac{4d}{\underset{1}{\cancel{3}}} = (3)(12)$$

$$4d = 36$$ Divide.

$$\frac{4d}{4} = \frac{36}{4}$$

$$\boldsymbol{d = 9}$$

Check: $\dfrac{4(\overset{3}{\cancel{9}})}{\underset{1}{\cancel{3}}} = 12$

$$12 = 12$$

(b)
$$2x + \frac{3}{4} = 1$$ Multiply each term by the denominator 4.

$$4(2x) + 4\left(\frac{3}{4}\right) = 4(1)$$ Reduce and multiply.

$$4(2x) + \overset{1}{\cancel{4}}\left(\frac{3}{\underset{1}{\cancel{4}}}\right) = 4(1)$$

$$8x + 3 = 4$$ Sort.

$$8x = 4 - 3$$ Combine.

$$8x = 1$$ Divide.

$$\frac{8x}{8} = \frac{1}{8}$$

$$x = \frac{1}{8}$$

Check: $2x + \frac{3}{4} = 1$ Substitute $\frac{1}{8}$ in place of x.

$$\overset{1}{2}\left(\frac{1}{\underset{4}{8}}\right) + \frac{3}{4} = 1 \qquad \text{Multiply.}$$

$$\frac{1}{4} + \frac{3}{4} = 1 \qquad \text{Add.}$$

$$1 = 1 \qquad\qquad\qquad \textbf{See Exercises 1–13.}$$

One advantage of clearing fractions is that the process can be applied to equations in which the variable is in the *denominator* of the fraction. Let's use this procedure to solve $\frac{10}{x} = 2$. Note that the term $\frac{10}{x}$ is *not* the product of 10 and x. Therefore, the coefficient of x is *not* 10.

STOP AND CHECK

Solve the equation.

1. $\dfrac{15}{x} = 5$

Answer:

1. $x = 3$

EXAMPLE 2

Solve $\dfrac{10}{x} = 2$.

$$\frac{10}{x} = 2 \qquad \text{Multiply both sides by denominator } x. \text{ Reduce where possible.}$$

$$(\overset{1}{x})\frac{10}{\underset{1}{x}} = (x)2$$

$$10 = 2x \qquad \text{Divide.}$$

$$\frac{10}{2} = \frac{2x}{2}$$

$$\mathbf{5 = x}$$

Check: $\dfrac{\overset{2}{10}}{\underset{1}{5}} = 2$

$$2 = 2 \qquad\qquad\qquad \textbf{See Exercises 14–15.}$$

Did You Know? The multiplication axiom applies only to multiplying (or dividing) by nonzero values. Therefore, when x is the denominator of a fraction, we must assume that x cannot be equal to zero. $x \neq 0$.

TIP Check for Extraneous Roots When you solve equations with a variable in the fraction's denominator, you may get a "root" that does not make a true statement when substituted in the original equation. Solutions or roots that do not make a true statement in the original equation are called **extraneous roots**. When solving an equation with the variable in the denominator, you *must* check the root to see if it results in a true statement in the original equation.

Extraneous roots: solutions or roots that do not result in a true statement in the original equation

Excluded value: a value that causes the denominator of any fraction to be zero. In an equation, an excluded value may generate an extraneous root

One situation that produces an extraneous root is a value that causes the denominator of any fraction to be zero. Such values are called **excluded values.**

> **To identify an excluded value that causes a denominator of zero:**
>
> 1. If a fraction has a variable in the denominator, write an equation by setting the denominator equal to zero.
>
> 2. Solve each resulting equation.
>
> 3. The solution for each equation from Step 1 is an excluded value.

STOP AND CHECK

Identify any excluded values. Solve the equation.

1. $-3 = \dfrac{0}{x}$

Answer:

1. Excluded value: 0; no solution

EXAMPLE 3

Identify any excluded values. Solve $\dfrac{0}{x} = 5$.

Excluded value:

$$x = 0$$

Set the denominator containing a variable equal to zero. When the denominator is 0, the fraction $\dfrac{0}{x}$ is indeterminate.

Solve the equation:

$$\frac{0}{x} = 5 \qquad \text{Multiply both sides by } x.$$

$$x\left(\frac{0}{x}\right) = 5(x)$$

$$0 = 5x$$

$$\frac{0}{5} = \frac{5x}{5}$$

$$0 = x \qquad \text{Possible solution}$$

Check: $\dfrac{0}{0} \neq 5$ Does not check. Zero is the excluded value.

The equation $\dfrac{0}{x} = 5$ has no solution.

See Exercises 16–17.

Equations in which the fraction contains more than one term in its numerator or denominator can also be cleared of fractions.

STOP AND CHECK

Identify any excluded values. Solve the equation.

1. $\dfrac{12}{T - 3} = 2$

Answer:

1. $T \neq 3; T = 9$

EXAMPLE 4

Identify any excluded values. Solve $\dfrac{12}{Q + 6} = 1$.

Excluded value:

$$Q + 6 = 0 \qquad \text{Set denominator equal to zero and solve for } Q.$$

$$Q = 0 - 6 \qquad \text{Sort.}$$

$$Q = -6 \qquad \text{Excluded value}$$

$$\frac{12}{Q + 6} = 1 \qquad \text{Multiply both sides of the equation by the denominator, the quantity } Q + 6. \text{ Reduce where possible.}$$

$$(Q + 6)\frac{12}{Q + 6} = (Q + 6)1 \qquad \text{Distribute.}$$

$$12 = Q + 6 \qquad \text{Sort.}$$

$$12 - 6 = Q \qquad \text{Combine like terms.}$$

$$\mathbf{6 = Q} \qquad \text{Because the numerical coefficient of } Q \text{ is already 1, the}$$
$$\text{equation is solved.}$$

Check: $\dfrac{12}{6 + 6} = 1 \qquad$ Substitute 6 for Q and evaluate.

$$\dfrac{12}{12} = 1$$

$$1 = 1 \qquad \text{Even though } -6 \text{ is an excluded value, 6 is a valid}$$
$$\text{solution.}$$

The solution of the equation is $Q = 6$.   See Exercises 18–21.

When an equation has more than one fractional term, we expand our process.

To solve an equation by clearing all fractions:

1. Multiply each term of the *entire* equation by the least common multiple (LCM) of the denominators of the equation.

2. Apply the distributive property to remove parentheses.

3. Combine like terms on each side of the equation.

4. Sort terms to collect the variable terms on one side and constants on the other (addition axiom).

5. Combine like terms on each side of the equation.

6. Solve the resulting equation by multiplying by the reciprocal of the coefficient of the variable term or dividing by the coefficient of the variable term (multiplication axiom).

STOP AND CHECK

Solve the equation by clearing all fractions first.

1. $-\dfrac{3}{4}x = 7 - \dfrac{1}{5}x$

Answer:

1. $x = -\dfrac{140}{11}$

EXAMPLE 5

Solve $-\dfrac{1}{4}x = 9 - \dfrac{2}{3}x$ by clearing all fractions first.

There are no excluded values since there are no variables in a denominator.

$$-\dfrac{1}{4}x = 9 - \dfrac{2}{3}x \qquad \text{Multiply each term by the LCM of the}$$
$$\text{denominators. LCM} = 12 \text{ or } 4(3).$$

$$(4)(3)\left(-\dfrac{1}{4}x\right) = (4)(3)(9) - (4)(3)\left(\dfrac{2}{3}x\right) \qquad \text{Reduce.}$$

$$(\overset{1}{4})(3)\left(-\dfrac{1}{\underset{1}{4}}x\right) = (4)(3)(9) - (4)(\overset{1}{3})\left(\dfrac{2}{\underset{1}{3}}x\right) \qquad \text{Multiply the remaining factors.}$$

$$-3x = 108 - 8x \qquad \text{This equation contains no fractions. Sort.}$$

$$-3x + 8x = 108 \qquad \text{Combine like terms.}$$

$$5x = 108 \qquad \text{Divide by the coefficient of } x.$$

$$\dfrac{5x}{5} = \dfrac{108}{5}$$

$$x = \dfrac{\mathbf{108}}{\mathbf{5}}$$

Check using a calculator:

$$-\frac{1}{4}x = 9 - \frac{2}{3}x$$

Substitute $\frac{108}{5}$ for x.

$$-\frac{1}{4}\left(\frac{108}{5}\right) = 9 - \frac{2}{3}\left(\frac{108}{5}\right)$$

Left Side: $(\ (-)\ 1\ \div\ 4\ )\ \times\ (\ 108\ \div\ 5\ )\ = \Rightarrow -5.4$
Right Side: $9\ -\ (\ 2\ \div\ 3\ )\ \times\ (\ 108\ \div\ 5\ )\ = \Rightarrow -5.4$ **See Exercises 21–38.**

Solutions of equations are not always whole numbers. Such solutions are usually written as *proper* or *improper fractions* in lowest terms. This representation is the *exact* solution. Solutions of applied problems are usually written in decimal or mixed-number form. If the decimal form is rounded, the rounded decimal is the *approximate* solution.

EXAMPLE 6

ELEC   Find the total resistance in a parallel DC circuit with three branches rated at 4 Ω, 10 Ω, and 20 Ω, respectively. Solve $\frac{1}{R} = \frac{1}{4} + \frac{1}{10} + \frac{1}{20}$ by clearing fractions first.

$R = 0$ is the excluded value.

$$\frac{1}{R} = \frac{1}{4} + \frac{1}{10} + \frac{1}{20}$$

The LCM is 20R because 20R is evenly divisible by R, 4, 10, and 20.

$$20R\left(\frac{1}{R}\right) = \overset{5}{20R}\left(\frac{1}{\underset{1}{4}}\right) + \overset{2}{20R}\left(\frac{1}{\underset{1}{10}}\right) + 20R\left(\frac{1}{\underset{1}{20}}\right)$$

Multiply each term in the *entire* equation by 20R and reduce.

$$20 = 5R + 2R + R$$ Combine like terms.

$$20 = 8R$$ Divide by the coefficient of R.

$$\frac{20}{8} = \frac{8R}{8}$$ Reduce.

$$\frac{5}{2} = R$$ $\frac{5}{2} = 2.5$

Interpretation
The resistance is 2.5 Ω.

Check: $\frac{1}{R} = \frac{1}{4} + \frac{1}{10} + \frac{1}{20}$ Substitute $\frac{5}{2}$ for R.

$$\frac{1}{\dfrac{5}{2}} = \frac{1}{4} + \frac{1}{10} + \frac{1}{20}$$

Perform calculations on each side of the equation.

$$\frac{1}{\dfrac{5}{2}} = \frac{5}{20} + \frac{2}{20} + \frac{1}{20}$$

$$1 \cdot \frac{2}{5} = \frac{8}{20}$$

$$\frac{2}{5} = \frac{2}{5}$$

See Exercises 39–42.

TIP **Fractions Versus Decimals** Sometimes applied problems that require fractions in their equations require that their solutions be expressed as decimal numbers. In these cases, we perform the division indicated by the fraction.

The equation in the preceding example is derived from the formula for finding total resistance in a parallel DC circuit with three branches rated at 4, 10, and 20 Ω. Ohms are expressed in decimal numbers, so in an application the solution should be converted to a decimal equivalent.

$$R = \frac{5}{2}\,\Omega \quad \text{or} \quad 2.5\,\Omega$$

2 **Solve Applied Problems Involving Rate, Time, and Work.** A *rate measure* often involves a unit of time. If car A travels 50 mi in 1 h, then car A's **rate of work** (travel) is 50 mi per 1 h, or $\frac{50\ \text{mi}}{1\ \text{h}}$ expressed as a fraction. If car A travels for 3 h, then the **amount of work** is $\frac{50\ \text{mi}}{\text{h}} \times 3\ \text{h} = 150\ \text{mi}$. Review rate measures in Chapter 4, Section 1, Outcome 5.

Rate of work: the ratio of the amount of work completed to the time worked

Amount of work: the quantity of work that is completed

> **To find the amount of work produced by one individual or machine:**
>
> 1. Identify the rate of work and the time worked.
>
> 2. Use the formula for amount of work.

> **Formula for amount of work:**
>
> amount of work = rate of work × time worked
>
> $$W = RT$$

EXAMPLE 7

CON A carpenter can install 1 door in 3 h. Find the number of doors the carpenter can install in 40 h.

Known facts

Rate of work = 1 door per 3 h, $\frac{1}{3}$ door per hour, or $\frac{1\ \text{door}}{3\ \text{h}}$

Time worked = 40 h

Unknown facts

$W =$ amount of work or number of doors installed

Relationships

Amount of work = rate of work × time worked

Estimation

It would take 30 h to install 10 doors. More than 10 doors can be installed in 40 h.

Dmitry Kalinovsky/123RF

Calculations

$$W = \left(\frac{1 \text{ door}}{3 \cancel{h}}\right)(40 \cancel{h}) \qquad \text{Multiply. Reduce dimensions.}$$

$$W = \frac{40}{3} \text{ doors}$$

$$W = 13\tfrac{1}{3} \text{ doors}$$

Interpretation

Thus, 13 doors can be installed in 40 h.   **See Exercises 43–44.**

If two workers or machines complete a job by working together, we can find the amount of work done by each worker or machine. Combined, the amounts equal 1 total job.

> **To find the amount of work each individual or machine produces when working together:**
>
> 1. Identify the rate of work for each individual or machine. If unknown, assign a letter to represent the unknown.
>
> 2. Identify the time worked for each individual or machine. If unknown, assign a letter to represent the unknown. *Note:* Only one letter should be used and other unknowns should be written in relationship to the one letter.
>
> 3. Use the formula for completing one job.

> **Formula for completing one job when A and B are working together:**
>
> $$\begin{pmatrix} \text{A's} \\ \text{amount of} \\ \text{work} \end{pmatrix} + \begin{pmatrix} \text{B's} \\ \text{amount of} \\ \text{work} \end{pmatrix} = 1 \text{ completed job}$$
>
> or
>
> $$\begin{pmatrix} \text{A's} \\ \text{rate of} & \times & \text{time} \\ \text{work} & & \text{worked} \end{pmatrix} + \begin{pmatrix} \text{B's} \\ \text{rate of} & \times & \text{time} \\ \text{work} & & \text{worked} \end{pmatrix} = 1 \text{ completed job}$$

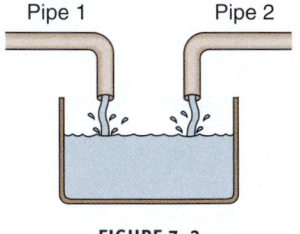

Pipe 1   Pipe 2

FIGURE 7–3

EXAMPLE 8

AG/H Pipe 1 fills a tank in 6 min and pipe 2 fills the same tank in 8 min (Fig. 7–3). How long does it take for both pipes together to fill the tank?

Known facts

Pipe 1 fills the tank at a rate of 1 tank per 6 min, $\frac{1}{6}$ tank per minute, or $\frac{1 \text{ tank}}{6 \text{ min}}$.

Pipe 2 fills the tank at a rate of 1 tank per 8 min, $\frac{1}{8}$ tank per minute, or $\frac{1 \text{ tank}}{8 \text{ min}}$.

Unknown facts

$T =$ time (in minutes) for both pipes together to fill the tank.

Relationships

Amount of work of pipe 1 $= \dfrac{1 \text{ tank}}{6 \text{ min}}(T)$. Amount of work of pipe 2 $= \dfrac{1 \text{ tank}}{8 \text{ min}}(T)$.

Amount of work together = pipe 1's work + pipe 2's work.

Estimation

Both pipes together should fill the tank more quickly than the faster rate, or in less than 6 min.

Calculations

$$\frac{1 \text{ tank}}{6 \text{ min}}(T \text{ min}) + \frac{1 \text{ tank}}{8 \text{ min}}(T \text{ min}) = 1 \text{ tank} \qquad \text{The LCM is 24.}$$

$$(24)\left(\frac{1}{6}T\right) + (24)\left(\frac{1}{8}T\right) = (24)(1) \qquad \text{Clear fractions.}$$

$$\overset{4}{(24)}\left(\frac{1}{\underset{1}{6}}T\right) + \overset{3}{(24)}\left(\frac{1}{\underset{1}{8}}T\right) = (24)(1) \qquad \text{Reduce and multiply.}$$

$$4T + 3T = 24 \qquad \text{Combine.}$$

$$7T = 24 \qquad \text{Divide.}$$

$$\frac{7T}{7} = \frac{24}{7}$$

$$T = \frac{24}{7}\left(\text{or } 3\frac{3}{7}\right) \text{min}$$

Interpretation

Both pipes together fill the tank in $3\frac{3}{7}$ min. **See Exercises 45–50.**

> **TIP** **Interpretation of Improper Fractions as Mixed Numbers, Decimal Equivalents, or Mixed Measurements** In applied problems, it is desirable to change improper fractions like $\frac{24}{7}$ into mixed numbers or decimal equivalents. The problem statement generally dictates the interpretation.
>
> In the preceding example, $\frac{24}{7}$ min can be interpreted as $3\frac{3}{7}$ min or 3.4 min (rounded). In some instances, you may need to change $\frac{3}{7}$ min to seconds $\left(\frac{3}{7} \text{ min} \times \frac{60 \text{ s}}{1 \text{ min}} \approx 25.7 \text{ s}\right)$. Then, $3\frac{3}{7}$ min becomes 3 min 26 s (rounded).

Two pipes, one a faucet and the other a drain, have opposite functions. Pipe 1 fills the tank. Pipe 2 is a drain and empties the tank. In this case, we subtract the work done by the drain from the work done by the faucet. This combined action, if the faucet fills at a faster rate than the drain empties, results in a full tank.

Formula for completing one job when A and B are working in opposition:

$$\begin{pmatrix} \text{A's} \\ \text{amount of} \\ \text{work} \end{pmatrix} - \begin{pmatrix} \text{B's} \\ \text{amount of} \\ \text{work} \end{pmatrix} = 1 \text{ completed job}$$

or

$$\begin{pmatrix} \text{A's} \\ \text{rate of} \quad \times \quad \text{time} \\ \text{work} \qquad \text{worked} \end{pmatrix} - \begin{pmatrix} \text{B's} \\ \text{rate of} \quad \times \quad \text{time} \\ \text{work} \qquad \text{worked} \end{pmatrix} = 1 \text{ completed job}$$

based on A's amount of work being greater than B's amount of work.

Faucet

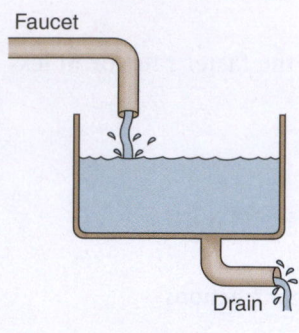

Drain

FIGURE 7–4

EXAMPLE 9

AG/H A faucet fills a tank in 6 min. A drain empties the tank in 8 min (Fig. 7–4). If both the faucet and drain are open, in how many minutes will the tank start to overflow if the faucet is not turned off?

Known facts

Faucet's rate of work = 1 tank filled per 6 min, $\frac{1}{6}$ tank per min, or $\dfrac{1 \text{ tank}}{6 \text{ min}}$.

Drain's rate of work = 1 tank emptied per 8 min, $\frac{1}{8}$ tank per min, or $\dfrac{1 \text{ tank}}{8 \text{ min}}$.

Unknown facts

T = time (min) until tank overflows with faucet and drain both working.

Relationships

Amount of work of faucet = $\dfrac{1 \text{ tank}}{6 \text{ min}} (T)$

Amount of work of drain = $\dfrac{1 \text{ tank}}{8 \text{ min}} (T)$

Amount of work when both faucet and drain are open = faucet's work − drain's work

Estimation

With both the faucet and drain open, it should take longer to fill the tank than if the faucet were open and the drain closed. It should take longer than 6 min.

Calculations

$$\frac{1 \text{ tank}}{6 \text{ min}} (T \text{ min}) - \frac{1 \text{ tank}}{8 \text{ min}} (T \text{ min}) = 1 \text{ tank} \qquad \text{Analyze dimensions.}$$

$$\frac{1}{6} T - \frac{1}{8} T = 1 \qquad \text{Clear fractions. LCM = 24}$$

$$(24)\left(\frac{1}{6} T\right) - (24)\left(\frac{1}{8} T\right) = (24)(1) \qquad \text{Reduce.}$$

$$\overset{4}{(24)}\left(\frac{1}{\underset{1}{6}} T\right) - \overset{3}{(24)}\left(\frac{1}{\underset{1}{8}} T\right) = (24)(1) \qquad \text{Multiply remaining factors.}$$

$$4T - 3T = 24 \qquad \text{Combine like terms.}$$
$$T = 24 \text{ min}$$

Interpretation

With the faucet and drain both open, the tank will be full in 24 min.

See Exercises 51–52.

3 **Solve Decimal Equations by Clearing the Decimals.** Since a decimal is a type of fraction, we can also solve equations containing decimals by clearing decimals. The place value of the last digit of the decimal determines the denominator of the fraction.

To solve a decimal equation by clearing decimals:
1. Multiply each term of the *entire* equation by the least common denominator (LCD) of the fractional amounts represented by the decimals.
2. Follow the same steps used in solving a linear equation.

> **TIP** **LCD for Decimals** Digits to the right of the decimal point represent fractions whose denominators are determined by the place value. To find the LCD for all the decimal numbers in an equation, find the decimal with the most digits after the decimal point. Use its denominator to clear the decimals.
>
> This procedure allows you to avoid dividing by a decimal, which can be a common source of error when performing calculations by hand.

EXAMPLE 10

Solve $0.38 + 1.1y = 0.6$ by first clearing the equation of decimals.

$0.38 + 1.1y = 0.6$	The LCD is **100**.
$100(0.38) + 100(1.1y) = 100(0.6)$	Multiply by 100.
$38 + 110y = 60$	Sort.
$110y = 60 - 38$	Combine.
$\dfrac{110y}{110} = \dfrac{22}{110}$	Divide.
$y = \mathbf{0.2}$	See Exercises 53–67.

The interest formula resembles the percentage formula, $P = RB$, but it includes the time period over which money was borrowed or invested. If we know three of the four elements, we can find the fourth.

> **Formula for simple interest:**
>
> $$I = PRT$$
>
> where I = **interest**, P = **principal**, R = **rate** or percent in decimal form, and T = **time**.

For comparison with the percentage formula, interest is the part or portion and the principal is the base.

EXAMPLE 11

PFIN A $1,000 investment is made for $2\frac{1}{2}$ years at 6.25%. Find the amount of interest.

When we use the interest formula, we change the percent to a decimal equivalent. When no time period is given for the rate, we assume the rate to be per year.

$I = PRT$ Substitute the values in the formula.

$\underset{\text{principal}}{} \quad \underset{\text{rate}}{} \quad \underset{\text{time}}{}$

$I = \$1,000(0.0625)(2.5 \text{ years})$ 6.25% = 0.0625; $2\frac{1}{2}$ years = 2.5 years

$I = \$156.25$

The interest for $2\frac{1}{2}$ years is $156.25.   See Exercises 68–72.

> **TIP** **Clearing Decimals Versus Using a Calculator** Working the preceding example by clearing decimals illustrates that in some cases the method is *not* the most efficient way to solve an equation with decimals. The numbers produced may be large and cumbersome.

$$I = \$1,000 \cdot 0.0625 \cdot 2.5$$

LCM is 10,000.

$$10,000I = \overset{1}{\cancel{10,000}}\left(\$1,\overset{100}{\cancel{000}} \cdot \frac{625}{\underset{1}{\cancel{10,000}}} \cdot \frac{25}{\underset{1}{\cancel{10}}}\right)$$

Multiply both sides of the equation by 10,000. Reduce.

$$10,000I = 100(625)(25)$$

Multiply.

$$10,000I = 1,562,500$$

Divide.

$$I = \frac{1,562,500}{10,000}$$

$$I = \$156.25$$

A calculator gives the solution more efficiently if we proceed using decimals.

EXAMPLE 12

ELEC The formula for voltage V is wattage W divided by amperage A: $V = \frac{W}{A}$. Find the voltage to the nearest hundredth needed for a circuit of 1,280 W with a current of 12.23 A.

Estimation

$$1,200 \div 12 = 100$$

$$V = \frac{W}{A}$$

Substitute values.

$$V = \frac{1,280}{12.23}$$

Divide.

$$V = 104.6606705, \quad \text{or} \quad 104.66 \text{ V}$$

Interpretation

The voltage is 104.66 V.

See Exercises 73–78.

7-3 EXERCISES

MyLab Math For additional practice go to your study plan in MyLab Math.

1 Solve the equations. *See Example 1.*

1. $\frac{2}{7}x = 8$ **2.** $-7 = \frac{21}{33}p$ **3.** $\frac{1}{3}r = \frac{6}{7}$ **4.** $0 = -\frac{2}{5}c$

5. $-\frac{5}{3}m = 9$ **6.** $-9m = \frac{5}{3}$ **7.** $\frac{5}{8}t = 1$ **8.** $\frac{5}{12}z = 20$

9. $10 = -\frac{1}{35}t$ **10.** $\frac{2x}{3} = 18$ **11.** $0 = \frac{x}{4}$ **12.** $\frac{P}{-8} = -72$

13. $\frac{3P}{7} = 12$

See Example 2.   Identify excluded values if appropriate and solve. *See Example 3.*

14. $\frac{7}{Q} = 21$ **15.** $-\frac{8}{P} = -72$ **16.** $\frac{0}{x} = 7$ **17.** $3 = \frac{0}{R}$

Identify excluded values if appropriate and solve. *See Example 4.*

18. $\frac{7}{p-4} = -8$ **19.** $-8 = \frac{4B}{B-6}$ **20.** $\frac{8-R}{76} = 1$ **21.** $\frac{y+1}{2} = 7$

See Example 5.

22. $-\dfrac{5}{7}p = -\dfrac{11}{21}$

23. $\dfrac{2}{9}c + \dfrac{1}{3}c = \dfrac{3}{7}$

24. $-\dfrac{1}{4}x = 9 - \dfrac{2}{3}x$

25. $\dfrac{2}{7}y + \dfrac{3}{8} = \dfrac{1}{7}y + \dfrac{5}{3}$

26. $\dfrac{1}{3}x + \dfrac{1}{2}x = \dfrac{20}{3}$

27. $\dfrac{7}{R} - \dfrac{2}{R} = -1$

28. $S = \dfrac{1}{15} + \dfrac{1}{5} + \dfrac{1}{30}$

29. $18 - \dfrac{1}{4}x = \dfrac{1}{2}$

30. $\dfrac{1}{7}H - \dfrac{1}{3}H = 0$

31. $\dfrac{7}{16}h + \dfrac{1}{9} = \dfrac{1}{3}$

32. $x + \dfrac{1}{4}x = 8$

33. $3y + 9 = \dfrac{1}{4}y$

34. $18 = \dfrac{4}{3x} - \dfrac{3}{2x}$

35. $\dfrac{2}{7}p + 1 = \dfrac{1}{3}p$

36. $S = \dfrac{1}{10} + \dfrac{1}{25} + \dfrac{1}{50}$

37. $-x + \dfrac{1}{7} = \dfrac{1}{2}x$

38. $0 = 1 + \dfrac{2}{9}c - c$

See Example 6.

39. **ELEC** Find the total resistance in a parallel DC circuit with two branches rated at 12 Ω and 30 Ω, respectively.

40. **ELEC** A parallel DC circuit has three branches rated at 12 Ω, 15 Ω, and 20 Ω. Find the total resistance.

41. **ELEC** Three branches of a parallel DC circuit are rated at 16 Ω, 24 Ω, and 32 Ω. Find the total resistance.

42. **ELEC** A parallel DC circuit has a total resistance of 3 Ω. One branch alone produces 18 Ω resistance. What is the resistance of the other branch?

2 Set up an equation and solve. *See Example 7.*

43. **INDTEC** A kiln fires 8 large vases in 2 h. How many vases can be fired in 20 h?

44. A baker can bake 48 cupcakes in 45 minutes. How many hours will it take to bake 240 cupcakes?

See Example 8.

45. **INDTEC** One machine packs 1 day's salmon catch in 8 h. A second machine packs 1 day's catch in 5 h. How much time does it require for 1 day's catch to be packed if both machines are used?

46. **CON** A painter can paint a house in 6 days. Another painter takes 8 days to paint the same house. If they work together, how much time will it take them to paint the house?

47. **INDTEC** One bottling machine can fill 400 bottles of water in 1 h and another machine can fill 400 bottles of water in $1\frac{1}{2}$ h. If both machines are working, how much time does it take to fill 400 bottles of water?

48. **AG/H** A tank has two pipes entering. Pipe 1 alone fills the tank in 4 min and pipe 2 takes 12 min to fill the tank. How much time does it take to fill the tank if both pipes are operating at the same time?

49. **AG/H** A tank has two pipes entering. Pipe 1 alone fills the tank in 30 min and together the pipes take 10 min to fill the tank. How much time does pipe 2 need to fill the tank alone?

50. **INDTR** A printing press produces 1 day's newspaper in 4 h. A higher-speed press does 1 day's newspaper in 2 h. How much time does it take both presses to produce 1 day's newspaper?

See Example 9.

51. One pipe can fill a tank in 1 hour. A second pipe can empty the tank in 1 hour and 15 minutes. If both pipes are operating at the same time, how long will it take for the tank to be full?

52. **AG/H** A tank has two pipes entering it and one leaving it. Pipe 1 fills the tank in 3 min. Pipe 2 takes 7 min to fill the same tank. Pipe 3, however, empties the tank in 21 min. How much time does it take to fill the tank with all three pipes operating at the same time?

3 Solve the equations. *See Example 10.*

53. $2.3x = 4.6$

54. $0.8R = 0.6$ (round to nearest tenth)

55. $0.33x + 0.25x = 3.5$ (round to nearest hundredth)

56. $0.3a = 4.8$

57. $1.5p = 7$ (round to nearest tenth)

58. $0.04x = 0.08 - x$ (round to nearest hundredth)

59. $0.4p = 0.014$

60. $0.47 = R + 0.4R$ (round to nearest hundredth)

61. $2.3 = 5.6 + y$

62. $4.3 = 0.3x - 7.34$

63. $2x + 3.7 = 10.3$

64. $0.16 + 2.3x = -0.3$

65. $1.5x + 2.1 = 3$

66. $3.82 - 2.5y = 1$

67. $0.15p = 2.4$

See Example 11.

68. **PFIN** Find the interest paid on a loan of $800 at $8\frac{1}{2}\%$ interest for 2 years.

69. **PFIN** Find the interest paid on a loan of $2,400 for 1 year at an interest rate of 11%.

Solve the problems using decimal equations.

70. **PFIN** Find the total amount of money (maturity value) that the borrower will pay back on a loan of $1,400 at $12\frac{1}{2}\%$ simple interest for 11 years.

71. **PFIN** Maddy Brown needed start-up money for her landscape service. She borrowed $12,000 for 30 months and paid $360 interest on the loan. What interest rate did she pay?

72. **PFIN** Find the rate of interest on an investment of $2,500 made by Nurse Honda for a period of 2 years if she received $612.50 in interest.

See Example 12.

73. **INDTR** The circumference of a circle equals π times the diameter. If a steel rod has a diameter of 1.5 in., what is the circumference of the rod to the nearest hundredth? (Circumference is the distance around a circle.) Use the calculator π key.

74. **AUTO** If the formula for force is force = pressure $\times$ area, how many pounds of force are produced by a pressure of 35 pounds per square inch on a piston whose surface area is 2.5 in^2? Express the pounds of force as a decimal number.

75. **AUTO** The distance formula is distance = rate $\times$ time If a tractor-trailer rig is driven 422.5 mi at 65 mi/h on interstate highways, how long does the trip take?

76. **AUTO** The distance formula is distance = rate $\times$ time If a trucker drove 682.5 mi at 55 mi per h (mi/h), how long did she drive? (Answer to the nearest whole number.)

77. **ELEC** Electrical resistance in ohms (Ω) is voltage V divided by amperage A. Find the resistance to the nearest tenth for a motor with a voltage of 12.4 V requiring 1.5 A.

78. **ELEC** The formula for electrical power is $V = \frac{W}{A}$: voltage (V) equals wattage (W) divided by amperage (A). Find the voltage needed for a circuit of 500 W with a current of 3.2 A.

7–4 Inequalities and Sets

LEARNING OUTCOMES

1 Use set terminology.

2 Show inequalities on a number line and write inequalities in interval notation.

LC LEARNING CATALYTICS

1. Arrange the numbers in order from smallest to largest: $-2.1, -3.5, 0.4$, and -2.09.

In Chapter 1 we defined an *inequality* as a mathematical statement that quantities *are not equal.* The symbol $\neq$ is read "is not equal to."

$$5 \neq 7 \qquad \text{5 is not equal to 7.}$$

More specific inequality symbols are $<$ (is less than) and $>$ (is greater than).

$$5 < 7 \qquad \text{5 is less than 7.} \qquad 7 > 5 \qquad \text{7 is greater than 5.}$$

The symbol $\leq$ indicates *is less than or equal to,* and the symbol $\geq$ indicates *is greater than or equal to.* As in equations, we represent missing amounts in inequalities by letters.

$$x \leq 7 \qquad \text{x is less than or equal to 7.} \qquad y \geq 5 \qquad \text{y is greater than or equal to 5.}$$

Solve an inequality: a process for finding the value or set of values of the unknown quantity that makes the statement true

To **solve an inequality**, we find the value or set of values of the unknown quantity that makes the statement true.

1   Use Set Terminology.   A *set* is a group or collection of items. For example, a set of days of the week that begin with the letter T includes Tuesday and Thursday. In this chapter, we examine sets of numbers. Numbers that belong to a set are called **members or elements of a set.** The description of a set clearly distinguishes between the elements that belong to the set and those that do not belong. This description can be given in words or by using **set notation.**

Members or elements of a set: numbers or items that belong to a set

Set notation: a notation used to describe the members of a set. It can take the form of a roster or set-builder notation

Roster of the elements of a set: a notation that lists the elements of a set, encloses them in braces, and separates them with commas

To illustrate set notation, we examine the set of whole numbers between 1 and 8. One notation is to make a list or **roster of the elements of a set.** These elements are enclosed in braces and separated with commas.

Set of whole numbers between 1 and 8 = $\{2, 3, 4, 5, 6, 7\}$

TIP   Common Symbols for Sets   The following capital letters are used to denote the indicated set of numbers:

N = natural numbers W = whole numbers
Z = integers Q = rational numbers
I = irrational numbers R = real numbers
M = imaginary numbers C = complex numbers

Symbols that substitute for phrases that are often used in describing sets are

| is read   "such that"
$\in$ is read   "is an element of"

Set-builder notation: a notation describing the elements of a set using the symbols for sets and a variable

Another method of illustrating a set is **set-builder notation**. The elements of the set are written in the form of an inequality using a variable to represent all the elements of the set.

Set of whole numbers between 1 and 8 = $\{x \mid x \in W \text{ and } 1 < x < 8\}$

This statement is read "the set of values of x *such that* each x *is an element of* the set of whole numbers and x is between 1 and 8."

EXAMPLE 1

Answer the statements as true or false.

(a) 5 is an element of the set of whole numbers.

(b) $\frac{3}{4}$ is an element of the set of Z.

(c) $-8 \in$ the set of real numbers with the property $\{x \mid x < -5\}$.

(d) 3.7 is an element of the set of rational numbers with the property $\{x \mid x > 3.7\}$.

(e) The set of prime numbers that are evenly divisible by 2 is an empty set.

(a) 5 is an element of the set of whole numbers. **True.**

(b) $\frac{3}{4}$ is an element of the set of integers. **False.** Integers include only whole numbers and their opposites.

(c) −8 is an element of the set of real numbers with the property $\{x \mid x < -5\}$.
True. −8 is a real number and it is less than −5. Both conditions are satisfied.

(d) 3.7 is an element of the set of rational numbers with the property $\{x \mid x > 3.7\}$.
False. 3.7 is a rational number. A number can equal itself, but a number cannot be greater than itself.

(e) The set of prime numbers evenly divisible by 2 is the empty set. **False.** 2 is a prime number, and 2 is divisible by 2. Therefore, the set contains the element 2.

See Exercises 1–4.

> **To identify the elements of a set using a roster or set-builder notation:**
>
> Roster
>
> **1.** List the elements of the set, separating the elements with commas.
>
> **2.** Enclose the list in braces $\{\}$.
>
> Set-Builder Notation
>
> **1.** Represent the elements of the set with a variable.
>
> **2.** Write symbolic statements that identify the elements of the set.
>
> **3.** Use the symbol $\mid$ to represent the phrase "such that."
>
> **4.** Enclose the results of Steps 1–3 in braces $\{\}$.

STOP AND CHECK

1. Write the set of natural numbers between 3 and 9 inclusive of the end points as a roster.

Answer:
1. $\{3, 4, 5, 6, 7, 8, 9\}$

EXAMPLE 2

Write the set of integers between −2 and 5 as a roster.

The word *between* does not include the stated boundaries. Therefore, the first element is −1 and the last is 4.

$$\{-1, 0, 1, 2, 3, 4\}$$

See Exercises 5–8.

STOP AND CHECK

1. Write the set of natural numbers that are less than or equal to −2 in set-builder notation.

Answer:
1. $\{x \mid x \in N \text{ and } x \le -2\}$

EXAMPLE 3

Write the set of rational numbers that are greater than 5 in set-builder notation.

Let x represent all of the elements of the set. The set of rational numbers is represented by the letter Q.

$\{x \mid x \in Q \text{ and } x > 5\}$ Read as the set of values of x such that x is an element of the set of rational numbers and $x > 5$.

See Exercises 9–10.

Empty set, ϕ, { }: a set containing no elements

A special set is the empty set. The **empty set** is a set containing no elements. Symbolically, the empty set is identified as $\{\ \}$ or ϕ (Greek letter phi, pronounced "fee"). An example of an empty set is the set of whole numbers between 1 and 2. The set of rational numbers between 1 and 2 includes numbers like $1\frac{1}{2}$ and 1.3, but there are no whole numbers between 1 and 2. Thus, the set of whole numbers between 1 and 2 is the empty set.

2 **Show Inequalities on a Number Line and Write Inequalities in Interval Notation.** The set of numbers that is represented by an inequality in one variable can be shown visually on a number line.

> **To graph an inequality in one variable:**
>
> 1. Determine the boundaries of the inequality, if they exist.
>
> 2. Locate the boundaries on a number line.
>
>   (a) If the inequality is $<$ or $>$, the boundary is *not* included. Represent this with a parenthesis,) or (.
>
>   (b) If the inequality is $\leq$ or $\geq$, the boundary *is* included. Represent this with a bracket,] or [.
>
> 3. Shade the portion of the number line between the boundaries. If there is no boundary in one or both directions, the graph continues indefinitely in that direction or both directions.

Interval notation: notation used to represent inequalities by separating the boundaries with a comma and enclosing them with a symbol indicating whether the boundary is included or not

Unbounded: a set is unbounded if it has no boundary in one or both directions

Another type of notation used to represent inequalities is **interval notation**. The two *boundaries* are separated by a comma and enclosed with a symbol that indicates whether the boundary is included or not. If there is no boundary in one or both directions, an infinity symbol, $-\infty$ or ∞, is used. We can say the set is **unbounded**.

> **To write an inequality in interval notation:**
>
> 1. Determine the boundaries of the interval, if they exist.
>
> 2. Write the left boundary or $-\infty$ first and write the right boundary or $+\infty$ second. Separate the two with a comma.
>
> 3. Enclose the boundaries with the appropriate grouping symbols. A parenthesis "(" or ")" indicates that the boundary is *not* included. A bracket "[" or "]" indicates that the boundary *is* included.

STOP AND CHECK

Represent the sets of numbers on a number line and by using interval notation.
1. $x < -2$
2. $x \geq 6$
3. $-1 < x \leq 3$

Answers:

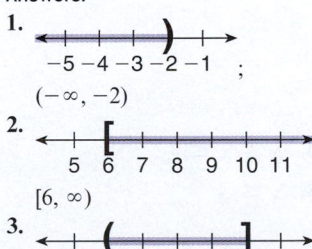

1.

$-5\ -4\ -3\ -2\ -1$; $(-\infty, -2)$

2.

$5\quad 6\quad 7\quad 8\quad 9\quad 10\quad 11$; $[6, \infty)$

3.

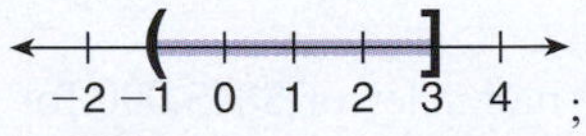

$-2\ -1\ \ 0\ \ 1\ \ 2\ \ 3\ \ 4$; $(-1, 3]$

EXAMPLE 4

Represent the sets of numbers on the number line and by using interval notation.

(a) $1 < x < 8$ (b) $1 \leq x \leq 8$ (c) $x < 3$

(d) $x \geq 3$ (e) all real numbers

(a) $1 < x < 8$

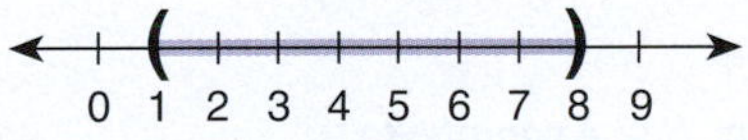

0 1 2 3 4 5 6 7 8 9

(1, 8)
Interval notation

FIGURE 7–5

Parentheses are used for both boundaries to indicate that 1 and 8 are *not* included in the solution set.

(b) $1 \leq x \leq 8$

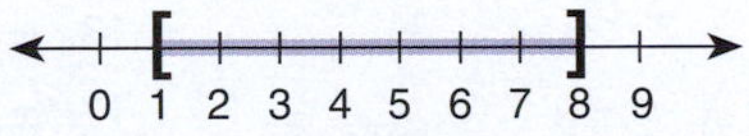

0 1 2 3 4 5 6 7 8 9

[1, 8]
Interval notation

FIGURE 7–6

Brackets are used for both boundaries to indicate that 1 and 8 *are* included in the solution set.

(c) $x < 3$

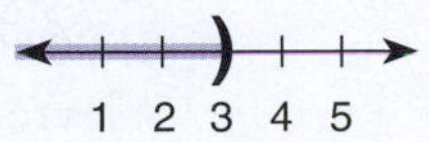

1 2 3 4 5

$(-\infty, 3)$
Interval notation

FIGURE 7–7

The left boundary is $-\infty$, and an open parenthesis is used as the left enclosure symbol. The right boundary is 3 but is not in the solution set so a close parenthesis is used as the right enclosure symbol.

(d) $x \geq 3$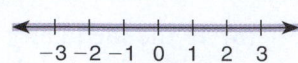

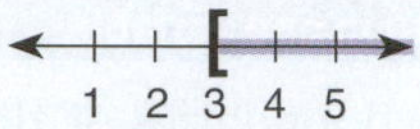

$[3, \infty)$
Interval notation

FIGURE 7–8

The left boundary is 3 and since 3 *is* included in the solution set, a bracket is used. The right boundary is ∞.

(e) All real numbers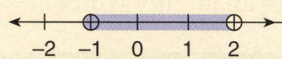

$(-\infty, \infty)$
Interval notation

FIGURE 7–9

All real numbers is an unbounded set of numbers, thus parentheses are used in the interval notation. The number line extends indefinitely in both directions.

See Exercises 11–25.

> **TIP** **Alternative Representation of an Inequality on a Number Line** Boundaries on a number line can also be indicated with open or shaded circles (Fig. 7–10).
>
> $-1 \leq x < 2$
>
> $[-1, 2)$
>
> **FIGURE 7–10**

7-4 EXERCISES   MyLab Math For additional practice go to your study plan in MyLab Math.

1 Answer the statements as true or false. *See Example 1.*

1. 0 is an element of the natural numbers.

2. $\frac{5}{8} \in N$

3. $-6 \in Z$

4. 5.3 is an element of the set of rational numbers with the property $\{x \mid x \leq 5.3\}$.

Write the sets as a roster. *See Example 2.*

5. The whole numbers between 5 and 12.

6. The even numbers between 3 and 8.

7. The whole numbers that are multiples of 5 and between 10 and 40.

8. The negative integers greater than -5.

Write the sets in set-builder notation. *See Example 3.*

9. $\{1, 3, 5, 7, 9\}$

10. $\{-7, -5, -3, -1\}$

2 Represent the sets on the number line and by using interval notation. *See Example 4.*

11. $5 < x < 9$

12. $-7 < x < -3$

13. $-5 \leq x \leq -3$

14. $8 \leq x \leq 12$

15. $x < 7$

16. $x < -2$

17. $x \geq 2$

18. $x \leq 3$

19. $x > 5$

20. $x \geq -3$

21. $x \leq -2$

22. All real numbers greater than -6.

23. University Trailer Company had sales of $843,000 for the previous year. The projected sales for the current year are more than the previous year, but less than $1,000,000, which is projected for the next year. Express sales for the current year using interval notation.

24. All real numbers with a minimum of 4 and a maximum of 18.

25. College classes generally must have a minimum number of students to avoid cancellation. If that number is 12, write and graph an inequality to represent the number of students that are in a canceled class.

7-5 Solving Linear Inequalities

LEARNING OUTCOME

1 Solve a linear inequality with one variable.

LC LEARNING CATALYTICS

1. List the set of integers between −1 and 3 as a roster.
2. List the set of integers between −1 and 3 inclusive as a roster.

1 Solve a Linear Inequality with One Variable. The procedures for solving inequalities are similar to the procedures for solving equations. The solution to an inequality is a *set* of numbers that satisfies the conditions of the statement.

The statement $x \leq 9$ means the *solution set* is 9 or any number less than 9.

Compare Figs. 7–11 and 7–12. In Fig. 7–11, the boundary 9 is *not* part of the solution set; thus, 9 is represented on the number line with a parenthesis. In Fig. 7–12, the boundary 9 *is* part of the solution set; thus, 9 is represented on the number line with a bracket.

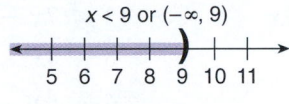

$x < 9$ or $(-\infty, 9)$

FIGURE 7–11

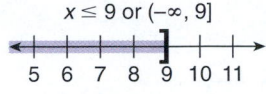

$x \leq 9$ or $(-\infty, 9]$

FIGURE 7–12

Look at the similarities in solving a linear equation and a linear inequality.

STOP AND CHECK

1. Solve the inequality $3x - 2 \leq x + 6$ and express the solution in set-builder notation, on a number line, and in interval notation.

Answer:

1. Set-builder notation: $\{x \mid x \leq 4\}$;

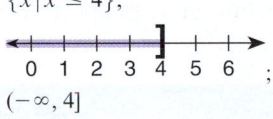

$(-\infty, 4]$

EXAMPLE 1

Solve the equation $4x + 6 = 2x - 2$ and the inequality $4x + 6 > 2x - 2$. Express the solution of the inequality in set-builder notation, on the number line, and in interval notation.

$4x + 6 = 2x - 2$	$4x + 6 > 2x - 2$	Sort.
$4x - 2x = -6 - 2$	$4x - 2x > -6 - 2$	Combine.
$2x = -8$	$2x > -8$	Divide.
$\dfrac{2x}{2} = \dfrac{-8}{2}$	$\dfrac{2x}{2} > \dfrac{-8}{2}$	
$x = -4$	$x > -4$ or $\{x \mid x > -4\}$	Set-builder notation
	$(-4, \infty)$	Interval notation

The solution of the equation is −4 (Fig. 7–13). The solution set of the inequality is any number greater than, but not including, −4 (Fig. 7–14).

$x = -4$

FIGURE 7–13

$x > -4$

FIGURE 7–14

See Exercises 1–9.

Symbolic representation of an inequality: a modified version of set-builder notation that includes just the inequality statement that represents the solution of an inequality

TIP **Three Ways to Represent the Solution Set of Inequalities: Symbolically, Graphically, and Interval Notation** In the preceding example, the solution to the inequality $4x + 6 > 2x - 2$ was written three ways.

▶ Symbolic representation: $x > -4$

▶ Graphical representation: Fig. 7–15

▶ Interval notation: $(-4, \infty)$

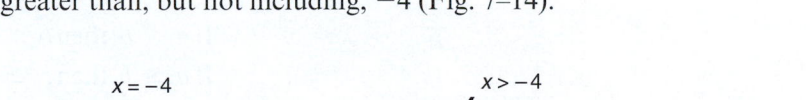

FIGURE 7–15

A modified version of set-builder notation that includes just the inequality statement that represents the solution of an inequality we will refer to as the **symbolic representation of an inequality**.

The symbolic representation evolves from the process of solving the inequality using the properties of inequalities and techniques for isolating the variable. Both the symbolic representation and the interval notation help us interpret and visualize the solution (Fig. 7–15) without having to see the solution represented on the number line.

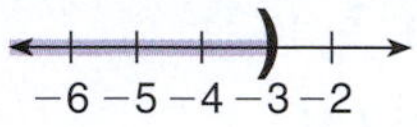

EXAMPLE 2

Solve the equation $3x + 5 = 7x + 13$ and the inequality $3x + 5 < 7x + 13$. Represent the solution of the inequality symbolically, in interval notation, and on a number line.

$3x + 5 = 7x + 13$	$3x + 5 < 7x + 13$	Sort.
$5 - 13 = 7x - 3x$	$5 - 13 < 7x - 3x$	Combine.
$-8 = 4x$	$-8 < 4x$	Divide.
$\dfrac{-8}{4} = \dfrac{4x}{4}$	$\dfrac{-8}{4} < \dfrac{4x}{4}$	
$-2 = x$	$-2 < x$	Interpret solution.
$x = -2$	$x > -2$ or $(-2, \infty)$	We prefer writing the variable on the left. Saying -2 *is less than x* is the same as saying *x is greater than* -2.

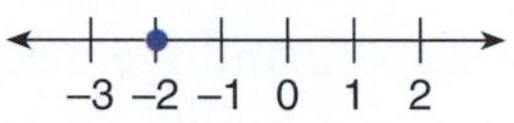

FIGURE 7–16

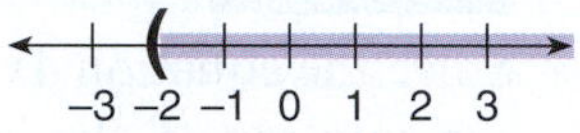

FIGURE 7–17

See Exercises 10–11.

The preceding example illustrates a very important difference in solving equations and solving inequalities. The sides of an equation can be interchanged without making any changes in the equal sign. This interchangeability is due to the *symmetric property of equality;* that is, if $a = b$, then $b = a$. The symmetric property does *not* apply to inequalities.

> **Interchanging the sides of an inequality:**
>
> When the sides of an inequality are interchanged, the sense of the inequality (direction of the inequality symbol) is reversed.
>
> $$\text{If } a < b, \text{ then } b > a. \qquad \text{If } a > b, \text{ then } b < a.$$
> $$\text{If } a \leq b, \text{ then } b \geq a. \qquad \text{If } a \geq b, \text{ then } b \leq a.$$

Sense of an inequality: the appropriate comparison symbol: less than, greater than, less than or equal to, and greater than or equal to

FIGURE 7–18

The **sense of an inequality** is the appropriate comparison symbol: less than, greater than, less than or equal to, and greater than or equal to.

To determine the effect of multiplying or dividing an inequality by a negative number, look at specific numbers and the number line (Fig. 7–18). We start with a true statement.

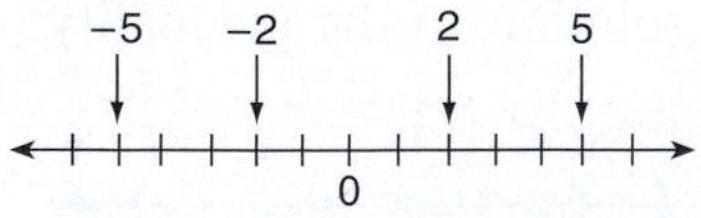

$$2 < 5 \qquad \text{2 is to the left of 5 on the number line.}$$

Then, multiply each side of the inequality by -1.

$$-1(2) \overset{?}{<} -1(5) \qquad \text{Is } -1(2) < -(5)?$$
$$-2 \overset{?}{<} -5 \qquad \text{This is a false statement: } -2 \text{ is to the right of } -5 \text{ and is greater than } -5.$$

To make the statement true, the sense of the inequality must be reversed.

$$-2 > -5$$

Multiplying or dividing an inequality by a negative number:

If both sides of an inequality are multiplied or divided by a negative number, the sense of the inequality is reversed.

$$\text{If } a < b, \text{ then } -a > -b. \qquad \text{If } a > b, \text{ then } -a < -b.$$
$$\text{If } a \leq b, \text{ then } -a \geq -b. \qquad \text{If } a \geq b, \text{ then } -a \leq -b.$$

To solve linear inequalities:

1. Follow the same sequence of steps that is normally used to solve a similar linear equation.

2. The sense of the inequality remains the same unless one of the following situations occurs:

   (a) The sides of the inequality are interchanged.

   (b) The steps used in solving the inequality require that the entire inequality (both sides) be multiplied or divided by a negative number.

3. If either situation **(a)** or **(b)** in Step 2 occurs in solving an inequality, *reverse* the sense of the inequality; that is, less than ($<$) becomes greater than ($>$), and vice versa.

STOP AND CHECK

Solve the inequality. Represent the solution symbolically, in interval notation, and on a number line.

1. $5x + 3 \leq 2(12 + x)$

Answer:

1. $x \leq 7; (-\infty, 7]$;

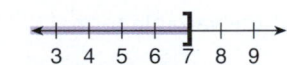

EXAMPLE 3

Solve the inequality, $4x - 2 \leq 3(25 - x)$. Represent the solution symbolically, in interval notation, and on a number line.

$4x - 2 \leq 3(25 - x)$	Distribute.
$4x - 2 \leq 75 - 3x$	Collect variable terms on the left and constants on the right (sort).
$4x + 3x \leq 75 + 2$	Combine like terms.
$7x \leq 77$	Divide by the coefficient of the variable. The coefficient is positive.
$\dfrac{7x}{7} \leq \dfrac{77}{7}$	

$x \leq 11 \quad$ or $\quad (-\infty, 11] \quad$ Solution set (See Fig. 7–19.)

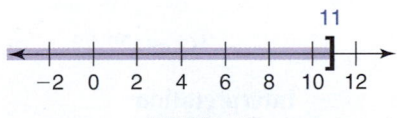

FIGURE 7–19

See Exercises 12–16.

STOP AND CHECK

Solve the inequality. Represent the solution symbolically, in interval notation, and on a number line.

1. $3x - (x + 5) < 7x - 2(x + 3)$

Answer:

1. $x > \frac{1}{3}; (\frac{1}{3}, \infty)$;

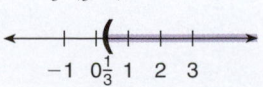

EXAMPLE 4

Solve the inequality, $2x - 3(x + 2) > 5x - (x - 5)$. Represent the solution symbolically, in interval notation, and on a number line.

$2x - 3(x + 2) > 5x - (x - 5)$	Distribute.
$2x - 3x - 6 > 5x - x + 5$	Combine like terms.
$-x - 6 > 4x + 5$	Sort.
$-x - 4x > 5 + 6$	Combine like terms.
$-5x > 11$	Divide by the coefficient of the variable. The coefficient is negative.

$$\frac{-5x}{-5} < \frac{11}{-5}$$

Both sides were divided by a negative number, so reverse the sense of the inequality.

$$x < -\frac{11}{5} \quad \text{or} \quad \left(-\infty, -\frac{11}{5}\right)$$

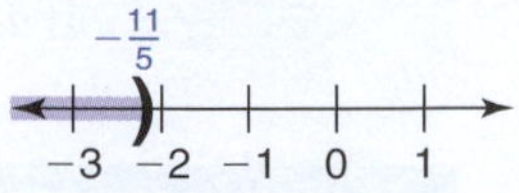

FIGURE 7–20

See Exercises 17–18.

EXAMPLE 5

An electrically controlled thermostat is set so the heating unit automatically comes on and continues to run when the temperature is equal to or below 72°F. At what Celsius temperatures to the nearest hundredth of a °C will the heating unit come on? One formula relating Celsius and Fahrenheit temperatures is $°F = \frac{9}{5}°C + 32$.

Using the formula $°F = \frac{9}{5}°C + 32$, the heating unit will operate when the expression $\frac{9}{5}°C + 32$ is less than or equal to 72.

$$\frac{9}{5}°C + 32 \leq 72$$

Estimation

The Celsius value will be less than the Fahrenheit value.

$$\frac{9}{5}°C + 32 \leq 72$$

Apply the addition axiom to isolate the variable term.

$$\frac{9}{5}°C \leq 72 - 32$$

Combine like terms.

$$\frac{9}{5}°C \leq 40$$

Multiply by the reciprocal of the coefficient of the variable term.

$$\frac{5}{9}\left(\frac{9}{5}\right)°C \leq \left(\frac{5}{9}\right)(40)$$

$$°C \leq \frac{200}{9}$$

Interpret solution.

$$°C \leq 22.22 \quad \text{or} \quad (-\infty, 22.22]$$

To nearest hundredth

Interpretation

The heating unit comes on and continues to run if the temperature is less than or equal to 22.22°C.

See Exercises 19–20.

7-5 EXERCISES

MyLab Math For additional practice go to your study plan in MyLab Math.

1 Solve the inequalities. Show the solution set on a number line and by using interval notation. *See Example 1.*

1. $y - 3 > 5$

2. $x + 7 < 8$

3. $x - 9 \leq -12$

4. $3x + 7x < 60$

5. $b + 6 > 5$

6. $5t - 18 < 12$

7. $2y + 7 < 17$

8. $2t + 6 \leq t + 13$

9. $4y - 8 > 2y + 14$

See Example 2.

10. $3a - 8 \geq 7a$

11. $x + 7 \leq 8x$

See Example 3.

12. $3(7 + x) \geq 30$

13. $6(3x - 1) < 12$

14. $2x > 7 - (x + 6)$

15. $4a + 5 \leq 3(2 + a) - 4$

16. $15 - 3(2x + 2) > 6$

See Example 4.

17. $5a - 6a \leq 3$

18. $y + 3y \leq 32$

See Example 5.

19. Kevin Presley sold $196 more than twice as much merchandise as Robyn Presley. If Kevin sold at least $52,800, how much did Robyn sell? Write an inequality to represent the facts and solve.

20. A supplier prices one brand of recordable CD at $3.60 and another brand of recordable CD at more than twice this price. Write an inequality to represent the facts and express the cost of the more expensive CD. Show the solution set on a number line.

| 7–6 | Solving Compound Inequalities |

LEARNING OUTCOMES

1 Identify subsets of sets and perform set operations.

2 Solve compound inequalities with the conjunction condition.

3 Solve compound inequalities with the disjunction condition.

LC LEARNING CATALYTICS

1. Make a roster of the set of integers between −2 and 2.

Compound inequality: a mathematical statement that combines two statements of inequality

Subset, ⊂: a set is a subset of a second set if every element of the first set is also an element of the second set

In many applications of inequalities, more than one condition is placed on the solution. As an example, when measurements are made, a range of acceptable values is specified. All measurements are approximations, and the range of acceptable values is generally stated as a tolerance. If an acceptable measurement is specified to be within ±0.005 in., then the range of acceptable values can be from 0.005 in. *less than* the ideal measurement to 0.005 in. *more than* the ideal measurement. This range can be stated symbolically as a compound inequality. A **compound inequality** is a mathematical statement that combines two statements of inequality.

The conditions placed on a compound inequality may use the connective *and* to indicate both conditions must be met simultaneously. Such compound inequalities may be written as a continuous statement. If an ideal measurement is 3 in. with a tolerance of ±0.005 in., the range of acceptable values would be

$$3 - 0.005 \leq x \leq 3 + 0.005$$
$$2.995 \leq x \leq 3.005$$

The conditions placed on a compound inequality may use the connective *or* to indicate that either condition may be met. Such compound inequalities *must* be written as two separate statements using the connective *or*.

1 Identify Subsets of Sets and Perform Set Operations. Before solving compound inequalities, let's look at additional properties of sets. A concept related to sets is the concept of subsets. The symbol ⊂ is read "is a subset of." A set is a **subset** of a second set if every element of the first set is also an element of the second set. If $A = \{1, 2, 3\}$ and $B = \{2\}$, then B is a subset of A. This is written in symbols as $B \subset A$ and is illustrated in Fig. 7–21.

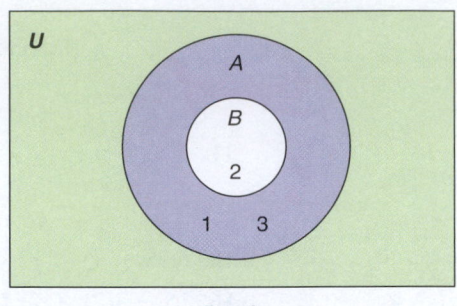

$B \subset A$

FIGURE 7–21

Universal set: a set that includes all the elements for a given description

Two special sets are the universal set and the empty set. The **universal set** includes all the elements for a given description. The universal set is sometimes identified as U. The *empty set* is a set containing no elements. The empty set is identified as $\{\ \}$ or ϕ.

> **Special set relationships:**
>
> Every set is a subset of itself: $A \subset A$
>
> The empty set is a subset of every set: $\phi \subset A$

STOP AND CHECK
Using the set definitions in Example 1, answer the statements as true or false.
1. $D \subset B$
2. $B \subset D$
3. $E \subset D$

Answers:
1. False **2.** False **3.** True

EXAMPLE 1

Using the given set definitions, answer the statements **(a)**–**(h)** as true or false.

$$U = \{0, 1, 2, 3, 4, 5, 6, 7, 8, 9\}; A = \{1, 3, 5, 7, 9\};$$
$$B = \{0, 2, 4, 6, 8\}; C = \{1, 2, 3\}; D = \{1\}; E = \{\ \}$$

(a) $A \subset U$ **(b)** $B \subset U$ **(c)** $A \subset B$ **(d)** $C \subset A$

(e) $C \subset U$ **(f)** $D \subset A$ **(g)** $E \subset U$ **(h)** $E \subset B$

(a) $A \subset U$ **True;** every element of A is also an element of U.

(b) $B \subset U$ **True;** every element of B is also an element of U.

(c) $A \subset B$ **False;** 1, 3, 5, 7, and 9 are not elements of B.

(d) $C \subset A$ **False;** 2 is not an element of A.

(e) $C \subset U$ **True;** every element of C is also an element of U.

(f) $D \subset A$ **True;** 1 is an element of A.

(g) $E \subset U$ **True;** the empty set is a subset of every set.

(h) $E \subset B$ **True;** the empty set is a subset of every set. **See Exercises 1–4.**

Union of two sets: a set that includes all elements that appear in *either* of the two sets; associated with the condition "or"

Intersection of two sets: a set that includes all elements that appear in *both* of the two sets; associated with the condition "and"

Two common set operations are union and intersection. The **union of two sets** is a set that includes all elements that appear in *either* of the two sets. Union is generally associated with the condition "or." The symbol for union is $\cup$. The colored portion in Fig. 7–22 represents $A \cup B$. The **intersection of two sets** is a set that includes all elements that appear in *both* of the two sets. Intersection is generally associated with the condition "and." The symbol for intersection is $\cap$. The colored portion of Fig. 7–23 represents $A \cap B$.

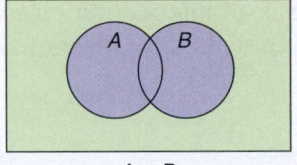

$A \cup B$

FIGURE 7-22

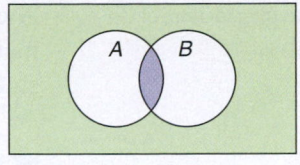

$A \cap B$

FIGURE 7-23

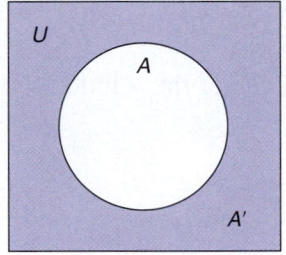

FIGURE 7-24

Complement of a set: a set that
includes every element of the
universal set that is *not* an element
of the given set

EXAMPLE 2

If $A = \{1, 2, 3, 4, 5\}$, $B = \{1, 3, 5, 7, 9\}$, and $C = \{2, 4, 6, 8, 10\}$, list the elements in the following sets.

(a) $A \cup B$ (b) $A \cup C$ (c) $B \cup C$ (d) $A \cap B$ (e) $A \cap C$ (f) $B \cap C$

(a) $A \cup B = \{1, 2, 3, 4, 5, 7, 9\}$ All elements in either A or B

(b) $A \cup C = \{1, 2, 3, 4, 5, 6, 8, 10\}$ All elements in either A or C

(c) $B \cup C = \{1, 2, 3, 4, 5, 6, 7, 8, 9, 10\}$ All elements in either B or C

(d) $A \cap B = \{1, 3, 5\}$ Elements common to both A and B

(e) $A \cap C = \{2, 4\}$ Elements common to both A and C

(f) $B \cap C = \{\ \}$ or ϕ No elements common to both B and C **See Exercises 5–7.**

Another set operation is complement. The **complement of a set** is a set that includes every element of the universal set that is *not* an element of the given set. The symbol for the complement of a set is $'$ and is read "prime." A' is represented by the colored portion of Fig. 7–24.

EXAMPLE 3

If $U = \{0, 1, 2, 3, 4, 5\}$, $A = \{1, 3, 5\}$, $B = \{0, 2, 4\}$, and $C = \{1, 2, 3\}$, list the elements in the following sets.

(a) A' (b) B' (c) C' (d) $(A \cup B)'$ (e) $(A \cap C)'$

(a) $A' = \{0, 2, 4\}$ Elements of U not in A

(b) $B' = \{1, 3, 5\}$ Elements of U not in B

(c) $C' = \{0, 4, 5\}$ Elements of U not in C

(d) $(A \cup B)'$

$A \cup B = \{0, 1, 2, 3, 4, 5\}$ Elements in either A or B

$(A \cup B)' = \{\ \}$ or ϕ Elements of U not in either A or B

(e) $(A \cap C)'$

$A \cap C = \{1, 3\}$ Elements common to both A and C

$(A \cap C)' = \{0, 2, 4, 5\}$ Elements of U not common to both A and C

See Exercises 8–10.

2 **Solve Compound Inequalities with the Conjunction Condition.** As we said before, a compound inequality is a statement that places more than one condition on the variable of the inequality. The solution set is the set of values for the variable that meets all conditions of the problem. One type of compound inequality is a *conjunction*.

Conjunction: an intersection, or *and,* set relationship

A **conjunction** is an intersection, or *and,* set relationship. Both conditions must be met simultaneously in a conjunction. A conjunction can be written as a continuous statement.

> **Property for conjunctions:**
>
> When a, b, and x are real numbers:
>
> $$\text{If } a < x \text{ and } x < b, \text{ then } a < x < b.$$
> $$\text{If } a > x \text{ and } x > b, \text{ then } a > x > b.$$
>
> Similar compound inequalities may also use $\leq$ and $\geq$.
>
> $$\text{If } a \leq x \text{ and } x \leq b, \text{ then } a \leq x \leq b.$$
> $$\text{If } a \geq x \text{ and } x \geq b, \text{ then } a \geq x \geq b.$$

> **To solve a compound inequality that is a conjunction:**
>
> 1. Separate the compound inequality into two simple inequalities using the conditions of the conjunction.
>
> 2. Solve each simple inequality.
>
> 3. Determine the solution set that includes the *intersection* of the solution sets of the two simple inequalities.

STOP AND CHECK

Find the solution set for each compound inequality. Graph the solution set on a number line and write the solution set in interval notation.
1. $-3 < x - 2 < 5$
2. $0 \leq x + 3 \leq 5$
3. $-8 < 3x - 2 \leq 10$

Answers:
1. $-1 < x < 7$;

![number line from -1 to 7 with open circles at -1 and 7]
$(-1, 7)$
2. $-3 \leq x \leq 2$;

![number line from -4 to 2 with brackets at -3 and 2]
$[-3, 2]$
3. $-2 < x \leq 4$;

![number line from -4 to 4 with open at -2 and bracket at 4]
$(-2, 4]$

EXAMPLE 4

Find the solution set for each compound inequality. Graph the solution set on a number line and write the solution set in interval notation.

(a) $5 < x + 3 < 12$ **(b)** $-7 \leq x - 7 \leq 1$

(c) $-6 < 3x < -3$ **(d)** $17 \leq 5x + 7 \leq 32$

(a) $5 < x + 3 < 12$ Separate into two inequalities.

$\qquad 5 < x + 3$ and $x + 3 < 12$ Solve each inequality.

$\qquad 5 - 3 < x \qquad\qquad\quad x < 12 - 3$

$\qquad\quad 2 < x \qquad\qquad\qquad x < 9$ Find the overlap.

Figure 7–25 shows the solution set graphically. The solution set as a continuous statement is **$2 < x < 9$** and in interval notation, **(2, 9).**

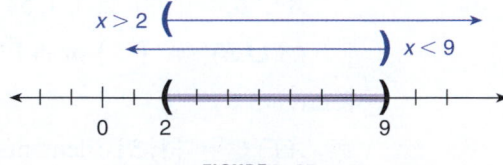

FIGURE 7–25

(b) $-7 \leq x - 7 \leq 1$ Separate into two inequalities.

$\qquad -7 \leq x - 7$ and $x - 7 \leq 1$ Solve each inequality.

$\qquad -7 + 7 \leq x \qquad\qquad\quad x \leq 1 + 7$

$\qquad\quad 0 \leq x \qquad\qquad\qquad x \leq 8$ Find the overlap.

Figure 7–26 shows the solution set graphically. The solution set as a continuous statement is **$0 \leq x \leq 8$** and in interval notation, **[0, 8].**

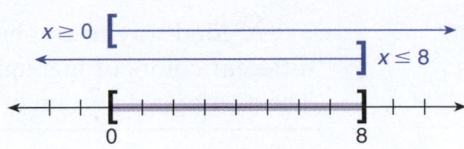

FIGURE 7-26

(c) $-6 < 3x < -3$ Separate into two inequalities.

$-6 < 3x$ and $3x < -3$ Solve each inequality.

$\dfrac{-6}{3} < \dfrac{3x}{3}$ $\dfrac{3x}{3} < \dfrac{-3}{3}$

$-2 < x$ $x < -1$ Find the overlap.

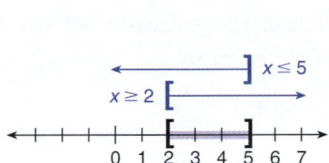

FIGURE 7-27

Figure 7–27 shows the solution set graphically. The solution set as a continuous statement is $-2 < x < -1$ and in interval notation is $(-2, -1)$.

(d) $17 \leq 5x + 7 \leq 32$ Separate into two inequalities.

$17 \leq 5x + 7$ and $5x + 7 \leq 32$ Solve each inequality.

$17 - 7 \leq 5x$ $5x \leq 32 - 7$

$10 \leq 5x$ $5x \leq 25$

$2 \leq x$ $x \leq 5$ Find the overlap.

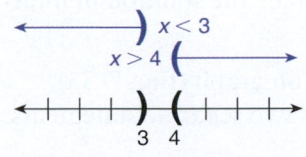

FIGURE 7-28

(e) Figure 7–28 shows the solution set graphically. The solution set as a continuous statement is $2 \leq x \leq 5$ and in interval notation, $[2, 5]$.   **See Exercises 11–20.**

TIP   **Greater Than Versus Less Than**   Even though continuous compound inequalities can be written using *greater than* or *less than* symbols, using *less than* symbols follows the natural positions of the boundaries on the number line. If $a > b > c$, then $c < b < a$. For instance, if $3 > 0 > -2$, then $-2 < 0 < 3$. Compound inequalities use the "less than" symbols more often. Do not mix symbols in a compound inequality.

As with equations and simple inequalities, a compound inequality may have no solution.

STOP AND CHECK

1. Find the solution set for the compound inequality, $-4 < 3x + 2 < -7$.

Answer:

1. No solution

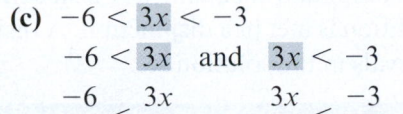

FIGURE 7-29

EXAMPLE 5

Find the solution set for the compound inequality, $7 < 2x - 1 < 5$.

Separate into two inequalities.

$7 < 2x - 1$ and $2x - 1 < 5$

$8 < 2x$ $2x < 6$

$4 < x$ $x < 3$

$x > 4$ Find the overlap.

As shown in Fig. 7–29, there is no overlap, so it is impossible for both conditions to be met at the same time. Also, note that the overall inequality $7 < 5$, is false. **The compound inequality has no solution.**   **See Exercises 21–22.**

TIP   **The Solution of a Conjunction Relationship Is the Overlap**   The importance of a graphical representation of the solution set of a conjunction is that you can visualize the overlap. Or, in the case of no solution to a conjunction, you can see there is no overlap.

A good way to emphasize the overlapped portion on a graph is to use two different colors of highlighters. Yellow and blue highlighters are good choices. The overlapped portion of the graph will be green.

Disjunction: a union, or *or*, set relationship

3   Solve Compound Inequalities with the Disjunction Condition. Disjunction is another type of compound inequality. A **disjunction** is a union, or *or*, set relationship. Either condition is met in a disjunction. A disjunction uses the union symbol to indicate the intervals in the solution set.

> **To solve a compound inequality that is a disjunction:**
>
> 1. Solve each simple inequality.
>
> 2. Determine the solution set that includes the *union* of the solution sets of the two simple inequalities.

STOP AND CHECK

1. Find the solution set for the compound inequality
$x - 1 \leq -5$ or
$x - 1 > 3$. Graph the solution set on a number line and write the solution in symbolic and interval notation.

Answer:
1.

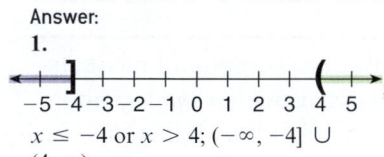

$x \leq -4$ or $x > 4$; $(-\infty, -4] \cup (4, \infty)$

EXAMPLE 6

Find the solution set for each compound inequality. Graph the solution set on a number line and write the solution in symbolic and interval notation.

(a) $x + 3 < -2$ or $x + 3 > 2$ **(b)** $x - 5 \leq -3$ or $x - 5 \geq 3$

(a) $x + 3 < -2$ or $x + 3 > 2$ Solve each inequality.

$\qquad x < -2 - 3 \qquad\qquad x > 2 - 3$

$\qquad \mathbf{x < -5}$ or $\mathbf{x > -1}$ Symbolic notation. See Fig. 7–30.

$\qquad \mathbf{(-\infty, -5)} \qquad \cup \qquad \mathbf{(-1, \infty)}$ Interval notation

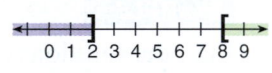

FIGURE 7–30

(b) $x - 5 \leq -3$ or $x - 5 \geq 3$ Solve each inequality.

$\qquad x \leq -3 + 5 \qquad\qquad x \geq 3 + 5$

$\qquad \mathbf{x \leq 2}$ or $\mathbf{x \geq 8}$ Symbolic notation. See Fig. 7–31.

$\qquad \mathbf{(-\infty, 2]} \qquad \cup \qquad \mathbf{[8, \infty)}$ Interval notation

FIGURE 7–31   **See Exercises 23–30.**

> **TIP   Let Your Graph Be Your Guide**   The graphical representation of your solution set can be a great guide for writing the symbolic solution or the solution in interval notation.
>
> Overlap on graph (Fig. 7–32)   Two intervals on graph (Fig. 7–33)
> The solution is one continuous   The solution is two separate statements.
> statement.
>
>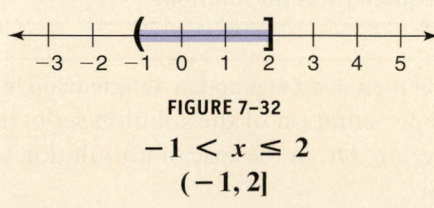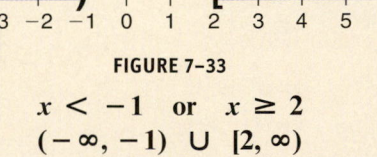
>
> **FIGURE 7–32**   **FIGURE 7–33**
>
> $\mathbf{-1 < x \leq 2}$ $\mathbf{x < -1}$ or $\mathbf{x \geq 2}$
> $\mathbf{(-1, 2]}$ $\mathbf{(-\infty, -1) \cup [2, \infty)}$

7-6 EXERCISES   MyLab Math For additional practice go to your study plan in MyLab Math.

1 Use the sets for Exercises 1–10.

$U = \{-5, -4, -3, -2, -1, 0, 1, 2, 3, 4, 5\}, A = \{-4, -2, 0\}, B = \{1, 2, 3, 4, 5\}, C = \{2, -2\}, D = \{0\}, E = \{\ \}$

Answer each statement as true or false and justify your answer. *See Example 1.*

1. $B \subset U$ **2.** $C \subset A$ **3.** $E \subset B$ **4.** $A \subset B$

List the elements in each set. *See Example 2.*

5. $A \cap B$ **6.** $C \cup D$ **7.** $A \cup B$

See Example 3.

8. A' **9.** $(A \cap D)'$ **10.** $(B \cup C)'$

2 Show the solution set of each inequality on a number line and by using symbolic and interval notation. *See Example 4.*

11. $6 \le 2x \le 8$ **12.** $15 \le 3x \le 21$ **13.** $-3 \le 3x \le 9$

14. $-3 \le 2x - 1 \le 3$ **15.** $0 < 5x < 15$ **16.** $4 \le 6 - x \le 8$

17. $6 < -2x < 12$ **18.** $1 < 6 - x < 3$ **19.** $-8 \le -3x - 2 \le 4$

20. $5 \le 5 + 7x \le 9$

See Example 5.

21. $-5 \le -x + 1 \le -7$ **22.** $1 < x - 2 < 5$

3 Show the solution set of each inequality on a number line and by using symbolic and interval notation. *See Example 6.*

23. $x + 3 < 5$ or $x - 7 > 2$ **24.** $x - 1 \le 2$ or $x \ge 7$ **25.** $2x - 1 \le 7$ or $3x \ge 15$

26. $5 - x \le 2$ or $x + 1 \le 2$ **27.** $5x - 2 \le 3x + 1$ or $2x \ge 7$ **28.** $-2x > 8$ or $3x \ge 27$

29. The blueprint specifications for a part show it has a measure of 5.27 cm with a tolerance of ±0.05 cm. Express the limit dimensions of the part with a compound inequality.

30. An automobile parts manufacturer makes a part that, according to the blueprint, measures 15.2 cm and has a tolerance of ±0.5 cm. Write an inequality that expresses the range of measures an acceptable part must have.

---

7 CHAPTER REVIEW OF KEY CONCEPTS

LEARNING OUTCOMES	KEY CONCEPTS AND EXAMPLES

Section 7–1

1 Identify equations, terms, factors, constants, variables, and coefficients (pp. 316–319).

Verify that an equation is true: 1. Find the value of the expression on each side of the equal sign. **2.** Compare the values for each side. **(a)** If the values for each side are equal, the equation is true. **(b)** If the values for each side are not equal, the equation is not true.

$x = 5 + 2$ is an *equation*. x is the *variable*. 7 is the *root* or *solution*.

Factors are expressions of multiplication.

$5a$ means 5 *times a*. 5 is a *factor* of $5a$. a is a *factor* of $5a$.

LEARNING OUTCOMES	KEY CONCEPTS AND EXAMPLES

Terms are algebraic expressions that are single quantities or products or quotients of quantities.

In the expression $5a + 3b + 7$, $5a$ is a *term* and $3b$ is a *term*. 7 is a *term*.

Constants are terms that contain only numbers.

In $5a + 3b + 7$, the *constant* term is 7.

Variable terms are terms that have at least one letter.

In $5a + 3b + 7$, $5a$ and $3b$ are *variable terms*.

A *coefficient* is one factor as it relates to the remaining factors of a term.

In $5a + 7$, 5 is the *coefficient* of a. 7 has no coefficient.

2 Write verbal interpretations of symbolic statements (p. 319).

1. Locate the variable or variable terms.   **2.** Examine the operations that link the factors and terms.   **3.** Write a verbal statement or statements that describe all the conditions of the symbolic statement.

$2x - 3 = 5$ can be translated into "3 less than twice a number is five."

3 Translate verbal statements into symbolic statements using variables (p. 320).

1. Assign a letter to represent the missing number.   **2.** Identify key words or phrases that imply or suggest specific operations.   **3.** Translate words into symbols.

The statement "A number increased by 13 results in 52" is translated into $x + 13 = 52$.

4 Simplify variable expressions (p. 321).

1. Change subtraction to addition if appropriate.   **2.** Add the constants using the appropriate rule for adding signed numbers. The sum is a signed number.   **3.** Add the numerical coefficients of the like variables using the appropriate rule for adding signed numbers. The sum has the same variable or variables as the like terms being added.

$5m + 4m = (5 + 4)m = 9m$

Apply the distributive principle by multiplying the factor outside the parentheses by each term inside the parentheses.

$-4(x + 2y - 5) = -4x - 8y + 20$

Distributing -1 changes only the sign of each term inside the parentheses.

$-(2x - 3) = -1(2x - 3) = -2x + 3$

Section 7–2

1 Solve linear equations using the addition axiom (pp. 323–325).

1. Locate the variable in the equation.   **2.** Identify the constant that is associated with the variable by addition (or subtraction).   **3.** Add the opposite of the constant to both sides of the equation.

LEARNING OUTCOMES	KEY CONCEPTS AND EXAMPLES

$$x + 3 = 2 \qquad \text{Add } -3 \text{ to both sides.}$$
$$x + 3 - 3 = 2 - 3 \qquad \text{Combine like terms.}$$
$$x + 0 = -1 \qquad x + 0 = x$$
$$x = -1$$

$$9 = x - 6 \qquad \text{Add 6 to both sides.}$$
$$9 + 6 = x - 6 + 6 \qquad \text{Combine like terms.}$$
$$15 = x + 0 \qquad x + 0 = x$$
$$15 = x$$

2 Solve linear equations using the multiplication axiom (pp. 325–327).

1. Multiply both sides of the equation by the reciprocal of the coefficient of the variable term; or **2.** Divide both sides of the equation by the coefficient of the variable term.

Solve:

$$\frac{x}{5} = 8 \qquad\qquad -8x = 24 \qquad\qquad \frac{1}{3}x = \frac{4}{9}$$

$$\left(\frac{5}{1}\right)\frac{x}{5} = \left(\frac{5}{1}\right)8 \qquad \frac{-8x}{-8} = \frac{24}{-8} \qquad \left(\frac{\overset{1}{\cancel{3}}}{1}\right)\left(\frac{1}{\cancel{3}}x\right) = \left(\frac{\cancel{3}}{1}\right)\left(\frac{4}{\cancel{9}}\right)$$

$$x = 40 \qquad\qquad x = -3 \qquad\qquad x = \frac{4}{3}$$

$$7.5 = 2.5x \qquad\qquad \frac{x}{0.6} = 2.9$$

$$\frac{7.5}{2.5} = \frac{2.5x}{2.5} \qquad 0.6\left(\frac{x}{0.6}\right) = 0.6(2.9)$$

$$3 = x \qquad\qquad x = 1.74$$

Verify or check the solution of an equation: 1. Substitute the solution in place of the variable in the original equation. **2.** Perform all indicated operations on each side of the equation. **3.** If the solution is correct, the value on the left side of the equation should equal the value on the right side.

Verify that $x = 3$ is the solution for the equation $2x - 1 = 5$.

$$2(3) - 1 = 5$$
$$6 - 1 = 5$$
$$5 = 5$$

3 Solve linear equations with like terms on the same side of the equation (p. 327).

1. Combine (add or subtract) like terms that are on the same side of the equal sign. **2.** *Multiply* both sides of the equation by the reciprocal of the coefficient of the variable or *divide* both sides of the equation by the coefficient of the variable.

Solve $3x - 5x = 12$ for x.

$$3x - 5x = 12$$
$$-2x = 12$$
$$\frac{-2x}{-2} = \frac{12}{-2}$$
$$x = -6$$

LEARNING OUTCOMES	KEY CONCEPTS AND EXAMPLES

4 Solve linear equations with like terms on opposite sides of the equation (pp. 327–330).

1. Identify the like terms and determine which term or terms should be moved to create an equation with like terms on each side.   **2.** Add to both sides of the equation the opposite of each term that is to be moved (addition axiom).   **3.** Combine like terms on each side of the equation.   **4.** Solve the resulting equation by using the multiplication axiom.

Solve $x - 5 = 7$ for x.

$$x - 5 = 7$$
$$x - 5 + 5 = 7 + 5$$
$$x = 12$$

Solve $7.2 = x - 3.5$ for x.

$$7.2 = x - 3.5$$
$$7.2 + 3.5 = x$$
$$10.7 = x$$

Solve $x - \dfrac{3}{8} = \dfrac{5}{8}$ for x.

$$x - \frac{3}{8} = \frac{5}{8}$$
$$x = \frac{5}{8} + \frac{3}{8}$$
$$x = \frac{8}{8}$$
$$x = 1$$

Solve $3x - 5 = 5x + 7$ for x.

$$3x - 5 = 5x + 7$$
$$3x - 5x = 7 + 5$$
$$-2x = 12$$
$$\frac{-2x}{-2} = \frac{12}{-2}$$
$$x = -6$$

5 Solve linear equations that contain parentheses (pp. 330–334).

1. Apply the distributive property to *remove parentheses.*   **2.** *Combine like terms* on each side of the equation.   **3.** *Sort terms* to collect the variable terms on one side and constants on the other (addition axiom).   **4.** *Combine like terms* on each side of the equation. **5.** *Multiply* by the reciprocal of the coefficient of the variable term or *divide* by the coefficient of the variable term (multiplication axiom).

Solve $3(2x - 5) = 8x + 7$ for x.

$3(2x - 5) = 8x + 7$	Distribute.
$6x - 15 = 8x + 7$	Sort terms (addition axiom).
$6x - 8x = 7 + 15$	Combine like terms.
$-2x = 22$	Divide by coefficient of x (multiplication axiom).
$\dfrac{-2x}{-2} = \dfrac{22}{-2}$	
$x = -11$	

Six-Step Problem-Solving Plan. See additional problem-solving strategies on pp. 333–334.

A shipment of college textbooks is sent in two boxes weighing 37 lb total. One box weighs 9 lb more than the other. What does each box weigh?

Unknown facts
Weight of each box.

Known facts
Total weight of two boxes is 37 lb.
One box weighs 9 lb more than the other.

Relationships

x = weight of one box $x + (x + 9) = 37$

$x + 9$ = weight of other box

LEARNING OUTCOMES	KEY CONCEPTS AND EXAMPLES

Estimation

If the boxes were the same weight, each would weigh $18\frac{1}{2}$ lb.

Thus, one box will weigh less than and one box will weigh more than $18\frac{1}{2}$ lb.

Calculations

$x + x + 9 = 37$	Combine like terms.
$2x + 9 = 37$	Sort terms (addition axiom).
$2x = 37 - 9$	Combine like terms.
$2x = 28$	Divide (multiplication axiom).
$\dfrac{2x}{2} = \dfrac{28}{2}$	
$x = 14\,\text{lb}$	Weight of one box
$x + 9 = 23\,\text{lb}$	Weight of the other box

Interpretation

The two boxes weigh 14 and 23 lb.

Section 7–3

1 Solve fractional equations by clearing the denominators (pp. 336–341).

Identify an excluded value that causes a denominator of zero: 1. If a fraction has a variable in the denominator, write an equation by setting the denominator equal to zero. **2.** Solve each resulting equation. **3.** The solution for each equation from Step 1 is an excluded value.

Solve an equation by clearing all fractions: 1. Multiply each term of the *entire* equation by the least common multiple (LCM) of the denominators of the equation. **2.** Apply the distributive property to remove parentheses. **3.** Combine like terms on each side of the equation. **4.** Sort terms to collect the variable terms on one side and constants on the other (addition axiom). **5.** Combine like terms on each side of the equation. **6.** Solve the resulting equation by multiplying by the reciprocal of the coefficient of the variable term or dividing by the coefficient of the variable term (multiplication axiom).

$$\frac{3}{a} + \frac{1}{3} = \frac{2}{3a} \qquad \text{LCM} = 3a. \text{ Excluded value} = 0.$$

$$(3a)\left(\frac{3}{a}\right) + (3a)\left(\frac{1}{3}\right) = (3a)\left(\frac{2}{3a}\right) \qquad \text{Multiply by } 3a.$$

$$(\overset{3}{3a})\left(\frac{3}{\underset{}{a}}\right) + (\overset{a}{3a})\left(\frac{1}{\underset{1}{3}}\right) = (\overset{1}{3a})\left(\frac{2}{\underset{1}{3a}}\right) \qquad \text{Reduce.}$$

$$9 + a = 2$$

$$a = 2 - 9$$

$$a = -7$$

If a variable is in a denominator, check the root to see if it makes a true statement and is not an *extraneous root*.

Check the example above:

$$\frac{3}{-7} + \frac{1}{3} = \frac{2}{3(-7)} \qquad \text{Substitute } -7 \text{ for } a.$$

$$\frac{3}{-7} + \frac{1}{3} = \frac{2}{-21} \qquad \text{LCM} = 21.$$

LEARNING OUTCOMES	KEY CONCEPTS AND EXAMPLES

$$-\frac{9}{21} + \frac{7}{21} = \frac{2}{-21}$$

$$-\frac{2}{21} = -\frac{2}{21} \qquad \text{The root makes a true statement.}$$

2 Solve applied problems involving rate, time, and work (pp. 341–344).

Use the formula for amount of work: rate × time = amount of work

A *rate* is a ratio of two measures such as 1 card per 15 min or $\frac{1 \text{ card}}{15 \text{ min}}$.

Lashonda can install a PC video card in 15 min. How many cards can she install in $1\frac{1}{2}$ hours?

amount of work (W) = rate × time

$$W = \frac{1 \text{ card}}{15 \text{ min}}(90 \text{ min}) \qquad 1\frac{1}{2}\text{h} = 90 \text{ min}$$

$$W = 6 \text{ cards}$$

Use the formula for completing one job when working together:

(A's rate × time) + (B's rate × time) = 1 job

Let T = time. Let 1 = 1 completed job.

Galenda assembles a product in 10 min and Marcus assembles the product in 15 min. How long will it take them together to assemble one product? Galenda's rate × time: $\frac{1}{10}T$. Marcus's rate × time: $\frac{1}{15}T$.

$$\frac{1}{10}T + \frac{1}{15}T = 1$$

Estimation: less than 10 min but more than half of 10, or 5 min for both to do one job.

$$(30)\left(\frac{1}{10}T\right) + (30)\left(\frac{1}{15}T\right) = (30)(1)$$

Galenda's rate: $\frac{1}{10}$ job per min.

$$(\overset{3}{30})\left(\frac{1}{\underset{1}{10}}T\right) + (\overset{2}{30})\left(\frac{1}{\underset{1}{15}}T\right) = (30)(1)$$

Marcus's rate: $\frac{1}{15}$ job per min.
Time in minutes: T. LCM = 30.
Multiply by 30.
Reduce.

$$3T + 2T = 30$$ Combine like terms.

$$5T = 30$$ Divide.

$$\frac{5T}{5} = \frac{30}{5}$$

$$T = 6 \text{ min}$$

Use the formula for completing one job when working in opposition:

(A's rate × time) − (B's rate × time) = 1 job

Solve as above but *subtract* the two amounts of work.

An inlet valve fills a vat in 2 h. A drain valve empties the vat in 5 h. With both valves open, how long does it take for the vat to fill?

Inlet valve's rate × time: $\frac{1}{2}T$. Drain valve's rate × time: $\frac{1}{5}T$.

$$\frac{1}{2}T - \frac{1}{5}T = 1$$ Estimation: more than 2 h.

LEARNING OUTCOMES	KEY CONCEPTS AND EXAMPLES

$$(10)\left(\frac{1}{2}T\right) - (10)\left(\frac{1}{5}T\right) = (10)1 \qquad \text{LCM = 10}$$
Multiply by 10.

$$(\overset{5}{\cancel{10}})\left(\frac{1}{\underset{1}{\cancel{2}}}T\right) - (\overset{2}{\cancel{10}})\left(\frac{1}{\underset{1}{\cancel{5}}}T\right) = (10)1 \qquad \text{Reduce.}$$

$$5T - 2T = 10$$

$$3T = 10$$

$$\frac{3T}{3} = \frac{10}{3}$$

$$T = 3\frac{1}{3}\text{ h}$$

3 Solve decimal equations by clearing the decimals (pp. 344–346).

1. Multiply each term of the *entire* equation by the least common denominator (LCD) of the fractional amounts represented by the decimals. (The place value of the decimal amount with the most places after the decimal point will be the LCD.)   **2.** Follow the same steps used in solving a linear equation.

$$3.5x + 2.75 = 10 \qquad \text{Place value of the LCD = 100.}$$
$$(100)(3.5x) + (100)(2.75) = (100)(10) \qquad \text{Multiply by 100.}$$
$$350x + 275 = 1{,}000 \qquad \text{Sort.}$$
$$350x = 1{,}000 - 275 \qquad \text{Combine like terms.}$$
$$350x = 725 \qquad \text{Divide.}$$
$$\frac{350x}{350} = \frac{725}{350}$$
$$x = 2.071428571$$
$$x = 2.07 \text{ (rounded)}$$

Section 7–4

1 Use set terminology (pp. 349–350).

Identify the elements of a set using a roster or set-builder notation:

Roster: 1. List the elements of the set, separating the elements with commas.   **2.** Enclose the list in braces { }.

Set-Builder Notation: 1. Represent the elements of the set with a variable.   **2.** Write symbolic statements that identify the elements of the set.   **3.** Use the symbol | to represent the phrase "such that."   **4.** Enclose the results of Steps 1–3 in braces { }.

Use a roster and set-builder notation to show the following set: the set of integers between -2 and 3, including 3.

Roster: $\{-1, 0, 1, 2, 3\}$

Set-builder notation: $\{x \mid x \in Z \text{ and } -2 < x \le 3\}$

2 Show inequalities on a number line and write inequalities in interval notation (pp. 350–352).

Graph an inequality in one variable: 1. Determine the boundaries of the inequality, if they exist.   **2.** Locate the boundaries on a number line. **(a)** If the inequality is $<$ or $>$, the boundary is *not* included. Represent this with a parenthesis,) or (. **(b)** If the inequality is $\le$ or $\ge$, the boundary *is* included. Represent this with a bracket,] or [. **3.** Shade the portion of the number line between the boundaries. If there is no boundary in one or both directions, the graph continues indefinitely in that direction or both directions.

LEARNING OUTCOMES

KEY CONCEPTS AND EXAMPLES

Write an inequality in interval notation: 1. Determine the boundaries of the interval, if they exist.   **2.** Write the left boundary or $-\infty$ first and write the right boundary or $+\infty$ second. Separate the two with a comma.   **3.** Enclose the boundaries with the appropriate grouping symbols. A parenthesis "(" or ")" indicates that the boundary is *not* included. A bracket "[" or "]" indicates that the boundary *is* included.

The solution of $-2 < x \le 3$ is shown on the number line in Fig. 7–34 and by using interval notation, $(-2, 3]$

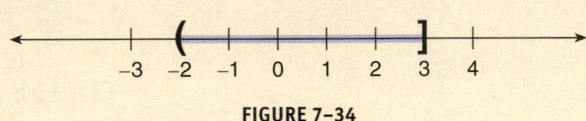

FIGURE 7–34

Section 7–5

1 Solve a linear inequality with one variable (pp. 353–356).

1. Follow the same sequence of steps that is normally used to solve a similar linear equation. **2.** The sense of the inequality remains the same unless one of the following situations occurs: **(a)** The sides of the inequality are interchanged. **(b)** The steps used in solving the inequality require that the entire inequality (both sides) be multiplied or divided by a negative number. **3.** If either situation **(a)** or **(b)** in Step 2 occurs in solving an inequality, *reverse* the sense of the inequality; that is, less than ($<$) becomes greater than ($>$), and vice versa. Show the solution set symbolically, graphically, and using interval notation.

Solve.

$$4x - 1 < 7$$
$$4x < 8$$
$$x < 2 \quad \text{or} \quad (-\infty, 2)$$

Symbolic   Interval
solution   notation

$$-2x + 3 \ge 9$$
$$-2x \ge 6 \qquad \text{Reverse sense.}$$
$$x \le -3 \quad \text{or} \quad (-\infty, -3]$$

Symbolic   Interval
solution   notation

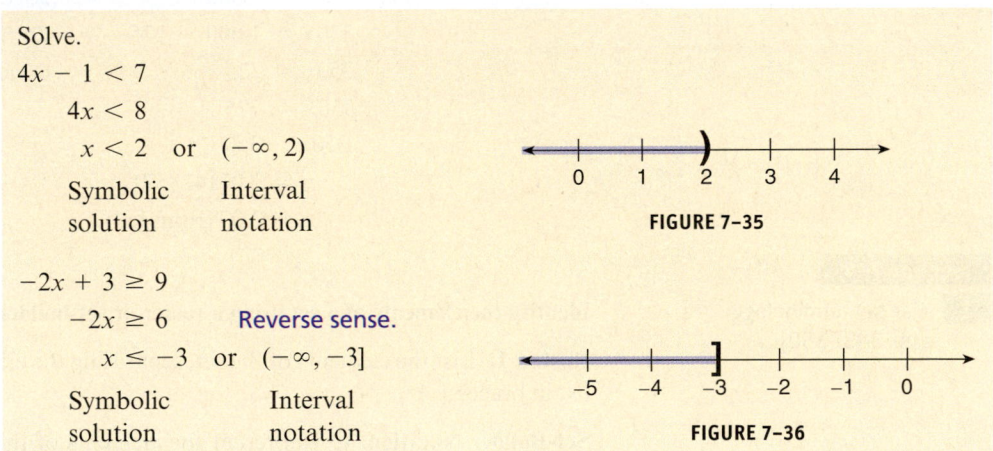

FIGURE 7–35

FIGURE 7–36

Section 7–6

1 Identify subsets of sets and perform set operations (pp. 357–359).

Every set has the set itself and the empty set (ϕ) as subsets. Set operations include union ($\cup$), intersection ($\cap$), and complements (A').

List the subsets of the set $\{3, 5, 8\}$:
$$\phi, \{3, 5, 8\} \text{ (the set itself)}, \{3\}, \{5\}, \{8\}, \{3, 5\}, \{3, 8\}, \{5, 8\}$$
If $U = \{1, 2, 3, 4, 5\}$, $A = \{1, 3\}$, and $B = \{2, 4\}$, find $A \cup B$, $A \cap B$, and A'.
$$A \cup B = \{1, 2, 3, 4\} \qquad A \cap B = \phi \qquad A' = \{2, 4, 5\}$$

2 Solve compound inequalities with the conjunction condition (pp. 359–362).

1. Separate the compound inequality into two simple inequalities using the conditions of the conjunction.   **2.** Solve each simple inequality.   **3.** Determine the solution set that includes the *intersection* of the solution sets of the two simple inequalities. *Note:* If there is no overlap in the sets, the solution is the empty set.

LEARNING OUTCOMES   **KEY CONCEPTS AND EXAMPLES**

Indicate the solution using a number line and using interval notation.

$$-3 < x + 2 \le 2$$

$$-3 < x + 2 \quad \text{and} \quad x + 2 \le 2$$

$$-3 - 2 < x \qquad\qquad x \le 2 - 2$$

$$-5 < x \qquad\qquad x \le 0$$

Solution: $-5 < x \le 0$ **Interval notation:** $(-5, 0]$

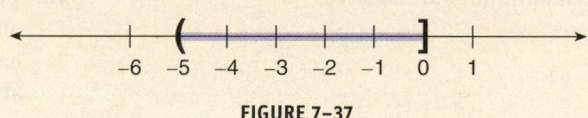

FIGURE 7–37

3 Solve compound inequalities with the disjunction condition (p. 362).

1. Solve each simple inequality.   **2.** Determine the solution set that includes the *union* of the solution sets of the two simple inequalities.

Solve and indicate the solution set on the number line and by using interval notation.

$$x - 1 < 3 \qquad \text{or} \qquad x + 2 > 8$$

$$x < 3 + 1 \qquad\qquad x > 8 - 2$$

$$x < 4 \qquad \text{or} \qquad x > 6$$

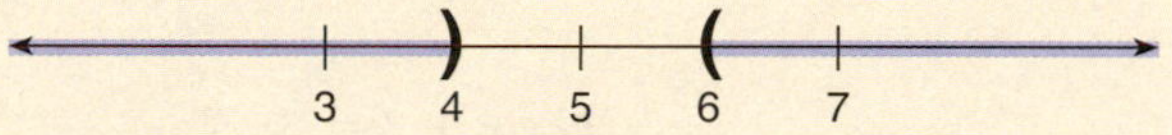

FIGURE 7–38

Interval notation: $(-\infty, 4) \cup (6, \infty)$

---

7   CHAPTER REVIEW EXERCISES

Section 7–1   MyLab Math   **For additional practice go to your study plan in MyLab Math.**

Verify that the statements are true equations.

1. $5 + 7 = 18 - 6$

2. $9(8) = 12 + 3(4) + 7^2 - 1$

3. $8 - 9 + 3^2 = -2^2 + 12$

4. $7 + 3 = 8 \cdot 1 - 5 + 4 + 3$

Find the value of the variable that makes the equation true.

5. $x = 5 + 9$

6. $8 - 5 = x$

7. $y = \dfrac{48}{-6}$

8. $7 + 3(5 - 2) = x$

Identify the terms in the expressions by drawing a box around each term.

9. $15x - \dfrac{3a}{7} + \dfrac{(x - 7)}{5}$

10. $5x - 8 + \dfrac{3}{y}$

State the equations in words.

11. $x + 5 = 2$

12. $x - 7 = 11$

13. $\dfrac{x}{8} = 7$

14. $3(x + 7) = -3$

Write the following statements in symbols.

15. 7 more than twice a number is 11.

16. A certain stock listed on the New York Stock Exchange closed at 42.375, a decrease of 3.125 points from the opening price.

17. Twice the sum of a number and 8 is 40.

18. The print shop used 31 cases of copy paper during one month. End-of-the-month inventory indicated 172 cases on hand. Write an equation to find the number of cases on hand at the beginning of the month. Then, solve the equation.

Simplify.

19. $3a + 2a$

20. $7a - 3b + 5 + 7a - 9b$

21. $3(2y - 4) + y$

22. $-(4y - 7)$

23. $-3(a + 2) - 5$

24. $5 - (a - 3)$

25. $11 - 2(x - 5)$

26. $-8(5x - 3y - 15)$

27. $3 - (2x - 4y - 8)$

28. $3(x - 2) - 2(5x - 7y + 1)$

29. $15a + 3(5a - 7b + 4)$

30. $12 - (x + y - 5)$

Section 7-2

Solve.

31. $x - 5 = 8$

32. $y + 7 = 3$

33. $x - 8 = -10$

34. $x - 15 = -7$

35. $x - 5 = 14$

36. $-2 = 8 - x$

37. $x + 7 = 10$

38. $x - 5 = 3$

39. $x - 3 = -4$

40. $3 + y = -5$

41. $1 = a - 4$

42. $t + 7 = 12$

43. $3x + 4 = 19$

44. $4x - 3 = 9$

45. $15 - 3x = -6$

46. $5 = 3x - 7$

47. $-7 = 6x - 31$

48. $4 = 7 - 4x$

49. $5x - 12 = 9x$

50. $7x = 8x + 4$

51. $3x = 21$

52. $4x = -28$

53. $-15 = 2b$

54. $-5 = -m$

55. $3 = \dfrac{1}{5}x$

56. $-\dfrac{2}{7}x = 8$

57. $-7y = -49$

58. $0.6 = -a$

59. $-\dfrac{3}{8}x = -24$

60. $\dfrac{1}{2}x = -5$

61. $42 = -\dfrac{6}{7}x$

62. $-\dfrac{5}{8}x = -10$

63. $\dfrac{1}{7}x = 12$

64. $5y - 7y = 14$

65. $2b - 7b = 10$

66. $0 = 4t - t$

67. $21 = x + 2x$

68. $8x - 3x = 6 + 9$

69. $4x + x = 25$

70. $36 = 9a - 5a$

71. $20 - 4 = 2x - 6x$

72. $13 - 27 = 3x - 10x$

73. $-12 = -8 - 2x$

74. $12x + 27 = 3x$

75. $3x + 9 = 10 + 3x + 1$

76. $10 + 4x = 5 - x$

77. $7 - 4y = y + 22$

78. $7x - 1 = 4x + 17 + 7x$

79. $7x - 5 + 2x = 3 - 4x + 12$

80. $2x - 3 + 15 = 7x - 8 - 6x$

81. $4y + 8 = 3y - 4$

82. $y - 5 = 6y + 30$

83. $8 - 2y = 15 - 3y$

84. $6 - 7x = 15 - x$

85. $5x - 12 = 2x + 15$

86. $18x - 21 = 15x + 33$

87. $3x - 5x + 2 = 6x - 5 + 12x$

88. $\dfrac{R}{7} - 6 = -R$

89. $0 = \dfrac{8}{9}c + \dfrac{1}{4}$

90. $0.9R = 0.3$ (round to tenths)

91. $0.86 = R + 0.4R$ (round to hundredths)

92. $0.04y = 0.02 - y$ (round to hundredths)

93. $18 = 6(2 - y)$

94. $4(6 + x) = 36$

95. $7x - 3(x - 8) = 28$

96. $3(x + 2) - 5 = 2x + 7$

97. $5(3 - 2x) = -5$

98. $3(2x + 1) = -3$

Solve the equations.

99. $3x = 3(9 + 2x)$

100. $4a = 8 - (a + 7)$

101. $5x = 7 + (x + 5)$

102. $3(x + 2) - 5 = 2x + 7$

103. $4(3 - x) = 2x$

104. $3(2x - 4) = 4x - 6$

105. $-2(4 - 2x) = -16 + 2x$

106. $-16 = -2(-2x + 4)$

107. $8 = 6 - 2(3x - 1)$

108. $4x - (x + 3) = 3x - 3$

109. $3(x - 1) = 18 - 2(x + 3)$

110. $-(x - 1) = 2(x + 7)$

111. $-(2x + 1) = -7$

112. $2 + 3(x - 4) = 2x - 5$

113. $7 = 3 + 4(x + 2)$

114. $7(x + 2) = -6 + 2x$

115. $3(4x + 3) = 3 - 4(x - 1)$

116. $3(2 - x) - 1 = 4(3 - x)$

Write the statements as equations and solve.

117. The difference between x and 6 is 8. Find x.

118. Twice a number increased by 5 is 17. Find the number.

119. 5 times the sum of x and 6 is 42 more than x. Find x.

120. How many gallons of water must be added to 46 gal of pure alcohol to make 100 gal of alcohol solution?

121. BUS   If one technician works 3 h less than another and their total hours worked are 51 h, how many hours has each technician worked?

122. CON   The shorter side of an L-shaped carpenter's square is 6 in. shorter than the longer side. If the total length of the carpenter's square is 24 in., what is the measure of each side?

123. AG/H   Ms. Galendez's backyard is a rectangle whose length is twice the width. If the perimeter of the yard is 720 ft, what are the dimensions of her yard?

124. BUS   Brubakers catered 32 chicken dinners for $409. This included a $25 delivery charge. Find the cost of each dinner excluding the delivery charge.

Section 7–3

Solve the equations.

125. $m + \dfrac{1}{4} = \dfrac{3}{4}$

126. $\dfrac{3}{5}y = 12$

127. $p = \dfrac{1}{2} + \dfrac{1}{3}$

128. $x + \dfrac{1}{7}x = 16$

129. $\dfrac{2}{5} - x = \dfrac{1}{2}x + \dfrac{4}{5}$

130. $\dfrac{R}{7} - 6 = -R$

131. $\dfrac{3}{7}m - \dfrac{1}{2} = \dfrac{2}{3}$

132. $\dfrac{1}{4}S = \dfrac{1}{4} + \dfrac{1}{10} + \dfrac{1}{20}$

133. $m = 2 + \dfrac{1}{4}m$

Solve the equations. Identify excluded values.

134. $\dfrac{1}{R} = \dfrac{1}{10} + \dfrac{1}{3} + \dfrac{1}{6}$

135. $\dfrac{2}{P} = \dfrac{1}{2} + \dfrac{1}{4} - \dfrac{5}{12}$

136. $\dfrac{3}{x} + 4 = \dfrac{1}{5} - 7$

Set up an equation with fractions and solve.

137. AG/H   Melissa can complete a landscape project in 3 h. Henry can complete the same project in 7 h. How long would it take Melissa and Henry working together to complete the landscape project?

138. COMP   One optical scanner reads a stack of sheets in 20 min. A second scanner reads the same stack in 12 min. How long does it take for both scanners together to process the one stack of sheets?

139. ELEC   An apprentice electrician can install 5 light fixtures in 2 h. How many light fixtures can be installed in 10 h?

140. CON   A brick mason can erect a retaining wall in 6 h. The brick mason's apprentice can do the same job in 10 h. How much time does it take both of them working together to erect the retaining wall?

141. **ELEC** A parallel DC circuit has 3 branches rated at $2\,\Omega$, $6\,\Omega$, and $12\,\Omega$. Find the total resistance to the nearest hundredth. *See Example 6 on p. 340.*

142. **ELEC** Find the total resistance of a parallel circuit with 2 branches rated at $12\,\Omega$ and $16\,\Omega$. Round to hundredths. *See Example 6 on p. 340.*

Solve the equations. Round to tenths if necessary.

143. $2.3x - 4.1 = 0.5$

144. $0.22 + 1.6x = -0.9$

145. $0.3x - 2.15 = 0.8x + 3.75$

146. **AUTO** The distance formula is distance = rate × time. If a portable MRI unit traveled 350.8 mi to and from a rural hospital at 50 mi/h, how long to the nearest hour did the trip to and from the hospital take?

147. **ELEC** Electrical resistance in ohms (Ω) is voltage (V) divided by amperage (A). Find the resistance of a small motor with a voltage of 8.5 V requiring 0.5 A.

Section 7–4

148. Describe the empty set and write the symbols used to indicate the empty set.

149. Use symbols and set-builder notation to write the statement: 5 is an element of the set of whole numbers.

Represent each set of numbers on the number line and with interval notation.

150. $x \le 3$

151. $x > -7$

152. $-3 < x < 5$

153. $-4 \le x < 2$

154. All real numbers

155. $-2 < x$

156. $-5 > x$

Section 7–5

Solve. Show the solution set on a number line and in symbolic and interval notation.

157. $42 > 8m - 2m$

158. $3m - 2m < 3$

159. $0 < 2x - x$

160. $1 \le x - 7$

161. $10 - 2x \ge 4$

162. $3x + 4 > x$

163. $10x + 18 > 8x$

164. $4x \le 5x + 8$

165. $12 + 5x > 6 - x$

166. $x - 7 < 2$

167. $x + 4 > 2$

168. $3x - 1 \le 8$

169. $3x - 2 \le 4x + 1$

170. $8 - 7y < y + 24$

171. $15 \ge 5(2 - y)$

172. $4t > 2(7 + 3t)$

173. $6x - 2(x - 3) \le 30$

174. $8x - (3x - 2) > 12$

175. A ream of 11×17 copy paper costs $5.60 more than a mechanical pencil. The total cost of two reams of paper and a dozen pencils must cost no more than $59.80. Write inequalities to find the cost range for a ream of paper and a single pencil.

176. A shirt costs $3 less than a certain tie. If the total cost for six ties and two shirts is less than $130, what is the most each shirt and tie can cost?

Use sets U, A, B, C, and D to give the elements in the sets in Exercises 177–181.

$U = \{-2, -1, 0, 1, 2, 3, 4, 5\}$, $\quad A = \{-2, -1\}$, $\quad B = \{0\}$, $\quad C = \{0, 1, 2, 3, 4\}$, $\quad D = \{-1, 0, 1\}$

177. $A \cap B$

178. $B \cup D$

179. $A \cap D$

180. A'

181. $(A \cup C)'$

Section 7–6

Solve each compound inequality. Show the solution set on a number line and in symbolic and interval notation.

182. $3x - 8 < x < 5x + 24$

183. $x + 1 < 5 < 2x + 1$

184. $x + 3 \le 7 \le 2x - 1$

185. $x + 2 < 7 < 2x - 15$

186. $2 < 3x - 4 < 8$

187. $2x + 3 < 15 < 3x + 9$

188. $5 < x - 3 < 8$

189. $-3 < 2x - 4 < 5$

190. $-7 < x - 5 < 7$

191. $-5 \le -3x + 1 < 10$

192. $2x + 3 \le 5x + 6 < -3x - 7$

193. $-3 \le 4x + 5 \le 2$

194. $4x - 2 < x + 8 < 9x + 1$

195. $x + 3 < 5$ or $x > 8$

196. $x - 7 < 4$ or $x + 3 > 8$

197. $x - 3 < -12$ or $x + 1 > 9$

198. $3x < -4$ or $2x - 1 > 9$

199. The price of a particular brand of PC whiteboard is $1,365, and the stand costs $199. A college will spend between $15,000 and $30,000 for the equipment. Assuming an equal number of boards and stands are purchased, what is the range for the number of boards and stands to be purchased?

200. The blueprint for a modular desk shows the front edge to be 48 in. with a tolerance of 0.25 in. If the width of the desk is 8 in. less than the length and has the same tolerance, what is the allowable range for the width of the desk?

7   TEAM PROBLEM-SOLVING EXERCISES

1. Your team has a travel agency and you have been asked to plan a trip for a high school math club. The math club has raised $8,700 for the trip. Your agency charges a one-time fee of $300 to set up the trip. The cost per person for the trip is $620.

   (a) Write an inequality that can be used to determine the maximum number of persons who can go on the trip with the money raised.
   (b) Solve the inequality created in part (a).
   (c) How much more money would need to be raised for one more person to be able to go on the trip?

7   CONCEPTS ANALYSIS

1. Describe like terms and explain how they are combined. Give an example and combine the terms.

2. Explain the difference between *factors* and *terms*. Give an example of each.

3. If you solve an equation and get $7 = x$, but want to write your root as $x = 7$, what mathematical property allows you to do so?

4. Write a paragraph explaining how the natural numbers, integers, rational numbers, and real numbers are related. Diagram the relationship of these numbers, and give examples of each type of number.

5. Write the following in the proper sequence for solving equations. An item can be used more than once.
   (a) Arrange the equation using the addition axiom so that number terms are on one side and letter terms are on the other.
   (b) Divide both sides by the coefficient of the variable term or multiply both sides by the reciprocal of the coefficient of the variable term (multiplication axiom).
   (c) Apply the distributive property to remove any parentheses.
   (d) Combine like terms that are on the same side of the equal sign.

6. Give an example of an equation that requires the use of the multiplication axiom. Solve the equation. If the solution requires several steps, identify the step that applies the multiplication axiom.

7. Write an example of an equation that requires the use of the addition axiom. Solve the equation. If several steps are required, identify the step that applies the addition axiom.

8. Solve the equation $\frac{3}{4}x = \frac{5}{8}$ in two different ways. Why are the two ways equivalent? What property of equality was used?

9. Write an example of an equation that requires the distributive property. Solve the equation. If several steps are required, identify the step that applies the distributive property.

10. Explain the estimation step we use to solve applied problems. Why is this step important in the problem-solving process?

Identify and explain the first mistake encountered in each problem. Then work the problem correctly.

11. $7(x - 3) + 2 = 4 + 2x$
    $7x - 3 + 2 = 4 + 2x$
    $7x - 1 = 4 + 2x$
    $5x - 1 = 4$
    $5x = 5$
    $x = 1$

12. $5 - 3(x + 2) = 3(x + 1)$
    $2(x + 2) = 3(x + 1)$
    $2x + 4 = 3x + 3$
    $-x = -1$
    $x = 1$

13. $5x - (3x - 6) = 18$
    $5x - 3x - 6 = 18$
    $2x - 6 = 18$
    $2x = 18 + 6$
    $2x = 24$
    $x = 12$

14. The symmetric property of equations states that if $a = b$ then $b = a$. Is an inequality symmetric? Illustrate your answer with an example.

15. In equations, if $a = b$, then $ac = bc$, if c equals any real number. Is the statement, if $a < b$ then $ac < bc$, true if c is any real number? For what values of c is the statement false? Illustrate your answer with an example.

16. Write in your own words two differences in the procedures for solving equations and for solving simple inequalities.

17. Explain the difference between the statements $x < 2$ and $x \leq 2$.

18. If $x < 2$ or $x > 10$, is it correct to write the statement as $2 > x > 10$? Why or why not? Explain your answer.

19. Explain the difference between an *and* relationship and an *or* relationship in inequalities.

20. In your own words, write a procedure for solving a compound inequality such as $5 < x + 2 < 9$.

21. Find the mistake in the inequality. Correct and briefly explain the mistake.

$$3x + 5 < 5x - 7$$
$$3x - 5x < -7 - 5$$
$$-2x < -12$$
$$x < 6$$

7 PRACTICE TEST

Solve. Round to hundredths if necessary.

1. $x + 5 = 19$

2. $-8y = 72$

3. $\dfrac{x}{2} = 5$

4. $5x + 2x = 49$

5. $5 - 2x = 3x - 10$

6. $5x + 3 - 7x = 2x + 4x - 11$

7. $3(x + 4) = 18$

8. $3x + 2 = 4(x - 1) - 1$

9. $\dfrac{8}{y + 2} = -7$

10. $\dfrac{4}{5}z + z = 8$

11. $5x + \dfrac{3}{5} = 2$

12. $3x + 2 = \dfrac{2}{3}$

13. $\dfrac{3}{5}x + \dfrac{1}{10}x = \dfrac{1}{3}$

14. $\dfrac{1}{x} = \dfrac{1}{3} + \dfrac{5}{6}$

Solve the equations. Round to hundredths when necessary.

15. $1.3x = 8.02$

16. $4.5y + 1.1 = 3.6$

17. $0.18x = 300 - x$

18. $7.9 = 0.5x - 8.35$

19. $0.23 + 7.1x = -0.8$

Write the statements in symbols and solve for the unknown amount.

20. The product of 5 and an unknown amount is equal to three times the amount plus 8.

21. Two times the difference of a number and 9 is 30.

Solve the problems involving fractions and decimal numbers.

22. A pipe fills 1 tank in 4 h. If a second pipe empties 1 tank in 6 h, how long does it take for the tank to fill with both pipes operating?

23. Find the interest paid on a loan of $4,500 at 9.3% interest for 3 years.

24. A pipe fills 1 tank in 8 h. A second pipe fills the tank in 12 h. How long does it take to fill the tank with both pipes filling the tank?

Represent each set of numbers on the number line and by using interval notation.

25. $x \geq -12$

26. $-3 < x \leq -2$

Graph the solution on a number line and write the solution in symbolic and interval notation.

27. $3x - 1 > 8$

28. $2 - 3x \geq 14$

29. $10 < 2 + 4x$

30. $-5b > -30$

31. $\frac{1}{3}x + 5 \leq 3$

32. $2(1 - y) + 3(2y - 2) \geq 12$

33. $5 - 3x < 3 - (2x - 4)$

34. $7 + 4x \leq 2x - 1$

35. $-5 < x + 3 < 7$

36. $\frac{1}{4}x + 2 < 3 < \frac{1}{3}x + 9$

37. $3x - 1 \leq 5 \leq x - 5$

38. $2(x - 1) < 3$ or $x + 7 > 15$

Use the following sets to give the elements in the indicated sets in Exercises 39–42.

$U = \{1, 2, 3, 4, 5, 6, 7, 8, 9\}$, $\qquad A = \{5, 8, 9\}$, $\qquad B = \{1, 2, 3, 4, 5, 6, 7, 8\}$

39. $A \cup B$

40. $A \cap B$

41. $A \cap B'$

42. List all the subsets of set A.

8

Formulas, Proportion, and Variation

Andre Nantel/Shutterstock

In Great Company

The Case of the Gimli Glider (July 23, 1983)

On July 23, 1983, Air Canada Flight 143, a Boeing 767, was about halfway through its flight from Montreal to Edmonton when this fully loaded commercial jet passenger plane *ran out of fuel*—at 41,000 feet above sea level.

The crew wisely refrained from asking the passengers to get out and push. Instead, the pilots successfully put the aircraft on a glide course to a safe emergency landing in Manitoba at Gimli Industrial Park Airport. This was no small feat, given that a Boeing 767 without engines loses all electricity, most of the instruments, and most of the hydraulics. And it was being landed on an abandoned airstrip that had been converted into a motor sports park—a park that was hosting an amateur sports car race as the plane attempted to land.

How does this happen? Was no one watching the fuel gauge?

Well, they were, but it didn't help. You see, the jetliner's fuel gauges had failed—before the crew even boarded the plane. But the Boeing 767 was a new plane and the pilots incorrectly thought this was acceptable.

With the fuel gauges gone, the ground crew needed to hand-calculate how much fuel to put in before takeoff. Canada had just switched to the metric system. The new 767 was calibrated according to the metric system. But the ground crew was used to gallons and pounds, not liters and kilograms. As a result, they calculated the fuel weight for the plane using a conversion factor of 1.77 instead of the correct factor: 0.803.

Instead of transferring 22,300 *kilograms* of fuel, they had put 22,300 *pounds* of fuel on board. That worked out to a little over 10,000 kg, less than half what was needed. The captain checked the ground crew calculations. The numbers were calculated correctly. He accepted the numbers. He didn't notice the conversion factor was wrong.

Reality crudely punished his poor math skills at an altitude of 41,000 feet—that's about 12,000 meters up if you're from Canada.

After the emergency landing, the plane was repaired, put back in service, and nicknamed the Gimli Glider. The captain and the first officer were both demoted for having failed to note the errors that led to the incident. They were both, however, also given the first Fédération Aéronautique Internationale Diploma for Outstanding Airmanship in landing the crippled craft. When the two officers returned to service after their suspension, their first flight turned out to be aboard . . . the Gimli Glider.

8-1   Formulas

LEARNING OUTCOMES

1 Evaluate formulas.

2 Rearrange formulas to solve for a specified variable.

LC LEARNING CATALYTICS

Evaluate:

1. $\dfrac{12(8)}{12 + 8}$

2. $\dfrac{\$35,000 - \$28,000}{\$35,000}$

Evaluate a formula: to substitute known values for some variables and perform the indicated operations to find the value of the variable in question

The most common use of formulas is for finding missing values. If we know values for all but one variable of a formula, we can find the unknown value.

1 **Evaluate Formulas.** To **evaluate a formula** is to substitute known values for some variables and perform the indicated operations to find the value of the variable in question. In Chapter 1, Section 3, Outcome 2, we evaluated formulas when the missing value was already isolated on the left. We can now evaluate formulas when the missing value is not isolated.

> **To evaluate a formula:**
>
> 1. Write the formula.
>
> 2. Rewrite the formula, substituting known values for variables of the formula.
>
> 3. Solve the equation from Step 2 for the missing variable.
>
> 4. Interpret the solution within the context of the formula.

EXAMPLE 1

AUTO Evaluate the formula for *engine efficiency* $E = \dfrac{I - P}{I}$ to the nearest thousandth if $I = 24,000$ calories (cal) and $P = 8,600$ cal. Round to thousandths.

$$E = \frac{I - P}{I}$$

Engine efficiency = difference between heat input and heat output divided by heat input.

$$E = \frac{24,000 - 8,600}{24,000}$$

Substitute given values. Perform calculations in numerator grouping.

$$E = \frac{15,400}{24,000}$$

Divide.

$$E = 0.6416666667$$

Round.

The engine efficiency is 0.642 (64.2% efficient).   See Exercises 1–7.

See Exercises 1–7.

STOP AND CHECK

1. Find the perimeter of a rectangle with a length of 9 m and a width of 7 m using the formula $P = 2(l + w)$.

Answer:

1. 32 m

EXAMPLE 2

ELEC   Evaluate the formula for *total resistance* $R_T = \dfrac{R_1 R_2}{R_1 + R_2}$ if $R_1 = 10\ \Omega$ and $R_2 = 6\ \Omega$.

$$R_T = \frac{R_1 R_2}{R_1 + R_2}$$

Total resistance = product of first resistance and second resistance divided by sum of first and second resistances.

$$R_T = \frac{10(6)}{10 + 6}$$

Substitute given values. Perform calculations in numerator and denominator groupings.

$$R_T = \frac{60}{16}$$

Divide.

$$R_T = 3.75$$

Dimension analysis: $\dfrac{\text{ohms (ohms)}}{\text{ohms}} = \text{ohms}$

The total resistance in the circuit is 3.75 Ω.   See Exercise 8.

See Exercise 8.

EXAMPLE 3

Solve the formula for the *perimeter of a rectangle* $P = 2(l + w)$ for w if $P = 12$ ft and $l = 4$ ft.

$$P = 2(l + w)$$

Perimeter of a rectangle = 2 times the sum of the length and the width. Substitute values.

$$12 = 2(4 + w)$$

Apply the distributive property.

$$12 = 8 + 2w$$

Isolate the term with the variable (addition axiom).

$$12 - 8 = 2w$$

Combine like terms.

$$4 = 2w$$

Divide.

$$\frac{4}{2} = \frac{2w}{2}$$

$$2 = w$$

Interpret solution.

The width is 2 ft.   See Exercise 9.

See Exercise 9.

EXAMPLE 4

BUS   Evaluate the formula for *simple interest* $I = PRT$ for principal (P) if interest (I) = \$94.50, rate (R) = 21%, and time (T) = $\frac{1}{2}$ year.

For convenience in using a calculator, convert $\frac{1}{2}$ year to 0.5 year. In this formula, the rate should be expressed as a decimal equivalent, 21% = 0.21.

$$I = \boxed{P}\,RT$$ Substitute values and solve for P.

$$94.50 = \boxed{P}\,(0.21)(0.5)$$ Multiply 0.21 and 0.5.

$$94.50 = 0.105\,\boxed{P}$$ Divide.

$$\frac{94.50}{0.105} = \frac{0.105\,\boxed{P}}{0.105}$$

$$900 = \boxed{P}$$ Interpret solution.

The principal is \$900.   **See Exercises 10–19.**

The next example has many steps and it is important to apply the order of operations. The formula is used to find the length of a belt connecting two pulleys.

STOP AND CHECK

1. Use the formula in Example 5 to find the belt length to the nearest hundredth when $C = 28$ in., $D = 12$ in., and $d = 8$ in.

Answer:
1. 87.58 in.

EXAMPLE 5

INDTEC   Evaluate the *belt length* using the formula $L = 2C + 1.57(D + d) + \dfrac{D + d}{4C}$

if $C = 24$ in., $D = 16$ in., and $d = 4$ in. Round to hundredths. See Fig. 8–1.

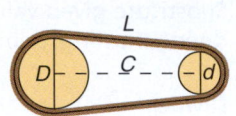

L = length of belt joining two pulleys
C = distance between centers of pulleys
D = diameter of large pulley
d = diameter of small pulley

FIGURE 8–1

$$L = 2C + 1.57(D + d) + \frac{D + d}{4C}$$ Substitute the given values.

$$L = 2(24) + 1.57(16 + 4) + \frac{16 + 4}{4(24)}$$ Work groupings in parentheses, numerator, and denominator.

$$L = 2(24) + 1.57(20) + \frac{20}{96}$$ Work multiplications and division.

$$L = 48 + 31.4 + 0.2083333333$$ Add.

$$L = 79.60833333$$ Round.

The pulley belt is 79.61 in. long.   **See Exercise 20.**

Formula rearrangement: the process of isolating a variable term other than the one already isolated (if any) in a formula

2   Rearrange Formulas to Solve for a Specified Variable.   **Formula rearrangement** generally refers to isolating a variable term other than the one already isolated in the formula. Solving formulas in this manner shortens our work when doing repeated formula evaluations. After we solve the formula for the desired variable, we rewrite the formula with the variable on the left side for convenience and for use in electronic spreadsheets.

The following formulas require applying the addition axiom.

EXAMPLE 6

BUS   Solve the *markup* formula, $M = S - C$, for S (selling price).

$$M = \boxed{S} - C$$ Markup = selling price − cost. Isolate S.

$$M + C = \boxed{S}$$ The coefficient of S is positive, so the formula is solved.

$$\boxed{S} = M + C$$ Rewrite S on the left for convenience.   **See Exercises 21–31.**

> **TIP** **Where Is the Variable or Unknown in a Formula?** It may help to think of the one letter we are solving for as the unknown or variable, and to think of the other letters as if they were *coefficients* or *constants* in an ordinary equation.

To rearrange a formula:

1. Determine which variable of the formula will be isolated (solved for).

2. Highlight or mentally locate all instances of the variable to be isolated.

3. Treat all other variables of the formula as you would treat numbers in an equation, and perform normal steps for solving an equation.

4. If the isolated variable is on the right side of the equation, interchange the sides so that it appears on the left side.

EXAMPLE 7

BUS Solve the formula $M = S - C$ for C (cost).

$M = S - C$ Markup = selling price − cost. Isolate C.

$M - S = -C$ The coefficient of C is negative, so divide both sides by − 1.

$\dfrac{M - S}{-1} = \dfrac{-C}{-1}$ Note effect on signs after division by − 1.

$-M + S = C$ Write the positive term first.

$S - M = C$ Rewrite with C on the left for convenience.

$C = S - M$ **See Exercises 32–35.**

When a formula contains a term of several factors and we need to solve for one of those factors, we treat the factor being solved for as the variable and the other factors as its coefficient.

STOP AND CHECK

1. Solve the formula in Example 8 for P.

Answer:

1. $P = \dfrac{I}{RT}$

EXAMPLE 8

BUS Solve for R in the formula $I = PRT$.

$I = PRT$ Interest = Principal × Rate × Time. Since we are solving for R, PT is its coefficient. Divide both sides by the coefficient of the variable.

$\dfrac{I}{PT} = \dfrac{PRT}{PT}$ Reduce.

$\dfrac{I}{PT} = R$ Rewrite with R on the left.

$R = \dfrac{I}{PT}$ **See Exercises 36–43.**

Sometimes formulas contain addition (or subtraction) and multiplication in which we must use the distributive property to solve for a particular letter. In the formula $A = P(1 + ni)$, we must use the distributive property to solve for either n or i because each appears inside the grouping.

STOP AND CHECK

1. Solve the formula in Example 9 for P.
2. Solve the formula in Example 9 for i.

Answers:

1. $P = \dfrac{A}{1 + ni}$ 2. $i = \dfrac{A - P}{Pn}$

EXAMPLE 9

BUS Solve for n in the formula for *compound amount*, $A = P(1 + ni)$.

$A = P(1 + ni)$	Identify the variable to be isolated. Use the distributive property to remove n from parentheses.
$A = P + Pni$	Use the addition axiom to isolate the term containing n.
$A - P = Pni$	Divide both sides by the coefficient of n, Pi.
$\dfrac{A - P}{Pi} = \dfrac{Pni}{Pi}$	Reduce.
$\dfrac{A - P}{Pi} = n$	Rewrite with n on the left.
$n = \dfrac{A - P}{Pi}$	See Exercises 44–45.

When the formula contains division, we clear the denominator before taking further steps.

EXAMPLE 10

ELEC The formula $V = \dfrac{P}{I}$ represents the relationship among the *voltage drop* (V), the electrical power (P), and the current (I). Rearrange the formula to solve for P.

$V = \dfrac{P}{I}$	Identify the variable to be isolated. Multiply both sides by the denominator I to clear it.
$(I)V = \dfrac{P}{I}(I)$	Reduce.
$IV = P$	Rewrite with P on the left.
$P = IV$	See Exercises 46–48.

Spreadsheets

Spreadsheet: information displayed in a table of rows and columns

Information is often displayed in a table of rows and columns called a **spreadsheet.** These rows and columns may show the results of calculations, such as totals or percents, in addition to the original data. Many software programs such as Excel help you build an **electronic spreadsheet** and draw appropriate graphs using formulas and equations. The calculations are made automatically once the formulas are defined in the spreadsheet.

Electronic spreadsheet: software program such as Excel that makes calculations for formulas defined in the spreadsheet

Address: a specific location on a spreadsheet

In a spreadsheet, key information is placed in a specific location called an **address,** and formulas are developed using these addresses as the variables. Spreadsheet programs are useful because key information can be changed while retaining the basic formulas of the spreadsheet. This process allows one to see quickly how various changes in the key data affect the results.

Let's examine a spreadsheet for the following situation. *The 7th Inning Sports Memorabilia Shop is developing an annual operating budget. The budget categories and the projected amount of expense for each category are shown in the spreadsheet in Fig. 8–2. The spreadsheet can be used to find the total operating budget for the year and the percent of the total annual budget for each category in the projected annual budget. Formulas must be developed to calculate this information.*

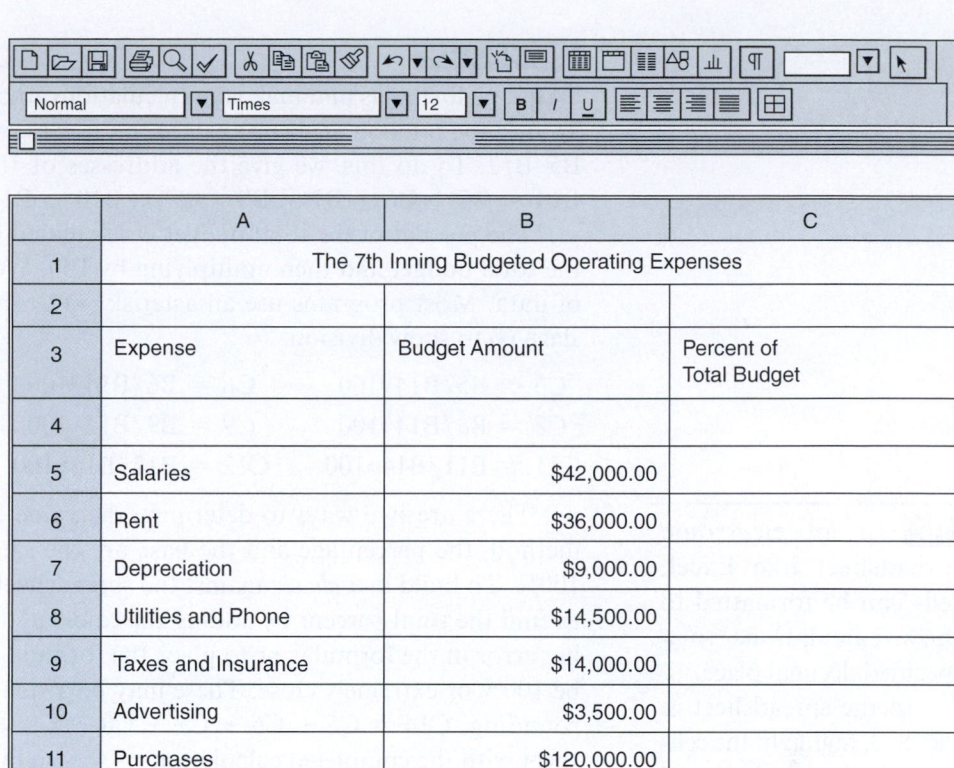

FIGURE 8-2

The spreadsheet program labels columns with letters like A, B, and C, and the rows with numbers. We'll use row 1 for the title of the spreadsheet, row 3 to label the columns of data, and column A to label the rows of data. Each position on the spreadsheet is a **cell,** and the program identifies each cell by its column letter and row number. For example, the amount budgeted for taxes and insurance is in cell B9 and is $14,000.00.

Cell: each position on a spreadsheet identified by its column letter and row number

> **TIP** **Spreadsheet Addresses** A spreadsheet address is an individual cell that is the intersection of a column and a row. This address is generally given in two parts: column and row.
>
> | Address | Interpretation |
> | --- | --- |
> | C 8 | Column C and Row 8 |
> | J 30 | Column J and Row 30 |
> | AA 5 | Column AA (follows Column Z) and Row 5 |

Now we develop formulas to make the calculations. Each spreadsheet program gives various shortcuts for writing formulas and formats for giving instructions unique to that program; however, we will write the basic concepts used and add program-specific conventions as appropriate.

EXAMPLE 11

Write the formulas and make the calculations to complete the Fig. 8–2 spreadsheet. To find the total budget to be placed in cell B14, we need to add the amounts in cells B5–B12. To do this, we give the addresses of the cells to be added in a formula: B14 = B5 + B6 + B7 + B8 + B9 + B10 + B11 + B12.

The percent of the total budget is calculated by dividing the specific amount by the total budget and then multiplying by 100. We will write a formula for each line of data. Most programs use an asterisk (∗) to show multiplication, and a forward slash (/) to show division.

C5 = B5/B14∗100	C6 = B6/B14∗100	C7 = B7/B14∗100
C8 = B8/B14∗100	C9 = B9/B14∗100	C10 = B10/B14∗100
C11 = B11/B14∗100	C12 = B12/B14∗100	

There are two ways to determine the value for cell C14. If we use the percent method, the percentage and the base are the same amount, so the total percent is 100%. To build in a *check* against the spreadsheet formulas, however, it is advisable to find the total percent by adding the calculated percents. It is easy to make a typing error in the formulas or to place the formula in the wrong cell. The total should be 100% or extremely close. There may be a small discrepancy due to the effects of rounding. C14 = C5 + C6 + C7 + C8 + C9 + C10 + C11 + C12. The spreadsheet with the completed calculations is shown in Fig. 8–3.

> **TIP** In an electronic spreadsheet like Excel, cells can be formatted to show calculations to a specified decimal place.
>
> In the spreadsheet in Fig. 8–3, highlight the cells c5–c14. Right click and select Format Cells. . . . Click on the Number tab and then Percentage. In the box following Decimal places: enter 1 (to have results displayed to nearest tenth of a percent). Then click on OK.

	A	B	C
1	The 7th Inning Budgeted Operating Expenses		
2			
3	Expense	Budget Amount	Percent of Total Budget
4			
5	Salaries	$42,000.00	17.5
6	Rent	$36,000.00	15.0
7	Depreciation	$9,000.00	3.8
8	Utilities and Phone	$14,500.00	6.1
9	Taxes and Insurance	$14,000.00	5.8
10	Advertising	$3,500.00	1.5
11	Purchases	$120,000.00	50.1
12	Other	$500.00	0.2
13			
14	Total	$239,500.00	100.0

FIGURE 8–3

See Exercises 49–50.

8–1 EXERCISES   MyLab Math   For additional practice go to your study plan in MyLab Math.

1 Evaluate the formulas. *See Example 1.*

1. $D = RT$ if $R = 40$ mi/h and $T = 7$ h

2. $S = C + M$ if $C = \$40$ and $M = \$80$

3. $A = 4\pi r^2$ if $\pi = 3.14$ and $r = 15$ in.

4. $A = bh$ if $b = 12$ m and $h = 9.8$ m

5. $P = a + b + c$ if $a = 50$ cm, $b = 43$ cm, and $c = 45$ cm

6. **AUTO** What is the percent efficiency (E) of an engine if the input (I) is 25,000 calories and the output (P) is 9,600 calories? Use the formula $E = \dfrac{I - P}{I}$.

7. **CON** Evaluate the formula for the area of a circle, $A = \pi r^2$, if $r = 7$ in., and $\pi = 3.14$. Round to the nearest tenth.

See Example 2.

8. **ELEC** Use the formula $R_T = \dfrac{R_1 R_2}{R_1 + R_2}$ to find the total resistance (R_T) if one resistance (R_1) is 12 Ω and the second resistance (R_2) is 8 Ω.

See Example 3.

9. **CAD/ARC** Find the length of a rectangular work area if the perimeter is 180 in. and the width is 24 in. Use the formula $P = 2(l + w)$.

See Example 4.

Evaluate the simple interest formula $I = PRT$ using the following values.

10. **BUS** Find the interest if $P = \$800$, $R = 15.5\%$, and $T = 2\frac{1}{2}$ years.

11. **BUS** Find the rate if $I = \$427.50$, $P = \$1,500$, and $T = 2$ years.

12. **BUS** Find the time if $I = \$236.25$, $P = \$750$, and $R = 10.5\%$.

13. **BUS** Find the principal if $I = \$838.50$, $R = 21\frac{1}{2}\%$, and $T = 1\frac{1}{2}$ years.

Evaluate Exercises 14–16 using the percentage formula, $P = RB$.

14. Find the portion if $R = 15\%$ and $B = 600$ lb.

15. Find the rate if $P = 24$ kg and $B = 300$ kg.

16. **BUS** Find the base if $P = \$250$ and $R = 7.4\%$. Round to hundredths.

17. Evaluate the rate formula, $R = \dfrac{P}{B}$ for P if $R = 16\%$ and $B = 85$.

18. Evaluate the base formula, $B = \dfrac{P}{R}$, for R if $B = \$2,200$ and $P = \$374$.

19. **INDTEC** The formula for the speed (s) of a driven pulley in revolutions per minute (rpm) is $s = \dfrac{DS}{d}$. Find the speed of a driven pulley with a diameter (d) of 5 in. if the diameter (D) of the driving pulley is 10 in. and its speed (S) is 800 rpm.

See Example 5.

20. **INDTEC** If the distance (C) between the centers of the pulleys in Exercise 19 is 24 in., find the length (L) to the nearest hundredth of the belt connecting them using the formula $L = 2C + 1.57(D + d) + \dfrac{D + d}{4C}$.

See Example 6.

21. **BUS** Using the markup formula, $M = S - C$, find the selling price (S) if the markup (M) is \$12.75 and the cost (C) is \$36.

22. **BUS** Find the cost (C) if the markup (M) on an item is \$5.25 and the selling price (S) is \$15.75. Use the formula $M = S - C$.

23. Using the formula for the side of a square, $s = \sqrt{A}$, find the length of a side of a square field whose area is 16 mi^2.

24. Evaluate the formula for the area of a square, $A = s^2$, if $s = 2.5$ km.

25. **ELEC** The formula for voltage (Ohm's law) is $E = IR$. Find the amperes of current (I) if the voltage (E) is 120 V and the resistance (R) is 80 Ω.

26. **INDTEC** According to Boyle's law, if temperature is constant, the volume of a gas is inversely proportional to the pressure on it. Find the final volume (V_2) of a gas using the formula $\dfrac{V_1}{V_2} = \dfrac{P_2}{P_1}$ if the original volume (V_1) is 15 ft^3, the original pressure (P_1) is 60 lb per square inch (psi), and the final pressure (P_2) is 150 psi.

27. **ELEC** The formula for power (P) in watts (W) is $P = I^2R$. Find the current (I) in amperes if a device draws 63 W and the resistance (R) is 7 Ω.

28. **AUTO** Use the formula $H = \dfrac{D^2N}{2.5}$ to find the number of cylinders (N) required in an engine of 3.2 hp (H) if the cylinder diameter (D) is 2 in.

29. Find the area of a circle whose radius is 54.5 cm using the formula $A = \pi r^2$ and using 3.14 for π. Round to the nearest tenth.

30. **AUTO** Distance is rate times time, or $D = RT$. Find the rate if the distance traveled is 140 mi and the time traveled is 4 h.

31. **ELEC** Find the impedance (Z) in ohms using the formula $Z = \sqrt{R^2 + X^2}$ if the reactance (X) is 15 Ω and the resistance (R) is 8 Ω. Round to tenths.

2 Rearrange the formulas. *See Example 7.*

32. **BUS** Solve $I_n = I - S$ for S, where I_n = new inventory, I = current inventory, and S = sales.

33. Solve $S = P - D$ for D.

34. Solve $y = mx + b$ for b.

35. Solve $S = C + M$ for C.

See Example 8.

36. Solve $V = \pi r^2 h$ for h.   37. Solve $S = 2\pi rh$ for r.   38. Solve $A = lw$ for l.   39. Solve $C = 2\pi r$ for r.

40. Solve $E = IR$ for R.   41. Solve $D = RT$ for R.   42. Solve $S = 2\pi rh$ for h.   43. Solve $C = \pi d$ for d.

See Example 9.

44. Solve $P = 2(b + s)$ for b.

45. Solve $R = AC - B$ for C.

See Example 10.

46. Solve $\dfrac{R}{100} = \dfrac{P}{B}$ for R.

47. **PFIN** The formula for finding the amount of a repayment on a loan is $A = I + P$, where A is the amount of the repayment, I is the interest, and P is the principal. Solve the formula for interest.

48. **PFIN** The formula for finding simple interest is $I = PRT$, where I represents interest, P represents principal, R represents rate, and T represents time. Rearrange the formula to find the time.

See Example 11.

49. Develop formulas to complete the spreadsheet in Fig. 8–4 to show data for the actual expenses for the 7th Inning Sports Memorabilia Shop.

	A	B	C	D	E	F
1	The 7th Inning Budgeted Operating Expenses And Actual Expenses					
2						
3	Expense	Budget Amount	Percent of Total Budget	Actual Expenses	% of Actual Total Expense	% Difference from Budget
4						
5	Salaries	$45,000.00		$42,000.00		
6	Rent	$37,000.00		$36,000.00		
7	Depreciation	$12,000.00		$14,000.00		
8	Utilities and Phone	$13,000.00		$10,862.56		
9	Taxes and Insurance	$15,000.00		$13,583.29		
10	Advertising	$2,000.00		$2,847.83		
11	Purchases	$125,000.00		$132,894.64		
12	Other	$2,000.00		$1,356.35		
13						
14	Total					

FIGURE 8–4

50. Use the formulas to complete the spreadsheet for the 7th Inning Sports Memorabilia Shop. The percent difference from the budget uses the budget as the base. Negative percents show percents under budget and positive percents show percents over budget.

8-2 Proportion

LEARNING OUTCOME

1 Solve equations that are proportions.

LC LEARNING CATALYTICS

1. Is $\frac{6}{8} = \frac{12}{32}$?

2. Multiply $3(2x - 7)$.

A common type of equation that contains fractions is a *proportion.* In a proportion, *each side* of the equation is a fraction or ratio. In Chapter 3, Section 2, Outcome 3, we used the percentage proportion to solve percentage problems.

1 Solve Equations That Are Proportions. An equation in the form of a proportion can be used to solve many types of applied problems. As with the percentage proportion, we use the property of proportions to solve equations in the form of a proportion.

> **Property of proportions:**
>
> The cross products in a proportion are equal. Symbolically, if $\frac{a}{b} = \frac{c}{d}$, then $a \cdot d = b \cdot c$, provided that b and d are not equal to zero. Also, if $a \cdot d = b \cdot c$, then $\frac{a}{b} = \frac{c}{d}$, a, b, c, and d represent real numbers and b and d are not zero.

STOP AND CHECK

Solve the proportions.

1. $\frac{x}{7} = \frac{9}{21}$

2. $\frac{x + 6}{x - 6} = \frac{9}{3}$

3. $\frac{5}{x} = 6$

Answers:

1. $x = 3$ **2.** $x = 12$ **3.** $x = \frac{5}{6}$

EXAMPLE 1

Solve the proportions.

(a) $\dfrac{x}{4} = \dfrac{9}{6}$ **(b)** $\dfrac{4x}{5} = \dfrac{17}{20}$ **(c)** $\dfrac{x - 2}{x + 8} = \dfrac{3}{5}$ **(d)** $\dfrac{3}{x} = 7$

(a) $\dfrac{x}{4} = \dfrac{9}{6}$ Cross multiply. $x(6) = 6x$; $4(9) = 36$

$6x = 36$ Solve for x.

$\dfrac{6x}{6} = \dfrac{36}{6}$

$x = 6$

(b) $\dfrac{4x}{5} = \dfrac{17}{20}$ Cross multiply. $4x(20) = 80x$; $5(17) = 85$

$80x = 85$ Solve for x.

$\dfrac{80x}{80} = \dfrac{85}{80}$

$x = \dfrac{85}{80}$ Simplify.

$x = \dfrac{17}{16}$

(c) $\dfrac{x - 2}{x + 8} = \dfrac{3}{5}$ Cross multiply. $3(1) = 3$; $x(7) = 7x$

$5(x - 2) = 3(x + 8)$ Distribute.

$5x - 10 = 3x + 24$ Sort.

$5x - 3x = 24 + 10$ Combine like terms.

$$2x = 34 \qquad \text{Solve for } x.$$

$$\frac{2x}{2} = \frac{34}{2}$$

$$x = 17$$

(d) $\dfrac{3}{x} = 7$ \qquad Write 7 as a fraction. $7 = \dfrac{7}{1}$

$$\frac{3}{x} = \frac{7}{1} \qquad \text{Cross multiply.}$$

$$3 = 7x \qquad \text{Solve for } x.$$

$$\frac{3}{7} = \frac{7x}{7}$$

$$\frac{3}{7} = x$$

See Exercises 1–40.

Did You Know? **Finding cross products is an application of the multiplication axiom.** The property of cross products is a shortcut application of the multiplication axiom for b and d not equal to zero.

$$\frac{a}{b} = \frac{c}{d} \qquad \text{Multiply both sides of the equation by both denominators.}$$

$$b(d)\left(\frac{a}{b}\right) = b(d)\left(\frac{c}{d}\right) \qquad \text{Simplify.}$$

$$\cancel{b}(d)\left(\frac{a}{\cancel{b}}\right) = b(\cancel{d})\left(\frac{c}{\cancel{d}}\right)$$

$$da = bc \qquad \text{Apply the commutative property of multiplication.}$$

$$ad = bc \qquad \text{Same as property of cross products}$$

A proportion is often used with measures. The measures are included in the multiplication and division processes.

STOP AND CHECK

1. Solve for x to the nearest tenth. $\dfrac{85 \text{ mi}}{3.5 \text{ gal}} = \dfrac{x \text{ mi}}{20 \text{ gal}}$.

Answer:
1. $x = 485.7$ mi

EXAMPLE 2

Solve for x to the nearest tenth. $\dfrac{450 \text{ ft}^2}{1.2 \text{ gal}} = \dfrac{1,250 \text{ ft}^2}{x \text{ gal}}$.

$$\frac{450 \text{ ft}^2}{1.2 \text{ gal}} = \frac{1,250 \text{ ft}^2}{x \text{ gal}} \qquad \text{Cross multiply.}$$

$$(450 \text{ ft}^2)(x \text{ gal}) = (1.2 \text{ gal})(1,250 \text{ ft}^2) \qquad \text{Divide both sides by } 450 \text{ ft}^2.$$

$$\frac{(450 \text{ ft}^2)(x \text{ gal})}{(450 \text{ ft}^2)} = \frac{(1.2 \text{ gal})(1,250 \text{ ft}^2)}{(450 \text{ ft}^2)} \qquad \text{Simplify dimensions.}$$

$$x \text{ gal} = \frac{(1.2)(1,250)}{450} \text{ gal} \qquad \text{Make numerical calculations.}$$

$$x = 3.333333333 \text{ gal}$$

$$x \approx 3.3 \text{ gal} \qquad \text{Rounded to nearest tenth}$$

See Exercises 41–54.

8-2 EXERCISES

MyLab Math For additional practice go to your study plan in MyLab Math.

1 Solve the proportions. Round to four significant digits if necessary. *See Example 1.*

1. $\dfrac{x}{5} = \dfrac{9}{15}$

2. $\dfrac{3x}{16} = \dfrac{3}{8}$

3. $\dfrac{x-1}{x+6} = \dfrac{4}{5}$

4. $\dfrac{5}{x} = 8$

5. $\dfrac{2}{7} = \dfrac{x-4}{x+3}$

6. $\dfrac{2x+1}{8} = \dfrac{3}{7}$

7. $\dfrac{3x-2}{3} = \dfrac{2x+1}{3}$

8. $\dfrac{5}{2x-2} = \dfrac{1}{8}$

9. $\dfrac{8}{3x+2} = \dfrac{8}{14}$

10. $\dfrac{2x}{8} = \dfrac{3x+1}{7}$

11. $\dfrac{-5}{x-2} = \dfrac{5}{x}$

12. $\dfrac{7}{x-6} = \dfrac{-3}{x}$

13. $\dfrac{4}{8} = \dfrac{x}{60}$

14. $\dfrac{5.2}{16} = \dfrac{x}{8.3}$

15. $\dfrac{5.2}{340} = \dfrac{12.8}{x}$

16. $\dfrac{3x+1}{4} = \dfrac{5x}{8}$

17. $\dfrac{7.35}{4} = \dfrac{5.21}{x}$

18. $\dfrac{9.12}{8} = \dfrac{x}{1.03}$

19. $\dfrac{2x}{7} = \dfrac{5.3}{6.7}$

20. $\dfrac{7x}{5} = \dfrac{91}{20}$

21. $\dfrac{0.107}{4x} = \dfrac{0.04}{321}$

22. $\dfrac{0.7x}{2.3} = \dfrac{5.6}{12.8}$

23. $\dfrac{0.04x}{5.2} = \dfrac{3.16}{14.08}$

24. $\dfrac{10^4}{x} = \dfrac{10^7}{10^5}$

25. $\dfrac{10^8}{10^5} = \dfrac{x}{10^9}$

26. $\dfrac{2x-3}{7} = \dfrac{3}{28}$

27. $\dfrac{32}{21} = \dfrac{8}{5x-1}$

28. $\dfrac{4}{3x-5} = \dfrac{18}{12}$

29. $\dfrac{12 \text{ in.}}{15 \text{ in.}} = \dfrac{8 \text{ in.}}{x}$

30. $\dfrac{15 \text{ ft}}{38 \text{ ft}} = \dfrac{x}{57 \text{ ft}}$

31. $\dfrac{12 \text{ k}\Omega}{8 \text{ k}\Omega} = \dfrac{x}{6 \text{ k}\Omega}$

32. $\dfrac{x}{15 \text{ W}} = \dfrac{60 \text{ W}}{3.5 \text{ W}}$

33. $\dfrac{150 \text{ V}}{0.6 \text{ V}} = \dfrac{65 \text{ V}}{x}$

34. $\dfrac{12 \text{ mA}}{8.2 \text{ mA}} = \dfrac{x}{0.5 \text{ mA}}$

35. $\dfrac{5.12}{14.87} = \dfrac{x}{3.91}$

36. $\dfrac{21.25}{3.2x} = \dfrac{212.5}{32}$

37. $\dfrac{\frac{3}{5}}{\frac{5}{8}} = \dfrac{\frac{4}{5}}{x}$

38. $\dfrac{x}{\frac{5}{9}} = \dfrac{\frac{3}{10}}{\frac{4}{5}}$

39. $\dfrac{2\frac{1}{4}}{x} = \dfrac{\frac{7}{10}}{\frac{4}{9}}$

40. $\dfrac{4\frac{3}{8}}{\frac{5}{8}} = \dfrac{x}{\frac{4}{5}}$

See Example 2.

41. $\dfrac{50 \text{ mi}}{1 \text{ h}} = \dfrac{400 \text{ mi}}{x \text{ h}}$

42. $\dfrac{2,500 \text{ mi}}{3.8 \text{ h}} = \dfrac{500 \text{ mi}}{x \text{ h}}$

43. $\dfrac{\$245,000}{2,500 \text{ ft}^2} = \dfrac{\$x}{1 \text{ ft}^2}$

44. $\dfrac{\$495,000}{5,200 \text{ ft}^2} = \dfrac{\$x}{1 \text{ ft}^2}$

45. $\dfrac{\$4.12}{20 \text{ oz}} = \dfrac{\$x}{1 \text{ oz}}$

46. $\dfrac{15 \text{ gal}}{40 \text{ acres}} = \dfrac{x \text{ gal}}{1 \text{ acres}}$

47. $\dfrac{350 \text{ mg}}{7.8 \text{ cm}^3} = \dfrac{x \text{ mg}}{1 \text{ cm}^3}$

48. $\dfrac{350 \text{ mg}}{60 \text{ cm}^3} = \dfrac{x \text{ mg}}{1 \text{ cm}^3}$

49. $\dfrac{5 \text{ gal}}{20 \text{ acres}} = \dfrac{x \text{ gal}}{1 \text{ acre}}$

50. $\dfrac{44 \text{ gal}}{1 \text{ min}} = \dfrac{x \text{ gal}}{3 \text{ min}}$

51. $\dfrac{1,500 \text{ mi}}{2.7 \text{ h}} = \dfrac{x \text{ mi}}{1 \text{ h}}$

52. $\dfrac{400 \text{ mg}}{20 \text{ cm}^3} = \dfrac{x \text{ mg}}{1 \text{ cm}^3}$

53. $\dfrac{1,000 \text{ mg}}{120 \text{ cm}^3} = \dfrac{x \text{ mg}}{1 \text{ cm}^3}$

54. $\dfrac{4,500 \text{ mg}}{560 \text{ cm}^3} = \dfrac{x \text{ mg}}{1 \text{ cm}^3}$

8-3	Direct and Joint Variation

LEARNING OUTCOMES

1 Solve problems of direct variation using proportions.

2 Solve problems of direct variation using a constant of variation.

3 Solve problems of joint variation using a constant of variation.

LC **LEARNING CATALYTICS**

1. Solve the proportion $\dfrac{15}{n} = \dfrac{12}{4}$.

Apples	Cost
4	$1
8	$2
12	$3
16	$4
independent variable	**dependent variable**

STOP AND CHECK

1. Find the cost of 14 tickets if 5 tickets cost $75.

Answer:
1. $210

Direct variation: a proportion in which the quantities being compared are directly related, so that as one quantity increases, the other quantity also increases. Similarly, as one quantity decreases, the other quantity decreases. This relationship is also called a direct proportion

Directly proportional: two ratios of quantities where an increase (or decrease) in one quantity causes an increase (or decrease) in the other; alternate terminology for direct variation

Independent variable: the variable that represents input values of a function

Dependent variable: the variable that is the result in a function of a relationship with the independent variable

1 Solve Problems of Direct Variation Using Proportions. Many problems in the workplace can be solved using proportions. The details of the problem can be grouped into two pairs of data that can be *directly* related.

A **direct variation** is one in which the quantities being compared are directly related, so that as one quantity increases (or decreases), the other quantity also increases (or decreases). We also say these quantities are **directly proportional.**

Data are often arranged in tables like the one on the left so these relationships can be examined. If 4 apples cost $1, set up a table to examine related costs of other amounts of apples.

In a pair of data, one item is identified as the *independent variable,* such as the *number of apples* purchased. The **independent variable** represents input values of a function. The other item depends on the first item and it is identified as the *dependent variable,* such as the *total cost* of the apples. The **dependent variable** is the result in a function of a relationship with the independent variable.

To set up a direct proportion:

1. Establish two pairs of related data.

2. Write one pair of data in the numerators of two ratios.

3. Write the other pair of data in the denominators of the two ratios.

4. Form a proportion using the two ratios.

EXAMPLE 1

PFIN   **(a)** Find the cost of 10 apples if 4 apples cost $1.   **(b)** How many apples can be purchased for $10?

(a) Pair 1: 4 apples cost $1.
     Pair 2: 10 apples cost c dollars.

Estimation

8 apples would cost $2. Therefore, 10 apples will cost more than $2.

$$\frac{4 \text{ apples}}{10 \text{ apples}} = \frac{\$1}{\$c}$$ Pair 1 is the numerator of each ratio.
     Pair 2 is the denominator of each ratio. Cross multiply.

$$4c = 10$$ Divide.

$$\frac{4c}{4} = \frac{10}{4}$$

$$c = 2.50$$ To the nearest cent

Interpretation

10 apples cost $2.50.

(b) Pair 1: 4 apples cost $1.
     Pair 2: a apples cost $10.

Estimation

If 10 apples cost $2.50, 4 times as many apples can be purchased for $10. That is, 40 apples can be purchased.

$$\frac{4 \text{ apples}}{a \text{ apples}} = \frac{\$1}{\$10}$$

Pair 1
Pair 2 Cross multiply.

$$40 = a$$

Interpretation

40 apples can be purchased for $10.

See Exercises 1–8.

Directly related data pairs can also be set up by making each pair a ratio. In the preceding example, we would write the ratio as

Pair 1 $\quad \dfrac{4 \text{ apples}}{\$1} = \dfrac{10 \text{ apples}}{\$c} \quad$ Pair 2

EXAMPLE 2

AUTO A truck travels 102 mi on 6 gal of gasoline. How far will it travel on 30 gal of gasoline?

Known facts

Pair 1: 102 mi uses 6 gal of gasoline.

Unknown facts

Pair 2: m mi uses 30 gal of gasoline.

Estimation

30 gal ÷ 6 gal = 5. Then, approximately 5 times 100 miles or 500 miles can be driven on 30 gal.

Calculations

$$\frac{102 \text{ mi}}{m \text{ mi}} = \frac{6 \text{ gal}}{30 \text{ gal}}$$

$$\frac{102}{m} = \frac{6}{30}$$

$$102(30) = 6m$$

$$\frac{3{,}060}{6} = \frac{6m}{6}$$

$$510 = m$$

Dimension Analysis

$$\frac{\text{distance}_1}{\text{distance}_2} = \frac{\text{gasoline}_1}{\text{gasoline}_2} \quad \begin{matrix} \text{Pair 1} \\ \text{Pair 2} \end{matrix}$$

Cross multiply. $\dfrac{\text{mi}}{\text{mi}} = \dfrac{\text{gal}}{\text{gal}}$

Divide. mi(gal) = gal(mi)

Reduce. $\dfrac{\text{mi(gal)}}{\text{gal}} = \dfrac{\text{gal(mi)}}{\text{gal}}$

m is expressed in miles.

Interpretation

The truck will travel 510 mi on 30 gal of gasoline.

See Exercises 9–12.

TIP **Analyze Dimensions** Even though we may analyze the dimensions of an equation separately, we *always* should analyze the dimensions to be sure we use the correct units in the solution.

EXAMPLE 3

INDTEC If a metal rod tapers 1 in. for every 24 in. of length, what is the amount of taper of a 30-in. piece of rod? (See Fig. 8–5.)

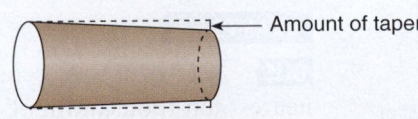

Amount of taper

FIGURE 8–5

Known facts

Pair 1: 1-in. taper for 24-in. length

Unknown facts

Pair 2: x-in. taper for 30-in. length

Estimation

A 30-in. rod will taper more than 1 in.

Calculations

$$\frac{\text{1-in. taper}}{x\text{-in. taper}} = \frac{\text{24-in. length}}{\text{30-in. length}} \qquad \text{Pair 1} \\ \text{Pair 2}$$

$$\frac{1}{x} = \frac{24}{30} \qquad \text{Cross multiply.}$$

$$1(30) = x(24)$$

$$30 = 24x$$

$$\frac{30}{24} = \frac{24x}{24} \qquad \text{Divide.}$$

$$\frac{5}{4} = x \qquad \text{or} \qquad x = 1\frac{1}{4}$$

Interpretation

The amount of taper for a 30-in. length of rod is $1\frac{1}{4}$ in.   **See Exercises 13–19.**

---

2   Solve Problems of Direct Variation Using a Constant of Variation. Another way to use directly related data is to identify a data pair in which both values are known then use that relationship to find a *conversion factor*. For example, if 4 apples cost $1, how much does 1 apple cost?

$$\$1 \div 4 \text{ apples} = \$0.25 \text{ cost per apple}$$

A conversion factor is used to multiply by the number of items. If each apple costs $0.25, then 10 apples cost $10 \times \$0.25$, or $2.50. A conversion factor is also called a *constant of direct variation*. The direct variation formula is $y = kx$, where x is the *independent variable, y* is the *dependent variable,* and k is the **constant of direct variation.** Another way to express this is $k = \dfrac{y}{x}$.

Constant of direct variation: a conversion factor for finding direct variation. The constant can be found by using the formula $k = \dfrac{y}{x}$.

> **To find the constant of direct variation:**
>
> 1. Identify a pair of related data in which both values are known.
>
> 2. Write the data pair as a ratio. The units in the denominator of the ratio should match the units of the independent variable. $k = \dfrac{y}{x}$
>
> 3. Leave the ratio as a fraction or change it to a decimal equivalent.

STOP AND CHECK

1. Find the unit cost to the nearest tenth of a cent (thousandth of a dollar) of a package of 225 paper plates if the package costs $7.95.

Answer:

1. $0.035/plate

Unit cost: the quotient of the cost of items and the number of items

EXAMPLE 4

BUS   A 15-oz box of cereal costs $2.29. Find the constant of direct variation to the nearest ten-thousandth of a dollar for the cost per ounce. (This is also referred to as the **unit cost.**)

$$k = \frac{y}{x}$$

Unit cost
or
Constant of direct variation

$$= \frac{cost}{oz} = \frac{\$}{oz} \frac{\text{dependent variable}}{\text{independent variable}}$$

$$k = \frac{\$2.29}{15 \text{ oz}}$$

Change to a decimal equivalent. Round to the nearest ten-thousandth.

$$k = \$0.1527/\text{oz}$$ Cost per ounce

The unit cost or constant of direct variation is $0.1527 \dfrac{\$}{oz}$.

See Exercises 20–22.

TIP **Making Connections with Terminology** Enhance your skills in interpreting applied problems by making connections among different terms. So far we have associated the terms *conversion factor* and *unit cost* with the *constant of direct variation.*

STOP AND CHECK

1. Compare the unit cost of the package of plates in the previous Stop and Check problem with the unit cost of a package of 500 paper plates costing $18.95. Which is the better buy?

Answer:

1. Since the unit cost of 500 paper plates costing $18.95 is $0.038, the package of 225 paper plates for $7.95 is the better buy.

EXAMPLE 5

BUS A 15-oz box of cereal costs $2.29 or a 36-oz box of cereal costs $4.89. Which is the better buy? Round the unit costs to the nearest ten-thousandth of a dollar.

In the preceding example we found the cost per ounce (unit cost) of the 15-oz box of cereal to be $0.1527.

Find the unit cost of the 36-oz box of cereal.

$$k = \frac{\$4.89}{36} = \$0.1358/\text{oz} \qquad \text{Ratio of } \frac{\$}{oz}$$

Compare the unit costs.

Unit cost of 15-oz box $= \$0.1527$ per oz; unit cost of 36-oz box $= \$0.1358$ per oz

The 36-oz box of cereal costs less per ounce and is the better buy.

See Exercises 23–29.

In the preceding example, we make a judgment solely on the basis of the mathematical facts. In reality, other factors are considered. Do you have the money to buy the larger box? Will you be able to use the larger amount of cereal before it gets stale? Are both boxes the same type of cereal?

To solve problems of direct variation using a constant of variation:

1. Find the constant of direct variation using known values for x and y and the formula $k = \dfrac{y}{x}$.

2. Evaluate the formula of direct variation

$$y = kx,$$

where x is the independent variable, y is the dependent variable, and k is the constant of direct variation.

STOP AND CHECK

1. If a map is scaled at 2 cm = 100 km, how far is a distance measuring 3.4 cm on the map?

Answer:

1. 170 km

EXAMPLE 6

CAD-ARC An architect's drawing is scaled at 0.25 in. = 5 ft. Find the actual measurements for the following measures shown on the drawing.

(a) 0.75 in. **(b)** 1.25 in.

Determine the scaled measure per foot of actual measurement (constant of direct variation). The actual measure is the independent variable and the scaled measure is the dependent variable.

$$\text{constant of direct variation} = \frac{\text{dependent variable}}{\text{independent variable}} \qquad k = \frac{y}{x}. \text{ Substitute known values.}$$

$$k = \frac{5 \text{ ft}}{0.25 \text{ in.}} \qquad\qquad\qquad \text{Divide.}$$

$$k = 20 \frac{\text{ft}}{\text{in.}} \qquad\qquad\qquad\qquad \frac{\text{ft}}{\text{in}}$$

Use the direct variation formula to find the respective values of the dependent variables.

(a) $y = kx$ Substitute $k = 20 \frac{\text{ft}}{\text{in.}}$ and $x = 0.75$ in.

$\quad y = 20 \frac{\text{ft}}{\text{in.}} (0.75 \text{ in.})$ Multiply and analyze the dimensions.

$\quad y = \mathbf{15 \ ft}$

(b) $y = kx$ Substitute $k = 20 \frac{\text{ft}}{\text{in.}}$ and $x = 1.25$ in.

$\quad y = 20 \frac{\text{ft}}{\text{in.}} (1.25 \text{ in.})$ Multiply.

$\quad y = \mathbf{25 \ ft}$ **See Exercises 30–46.**

> **TIP** **Which Method for Direct Variation Is Preferred?** Personal opinions vary in selecting a preferred method for solving direct proportion problems. The proportion method is very versatile for a variety of situations. Using the constant of direct variation is useful in computer programs and electronic spreadsheets.

3 **Solve Problems of Joint Variation Using a Constant of Variation.** One quantity may vary directly as a product of two or more other quantities. This variation is called **joint variation.**

Joint variation: a quantity varying directly as a product of two or more other quantities

> **To solve problems of joint variation using a constant of variation:**
>
> 1. Symbolically represent the relationships among the variables.
>
> 2. Find the constant of joint variation by substituting known values into the equation from Step 1 and solving for k.
>
> 3. Evaluate the equation from Step 1 using the constant of joint variation from Step 2 and other known values.

An example of an equation of joint variation is $y = kxz$. It can be read as "y varies jointly as x and z" or "y is jointly proportional to x and z." A third statement is "$y = kxz$ for some constant k."

STOP AND CHECK

1. Find m if m varies jointly as n and p, and $n = 38.5$ and $p = 4$. The constant of variation is 0.2.

Answer:

1. $m = 30.8$

EXAMPLE 7

Find w if w varies jointly as x and y, and $x = 140$, $y = 6$, and the constant of joint variation is 0.5.

Write the equation using the joint variation model.

$w = kxy$ Substitute known values and solve for w.

$w = 0.5(140)(6)$ Evaluate.

$w = \mathbf{420}$ **See Exercises 47–49.**

EXAMPLE 8

The *simple interest* for an account is jointly proportional to the time and principal. If the quarterly interest for an account balance of $7,000 is $87.50, find the interest on a balance of $8,500 for 9 months.

In the joint variation equation, let y represent the interest, x represent the time, and z represent the principal.

$$y = k(\text{time})(\text{principal})$$
$$y = kxz$$

Find k. Substitute known values for interest, time, and principal.

Quarterly means 3 months or $\frac{3}{12}$ year, or 0.25 year.

$$\$87.50 = k(0.25)(\$7,000)$$

Solve for k.

$$\frac{\$87.50}{(0.25)\$7,000} = k$$

Evaluate.

$$k = 0.05$$

Constant of joint variation

Use the constant of joint variation to find the interest.

$$y = k(\text{time})(\text{principal})$$
$$y = kxz$$
$$y = (0.05)(0.75)(\$8,500)$$
$$y = \$318.75$$

Find the interest. Substitute known values for k, the time, and the principal. 9 months = 0.75 year.

Evaluate.

The interest for $8,500 for 9 months is $318.75.　　　　　**See Exercises 50–52.**

In the preceding example, note the similarity between the joint variation model and the simple interest formula—Interest = principal(rate)(time) or $I = PRT$. The constant of joint variation is represented by the rate R in the simple interest formula.

8-3 EXERCISES

MyLab Math For additional practice go to your study plan in MyLab Math.

1 Solve using proportions. *See Example 1.*

1. **PFIN** If 7 cans of dog food sell for $4.13, how much will 10 cans sell for?

2. **BUS** If 6 cans of coffee sell for $22.24, how much will 20 cans sell for?

3. **AUTO** A mechanic took 7 h to tune up 9 fuel-injected engines. At this rate, how many fuel-injected engines can be tuned up in 37.5 h? Round to the nearest whole number.

4. **BUS** A costume maker took 9 h to make 4 headpieces for a Mardi Gras ball. At this rate, how many complete headpieces can be made in 35 h?

5. How far can a family travel in 5 days if it travels at the rate of 855 mi in 3 days?

6. **AUTO** How far can a tractor-trailer rig travel in 8 days if it travels at the rate of 1,680 mi in 4 days?

7. **AG/H** How much crystallized insecticide does 275 acres of farmland need if the insecticide treats 50 acres per 100 lb?

8. **AG/H** How much fertilizer does 2,625 ft^2 of lawn need if the fertilizer treats 1,575 ft^2 per gallon? Express the answer to the nearest tenth of a gallon.

See Example 2.

9. **INDTEC** Two gears have a ratio of 8 to 2. If the larger gear has 48 teeth, how many teeth does the smaller gear have?

10. **INDTEC** Two gears have a ratio of 9 to 4. If the larger gear has 72 teeth, how many teeth does the smaller gear have?

11. **INDTR** The diameter of the larger of two pulleys connected with a belt is 36 cm. If the ratio is 9:1, what is the diameter of the smaller pulley?

12. **INDTR** The diameter of the smaller of two pulleys connected with a belt is 48 cm. If the ratio is 8:1, what is the diameter of the larger pulley?

See Example 3.

13. **CON**   The slope of a roof is the ratio of the rise to the run $\left(\dfrac{\text{rise}}{\text{run}}\right)$ (Fig. 8–6). A roof has a slope of $\dfrac{3}{12}$. What is the rise for this roof if it has a run of 28 ft?

14. **CON**   A roof has a slope of $\dfrac{1}{6}$. What is the rise for this roof if it has a run of 36 ft?

15. **CON**   The *pitch* of a roof is the ratio of the rise of a roof to its span (Fig. 8–7). A roof has a pitch of 1:4. If the span is 40 ft, what is the rise?

16. **CON**   A roof has a pitch of 1:3. If the span is 42 ft, what is the rise?

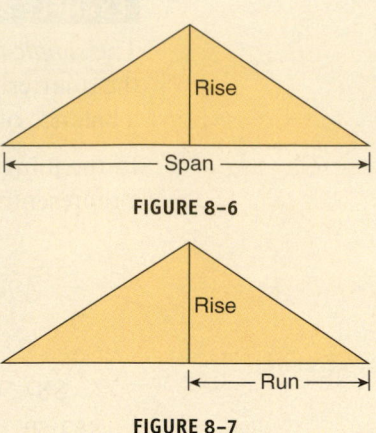

FIGURE 8–6

FIGURE 8–7

18. **HLTH/N**   A person who weighs 185 lb should be given how many milligrams of medication if the dosage is 15 mg for every 10 lb?

17. **HLTH/N**   A person who weighs 142 lb should be given how many milligrams of medication if the dosage is 25 mg for every 10 lb?

19. **HLTH/N**   The pediatric dosage for chlorpromazine hydrochloride is 0.25 mg/lb. What is the dosage for a child that weighs 40 lb?

2   Solve using a constant of variation. *See Example 4.*

20. **HLTH/N**   Chlorpromazine injection strength contains 50 mg per 2 mL. A patient is prescribed 0.5 g. How many milliliters should be administered (1 g = 1,000 mg)?

21. **BUS**   A can of tomatoes contains 793 g and costs $1.49. Find the constant of direct variation for the cost per gram. Round to the nearest ten-thousandth.

22. **PFIN**   A can of tomatoes contains 411 g and costs $0.89. Find the constant of direct variation for the cost per gram. Round to the nearest ten-thousandth.

See Example 5.

23. From Exercises 21 and 22 compare the unit cost of the small can of tomatoes with the unit cost of the large can to determine which size can is more economical.

24. **INDTEC**   A bottling machine can fill 800 bottles in 2.5 h. Find the constant of direct variation for the number of bottles filled per hour.

25. **HLTH/N**   A patient is prescribed 50 mg of Librium® IM. The medication is available as 0.4 mg per 2 mL. How many milliliters should be injected?

26. **HLTH/N**   A patient is prescribed an injection of 0.2 mg of Atropine® IM. The drug is available as 0.5 mg per milliliter. How many milliliters should be injected?

27. **AG/H**   Instructions for mixing a chemical pesticide state that 6 oz of chemical should be mixed with 64 oz of water. How many ounces of chemical should be used with 160 oz of water?

28. **ELEC**   An electrical transformer shown in Fig. 8–8 is constructed by winding a number of turns of wire into a primary. Another set of turns is constructed by winding to form one or more secondaries. The ratio of the primary voltage (E_p) to the secondary voltage (E_s) is the same as the ratio of the number of primary turns (T_p) to the number of secondary turns (E_s).

$$\frac{E_p}{E_s} = \frac{T_p}{T_s}$$

Find the secondary voltage if the transformer has a 360-turn primary and a 30-turn secondary and 150 V are applied.

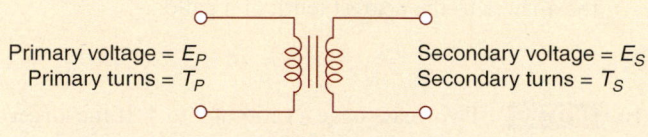

Primary voltage = E_P
Primary turns = T_P

Secondary voltage = E_S
Secondary turns = T_S

FIGURE 8–8

29. **HLTH/N**   A doctor ordered streptomycin 250 mg IM for a patient. The dosage available for use contains 1 g per 4 mL. How many milliliters should be injected (1 g = 1,000 mg)?

See Example 6.

30. **CAD/ARC** A blueprint has a scale of $\frac{1}{2}$ in. = 1 ft.

    On a blueprint a wall is drawn $7\frac{1}{2}$ in. long. What is the actual measure of the wall?

31. **CAD/ARC** A blueprint has a scale of $\frac{3}{4}$ in. = 1 ft.

    On a blueprint a wall is drawn $8\frac{3}{4}$ in. long. What is the actual measure of the wall?

32. **CAD/ARC** A building that is 120 ft long would be shown as what length on a blueprint if the scale were $\frac{1}{4}$ in. = 1 ft?

33. **CAD/ARC** A building that is 320 ft long would be shown as what length on a blueprint if the scale were $\frac{1}{8}$ in. = 1 ft?

34. **ELEC** The length of a wire is proportional to its resistance. The resistance of 200 ft of a certain wire is 0.0062 Ω. What is the resistance of 750 ft of the same wire?

35. **ELEC** The length of a wire is proportional to its resistance. The resistance of 100 ft of a certain wire is 0.048 Ω. What is the resistance of 1,200 ft of the same wire?

Use the map in Fig. 8–9 and a U.S. customary rule in Exercises 36–38.

36. **AVIA** Find the air distance to the nearest mile from Upper Sandusky to Lima.

37. **AVIA** Find the air distance to the nearest mile from Sidney to Bellefontaine.

38. **AVIA** Find the air distance to the nearest mile from Lima to Bellefontaine.

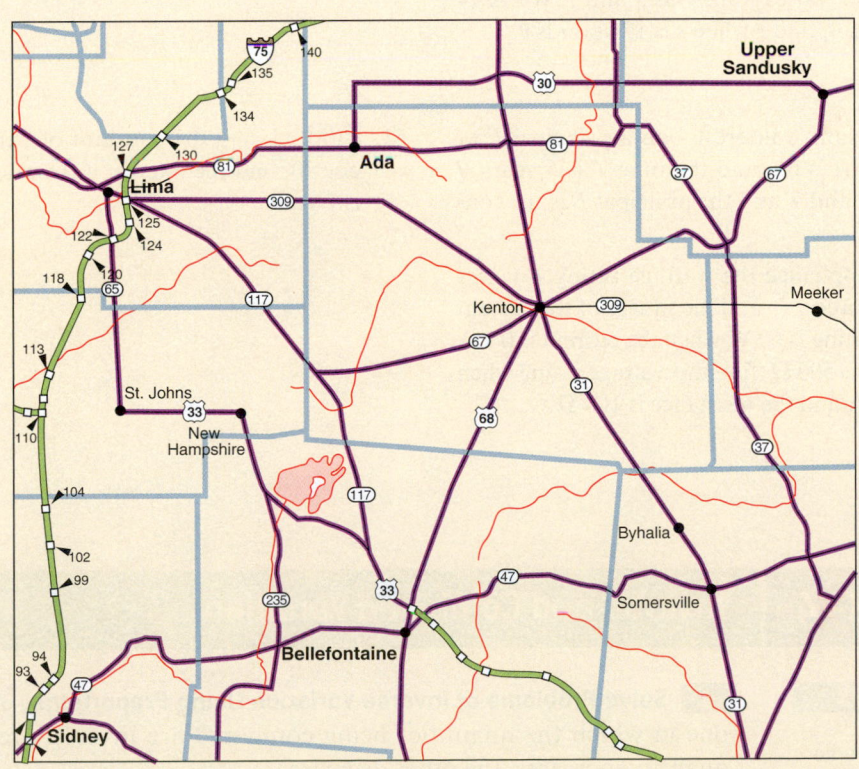

$\frac{13}{16}$ inch = 10 miles

FIGURE 8–9

39. **INDTR** What is the overall width of the actual plate represented in Fig. 8–10?

40. **INDTR** A machinist is to make the metal plate represented in Fig. 8–10. What is the overall length of the actual plate?

41. **INDTR** What is the diameter in inches of the semicircle in Fig. 8–10? The diameter is the distance across the center of a circle.

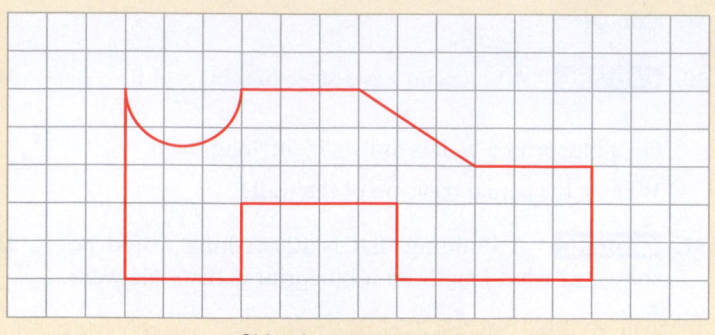

Side of 1 square = 2.5 in.

FIGURE 8–10

Use square-ruled paper to make a line drawing of objects described in Exercises 42–46.

42. A rectangle 10 ft by 7.5 ft, using a scale of 1 square = 2.5 ft

43. A square 10 cm on a side, using a scale of 1 square = 1 cm

44. A circle with a 24-in. diameter, using a scale of 1 square = 4 in.

45. A rectangle 15 m by 12 m, using a scale of 1 square = 3 m

46. A rectangle 4.5 cm by 3.5 cm, using a scale of 1 square = 0.5 cm

3 Solve problems of joint variation using a constant of variation. *See Example 7.*

47. If x varies jointly with y and z, find x when $y = 18$ and $z = 20$ if the constant of variation is $k = 3$.

48. If I varies jointly with R and T, find I when $R = 0.05$, $T = 3$, and the constant of variation is $P = 5,000$.

49. In a relationship, r varies jointly as s and t. If r is 84 when s is 7 and t is 4, find r when s is 12 and t is 9.

See Example 8.

50. **PFIN** Find the simple interest I on a principal P of $8,000 if the rate R is 6% and the time T is 5 years. I varies jointly as R and T and the principal P is the constant of variation.

51. **BUS** Using the constant of variation given in Exercise 50, find the interest if the rate is 3.5% and the time is 7 years.

52. **ELEC** For an appliance the wattage rating P varies jointly as the resistance, R, and the square of the current I. If the wattage rating is 60 W when the current is 0.1 A and the resistance is 500 Ω, find the wattage rating when the current is 0.3 A and the resistance is 100 Ω.

8–4 Inverse and Combined Variation

LEARNING OUTCOMES

1 Solve problems of inverse variation using proportions.

2 Solve problems of inverse variation using a constant of variation.

1 Solve Problems of Inverse Variation Using Proportions. An **inverse variation** is one in which the quantities being compared are inversely related. That is, as one quantity increases, the other decreases.

For example, as we *increase* pressure on a gas, the gas compresses and so *decreases* in volume. Or, as we *decrease* pressure on the gas, it *increases* in volume as it expands. This relationship is the opposite, or inverse, of direct variation.

3 Solve problems of combined variation using a constant of variation.

LC LEARNING CATALYTICS
1. If it takes 5 hours for 3 workers to load a truck, should 2 workers be able to load the truck in more or less time?

Inverse variation: a proportion in which the quantities being compared are inversely related, so that as one quantity increases (or decreases), the other quantity decreases (or increases). This relationship is also called an inverse proportion

Inversely proportional: two ratios of quantities where an increase in one quantity causes a decrease in the other; alternate terminology for inverse variation

STOP AND CHECK

1. If it takes 3 people to complete a job in 4 days, how long will it take 8 people to complete the job?

Answer:

1. 1.5 days

If 3 workers frame a house in 2 weeks and the contractor *increases* the number of workers to 6, the framing time *decreases* to 1 week, assuming the workers work at the same rate. In other words, the framing time is **inversely** **proportional** to the number of workers on the job.

Unlike the directly related ratios in a proportion, inversely related ratios do not allow us the flexibility we had in setting up ratios of unlike measures.

To set up an inverse proportion:

1. Establish two pairs of related data.

2. Arrange one pair as the numerator of one ratio and the denominator of the other ratio.

3. Arrange the other pair so that each ratio contains like measures.

4. Form a proportion using the two ratios.

EXAMPLE 1

INDTEC If 5 machines take 12 days to complete a job, how long will it take for 8 machines to do the job?

As the number of machines *increases,* the amount of time required to do the job *decreases.* Thus, the quantities are *inversely proportional.*

Pair 1: 5 machines finish in 12 days.

Pair 2: 8 machines finish in x days.

Estimation

We expect more machines to do the job in fewer than 12 days. Also, we did not double the number of machines, so it will take more than half of the time (more than 6 days).

$$\frac{5 \text{ machines}}{8 \text{ machines}} = \frac{x \text{ days for 8 machines}}{12 \text{ days for 5 machines}}$$ Each ratio uses like measures. The pairs are arranged inversely.

Dimension Analysis

$$\frac{5}{8} = \frac{x}{12}$$ $$\frac{\text{machines}}{\text{machines}} = \frac{\text{days}}{\text{days}}$$

$$5(12) = 8x$$ Cross multiply. machines(days) = machines(days)

$$60 = 8x$$ Divide by machines.

$$\frac{60}{8} = x$$ $$\frac{\text{machines (days)}}{\text{machines}} = \text{days}$$

$$x = \frac{15}{2} \quad \text{or} \quad 7\frac{1}{2} \text{ days}$$

Interpretation

It will take $7\frac{1}{2}$ days for 8 machines to do the job.   See Exercises 1–6.

TIP **Why Is Estimation So Important?** Suppose we arrange the data so that each pair forms a ratio of unlike measures and then we invert one of the ratios. Look at the value to see if the answer conforms to what we expect the answer to be.

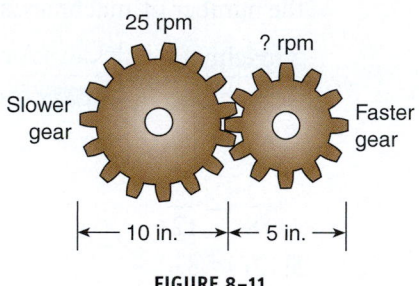

$$\frac{5 \text{ machines}}{12 \text{ days}} = \frac{x \text{ days}}{8 \text{ machines}}$$

$$12x = 40$$

$$x = 3.33 \text{ days}$$

How does this compare to our estimate? We expected our answer to be less than 12 days but more than 6 days. The answer above, 3.33 days, is not consistent with our estimate.

$$\frac{\text{machines}}{\text{days}} = \frac{\text{days}}{\text{machines}}$$

Analyzing the dimensions, we produce an incorrect statement.

$$\text{machines(machines)} = \text{days(days)}$$

These measures are not equal. Estimation and dimension analysis can identify proportions that are set up incorrectly!

The speed and size of gears and pulleys involve *inverse* relationships. Suppose a larger gear and a smaller gear are in mesh, or a larger pulley is connected by a belt to a smaller pulley. The larger gear or pulley has the *slower* speed, and the smaller gear or pulley has the *faster* speed.

EXAMPLE 2

INDTEC   A 10-in.-diameter gear is in mesh with a 5-in.-diameter gear (Fig. 8–11). If the larger gear has a speed of 25 rpm, at how many rpm does the smaller gear turn?

Umberto Shtanzman/Shutterstock

FIGURE 8–11

Because gears in mesh are *inversely* related, we set up an inverse proportion. Each ratio uses like measures and the ratios are in inverse order.

Pair 1: 25 rpm of larger gear for 10-in. size of larger gear.

Pair 2: x rpm of smaller gear for 5-in. size of smaller gear.

Estimation

We expect the speed of the smaller gear to be faster or greater than the 25-rpm speed of the larger gear.

$$\frac{25 \text{ rpm of larger gear}}{x \text{ rpm of smaller gear}} = \frac{5 \text{ in. of smaller gear}}{10 \text{ in. of larger gear}} \quad \begin{array}{l} \text{Pair 2} \\ \text{Pair 1} \end{array}$$

Set up ratios in inverse order.

$$\frac{25}{x} = \frac{5}{10}$$

Cross multiply.

$$25(10) = x(5)$$

$$250 = 5x \qquad \text{Divide.}$$

$$\frac{250}{5} = x$$

$$50 = x$$

Interpretation

The smaller gear turns at the faster, or greater, speed of 50 rpm. **See Exercises 7–12.**

Constant of inverse variation: a conversion factor for finding inverse variation. The formula can be written as $k = xy$

2 **Solve Problems of Inverse Variation Using a Constant of Variation.** Inverse variation has a conversion factor similar to the constant of direct variation. The inverse variation formula is $y = \dfrac{k}{x}$, where x is the nonzero independent variable, y is the dependent variable, and k is the **constant of inverse variation.** This formula can also be written as $k = xy$.

To find the constant of inverse variation:

1. Identify a pair of inversely related data in which both values are known.

2. Write the related pair as a product. $k = xy$

3. Multiply to find the constant of inverse variation.

To solve problems of inverse variation using a constant of variation:

1. Find the constant of inverse variation using known values for x and y and the formula $k = xy$.

2. Evaluate the formula of inverse variation

$$y = \frac{k}{x}$$

where x is the independent variable, y is the dependent variable, and k is the constant of inverse variation.

EXAMPLE 3

AG/H A landscaper needs a five-person crew to complete a commercial project in 30 h. How many hours will it take to complete the project for the following number of workers?

(a) 3 workers **(b)** 8 workers

Determine the number of hours necessary to complete the job per worker (constant of inverse variation). The number of workers is the independent variable and the number of hours to complete the job is the dependent variable.

constant of inverse variation = (independent variable)(dependent variable)

$$k = xy \qquad \text{Substitute known values.}$$

$$k = (5 \text{ workers})(30 \text{ h}) \qquad \text{Multiply.}$$

$$k = 150 \text{ worker-hours}$$

Use the inverse variation formula to find the respective values of the dependent variables.

(a) $y = \dfrac{k}{x}$ Substitute $k = 150$ worker-hours and $x = 3$ workers.

$y = \dfrac{150 \text{ worker-hours}}{3 \text{ workers}}$ Divide.

$y = \mathbf{50\ h}$

(b) $y = \dfrac{k}{x}$ Substitute $k = 150$ worker-hours and $x = 8$ workers.

$y = \dfrac{150 \text{ worker-hours}}{8 \text{ workers}}$ Divide.

$y = \mathbf{18.75\ h}$ **See Exercises 13–30.**

TIP **How to Distinguish Between Direct and Inverse Variation** We can distinguish between direct and inverse variation by anticipating cause-and-effect situations.

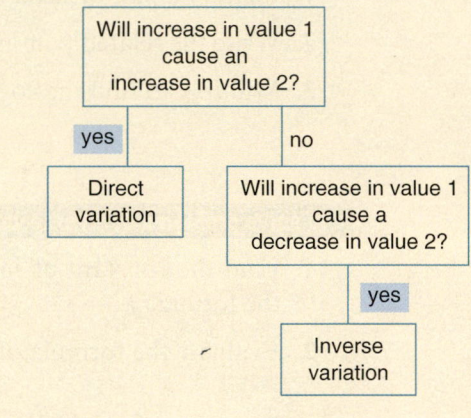

Similarly,

$$\text{decrease causes decrease} \Rightarrow \text{direct variation}$$
$$\text{decrease causes increase} \Rightarrow \text{inverse variation}$$

---

3 **Solve Problems of Combined Variation Using a Constant of Variation.** In certain real-life situations the variables may be related through both direct and inverse variations. Such variations are called **combined variation.**

Combined variation: variables are related through both direct and inverse variations

To solve problems of combined variation using a constant of variation:

1. Symbolically represent the relationships among the variables.

2. Find the constant of combined variation by substituting known values into the equation from Step 1 and solving for k.

3. Evaluate the equation from Step 1 for the established constant of combined variation (Step 2) and other known values.

An example of an equation of combined variation is $y = \dfrac{kx}{z}$. It can be read as "y varies directly as x and inversely as z" or "y is directly proportional to x and inversely proportional to z."

EXAMPLE 4

z varies directly as the square of x and inversely as y. Find z when $x = 8$, $y = 15$, and the constant of combined variation is 12.

Write an equation of combined variation.

$z = \dfrac{kx^2}{y}$ Substitute known values.

$z = \dfrac{12(8^2)}{15}$ Evaluate.

$z = \dfrac{12(64)}{15}$

$z = \dfrac{768}{15}$

$z = \mathbf{51.2}$

See Exercises 31–36.

Dmitry Kalinovsky/Shutterstock

EXAMPLE 5

The number of hours h that it takes p persons to assemble m machines varies directly as the number of machines and inversely as the number of persons. Three persons can assemble 10 machines in 5 h. How many persons would it take to assemble 36 machines in 8 h?

Write an equation of combined variation.

$h = \dfrac{km}{p}$ The number of hours h varies directly as the number of machines m and inversely as the number of persons p.

Find the constant of combined variation.

$h = \dfrac{km}{p}$ Substitute known values and solve for k.

$5 = \dfrac{k(10)}{3}$ Clear fraction.

$15 = 10k$

$\dfrac{15}{10} = k$ Reduce and apply symmetric property of equality.

$k = \dfrac{3}{2}$, or 1.5 Constant of combined variation

Find p for $h = 8$ and $m = 36$.

$$h = \frac{1.5m}{p}$$ Substitute $h = 8$ and $m = 36$. Solve for p.

$$8 = \frac{1.5(36)}{p}$$

$$8p = 1.5(36)$$

$$8p = 54$$

$$p = \frac{54}{8} \quad \text{or} \quad 6.75$$ Round up to the next whole number.

Seven persons are required to complete the job in the allotted time.

See Exercises 37–38.

8-4 EXERCISES

MyLab Math   For additional practice go to your study plan in MyLab Math.

1 Solve using proportions. *See Example 1.*

1. **AG/H**   Two groundskeepers take 25 h to prepare a golf course for a tournament. How long would it take five groundskeepers to prepare the golf course?

2. **INDTR**   The volume of a certain gas is inversely proportional to the pressure on it. If the gas has a volume of 160 in^3 under a pressure of 20 lb per square inch (psi), what is the volume if the pressure is decreased to 16 psi?

3. **INDTEC**   Three machines complete a printing project in 5 h. How many machines are needed to finish the same project in 3 h?

4. **CON**   Six painters can trim the exterior of all the new brick homes in a subdivision in 9 weeks. The contractor wants to have the homes ready in just 3 weeks and so needs more painters. How many painters are needed for the job if they all work at the same rate?

5. **HLTH/N**   Nurse Lee prepares dosages for her patients in 30 min. If she gets help from assistants, who also work at her rate, and together they complete the preparation in 6 min, how many helpers did she get?

6. **CON**   Two painters working at the same speed can paint 800 ft^2 of wall space in 6 h. If a third painter paints at the same speed, how long will it take all three to paint the same wall space?

See Example 2.

7. **INDTR**   The fan pulley and alternator pulley are connected by a fan belt on an automobile engine. The fan pulley is 225 cm in diameter and the alternator pulley is 125 cm in diameter. If the fan pulley turns at 500 rpm, how many revolutions per minute does the faster alternator pulley turn?

8. **INDTEC**   A gear measures 5 in. across. It turns another gear 2.5 in. across. If the larger gear has a speed of 25 rpm, what is the rpm of the smaller gear?

9. **INDTR**   A pulley that measures 15 in. across (diameter) turns at 1,600 rpm and drives a larger pulley at the rate of 1,200 rpm. What is the diameter of the larger pulley in this inverse relationship?

10. **INDTEC**   A gear measures 6 in. across. It is in mesh with another gear with a diameter of 3 in. If the larger gear has a speed of 60 rpm, what is the rpm of the smaller gear?

11. **INDTR**   A small pulley 3 in. in diameter turns 250 rpm and drives a larger pulley at 150 rpm. What is the diameter (distance across) of the larger pulley?

12. **INDTR**   A 4.5-in. pulley turning at 1,000 rpm is belted to a larger pulley turning at 500 rpm. What is the size of the larger pulley?

2 *See Example 3.*

13. **INDTR**   A 12-in. pulley turns at 30 rpm. Find the constant of inverse variation.

14. **INDTR**   A 16-in. gear makes 60 rpm. Find the constant of inverse variation.

The diameter of two gears in mesh is inversely proportional to the number of revolutions per minute each turns. Complete the table for two gears in mesh using a constant of variation.

	LARGE GEAR		SMALL GEAR	
	DIAMETER	RPM	DIAMETER	RPM
15.	30 in.	120	15 in.	
16.	24 in.		18 in.	300
17.		70	14 cm	3,500
18.	20 cm	100		500
19.		180	12 cm	540
20.	32 in.	225	12 in.	

The number of teeth of two gears in mesh is inversely proportional to the number of revolutions per minute each turns. Complete the table for two gears in mesh using a constant of variation.

	LARGE GEAR		SMALL GEAR	
	NUMBER OF TEETH	RPM	NUMBER OF TEETH	RPM
21.	60	300	45	
22.	250	28	50	
23.		1.7	60	8.5
24.	300	70		600
25.	448		56	300
26.	84	24		96

27. INDTEC A gear with 40 teeth turns 30 rpm on another gear with 120 teeth. At how many rpm does the second gear turn?

28. INDTEC A gear turning at 19 rpm with 150 teeth turns a smaller gear with 75 teeth. At how many rpm does the smaller gear turn?

29. INDTEC A large gear turns at 40 rpm in mesh with a gear that has 60 teeth and turns at 120 rpm. How many teeth does the larger gear have?

30. INDTEC A small gear turns at 150 rpm in mesh with a large gear that has 180 teeth and turns at 90 rpm. How many teeth has the small gear?

3 Solve problems of combined variation using a constant of variation. *See Example 4.*

31. In a relationship, P varies directly as m and inversely as n and the constant of variation is 18. Find P if $m = 14$ and $n = 28$.

32. If x varies directly as y and inversely as z and $x = 18$ when $y = 12$ and $z = 2$, find x when $y = 75$ and $z = 15$.

33. If x varies directly as y and z and inversely as the square of r, and $x = 16$ when $y = 3$, $z = 12$, and $r = 3$, find x when $y = 5$, $z = 30$, and $r = 10$.

34. J varies directly as the cube of R and inversely with the square of S. If $J = 972$ when $R = 3$ and $S = 6$, find J when $R = 4$ and $S = 8$.

35. A varies directly as the square of B and inversely with C, and $A = 5,000$ when $B = 40$ and $C = 8$. Write an equation that relates the variables. Find the constant of variation.

36. Find A in Exercise 35 if $B = 20$ and $C = 2$.

See Example 5.

37. A horizontal beam safely supports a load P that varies jointly as the product of the width W of the beam and the square of the depth D and inversely as the length L. A beam that has a width of 5 in., depth of 12 in., and length of 8 in. can safely support 1,260 lb. Determine the safe load in pounds of a beam made from the same material if the beam is 20 in. long.

38. Use the relationship in Exercise 37 to find the load P of a horizontal beam that has the same constant of variation and has a width of 6 in., a depth of 14 in., and a length of 12 in.

8 | CHAPTER REVIEW OF KEY CONCEPTS

LEARNING OUTCOMES	KEY CONCEPTS AND EXAMPLES

Section 8-1

1 Evaluate formulas (pp. 378–380).

1. Write the formula. **2.** Rewrite the formula, substituting known values for variables of the formula. **3.** Solve the equation from Step 2 for the missing variable. **4.** Interpret the solution within the context of the formula.

Find the length (l) if the area (A) is 8 ft^2 and the width (w) is 2 ft.

$A = lw$ — Substitute values.

$8 = (l)(2)$ — Divide.

$l = 4$ ft

If the perimeter (P) of a square is 12 in., find the length of a side.

$P = 4s$ — Substitute value for P.

$12 = 4s$ — Divide.

$3 = s$ — Interchange sides.

$s = 3$ in.

2 Rearrange formulas to solve for a specified variable (pp. 380–384).

1. Determine which variable of the formula will be isolated (solved for). **2.** Highlight or mentally locate all instances of the variable to be isolated. **3.** Treat all other variables of the formula as you would treat numbers in an equation, and perform normal steps for solving an equation. **4.** If the isolated variable is on the right side of the equation, interchange the sides so that it appears on the left side.

Solve the formula $S = \dfrac{R + P}{2}$ for R.

$(2)S = (2)\dfrac{R + P}{2}$ — Clear the denominator.

$2S = R + P$ — Isolate R (sort terms).

$2S - P = R$ — Interchange sides so the variable appears on the left.

$R = 2S - P$

Section 8-2

1 Solve equations that are proportions (pp. 387–388).

Use the property of proportions: If $\dfrac{a}{b} = \dfrac{c}{d}$, then $ad = bc$ ($b, d \neq 0$).

Solve a proportion: 1. Find the cross products. **2.** Divide both sides by the coefficient of the variable.

$\dfrac{x}{6} = \dfrac{4}{2}$

$(4)(6) = (2)(x)$ — Cross products.

$24 = 2x$ — Divide by the coefficient of x.

$\dfrac{24}{2} = \dfrac{2x}{2}$

$12 = x$

LEARNING OUTCOMES	KEY CONCEPTS AND EXAMPLES

Section 8-3

1 Solve problems of direct variation using proportions (pp. 390–392).

Verify that the problem involves direct variation: As one amount increases, the other amount increases. Similarly, as one amount decreases, the other amount decreases.

Set up a direct proportion: 1. Establish two pairs of related data. **2.** Write one pair of data in the numerators of two ratios. **3.** Write the other pair of data in the denominators of the two ratios. **4.** Form a proportion using the two ratios.

Kinta makes $72 for 8 h of work in a hospital business office. How many hours must he work to earn $150?

Pairs are directly related: As hours increase, pay increases.

$$\frac{\$72}{\$150} = \frac{8\,h}{x\,h}$$ Pair 1. $72; 8 h.

Pair 2. $150; x h.   Estimation: $x\,h > 8\,h$
because $150 > $72.

$$\frac{72}{150} = \frac{8}{x}$$ Cross multiply.

$(72)(x) = (150)(8)$

$72x = 1{,}200$ Divide.

$$\frac{72x}{72} = \frac{1{,}200}{72}$$

$x = 16\dfrac{2}{3}\,h$ Kinta must work $16\dfrac{2}{3}$ h to earn $150.

2 Solve problems of direct variation using a constant of variation (pp. 392–394).

1. Find the constant of direct variation using known values for x and y and the formula $k = \dfrac{y}{x}$. **2.** Evaluate the formula of direct variation $y = kx$, where x is the independent variable, y is the dependent variable, and k is the constant of direct variation.

A caterer can purchase 4 cartons of Coca-Cola® for $10. What is the cost of 5 cartons?

$$\text{Constant of direct variation} = \frac{\text{dependent variable}}{\text{independent variable}}$$

$$k = \frac{y}{x}$$ Substitute known values.

$$k = \frac{\$10}{4\ \text{cartons}}$$ Divide to simplify ratio.

$$k = \frac{\$2.50}{\text{carton}}$$ $2.50 per carton

$y = kx$ Substitute k = $2.50 per carton and x = 5 cartons.

$$y = \left(\frac{\$2.50}{\text{carton}}\right)(5\ \text{cartons})$$ Multiply.

$y = \$12.50$

Five cartons of Coca-Cola® cost $12.50.

LEARNING OUTCOMES	KEY CONCEPTS AND EXAMPLES

3 Solve problems of joint variation using a constant of variation (pp. 394–395).

1. Symbolically represent the relationships among the variables. **2.** Find the constant of joint variation by substituting known values into the equation from Step 1 and solving for k. **3.** Evaluate the equation from Step 1 for the established constant of joint variation (Step 2) and other known values.

p varies jointly as q and r. If $p = 42$ when $q = 5$ and $r = 7$, find p when $q = 9$ and $r = 12$.

Find the constant of variation:

$p = kqr$ Substitute known values and solve for k.

$42 = k(5)(7)$

$42 = k(35)$

$\dfrac{42}{35} = k$

$1.2 = k$ Constant of variation

Find p when $q = 9$ and $r = 12$:

$p = kqr$ Substitute known values and solve for p.

$p = 1.2(9)(12)$

$p = 129.6$

Section 8–4

1 Solve problems of inverse variation using proportions (pp. 398–401).

Verify that the problem involves inverse variation: As one amount increases, the other amount decreases. Similarly, as one amount decreases, the other amount increases.

Set up an inverse proportion: 1. Establish two pairs of related data. **2.** Arrange one pair as the numerator of one ratio and the denominator of the other. **3.** Arrange the other pair so that each ratio contains like measures. **4.** Form a proportion using the two ratios.

Jane can paint a room in 4 h. If she has two helpers who also can paint a room in 4 h, how long will it take all three to paint the same room?

Pairs are inversely related: As the number of painters increases, time decreases.

$\dfrac{1 \text{ painter}}{3 \text{ painters}} = \dfrac{x \text{ h}}{4 \text{ h}}$ Pair 1: 1 painter, 4 h

Pair 2: 3 painters, x h   Estimation: x h < 4 h since 1 painter < 3 painters.

$\dfrac{1}{3} = \dfrac{x}{4}$

$1(4) = 3(x)$

$4 = 3x$

$\dfrac{4}{3} = \dfrac{3x}{3}$

$x = 1\dfrac{1}{3} \text{ h}$ for 3 painters to paint the room.

LEARNING OUTCOMES	KEY CONCEPTS AND EXAMPLES

2 Solve problems of inverse variation using a constant of variation (pp. 401–402).

1. Find the constant of inverse variation using known values for x and y and the formula $k = xy$. **2.** Evaluate the formula of inverse variation $y = \dfrac{k}{x}$ where x is the independent variable, y is the dependent variable, and k is the constant of inverse variation.

The number of teeth on two gears in mesh is inversely proportional with the speed at which the gears turn. If a gear with 15 teeth turns at 600 revolutions per minute $\left(\dfrac{r}{min}\right)$, how many teeth are on a gear in mesh that turns at $750\,\dfrac{r}{min}$?

constant of inverse variation = (independent variable)(dependent variable).

$$k = xy \qquad \text{Substitute known values.}$$

$$k = \left(600\,\frac{r}{min}\right)(15\text{ teeth}) \qquad \text{Multiply.}$$

$$k = 9{,}000\,\frac{r}{min}\text{-teeth}$$

$$y = \frac{k}{x} \qquad \text{Substitute.}$$

$$y = \frac{9{,}000\,\dfrac{r}{min}\text{-teeth}}{750\,\dfrac{r}{min}} \qquad\qquad k = 9{,}000\,\frac{r}{min}\text{-teeth and}$$

$$x = 750\,\frac{r}{min}.$$

Divide.

$$y = 12\text{ teeth}$$

The gear that turns at $750\,\dfrac{r}{min}$ has 12 teeth.

3 Solve problems of combined variation using a constant of variation (pp. 402–404).

1. Symbolically represent the relationships among the variables. **2.** Find the constant of combined variation by substituting known values into the equation from Step 1 and solving for k. **3.** Evaluate the equation from Step 1 for the established constant of combined variation (Step 2) and other known values.

Q varies jointly as S and T and inversely as the square of R. Find the constant of combined variation when $S = 5$, $T = 7$, $R = 2$, and $Q = 4.375$.

$$Q = \frac{kST}{R^2} \qquad \begin{array}{l}\text{Symbolic relationships}\\ \text{Substitute known values.}\end{array}$$

$$4.375 = \frac{k(5)(7)}{2^2} \qquad \text{Solve for } k.$$

$$4.375 = \frac{k(35)}{4}$$

$$4.375(4) = 35k$$

$$17.5 = 35k$$

$$\frac{17.5}{35} = k$$

$$0.5 = k \qquad \text{Constant of variation}$$

LEARNING OUTCOMES	KEY CONCEPTS AND EXAMPLES

Find Q to the nearest hundredth when $S = 3$, $T = 8$, $R = 3$, and $k = 0.5$.

$Q = \dfrac{kST}{R^2}$ Substitute known values.

$Q = \dfrac{0.5(3)(8)}{3^2}$ Evaluate.

$Q = \dfrac{12}{9}$

$Q = 1.33$ Rounded

8   CHAPTER REVIEW EXERCISES

Section 8–1   MyLab Math   **For additional practice go to your study plan in MyLab Math.**

1. **PFIN**   Use the simple interest formula to find the time if $I = \$387.50$, $P = \$1,550$, and $R = 12.5\%$.

2. **PFIN**   Find the rate if $I = \$2,484$, $P = \$4,600$, and $T = 3$ years.

3. **ELEC**   Ohm's law is $E = IR$. Find the amperes of current (I) if the voltage (E) is 220 V and the resistance (R) is 80 Ω.

4. **ELEC**   Use the formula $R_T = \dfrac{R_1 R_2}{R_1 + R_2}$ to find the total resistance (R_T) if one resistance (R_1) is 10 Ω and the second resistance (R_2) is 9 Ω. Round to tenths.

5. **ELEC**   The formula for power (P) in watts (W) is $P = I^2 R$. Find the current (I) in amperes if a device draws 392 W and the resistance (R) is 8 Ω.

6. **AUTO**   Use the formula $E = \dfrac{I - P}{I}$ to find the percent efficiency (E) of an engine if the input (I) is 22,600 cal and the output (P) is 5,600 cal. Round to the nearest tenth of a percent.

Solve the formulas for the indicated variable.

7. $V = lwh$ for w

8. $s = c + m$ for c

9. $s = r - d$ for r

10. $s = r - d$ for d

11. $v = v_0 - 32t$ for t

12. $V = \frac{1}{3} Bh$ for h

13. **BUS**   The formula for finding the sale price on an item is $S = P - D$, where S is the sale price, P is the original price, and D is the discount. Solve the formula for the original price.

14. **BUS**   The formula for finding sales tax is $T = RM$, where T represents sales tax, R represents the tax rate, and M represents the marked price. Rearrange the formula to find the marked price.

Section 8–2

Solve the proportions.

15. $\dfrac{7}{x} = 6$

16. $\dfrac{2}{3} = \dfrac{x + 3}{x - 7}$

17. $\dfrac{4x + 3}{15} = \dfrac{1}{3}$

18. $\dfrac{3x - 2}{3} = \dfrac{2x + 1}{4}$

19. $\dfrac{5}{4x - 3} = \dfrac{3}{8}$

20. $\dfrac{8}{3x - 2} = \dfrac{2}{3}$

21. $\dfrac{4x}{7} = \dfrac{2x + 3}{3}$

22. $\dfrac{2x}{3x - 2} = \dfrac{5}{8}$

23. $\dfrac{5x}{3} = \dfrac{2x + 1}{4}$

24. $\dfrac{5}{9} = \dfrac{x}{2x - 1}$

25. $\dfrac{7}{x} = \dfrac{5}{4x + 3}$

26. $\dfrac{3}{5} = \dfrac{2x - 3}{7x + 4}$

Section 8–3

Solve using proportions.

27. There are 25 women in a class of 35 students. If this is typical of all classes in the college, how many women are enrolled in the college if it has 6,300 students altogether?

28. **HOSP** In preparing a banquet for 30 people, Kristin Bennett uses 9 lb of potatoes. How many pounds of potatoes will be needed for a banquet for 175 people?

29. **CAD/ARC** A CAD program scales a blueprint so that $\frac{5}{8}$ in. = 2 ft. What is the actual measure of a wall that is shown as $1\frac{5}{16}$ in. on the blueprint?

30. **AG/H** A 6-ft landscape engineer casts a 5-ft shadow on the ground. How tall is a nearby tree that casts a 30-ft shadow?

31. **HLTH/N** On recent trips to rural health centers, a portable mammography unit used 81.2 gal of unleaded gasoline. If the travel involved a total of 845 mi, how many gallons of gasoline would be used for travel of 1,350 mi? Round to tenths.

32. **BUS** A certain fabric sells at a rate of 3 yd for $7.00. How many yards can Emily Bennett buy for $35.00?

Solve using a constant of variation.

33. **CON** A contractor estimates that a painter can paint 300 ft^2 of wall space in 3.5 h. How many hours should the contractor estimate for the painter to paint 425 ft^2? Express your answer to the nearest tenth.

34. **CAD/ARC** An architect's drawing is scaled at $\frac{3}{4}$ in. = 6 ft. What is the actual height of a door that measures $\frac{7}{8}$ in. on the drawing?

35. **ELEC** A wire 825 ft long has a resistance of 1.983 Ω. How long is a wire of the same diameter if the resistance is 3.247 Ω? Round to the nearest whole foot.

36. **BUS** A coffee company mixes 1.6 lb of chicory with every 3.5 lb of coffee. At this ratio, how many pounds of chicory are needed to mix with 2,500 lb of coffee? Round to the nearest whole number.

37. If b varies jointly with d and e and the constant of variation is 0.8, find b when $d = 12$ and $e = 25$.

38. If p varies jointly as the square of q and the cube of r and $p = 108$ when $q = 3$ and $r = 2$, find p when $q = 4$ and $r = 3$.

39. The area of a triangle A varies jointly as b and h. If $A = 225$ in^2 when $b = 18$ in. and $h = 25$ in., find A when $b = 35$ in. and $h = 40$ in.

Section 8–4

Solve using proportions.

40. **AG/H** It takes five people 7 days to clear an acre of land of debris left by a tornado; inversely, more people can do the job in less time. How long would it take seven people all working at the same rate?

41. **INDTEC** If 15 machines can complete a job in 6 weeks, how many machines are needed to complete the job in 4 weeks?

42. **INDTR** A 10-in. pulley makes 900 revolutions every minute. It drives a larger pulley at 500 rpm. What is the diameter of the larger pulley in this inverse relationship?

43. A car with a speed control device travels 100 mi at 50 mi/h. The trip takes 2 h. If the car traveled at 40 mi/h, how much time would the driver need to reach the same destination?

44. **INDTR** A pulley whose diameter is 3.5 in. is belted to a pulley whose diameter is 8.5 in. In this inverse relationship, if the smaller, faster pulley turns at the rate of 1,200 rpm, what is the rpm of the slower pulley? Round to the nearest whole number.

45. **INDTR** A gear with a diameter of 45 cm is in mesh with a gear that has a diameter of 30 cm. If the larger gear turns at 1,000 rpm, how many revolutions per minute does the smaller gear make?

Solve using a constant of variation.

46. **INDTR** A small pulley with a 6-in. diameter turns 350 rpm and drives a larger pulley at 150 rpm. What is the diameter (distance across) of the larger pulley?

47. **INDTEC** Three workers take 5 days to assemble a shipment of microwave ovens; inversely, more workers can do the job in less time. How long will it take five workers to do the same job?

48. **INDTEC**   Matthew Bennett can install a hard drive in 10 PCs in 6 h. He gets help from assistants who work at his rate and together they complete the installations in 2 h. How many helpers did he get?

49. **INDTR**   A gear is 4 in. across. It turns a smaller gear 2 in. across. If the larger gear has a speed of 30 rpm, what is the rpm of the smaller gear?

50. If x varies directly as the square of y and inversely with z and the constant of variation is 3.2, find x when $y = 14$ and $z = 16$.

51. P varies directly as r and inversely as the square of s. If $P = 175$, $r = 252$, and $s = 6$, find P when $r = 196$ and $s = 7$.

52. The load P that a beam can support varies jointly as the product of the beam width W and the square of the beam depth D and inversely as the beam length L. A beam 8 in. wide, 15 in. deep, and 40 in. long can safely support 1,116 lb. How many pounds can be supported by a beam that is 12 in. wide, 6 in. deep, and 20 in. long?

---

8 | TEAM PROBLEM-SOLVING EXERCISES

1. A formula is an equation that gives a model for solving a certain type of application. Devise formulas for the following relationships.
  (a) An electrical power company computes the monthly charges by multiplying the kilowatts of power used times the cost per kilowatt and adds to that a fixed monthly fee.
  (b) A store calculates the ending balance on a charge account by multiplying the interest rate times the previous unpaid balance and then adding the previous balance, the interest, and purchases and subtracting payments.
  (c) Profit on the sale of a particular item is the product of the number of items sold and the difference between the selling price of the item and its cost to the seller.

2. Formula rearrangement is a way of devising variations of formulas.
  (a) Explain the usefulness of formula rearrangement.
  (b) Select a formula that has at least three variables. Find a variation of the formula for each variable of the formula.
  (c) Describe at least two occasions when it is desirable to rearrange a formula.

3. An effective measure of your understanding of a concept is your ability to apply the concept to real-world situations.
  (a) Develop and solve a word problem that can be solved using a direct proportion. Include in your storyline a lawn mower, tanks of gasoline, and acres to be mowed.
  (b) Develop and solve a word problem that can be solved using a direct proportion.

4. Inverse proportions can be used to solve certain types of real-world applications.
  (a) Develop and solve a word problem that can be solved using an inverse proportion. Include in your storyline a belt, pulleys, rpms, and diameters of pulleys.
  (b) Develop and solve a word problem that can be solved using an inverse proportion.

---

8 | CONCEPTS ANALYSIS

1. Illustrate the property of proportions using two equivalent fractions.

2. Explain the difference between a direct proportion and an inverse proportion.

3. Explain how to set up a direct proportion.

4. Give some examples of situations that are directly proportional.

5. Explain how to set up an inverse proportion.

6. Give some examples of situations that are inversely proportional.

7. If the constant of variation is positive and y varies directly as x, how will y change when x increases? Use an example to support your answer.

8. If the constant of variation is positive and y varies inversely as x, how will y change when x increases? Use an example to support your answer.

9. Suppose y varies directly as the square of x. How will y change when x is doubled? Illustrate your answer with an example.

10. Suppose y varies inversely as the square of x. How will y change when x is doubled? Illustrate your answer with an example.

8 | PRACTICE TEST

1. Ohm's law is $E = IR$. Find the amperes of current (I) if the voltage (E) is 110 V and the resistance (R) is 50 Ω?

2. The formula for the volume (V) of a solid rectangular figure is $V = lwh$ (length $\times$ width $\times$ height). If the volume of a mailing container is 7.5 m^3, its length is 1.5 m, and its width is 0.5 m, what is its height?

3. The electrical resistance of a wire is found from the formula $R = \dfrac{PL}{A}$. Rearrange the formula to find the length L of the wire.

4. Use the formula $R_T = \dfrac{R_1 R_2}{R_1 + R_2}$ to find the total resistance (R_T) if one resistance (R_1) is 9 Ω and the second resistance (R_2) is 8 Ω. Round to tenths.

5. Engine displacement d is found using the formula $d = \pi r^2 sn$. Solve to find s.

6. If the efficiency (E) of an engine is 70% and the input (I) is 40,000 cal, find the output (P) in calories.

Use the formula $E = \dfrac{I - P}{I}$

Solve the proportions.

7. $\dfrac{x}{12} = \dfrac{5}{8}$

8. $\dfrac{x}{6} = \dfrac{12}{8}$

9. $\dfrac{640}{24} = \dfrac{x}{360}$

10. $\dfrac{45 \text{ cm}}{18 \text{ cm}} = \dfrac{9 \text{ cm}}{x}$

11. $\dfrac{2\frac{1}{2}}{x} = \dfrac{1\frac{1}{4}}{3\frac{1}{5}}$

12. $\dfrac{3.9}{5.4} = \dfrac{x}{8.1}$

13. A large gear with 300 teeth turns at 40 rpm. Find the rpm of a small gear that has 60 teeth.

Solve the problems involving fractions, decimal numbers, and proportions.

14. A pipe fills a 120-gal tank in 4 h. How long does it take for the pipe to fill a 420-gal tank?

15. A 9-in. gear is in mesh with a 4-in. gear. If the larger gear makes 75 rpm, how many revolutions per minute does the smaller gear make in this inverse relationship?

16. One employee can wallpaper a room in 12 h. Four employees working at the same rate can wallpaper the same room in how many hours?

17. A force of 32.75 lb exerts pressure on a surface of 24.65 in^2 in a hydraulic system. A force of 117.9 lb exerts the same pressure on what area? Force and area are directly proportional when the pressure is constant.

18. A $34,475 loan for the purchase of an electronically controlled assembly machine in a factory cost the management $2,758 in simple interest. How much would a machine cost if the interest were $1,879? Round to the nearest dollar.

19. If a compact car used 62.5 L of unleaded gasoline to travel 400 mi, how many liters of gasoline would the driver use to travel 350 mi? Round to tenths.

20. If three workers take 8 days to complete a job, how many workers would be needed to finish the same job in only 6 days if each worked at the same rate? (More workers take fewer days.)

21. If voltage is 40 V and amperage is 3.5 A, find the equivalent amperage for a voltage of 100 V. Voltage and amperage are directly proportional. Express the answer as a decimal rounded to tenths.

22. The ratio of men to women in technical and trade occupations is estimated to be 3 to 1, that is, $\frac{3}{1}$. If 56,250 men are employed in such occupations in a certain city, how many employees are women?

23. If an ice maker produces 75 lb of ice in $3\frac{1}{2}$ h, how many pounds of ice would it produce in 5 h? Round to the nearest whole number.

24. If D varies jointly with E and F and $D = 2,100$ when $E = 12$ and $F = 35$, find D when $E = 18$ and $F = 40$.

25. Simple interest I varies jointly as the interest rate R and the time T in years. The simple interest for a loan at 8% for 4 years requires $320. How much interest is required for a loan of the same amount of money if it is made at 9% for 5 years?

26. H varies directly as J and inversely with M. If the constant of variation is $\frac{3}{4}$, find H when $J = 16$ and $M = 24$.

9

Linear Equations, Functions, and Inequalities in Two Variables

Chris Willson/Alamy Stock Photo

In Great Company

Floating Point Fiascos (1993)

In 1993, Intel Corporation was flying high on the back of a brand new Pentium P5 processor. The chip used a new design that was being heavily promoted for the speed it brought to scientific calculations. But on June 13, 1994, a small hiccup developed. As Professor Thomas R. Nicely of Lynchburg College, Virginia, was researching prime numbers, he discovered some strange errors in his solution sets. It took him four months to trace the problem to the new Pentium processor. For a few certain numbers, the chip didn't do long division correctly, getting anything past the ten-thousandths place wrong. On October 24, 1994, five days after Nicely was sure he had found the culprit, he reported the problem to Intel.

It turned out Intel already knew about it. It had discovered the same error back in May, but sat on it in the hope that no one else would notice. Professor Nicely thought such silence wasn't . . . nice. He wrote various contacts, asking them to verify the problem, which they quickly did. In fact, Nicely's contacts found the error could go as high as 61 parts per million. The computer press soon caught wind of it, and by November 7, the science geeks were having a field day with it. The noise got so loud that by November 21, 1994, CNN began reporting it to the general public.

That's when things got ugly. Intel had had six months to work on the problem, but had done nothing because the flaw would only crop up under special conditions that most users wouldn't ever see. The general public, however, would not accept that line of reasoning. Everyone from secretaries to scientists were asking for a fix, even though Intel was correct—people who were not doing scientific or engineering calculations really weren't affected by the problem.

Although only about 1 in 9 billion calculations were affected, Intel's competitors quickly put out public condemnations of the error, turning it into a public relations nightmare. With an estimated 5 million defective chips in circulation, Intel eventually had to replace the chip for anyone who complained.

How did this happen? The divider in the Pentium floating point unit looked up answers in a division table. The table was missing about five out of a thousand entries. Although that means the table had an accuracy rate of 99.5%, an "A" in most instructors' gradebooks, that wasn't good enough. The company lost nearly $475 million. It took years to get its reputation back.

9–1   Graphical Representation of Linear Equations and Functions

LEARNING OUTCOMES

1. Locate points on a rectangular coordinate system.

2. Represent an equation in two variables as a function.

3. Make a table of solutions for a linear equation or function.

4. Graph a linear equation or function using a table of solutions.

LC LEARNING CATALYTICS

1. Find y if $x = 4$ in the equation $3x + 2y = 6$.

One-dimensional graph: a visual representation of the distance and direction that a value is from zero; for example, a number line

Rectangular coordinate system: a graphical representation of two-dimensional values. It is two number lines positioned to form a right angle or square corner

Linear equations in one variable have at most one solution. The solution can be represented graphically as a point on a number line. A linear equation in two variables has an unlimited number of solutions. The solutions of a linear equation in two variables can be represented graphically as a line on a rectangular coordinate system.

1 Locate Points on a Rectangular Coordinate System. The number line shows values pictorially. This kind of visual representation is called a **one-dimensional graph.** The one dimension represents the distance and direction that a value is from zero.

The **rectangular coordinate system** gives a graphical representation of two-dimensional values. In the rectangular coordinate system, two number lines are positioned to form a right angle or square corner (see Fig. 9–1). The number line that runs from left to right is the **horizontal axis or x-axis,** and is represented by the letter x. The number line that runs from top to bottom is the **vertical axis or y-axis,** and is represented by the letter y. The location of points or lines on the rectangular coordinate system is also referred to as a **two-dimensional graph**.

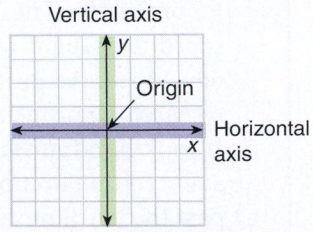

FIGURE 9–1

Horizontal axis (x-axis): the number line that runs from left to right in the rectangular coordinate system. It is represented by the letter x

Describe the location of the points on the given rectangular coordinate system.
1. Point E
2. Point F

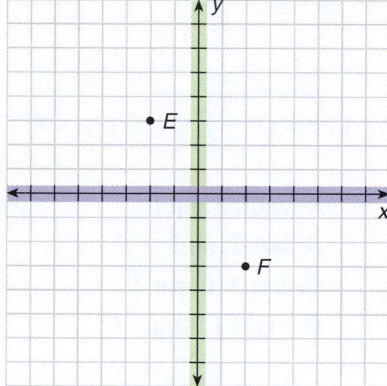

Answers:
1. Point E: horizontal movement, -2; vertical movement, $+3$
2. Point F: horizontal movement, $+2$; vertical movement, -3

Vertical axis (y-axis): the number line that runs from top to bottom in the rectangular coordinate system. It is represented by the letter y

Two-dimensional graph: a visual representation of points and lines using two number lines at right angles to each other and in the vertical and horizontal directions. The rectangular coordinate system with an x- and y-axis provides the structure for creating a two-dimensional graph

Origin: the point where the two number lines of the rectangular coordinate system cross. It is the zero point on both number lines

Write the points in the previous Stop and Check using point notation.
1. Point E
2. Point F

Answers:
1. $(-2, +3)$
2. $(+2, -3)$

Zero on both number lines is located at the point where the two number lines cross. This point is called the **origin**. Points are located by two dimensions, the horizontal distance from the origin and the vertical distance from the x-axis.

EXAMPLE 1

Describe the location of the points in Fig. 9–2 by giving the amount of horizontal and vertical movement from the origin and vertical movement from the x-axis.

Point A: horizontal movement, $+3$; vertical movement, $+2$
Point B: horizontal movement, -2; vertical movement, -1
Point C: horizontal movement, -4; vertical movement, $+3$
Point D: horizontal movement, $+1$; vertical movement, -4

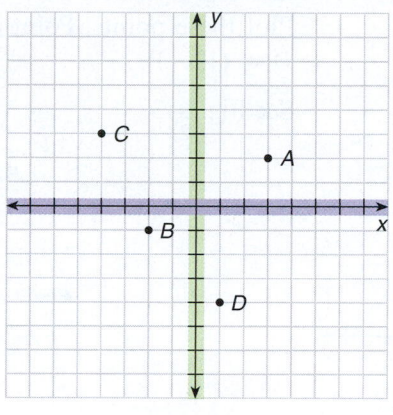

FIGURE 9–2

See Exercises 1–8.

The location of points on a rectangular coordinate system can be written in an abbreviated, symbolic form. **Point notation** uses two signed numbers to show the horizontal movement from the origin and the vertical movement from the x-axis to a given point. Horizontal movement is represented by the **x-coordinate** and is written first. Vertical movement is represented by the **y-coordinate** and is written second. These signed numbers are separated with a comma and enclosed in parentheses.

Symbolically, we represent the location of a point as an **ordered pair** of numbers

$$(x, y)$$

where x is the horizontal movement from the origin, indicated by the x-coordinate, and y is the vertical movement from the x-axis, indicated by the y-coordinate.

EXAMPLE 2

Write the points in Example 1 (Fig. 9–2) using point notation.

Point $A = $ (horizontal movement of $+3$, vertical movement of $+2$) or $(+3, +2)$
Point $B = $ (horizontal movement of -2, vertical movement of -1) or $(-2, -1)$
Point $C = $ (horizontal movement of -4, vertical movement of $+3$) or $(-4, +3)$
Point $D = $ (horizontal movement of $+1$, vertical movement of -4) or $(+1, -4)$

See Exercises 9–17.

Point notation: two signed numbers enclosed in parentheses and separated with a comma to show the horizontal movement from the origin and the vertical movement from the *x*-axis to a given point

x-coordinate: the value of the horizontal movement of a point on a rectangular coordinate system. This coordinate is written first in point notation

To **plot a point** means to show its location on the rectangular coordinate system.

> **To plot a point on the rectangular coordinate system:**
>
> 1. Start at the origin.
> 2. Count to the left or right the number of units of the first signed number (*x*-coordinate) in the ordered pair.
> 3. From the ending point of Step 2, count up or down the number of units of the second signed number (*y*-coordinate).
> 4. Place a dot to show the point and write the coordinates beside the point.

STOP AND CHECK

Plot these points.
1. Point $E = (-3, 5)$
2. Point $F = (5, -3)$

Answers:

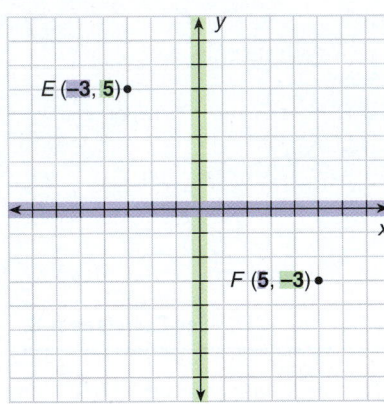

EXAMPLE 3

Plot these points: Point $A = (3, 1)$, point $B = (-2, 5)$, point $C = (-3, -2)$, point $D = (1, -3)$.

See Fig. 9–3.

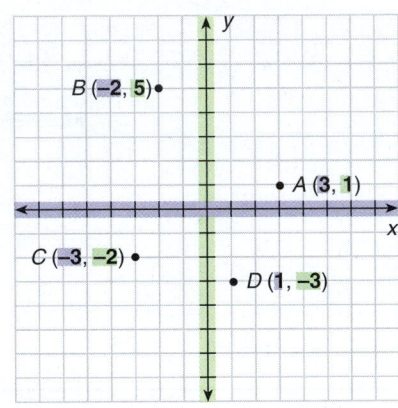

FIGURE 9–3

A: **right 3, up 1**
B: **left 2, up 5**
C: **left 3, down 2**
D: **right 1, down 3**

See Exercises 18–27.

y-coordinate: the value of the vertical movement of a point on a rectangular coordinate system. This coordinate is written second in point notation

Ordered pair: a notation, such as point notation, in which the order of the values is defined. In point notation the order is (x, y)

Plot a point: to show the location of a point on the rectangular coordinate system

Quadrant: one of four regions of the rectangular coordinate system. The quadrants are numbered I, II, III, and IV starting with the upper right quadrant and moving in a counterclockwise direction

> **TIP**   **Develop Your Spatial Sense** Moving in the wrong direction is a common mistake when plotting points. To develop your spatial sense, consider the signs of the coordinates of points being plotted. Then, visualize in which part of the graph the point will fall. The four regions of the rectangular coordinate system are called **quadrants.**
>
> Refer to the points $A = (3, 1)$; $B = (-2, 5)$; $C = (-3, -2)$; and $D = (1, -3)$.
>
> Point A (3, 1): move *right* and *up,* as shown in Fig. 9–4. This is quadrant I.
>
> Point B (−2, 5): move *left* and *up,* as shown in Fig. 9–5. This is quadrant II.
>
>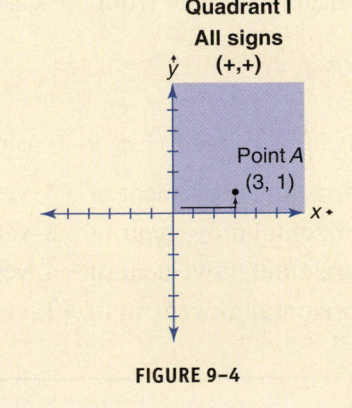
>
> **FIGURE 9–4**
>
>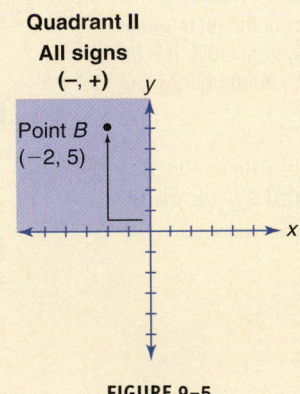
>
> **FIGURE 9–5**

Point $C(-3, -2)$: move *left* and *down*, as shown in Fig. 9–6. This is quadrant III.

Point $D(1, -3)$: move *right* and *down,* as shown in Fig. 9–7. This is quadrant IV.

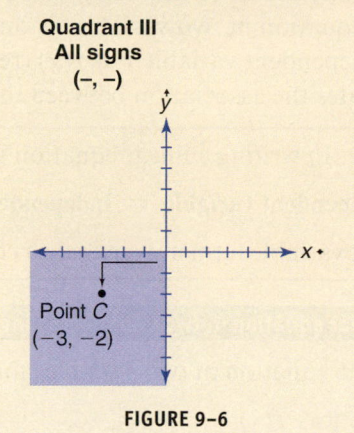

FIGURE 9–6

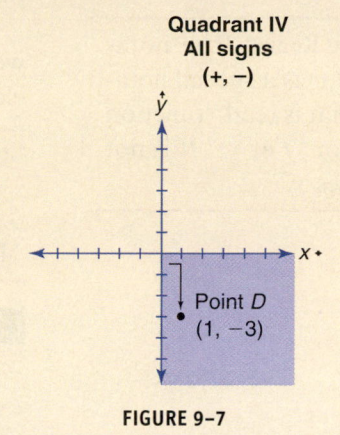

FIGURE 9–7

A point that shows movement in only one direction is written in point notation by using 0 as the coordinate that represents no movement. Points on the *x*-axis have no vertical movement. Points on the *y*-axis have no horizontal movement.

STOP AND CHECK
Plot the points.
1. $E = (-3, 0)$
2. $F = (0, -3)$

Answers:

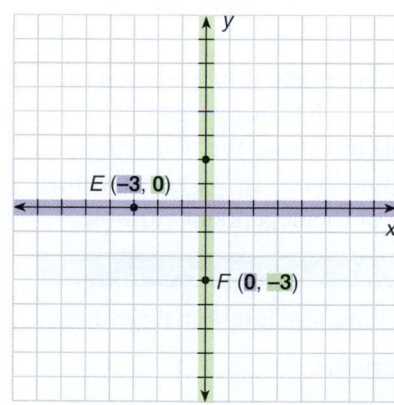

EXAMPLE 4

Plot the points: $A = (2, 0)$, $B = (0, 2)$, $C = (-2, 0)$, $D = (0, -2)$.

See Fig. 9–8.

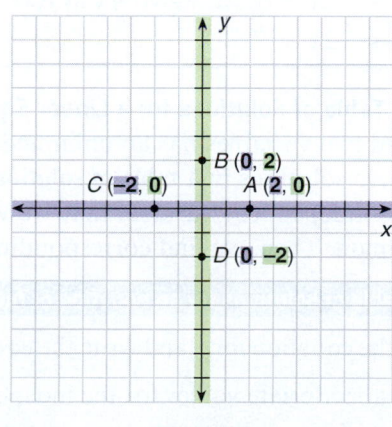

FIGURE 9–8

Point A $= (2, 0)$	right 2, then stop
Point B $= (0, 2)$	up 2
Point C $= (-2, 0)$	left 2, then stop
Point D $= (0, -2)$	down 2

See Exercises 18–27.

2 **Represent an Equation in Two Variables as a Function.** The equations that we have examined so far have been linear equations in one variable. We will expand our study of linear equations to include equations with two variables. In general, the variable that is the *independent variable* will be represented with the letter *x*. The *dependent variable* will be represented with the letter *y*.

An example of a linear equation in two variables is $y = x + 3$. This type of equation does not have just one solution. It has an unlimited number of solutions. For instance, if *x* is 2, then *y* is $2 + 3$, or 5. If *x* is 3, then *y* is $3 + 3$, or 6. Each solution of an equation in two variables has a value for each variable. We write these solutions as *ordered pairs* (independent variable, dependent variable).

The mathematical term for a set of ordered pairs is a **relation**. The first component of each ordered pair or the set of values for the independent variable is called the **domain**. The second component of each ordered pair or the set of values for the dependent variable is called the **range**. When each value of the domain corresponds

Relation: a set of ordered pairs

Domain: the first component of an ordered pair; the set of values for the independent variable

Range: the second component of an ordered pair; the set of values for the dependent variable

Function: when each value of the domain corresponds to exactly one value in the range

> **Did You Know?** The notation $f(x)$ is a special notation that is read "function of x" or "f of x." It is not "f times x."

to exactly one value in the range, the relation is called a **function.** A function shows the association between the values from the domain and the range. The domain, range, and relations will be examined in more detail in Chapter 15.

A linear equation in two variables is one type of function. In function notation we write the dependent variable y as $f(x)$ (read "function of x"). This notation more clearly illustrates the association between the independent and dependent variables.

> **Did You Know?** In writing a linear equation in function notation, we use the pattern
>
> **Dependent variable = Independent variable term + Constant**
>
> Independent variable term may include a coefficient and the independent variable.

> **To write a linear equation in two variables using function notation:**
>
> 1. Solve the equation in two variables for the dependent variable y.
>
> 2. Rewrite y as $f(x)$.

STOP AND CHECK

Rewrite the equations in function notation.
1. $y = -5x + 2$
2. $3x - y = -5$

Answers:
1. $f(x) = -5x + 2$
2. $f(x) = 3x + 5$

EXAMPLE 5

Rewrite the equations in function notation.

(a) $y = 2x - 3$ (b) $2x + y = 7$

(a) $y = 2x - 3$ Rewrite y as $f(x)$.
     $f(x) = 2x - 3$

(b) $2x + y = 7$ Solve for y.
      $y = -2x + 7$ Rewrite y as $f(x)$.
      $f(x) = -2x + 7$ See Exercises 28–36.

3 **Make a Table of Solutions for a Linear Equation or Function.** Since an equation or function in two variables has an unlimited number of solutions, we can organize several of these solutions in a **table of solutions** or **table of values.** We determine the values of the independent variable that we will examine. That is, we select values from the domain. Then, we find corresponding values of the dependent variable.

Table of solutions or table of values: a table that shows two or more corresponding values for the independent and dependent variables

> **To make a table of solutions for an equation or function:**
>
> 1. Solve the equation for y and write the equation in function notation, if desired.
>
> 2. Select appropriate values for the independent variable.
>
> 3. Perform the calculations associated with the function for each selected value of the independent variable.
>
> 4. Write the results of each calculation as an ordered pair.

STOP AND CHECK

Make a table of solutions for the equations.
1. $y = -5x + 2$
2. $3x - y = -5$

Answers:
Answers may vary.

1.

x	y	(x, y)
-1	7	$(-1, 7)$
0	2	$(0, 2)$
1	-3	$(1, -3)$

2.

x	y	(x, y)
-2	-1	$(-2, -1)$
0	5	$(0, 5)$
2	11	$(2, 11)$

EXAMPLE 6

Make a table of solutions for the equations.

(a) $y = 2x - 3$ (b) $2x + y = 7$

(a) $y = 2x - 3$ Rewrite as $f(x)$.
     $f(x) = 2x - 3$ Select appropriate values for the independent variable x.

The domain of the function is all real numbers. Select values in the domain that are easy to work with and reasonable to graph. Any number of values can be selected but we will select at least three that are not too close together. Let's select $-3, 0,$ and 3.

Table of Solutions

x	f(x)	y	(x, y)
−3	$f(x) = 2x - 3$ $f(-3) = 2(-3) - 3$ $f(-3) = -6 - 3$ $f(-3) = -9$	−9	(−3, −9)
0	$f(x) = 2x - 3$ $f(0) = 2(0) - 3$ $f(0) = 0 - 3$ $f(0) = -3$	−3	(0, −3)
3	$f(x) = 2x - 3$ $f(3) = 2(3) - 3$ $f(3) = 6 - 3$ $f(3) = 3$	3	(3, 3)

(b) $2x + y = 7$ Solve for y.

$\quad\quad y = -2x + 7$ Rewrite y as f(x).

$\quad f(x) = -2x + 7$

Select appropriate values of x. Let's select $-2, 0,$ and 2. In making our table, we will perform some of the steps mentally.

x	f(x)	y	(x, y)
−2	$f(-2) = -2(-2)+7$	11	(−2, 11)
0	$f(0)\ \ = -2(0) + 7$	7	(0, 7)
2	$f(2)\ \ = -2(2) + 7$	3	(2, 3)

See Exercises 37–50.

TIP **Make a Table of Solutions on a Calculator** A graphing calculator can be used to build a table of solutions for a function having two variables if the equation is solved for the dependent variable.

Set the desired characteristics of the table (starting point of the desired domain and the desired increment of the table values of the independent variable):

TI-84

To access features that are written in blue type above a key, use the blue key labeled 2ND. For illustration in this text when the feature is written in blue type above a key, we will highlight the key with blue type.

Access 2ND TBLSET feature. Enter starting value of desired domain after TblStart =.

Enter the desired increment after △Tbl =.

Be sure that both the independent and dependent variables are set to Auto.

Enter the function: Y = CLEAR. Then, enter the right side of the function at \Y₁ =. Enter a variable using the X, T, θ, n key.

View the table by pressing 2ND TABLE. The arrow keys allow you to move up and down the table.

Use Part (b) of the previous example ($y = -2x + 7$) to build a table of solutions.
2ND TBLSET TblStart = −3 △Tbl = 1 Auto Auto 2ND QUIT
Y = CLEAR (−) 2 X, T, θ, n + 7 2ND TABLE. See Fig. 9–9.

X	Y1
−3	13
−2	11
−1	9
0	7
1	5
2	3
3	1

X = −3

FIGURE 9–9

4 **Graph a Linear Equation or Function Using a Table of Solutions.** The *table-of-solutions* or *table-of-values* procedure for graphing an equation works for *all* types of equations. Other methods for graphing equations focus on specific properties of each type of equation; these methods are generally less time-consuming and more efficient than the table-of-solutions procedure.

To graph a linear equation or function using a table of solutions:

1. Prepare a table of solutions by evaluating the equation or function using different values (at least three) in the domain of the independent variable.

2. Plot the points from the table of solutions on a rectangular coordinate system.

3. Connect the points with a straight line. Extend the graph beyond the three points and place an arrow on each end as appropriate to indicate that the line extends indefinitely.

STOP AND CHECK

1. Graph the equations in the previous Stop and Check using the table-of-solutions method.

Answers:

1.

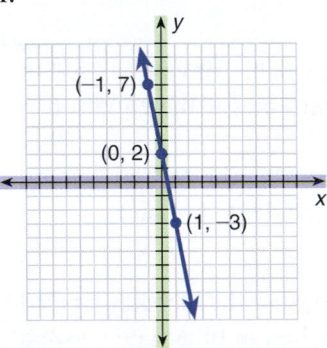

2.

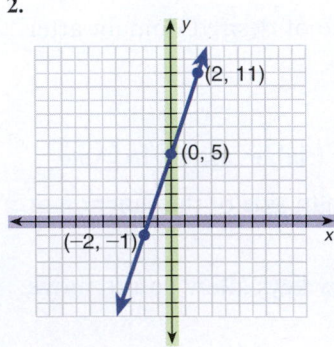

EXAMPLE 7

Graph the equations $y = 2x - 3$ and $2x + y = 7$ using the table-of-solutions method.

Use the tables of solutions from Example 6.
$y = 2x - 3$ (see Fig. 9–10)

x	y
−3	−9
0	−3
3	3

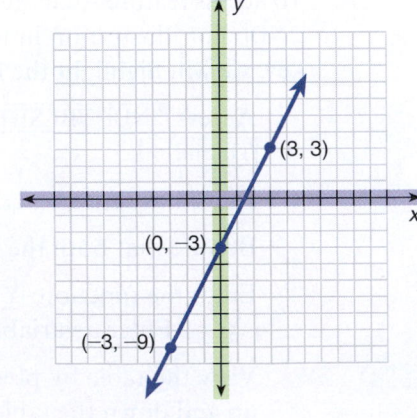

FIGURE 9–10

$2x + y = 7$ (see Fig. 9–11)   Solve the equation for y.
$y = -2x + 7$ Since the solution $(-2, 11)$ is off the graph, another point
                                 was selected. $f(5) = -2(5) + 7 = -10 + 7 = -3$.

x	y
-2	11
0	7
2	3
5	-3

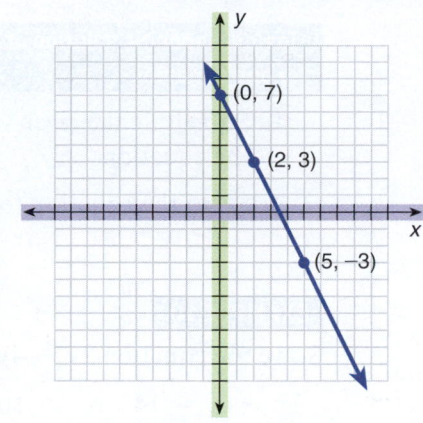

FIGURE 9–11   **See Exercises 51–58.**

Linear equation: an equation for which the graph is a straight line

A **linear equation** is an equation for which the graph is a straight line. We can *check* each solution of an equation in two variables by substituting the values for the respective variables and performing the operations indicated.

> **To check a solution of an equation in two variables:**
>
> 1. Substitute the values from the ordered pair for the respective variables in the equation.
>
> 2. Evaluate each side of the equation using the order of operations.
>
> 3. The two sides of the equation will be equal, if the ordered pair is a solution of the equation.

EXAMPLE 8

Check to see if the ordered pair $(-2, 16)$ is a solution for the equation $2x + y = 12$.

$2x + y = 12$ Substitute -2 for x and 16 for y.
$2(-2) + 16 = 12$ Simplify the left side.
$-4 + 16 = 12$
$12 = 12$ True

The ordered pair makes the equation true. So $(-2, 16)$ is a solution of the equation.

See Exercises 59–64.

EXAMPLE 9

Is the point $(3, -2)$ on the graph of the equation $y = 2x - 5$?

$y = 2x - 5$ Substitute 3 for x and -2 for y.
$-2 = 2(3) - 5$ Simplify the right side.
$-2 = 6 - 5$
$-2 = 1$ False

The ordered pair does not make the equation true. So $(3, -2)$ is not a solution.

No, the point $(3, -2)$ is not on the graph of the equation $y = 2x - 5$.

See Exercises 65–70.

If we are given the value of one variable for an equation with two variables, we can substitute the given value in the equation and solve the equation for the other variable.

> **To find a specific solution of an equation in two variables when given the value of one variable:**
>
> 1. Substitute the given value of the variable in each place that variable occurs in the equation.
>
> 2. Solve the equation for the other variable.

STOP AND CHECK

1. Find y in the equation $-3x + 2y = 13$, if $x = -3$.

Answer:

1. $y = 2$

EXAMPLE 10

Solve the equation $2x - 4y = 14$, if $x = 3$.

$$2x - 4y = 14 \qquad \text{Substitute 3 for } x.$$
$$2(3) - 4y = 14$$
$$6 - 4y = 14 \qquad \text{Sort terms.}$$
$$-4y = 14 - 6$$
$$-4y = 8 \qquad \text{Divide both sides by } -4.$$
$$\frac{-4y}{-4} = \frac{8}{-4}$$
$$y = -2$$

The solution is $(3, -2)$; $x = 3$ and $y = -2$.   See Exercises 71–76.

Standard form of a linear equation in two variables: a linear equation written in standard form $ax + by = c$. The leading term has the x variable and is positive and the equation contains no fractions

The **standard form of a linear equation in two variables** is $ax + by = c$ for all real numbers a, b, and c. Both a and b are not zero; however, either a or b may be zero.

When a or b is zero, the variable for which zero is the coefficient is eliminated from the equation. For example, consider $b = 0$ in the following equation.

$$ax + by = 15 \qquad \text{Let } a = 5 \text{ and } b = 0.$$
$$5x + 0(y) = 15 \qquad 0(y) = 0.$$
$$5x + 0 = 15 \qquad \text{Linear equation in one variable}$$
$$5x = 15 \qquad \text{Divide both sides by 5.}$$
$$\frac{5x}{5} = \frac{15}{5}$$
$$x = 3$$

An equation in one variable like $x = 3$ is a linear equation, and we think of the coefficient of the missing variable term as zero:

$$x = 3 \qquad \text{is considered as} \qquad x + 0y = 3 \qquad \text{in standard form}$$

This means that x is 3 for any value of y.

STOP AND CHECK

1. For the equation $y = -3$, complete the ordered pairs:
2. $(5, \ )$; $(-1, \ )$; $(0, \ )$.

Answer:

1. $(5, -3)$; $(-1, -3)$; $(0, -3)$

EXAMPLE 11

For the equation $x = 6$, complete the given ordered pairs: $(\ , 2)$; $(\ , -3)$; $(\ , 1)$.

Equation	Ordered Pairs			
$x = 6$	$(\ , 2)$	$(\ , -3)$	$(\ , 1)$	x is 6 for *any* value of y.
	$(6, 2)$	$(6, -3)$	$(6, 1)$	See Exercises 77–78.

Stephen Coburn/123RF

Model: a real-life situation written as an equation

Numerous application problems can be solved with equations in two variables.

EXAMPLE 12

BUS A small business photocopied a report that included 12 black-and-white pages and 25 color pages. The cost was $23.70. Letting b equal the cost for a black-and-white page and c equal the cost for a color page, write an equation with two variables.

Cost of black-and-white copies: $12b$ (12 times cost per page)

Cost of color copies: $25c$ (25 times cost per page)

Total cost: cost of black-and-white plus cost of color = $23.70

Equation: $12b + 25c = \$23.70$ See Exercises 79–82.

This application of an equation in two variables can be called a **model** for a real-life situation. What if we knew the cost for a color copy but not for the black-and-white copy? Example 13 shows how a model can be used to find missing information.

An equation containing two variables shows the relationship between the two unknown amounts. If a value is known for either amount, the other amount can be found by solving the equation for the missing amount.

EXAMPLE 13

Solve the equation from Example 12 to find the cost for each black-and-white copy if color copies cost $0.90 each.

$$12b + 25c = 23.70 \qquad \text{Substitute 0.90 for } c.$$
$$12b + 25(0.90) = 23.70$$
$$12b + 22.50 = 23.70 \qquad \text{Sort terms.}$$
$$12b = 23.70 - 22.50$$
$$12b = 1.20 \qquad \text{Divide both sides by 12.}$$
$$\frac{12b}{12} = \frac{1.20}{12}$$
$$b = 0.10$$

Each black-and-white copy costs $0.10.   See Exercises 79–82.

9-1 EXERCISES

MyLab Math For additional practice go to your study plan in MyLab Math.

1 Locate the points in Fig. 9–12 by giving the amount of horizontal and vertical movement from the origin. *See Example 1.*

1. R 2. S

3. T 4. U

5. V 6. W

7. X 8. Y

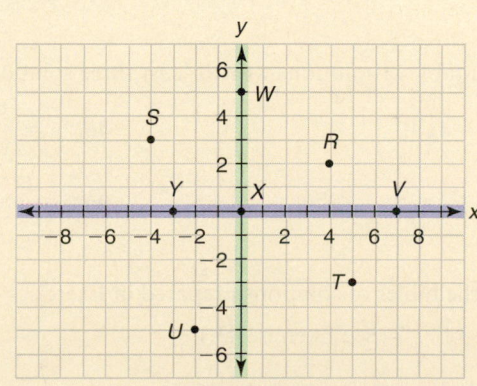

FIGURE 9-12

See Example 2.

Write the points in Fig. 9–12 in point notation.

9. *R*   **10.** *S*   **11.** *T*   **12.** *U*

13. *V*   **14.** *W*   **15.** *X*   **16.** *Y*

17. Write the coordinates for the points *A, B, C,* and *D* on the graph in Fig. 9–13.

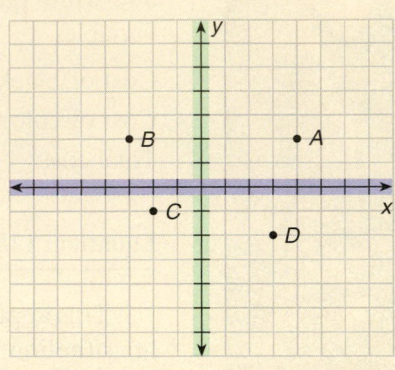

FIGURE 9–13

Draw a rectangular coordinate system, plot the points, and give the quadrant for each point in Exercises 18–23. *See Examples 3 and 4.*

18. $A = (-7, 4)$ **19.** $B = (0, -10)$ **20.** $C = (-2, 0)$

21. $D = (-7, -7)$ **22.** $E = (3, 0)$ **23.** $F = (4, -3)$

24. What common property describes the coordinates of the points on the *x*-axis?

25. What common property describes the coordinates of the points on the *y*-axis?

26. Describe the signs of the coordinates of points that lie in the upper-left quadrant of a graph.

27. Describe the signs of the coordinates of points that lie in the lower-right quadrant of a graph.

2 Rewrite the equation in function notation. *See Example 5.*

28. $y = 5x - 7$ **29.** $y = -3x + 2$ **30.** $2y = 8x - 10$

31. $4y = -7x - 3$ **32.** $3x + y = 12$ **33.** $6x - 2y = 8$

34. $3x + 2y = 9$ **35.** $x - 7y = 14$ **36.** $5x - 3y = 7$

3 Make a table of solutions to obtain at least three sets of coordinates that satisfy each equation. *See Example 6.*

37. $y = 5x - 7$ **38.** $y = -3x + 2$ **39.** $2y = 8x - 10$

40. $4y = -7x - 3$ **41.** $3x + y = 12$ **42.** $6x - 2y = 8$

43. Make a table of at least three solutions for the equation $y = 3x - 7$.

44. Make a table of three solutions for the equation $y = -2x + 3$.

Prepare a table of solutions with at least three solutions for the equations and functions.

45. $y = -4x + 5$ **46.** $y = \frac{1}{2}x + 3$ **47.** $y = -\frac{2}{3}x - 1$

48. $y = 5x + 2$ **49.** $f(x) = -\frac{1}{2}x + 2$ **50.** $y = \frac{1}{3}x + 2$

4 Use the tables of solutions prepared in Exercises 43–50 to graph the following solutions. *See Example 7.*

51. $y = 3x - 7$ (See Exercise 43.)   **52.** $y = -2x + 3$ (See Exercise 44.)

53. $y = -4x + 5$ (See Exercise 45.)   **54.** $y = \frac{1}{2}x + 3$ (See Exercise 46.)

55. $y = -\dfrac{2}{3}x - 1$ (See Exercise 47.)

56. $y = 5x + 2$ (See Exercise 48.)

57. $f(x) = -\dfrac{1}{2}x + 2$ (See Exercise 49.)

58. $y = \dfrac{1}{3}x + 2$ (See Exercise 50.)

Determine which of the ordered pairs are solutions for the equation $2x - 5y = 9$. *See Example 8.*

59. $(4, 5)$ **60.** $(17, 5)$ **61.** $(12, 3)$ **62.** $(6, 1)$ **63.** $(7, 1)$ **64.** $(4.5, 0)$

Determine which of the ordered pairs are solutions for the equation $3y = 2 - 4x$. *See Example 9.*

65. $(2, -2)$ **66.** $(-2, 2)$ **67.** $(5, -6)$ **68.** $(-6, 5)$ **69.** $\left(0, \dfrac{2}{3}\right)$ **70.** $\left(6, -7\dfrac{1}{3}\right)$

Write the specific solution for each equation in ordered pair form. *See Example 10.*

71. $x - y = 3$, if $x = 8$ **72.** $x + y = 7$, if $x = 2$ **73.** $x + 2y = -4$, if $y = 1$

74. $x - 3y = 12$, if $y = 3$ **75.** $5x - y = 10$, if $x = 5$ **76.** $4x + y = 8$, if $x = 2$

Complete the ordered pairs for each equation. *See Example 11.*

77. $x = 1$: $(\ , 3)$; $(\ , -2)$; $(\ , 0)$; $(\ , 1)$

78. $y = 4$: $(-1, \ )$; $(3, \ )$; $(0, \ )$; $(2, \ )$

Solve using equations with two variables. *See Examples 12 and 13.*

79. **BUS** Anne Richards purchased three shirts at one price and five belts at another price. Let $s =$ the price of each shirt and $b =$ the price of each belt. Each shirt cost \$22. How much did each belt cost if the total purchase price was \$126?

80. **AUTO** Allen and Cecile Bell paid \$13 for 4 spark plugs and 5 quarts of motor oil. Let $p =$ the cost of each plug and $q =$ the cost of each quart of oil. If the oil cost \$1 per quart, how much did each spark plug cost?

81. **BUS** Lola Bowers charged \$160 for four pairs of pants and two shirts. If the shirts cost \$12 each, how much did each pair of pants cost? Let $p =$ the cost of each pair of pants and $s =$ the cost of each shirt.

82. **AUTO** Coach Barton paid \$15,000 for an automobile. The purchase price included three extended-warranty payments of \$400 each. How much did the car cost without the extended warranty? Let $c =$ the cost of the car and $w =$ the cost of each warranty payment.

9–2   Graphing Linear Equations and Inequalities in Two Variables Using Alternative Methods

LEARNING OUTCOMES

1. Graph a linear equation using intercepts.
2. Graph a linear equation using the slope and y-intercept.
3. Graph a linear equation using a graphing calculator.
4. Solve a linear equation in one variable using a graph.
5. Graph a linear inequality in two variables.

LC LEARNING CATALYTICS

1. Find y when $x = 0$ for the equation $5x - 2y = 10$.

We have graphed linear equations using a table of solutions. Now let's examine other graphing procedures and specific properties of linear equations in two variables.

1 Graph a Linear Equation Using Intercepts. Two important points on the graph of an equation are the points where the graph crosses the x- and y-axis. These points are called **intercepts**. The **x-intercept** is the point on the x-axis through which the line passes; that is, the y-value is zero $(x, 0)$. The **y-intercept** is the point on the y-axis through which the line passes; that is, the x-value is zero $(0, y)$.

To find the intercepts of a linear equation:

1. Find the x-intercept; let $y = 0$ and solve for x.
2. Find the y-intercept; let $x = 0$ and solve for y.

Intercept: a point where a graph crosses one of the axes in the rectangular coordinate system

x-intercept: the point on the x-axis through which the line passes; that is, the y-value is zero $(x, 0)$

y-intercept: the point on the y-axis through which the line passes; that is, the x-value is zero $(0, y)$

STOP AND CHECK

1. Find the intercepts of the equation $-3x + y = 12$.

Answer:

1. x-intercept: $(-4, 0)$; y-intercept: $(0, 12)$

EXAMPLE 1

Find the intercepts of the equation $3x - y = 5$.

$3x - y = 5$ For the x-intercept, let $y = 0$.

$3x - 0 = 5$ Solve for x.

$3x = 5$

$x = \dfrac{5}{3}$ A mixed number or decimal is easier to plot than an improper fraction.

$x = 1\dfrac{2}{3}$,   or 1.67 $1\frac{2}{3}$ is the exact value and 1.67 is an approximate value.

The x-intercept is $\left(1\frac{2}{3}, 0\right)$.

$3x - y = 5$ For the y-intercept, let $x = 0$.

$3(0) - y = 5$ Solve for y.

$-y = 5$

$y = -5$

The y-intercept is $(0, -5)$.   See Exercises 1–2.

To graph a linear equation by the intercepts method:

1. Find the x- and y-intercepts.

2. Plot the intercepts on a rectangular coordinate system.

3. Draw the line through the two points and extend it beyond each point.

4. Check by examining one additional solution of the equation.

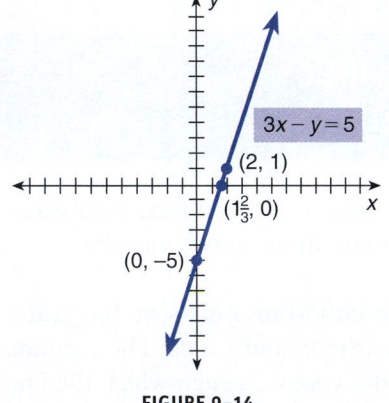

FIGURE 9–14

EXAMPLE 2

Graph the equation $3x - y = 5$ by using the intercepts method.

Plot the two intercepts found in the preceding example, $\left(1\frac{2}{3}, 0\right)$ and $(0, -5)$. Draw the line connecting these points and extending beyond. See Fig. 9–14. We can check by finding the coordinates of one other point using the table-of-solutions method and plotting the point. If the point is on the line, the graph is correct. Find y when $x = 2$:

$3x - y = 5$ Substitute 2 for x.

$3(2) - y = 5$ Solve for y.

$6 - 5 = y$

$1 = y$ The point (2, 1) is on the graph.   See Exercises 3–8.

TIP   Graphing an Equation When Both Intercepts Are (0, 0) If both intercepts are (0, 0), the intercepts coincide at the origin and form only *one point*. An additional point must be found by the table-of-solutions method to have two distinct points to draw the line. A third point is still useful to check your work.

Graph the equations using the intercepts method.

1. $x - 3y = 9$

2. $-2y = -6x$

Answers:

1.

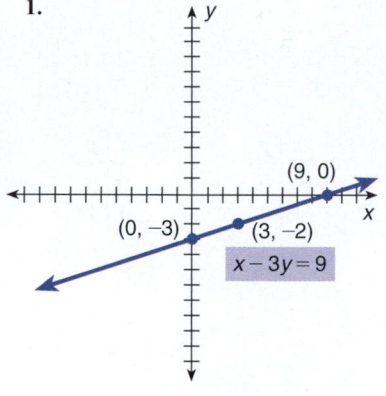

2.

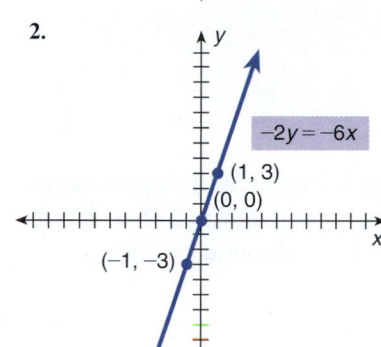

EXAMPLE 3

Graph $y = 7x$ using the intercepts method.

Find the intercepts.

$y = 7x$ Let $y = 0$.

$0 = 7x$

$\dfrac{0}{7} = \dfrac{7x}{7}$

$0 = x$ Coordinates of the x-intercept are $(0, 0)$.

$y = 7x$ Let $x = 0$.

$y = 7(0)$

$y = 0$ Coordinates of the y-intercept are $(0, 0)$.

Find the coordinates of two other points.

$y = 7x$ Let $x = 1$. $y = 7x$ Let $y = -7$.

$y = 7(1)$ $-7 = 7x$

$y = 7$ $-1 = x$

Additional points $= (1, 7)$ and $(-1, 7)$

Plot the points $(0, 0)$ and $(1, 7)$. Draw the line through the two points and extend it beyond each point, as shown in Fig. 9–15.   **See Exercises 9–10.**

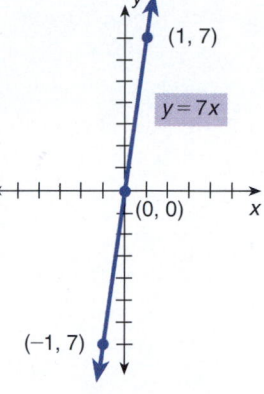

FIGURE 9–15

2 Graph a Linear Equation Using the Slope and y-Intercept. We can identify characteristics of the graph of an equation by inspection. To see these characteristics, we solve the equation for y. The equation will be in the form $y = mx + b$, where m is the coefficient of x and b is a constant.

First, let's examine the characteristics identified by the constant. Let the coefficient of x equal 1 and look at the graphs of the equations in Fig. 9–16.

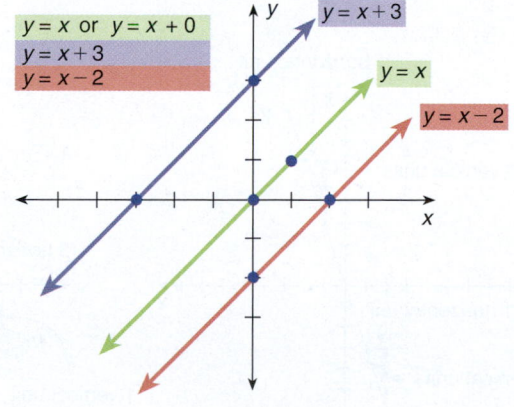

$y = x$		$y = x + 3$		$y = x - 2$	
x	y	x	y	x	y
0	0	0	3	0	-2
1	1	-3	0	2	0

FIGURE 9–16

The three graphs have the same slope (steepness or slant), but they have different y-intercepts. That is, they cross the y-axis at different points.

$y = x + 0$ crosses the y-axis at 0; y-intercept $= (0, 0)$

$y = x + 3$ crosses the y-axis at $+3$; y-intercept $= (0, 3)$

$y = x - 2$ crosses the y-axis at -2; y-intercept $= (0, -2)$

The y-coordinate of the point where the graph crosses the y-axis is the same as the constant in the equation. The constant (b) in an equation in the form $y = mx + b$ is the y-coordinate of the y-intercept. The y-intercept can be written $(0, b)$.

Let's examine equations that have graphs with a different slant or steepness but the same y-intercept. The x- and y-intercepts in each of the three equations in Fig. 9–17 are $(0, 0)$. We will find one additional point for each equation.

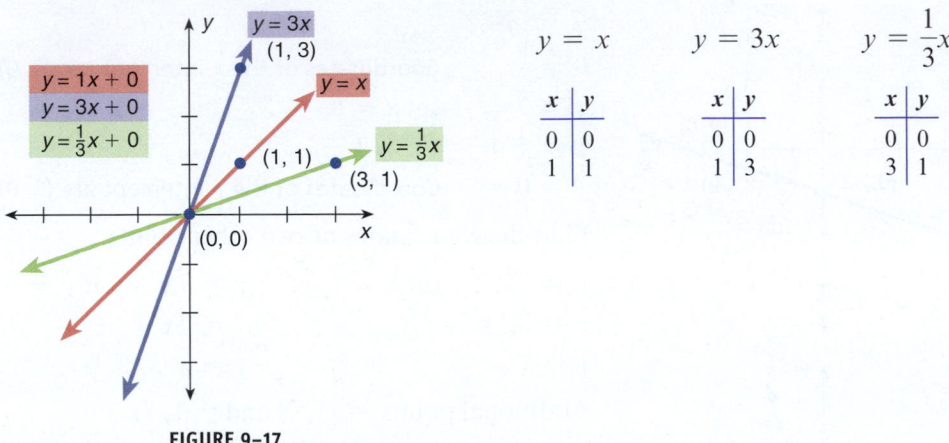

FIGURE 9–17

The three graphs have the same y-intercepts, but they have different slopes (steepness or slants). We define this slope or steepness so that a numerical value identifies it. The **slope** of a line is the ratio of the vertical change to the horizontal change.

Slope: the ratio of the vertical change (rise) of a line to the horizontal change (run) of a line

$$\text{slope} = \frac{\text{vertical change}}{\text{horizontal change}}$$

A slope of 1 (written as $\frac{1}{1}$) means a change of 1 vertical unit for every 1 horizontal unit of change. A slope of 3 (written as the ratio $\frac{3}{1}$) means a change of 3 vertical units for every 1 horizontal unit. A slope of $\frac{1}{3}$ means a change of 1 vertical unit for every 3 horizontal units (see Fig. 9–18).

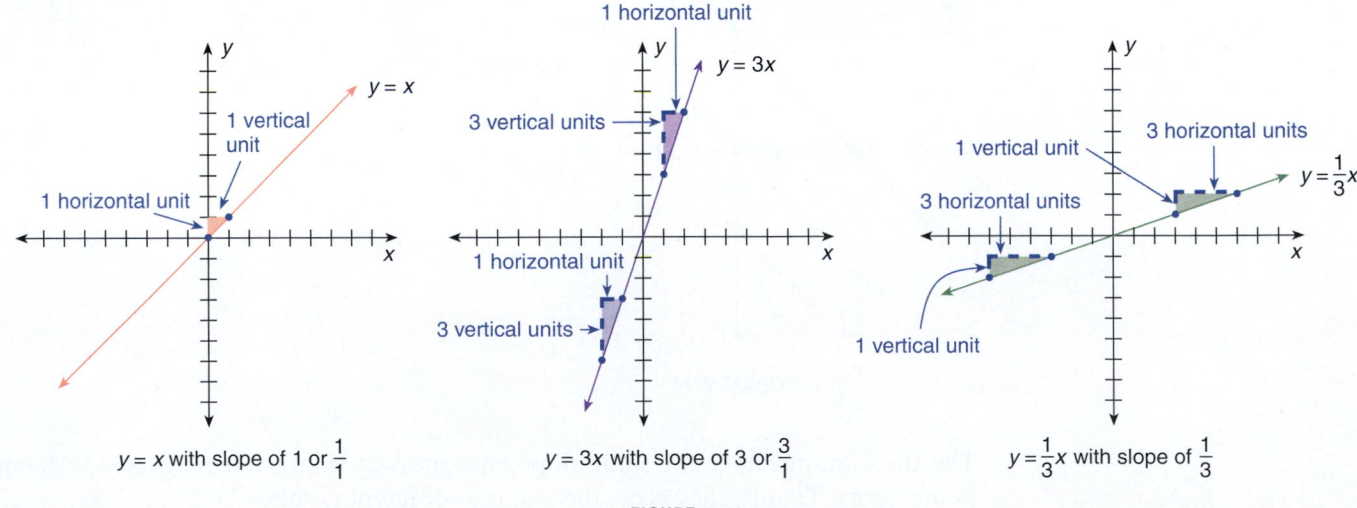

FIGURE 9–18

Notice that the slope of the line is the same as the fractional form of the coefficient of the x-term in the equation. Look again at the equations in Figs. 9–16 and 9–17.

From Fig. 9–16	Slope-Intercept Form	From Fig. 9–17	Slope-Intercept Form
$y = x$	$y = 1x + 0$	$y = x$	$y = 1x + 0$
$y = x + 3$	$y = 1x + 3$	$y = 3x$	$y = 3x + 0$
$y = x - 2$	$y = 1x - 2$	$y = \dfrac{1}{3}x$	$y = \dfrac{1}{3}x + 0$

Slope-intercept form of a linear equation: the linear equation $y = mx + b$, where m is the slope and b is the y-coordinate of the y-intercept

In each case, the equation is solved for y. This form of equation is called the **slope-intercept form of a linear equation,** or $y = mx + b$, where m is the slope and b is the y-coordinate of the y-intercept. When equations are written in this form, the slope and y-intercept can be identified by inspection.

> **Slope-intercept form of an equation:**
>
> $$y = mx + b$$
>
> where m and b are real numbers and $m =$ slope and $b = y$-coordinate of the y-intercept

> **To determine the slope and y-intercept of a linear equation by inspection:**
>
> 1. Solve the equation for y and write it in the form $y = mx + b$.
>
> 2. The *slope* is m, which is the coefficient of x.
>
> 3. The y-coordinate of the y-intercept is b. The y-intercept is $(0, b)$.

STOP AND CHECK

1. Write the equation $6x - 5y = 15$ in slope-intercept form and identify the slope and y-intercept.

Answer:

1. $y = \dfrac{6}{5}x - 3$; slope $= \dfrac{6}{5}$; y-intercept $= (0, -3)$

EXAMPLE 4

Write the equations in slope-intercept form and identify the slope and y-intercept.

(a) $2x + y = 4$ **(b)** $5x - y = -2$ **(c)** $3x + 4y = -12$

(a) $2x + y = 4$ Solve for y.

$$y = -2x + 4$$ Slope-intercept form

$$\text{slope} = -2 \text{ or } \frac{-2}{1} \text{ or } \frac{2}{-1}$$ Coefficient of x

y-coordinate of y-intercept $= 4$ Constant

y-intercept $= (0, 4)$

(b) $5x - y = -2$ Solve for y. Divide by -1.

$$\frac{-y}{-1} = \frac{-5x - 2}{-1}$$ Write the right side as separate terms and simplify.

$$y = 5x + 2$$ Slope-intercept form

$$\text{slope} = 5 \text{ or } \frac{5}{1} \text{ or } \frac{-5}{-1}$$ Coefficient of x

y-coordinate of y-intercept $= 2$ Constant

y-intercept $= (0, 2)$

(c) $3x + 4y = -12$ Solve for y. Sort terms to isolate $4y$. Divide by 4.

$$\frac{4y}{4} = \frac{-3x - 12}{4}$$ Write the right side as separate terms and reduce.

$$y = -\frac{3}{4}x - 3 \qquad \text{Slope-intercept form}$$

$$\text{slope} = -\frac{3}{4} \text{ or } \frac{-3}{4} \text{ or } \frac{3}{-4} \qquad \text{Coefficient of } x$$

y-coordinate of y-intercept $= -3$ Constant

y-intercept $= (0, -3)$ See Exercises 11–20.

An equation written in the slope-intercept form can be graphed using just the slope and y-intercept.

To graph a linear equation in the form $y = mx + b$ using the slope-intercept method:

1. Locate the y-intercept on the y-axis.

2. Using the slope in fractional form, determine the amount of vertical and horizontal movement indicated.

3. From the y-intercept, locate additional points on the graph of the equation by counting the indicated vertical and horizontal movements of the slope.

4. Draw the line connecting the points and extending beyond the points.

STOP AND CHECK

1. Graph the equation $3x - 2y = -12$ using the slope-intercept method.

Answer:

1.

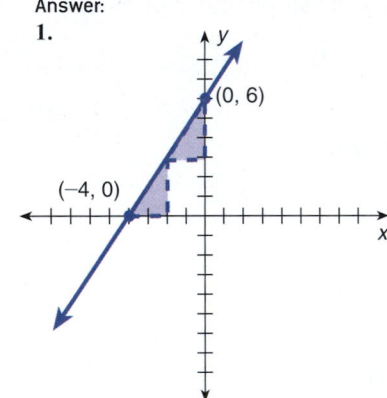

EXAMPLE 5

Graph the equations using the slope-intercept method.

(a) $2x + y = 4$ (b) $5x - y = -2$ (c) $3x + 4y = -12$

(a) $2x + y = 4$ Solve for y.

$$y = -2x + 4$$

y-intercept $= (0, 4)$ Locate this point on the y-axis.

$$\text{slope} = \frac{-2}{1} \text{ or } \frac{2}{-1}$$ Write the slope in fractional form as two equivalent fractions by manipulating the signs of the fraction.

$\frac{-2}{1}$ indicates vertical movement of -2 and horizontal movement of $+1$ from the y-intercept.

$\frac{2}{-1}$ indicates vertical movement of $+2$ and horizontal movement of -1. See Fig. 9–19.

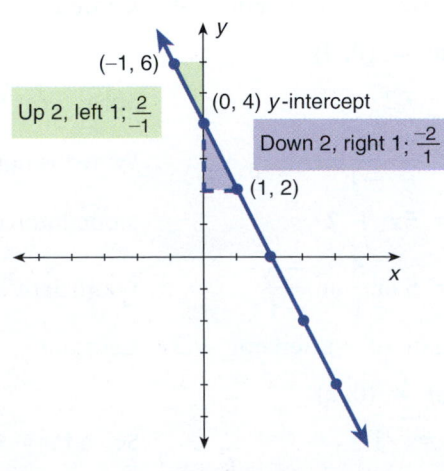

FIGURE 9–19

(b) $5x - y = -2$ Solve for y.

$y = 5x + 2$ Slope-intercept form

y-intercept $= (0, 2)$ Locate this point on the y-axis.

$$\text{slope} = \frac{5}{1} \text{ or } \frac{-5}{-1}$$

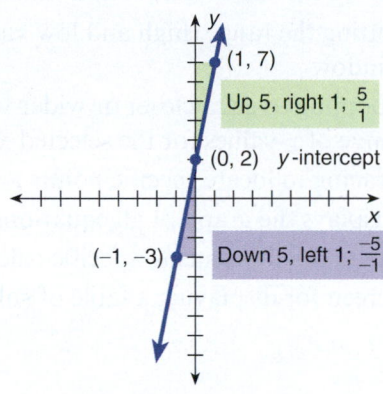

$\frac{5}{1}$ indicates vertical movement of $+5$ and horizontal movement of $+1$ from the y-intercept.

$\frac{-5}{-1}$ indicates vertical movement of -5 and horizontal movement of -1. See Fig. 9–20.

FIGURE 9–20

(c) $3x + 4y = -12$ Solve for y.

$$y = -\frac{3}{4}x - \frac{12}{4}$$ Slope-intercept form

$$y = -\frac{3}{4}x - 3$$

y-intercept $= (0, -3)$ Locate this point on the y-axis.

$$\text{slope} = \frac{-3}{4} \text{ or } \frac{3}{-4}$$

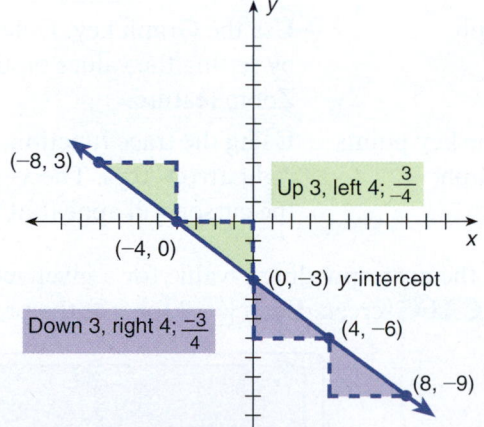

$\frac{-3}{4}$ indicates vertical movement of -3 and horizontal movement of $+4$ from the y-intercept.

$\frac{3}{-4}$ indicates vertical movement of $+3$ and horizontal movement of -4. See Fig. 9–21.

FIGURE 9–21

See Exercises 21–26.

3 Graph Linear Equations Using a Graphing Calculator. A graphing calculator can be used to graph linear equations written in the form $y = mx + b$. Also, computer software can be used to graph linear equations. These tools free us from the tedious task of graphing equations and allow us to focus on the patterns and properties of graphs.

Every model of graphing calculator and every brand of computer software may have a different set of keystrokes for graphing equations. You will need to refer to the owner's manual to adapt to a particular model or brand. However, we will illustrate the usefulness of these tools by showing a few features common to most calculators.

Feature	Purpose
Y =	Screen for entering equation that is solved for y
Window	Setting the range (high and low values of each axis) for the viewing window
Zoom	Zooming to get a closer or wider view or a view to "fit" the full range of y-values for the selected x-values
Trace	Tracing to locate specific points and determine their coordinates
Graph	Displays the graph of all equations that have been entered
Calc	Menu for shortcuts to specific calculations
Table	Screen for displaying a table of solutions

EXAMPLE 6

Graph $y = 5x + 2$ using a graphing calculator. Find the y-value for $x = -2$.

Perform the appropriate function on the graphing calculator of your choice.

Clear graphing screen.	Erase previous equations from the Y = screen.
Set or initialize range.	Define the viewing range of the graph window. One option is to select the standard option from the Zoom screen. That means the range will be reset at a predetermined, factory-set domain and range.
Enter equation.	Equation must be solved for y. On the Y = screen enter the equation using the appropriate key for the variable. This key is often labeled $\boxed{X, T, \theta, n}$.
View graph.	Use the Graph key. Determine a different viewing window by setting the values on the Window screen or using the Zoom feature.
Determine key points on the graph.	Using the trace function, move the cursor with the left or right arrow keys. The x- and y-coordinates of the point at the cursor will appear at the bottom of the screen.

Find the corresponding y-value for a given x-value by using the *value* function from the **CALC** screen. Enter $\boxed{(-)}$ 2 for x at the prompt.

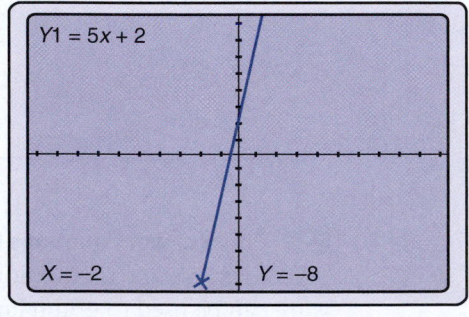

FIGURE 9–22

When $x = -2$, $y = -8$.

See Exercises 27–38.

4 Solve a Linear Equation in One Variable Using a Graph. A linear equation written in slope-intercept form is also referred to as a *function* of *x*, the independent variable. An equation in one variable, such as $2x + 5 = 3$, can be written in the form of a function by first rewriting all terms on either side of the equation.

$$2x + 5 = 3 \qquad \text{or} \qquad 2x + 5 = 3$$
$$2x + 5 - 3 = 0 \qquad\qquad 0 = 3 - 2x - 5$$
$$2x + 2 = 0 \qquad\qquad 0 = -2x - 2$$

Then, write the nonzero side of each equation as a function of *x*.

$$y = 2x + 2 \qquad \text{or} \qquad y = -2x - 2$$

The equation can also be written as $f(x) = 2x + 2$ or $f(x) = -2x - 2$.

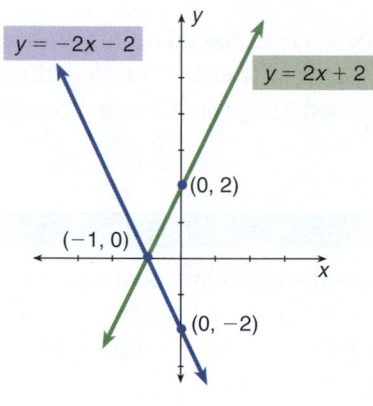

FIGURE 9–23

> **To write an equation in one variable as a function:**
>
> 1. Rewrite the equation so that all terms are on one side of the equation.
>
> 2. Write the equation in function notation using $y =$ the nonzero side of the equation from Step 1.

The solution of the original equation in one variable is the value of *x* that makes the value of either function zero.

If an equation in one variable is written in function notation, the solution of the equation is shown on the graph at the point where the graph crosses the *x*-axis. In Fig. 9–23, both functions cross the *x*-axis at $(-1, 0)$ and the solution of the equation $2x + 5 = 3$ is $x = -1$. Thus, we can solve equations in one variable by using the graphing function on a calculator or computer.

> **To solve an equation in one variable using a graph:**
>
> 1. Write the equation as a function.
>
> 2. Graph the function using a calculator or computer.
>
> 3. The solution is the *x*-intercept. This solution is also referred to as the **zero** ($y = 0$) **of the function**.

Zero of a function: the point or points where a graph crosses the *x*-axis and it is the solution of an equation in one variable

> **TIP** **Find the Zero of a Function Using a Calculator** On most graphing calculators there is a feature to find the zero point or points of a function. On one popular model, the TI-84 Plus Silver Edition™, the feature is a selection on the CALC screen.
>
> The option for finding the zero point requires several steps that are illustrated in the next example.

STOP AND CHECK

1. Solve the equation $5x - 6 = 19$ using a graphing calculator.

Answer:

1. $x = 5$

EXAMPLE 7

Solve the equation $2x + 5 = 3$ using the TI-84 Plus Silver Edition™.

$$2x + 5 = 3$$
$$2x + 5 - 3 = 0$$
$$2x + 2 = 0$$
$$y = 2x + 2 \qquad \text{Function form of the equation}$$
$$\text{Y} = \qquad\qquad \text{Enter right side of equation: } 2\,\boxed{\text{X, T, }\theta, n}+2$$
$$\text{CALC} \qquad\qquad \boxed{\text{CALC}}$$

Zero	Option 2. Press ENTER.
Left bound?	Move cursor to any point to the left of the x-intercept. Press ENTER.
Right bound?	Move cursor to any point to the right of the x-intercept. Press ENTER.
Guess?	Move cursor close to the x-intercept. Press ENTER.
$x = -1$	Read solution from x value at the bottom left of the screen.

See Exercises 39–44.

5 **Graph a Linear Inequality in Two Variables.** When we solved inequalities in one variable the solution was a set of numbers that makes the inequality true that we can represent graphically. The **solution of a linear inequality in two variables** is a set of coordinate points that makes the inequality true.

The graphical solution for a linear inequality ($<$ or $>$) includes all points on one side of the line representing the graph of the similar equation. If either the "less than or equal" ($\leq$) or "greater than or equal" ($\geq$) symbol is used, the points on the boundary line are also included in the solution set.

Solution of a linear inequality in two variables: a set or collection of coordinate points that makes the inequality true

---

To graph a linear inequality in two variables:

1. Find the boundary by graphing an equation that substitutes an equal sign for the inequality symbol. Make a solid line if the boundary is included ($\leq$ or $\geq$) or a dashed line if the boundary is not included ($<$ or $>$) in the solution set.

2. Test any point that is *not* on the boundary line.

3. If the test point makes a true statement with the original inequality, shade the side containing the test point.

4. If the test point makes a false statement with the original inequality, shade the side opposite the side containing the test point.

STOP AND CHECK

1. Graph the inequality $3x - y \geq 6$.

Answer:

1.

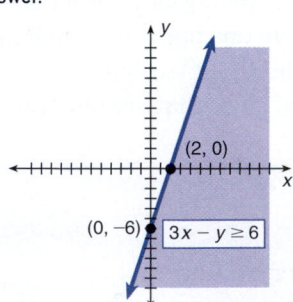

$3x - y \geq 6$

EXAMPLE 8

Graph the inequality $2x + y < 3$.

Graph the equation $2x + y = 3$ to establish the boundary. The boundary will not be included in the solution set since we have a strict inequality $<$, and is represented with a dashed line (Fig. 9–24).

$2x + y = 3$	Solve the equation for y.
$y = -2x + 3$	
y-intercept $= 3$ or $(0, 3)$	Constant term
slope $= -2$ or $\dfrac{-2}{1}$ or $\dfrac{2}{-1}$	Coefficient of x

Select one point, on either side of the boundary, to use as a test point. Suppose that we choose $(2, 2)$ as the point above the boundary.

Is $2x + y < 3$ when $x = 2$ and $y = 2$?	Substitute values.
$2(2) + 2 < 3$	Simplify.
$4 + 2 < 3$	
$6 < 3$	False statement

Thus, the side of the line including the point (2, 2) is *not* in the solution set. Shade the opposite side. See Fig. 9–25.

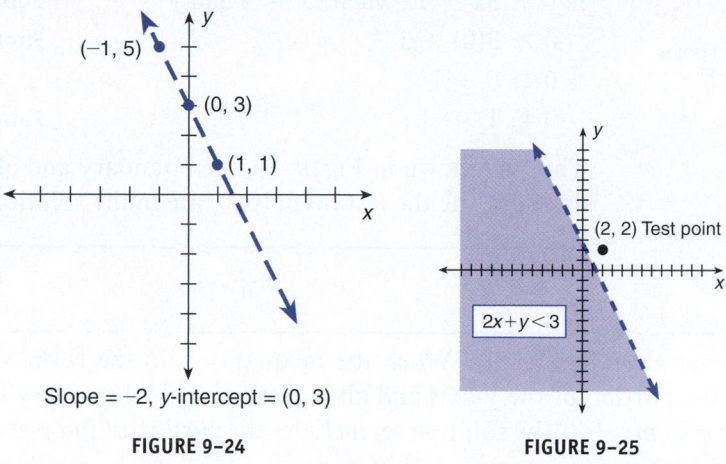

Slope = −2, *y*-intercept = (0, 3)

FIGURE 9–24   **FIGURE 9–25**   See Exercises 45–50.

TIP **Solid or Dashed Boundary Lines** Examine the boundaries and shaded regions of the graphs in Figs. 9–26, 9–27, and 9–28.

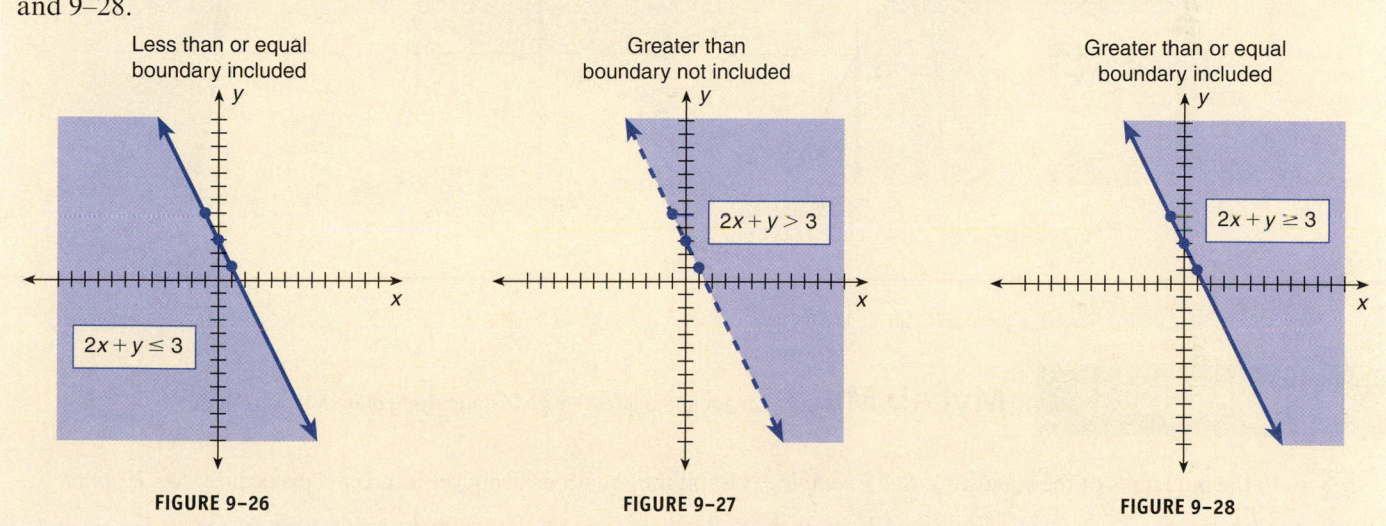

FIGURE 9–26   **FIGURE 9–27**   **FIGURE 9–28**

TIP **Selecting Test Points** Choose numbers that are easy to work with when selecting a test point. If the boundary line does not pass through the origin, a good point to use is (0, 0).

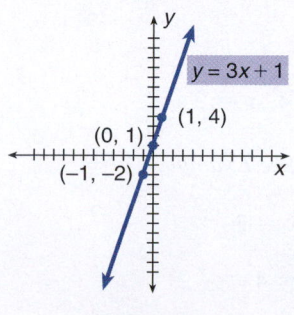

FIGURE 9–29

EXAMPLE 9

Graph the inequality $y \geq 3x + 1$.
    Graph the line $y = 3x + 1$.
    y-intercept $= (0, 1)$ **Constant term**
    slope $= 3$ or $\dfrac{3}{1}$ or $\dfrac{-3}{-1}$ **Coefficient of *x***

The boundary line is shown in Fig. 9–29 as a solid line because the inequality $y \geq 3x + 1$ *does* include the boundary.

The point $(0, 0)$ is not on the boundary line and it is on the right side of the boundary. We use $(0, 0)$ as our test point.

Is $y \geq 3x + 1$ when $x = 0$ and $y = 0$?   Substitute values.

$0 \geq 3(0) + 1$ Simplify.

$0 \geq 0 + 1$

$0 \geq 1$ False statement

Thus, as shown in Fig. 9–30, the boundary and all points on the side of the boundary opposite the test point $(0, 0)$ are in the solution set.

See Exercises 51–54.

FIGURE 9–30

TIP **General Observation About Inequalities** When the inequality is in the form $y < mx + b$ or $y \leq mx + b$, the solution set includes the portion of the y-axis and all points *below* the boundary line. When an inequality is in the form $y > mx + b$ or $y \geq mx + b$, the solution set includes the portion of the y-axis and all points *above* the boundary line (Fig. 9–31).

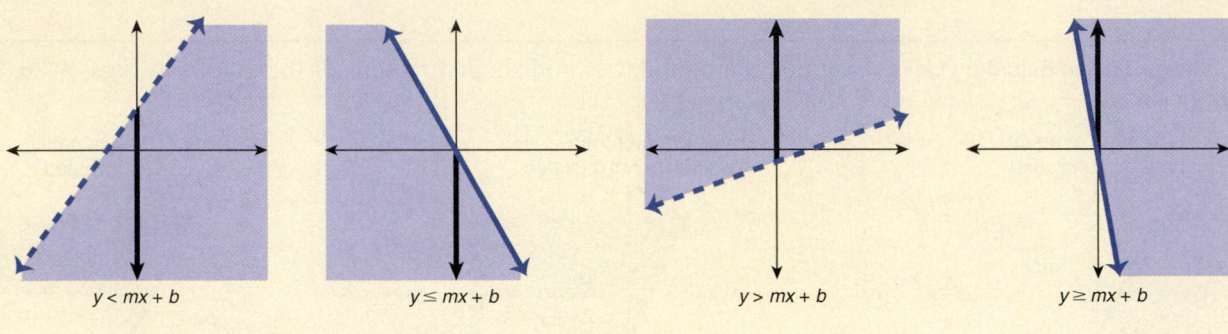

| $y < mx + b$ | $y \leq mx + b$ | $y > mx + b$ | $y \geq mx + b$ |

FIGURE 9–31

9–2 EXERCISES   **MyLab Math** For additional practice go to your study plan in MyLab Math.

1 Find the intercepts of the equations. *See Example 1.* Graph the equations using the intercepts procedure. *See Example 2.*

1. $5x - y = 15$

2. $2x + 3y = 9$

3. $x + y = 5$

4. $x + 3y = 5$

5. $\dfrac{1}{2}y = 4 + x$

6. $y = 3x - 1$

7. $5x = y + 2$

8. $x = -2y + 3$

See Example 3.

9. $y = -4x$

10. $y = -x$

2 Determine the slope and y-intercept by inspection. *See Example 4.*

11. $y = 4x + 3$

12. $y = -5x + 6$

13. $y = -\dfrac{7}{8}x - 3$

14. $y = 3$

Rewrite the following equations in slope-intercept form and determine the slope and y-intercept. *See Example 4.*

15. $4x - 2y = 10$

16. $2y = 5$

17. $\dfrac{1}{2}y + x = 3$

18. $2.1y - 4.2x = 10.5$

19. $2x = 8$

20. $x - 5 = 4$

Graph the equations using the slope-intercept procedure. *See Example 5.*

21. $y = 2x - 3$

22. $y = -\dfrac{1}{2}x - 2$

23. $y = -\dfrac{3}{5}x$

24. $x - 2y = 3$

25. $2x + y = 1$

26. $y = \dfrac{3}{4}x + 2$

3 Graph the equations using a graphing calculator and find the value of y as indicated. Use the appropriate viewing window. *See Example 6.*

27. $y = 0.3x + 0.5$

28. Find y when $x = -0.4$ in $y = 0.3x + 0.5$.

29. $y = -2.6x - 1.7$

30. Find y when $x = 1.3$ in $y = -2.6x - 1.7$.

31. $y = 212x + 757$

32. Find y when $x = 5$ in $y = 212x + 757$.

33. $y = 48x - 200$

34. Find y when $x = 7$ in $y = 48x - 200$.

35. **BUS** The cost of printing a magazine is $5,000 to typeset and prepare copy for printing and $3 per copy to print. This is expressed as an equation $y = 3x + 5,000$, where x is the number of copies printed and y is the total cost of printing x copies. Find the cost of printing 10,000 copies. Find the cost of printing 30,000 copies.

36. **BUS** Income is expressed by the equation $y = 8.30x$, where x is the number of hours per week an employee works. Find the income for an employee who works 40 hours in a week. Find the annual income if a person works for an entire year (52 weeks).

37. **HLTH/N** Some researchers say a person's maximum heart rate in beats per minute can be found by subtracting age from 220. This can be expressed as an equation $y = 220 - x$. What is the maximum heart rate for a person 25 years old? What is the maximum heart rate for a person 65 years old?

38. **BUS** An inventory shows 196 jigsaw puzzles in stock. These puzzles can be produced at a rate of 15 per hour. The number of puzzles on hand can be expressed as an equation $y = 15x + 196$, where x represents the number of hours of production and y represents the total inventory. How many puzzles are in inventory if they have been produced for 60 h?

4 Solve the following equations using a graphing calculator. *See Example 7.*

39. $3x + 2 = 7$

40. $5x - 4 = 2$

41. $4 = 6x - 1$

42. $3x + 7 = 4x + 6$

43. $3x - 1 = 4(x + 2)$

44. $2x - 8 = 4(x + 2)$

5 Graph the following linear inequalities using test points. *See Example 8.*

45. $x - y < 8$

46. $x + y > 4$

47. $3x + y < 2$

48. $x + 2y < 3$

49. $y > \dfrac{1}{2}x - 3$

50. $y > -\dfrac{3}{2}x + \dfrac{1}{2}$

See Example 9.

51. $2x + y \le 1$

52. $2x + 2y \ge 3$

53. $y \ge 2x - 3$

54. $y \ge -3x + 1$

9–3 Slope

LEARNING OUTCOMES

1 Calculate the slope of a line, given two points on the line.

2 Determine the slope of a horizontal or vertical line.

The concept of the *rate of change* is important in real-world applications. This concept is often referred to as *slope* and is used when we consider the slant of a roof or the grade of a roadway. We use the rate of change in applications, like changes in temperature, in the quality of a product, or in sales.

1 **Calculate the Slope of a Line, Given Two Points on the Line.** The *slope* of a straight line is the **rate of change** of the line or the ratio of the vertical rise of a line to the horizontal run of the line (see Fig. 9–32).

1. Evaluate $\dfrac{-2 - (-5)}{3 - (-2)}$.

Rate of change: the ratio of the vertical rise of a straight line to the horizontal run of the line. Also called *slope*

Rise: the vertical change; the difference in the y-coordinates

Run: the horizontal change; the difference in the x-coordinates

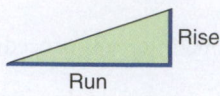

FIGURE 9–32

$$\text{slope} = \frac{\text{rise}}{\text{run}} \quad \text{or} \quad \frac{\text{vertical change}}{\text{horizontal change}}$$

> **To find the slope of a line from two given points on the line:**
>
> 1. Designate either point as point 1 with coordinates (x_1, y_1). Designate the other point as point 2 with coordinates (x_2, y_2).
>
> 2. Calculate the change (*difference*) in the y-coordinates to find the vertical **rise** $(y_2 - y_1)$ and the change in the x-coordinates to find the horizontal **run** $(x_2 - x_1)$.
>
> 3. Write a ratio of the rise to the run $\left(\frac{\text{rise}}{\text{run}}\right)$ and reduce the ratio to lowest terms. The slope can also be written in decimal notation if desired.
>
> Symbolically, if P_1 and P_2 are any two points on the line,
>
> $$\text{slope} = \frac{\Delta y}{\Delta x} = \frac{y_2 - y_1}{x_2 - x_1}$$
>
> where $P_1 = (x_1, y_1)$ and $P_2 = (x_2, y_2)$.
>   The Greek capital letter delta (Δ) is used to indicate a change and to write the slope definition symbolically.

STOP AND CHECK

1. Find the slope of a line if the points $(4, -3)$ and $(1, -2)$ are on the line.

Answer:

1. $-\dfrac{1}{3}$

EXAMPLE 1

Find the slope of a line if the points $(2, -1)$ and $(5, 3)$ are on the line (see Fig. 9–33).

Identify $(2, -1)$ as point 1 (P_1) and $(5, 3)$ as point 2 (P_2). The coordinates of P_1 are (x_1, y_1) and the coordinates of P_2 are (x_2, y_2). Write the slope definition symbolically and substitute known values.

$$\text{change in } y = \Delta y = \text{rise} = y_2 - y_1$$
$$= 3 - (-1)$$
$$\text{change in } x = \Delta x = \text{run} = x_2 - x_1$$
$$= 5 - 2$$

$$\text{slope} = \frac{\Delta y}{\Delta x} = \frac{\text{rise}}{\text{run}} = \frac{y_2 - y_1}{x_2 - x_1}$$

$$= \frac{3 - (-1)}{5 - 2} = \frac{4}{3}$$

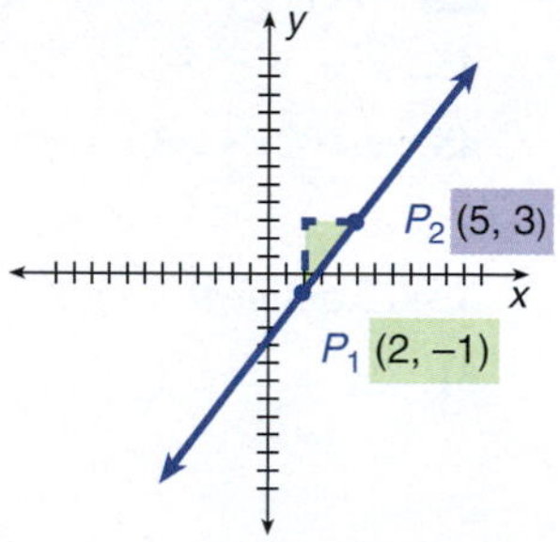

FIGURE 9–33

The slope of the line through points $(2, -1)$ and $(5, 3)$ is $\dfrac{4}{3}$.

See Exercises 1–11.

Any line has an infinite number of points. *Any* two points on the line can be used to find the slope. Also, the designation of P_1 and P_2 is not critical. Let's find the slope of the line in the preceding example by designating P_1 as $(5, 3)$ and P_2 as $(2, -1)$.

$$\Delta y = \text{rise} = y_2 - y_1 = -1 - 3 = -4$$
$$\Delta y = \text{run} = x_2 - x_1 = 2 - 5 = -3$$
$$\frac{\Delta y}{\Delta x} = \frac{\text{rise}}{\text{run}} = \frac{-4}{-3} = \frac{4}{3} \qquad \text{Note that the slope is the same.}$$

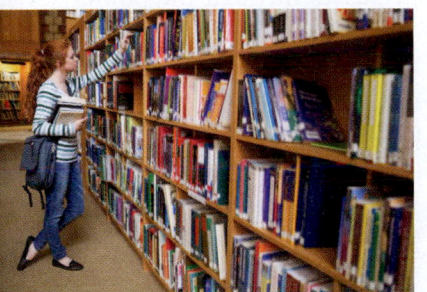

FIGURE 9-34

EXAMPLE 2

Find the slope of the line passing through the pair of points.

$(-5, -1)$ and $(3, -2)$

Visualize the line on a coordinate system (Fig. 9–34).

Let $P_1 = (-5, -1)$ and $P_2 = (3, -2)$.

$$\text{slope} = \frac{\Delta y}{\Delta x} = \frac{y_2 - y_1}{x_2 - x_1} = \frac{-2 - (-1)}{3 - (-5)} = \frac{-2 + 1}{3 + 5} = \frac{-1}{8} = -\frac{1}{8}$$

The slope is $-\dfrac{1}{8}$.

See Exercises 1–11.

Examine the slopes found in Examples 1 and 2 and the corresponding figures (Fig. 9–33 and Fig. 9–34). The first example has a positive slope of $\dfrac{4}{3}$. The graph of the line in Fig. 9–33 *rises* from left to right. The second example has a negative slope of $-\dfrac{1}{8}$. The graph of the line in Fig. 9–34 *falls* from left to right.

> **Lines with positive or negative slope:**
>
> ▶ A line that has a positive slope will *rise* as the line moves from left to right.
>
> ▶ A line that has a negative slope will *fall* as the line moves from left to right.

EXAMPLE 3

Table 9–1 on this page lists the annual cost in dollars of tuition and fees at public 2-year colleges for selected years. Find the rate of change in tuition and fees from the 2006–2007 academic year to the 2007–2008 academic year. Also, find the rate of change from 2007–2008 to 2012–2013. Use the initial year of the academic year as the reference.

Find the rate of change of tuition and fees from 2006–2007 to 2007–2008.

$(2006, 2,554)$ and $(2007, 2,649)$ Write the data as ordered pairs. Use the initial year of the academic year as the reference.

$$m = \frac{y_2 - y_1}{x_2 - x_1}$$ Use the slope formula to find the rate of change.

$$\text{rate of change} = \frac{2,649 - 2,554}{2007 - 2006}$$

$$\text{rate of change} = \frac{95}{1} = \textbf{\$95 per year}$$

Find the rate of change from 2007–2008 to 2012–2013.

$(2007, 2,649)$ and $(2012, 3,076)$

$$m = \frac{y_2 - y_1}{x_2 - x_1}$$

$$\text{rate of change} = \frac{3,076 - 2,649}{2012 - 2007}$$

$$\text{rate of change} = \frac{427}{5} = \textbf{\$85.40 per year}$$

Wavebreakmedia/Shutterstock

Table 9-1 Tuition and Fees at 2-Year Public Colleges Current Dollars

Academic Year	Tuition and Fees
02–03	$2,310
03–04	$2,330
04–05	$2,465
05–06	$2,567
06–07	$2,584
07–08	$2,649
08–09	$2,662
09–10	$2,668
10–11	$2,674
11–12	$2,909
12–13	$3,076

The rate of change in tuition for one year from 2006–2007 to 2007–2008 was $95. But the rate of change from 2007–2008 to 2012–2013 was $85.40 per year. This rate represents an average change per year over the 5-year period.   **See Exercises 12–18.**

2 **Determine the Slope of a Horizontal or Vertical Line.** Let's examine the slope of a horizontal line and a vertical line (Figs. 9–35 and 9–36). First look at the horizontal line that passes through the points (3, 2) and (−3, 2).

EXAMPLE 4

Find the slope of the horizontal line that passes through (3, 2) and (−3, 2).

Let $P_1 = $ (3, 2) and $P_2 = $ (−3, 2). See Fig. 9–35.

$$\text{slope} = \frac{\Delta y}{\Delta x} = \frac{y_2 - y_1}{x_2 - x_1} = \frac{2 - 2}{-3 - 3} = \frac{0}{-6} = 0$$

The horizontal line has a slope of 0.   **See Exercises 19–20.**

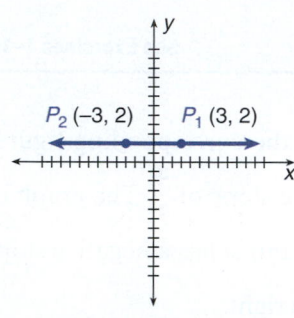

FIGURE 9–35

> **Slope of horizontal lines:**
>
> The slope of a horizontal line is zero.
>
> ▶ Two points are on the same horizontal line if their *y*-coordinates are equal.
>
> ▶ A horizontal line has no rise or no change in *y*-values.

Next, let's look at the slope of the vertical line that passes through the points (3, 2) and (3, −2).

EXAMPLE 5

Find the slope of the vertical line that passes through the points (3, 2) and (3, −2).

Let $P_1 = $ (3, 2) and $P_2 = $ (3, −2). See Fig. 9–36.

$$\text{slope} = \frac{\Delta y}{\Delta x} = \frac{y_2 - y_1}{x_2 - x_1} = \frac{-2 - 2}{3 - 3} = \frac{-4}{0}$$

Division by zero is undefined; **therefore, the slope of the vertical line passing through (3, 2) and (3, −2) is undefined.**

The vertical line has undefined slope.   **See Exercises 21–22.**

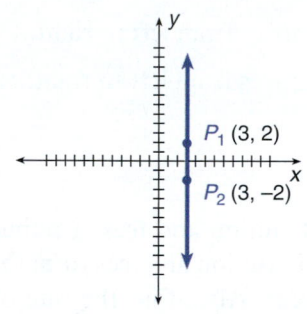

FIGURE 9–36

> **Slope of vertical lines:**
>
> The slope of a vertical line is undefined.
>
> ▶ Two points are on the same vertical line if their *x*-coordinates are equal.
>
> ▶ A vertical line has no run or no change in *x*-values.

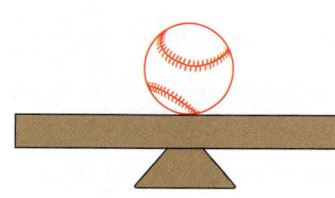

Zero slope

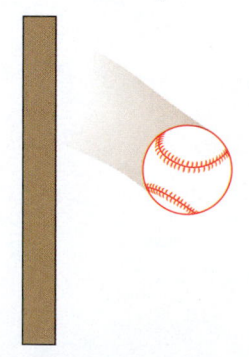

Undefined slope

FIGURE 9–37

TIP   **Zero Slope Versus Undefined Slope** A slope of zero and an undefined slope are *not* the same. Zero is a real number. A slope of zero is a real number. On the other hand, when zero is the denominator of a fraction, the number is undefined. As a memory aid, think of the number zero as a ball. It can be balanced on a horizontal line. It will not balance on a vertical line (Fig. 9–37).

Sketch the line passing through the pair of points and identify the line as having a positive slope, a negative slope, a zero slope, or an undefined slope.

1. $(-2, 1)$ and $(5, 1)$
2. $(5, 4)$ and $(6, -1)$
3. $(4, -2)$ and $(4, 3)$

Answers:
1.

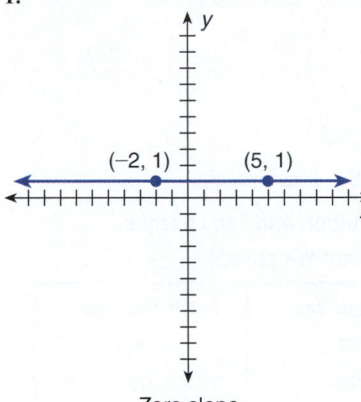

Zero slope

2.

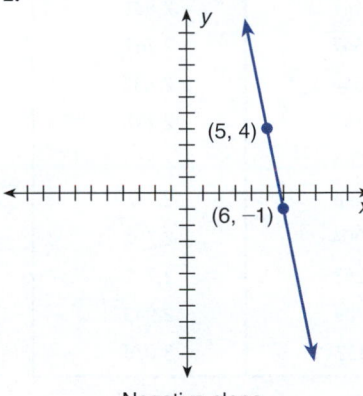

Negative slope

3.

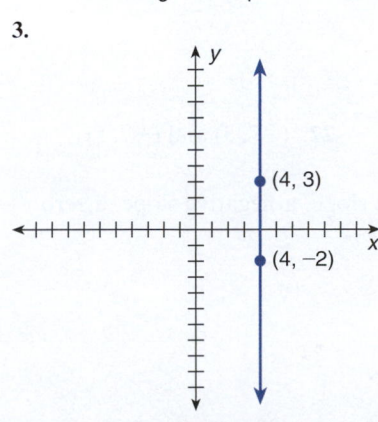

Undefined slope

To identify a horizontal or vertical line:

1. Examine the y-coordinates of two points on the line. If the y-coordinates are the same, the line is horizontal and the slope is 0.

2. Examine the x-coordinates of two points on the line. If the x-coordinates are the same, the line is vertical and the slope is undefined.

EXAMPLE 6

Sketch the line passing through the pair of points and identify the line as having a positive slope, a negative slope, a zero slope, or an undefined slope.

(a) $(-4, 2)$ and $(-4, -5)$　　(b) $(-2, -3)$ and $(1, 6)$　　(c) $(0, 7)$ and $(2, 5)$

(d) $(-6, -3)$ and $(5, -3)$

(a)

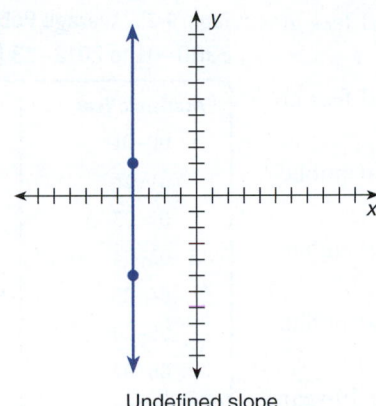

Undefined slope

FIGURE 9-38

(b)

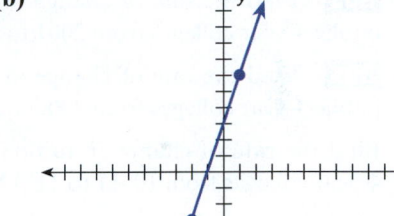

Positive slope

FIGURE 9-39

(c)

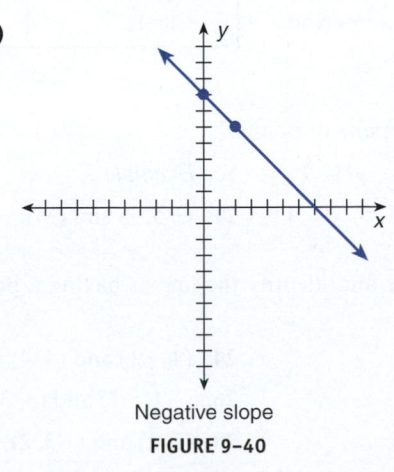

Negative slope

FIGURE 9-40

(d)

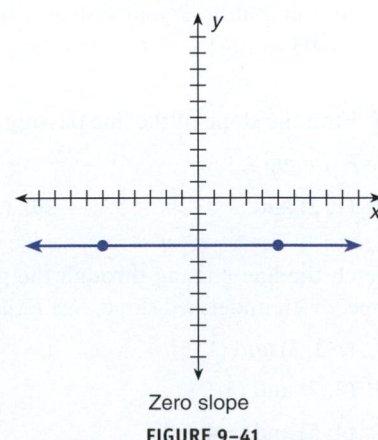

Zero slope

FIGURE 9-41

See Exercises 23–34.

9–3 EXERCISES   MyLab Math For additional practice go to your study plan in MyLab Math.

1 Find the slope of the line passing through the pairs of points. *See Examples 1 and 2.*

1. $(-3, 3)$ and $(1, 5)$

2. $(4, -1)$ and $(1, 4)$

3. $(3, 1)$ and $(5, 7)$

4. $(-1, -1)$ and $(3, 3)$

5. $(4, 5)$ and $(-4, -1)$

6. $(4, -5)$ and $(0, 0)$

7. $(2, -2)$ and $(5, -5)$

8. $(4, -4)$ and $(1, 5)$

9. **BUS** Dee Wallace's salary was $62,000 in 2005 and increased to $64,800 in 2007. What was the annual rate of change in his salary?

10. **BUS** If 100 backpacks cost $2,000 to produce and 800 backpacks cost $9,000 to produce, find the unit cost (slope) of producing the backpacks.

11. **AVIA** An airplane takes off from the ground (altitude is 0 ft and time is 0 min) and after 3 min it has an altitude of 6,000 ft. Find the rate of change in feet per minute.

Use Table 9–2 for Exercises 12–18. Use the initial year of the academic year as the reference. *See Example 3.*

12. **BUS** Find the rate of change in tuition and fees at public 4-year colleges from 2001 to 2011.

13. **BUS** Find the rate of change in tuition and fees at public 4-year colleges from 2006 to 2011.

14. Find the rate of change in tuition and fees at public 4-year colleges from 10–11 to 11–12.

15. Find the rate of change in tuition and fees at public 2-year colleges from 01–02 to 11–12.

16. Find the rate of change in tuition and fees at public 2-year colleges from 06–07 to 10–11.

17. The rate of change in tuition and fees for the 10-year period 01–02 to 11–12 is not the same as the rate of change from 06–07 to 10–11 for public 4-year colleges. Will the data graph to form a straight line? Explain.

18. Compare the rate of change in tuition and fees at public 4-year colleges with the rate of change in tuition and fees at public 2-year colleges for the 10-year period 02–03 to 12–13.

Table 9-2 Average Published Tuition and Fee Charges, 2000–01 to 2012–13 (Enrollment-Weighted)

Academic Year	Public Four-Year	Public Two-Year
00–01	4,102	2,483
01–02	4,285	2,613
02–03	4,407	2,650
03–04	4,447	2,631
04–05	4,547	2,747
05–06	4,644	2,835
06–07	4,742	2,791
07–08	4,766	2,847
08–09	4,796	2,777
09–10	5,004	2,710
10–11	5,263	2,741
11–12	5,729	2,943
12–13	6,132	3,076

2 Find the slope of the line passing through the pairs of points.

See Example 4.

19. $(7, 2)$ and $(-3, 2)$

20. $(3, 5)$ and $(-4, 5)$

See Example 5.

21. $(-5, 2)$ and $(-5, -4)$

22. $(-7, 3)$ and $(-7, 5)$

Sketch the line passing through the pair of points and identify the line as having a positive slope, a negative slope, a zero slope, or an undefined slope. *See Example 6.*

23. $(-3, 3)$ and $(1, 5)$

24. $(4, -1)$ and $(4, 4)$

25. $(3, 7)$ and $(5, 7)$

26. $(-1, -1)$ and $(3, 3)$

27. $(4, 5)$ and $(-4, -1)$

28. $(7, 2)$ and $(-3, 2)$

29. $(4, -5)$ and $(0, 0)$

30. $(5, -2)$ and $(5, -5)$

31. $(-5, 2)$ and $(-5, -4)$

32. $(4, -4)$ and $(1, 5)$

33. $(3, 5)$ and $(-4, 5)$

34. $(-7, 3)$ and $(-7, 5)$

9-4   Linear Equation of a Line

LEARNING OUTCOMES

1. Find the equation of a line, given the slope and one point.

2. Find the equation of a line, given two points on the line.

3. Find the equation of a line, given the slope and y-intercept.

4. Find the equation of a line, given a point on the line and an equation of a line parallel to that line.

5. Find the equation of a line, given a point on the line and an equation of a line perpendicular to that line.

LC LEARNING CATALYTICS

1. Identify the slope and y-intercept in the equation $y = -\dfrac{3}{8}x - 5$.

Point-slope form of a linear equation: a linear equation written in the form of $y - y_1 = m(x - x_1)$

STOP AND CHECK

1. Find the equation of the line passing through the point $(-1, 4)$ with a slope of $-\dfrac{3}{4}$. Write the equation in slope-intercept form.

Answer:

1. $y = -\dfrac{3}{4}x + \dfrac{13}{4}$

1 Find the Equation of a Line, Given the Slope and One Point. It is often desirable to know the equation of the line. The equation of the line can serve as a model for finding any point on the line and for solving various applications.

> **To find the equation of a line if the slope and one point on the line are known:**
>
> 1. Use the following version of the **point-slope form of a linear equation** of a straight line.
>
> $$y - y_1 = m(x - x_1)$$
>
> where x_1 and y_1 are coordinates of the known point, m is the slope of the line, and x and y are the variables of the equation.
>
> 2. Substitute known values for x_1, y_1, and m.
>
> 3. Rearrange the equation to be in standard form ($ax + by = c$) or slope-intercept form ($y = mx + b$).

EXAMPLE 1

Find the equation of the line passing through the point $(3, -2)$ with a slope of $\frac{2}{3}$. Write the equation in slope-intercept form.

$y - y_1 = m(x - x_1)$	Substitute into the point-slope form.
	$m = \dfrac{2}{3}; x_1 = 3; y_1 = -2$
$y - (-2) = \dfrac{2}{3}(x - 3)$	Simplify left side.
$y + 2 = \dfrac{2}{3}(x - 3)$	Distribute $\dfrac{2}{\cancel{3}} \cdot \dfrac{\overset{-1}{\cancel{-3}}}{1} = -2$.
$y + 2 = \dfrac{2}{3}x - 2$	Sort terms.
$y = \dfrac{2}{3}x - 2 - 2$	Combine like terms.
$y = \dfrac{2}{3}x - 4$	Slope-intercept form

See Exercises 1–4.

> **Standard form of a linear equation:**
>
> $$ax + by = c$$
>
> The characteristics of an equation in standard form are
>
> ▶ Both variable terms are on the left.
>
> ▶ The leading term has the x variable and is positive.
>
> ▶ The equation contains no fractions.

STOP AND CHECK

1. Write $y = -\dfrac{3}{4}x + \dfrac{13}{4}$ in standard form.

Answer:
1. $3x + 4y = 13$

EXAMPLE 2

Write $y = \dfrac{2}{3}x - 4$ in standard form.

$$y = \dfrac{2}{3}x - 4$$ Rearrange with variable terms on the left with the x-variable as the first term.

$$-\dfrac{2}{3}x + y = -4$$ Clear the fraction by multiplying by the denominator 3.

$$3\left(-\dfrac{2}{3}x + y\right) = 3(-4)$$ Distribute.

$$-2x + 3y = -12$$ Multiply by -1 to make the leading term positive.

$$\mathbf{2x - 3y = 12}$$ Standard form   **See Exercises 1–4.**

---

2 **Find the Equation of a Line, Given Two Points on the Line.** To find the equation of a line when two points on the line are known, we use both the slope and the point-slope formulas.

> **To find the equation of a line if two points on the line are known:**
>
> 1. Use the slope formula to find the slope, given two points.
>
> $$m = \dfrac{y_2 - y_1}{x_2 - x_1}$$
>
> 2. Use the point-slope form of an equation of a straight line, the calculated slope from Step 1, and the coordinates of either one of the given points.
>
> $$y - y_1 = m(x - x_1)$$
>
> 3. Write the equation in slope-intercept or standard form.

STOP AND CHECK

1. Find the equation of the line that passes through the points $(0, -3)$ and $(4, 0)$. Write the equation in slope-intercept form.

Answer:

1. $y = \dfrac{3}{4}x - 3$

EXAMPLE 3

Find the equation of the line that passes through the points $(0, 8)$ and $(5, 0)$. Write the equation in slope-intercept form.

$$\text{slope} = \dfrac{\Delta y}{\Delta x} = \dfrac{y_2 - y_1}{x_2 - x_1} = \dfrac{0 - 8}{5 - 0} = -\dfrac{8}{5}$$ Substitute known values into the slope formula. Let $P_1 = (0, 8)$ and $P_2 = (5, 0)$.

$$y - y_1 = m(x - x_1)$$ Substitute the slope and values for one point into the point-slope form. Use P_1.

$$y - 8 = -\dfrac{8}{5}(x - 0)$$ Simplify.

$$y - 8 = -\dfrac{8}{5}x$$ Solve for y.

$$y = -\dfrac{8}{5}x + 8$$ Slope-intercept form

See Exercises 5–6.

EXAMPLE 4

Find the equation of the line that passes through the points $(-3, 4)$ and $(7, 4)$.

We let $P_1 = (-3, 4)$ and $P_2 = (7, 4)$.

$$\text{slope} = \dfrac{\Delta y}{\Delta x} = \dfrac{y_2 - y_1}{x_2 - x_1} = \dfrac{4 - 4}{7 - (-3)} = \dfrac{0}{7 + 3} = \dfrac{0}{10} = 0$$ Substitute values.

Use the point-slope form of an equation and P_1.

$y - y_1 = m(x - x_1)$	Substitute values.
$y - 4 = 0[x - (-3)]$	Simplify.
$y - 4 = 0$	Solve for y.
$y = 4$	**See Exercises 7–8.**

The situation in the preceding example is generalized in the equation of a horizontal line.

> **Equation of a horizontal line:**
>
> The equation of a line with a slope of zero (a horizontal line) is
>
> $$y = k$$
>
> where k is the common y-coordinate for all points on the line.

There is one special situation in which the point-slope form of an equation *cannot* be used to determine the equation of a line, that is, when the slope is undefined.

> **Equation of a vertical line:**
>
> The equation of a line with a slope that is undefined (a vertical line) is
>
> $$x = k$$
>
> where k is the common x-coordinate for all points on the line.

STOP AND CHECK

Find the equation of the line that passes through the given points.
1. $(-2, 4)$ and $(3, 4)$
2. $(-5, 3)$ and $(-5, -1)$

Answers:
1. $y = 4$ 2. $x = -5$

EXAMPLE 5

Find the equation of the line that passes through the points $(5, 3)$ and $(5, 7)$.

Since the x-coordinate is the same in both points, the slope of the line is undefined. The equation of the line is $x = k$, where k is the common x-coordinate. In this example, $k = 5$.

The equation of the line through $(5, 3)$ and $(5, 7)$ is $x = 5$.   **See Exercises 9–10.**

3 **Find the Equation of a Line, Given the Slope and y-Intercept.** So far we have written the equation of a line when we know either two points on the line or the slope and at least one point on the line. If the one point is the y-intercept $(0, b)$, then we use the slope-intercept form of an equation, $y = mx + b$, and we can write the equation by inspection.

> **To find the equation of a line if the slope and y-intercept are known:**
>
> 1. Use the slope-intercept form of an equation of a straight line.
>
> $$y = mx + b$$
>
> where $m =$ slope and $b =$ the y-coordinate of the y-intercept.
>
> 2. Substitute known values for m and b.
>
> 3. Write the equation in standard form or leave the equation in slope-intercept form.

1. Write the equation for a line with a slope of $\frac{2}{3}$ and a y-intercept of $(0, 4)$.

Answer:

1. $y = \frac{2}{3}x + 4$

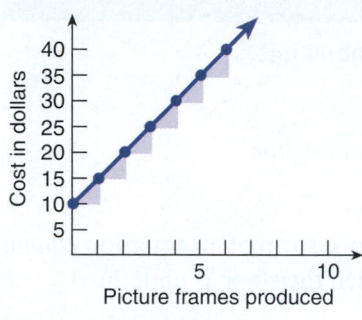

FIGURE 9–42

EXAMPLE 6

Write the equation for a line with a slope of -3 and a y-intercept of $(0, 5)$.

Slope $= m = -3$; y-intercept $= b = 5$

$y = mx + b$ Substitute values.

$y = -3x + 5$ Slope-intercept form   **See Exercises 13–20.**

The necessary facts for writing the equation of a line are often obtained from the graph of the equation.

EXAMPLE 7

Fig. 9–42 shows the cost of producing picture frames.

(a) Write an equation that represents the graph.

(b) Use the equation to find the cost of producing 20 picture frames.

(a) The y-intercept represents the fixed cost of producing picture frames. $b = \$10$. The slope m is 5 units vertical change for every 1 unit horizontal change, which is $\frac{5}{1} = 5$. Notice the different scales used for the x- and y-axis.

$$y = 5x + 10 \quad \text{Equation of the line}$$

This type of equation is a *cost function*.

(b) Use the equation of the line found in part (a) to find y when $x = 20$.

$y = 5x + 10$ Substitute $x = 20$.

$y = 5(20) + 10$ Evaluate.

$y = 100 + 10$

$y = 110$ Interpret.

The cost of producing 20 frames is \$110.   **See Exercises 21–25.**

1. Identify the slope and y-intercept and write the equation of the line on the given graph.

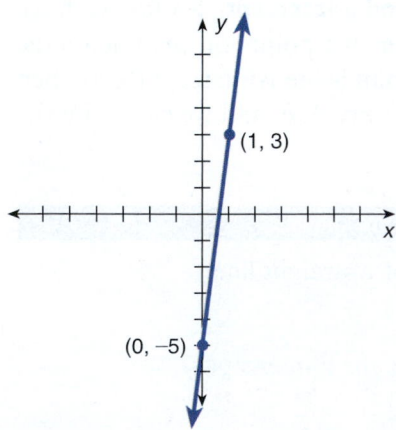

Answer:

1. Slope $= 8$; y-intercept $= (0, -5)$; $y = 8x - 5$

EXAMPLE 8

Identify the slope and y-intercept and write the equation of the line that is graphed in Figure 9–43.

y-intercept $= \mathbf{(0, -3)}$

Another point $= (-4, 0)$

Slope $= \dfrac{-3 - 0}{0 - (-4)} = \dfrac{-3}{4} = -\dfrac{3}{4}$

$y = mx + b$

Substitute values into slope-intercept form.

$y = -\dfrac{3}{4}x - 3$

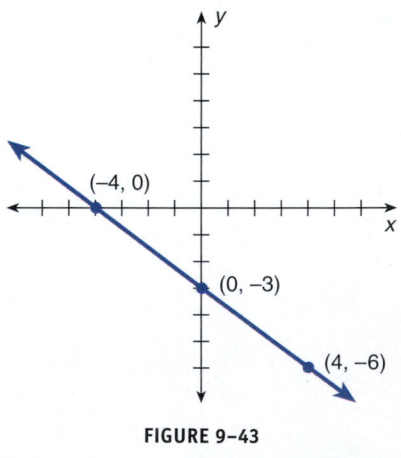

FIGURE 9–43

See Exercises 26–28.

4   **Find the Equation of a Line, Given a Point on the Line and an Equation of a Line Parallel to That Line.** **Parallel lines** are two or more lines that are the same distance apart everywhere. They have no points in common. See Fig. 9–44.

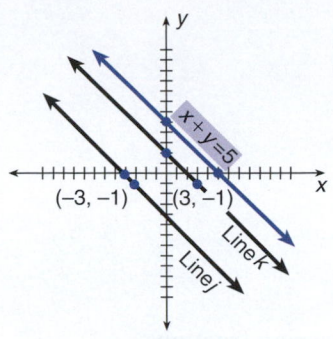

FIGURE 9-44

Parallel lines: two or more lines in the same plane that are the same distance apart everywhere. They have no points in common

Coincides: when two lines are positioned one on top of the other. They are the same line

If we write $x + y = 5$ in slope-intercept form, we have $y = -x + 5$ and the slope is -1. *Any* equation that has a slope of -1 either is parallel to or coincides with the line formed by the equation $x + y = 5$. **Coincides** means one line lies on top of the other; that is, they are the same line.

Although parallel lines have the same slopes, they have different x- and y-intercepts.

> **Slopes of parallel lines:**
>
> The slopes of parallel lines are equal.

EXAMPLE 9

Write the equations for lines j and k in Fig. 9-44.

$x + y = 5$ Solve the given equation for y to determine the slope of all the parallel lines.

$y = -x + 5$ Slope $= -1$.

The y-intercepts for lines j and k can be determined from the graph.

Line j: $y = -x - 4$ Substitute $m = -1$ and $b = -4$.

Line k: $y = -x + 2$ Substitute $m = -1$ and $b = 2$.   **See Exercises 29–40.**

> **To find the equation of a line that is parallel to a given line when at least one point on the parallel line is known:**
>
> 1. Solve the equation of the given line for y.
>
> 2. Determine the slope m from the equation in Step 1.
>
> 3. The slope of the parallel line is the same as the slope of the given line.
>
> 4. Use the point-slope form of a straight line $y - y_1 = m(x - x_1)$ and substitute values for m, x_1, and y_1.
>
> 5. Rearrange the equation to be in standard form ($ax + by = c$) or slope-intercept form ($y = mx + b$).

EXAMPLE 10

Write the equation of a line that is parallel to $2y = 3x + 8$ and passes through the point (2, 1). Write the new equation in slope-intercept form and in standard form.

$2y = 3x + 8$ Write the original equation in slope-intercept form.

$\dfrac{2y}{2} = \dfrac{3x + 8}{2}$

$y = \dfrac{3}{2}x + 4$ Identify the slope.

The slope of $2y = 3x + 8$ is $\dfrac{3}{2}$.

$y - y_1 = m(x - x_1)$ Substitute into the point-slope form. The slope of a parallel line is equal to the slope of the given line.

$m = \dfrac{3}{2}; x_1 = 2; y_1 = 1$

$$y - 1 = \frac{3}{2}(x - 2) \qquad \text{Distribute.}$$

$$y - 1 = \frac{3}{2}x - 3 \qquad \text{Solve for } y.$$

$$y = \frac{3}{2}x - 3 + 1 \qquad \text{Combine like terms.}$$

$$y = \frac{3}{2}x - 2 \qquad \text{Slope-intercept form}$$

To write the equation in standard form, begin by clearing fractions.

$$2y = 2\left(\frac{3}{2}x - 2\right) \qquad \text{To clear fractions, multiply by 2.}$$

$$2y = 3x - 4 \qquad \text{Move variable term 3x to the left side of the equation.}$$

$$-3x + 2y = -4 \qquad \text{Multiply each term by } -1 \text{ so that the coefficient of } x \text{ is positive.}$$

$$3x - 2y = 4 \qquad \text{Standard form} \qquad \qquad \text{See Exercises 29–40.}$$

5 **Find the Equation of a Line, Given a Point on the Line and an Equation of a Line Perpendicular to That Line.** **Perpendicular lines** are two lines that intersect to form right angles (90° angles). Another term for *perpendicular* is **normal.** In Fig. 9–45, the perpendicular (or normal) line that passes through the given point is indicated by the symbol ⌐, which means that a right angle is formed by the lines.

Perpendicular lines: two lines that intersect to form right angles (90° angles). Another term for *perpendicular* is *normal*

Normal: another term for *perpendicular*

To explore the relationship of the slopes of perpendicular lines, examine the graphs and equations in Fig. 9–45. The slopes of the two lines on the left are 2 and $-\frac{1}{2}$, whereas the slopes of the two lines on the right are $\frac{2}{3}$ and $-\frac{3}{2}$.

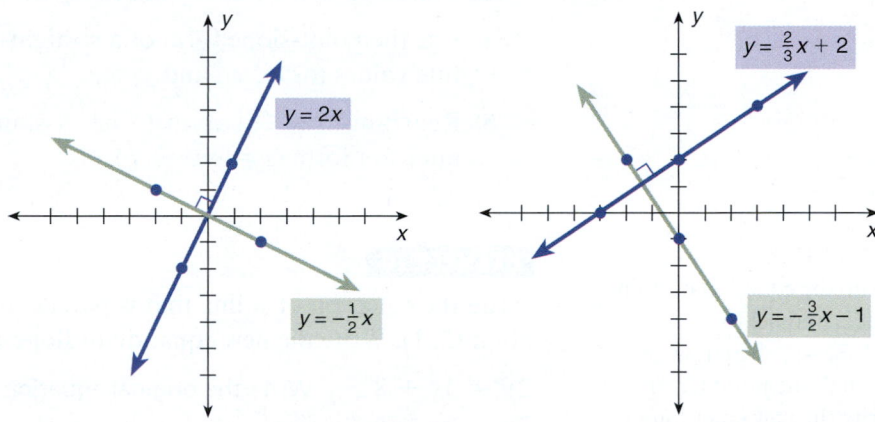

FIGURE 9–45

Negative reciprocal: the opposite and the reciprocal of a number

The slope of *any* line that is perpendicular to a given line is the **negative reciprocal** of the slope of the given line.

> **TIP** **Negative Reciprocals** The negative reciprocal of a given number is not necessarily a negative value. It has the *opposite* sign of the given number. That is, it is the opposite *and* the reciprocal.
>
> To find the negative reciprocal of a number:
>
> 1. Interchange the numerator and denominator to form the reciprocal.
>
> 2. Give the reciprocal the sign opposite the sign of the original number.

The negative reciprocal of -5 is $+\dfrac{1}{5}$. The negative reciprocal of $-\dfrac{3}{4}$ is $+\dfrac{4}{3}$.

The negative reciprocal of $\dfrac{4}{5}$ is $-\dfrac{5}{4}$. The negative reciprocal of 3 is $-\dfrac{1}{3}$.

To find the equation of a line that is perpendicular (normal) to a given line and passes through a given point:

1. Determine the slope of the *given* line.

2. Find the *negative reciprocal* of this slope. The negative reciprocal is the slope of the perpendicular line.

3. Write the equation for the perpendicular line by substituting the coordinates of the *given* point for x_1 and y_1 and the slope of the *perpendicular* line for m into the point-slope form of the equation $y - y_1 = m(x - x_1)$.

4. Write the equation in slope-intercept or standard form.

TIP **Words as Subscripts** In the next example, we clarify the notation with subscripts. The phrase *slope of a given line* is notated as "slope$_{given}$." The *slope of the perpendicular line* is notated as "slope$_{perpendicular}$."

STOP AND CHECK

1. Find the equation of a line that is perpendicular to $x - 3y = -9$ and passes through the point $(2, 5)$. Write the new equation in slope-intercept form and in standard form.

Answer:

1. $y = -3x + 11; 3x + y = 11$

EXAMPLE 11

Find the equation of the line that is perpendicular to $4x + y = -3$ and passes through $(0, -3)$. Write the new equation in slope-intercept form and in standard form.

$$4x + y = -3 \quad \text{Given equation. Solve for } y.$$
$$y = -4x - 3 \quad \text{Identify the slope.}$$
$$\text{slope}_{given} = -4$$
$$\text{slope}_{perpendicular} = +\frac{1}{4} \quad \text{Negative reciprocal of } -4$$
$$y - y_1 = m(x - x_1) \quad \text{Point-slope form. Substitute } m = \tfrac{1}{4}, P_1 = (0, -3).$$
$$y - (-3) = \frac{1}{4}(x - 0) \quad \text{Simplify.}$$
$$y + 3 = \frac{1}{4}x \quad \text{Solve for } y.$$
$$y = \frac{1}{4}x - 3 \quad \text{Slope-intercept form. For standard form clear the fraction and rearrange.}$$

or

$$4y = x - 12 \quad \text{Move the } x\text{-variable term to the left and make the leading coefficient positive.}$$
$$x - 4y = 12 \quad \text{Standard form} \qquad \text{See Exercises 41–50.}$$

EXAMPLE 12

Which of the equations represents a line that is perpendicular to $2x - 3y = 1$ passing through $(2, -1)$?

(a) $y = \dfrac{2}{3}x - \dfrac{1}{3}$ (b) $y = \dfrac{2}{3}x - \dfrac{7}{3}$

(c) $y = -\dfrac{3}{2}x - \dfrac{1}{3}$ **(d)** $y = -\dfrac{3}{2}x + 2$

$$2x - 3y = 1$$ Rewrite the given equation in slope-intercept form, that is, solve for y.

$$-3y = -2x + 1$$

$$\frac{-3y}{-3} = \frac{-2x}{-3} + \frac{1}{-3}$$

$$y = \frac{2}{3}x - \frac{1}{3}$$ Slope-intercept form of given equation

$$\text{slope}_{\text{given}} = \frac{2}{3}$$ From $y = \dfrac{2}{3}x - \dfrac{1}{3}$

$$\text{slope}_{\text{perpendicular}} = -\frac{3}{2}$$ Negative reciprocal of given slope

The choice is now limited to (c) or (d) since (a) and (b) have slopes of $\dfrac{2}{3}$.

$$y - y_1 = m(x - x_1)$$ Substitute $m = -\dfrac{3}{2}$, $x_1 = 2$, $y_1 = -1$.

$$y - (-1) = -\frac{3}{2}(x - 2)$$ Distribute and simplify.

$$y + 1 = -\frac{3}{2}x + 3$$ Solve for y.

$$y = -\frac{3}{2}x + 3 - 1$$

$$y = -\frac{3}{2}x + 2$$

The correct equation is $y = -\dfrac{3}{2}x + 2$**, or choice (d).**

See Exercises 41–50.

9–4 EXERCISES

MyLab Math For additional practice go to your study plan in MyLab Math.

1 Find the equation of a line passing through the given point with the given slope. Solve the equation for y when necessary. Write in both slope-intercept and standard form. *See Examples 1 and 2.*

1. $(-8, 3)$, $m = \dfrac{2}{3}$ **2.** $(4, 1)$, $m = -\dfrac{1}{2}$ **3.** $(-3, -5)$, $m = 2$ **4.** $(0, -1)$, $m = 1$

2 Find the equation of a line passing through the given pairs of points. Solve the equation for y. *See Example 3.*

5. $(4, 6)$ and $(7, 1)$ **6.** $(-2, -4)$ and $(5, 10)$

See Example 4.   *See Example 5.*

7. $(-1, -3)$ and $(3, -3)$ **8.** $(-4, 0)$ and $(6, 0)$ **9.** $(-1, 6)$ and $(-1, 4)$ **10.** $(-4, 4)$ and $(-4, -2)$

11. **BUS** If 50 DVDs cost $2,185 to produce and 2,000 DVDs cost $11,935 to produce, write a cost function for producing DVDs.

12. **BUS** The first 5,000 leather bags can be produced at a cost of $160,000, and 50,000 bags can be produced at a cost of $1,150,000. Write a cost function for producing these bags.

3 Write the equation of the line with the given slope and *y*-intercept. *See Example 6.*

13. $m = \dfrac{1}{4}, b = 7$

14. $m = -8, b = -4$

15. $m = -2, b = 3$

16. $m = \dfrac{3}{5}, b = -2$

17. $m = 1, b = 0$

18. $m = 5, b = -\dfrac{1}{5}$

19. $m = 2, b = -2$

20. $m = -\dfrac{3}{4}, b = 0$

See Example 7.

21. **BUS**   A local business rents computer time for a $3 setup charge and $0.20 for every minute the computer is used. Write an equation to represent the cost *y* of using the computer for *x* minutes. How much does 45 min of computer time cost?

22. **BUS**   The cost of producing fine china plates is $8 per plate plus a one-time equipment charge of $12,000. Write an equation that represents the total cost *y* of producing *x* plates. What is the total cost of producing 8,000 plates?

Solve Exercises 23 and 24 using the information given in Fig. 9–46, which shows the cost of producing widgets in dollars.

23. **INDTEC**   Write an equation that represents the cost of producing widgets shown in the graph.

24. **INDTEC**   Use the equation to find the cost of producing 10 widgets.

25. **AUTO**   A snowplow has a maximum speed of 40 mi/h per hour on a dry, flat road surface. Its speed decreases 0.9 mi/h per hour for every inch of snow on the highway. Write an equation that represents the speed *y* of the snowplow when moving *x* inches of snow. What is the maximum speed of the plow in moving 10 in. of snow?

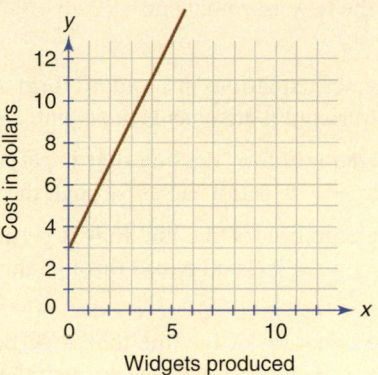

FIGURE 9–46

Identify the slope and *y*-intercept to write the equations of the lines that are graphed in Figs. 9–47 through 9–49. *See Example 8.*

26.

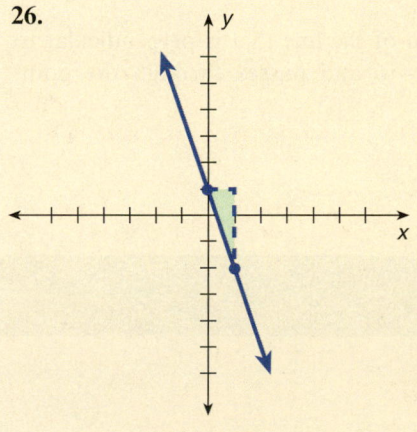

FIGURE 9–47

27.

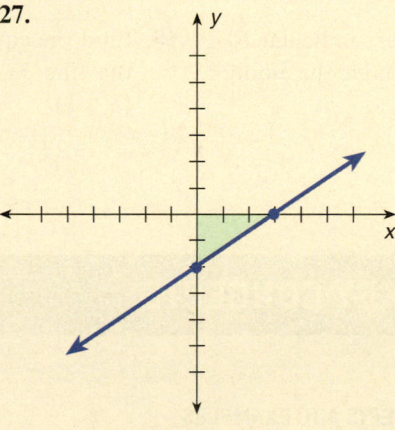

FIGURE 9–48

28.

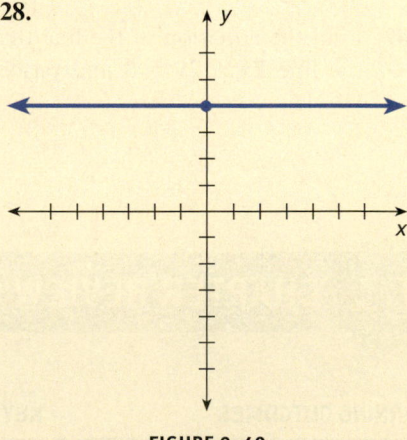

FIGURE 9–49

4 Write the new equations in standard form. Verify your answers using a graphing calculator. *See Examples 9 and 10.*

29. Find the equation of the line that is parallel to the line $x + y = 6$ and passes through the point $(2, -3)$.

30. Find the equation of the line that is parallel to the line $2x + y = 5$ and passes through the point $(1, 7)$.

31. Find the equation of the line that is parallel to the line $3y = x - 2$ and passes through the point $(4, 0)$.

32. Find the equation of the line that is parallel to the line $3x - y = -2$ and passes through the point $(-3, -2)$.

33. Find the equation of the line that is parallel to the line $x + 3y = 7$ and passes through the point $(4, 1)$.

34. Find the equation of the line that is parallel to the line $3x - y = 4$ and passes through the point $(0, 3)$.

35. Find the equation of the line that is parallel to the line $2x + 3y = 5$ and passes through the point $(1, 1)$.

36. Find the equation of the line that is parallel to the line $3x + 2y = 1$ and passes through the point $(2, 0)$.

37. Find the equation of the line that is parallel to the line $2x - 5y = 0$ and passes through the point $(3, -1)$.

38. Find the equation of the line that is parallel to the line $-3x + 4y = -1$ and passes through the point $\left(\frac{1}{2}, 0\right)$.

39. **BUS**   The cost of producing tires on an assembly line includes a fixed cost of \$45,000 and a variable cost of \$12 per tire. This can be written as a cost function $y = 12x + 45{,}000$, where x is the number of tires produced and y is the total cost of x tires. Research shows that 10,000 tires can be produced on a newly installed assembly line at a total cost of \$140,000 with the same variable cost per tire as the original assembly line. Write an equation that represents the total cost of producing tires on the new assembly line.

40. **BUS**   A car rental agency uses the function $y = 0.5x + 25$, where x is the number of miles driven and y is the total cost of car rental. Another car rental agency has the same variable cost of \$0.50 per mile and quotes the total cost of driving 500 mi as \$295. Write an equation to represent the second agency's rental charges.

5 Write the new equations in standard form. Verify your results using a graphing calculator. Be sure the range is set so the vertical and horizontal increments are equal. *See Examples 11 and 12.*

41. Find the equation of the line that is perpendicular to the line $x + y = 6$ and passes through the point $(2, 3)$.

42. Find the equation of the line that is perpendicular to the line $2x + y = 5$ and passes through the point $(1, 7)$.

43. Find the equation of the line that is perpendicular to the line $3y = x - 2$ and passes through the point $(4, 0)$.

44. Find the equation of the line that is perpendicular to the line $3x - y = 2$ and passes through the point $(-3, -2)$.

45. Find the equation of the line that is perpendicular to the line $x + 3y = 7$ and passes through the point $(4, 1)$.

46. Find the equation of the line that is perpendicular to the line $x + 2y = 7$ and passes through the point $(-2, 3)$.

47. Find the equation of the line that is perpendicular to the line $2x + 3y = 4$ and passes through the point $(3, -1)$.

48. Find the equation of the line that is perpendicular to the line $4x + y = 1$ and passes through the point $(0, 0)$.

49. Find the equation of the line that is perpendicular to the line $2x + 2y = 3$ and passes through the point $\left(\frac{1}{2}, 2\right)$.

50. Find the equation of the line that is perpendicular to the line $5x - y = 6$ and passes through the point $\left(5, -\frac{1}{5}\right)$.

---

9 | CHAPTER REVIEW OF KEY CONCEPTS

LEARNING OUTCOMES	KEY CONCEPTS AND EXAMPLES

Section 9–1

1 Locate points on a rectangular coordinate system (pp. 416–419).

Plot a point on the rectangular coordinate system: 1. Start at the origin.   **2.** Count to the left or right the number of units of the first signed number (x-coordinate) in the ordered pair. **3.** From the ending point of Step 2, count up or down the number of units of the second signed number (y-coordinate).   **4.** Place a dot to show the point and write the coordinates beside the point.

LEARNING OUTCOMES	KEY CONCEPTS AND EXAMPLES

Draw a coordinate system and locate the following points: $A(-3, -2)$, $B(4, -1)$, $C(0, -3)$, $D(2, 3)$. See Fig. 9–50.

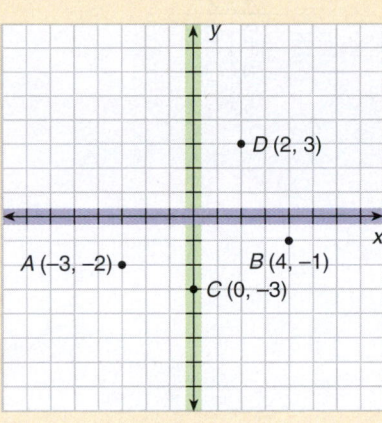

FIGURE 9–50

2 Represent an equation in two variables as a function (pp. 419–420).

1. Solve the equation in two variables for the dependent variable y.   **2.** Rewrite y as $f(x)$.

Rewrite $3x - y = -2$ as a function.

$3x - y = -2$ Solve for y.

$-y = -3x - 2$ Multiply entire equation by -1.

$y = 3x + 2$ Rewrite y as $f(x)$.

$f(x) = 3x + 2$

3 Make a table of solutions for a linear equation or function (pp. 420–422).

1. Solve the equation for y and write the equation in function notation, if desired.   **2.** Select appropriate values for the independent variable.   **3.** Perform the calculations associated with the function for each selected value of the independent variable.   **4.** Write the results of each calculation as an ordered pair.

Make a table of solutions for the equation $y = -x + 3$. The domain is all real numbers.

When $x = -2$ When $x = 0$ When $x = 2$

$y = -(-2) + 3$ $y = -0 + 3$ $y = -2 + 3$

$y = 2 + 3$ $y = 3$ $y = 1$

$y = 5$

x	y
-2	5
0	3
2	1

Each solution is an ordered pair that can be written in point notation: $(-2, 5)$, $(0, 3)$, and $(2, 1)$. The range is all real numbers.

4 Graph a linear equation or function using a table of solutions (pp. 422–425).

1. Prepare a table of solutions by evaluating the equation or function using different values (at least three) in the domain of the independent variable.   **2.** Plot the points from the table of solutions on a rectangular coordinate system.   **3.** Connect the points with a straight line. Extend the graph beyond the three points and place an arrow on each end as appropriate to indicate that the line extends indefinitely.

LEARNING OUTCOMES   **KEY CONCEPTS AND EXAMPLES**

Prepare a table of values and graph $y = -x + 3$ (Fig. 9–51).

x	y
-2	5
0	3
2	1

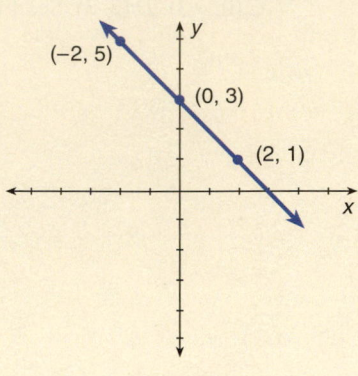

FIGURE 9–51

Check a solution of an equation in two variables: 1. Substitute the values from the ordered pair for the respective variables in the equation.   **2.** Evaluate each side of the equation using the order of operations.   **3.** The two sides of the equation will be equal if the ordered pair is a solution of the equation.

A solution for the equation $3x - y = 1$ is (1, 2). Check to verify the solution.

$3x - y = 1$ Substitute values.
$3(1) - 2 = 1$ Evaluate.
$3 - 2 = 1$
$1 = 1$ True. Solution checks.

Find a specific solution of an equation in two variables when given the value of one variable: 1. Substitute the given value of the variable in each place that variable occurs in the equation.
2. Solve the equation for the other variable.

Evaluate the equation $4x - y = 6$ for x, if $y = 14$.

$4x - y = 6$ Substitute 14 for y.
$4x - 14 = 6$ Solve for x.
$4x = 20$
$x = 5$
Solution: (5, 14)

Section 9–2

1 Graph linear equations using intercepts (pp. 427–429).

Find the intercepts of a linear equation: 1. Find the x-intercept; let $y = 0$ and solve for x.
2. Find the y-intercept; let $x = 0$ and solve for y.

Graph linear equations by the intercepts method: 1. Find the x- and y-intercepts.   **2.** Plot the intercepts on a rectangular coordinate system.   **3.** Draw the line through the two points and extend it beyond each point.   **4.** Check by examining one additional solution of the equation.

LEARNING OUTCOMES	KEY CONCEPTS AND EXAMPLES

Graph the equation $y = 2x - 1$ by using the intercepts method.

x-intercept: $0 = 2x - 1$ **Substitute $y = 0$.** y-intercept: $y = 2(0) - 1$ **Substitute $x = 0$.**

$\qquad\qquad 1 = 2x$ **Solve for x.** $\qquad\qquad\qquad\quad y = 0 - 1$ **Solve for y.**

$\qquad\qquad \dfrac{1}{2} = x$ $\qquad\qquad\qquad\qquad\qquad\qquad\quad y = -1$

$\left(\dfrac{1}{2}, 0 \right)$ $\qquad\qquad\qquad\qquad\qquad\qquad\qquad\qquad (0, -1)$

Plot the two intercepts: then draw the graph (Fig. 9–52).

Check point: For $x = 3$

$y = 2x - 1$

$y = 2(3) - 1$

$y = 6 - 1$

$y = 5$

$(3, 5)$

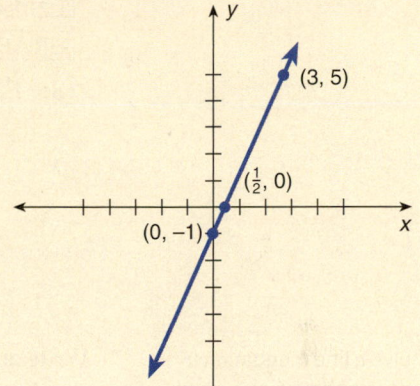

FIGURE 9–52

2 Graph linear equations using the slope and y-intercept (pp. 429–433).

Determine the slope and y-intercept of a linear equation by inspection: 1. Solve the equation for y and write it in the form $y = mx + b$.   **2.** The *slope* is m, which is the coefficient of x.   **3.** The y-coordinate of the y-intercept is b. The y-intercept is $(0, b)$.

Graph a linear equation in the form $y = mx + b$ using the slope-intercept method: 1. Locate the y-intercept on the y-axis.   **2.** Using the slope in fractional form, determine the amount of vertical and horizontal movement indicated.   **3.** From the y-intercept, locate additional points on the graph of the equation by counting the indicated vertical and horizontal movements of the slope.   **4.** Draw the line connecting the points and extending beyond the points.

Use the slope-intercept method to graph $y = 2x - 1$. The y-coordinate of the y-intercept is -1, so count down 1 from the origin. The coordinates of this point are $(0, -1)$. From this point, move $+2$ vertically and $+1$ horizontally (or -2 vertically and -1 horizontally). The coordinates of the second point are $(1, 1)$. Connect the two points (Fig. 9–53).

y-intercept $= (0, -1)$

$m = 2 = \dfrac{2}{1}$ or $\dfrac{-2}{-1}$

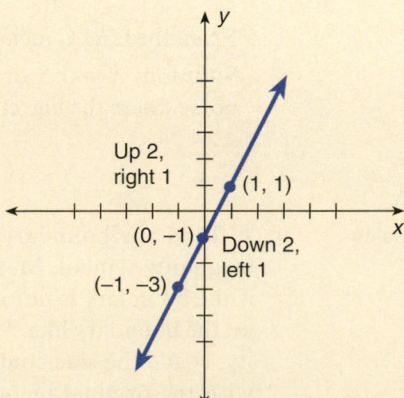

FIGURE 9–53

LEARNING OUTCOMES	KEY CONCEPTS AND EXAMPLES

3 Graph linear equations using a graphing calculator (pp. 433–434).

1. Solve the equation for y. **2.** Press the function key $\boxed{Y=}$; then enter the right side of the equation. **3.** Press the Graph key to show the graph on the screen. Adjust the view with the Window or Zoom key if appropriate.

Graph the equation $y = 2x - 1$.

Using a TI-84 Plus Silver Edition™:

$\boxed{Y=}$ $\boxed{\text{CLEAR}}$ 2 $\boxed{X, T, \theta, n}$ $\boxed{-}$ 1

$\boxed{\text{GRAPH}}$

(See Fig. 9–54.)

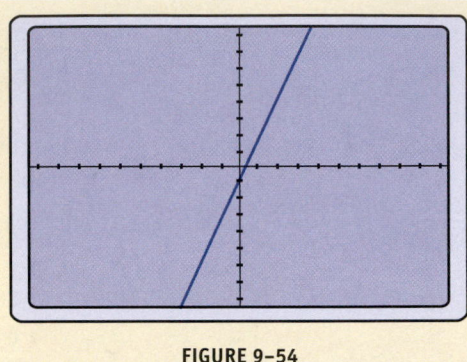

FIGURE 9–54

4 Solve a linear equation in one variable using a graph (pp. 435–436).

Write an equation in one variable as a function: 1. Rewrite the equation so that all terms are on one side of the equation. **2.** Write the equation in function notation using $y =$ the non-zero side of the equation from Step 1.

Solve an equation in one variable using a graph: 1. Write the equation as a function. **2.** Graph the function using a calculator or computer. **3.** The solution is the x-intercept. This solution is also referred to as the *zero* ($y = 0$) of the function.

Write $2x + 3 = 6$ as a function, graph the function, and find the solution from the graph.

$$2x + 3 = 6$$
$$2x + 3 - 6 = 0$$
$$2x - 3 = 0$$

$y = 2x - 3$ Enter $y = 2x - 3$ in calculator.

$y = 2(0) - 3$ At $x = 0$

$y = -3$

$y = 2(1) - 3$ At $x = 1$

$y = -1$

From the CALC menu, choose the zero option.

Solution: $x = 1.5$ or $1\frac{1}{2}$, the x-coordinate of the point where the line crosses the x-axis (Fig. 9–55).

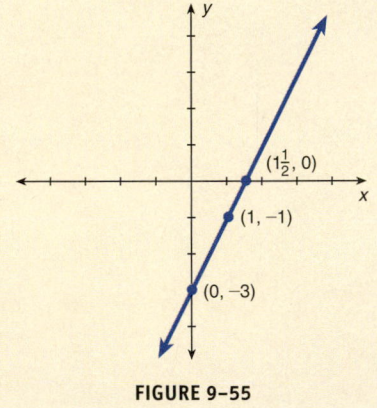

FIGURE 9–55

5 Graph a linear inequality in two variables (pp. 436–438).

1. Find the boundary by graphing an equation that substitutes an equal sign for the inequality symbol. Make a solid line if the boundary is included ($\leq$ or $\geq$) or a dashed line if the boundary is not included ($<$ or $>$) in the solution set. **2.** Test any point that is *not* on the boundary line. **3.** If the test point makes a true statement with the original inequality, shade the side containing the test point. **4.** If the test point makes a false statement with the original inequality, shade the side opposite the side containing the test point. For $>$ or $<$ inequalities, make the boundary line dashed. For $\geq$ or $\leq$ inequalities, make the boundary line solid.

LEARNING OUTCOMES   **KEY CONCEPTS AND EXAMPLES**

Graph $y \geq 2x - 1$.

First graph $y = 2x - 1$. Make the boundary line solid.

Test $(1, 3)$:

$3 \geq 2(1) - 1$ Substitute $x = 1$ and $y = 3$.

$3 \geq 2 - 1$ Simplify.

$3 \geq 1$ True

Shade the side of the boundary line that contains the point since the coordinates of the test point made a true statement in the inequality (Fig. 9–56).

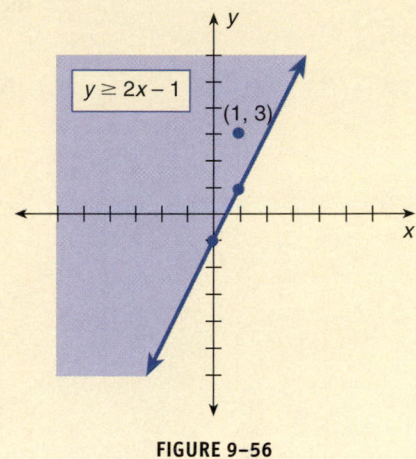

FIGURE 9–56

Section 9-3

1 Calculate the slope of a line, given two points on the line (pp. 439–442).

The slope of a line joining two points is the difference of the y-coordinates divided by the difference of the x-coordinates, that is, $\frac{\text{rise}}{\text{run}}$.

Find the slope of a line from two given points on the line: 1. Designate either point as point 1 with coordinates (x_1, y_1). Designate the other point as point 2 with coordinates (x_2, y_2). **2.** Calculate the change (*difference*) in the y-coordinates to find the vertical *rise* $(y_2 - y_1)$ and the change in the x-coordinates to find the horizontal *run* $(x_2 - x_1)$. **3.** Write the ratio of the rise to the run $\left(\frac{\text{rise}}{\text{run}}\right)$ and reduce the ratio to lowest terms. The slope can also be written in decimal notation if desired.

$$m = \frac{\text{rise}}{\text{run}} = \frac{\Delta y}{\Delta x} = \frac{y_2 - y_1}{x_2 - x_1}$$

Find the slope of the line passing through the points $(3, 1)$ and $(-2, 5)$.

$$\frac{y_2 - y_1}{x_2 - x_1} \qquad \frac{5 - 1}{-2 - 3} = \frac{4}{-5} \text{ or } -\frac{4}{5} \qquad P_1 = (3, 1), P_2 = (-2, 5)$$

2 Determine the slope of a horizontal or vertical line (pp. 442–443).

The slope of a horizontal line is zero, and all points on the same horizontal line have the same y-coordinate. The slope of a vertical line is not defined, and all points on the same vertical line have the same x-coordinate.

Identify a horizontal or vertical line: 1. Examine the y-coordinates of two points on the line. If the y-coordinates are the same, the line is horizontal and the slope is 0. **2.** Examine the x-coordinates of two points on the line. If the x-coordinates are the same, the line is vertical and the slope is undefined.

Examine the coordinate pairs and indicate which points lie on the same horizontal line and which lie on the same vertical line: $A\,(4, -3)$, $B\,(2, 5)$, $C\,(4, 5)$, $D\,(2, -3)$.

Points A and D and points B and C lie on the same horizontal lines, respectively.

Points A and C and points B and D lie on the same vertical lines, respectively.

LEARNING OUTCOMES	KEY CONCEPTS AND EXAMPLES

Section 9-4

1 Find the equation of a line, given the slope and one point (pp. 445–446).

1. Use the point-slope form of a linear equation of a straight line. $y - y_1 = m(x - x_1)$

where x_1 and y_1 are coordinates of the known point, m is the slope of the line, and x and y are the variables of the equation.   **2.** Substitute known values for x_1, y_1, and m.   **3.** Rearrange the equation to be in standard form ($ax + by = c$) or slope-intercept form ($y = mx + b$).

A line has slope 2 and passes through the point (3, 4). Find the equation of the line.

$y - y_1 = m(x - x_1)$	Substitute values. $m = 2, x_1 = 3, y_1 = 4$.
$y - 4 = 2(x - 3)$	Distribute.
$y - 4 = 2x - 6$	Solve for y. $-6 + 4 = -2$
$y = 2x - 2$	

or

$-2x + y = -2$	Put variables on same side of the equation. Then multiply by -1.
$2x - y = 2$	Standard form

2 Find the equation of a line, given two points on the line (pp. 446–447).

1. Use the slope formula to find the slope, given two points. $m = \dfrac{y_2 - y_1}{x_2 - x_1}$.   **2.** Use the point-slope form of an equation of a straight line, the calculated slope from Step 1, and the coordinates of either one of the given points. $y - y_1 = m(x - x_1)$.   **3.** Write the equation in slope-intercept or standard form.

Find the equation of a line that passes through the two points (3, 1) and (−3, 2).

First, find the slope.

$m = \dfrac{2 - 1}{-3 - 3} = \dfrac{1}{-6}$ or $-\dfrac{1}{6}$	Find the slope. Substitute values from the two points.
$y - y_1 = m(x - x_1)$	Substitute $m = -\dfrac{1}{6}, x_1 = 3$, and $y_1 = 1$.
$y - 1 = -\dfrac{1}{6}(x - 3)$	Distribute.
$y - 1 = -\dfrac{1}{6}x + \dfrac{1}{2}$	Solve for y. $\dfrac{1}{2} + 1 = \dfrac{1}{2} + \dfrac{2}{2} = \dfrac{3}{2}$
$y = -\dfrac{1}{6}x + \dfrac{3}{2}$	Slope-intercept form

or

$6y = -1x + 9$	Clear fractions and rearrange.
$x + 6y = 9$	Standard form

The equation of a line with a slope of zero (a horizontal line) is $y = k$, where k is the common y-coordinate for all points on the line. The equation of a line with a slope that is undefined (a vertical line) is $x = k$, where k is the common x-coordinate for all points on the line.

3 Find the equation of a line, given the slope and y-intercept (pp. 447–448).

1. Use the slope-intercept form of an equation of a straight line $y = mx + b$, where $m = $ slope and $b = $ the y-coordinate of the y-intercept.   **2.** Substitute known values for m and b.   **3.** Write the equation in standard form or leave the equation in slope-intercept form.

LEARNING OUTCOMES	KEY CONCEPTS AND EXAMPLES

Write the equation of a line that has slope $\frac{2}{3}$ and the y-intercept $(0, -1)$. Use the slope-intercept form of the equation:

$$y = mx + b \qquad \text{Substitute values, } m = \frac{2}{3}, b = -1.$$

$$y = \frac{2}{3}x - 1$$

4 Find the equation of a line, given a point on the line and an equation of a line parallel to that line (pp. 448–450).

The slopes of parallel lines are equal; that is, lines that have the same slope are parallel lines. Equations that have equal slopes have graphs that are parallel lines.

Determine which two of the three given equations have graphs that are parallel lines:

(a) $y = 3x - 5$ **(b)** $3x - 2y = 10$ **(c)** $6x - 2y = 8$

Write each of the three equations in slope-intercept form and compare the slopes (coefficients of x).

(a) $y = 3x - 5$ **(b)** $y = \frac{3}{2}x - 5$ **(c)** $y = 3x - 4$

(a) and **(c)** have the same slope and their graphs are parallel.

Find the equation of a line that is parallel to a given line when at least one point on the parallel line is known: 1. Solve the equation of the given line for y. **2.** Determine the slope m from the equation in Step 1. **3.** The slope of the parallel line is the same as the slope of the given line. **4.** Use the point-slope form of a straight line $y - y_1 = m(x - x_1)$ and substitute values for m, x_1, and y_1. **5.** Rearrange the equation to be in standard form $(ax + by = c)$ or slope-intercept form $(y = mx + b)$.

Find the equation of a line that passes through the point $(2, 3)$ and is parallel to the line that has equation $y = 4x - 1$. Write the new equation in slope-intercept form.

Use the slope of the given equation, 4, for the slope of the new equation.

$$y - y_1 = m(x - x_1) \qquad \text{Substitute } m = 4, x_1 = 2, \text{ and } y_1 = 3.$$
$$y - 3 = 4(x - 2) \qquad \text{Distribute.}$$
$$y - 3 = 4x - 8 \qquad \text{Solve for } y. -8 + 3 = -5$$
$$y = 4x - 5 \qquad \text{Slope-intercept form}$$

5 Find the equation of a line, given a point on the line and an equation of a line perpendicular to that line (pp. 450–452).

The slopes of perpendicular lines are negative reciprocals. Equations with slopes that are negative reciprocals have graphs that are perpendicular lines.

Determine which two of the three given equations have graphs that are perpendicular lines:

(a) $y = 3x - 5$ **(b)** $3x + 9y = 10$ **(c)** $9x + 3y = 12$

Write each of the three equations in slope-intercept form and compare the slopes (coefficients of x).

(a) $y = 3x - 5$ **(b)** $y = -\frac{1}{3}x + \frac{10}{9}$ **(c)** $y = -3x + 4$

(a) and **(b)** have slopes that are negative reciprocals and their graphs are perpendicular lines. Note that **(a)** and **(c)** are not perpendicular. Their slopes are opposites but *not* reciprocals. The slopes of **(b)** and **(c)** are reciprocals but *not* opposites.

LEARNING OUTCOMES	KEY CONCEPTS AND EXAMPLES
	Find the equation of a line that is perpendicular (normal) to a given line and passes through a given point: 1. Determine the slope of the *given* line. **2.** Find the *negative reciprocal* of this slope. The negative reciprocal is the slope of the perpendicular line. **3.** Write the equation for the perpendicular line by substituting the coordinates of the *given* point for the x_1 and y_1 and the slope of the *perpendicular* line for m into the point-slope form of the equation $y - y_1 = m(x - x_1)$. **4.** Write the equation in slope-intercept or standard form.

Find the equation of a line that passes through the point $(-1, 2)$ and is perpendicular to the line represented by the equation $y = \frac{1}{3}x - 5$. Write the equation in standard form.

The slope for the new equation is the negative reciprocal of $\frac{1}{3}$, which is -3.

$y - y_1 = m(x - x_1)$	Substitute $m = -3$, $x_1 = -1$, and $y_1 = 2$.
$y - 2 = -3[x - (-1)]$	Simplify and distribute.
$y - 2 = -3x - 3$	Rearrange with variables on left.
$3x + y = -3 + 2$	Combine like terms.
$3x + y = -1$	Standard form

9 | CHAPTER REVIEW EXERCISES

Section 9-1 MyLab Math For additional practice go to your study plan in MyLab Math.

Draw a rectangular coordinate system and locate the points.

1. $A = (5, -2)$

2. $B = (-8, -3)$

3. $C = (0, -4)$

4. $D = (3, 7)$

5. $E = (-3, 2)$

6. $F = (-3, 0)$

7. What are the coordinates of the origin?

8. Which of the four quadrants of the coordinate system is used to plot points with coordinates that are both negative?

9. Write the coordinates for points A through E on the graph in Fig. 9–57.

10. Write the coordinates of a point that lies on the y-axis and is four units below the x-axis.

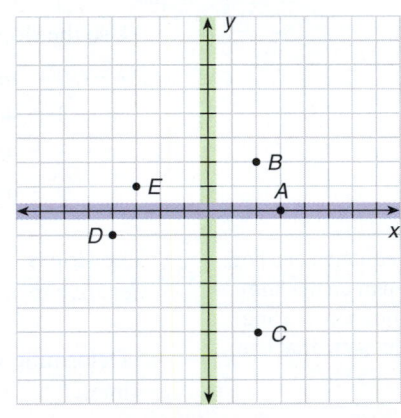

FIGURE 9–57

Represent the solutions of the equations in a table of solutions and on a graph.

11. $y = 2x - 3$

12. $y = -4x + 1$

13. $y = 3x$

14. $y = 4x$

15. $y = -3x$

16. $y = -4x$

17. $y = 2x + 1$

18. $y = 2x + 5$

19. $y = 4x - 2$

20. $y = -5x + 1$

21. $y = \frac{1}{2}x - 2$

22. $y = -\frac{1}{4}x + 1$

Which of the coordinate pairs are solutions for the equation $2x - 3y = 12$?

23. $(-2, -3)$

24. $(1, -3)$

25. $(3, -2)$

26. $(0, 4)$

27. $(6, 0)$

28. $(3, 4)$

29. In the equation $y = 3x - 1$, find y when $x = -2$.

30. In the equation $2x + y = 8$, find x when $y = 10$.

31. Find y in the equation $x - 3y = 5$ when $x = 8$.

32. Find y if $x = 7$ in the equation $3x - 4y = -2$.

Section 9–2

Graph the equations using the intercepts procedure.

33. $x = -4y - 1$

34. $x + y = -4$

35. $3x - y = 1$

36. $x = -4y$

37. $\frac{1}{2}x + \frac{1}{3}y = 1$

38. $2x + 3y = 6$

Graph using the slope and y-intercept procedure.

39. $y = 5x - 2$

40. $y = -x$

41. $y = -3x - 1$

42. $y = \frac{1}{2}x + 3$

43. $x - y = 4$

44. $2y + 4 = -3$

45. $x - 2y = -1$

46. $x + y = 5$

47. $y - 2x = -2$

48. $y + 3x = 4$

Graph the equations using a graphing calculator. Use the appropriate viewing window.

49. $y = 0.5x - 3$

50. $y = 2.1x + 0.5$

51. $y = 20x - 15$

52. $y = -30x + 10$

53. The function $y = 5x + 8,000$ is used to express the cost of manufacturing a widget, where $8,000 is the fixed cost of constructing molds and $5 is the cost of materials to make each widget. What is the cost of manufacturing 10,000 widgets?

54. A salesperson earns a salary of $200 weekly plus $12.50 commission for each photo sitting. Weekly income is expressed by the function $y = 12.50x + 200$, where x represents the number of sittings in a week. Find the weekly income if 28 sittings are sold.

Write each equation as a function and find the solution graphically.

55. $x + 2 = 8$

56. $3x - 7 = 1$

57. $2x + 1 = 5x + 7$

58. $-14 = 2(x - 7)$

59. $5(x + 2) = 3(x + 4)$

60. $2x + 1 = 7x + 2 - 8$

Determine the slope and y-intercept of the given equations by inspection.

61. $y = 3x + \dfrac{1}{4}$

62. $y = \dfrac{2}{3}x - \dfrac{3}{5}$

63. $y = -5x + 4$

64. $y = 7$

65. $x = 8$

66. $y = \dfrac{1}{3}x - \dfrac{5}{8}$

67. $y = \dfrac{x}{8} - 5$

68. $y = -\dfrac{x}{5} + 2$

Write the given equations in slope-intercept form and determine the slope and y-intercept.

69. $2x + y = 8$

70. $4x + y = 5$

71. $3x - 2y = 6$

72. $5x - 3y = 15$

73. $\dfrac{3}{5}x - y = 4$

74. $2.2y - 6.6x = 4.4$

75. $3y = 5$

76. $3x - 6y = 12$

Graph the linear inequalities using test points and verify with a graphing calculator.

77. $4x + y < 2$

78. $x + y > 6$

79. $3x + y \leq 2$

80. $x + 3y > 4$

81. $x - 2y < 8$

82. $3x + 2y \geq 4$

83. $y \geq 3x - 2$

84. $y \leq -2x + 1$

85. $y > \frac{2}{3}x - 2$

Section 9–3

Find the slope of the line passing through the given pairs of points.

86. $(3, -1)$ and $(1, 3)$

87. $(3, 2)$ and $(5, 6)$

88. $(-1, -1)$ and $(2, 2)$

89. $(4, 3)$ and $(-4, -2)$

90. $(6, 2)$ and $(-3, 2)$

91. $(3, -4)$ and $(0, 0)$

92. $(1, -1)$ and $(5, -5)$

93. $(-4, 1)$ and $(-4, 3)$

94. $(4, -4)$ and $(1, 3)$

95. $(5, 0)$ and $(-2, 4)$

96. $(-2, 1)$ and $(0, 3)$

97. $(-4, -8)$ and $(-2, -1)$

98. (3, 3) and (3, 0)

99. (5, −3) and (−1, −3)

100. (−5, −1) and (−7, −3)

101. (−7, 0) and (−7, 5)

102. (3, 5) and (2, 5)

103. Write the coordinates of two points that lie on the same horizontal line.

104. Write the coordinates of two points that lie on the same vertical line.

Use Table 9–2 on page 444 for Exercises 105–110. Use the initial year of the academic year as the reference.

105. What is the rate of change in tuition and fees at public 2-year colleges from 2006 to 2007?

106. What is the rate of change in tuition and fees at public 2-year colleges from 2010 to 2011?

107. What is the rate of change in tuition and fees at public 2-year colleges from 2001 to 2002?

108. What is the rate of change in tuition and fees at public 2-year colleges from 2000 to 2001?

109. Explain why the rates of change found in Exercises 106, 107, and 108 are different.

110. If the data in Table 9–2 formed a straight line when graphed, what would you expect to find for the rates of change for the data in Exercises 106, 107, and 108?

Section 9–4

Find the equation of a line passing through the given point with the given slope. Write the equation in slope-intercept form.

111. (−6, 2), $m = \dfrac{1}{3}$

112. (3, 2), $m = -\dfrac{2}{5}$

113. (4, 0), $m = \dfrac{3}{4}$

Find the equation of a line passing through the given point with the given slope. Write the equation in standard form.

114. (0, −2), $m = 2$

115. (2, 3), $m = 4$

116. (6, 0), $m = -1$

Find the equation of a line passing through the given pairs of points. Solve the equation for y if necessary.

117. (−5, 2) and (6, 1)

118. (1, 4) and (−1, 3)

119. (−1, −3) and (3, 4)

120. (−3, 0) and (4, 0)

121. (−2, −3) and (3, 6)

122. (2, −4) and (3, −4)

123. (5, 2) and (6, 3)

124. (4, 6) and (1, −1)

125. (−1, −2) and (−3, −4)

126. (4, 0) and (4, −3)

127. (5, −2) and (3, −2)

128. (5, 4) and (0, 4)

129. A salesperson sells 80 items and earns $3,800 in one month, and in another month 120 items are sold resulting in $4,200 earnings. If we assume this is a linear function, write a function that expresses the salesperson's monthly salary plus commission, S, as a function of the number of items sold, x. Salary is the y-intercept.

130. Simple interest is a linear function. Write an equation to express interest earned if $500 is earned in 2 years and $2,000 is earned in 8 years.

Write the equations using the given slope and y-intercept.

131. $m = 3, b = -2$

132. Slope $= \dfrac{3}{5}$; y-intercept $= -7$

Identify the slope and y-intercept to write the equations using information from the graphs in Figs. 9–58 and 9–59.

133.

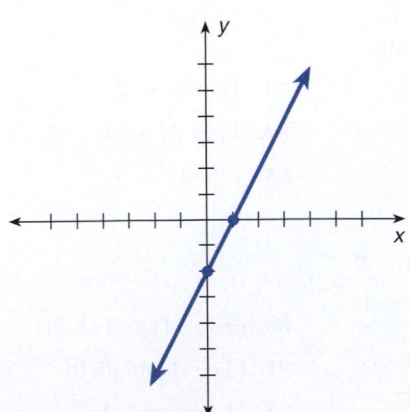

FIGURE 9–58

134.

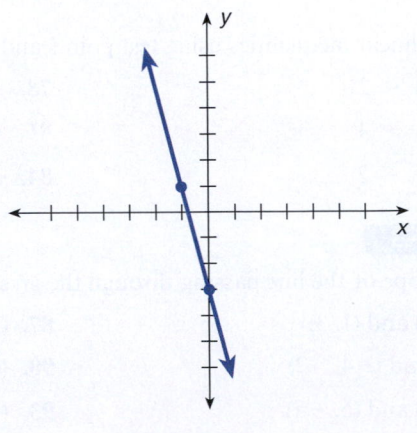

FIGURE 9–59

Write the equations in standard form. Graph both equations for each exercise on a graphing calculator to verify parallelism.

135. Find the equation of the line that is parallel to the line $x + y = 4$ and passes through the point $(2, 5)$.

136. Find the equation of the line that is parallel to the line $3x + y = 6$ and passes through the point $(1, 0)$.

137. Find the equation of the line that is parallel to the line $2y = x - 3$ and passes through the point $(2, -3)$.

138. Find the equation of the line that is parallel to the line $4x - y = -1$ and passes through the point $(0, -2)$.

139. Find the equation of the line that is parallel to the line $x - 3y = 5$ and passes through the point $(5, -5)$.

140. Find the equation of the line that is parallel to the line $3x - 2y = 2$ and passes through the point $(0, -3)$.

141. Find the equation of the line that is parallel to the line $x + 3y = 6$ and passes through the point $(-4, -2)$.

142. Find the equation of the line that is parallel to the line $4x + 3y = 1$ and passes through the point $\left(3, -\frac{1}{2}\right)$.

143. Find the equation of the line that is parallel to the line $3x - 4y = 0$ and passes through the point $\left(\frac{1}{3}, 2\right)$.

144. Find the equation of the line that is parallel to the line $-2x + 3y = 2$ and passes through the point $(-1, -1)$.

Write the equations in standard form. Use a graphing calculator to verify that the two lines in each exercise are perpendicular.

145. Find the equation of the line that is perpendicular to the line $x + y = 4$ and passes through the point $(-3, 1)$.

146. Find the equation of the line that is normal to the line $3x + y = 6$ and passes through the point $(1, 4)$.

147. Find the equation of the line that is perpendicular to the line $x + 2y = 5$ and passes through the point $(-2, 0)$.

148. Find the equation of the line that is normal to the line $2x + 2y = 4$ and passes through the point $(0, 0)$.

149. Find the equation of the line that is normal to the line $5x + y = 8$ and passes through the point $(-1, 2)$.

150. Find the equation of the line that is perpendicular to the line $3y = x - 4$ and passes through the point $(2, 3)$.

151. Find the equation of the line that is perpendicular to the line $5x - y = 10$ and passes through the point $\left(\frac{1}{2}, 3\right)$.

152. Find the equation of the line that is normal to the line $x - 3y = 6$ and passes through the point $(-2, 4)$.

153. Find the equation of the line that is perpendicular to the line $4x - y = 8$ and passes through the point $\left(4, -\frac{1}{2}\right)$.

154. Find the equation of the line that is perpendicular to the line $4y = 2x + 1$ and passes through the point $(3, -1)$.

9   TEAM PROBLEM-SOLVING EXERCISES

1. A 1-gal can of indoor house paint is advertised to cover 400 ft^2 of wall surface.
  (a) Make a table to show the amount of wall surface area that can be covered by 1, 2, 3, . . . , 10 gal of paint.
  (b) Graph these data and extend the line beyond the data to 15 gal of paint.
  (c) Use the graph to decide how many 1-gal cans of paint it would take to cover 5,500 ft^2 of wall surface.
  (d) The paint being used for this job can be purchased for $19.05 a gallon. Find the cost of the paint for the 5,500 ft^2 of wall surface.
  (e) The sales tax rate is 8.25%. Calculate the total cost of the paint.

2. Linda Kodama is introducing a new lipstick in Bright-glow's product line. As product development manager, she estimates the cost of the new lipstick to be $4.53 per item, plus an additional cost of $5,000 for product development.
  (a) Make a table of the estimated cost for producing 0 lipsticks, 100 lipsticks, 1,000 lipsticks, and 2,000 lipsticks.
  (b) Represent these costs as ordered pairs (number of lipsticks, cost).
  (c) Plot the ordered pairs and graph a line that fits the ordered pairs.
  (d) Write an equation that represents the cost of the new lipstick as a function of the number of lipsticks produced.
  (e) Linda projects the selling price of the new lipstick to be $8.99. How many lipsticks must the company sell to recover the cost of producing the new items?

9   CONCEPTS ANALYSIS

1. Describe the graph of a line with slope that is positive.

2. Describe the graph of a line with slope that is negative.

3. Describe the graph of a line with slope that is a fraction between 0 and 1.

4. Describe the graph of a line with slope that is a number greater than 1.

5. Describe the graph of a line with slope that is a fraction between -1 and 0.

6. Describe the graph of a line with slope that is a number less than -1.

7. What is the slope of a horizontal line? Why?

8. What is the slope of a vertical line? Why?

9. If the standard form of an equation of a horizontal line is $y = k$, what does k represent?

10. If the standard form of an equation of a vertical line is $x = k$, what does k represent?

11. If an equation is solved for y, what information about the graph can be read from the equation?

12. Give two basic characteristics of an equation that graphs into a straight line.

13. How do the slopes of parallel lines compare?

14. How do the slopes of perpendicular lines compare?

15. List the steps to find the x-intercept of a line if you are given two points on the line.

16. What do we mean when we say a point is represented by an ordered pair of numbers?

17. What does the graph of an equation represent?

18. Explain the procedure for plotting the point $(5, -2)$ on a grid representing the rectangular coordinate system.

19. Discuss the similarities and differences in the table-of-solutions method, the intercepts method, and the slope-intercept method for graphing a linear equation.

20. Why is it generally helpful to solve an equation for y before using the table-of-solutions method for graphing?

9 | PRACTICE TEST

Make a table of values to represent the solutions to the following equations and show the solutions on a graph.

1. $y = \dfrac{1}{2}x$

2. $y = \dfrac{1}{2}x + 1$

3. $y = 2x - 4$

4. $y = 5 - x$

Write as functions and find the solutions graphically.

5. $3x + 2 = 5$

6. $2(x + 1) = 3x$

Write the specific solution in ordered-pair form for each equation.

7. $x + y = 7$, if $x = 2$

8. $2x - y = 1$, if $y = -3$

9. A cake bakery estimates the profit function to be $y = 3x - 800$, where x is the number of cakes produced in a month and $800 is cost of overhead such as utilities. Find the profit if 8,000 cakes are produced.

10. A cost function is known to be $y = 10x + 250$, where x is the number of units produced and $250 is the fixed cost. Find the total cost of producing 5,000 units.

Graph using the intercepts method.

11. $x + y = -5$

12. $2x - 3y = 6$

13. $x + 2y = 8$

Graph using the slope-intercept method.

14. $y = -3x + 1$

15. $2x + y = -3$

16. $x + 2y = 1$

Graph the linear inequalities.

17. $2x - y \le 2$

18. $x + 2y < 4$

19. $x + y < 1$

20. $y \le 2x + 2$

Find the slope of the line passing through the given pairs of points.

21. $(-3, 6)$ and $(3, 2)$

22. $(0, 4)$ and $(-1, 6)$

Write the equations in slope-intercept form and determine the slope and y-intercept.

23. $-2x + y = 34$

24. $x - y = 4$

25. $x = 4y$

26. $2y - x = 3$

Find the equation of the line passing through the given point with the given slope. Solve the equation for y.

27. $(3, -5), m = \dfrac{2}{3}$

28. $(5, 1), m = -2$

Find the equation of the line passing through the given pairs of points. Solve the equation for y.

29. $(1, 3)$ and $(4, 5)$

30. $(-1, 1)$ and $(4, -4)$

31. $(5, 2)$ and $(-1, 2)$

32. Write the equation in slope-intercept form of the line that has a slope $= -2$ and y-intercept $= -3$.

33. Write the equation of the line shown in Fig. 9–60 in slope-intercept form.

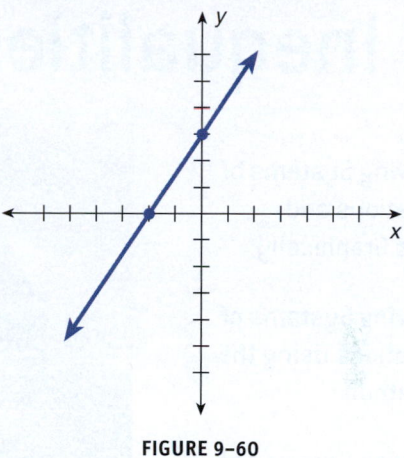

FIGURE 9–60

Write the equations in standard form.

34. Find the equation of the line that is parallel to $y - 2x = 3$ and passes through the point $(2, 5)$.

35. Find the equation of the line that is parallel to $2x + y = 4$ and passes through the point $(4, -3)$.

36. Find the equation of the line perpendicular to $y - 2x = 3$ and passing through the point $(2, 5)$.

37. Find the equation of the line perpendicular to $2x + y = 4$ and passing through the point $(4, -3)$.

10

Systems of Linear Equations and Inequalities

Konstantin32/123RF

In Great Company

It Don't Mean a Thing if It Ain't Got That Swing (June 2000)

The morning of June 10, 2000, was not just the opening of a new day, but the opening of a new bridge. The Millennium Bridge is a 1,066-foot-long suspension footbridge across the Thames River in the heart of London. Costing $30 million, it was designed to support the weight of 5,000 people at once—a total weight of 500 tons. Although the opening was two months late, the new design promised to give unparalleled views of the London skyline.

Within 48 h, the bridge was closed. It would stay closed for the next two years, as engineers desperately tried to figure out how to fix what had gone wrong.

As to what had gone wrong, that was obvious. Any attempt to walk across the bridge caused such violent swaying that pedestrians were often knocked off their feet. The bridge's sway was caused by a positive feedback loop called "synchronous lateral excitation." What that means is this: As someone steps onto the bridge, it sways slightly. To keep their footing, pedestrians unconsciously match their next step to the sway of the bridge. This increases the sway. And so on.

This phenomenon has been known since 1831. Soldiers marching in lockstep across the solid roadway of the Broughton Suspension bridge created enough mechanical resonance to destroy the bridge as they marched across. Since then, troops have typically been ordered to break step when marching across solid roadway bridges to prevent another occurrence.

In this case, it took two years, 37 fluid dampers, 52 tuned mass dampers, and roughly $9 million to control the bridge's wobbles. But this isn't the first or last bridge to have this problem.

The most famous is probably the Tacoma Narrows Bridge Collapse of 1940. The third largest suspension bridge in the world when it opened on July 1, it stretched across Puget Sound like a steel ribbon. Four months later, it tore itself apart as the wind hit the bridge at just the right speed and angle to set up a resonant vibration. All of it was caught on color film and can be found on YouTube—search for Galloping Gertie and watch for the man who tries to save a dog from a car.

More recently, a May 2010 YouTube video shows a Volgograd bridge over the Volga River suffering the same problem. The film is available at http://www.youtube.com/watch?v=uWP5d2t2JVE&feature=player_embedded.

10-1 Solving Systems of Linear Equations and Inequalities Graphically

LEARNING OUTCOMES

1. Solve a system of linear equations by graphing.
2. Solve a system of linear inequalities by graphing.

LC LEARNING CATALYTICS

1. Is the point $(-1, 6)$ on the graph of the equation $5x + y = 1$?

System of two linear equations: a system that has two linear equations with one or two variables in each equation

STOP AND CHECK

1. Graph the two equations and find the point of intersection:
   $x + y = 15$
   $x - y = 3$

Answer:
1. $(9, 6)$, or $x = 9$ and $y = 6$

1 Solve a System of Linear Equations by Graphing. A **system of two linear equations,** is two linear equations with one or two variables in each equation. The system is solved when we find the one ordered pair of solutions that satisfies *both* equations. There are also systems that have no solutions or for which any solution of one equation is also a solution of the other. One method of solving systems of two equations is to graph each equation and find the intersection of these graphs. The point where the two graphs intersect represents the ordered pair of solutions that the two graphs have in common.

To solve a system of two linear equations with two variables by graphing:

1. Graph each equation on the same pair of axes.
2. The solution will be the common point.

EXAMPLE 1

CON A board is 20 ft long. It needs to be cut so that one piece is 2 ft longer than the other. What should be the length of each piece?

Write two equations to describe all the conditions of the problem.

Since the board is not cut into equal pieces, we let the letter *l* represent the *longer* piece and the letter *s* represent the *shorter* piece.

Condition 1:	The total length of the board is 20 ft. Thus, the two pieces (*l* and *s*) total 20 ft: $l + s = 20$.
Condition 2:	One piece is 2 ft longer than the other. Thus, the shorter piece plus 2 ft equals the longer piece: $s + 2 = l$.

The two equations become a *system of equations.*

$l + s = 20$ Condition 1
$s + 2 = l$ Condition 2

Graph each equation on the same set of axes and examine the intersection of the graphs. Let *s* be the independent variable and *l,* the dependent variable.

Condition 1	**Condition 2**
$l + s = 20$	$s + 2 = l$
or	or
$l = 20 - s$ Domain: [0, 20]	$l = s + 2$ Domain: [0, 20]

Make a table of solutions for each equation.

s	l	
8	12	Choose values that are near
10	10	the middle of the domain.
12	8	

s	l	
0	2	Choose only zero and positive
5	7	values.
10	12	

The point of intersection (Fig. 10–1) is (9, 11), which means that $s = 9$ and $l = 11$. **The shorter length is 9 ft and the longer length is 11 ft.**

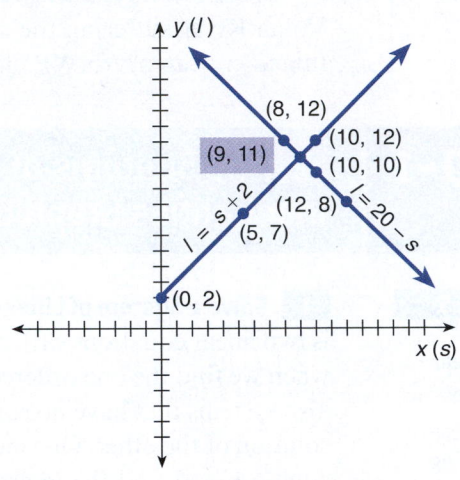

FIGURE 10–1

Check to see if the solution ($s = 9, l = 11$) satisfies both equations.

$l + s = 20$	$s + 2 = l$	
$11 + 9 = 20$	$9 + 2 = 11$	The ordered pair checks
$20 = 20$	$11 = 11$	in both equations.

See Exercises 1–21.

Independent system: a system of linear equations that results in the graphs of the equations intersecting at just one point. There is one pair of values that satisfies both equations

Inconsistent system: a system of linear equations that results in the graphs of the equations not intersecting at all. There is no solution

Dependent system: a system of linear equations that results in the graphs of the equations coinciding or falling exactly in the same place. Any pair of values that satisfies one equation satisfies both equations

Line relationships:

When graphing two straight lines, three possibilities can occur.

1. The two lines can intersect at *just one point*. This means the system is **independent** and one pair of values satisfies both equations.
2. The two lines *do not intersect* at all. This means the system is **inconsistent** and no pair of values satisfies both equations.
3. The two lines *coincide* or fall exactly in the same place. This means the system is **dependent** and the equations are identical or are multiples, and any pair of values that satisfies one equation satisfies both equations.

To solve a system of equations using a graphing calculator:

1. Enter both equations in the $\boxed{Y=}$ screen.
2. Graph both equations on the same screen.

3. Use the Trace feature or the CALC then 5: Intersect feature to determine the approximate coordinates of the intersection of the graphs.

4. Use the Zoom or Box (Window) feature to get a closer view and thus a more accurate approximation of the intersection of the graphs.

2 **Solve a System of Linear Inequalities by Graphing.** When two intersecting lines are plotted on the same set of axes, the rectangular coordinate system is separated into four regions (Fig. 10–2). The solution set of a system of linear inequalities will include all the points in one of the regions.

To solve a system of two linear inequalities with two variables by graphing:

1. Graph each inequality on the same pair of axes.

2. The solution set of the system will be the *overlapping* region of the solution sets of the two inequalities.

FIGURE 10-2

STOP AND CHECK

1. Shade in yellow the portion of the graph in Fig. 10–5 that corresponds to $y \geq 3x + 5$ and $x + y > 7$.

2. Shade in red the portion of the graph in Fig. 10–5 that corresponds to $y \geq 3x + 5$ and $x + y < 7$.

3. Shade in dark blue the portion of the graph in Fig. 10–5 that corresponds to $y \leq 3x + 5$ and $x + y < 7$.

Answers:

1.

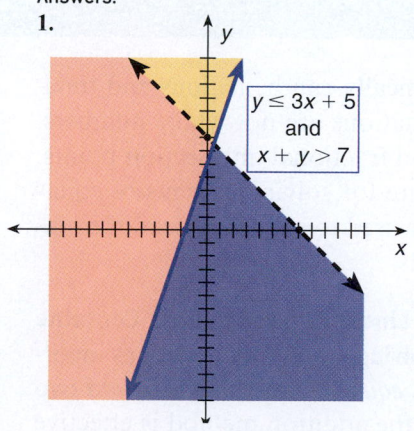

$y \leq 3x + 5$
and
$x + y > 7$

EXAMPLE 2

Shade the portion on the graph that is represented by the following conditions: $y \leq 3x + 5$ and $x + y > 7$.

Graph the inequality $y \leq 3x + 5$:

The boundary $y = 3x + 5$ has a slope of 3 and a y-intercept of 5. The boundary will be included in the solution set.

For the *less than* relationship, shade the region that includes the part of the y-axis that is *below* the boundary (Fig. 10–3).

Graph the inequality, $x + y > 7$:

The boundary $x + y = 7$ in slope-intercept form is $y = -x + 7$. The slope is −1 and the y-intercept is 7. The boundary will not be included.

For the *greater than* relationship, shade the region that includes the part of the y-axis that is *above* the graph (Fig. 10–4).

Find the solution set common to both inequalities. Visualize both graphs on the same axes. The region that has overlapping shading represents the points that satisfy *both* conditions and forms the solution set for the system of inequalities (see Fig. 10–5).

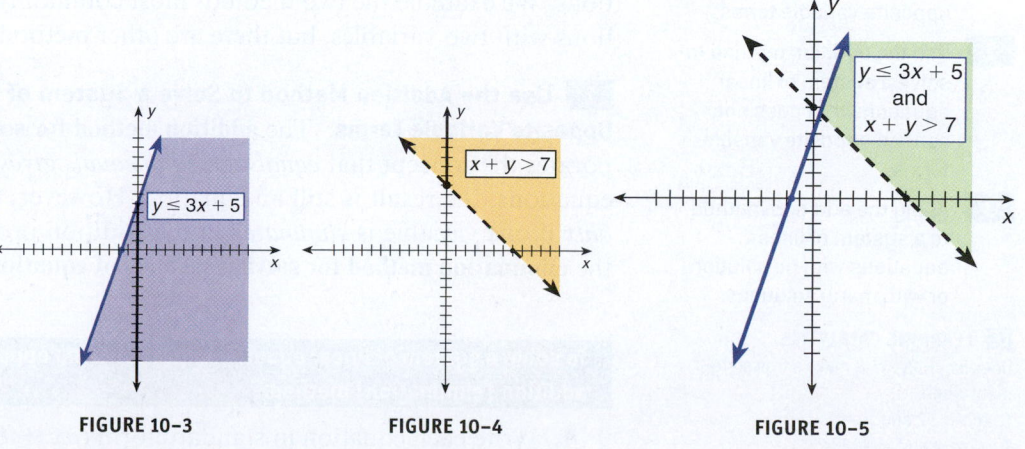

$y \leq 3x + 5$

$x + y > 7$

$y \leq 3x + 5$
and
$x + y > 7$

FIGURE 10-3   **FIGURE 10-4**   **FIGURE 10-5**

See Exercises 22–27.

10–1 EXERCISES

MyLab Math For additional practice go to your study plan in MyLab Math.

1 Solve the systems of equations by graphing. *See Example 1.*

1. $x + y = 12$
$x - y = 2$

2. $2x + y = 9$
$3x - y = 6$

3. $2x - y = 5$
$4x - 2y = 8$

4. $x - y = 9$
$3x - 3y = 27$

5. $3x + 2y = 8$
$x + y = 2$

6. $y = x - 2$
$y = -2x + 7$

7. $y = -3$
$y = x - 7$

8. $y = 5x$
$y = -2x - 7$

9. $y = -2x$
$y = x - 3$

10. $x - y = 3$
$x + 2y = 6$

11. $5x + y = -2$
$x - y = 8$

12. $2x - y = -3$
$x + 2y = 6$

13. $4x + y = -1$
$x + y = -4$

14. $x + y = 3$
$2y = 3x - 4$

15. $y = 2x - 5$
$y = x + 2$

16. $x + 2y = 6$
$x + y = 2$

17. $2x - y = 7$
$x + y = 5$

18. $3x + y = 5$
$x - y = 7$

19. $2x - y = 6$
$5x + y = 8$

20. $5x + y = 0$
$x = y$

21. $-x + y = -2$
$2x + y = 7$

2 Graph each system of inequalities and shade the portion on the graph represented by the solution set. *See Example 2.*

22. $y < 2x + 1$ and $x + y > 5$

23. $x + y \geq 3$ and $x - y \geq 2$

24. $x - 2y \leq -1$ and $x + 2y \geq 3$

25. $x + y > 4$ and $x - y > -3$

26. $3x - 2y < 8$ and $2x + y \leq -4$

27. $x - 2y < -6$ and $2x + y \leq 5$

| **10–2** | **Solving Systems of Linear Equations Using the Addition Method** |

LEARNING OUTCOMES

1 Use the addition method to solve a system of linear equations that contains opposite variable terms.

2 Use the addition method to solve a system of linear equations that does not contain opposite variable terms.

3 Apply the addition method to a system of linear equations with no solution or with many solutions.

LC LEARNING CATALYTICS

Does $5x - 2y = 8$ make a true statement when

1. $x = -2$ and $y = -1$?

2. $x = 2$ and $y = 1$?

We can see that solving systems of equations graphically can be tedious and time-consuming. Also, many solutions to systems of equations are not whole numbers. Graphically, it is difficult to plot fractions or to read fractional intersection points. Therefore, we need a convenient algebraic procedure for solving systems of equations. We examine the two methods most commonly used to solve a system of equations with two variables, but there are other methods.

1 Use the Addition Method to Solve a System of Linear Equations That Contains Opposite Variable Terms. The **addition method for solving systems of equations** incorporates the concept that *equals added to equals gives equals.* Thus, when we add two equations, the result is still an equation. However, the addition method is effective *only* if one variable is *eliminated* in the addition process. This method is also called the **elimination method for solving systems of equations.**

To solve a system of linear equations that contains opposite variable terms by the addition (elimination) method:

1. Write each equation in standard form $(ax + by = c)$.

2. Add two equations.

Addition method for solving systems of equations: a method for solving systems of equations that incorporates the concept *equals added to equals gives equals*

Elimination method for solving systems of equations: another name for the addition method

STOP AND CHECK

Solve the systems of equations by the addition method.

1. $x + y = 10$
   $x - y = 6$
2. $2x - 3y = 9$
   $4x + 3y = 15$

Answers:

1. $x = 8$ and $y = 2$, or $(8, 2)$
2. $x = 4$ and $y = -\dfrac{1}{3}$, or $\left(4, -\dfrac{1}{3}\right)$

---

3. Solve the equation from Step 2. The result is one part of the ordered pair solution.

4. Substitute the solution from Step 3 in either equation and solve for the remaining variable. This completes the solution ordered pair.

5. Check the solution in both original equations.

EXAMPLE 1

Solve the systems of equations by the addition method:

(a) $x + y = 7$ (b) $3x + 8y = 7$
     $x - y = 5$ $-3x - 3y = 8$

(a) $x + y = 7$ Equation 1. Add the equations to eliminate the *y*-variable.
$\underline{x - y = 5}$ Equation 2
$2x = 12$ Solve for *x*.
$\boxed{x = 6}$ Substitute 6 in place of *x* in *either* of the given equations.
$x + y = 7$ Substitute $x = 6$ in Equation 1.
$\boxed{6} + y = 7$ Solve for *y*.
$y = 7 - 6$ Simplify.
$\boxed{y = 1}$

The solution is $x = 6$ and $y = 1$, or (6, 1).

Check in the *other* equation:

$\boxed{x} - \boxed{y} = 5$ Equation 2. Substitute values from the solution ordered pair.
$\boxed{6} - \boxed{1} = 5$ Simplify.
$5 = 5$ The solution checks.

Checking in both equations helps ensure that no errors have been made in the addition process or in the substitution process.

(b) $3x + 8y = 7$ Equation 1. Add the equations to eliminate the *x*-variable.
$\underline{-3x - 3y = 8}$ Equation 2
$5y = 15$ Solve for *y*.
$\boxed{y = 3}$
$3x + 8y = 7$ Substitute 3 for *y* in Equation 1.
$3x + 8(\boxed{3}) = 7$ Solve for *x*.
$3x + 24 = 7$
$3x = 7 - 24$
$3x = -17$
$\boxed{x = \dfrac{-17}{3}}$

The solution is $\left(-\dfrac{17}{3}, 3\right)$.

Check in Equation 2:

$-3x - 3y = 8$ Equation 2. Substitute $x = -\dfrac{17}{3}$ and $y = 3$.
$-3\left(\dfrac{-17}{3}\right) - 3(\boxed{3}) = 8$ Multiply.
$17 - 9 = 8$ Simplify.
$8 = 8$ The solution checks.   See Exercises 1–9.

2 **Use the Addition Method to Solve a System of Linear Equations That Does Not Contain Opposite Variable Terms.** In the addition method one unknown must be eliminated; that is, the terms must add to 0. Thus, if neither pair of variable terms in a system of equations is opposite terms that will add to 0, we multiply one or both of the equations by numbers that *cause* the terms of one variable to add to 0. This is an application of the multiplication axiom (see Chapter 7, Section 2, Outcome 2).

> **To solve a system of linear equations that does not contain opposite variable terms using the addition (elimination) method:**
>
> 1. Write each equation in standard form ($ax + by = c$).
>
> 2. If necessary, multiply one or both equations by numbers that cause the terms of one variable to add to zero.
>
> 3. Add the two equations to eliminate a variable.
>
> 4. Solve the equation from Step 3 for the remaining variable.
>
> 5. Substitute the solution from Step 4 in either equation and solve for the remaining variable.
>
> 6. Check the solution in both original equations.

STOP AND CHECK

1. Solve the system of equations: $3x - y = 17$ and $x - y = 7$.

Answer:

1. $(5, -2)$

EXAMPLE 2

Solve the system of equations: $2x + y = 7$ and $x + y = 3$.

There are several possibilities for eliminating one variable by applying the multiplication axiom. We will examine three choices.

Choice 1.
$$-2x - y = -7$$
$$\underline{x + y = 3}$$
$$-x = -4$$
$$x = 4$$

Multiply the first equation by -1 and add the equations to eliminate the y-terms.

Solve for x.

Choice 2.
$$2x + y = 7$$
$$\underline{-x - y = -3}$$
$$x = 4$$

Multiply the second equation by -1 and add the equations to eliminate the y-terms.

Choice 3.
$$2x + y = 7$$
$$\underline{-2x - 2y = -6}$$
$$-y = 1$$
$$y = -1$$

Multiply the second equation by -2 and add the equations to eliminate the x-terms.

Solve for y.

For choices 1 and 2, substitute $x = 4$:

$$2x + y = 7$$
$$2(4) + y = 7$$
$$8 + y = 7$$
$$y = 7 - 8$$
$$y = -1$$

For choice 3, substitute $y = -1$:

$$2x + y = 7$$
$$2x + (-1) = 7$$
$$2x - 1 = 7$$
$$2x = 7 + 1$$
$$2x = 8$$
$$x = 4$$

The solution is $(4, -1)$.

Check the solution in the original equations:

$$2x + y = 7 \qquad\qquad x + y = 3 \qquad \text{Substitute } x = 4 \text{ and } y = -1.$$
$$2(4) + (-1) = 7 \qquad 4 + (-1) = 3$$
$$8 + (-1) = 7 \qquad\qquad 3 = 3 \qquad \text{Solution checks.}$$
$$7 = 7$$

See Exercises 10–12.

When an original equation is altered, it is important to check your solution in *both* original equations. This enables you to identify mistakes such as forgetting to multiply *each* term in the equation by a number.

STOP AND CHECK

1. Solve the system of equations: $4x - 3y = 17$ and $5x + 2y = 4$.

Answer:

1. $(2, -3)$

EXAMPLE 3

Solve the system of equations: $2x + 3y = 1$ and $3x + 4y = 2$.

In this system, no integer can be multiplied by just one equation to eliminate a letter. Therefore, we need to multiply each equation by some number. There are several possibilities; we examine just two.

Choice 1.

$-3(2x + 3y) = -3(1)$		Multiply Equation 1 by -3.
$2(3x + 4y) = 2(2)$		Multiply Equation 2 by $+2$.
$-6x - 9y = -3$		Add equations to eliminate the x-variable.
$6x + 8y = 4$		
$-y = 1$		Solve for y.
$y = -1$		y-value of solution
$2x + 3y = 1$		Substitute $y = -1$.
$2x + 3(-1) = 1$		Solve for x.
$2x - 3 = 1$		
$2x = 1 + 3$		
$2x = 4$		
$x = 2$		x-value of solution

The solution is $(2, -1)$.

Choice 2.

$4(2x + 3y) = 4(1)$		Multiply Equation 1 by 4.
$-3(3x + 4y) = -3(2)$		Multiply Equation 2 by -3.
$8x + 12y = 4$		Add equations to eliminate the y-variable.
$-9x - 12y = -6$		
$-x = -2$		Solve for x.
$x = 2$		x-value of solution
$2x + 3y = 1$		Substitute $x = 2$.
$2(2) + 3y = 1$		Solve for y.
$3y = 1 - 4$		
$3y = -3$		
$y = -1$		y-value of solution

The solution is $(2, -1)$.

Check the solution in the original equations:

$$2x + 3y = 1 \qquad 3x + 4y = 2 \qquad \text{Substitute } x = 2 \text{ and } y = -1.$$
$$2(2) + 3(-1) = 1 \quad 3(2) + 4(-1) = 2$$
$$4 + (-3) = 1 \qquad 6 + (-4) = 2$$
$$1 = 1 \qquad\qquad 2 = 2 \qquad \text{Solution checks.} \qquad \textbf{See Exercises 13–21.}$$

3 **Apply the Addition Method to a System of Linear Equations with No Solution or with Many Solutions.** Sometimes a system of equations has no solution; the graphs of the two equations do not intersect. There are also instances when a system of equations has all solutions in common; the graphs of the two equations coincide.

EXAMPLE 4

Solve the system: $x + y = 7$ and $x + y = 5$.

$$x + y = \quad 7 \qquad \text{Multiply Equation 2 by } -1.$$
$$-1(x + y) = -1(5)$$
$$x + y = \quad 7 \qquad \text{Add the equations to eliminate a variable.}$$
$$\underline{-x - y = -5}$$
$$0 = \quad 2 \qquad \text{Both variables are eliminated.}$$

Notice, both variables are eliminated and the resulting equation, $0 = 2$, is *false*.

There are no solutions to this system. $\qquad\qquad$ **See Exercises 22–23.**

If both variables in a system of equations are eliminated and the resulting statement is false, the system is *inconsistent* and has no solution. The graphs of the equations are parallel lines.

EXAMPLE 5

Solve the system: $2x - y = 7$ and $4x - 2y = 14$.

$$-2(2x - y) = -2(7) \qquad \text{Multiply Equation 1 by } -2.$$
$$-4x + 2y = -14 \qquad \text{Add the equations to eliminate a variable.}$$
$$\underline{4x - 2y = \quad 14}$$
$$0 = \quad 0 \qquad \text{Both variables are eliminated.}$$

Both variables are eliminated; however, the result is a *true* statement ($0 = 0$). In this situation, **all solutions of one equation are also solutions of the other equation.** For example, $x = 4$ and $y = 1$ is a solution of both equations as is $(-2, -11)$, and infinitely many other points. $\qquad$ **See Exercises 24–30.**

If both variables in a system of equations are eliminated and the resulting statement is true, then the system is *dependent* and has many solutions. The graphs of the equations coincide.

10-2 EXERCISES   MyLab Math   For additional practice go to your study plan in MyLab Math.

1 Solve the systems of equations using the addition method. *See Example 1.*

1. $a - 2b = 7$
$3a + 2b = 13$

2. $3m + 4n = 8$
$2m - 4n = 12$

3. $x - 4y = 5$
$-x - 3y = 2$

4. $a - b = 6$
$2a + b = 3$

5. $x + 2y = 5$
$3x - 2y = 3$

6. $x - 5y = 7$
$2x + 5y = 5$

7. $5x + 2y = -3$
$-5x - 4y = -7$

8. $-8x - 3y = 7$
$8x + 5y = 3$

9. $x + 3y = 5$
$-x + 3y = 13$

2 Solve the systems of equations using the addition method. *See Example 2.*

10. $3x + y = 9$
$x + y = 3$

11. $5x - 3y = -4$
$2x - 3y = -7$

12. $a + 6b = 18$
$4a - 3b = 0$

See Example 3.

13. $7x + 2y = 17$
$y = 3x + 2$

14. $a = 6y$
$2a - y = 11$

15. $3x + y = -1$
$4x - 2y = -8$

16. $2x - 3y = 10$
$3x + 2y = 2$

17. $4x - 3y = -19$
$-2x - 4y = 4$

18. $2x + 5y = 0$
$-3x - 2y = -11$

19. $x + 2y = 6$
$2x + 5y = 15$

20. $5x - y = -8$
$x - 2y = -7$

21. $-x + 2y = -7$
$x - 6y = -5$

3 Solve the systems of equations using the addition method. *See Example 4.*

22. $x + y = 8$
$x + y = 3$

23. $x + y = 9$
$x + y = 3$

24. $3a - 2b = 14$
$3a = 2b + 2$

25. $-3x + 2y = 3$
$9x - 6y = -8$

26. $2x + 5y = 3$
$16x + 40y = 38$

27. $-x - 2y = 3$
$4x + 8y = 15$

See Example 5.

28. $3x - 2y = 6$
$9x - 6y = 18$

29. $2a + 4b = 10$
$a + 2b = 5$

30. $5x - 7y = 8$
$10x - 14y = 16$

10-3 | **Solving Systems of Linear Equations Using the Substitution Method**

LEARNING OUTCOME

1 Use the substitution method to solve a system of linear equations.

LC LEARNING CATALYTICS

Multiply.
1. $5(12 - 3y)$ **2.** $-2(-11 + 2x)$

1 **Use the Substitution Method to Solve a System of Linear Equations.** Another method for solving systems of equations is by substitution. Recall that in formula rearrangement (Chapter 8, Section 1, Outcome 2), whenever more than one variable is used in an equation or formula, we can rearrange the equation to solve for a particular variable. In the **substitution method for solving systems of equations**, we solve one equation for one variable and then substitute the equivalent expression in place of the variable in the other equation.

Substitution method for solving systems of equations: a method for solving systems of equations by solving one equation for one variable and then substituting the equivalent expression in place of the variable in the other equation

STOP AND CHECK

1. Solve the system of equations $x - 4y = 5$ and $y = -4x - 14$ using the substitution method.

Answer:

1. $(-3, -2)$

To solve a system of equations by substitution:

1. Rearrange either equation to isolate one variable.

2. Substitute the equivalent expression from Step 1 into the *other* equation and solve for the remaining variable.

3. Substitute the value of the variable found in Step 2 into the equation from Step 1 to find the remaining value of the solution.

4. Check the solution in both original equations.

EXAMPLE 1

Solve the system of equations $x + y = 15$ and $y = 2x$ using the substitution method.

$$x + y = 15 \qquad \text{Equation 1}$$
$$y = 2x \qquad \text{Equation 2 is already solved for } y.$$
$$x + y = 15 \qquad \text{Substitute } 2x \text{ for } y \text{ in Equation 1 and solve.}$$
$$x + 2x = 15 \qquad \text{Solve for } x.$$
$$3x = 15$$
$$x = 5 \qquad x\text{-value of solution}$$
$$y = 2x \qquad \text{Substitute the solution for } x \text{ in Equation 2 to find } y.$$
$$y = 2(5) \qquad \text{Simplify.}$$
$$y = 10 \qquad y\text{-value of solution}$$

The solution is (5, 10).

Check:

$$x + y = 15 \qquad\qquad y = 2x \qquad \text{Substitute } x = 5 \text{ and } y = 10 \text{ in both equations.}$$
$$5 + 10 = 15 \qquad\qquad 10 = 2(5) \qquad \text{Simplify.}$$
$$15 = 15 \qquad\qquad 10 = 10 \qquad \text{The solution checks in both equations.}$$

See Exercises 1–9.

EXAMPLE 2

Solve the following system of equations using the substitution method.

$$2x - 3y = -14$$
$$x + 5y = 19$$

Either equation can be solved for either unknown. In this example, the x-term in the second equation has a coefficient of 1, so the simplest choice would be to solve the second equation for x.

Step 1	**Step 2**	**Step 3**
$x + 5y = 19$	$2x - 3y = -14$	$x = 19 - 5y$
$x = 19 - 5y$	$2(19 - 5y) - 3y = -14$	$x = 19 - 5(4)$
	$38 - 10y - 3y = -14$	$x = 19 - 20$
	$38 - 13y = -14$	$x = -1$
	$-13y = -14 - 38$	
	$-13y = -52$	
	$\dfrac{-13y}{-13} = \dfrac{-52}{-13}$	
	$y = 4$	

The solution is $(-1, 4)$.

Check the solution $x = -1$, $y = 4$ in both original equations:

Step 4

$$
\begin{array}{ll}
2x - 3y = -14 & x + 5y = 19 \\
2(-1) - 3(4) = -14 & -1 + 5(4) = 19 \\
-2 - 12 = -14 & -1 + 20 = 19 \\
-14 = -14 & 19 = 19
\end{array}
$$

See Exercises 10–21.

TIP **Long Problems Don't Have to Be Difficult** Sometimes we let ourselves become overwhelmed by the mere length of a problem. Look at the previous example. Each step of the solution involves skills we have previously used many times. Here are some tips to help you manage longer problems.

▶ Get a global or overall understanding of the problem you are solving.

▶ Make a prediction or estimate of the solution if appropriate.

▶ Get a global or overall understanding of the process you are using to solve the problem.

▶ List in your own words (as briefly as possible) the steps of the process.

▶ Focus on one step at a time.

▶ Examine the solution to see if it matches your prediction or estimate. Check if appropriate.

10-3 EXERCISES

MyLab Math For additional practice go to your study plan in MyLab Math.

1 Solve the systems of equations using the substitution method. *See Example 1.*

1. $2a + 2b = 60$
$\quad a = 10 + b$

2. $7r + c = 42$
$\quad 3r - 8 = c$

3. $x - 3 = 2y$
$\quad x = 3y - 2$

4. $x + y = 12$
$\quad x = 2 + y$

5. $a = 3x - 1$
$\quad x = a + 5$

6. $x + 2y = 5$
$\quad x = 3y$

7. $5x + 2y = 7$
$\quad x = 2y - 1$

8. $x + y = 7$
$\quad x = y - 5$

9. $y = 3x - 2$
$\quad y = x + 6$

See Example 2.

10. $x - 35 = -2y$
$\quad 3x - 2y = 17$

11. $2p + 3k = 2$
$\quad 2p - 3k = 0$

12. $x - 2y = 6$
$\quad 4x + 3y = 35$

13. $2x - 3y = 0$
$\quad 3x - y = 7$

14. $2x - 3y = -3$
$\quad 4x + 2y = 18$

15. $6x - 2y = 3$
$\quad 3x + 4y = 9$

16. $6x + 2y = 3$
$\quad 3x - 4y = -1$

17. $8x - 6y = -2$
$\quad 4x + 9y = 7$

18. $4x + 6y = 1$
$\quad 8x - 4y = -6$

19. $x - y = -4$
$\quad 5x + 9y = 8$

20. $x - 6y = 8$
$\quad x + 3y = -1$

21. $x + 3y = 1$
$\quad 2x + 4y = -4$

10–4 | Problem Solving Using Systems of Linear Equations

LEARNING OUTCOME

1   Use a system of linear equations to solve application problems.

LC **LEARNING CATALYTICS**
Is (3.7, 5.2) a solution for the equation $0.4x = 0.2(2y - 3)$?

1 **Use a System of Linear Equations to Solve Application Problems.** Many job-related problems can be solved by setting up and solving systems of equations. The addition method or the substitution method can be used to solve application problems.

EXAMPLE 1

ELEC   Two dry cells connected in series have a total internal resistance of 0.09 Ω. The difference between the internal resistances of the individual dry cells is 0.03 Ω. How much is each internal resistance?

Known facts

The total internal resistance of the two dry cells is 0.09 Ω.
The difference in the internal resistances of the two dry cells is 0.03 Ω.

Unknown facts

What is the resistance of dry cell 1 (r_1)?
What is the resistance of dry cell 2 (r_2)?

Relationships

$r_1 + r_2 = 0.09$ Equation 1
$r_1 - r_2 = 0.03$ Equation 2

Estimation

If resistances were the same, they would each be 0.045 Ω. Because they are not the same, one will be more than 0.045 Ω and one will be less than 0.045 Ω.

Calculations

$$r_1 + r_2 = 0.09 \qquad \text{Add the equations to eliminate } r_2.$$
$$\underline{r_1 - r_2 = 0.03}$$
$$2r_1 \quad\;\; = 0.12 \qquad \text{Solve for } r_1.$$
$$r_1 = \frac{0.12}{2}$$
$$r_1 = 0.06 \qquad r_1\text{-value of solution}$$
$$r_1 + r_2 = 0.09 \qquad \text{Substitute for } r_1 \text{ in Equation 1.}$$
$$0.06 + r_2 = 0.09 \qquad \text{Solve for } r_2.$$
$$r_2 = 0.09 - 0.06$$
$$r_2 = 0.03 \qquad r_2\text{-value of solution}$$

Interpretation

The larger internal resistance is 0.06 Ω and the smaller internal resistance is 0.03 Ω.

See Exercises 1–2.

EXAMPLE 2

AG/H   A tank holds a solution that is 10% herbicide. Another tank holds a solution that is 50% herbicide. If a farmer wants to mix the two solutions to get 200 gal of a solution that is 25% herbicide, how many gallons of each solution should be mixed?

Chaiyaporn Baokaew/Shutterstock

Known facts

There are two strengths of herbicide, 10% and 50%.

200 gal of 25% herbicide are needed.

Unknown facts

How many gallons of 10% herbicide (h) are needed?
How many gallons of 50% herbicide (H) are needed?

Relationships

200 gal of the new herbicide are needed: $h + H = 200$
Amount of pure herbicide in h gal of 10% herbicide: $0.1h$
Amount of pure herbicide in H gal of 50% herbicide: $0.5H$
Amount of pure herbicide in 200 gal of 25% herbicide: $0.25(200)$

$$h + H = 200 \qquad \text{Equation 1 (total gallons)}$$
$$0.1h + 0.5H = 0.25(200) \qquad \text{Equation 2 (gallons of pure herbicide)}$$

Estimation

If equal amounts of herbicide were needed, we would need 100 gal of each solution. However, since the desired solution strength is not exactly halfway between the two original herbicide strengths, we will need unequal amounts of herbicide. One amount will be less than 100 gal and the other will be more than 100 gal.

Calculations

Solve by the substitution method.

$$h + H = 200 \qquad \text{Solve Equation 1 for } h.$$
$$h = 200 - H \qquad \text{Equivalent expression for } h$$
$$0.1h + 0.5H = 0.25(200) \qquad \text{Substitute } (200 - H) \text{ for } h \text{ in Equation 2.}$$
$$0.1(200 - H) + 0.5H = 0.25(200) \qquad \text{Solve for } H.$$
$$20 - 0.1H + 0.5H = 50$$
$$20 + 0.4H = 50$$
$$0.4H = 50 - 20$$
$$0.4H = 30$$
$$H = \frac{30}{0.4}$$
$$H = 75 \text{ gal} \qquad H\text{-value of solution}$$
$$h + H = 200 \qquad \text{Substitute 75 for } H \text{ in Equation 1.}$$
$$h + 75 = 200 \qquad \text{Solve for } h.$$
$$h = 200 - 75$$
$$h = 125 \text{ gal} \qquad h\text{-value of solution}$$

Interpretation

The farmer must mix 75 gal of the 50% herbicide and 125 gal of the 10% herbicide to make 200 gal of a 25% herbicide.   **See Exercises 3–6.**

STOP AND CHECK

1. A mixture of cheddar and caramel popcorn is called the Chicago mix. If 1 lb of cheddar popcorn costs \$3.25 and 1 lb of caramel popcorn costs \$4.50, how much of each type of popcorn should be used for a mixture of 10 pounds of popcorn costing \$4.00 per pound?

Answer:

1. 4 lb of cheddar popcorn and 6 lb of caramel popcorn

EXAMPLE 3

PFIN Rosita has \$5,500 to invest and for tax purposes wants to earn exactly \$500 interest for 1 year. She wants to invest part at 10% and the remainder at 5%. How much must she invest at each interest rate to earn exactly \$500 interest in 1 year?

Let $x =$ the amount invested at 10%. Let $y =$ the amount invested at 5%. Interest for 1 year = rate × amount invested. Convert percents to decimals. Using these relationships, we derive a system of equations.

Known facts

Total of $5,500 to be invested

$500 interest to be earned in one year

Unknown facts

How much should be invested at 10%?

How much should be invested at 5%?

Relationships

Amount invested at 10%: x

Interest earned at 10%: $0.1x$

Amount invested at 5%: y

Interest earned at 5%: $0.05y$

$x + y = 5,500$	Equation 1 (total investment)
$0.1x + 0.05y = 500$	Equation 2 (total interest in 1 year)

Estimation

If the total amount were invested at 10%, the interest (in 1 year) would be $550 (0.1 × $5,500). Since we want $500 in interest, most of the money will need to be invested at 10%.

Calculations

Solve by the substitution method.

$x + y = 5,500$	Solve Equation 1 for x.
$x = 5,500 - y$	Equivalent expression for x
$0.1x + 0.05y = 500$	Substitute $(5,500 - y)$ for x in Equation 2.
$0.1(5,500 - y) + 0.05y = 500$	Solve for x.
$550 - 0.1y + 0.05y = 500$	$-0.1y + 0.05y = -0.05y$
$550 - 0.05y = 500$	
$-0.05y = 500 - 550$	
$-0.05y = -50$	
$y = \dfrac{-50}{-0.05}$	
$y = \$1,000$	Amount invested at 5%
$x + y = \$5,500$	Substitute $1,000 for y in Equation 1.
$x + \$1,000 = \$5,500$	
$x = \$5,500 - \$1,000$	
$x = \$4,500$	Amount invested at 10%

Interpretation

Rosita must invest $4,500 at 10% and $1,000 at 5% for 1 year to earn $500 interest.

See Exercises 7–9.

10-4 EXERCISES   MyLab Math For additional practice go to your study plan in MyLab Math.

1 Solve the problems using systems of equations with two unknowns. *See Example 1.*

1. **CON** Two lengths of board total 48 in. If one board is 17 in. shorter than the other, find the length of each board.

2. **ELEC** Two resistances have a sum of 21 Ω. Their difference is 13 Ω. Write two equations using R_1 as the first resistance and R_2 as the second resistance. Find each resistance.

See Example 2.

3. **AG/H** A lawn-care technician wants to spread a 200-lb seed mixture that is 50% bluegrass. If the technician has on hand a mixture that is 75% bluegrass and a mixture that is 10% bluegrass, how many pounds of each mixture are needed to make 200 lb of the 50% mixture? Round to the nearest whole pound.

4. A photographer has a container with a solution of 75% developer and a container with a solution of 25% developer. If she wants to mix the solutions to get 8 pt of solution with 50% developer, how many pints of each solution does she need to mix?

5. **BUS** A college bookstore received a partial shipment of 50 scientific calculators and 25 graphing calculators at a total cost of $2,200. Later the bookstore received the balance of the calculators: 25 scientific and 50 graphing at a cost of $3,800. Find the cost of each calculator.

6. **HOSP** For the first performance at the Overton Park Shell, 40 reserved seats and 80 general admission seats were sold for $2,000. For the second performance, 50 reserved seats and 90 general admission seats were sold for $2,350. What was the cost for a reserved seat and for a general admission seat?

See Example 3.

7. **BUS** A broker invested $35,000 in two different stocks. One earned dividends at 4% and the other at 5%. If a $1,570 dividend was earned on both stocks together, how much was invested in each? (*Reminder:* Change 4% and 5% to decimals.)

8. **BUS** A consumer received two 1-yr loans totaling $10,000 at interest rates of 10% and 15%. If the consumer paid $1,300 interest, how much money was borrowed at each rate?

9. **PFIN** In 1 year, Dee Wallace earned $660 in interest on two investments totaling $8,000. If he received 7% and 9% rates of return, how much did he invest at each rate?

10. **AUTO** A mechanic makes $105 on each 8-cylinder engine tune-up and $85 on each 4-cylinder engine tune-up. If the mechanic did 10 tune-ups and made a total of $990, how many 8-cylinder jobs and how many 4-cylinder jobs were completed?

11. **ELEC** 30 resistors and 15 capacitors cost $12. And 10 resistors and 20 capacitors cost $8.50. How much does each capacitor and resistor cost?

12. **AVIA** A private airplane flew 420 mi in 3 h with the wind. The return trip against the wind took 3.5 h. Find the rate of the plane in calm air and the rate of the wind.

13. **BUS** A department store buyer ordered 12 shirts and 8 hats for $380 one month and 24 shirts and 10 hats for $664 the following month. What was the cost of each shirt and each hat?

14. A motorboat went 40 mi with the current in 3 h. The return trip against the current took 4 h. How fast was the current? What would have been the speed of the boat in calm water? Round to the nearest hundredth mile per hour.

15. **PFIN** A visitor to south Louisiana purchased 3 lb of dark-roast pure coffee and 4 lb of coffee with chicory for $27.30 in a local supermarket. Another visitor at the same store purchased 2 lb of coffee with chicory and 5 lb of dark-roast pure coffee for $28. How much did each coffee cost per pound?

16. **AG/H** The total weight of a fertilizer composed of nitrogen and potassium is 480 lb. The fertilizer has three times as much nitrogen as potassium. How many pounds of each chemical are in the fertilizer?

17. **AG/H** A plant nursery purchased holly shrubs that cost $4 each and nandinas that cost $5 each. The total cost was $260 for 60 shrubs. How many of each type of shrub did the nursery purchase?

18. **CON** A mortar mix contains five times as much sand as water. The total volume is 12 ft^3. How much of each ingredient is in the mix?

10 | CHAPTER REVIEW OF KEY CONCEPTS

LEARNING OUTCOMES

KEY CONCEPTS AND EXAMPLES

Section 10–1

1 Solve a system of linear equations by graphing (pp. 469–471).

Solve a system of two linear equations with two variables by graphing: 1. Graph each equation on the same pair of axes.   **2.** The solution will be the common point. (The table-of-solutions, intercepts, or slope-intercept method may also be used to graph each equation.)

Line relationships: When graphing two straight lines, three possibilities can occur.   **1.** The two lines can intersect at *just one point*. This means the system is **independent** and one pair of values satisfies both equations.   **2.** The two lines *do not intersect* at all. This means the system is **inconsistent** and no pair of values satisfies both equations.   **3.** The two lines *coincide* or fall exactly in the same place. This means the system is **dependent** and the equations are identical or are multiples, and any pair of values that satisfies one equation satisfies both equations.

Graph $x + y = 2$ and $x - y = 4$ (Fig. 10–6).

$x + y = 2$ $\qquad\qquad$ $x - y = 4$

$\quad y = -x + 2$ $\qquad\quad$ $\quad y = x - 4$

x	y
0	2
1	1
2	0

x	y
0	−4
1	−3
2	−2

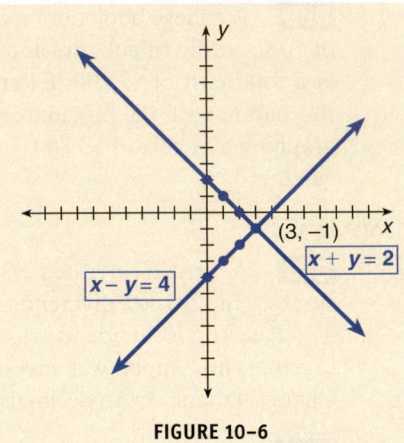

FIGURE 10–6

Solution: $x = 3,\; y = -1.$

2 Solve a system of linear inequalities by graphing (p. 471).

Solve a system of equations using a graphing calculator: 1. Enter both equations in the $\boxed{Y=}$ screen.   **2.** Graph both equations on the same screen.   **3.** Use the Trace feature or the CALC then 5: Intersect feature to determine the approximate coordinates of the intersection of the graphs.   **4.** Use the Zoom or Box (Window) feature to get a closer view and thus a more accurate approximation of the intersection of the graphs.

Solve a system of two linear inequalities with two variables by graphing: 1. Graph each inequality on the same pair of axes.   **2.** The solution set of the system will be the *overlapping* region of the solution sets of the two inequalities.

Graph $x + y \le -2$ and $x - y > 3$.
For $x + y = -2$,

when $x = 0, y = -2$
when $y = 0, x = -2$
when $x = 1, y = -3$
Note the line is solid (is included). Shade *below* the line since the test point (0, 0) makes a false statement.

For $x - y = 3$,

when $x = 0, y = -3$
when $y = 0, x = 3$
when $x = 1, y = -2$
Note the line is dashed (is not included). Shade *below* the line since the test point (0, 0) makes a false statement.

FIGURE 10–7

The overlapping shaded area meets both conditions of the system of inequalities (Fig. 10–7).

LEARNING OUTCOMES	KEY CONCEPTS AND EXAMPLES

Section 10–2

1 Use the addition method to solve a system of linear equations that contains opposite variable terms (pp. 472–473).

Solve a system of linear equations that contains opposite variable terms by the addition (elimination) method: 1. Write each equation in standard form ($ax + by = c$). **2.** Add two equations. **3.** Solve the equation from Step 2. The result is one part of the ordered pair solution. **4.** Substitute the solution from Step 3 in either equation and solve for the remaining variable. This completes the solution ordered pair. **5.** Check the solution in both original equations.

Solve $x - y = 6$ and $x + y = 4$ ($-y$ and $+y$ are opposites).

$$\begin{array}{ll} x - y = 6 & \text{Equation 1. Add the equations to eliminate } y. \\ \underline{x + y = 4} & \text{Equation 2} \\ 2x \quad\;\; = 10 & \text{Solve for } x. \end{array}$$

$$\frac{2x}{2} = \frac{10}{2}$$

$x = 5$ *x*-value of solution

$$\begin{array}{ll} x - y = 6 & \text{Equation 1. Substitute } x = 5. \\ 5 - y = 6 & \text{Solve for } y. \\ -y = 6 - 5 & \\ -y = 1 & \\ y = -1 & \text{\textit{y}-value of solution} \end{array}$$

The solution is (5, −1).

Check:

$$\begin{array}{l} x - y = 6 \\ 5 - (-1) = 6 \\ 5 + 1 = 6 \\ 6 = 6 \\ \\ x + y = 4 \\ 5 + (-1) = 4 \\ 4 = 4 \\ \\ \text{Solution checks.} \end{array}$$

2 Use the addition method to solve a system of linear equations that does not contain opposite variable terms (pp. 474–476).

Solve a system of linear equations that does not contain opposite variable terms using the addition (elimination) method: 1. Write each equation in standard form ($ax + by = c$). **2.** If necessary, multiply one or both equations by numbers that cause the terms of one variable to add to zero. **3.** Add the two equations to eliminate a variable. **4.** Solve the equation from Step 3 for the remaining variable. **5.** Substitute the solution from Step 4 in either equation and solve for the remaining variable. **6.** Check the solution in both original equations.

Solve $2x - y = 6$ and $4x + 2y = 4$.

$$\begin{array}{ll} 2(2x - y) = 2(6) & \text{Multiply Equation 1 by 2. Add equations to eliminate } y. \\ 4x - 2y = 12 & \\ \underline{4x + 2y = 4} & \\ 8x \quad\;\; = 16 & \text{Solve for } x. \end{array}$$

$$\frac{8x}{8} = \frac{16}{8}$$

$x = 2$ *x*-value of solution

$$\begin{array}{ll} 4x + 2y = 4 & \text{Equation 2. Substitute } x = 2. \\ 4(2) + 2y = 4 & \\ 8 + 2y = 4 & \text{Solve for } y. \\ 2y = 4 - 8 & \\ 2y = -4 & \end{array}$$

$$\frac{2y}{2} = \frac{-4}{2}$$

$y = -2$ *y*-value of solution

Solution: (2, −2)

Check:

$$\begin{array}{l} 2x - y = 6 \\ 2(2) - (-2) = 6 \\ 4 + 2 = 6 \\ 6 = 6 \\ \\ 4x + 2y = 4 \\ 4(2) + 2(-2) = 4 \\ 8 - 4 = 4 \\ 4 = 4 \\ \\ \text{Solution checks.} \end{array}$$

LEARNING OUTCOMES	**KEY CONCEPTS AND EXAMPLES**

3 Apply the addition method to a system of linear equations with no solution or with many solutions (p. 476).

Apply the addition rule. But note that *both* variables are eliminated. If the resulting statement is false, there is no solution. The system is *inconsistent*. If the resulting statement is true, there are many solutions. The system is *dependent*.

Solve $a + b = 2$ and $a + b = 4$.

$$-1(a + b) = -1(2) \qquad \text{Multiply the first equation by } -1. \text{ Add to eliminate variables.}$$

$$-a - b = -2$$
$$\underline{a + b = 4}$$
$$0 = 2 \qquad \text{False. No solution. Inconsistent system}$$

Section 10-3

1 Use the substitution method to solve a system of linear equations (pp. 477–479).

Solve a system of equations by substitution: 1. Rearrange either equation to isolate one variable. **2.** Substitute the equivalent expression from Step 1 into the *other* equation and solve for the remaining variable. **3.** Substitute the value of the variable found in Step 2 into the equation from Step 1 to find the remaining value of the solution. **4.** Check the solution in both original equations.

Solve $2x + y = 6$ and $2x - 2y = 8$.

$$y = 6 - 2x \qquad \text{Isolate } y \text{ in the first equation.}$$
$$2x - 2y = 8 \qquad \text{Substitute } 6 - 2x \text{ for } y \text{ in the second equation.}$$
$$2x - 2(6 - 2x) = 8 \qquad \text{Solve for } x.$$
$$2x - 12 + 4x = 8$$
$$6x - 12 = 8$$
$$6x = 8 + 12$$
$$6x = 20$$
$$\frac{6x}{6} = \frac{20}{6}$$
$$x = \frac{10}{3} \qquad x\text{-value of solution}$$

$$\text{Check: } 2x + y = 6 \qquad\qquad 2x - 2y = 8$$

$$y = 6 - 2\left(\frac{10}{3}\right) \qquad 2\left(\frac{10}{3}\right) + \left(\frac{-2}{3}\right) = 6 \qquad 2\left(\frac{10}{3}\right) - 2\left(-\frac{2}{3}\right) = 8$$

$$y = \frac{18}{3} - \frac{20}{3} \qquad \frac{20}{3} + \frac{-2}{3} = 6 \qquad\qquad \frac{20}{3} + \frac{4}{3} = 8$$

$$y = -\frac{2}{3} \quad \begin{array}{l} y\text{-value} \\ \text{of solution} \end{array} \qquad \frac{18}{3} = 6 \qquad\qquad \frac{24}{3} = 8$$

$$6 = 6 \qquad\qquad 8 = 8$$

Solution: $\left(\frac{10}{3}, -\frac{2}{3}\right)$ \qquad Solution checks.

Section 10-4

1 Use a system of linear equations to solve application problems (pp. 480–482).

Let two variables represent the unknown values. Use numbers and the variables to represent the conditions of the problem. Set up a system of equations and solve.

LEARNING OUTCOMES	KEY CONCEPTS AND EXAMPLES

A taxidermist bought two pairs of glass deer eyes and five pairs of glass duck eyes for $15. She later purchased three pairs of deer eyes and five pairs of duck eyes for $20. Find the price per pair of each type of glass eye.

Let x = the price of a pair of deer eyes. Let y = the price of a pair of duck eyes. Price = the number of pairs of eyes times price of eye type:

$2x + 5y = 15$ 2 pairs of deer eyes plus 5 pairs of duck eyes cost $15.

$3x + 5y = 20$ 3 pairs of deer eyes plus 5 pairs of duck eyes cost $20.

$-1(2x + 5y) = -1(15)$ Multiply first equation by -1. Add equations to eliminate y.

$-2x - 5y = -15$

$3x + 5y = 20$

$x = 5$ x-value of solution

$2x + 5y = 15$ Use first equation to find y. Substitute $x = 5$.

$2(5) + 5y = 15$ Solve for y.

$10 + 5y = 15$

$5y = 15 - 10$

$5y = 5$

$\dfrac{5y}{5} = \dfrac{5}{5}$

$y = 1$ y-value of solution

The deer eyes cost $5 a pair and the duck eyes cost $1 a pair.

10   CHAPTER REVIEW EXERCISES

Section 10–1 MyLab Math For additional practice go to your study plan in MyLab Math.

Solve the systems of equations by graphing.

1. $y = 5x - 1$
$y = 3x + 3$

2. $y = -x + 5$
$y = 3x - 7$

3. $y = 5x + 1$
$y = 3x - 1$

4. $2x - y = 5$
$4x - 2y = 2$

5. $x + y = 8$
$x - y = 2$

6. $3x + 2y = 13$
$x - 2y = 7$

7. $2x + 2y = 10$
$3x + 3y = 15$

8. $x + y = 1$
$3x - 4y = 10$

Solve the systems of inequalities by graphing.

9. $2x + y < 6$ and $x - y < 1$

10. $x + 2y < -1$ and $x + 2y > 3$

11. $2x + y > 3$ and $x - y \le 1$

12. $x + 2y > -4$ and $x - 2y > -1$

Section 10–2

Solve the systems of equations using the addition method.

13. $3x + y = 9$
$2x - y = 6$

14. $7x - y = 4$
$8x - y = 3$

15. $5x - 3y = 8$
$-5x + 2y = -7$

16. $2a + 3b = 8$
$a - b = 4$

17. $Q = 2P + 8$
$2Q + 3P = 2$

18. $4j + k = 3$
$8j + 2k = 6$

19. $r = 2y + 6$
$2r + y = 2$

20. $3a + 3b = 3$
$2a - 2b = 6$

21. $c = 2y$
$2c + 3y = 21$

22. $2x + 4y = 9$
$x + 2y = 3$

23. $3R - 2S = 7$
$-14 = -6R + 4S$

24. $x - 3 = -y$
$2y = 9 - x$

25. $c = 2 + 3d$
$3c - 14 = d$

26. $Q - 10 = T$
$T = 2 - 2Q$

27. $x - 18 = -6y$
$4x - 0 = 3y$

28. $R + S = 3$
$S - 9 = -3R$

29. $3a - 2b = 6$
$6a - 12 = b$

30. $a - b = 2$
$a + b = 12$

31. $x + 2y = 7$
$x - y = 1$

32. $2c + 3b = 2$
$2c - 3b = 0$

33. $x + 2r = 5.5$
$2x = 1.5r$

34. $7x + 2y = 6$
$4y = 12 - x$

Section 10–3

Solve the systems of equations using either the addition or the substitution method.

35. $a + 7b = 32$
$3a - b = 8$

36. $x + y = 1$
$4x + 3y = 0$

37. $c - d = 2$
$c = 12 - d$

38. $3a + 4b = 0$
$a + 3b = 5$

39. $7x - 4 = -4y$
$3x + y = 6$

40. $5Q - 4R = -1$
$R + 3Q = -38$

41. $a = 2b + 11$
$3a + 11 = -5b$

42. $y = 5 - 2x$
$3x - 2y = 4$

43. $c = 2q$
$2c + q = 2$

44. $3x + 2y = 10$
$y = 6 - x$

45. $4x - 2.5y = 2$
$2x - 1.5y = -10$

46. $2a - c = 4$
$a = 2 + c$

47. $4d - 7 = -c$
$3c - 6 = -6d$

48. $x = 10 - y$
$5x + 2y = 11$

49. $3.5a + 2b = 2$
$0.5b = 3 - 1.5a$

50. $c + d = 12$
$c - d = 2$

51. $x + 4y = 20$
$4x + 5y = 58$

52. $a + 5y = 7$
$a + 4y = 8$

53. $3a + 1 = -2b$
$4b + 23 = 15a$

54. $6y + 0 = 5p$
$4y - 3p = 38$

Section 10–4

Solve the problems using systems of equations with two unknowns.

55. **ELEC** Three electricians and four apprentices earned a total of $365 on one job. At the same rate of pay, one electrician and two apprentices earned a total of $145. How much pay did each apprentice and electrician receive?

56. **AG/H** Six bushels of bran and 2 bushels of corn weigh 182 lb. If 2 bushels of bran and 4 bushels of corn weigh 154 lb, how much do 1 bushel of bran and 1 bushel of corn each weigh?

57. **CON** A painter paid $22.50 for 2 qt of white shellac and 5 qt of thinner. If 3 qt of shellac and 2 qt of thinner cost the painter's helper $14.50, what is the cost of each quart of shellac and thinner?

58. **ELEC** A main current of electricity is the sum of two smaller currents whose difference is 0.8 A. What are the two smaller currents if the main current is 10 A?

59. The sum of two angles is 175°. Their difference is 63°. What is the measure of each angle?

60. **HOSP** A tour boat traveled 20 mi in 2 h with the current. The return trip took 3 h. Find the rate of the boat in calm water and the rate of the current.

61. **PFIN**   In 1 year, Sholanda Brown earned $280 on two investments totaling $5,000. If she received 5% and 6% rates of return, how much did she invest at each rate?

62. **AVIA**   A plane flew 300 km against the wind in 4 h. The return trip with the wind took 3 h. How fast was the wind? What would have been the speed of the plane in calm air?

63. **HOSP**   A restaurant purchased 30 lb of Colombian coffee and 10 lb of blended coffee for $190. In a second purchase, the same restaurant paid $120 for 20 lb of Colombian coffee and 5 lb of blended coffee. How much did each coffee cost per lb?

64. **AG/H**   A rancher wants to spread a 300-lb grass seed mixture that is 50% tall fescue. If the rancher has on hand a seed mixture that is 80% tall fescue and a mixture that is 20% tall fescue, how many pounds of each mixture are needed to make 300 lb of the 50% mixture?

65. **AUTO**   An automotive service station purchased 25 maps of Ohio and 8 maps of Alaska at a total cost of $65.55. Later the station purchased 20 maps of Ohio and 5 maps of Alaska for $49.50. Find the cost of each map.

66. **PFIN**   Emily Harrington made two 1-year investments totaling $7,000 at interest rates of 4% and 7%. If she received $415 in return, how much money was invested at each rate?

67. **BUS**   At the first of the month, store buyer Selena Henson placed a $12,525 order for 20 name-brand suits and 35 suits with generic labels. At the end of the month, she placed a $15,725 order for 30 name-brand suits and 35 generic-label suits. How much did she pay for each type of suit?

68. **BUS**   A taxidermist has a container with a solution of 10% tanning chemical and a container with a solution of 50% tanning chemical. If the taxidermist wants to mix the solutions to get 10 gal of solution with 25% tanning chemical, how many gallons of each solution should be mixed?

69. **BUS**   Jorge makes 5% commission on telephone sales and 6% commission on showroom sales. If his sales totaled $40,000 and his commission was $2,250, how much did he sell by telephone? How much did he sell on the showroom floor?

70. **BUS**   Sing-Fong has 60 coins in nickels and quarters. The total value of the coins is $12. How many coins of each type does she have?

10   TEAM PROBLEM-SOLVING EXERCISES

1. A true measure of your understanding of a concept is the ability to apply this understanding to real-world situations.
   (a) Write a story line for a mixture problem that can be solved using the system

$$0.5x + 0.3y = 8$$
$$x + y = 20$$

   (b) Solve the system and interpret the results within the context of your story line.

2. Select a career or business situation of interest to your team.
   (a) Write a story line for a mixture problem that can be solved using the system

$$3x + 5y = 28$$
$$x - y = 4$$

   (b) Solve the system and interpret the results within the context of your story line.

10   CONCEPTS ANALYSIS

1. What does the graphical solution of a system of two equations with two unknowns represent?

2. Describe the solution of a system of two equations if the graphs of the two equations are parallel.

3. What can be said about the solution of a system of two equations if the graphs of the two coincide?

4. Describe the solution of a system of two equations if the graphs intersect in exactly one point.

5. Identify the first error that occurs in the solution. Solve the system correctly.

$$x + 2y = 3$$
$$2x - 3y = -1$$
$$2x + 4y = 6$$
$$2x - 3y = -1$$
$$y = 5$$

$$x + 2y = 3$$
$$x + 2(5) = 3$$
$$x + 10 = 3$$
$$x = -7$$
$$\text{Solution: } (-7, 5)$$

6. Explain the addition or elimination method of solving a system of two equations with two unknowns.

7. Explain the substitution method of solving a system of two equations with two unknowns.

8. What does it mean when solving a system of two equations with two unknowns if the system produces a true statement like $0 = 0$?

9. What does it mean when solving a system of two equations with two unknowns if the system produces a false statement like $3 = 0$?

10. Can a system of two linear equations in two unknowns have exactly two ordered pairs as solutions? Explain your answer.

10   PRACTICE TEST

Solve the systems of equations and inequalities graphically.

1. $2a + b = 10$
   $a - b = 5$

2. $x + y = 5$
   $2x - y = 4$

3. $3x + 4y = 6$
   $x + y = 5$

4. $a - 3b = 7$
   $a - 5 = b$

5. $2c - 3d = 6$
   $c - 12 = 3d$

6. $x + y < 4$
   $y > 3x + 2$

Solve the systems of equations using the addition method.

7. $2x - y = 12$
   $y = 4$

8. $x + y = 6$
   $x - y = 2$

9. $p + 2m = 0$
   $2p = -m$

10. $6p + 5t = -16$
    $3p - 3 = 3t$

11. $3x + y = 5$
    $2x - y = 0$

12. $7c - 2b = -2$
    $c - 4b = -4$

Solve the systems of equations with two variables using the substitution method.

13. $4x + 3y = 14$
    $x - y = 0$

14. $a + 2y = 6$
    $a + 3y = 3$

15. $7p + r = -6$
    $3p + r = 6$

Solve the systems of equations with two variables using either the addition or the substitution method.

16. $4x + 4 = -4y$
    $6 + y = -6x$

17. $38 + d = -3a$
    $5a + 1 = 4d$

Solve the problems using systems of equations with two variables.

18. Two lengths of stereo speaker wire total 32.5 ft. One length is 2.9 ft longer than the other. How long is each length of speaker wire?

19. Two currents add to 35 A and their difference is 5 A. How many amperes are in each current?

20. Six packages of common nails and four packages of finishing nails weigh 6.5 lb. If two packages of common nails and three packages of finishing nails weigh 3.0 lb, how much does one package of each kind of nail weigh?

21. The length of a piece of sheet metal is $1\frac{1}{2}$ times the width. The difference between the length and the width is 17 in. Find the length and the width.

22. A mixture of fieldstone is needed for a construction job and will cost $954 for the 27 tons of stone. The stone is of two types, one costing $38 per ton and one costing $32 per ton. How many tons of each are required?

23. A broker invested $25,000 in two different stocks. One stock earned dividends at 3.5% and the other at 4%. If a dividend of $900 was earned on both stocks together, how much was invested in each stock?

24. A mason purchased 2-in. cold-rolled channels and $\frac{3}{4}$-in. cold-rolled channels whose total weight was 820 lb. The difference in weight between the heavier 2-in. and lighter $\frac{3}{4}$-in. channels was 280 lb. How many pounds of each type of channel did the mason purchase?

25. A total capacitance in parallel is the sum of two capacitances. If the capacitances total 0.00027 farad (F) and the difference between the two capacitances is 0.00016 F, what is the value of each capacitance in the system?

7–10   CUMULATIVE PRACTICE TEST

Simplify.

1. $4 - 3(2x - 5)$

2. $5x - (3x - 2)$

Solve.

3. $7x - 3(x - 8) = 28$

4. $18 - 6(2 - y) = 24$

5. $\dfrac{5}{12}x - \dfrac{3}{4} = \dfrac{1}{9} - \dfrac{2}{3}x$

6. $4x - 3.2 + x \le 3.3 - 2.4x$

7. Find the impedance (Z) in ohms using the formula $Z = \sqrt{R^2 + X^2}$ if the resistance (R) is 4 Ω and the reactance (X) is 7 Ω. Round to tenths.

Solve.

8. $\dfrac{3x}{8} = \dfrac{3}{4}$

9. $\dfrac{2x}{7} = \dfrac{3}{5}$

10. $\dfrac{2}{12} = \dfrac{7}{4x}$

11. A large gear has 400 teeth and turns at 30 rpm. Find the number of teeth of a small gear in mesh that turns at 75 rpm.

12. A blueprint has a scale of $\frac{1}{4}$ in. = 1 ft. What is the actual measurement of a wall that is five inches on the blueprint?

13. Graph using a table of solutions: $y = -2x + 4$

14. Graph using the intercepts procedure: $2x - y = 6$

15. Graph using the slope and y-intercept procedure:

$y - \dfrac{3}{4}x = 2$

Determine the slope and y-intercept.

16. $y = \dfrac{3}{4}x + 2$

17. $3x - 2y = 10$

18. Find the slope of the line passing through the points (5, 2) and (−2, 3).

Find the equation of the line passing through the given point with the given slope. Write the equation in slope y-intercept form.

19. $(5, -4), m = -\dfrac{2}{3}$

20. $(-1, -5), m = 3$

Find the equation of the line passing through the given pairs of points. Solve the equation for y.

21. (3, 2) and (5, 6)

22. (−2, 1) and (4, −2)

Write the equation with the given slope and y-intercept. Solve the equation for y.

23. $m = 4, b = 3$

24. slope $= -2$, y-intercept $= \dfrac{1}{3}$

25. Write the equation in standard form of the line that is parallel to the line $x - y = 3$ and contains the point (5, 4).

26. Write the equation in standard form of the line that is perpendicular to the line $x - 2y = 6$ and passes through the point (3, 0).

Solve the systems of equations using the addition method.

27. $2x - y = -1$

$x + y = 4$

28. $3x - 2y = -1$

$2x + 4y = 10$

29. Two resistances have a sum of 32 Ω and a difference of 8 Ω. Find each resistance.

30. The total liquid in a herbicide mixture is 20 L. If the mixture has 4 times as much water as herbicide, how many liters of each substance are in the mixture?

11

Powers and Polynomials

RGB Ventures/SuperStock/Alamy Stock Photo

In Great Company

Castle Bravo (1954)

To this day, the Castle Bravo explosion of March 1, 1954, is not just the largest nuclear explosion in U.S. history, it is the fifth largest nuclear explosion the world has ever seen. It was also a complete surprise, even to the bomb designers.

Castle Bravo was designed to be a 5-megaton fusion bomb. Early nuclear weapons fissioned, or split, uranium atoms. Fusion bombs force together, or fuse, hydrogen atoms. Splitting atoms is easy. Fusing them is hard. A hydrogen bomb is really two bombs in one. The hydrogen bomb trigger is actually a fission bomb, a little nuclear weapon. When the uranium atoms split, the explosive power forces nearby hydrogen atoms to fuse. The hydrogen fusion releases massive energy: the fusion explosion. The fission weapon is the match; the hydrogen is the fuel.

Hydrogen is usually hard to hold in place, but if hydrogen is attached to lithium, it stays in place. Lithium has two isotopes, or variants: lithium-6 and lithium-7. The bomb designers knew lithium-6 would break down and contribute to the explosive force of the fusion bomb. Calculations showed lithium-7 would not do this; it was inert.

Lithium-6 is difficult to separate from lithium-7. But, the designers didn't try very hard to separate the two. As long as they knew how much hydrogen and lithium-6 were in the bomb, they could correctly calculate its explosive force. It would be about 5 megatons.

They planned to detonate the bomb on Bikini Atoll, an empty Pacific island. They placed recording devices, built safety bunkers, made evacuation plans. They designed everything for a 5-megaton detonation.

The day came. The men in the safety bunker pushed the button. Less than a second later, they found out their calculations were wrong. Lithium-7 was not inert. It burned just as well as lithium-6. Castle Bravo had a lot more lithium-7 than lithium-6.

The detonation wasn't 5 megatons. It was 15 megatons, 200% more than expected. Radioactive fallout was a thousandfold higher than expected. Most recording instruments were destroyed; men in observation bunkers were trapped by fallout. The crew of a nearby Japanese tuna boat, *Lucky Dragon #5,* received massive radiation overdoses—the radio operator died. Nearby atolls had to be evacuated. Fallout reached Japan, Australia, India, the United States, and eventually parts of Europe. Cancer rates for Pacific Island natives dramatically increased. It was the biggest nuclear contamination in U.S. history. The Japanese considered it a "second Hiroshima." That calculation mistake eventually led to a ban on atmospheric testing of nuclear devices.

11–1 Laws of Exponents

LEARNING OUTCOMES

1. Multiply powers with like bases.

2. Divide powers with like bases.

3. Find a power of a power.

LC LEARNING CATALYTICS

Perform the operations.
1. $-3 - (-5)$
2. $-3 - 5$

Laws of exponents: laws that allow the shortening of processes for performing operations with exponents

We have seen in various equations and formulas that a variable can represent all types of numbers: natural numbers, whole numbers, integers, fractions, decimals, and signed numbers. When raising a variable to a power, we must consider all the types of numbers the variable can be.

The **laws of exponents** have evolved from the patterns that formed when interpreting powers as repeated multiplication. These laws allow us to shorten our processes and to take advantage of the accessibility of technology. Be sure to notice the circumstances in which a law is applicable. For example, many of the laws apply to factors with *like* bases.

1 Multiply Powers with Like Bases. To multiply 2^3 by 2^2 using repeated multiplication, we have $2(2)(2)$ times $2(2)$ or $2(2)(2)(2)(2)$. Another way of writing this is $2^3 \times 2^2 = 2^5$. We can multiply x^4 and x^2. Even though the value of x is not known, $x^4(x^2)$ is $x(x)(x)(x)$ times $x(x)$ or $x(x)(x)(x)(x)(x)$.

$$x^4(x^2) = x^6$$

The product contains $4 + 2$ or 6 factors of x. The following law is a shortcut for using repeated multiplication.

To multiply powers that have like bases:

1. Verify that the bases are the same. Use this base as the base of the *product*.

2. Add the exponents for the exponent of the product.

This rule can be stated symbolically as

$$a^m(a^n) = a^{m+n} \qquad \text{where } a, m, \text{ and } n \text{ are real numbers and } a \neq 0.$$

STOP AND CHECK

Write the products.

1. $x^7(x^2)$
2. $y(y^4)$
3. $-2x^2(3x^3)$

Answers:

1. x^9　2. y^5　3. $-6x^5$

EXAMPLE 1

Write the products.

(a) $y^4(y^3)$　**(b)** $a(a^2)$　**(c)** $b(b)$　**(d)** $x(x^3)(x^2)$　**(e)** $x^2(y^3)$　**(f)** $5x^3(2x^4)$

(a) $y^4(y^3) = y^{4+3} = \boldsymbol{y^7}$　　　The bases are the same, so add the exponents.

(b) $a(a^2) = a^{1+2} = \boldsymbol{a^3}$　　　When the exponent for a base is not written, it is 1.

(c) $b(b) = b^{1+1} = \boldsymbol{b^2}$

(d) $x(x^3)(x^2) = x^{1+3+2} = \boldsymbol{x^6}$

(e) $x^2(y^3) = \boldsymbol{x^2y^3}$　　　Bases are unlike.

(f) $5x^3(2x^4) = \boldsymbol{10x^7}$　　　Multiply coefficients. Add exponents of like bases.

See Exercises 1–16.

> **TIP**　**Do You Always Add Exponents When Multiplying?**　The previous law applies *only* to expressions with *like* bases. Thus, $x^2(y^3)$ can be written only as x^2y^3. No other simplification can be made.

2　**Divide Powers with Like Bases.**　When reducing fractions, we can reduce to a factor of 1 any factors common to both the numerator and denominator. We use this concept when dividing powers that have like bases.

STOP AND CHECK

Perform the division indicated by each fraction.

1. $\dfrac{x^6}{x}$　2. $\dfrac{y^2}{y^2}$　3. $\dfrac{a^3}{a^5}$

Answers:

1. x^5　2. 1　3. $\dfrac{1}{a^2}$

EXAMPLE 2

Reduce or simplify the fractions; that is, perform the division indicated by each fraction.

(a) $\dfrac{x^5}{x^2}$　**(b)** $\dfrac{a^2}{a}$　**(c)** $\dfrac{b^7}{c^5}$　**(d)** $\dfrac{m^3}{m^3}$　**(e)** $\dfrac{y^3}{y^4}$

(a) $\dfrac{x^5}{x^2} = \dfrac{x(x)(x)(x)(x)}{x(x)} = \boldsymbol{x^3}$　　　The quotient contains $5 - 2 = 3$ factors of x, or x^3.

(b) $\dfrac{a^2}{a} = \dfrac{a(a)}{a} = a$ **(or a^1)**　　　The quotient contains $2 - 1 = 1$ factor of a.

(c) $\dfrac{b^7}{c^5} = \dfrac{b^7}{c^5}$　　　Bases are unlike, so no simplification can be made.

(d) $\dfrac{m^3}{m^3} = \dfrac{(m)(m)(m)}{(m)(m)(m)} = \boldsymbol{1}$　　　All factors of m reduce to 1.

(e) $\dfrac{y^3}{y^4} = \dfrac{(y)(y)(y)}{(y)(y)(y)(y)} = \dfrac{1}{y}$　　　The denominator has more factors of y than the numerator.　　**See Exercises 17–20.**

In Example 2, examine the exponents in the division and the exponent of the quotient. In parts a and b, the exponent of the quotient is the difference of the exponents of the dividend and the divisor.

(a) $\dfrac{x^5}{x^2} = x^{5-2} = x^3$　　　　　　(b) $\dfrac{a^2}{a} = a^{2-1} = a^1 = a$

In part c, the bases are unlike and no simplification can be done.

The results in parts d and e illustrate that the interpretation of exponents goes beyond just repeated multiplication; a new notation for 1 and for reciprocals is needed.

(d) $\dfrac{m^3}{m^3} = m^{3-3} = m^0 = 1$ Any nonzero base raised to the zero power equals 1.

(e) $\dfrac{y^3}{y^4} = y^{3-4} = y^{-1} = \dfrac{1}{y^1} = \dfrac{1}{y}$ An expression with a negative exponent equals the reciprocal of the expression written with a positive exponent having the same absolute value.

> **TIP** **Zero, Reciprocals, and Negative Exponents** Any nonzero number raised to the zero power is equal to 1. The notation for the zero power is a **zero exponent.**
>
> $$n^0 = 1 \qquad \text{where } n \text{ is a real number and } n \neq 0$$
>
> An expression with a **negative exponent** can be written as an equivalent expression with a positive exponent.
>
> $$n^{-1} = \dfrac{1}{n} \qquad \dfrac{1}{n^{-1}} = n \qquad \text{where } n \text{ is a real number and } n \neq 0.$$
>
> A nonzero number times its reciprocal equals 1.
>
> $$n \cdot \dfrac{1}{n} = 1 \qquad n^1 \cdot n^{-1} = n^0 = 1 \qquad \text{where } n \text{ is a real number and } n \neq 0.$$

Zero exponent: a notation for raising a nonzero number to the zero power; $n^0 = 1$ where n is a real number and $n \neq 0$

Negative exponent: a notation for showing the reciprocal as an equivalent expression with a positive exponent; $n^{-1} = \dfrac{1}{n}$ and $\dfrac{1}{n^{-1}} = n$ where n is a real number and $n \neq 0$

> **To divide powers that have like bases:**
>
> 1. Verify that the bases are the same. Use this base as the base of the *quotient.*
>
> 2. Subtract the exponents for the exponent of the quotient.
>
> This rule can be stated symbolically as
>
> $$\dfrac{a^m}{a^n} = a^{m-n} \qquad \text{where } a, m, \text{ and } n \text{ are real numbers except that } a \neq 0.$$

Because expressions with positive integral exponents are evaluated by using repeated multiplication, it is preferable to rewrite expressions with negative exponents as equivalent expressions with positive exponents. This manipulation is accomplished by applying the definition of negative exponents and exponents of zero.

STOP AND CHECK

Write the quotients using positive exponents.

1. $\dfrac{x^6}{x^{10}}$ 2. $\dfrac{y^{-2}}{y^2}$

3. $\dfrac{a^2}{a^{-1}}$ 4. $\dfrac{15x^4}{6x}$

Answers:

1. $\dfrac{1}{x^4}$ 2. $\dfrac{1}{y^4}$ 3. a^3 4. $\dfrac{5x^3}{2}$

EXAMPLE 3

Write the quotients using positive exponents.

(a) $\dfrac{x^5}{x^8}$ (b) $\dfrac{a}{a^4}$ (c) $\dfrac{y^{-3}}{y^2}$ (d) $\dfrac{x^3}{x^{-5}}$ (e) $\dfrac{12x^7}{8x^5}$

(a) $\dfrac{x^5}{x^8} = x^{5-8} = x^{-3} = \dfrac{1}{x^3}$ (b) $\dfrac{a}{a^4} = a^{1-4} = a^{-3} = \dfrac{1}{a^3}$

(c) $\dfrac{y^{-3}}{y^2} = y^{-3-2} = y^{-5} = \dfrac{1}{y^5}$ (d) $\dfrac{x^3}{x^{-5}} = x^{3-(-5)} = x^{3+5} = x^8$

(e) $\dfrac{12x^7}{8x^5} = \dfrac{3x^{7-5}}{2} = \dfrac{3x^2}{2}$ or $\dfrac{3}{2}x^2$

See Exercises 21–28.

TIP   **Manipulating Negative and Positive Exponents** The inverse relationship of multiplication and division allows us flexibility in applying the laws of exponents.
  Look at part a of the previous example.

$$\frac{x^5}{x^8} \quad \text{is} \quad x^5 \div x^8 \quad \text{or} \quad x^5 \cdot \frac{1}{x^8} \quad \text{or} \quad x^5 \cdot x^{-8}$$

The multiplication law of exponents can be used.

$$x^5 \cdot x^{-8} = x^{5+(-8)} = x^{-3}$$

A *factor* can be moved from a numerator to a denominator (or vice versa) by changing the sign of its exponent.

$$x^{-2} = \frac{x^{-2}}{1} = \frac{1}{x^2}, \qquad \frac{1}{x^{-3}} = \frac{x^3}{1} = x^3$$

This property *does not apply* to a *term* that is part of a numerator or denominator that has two or more terms.

$$\frac{x^{-2} + 1}{3} \qquad \textit{does not equal} \qquad \frac{1}{3x^2}.$$

In other words, *only factors* of the entire numerator or denominator can be moved. If more than one term is in the numerator, each term is divided by the denominator.

$$\frac{x^{-2} + 1}{3} = \frac{x^{-2}}{3} + \frac{1}{3} = \frac{1}{3x^2} + \frac{1}{3}$$

STOP AND CHECK

Simplify the expressions and make all exponents positive.

1. $\dfrac{xy^3}{x^2y^{-1}}$ 2. $\dfrac{a^2 - b^3}{ab}$

3. $\dfrac{a^{-3}b}{a^2b^{-1}}$

Answers:

1. $\dfrac{y^4}{x}$ 2. $\dfrac{a^2 - b^3}{ab}$ or $\dfrac{a}{b} - \dfrac{b^2}{a}$

3. $\dfrac{b^2}{a^5}$

EXAMPLE 4

Simplify the expressions and make all exponents positive.

(a) $\dfrac{a^2b^{-3}}{ab^{-1}}$ (b) $\dfrac{xy^{-1}}{xy^2}$ (c) $\dfrac{x^3 + y^2}{xy}$

  There is more than one way to simplify the expressions.

(a)

Option 1

$$\frac{a^2b^{-3}}{ab^{-1}} = a^{2-1}\, b^{-3-(-1)}$$
Apply the division law of exponents. $2 - 1 = 1$; $-3 - (-1) = -3 + 1 = -2$. Be careful with the signs.

$$= ab^{-2}$$
Make all exponents positive.

$$= \frac{a}{b^2}$$

Option 2

$$\frac{a^2\, b^{-3}}{ab^{-1}} = \frac{a^2\, a^{-1}\, b^{-3}\, b^1}{1}$$
Apply the property of negative exponents to write all factors in the numerator. Apply the multiplication law of exponents.

$$= ab^{-2} = \frac{a}{b^2}$$
Make all exponents positive.

(b) $\dfrac{xy^{-1}}{xy^2} = x^{1-1}y^{-1-2}$ Apply the division law of exponents. $x^0 = 1$.

$= y^{-3}$ Make the exponent positive.

$= \dfrac{1}{y^3}$

(c) $\dfrac{x^3 + y^2}{xy} = \dfrac{x^3}{xy} + \dfrac{y^2}{xy}$ The fraction is already simplified. It can be written as two separate fractions if desired. Separate into two terms and apply the division law to like bases.

$= \dfrac{x^{3-1}}{y} + \dfrac{y^{2-1}}{x}$

$= \dfrac{x^2}{y} + \dfrac{y}{x}$ See Exercises 29–36.

---

3 **Find a Power of a Power.** In the term $(2^3)^2$, we have a power raised to a power. 2^3 is the base of the expression and 2 is the exponent.

$$(2^3)^2 \quad \text{is} \quad (2^3)(2^3) = 2^{3+3} = 2^6 \quad \text{or} \quad 64$$

Let's look at some examples of raising numerical and variable powers to a power.

STOP AND CHECK
Find the power of powers by first writing each expression as repeated multiplication and then applying the multiplication law of exponents.
1. $(2^4)^2$
2. $(x^2)^5$

Answers:
1. $2^4(2^4) = 256$ and $2^{4(2)} = 2^8 = 256$
2. $x^2(x^2)(x^2)(x^2)(x^2) = x^{10}$ and $x^{2(5)} = x^{10}$

EXAMPLE 5

Find the powers of powers by first writing each expression as repeated multiplication and then applying the multiplication law of exponents.

(a) $(3^2)^3$ **(b)** $(x^3)^4$ **(c)** $(a^2)^2$ **(d)** $(n^3)^5$

(a) $(3^2)^3 = (3^2)(3^2)(3^2) = 3^{2+2+2} = 3^6 = \mathbf{729}$

(b) $(x^3)^4 = (x^3)(x^3)(x^3)(x^3) = x^{3+3+3+3} = \mathbf{x^{12}}$

(c) $(a^2)^2 = (a^2)(a^2) = a^{2+2} = \mathbf{a^4}$

(d) $(n^3)^5 = (n^3)(n^3)(n^3)(n^3)(n^3) = n^{3+3+3+3+3} = \mathbf{n^{15}}$ See Exercises 37–41.

From Example 5 we can see a pattern developing. In each case, if we multiply the exponents, we get the exponent of the new power.

> **To raise a power to a power:**
>
> 1. Multiply the exponents.
>
> 2. Keep the same base.
>
> This rule can be stated symbolically as
>
> $$(a^m)^n = a^{mn} \quad \text{where } a, m, \text{ and } n \text{ are real numbers and } a \neq 0.$$

Applying this rule to the problems in Example 5, we have

(a) $(3^2)^3 = 3^{2(3)} = 3^6 = 729$ **(b)** $(x^3)^4 = x^{3(4)} = x^{12}$

(c) $(a^2)^2 = a^{2(2)} = a^4$ **(d)** $(n^3)^5 = n^{3(5)} = n^{15}$

Other laws of exponents are extensions or combinations of the three laws we have already examined. These additional laws are useful tools for simplifying expressions containing exponents.

> **To raise a fraction or quotient to a power:**
>
> 1. Raise the numerator (dividend) to the power.
> 2. Raise the denominator (divisor) to the power.
>
> This rule can be stated symbolically as
>
> $$\left(\frac{a}{b}\right)^n = \frac{a^n}{b^n} \quad \text{where } a, b, \text{ and } n \text{ are real numbers and } b \neq 0.$$

EXAMPLE 6

Raise the fractions to the indicated powers.

(a) $\left(\dfrac{2}{3}\right)^2$ **(b)** $\left(\dfrac{-3}{4}\right)^2$ **(c)** $\left(\dfrac{-1}{3}\right)^3$ **(d)** $\left(\dfrac{x}{y^3}\right)^3$ **(e)** $\left(\dfrac{x^2}{y^3}\right)^4$

(a) $\left(\dfrac{2}{3}\right)^2 = \dfrac{2^2}{3^2} = \dfrac{4}{9}$

(b) $\left(\dfrac{-3}{4}\right)^2 = \dfrac{(-3)^2}{4^2} = \dfrac{9}{16}$ $(-3)(-3) = +9$

(c) $\left(\dfrac{-1}{3}\right)^3 = \dfrac{(-1)^3}{3^3} = \dfrac{-1}{27}$ or $-\dfrac{1}{27}$ $(-1)(-1)(-1) = -1$

(d) $\left(\dfrac{x}{y^3}\right)^3 = \dfrac{x^3}{(y^3)^3} = \dfrac{x^3}{y^9}$

(e) $\left(\dfrac{x^2}{y^3}\right)^4 = \dfrac{(x^2)^4}{(y^3)^4} = \dfrac{x^8}{y^{12}}$

See Exercises 42–45.

> **To raise a product to a power:**
>
> Raise each factor to the indicated power.
>
> This rule can be stated symbolically as
>
> $$(ab)^n = a^n b^n \quad \text{where } a, b, \text{ and } n \text{ are real numbers.}$$

EXAMPLE 7

Raise the products to the indicated powers.

(a) $(ab)^2$ **(b)** $(a^2b)^3$ **(c)** $(xy^2)^2$ **(d)** $(3x)^2$ **(e)** $(2x^2y)^3$ **(f)** $(-5xy)^2$

(a) $(ab)^2 = a^{1(2)}\,b^{1(2)} = a^2\,b^2$ Unwritten exponents are understood to be 1.

(b) $(a^2b)^3 = a^{2(3)}b^{1(3)} = a^6\,b^3$

(c) $(xy^2)^2 = x^{1(2)}y^{2(2)} = x^2\,y^4$

(d) $(3x)^2 = 3^{1(2)}\,x^{1(2)} = 3^2 x^2 = 9x^2$ Evaluate numerical factors.

(e) $(2x^2y)^3 = 2^{1(3)}x^{2(3)}y^{1(3)} = 2^3x^6y^3 = 8x^6\,y^3$

(f) $(-5xy)^2 = (-5)^{1(2)}x^{1(2)}y^{1(2)} = (-5)^2x^2y^2 = 25x^2y^2$ **See Exercises 46–60.**

TIP   **Limitations of the Laws of Exponents**   It is very important to understand what the laws of exponents *do not* include.

▶ The product-raised-to-power law applies to factors, not terms.

$$(a + b)^3 \quad \textbf{does not equal} \quad a^3 + b^3$$

▶ The multiplication-of-powers law applies to *like* bases.

$$a^2(b^3) \quad \textbf{does not equal} \quad ab^5 \text{ or } (ab)^5$$

▶ An exponent affects only the one factor or grouping immediately to the left.

$$3x^2 \quad \text{and} \quad (3x)^2 \quad \textbf{are not equal}$$

In the term $3x^2$, the numerical coefficient 3 is multiplied times the square of x. In the term $(3x)^2$, $3x$ is squared. Thus, $(3x)^2 = 3^2x^2 = 9x^2$.

▶ A negative coefficient of a base is not affected by the exponent.

$$-x^3 \quad \textbf{means} \quad -(x)(x)(x)$$

If $x = 4$, $-x^3 = -(4)^3$ or $-(64) = -64$. If $x = -4$, $-x^3 = -(-4)^3$ or $-(-64) = 64$.

STOP AND CHECK

Simplify the expressions and write with positive exponents.

1. $\dfrac{18x^2(x^3)}{24x^{-2}}$ **2.** $\dfrac{(4^{-3}x)^2}{4^{-5}x^{-3}}$

Answers:

1. $\dfrac{3x^7}{4}$ **2.** $\dfrac{x^5}{4}$

EXAMPLE 8

Simplify the expressions and write with positive exponents

(a) $\dfrac{(12x^3)(x^2)}{18x^{-5}}$ and **(b)** $\dfrac{(5^{-2}d)^3}{5^{-4}d^{-2}}$.

(a) $\dfrac{(12x^3)(x^2)}{18x^{-5}}$ Perform operations in numerator.

$\dfrac{12x^5}{18x^{-5}}$ Reduce coefficients. Subtract exponents.

$\dfrac{2x^{5-(-5)}}{3} = \dfrac{2x^{10}}{3}$

(b) $\dfrac{(5^{-2}d)^3}{5^{-4}d^{-2}}$ Raise to power in numerator.

$\dfrac{5^{-6}d^3}{5^{-4}d^{-2}}$ Subtract exponents for like bases.

$5^{-6-(-4)}d^{3-(-2)}$

$5^{-2}d^5$ Write expression with positive exponents.

$\dfrac{d^5}{5^2} = \dfrac{d^5}{25}$

See Exercises 61–64.

11–1 EXERCISES

MyLab Math For additional practice go to your study plan in MyLab Math.

1 Write the products. *See Example 1.*

1. $x^3(x^4)$

2. $m(m^3)$

3. $a(a)$

4. $x^2(x^3)(x^5)$

5. $y(y^2)(y^3)$

6. $a^2(b)$

7. $3a^5(4b^7)$

8. $2x^3(5x^8)$

9. $(5x^2yz)(-2xy^2z^3)$

10. $(-8x^2y)(-3xy^3)$

11. $\left(\dfrac{3}{5}a^2b\right)\left(\dfrac{10}{21}ab^4\right)$

12. $\left(\dfrac{2}{3}a^3b^2\right)\left(\dfrac{9}{16}ab\right)$

13. $(-1.2m^2n^7)(3.5m^{-4}n^3)$ **14.** $(-3a^2)(-4a^3)(-6a)$ **15.** $(32x^3)\left(\dfrac{3}{4}x^4\right)$

16. $(4a^2b)(-3ab)(-2a^{-2}b^2)$

2 Write the quotients with positive exponents. *See Example 2.*

17. $\dfrac{y^7}{y^2}$ **18.** $\dfrac{x^5}{x}$ **19.** $\dfrac{a^3}{a^4}$ **20.** $\dfrac{b^6}{b^5}$

See Example 3.

21. $\dfrac{m^2}{m^2}$ **22.** $\dfrac{x}{x^3}$ **23.** $\dfrac{y^5}{y}$ **24.** $\dfrac{n^2}{n^{-5}}$

25. $\dfrac{x^{-4}}{x^6}$ **26.** $\dfrac{8n^2}{4n^3}$ **27.** $\dfrac{18x^7}{12x^0}$ **28.** $\dfrac{x^{-3}}{x^2}$

See Example 4.

29. $\dfrac{x^2y^{-2}}{xy^4}$ **30.** $\dfrac{ab^2}{a^{-2}b^4}$ **31.** $\dfrac{x^2y^4}{xy^2}$ **32.** $\dfrac{a^3b^2}{a^2b^3}$

33. $\dfrac{21a^3b^{-4}}{7a^2b}$ **34.** $\dfrac{39r^2s^7}{26rs^{-5}}$ **35.** $\dfrac{12x^3y^5}{18x^7y^3}$ **36.** $\dfrac{51m^2n^5}{34m^{-4}n^2}$

3 Simplify the expressions. Make all exponents positive. *See Example 5.*

37. $(4^2)^3$ **38.** $(x^4)^2$ **39.** $(y^7)^0$ **40.** $(x^{10})^4$ **41.** $(a^7)^3$

See Example 6.

42. $\left(-\dfrac{1}{2}\right)^3$ **43.** $\left(-\dfrac{2}{7}\right)^2$ **44.** $\left(\dfrac{a}{b}\right)^4$ **45.** $\left(\dfrac{x^2}{y}\right)^3$

See Example 7.

46. $(2m^2n)^3$ **47.** $(x^2y^4)^3$ **48.** $(-2a)^2$ **49.** $(x^2y)^3$ **50.** $(-3ab^2)^3$

51. $(-7x^2)^5$ **52.** $(4x^2y^8)^2$ **53.** $-7x^3(x^{-5})^4$ **54.** $-x^4$ **55.** $(-x)^4$

56. $-6(x^2)^4$ **57.** $-x^3y(x^2)^5$ **58.** $-3x^9(x^{-2})^3$ **59.** $-x^2y^{-7}$ **60.** $(-3a^2bc)^3$

See Example 8.

61. $\dfrac{(-9x^5)(x^{-2})}{15x^3}$ **62.** $\dfrac{(20y^3)(y^4)}{65y^{-6}}$ **63.** $\dfrac{(6^{-3}d^3)^2}{6^{-6}d^{-3}}$ **64.** $\dfrac{45x^{-3}yx^2}{9xyz}$

11–2   Polynomials

LEARNING OUTCOMES

1 Identify polynomials, monomials, binomials, and trinomials.

1   **Identify Polynomials, Monomials, Binomials, and Trinomials.** A **polynomial** is an algebraic expression in which the exponents of the variables are nonnegative integers and there are no variables in a denominator.

2 Identify the degree of terms and polynomials.

3 Arrange polynomials in descending order.

STOP AND CHECK

Identify which expressions are polynomials. If an expression is not a polynomial, explain why.

1. $3xy + 5x - \dfrac{1}{x}$

2. $2x^2 + 3x^{-1}$

3. $6x^2 + 5$

Answers:

1. Not a polynomial; variable in denominator
2. Not a polynomial; negative exponent of variable
3. Polynomial

Polynomial: an algebraic expression in which the exponents of the variables are nonnegative integers and there are no variables in a denominator

Monomial: a polynomial containing one term. A term may have more than one factor

Binomial: a polynomial containing two terms. Examples include $2x + 4$ and $7xy - 21a$

Trinomial: a polynomial containing three terms. Examples include $a + b + c$ and $x^2 + 2x - 1$

STOP AND CHECK

Identify which of the expressions are polynomials; then state whether each polynomial is a monomial, binomial, or trinomial.

1. $5(2x - 3)$
2. $5x - 2$
3. $10x^2 - 15x + 3$

Answers:

1. Monomial
2. Binomial
3. Trinomial

EXAMPLE 1

Identify which expressions are polynomials. If an expression is not a polynomial, explain why.

(a) $5x^2 + 3x + 2$ **(b)** $5x - \dfrac{3}{x}$ **(c)** 9 **(d)** $-\dfrac{1}{2}x + 3x^{-2}$

(a) $5x^2 + 3x + 2$ **is a polynomial.**

(b) $5x - \dfrac{3}{x}$ **is not a polynomial** because the term $-\dfrac{3}{x}$ has a variable in the denominator.

(c) **9 is a polynomial** because it is equivalent to $9x^0$ and the exponent, zero, is a non-negative integer.

(d) $-\dfrac{1}{2}x + 3x^{-2}$ **is not a polynomial** because $3x^{-2}$ has a negative exponent.

See Exercises 1–4.

Some polynomials have special names depending on the number of terms contained in the polynomial.

A **monomial** is a polynomial containing one term. A term may have more than one factor.

$$3, -2x, 5ab, 7xy^2, \frac{3a^2}{4} \text{ are monomials.}$$

A **binomial** is a polynomial containing two terms.

$$x + 3, 2x^2 - 5x, x + \frac{y}{4} \text{ are binomials.}$$

A **trinomial** is a polynomial containing three terms.

$$a + b + c, x^2 - 3x + 4, x + \frac{2a}{7} - 5 \text{ are trinomials.}$$

To identify polynomials, monomials, binomials, and trinomials:

1. Write all variables in the numerator if necessary.

2. Expressions that have a negative exponent in *any* term or a variable in a denominator are not polynomials.

3. Identify the expression based on the number of terms it contains. A monomial has one term, a binominal has two terms, and a trinomial has three terms.

EXAMPLE 2

Identify which of the expressions are polynomials; then state whether each polynomial is a monomial, binomial, or trinomial.

(a) $x^2y - 1$ **(b)** $4(x - 2)$ **(c)** $\dfrac{2x + 5}{2y}$ **(d)** $3x^2 - x + 1$

(a) $x^2y - 1$ — **Binomial**

(b) $4(x - 2)$ — **Monomial**

(c) $\dfrac{2x + 5}{2y}$ — This one-term expression is **not a monomial because it is not a polynomial** (there is a variable in the denominator).

(d) $3x^2 - x + 1$ — **Trinomial**

See Exercises 5–16.

Degree of a term: the degree of a term that has only one variable with a nonnegative exponent is the same as the exponent of the variable; the degree of a *constant* term is zero; the degree of a term that has more than one variable factor is the *sum* of the exponents of all the variable factors

2 Identify the Degree of Terms and Polynomials. The **degree of a term** that has only one variable with a nonnegative exponent is the same as the exponent of the variable.

$$3x^4, \quad \text{fourth degree} \qquad -5x, \quad \text{first degree}$$

The degree of a *constant* is 0. A variable to the zero power is implied.

$$5 = 5x^0, \quad \text{degree zero}$$

If a term has more than one variable, the degree of the term is the *sum* of the exponents of all the variable factors.

$$2xy = 2x^1y^1, \quad \text{second degree} \qquad -2ab^2 = -2a^1b^2, \quad \text{third degree}$$

To identify the degree of a term:

1. Exponents of variable factors must be integers greater than zero.

2. For a term that is a constant, the degree is zero.

3. For a term that has only one variable factor, the degree is the same as the exponent of the variable.

4. For a term that has more than one variable factor, the degree is the sum of the exponents of the variables.

STOP AND CHECK
Identify the degree of each term in the polynomial.
1. $3x^4 - 2x^2 - 4$

Answer:
1. $3x^4$, degree 4; $-2x^2$, degree 2; -4, degree 0

EXAMPLE 3

Identify the degree of each term in the polynomial $5x^3 + 2x^2 - 3x + 3$.

$5x^3$,	degree 3 (or third degree)
$2x^2$,	degree 2 (or second degree)
$-3x$,	degree 1 (or first degree)
3,	degree 0

See Exercises 17–28.

Linear term: a term that has degree 1

Quadratic term: a term that has degree 2

Cubic term: a term that has a degree 3

Degree of polynomial: the degree of a polynomial that has only one variable and only positive integral exponents is the degree of the term with the largest exponent

Linear polynomial: a polynomial that has degree 1

Quadratic polynomial: a polynomial that has degree 2

Cubic polynomial: a polynomial that has degree 3

Special names are associated with terms of degree 0, 1, 2, and 3. A *constant term* has degree 0. A **linear term** has degree 1. A **quadratic term** has degree 2. A **cubic term** has degree 3.

The **degree of a polynomial** that has only one variable and only positive integral exponents is the degree of the term with the largest exponent.

A **linear polynomial** has degree 1. A **quadratic polynomial** has degree 2. A **cubic polynomial** has degree 3.

$$5x^3 - 2 \text{ has a degree of 3 and is a cubic polynomial.}$$
$$x + 7 \text{ has a degree of 1 and is a linear polynomial.}$$
$$7x^2 - 4x + 5 \text{ has a degree of 2 and is a quadratic polynomial.}$$

To identify the degree of a polynomial:

1. Identify the degree of each term of the polynomial.

2. Compare the degrees of each term of the polynomial and select the greatest degree as the degree of the polynomial.

STOP AND CHECK

Identify the degree of the polynomials.

1. $2x^3 - 16$

2. $3x - \dfrac{4}{5}$

3. $x^2 - 5x + 9$

Answers:

1. Degree 3, cubic polynomial
2. Degree 1, linear polynomial
3. Degree 2, quadratic polynomial

EXAMPLE 4

Identify the degree of the polynomials.

(a) $5x^4 + 2x - 1$ **(b)** $3x^3 - 4x^2 + x - 5$ **(c)** 7   **(d)** $x - \dfrac{1}{2}$

(a) $5x^4 + 2x - 1$ has a **degree of 4.**

(b) $3x^3 - 4x^2 + x - 5$ has a **degree of 3 and is a cubic polynomial.**

(c) 7 has a **degree of 0 and is a constant.**

(d) $x - \dfrac{1}{2}$ has a **degree of 1 and is a linear polynomial.**

See Exercises 29–36.

3 **Arrange Polynomials in Descending Order.** The terms of a polynomial in one variable are customarily arranged in order based on the degree of each term of the polynomial. The terms can be arranged beginning with the term with the highest degree and then in order of descending degrees (**descending order**) or beginning with the term with the lowest degree and then in order of ascending degrees (**ascending order**).

Descending order: arranging the terms of the polynomial beginning with the term with the highest degree

Ascending order: arranging the terms of the polynomial beginning with the term with the lowest degree

Leading term: the first term of a polynomial arranged in descending order

Leading coefficient: the coefficient of the leading term of a polynomial

> **TIP** **Most Common Arrangement of Polynomials** Polynomials are most often arranged in *descending* order so that the degree of the polynomial is the degree of the first term.

The first term of a polynomial arranged in descending order is called the **leading term** of the polynomial. The coefficient of the leading term of a polynomial is called the **leading coefficient.**

> **To arrange polynomials in descending order of a variable:**
>
> 1. Identify the variable on which the terms of the polynomial will be arranged if the polynomial has more than one variable.
>
> 2. Compare the degrees of the selected variable for each term.
>
> 3. List the term with highest degree of the specified variable first.
>
> 4. Continue to list the terms of the polynomial in descending order of the selected variable.

STOP AND CHECK

Arrange each polynomial in descending order and identify the degree, the leading term, and the leading coefficient of the polynomial.

1. $3x^2 - 5 + 4x - x^3 + 2x^4$

Answer:

1. $2x^4 - x^3 + 3x^2 + 4x - 5$; degree 4; leading term $2x^4$; leading coefficient 2

EXAMPLE 5

Arrange each polynomial in descending order and identify the degree, the leading term, and the leading coefficient of the polynomial.

(a) $5x + 3x^3 - 7 + 6x^2$ **(b)** $x^4 - 2x + 3$ **(c)** $4 + 2x$ **(d)** $x^2 + 5$

(a) $5x + 3x^3 - 7 + 6x^2 = \mathbf{3x^3 + 6x^2 + 5x - 7.}$
Third degree, leading term is $3x^3$, leading coefficient is 3.

(b) $\mathbf{x^4 - 2x + 3}$ is already in descending order.
Fourth degree, leading term is x^4, leading coefficient is 1.

(c) $4 + 2x = \mathbf{2x + 4.}$
First degree or linear polynomial, leading term is x, leading coefficient is 2.

(d) $\mathbf{x^2 + 5}$ is already in descending order.
Second degree or quadratic polynomial, leading term is x^2, leading coefficient is 1.

See Exercises 37–48.

A polynomial arranged in descending order can be written with every successive degree represented. Missing terms have a coefficient of 0.

$$x^4 - 2x + 3 \text{ is the same as } x^4 + 0x^3 + 0x^2 - 2x^1 + 3x^0.$$

11–2 EXERCISES

MyLab Math For additional practice go to your study plan in MyLab Math.

1 Identify which expressions are polynomials. If an expression is not a polynomial, explain why. *See Example 1.*

1. $3x^2 - 5 + \dfrac{1}{2x}$

2. 245

3. $-3x^3 + \dfrac{3}{4}$

4. $-8x^4 + 2x^2 - 3x^{-1}$

Identify each of the following expressions as a monomial, binomial, or trinomial. *See Example 2.*

5. $2x^3y - 7x$

6. $5xy^2 + 8y$

7. $3xy$

8. $7ab$

9. $5(3x - y)$

10. $(x - 5)(2x + 4)$

11. $5x^2 - 8x + 3$

12. $7y^2 + 5y - 1$

13. $\dfrac{4x - 1}{5}$

14. $\dfrac{x}{6} - 5$

15. $4(x^2 - 2) + x^3$

16. $3x^2 - 7(x - 8)$

2 Identify the degree of each term in each polynomial. *See Example 3.*

17. $6x$

18. $8x^2$

19. $6x^2 - 8x + 12$

20. $7x^3 - 8x + 12$

21. $x - 12$

22. $3x^2 - 8$

23. 15

24. 21

25. $2x - \dfrac{1}{4}$

26. $8x^2 + \dfrac{5}{6}$

27. $5x^2y^3 + 7xy$

28. $8r^4s - 5r^3s^5$

Identify the degree of each polynomial. *See Example 4.*

29. $5x^2 + 8x - 14$

30. $x^3 - 8x^2 + 5$

31. $9 - x^3 + x^6$

32. $12 - 15x^2 - 7x^5$

33. $2x - \dfrac{4}{5}x^2$

34. $\dfrac{7}{8} - x$

35. $5x^3y + 3x^2y^2 + 5xy^5$

36. $7x^2y - 4xy^5$

3 Arrange each polynomial in descending powers of x, and identify the degree, leading term, and leading coefficient of the polynomial. *See Example 5.*

37. $5x - 3x^2$

38. $7 - x^3$

39. $4x - 8 + 9x^2$

40. $5x^2 + 8 - 3x$

41. $7x^3 - x + 8x^2 - 12$

42. $7 - 15x^4 + 12x$

43. $-7x + 8x^6 - 7x^3$

44. $15 - 14x^8 + x$

45. $12x + 8x^4 + 15x^3$

46. $5x^3 - 7x^4 + 3x - 8$

47. $3x - 5x^4 + 2x^2 - 5$

48. $-3 + 7x - 8x^4 - x^3$

11–3 Basic Operations with Polynomials

LEARNING OUTCOMES

1 Add and subtract polynomials.

2 Multiply polynomials.

3 Use the FOIL method to multiply two binomials.

1 Add and Subtract Polynomials. Now that variables with exponents have been introduced, we can broaden our concept of *like terms*. For variable terms to be like terms, the variables as well as the exponents of the variables must be exactly the same; that is, $2x^2$ and $-4x^2$ are like terms because the x's are both squared. But $2x^2$ and $4x$ are not like terms because the x terms do not have the same exponent.

4 Multiply polynomials that result in special products.

5 Divide polynomials.

LC LEARNING CATALYTICS
Perform the operations.
1. $-3x(2x - 5)$
2. $\left(\dfrac{-3}{2}\right)^3$

STOP AND CHECK

1. Are $3xy$ and $-\dfrac{2}{5}xy$ like terms?

2. Are $-6x^3y$ and $5xy^3$ like terms?

Answers:
1. Yes 2. No

EXAMPLE 1

Which pairs of terms are like terms?

(a) $3a^2b$ and $-\frac{2}{3}a^2b$ **(b)** $-8xy^2$ and $7x^2y$ **(c)** $10ab^2$ and $(-2ab)^2$

(d) $5x^4y^3$ and $-2y^3x^4$ **(e)** $2x^2$ and $3x^3$

(a) $3a^2b$ and $-\frac{2}{3}a^2b$ **are like terms.** All variables and their corresponding exponents are the same.

(b) $-8xy^2$ and $7x^2y$ are *not* **like terms.** In $-8xy^2$ the exponent of x is 1, and in $7x^2y$ the exponent of x is 2. Also, the exponents of y are not the same.

(c) $10ab^2$ and $(-2ab)^2$ are *not* **like terms.** In the term $10ab^2$, only the b is squared. In $(-2ab)^2$, the entire term is squared, and the result of the squaring is $4a^2b^2$.

(d) $5x^4y^3$ and $-2y^3x^4$ **are like terms.** Both x factors have an exponent of 4, and both y factors have an exponent of 3. Because multiplication is commutative, the order of factors does not matter.

(e) $2x^2$ and $3x^3$ are *not* **like terms.** The exponents of x are not the same.

See Exercises 1–4.

Algebraic expressions containing several terms are simplified as much as possible by combining like terms.

> **To add or subtract (combine) like terms:**
>
> 1. Combine the coefficients of the like terms using the rules for adding or subtracting signed numbers.
>
> 2. The variable factors and their exponents do not change in the sum or difference.

TIP **What Does *Simplify* Mean?** The instructions to "simplify the expression" are vague but often used in mathematics exercises. In general, to simplify an expression means to write the expression using fewer terms or reduced or with lower coefficients or exponents. When the instructions say "simplify," the intent is for you to examine the expression and see which laws or operations allow you to rewrite the expression in a simpler form.

STOP AND CHECK

Simplify.
1. $3x^2 - x^2$
2. $2x^3 - 7x^3$
3. $5a + 2b - 3a - 6b$

Answers:
1. $2x^2$ 2. $-5x^3$ 3. $2a - 4b$

EXAMPLE 2

Simplify the algebraic expressions by combining like terms.

(a) $5x^3 + 2x^3$ **(b)** $x^5 - 4x^5$ **(c)** $a^3 + 4a^2 + 3a^3 - 6a^2$

(d) $3m + 5n - m$ **(e)** $y + y - 5y^2 + y^3$

(a) $5\ x^3 + 2\ x^3 = (5 + 2)\ x^3 = 7\ x^3$ $5x^3$ and $2x^3$ are like terms; add the coefficients 5 and 2. The sum has the same variable factor and exponent as the like terms.

(b) $x^5 - 4\ x^5 = (1 - 4)\ x^5 = -3\ x^5$ x^5 and $-4x^5$ are like terms. $1 - 4 = -3$. The difference has the same variable factor and exponent as the like terms.

(c) $a^3 + 4a^2 + 3a^3 - 6a^2 =$ **$4a^3 - 2a^2$** a^3 and $3a^3$ are like terms. $4a^2$ and $-6a^2$ are like terms. Combine coefficients mentally.

(d) $3m + 5n - m =$ **$2m + 5n$** $3m$ and $-m$ are like terms.

(e) $y + y - 5y^2 + y^3 =$ **$2y - 5y^2 + y^3$** y and y are like terms. Coefficients are 1.

See Exercises 5–10.

When an algebraic expression contains a grouping preceded immediately by a negative sign, we subtract the entire grouping. Another interpretation is that the grouping is multiplied by -1, the implied coefficient of the grouping. Either interpretation causes *each* sign within the grouping to be changed to its opposite and at the same time removes the parentheses. If the grouping is preceded by a positive sign or an unexpressed positive sign, parentheses are removed without changing signs, as if each term in the grouping were multiplied by $+1$, the implied coefficient.

STOP AND CHECK
Simplify the expression.
1. $x^2 - 4y + 3 - (2x^2 - 5y - 2)$

Answer:
1. $-x^2 + y + 5$

EXAMPLE 3

Simplify the expressions.

(a) $y^2 + 2y - (3y^2 + 5y)$

(b) $(m^3 - 3m^2 - 5m + 4) - (4m^3 - 2m^2 - 5m + 2)$

(a) $y^2 + 2y - (3y^2 + 5y) =$

$y^2 + 2y - 1(3y^2 + 5y) =$ Distribute the implied coefficient of -1.

$y^2 + 2y - 3y^2 - 5y =$ Combine like terms.

$-2y^2 - 3y$

(b) $(m^3 - 3m^2 - 5m + 4) - (4m^3 - 2m^2 - 5m + 2) =$ Distribute the implied coefficient of -1.

$(m^3 - 3m^2 - 5m + 4) - 1(4m^3 - 2m^2 - 5m + 2) =$ Combine like terms.

$m^3 - 3m^2 - 5m + 4 - 4m^3 + 2m^2 + 5m - 2 =$

$-3m^3 - m^2 + 2$ **See Exercises 11–16.**

2 **Multiply Polynomials.** Simplifying algebraic expressions may also involve multiplication.

> **To multiply by a monomial:**
>
> 1. Multiply the coefficients using the rules for signed numbers.
>
> 2. Multiply the variable factors using the laws of exponents for factors with like bases.
>
> 3. Distribute if multiplying a polynomial by a monomial.

STOP AND CHECK
Multiply.
1. $-5x^2(-x^3)$
2. $3x(15x^3)$

Answers:
1. $5x^5$ **2.** $45x^4$

EXAMPLE 4

Multiply.

(a) $4x(3x^2)$ **(b)** $-6y^2(2y^3)$ **(c)** $-a(-3a)$

(a) $4x(3x^2) = 4(3)(x^{1+2}) =$ **$12x^3$**   Multiply coefficients. Add exponents.

(b) $-6y^2(2y^3) = -6(2)(y^{2+3}) =$ **$-12y^5$**   Multiply coefficients. Add exponents.

(c) $-a(-3a) = \boxed{-1(-3)}\,(a^{1+1}) = \boxed{3\,a^2}$ Multiply coefficients. Add exponents. The coefficient of $-a$ is -1. The exponent of $-a$ is 1.   **See Exercises 17–20.**

If factors are to be multiplied times a factor with more than one term, apply the distributive property.

STOP AND CHECK

Perform the multiplications.
1. $x(5x^2 - 3x + 2)$
2. $-3x^2(x - 7)$

Answers:
1. $5x^3 - 3x^2 + 2x$
2. $-3x^3 + 21x^2$

EXAMPLE 5

Perform the multiplications.

(a) $2x(x^2 - 4x)$ **(b)** $-2y^3(2y^2 + 5y - 6)$ **(c)** $4a(3a^3 - 2a^2 - a)$

(a) $2x\,(x^2 - 4x) = \mathbf{2x^3 - 8x^2}$ Distribute.

(b) $-2y^3\,(2y^2 + 5y - 6) = \mathbf{-4y^5 - 10y^4 + 12y^3}$ Distribute.

(c) $4a\,(3a^3 - 2a^2 - a) = \mathbf{12a^4 - 8a^3 - 4a^2}$ Distribute.

    See Exercises 21–28.

We multiply two binomials by applying the distributive property more than once. According to the distributive property, each term of the first factor is multiplied by each term of the second factor. This means we are required to use the distributive property more than once.

STOP AND CHECK

Multiply.
1. $(x + 1)(x + 3)$
2. $(2x + 3)(x - 4)$

Answers:
1. $x^2 + 4x + 3$
2. $2x^2 - 5x - 12$

EXAMPLE 6

Multiply $(x + 4)(x + 2)$.

$$
\begin{aligned}
(x + 4)\,(x + 2) &= x\,(x + 2) + 4\,(x + 2) && \text{Apply the distributive property.}\\
&= x^2 + 2x + 4x + 8 && \text{Combine like terms.}\\
&= x^2 + 6x + 8 && \textbf{See Exercises 29–32.}
\end{aligned}
$$

To multiply two polynomials:

1. Use the distributive property to multiply each term of the first polynomial times the entire second polynomial.

2. Combine like terms.

3. Symbolically,

$$(a + b)\,(c + d) = a\,(c + d) + b\,(c + d) = ac + ad + bc + bd$$

$$(a + b + c)\,(d + e + f) = a\,(d + e + f) + b\,(d + e + f) + c\,(d + e + f)$$

$$= ad + ae + af + bd + be + bf + cd + ce + cf$$

STOP AND CHECK

Multiply.
1. $(x - 7)(2x^2 - x + 1)$
2. $(3x^2 + 2x - 1)(x^2 - 5x + 2)$

Answers:
1. $2x^3 - 15x^2 + 8x - 7$
2. $3x^4 - 13x^3 - 5x^2 + 9x - 2$

EXAMPLE 7

Multiply **(a)** $(3x - 5)(2x^2 - x - 1)$ and **(b)** $(2x^2 + 3x - 2)(3x^2 - 5x + 6)$.

(a)
$$
\begin{aligned}
(3x - 5)(2x^2 - x - 1) &=\\
3x(2x^2 - x - 1) - 5(2x^2 - x - 1) &= && \text{Distribute.}\\
6x^3 - 3x^2 - 3x - 10x^2 + 5x + 5 &= && \text{Combine like terms.}\\
6x^3 - 13x^2 + 2x + 5 &&&
\end{aligned}
$$

(b)

$$(2x^2 + 3x - 2)(3x^2 - 5x + 6) =$$

$$2x^2(3x^2 - 5x + 6) + 3x(3x^2 - 5x + 6) - 2(3x^2 - 5x + 6) = \qquad \text{Distribute.}$$

$$6x^4 - 10x^3 + 12x^2 + 9x^3 - 15x^2 + 18x - 6x^2 + 10x - 12 = \qquad \text{Combine like terms.}$$

$$6x^4 - x^3 - 9x^2 + 28x - 12 \qquad \text{See Exercises 33–48.}$$

TIP   **Use Long Multiplication to Multiply Polynomials**   The preceding problem can be organized using a procedure similar to the long-multiplication procedure in arithmetic.

1. Multiply each term in the multiplier times the entire multiplicand.

2. Align partial products so that like terms are in columns.

3. Combine like terms.

$$
\begin{array}{r}
3x^2 - 5x + 6 \\
2x^2 + 3x - 2 \\
\hline
\end{array}
$$

$$-6x^2 + 10x - 12 \qquad \text{Multiply by } -2.$$
$$9x^3 - 15x^2 + 18x \qquad \text{Multiply by } 3x.$$
$$6x^4 - 10x^3 + 12x^2 \qquad \text{Multiply by } 2x^2.$$
$$\overline{\;6x^4 - x^3 - 9x^2 + 28x - 12\;} \qquad \text{Combine like terms.}$$

Since multiplication is commutative, the polynomials can be interchanged.

$$
\begin{array}{r}
2x^2 + 3x - 2 \\
3x^2 - 5x + 6 \\
\hline
\end{array}
$$

$$+12x^2 + 18x - 12 \qquad \text{Multiply by } 6.$$
$$-10x^3 - 15x^2 + 10x \qquad \text{Multiply by } -5x.$$
$$6x^4 + 9x^3 - 6x^2 \qquad \text{Multiply by } 3x^2.$$
$$\overline{\;6x^4 - x^3 - 9x^2 + 28x - 12\;} \qquad \text{Combine like terms.}$$

---

3   **Use the FOIL Method to Multiply Two Binomials.**   When multiplying two binomials, we can guide ourselves through the repeated applications of the distributive property by using the acronym **FOIL**.

FOIL: a systematic method for multiplying two binomials, where F refers to the product of the first term of each factor. O refers to the product of the two outer terms, I refers to the product of the two inner terms, and L refers to the product of the last term of each factor

To multiply two binomials using the FOIL method:

1. Write the product of the *first* term of each factor.

2. Write the product of the two *outer* terms of the factors.

3. Write the product of the two *inner* terms of the factors.

4. Write the product of the *last* term of each factor.

5. Combine like terms.

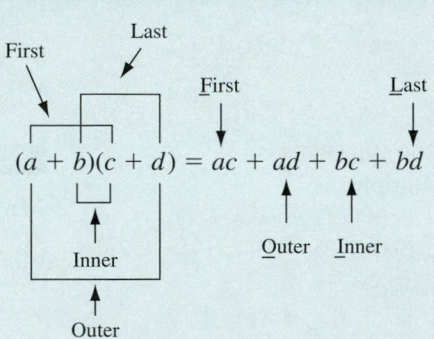

EXAMPLE 8

Use the FOIL method to multiply $(2x - 3)(x + 1)$.

$$(2x - 3)(x + 1) = 2x^2 + 2x - 3x - 3$$
$$= 2x^2 - x - 3$$

See Exercises 49–75.

4 Multiply Polynomials That Result in Special Products. When the factors being multiplied have certain characteristics or conditions, we can anticipate the product without going through all the steps. Identifying and using these special characteristics allows us to do many multiplications mentally.

When we multiply the sum of two terms by the difference of the same two terms, we make some special observations about the product.

$$(x + 3)(x - 3) = x^2 - 3x + 3x - 9 = x^2 - 9$$

Examine the final product. Notice that the product has only *two* terms and is a *difference*. The sum of the outer and inner products is zero $(-3x + 3x = 0)$. The absolute values of the terms of the product are *perfect squares*.

Compare the product with its factors. The first term of the product is the *square of the first term* in either of its factors. The second term in the product is the *square of the absolute value of the second term* in either of its factors.

We call pairs of factors like $(a + b)(a - b)$ the **sum and difference of the same two terms,** or **conjugate pairs.** The product, $a^2 - b^2$, is called the **difference of two perfect squares.**

Sum and difference of the same two terms: pairs of factors like $(a + b)(a - b)$; also called conjugate pairs

Conjugate pairs: pairs of factors like $(a + b)(a - b)$

Difference of two perfect squares: an algebraic expression in the form $a^2 - b^2$

To mentally multiply the sum and difference of the same two terms (conjugate pairs):

1. Square the first term of either binomial.
2. Insert a minus sign.
3. Square the second term of either binomial.

Symbolically, $(a + b)(a - b) = a^2 - b^2$

EXAMPLE 9

Find the products mentally.

(a) $(x + 2)(x - 2)$ (b) $(2a + y)(2a - y)$

Since each of these is the sum and difference of the same two terms, the product is the difference of two perfect squares.

(a) $(x + 2)(x - 2) = x^2 - 4$ Square x. Square 2.

(b) $(2a + y)(2a - y) = 4a^2 - y^2$ Square 2a. Square y. See Exercises 76–87.

Binomial square: a grouping including a binomial that is squared

Square of a binomial: a grouping including a binomial that is squared

When we multiply the same two binomials, we make some special observations. Because $x \cdot x = x^2$, $(x + 3)(x + 3) = (x + 3)^2$. The quantity $(x + 3)^2$ is called a **binomial square** or the **square of a binomial.**

EXAMPLE 10

Square the binomials using the FOIL method.

(a) $(x + 2)^2$ **(b)** $(2x - 5)^2$ **(c)** $(3a - 2b)^2$

To square the binomials, write each binomial square as two factors; then use the FOIL method to multiply.

(a) $(x + 2)^2 = (x + 2)(x + 2)$
$$= x^2 + 4x + 4 \qquad \text{Outer + Inner: } 2x + 2x = 4x$$

(b) $(2x - 5)^2 = (2x - 5)(2x - 5)$
$$= 4x^2 - 20x + 25 \qquad \text{Outer + Inner: } -10x - 10x = -20x$$

(c) $(3a - 2b)^2 = (3a - 2b)(3a - 2b)$
$$= 9a^2 - 12ab + 4b^2 \qquad \text{Outer + Inner: } -6ab - 6ab = -12ab$$

See Exercises 88–102.

Perfect-square trinomial: a trinomial whose first and last terms are positive perfect squares and whose middle term is twice the product of the square root of the first and last terms. The middle term can be positive or negative.

Notice that each product is a trinomial (three terms). These are special trinomials called **perfect-square trinomials.**

To mentally square a binomial:

1. Square the first term of the binomial for the first term of the perfect-square trinomial.

2. Double the product of the two terms of the binomial for the middle term of the trinomial.

3. Square the second term of the binomial for the third term of the trinomial.

     Symbolically, $(a + b)^2 \quad = \quad a^2 + 2ab + b^2$
                            $(a - b)^2 \quad = \quad a^2 - 2ab + b^2$

            binomial square   perfect-square trinomial

EXAMPLE 11

Square the binomials mentally.

(a) $(x + 2)^2$ **(b)** $(2x - 3)^2$ **(c)** $(5x + 1)^2$

	Square First Term	Double Product of Terms	Square Second Term
(a) $(x + 2)^2 =$	$(x)^2$	$2 \cdot (2x)$	$(2)^2$
	x^2	$+4x$	$+4$
(b) $(2x - 3)^2 =$	$(2x)^2$	$2 \cdot (-6x)$	$(-3)^2$
	$4x^2$	$-12x$	$+9$
(c) $(5x + 1)^2 =$	$(5x)^2$	$2 \cdot (5x)$	$(1)^2$
	$25x^2$	$+10x$	$+1$

See Exercises 88–102.

Many patterns emerge when we multiply polynomials. However, it is not practical to develop a special process for every pattern. We will examine only the most common special products.

Let's look at the special products of the following two polynomials:

$$(a + b)(a^2 - ab + b^2)$$
$$(a - b)(a^2 + ab + b^2)$$

> **TIP** **Recognizing Patterns** Special products are products that fit a specific pattern. If the factors fit the pattern, then the product is predictable and can be done mentally. Look carefully at the pattern developed by the special products:
>
> $$(a + b)(a^2 - ab + b^2) \quad \text{and} \quad (a - b)(a^2 + ab + b^2)$$
>
> ▶ The first and last terms of the trinomial are the squares of the two terms of the binomial.
>
> ▶ The absolute value of the middle term of the trinomial is the product of the two terms of the binomial.
>
> ▶ When the binomial is a *sum* (+), the middle term of the trinomial is *negative* (−).
>
> ▶ When the binomial is a *difference* (−), the middle term of the trinomial is *positive* (+).

Now, let's multiply the binomial and trinomial by an extension of the FOIL method or repeated applications of the distributive property.

EXAMPLE 12

Multiply **(a)** $(a + b)(a^2 - ab + b^2)$ and **(b)** $(a - b)(a^2 + ab + b^2)$.

(a) Multiply each term in the trinomial by each term in the binomial.

Thus, $(a + b)(a^2 - ab + b^2) = a^3 + b^3$, the **sum of two perfect cubes.**

Sum of two perfect cubes: a binomial in the form $a^3 + b^3$

(b) Multiply each term in the trinomial by each term in the binomial.

Thus, $(a - b)(a^2 + ab + b^2) = a^3 - b^3$, the **difference of two perfect cubes.**

Difference of two perfect cubes: a binomial in the form $a^3 - b^3$

See Exercises 103–114.

> **To mentally multiply a binomial and a trinomial of the types $(a + b)(a^2 - ab + b^2)$ and $(a - b)(a^2 + ab + b^2)$:**
>
> 1. Cube the first term of the binomial.
>
> 2. Cube the second term of the binomial.
>
> Symbolically,
>
> $$(a + b)(a^2 - ab + b^2) = a^3 + b^3$$
> $$(a - b)(a^2 + ab + b^2) = a^3 - b^3$$

> **TIP** **Using Symbolic Patterns** The pattern $(a + b)(a^2 - ab + b^2)$ is almost impossible to describe in sentence form. However, we can identify the pattern at a glance when it is written symbolically.
>
> When we apply this pattern to different polynomials, we substitute for the letters a and b in the pattern.
>
> $$\text{Multiply } (2x + 3)(4x^2 - 6x + 9).$$
>
> Do these factors match the pattern? Let $2x$ represent a and 3 represent b.
>
> $a^2 = (2x)^2 = 4x^2$ Matches pattern.
> $b^2 = 3^2 = 9$ Matches pattern.
> $ab = (2x)(3) = 6x$ Matches pattern.
> Sign in binomial is positive.
> Middle sign in trinomial is negative. $\Big\}$ Matches pattern.
>
> Now, apply the pattern of the product.
>
> $$(a + b)(a^2 - ab + b^2) = a^3 + b^3$$
> $$a^3 = (2x)^3 = 8x^3$$
> $$b^3 = 3^3 = 27$$
>
> Then, $(2x + 3)(4x^2 - 6x + 9) = 8x^3 + 27$.

STOP AND CHECK
1. Mentally multiply
$(3x - 2)(9x^2 + 6x + 4)$.

Answer:
1. $27x^3 - 8$

EXAMPLE 13

Mentally multiply **(a)** $(m + p)(m^2 - mp + p^2)$ and **(b)** $(4c - d)(16c^2 + 4cd + d^2)$.

(a) $(a + b)(a^2 - ab + b^2) = a^3 + b^3$ Use the pattern and substitute m for a and p for b.

$$(m + p)(m^2 - mp + p^2) = \mathbf{m^3 + p^3}$$

(b) $(a - b)(a^2 + ab + b^2) = a^3 - b^3$ Substitute $4c$ for a and d for b.

$(4c - d)(16c^2 + 4cd + d^2) = \mathbf{64c^3 - d^3}$ $4^3 = 64$ **See Exercises 103–114.**

5 **Divide Polynomials.** Simplifying algebraic expressions may also involve division.

> **To divide a monomial by a monomial:**
>
> 1. Divide the coefficients using the rules for signed numbers.
>
> 2. Divide the variable factors using the laws of exponents for factors with like bases.

STOP AND CHECK
Divide. Express answers with
positive exponents.

1. $\dfrac{5x^3}{x^2}$ 2. $\dfrac{-3x^2}{9x}$

Answers:

1. $5x$ 2. $-\dfrac{x}{3}$

EXAMPLE 14

Divide. Express answers with positive exponents.

(a) $\dfrac{2x^4}{x}$ (b) $\dfrac{-6y^5}{2y^3}$ (c) $\dfrac{-4x}{4x^2}$ (d) $\dfrac{-5x^4}{15x^2}$ (e) $\dfrac{3x}{12x^3}$

(a) $\dfrac{2x^4}{x} = \dfrac{2}{1}(x^{4-1}) = \mathbf{2\,x^3}$ Divide coefficients. Subtract exponents. The coefficient of x in the denominator is 1.

(b) $\dfrac{-6y^5}{2y^3} = \dfrac{-6}{2}(y^{5-3}) = \mathbf{-3\,y^2}$ Divide coefficients. Subtract exponents.

(c) $\dfrac{-4x}{4x^2} = \dfrac{-4}{4}(x^{1-2}) = \mathbf{-1\,x^{-1}}$ **or** $-\dfrac{1}{x}$ Divide coefficients. Subtract exponents. Write factors with negative exponents as equivalent positive exponents.

This can also be written as $\dfrac{1}{-x}$ or $\dfrac{-1}{x}$.

(d) $\dfrac{-5x^4}{15x^2} = \dfrac{-5}{15}(x^{4-2}) = \mathbf{-\dfrac{1}{3}\,x^2}$ **or** $-\dfrac{x^2}{3}$ Reduce coefficients. Subtract exponents.

$\dfrac{-1}{3}x^2$ is same as $-\dfrac{1}{3}\left(\dfrac{x^2}{1}\right)$ or $-\dfrac{x^2}{3}$.

(e) $\dfrac{3x}{12x^3} = \dfrac{3}{12}(x^{1-3}) = \dfrac{1}{4}x^{-2} = \mathbf{\dfrac{1}{4\,x^2}}$ Reduce coefficients. Subtract exponents. Make the exponent positive.

Since $\dfrac{1}{4}x^{-2} = \dfrac{1}{4}\left(\dfrac{1}{x^2}\right)$, this can be written as $\dfrac{1}{4x^2}$.   **See Exercises 115–119.**

When a polynomial is divided by a monomial, *each* term in the dividend (numerator) is divided by the divisor (denominator).

STOP AND CHECK
Perform the divisions.

1. $\dfrac{18x^3 - 12x^2 - 6x}{3x}$

2. $\dfrac{4x^2 - 3x}{2x^2}$

Answers:

1. $6x^2 - 4x - 2$ 2. $2 - \dfrac{3}{2x}$

EXAMPLE 15

Perform the divisions.

(a) $\dfrac{18a^4 + 15a^3 - 9a^2 - 12a}{3a}$ (b) $\dfrac{3x^3 - x^2}{x^3}$ (c) $\dfrac{6x^3 + 2x^2}{2x^2}$

(a) $\dfrac{18a^4 + 15a^3 - 9a^2 - 12a}{3a} = \dfrac{18a^4}{3a} + \dfrac{15a^3}{3a} - \dfrac{9a^2}{3a} - \dfrac{12a}{3a}$ Write as separate terms.

$\qquad\qquad = \mathbf{6a^3 + 5a^2 - 3a - 4}$ Simplify each term.

(b) $\dfrac{3x^3 - x^2}{x^3} = \dfrac{3x^3}{x^3} - \dfrac{x^2}{x^3} = \mathbf{3 - x^{-1}}$ **or** $3 - \dfrac{1}{x}$ Write as separate terms and simplify each term. Make the exponent positive.

(c) $\dfrac{6x^3 + 2x^2}{2x^2} = \dfrac{6x^3}{2x^2} + \dfrac{2x^2}{2x^2} = \mathbf{3x + 1}$ Write as separate terms and simplify each term.

See Exercises 120–128.

> **TIP**   **Why Can't We Reduce Terms?** **You reduce *factors*, but not *terms*.** Look at part c of the previous example, $\dfrac{6x^3 + 2x^2}{2x^2}$. A common *mistake* is to reduce the terms.
>
>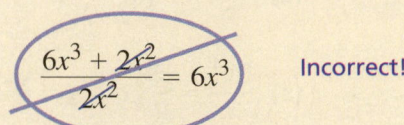
>
> $$\dfrac{6x^3 + 2x^2}{2x^2} = 6x^3 \qquad \text{Incorrect!}$$
>
> Why is this not correct? To check a division, multiply the quotient by the divisor (denominator). The result will be the dividend (numerator).
>
> Does $6x^3(2x^2) = 6x^3 + 2x^2$? NO!

Expressions that have several operations can be simplified.

STOP AND CHECK

Simplify and write all exponents as positive exponents.

1. $\dfrac{9x^4}{3x^2} + 2x(5x)$

2. $3x(-2x^2y)^2$

3. $\dfrac{4xy^2 - 2x^2y + 1}{xy}$

Answers:
1. $13x^2$ 2. $12x^5y^2$

3. $4y - 2x + \dfrac{1}{xy}$

EXAMPLE 16

Simplify and write all exponents as positive exponents.

(a) $\dfrac{5x^5}{10x^2} + 3x(2x^2)$ (b) $\dfrac{3ab^2 - a^2b + 4}{ab}$ (c) $3x(4xy^2)^2$

(a) $\dfrac{5x^5}{10x^2} + 3x(2x^2) =$ Simplify each term.

     $\dfrac{1}{2}x^3 + 6x^3 =$ Combine terms. $\dfrac{1}{2} + 6 = \dfrac{1}{2} + \dfrac{12}{2} = \dfrac{13}{2}$

     $\dfrac{13}{2}x^3$ or $\dfrac{13x^3}{2}$

(b) $\dfrac{3ab^2 - a^2b + 4}{ab} =$ Write or mentally visualize as separate terms.

     $\dfrac{3ab^2}{ab} - \dfrac{a^2b}{ab} + \dfrac{4}{ab} =$ Simplify each term.

     $3b - a + \dfrac{4}{ab}$

(c) $3x(4xy^2)^2 =$ Follow the order of operations. Raise to a power first.

     $3x(16x^2y^4) =$ Multiply.

     $48x^3y^4$ **See Exercises 129–130.**

A polynomial can be divided by a polynomial by using a long-division procedure. If the remainder is 0, both the divisor and the quotient are factors of the dividend.

> **To divide a polynomial by a polynomial using long division:**
>
> 1. Divide the first term of the dividend by the first term of the divisor. The partial quotient is placed above the first term of the dividend.
>
> 2. Multiply the partial quotient times the divisor and align the product under like terms of the dividend.

3. Subtract (change subtrahend to opposite and use addition rules).

4. Bring down the next term of the dividend and repeat Steps 1–3.

5. Repeat Steps 1– 4 until all terms of the dividend have been brought down. The result of the last subtraction is the remainder.

6. Write the remainder as a fraction with the remainder as the numerator and the divisor as the denominator.

STOP AND CHECK

1. Divide $3x^3 + 10x^2 - 22x + 15$ by $x + 5$.

2. Divide $x^3 - 5$ by $x - 2$.

Answers:

1. $3x^2 - 5x + 3$

2. $x^2 + 2x + 4 + \dfrac{3}{x - 2}$

EXAMPLE 17

Divide **(a)** $2x^3 - 9x^2 + 7x - 12$ by $x - 4$ and **(b)** $x^3 + 1$ by $x - 1$.

(a)
$$
\begin{array}{r}
2x^2 - x + 3 \\
x - 4 \,\overline{\big)\, 2x^3 - 9x^2 + 7x - 12} \\
\underline{2x^3 - 8x^2} \\
-x^2 + 7x \\
\underline{-x^2 + 4x} \\
3x - 12 \\
\underline{3x - 12} \\
0
\end{array}
$$

Divide: $\dfrac{2x^3}{x} = 2x^2$

Multiply: $2x^2(x - 4) = 2x^3 - 8x^2$

Subtract: $2x^3 - 2x^3 = 0$; $-9x^2 - (-8x^2) = -x^2$

Bring down next term. Divide: $\dfrac{-x^2}{x} = -x$.

Multiply: $-x(x - 4) = -x^2 + 4x$. Subtract.

Divide: $\dfrac{3x}{x} = 3$.

Multiply: $3(x - 4) = 3x - 12$. Subtract.

Remainder = 0.

The quotient is $2x^2 - x + 3$.

$x - 4$ is a factor of $2x^3 - 9x^2 + 7x - 12$.

(b)
$$
\begin{array}{r}
x^2 + x + 1 + \dfrac{2}{x - 1} \\
x - 1 \,\overline{\big)\, x^3 + 0x^2 + 0x + 1} \\
\underline{x^3 - x^2} \\
x^2 + 0x \\
\underline{x^2 - x} \\
x + 1 \\
\underline{x - 1} \\
2
\end{array}
$$

Represent missing powers of x with terms having a coefficient of 0.

Divide $\dfrac{x^3}{x} = x^2$, then multiply by x^2.

Subtract.

Bring down next term. Divide and then multiply.

Subtract.

Bring down next term. Divide and then multiply.

Subtract.

Express remainder as a fraction using the divisor as the denominator.

The quotient is $x^2 + x + 1 + \dfrac{2}{x - 1}$.

$x - 1$ is not a factor of $x^3 + 1$.

See Exercises 131–140.

11-3 EXERCISES

MyLab Math For additional practice go to your study plan in MyLab Math.

1 Determine if the pairs of terms are like terms. If they are not like terms, explain why. *See Example 1.*

1. $4a^2b^3$ and $-0.5a^2b^3$

2. $-2a$ and 7

3. $-3ab^2$ and $5(ab)^2$

4. $9x^2$ and $2x^3$

Simplify. *See Example 2.*

5. $3a^2 + 4a^2$

6. $5x^3 - 2x^3$

7. $b^2 + 3a^2 + 2b^2 - 5a^2$

8. $3a - 2b - a$

9. $x - 3x - 2x^2 - 3x^2$

10. $3a^2 - 2a^2 + 4a^2$

See Example 3.

11. $x^2 + 3y - (2x^2 + 5y)$ **12.** $4m^2 - 2n^2 - (2m^2 - 3n^2)$ **13.** $7a + 3b + 8c + 2a - (b - 2c)$

14. $5x + 3y - (7x - 2z)$ **15.** $(4x^2 - 3) - (3x^2 + 2) - (x^2 - 1)$ **16.** $5x - (3x + 7) - (-x - 8)$

2 Multiply. *See Example 4.*

17. $7x(2x^2)$ **18.** $(-2m)(-m^2)$ **19.** $(-3m)(7m)$ **20.** $(-y^3)(2y^3)$

See Example 5.

21. $4x^2(2x - 7)$ **22.** $-3ab(2a^2b - 5ab^2)$ **23.** $7x(5x^3 - 3x^2 - 7)$ **24.** $2xy(3x^2y - 5xy^2)$

25. $3x(x - 6)$ **26.** $4x(3x^2 - 7x + 8)$ **27.** $-4x(2x - 3)$ **28.** $2x^2(5 + 2x)$

Multiply. *See Example 6.*

29. $(x + 7)(x + 3)$ **30.** $(x + 8)(x + 5)$ **31.** $(2x - 1)(x + 2)$ **32.** $(3x + 7)(x - 5)$

See Example 7.

33. $(x + 5)(x^2 + 2x - 1)$ **34.** $(x + 7)(x^2 - 3x + 2)$ **35.** $(x - 7)(x^2 - 5x + 2)$

36. $(x - 3)(x^2 - 8x + 1)$ **37.** $(2x - 5)(x^2 - x + 1)$ **38.** $(3x - 2)(x^2 + x - 3)$

39. $(2x + 7)(3x^2 - 5x + 2)$ **40.** $(5x - 3)(4x^2 - 2x - 3)$ **41.** $(x - 2)(x^2 + 2x + 4)$

42. $(2x - 3)(4x^2 + 6x + 9)$ **43.** $(x + 3)(x^2 - 3x + 9)$ **44.** $(3x + 2)(9x^2 - 6x + 4)$

45. $(x + 5)(x^2 - 5x + 25)$ **46.** $(11x - 3)(121x^2 + 33x + 9)$ **47.** $(x^2 - 2x + 1)(3x^2 + 4x - 2)$

48. $(4x^2 + x + 3)(2x^2 - 3x + 1)$

3 Use the FOIL method to find the products. Practice combining the outer and inner products mentally. *See Example 8.*

49. $(a + 3)(a + 8)$ **50.** $(x - 4)(x + 5)$ **51.** $(y - 2)(y - 9)$

52. $(y - 7)(y - 3)$ **53.** $(2a + 3)(a + 4)$ **54.** $(3a - 5)(a + 1)$

55. $(3a - 2b)(a - 2b)$ **56.** $(5x - y)(x - 5y)$ **57.** $(3x - 4)(2x - 3)$

58. $(a - b)(2a - 5b)$ **59.** $(7 - m)(3 - 7m)$ **60.** $(5 - 2x)(8 - x)$

61. $(x + 7)(x + 4)$ **62.** $(y - 7)(y - 5)$ **63.** $(m + 3)(m - 7)$

64. $(3b - 2)(x + 6)$ **65.** $(4r - 5)(3r + 2)$ **66.** $(5 - x)(7 - 3x)$

67. $(4 - 2m)(1 - 3m)$ **68.** $(2 + 3x)(3 + 2x)$ **69.** $(x + 3)(2x - 5)$

70. $(5x - 7y)(4x + 3y)$ **71.** $(2a + 3b)(7a - b)$ **72.** $(5a + 2b)(6a - 5b)$

73. $(9x - 2y)(3x + 4y)$ **74.** $(5x - 8y)(4x - 3y)$ **75.** $(7m - 2n)(3m + 5n)$

4 Find the special products using patterns. *See Example 9.*

76. $(a + 3)(a - 3)$ **77.** $(2x + 3)(2x - 3)$ **78.** $(a - y)(a + y)$

79. $(4r + 5)(4r - 5)$ **80.** $(5x + 2)(5x - 2)$ **81.** $(7 + m)(7 - m)$

82. $(3y - 5)(3y + 5)$ **83.** $(8y + 3)(8y - 3)$ **84.** $(3a - 11b)(3a + 11b)$

85. $(5y - 3)(5y + 3)$ **86.** $(x - 7)(x + 7)$ **87.** $(x - 11)(x + 11)$

See Examples 10 and 11.

88. $(2 - 3x)^2$ **89.** $(3x + 4)^2$ **90.** $(Q + L)^2$

91. $(a^2 + 1)^2$ **92.** $(2d - 5)^2$ **93.** $(3a + 2x)^2$

94. $(3x - 7)^2$ **95.** $(6 + Q)^2$ **96.** $(y + 5x)^2$

97. $(4 - 3j)^2$ **98.** $(3m - 2p)^2$ **99.** $(m^2 + p^2)^2$

100. $(2a - 7c)^2$ **101.** $(9 - 13a)^2$ **102.** $(2x - 5y)^2$

Use the extension of the FOIL method or the special products patterns to find the products. *See Examples 12 and 13.*

103. $(x + p)(x^2 - xp + p^2)$

104. $(Q + L)(Q^2 - QL + L^2)$

105. $(3 + a)(9 - 3a + a^2)$

106. $(2x + 5p)(4x^2 - 10xp + 25p^2)$

107. $(3m + 2)(9m^2 - 6m + 4)$

108. $(6 - p)(36 + 6p + p^2)$

109. $(5y - p)(25y^2 + 5yp + p^2)$

110. $(x - 2y)(x^2 + 2xy + 4y^2)$

111. $(Q - 6)(Q^2 + 6Q + 36)$

112. $(3t - 2)(9t^2 + 6t + 4)$

113. $(2x + 3y)(4x^2 - 6xy + 9y^2)$

114. $(3x - 4y)(9x^2 + 12xy + 16y^2)$

5 Divide. *See Example 14.*

115. $\dfrac{6x^4}{3x^2}$

116. $\dfrac{-5a^2}{10a}$

117. $\dfrac{-7x}{-14x^3}$

118. $\dfrac{-9x^5}{12x^2}$

119. $\dfrac{6x^2 - 4x}{2x}$

See Example 15.

120. $\dfrac{12x^5 - 6x^3 - 3x^2}{3x^2}$

121. $\dfrac{7x^4 - x^2}{x^3}$

122. $\dfrac{8x^4 + 6x^3}{2x^2}$

123. $\dfrac{6x^4 + 8x^2}{18x^2}$

124. $\dfrac{5a^2b^3 - 3ab^2 - 7}{ab}$

125. $\dfrac{15a^2b^3 - 3ab^2}{6ab}$

126. $\dfrac{18x^2 - 12y^2 - 6xy}{6xy}$

127. $\dfrac{-3x - (5x + 8x^3)}{2x}$

128. $\dfrac{6x(2x^3y^2)^3 + 8x}{2x}$

See Example 16.

129. $\dfrac{15x^2}{3x} - 2x(7x^3)$

130. $\dfrac{3x^2 - 5y^2}{15xy} - \dfrac{8x^3}{4x}$

Perform the indicated division and determine if the binomial is a factor of the dividend. *See Example 17.*

131. $x - 5 \overline{\smash{)}\,x^2 - x - 20}$

132. $x - 3 \overline{\smash{)}\,x^2 + 3x - 18}$

133. $x + 1 \overline{\smash{)}\,x^3 - 2x^2 - x + 2}$

134. $x + 2 \overline{\smash{)}\,x^3 + 3x^2 - 3x - 10}$

135. $x + 3 \overline{\smash{)}\,3x^3 + 7x^2 - 3x + 8}$

136. $x - 5 \overline{\smash{)}\,2x^3 - 3x^2 - 33x + 8}$

137. $x + 1 \overline{\smash{)}\,x^3 + 2x^2 - 5x - 6}$

138. $x + 3 \overline{\smash{)}\,x^3 + 2x^2 - 5x - 6}$

139. $x - 3 \overline{\smash{)}\,x^3 - 27}$

140. $x + 5 \overline{\smash{)}\,x^3 + 125}$

---

11 | **CHAPTER REVIEW OF KEY CONCEPTS**

LEARNING OUTCOMES	KEY CONCEPTS AND EXAMPLES

Section 11–1

1 Multiply powers with like bases (pp. 493–494).

1. Verify that the bases are the same. Use this base as the base of the *product*. **2.** Add the exponents for the exponent of the product. Symbolically, $a^m(a^n) = a^{m+n}$, where a, m, and n are real numbers and $a \neq 0$.

Multiply. $x^5(x^{-7})$

$$x^5(x^{-7}) = x^{5+(-7)} = x^{-2} = \frac{1}{x^2} \qquad \text{Add exponents.}$$

LEARNING OUTCOMES	KEY CONCEPTS AND EXAMPLES

2 Divide powers with like bases (pp. 494–497).

1. Verify that the bases are the same. Use this base as the base of the *quotient*.   **2.** Subtract the exponents for the exponent of the quotient. Symbolically, $\dfrac{a^m}{a^n} = a^{m-n}$, where a, m, and n are real numbers except that $a \neq 0$.

> Divide. $\dfrac{a^7}{a^4}$
>
> $$\dfrac{a^7}{a^4} = a^{7-4} = a^3 \qquad \text{Subtract exponents.}$$

3 Find a power of a power (pp. 497–499).

1. Multiply the exponents.   **2.** Keep the same base. Symbolically, $(a^m)^n = a^{mn}$, where a, m, and n are real numbers and $a \neq 0$.

> Simplify. $(y^5)^4$
>
> $$y^5(y^4) = y^{5(4)} = y^{20} \qquad \text{Multiply exponents.}$$

Raise a fraction or quotient to a power: 1. Raise the numerator (dividend) to the power. **2.** Raise the denominator (divisor) to the power. Symbolically, $\left(\dfrac{a}{b}\right)^n = \dfrac{a^n}{b^n}$ are real numbers and $b \neq 0$.

> $$\left(\dfrac{-2}{x^2}\right)^3 = \dfrac{(-2)^3}{(x^2)^3} = \dfrac{-8}{x^6} \qquad \begin{array}{l}\text{Cube numerator.}\\ \text{Cube denominator.}\end{array}$$

Raise a product to a power: Raise each factor to the indicated power. Symbolically, $(ab)^n = a^n b^n$, where a, b, and n are real numbers.

> $$(5x^2y)^2 = 5^2 x^4 y^2 = 25 x^4 y^2 \qquad \text{Raise each factor in the grouping to the indicated power.}$$

Section 11–2

1 Identify polynomials, monomials, binomials, and trinomials (pp. 500–501).

1. Write all variables in the numerator if necessary.   **2.** Expressions that have a negative exponent in *any* term or a variable in a denominator are not polynomials.   **3.** Identify the expression based on the number of terms it contains. A monomial has one term, a binomial has two terms, and a trinomial has three terms.

> Give an example of a polynomial, a monomial, a binomial, and a trinomial.
>
> Polynomial: $4x^3 + 6x^2 - x + 3$ 　　　　Monomial: $8x$
>
> Binomial: $6x + 3$ 　　　　Trinomial: $8x^2 - 4x - 3$

2 Identify the degree of terms and polynomials (pp. 502–503).

Identify the degree of a term: 1. Exponents of variable factors must be integers greater than zero.   **2.** For a term that is a constant, the degree is zero.   **3.** For a term that has only one variable factor, the degree is the same as the exponent of the variable.   **4.** For a term that has more than one variable factor, the degree is the sum of the exponents of the variables.

Identify the degree of a polynomial: 1. Identify the degree of each term of the polynomial. **2.** Compare the degrees of each term of the polynomial and select the greatest degree as the degree of the polynomial.

LEARNING OUTCOMES	KEY CONCEPTS AND EXAMPLES

Identify the degree of each term and the degree of the polynomial.

$$4x^3 \quad + \quad 6x^2 \quad - \quad x \quad + \quad 3 \qquad \text{The polynomial has a degree of 3.}$$
$$\text{degree 3} \quad \text{degree 2} \quad \text{degree 1} \quad \text{degree 0}$$

3 Arrange polynomials in descending order (pp. 503–504).

Arrange polynomials in descending order of a variable: 1. Identify the variable on which the terms of the polynomial will be arranged if the polynomial has more than one variable. **2.** Compare the degrees of the selected variable for each term. **3.** List the term with the highest degree of the specified variable first. **4.** Continue to list the terms of the polynomial in descending order of the selected variable.

Arrange the polynomial in descending order: $4 - 2x + 7x^5 - 3x^2 + x^3$

Descending order:

$$7x^5 + x^3 - 3x^2 - 2x + 4 \qquad \text{Begin with the term with the highest degree.}$$

Section 11–3

1 Add and subtract polynomials (pp. 504–506).

Add or subtract (combine) like terms: 1. Combine the coefficients of the like terms using the rules for adding or subtracting signed numbers. **2.** The variable factors and their exponents do not change in the sum or difference.

Simplify. $3x^4 + 8x^2 - 7x^2 + 2x^4 = 5x^4 + x^2$

2 Multiply polynomials (pp. 506–508).

Multiply by a monomial: 1. Multiply the coefficients using the rules for signed numbers. **2.** Multiply the variable factors using the laws of exponents for factors with like bases. **3.** Distribute if multiplying a polynomial by a monomial.

Simplify. $(3x^4y^5)(7x^2yz) = 21x^6y^6z$

Multiply two polynomials: 1. Use the distributive property to multiply each term of the first polynomial times the entire second polynomial. **2.** Combine like terms. **3.** Symbolically,

$$(a + b)(c + d) = a(c + d) + b(c + d) = ac + ad + bc + bd$$
$$(a + b + c)(d + e + f) = a(d + e + f) + b(d + e + f) + c(d + e + f)$$
$$= ad + ae + af + bd + be + bf + cd + ce + cf$$

Multiply $(x^2 + 3x + 1)(x^2 - 3x + 2)$.

$$x^2(x^2 - 3x + 2) + 3x(x^2 - 3x + 2) + 1(x^2 - 3x + 2) =$$
$$x^4 - 3x^3 + 2x^2 + 3x^3 - 9x^2 + 6x + x^2 - 3x + 2 =$$
$$x^4 - 6x^2 + 3x + 2$$

3 Use the FOIL method to multiply two binomials (pp. 508–509).

Multiply two binomials using the FOIL method: 1. Write the product of the *first* term of each factor. **2.** Write the product of the two *outer* terms of the factors. **3.** Write the product of the two *inner* terms of the factors. **4.** Write the product of the *last* term of each factor. **5.** Combine like terms. (Multiply **F**irst terms, **O**uter terms, **I**nner terms, and **L**ast terms.)

Multiply $(3a - 2)(a + 3)$.

$$\begin{array}{cccc} \text{F} & \text{O} & \text{I} & \text{L} \end{array}$$
$$3a^2 + 9a - 2a - 6 \qquad \text{Combine like terms.}$$
$$3a^2 + 7a - 6$$

LEARNING OUTCOMES	KEY CONCEPTS AND EXAMPLES

4 Multiply polynomials that result in special products (pp. 509–512).

Mentally multiply the sum and difference of the same two terms (conjugate pairs): 1. Square the first term of either binomial. **2.** Insert a minus sign. **3.** Square the second term of either binomial. Symbolically, $(a + b)(a - b) = a^2 - b^2$.

Multiply $(3b - 2)(3b + 2)$.

$9b^2 - 4$ $(3b)^2 = 9b^2; (2)^2 = 4$

Mentally square a binomial: 1. Square the first term of the binomial for the first term of the perfect-square trinomial. **2.** Double the product of the two terms of the binomial for the middle term of the trinomial. **3.** Square the second term of the binomial for the third term of the trinomial.

Symbolically, $(a + b)^2 = a^2 + 2ab + b^2$
$(a - b)^2 = a^2 - 2ab + b^2$

Multiply $(2a - 3)^2$.

$4a^2 - 12a + 9$ $(2a)^2 = 4a^2; 2(2a)(-3) = -12a; (-3)^2 = 9$

Mentally multiply a binomial and a trinomial of the types $(a + b)(a^2 - ab + b^2)$ and $(a - b)(a^2 + ab + b^2)$: 1. Cube the first term of the binomial. **2.** Cube the second term of the binomial. Symbolically,

$$(a + b)(a^2 - ab + b^2) = a^3 + b^3$$
$$(a - b)(a^2 + ab + b^2) = a^3 - b^3$$

Multiply $(8b - 3)(64b^2 + 24b + 9)$.

$512b^3 - 27$ $(8b)^3 = 512b^3; 3^3 = 27$

Multiply $(2x + 5)(4x^2 - 10x + 25)$.

$8x^3 + 125$ $(2x)^3 = 8x^3; 5^3 = 125$

5 Divide polynomials (pp. 512–515).

Divide a monomial by a monomial: 1. Divide the coefficients using the rules for signed numbers. **2.** Divide the variable factors using the laws of exponents for factors with like bases.

Simplify. $\dfrac{10x^7y^4}{5x^8y^2} = 2x^{-1}y^2 = \dfrac{2y^2}{x}$ $\dfrac{3x^2 + 5x}{x} = \dfrac{3x^2}{x} + \dfrac{5x}{x} = 3x + 5$

Divide a polynomial by a polynomial using long division: 1. Divide the first term of the dividend by the first term of the divisor. The partial quotient is placed above the first term of the dividend. **2.** Multiply the partial quotient times the divisor and align the product under like terms of the dividend. **3.** Subtract (change subtrahend to opposite and use addition rules). **4.** Bring down the next term of the dividend and repeat Steps 1–3. **5.** Repeat Steps 1– 4 until all terms of the dividend have been brought down. The result of the last subtraction is the remainder. **6.** Write the remainder as a fraction with the remainder as the numerator and the divisor as the denominator.

Divide $(3x^2 + 13x - 10)$ by $(x + 5)$.

$$
\begin{array}{r}
3x - 2 \\
x + 5 \overline{\smash{)}\ 3x^2 + 13x - 10} \\
\underline{3x^2 + 15x} \\
-2x - 10 \\
\underline{-2x - 10} \\
0
\end{array}
$$

11 | CHAPTER REVIEW EXERCISES

Section 11–1 MyLab Math For additional practice go to your study plan in MyLab Math.

Perform the indicated operations. Write the answers with positive exponents.

1. $x^5 \cdot x^5$

2. $x^2(x^4)$

3. $3x^4 \cdot 7x^5$

4. $5x^3 \cdot 8x$

5. $\dfrac{x^8}{x^5}$

6. $\dfrac{x^3}{x^5}$

7. $\dfrac{21x^4}{3x}$

8. $\dfrac{24y^7}{18y^{10}}$

9. $\dfrac{x^3 y^{-1}}{x^2 y^2}$

10. $\dfrac{x^3 y^2}{x^2 y}$

11. $(x^3)^4$

12. $(-x^3)^3$

13. $(x^{-3})^{-5}$

14. $(x^2 y^5)^2$

15. $(-3x^2)^3$

16. $\dfrac{a^2 b^7}{ab^2}$

17. $\dfrac{xy^3}{xy^5}$

18. $(-x^4)^3$

19. $5x^{-2} \cdot 8x^4$

20. $\dfrac{28x^3 y^{-3}}{7xy^{-4}}$

Section 11–2

Identify the degree of each polynomial.

21. $7m^2 - 8m + 12m^4$

22. $5a^4 - 7a^3 + 12a^2 - 38$

23. $2x^3 y^2 - 15xy^3 + 21y^4$

24. Is the expression $5x^3 - 3x^{-2}$ a polynomial? Why or why not?

Arrange the following polynomials in descending order and identify the degree of each polynomial, the leading term, and the leading coefficient.

25. $5x + 3x^3 - 8 + x^2$

26. $3y^5 - 7y - 8y^4 + 12$

27. $5x^2 - 12x + 2x^4 - 32$

Section 11–3

Simplify the following:

28. $4x^3 + 7x - 3x^3 - 5x$

29. $8x - 2x^4 - (3x^3 + 5x - x^3)$

30. $5x - 3x + (7x^2 - 8x)$

31. $4x^2 - (3y^2 + 7x^2 - 8y^2)$

32. $4x^3(-3x^4)$

33. $-7x^8(-3x^{-2})$

34. $5a(a^2 - 7)$

35. $2x(x^2 + 3x - 5)$

36. $-2y(3y^2 - 7y - 12)$

37. $-2x(x^3 - 7x^2 + 15)$

38. $(y - 7)(y - 5)$

39. $(m + 3)(m - 7)$

40. $(3b - 2)(x + 6)$

41. $(4r - 5)(3r + 2)$

42. $(5 - x)(7 - 3x)$

43. $(4 - 2m)(1 - 3m)$

44. $(2 + 3x)(3 + 2x)$

45. $(x + 3)(2x - 5)$

46. $(5x - 7y)(4x + 3y)$

47. $(2a + 3b)(7a - b)$

48. $(5a + 2b)(6a - 5b)$

49. $(9x - 2y)(3x + 4y)$

50. $(5x - 8y)(4x - 3y)$

51. $(7m - 2n)(3m + 5n)$

52. $(x + 3)(2x^2 + 3x + 2)$

53. $(5x + 3)(x^2 - 3x + 1)$

Find the special products mentally.

54. $(y - 4)(y + 4)$

55. $(6x - 5)(6x + 5)$

56. $(3m + 4)(3m - 4)$

57. $(7y + 11)(7y - 11)$

58. $(5x - 2y)(5x + 2y)$

59. $(8a - 5b)(8a + 5b)$

60. $(12r - 7s)(12r + 7s)$

61. $(\sqrt{8} + 2)(\sqrt{8} - 2)$

62. $(\sqrt{5} + \sqrt{2})(\sqrt{5} - \sqrt{2})$

63. $(3 + x)(3 - x)$

64. $(7 - x)(7 + x)$

65. $(x + 9)^2$

66. $(x - 7)^2$

67. $(x - 3)^2$

68. $(2x - 3)^2$

69. $(4x - 15)^2$ **70.** $(5 + 3m)^2$ **71.** $(8 + 7m)^2$

72. $(5x - 13)^2$ **73.** $(4x - 11)^2$ **74.** $(K + L)(K^2 - KL + L^2)$

75. $(g - h)(g^2 + gh + h^2)$ **76.** $(4 + a)(16 - 4a + a^2)$ **77.** $(2H - 3T)(4H^2 + 6HT + 9T^2)$

78. $(3a - 5)(9a^2 + 15a + 25)$ **79.** $(6 + x)(36 - 6x + x^2)$ **80.** $(9y - p)(81y^2 + 9yp + p^2)$

81. $(z + 2t)(z^2 - 2zt + 4t^2)$ **82.** $(g - 2)(g^2 + 2g + 4)$ **83.** $(7T + 2)(49T^2 - 14T + 4)$

Find the quotients.

84. $\dfrac{12x^5}{6x^3}$ **85.** $\dfrac{12x^7}{-18x^4}$ **86.** $\dfrac{11x^4}{22x^7}$

87. $\dfrac{42x^3 y}{-15x^3 y^3}$ **88.** $\dfrac{-8x^3 y^5}{20x^4 y}$ **89.** $\dfrac{6x^3 - 12x^2 + 21x}{3x}$

90. $\dfrac{25y^5 - 85y^3 + 70y^2}{-5y}$ **91.** $\dfrac{4x^5}{8x^2} - 3x^2(2x^4)$ **92.** $2x(3x^2 y)^3 + \dfrac{7x^3}{21x^2}$

93. $\dfrac{16x^3 + 12x^4 - 20x^5}{-4x^2}$ **94.** $\dfrac{9x^2 y^5 - 15xy^3}{3xy^3}$ **95.** $x - 9 \overline{\smash{\big)}\, x^2 - 11x + 18}$

96. $2x - 3 \overline{\smash{\big)}\, 6x^2 - 13x + 21}$ **97.** $5x + 2 \overline{\smash{\big)}\, 15x^2 - 4x - 4}$ **98.** $x + 2 \overline{\smash{\big)}\, x^3 - x^2 - 4x + 4}$

99. $x - 5 \overline{\smash{\big)}\, x^3 - 4x^2 - 10x + 25}$ **100.** $2x - 1 \overline{\smash{\big)}\, 6x^3 - 7x^2 + 4x - 3}$ **101.** $2x^2 - x - 1 \overline{\smash{\big)}\, 6x^3 - x^2 - 4x - 1}$

11 | TEAM PROBLEM-SOLVING EXERCISES

1. The symbolic representation of the laws of exponents illustrates many properties and restrictions. Explain these properties and restrictions in words.

(a) $a^m \cdot a^n = a^{m+n}$

(b) $\dfrac{a^m}{a^n} = a^{m-n}, a \neq 0$

(c) $(a^m)^n = a^{mn}$

(d) $\left(\dfrac{a}{b}\right)^n = \dfrac{a^n}{b^n}, b \neq 0$

(e) $(ab)^n = a^n b^n$

2. Give an example to illustrate each of the laws of exponents.

(a) $a^m \cdot a^n = a^{m+n}$

(b) $\dfrac{a^m}{a^n} = a^{m-n}, a \neq 0$

(c) $(a^m)^n = a^{mn}$

(d) $\left(\dfrac{a}{b}\right)^n = \dfrac{a^n}{b^n}, b \neq 0$

(e) $(ab)^n = a^n b^n$

11 | CONCEPTS ANALYSIS

Find the mistake in the examples. Explain the mistake and correct it.

1. $\dfrac{9x^3 - 12x^2 + 3x}{3x} = 3x^2 - 4x$

2. $x^5(x^3) = x^{15}$

3. $4x^3 + 3x^3 = 7x^6$

4. Explain how the FOIL process for multiplying two binomials is an application of the distributive property.

5. (a) How is the distributive property used to multiply a binomial and a trinomial?

(b) How is the distributive property used to multiply two trinomials?

6. Give an example of the product of a binomial and a trinomial to illustrate your answer to Question 5a.

7. Give an example of the product of two trinomials to illustrate your answer to Question 5b.

8. List the properties of the two binomials that result in the product of the difference of two perfect squares.

9. List the properties of a perfect-square trinomial that result from squaring a binomial.

10. Explain the difference in the processes used to divide a polynomial by a monomial and to divide a polynomial by a binomial.

11   PRACTICE TEST

Perform the indicated operations. Write the answers with positive exponents.

1. $(x^4)(x)$

2. $\dfrac{x^0}{x^2}$

3. $\left(\dfrac{4}{7}\right)^2$

4. $\dfrac{24x^2y^{-1}}{16xy^3}$

5. $\dfrac{x^{-7}}{x^3}$

6. $(6a^2b)^2$

7. $\left(\dfrac{x^2}{y}\right)^2$

8. $\dfrac{12x^2}{4x^3}$

9. $4a(3a^2 - 2a + 5)$

10. $\dfrac{60x^3 - 45x^2 - 5x}{5x}$

11. $5x^2 - 3x - (2x + 4x^2)$

12. $4x^3 + 2x - 7 + (3x^3 - 7x - 5)$

13. $-7x^3(-8x^4)$

14. $-(2x - 8) + 12$

15. $4xy - (3x - 2) + 4$

What is the degree of the polynomial?

16. $5x^3 - 4x^3 + x^2$

17. $14x - 3x + 21x$

Arrange each polynomial in decreasing powers of x.

18. $4 - 12x + 15x^3$

19. $6 - 3x^4 - 2x^3$

20. $6x^4 - 2x^5 - 5$

Find the products.

21. $(m - 7)(m + 7)$

22. $(3x - 2)(3x + 2)$

23. $(a + 3)^2$

24. $(2x - 7)^2$

25. $(x - 3)(2x - 5)$

26. $(7x - 3)(2x + 1)$

27. $(x - 2)(x^2 + 2x + 4)$

28. $(3y + 4)(9y^2 - 12y + 16)$

29. $(5a - 3)(25a^2 + 15a + 9)$

30. $(a + 6)(a^2 - 6a + 36)$

Divide using long division.

31. $x - 3 \overline{\smash{\big)}\ 2x^2 + x - 21}$

32. $x + 3 \overline{\smash{\big)}\ 2x^3 - 3x^2 + 7x - 6}$

12

Roots and Radicals

3Dsculptor/Shutterstock

In Great Company

All the Time In the World (1983)

On September 26, 1983, Ronald Reagan had been president of the United States for almost three years. For those three years, he had been working with Margaret Thatcher, Pope John Paul II, and the Polish Solidarity movement to seriously undermine the stability of the Soviet Union. Reagan had announced the Strategic Defense Initiative in March 1983, and NATO had begun to deploy Pershing II missiles in Germany.

Yuri Andropov, a former KGB agent, feared the United States was positioning itself to make a first strike. Just 25 days earlier, on September 1, a trigger-happy Soviet pilot shot down a South Korean commercial airliner that had inadvertently strayed into Soviet airspace. Two hundred sixty-nine people died. Each side accused the other of responsibility for the deaths.

Needless to say, tensions between the Soviets and the Americans were at one of the highest points since World War II. It was about to get a whole lot worse.

The clock had just ticked past midnight when Stanislav Yevgrafovich Petrov settled into the commander's chair inside the Oko bunker at Serpukhov-15. It was the secret bunker that controlled the Soviet nuclear early warning command center.

As the second hand ticked round, alarms began to sound. Satellite readings indicated that the United States had just launched one . . . two . . . three . . . four . . . five nuclear ballistic missiles: a nuclear first strike aimed right at the heart of the Soviet Union.

As the clock ticked the seconds away, Petrov stared at the flashing red button labeled "Start." He had only minutes to verify the stream of incoming data. Once he punched that button, he would be verifying to his commanders that the attack was real, that the Soviets needed to launch an immediate, massive, nuclear response.

He knew the computer system had just been upgraded. He couldn't find any other data indicating that the ballistic missiles were being tracked. He "had a funny feeling in my gut" that the United States would never launch just five missiles. As he later said, "I didn't want to make a mistake." After tense minutes with a phone in one hand, intercom in the other, he told his men to stand down. He turned the alarms off.

Later investigation proved his call correct. The software upgrade had introduced a math error that caused the satellite to mistake sunlight reflecting off high-altitude clouds for ICBM missile launches. After intensive interrogations, Petrov was neither rewarded nor punished for his actions. He took early retirement and soon suffered a nervous breakdown, thinking about how close he had come to being wrong.

12–1 Irrational Numbers and Real Numbers

LEARNING OUTCOMES

1 Write roots using radical and exponential notation.

2 Approximate an irrational number.

3 Write powers and roots using rational exponents and radical notation.

LC LEARNING CATALYTICS

1. Write the principal square root of 9.
2. Write the square of 9.

In Chapter 1 we saw that the opposite or inverse operation for raising to powers was finding roots. The *square root* of a number is the number that is used as a factor 2 times to equal the square. The *radical notation* for square root is $\sqrt{}$, $\sqrt{25} = 5$. The *index* of a square root is 2, but is not required in radical notation.

A *cube root* is the number that is used as a factor 3 times to give the cube. The index of a cube root is 3. For cube roots, the index 3 is written in the $\sqrt{}$ portion of the radical sign: $\sqrt[3]{8} = 2$. For *fourth roots*, the index 4 is written in the $\sqrt{}$ portion of the radical sign: $\sqrt[4]{81} = 3$.

The results when natural numbers are squared, cubed, raised to the fourth power, and so on are called *perfect powers*. The principal (or positive) root of a perfect power is a natural number.

Powers of 2	Powers of 3	Powers of 4	Powers of 5
$2^1 = 2$	$3^1 = 3$	$4^1 = 4$	$5^1 = 5$
$2^2 = 4$	$3^2 = 9$	$4^2 = 16$	$5^2 = 25$
$2^3 = 8$	$3^3 = 27$	$4^3 = 64$	$5^3 = 125$
$2^4 = 16$	$3^4 = 81$	$4^4 = 256$	$5^4 = 625$
$2^5 = 32$	$3^5 = 243$	$4^5 = 1,024$	$5^5 = 3,125$
$2^6 = 64$	$3^6 = 729$	$4^6 = 4,096$	$5^6 = 15,625$
Selected roots			
$\sqrt[4]{16} = 2$	$\sqrt[5]{243} = 3$	$\sqrt[3]{64} = 4$	$\sqrt[6]{15,625} = 5$

1 Write Roots Using Radical and Exponential Notation. A rational or fractional exponent is another notation used to indicate a root. To indicate a square root, use the exponent $\frac{1}{2}$. To indicate a cube root, use the exponent $\frac{1}{3}$. For a fourth root, use the exponent $\frac{1}{4}$. To generalize, the exponential notation for the root of a number is the number written with the fractional exponent. The index of the root is the denominator of the fractional exponent and 1 is the numerator.

$$\sqrt[n]{a} = a^{\frac{1}{n}}$$ where n is a natural number greater than 1 and a is a positive number unless otherwise indicated.

To write roots using radical and exponential notation:

1. Identify the index of the root.

2. For radical notation, place the index of the root in the $\sqrt{}$ portion of the radical sign with the radicand under the bar portion of the sign. The index 2 for square roots does not have to be written.

3. For exponential notation, write the radicand as the base and the index of the root as the denominator of a fractional exponent that has a numerator of 1. The decimal equivalent of the fractional exponent can also be used.

> **TIP** **Equivalent Notations for Roots** Just as $\frac{1}{2}$, 0.5, and 50% are three notations for writing equivalent values, radical notation and exponential notation are notations for writing equivalent values for roots.
>
> | In Words | Radical Notation | Exponential Notation |
> |---|---|---|
> | Square root of n | $\sqrt{n}$ | $n^{1/2}$ or $n^{\frac{1}{2}}$ or $n^{0.5}$ |
> | Cube root of n | $\sqrt[3]{n}$ | $n^{1/3}$ or $n^{\frac{1}{3}}$ or $n^{0.\overline{3}}$ |
> | Fourth root of n | $\sqrt[4]{n}$ | $n^{1/4}$ or $n^{\frac{1}{4}}$ or $n^{0.25}$ |

STOP AND CHECK

Write the roots using radical notation and exponential notation.

1. Cube root of 8
2. Square root of 121

Answers:
1. $\sqrt[3]{8}$; $8^{1/3}$ or $8^{0.3}$
2. $\sqrt{121}$; $121^{1/2}$ or $121^{0.5}$

EXAMPLE 1

Write the roots using radical notation and exponential notation.

(a) Square root of 25　　(b) Cube root of 125　　(c) Fourth root of 625

(a) Square root of 25 = $\sqrt{25}$ = $\mathbf{25^{1/2}}$ **or** $\mathbf{25^{0.5}}$

(b) Cube root of 125 = $\sqrt[3]{125}$ = $\mathbf{125^{1/3}}$ **or** $\mathbf{125^{0.\overline{3}}}$

(c) Fourth root of 625 = $\sqrt[4]{625}$ = $\mathbf{625^{1/4}}$ **or** $\mathbf{625^{0.25}}$ 　　**See Exercises 1–6.**

2 Approximate an Irrational Number. Not all numbers are perfect powers. The root of a nonperfect power is an **irrational number.**

Irrational number: the root of a nonperfect power is an example of an irrational number. The decimal equivalent is nonterminating and nonrepeating

To find the two whole numbers that are closest to the value of an irrational number expressed as a radical:

1. Make a list of perfect powers that go beyond the given radicand.

2. Identify the two perfect powers that the radicand is between.

3. Find the principal roots of the two perfect powers from Step 2.

4. The root of the radicand is between the roots found in Step 3.

STOP AND CHECK
Find two whole numbers
that are closest to each given
irrational number.
1. $\sqrt{60}$
2. $\sqrt[3]{55}$

Answers:
1. 7 and 8   2. 3 and 4

EXAMPLE 2

Find two whole numbers that are closest to **(a)** $\sqrt{75}$ and **(b)** $\sqrt[3]{120}$.

(a) 1, 4, 9, 16, 25, 36, 49, 64, 81   List perfect squares. 75 is between 64 and 81.

$\sqrt{64} = 8$ $\sqrt{81} = 9$ Square roots of 64 and 81

$\sqrt{75}$ **is between 8 and 9.**

(b) 1, 8, 27, 64, 125   List perfect cubes. 120 is between 64 and 125.

$\sqrt[3]{64} = 4$ $\sqrt[3]{125} = 5$ Cube roots of 64 and 125

$\sqrt{120}$ **is between 4 and 5.**   **See Exercises 7–14.**

Real numbers include both rational and irrational numbers (Fig. 12–1).

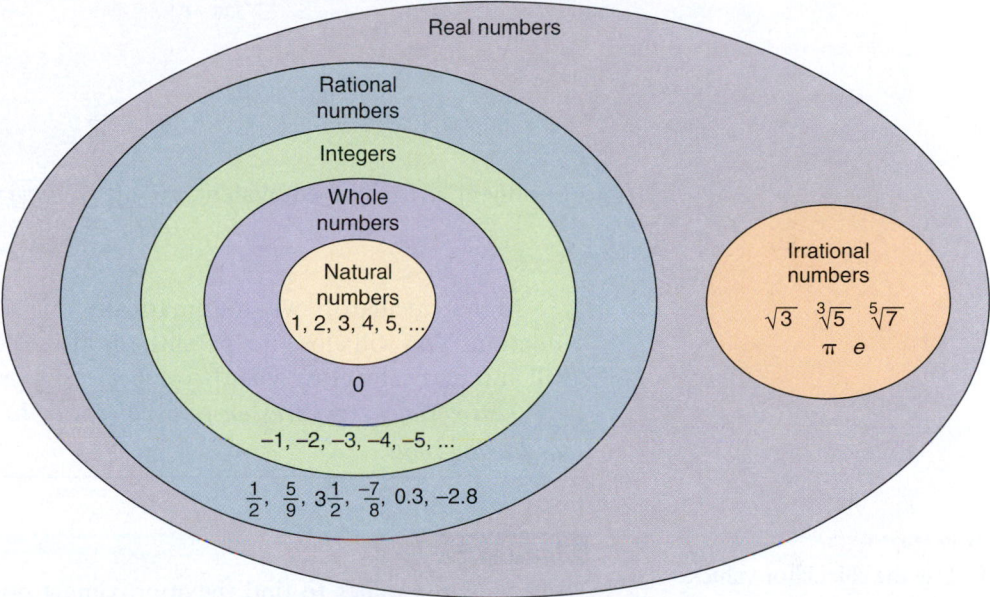

FIGURE 12-1 Real-number system.

Irrational numbers are ordered and have a position on the number line similar to the ordering of rational numbers.

Because of the availability of scientific and graphing calculators, radical notation, which once was the most popular notation for roots, is now being replaced by the rational exponent notation for roots.

TIP **Various Options for Roots on Your Calculator** Scientific and graphing calculators use a variety of symbols and processes to find roots.

Special root keys:
Some calculators have special function keys or menu options for the most common roots: square roots $\boxed{\sqrt{}}$ and cube roots $\boxed{\sqrt[3]{}}$. To use these functions, select the special root option, enter the base or radicand, and then press the $\boxed{=}$ or $\boxed{\text{ENTER}}$ key.

To find roots in general, other calculator functions are necessary.

General root key:
Use the general root key, $\boxed{x^{1/y}}$, on your calculator, to find the fourth root of 16. Enter the base or radicand, press the general root key, enter only the index of the

Special root key: a key or function on a scientific or graphing calculator for finding a specific root, such as square root or cube root

General root key: a key or function on a scientific or graphing calculator for finding any root

root (denominator of exponent), and then press $\boxed{=}$. Some calculators have a general root key in radical notation, $\boxed{\sqrt[y]{x}}$. Some typical sequences of keys are

$16\,\boxed{x^{1/y}}\,4\,\boxed{=} \Rightarrow 2$

$16\,\boxed{\sqrt[y]{x}}\,4\,\boxed{=} \Rightarrow 2$ Enter index second.

$4\,\boxed{\sqrt[y]{x}}\,16\,\boxed{=} \Rightarrow 2$ Enter index first.

General power key: a key or function on a scientific or graphing calculator for any power, including any type of rational number

General power key:

Using the general power key, $\boxed{\wedge}$ or $\boxed{x^y}$, on a calculator, find the fourth root of 16.

Exponent as indicated division: $16\,\boxed{\wedge}\,\boxed{(}\,1\,\boxed{\div}\,4\,\boxed{)}\,\boxed{=} \Rightarrow 2$

$$16^{1/4} = 2$$

Exponent as fraction: $16\,\boxed{\wedge}\,1\,\boxed{a\dfrac{b}{c}}\,4\,\boxed{=} \Rightarrow 2$ Some calculators with a fraction key require the fraction be placed in parentheses; others do not.

$$16^{1/4} = 2$$

Exponent as decimal equivalent: $16\,\boxed{\wedge}\,\boxed{\cdot}\,25\,\boxed{=} \Rightarrow 2$ $\dfrac{1}{4} = 0.25$

$$16^{0.25} = 2$$

Many calculators will automatically open a parenthesis when using a root function. You will close the parenthesis after the radicand has been entered or it will automatically close when $\boxed{=}$ or $\boxed{\text{ENTER}}$ is pressed.

Test various sequences on your calculator to determine what options you have for finding roots.

STOP AND CHECK

1. Use the calculator value to find the approximate position of $\sqrt{8}$.

Answer:
1.

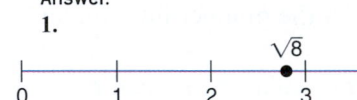

EXAMPLE 3

Use calculator values to find the approximate positions of the square roots on the number line (Fig. 12–2).

(a) $\sqrt{2}$ **(b)** $\sqrt{3}$ **(c)** $\sqrt{5}$ **(d)** $\sqrt{6}$

(a) $\sqrt{2} \approx 1.414$ **(b)** $\sqrt{3} \approx 1.732$

(c) $\sqrt{5} \approx 2.236$ **(d)** $\sqrt{6} \approx 2.449$

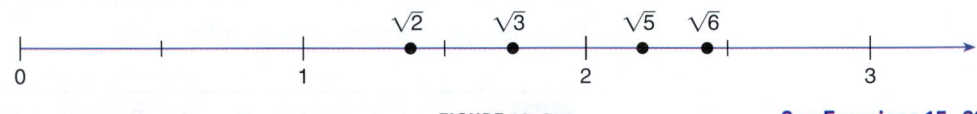

FIGURE 12–2   See Exercises 15–22.

3 **Write Powers and Roots Using Rational Exponents and Radical Notation.**
Powers and roots are inverse operations for nonnegative values. In the order of operations, powers and roots (Exponents in "Please Excuse My Dear Aunt Sally") have the same priority. The order in which the operations of powers and roots are performed does not matter.

	Rational Exponent Notation	**Radical Notation**
For $x \geq 0$ and n a natural number greater than 1:	$(x^{1/n})^n = x$	$(\sqrt[n]{x})^n = x$
	$(x^n)^{1/n} = x$	$\sqrt[n]{x^n} = x$

STOP AND CHECK

Write both the rational exponent and radical notations.

1. The cube root of the fourth power of x
2. The square of the cube root of x

Answers:

1. $\sqrt[3]{x^4}$; $x^{4/3}$ 2. $(\sqrt[3]{x})^2$; $x^{2/3}$

EXAMPLE 4

Write both the rational exponent and radical notations.

(a) The fourth root of the square of x

(b) The square of the fourth root of x

(c) The square root of the cube of x

(d) The cube of the square root of x

Rational Exponent Notation	Radical Notation	
(a) $(x^2)^{1/4}$	$\sqrt[4]{x^2}$	
(b) $(x^{1/4})^2$	$(\sqrt[4]{x})^2$	
(c) $(x^3)^{1/2}$	$\sqrt{x^3}$	
(d) $(x^{1/2})^3$	$(\sqrt{x})^3$	**See Exercises 23–28.**

To write powers and roots using rational exponents and radical notation:

Rational exponent to radical notation:

1. Write a decimal exponent as an equivalent fractional exponent.

2. Write the numerator of a rational exponent as the power.

3. Write the denominator of the rational exponent as the index of the root.

Symbolically, $x^{\text{power/root}} = \sqrt[\text{root}]{x^{\text{power}}}$ or $\left(\sqrt[\text{root}]{x}\right)^{\text{power}}$

Radical to rational exponent:

1. Write the power as the numerator of the rational exponent.

2. Write the index of the root as the denominator of the rational exponent.

3. Write the fraction in lowest terms or as a decimal equivalent.

Symbolically, $\sqrt[\text{root}]{x^{\text{power}}} = x^{\text{power/root}}$ or $x^{\frac{\text{power}}{\text{root}}}$

STOP AND CHECK

1. Write $x^{5/6}$ in radical form.

Answer:

1. $\sqrt[6]{x^5}$

EXAMPLE 5

Write $x^{2/3}$ in radical form.

$x^{2/3}$ Write the numerator of the rational exponent as the power.
   Write the denominator of the rational exponent as the index of the root.

$\sqrt[3]{x^2}$

See Exercises 29–36.

STOP AND CHECK

1. Write $\sqrt[3]{x^5}$ in rational exponent form.

Answer:

1. $x^{5/3}$ or $x^{1.\overline{6}}$

EXAMPLE 6

Write $\sqrt[4]{x^3}$ in rational exponent notation.

$\sqrt[4]{x^3}$ Write the power of the radicand as the numerator of the fractional exponent.
   Write the index as the denominator of the fractional exponent.

$x^{3/4}$ or $x^{0.75}$

See Exercises 37–44.

Interpreting an exponent depends on the type of number used as the exponent. Let's summarize the different types of numbers we have used as exponents.

TIP	**Different Types of Numbers Used as Exponents Indicate Different Interpretations**	
Exponent Type		**Means**
Natural number exponents greater than 1		Repeated multiplication
$4^3 = 4 \times 4 \times 4 = 64$		
Exponent of 1		Expression equals the base
$7^1 = 7$		
Exponent of 0 when base is not equal to 0		Expression equals 1
$5^0 = 1$		
Negative integral exponents		Reciprocal of power
$3^{-2} = \dfrac{1}{3^2} = \dfrac{1}{9}$		
Fractional or decimal exponents		Roots
$9^{1/2} = 9^{0.5} = \sqrt{9} = 3$		
$8^{1/3} = 8^{0.\overline{3}} = \sqrt[3]{8} = 2$		

12–1 EXERCISES   MyLab Math For additional practice go to your study plan in MyLab Math.

1 Write the roots in radical notation and exponential notation. *See Example 1.*

1. Square root of 36

2. Cube root of 27

3. Fourth root of 81

4. Fifth root of 32

5. Square root of 81

6. Cube root of 1,331

2 Find the two whole numbers that are closest to each root. *See Example 2.*

7. $\sqrt{40}$

8. $\sqrt{68}$

9. $\sqrt{12}$

10. $\sqrt{120}$

11. $\sqrt[3]{12}$

12. $\sqrt[3]{36}$

13. $\sqrt[3]{50}$

14. $\sqrt[3]{65}$

Use calculator values to find the approximate position. Then position each root on a number line. Label the points. *See Example 3.*

15. $\sqrt{25}$

16. $\sqrt{17}$

17. $\sqrt{20}$

18. $\sqrt{10}$

19. $\sqrt{18}$

20. $\sqrt{32}$

21. $\sqrt{7}$

22. $\sqrt{40}$

3 Write each expression in rational exponent notation and radical notation. *See Example 4.*

23. Square root of x^3

24. Cube root of the square of x

25. Fourth power of cube root of x

26. Square of cube root of x

27. Cube of fourth root of x

28. Fifth power of cube root of x

Write in radical notation. *See Example 5.*

29. $x^{3/5}$

30. $x^{2/5}$

31. $x^{3/4}$

32. $x^{5/8}$

33. $y^{1/5}$

34. $y^{7/8}$

35. $y^{3/8}$

36. $y^{4/5}$

Write in rational exponent notation in simplest form. *See Example 6.*

37. $\sqrt[3]{x^2}$

38. $\sqrt[4]{x^2}$

39. $\sqrt[5]{x^2}$

40. $\sqrt[5]{x^3}$

41. $\sqrt[6]{x^5}$

42. $\sqrt[6]{x^3}$

43. $\sqrt[6]{x^4}$

44. $\sqrt[7]{x^4}$

12-2 Simplifying Irrational Expressions

LEARNING OUTCOMES

1 Find the square root of variables.

2 Simplify square-root radicals using rational exponents and the laws of exponents.

3 Simplify square-root radical expressions containing perfect-square factors.

LC LEARNING CATALYTICS

1. Subtract $\dfrac{1}{3} - \dfrac{1}{4}$.

STOP AND CHECK

Find the square root of the positive variables.

1. $\sqrt{x^6}$

2. $\pm\sqrt{\dfrac{a^4}{b^{10}}}$

Answers:

1. x^3 2. $\pm\dfrac{a^2}{b^5}$

1 Find the Square Root of Variables. The square root of a variable is best understood by using the rational exponent notation and the laws of exponents. We will continue to consider only real numbers, and we will assume that the variables represent *positive values*.

$$\sqrt{x^2} = (x^2)^{1/2} = x^1 = x$$
$$\sqrt{x^4} = (x^4)^{1/2} = x^2$$
$$\sqrt{x^6} = (x^6)^{1/2} = x^3$$

> **TIP** **Perfect-Square Variable Factors** A positive variable with an even-number exponent is a perfect square. To find the square root of a variable factor, take $\frac{1}{2}$ of the exponent.

EXAMPLE 1

Find the square root of the positive variables.

(a) $\sqrt{x^8}$ **(b)** $-\sqrt{x^{12}}$ **(c)** $\pm\sqrt{x^{18}}$ **(d)** $\sqrt{\dfrac{a^2}{b^4}}$

(a) $\sqrt{x^8} = (x^8)^{1/2} = \mathbf{x^4}$ **(b)** $-\sqrt{x^{12}} = -(x^{12})^{1/2} = \mathbf{-x^6}$

(c) $\pm\sqrt{x^{18}} = \pm(x^{18})^{1/2} = \mathbf{\pm x^9}$ **(d)** $\sqrt{\dfrac{a^2}{b^4}} = \dfrac{(a^2)^{1/2}}{(b^4)^{1/2}} = \dfrac{\mathbf{a}}{\mathbf{b^2}}$

See Exercises 1–8.

> **TIP** **Principal and Negative Square Roots** There are two square roots for a positive radicand. It is customary to indicate whether the positive (principal) square root, the negative square root, or both square roots are to be considered by putting the appropriate sign in front of the radicand or its coefficient.
>
> $$\sqrt{4} = 2 \qquad -\sqrt{4} = -2 \qquad \pm\sqrt{4} = \pm 2$$

2 Simplify Square-Root Radicals Using Rational Exponents and the Laws of Exponents. Even though the calculator makes the pencil-and-paper method of taking roots outmoded, it is still helpful to perform some mental manipulations on expressions containing powers and roots before evaluating these expressions using the calculator. One common manipulation is to simplify rational exponents and radical expressions by applying the laws of exponents.

To simplify radicals using rational exponents and the laws of exponents:

1. Convert the radicals to equivalent expressions using rational exponents.

2. Apply the laws of exponents and the arithmetic of fractions.

3. Convert simplified expressions back to radical notation if desired.

EXAMPLE 2

Convert the radical expressions to equivalent expressions using rational exponents and simplify if appropriate.

(a) $\sqrt[3]{x}$ **(b)** $\sqrt[5]{2y}$ **(c)** $(\sqrt{ab})^3$ **(d)** $\sqrt[4]{16b^8}$ **(e)** $(\sqrt[3]{27xy^5})^4$

(a) $\sqrt[3]{x} = x^{1/3}$ **(b)** $\sqrt[5]{2y} = (2y)^{1/5}$ or $2^{1/5}y^{1/5}$

(c) $(\sqrt{ab})^3 = (ab)^{3/2}$ or $a^{3/2}b^{3/2}$

(d) $\sqrt[4]{16b^8} =$ 16 is a perfect fourth power. $2^4 = 16$

      $(2^4 b^8)^{1/4} =$ Raise each factor to the $\frac{1}{4}$ power.

      $(2^4)^{1/4}(b^8)^{1/4} =$ $(2^4)^{1/4} = 2^1 = 2; (b^8)^{1/4} = b^2$

      $2b^2$

(e) $(\sqrt[3]{27xy^5})^4 =$ 27 is a perfect cube. $3^3 = 27$

      $(3^3 xy^5)^{4/3} =$ Raise each factor to the $\frac{4}{3}$ power.

      $(3^3)^{4/3} x^{4/3}(y^5)^{4/3} =$ $(3^3)^{4/3} = 3^4 = 81; (y^5)^{4/3} = y^{20/3}$

      $81x^{4/3}y^{20/3}$ See Exercises 9–21.

> **TIP**   **Simplifying Coefficients**   When coefficients of variable terms have rational exponents, the numerical equivalent can be determined if desired. Usually, if the coefficient is a perfect power of the indicated root, we will evaluate the coefficient. Otherwise, we may leave the coefficient with the rational exponent.
>
> $$8^{2/3} = (8^{1/3})^2 = ((2^3)^{1/3})^2 = 2^2 = 4 \qquad 5^{2/3} \text{ does not simplify}$$

The laws of exponents are applied to rational exponents of factors having *like bases*. The following example illustrates the laws of exponents applied to rational exponents.

EXAMPLE 3

Perform the following operations and simplify. Express answers with positive exponents in lowest terms.

(a) $(x^{3/2})(x^{1/2})$ **(b)** $(3a^{1/2}b^3)^2$ **(c)** $\dfrac{x^{1/2}}{x^{1/3}}$ **(d)** $\dfrac{10a^3}{2a^{1/2}}$

(a) $(x^{3/2})(x^{1/2}) = x^{3/2+1/2} = x^{4/2} = \boldsymbol{x^2}$ Add exponents: $\frac{3}{2} + \frac{1}{2} = \frac{4}{2} = 2$

(b) $(3a^{1/2}b^3)^2 = 3^2 ab^6 = \boldsymbol{9ab^6}$ Multiply exponents: $\frac{1}{2} \cdot 2 = 1; 3 \cdot 2 = 6$

(c) $\dfrac{x^{1/2}}{x^{1/3}} = x^{1/2-1/3} = \boldsymbol{x^{1/6}}$ Subtract exponents: $\frac{1}{2} - \frac{1}{3} = \frac{3}{6} - \frac{2}{6} = \frac{1}{6}$

(d) $\dfrac{10a^3}{2a^{1/2}} = 5a^{3-1/2} = \boldsymbol{5a^{5/2}}$ Reduce coefficients. Subtract exponents: $3 - \frac{1}{2} = \frac{3}{1} - \frac{1}{2} = \frac{6}{2} - \frac{1}{2} = \frac{5}{2}$

           See Exercises 22–31.

EXAMPLE 4

Evaluate the expressions **(a)** and **(d)** from Example 3 for $x = 2$ and $a = 3$ both before and after simplifying.

(a) $(x^{3/2})(x^{1/2}) =$

Before simplifying: $(x^{3/2})(x^{1/2}) = (2^{3/2})(2^{1/2})$ Substitute $x = 2$.

$2\boxed{\wedge}\boxed{(}3\boxed{\div}2\boxed{)}\boxed{\times}$ Enter into the calculator using the general power key $\boxed{\wedge}$ or $\boxed{x^y}$.

$2\boxed{\wedge}\boxed{(}1\boxed{\div}2\boxed{)}\boxed{=}\Rightarrow \mathbf{4}$

After simplifying: $(x^{3/2})(x^{1/2}) = x^{3/2+1/2}$ Add exponents.

$\qquad\qquad\qquad\qquad = x^2$ Substitute $x = 2$.

$\qquad\qquad\qquad\quad 2^2 = \mathbf{4}$ Evaluate.

(d) $\dfrac{10a^3}{2a^{1/2}} =$

Before simplifying: $\dfrac{10a^3}{2a^{1/2}} = \dfrac{10 \cdot 3^3}{2 \cdot 3^{1/2}}$ Substitute $a = 3$.

$\boxed{(}10\boxed{\times}3\boxed{\wedge}3\boxed{)}\boxed{\div}$ Both the numerators and denominators are groupings.

$\boxed{(}2\boxed{\times}3\boxed{\wedge}\boxed{(}1\boxed{\div}2\boxed{)}\boxed{)}\boxed{=}$

$\Rightarrow \mathbf{77.94228634}$

After simplifying: $\dfrac{10\,a^3}{2a^{1/2}} = \dfrac{10}{2} \cdot \dfrac{a^3}{a^{1/2}}$ Reduce coefficients. Subtract exponents.

$\qquad\qquad\qquad = 5a^{3-\frac{1}{2}}$ $3 - \dfrac{1}{2} = \dfrac{6}{2} - \dfrac{1}{2} = \dfrac{5}{2}$

$\qquad\qquad\qquad = 5a^{5/2}$ or $5a^{2.5}$ Substitute $a = 3$.

$\qquad\qquad\qquad = 5 \cdot 3^{2.5}$ Evaluate.

$5\boxed{\times}3\boxed{\wedge}2.5\boxed{=}$

$\Rightarrow \mathbf{77.94228634}$ **See Exercises 32–33.**

Is it really worth the effort to simplify an expression before evaluating? Yes, in most cases. Long sequences of calculator steps are very tedious to enter.

As we saw in part (d) of the preceding example, *decimal exponents* are often desirable in finding roots.

STOP AND CHECK

Perform the operations. Express the answers with positive exponents.

1. $x^{1.2}(x^3)$

2. $(2x^{2.5}y^{0.3})^3$

Answers:

1. $x^{4.2}$ **2.** $8x^{7.5}y^{0.9}$

Factor an algebraic expression: to write the expression as the indicated product of two or more factors, that is, as a multiplication

EXAMPLE 5

Perform the operations. Express the answers with positive exponents.

(a) $a^{2.3}(a^3)$ **(b)** $(5a^{3.5}b^{0.5})^2$

(a) $a^{2.3}(a^3) = a^{2.3+3} = \mathbf{a^{5.3}}$

What does $a^{5.3}$ mean? If written as an improper fraction, $a^{5.3} = a^{5\frac{3}{10}} = a^{53/10}$. This means we take the tenth root of a to the 53rd power or $\sqrt[10]{a^{53}}$.

(b) $(5a^{3.5}b^{0.5})^2 = 5^2 a^{3.5(2)} b^{0.5(2)}$

$\qquad\qquad\qquad = \mathbf{25a^7 b}$ **See Exercises 34–38.**

3 **Simplify Square-Root Radical Expressions Containing Perfect-Square Factors.** Some radicands are perfect squares, whereas others are not perfect squares. In the latter case, it may be useful to simplify the radicand. This is usually done by first writing the radicand as appropriate factors.

To **factor an algebraic expression** is to write it as the indicated product of two or more factors, that is, as a multiplication.

Some radicands that are not perfect squares can be *factored* into a perfect square times other factors. *We simplify radicals by taking the square root of all the perfect-square factors.*

> **To simplify square-root radicals containing perfect-square factors:**
>
> 1. If the radicand is a perfect square, express it as a square root without the radical sign.
>
> 2. If the radicand is *not* a perfect square, factor the radicand so that one factor is the largest possible perfect-square factor. The square roots of the perfect-square factors are written *outside* the radical and the other factors stay *inside* (under) the radical sign.
>
> If the radicand is *not* a perfect square and *cannot* be factored into one or more perfect-square factors, it is in simplest radical form.

> **Did You Know?**
>
> ▶ A variable factor with an even exponent is a perfect-square factor.
>
> $$\sqrt{x^4} = x^2 \quad \sqrt{x^6} = x^3 \quad \sqrt{x^8} = x^4$$
>
> ▶ A term with variable factors with an odd exponent that is greater than 1 has a perfect-square factor.
>
> $$\sqrt{x^5} = \sqrt{x^4 x} = x^2\sqrt{x} \quad \sqrt{x^7} = \sqrt{x^6 x} = x^3\sqrt{x} \quad \sqrt{x^9} = \sqrt{x^8 x} = x^4\sqrt{x}$$

STOP AND CHECK

Simplify the radicals.
1. $\sqrt{50}$
2. $\sqrt{8x^3}$
3. $\sqrt{54x^4y^7}$

Answers:
1. $5\sqrt{2}$ 2. $2x\sqrt{2x}$
3. $3x^2y^3\sqrt{6y}$

TIP **In part c of Example 6, is x^2 a coefficient?** Yes. It is just not a numerical coefficient. It is a factor outside of a radical and can be called a coefficient. Another way of describing the process is to multiply the factors that are outside the radicals.

EXAMPLE 6

Simplify the radicals.

(a) $\sqrt{32}$ **(b)** $\sqrt{y^7}$ **(c)** $\sqrt{18x^5}$ **(d)** $\sqrt{75xy^3z^5}$ **(e)** $\sqrt{7x}$

When factoring coefficients, some perfect squares that can be used are 4, 9, 16, 25, 36, 49, 64, and 81.

(a) $\sqrt{32} = \sqrt{16 \cdot 2}$ What is the *largest* perfect-square factor of 32? 4 is a factor of 32, but 16 is also a factor of 32. Use the *largest* perfect-square factor.

$$= \sqrt{16}(\sqrt{2}) = 4\sqrt{2}$$

(b) $\sqrt{y^7} = \sqrt{y^6 \cdot y^1} = \sqrt{y^6}(\sqrt{y})$ The largest perfect-square factor of y^7 is y^{7-1}, or y^6. For variables, perfect-square factors have *even-number* exponents.

$$= y^3\sqrt{y}$$

(c) $\sqrt{18x^5}$ Factor the radicand into as many perfect-square factors as possible.

$$= \sqrt{9 \cdot 2 \cdot x^4 \cdot x^1}$$ Write each factor as a separate radical.

$$= \sqrt{9}(\sqrt{2})(\sqrt{x^4})(\sqrt{x})$$ Take the square root of the two perfect squares.

$$= 3(\sqrt{2})(x^2)\sqrt{x}$$ Multiply coefficients. Multiply radicands.

$$= 3x^2\sqrt{2x}$$

In the previous example we showed each step in the simplifying process. However, we customarily do most of these steps mentally.

(d) $\sqrt{75xy^3z^5} = \sqrt{\boxed{25} \cdot 3 \cdot x \cdot \boxed{y^2} \cdot y \cdot \boxed{z^4} \cdot z}$

Write the square roots of the perfect-square factors outside the radical sign. The remaining factors are written under the radical sign.

$= \mathbf{5yz^2} \sqrt{\mathbf{3xyz}}$

(e) $\sqrt{7x} = \sqrt{7x}$

7 and x contain no perfect-square factors. The radical is in simplest form.

See Exercises 39–47.

> **TIP** **Finding Perfect-Square Factors**
>
> ▶ Natural-number perfect squares: 1, 4, 9, 16, 25, 36, 49, 64, 81, 100, 121, 144, ...
>
> ▶ 1 is a factor of any number: $8 = 8 \cdot 1$. To factor using the perfect square 1 does not simplify a radicand.
>
> ▶ Any variable with an exponent greater than 1 is a perfect square or has a perfect-square factor.
>
> ▶ Perfect-square variables have even numbers as exponents:
> $$x^2, x^4, x^6, x^8, x^{10}, \ldots$$
>
> ▶ Variables with a perfect-square factor:
> $$x^3 = x^2 \cdot x^1, \qquad x^5 = x^4 \cdot x^1$$
> $$x^7 = x^6 \cdot x^1, \qquad x^9 = x^8 \cdot x^1$$
>
> ▶ A convenient way to keep track of perfect-square factors is to circle them. The square roots of circled factors are written outside the radical sign. The uncircled factors stay in the radicand as is.
> $$\sqrt{75ab^4c^3} = \sqrt{\boxed{25} \cdot 3 \cdot a \cdot \boxed{b^4} \cdot \boxed{c^2} \cdot c^1}$$
> $$= 5b^2c\sqrt{3ac}$$

STOP AND CHECK

Simplify.

1. $\sqrt{6x^4}$

2. $\sqrt{12x^3y^8}$

Answers:

1. $x^2\sqrt{6}$ **2.** $2xy^4\sqrt{3x}$

EXAMPLE 7

Simplify.

(a) $\sqrt{7x^2}$ **(b)** $\sqrt{9a}$ **(c)** $\sqrt{32m^5n^6}$

(a) $\sqrt{7\boxed{x^2}} = x\sqrt{7}$ **(b)** $\sqrt{\boxed{9}a} = 3\sqrt{a}$

(c) $\sqrt{32m^5n^6} = \sqrt{\boxed{16} \cdot 2 \cdot \boxed{m^4} \cdot m \cdot \boxed{n^6}} = 4m^2n^3\sqrt{2m}$

See Exercises 39–47.

12–2 EXERCISES

MyLab Math For additional practice go to your study plan in MyLab Math.

1 Find the square root of the positive variables. *See Example 1.*

1. $\sqrt{x^{10}}$

2. $\sqrt{x^6}$

3. $\pm\sqrt{x^{16}}$

4. $\sqrt{\dfrac{x^{14}}{y^{24}}}$

5. $\sqrt{a^6b^{10}}$

6. $-\sqrt{a^2b^4c^{12}}$

7. $\sqrt{\dfrac{x^2y^4}{z^{10}}}$

8. $\sqrt{\dfrac{a^4}{b^{10}c^{12}}}$

2 Convert the radical expressions to equivalent expressions using rational exponents, and simplify if appropriate. *See Example 2.*

9. $\sqrt[3]{y}$

10. $\sqrt[3]{5x}$

11. $(\sqrt{xy})^5$

12. $(\sqrt{r})^7$

13. $\sqrt[4]{81x^8}$

14. $\sqrt[4]{16r^{12}}$

15. $(\sqrt[3]{8x^3y^6})^4$

16. $(\sqrt[4]{16xy^5})^5$

17. $\sqrt[4]{x}$

18. $\sqrt[5]{7x}$

19. $(\sqrt[4]{xy})^4$

20. $\sqrt[4]{81b^{20}}$

21. $(\sqrt[3]{8xy^7})^2$

Perform the operations. Express answers with positive exponents in lowest terms. *See Example 3.*

22. $x^{5/2} \cdot x^{3/2}$

23. $y^{4/3} \cdot y^{5/3}$

24. $(4a^{1/3}b^2)^3$

25. $(2x^{2/3}y^{5/6})^3$

26. $\dfrac{x^{1/3}}{x^{2/3}}$

27. $\dfrac{x^{4/5}}{x^{1/5}}$

28. $\dfrac{y^{1/3}}{y^{1/2}}$

29. $\dfrac{b^{2/3}}{b^{1/2}}$

30. $\dfrac{12x^4}{3x^{1/2}}$

31. $\dfrac{20x^3}{4x^{2/3}}$

See Example 4.

32. Evaluate $(x^{1/3})(x^{2/3})$ if $x = 2$.

33. Evaluate $\dfrac{12x^2}{2x^{1/3}}$ if $x = 8$.

Perform the operations. *See Example 5.*

34. $a^{1.2}(a^4)$

35. $(7a^{1.7} \cdot a^{2.3})^2$

36. $\dfrac{x^{3.7}}{x^{1.7}}$

37. $\dfrac{12x^{2.9}}{4x^{0.9}}$

38. $2x^{2.1}(3x^{3.9})$

3 Simplify. *See Examples 6 and 7.*

39. $\sqrt{24}$

40. $\sqrt{98}$

41. $\sqrt{48}$

42. $\sqrt{x^9}$

43. $\sqrt{y^{15}}$

44. $\sqrt{12x^3}$

45. $\sqrt{56a^5}$

46. $\sqrt{72a^3x^4}$

47. $\sqrt{44x^5y^2z^7}$

48. Create a square-root radical with the product of a constant factor that is not a perfect square but has a perfect-square factor and a variable factor with an odd exponent greater than 1. Then simplify.

12–3 Basic Operations with Square-Root Radicals

LEARNING OUTCOMES

1 Add or subtract square-root radicals.

2 Multiply or divide square-root radicals.

1 **Add or Subtract Square-Root Radicals.** As you recall, when adding or subtracting measures or algebraic terms, we only add or subtract like quantities. Similarly, only *like* radicals are added or subtracted.

Like radicals are radical expressions with radicands that are identical and have the same order or index.

LC LEARNING CATALYTICS

1. Use the FOIL method to multiply $(2x + 3)(3x - 5)$.

Like radicals: radical expressions with radicands that are identical and have the same order or index

To add or subtract (combine) like square-root radicals:
1. Simplify all radicals.
2. Add the coefficients of like radicals.
3. Use the common radical as a factor in the solution.
Symbolically, $a\sqrt{b} + c\sqrt{b} = (a + c)\sqrt{b}, b > 0$

STOP AND CHECK

Add or subtract the radicals when possible.

1. $5\sqrt{3} - 2\sqrt{3}$
2. $3\sqrt{2} - 5\sqrt{2} + \sqrt{2}$
3. $4\sqrt{3} - \sqrt{5}$

Answers:
1. $3\sqrt{3}$ 2. $-\sqrt{2}$
3. $4\sqrt{3} - \sqrt{5}$

EXAMPLE 1

Add or subtract the radicals when possible.

(a) $3\sqrt{7} + 2\sqrt{7}$ (b) $4\sqrt{2} + \sqrt{2}$ (c) $5\sqrt{3} + 7\sqrt{5} + 2\sqrt{3} - 4\sqrt{5}$
(d) $3 + \sqrt{3}$ (e) $3\sqrt{11} - 3\sqrt{11}$ (f) $2\sqrt{2} - \sqrt{3}$

(a) $3\sqrt{7} + 2\sqrt{7} = \mathbf{5\sqrt{7}}$ — Radicals are like. Add coefficients.

(b) $4\sqrt{2} + \sqrt{2} = \mathbf{5\sqrt{2}}$ — When no coefficient is written in front of a radical, the coefficient is 1.

(c) $5\sqrt{3} + 7\sqrt{5} + 2\sqrt{3} - 4\sqrt{5} =$
$\mathbf{7\sqrt{3} + 3\sqrt{5}}$ — Combine like radicals by adding appropriate coefficients.

(d) $\mathbf{3 + \sqrt{3}}$ — No addition can be performed. These are not like radicals.

(e) $3\sqrt{11} - 3\sqrt{11} = 0\sqrt{11} = \mathbf{0}$ — Zero times any number is zero.

(f) $\mathbf{2\sqrt{2} - \sqrt{3}}$ — The terms cannot be combined. These are not like radicals.

See Exercises 1–6.

We add or subtract square-root radical expressions only if they have *like* radicands. However, when radicals are not in simplest form, we simplify the radical expressions. If we obtain like radicands after simplifying, then we add or subtract.

STOP AND CHECK

Add or subtract the radical expressions when possible.

1. $7\sqrt{6} + 2\sqrt{24}$
2. $3\sqrt{2} - \sqrt{18}$
3. $6\sqrt{5} + 2\sqrt{32}$

Answers:
1. $11\sqrt{6}$ 2. 0
3. $6\sqrt{5} + 8\sqrt{2}$

EXAMPLE 2

Add or subtract the radical expressions when possible.

(a) $12\sqrt{5} + 3\sqrt{20}$ (b) $\sqrt{3} - \sqrt{27}$ (c) $6\sqrt{3} + 2\sqrt{8}$

(a) $12\sqrt{5} + 3\sqrt{20}$ — $\sqrt{20}$ can be simplified. $\sqrt{20} = \sqrt{4 \cdot 5}$
$12\sqrt{5} + 3\sqrt{4 \cdot 5}$ — $\sqrt{4} = 2$. The 2 becomes a coefficient. Now, 3 and 2 are both coefficients of $\sqrt{5}$.
$12\sqrt{5} + 3 \cdot 2\sqrt{5}$ — Multiply 3 and 2.
$12\sqrt{5} + 6\sqrt{5}$ — Now we have like radicands. Add coefficients.
$\mathbf{18\sqrt{5}}$ — Keep like radicand.

(b) $\sqrt{3} - \sqrt{27}$ — $\sqrt{27}$ can be simplified. $\sqrt{27} = \sqrt{9 \cdot 3}$
$\sqrt{3} - \sqrt{9 \cdot 3}$ — $\sqrt{9} = 3$. The 3 becomes a coefficient.
$\sqrt{3} - 3\sqrt{3}$ — Now we have like radicals. Add coefficients. $1 - 3 = -2$
$\mathbf{-2\sqrt{3}}$ — Keep like radicand.

(c) $6\sqrt{3} + 2\sqrt{8}$ $\sqrt{8}$ can be simplified. $\sqrt{8} = \sqrt{4\cdot2}$

$6\sqrt{3} + 2\sqrt{4\cdot2}$ $\sqrt{4} = 2$. The 2 becomes a coefficient.

$6\sqrt{3} + 2\cdot2\sqrt{2}$ $2\cdot2\sqrt{2} = 4\sqrt{2}$

$6\sqrt{3} + 4\sqrt{2}$ Terms cannot be combined. Radicals are still unlike.

See Exercises 7–18.

---

2 Multiply or Divide Square-Root Radicals. When multiplying two square-root radicals, the expressions under the radical signs (radicands) are multiplied together. Numbers in front of the radical signs are coefficients of the radicals and are multiplied separately.

> **To multiply square-root radicals:**
>
> 1. Multiply coefficients to give the coefficient of the product.
>
> 2. Multiply radicands to give the radicand of the product.
>
> 3. Simplify if possible.
>
> Symbolically, $a\sqrt{b} \cdot c\sqrt{d} = ac\sqrt{bd}$.

> **TIP Express Procedures in Your Own Words** The precise details of a procedure can sometimes be overwhelming. The procedure written symbolically often guides you through the process. However, it is desirable to describe the procedure in your own words.
>
> For example, the procedure for multiplying square-root radicals could be casually phrased in various ways.
>
> Outside times outside $\Rightarrow$ Stays outside
>
> Inside times inside $\Rightarrow$ Stays inside
>
> $(2\sqrt{3})(5\sqrt{2}) = 10\sqrt{6}$
>
> Another example might be used with the procedure for simplifying perfect-square radicands. The radical symbol defines a confined area. "Perfect" factors get out; "imperfect factors" stay in. As perfect factors come out, they are different (square root is taken).
>
> $\sqrt{9x^2y} =$ 9 is "perfect." It comes out as 3.
>
> x^2 is "perfect." It comes out as x.
>
> y is "imperfect." It stays in as y.
>
> $3x\sqrt{y}$

STOP AND CHECK

Multiply the following radicals.

1. $\sqrt{2}\cdot\sqrt{7}$

2. $\sqrt{\dfrac{3}{5}}\cdot\sqrt{\dfrac{7}{10}}$

3. $5\sqrt{3}\cdot3\sqrt{5}$

Answers:

1. $\sqrt{14}$ 2. $\sqrt{\dfrac{21}{50}}$

3. $15\sqrt{15}$

EXAMPLE 3

Multiply the following radicals.

(a) $\sqrt{3}\cdot\sqrt{5}$ **(b)** $\sqrt{\dfrac{7}{8}}\cdot\sqrt{\dfrac{2}{3}}$ **(c)** $3\sqrt{2}\cdot4\sqrt{3}$ **(d)** $\sqrt{3}\cdot\sqrt{12}$

(a) $\sqrt{3}\cdot\sqrt{5} = \sqrt{15}$ $\sqrt{15}$ will not simplify.

(b) $\sqrt{\dfrac{7}{8}}\cdot\sqrt{\dfrac{2}{3}} = \sqrt{\dfrac{14}{24}} = \sqrt{\dfrac{7}{12}}$ Reduce.

(c) $3\sqrt{2}\cdot4\sqrt{3} = 12\sqrt{6}$ — Multiply coefficients 3 and 4. Then multiply radicands 2 and 3.

(d) $\sqrt{3}\cdot\sqrt{12} = \sqrt{36} = 6$ — 36 is a perfect square. Take the square root of 36.

See Exercises 19–24.

The distributive property can be used to multiply radical expressions.

EXAMPLE 4

Multiply by distributing.

(a) $5(\sqrt{7} - 2)$

(b) $\sqrt{3}(\sqrt{11} - 5)$

(c) $\sqrt{2}(\sqrt{5} - \sqrt{7})$

(a) $5(\sqrt{7} - 2) = 5\sqrt{7} - 5\cdot2$ — Distribute. Multiply $5 \cdot 2$.
$= 5\sqrt{7} - 10$

(b) $\sqrt{3}(\sqrt{11} - 5) = \sqrt{3}\cdot\sqrt{11} - 5\sqrt{3}$ — Distribute. Multiply $\sqrt{3}\cdot\sqrt{11}$.
$= \sqrt{33} - 5\sqrt{3}$

(c) $\sqrt{2}(\sqrt{5} - \sqrt{7}) = \sqrt{2}\cdot\sqrt{5} - \sqrt{2}\cdot\sqrt{7}$ — Distribute. Multiply radicals.
$= \sqrt{10} - \sqrt{14}$

See Exercises 25–29.

The FOIL method works for all types of numbers and expressions.

EXAMPLE 5

Find the product $(2 + \sqrt{3})(5 + 2\sqrt{3})$.

$(2 + \sqrt{3})(5 + 2\sqrt{3})$ — Multiply applying the FOIL method.

F O I L
$2\cdot5 + 2\cdot2\sqrt{3} + \sqrt{3}\cdot5 + \sqrt{3}\cdot2\sqrt{3} =$ — $\sqrt{3}\cdot2\sqrt{3} = 2\sqrt{9} = 6$
$10 + 4\sqrt{3} + 5\sqrt{3} + 6 =$ — Combine like terms.

$16 + 9\sqrt{3}$

See Exercises 30–32.

Factor pairs like $(a + b)(a - b)$ are called the *sum and difference of the same two terms,* or *conjugate pairs.* Their product, $a^2 - b^2$, is called the *difference of two perfect squares.*

To multiply the sum and difference of two terms or conjugate pairs:

1. Square the first term of either binomial.

2. Insert a minus sign.

3. Square the second term of either binomial.

Symbolically, $(a + b)(a - b) = a^2 - b^2$, where a and b are real numbers.

The product of conjugate pairs is useful when expressions have irrational or imaginary terms or factors.

STOP AND CHECK
Multiply.
1. $(2 + \sqrt{7})(2 - \sqrt{7})$
2. $(3\sqrt{5} + \sqrt{2})(3\sqrt{5} - \sqrt{2})$

Answers:
1. -3 2. 43

EXAMPLE 6

Multiply **(a)** $(3 + \sqrt{5})(3 - \sqrt{5})$ **(b)** $(2\sqrt{3} + 4)(2\sqrt{3} - 4)$

(c) $(\sqrt{2} + \sqrt{3})(\sqrt{2} - \sqrt{3})$

(a) $(3 + \sqrt{5})(3 - \sqrt{5})$ These are conjugate pairs.

$(a + b)(a - b) = a^2 - b^2$ Apply the special product pattern.

$(3 + \sqrt{5})(3 - \sqrt{5}) = 9 - 5$ $3^2 = 9; (\sqrt{5})^2 = 5.$ Combine like terms.

$= 4$

(b) $(2\sqrt{3} + 4)(2\sqrt{3} - 4)$ These are conjugate pairs.

$(a + b)(a - b) = a^2 - b^2$ Apply the special product pattern.

$(2\sqrt{3} + 4)(2\sqrt{3} - 4) = 12 - 16$ $(2\sqrt{3})^2 = 4 \cdot 3 = 12; 4^2 = 16.$ Combine

$= -4$ like terms.

(c) $(\sqrt{2} + \sqrt{3})(\sqrt{2} - \sqrt{3})$ These are conjugate pairs.

$(a + b)(a - b) = a^2 - b^2$ Apply the special product pattern.

$(\sqrt{2} + \sqrt{3})(\sqrt{2} - \sqrt{3}) = 2 - 3$ $(\sqrt{2})^2 = 2; (\sqrt{3})^2 = 3.$ Combine like

$= -1$ terms.   **See Exercises 33–35.**

The next example examines the process of squaring a binomial when a radical is involved.

STOP AND CHECK
1. Multiply $(6 - \sqrt{3})^2$.

Answer:
1. $39 - 12\sqrt{3}$

EXAMPLE 7

Multiply $(3 + \sqrt{5})^2$.

$(3 + \sqrt{5})^2 =$ Apply the squaring process or FOIL.

$9 + 2 \cdot 3\sqrt{5} + (\sqrt{5})^2 =$ Simplify.

$9 + 6\sqrt{5} + 5 =$ Combine like terms.

$\mathbf{14 + 6\sqrt{5}}$ **See Exercises 36–38.**

When dividing square-root radicals, we follow a similar procedure as with multiplication. Coefficients and radicands are divided (or reduced) separately.

> **To divide square-root radicals:**
>
> 1. Divide coefficients to give the coefficient of the quotient.
>
> 2. Divide radicands to give the radicand of the quotient.
>
> 3. Simplify if possible.
>
> $$\frac{a\sqrt{b}}{c\sqrt{d}} = \frac{a}{c}\sqrt{\frac{b}{d}} \qquad c \text{ and } d \neq 0$$

STOP AND CHECK
Divide the following radicals.

1. $\dfrac{\sqrt{20}}{\sqrt{5}}$ 2. $\dfrac{\sqrt{\frac{3}{5}}}{\sqrt{\frac{2}{7}}}$

3. $\dfrac{12x^4\sqrt{4x^2}}{4x^2\sqrt{25x^3}}$

Answers:
1. 2 2. $\sqrt{\dfrac{21}{10}}$ 3. $\dfrac{6x^2}{5\sqrt{x}}$

EXAMPLE 8

Divide the following radicals.

(a) $\dfrac{\sqrt{12}}{\sqrt{4}}$ **(b)** $\dfrac{\sqrt{\frac{2}{3}}}{\sqrt{\frac{7}{4}}}$ **(c)** $\dfrac{3\sqrt{6}}{6}$ **(d)** $\dfrac{5\sqrt{20}}{\sqrt{10}}$ **(e)** $\dfrac{8x^3\sqrt{9x^2}}{2x^2\sqrt{16x^3}}$

(a) $\dfrac{\sqrt{12}}{\sqrt{4}} = \sqrt{\dfrac{12}{4}} = \sqrt{3}$ Divide or reduce.

(b) $\dfrac{\sqrt{\dfrac{2}{3}}}{\sqrt{\dfrac{7}{4}}} = \sqrt{\dfrac{\dfrac{2}{3}}{\dfrac{7}{4}}} = \sqrt{\dfrac{2}{3}\left(\dfrac{4}{7}\right)} =$ Multiply $\frac{2}{3}$ by the reciprocal of $\frac{7}{4} \cdot \left(\frac{2}{3} \div \frac{7}{4} = \frac{2}{3} \cdot \frac{4}{7}\right)$

$\sqrt{\dfrac{8}{21}} = \sqrt{\dfrac{4 \cdot 2}{21}} = \dfrac{2\sqrt{2}}{\sqrt{21}}$ Factor and take the square root of perfect-square factors.

(c) $\dfrac{3\sqrt{6}}{6}$ The coefficient 3 and the denominator 6 can be divided (or reduced) because they are both outside the radical.

$\dfrac{\overset{1}{\cancel{3}}\sqrt{6}}{\underset{2}{\cancel{6}}} = \dfrac{\sqrt{6}}{2}$ A coefficient of 1 does not have to be written in front of the radical.

(d) $\dfrac{5\sqrt{20}}{\sqrt{10}} = 5\sqrt{2}$ The 20 and 10 are divided because they are both square-root radicands.

(e) $\dfrac{8x^3\sqrt{9x^2}}{2x^2\sqrt{16x^3}} =$ Reduce the factors outside the radical.

$\dfrac{4x\sqrt{9x^2}}{\sqrt{16x^3}} =$ Reduce the factors inside the radical.

$\dfrac{4x\sqrt{9}}{\sqrt{16x}} =$ Remove perfect-square factors from all radicands.

$\dfrac{4x(3)}{4\sqrt{x}} =$ Reduce the resulting coefficients.

$\dfrac{3x}{\sqrt{x}}$ **See Exercises 39–56.**

Fractions with radical factors in the denominator are often rewritten in an equivalent form. Before the common use of calculators, this was a popular manipulation. Dividing by a rational number, and most often a whole number, was easier and more accurate than dividing by a rounded decimal approximation of the irrational number. Even now, finding common denominators and other procedures are easier if all denominators contain only rational numbers. Thus, $\frac{1}{\sqrt{3}}$ and $\sqrt{\frac{2}{5}}$ are generally rewritten so that the denominator is a rational number. This procedure is called *rationalizing the denominator*.

To rationalize a denominator:

1. If the denominator has an irrational factor, multiply the denominator by another irrational factor so that the resulting radicand is a perfect square.

2. To preserve the value of the fraction, multiply the numerator by the same radical factor used in Step 1. Thus, in Steps 1 and 2 together, we have multiplied by an equivalent of 1.

3. Simplify all radicals and reduce the resulting fraction, if possible.

STOP AND CHECK
Rationalize all denominators. Simplify the answers if possible.

1. $\sqrt{\dfrac{21}{10}}$ 2. $\dfrac{3}{5\sqrt{x}}$

3. $\dfrac{4\sqrt{3}}{x^3\sqrt{2x}}$

Answers:
1. $\dfrac{\sqrt{210}}{10}$ 2. $\dfrac{3\sqrt{x}}{5x}$ 3. $\dfrac{2\sqrt{6x}}{x^4}$

EXAMPLE 9

Rationalize all denominators. Simplify the answers if possible.

(a) $\dfrac{5}{\sqrt{7}}$ (b) $\dfrac{2}{\sqrt{x}}$ (c) $\dfrac{4}{\sqrt{8}}$ (d) $\dfrac{5\sqrt{2}}{x^2\sqrt{3x}}$ (e) $\sqrt{\dfrac{2}{3}}$

(a) $\dfrac{5}{\sqrt{7}} = \dfrac{5}{\sqrt{7}} \cdot \dfrac{\sqrt{7}}{\sqrt{7}} = \dfrac{5\sqrt{7}}{7}$ $\sqrt{7}\cdot\sqrt{7} = (\sqrt{7})^2 = 7$. Denominator now has no irrational factor.

(b) $\dfrac{2}{\sqrt{x}} = \dfrac{2}{\sqrt{x}} \cdot \dfrac{\sqrt{x}}{\sqrt{x}} = \dfrac{2\sqrt{x}}{x}$ $\sqrt{x}\cdot\sqrt{x} = (\sqrt{x})^2 = x$. Denominator now has no irrational factor.

(c) $\dfrac{4}{\sqrt{8}} = \dfrac{4}{\sqrt{4\cdot2}} = \dfrac{4}{2\sqrt{2}} = \dfrac{2}{\sqrt{2}}$ Simplify perfect-square factors in radicands and reduce.

$= \dfrac{2}{\sqrt{2}} \cdot \dfrac{\sqrt{2}}{\sqrt{2}} = \dfrac{2\sqrt{2}}{2} = \sqrt{2}$ Rationalize the denominator by multiplying by $\dfrac{\sqrt{2}}{\sqrt{2}}$. Reduce coefficients.

(d) $\dfrac{5\sqrt{2}}{x^2\sqrt{3x}} = \dfrac{5\sqrt{2}}{x^2\sqrt{3x}} \cdot \dfrac{\sqrt{3x}}{\sqrt{3x}} =$

$\dfrac{5\sqrt{6x}}{x^2\cdot3x} = \dfrac{5\sqrt{6x}}{3x^3}$

(e) $\sqrt{\dfrac{2}{3}} = \dfrac{\sqrt{2}}{\sqrt{3}} \cdot \dfrac{\sqrt{3}}{\sqrt{3}} = \dfrac{\sqrt{6}}{3}$ Separate fractional radicand into a fraction with a radical factor in both the numerator and denominator.

See Exercises 57–64.

TIP   Are Calculator Values Less Accurate if the Denominator Is Not Rationalized? Compare the approximate value of the expressions:

$$\dfrac{1}{\sqrt{2}} \quad \text{and} \quad \dfrac{\sqrt{2}}{2}$$

$\dfrac{1}{\sqrt{2}} = 1 \div \boxed{\sqrt{}}\, 2 \boxed{=} \Rightarrow 0.7071067812$

$\dfrac{\sqrt{2}}{2} = \boxed{\sqrt{}}\, 2 \boxed{)} \div 2 \boxed{=} \Rightarrow 0.7071067812$

The approximate values of the radical expressions are equivalent; so, no, expressions with rationalized denominators do not produce more accurate calculator equivalents.

When calculators are used to convert irrational numbers from radical notation to decimal notation, they become *approximate values*. That is, a decimal notation is a rounded value. In applications we are most often interested in approximate numbers. *Exact values* in simplest form are useful in minimizing the calculator steps when finding approximate values.

Did You Know? Radical expressions are in simplest form if:

1. There are no perfect-square factors in any radicand.

2. There are no fractional radicands.

3. The denominator of a radical expression is rational (contains no radicals).

A logical sequence for simplifying radical expressions follows.

> **To simplify a radical expression:**
>
> 1. Reduce coefficients and radicands whenever possible.
> 2. Simplify expressions by removing perfect-square factors from all radicands.
> 3. Reduce coefficients and radicands again whenever possible.
> 4. Rationalize denominators that contain radical factors.
> 5. Reduce coefficients and radicands again whenever possible.

STOP AND CHECK

Perform the operation and simplify if possible.

1. $\sqrt{2x^5} \cdot \sqrt{6x^2}$

2. $\sqrt{\dfrac{3}{5}} \cdot \sqrt{\dfrac{15}{2}}$

Answers:

1. $2x^3\sqrt{3x}$ 2. $\dfrac{3\sqrt{2}}{2}$

EXAMPLE 10

Perform the operation and simplify if possible.

(a) $\sqrt{y^3} \cdot \sqrt{8y^2}$ 　　(b) $\sqrt{\dfrac{1}{3}} \cdot \sqrt{\dfrac{8}{3x}}$

(a) $\sqrt{y^3} \cdot \sqrt{8y^2} = \sqrt{8y^5} = \sqrt{4 \cdot 2 \cdot y^4 \cdot y} = \mathbf{2y^2\sqrt{2y}}$

(b) $\sqrt{\dfrac{1}{3}} \cdot \sqrt{\dfrac{8}{3x}} = \sqrt{\dfrac{8}{9x}} = \dfrac{\sqrt{8}}{\sqrt{9x}} = \dfrac{\sqrt{4\cdot2}}{\sqrt{9\cdot x}} = \dfrac{2\sqrt{2}}{3\sqrt{x}} \cdot \dfrac{\sqrt{x}}{\sqrt{x}} = \dfrac{\mathbf{2\sqrt{2x}}}{\mathbf{3x}}$

See Exercises 65–66.

12-3 EXERCISES

MyLab Math For additional practice go to your study plan in MyLab Math.

1 Add or subtract. Simplify radicals where necessary. *See Example 1.*

1. $5\sqrt{3} + 7\sqrt{3}$
2. $8\sqrt{5} - 12\sqrt{5}$
3. $4\sqrt{7} + 3\sqrt{7} - 5\sqrt{7}$
4. $2\sqrt{3} - 8\sqrt{5} + 7\sqrt{3}$
5. $9\sqrt{11} - 3\sqrt{6} + 4\sqrt{6} - 12\sqrt{11}$
6. $44\sqrt{2} + \sqrt{3} - \sqrt{2} + 5\sqrt{3}$

See Example 2.

7. $2\sqrt{3} + 5\sqrt{12}$
8. $7\sqrt{5} + 2\sqrt{45}$
9. $4\sqrt{63} - \sqrt{7}$
10. $3\sqrt{6} - 2\sqrt{54}$
11. $8\sqrt{2} - 3\sqrt{28}$
12. $2\sqrt{3} + \sqrt{48}$
13. $3\sqrt{5} + 4\sqrt{180}$
14. $7\sqrt{98} - 2\sqrt{2}$
15. $6\sqrt{40} - 2\sqrt{90}$
16. $\sqrt{12} - \sqrt{27}$
17. $3\sqrt{112} + 5\sqrt{7}$
18. $2\sqrt{32} - 3\sqrt{8}$

2 Multiply and simplify if possible. *See Example 3.*

19. $\sqrt{6} \cdot \sqrt{7}$
20. $\sqrt{18} \cdot \sqrt{2}$
21. $5\sqrt{2} \cdot 3\sqrt{5}$
22. $8\sqrt{3} \cdot 5\sqrt{12}$
23. $5\sqrt{3x} \cdot 4\sqrt{5x^2}$
24. $2x\sqrt{3x^4} \cdot 7x^2\sqrt{8x}$

See Example 4.

25. $7(\sqrt{2} + 5)$
26. $4(12 - \sqrt{3})$
27. $\sqrt{5}(\sqrt{2} - 3)$
28. $\sqrt{7}(4 + \sqrt{3})$
29. $\sqrt{3}(\sqrt{5} + \sqrt{2})$

See Example 5.

30. $(\sqrt{11} - 5)(\sqrt{2} + \sqrt{3})$
31. $(2\sqrt{3} - 1)(\sqrt{2} + 3)$
32. $(\sqrt{5} - 4)(\sqrt{2} - 3)$

See Example 6.

33. $(\sqrt{7} + \sqrt{2})(\sqrt{7} + 3\sqrt{5})$
34. $(\sqrt{3} - 2)(\sqrt{3} + 2)$
35. $(\sqrt{6} - 1)(\sqrt{6} + 1)$

See Example 7.

36. $(2\sqrt{3} - 5)^2$
37. $(6\sqrt{2} + 2)^2$
38. $(7 - 3\sqrt{3})^2$

Perform the indicated operations and simplify if possible. *See Example 8.*

39. $\sqrt{x^2}\cdot\sqrt{3x}$

40. $\sqrt{\dfrac{2}{3}}\cdot\sqrt{\dfrac{4}{5y}}$

41. $\sqrt{\dfrac{1}{x}}\cdot\sqrt{\dfrac{8}{7x}}$

42. $\sqrt{\dfrac{4}{9y^2}}\cdot\sqrt{\dfrac{1}{2y}}$

43. $\sqrt{\dfrac{8x}{3}}\cdot\sqrt{\dfrac{2x^2}{3}}$

44. $\sqrt{7}\cdot\sqrt{x^2}$

45. $\sqrt{\dfrac{4x^2}{x^3}}\cdot\sqrt{12x}$

46. $\sqrt{\dfrac{1}{2}}\cdot\sqrt{\dfrac{x^2}{3}}$

47. $\sqrt{\dfrac{y^3}{2}}\cdot\sqrt{\dfrac{2}{7y}}$

48. $\sqrt{8x}\cdot\sqrt{x^2}$

49. $\sqrt{3x}\cdot\sqrt{18x^2}$

50. $\sqrt{12x^3}\cdot\sqrt{9y^4}$

Divide and simplify.

51. $\dfrac{\sqrt{18}}{\sqrt{2}}$

52. $\dfrac{5\sqrt{10}}{10}$

53. $\dfrac{4\sqrt{12}}{2\sqrt{6}}$

54. $\dfrac{15\sqrt{24}}{9\sqrt{2}}$

55. $\dfrac{12x^2\sqrt{8x}}{15x\sqrt{6x^3}}$

56. $\dfrac{6x^4\sqrt{25x^3}}{2x\sqrt{16x^2}}$

Rationalize the denominator and simplify. *See Example 9.*

57. $\dfrac{5}{\sqrt{3}}$

58. $\dfrac{6}{\sqrt{5}}$

59. $\dfrac{1}{\sqrt{8}}$

60. $\dfrac{\sqrt{5}}{\sqrt{11}}$

61. $\dfrac{\sqrt{8}}{\sqrt{12}}$

62. $\dfrac{5x}{\sqrt{3x}}$

63. $\dfrac{2x^2}{\sqrt{7x^2}}$

64. $\dfrac{4x^5}{\sqrt{12x^3}}$

See Example 10.

65. $\sqrt{x^5}\cdot\sqrt{12x^4}$

66. $\sqrt{\dfrac{1}{5}}\cdot\sqrt{\dfrac{18}{5x}}$

12–4   Complex and Imaginary Numbers

LEARNING OUTCOMES

1 Write imaginary numbers using the letter *i*.

2 Simplify powers of imaginary numbers.

3 Write real and imaginary numbers in complex form, $a + bi$.

4 Combine complex numbers.

5 Multiply complex numbers.

LC LEARNING CATALYTICS

1. Multiply $(5 + 2\sqrt{3})(5 - 2\sqrt{3})$.
2. Simplify $(x^3y^4)(-2xy)^3$.

Imaginary number: an imaginary number has a factor of $\sqrt{-1}$, which is represented by *i*

1 Write Imaginary Numbers Using the Letter *i*. Taking the square root of a negative number introduces a new type of number. This new type of number is called an *imaginary number*. An **imaginary number** has a factor of $\sqrt{-1}$, which is represented by $i\,(i = \sqrt{-1})$.

Imaginary numbers and real numbers combine to form the set of complex numbers (Fig. 12–3). A **complex number** is a number that can be written in the form $a + bi$, where a and b are real numbers and i is $\sqrt{-1}$.

The square root of a negative number, say, $\sqrt{-16}$, can be simplified as $\sqrt{-1\cdot 16}$ or $4\sqrt{-1}$. Because the square root of -1 is an *imaginary-number* factor and 4 is a real-number factor, we will use the letter i to represent $\sqrt{-1}$ and rewrite $4\sqrt{-1}$ as $4i$. Similarly, $\sqrt{-4} = \sqrt{-1\cdot 4} = 2\sqrt{-1} = 2i$.

TIP Is *i* Different from *j*? Electronics and other applications of imaginary numbers may use j rather than i to represent $\sqrt{-1}$. Thus, $3 + 4i$ and $3 + 4j$ represent the same quantity.

The **j-factor** is another representation for the imaginary number i.

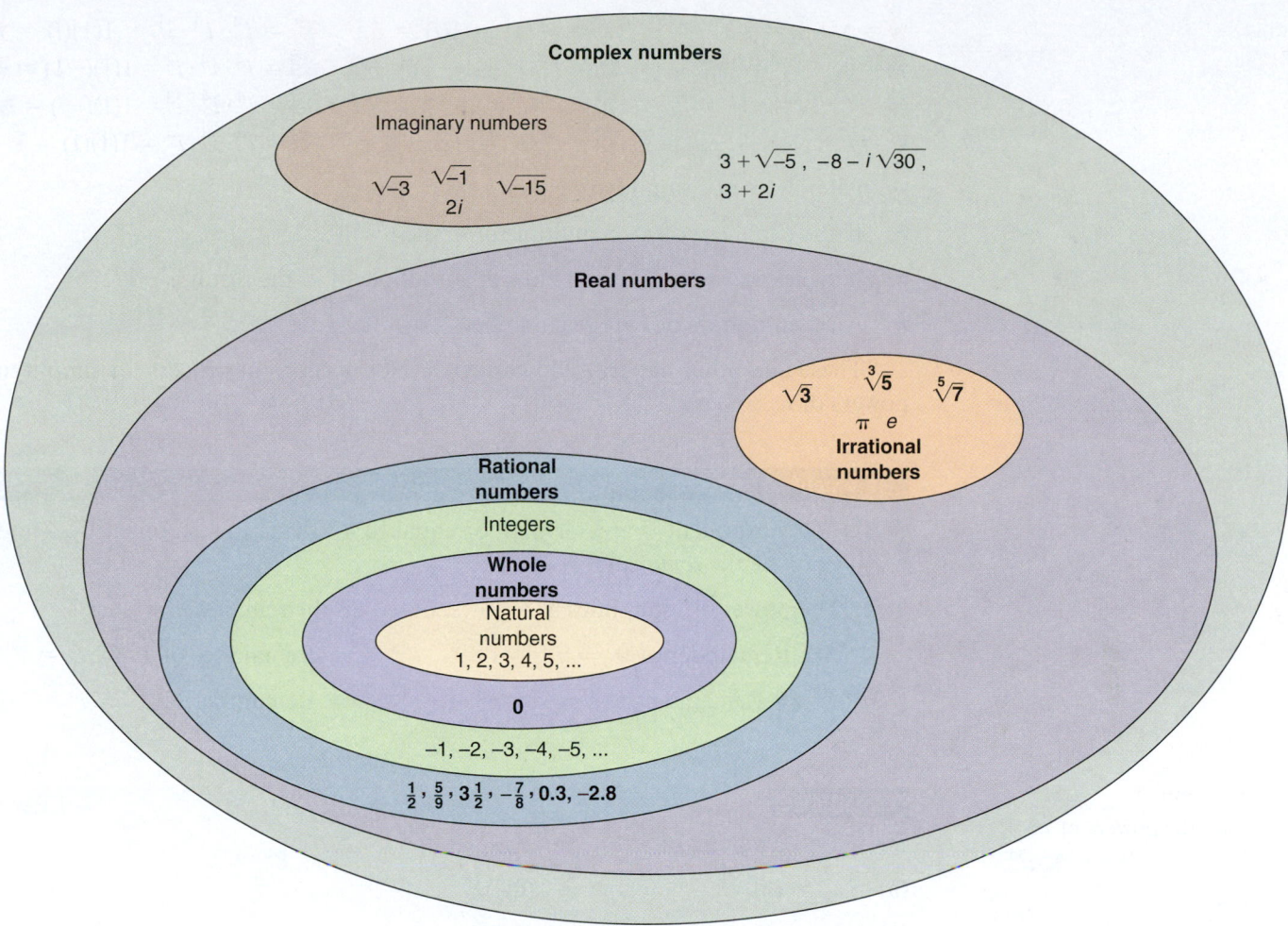

FIGURE 12-3 Complex-number system.

Complex number: a number that can be written in the form $a + bi$, where a and b are real numbers and i is $\sqrt{-1}$

j-factor: another representation for the imaginary number i

EXAMPLE 1

Rewrite the imaginary numbers using the letter i for $\sqrt{-1}$.

(a) $\sqrt{-9}$ **(b)** $\sqrt{-25}$ **(c)** $\sqrt{-7}$

(a) $\sqrt{-9} = \sqrt{-1 \cdot 9} = 3\sqrt{-1} = 3i$

(b) $\sqrt{-25} = \sqrt{-1 \cdot 25} = 5\sqrt{-1} = 5i$

(c) $\sqrt{-7} = \sqrt{-1 \cdot 7} = \sqrt{7} \cdot \sqrt{-1} = \sqrt{7}i$ or $i\sqrt{7}$.

See Exercises 1–4.

TIP **Are Numerical Coefficients Always First?** $\sqrt{7}i$ and $\sqrt{7i}$ do not represent the same amount. In the first term, $\sqrt{7}i$, i is not under the radical symbol. In the second term, $\sqrt{7i}$, i is under the radical symbol. Because it is easy to confuse the two terms, when the coefficient of an imaginary number is an irrational number, we write the i factor first: $\sqrt{7}i = i\sqrt{7}$.

2 **Simplify Powers of Imaginary Numbers.** Some powers of imaginary numbers are real numbers. Examine the pattern that develops with powers of i.

$$i = \sqrt{-1} = i$$

$$i^2 = (\sqrt{-1})^2 = -1$$

$$i^3 = i^2 \cdot i^1 = -1i = -i$$

$$i^4 = i^2 \cdot i^2 = -1(-1) = 1$$

$$i^5 = i^4 \cdot i^1 = 1(i) = i$$

$$i^6 = i^4 \cdot i^2 = 1(-1) = -1$$

$$i^7 = i^4 \cdot i^3 = 1(-i) = -i$$

$$i^8 = i^4 \cdot i^4 = 1(1) = 1$$

$$i^9 = i^4 \cdot i^4 \cdot i^1 = 1(1)(i) = i$$

$$i^{10} = i^4 \cdot i^4 \cdot i^2 = 1(1)(-1) = -1$$

$$i^{11} = i^4 \cdot i^4 \cdot i^3 = 1(1)(-i) = -i$$

$$i^{12} = i^4 \cdot i^4 \cdot i^4 = 1(1)(1) = 1$$

▶ All powers of i simplify to either i, -1, $-i$, or 1.

▶ If the exponent of i is a multiple of 4, the result is 1.

▶ If the exponent of i is even but not a multiple of 4, the result is -1.

▶ All even powers of i are real numbers.

These and other patterns allow us to develop a shortcut process for simplifying powers of i.

To simplify a power of i.

1. If the exponent is greater than or equal to 4, divide the exponent by 4 and examine the *remainder*.

2. The power of i simplifies as follows, based on the remainder in Step 1.

$i^{4x} \Rightarrow$ Remainder of 0 $\Rightarrow i^0 = 1$ $i^{4x+1} \Rightarrow$ Remainder of 1 $\Rightarrow i^1 = i$

$i^{4x+2} \Rightarrow$ Remainder of 2 $\Rightarrow i^2 = -1$ $i^{4x+3} \Rightarrow$ Remainder of 3 $\Rightarrow i^3 = -i$

STOP AND CHECK

Simplify the powers of i.

1. i^{19} 2. i^{24} 3. i^{30}

Answers:

1. $-i$ 2. 1   3. -1

EXAMPLE 2

Simplify the powers of i.

(a) i^{15} (b) i^{20} (c) i^{33} (d) i^{18}

(a) $i^{15} = -i$ $15 \div 4 = 3\ R3$. Remainder of 3 $\Rightarrow i^3 = -i$.

(b) $i^{20} = 1$ $20 \div 4 = 5$. Remainder 0 $\Rightarrow i^0 = 1$.

(c) $i^{33} = i$ $33 \div 4 = 8\ R1$. Remainder 1 $\Rightarrow i^1 = i$.

(d) $i^{18} = -1$ $18 \div 4 = 4\ R2$. Remainder 2 $\Rightarrow i^2 = -1$.   **See Exercises 5–12.**

Real part of a complex number: in a complex number of the form $a + bi$, the real part is a

Imaginary part of a complex number: in a complex number of the form $a + bi$, the imaginary part is bi

3 Write Real and Imaginary Numbers in Complex Form, $a + bi$. A complex number has two parts, a **real part** and an **imaginary part**. In $a + bi$, if $a = 0$, then $a + bi$ is the same as $0 + bi$ or bi, which is an imaginary number. If $b = 0$, then $a + bi$ is the same as $a + 0 \cdot i$ or a, which is a real number. Thus, real numbers and imaginary numbers are also complex numbers.

STOP AND CHECK

Rewrite in the form $a + bi$.

1. 3

2. $\sqrt{-25}$

3. $4 + \sqrt{-17}$

Answers:

1. $3 + 0i$ 2. $0 + 5i$

3. $4 + i\sqrt{17}$

EXAMPLE 3

Rewrite in the form $a + bi$.

(a) 5   (b) $\sqrt{-36}$ (c) $-3i^2$ (d) $6 - \sqrt{-5}$

(a) $5 = 5 + 0i$ $a = 5, b = 0$

(b) $\sqrt{-36} = 6i = 0 + 6i$ $a = 0, b = 6$

(c) $-3i^2 = -3(-1) = 3 = 3 + 0i$ $a = 3, b = 0$

(d) $6 - \sqrt{-5} = 6 - \sqrt{-1(5)} = 6 - i\sqrt{5}$ $a = 6, b = -\sqrt{5}$

See Exercises 13–22.

4 **Combine Complex Numbers.** Complex numbers are combined by adding *like* parts. Real parts are added together and imaginary parts are added together.

> **To add or subtract (combine) complex numbers:**
>
> 1. Add or subtract real parts for the real part of the answer.
>
> 2. Add or subtract imaginary parts for the imaginary part of the answer.
>
> Symbolically, $(a + bi) + (c + di) = (a + c) + (b + d)i$

STOP AND CHECK

Combine the complex numbers.
1. $(5 + 3i) + (-3 - 6i)$
2. $6 - (4 - 5i)$

Answers:
1. $2 - 3i$ 2. $2 + 5i$

EXAMPLE 4

Combine the complex numbers.

(a) $(3 + 5i) + (8 - 2i)$ (b) $(-3 + i) - (7 - 4i)$ (c) $-2 + (5 + 6i)$

(a) $(3 + 5i) + (8 - 2i) = \mathbf{11 + 3i}$ $3 + 8 = 11; 5i - 2i = 3i.$

(b) $(-3 + i) - (7 - 4i) =$ Distribute $-1.$

$(-3 + i) + (-7 + 4i) = \mathbf{-10 + 5i}$ $-3 - 7 = -10; i + 4i = 5i.$

(c) $-2 + (5 + 6i) = \mathbf{3 + 6i}$ $-2 + 5 = 3.$ **See Exercises 23–28.**

5 **Multiply Complex Numbers.** Multiplying complex numbers uses procedures similar to multiplying polynomials.

Use the FOIL method to multiply complex numbers.

STOP AND CHECK

1. Multiply $(4 + 7i)(2 - 3i)$.

Answer:
1. $29 + 2i$

EXAMPLE 5

Find the product $(5 + 3i)(6 - 7i)$.

$(5 + 3i)(6 - 7i)$ Multiply applying the FOIL method.

$$\underset{\mathbf{F}}{5 \cdot 6} + \underset{\mathbf{O}}{5 \cdot (-7i)} + \underset{\mathbf{I}}{3i \cdot 6} + \underset{\mathbf{L}}{3i(-7i)} =$$

$30 - 35i + 18i - 21i^2 =$ $i^2 = -1; -21i^2 = -21(-1) = 21$

$\mathbf{51 - 17i}$ **See Exercises 29–30.**

The product of conjugate pairs of complex numbers is useful when expressions have imaginary terms or factors.

STOP AND CHECK

Multiply.
1. $(5 + 2i)(5 - 2i)$
2. $(9i + \sqrt{3})(9i - \sqrt{3})$

Answers:
1. 29 2. -84

EXAMPLE 6

Multiply.

(a) $(7 + 3i)(7 - 3i)$ (b) $(5i - \sqrt{2})(5i + \sqrt{2})$

(a) $(7 + 3i)(7 - 3i)$ These are conjugate pairs.

$(a + b)(a - b) = a^2 - b^2$ Apply the special product pattern.

$(7 + 3i)(7 - 3i) = 49 - (-9)$ $(3i)^2 = 9i^2 = 9(-1) = -9$

$= \mathbf{58}$

(b) $(5i - \sqrt{2})(5i + \sqrt{2})$ These are conjugate pairs.

$(5i - \sqrt{2})(5i + \sqrt{2}) = (5i)^2 - (\sqrt{2})^2$ $(5i)^2 = 25i^2 = 25(-1) = -25$

$= -25 - 2$

$= \mathbf{-27}$ **See Exercises 31–37.**

Squaring a binomial with a complex term uses the same procedures as squaring other binomials.

EXAMPLE 7

Multiply.

(a) $(3 + 5i)^2$ **(b)** $(5 - 2i)^2$

(a)
$$(3 + 5i)^2 =$$ Apply the squaring process.
$$9 + 2 \cdot 3\,(5i) + (5i)^2 =$$ Simplify: $2 \cdot 3\,(5i) = 30i$; $(5i^2) = -25$
$$9 + 30i - 25 =$$ Combine like terms.
$$\mathbf{-16 + 30i}$$

(b)
$$(5 - 2i)^2 =$$ Apply the squaring process.
$$25 + 2 \cdot 5(-2i) + (-2i)^2 =$$ Simplify.
$$25 - 20i + 4i^2 =$$ $4i^2 = 4(-1) = -4$
$$25 - 20i - 4 =$$ Combine like terms.
$$\mathbf{21 - 20i}$$ See Exercises 38–40.

12-4 EXERCISES   MyLab Math   For additional practice go to your study plan in MyLab Math.

1 Write the imaginary numbers using the letter i. Simplify if possible. *See Example 1.*

1. $\sqrt{-25}$ **2.** $\sqrt{-36}$ **3.** $\sqrt{-64x^2}$ **4.** $\sqrt{-32y^5}$

2 Simplify the powers of i. *See Example 2.*

5. i^{17} **6.** i^5 **7.** i^{28} **8.** i^{10}

9. i^{24} **10.** i^9 **11.** i^{32} **12.** i^{23}

3 Write the real and imaginary numbers in simplified complex form. *See Example 3.*

13. 15 **14.** 17 **15.** $\sqrt{-49}$ **16.** $\sqrt{-81}$

17. $33i$ **18.** $-7i^3$ **19.** $4i^6$ **20.** $5 + \sqrt{-4}$

21. $8 + \sqrt{-32}$ **22.** $7 - \sqrt{-3}$

4 Combine the complex numbers. *See Example 4.*

23. $(4 + 3i) + (7 + 2i)$ **24.** $\sqrt{12} - 5\sqrt{-3} + \left(\sqrt{8} - \sqrt{-27}\right)$

25. $12 - 3i - (8 - 4i)$ **26.** $15 + 8i - (3 - 12i)$

27. $(4 + 7i) - (3 - 2i)$ **28.** $(8 - 5i) + (4 - 3i)$

5 Multiply the complex numbers and simplify powers of i. *See Example 5.*

29. $(3i - 1)(2i + 3)$ **30.** $(7i + 2)(2i - 1)$

See Example 6.

31. $(8 - i)(8 + i)$ **32.** $(5 - i)(5 + i)$ **33.** $(i - 1)(i + 1)$ **34.** $(4 - i)(4 + i)$

35. $(7i - 5)\,(7i + 5)$ **36.** $\left(2i - \sqrt{3}\right)\left(2i + \sqrt{3}\right)$ **37.** $(i - 5)(2i - 3)$

See Example 7.

38. $(i - 3)^2$ **39.** $(2i - 7)^2$ **40.** $(4 + 3i)^2$

12 | CHAPTER REVIEW OF KEY CONCEPTS

| **LEARNING OUTCOMES** | **KEY CONCEPTS AND EXAMPLES** |

Section 12–1

1 Write roots using radical and exponential notation (p. 526).

1. Identify the index of the root.   **2.** For radical notation, place the index of the root in the $\sqrt{}$ portion of the radical sign with the radicand under the bar portion of the sign. The index 2 for square roots does not have to be written.   **3.** For exponential notation, write the radicand as the base and the index of the root as the denominator of a fractional exponent that has a numerator of 1. The decimal equivalent of the fractional exponent can also be used.

The square root of 16	The cube root of 125	The fifth root of 32
$16^{1/2} = \sqrt{16} = 4$	$125^{1/3} = \sqrt[3]{125} = 5$	$32^{1/5} = \sqrt[5]{32} = 2$

2 Approximate an irrational number (pp. 526–528).

Find the two whole numbers that are closest to the value of an irrational number expressed as a radical: 1. Make a list of perfect powers that go beyond the given radicand.   **2.** Identify the two perfect powers that the radicand is between.   **3.** Find the principal roots of the two perfect powers from Step 2.   **4.** The root of the radicand is between the roots found in Step 3.

15 is between 9 and 16, so $\sqrt{15}$ or $15^{1/2}$ is between 3 and 4.

38 is between 36 and 49, so $\sqrt{38}$ or $38^{1/2}$ is between 6 and 7.

Position irrational numbers on the number line (Fig. 12–4) by approximating the value of the irrational number mentally or with a calculator.

$\sqrt{6} = 6^{1/2} \approx 2.449489743$ $\quad$ $\sqrt{7} = 7^{1/2} \approx 2.645751311$ $\quad$ $\sqrt{8} = 8^{1/2} \approx 2.828427125$
$\sqrt{10} = 10^{1/2} \approx 3.16227766$

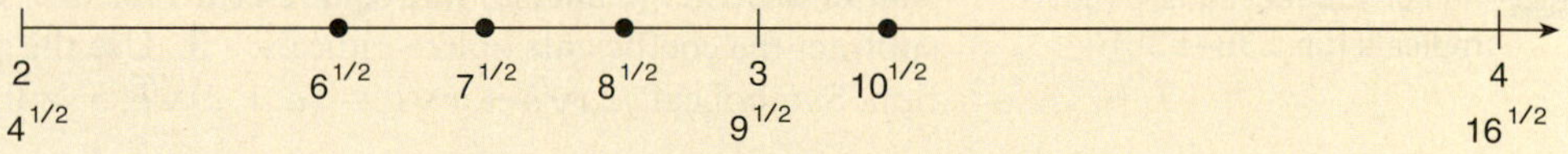

FIGURE 12–4

3 Write powers and roots using rational exponents and radical notation (pp. 528–530).

Change rational exponent to radical notation: 1. Write a decimal exponent as an equivalent fractional exponent.   **2.** Write the numerator of a rational exponent as the power.   **3.** Write the denominator of the rational exponent as the index of the root. Symbolically, $x^{\text{power/root}} = \sqrt[\text{root}]{x^{\text{power}}}$ or $\left(\sqrt[\text{root}]{x}\right)^{\text{power}}$.

Change radical to rational exponent: 1. Write the power as the numerator of the rational exponent.   **2.** Write the index of the root as the denominator of the rational exponent.   **3.** Write the fraction in lowest terms or as a decimal equivalent. Symbolically, $\sqrt[\text{root}]{x^{\text{power}}} = x^{\text{power/root}}$ or $x^{\frac{\text{power}}{\text{root}}}$.

Write in rational exponent notation.

$$\sqrt[5]{3^2} = 3^{2/5}$$

Write in radical notation:

$$5^{1/2} = \sqrt{5}; \qquad 7^{3/5} = \sqrt[5]{7^3} \text{ or } \left(\sqrt[5]{7}\right)^3$$

LEARNING OUTCOMES	KEY CONCEPTS AND EXAMPLES

Section 12-2

1 Find the square root of variables (p. 531).

Variable factors are perfect squares if the exponent is divisible by 2. To find the square root of a perfect-square variable with an even-number exponent, take one-half of the exponent and keep the same base.

> Give the square root of the following: x^6, x^{10}, x^{24}.
> $$\sqrt{x^6} = x^3, \qquad \sqrt{x^{10}} = x^5, \qquad \sqrt{x^{24}} = x^{12}$$

2 Simplify square-root radicals using rational exponents and the laws of exponents (pp. 531–533).

1. Convert the radicals to equivalent expressions using rational exponents. **2.** Apply the laws of exponents and the arithmetic of fractions. **3.** Convert simplified expressions back to radical notation if desired.

> Simplify.
> $$x^{1/3} \cdot x^{2/3} = x^{1/3+2/3} = x^{3/3} = x; \qquad \frac{x^{7/8}}{x^{3/4}} = x^{7/8-3/4} = x^{7/8-6/8} = x^{1/8}$$

3 Simplify square-root radical expressions containing perfect-square factors (pp. 533–535).

1. If the radicand is a perfect square, express it as a square root without the radical sign. **2.** If the radicand is *not* a perfect square, factor the radicand so that one factor is the largest possible perfect-square factor. The square roots of the perfect-square factors are written *outside* the radical and the other factors stay *inside* (under) the radical sign.

> Simplify the following radical expressions: $\sqrt{98}, \sqrt{x^{13}}, \sqrt{72y^9}, 5\sqrt{12}$.
> $$\sqrt{98} = \sqrt{49(2)} = 7\sqrt{2} \qquad\qquad \sqrt{x^{13}} = \sqrt{x^{12}(x)} = x^6\sqrt{x}$$
> $$\sqrt{72y^9} = \sqrt{36 \cdot 2 \cdot y^8 \cdot y} = 6y^4\sqrt{2y} \qquad 5\sqrt{12} = 5\sqrt{4 \cdot 3} = 5 \cdot 2\sqrt{3} = 10\sqrt{3}$$

Section 12-3

1 Add or subtract square-root radicals (pp. 536–538).

Add or subtract (combine) like square-root radicals: 1. Simplify all radicands. **2.** Add or subtract the coefficients of like radicals. **3.** Use the common radical as a factor in the solution. Symbolically, $a\sqrt{b} + c\sqrt{b} = (a + c)\sqrt{b}, b > 0$.

> Add or subtract. $3\sqrt{5x} + 7\sqrt{5x}$; $2\sqrt{18} - 5\sqrt{8}$.
> $$3\sqrt{5x} + 7\sqrt{5x} = 10\sqrt{5x}$$
> $$2\sqrt{18} - 5\sqrt{8} = 2\sqrt{9 \cdot 2} - 5\sqrt{4 \cdot 2} = 2 \cdot 3\sqrt{2} - 5 \cdot 2\sqrt{2} =$$
> $$6\sqrt{2} - 10\sqrt{2} = -4\sqrt{2}$$

2 Multiply or divide square-root radicals (pp. 538–543).

Multiply square-root radicals: 1. Multiply coefficients to give the coefficient of the product. **2.** Multiply radicands to give the radicand of the product. **3.** Simplify if possible. Symbolically, $a\sqrt{b} \cdot c\sqrt{d} = ac\sqrt{bd}$.

> Multiply. $5\sqrt{7} \cdot 8\sqrt{14}$; $\sqrt{5}(\sqrt{7} - \sqrt{2})$.
> $$5\sqrt{7} \cdot 8\sqrt{14} = 40\sqrt{98} = 40\sqrt{49 \cdot 2} = 40 \cdot 7\sqrt{2} = 280\sqrt{2};$$
> $$\sqrt{5}(\sqrt{7} - \sqrt{2}) = \sqrt{5} \cdot \sqrt{7} - \sqrt{5} \cdot \sqrt{2} = \sqrt{35} - \sqrt{10}$$

Multiply the sum and difference of two terms or conjugate pairs: 1. Square the first term of either binomial. **2.** Insert a minus sign. **3.** Square the second term of either binomial. Symbolically, $(a + b)(a - b) = a^2 - b^2$, where a and b are real numbers.

LEARNING OUTCOMES	KEY CONCEPTS AND EXAMPLES

Divide square-root radicals: **1.** Divide coefficients to give the coefficient of the quotient. **2.** Divide radicands to give the radicand of the quotient. **3.** Simplify if possible.

Divide. $\dfrac{12\sqrt{75}}{8\sqrt{6}} = \dfrac{3\sqrt{3\cdot 25}}{2\sqrt{3\cdot 2}} = \dfrac{3\sqrt{25}}{2\sqrt{2}} = \dfrac{3\cdot 5}{2\sqrt{2}} = \dfrac{15}{2\sqrt{2}}$ not rationalized

To rationalize a denominator: **1.** If the denominator has an irrational factor, multiply the denominator by another irrational factor so that the resulting radicand is a perfect square. **2.** To preserve the value of the fraction, multiply the numerator by the same radical factor used in Step 1. Thus, in Steps 1 and 2 together, we have multiplied by an equivalent of 1. **3.** Simplify all radicals and reduce the resulting fraction, if possible.

Rationalize the denominator. $\sqrt{\dfrac{3}{5}} = \dfrac{\sqrt{3}}{\sqrt{5}}\cdot\dfrac{\sqrt{5}}{\sqrt{5}} = \dfrac{\sqrt{15}}{5}$

Simplify a radical expression: **1.** Reduce coefficients and radicands whenever possible. **2.** Simplify expressions by removing perfect-square factors from all radicands. **3.** Reduce coefficients and radicands again whenever possible. **4.** Rationalize denominators that contain radical factors. **5.** Reduce coefficients and radicands again whenever possible.

Section 12–4

1 Write imaginary numbers using the letter i (pp. 544–545).

The letter i is used to represent $\sqrt{-1}$. Thus, the square root of negative numbers can be expressed as imaginary numbers and simplified using the letter i. Be careful to distinguish when the negative is *outside* the radical and when it is *under* the radical.

Simplify. $\sqrt{-48} = \sqrt{-1\cdot 16\cdot 3} = 4i\sqrt{3}$

2 Simplify powers of imaginary numbers (pp. 545–546).

1. If the exponent is greater than or equal to 4, divide the exponent by 4 and examine the *remainder*. **2.** The power of i simplifies as follows, based on the remainder in Step 1.

Remainder of $0 \Rightarrow i^{4x} = 1$ Remainder of $1 \Rightarrow i^{4x+1} = i$

Remainder of $2 \Rightarrow i^{4x+2} = -1$ Remainder of $3 \Rightarrow i^{4x+3} = -i$

Simplify. $i^{17} = i^{16}\cdot i^1 = i^1 = i;\ i^{42} = i^{40}\cdot i^2 = i^2 = -1$

3 Write real and imaginary numbers in complex form, $a + bi$ (p. 546).

A complex number is a number that can be written in the form $a + bi$, where a and b are real numbers and i is $\sqrt{-1}$. Either a or b can be zero. If a is zero, the number is an imaginary number; if b is zero, the number is a real number.

Rewrite as complex numbers: $\sqrt{-81}, 38, \sqrt{9}, \sqrt{-19}$.

$\sqrt{-81} = 9i = 0 + 9i$ $38 = 38 + 0i$

$\sqrt{9} = 3 = 3 + 0i$ $\sqrt{-19} = i\sqrt{19} = 0 + i\sqrt{19}$

4 Combine complex numbers (p. 547).

1. Add or subtract real parts for the real part of the answer. **2.** Add or subtract imaginary parts for the imaginary part of the answer. Symbolically, $(a + bi) + (c + di) = (a + c) + (b + d)i$.

Simplify. $(5 + 3i) + (7 - 8i) = (5 + 7) + (3 - 8)i = 12 - 5i;$

$(3 - 8i) - (4 + 6i) = (3 - 4) + (-8 - 6)i = -1 - 14i$

LEARNING OUTCOMES	KEY CONCEPTS AND EXAMPLES
5 Multiply complex numbers (pp. 547–548).	Multiplying complex numbers applies similar procedures to multiplying polynomials.

Multiply $(3 + \sqrt{5})(2 - 3\sqrt{5})$.

$$
\begin{array}{cccc}
\mathbf{F} & \mathbf{O} & \mathbf{I} & \mathbf{L}
\end{array}
$$

$$
(3 + \sqrt{5})(2 - 3\sqrt{5}) = 3(2) + 3(-3\sqrt{5}) + 2\sqrt{5} + \sqrt{5}(-3\sqrt{5})
$$
$$
= 6 - 9\sqrt{5} + 2\sqrt{5} - 3(5)
$$
$$
= -9 - 7\sqrt{5}
$$

Multiply $(2 + 5i)(2 - 5i)$.

$$
(2 + 5i)(2 - 5i) = 4 - 25i^2 = 4 - 25(-1) = 4 + 25 = 29
$$

12 | CHAPTER REVIEW EXERCISES

Section 12–1   MyLab Math   For additional practice go to your study plan in MyLab Math.

Write the roots in radical notation and exponential notation.

1. Square root of 49

2. Cube root of 125

3. Fourth root of 16

4. Fifth root of 3,125

5. Square root of 121

6. Cube root of 343

Find the two whole numbers that are closest to each root.

7. $\sqrt{38}$

8. $\sqrt{15}$

9. $\sqrt{135}$

10. $\sqrt{75}$

11. $\sqrt[3]{60}$

12. $\sqrt[3]{135}$

Position each square root on a number line. Label the points. Estimate the roots and check estimation with a calculator.

13. $\sqrt{15}$

14. $\sqrt{27}$

15. $\sqrt{5}$

16. $\sqrt{38}$

Write in both rational exponent and radical notations.

17. The square root of the seventh power of x

18. The cube root of the fourth power of x

19. The square of the cube root of x

20. The fifth power of the fifth root of x

Write in rational exponent notation in simplest form.

21. $\sqrt{x}$

22. $\sqrt[3]{x^5}$

23. $\sqrt[5]{x^4}$

24. $\sqrt[4]{9x}$

25. $(\sqrt[3]{xy})^4$

26. $\sqrt[3]{64x^{10}}$

27. $\sqrt{7}$

Write in radical notation.

28. $x^{5/8}$

29. $y^{3/5}$

30. $a^{1/4}$

Section 12-2

Find the square roots if all variables represent positive numbers.

31. $\sqrt{y^{12}}$ **32.** $\sqrt{a^{10}}$ **33.** $-\sqrt{b^{18}}$ **34.** $\pm\sqrt{\dfrac{x^4}{y^6}}$

Convert the radicals to equivalent expressions using rational exponents and simplify if appropriate.

35. $\sqrt[3]{x}$ **36.** $\sqrt[4]{p}$ **37.** $\sqrt[5]{4y}$

38. $\left(\sqrt{ab}\right)^6$ **39.** $\sqrt[3]{8b^{12}}$ **40.** $\left(\sqrt{49x^2y^3}\right)^4$

Perform the operations. Express the answers with positive exponents in lowest terms.

41. $(a^{1/2})(a^{3/2})$ **42.** $(a^{4/3})(a^{2/3})$ **43.** $y^{3/4} \cdot y^{1/4}$ **44.** $y^{5/8} \cdot y^{1/8}$ **45.** $(3x^{1/4}y^2)^3$

46. $(2x^{3/4}y)^2$ **47.** $(4ax^{1/2})^3$ **48.** $(x^{1/2})^{1/3}$ **49.** $\dfrac{x^{3/4}}{x^{1/4}}$ **50.** $\dfrac{x^{1/6}}{x^{5/6}}$

51. $\dfrac{a^{5/6}}{a^{-1/3}}$ **52.** $\dfrac{a^{7/10}}{a^{2/5}}$ **53.** $\dfrac{x^{5/8}}{x^{3/4}}$ **54.** $\dfrac{y^{1/3}}{y^{5/6}}$ **55.** $\dfrac{a^3}{a^{1/3}}$

Use a calculator to evaluate the expressions in Exercises 56 to 63 for $a = 2$, $b = 1$, and $x = 3$ both before and after simplifying.

56. $\dfrac{a^2}{a^{3/5}}$ **57.** $\dfrac{12a^4}{6a^{1/2}}$ **58.** $\dfrac{27x^3}{9x^{2/3}}$ **59.** $\dfrac{15a^{3/5}}{10a^5}$

60. $\dfrac{14a^{5/6}}{24a^2}$ **61.** $a^{2.3}(a^4)$ **62.** $(3a^{1.2}b^2)^3$ **63.** $(4a^6b^8)^{1/2}$

Simplify the expressions.

64. $\left(\sqrt{x^5}\right)^2$ **65.** $\sqrt{x^2}$ **66.** $\left(\sqrt{8}\right)^2$ **67.** $\sqrt{9P^3}$

68. $\sqrt{8^2}$ **69.** $\sqrt{18a^2b}$ **70.** $\sqrt{12x^2y^3}$ **71.** $\sqrt{32x^5y^2}$

72. $\sqrt{63x^4y^7}$ **73.** $\sqrt{75x^{10}y^9}$ **74.** $\sqrt{125xy^3}$ **75.** $\sqrt{147xy^8}$

Section 12-3

Add or subtract the radicals.

76. $12\sqrt{11} - 5\sqrt{11}$ **77.** $5\sqrt{3} - 7\sqrt{3}$

78. $4\sqrt{2} + 3\sqrt{5} - 8\sqrt{2} + 6\sqrt{5}$ **79.** $3\sqrt{7} - 2\sqrt{28}$

80. $\sqrt{2} - \sqrt{8}$ **81.** $2\sqrt{6} + 3\sqrt{54}$

82. $3\sqrt{5} - 2\sqrt{45}$ **83.** $4\sqrt{3} - 8\sqrt{48}$

84. $\sqrt{40} + \sqrt{90}$ **85.** $5\sqrt{8} - 3\sqrt{50}$

86. $5\sqrt{7} - 4\sqrt{63}$ **87.** $3\sqrt{2} - 5\sqrt{32}$

Multiply the radicals and simplify.

88. $\sqrt{6} \cdot \sqrt{3}$ **89.** $2\sqrt{8} \cdot 3\sqrt{6}$ **90.** $2\sqrt{a} \cdot \sqrt{b}$

91. $5\sqrt{3} \cdot 8\sqrt{7}$ **92.** $2\sqrt{3} \cdot 5\sqrt{18}$ **93.** $-8\sqrt{5} \cdot 4\sqrt{30}$

94. $5(\sqrt{3} - 2)$ **95.** $\sqrt{3}(\sqrt{12} - 5)$ **96.** $\sqrt{2}(\sqrt{6} - \sqrt{10})$

97. $\sqrt{3}(\sqrt{6} - \sqrt{15})$ **98.** $(\sqrt{7} - 5)(\sqrt{5} - 3)$ **99.** $(\sqrt{5} - 8)(\sqrt{5} + 8)$

Divide the radicals and simplify.

100. $\dfrac{4\sqrt{8}}{2}$

101. $\dfrac{3\sqrt{5}}{2\sqrt{20}}$

102. $\dfrac{2\sqrt{90}}{\sqrt{5}}$

103. $\dfrac{6\sqrt{18}}{8\sqrt{12}}$

104. $\dfrac{14\sqrt{56}}{7\sqrt{7}}$

105. $\dfrac{5\sqrt{48}}{20\sqrt{20}}$

106. $\dfrac{\sqrt{9x}}{\sqrt{3x}}$

107. $\dfrac{\sqrt{3y^3}}{\sqrt{y^3}}$

108. $\left(\sqrt{\dfrac{25}{36}}\right)^2$

109. $\left(\sqrt{\dfrac{9}{16}}\right)^2$

110. $\sqrt{\dfrac{9c^4}{25y^6}}$

111. $\sqrt{\dfrac{36x^8}{81y^{10}}}$

Rationalize the denominator and simplify.

112. $\dfrac{1}{\sqrt{8}}$

113. $\dfrac{\sqrt{7}}{\sqrt{12}}$

114. $\dfrac{\sqrt{3}}{\sqrt{7x}}$

115. $\dfrac{\sqrt{3}}{\sqrt{8}}$

116. $\dfrac{\sqrt{7}}{5\sqrt{18}}$

117. $\dfrac{5\sqrt{3}}{\sqrt{24}}$

118. $\dfrac{\sqrt{15}}{5\sqrt{7}}$

119. $\dfrac{2\sqrt{5}}{\sqrt{18}}$

Section 12–4

Write the numbers using the letter i. Simplify if possible.

120. $\sqrt{-144}$

121. $\sqrt{-100}$

122. $-\sqrt{-16x^2}$

123. $\pm\sqrt{-24y^7}$

Simplify the powers of i.

124. i^6

125. i^{14}

126. i^{98}

127. i^{77}

Write as complex numbers in simplified form.

128. 5

129. $15i$

130. $3 + \sqrt{-9}$

131. $-12i^5$

132. $-6i^{11}$

Perform the indicated operation and simplify.

133. $(5 + 3i) + (2 - 7i)$

134. $(4 - i) - (3 - 2i)$

135. $(7 - \sqrt{-9}) + (4 + \sqrt{-16})$

136. $(5i - 3)(2i - 3)$

137. $(4i + 3)(4i - 3)$

138. $(7i - 4)^2$

---

12   TEAM PROBLEM-SOLVING EXERCISES

1. Laws of radicals that pattern the laws of exponents can be used to simplify radicals that are not square-root radicals. For example, $\sqrt[n]{xy} = \sqrt[n]{x} \cdot \sqrt[n]{y}$, for positive values of x and y and a natural number n greater than 1.
   (a) Illustrate this property with a numerical example for a natural-number value of n that is greater than 2.
   (b) Illustrate with a numerical example the property $\sqrt[m]{\sqrt[n]{x}} = \sqrt[n]{\sqrt[m]{x}} = \sqrt[mn]{x}$ for a positive value of x and a natural-number values of m and n that is greater than 2.

2. Illustrate with a numerical example the following properties of radicals and rational exponents.
   (a) $\sqrt[n]{\dfrac{x}{y}} = \dfrac{\sqrt[n]{x}}{\sqrt[n]{y}}$ for positive values of x and y and a natural-number value of n that is greater than 2.
   (b) $x^{-m/n} = \dfrac{1}{x^{m/n}}$ for a positive value of x and values of m and n that are natural numbers greater than 2.

---

12   CONCEPTS ANALYSIS

For 1–4, write the rules in words. Assume that all radicands represent positive values.

1. $a\sqrt{b} \cdot c\sqrt{d} = ac\sqrt{bd}$

2. $\dfrac{a\sqrt{b}}{c\sqrt{d}} = \dfrac{a}{c}\sqrt{\dfrac{b}{d}}; c, d \neq 0$

3. $a\sqrt{b} + c\sqrt{b} = (a + c)\sqrt{b}$

4. $(\sqrt{x})^2 = x$ or $\sqrt{x^2} = x$, for positive values of x.

5. List the conditions for a radical expression to be in simplest form.

6. What does it mean to *rationalize* a denominator? What calculations or manipulations with fractions are easier if the denominator is a rational number?

7. Write the property in words.
$$x^{1/n} = \sqrt[n]{x}$$

8. Write the following property in words:
$$x^{m/n} = \sqrt[n]{x^m} \text{ or } (\sqrt[n]{x})^m$$

9. Explain how to analyze a power of i to determine if the power simplifies to be i, -1, $-i$, or 1.

10. Write an example of two binomials that are conjugates and contain radicals. Multiply the conjugates to eliminate the radical.

---

12 PRACTICE TEST

Perform the indicated operations. Simplify if possible. Rationalize the denominators if needed.

1. $2\sqrt{7} \cdot 3\sqrt{2}$

2. $\dfrac{\sqrt{8}}{\sqrt{2}}$

3. $4\sqrt{3} + 2\sqrt{3}$

4. $3\sqrt{8} - 4\sqrt{8}$

5. $\dfrac{4\sqrt{2}}{\sqrt{3}}$

6. $\sqrt{3y^3} \cdot \sqrt{15y^2}$

7. $\dfrac{6\sqrt{8}}{2\sqrt{3}}$

8. $\dfrac{3\sqrt{a}}{\sqrt{b}}$

9. $\dfrac{3\sqrt{5}}{2} \cdot \dfrac{7}{\sqrt{3x}}$

Convert the radical expressions to equivalent expressions using rational exponents and simplify.

10. $\sqrt[6]{x}$

11. $\sqrt[3]{27x^{15}}$

12. $(\sqrt{5x^4})^6$

13. $\sqrt[5]{x^{10}y^{15}z^{30}}$

Perform the operations. Express answers with positive exponents in lowest terms.

14. $a^{4/5} \cdot a^{1/5}$

15. $(125x^{1/2}y^6)^{1/3}$

16. $\dfrac{b^{3/4}}{b^{1/4}}$

17. $\dfrac{12x^{3/5}}{6x^{-2/5}}$

18. $\dfrac{r^{-1/5}s^{1/3}}{r^{3/5}s^{-5/3}}$

Simplify.

19. i^{23}

20. i^{88}

Add or subtract.

21. $(5 + 3i) - (8 - 2i)$

22. $(7 - i) + (4 - 3i)$

23. $5\sqrt{3} + 8\sqrt{5} - 7\sqrt{3}$

24. $2\sqrt{24} - 7\sqrt{54} + 3\sqrt{96}$

Multiply and simplify.

25. $\sqrt{7}(\sqrt{5} - 4)$

26. $\sqrt{3}(\sqrt{6} - \sqrt{12})$

27. $(5\sqrt{2} - 3)(5\sqrt{2} + 3)$

28. $(3i - 8)(3i + 8)$

13

Factoring

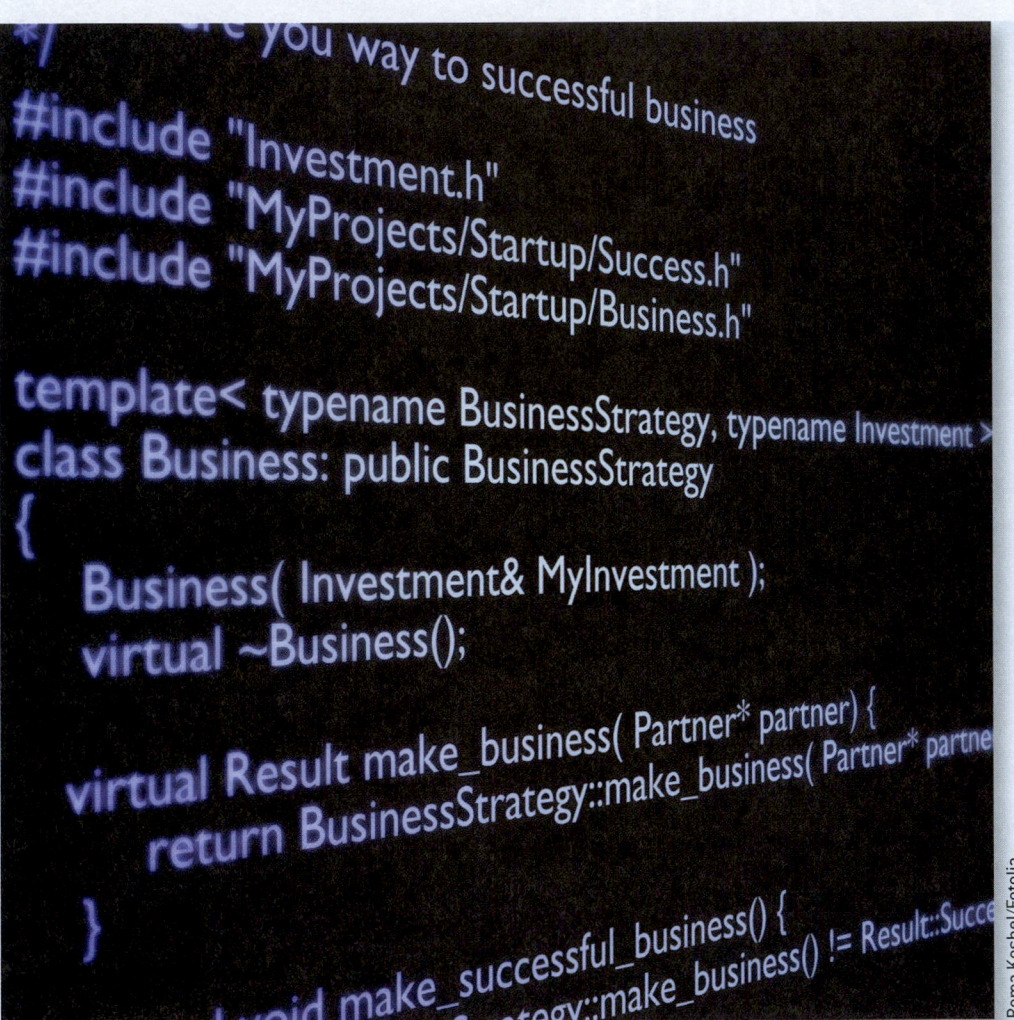

Roma Koshel/Fotolia

In Great Company

Double-Check Your Calculations (1990)

In 1990, AT&T had a long-distance calling network that was renowned, robust, and reliable. It carried 70% of the nation's long-distance traffic, routing over 115 million calls a day. It balanced the long-distance load between 114 computer-controlled electronic switches constructed in key locations across the country; together they handled 700,000 calls an hour.

Each switch automatically transferred each long-distance call to the appropriate remote switch in roughly 5 s. On Monday, January 15, 1990, at 2:25 P.M., half of AT&T's renowned nationwide long-distance calling network disintegrated. It took roughly 5 s.

In math, getting things wrong is easy. Right takes longer. It took over 100 technicians over 9 h to discover the cause of the error, repair it, and rebuild the network. That 9-h outage cost AT&T at least 750,000 unconnected phone calls, and $60 million in lost revenue. An estimated 200,000 airline reservations were lost, and untold numbers of hotel and car rentals were all gone.

A simple mechanical problem and a one-line (actually, one-word) mistake in a computer program accomplished the deed. A switch at one of AT&T's 114 switching centers suffered a minor mechanical problem and shut down. When it was repaired and reset, it activated brand-new software that had just been put in place in December, less than a month before, as part of a general "switch upgrade" package.

The software was supposed to speed up calling. It was designed to "self-heal" the network by monitoring switches to make sure they were all operating correctly. It automatically isolated defective switches to limit network damage. It was put in place on a physical network that was specifically designed to have no single point of failure.

And it didn't have a single point of failure. The same code was running on all the switches. Given the right conditions, the software could force any switch to reset. When it did, the switch passed its work to its backup in such a way that the backup was also forced to reset. Because all the code on the switches was identical, everything failed simultaneously.

What, you may ask, was the word that caused all the trouble? You must keep in mind that the C programming language has special technical meanings for the words it uses. The C programming command doesn't actually mean what the English word means. Still, there is some irony here. The computer command which brought down the network was *break*.

13-1 The Distributive Property and Common Factors

LEARNING OUTCOME

1 Factor an expression containing a common factor.

LC LEARNING CATALYTICS
1. Are $4a$ and $3a^3$ like terms?
2. Name any factor that $4a$ and $3a^3$ have in common.

Factored form: an expression written as the indicated product of two or more factors, that is, as a multiplication

Expanded form: an expression written as one or more terms with the multiplication of any instances of the distributive property

Throughout our study of mathematics, we have examined products and factors. To reduce fractions, we looked for factors common to both the numerator and the denominator. In this chapter, we again find it useful to examine products and factors.

We discussed the distributive property and finding common factors earlier in the text and applied them in different contexts. In this section, rather than use the distributive property to multiply and obtain a product, we start with a product and regenerate the factors that produce the product. In other words, we want to undo the multiplication. Factoring resembles division, which is the inverse operation of multiplication.

1 Factor an Expression Containing a Common Factor. The multiplication problem $7a(3a + 2)$ is written in **factored form.** It is the indicated product of $7a$ and the grouped quantity $3a + 2$. After the expression is multiplied, we have two terms written as the sum $21a^2 + 14a$. This is the **expanded form.** To rewrite the expression $21a^2 + 14a$ as the indicated product $7a(3a + 2)$ is to *factor* it.

Let's look at a general example of the distributive property:

$$a(x + y) \quad = \quad ax + ay$$

factored form   expanded form

Notice that *a* appears as a factor in both terms in the expanded form. When a factor appears in each of several terms, it is called a *common factor* of the terms. The distributive property in reverse can be used to write the addition as a multiplication. In other words, we can *factor* the expression.

$$ax + ay = a(x + y)$$

To factor an expression containing a common factor:

1. Find the *greatest* factor common to *each* term of the expression.

2. Divide each term by the common factor. Divide mentally if practical.

3. Rewrite the expression as the indicated product of the greatest common factor (GCF) and the quotients in Step 2.

STOP AND CHECK

1. Write $7x - 7y$ in factored form.

Answer:
1. $7(x - y)$

EXAMPLE 1

Write $4a + 4b$ in factored form.

We can use the distributive property to factor the expression.

$4a + 4b =$	Write 4 as a factor and divide each term by 4.
$4\left(\dfrac{4a}{4} + \dfrac{4b}{4}\right) =$	Simplify each fraction in parentheses.
$4(a + b)$	Factored form See Exercises 1–8.

The distributive property also applies if we have more than two terms.

STOP AND CHECK

1. Write $10xy + 15x - 20y$ in factored form.

Answer:
1. $5(2xy + 3x - 4y)$

EXAMPLE 2

Write $3ab + 9a + 12b$ in factored form.

$3ab + 9a + 12b$	3 is the common factor. Divide.
$3 \cdot 3 \quad 3 \cdot 4$	
$3\left(\dfrac{3 \cdot ab}{3} + \dfrac{3 \cdot 3a}{3} + \dfrac{3 \cdot 4b}{3}\right)$	Simplify each fraction in parentheses.
$3(ab + 3a + 4b)$	Factored form See Exercises 9–11.

When looking for common factors, we always look for *all* common factors.

STOP AND CHECK

Factor completely.
1. $12x^3 - 10x$
2. $3y + 6y^3$

Answers:
1. $2x(6x^2 - 5)$ 2. $3y(1 + 2y^2)$

EXAMPLE 3

Factor completely. **(a)** $10a^2 + 6a$ **(b)** $2x^2 + 4x^3$

(a)

$10a^2 + 6a =$	The GCF is $2a$. Write $2a$ as a factor and divide each term by $2a$.
$2a\left(\dfrac{10a^2}{2a} + \dfrac{6a}{2a}\right) =$	Simplify.
$2a(5a + 3)$	Factored form

(b)

$2x^2 + 4x^3 =$	The GCF is $2x^2$. Write $2x^2$ as a factor and divide each term by $2x^2$.
$2x^2\left(\dfrac{2x^2}{2x^2} + \dfrac{4x^3}{2x^2}\right) =$	Simplify.
$2x^2(1 + 2x)$	Term of 1 must be written. See Exercises 12–19.

Prime polynomial: an expression for which the greatest common factor of all terms is 1

EXAMPLE 4

Write $2x + 3y$ in factored form.

$2x + 3y =$ The GCF is 1. The expression can be written in factored form only as $1(2x + 3y)$.

$\mathbf{1(2x + 3y)}$ Factored form   See Exercises 20–21.

When 1 is the greatest common factor of an expression, the expression is a **prime polynomial**. An example of a prime polynomial is $5x^2 + 7x + 11$.

TIP   **When Is It Necessary to Write a 1?**   We have found that it is not always necessary to write the number 1. When is it necessary?

When 1 is a *term*, it must be written.

$$2x^2 - x = x\left(\frac{2x^2}{x} - \frac{x}{x}\right) = x(2x - 1)$$

When 1 is a *factor*, writing the 1 is optional: $1 \cdot n = n$.

$$2a + 2b = 2\left(\frac{2a}{2} + \frac{2b}{2}\right) = 2(1a + 1b) \qquad \text{or} \qquad 2(a + b)$$

Also, in Example 4, the factor of 1 is optional. The factored form of the fraction could be written as $2x + 3y$.

When 1 is an *exponent*, writing the 1 is optional: $n^1 = n$.

$$2x^3 - 5x^2 = x^2\left(\frac{2x^3}{x^2} - \frac{5x^2}{x^2}\right) = x^2(2x^1 - 5x^0) = x^2(2x - 5)$$

Also, recall that $n^0 = 1$ for any real number n, $n \neq 0$.

Sometimes a binomial factor or a grouping is the common factor. Example 5 is an example of an expression that has a binomial factor as its GCF.

EXAMPLE 5

Factor $7y(2y - 5) + 3(2y - 5)$.

$7y(2y - 5) + 3(2y - 5) =$ Common factor is $(2y - 5)$.

$(2y - 5)\left[\dfrac{7y(2y - 5)}{(2y - 5)} + \dfrac{3(2y - 5)}{(2y - 5)}\right] =$

$\mathbf{(2y - 5)(7y + 3)}$ See Exercises 22–29.

If the leading coefficient of a polynomial is negative, it is often helpful to factor a common factor of -1.

EXAMPLE 6

Factor $-3x^2 + 2x - 5$.

$-3x^2 + 2x - 5 =$ Common factor is -1.

$-1\left(\dfrac{-3x^2}{-1} + \dfrac{2x}{-1} - \dfrac{5}{-1}\right) =$

$\mathbf{-1(3x^2 - 2x + 5)}$ See Exercises 30–37.

13–1 EXERCISES

MyLab Math For additional practice go to your study plan in MyLab Math.

1 Factor completely. Check.

See Example 1.

1. $7a + 7b$
2. $12x + 12y$
3. $m^2 + 2m$
4. $5y^3 + 8y^2$
5. $6x^2 + 3x$
6. $12y^3 + 18y^4$
7. $12x^5 - 6x^4$
8. $5x - 15xy$

See Example 2.

9. $5ab + 10a + 20b$
10. $4ax + 6x + 10a$
11. $5a - 7ab + 35b$

See Example 3.

12. $12a^2 - 15b^2 + 6a$
13. $3x^3 - 9x^2 - 6x$
14. $8a^2b + 14ab^3 + 28a^3b^3$
15. $3m^2 - 6m^3 + 12m^4$
16. $12x^2y - 18xy^3 + 24x^2y^2$
17. $15a^2bc + 18a^3b^2c^3 - 21a^4bc^5$
18. $20x^2y^3z - 35x^3y^2z - 40x^2y^2z$
19. $8x^4y^2 - 12x^2y^4 - 4x^2y^2$

Write in factored form. *See Example 4.*

20. $5y + 3z$
21. $18a - 7b$

See Example 5.

22. $5x(x + 3) + 8y(x + 3)$
23. $3x(2x - 1) + 5(2x - 1)$
24. $4y(3y - 5) + 7(3y - 5)$
25. $7a(a - b) + 2b(a - b)$
26. $5m(2m - 3n) - 7n(2m - 3n)$
27. $y(y - 2) - 3(y - 2)$
28. $3x(2x - 7) - 8(2x - 7)$
29. $7y(9y - 2) - 5(9y - 2)$

Write in factored form so the leading coefficient of the polynomial factor is positive. *See Example 6.*

30. $-5x + 2$
31. $-12x + 7$
32. $-x^2 + 3x - 8$
33. $-2x^2 - 7x - 11$
34. $-2x^2 + 6x - 8$
35. $-3x^2 - 9x + 15$
36. $-7x^2 - 21x + 14$
37. $-12x^2 + 18x + 6$

13–2 Factoring Special Products

LEARNING OUTCOMES

1 Recognize and factor the difference of two perfect squares.

2 Recognize and factor a perfect-square trinomial.

3 Recognize and factor the sum or difference of two perfect cubes.

LC LEARNING CATALYTICS

1. Multiply $(2x + 3)(4x^2 - 6x + 9)$.

To factor any of the special products we used in Chapter 11, we apply the inverse of the process. That is, we start with the product and "work back" to the factors that produce these special products.

To rewrite a special product in factored form, we must recognize the product as a pattern. Once we identify the special product, then we must know the pattern of the product in factored form.

1 Recognize and Factor the Difference of Two Perfect Squares. Before we can factor such a special product, we must be able to recognize an expression as a special product. First, we examine the *sum and difference of the same two terms*. The product is the difference of two perfect squares.

To identify a binomial as the difference of two perfect squares:

1. Verify that the expression is a binomial (two terms).
2. Verify that the absolute value of each term is a perfect square.
3. Verify that the second term is negative.

The pattern is $a^2 - b^2$.

STOP AND CHECK
Identify the special products
that are the difference of two
perfect squares.
1. $x^2 - 8$
2. $x^2 - 2x - 1$
3. $4x^2 - 25$

Answers:
1. Not a difference of two perfect
   squares
2. Not a difference of two perfect
   squares
3. Difference of two perfect
   squares

EXAMPLE 1

Identify the special products that are the difference of two perfect squares.

(a) $x^2 - 9$ (b) $a^2 + 49$ (c) $m^2 - 27$

(d) $3y^2 - 25$ (e) $9x^2 - 4$ (f) $4x^2 - 4x + 1$

(a) Difference of two perfect squares.

(b) Not the difference of two perfect squares; this is a *sum,* not a difference.

(c) Not the difference of two perfect squares; 27 is not a perfect square.

(d) Not the difference of two perfect squares; in $3y^2$, 3 is not a perfect square.

(e) Difference of two perfect squares.

(f) Not the difference of two perfect squares; this is a trinomial, not a binomial.

See Exercises 1–6.

To factor the difference of two perfect squares:

1. Take the square root of the first term.

2. Take the square root of the absolute value of the second term.

3. Write one factor as the *sum* of the square roots found in Steps 1 and 2, and write the other factor as the *difference* of the square roots from Steps 1 and 2.

Symbolically,

$$a^2 - b^2 = (a + b)(a - b)$$

STOP AND CHECK
Factor the special products,
which are the differences of
two perfect squares.
1. $9x^2 - 1$
2. $-16 + 25x^2$
3. $x^2 - 4$

Answers:
1. $(3x + 1)(3x - 1)$
2. $(5x + 4)(5x - 4)$
3. $(x + 2)(x - 2)$

EXAMPLE 2

Factor the special products, which are the differences of two perfect squares.

(a) $a^2 - 9$ (b) $x^2 - 36$ (c) $4x^2 - 1$ (d) $-49 + 16m^2$

(a) $a^2 - 9 = (a + 3)(a - 3)$

(b) $x^2 - 36 = (x + 6)(x - 6)$

(c) $4x^2 - 1 = (2x + 1)(2x - 1)$

(d) $-49 + 16m^2 = 16m^2 - 49 = (4m + 7)(4m - 7)$

See Exercises 7–21.

TIP **Order of Factors** Because multiplication is commutative, the factors given as answers in the preceding example may be expressed in any order, such as $(a + 3)(a - 3)$ or $(a - 3)(a + 3)$, $(x + 6)(x - 6)$ or $(x - 6)(x + 6)$.

2 **Recognize and Factor a Perfect-Square Trinomial.** A trinomial is a *perfect-square trinomial* if the first and last terms are positive perfect squares and the absolute value of the middle term is *twice* the product of the square roots of the first and last terms. We need to be able to distinguish these special products from other expressions before we factor them.

> **To identify a perfect-square trinomial:**
>
> 1. Verify that the expression is a trinomial.
>
> 2. Verify that the first and last terms are positive and perfect squares.
>
> 3. Mentally take the square root of the first and last terms and multiply the results. Two times this product should equal the absolute value of the middle term of the original trinomial.
>
> The pattern is $a^2 + 2ab + b^2$ or $a^2 - 2ab + b^2$.

STOP AND CHECK

1. Verify that $x^2 - 12x + 36$ is a perfect-square trinomial.

Answer:

1. The first and last terms are positive perfect squares, and the middle term of $-12x$ has an absolute value that is twice the product of the square roots of the first and last terms, x^2 and 36.

EXAMPLE 3

Verify that the trinomials are perfect-square trinomials.

(a) $x^2 + 14x + 49$ **(b)** $4m^2 - 12m + 9$ **(c)** $9x^2 + 24xy + 16y^2$

(a) The first and last terms, x^2 and 49, are positive perfect squares. The middle term, $14x$, has an absolute value that is twice the product of the square roots of x^2 and 49. That is, $2(7 \cdot x) = 14x$.

(b) The first and last terms, $4m^2$ and 9, are positive perfect squares. The middle term, $-12m$, has an absolute value that is twice the product of the square roots of $4m^2$ and 9. That is, $2(2m \cdot 3) = 12m$.

(c) The first and last terms, $9x^2$ and $16y^2$, are positive perfect squares. The middle term, $24xy$, has an absolute value that is twice the product of the square roots of $9x^2$ and $16y^2$. That is, $2(3x \cdot 4y) = 24xy$. **See Exercises 22–27.**

STOP AND CHECK

Explain why the trinomials are *not* perfect-square trinomials.
1. $4x^2 + 12x - 9$
2. $x^2 + 10x + 16$

Answers:
1. The last term, -9, is negative.
2. The middle term, $10x$, is not *twice* the product of x and 4.

EXAMPLE 4

Explain why the trinomials are *not* perfect-square trinomials.

(a) $x^2 + 2x - 1$ **(b)** $4x^2 + 6x + 9$

(c) $x^2 - 5x + 4$ **(d)** $-4x^2 - 4x + 1$

(a) The last term, -1, is negative. This term must be positive in a perfect-square trinomial.

(b) The middle term, $6x$, is not *twice* the product of $2x$ and 3.

(c) The middle term, $-5x$, is not *twice* the product of x and 2.

(d) The first term, $-4x^2$, is negative. The first and last terms of a perfect-square trinomial are positive. **See Exercises 22–27.**

> **To factor a perfect-square trinomial:**
>
> 1. Write the square root of the first term.
>
> 2. Write the sign of the middle term.
>
> 3. Write the square root of the last term.
>
> 4. Indicate the square of this binomial quantity.
>
> Symbolically,
>
> $$a^2 + 2ab + b^2 = (a + b)^2 \text{ and } a^2 - 2ab + b^2 = (a - b)^2$$

EXAMPLE 5

Factor the perfect-square trinomials.

(a) $x^2 + 14x + 49$ **(b)** $4m^2 - 12m + 9$ **(c)** $9x^2 + 24xy + 16y^2$

	Square root of *first* term	Sign of *middle* term	Square root of *last* term	*Square the quantity*
(a) $x^2 + 14x + 49$	x	$+$	7	$(x + 7)^2$
(b) $4m^2 - 12m + 9$	$2m$	$-$	3	$(2m - 3)^2$
(c) $9x^2 + 24xy + 16y^2$	$3x$	$+$	$4y$	$(3x + 4y)^2$

See Exercises 28–42.

3 **Recognize and Factor the Sum or Difference of Two Perfect Cubes.** Before we factor the sum or difference of two perfect cubes, we should be able to recognize an expression that matches the pattern.

To identify the sum or difference of two perfect cubes:

1. Verify that the expression is a binomial.

2. Verify that each term is a perfect cube.

The pattern is $a^3 + b^3$ or $a^3 - b^3$.

EXAMPLE 6

Identify the expressions that are the sum or difference of two perfect cubes.

(a) $a^3 - 27$ **(b)** $8x^3 + 9$ **(c)** $8y^3 - 125$ **(d)** $64 - 8a^3$

(e) $a^3 + 16b^3$ **(f)** $c^3 + y^3 - 27$ **(g)** $27x^3 + 8$ **(h)** $3a^3 - 64$

(a) Difference of two perfect cubes.

(b) Not the sum or difference of two perfect cubes; 9 is not a perfect cube.

(c) Difference of two perfect cubes.

(d) Difference of two perfect cubes.

(e) Not the sum or difference of two perfect cubes; 16 is not a perfect cube.

(f) Not the sum or difference of two perfect cubes; this is a trinomial.

(g) Sum of two perfect cubes.

(h) Not the sum or difference of two perfect cubes; 3 is not a perfect cube.

See Exercises 43–48.

The sum or difference of two perfect cubes is factored as the product of a binomial and a trinomial.

To factor the sum of two perfect cubes:

1. Write the binomial factor as the *sum* of the cube roots of the two terms.

2. Write the trinomial factor as the square of the first term from Step 1, *minus* the product of the two terms from Step 1, *plus* the square of the second term from Step 1.

Symbolically,

$$a^3 + b^3 = (a + b)(a^2 - ab + b^2)$$

> **To factor the difference of two perfect cubes:**
>
> 1. Write the binomial factor as the *difference* of the cube roots of the two terms.
> 2. Write the trinomial factor as the square of the first term from Step 1, *plus* the product of the two terms from Step 1, *plus* the square of the second term from Step 1.
>
> Symbolically,
>
> $$a^3 - b^3 = (a - b)(a^2 + ab + b^2)$$

Notice that the sign of the third term of the trinomial is positive when factoring either the sum or the difference of two cubes.

EXAMPLE 7

Factor the sum or difference of two perfect cubes.

(a) $27a^3 - 8$ (b) $m^3 + 125n^3$

(a) $27a^3 - 8 =$
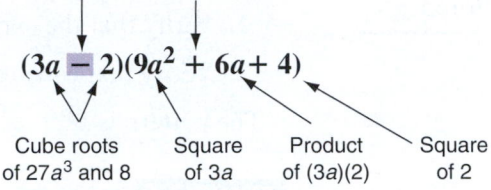

(b) $m^3 + 125n^3 =$
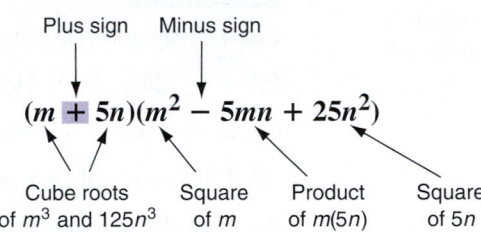

See Exercises 49–60.

13-2 EXERCISES

MyLab Math For additional practice go to your study plan in MyLab Math.

1 Identify the special products that are the difference of two perfect squares. *See Example 1.*

1. $r^2 - s^2$
2. $d^2 - 4d + 10$
3. $4y^2 - 16$
4. $25m^2 - 9n^2$
5. $9 + 4a^2$
6. $64p^2 - q^2$

Factor the special products. *See Example 2.*

7. $y^2 - 49$
8. $16x^2 - 1$
9. $9a^2 - 100$
10. $4m^2 - 81n^2$
11. $9x^2 - 64y^2$
12. $25x^2 - 64$
13. $100 - 49x^2$
14. $4x^2 - 49y^2$
15. $121m^2 - 49n^2$
16. $81x^2 - 169$
17. $-9 + 4a^2$
18. $-16 + 25r^2$
19. $36x^2 - 49y^2$
20. $49 - 144x^2$
21. $16x^2 - 81y^2$

2 Identify the trinomials that are perfect-square trinomials. If a trinomial is not a perfect-square trinomial, state why. *See Examples 3 and 4.*

22. $4y^2 + 2y + 16$
23. $9m^2 - 24mn + 16n^2$
24. $16a^2 + 8a - 1$
25. $-9r^2 + 12r + 4$
26. $y^2 - 14y + 49$
27. $p^2 + 10p + 25$

Factor the special products. *See Example 5.*

28. $x^2 + 6x + 9$

29. $x^2 + 14x + 49$

30. $x^2 - 12x + 36$

31. $x^2 - 16x + 64$

32. $4a^2 + 4a + 1$

33. $25x^2 - 10x + 1$

34. $9m^2 - 48m + 64$

35. $4x^2 - 36x + 81$

36. $x^2 - 12xy + 36y^2$

37. $4a^2 - 20ab + 25b^2$

38. $y^2 - 10y + 25$

39. $9x^2 + 60xy + 100y^2$

40. $-x^2 - 12x - 36$

41. $-9x^2 + 6x - 1$

42. $-x^2 - 8x - 16$

3 Identify the special products that are the sum or difference of two perfect cubes. *See Example 6.*

43. $8b^3 - 125$

44. $y^2 - 14y + 49$

45. $T^3 + 27$

46. $c^3 + 16$

47. $125x^3 - 8y^3$

48. $64 + a^3$

Factor the special products, which are the sum or difference of two perfect cubes. *See Example 7.*

49. $m^3 - 8$

50. $y^3 - 125$

51. $Q^3 + 27$

52. $c^3 + 1$

53. $125d^3 - 8p^3$

54. $a^3 + 64$

55. $216a^3 - b^3$

56. $x^3 + Q^3$

57. $8p^3 - 125$

58. $27 - 8y^3$

59. $-a^3 - 8$

60. $-x^3 - 27$

13–3	Factoring General Trinomials

LEARNING OUTCOMES

1 Factor general trinomials whose squared term has a coefficient of 1.

2 Remove common factors after grouping an expression.

3 Factor a general trinomial by grouping.

4 Factor any binomial or trinomial that is not prime.

LC LEARNING CATALYTICS

1. Multiply $(3x - 7)(5x + 3)$.

General trinomial: a trinomial that is not a perfect-square trinomial

Many trinomials do not have a common factor and do not match the pattern of a special product. Some will still factor as the product of two binomials. First, let's examine a trinomial whose squared term has a coefficient of 1.

1 **Factor General Trinomials Whose Squared Term Has a Coefficient of 1.** Trinomials that are not perfect-square trinomials are **general trinomials**.

To factor a trinomial with a squared term that has a coefficient of 1:

1. Ensure the trinomial is arranged in descending powers of one variable.

2. Determine the signs of the second term of each binomial factor by examining the sign of the *third* term of the trinomial.

   $+ \Rightarrow$ like signs (both $+$ or both $-$, matching the middle sign)
   $- \Rightarrow$ unlike signs (one $+$ and one $-$, with the sign of the larger absolute value of the factor pair matching the sign of the middle term of the trinomial)

3. Write all factor pairs of the coefficient of the third term.

4. Select the factor pair that adds (algebraically) to the coefficient of the middle term of the trinomial. Include appropriate signs.

5. Write the binomial factors of the trinomial with the square root of the first-term variable as the first term of each factor and the two factors from Step 4 as the second terms of the binomials.

EXAMPLE 1

Factor $x^2 + 6x + 5$.

$x^2 + 6x + 5$	Terms are already in descending order.
$(\ +\ )(\ +\ )$	Last term $+$ and middle term $+$ means signs are alike and both positive.
	Only factor pair of 5 is 1(5). The algebraic sum is $+1 + (+5) = +6$.
$(x + 1)(x + 5)$	First term of each factor is x. Second terms are $+1$ and $+5$.

Check by using FOIL to multiply.   **See Exercises 1–14.**

EXAMPLE 2

Factor $x^2 + 4x - 12$.

$x^2 + 4x - 12$	Terms are already in descending order. Last term $-$ means signs of the binomials are unlike.
$(\ +\ )(\ -\ )$	Middle term $+$ means the factor with the larger absolute value in the selected factor pair will be $+$.
	Factor pairs of 12:
	$1 \cdot 12$
	$\boxed{2 \cdot 6}$ Algebraic sum of -2 and $+6$ is $+4$.
	$3 \cdot 4$
$(x - 2)(x + 6)$	First term of each factor is x. Second terms are -2 and 6.

See Exercises 15–28.

EXAMPLE 3

Factor $x^2 - 20 - x$.

$x^2 - x - 20 =$	Arrange in descending order of x. Last sign $-$ means signs of the binomials are unlike.
	Middle sign $-$ means the factor with the larger absolute value in the selected factor pair will be $-$.
	Factor pairs of 20:
	$1 \cdot 20$
	$2 \cdot 10$
	$\boxed{4 \cdot 5}$ Algebraic sum of $+4$ and -5 is -1.
$(x + 4)(x - 5)$	First term of each factor is x. Second terms are $+4$ and -5.

See Exercises 29–40.

2 **Remove Common Factors After Grouping an Expression.** Algebraic expressions that have more than three terms and have no factors common to every term in the expression may have common factors for some groups of terms. In these cases, the expression may be written as groupings.

> **To remove common factors after grouping an expression:**
>
> **1.** Identify groupings.
>
> **2.** Factor all common factors from each grouping.
>
> **3.** Examine each grouping to see if there are common groupings. If so, factor out the common grouping.

STOP AND CHECK

1. Write
$2x^2 - 3xy + 2bx - 2ab$ as
an alternate expression by
grouping pairs of terms.
Factor common factors
from each grouping.

Answer:

1. $x(2x - 3y) + 2b(x - a)$

EXAMPLE 4

Write the expression $2x^2 - 2xb + ax - ay$ as an alternate expression by grouping pairs of terms. Factor common factors from each grouping.

$2x^2 - 2xb + ax - ay =$	Group two terms in each grouping.
$(2x^2 - 2xb) + (ax - ay) =$	Factor common factors from each grouping.
$\mathbf{2x(x - b) + a(x - y)}$	There is no common grouping in the terms.

See Exercises 41–44.

> **TIP** **Groupings Are Not Necessarily Factors** In the preceding example the groupings $(2x^2 - 2xb)$ and $(ax - ay)$ are not factors. They are groupings that are added. To rewrite the expression as $2x(x - b) + a(x - y)$ is not in factored form. The expression has two terms and each term has two or more factors. *An expression in factored form is only one term.*

STOP AND CHECK

Write the expressions in factored form by using grouping.

1. $xy + 2x - 3y - 6$
2. $6x^2 - 4x - 15x + 10$
3. $2x^2 - 3x + 7x + 21$

Answers:

1. $(x - 3)(y + 2)$
2. $(2x - 5)(3x - 2)$
3. Prime

EXAMPLE 5

Write the expressions in factored form by using grouping.

(a) $mx + 2m - 4x - 8$ **(b)** $y^2 + 2xy + 3y + 6x$

(c) $3x^2 - 9x - 7x + 21$ **(d)** $2x^2 + 8x + 5y - 15$

(a)

$mx + 2m - 4x - 8 =$	Group the four-termed expression into two groups.
$(mx + 2m) + (-4x - 8) =$	Factor out common factors in each of the two groups.
$m(x + 2) + -4(x + 2) =$	Convert double signs to an equivalent single sign.
$m(x + 2) - 4(x + 2) =$	Factor out the common binomial factor $(x + 2)$.
$\mathbf{(x + 2)(m - 4)}$ or $\mathbf{(m - 4)(x + 2)}$	Check the result by using the FOIL method.
$(x + 2)(m - 4) = xm - 4x + 2m - 8$	Rearrange terms and factors (commutative properties of multiplication and addition).
$= mx + 2m - 4x - 8$	The factoring checks.

(b)

$y^2 + 2xy + 3y + 6x =$	Group into two groups.
$(y^2 + 2xy) + (3y + 6x) =$	Factor out common factors in each group.
$y(y + 2x) + 3(y + 2x) =$	Factor out the common binomial factor.
$\mathbf{(y + 2x)(y + 3)}$ or $\mathbf{(y + 3)(y + 2x)}$	

(c)

$3x^2 - 9x - 7x + 21 =$	Group into two groups.
$(3x^2 - 9x) + (-7x + 21) =$	Factor each group.
$3x(x - 3) + -7(x - 3) =$	Convert double signs to a single sign.
$3x(x - 3) - 7(x - 3) =$	Factor out the common binomial factor.
$\mathbf{(x - 3)(3x - 7)}$ or $\mathbf{(3x - 7)(x - 3)}$	

(d) $2x^2 + 8x + 5y - 15 =$ Group into two groups.

$(2x^2 + 8x) + (5y - 15) =$ Factor each group.

$2x(x + 4) + 5(y - 3)$

In this example, the two groups do not have a common factor. Thus, the expression cannot be factored. Even if we rearrange the terms, we will not be able to write the expression as a single term in factored form.

$2x^2 + 8x + 5y - 15$ **is prime.**   See Exercises 45–55.

> **TIP** **Make Leading Coefficient of Binomial Factor Positive** When factoring by grouping, manipulate the signs of the common factor so that the leading coefficient of the binomial factor is positive.
>
> $$-3x + 9$$
>
> Factor as $-3(x - 3)$, *not* $3(-x + 3)$. Parts a and c of the preceding example illustrate this tip.

---

3 **Factor a General Trinomial by Grouping.** We can now use a systematic method to factor general trinomials with leading integral coefficients that are positive and not equal to 1.

> **To factor a general trinomial of the form $ax^2 + bx + c$ by grouping:**
>
> 1. Arrange terms in descending order and verify that the first term is positive or rewrite with -1 as the common factor.
> 2. Multiply the coefficient of the first term of the trinomial by the coefficient of the last term.
> 3. Factor the product from Step 2 into a pair of factors whose *sum* is the coefficient of the middle term.
>   **(a)** If the sign of the last term is positive, the factors will have like signs, either both positive or both negative, depending on the sign of the middle term.
>   **(b)** If the sign of the last term is negative, the factors will have unlike signs with the factor with the larger absolute value matching the sign of the middle term.
>   If there is no factor pair that meets one of these conditions, the original trinomial is prime.
> 4. Rewrite the trinomial as a polynomial with four terms by replacing the middle term with two terms that have the coefficients identified in Step 3.
> 5. Group the polynomial with four terms from Step 4 into two groups of two terms.
> 6. Factor the common factors from each of the two groups.
> 7. Factor out the common binomial factor.

STOP AND CHECK

1. Factor $12x^2 + 23x + 5$ by grouping.

Answer:

1. $(4x + 1)(3x + 5)$

EXAMPLE 6

Factor $6x^2 + 19x + 10$ by grouping.

$6 \cdot 10 = 60$ Multiply the coefficients of the first and third terms.

$60 = 1 \cdot 60$ List all factor pairs of 60.
$ 2 \cdot 30$
$ 3 \cdot 20$
$ 4 \cdot 15$ Identify the pair whose sum is 19.
$ 5 \cdot 12$
$ 6 \cdot 10$

$$6x^2 + 19x + 10 =$$ Separate $+19x$ into two terms using the coefficients 4 and 15.

$$6x^2 + 4x + 15x + 10 =$$ Group into two groups.

$$(6x^2 + 4x) + (15x + 10) =$$ Factor common factors in each group.

$$2x(3x + 2) + 5(3x + 2) =$$ Factor out the common binomial factor.

$$(3x + 2)(2x + 5) \text{ or } (2x + 5)(3x + 2)$$ Check using the FOIL method.

See Exercises 56–62.

STOP AND CHECK

1. Factor $15x^2 - 41x + 14$ by grouping.

Answer:

1. $(3x - 7)(5x - 2)$

EXAMPLE 7

Factor $20x^2 - 23x + 6$ by grouping.

$$20 \cdot 6 = 120$$ Find the product of 20 and 6.

List the factor pairs of 120 and select the pair that has a sum of 23.

$$120 = 1 \cdot 120$$ List all factor pairs of 120.
$$2 \cdot 60$$
$$3 \cdot 40$$
$$4 \cdot 30$$
$$5 \cdot 24$$
$$6 \cdot 20$$
$$8 \cdot 15$$ Identify the pair that *adds* to -23. Since the sign of
$$10 \cdot 12$$ the last term is *positive,* both factors will be negative.

$$20x^2 - 23x + 6 =$$ Separate $-23x$ into two terms using the coefficients -8 and -15.

$$20x^2 - 8x - 15x + 6 =$$ Group into two groups. Be sure the second grouping keeps the negative sign with $15x$.

$$(20x^2 - 8x) + (-15x + 6) =$$ Factor common factors from each group so that the leading coefficient in each binomial factor is positive.

$$4x(5x - 2) - 3(5x - 2) =$$ Factor out the common binomial factor.

$$(5x - 2)(4x - 3)$$ Check using the FOIL method.   **See Exercises 63–69.**

If the third term of a trinomial is negative, we use the same procedure but the two factors will have unlike signs and the sign of the sum will match the sign of the coefficient of the middle term.

STOP AND CHECK

1. Factor $9x^2 + 9x - 28$ by grouping.

Answer:

1. $(3x + 7)(3x - 4)$

EXAMPLE 8

Factor $10x^2 + 19x - 15$ by grouping.

$$10 \cdot 15 = 150$$ Find the product of 10 and 15.

List the factor pairs of 150 and select the pair that has a difference of 19.

$$150 = 1 \cdot 150$$ List all factor pairs of 150.
$$2 \cdot 75$$
$$3 \cdot 50$$
$$5 \cdot 30$$
$$6 \cdot 25$$ Identify the pair whose difference is 19,
$$10 \cdot 15$$ since the last term is *negative.*

The factors 25 and 6 have a difference of 19. When we rewrite the trinomial, we write $19x$ as $25x - 6x$.

$$10x^2 + 19x - 15 =$$ Separate $+19x$ into two terms using the coefficients 25 and -6. The signs will be different and the larger coefficient will be positive since the sign of $19x$ is positive.

$$10x^2 + 25x - 6x - 15 =$$ Group into two terms.

$$(10x^2 + 25x) + (-6x - 15) =$$ Factor common factors in each term so that the leading coefficient in each term is positive.

$$5x(2x + 5) - 3(2x + 5) =$$ Factor the common binomial.

$$(2x + 5)(5x - 3)$$ Check using the FOIL method.

See Exercises 70–82.

TIP **Practice Moves You from Systematic Processes to Intuitive Processes** Systematic processes help you understand the concepts and build confidence. They also help you develop your mathematical senses such as your number sense and your spatial sense. The more you practice, the more you develop your mathematical senses.

Many students instinctively and automatically move to shortened and mental processes. Many times you can test a few combinations of coefficients and find the correct factors without going through the entire systematic process. With practice, you can write expressions in factored form more readily.

---

4 **Factor Any Binomial or Trinomial That Is Not Prime.** Now let's develop a strategy for factoring any binomial or trinomial that can be factored. This strategy allows us to determine with confidence if a particular binomial or trinomial is prime and cannot be factored.

To factor any binomial or trinomial or polynomial:

Perform the following steps in order.

1. Write terms in decending order and factor the greatest common factor (if any).

2. Check the binomial or trinomial to see if it is a special product.

   (a) If it is the difference of two perfect squares, use the pattern.

   (b) If it is a perfect-square trinomial, use the pattern.

   (c) If it is the sum or difference of two perfect cubes, use the appropriate pattern.

3. If it is not a special product, factor by grouping or by any appropriate process.

4. Examine each factor to see if it can be factored further.

5. Check factoring by multiplying.

The steps for factoring any binomial or trinomial are presented visually in the flowchart (Fig. 13–1).

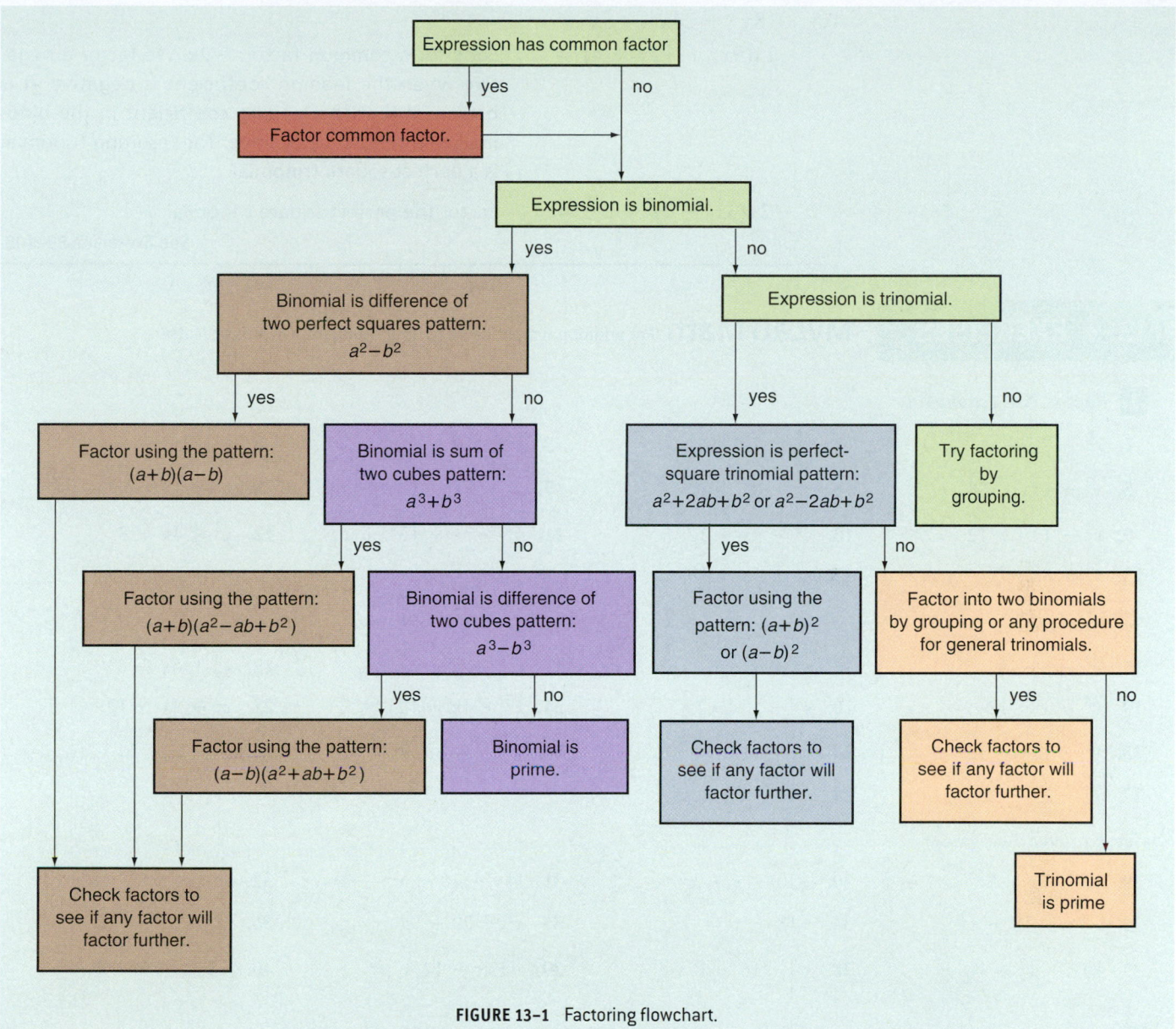

FIGURE 13-1 Factoring flowchart.

STOP AND CHECK

Factor completely.
1. $18x^2 - 8$
2. $6x^3 + 3x^2 - 9x$
3. $-75x^3 + 30x^2 - 3x$

Answers:
1. $2(3x + 2)(3x - 2)$
2. $3x(2x + 3)(x - 1)$
3. $-3x(5x - 1)^2$

EXAMPLE 9

Factor completely **(a)** $4x^3 - 2x^2 - 6x$ **(b)** $12x^2 - 27$ **(c)** $-18x^3 + 24x^2 - 8x$

(a) $4x^3 - 2x^2 - 6x =$

$2x(2x^2 - x - 3) =$ 2x is the common factor. The trinomial is not a special product. Factor the trinomial using any appropriate process and *keep* the 2x factor.

$2x(2x - 3)(x + 1) =$ Check by multiplying.

(b) $12x^2 - 27 =$

$3(4x^2 - 9) =$ 3 is the common factor. The binomial is the difference of two perfect squares.

$3(2x - 3)(2x + 3) =$ Factor the difference of two perfect squares. Keep the common factor of 3. Check.

(c) $-18x^3 + 24x^2 - 8x =$
$-2x(9x^2 - 12x + 4) =$ Look for a common factor: $-2x$. We factor a negative when the leading coefficient is negative. It is preferred that the leading coefficient in the binomial or trinomial be positive. The resulting trinomial is a perfect-square trinomial.

$$-2x(3x - 2)^2 \qquad = \qquad$$ Factor the perfect-square trinomial.

See Exercises 83–106.

13–3 EXERCISES

MyLab Math For additional practice go to your study plan in MyLab Math.

1 Factor. *See Example 1.*

1. $x^2 + 7x + 6$ **2.** $x^2 - 7x + 6$ **3.** $x^2 - 5x + 6$ **4.** $x^2 + 5x + 6$

5. $x^2 - 11x + 28$ **6.** $x^2 + 8x + 12$ **7.** $x^2 - 8x + 12$ **8.** $x^2 + 13x + 12$

9. $x^2 - 13x + 12$ **10.** $x^2 + 7x + 12$ **11.** $x^2 - 7x + 12$ **12.** $x^2 - 4x + 3$

13. $x^2 + 8x + 7$ **14.** $x^2 + 7x + 10$

See Example 2.

15. $x^2 - x - 6$ **16.** $x^2 + x - 6$ **17.** $x^2 - 5x - 6$ **18.** $x^2 + 5x - 6$

19. $x^2 - x - 12$ **20.** $x^2 + x - 12$ **21.** $x^2 + 4x - 12$ **22.** $x^2 - 4x - 12$

23. $x^2 - 11x - 12$ **24.** $x^2 + 11x - 12$ **25.** $y^2 - 3y - 10$ **26.** $y^2 - y - 20$

27. $b^2 + 2b - 3$ **28.** $-14 - 5b + b^2$

See Example 3.

29. $12 - 7x + x^2$ **30.** $-30 - x + x^2$ **31.** $11x + 18 + x^2$ **32.** $-9x + x^2 + 18$

33. $-7x - 18 + x^2$ **34.** $-18 + x^2 + 17x$ **35.** $x^2 + 20 + 9x$ **36.** $12x - x^2 + 20$

37. $10x - x^2 + 16$ **38.** $x^2 - 16 + 17x$ **39.** $-13x - 14 + x^2$ **40.** $-5x - 14 + x^2$

2 Factor the polynomials by removing the common factors after grouping. *See Example 4.*

41. $x^2 + xy + 4x + 4y$ **42.** $6x^2 + 4x - 3xy - 2y$ **43.** $3mx + 5m - 6nx - 10n$

44. $30xy - 35y - 36x + 42$

See Example 5.

45. $x^2 - 2x + 8x - 16$ **46.** $6x^2 - 2x - 21x + 7$ **47.** $x^2 - 4x + x - 4$

48. $8x^2 - 4x + 6x - 3$ **49.** $x^2 - 5x + 4x - 20$ **50.** $3x^2 - 6x + 5x - 10$

51. $4x^2 + 8x - 3x - 6$ **52.** $4x^2 - 8x + 3x - 6$ **53.** $4x^2 + 8x + 3x + 6$

54. $4x^2 - 8x - 3x + 6$ **55.** $8x^2 + 4x - 6x - 3$

3 Factor the trinomials by grouping. *See Example 6.*

56. $3x^2 + 7x + 2$ **57.** $3x^2 + 14x + 8$ **58.** $6x^2 + 13x + 6$ **59.** $8x^2 + 10x + 3$

60. $8x^2 + 26x + 15$ **61.** $12x^2 + 11x + 2$ **62.** $2a^2 + 21a + 19$

See Example 7.

63. $2x^2 - 9x + 10$ **64.** $6x^2 - 13x + 5$ **65.** $8x^2 - 10x + 3$ **66.** $6x^2 - 17x + 12$

67. $24x^2 - 11x + 1$ **68.** $6x^2 - 11x + 5$ **69.** $12x^2 - 11x + 2$

See Example 8.

70. $12x^2 + 5x - 2$ **71.** $24x^2 + 5x - 1$ **72.** $6x^2 + 7xy - 10y^2$ **73.** $7x^2 + x - 8$

74. $12x^2 + 8x - 15$ **75.** $8x^2 + 2x - 3$ **76.** $10x^2 + x - 3$ **77.** $2x^2 - 5x - 7$

78. $15x^2 - 22x - 5$ **79.** $12x^2 - 5x - 2$ **80.** $20x^2 - xy - 12y^2$ **81.** $18x^2 - 3x - 10$

82. $6a^2 - 17ab - 14b^2$

4 Factor the polynomials completely. Identify common factors first, then identify special cases. *See Example 9.*

83. $4x - 4$ **84.** $x^2 + x - 56$ **85.** $2x^2 + x - 3$ **86.** $x^2 - 9$

87. $4x^2 - 16$ **88.** $m^2 + 2m - 15$ **89.** $2a^2 + 6a + 4$ **90.** $b^2 + 6b + 9$

91. $16m^2 - 8m + 1$ **92.** $x^2 - 8x + 7$ **93.** $2m^2 + 5m + 2$ **94.** $2m^2 - 5m - 3$

95. $2a^2 - 3a - 5$ **96.** $3x^2 + 10x - 8$ **97.** $6x^2 + x - 15$ **98.** $8x^2 + 10x - 3$

99. $-2x^2 + 6x - 4$ **100.** $x^4 - 16$ **101.** $x^6 - 81$ **102.** $x^6 - 27$

103. $3x^4 - 48$ **104.** $6x^3 - 48$ **105.** $3x^2 - 6x - 24$ **106.** $64a^9 + b^3$

---

13   CHAPTER REVIEW OF KEY CONCEPTS

LEARNING OUTCOMES	**KEY CONCEPTS AND EXAMPLES**

Section 13–1

1 Factor an expression containing a common factor (pp. 557–559).

1. Find the *greatest* factor common to *each* term of the expression.   **2.** Divide each term by the common factor. Divide mentally if practical.   **3.** Rewrite the expression as the indicated product of the greatest common factor (GCF) and the quotients in Step 2.

Factor $3ab + 9b^2$.

$$3b\left(\frac{3ab}{3b} + \frac{9b^2}{3b}\right) = 3b(a + 3b)$$

Section 13–2

1 Recognize and factor the difference of two perfect squares (pp. 560–561).

Identify a binomial as the difference of two perfect squares: 1. Verify that the expression is a binomial (two terms).   **2.** Verify that the absolute value of each term is a perfect square.   **3.** Verify that the second term is negative. The pattern is $a^2 - b^2$.

Identify which expression is the special product, the difference of two squares.

(a) $36 - 27a^2$; no, 27 is not a perfect square.

(b) $9c^2 - 4$; yes, both terms are perfect squares. The terms are subtracted.

LEARNING OUTCOMES	KEY CONCEPTS AND EXAMPLES

Factor the difference of two perfect squares: 1. Take the square root of the first term. **2.** Take the square root of the absolute value of the second term.   **3.** Write one factor as the *sum* of the square roots found in Steps 1 and 2, and one factor as the *difference* of the square roots from Steps 1 and 2. Symbolically, $a^2 - b^2 = (a + b)(a - b)$.

Factor $9c^2 - 4$.

$$(3c + 2)(3c - 2)$$

2   Recognize and factor a perfect-square trinomial (pp. 561–563).

Identify a perfect-square trinomial: 1. Verify that the expression is a trinomial.   **2.** Verify that the first and last terms are positive and perfect squares.   **3.** Mentally take the square root of the first and last terms and multiply the results. Two times this product should equal the absolute value of the middle term of the original trinomial. The pattern is $a^2 + 2ab + b^2$ or $a^2 - 2ab + b^2$.

Identify which expression is a perfect-square trinomial.

(a) $4x^2 + 18x + 9$; not a perfect-square trinomial. The middle term is not twice the product of the square roots of the first and last terms. $2(2x \cdot 3) = 12x$

(b) $9a^2 + 12a + 4$; the first and last terms are perfect squares. The middle term is twice the product of the square roots of the first and last terms, $2(3a \cdot 2) = 12a$, so it is a perfect-square trinomial.

Factor a perfect-square trinomial: 1. Write the square root of the first term.   **2.** Write the sign of the middle term.   **3.** Write the square root of the last term.   **4.** Indicate the square of this binomial quantity.

Symbolically, $a^2 + 2ab + b^2 = (a + b)^2$ and $a^2 - 2ab + b^2 = (a - b)^2$.

Factor $9a^2 + 12a + 4$.

$$(3a + 2)^2$$

3   Recognize and factor the sum or difference of two perfect cubes (pp. 563–564).

Identify the sum or difference of two perfect cubes: 1. Verify that the expression is a binomial. **2.** Verify that each term is a perfect cube. The pattern is $a^3 + b^3$ or $a^3 - b^3$.

Identify the sum or difference of two perfect cubes:

(a) $b^3 - 27$; yes, both terms are perfect cubes; difference of two cubes.

(b) $x^3 - 6$; no, 6 is not a perfect cube.

(c) $8 + y^3$; yes, both terms are perfect cubes; sum of two cubes.

Factor the sum (difference) of two perfect cubes: 1. Write the binomial factor as the sum (difference) of the cube roots of the two terms.   **2.** Write the trinomial factor as the square of the first term from Step 1, insert a minus sign if factoring a sum (a plus sign if factoring a difference). Write the product of the two terms from Step 1. The third term of the trinomial is the square of the second term from Step 1.

Symbolically, $a^3 + b^3 = (a + b)(a^2 - ab + b^2)$ and $a^3 - b^3 = (a - b)(a^2 + ab + b^2)$.

Factor $8 + y^3$.

$$(2 + y)(4 - 2y + y^2)$$

Factor $b^3 - 27$.

$$(b - 3)(b^2 + 3b + 9)$$

LEARNING OUTCOMES	KEY CONCEPTS AND EXAMPLES

Section 13–3

1 Factor general trinomials whose squared term has a coefficient of 1 (pp. 565–566).

Factor a trinomial with a squared term that has a coefficient of 1: 1. Ensure the trinomial is arranged in descending powers of one variable.   **2.** Determine the signs of the second term of each binomial factor by examining the sign of the *third* term of the trinomial.

$+ \Rightarrow$ like signs (both $+$ or both $-$, matching the middle sign)
$- \Rightarrow$ unlike signs (one $+$ and one $-$, with the sign of the larger absolute value of the factor pair matching the sign of the middle term of the trinomial)

3. Write all factor pairs of the coefficient of the third term.   **4.** Select the factor pair that adds (algebraically) to the coefficient of the middle term of the trinomial. Include appropriate signs.   **5.** Write the binomial factors of the trinomial with the square root of the first-term variable as the first term of each factor and the two factors from Step 4 as the second terms of the binomials.

> Factor $a^2 - 5a + 6$.
>
> $(\quad)(\quad)$ Factors of 6 that add to 5 are 3 and 2.
> $(a \quad 3)(a \quad 2)$ Use terms with like signs that are negative and add to -5.
> $(a - 3)(a - 2)$ These factors of 6, -3 and -2, give $-5a$ as the middle term.

2 Remove common factors after grouping an expression (pp. 566–568).

1. Identify groupings.   **2.** Factor all common factors from each grouping.   **3.** Examine each grouping to see if there are common groupings. If so, factor out the common grouping.

> Factor $2x^2 - 4x + xy - 2y$ completely.
>
> $(2x^2 - 4x) + (xy - 2y)$ Arrange into groups of two terms. Factor common factors.
> $2x(x - 2) + y(x - 2)$ Factor common binomial.
> $(x - 2)(2x + y)$

3 Factor a general trinomial by grouping (pp. 568–572).

1. Arrange terms in descending order and verify that the first term is positive or rewrite with -1 as the common factor.   **2.** Multiply the coefficient of the first term of the trinomial by the coefficient of the last term.   **3.** Factor the product from Step 2 into a pair of factors whose *sum* is the coefficient of the middle term. **(a)** If the sign of the last term is positive, the factors will have like signs, either both positive or both negative, depending on the sign of the middle term. **(b)** If the sign of the last term is negative, the factors will have unlike signs with the factor with the larger absolute value matching the sign of the middle term.

If there is no factor pair that meets one of these conditions, the original trinomial is prime.   **4.** Rewrite the trinomial as a polynomial with four terms by replacing the middle term with two terms that have the coefficients identified in Step 3.   **5.** Group the polynomial with four terms from Step 4 into two groups of two terms.   **6.** Factor the common factors from each of the two groups.   **7.** Factor out the common binomial factor.

> Factor $6x^2 - 17x + 12$.
>
> $6 \cdot 12 = 72$ Factor pairs of 72.
> $1 \cdot 72$ Find the factor pair that has a sum of 17.
> $2 \cdot 36$
> $3 \cdot 24$
> $4 \cdot 18$
> $6 \cdot 12$
> $8 \cdot 9$ Selected pair: $-8 + (-9) = -17$
> $6x^2 - 8x - 9x + 12$ Rewrite the middle term. $-17x = -8x + (-9x)$

LEARNING OUTCOMES	KEY CONCEPTS AND EXAMPLES

$(6x^2 - 8x) + (-9x + 12)$ Factor common factor from each binomial.
$2x(3x - 4) - 3(3x - 4)$ Factor common binomial.
$(3x - 4)(2x - 3)$

4 Factor any binomial or trinomial that is not prime (pp. 570–572).

1. Factor the greatest common factor (if any).   2. Check the binomial or trinomial to see if it is a special product.   3. If there is no special product, factor by grouping or by any appropriate process.   4. Examine each factor to see if it can be factored further.   5. Check factoring by multiplying.

Factor $12x^3 + 6x^2 - 18x$ completely.

$$6x(2x^2 + x - 3) =$$ Factor common factors.
$$6x(2x^2 + 3x - 2x - 3) =$$ Separate x into two terms.
$$6x[(2x^2 + 3x) + (-2x - 3)] =$$ Group and factor common factors from each grouping.
$$6x[x(2x + 3) - 1(2x + 3)] =$$ Factor common grouping.
$$6x(2x + 3)(x - 1)$$

(After removing the common factor, procedures for factoring general trinomials can be used to factor the remaining trinomial.)

13 CHAPTER REVIEW EXERCISES

Section 13–1   MyLab Math   For additional practice go to your study plan in MyLab Math.

Factor by removing the greatest common factor.

1. $5x + 5y$
2. $2x + 5x^2$
3. $12m^2 - 8n^2$
4. $25x^2y - 10xy^3 + 5xy$
5. $2a^3 - 14a^2 - 2a$
6. $30a^3 - 18a^2 - 12a$
7. $15x^3 - 5x^2 - 20x$
8. $9a^2b^3 - 6a^3b^2$
9. $18a^3 - 12a^2$
10. $a^2b + ab^2$
11. $-6x^2 - 10x$
12. $-12n^2 - 18$
13. $5\sqrt{3} + 15\sqrt{7}$
14. $12\sqrt{5} - 21a\sqrt{5}$
15. $\sqrt{12} - 10\sqrt{7}$

Section 13–2

Identify the special products that are the difference of two perfect squares.

16. $25m^2 - 4n^2$
17. $64 + 4a^2$
18. $4f^2 - 9g^2$
19. $H^2 - G^2$
20. $s^2 - 4s + 10$
21. $64b^2 - 49$

Verify which of the following trinomials are perfect-square trinomials.

22. $16c^2 + 8c - 1$
23. $-9x^2 + 12x + 4$
24. $4d^2 + 2d + 16$
25. $9t^2 - 24tp + 16p^2$
26. $a^2 - 14a + 49$
27. $j^2 + 10j + 25$

Identify the special products that are the sum or difference of two perfect cubes.

28. $R^3 + 81$
29. $125a^3 - 8b^3$
30. $64 + m^3$
31. $8z^3 - 125$
32. $d^3 - 12d + 36$
33. $64W^3 + 27$

Factor the expressions that are special products. Explain why the other expressions are not special products.

34. $x^2 - 81$

35. $25y^2 - 4$

36. $100a^2 - 8ab^2$

37. $a^2b^2 + 49$

38. $121 - 9m^2$

39. $a^2 + 2a + 1$

40. $4x^2 + 12x + 9$

41. $16c^2 - 24bc + 9b^2$

42. $y^2 - 12y - 4$

43. $n^2 + 169 - 26n$

44. $16d^2 - 20d + 25$

45. $36a^2 + 84ab + 49b^2$

46. $4x^2 - 25y^2$

47. $49 - 14x + x^2$

48. $9x^2y^2 - 49z^2$

49. $64 + 25x^2$

50. $4x^2 + 12xy + 9y^2$

51. $16x^2 + 24x + 9y^2$

52. $36 - x^2$

53. $49 - 81y^2$

54. $64x^2 - 25y^2$

55. $9x^2 - 100y^2$

56. $a^2 - 10a + 25$

57. $9x^2 - 6xy + y^2$

58. $25 - 16a^2b^2$

59. $9x^2y^2 - z^2$

60. $x^2 - y^2$

61. $x^2 + 4x + 4$

62. $\frac{1}{4}x^2 - \frac{1}{9}y^2$

63. $\frac{4}{25}x^2 - \frac{1}{16}y^2$

64. $27v^3 - 8$

65. $T^3 - 8$

66. $8r^3 + 27$

67. $d^3 + 729$

68. $125c^3 - 216d^3$

69. $27K^3 + 64$

Section 13-3

Factor the trinomials.

70. $x^2 + 10x + 21$

71. $x^2 + 11x + 24$

72. $x^2 + 29x + 28$

73. $x^2 + 13x + 30$

74. $x^2 - 13x + 40$

75. $x^2 - 9x + 8$

76. $x^2 - 17x - 18$

77. $x^2 - 11x - 26$

78. $x^2 + x - 30$

79. $x^2 + 5x - 24$

80. $6x^2 + 25x + 14$

81. $6x^2 + 25x + 4$

82. $4x^2 - 23x + 15$

83. $5x^2 - 34x + 24$

84. $3x^2 - x - 14$

85. $6x^2 - x - 35$

86. $3x^2 + 11x - 4$

87. $7x^2 - 13x - 24$

Factor the polynomials. Look for common factors and special cases.

88. $5mn - 25m$

89. $9a^2 - 100$

90. $a^2 - b^2$

91. $2x^2 - 3x - 2$

92. $b^3 - 8b^2 - b$

93. $a^2 - 81$

94. $x^2 - 14x + 13$

95. $y^2 - 14y + 49$

96. $m^2 - 3m + 2$

97. $b^2 + 8b + 15$

98. $x^2 - 13x + 30$

99. $169 - m^2$

100. $5x^2 + 13x + 6$

101. $x^2 - 4x - 32$

102. $x^2 + 9x + 14$

103. $x^2 + 19x - 20$

104. $x^2 + 8x - 20$

105. $2x^2 - 4x - 16$

106. $5x^2 - 20$

107. $2x^3 - 10x^2 - 12x$

108. $25m^2 - 121n^2$

13 TEAM PROBLEM-SOLVING EXERCISES

1. A standard-sized rectangular swimming pool is 25 ft long and 15 ft wide. This gives a water-surface area of 375 ft². A customer wants to examine some options for varying the size of the pool.

   (a) Write an expression in both factored and expanded form for finding the water-surface area of a pool that is changed by x feet in length and y feet in width.

(b) Write an expression using one variable in both factored and expanded form for the water-surface area of a pool when the length and width increase by the same amount.

2. Use the expressions written in Exercise 1 to answer the following.

   (a) Will the expressions work for both increasing and decreasing the size of the pool?

   (b) Illustrate your answer to part (a) with numerical examples.

13 CONCEPTS ANALYSIS

1. List the properties of a binomial that is the difference of two perfect squares.

2. List the properties of a perfect-square trinomial.

3. Explain how the sign of the third term of a general trinomial affects the signs between the terms of its binomial factors when the trinomial is written in factored form.

4. What do we mean if we say that a polynomial is prime?

5. Write a brief comment explaining each lettered step of the following example.

   Factor $10x^2 - 19x - 12$ using grouping.

   (a)
   $$\frac{120}{\begin{array}{l} 1, 120 \\ 2, 60 \\ 3, 40 \\ 4, 30 \\ 5, 24* \\ 6, 20 \\ 8, 15 \\ 10, 12 \end{array}}$$
       5, 24*   Factors with difference of 19.

   $10x^2 - 19x - 12$

   (b) $10x^2 + 5x - 24x - 12$

   (c) $5x(2x + 1) - 12(2x + 1)$

   (d) $(2x + 1)(5x - 12)$

6. Find a mistake in each of the following. Briefly explain the mistake, then work the problem correctly.

   (a) $(3x - 2y)^2 = 9x^2 + 4y^2$

   (b) $5xy^2 - 45x = 5x(y^2 - 9)$

   (c) $2x^2 - 28x + 96 = (2x - 12)(x - 8)$

   (d) $x^2 - 5x - 6 = (x - 2)(x - 3)$

7. Write the pattern for factoring the difference of two perfect cubes and illustrate it with an example.

8. What is the value of being able to recognize the special products when we factor algebraic expressions?

9. Write the pattern for factoring the difference of two perfect squares and illustrate it with an example.

10. Write the pattern for factoring a perfect-square trinomial and illustrate it with an example.

13 PRACTICE TEST

Factor by removing the greatest common factor.

1. $7x^2 + 8x$

2. $6ax + 15bx$

3. $7a^2b - 14ab$

Factor into special products.

4. $a^2 - 25$

5. $9x^2 - 25$

6. $x^2 + 4x + 4$

7. $x^2 - 18x + 81$

8. $125m^3 - 8n^3$

9. $27r^3 + 64s^3$

Factor into two binomials.

10. $x^2 + 5x - 24$

11. $6x^2 - 5x - 6$

12. $y^2 - y - 6$

13. $a^2 + 16ab + 64b^2$

14. $x^2 - 7x + 10$

15. $b^2 - 3b - 10$

16. $3x^2 + x - 4$

17. $3m^2 - 5m + 2$

18. $3x^2 + 11xy + 3y^2$

Factor completely.

19. $3x^2 - 12$

20. $3x^2 - 3x - 18$

21. $5x^2 - 20$

22. $27a^2 - 48$

23. $3x^2 + 12x + 12$

24. $3x^2 - 15x - 18$

14

Rational Expressions, Equations, and Inequalities

Richard Laschon/Shutterstock

In Great Company

Do Your Work Neatly! (July 22, 1962)

After months of effort by hundreds of people, at a combined cost of $18.5 million, the Jet Propulsion Laboratory, the U.S. Air Force, and NASA collaborated to launch the flagship of its new Mariner program. When the rocket carrying *Mariner 1* lifted off on the bright, clear morning of July 22, 1962, everyone looked forward to its flyby of Venus and all the scientific knowledge that would be gained. What they got, 4 minutes and 25 seconds after lift-off, was a missile detonation and a rather short fireworks show. What happened?

579

Well, accounts aren't entirely clear, but what seems to have happened was a typing error. The persons entering the equations for the guidance control system appear to have copied the equation incorrectly. To be specific, they missed a single overbar in a single equation. In congressional testimony, the NASA officials referred to the overbar as a hyphen, on the theory that the congressman wouldn't understand what an overbar was—which was probably true.

The overbar represented the need to smooth the value of the time derivative when calculating velocity. Because rocket fuel burns a little unevenly, the guidance computer was supposed to ignore minor variations in velocity. But since the smoothing function represented by the smoothing bar wasn't in the software, the guidance computer didn't know that. It treated those normal minor fuel-burn variations as if they were major problems. It kept "correcting" course, even though there was nothing wrong with the course.

Later investigation discovered this error had been in previous "successful" versions of the software. But that portion of the software executed only if radio guidance was lost, which had never happened before, so it had not been noticed earlier.

It was noticed this time. This time, ground communications failed. The code executed. The rocket nosed down hard left. The ground control crew had only seconds to decide what to do. They triggered self-destruct 6 s before they would have lost complete control of the missile. Sigh.

When working a math problem, it is really important to make sure you copy everything correctly from step to step. Dropping a single coefficient, a single exponent, or a single smoothing bar, can destroy all the hard work you did up to that point.

Now, this kind of error is going to happen. Everybody does it. There's no way to stop it. When it happens, be happy you lose only a few points on an assignment. It sure beats losing nearly $19 million and a satellite.

14–1   Simplifying Rational Expressions

LEARNING OUTCOME

1   Simplify or reduce rational expressions.

LC LEARNING CATALYTICS
1. Write $x^2 - 16$ in factored form.
2. Which pair of factors is correct for the factored form of $x^2 - 5x - 36$: $(x - 4)(x + 9)$ or $(x + 4)(x - 9)$?

STOP AND CHECK

Simplify or reduce the fractions. Write answers with positive exponents only.

1. $\dfrac{15}{35}$

2. $\dfrac{3xy^2z}{9x^2yz}$

3. $\dfrac{12a^3b}{4ab}$

Answers:

1. $\dfrac{3}{7}$ 2. $\dfrac{y}{3x}$ 3. $3a^2$

Once we become proficient in factoring algebraic expressions, we can use this skill to simplify *rational expressions,* also called *algebraic fractions.*

1 Simplify or Reduce Rational Expressions. In arithmetic, we learned to reduce fractions by reducing or dividing *factors* that are common to both the numerator and denominator. In our study of the laws of exponents, we also *reduced* or *simplified* algebraic fractions by reducing common factors. This process is accomplished whenever the numerator and denominator of the fraction are written in factored form. In the following example, we review the arithmetic process with fractions and the division laws of exponents. To focus on key aspects of the process, let's write steps that we often do mentally.

EXAMPLE 1

Simplify or reduce the following fractions. Write answers with positive exponents only.

(a) $\dfrac{12}{15}$ (b) $\dfrac{24}{36}$ (c) $\dfrac{2x^2yz^3}{4xy^2z}$ (d) $\dfrac{9a^2b^3}{3ab}$

(a) $\dfrac{12}{15} = \dfrac{2(2)(3)}{(3)(5)} =$ Write in factored form using prime factorization.

$\dfrac{2(2)(\cancel{3})}{\cancel{3}(5)} = \dfrac{4}{5}$ Reduce the common factors and multiply the remaining factors.

(b) $\dfrac{24}{36} = \dfrac{2(2)(2)(3)}{2(2)(3)(3)} =$ Write in factored form using prime factorization.

$\dfrac{\cancel{2}(\cancel{2})(2)(\cancel{3})}{\cancel{2}(\cancel{2})(\cancel{3})(3)} = \dfrac{2}{3}$ Reduce the common factors.

(c) $\dfrac{2x^2yz^3}{4xy^2z} = \dfrac{\cancel{2}x^{2-1}y^{1-2}z^{3-1}}{\cancel{2}(2)} =$ Write coefficients in factored form and reduce. Apply the laws of exponents to variable factors with like bases.

$\dfrac{xy^{-1}z^2}{2} = \dfrac{xz^2}{2y}$ Express all variable factors with positive exponents.

(d) $\dfrac{9a^2b^3}{3ab} = \dfrac{\cancel{3}(3)a^{2-1}b^{3-1}}{\cancel{3}}$ Write coefficients in factored form and reduce. Apply the laws of exponents to variable factors with like bases.

$= 3ab^2$ **See Exercises 1–4.**

Before continuing, we must understand that we are always reducing common *factors*. This procedure *does not* apply to *addends* or *terms*.

TIP **Can Common Addends Be Reduced?** **Does** $\frac{5}{10}$ **reduce to** $\frac{3}{8}$**?** We know from previous experience that $\frac{5}{10}$ reduces to $\frac{1}{2}$, or 0.5. What is wrong with the following argument?

Both the numerator and the denominator of the fraction $\frac{5}{10}$ can be rewritten as addends or terms:

Does $\dfrac{5}{10} = \dfrac{\cancel{2}+3}{\cancel{2}+8} = \dfrac{3}{8}$? Can common addends be reduced? No!

We can verify with a calculator that this is an incorrect statement. Common *addends* cannot be reduced. To correct the process, we reduce *only factors*. Thus, we rewrite $\frac{5}{10}$ in factored form:

$$\frac{5}{10} = \frac{1(\cancel{5})}{2(\cancel{5})} = \frac{1}{2}$$

If common factors are reduced,

$$\frac{5}{10} = \frac{1}{2}$$

Rational expression: an algebraic fraction in which the numerator or denominator or both are polynomials

STOP AND CHECK

Simplify the rational expressions that are already in factored form.

1. $\dfrac{5(2x - 3)}{(x + 3)(2x - 3)}$

2. $\dfrac{5xy(x - 7)}{2x^2(x - 2)}$

3. $\dfrac{(x - 2)(2x + 3)}{(x + 2)(2x - 3)}$

Answers:

1. $\dfrac{5}{x + 3}$

2. $\dfrac{5y(x - 7)}{2x(x - 2)}$ or $\dfrac{5xy - 35y}{2x^2 - 4x}$

3. $\dfrac{(x - 2)(2x + 3)}{(x + 2)(2x - 3)}$ or $\dfrac{2x^2 - x - 6}{2x^2 + x - 6}$

Now let's use our knowledge of factoring *polynomials* to simplify rational expressions. A **rational expression** is an algebraic fraction in which the numerator or denominator or both are polynomials.

First, let's look at some examples that are written in factored form.

EXAMPLE 2

Simplify the rational expressions that are already in factored form.

(a) $\dfrac{x(x + 2)}{(x + 2)(x + 3)}$ **(b)** $\dfrac{3ab(a + 4)}{6ab(a + 3)}$

(c) $\dfrac{2x^2(2x - 1)}{x^3(x - 1)}$ **(d)** $\dfrac{(x + 3)(x - 4)}{(x - 3)(x + 4)}$

(a) $\dfrac{x(x+2)}{(x+2)(x+3)} = \dfrac{x(\cancel{x+2})}{(\cancel{x+2})(x+3)}$ Reduce common binomial factors.

$\qquad\qquad\qquad = \dfrac{x}{x+3}$

The remaining x and $x+3$ cannot be reduced. The x in the numerator is a factor ($x = 1 \cdot x$); however, the x in the denominator is an *addend* or *term*.

(b) $\dfrac{3ab(a+4)}{6ab(a+3)} = \dfrac{\overset{1}{\cancel{3ab}}(a+4)}{\underset{2}{\cancel{6ab}}(a+3)} =$ Reduce common factors:
$1(a+4) = a+4$

$\qquad \dfrac{a+4}{2(a+3)}$ or $\dfrac{a+4}{2a+6}$ Reduced rational expressions can be written in *either* factored or expanded form.

(c) $\dfrac{2x^2(2x-1)}{x^3(x-1)} =$ Reduce common factor:
$\dfrac{x^2}{x^3} = x^{2-3} = x^{-1} = \dfrac{1}{x}$

$\qquad \dfrac{2(2x-1)}{x(x-1)}$ or $\dfrac{4x-2}{x^2-x}$ Reduced rational expressions can be written in *either* factored or expanded form.

(d) $\dfrac{(x+3)(x-4)}{(x-3)(x+4)}$ or $\dfrac{x^2-x-12}{x^2+x-12}$ There are no common factors; therefore, the fraction is already in lowest terms. Reduced rational expressions can be written in *either* factored or expanded form.   **See Exercises 5–8.**

To simplify rational expressions:

1. Factor *completely* both the numerator and denominator.

2. Reduce factors common to both the numerator and denominator.

3. Write the simplified expression in either factored or expanded form.

STOP AND CHECK

Reduce each rational expression to its simplest form.

1. $\dfrac{x-y}{3x^2-3xy}$

2. $\dfrac{4x^2+4y^2}{4x^2-4y^2}$

3. $\dfrac{x^2-8x+16}{x^2-16}$

4. $\dfrac{2x-5}{5-2x}$

Answers:

1. $\dfrac{1}{3x}$ 2. $\dfrac{x^2+y^2}{x^2-y^2}$ or
$\dfrac{x^2+y^2}{(x+y)(x-y)}$ 3. $\dfrac{x-4}{x+4}$ 4. -1

EXAMPLE 3

Reduce each rational expression to its simplest form.

(a) $\dfrac{x+y}{4x^2+4xy}$ (b) $\dfrac{a^2+b^2}{a^2-b^2}$ (c) $\dfrac{x^2-6x+9}{x^2-9}$

(d) $\dfrac{3x^2-12}{6x+12}$ (e) $\dfrac{a-b}{b-a}$

(a) $\dfrac{x+y}{4x^2+4xy} = \dfrac{(x+y)}{4x(x+y)} =$ Factor the common factor in the denominator and recall that the numerator is a grouping. Reduce common binomial factor.

$\qquad \dfrac{(\cancel{x+y})}{4x(\cancel{x+y})} = \dfrac{1}{4x}$ Numerator of 1 is necessary: $x+y = 1(x+y)$

(b) $\dfrac{a^2+b^2}{a^2-b^2} = \dfrac{a^2+b^2}{(a+b)(a-b)}$ Factor the difference of the squares in the denominator.

The numerator, which is the *sum* of the squares, will not factor. There are no factors common to both the numerator and denominator. Thus, the fraction is in simplest form:

$$\frac{a^2 + b^2}{(a + b)(a - b)} \quad \text{or} \quad \frac{a^2 + b^2}{a^2 - b^2}$$

(c) $\dfrac{x^2 - 6x + 9}{x^2 - 9} = \dfrac{(x - 3)(x - 3)}{(x + 3)(x - 3)}$ Write both the numerator and the denominator in factored form.

$= \dfrac{(x - 3)(x \!\!\!\!\diagdown\!\!\!\! 3)}{(x + 3)(x \!\!\!\!\diagdown\!\!\!\! 3)} = \dfrac{x - 3}{x + 3}$ Reduce common binomial factors.

(d) $\dfrac{3x^2 - 12}{6x + 12} = \dfrac{3(x^2 - 4)}{6(x + 2)} = \dfrac{3(x + 2)(x - 2)}{3(2)(x + 2)}$ Factor the numerator and the denominator completely.

$= \dfrac{\cancel{3}(\cancel{x + 2})(x - 2)}{\cancel{3}(2)(\cancel{x + 2})} = \dfrac{x - 2}{2}$ Reduce common numerical and binomial factors.

(e) $\dfrac{a - b}{b - a} =$ Write terms in numerator and denominator in same order.

$\dfrac{a - b}{-a + b} =$ Factor the denominator so that the leading coefficient of the binomial is positive.

$\dfrac{a \!\!\!\!\diagdown\!\!\!\! b}{-1(a \!\!\!\!\diagdown\!\!\!\! b)} =$ Reduce the common binomial factor.

$\dfrac{1}{-1} = -1$

See Exercises 9–24.

TIP **Don't Forget 1 or −1** In the preceding example, part (e), we have some very important observations to make.

▶ When all other factors of a numerator or denominator reduce, a factor of 1 remains.

▶ Polynomial factors are opposites when every term of one grouping has an opposite in the other grouping.

$(a - b)$ and $(b - a)$ are opposites.   From part (e)
$(2m - 5)$ and $(5 - 2m)$ are opposites.

▶ Opposites differ by a factor of −1.

$b - a = -a + b = -1(a - b)$ From part (e)
$5 - 2m = -2m + 5 = -1(2m - 5)$

▶ When a numerator and denominator are opposites, the fraction reduces to −1.

$$\frac{a - b}{b - a} = \frac{1(a - b)}{-1(a - b)} = -1 \qquad \frac{2m - 5}{5 - 2m} = \frac{1(2m - 5)}{-1(2m - 5)} = -1$$

▶ It is helpful in recognizing common factors to rearrange so that like terms are in the same order in the factors and factor with −1 as a common factor so that leading coefficients within a grouping are positive: $(5 - x) = (-x + 5) = -1(x - 5)$.

▶ It is helpful in recognizing common factors for the terms in both the numerator and denominator to be arranged in the same order.

$$\frac{x + 5}{5 + x} = \frac{x + 5}{x + 5} \qquad \frac{x - 2}{2 - x} = \frac{x - 2}{-x + 2} = \frac{x - 2}{-1(x - 2)}$$

14-1 EXERCISES

MyLab Math For additional practice go to your study plan in MyLab Math.

1 Simplify the rational expressions. *See Exercise 1.*

1. $\dfrac{8}{18}$

2. $\dfrac{9}{24}$

3. $\dfrac{4a^2b^3}{2ab}$

4. $\dfrac{27a^3bc^2}{18a^2b^4c^2}$

See Example 2.

5. $\dfrac{x(x+3)}{(x+3)(x+2)}$

6. $\dfrac{3x^2(3x+2)}{x^3(2x-1)}$

7. $\dfrac{(x+2)(x-5)}{(x+5)(x-2)}$

8. $\dfrac{x+y}{2x+2y}$

See Example 3.

9. $\dfrac{a+b}{a^2-b^2}$

10. $\dfrac{x^2-4x+4}{x^2-4}$

11. $\dfrac{4x^2-16}{6x+12}$

12. $\dfrac{3m-11}{11-3m}$

13. $\dfrac{x^2-x-6}{3-x}$

14. $\dfrac{b-a}{2a-2b}$

15. $\dfrac{3x-6}{2-x}$

16. $\dfrac{2x-3y}{6y-4x}$

17. $\dfrac{5m-10n}{2n-m}$

18. $\dfrac{2x-6}{-x^2+5x-6}$

19. $\dfrac{x^2-3x-10}{-x^2-3x-2}$

20. $\dfrac{-x^2-9x-14}{x^2+9x+14}$

21. $\dfrac{x^2-4x-21}{x^2-5x-14}$

22. $\dfrac{2x^2-7x+3}{2x^2+5x-3}$

23. $\dfrac{3x^2-4x-7}{3x^2+5x-28}$

24. $\dfrac{6x^2+x-2}{9x^2-4}$

14-2   Multiplying and Dividing Rational Expressions

LEARNING OUTCOMES

1 Multiply and divide rational expressions.

2 Use multiplication and conjugates to rationalize a numerator or denominator of a fraction that has a binomial with an irrational term.

LC **LEARNING CATALYTICS**

1. Multiply and reduce to lowest terms
$\dfrac{10}{21} \cdot \dfrac{14}{15}$.

2. Divide and reduce to lowest terms
$\dfrac{8}{15} \div 3$.

STOP AND CHECK

Multiply or divide as indicated. Express answers in simplest form.

1. $\dfrac{9}{16} \div 3$

2. $\dfrac{12a^2b}{ab^2} \cdot \dfrac{(ab)^2}{4a}$

Answers:

1. $\dfrac{3}{16}$ **2.** $3a^2b$

1 **Multiply and Divide Rational Expressions.** We can connect our knowledge of arithmetic of multiplying and dividing fractions and the laws of exponents to expand our mathematical experience. We express results in lowest terms or in simplified form.

EXAMPLE 1

Multiply or divide as indicated. Express answers in simplest form.

(a) $\dfrac{5}{8} \cdot \dfrac{16}{25}$ **(b)** $\dfrac{7}{16} \div 2$ **(c)** $\dfrac{15xy^2}{xy^3} \cdot \dfrac{(xy)^3}{5x}$

(a) $\dfrac{5}{8} \cdot \dfrac{16}{25} = \dfrac{\overset{1}{\cancel{5}}}{\underset{1}{\cancel{8}}} \cdot \dfrac{\overset{2}{\cancel{16}}}{\underset{5}{\cancel{25}}}$ Reduce factors common to a numerator and denominator. Multiply remaining factors.

$= \dfrac{2}{5}$

If all factors that are common to a numerator and denominator are reduced or canceled before multiplying, the product will be in lowest terms.

(b) $\dfrac{7}{16} \div 2$ Rewrite 2 as a fraction.

$\dfrac{7}{16} \div 2 = \dfrac{7}{16} \div \dfrac{2}{1}$ Rewrite division as multiplication and multiply.

$= \dfrac{7}{16} \cdot \dfrac{\overset{1}{\cancel{1}}}{\underset{2}{\cancel{2}}} = \dfrac{7}{32}$

(c) $\dfrac{15xy^2}{xy^3} \cdot \dfrac{(xy)^3}{5x}$ Raise grouping to power.

$= \dfrac{15xy^2}{xy^3} \cdot \dfrac{x^3y^3}{5x}$ Apply the laws of exponents.

$= \dfrac{15x^4y^5}{5x^2y^3}$ Reduce. (Reducing could have occurred before multiplication.)

$= \dfrac{15(x^{4-2})(y^{5-3})}{5}$

$= 3x^2y^2$ See Exercises 1–4.

When applying the process of multiplying or dividing fractions to rational expressions, the key fact to remember is that the numerators and denominators should be written in *factored* form whenever possible.

> **To multiply or divide rational expressions:**
>
> 1. Convert any division to an equivalent multiplication.
> 2. Factor completely every numerator and denominator.
> 3. Reduce factors that are common to a numerator and denominator.
> 4. Multiply remaining factors.
> 5. The result can be written in factored or expanded form.

STOP AND CHECK

Perform the operation and simplify.

1. $\dfrac{x^2 + 3x - 10}{5x - 10} \cdot \dfrac{x - 5}{x^2 - 10x + 25}$

2. $\dfrac{3x + 6}{3 - x} \div \dfrac{x^2 + 5x + 6}{x^2 - 9}$

Answers:

1. $\dfrac{x + 5}{5(x - 5)}$ or $\dfrac{x + 5}{5x - 25}$

2. -3

EXAMPLE 2

Perform the operation and simplify.

(a) $\dfrac{x^2 - 4x - 12}{2x - 12} \cdot \dfrac{x - 4}{x^2 + 4x + 4}$

(b) $\dfrac{4y^2 - 9}{2y^2} \div \dfrac{y^2 - 2y - 15}{4y^2 + 12y}$

(c) $\dfrac{2x - 6}{1 - x} \div \dfrac{x^2 - 2x - 3}{x^2 - 1}$

(a) $\dfrac{x^2 - 4x - 12}{2x - 12} \cdot \dfrac{x - 4}{x^2 + 4x + 4}$ Factor each numerator and denominator.

$= \dfrac{(x + 2)(x - 6)}{2(x - 6)} \cdot \dfrac{x - 4}{(x + 2)(x + 2)}$ Reduce factors common to a numerator and denominator.

$= \dfrac{(x + 2)(x - 6)}{2(x - 6)} \cdot \dfrac{x - 4}{(x + 2)(x + 2)}$ Multiply remaining factors.

$= \dfrac{x - 4}{2(x + 2)}$ or $\dfrac{x - 4}{2x + 4}$ Write in factored or expanded form.

(b) $\dfrac{4y^2 - 9}{2y^2} \div \dfrac{y^2 - 2y - 15}{4y^2 + 12y}$ Convert to multiplication.

$= \dfrac{4y^2 - 9}{2y^2} \cdot \dfrac{4y^2 + 12y}{y^2 - 2y - 15}$ Factor each numerator and denominator.

$$= \frac{(2y + 3)(2y - 3)}{2y^2} \cdot \frac{4y(y + 3)}{(y + 3)(y - 5)}$$

Reduce factors common to a numerator and denominator.

$$= \frac{(2y + 3)(2y - 3)}{2\overset{}{\underset{y}{y^2}}} \cdot \frac{\overset{2}{4y}(y + 3)}{(y + 3)(y - 5)}$$

Multiply remaining factors.

$$= \frac{2(2y + 3)(2y - 3)}{y(y - 5)} \quad \text{or} \quad \frac{2(4y^2 - 9)}{y^2 - 5y} \quad \text{or} \quad \frac{8y^2 - 18}{y^2 - 5y} \quad \text{or} \quad \frac{2(4y^2 - 9)}{y(y - 5)}$$

(c) $\dfrac{2x - 6}{1 - x} \div \dfrac{x^2 - 2x - 3}{x^2 - 1}$

Convert to multiplication.

$$= \frac{2x - 6}{1 - x} \cdot \frac{x^2 - 1}{x^2 - 2x - 3}$$

Factor each numerator and denominator. Note the effect of factoring -1 in the denominator of the first fraction so the $x - 1$ will reduce with $x - 1$ in the numerator of the second fraction.

$$= \frac{2(x - 3)}{-1(x - 1)} \cdot \frac{(x + 1)(x - 1)}{(x + 1)(x - 3)}$$

Reduce factors common to a numerator and denominator.

$$= \frac{2(x - 3)}{-1(x - 1)} \cdot \frac{(x + 1)(x - 1)}{(x + 1)(x - 3)}$$

Multiply remaining factors.

$$= \frac{2}{-1} = -2$$

See Exercises 5–21.

Complex rational expression: a rational expression that has a rational expression in its numerator or its denominator or both

A **complex rational expression** is a rational expression that has a rational expression in its numerator or its denominator or both. Some examples of complex rational expressions are

$$\frac{\dfrac{4xy}{2x}}{5} \qquad \frac{\dfrac{x^2 - y^2}{2x}}{\dfrac{x - y}{3x^2}} \qquad \frac{1 + \dfrac{1}{x}}{\dfrac{2}{3}} \qquad \frac{\dfrac{2}{x} - \dfrac{5}{2x}}{\dfrac{3}{4x} + \dfrac{3}{x}}$$

To simplify a complex rational expression:

1. Rewrite the complex rational expression as a division then convert to an equivalent multiplication of rational expressions.

2. Multiply the rational expressions.

Before we look at examples involving complex rational expressions, the following Tip will eliminate some of our written steps.

TIP **Mentally Converting from Complex Form to Division and Then to Multiplication** Examine the symbolic representation for converting a complex expression to multiplication. The numerator is multiplied by the reciprocal of the denominator.

$$\frac{\dfrac{a}{b}}{\dfrac{c}{d}} = \frac{a}{b} \div \frac{c}{d} = \frac{a}{b} \cdot \frac{d}{c} \qquad b, c, \text{ and } d \neq 0$$

When simplifying a complex rational expression, we can eliminate some written steps by making this conversion mentally; for example,

$$\dfrac{\dfrac{4xy}{2x}}{5} \qquad \text{becomes} \qquad \dfrac{4xy}{1} \cdot \dfrac{5}{2x}$$

STOP AND CHECK

Simplify.

1. $\dfrac{\dfrac{6ab}{3a}}{2}$

2. $\dfrac{\dfrac{x^2 + 2xy + y^2}{3x^2}}{\dfrac{x + y}{6x}}$

Answers:
1. $4b$
2. $\dfrac{2(x + y)}{x}$ or $\dfrac{2x + 2y}{x}$

EXAMPLE 3

Simplify.

(a) $\dfrac{\dfrac{4xy}{2x}}{5}$ (b) $\dfrac{\dfrac{x^2 - y^2}{2x}}{\dfrac{x - y}{3x^2}}$

(a) $\dfrac{\dfrac{4xy}{2x}}{5} = \dfrac{4xy}{1} \cdot \dfrac{5}{2x} =$ Multiply the numerator by the reciprocal of the denominator.

$\dfrac{\overset{2}{4}xy}{1} \cdot \dfrac{5}{\cancel{2x}} =$ Reduce. Multiply remaining factors.

$\dfrac{10y}{1} = \mathbf{10y}$ Simplify.

(b) $\dfrac{\dfrac{x^2 - y^2}{2x}}{\dfrac{x - y}{3x^2}} = \dfrac{x^2 - y^2}{2x} \cdot \dfrac{3x^2}{x - y} =$ Multiply the numerator by the reciprocal of the denominator. Factor the numerator of the first fraction.

$\dfrac{(x + y)(x - y)}{2x} \cdot \dfrac{3x^2}{x - y} =$ Reduce.

$\dfrac{(x + y)(x - y)}{2\cancel{x}} \cdot \dfrac{3\overset{x}{\cancel{x^2}}}{x - y} =$ Multiply remaining factors.

$= \dfrac{3x(x + y)}{2}$ or $\dfrac{3x^2 + 3xy}{2}$ Write in factored or expanded form.

See Exercises 22–31.

2 **Use Multiplication and Conjugates to Rationalize a Numerator or Denominator of a Fraction That Has a Binomial with an Irrational Term.** Our understanding of fractions and rational expressions can be used to manipulate fractional expressions that have irrational terms. In an advanced study of mathematics, there are occasions when you may want a numerator or denominator clear of irrational terms.

To use multiplication and conjugates to rationalize a numerator or denominator of a fraction that has a binomial with an irrational term:

1. Determine which part of the fraction (numerator or denominator) is to be cleared of a binomial with an irrational term.

2. Multiply by 1 in the form of $\dfrac{n}{n}$, where n is the conjugate of the binomial with the irrational term.

3. Simplify the resulting expression.

STOP AND CHECK

1. Rationalize the denominator of the fraction $\dfrac{2 + 3\sqrt{2}}{3 - 2\sqrt{3}}$.

Answer:

1. $\dfrac{6 + 4\sqrt{3} + 9\sqrt{2} + 6\sqrt{6}}{-3}$ or

$-\dfrac{6 + 4\sqrt{3} + 9\sqrt{2} + 6\sqrt{6}}{3}$

EXAMPLE 4

Rationalize the denominator of the fraction $\dfrac{1 + 2\sqrt{3}}{2 - 5\sqrt{2}}$.

$\dfrac{1 + 2\sqrt{3}}{2 - 5\sqrt{2}}$

The conjugate of $2 - 5\sqrt{2}$ is $2 + 5\sqrt{2}$.

Multiply by 1 in the form of $\dfrac{2 + 5\sqrt{2}}{2 + 5\sqrt{2}}$.

$\dfrac{1 + 2\sqrt{3}}{2 - 5\sqrt{2}} \cdot \dfrac{2 + 5\sqrt{2}}{2 + 5\sqrt{2}}$

FOIL the numerators and use the special product for the sum and difference of two terms for the denominator.

$\dfrac{2 + 1(5\sqrt{2}) + 2(2\sqrt{3}) + 2\sqrt{3}(5\sqrt{2})}{2^2 - (5\sqrt{2})^2}$

Simplify by multiplying coefficients.

$\dfrac{2 + 5\sqrt{2} + 4\sqrt{3} + 10\sqrt{6}}{4 - 25(2)} =$

Simplify the denominator.

$\dfrac{2 + 5\sqrt{2} + 4\sqrt{3} + 10\sqrt{6}}{4 - 50} =$

Combine like terms in the denominator.

$\dfrac{2 + 5\sqrt{2} + 4\sqrt{3} + 10\sqrt{6}}{-46}$

Manipulate the signs of the fraction to make the denominator positive.

or

$-\dfrac{2 + 5\sqrt{2} + 4\sqrt{3} + 10\sqrt{6}}{46}$

See Exercises 32–37.

STOP AND CHECK

1. Rationalize the numerator of the fraction $\dfrac{2 + 3\sqrt{2}}{3 - 2\sqrt{3}}$.

Answer:

1. $\dfrac{-14}{6 - 9\sqrt{2} - 4\sqrt{3} + 6\sqrt{6}}$ or

$-\dfrac{14}{6 - 9\sqrt{2} - 4\sqrt{3} + 6\sqrt{6}}$

EXAMPLE 5

Rationalize the numerator of the fraction $\dfrac{1 + 2\sqrt{3}}{2 - 5\sqrt{2}}$.

$\dfrac{1 + 2\sqrt{3}}{2 - 5\sqrt{2}} =$

The conjugate of $1 + 2\sqrt{3}$ is $1 - 2\sqrt{3}$. Multiply by 1 in the form of $\dfrac{1 - 2\sqrt{3}}{1 - 2\sqrt{3}}$.

$\dfrac{1 + 2\sqrt{3}}{2 - 5\sqrt{2}} \cdot \dfrac{1 - 2\sqrt{3}}{1 - 2\sqrt{3}} =$

Use the special product for the sum and difference of two terms for the numerator and FOIL the denominator.

$\dfrac{1^2 - (2\sqrt{3})^2}{2 + 2(-2\sqrt{3}) + 1(-5\sqrt{2}) - 5\sqrt{2}(-2\sqrt{3})} =$

Simplify.

$\dfrac{1 - 4(3)}{2 - 4\sqrt{3} - 5\sqrt{2} + 10\sqrt{6}} =$

$\dfrac{-11}{2 - 4\sqrt{3} - 5\sqrt{2} + 10\sqrt{6}}$ or $-\dfrac{11}{2 - 4\sqrt{3} - 5\sqrt{2} + 10\sqrt{6}}$

See Exercises 38–43.

TIP **Does Rationalizing a Numerator or Denominator of a Fraction Make the Expression Easier to Evaluate?** Probably not. Rationalizing is done whenever you need to manipulate a fraction so that either the numerator or denominator has only rational numbers. In more advanced studies of mathematics, there are specific situations when it is desirable to perform this manipulation.

14–2 EXERCISES

MyLab Math For additional practice go to your study plan in MyLab Math.

1 Multiply or divide the fractions. Reduce to simplest form. *See Example 1.*

1. $\dfrac{7}{12} \cdot \dfrac{18}{21}$

2. $\dfrac{3}{8} \cdot \dfrac{4}{15}$

3. $\dfrac{5x^2}{3y} \cdot \dfrac{2x}{3y}$

4. $\dfrac{8a^2}{15ab} \cdot \dfrac{21ab}{24a^2}$

See Example 2.

5. $\dfrac{13r^2}{20a^2} \div \dfrac{39r^2}{5a}$

6. $\dfrac{4}{5} \div \dfrac{8}{15}$

7. $\dfrac{3}{8} \div \dfrac{9}{16}$

8. $\dfrac{5x^2y}{3y} \div \dfrac{10x^3}{9y^3}$

9. $\dfrac{7a^2b}{15b} \div \dfrac{14a}{9b}$

10. $\dfrac{(x+2)(x-7)}{(x+1)(x-2)} \cdot \dfrac{3(x+1)}{(x-7)(x+2)}$

11. $\dfrac{-4(x-6)}{(x-3)(x+5)} \cdot \dfrac{(3-x)(x+5)}{(x-6)(x-2)}$

12. $\dfrac{a^2-b^2}{4} \cdot \dfrac{12}{a+b}$

13. $\dfrac{2u+2v}{5} \cdot \dfrac{10}{u+v}$

14. $\dfrac{a^2-49}{b^2-25} \cdot \dfrac{b-5}{a+7}$

15. $\dfrac{x^2+2x+1}{5x-5} \cdot \dfrac{15}{x+1}$

16. $\dfrac{a-b}{4} \div \dfrac{a-b}{2}$

17. $\dfrac{b-a}{7} \div \dfrac{a-b}{14}$

18. $\dfrac{2x-y}{x+y} \div \dfrac{y-2x}{-x-y}$

19. $\dfrac{x}{x^2-4x+4} \div \dfrac{1}{x-2}$

20. $\dfrac{5a^2-5b^2}{a^2b^2} \div \dfrac{a+b}{10ab}$

21. $\dfrac{x^2+4x+3}{x^2-4x-5} \div \dfrac{x+3}{x-5}$

Simplify. *See Example 3.*

22. $\dfrac{\frac{5}{9}}{\frac{3}{5}}$

23. $\dfrac{\frac{7}{8}}{\frac{5}{6}}$

24. $\dfrac{\frac{x-5}{4}}{3x}$

25. $\dfrac{\frac{6}{x-2}}{9x}$

26. $\dfrac{\frac{4x}{8x}}{x+3}$

27. $\dfrac{\frac{7x}{21x}}{x-3}$

28. $\dfrac{\frac{x^2-y^2}{3x+y}}{\frac{y-x}{-3x-y}}$

29. $\dfrac{\frac{x^2-5x+4}{x^2+4x+3}}{\frac{4-x}{x+3}}$

30. $\dfrac{\frac{x^2-36}{5x}}{\frac{x+6}{15x}}$

31. $\dfrac{\frac{x^2-5x}{8}}{\frac{2x-10}{12x}}$

2 Rationalize the denominator of each fraction. *See Example 4.*

32. $\dfrac{3}{2-\sqrt{5}}$

33. $\dfrac{13}{5-2\sqrt{3}}$

34. $\dfrac{5-\sqrt{2}}{4+\sqrt{2}}$

35. $\dfrac{7+\sqrt{7}}{5+\sqrt{7}}$

36. $\dfrac{2+3\sqrt{5}}{7-2\sqrt{3}}$

37. $\dfrac{5-\sqrt{6}}{5-2\sqrt{6}}$

Rationalize the numerator of each fraction. *See Example 5.*

38. $\dfrac{2-\sqrt{5}}{7}$

39. $\dfrac{5-\sqrt{11}}{15}$

40. $\dfrac{4+\sqrt{6}}{20}$

41. $\dfrac{3+\sqrt{2}}{7}$

42. $\dfrac{4+3\sqrt{2}}{7}$

43. $\dfrac{5+2\sqrt{3}}{8}$

14-3 Adding and Subtracting Rational Expressions

1 Add and subtract rational expressions.

2 Use addition and subtraction of rational expressions to simplify complex fractions.

LC LEARNING CATALYTICS

1. Add $\dfrac{5}{16} + \dfrac{1}{16}$ and reduce to lowest terms.

2. Subtract $\dfrac{4}{5} - \dfrac{2}{3}$.

STOP AND CHECK

Add or subtract as indicated. Write answers in lowest terms.

1. $\dfrac{5}{9} + \dfrac{2}{3}$

2. $6 - \dfrac{4}{5}$

3. $\dfrac{3}{x} + \dfrac{2}{3x} + \dfrac{5}{6}$

Answers:

1. $\dfrac{11}{9}$ or $1\dfrac{2}{9}$ 2. $\dfrac{26}{5}$ or $5\dfrac{1}{5}$

3. $\dfrac{22 + 5x}{6x}$

1 Add and Subtract Rational Expressions. In adding and subtracting fractions, we can add or subtract only fractions with like denominators. Whenever we have unlike fractions, we must first find a common denominator for the fractions. We then convert each fraction to an equivalent fraction with the common denominator and add or subtract the numerators. Let's refresh our memory of the procedure for adding and subtracting fractions.

EXAMPLE 1

Add or subtract as indicated. Write answers in lowest terms.

(a) $\dfrac{3}{8} + \dfrac{1}{8}$ **(b)** $\dfrac{7}{12} - \dfrac{1}{3}$ **(c)** $5 - \dfrac{2}{3}$ **(d)** $\dfrac{5}{2x} + \dfrac{3}{x} + \dfrac{9}{4}$

(a) $\dfrac{3}{8} + \dfrac{1}{8} = \dfrac{4}{8} = \dfrac{1}{2}$ Add the numerators and *keep* the *like* denominator. Reduce.

(b) $\dfrac{7}{12} - \dfrac{1}{3} =$ Select a common denominator and change to equivalent fractions with the common denominator of 12:

$\dfrac{1}{3} = \dfrac{1}{3}\left(\dfrac{4}{4}\right) = \dfrac{4}{12}$

$\dfrac{7}{12} - \dfrac{4}{12} = \dfrac{3}{12} = \dfrac{1}{4}$ Subtract and reduce.

(c) $5 - \dfrac{2}{3} =$ Convert 5 to a fraction with a denominator of 3:

$\dfrac{5}{1}\left(\dfrac{3}{3}\right) = \dfrac{15}{3}$

$\dfrac{15}{3} - \dfrac{2}{3} = \dfrac{13}{3}$ or $4\dfrac{1}{3}$ Subtract. Write improper fraction as a mixed number if desired.

(d) $\dfrac{5}{2x} + \dfrac{3}{x} + \dfrac{9}{4} =$ Convert to equivalent fractions with a common denominator of $4x$.

$\dfrac{5(2)}{2x(2)} + \dfrac{3(4)}{x(4)} + \dfrac{9(x)}{4(x)} =$ Multiply by 1 in the form of $\dfrac{n}{n}$ to get an equivalent fraction with a denominator of $4x$.

$\dfrac{10}{4x} + \dfrac{12}{4x} + \dfrac{9x}{4x} =$ Add numerators and *keep* the like (common) denominator.

$\dfrac{22 + 9x}{4x}$

See Exercises 1–4.

TIP **Writing Improper Fractions Versus Mixed Numbers** In arithmetic, we often write an improper fraction as a mixed number. In algebra, we use the mixed-number form only when the final result contains only numbers and when we are interpreting the result within the context of an applied problem. A rational expression like the solution in the preceding example, part (d), can be written in an alternative form.

$$\dfrac{22 + 9x}{4x} = \dfrac{22}{4x} + \dfrac{9x}{4x} = \dfrac{11}{2x} + \dfrac{9}{4}$$

Unless the context of the problem requires the step, there's usually no reason to do the extra work.

> **To add or subtract rational expressions with unlike denominators:**
>
> 1. Find the *least common denominator (LCD)*.
>
> 2. Change *each* fraction to an equivalent fraction with the least common denominator.
>
> 3. Add or subtract numerators.
>
> 4. Keep the same (common) denominator.
>
> 5. Reduce (or simplify) if possible.

STOP AND CHECK

1. Add $\dfrac{2}{x + 5} + \dfrac{3}{x - 5}$.

Answer:

1. $\dfrac{5x + 5}{(x + 5)(x - 5)}$ or

$\dfrac{5(x + 1)}{(x + 5)(x - 5)}$

EXAMPLE 2

Add $\dfrac{4}{x + 3} + \dfrac{3}{x - 3}$.

First, convert to equivalent expressions with a common denominator. The LCD is the product $(x + 3)(x - 3)$.

$$\frac{4}{x + 3} = \frac{4(x - 3)}{(x + 3)(x - 3)} = \frac{4x - 12}{(x + 3)(x - 3)}$$ Multiply the numerator and denominator by $(x - 3)$.

$$\frac{3}{x - 3} = \frac{3(x + 3)}{(x - 3)(x + 3)} = \frac{3x + 9}{(x + 3)(x - 3)}$$ Multiply the numerator and denominator by $(x + 3)$. The factors in the denominator can be written in any order.

Next, use the equivalent expressions to proceed.

$$\frac{4x - 12}{(x + 3)(x - 3)} + \frac{3x + 9}{(x + 3)(x - 3)} =$$ Add numerators.

$$\frac{4x - 12 + 3x + 9}{(x + 3)(x - 3)} =$$ Combine like terms in the numerator.

$$\frac{7x - 3}{(x + 3)(x - 3)} \quad \text{or} \quad \frac{7x - 3}{x^2 - 9}$$ Write in factored or expanded form.
See Exercises 5–8.

The preceding example could be written with fewer steps, but the example illustrates each step that must be performed, either mentally or on paper, to add the rational expressions. Look at the next example.

STOP AND CHECK

1. Subtract $\dfrac{2x}{x + 6} - \dfrac{7}{x - 2}$.

Answer:

1. $\dfrac{2x^2 - 11x - 42}{(x + 6)(x - 2)}$ or

$\dfrac{2x^2 - 11x - 42}{x^2 + 4x - 12}$

EXAMPLE 3

Subtract $\dfrac{x}{x - 2} - \dfrac{5}{x + 4}$.

Change to equivalent expressions with a common denominator. The LCD is the product $(x - 2)(x + 4)$.

$$\frac{x}{x - 2} = \frac{x(x + 4)}{(x - 2)(x + 4)} = \frac{x^2 + 4x}{(x - 2)(x + 4)}$$ Multiply the numerator and denominator by $(x + 4)$.

$$\frac{5}{x + 4} = \frac{5(x - 2)}{(x + 4)(x - 2)} = \frac{5x - 10}{(x - 2)(x + 4)}$$ Multiply the numerator and denominator by $(x - 2)$.

Use equivalent expressions and proceed.

$$\frac{x^2 + 4x}{(x - 2)(x + 4)} - \frac{5x - 10}{(x - 2)(x + 4)} =$$ Subtract numerators.

$$\frac{x^2 + 4x - (5x - 10)}{(x - 2)(x + 4)} =$$ *Note:* The *entire* numerator following the subtraction sign is subtracted.

$$\frac{x^2 + 4x - 5x + 10}{(x - 2)(x + 4)} =$$ Be careful with the signs when the parentheses are removed.

$$\frac{x^2 - x + 10}{(x - 2)(x + 4)} \quad \text{or} \quad \frac{x^2 - x + 10}{x^2 + 2x - 8}$$ $x^2 - x + 10$ will not factor, and the fraction cannot be reduced.

See Exercises 9–10.

---

2 **Use Addition and Subtraction of Rational Expressions to Simplify Complex Fractions.** Recall that a complex rational expression may have rational expressions in the numerator or denominator or both. Thus, there may be complex expressions that require the addition or subtraction of rational expressions before conversion to an equivalent multiplication.

> **To simplify a complex fraction that requires adding or subtracting rational terms:**
>
> 1. Use addition or subtraction to combine terms in the numerator or denominator so that each contains a single fraction.
>
> 2. Rewrite the complex expression as a multiplication of the reciprocal of the denominator.
>
> 3. Proceed using the procedure for multiplying rational expressions.

STOP AND CHECK

Simplify.

1. $$\dfrac{2 + \dfrac{3}{x}}{\dfrac{1}{4}}$$

2. $$\dfrac{\dfrac{3}{x} - \dfrac{4}{3x}}{\dfrac{1}{6x} - \dfrac{2}{x}}$$

Answers:

1. $\dfrac{4(2x + 3)}{x}$ or $\dfrac{8x + 12}{x}$

2. $-\dfrac{10}{11}$

EXAMPLE 4

Simplify.

(a) $$\dfrac{1 + \dfrac{1}{x}}{\dfrac{2}{3}}$$

(b) $$\dfrac{\dfrac{2}{x} - \dfrac{5}{2x}}{\dfrac{3}{4x} + \dfrac{3}{x}}$$

(a) $$\dfrac{1 + \dfrac{1}{x}}{\dfrac{2}{3}} = \dfrac{\dfrac{x}{x} + \dfrac{1}{x}}{\dfrac{2}{3}} =$$ Add the terms in the numerator. The common denominator for the numerator is x: $1 = \dfrac{x}{x}$

$$\dfrac{\dfrac{x + 1}{x}}{\dfrac{2}{3}} =$$ Multiply the numerator by the reciprocal of the denominator.

$$\dfrac{x + 1}{x} \cdot \dfrac{3}{2} =$$

$$\dfrac{3(x + 1)}{2x} \quad \text{or} \quad \dfrac{3x + 3}{2x}$$ Write in factored or expanded form.

(b) $\dfrac{\dfrac{2}{x} - \dfrac{5}{2x}}{\dfrac{3}{4x} + \dfrac{3}{x}} = \dfrac{\dfrac{4}{2x} - \dfrac{5}{2x}}{\dfrac{3}{4x} + \dfrac{12}{4x}} =$

Add or subtract fractions in the numerator and in the denominator. Use the common denominator for the numerator, $2x$, and the denominator, $4x$, and convert to equivalent fractions.

$= \dfrac{\dfrac{-1}{2x}}{\dfrac{15}{4x}} =$

Multiply the numerator by the reciprocal of the denominator.

$= \dfrac{-1}{2x} \cdot \dfrac{4x}{15} =$

Reduce and multiply.

$= \dfrac{-1}{2\overset{1}{\cancel{x}}} \cdot \dfrac{\overset{2}{\cancel{4x}}}{15} =$

$= \dfrac{-2}{15} = -\dfrac{2}{15}$

See Exercises 11–18.

14–3 EXERCISES

MyLab Math For additional practice go to your study plan in MyLab Math.

1 Add or subtract the rational expressions. Reduce to simplest terms. *See Example 1.*

1. $\dfrac{3}{7} + \dfrac{2}{7}$

2. $\dfrac{3}{8} + \dfrac{7}{16}$

3. $\dfrac{2x}{3} + \dfrac{5x}{6}$

4. $\dfrac{3x}{8} + \dfrac{7x}{12}$

See Example 2.

5. $\dfrac{2}{x} + \dfrac{3}{2}$

6. $\dfrac{5}{2x} + \dfrac{6}{x} + \dfrac{2}{3x}$

7. $\dfrac{5}{x + 1} + \dfrac{4}{x - 1}$

8. $\dfrac{6}{x + 3} + \dfrac{4}{x - 1}$

See Example 3.

9. $\dfrac{5}{2x + 1} - \dfrac{2}{x - 1}$

10. $\dfrac{7}{x - 6} - \dfrac{6}{x - 5}$

2 Simplify. *See Example 4.*

11. $\dfrac{2 + \dfrac{1}{x}}{\dfrac{3}{5}}$

12. $\dfrac{3 + \dfrac{2}{x}}{\dfrac{5}{8}}$

13. $\dfrac{\dfrac{3}{x} + \dfrac{5}{3x}}{\dfrac{3}{6x} - \dfrac{2}{x}}$

14. $\dfrac{\dfrac{x}{6} - \dfrac{3x}{9}}{\dfrac{5x}{12} + \dfrac{3x}{4}}$

15. $\dfrac{\dfrac{4}{x} + \dfrac{3}{2x}}{\dfrac{7x}{5} + \dfrac{x}{10}}$

16. $\dfrac{\dfrac{2}{3x} + \dfrac{3}{5}}{\dfrac{1}{5x} + \dfrac{2}{3}}$

17. $\dfrac{\dfrac{7}{x + 2} - \dfrac{5}{x + 2}}{\dfrac{3}{2 - x} + \dfrac{2}{x - 2}}$

18. $\dfrac{\dfrac{5}{x - 3} + \dfrac{2}{3 - x}}{\dfrac{4}{2x - 6} - \dfrac{5}{3x - 9}}$

LC LEARNING CATALYTICS

1. Find the least common denominator for the fractions $\frac{5}{8}$ and $\frac{7}{18}$.

Rational equation: an equation that contains one or more rational expressions

STOP AND CHECK

Determine the value or values that must be excluded as possible solutions.

1. $\frac{4}{y} - \frac{1}{3} = 4$

2. $\frac{1}{x + 3} + \frac{1}{x + 1} = \frac{1}{5}$

Answers:
1. $y = 0$
2. $x = -3$ and $x = -1$

1 **Exclude Certain Values as Solutions of Rational Equations.** A **rational equation** is an equation that contains one or more rational expressions. Some examples of rational equations are

$$\frac{1}{2} + \frac{1}{x} = \frac{1}{3}, \qquad x - \frac{3}{x} = 4, \qquad \frac{x}{x - 2} = \frac{1}{x + 3}$$

We solved some types of rational equations in Chapter 7 by first clearing the equation of all fractions. An important caution is necessary when a variable is in the denominator of a rational expression. Division by zero is impossible, and any value of the variable that makes the value of any denominator zero is excluded as a possible solution or root of an equation. These values are called *excluded values*.

> **To find excluded values of rational equations:**
>
> 1. Set each denominator containing a variable equal to zero and solve for the variable.
>
> 2. Each equation in Step 1 produces an excluded value; however, some values may be repeats.

EXAMPLE 1

Determine the value or values that must be excluded as possible solutions.

(a) $\frac{1}{x} + \frac{1}{2} = 5$ **(b)** $\frac{1}{x} = \frac{1}{x - 3}$

(c) $\frac{1}{x + 2} + \frac{1}{x - 2} = \frac{1}{4}$ **(d)** $\frac{3x}{x + 2} = 3 - \frac{5}{2x}$

(a) $\frac{1}{x} + \frac{1}{2} = 5$ Set the denominator containing a variable equal to 0.

$x = 0$

Excluded value is 0.

(b) $\frac{1}{x} = \frac{1}{x - 3}$ Set each denominator containing a variable equal to 0.

$x = 0, \qquad x - 3 = 0$ Solve each equation.

$x = 3$

Excluded values are **0** and **3**.

(c) $\frac{1}{x + 2} + \frac{1}{x - 2} = \frac{1}{4}$ Set each denominator containing a variable equal to 0.

$x + 2 = 0 \qquad x - 2 = 0$ Solve each equation.

$x = -2 \qquad x = 2$

Excluded values are **−2** and **2**.

(d) $\dfrac{3x}{x+2} = 3 - \dfrac{5}{2x}$ Set each denominator containing a variable equal to 0.

$\quad\quad x+2 = 0 \quad\quad 2x = 0$ Solve each equation.

$\quad\quad\quad x = -2 \quad\quad\quad x = 0$

Excluded values are −2 and 0.   **See Exercises 1–8.**

Finding excluded values is another way to check whether roots are *extraneous roots*.

2 **Solve Rational Equations with Variable Denominators.**

> **To solve rational equations:**
>
> 1. Determine the excluded values.
>
> 2. Clear the equation of all denominators by multiplying the entire equation by the least common multiple (LCM) of the denominators.
>
> 3. Complete the solution using previously learned strategies.
>
> 4. Eliminate any excluded values as solutions.
>
> 5. Check the solutions.

STOP AND CHECK

Solve the rational equations. Check.

1. $\dfrac{4}{y} - \dfrac{1}{3} = 4$

2. $\dfrac{3}{x-5} - \dfrac{4x}{x-5} = -\dfrac{17}{x-5}$

Answers:

1. $\dfrac{12}{13}$ **2.** No solution

EXAMPLE 2

Solve the rational equations. Check.

(a) $\dfrac{2}{y} + \dfrac{1}{3} = 1$ **(b)** $\dfrac{2}{y-2} = \dfrac{4}{y+1}$ **(c)** $\dfrac{5}{x-2} - \dfrac{3x}{x-2} = -\dfrac{1}{x-2}$

(a) $\dfrac{2}{y} + \dfrac{1}{3} = 1$ Excluded value: $y = 0$

$3y\left(\dfrac{2}{y}\right) + 3y\left(\dfrac{1}{3}\right) = 3y(1)$ Multiply by LCM, 3y, and reduce.

$(3\cancel{y})\left(\dfrac{2}{\cancel{y}}\right) + (\cancel{3}y)\left(\dfrac{1}{\cancel{3}}\right) = 3y(1)$ Multiply remaining factors.

$6 + y = 3y$ Sort.

$6 = 3y - y$ Combine.

$6 = 2y$ Divide.

$\dfrac{6}{2} = \dfrac{2y}{2}$ Simplify.

$3 = y$ 3 is not an excluded value.

Check: $\dfrac{2}{y} + \dfrac{1}{3} = 1$ Substitute 3 for y.

$\dfrac{2}{3} + \dfrac{1}{3} = 1$ Combine like terms.

$\dfrac{3}{3} = 1$ Simplify.

$1 = 1$ Solution $y = 3$ checks.

(b) $\dfrac{2}{y-2} = \dfrac{4}{y+1}$

Excluded values:
$y = 2, y = -1$

$(y-2)(y+1)\left(\dfrac{2}{y-2}\right) = (y-2)(y+1)\left(\dfrac{4}{y+1}\right)$

Multiply by LCM, $(y-2)(y+1)$, and reduce.

$(\cancel{y-2})(y+1)\dfrac{2}{\cancel{y-2}} = (y-2)(\cancel{y+1})\dfrac{4}{\cancel{y+1}}$

Multiply remaining factors.

$2(y+1) = 4(y-2)$

Distribute.

$2y + 2 = 4y - 8$

Sort.

$2y - 4y = -8 - 2$

Combine like terms.

$-2y = -10$

Divide.

$\dfrac{-2y}{-2} = \dfrac{-10}{-2}$

$y = 5$

5 is not an excluded value.

Check: $\dfrac{2}{y-2} = \dfrac{4}{y+1}$

Substitute 5 for y.

$\dfrac{2}{5-2} = \dfrac{4}{5+1}$

Simplify.

$\dfrac{2}{3} = \dfrac{4}{6}$

Reduce $\frac{4}{6}$.

$\dfrac{2}{3} = \dfrac{2}{3}$

Solution $y = 5$ checks.

(c) $\dfrac{5}{x-2} - \dfrac{3x}{x-2} = -\dfrac{1}{x-2}$

Excluded value: $x = 2$

$(x-2)\dfrac{5}{x-2} - (x-2)\dfrac{3x}{x-2} = (x-2)\left(-\dfrac{1}{(x-2)}\right)$

Multiply by LCM, $x - 2$, and reduce.

$(\cancel{x-2})\dfrac{5}{\cancel{x-2}} - (\cancel{x-2})\dfrac{3x}{\cancel{x-2}} = (\cancel{x-2})\left(-\dfrac{1}{(\cancel{x-2})}\right)$

$5 - 3x = -1$

Sort.

$-3x = -1 - 5$

Combine.

$-3x = -6$

Divide.

$\dfrac{-3x}{-3} = \dfrac{-6}{-3}$

$x = 2$

This is an excluded value. The root will not check.

There is no solution.

Check: $\dfrac{5}{x-2} - \dfrac{3x}{x-2} = -\dfrac{1}{x-2}$

Substitute 2 for x.

$\dfrac{5}{2-2} - \dfrac{3(2)}{2-2} = -\dfrac{1}{2-2}$

$\dfrac{5}{0} - \dfrac{6}{0} = -\dfrac{1}{0}$

Division by zero is impossible. The equation has no solution.

See Exercises 9–14.

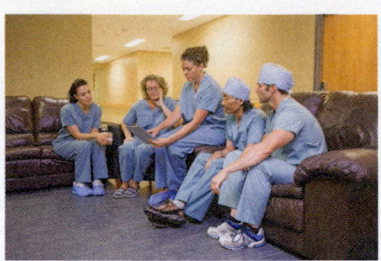

Tyler Olson/Shutterstock

EXAMPLE 3

HLTH/N   Nurse Vance can prepare dosages for her patients in 30 min. If she gets help from assistants who work at the same rate, together they can complete the preparation in 5 min. How many assistants helped her?

Unknown facts

a = number of assistants

Known facts

30 min = time it took Nurse Vance to complete the task alone
 5 min = time it took Nurse Vance and assistants to complete the task
Assistants worked at the same rate as Nurse Vance.

Relationships

$a + 1$ = total number of assistants (Nurse Vance and a assistants)

$\dfrac{30 \text{ min}}{a + 1}$ = amount of time that it would take to complete the task with $a + 1$ assistants

$\dfrac{30 \text{ min}}{a + 1} = 5$ Both sides of the equation represent the amount of time to complete the task with $a + 1$ assistants.

Estimation

Two people (one additional assistant) could do the job in 15 min. Four people would take 7.5 min. Eight people would take less than 4 min. The total number of people will be between 4 and 8 and the number of assistants will be between 3 and 7.

Calculations

$$\frac{30 \text{ min}}{a + 1} = 5$$ Excluded value is -1, which is not appropriate within the context of the problem. Clear the equation of fractions.

$$(a + 1)\left(\frac{30 \text{ min}}{a + 1}\right) = (a + 1)5$$ Simplify.

$$30 = 5a + 5$$ Solve for a.

$$30 - 5 = 5a$$ Combine.

$$25 = 5a$$ Divide.

$$5 = a$$ Number of additional assistants

Interpretation

Nurse Vance got 5 assistants to help her.   See Exercises 15–20.

3   Solve Rational Inequalities in One Variable. When is a rational expression greater than zero? To solve a **rational inequality** like $\dfrac{x + 2}{x - 5} > 0$, treat the numerator and denominator as factors and find the critical values by setting each factor equal to zero.

Rational inequality: an inequality that contains one or more rational expressions

> **To solve a rational inequality like $\frac{x + a}{x + b} < 0$ or $\frac{x + a}{x + b} > 0$:**
>
> 1. Find critical values by setting both the numerator and denominator equal to zero.
>
> 2. Solve each equation from Step 1 and use the solutions (critical values) to divide the number line into three regions.
>
> 3. Evaluate the numerator and denominator for a test point in *each* region. Examine the signs of the results.
>
> 4. Regions that have like signs for both the numerator and denominator are in the solution set of $\frac{x + a}{x + b} > 0$ for $x \neq -b$ ($x = -b$ is an excluded value).
>
> 5. Regions that have unlike signs for the numerator and denominator are in the solution set of $\frac{x + a}{x + b} < 0$ for $x \neq -b$ ($x = -b$ is an excluded value).

TIP **Make Connections to Previously Learned Concepts** Our strategy for solving a rational inequality is based on the rule for dividing signed numbers.

Like signs: $\dfrac{\text{positive}}{\text{positive}} = \text{positive} \quad x > 0 \qquad \dfrac{\text{negative}}{\text{negative}} = \text{positive} \quad x > 0$

Unlike signs: $\dfrac{\text{positive}}{\text{negative}} = \text{negative} \quad x < 0 \qquad \dfrac{\text{negative}}{\text{positive}} = \text{negative} \quad x < 0$

STOP AND CHECK

Solve the inequalities. Write the solutions symbolically, on the number line, and in interval notation.

1. $\dfrac{x - 6}{x + 7} > 0$

2. $\dfrac{x - 6}{x + 7} \leq 0$

Answers:

1. $x < -7$ or $x > 6$;
$(\infty, -7)$ or $(6, \infty)$

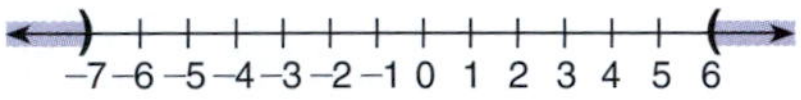

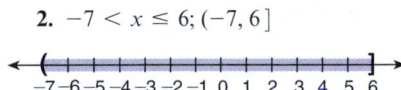

2. $-7 < x \leq 6; (-7, 6]$

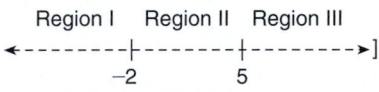

EXAMPLE 4

Solve the following inequalities. Write the solutions symbolically, on the number line, and in interval notation.

(a) $\dfrac{x + 2}{x - 5} < 0$ **(b)** $\dfrac{x + 2}{x - 5} \geq 0$

$x + 2 = 0 \qquad x - 5 = 0$ Set numerator and denominator equal to zero.

$\quad x = -2 \qquad \quad x = 5$ Critical values for inequalities in both (a) and

Excluded value: $x = 5$ (b). The critical value for the denominator is also an excluded value.

Evaluate the numerator and denominator for one point in each region (See Fig. 14–1).

Region I   Region II   Region III

$$\longleftarrow \!\text{-}\text{-}\text{-}\text{-}\text{-}\text{-}\text{-}\text{-}\!\!\overset{\big|}{\underset{-2}{}}\!\text{-}\text{-}\text{-}\text{-}\text{-}\text{-}\text{-}\!\!\overset{\big|}{\underset{5}{}}\!\text{-}\text{-}\text{-}\text{-}\text{-}\text{-}\text{-}\!\!\rightarrow]$$

FIGURE 14–1

Region I:
For $x < -2$, test $x = -3$.

Region II:
For $-2 < x < 5$, test $x = 0$.

Region III:
For $x > 5$, test $x = 6$.

Numerator:

$\quad x + 2 = -3 + 2$
$\qquad \quad = -1 \text{ nagative}$

$\quad x + 2 = 0 + 2$
$\qquad \quad = 2 \text{ positive}$

$\quad x + 2 = 6 + 2$
$\qquad \quad = 8 \text{ positive}$

Critical values for numerator

Denominator:

$\quad x - 5 = -3 - 5$
$\qquad \quad = -8 \text{ nagative}$

$\quad x - 5 = 0 - 5$
$\qquad \quad = -5 \text{ nagative}$

$\quad x - 5 = 6 - 5$
$\qquad \quad = 1 \text{ positive}$

Critical values for denominator

Examine the signs of the results for the numerator and denominator in each region:

Region I:
For $x < -2$, like signs.

Region II:
For $-2 < x < 5$, unlike signs.

Region III:
For $x > 5$, like signs.

See Fig. 14–2. The quotient in Region I is positive (like signs). The quotient in Region II is negative (unlike signs). The quotient in Region III is positive (like signs).

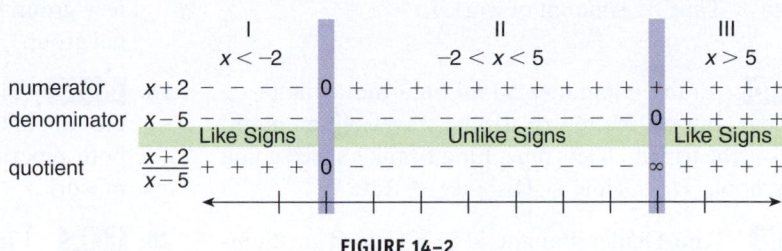

FIGURE 14-2

(a) Solution set (see Fig. 14–3):

For $\dfrac{x+2}{x-5} < 0$,

Unlike signs

$-2 < x < 5$

$(-2, 5)$

(b) Solution set (see Fig. 14–4):

For $\dfrac{x+2}{x-5} \geq 0$,

Like signs

$x \leq -2$ or $x > 5$

$(-\infty, -2] \cup (5, \infty)$

The boundary point, 5, is an excluded value. Therefore, it is not included in the solution set.

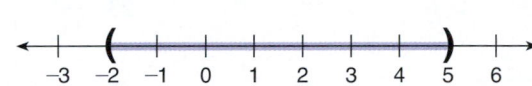

FIGURE 14-3

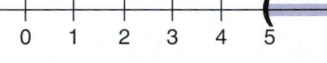

FIGURE 14-4

See Exercises 21–29.

14-4 EXERCISES

MyLab Math For additional practice go to your study plan in MyLab Math.

1 Determine the value or values that must be excluded as possible solutions. *See Example 1.*

1. $\dfrac{5}{x} + \dfrac{3}{5} = 7$

2. $\dfrac{6}{x} + 5 = \dfrac{5}{12}$

3. $\dfrac{9}{x-5} = \dfrac{7}{x}$

4. $\dfrac{15}{2x} = \dfrac{6}{x+9}$

5. $\dfrac{7}{x+8} + \dfrac{2}{x-8} = \dfrac{5}{x^2-64}$

6. $\dfrac{9}{x-2} + \dfrac{5}{x+2} = \dfrac{1}{x^2-4}$

7. $\dfrac{2x}{4x-3} - 4 = \dfrac{5}{6x}$

8. $\dfrac{1}{9x} + \dfrac{8x}{5x+15} = 11$

2 Solve the equations. Check for extraneous roots. *See Example 2.*

9. $\dfrac{6}{7} + \dfrac{5}{x} = 1$

10. $\dfrac{4}{x} - \dfrac{3}{4} = \dfrac{1}{20}$

11. $\dfrac{2}{x} + \dfrac{5}{2x} - \dfrac{1}{2} = 4$

12. $\dfrac{5}{2x} - \dfrac{1}{x} = 6$

13. $\dfrac{2}{x-3} + \dfrac{7}{x+3} = \dfrac{12}{x-3}$

14. $\dfrac{10}{x+5} = \dfrac{1}{x-5} - \dfrac{2}{x+5}$

Use rational equations to solve the problems. Check for extraneous roots. *See Example 3.*

15. **BUS** Shawna can complete an electronic lab project in 3 days. Tom can complete the same project in 4 days. How many days will it take to complete the project if Shawna and Tom work together? (*Hint:* Rate × Time = Amount of work.)

16. **BUS** A group of investors purchased land in Maine for $12,000. More investors joined the group and the cost dropped $500 per person in the original group. How many investors were in the original group if the new group had $1\frac{1}{2}$ times as many investors as the original group?

17. **AUTO** Frank commutes 50 mi on a motorbike to a college, but on the return trip he travels 40 mi/h and makes the trip in $\frac{3}{4}$ h less time. Find Frank's speed going to school. (*Hint:* Time = Distance ÷ Rate.)

18. **INDTR** Pipe 1 fills a tank in 2 min and pipe 2 empties the tank in 6 min. How long does it take to fill the tank if both pipes are open? (*Hint:* Rate × Time = Amount of work.)

19. **PFIN** Kim Denley bought $150 worth of medium-roast coffee and, at the same cost per pound, $100 worth of dark-roast coffee. If she bought 25 pounds more of the medium-roast coffee than the dark-roast coffee, how many pounds of each did she buy and what was the cost per pound?

20. **ELEC** Find resistance$_2$ if resistance$_1$ is 10 Ω and resistance$_t$ is 3.75 Ω, using the formula

$$R_t = \frac{R_1 R_2}{R_1 + R_2}.$$

3 Graph the solution of the rational inequalities on a number line and write the solution in symbolic and interval notation. *See Example 4.*

21. $\dfrac{x-3}{x+7} < 0$

22. $\dfrac{x+8}{x-3} > 0$

23. $\dfrac{2x-6}{3x-1} > 0$

24. $\dfrac{x-7}{x+7} \le 0$

25. $\dfrac{x}{x-1} < 0$

26. $\dfrac{5x-10}{6x-18} \ge 0$

27. $\dfrac{3x-9}{6x-12} \le 0$

28. $\dfrac{3x}{2x-8} \ge 0$

29. $\dfrac{5x-15}{4x+4} \ge 0$

14 | CHAPTER REVIEW OF KEY CONCEPTS

LEARNING OUTCOMES	KEY CONCEPTS AND EXAMPLES

Section 14–1

1 Simplify or reduce rational expressions (pp. 580–583).

Simplify rational expressions: 1. Factor *completely* both the numerator and denominator. **2.** Reduce factors common to both the numerator and denominator. **3.** Write the simplified expression in either factored or expanded form.

Simplify $\dfrac{a+b}{16a^2 + 16ab}$.   Factor common factor of 16a in the denominator.

$\dfrac{a+b}{16a(a+b)}$ Reduce the common binomial factor.

$\dfrac{a+\!\!\!\!/\,b}{16a(a+\!\!\!\!/\,b)} = \dfrac{1}{16a}$

Section 14–2

1 Multiply and divide rational expressions (pp. 584–587).

Multiply or divide rational expressions: 1. Convert any division to an equivalent multiplication. **2.** Factor completely every numerator and denominator. **3.** Reduce factors that are common to a numerator and denominator. **4.** Multiply remaining factors. **5.** The result can be written in factored or expanded form.

LEARNING OUTCOMES	KEY CONCEPTS AND EXAMPLES

Multiply $\dfrac{x}{2x+8} \cdot \dfrac{2}{x+1}$. — Write each numerator and denominator in factored form.

$\dfrac{x}{2(x+4)} \cdot \dfrac{2}{x+1} =$ — Reduce and multiply.

$\dfrac{x}{(x+4)(x+1)}$ or $\dfrac{x}{x^2+5x+4}$ — Factored or expanded form

Divide $\dfrac{x}{x+2} \div \dfrac{2}{x+2}$. — Multiply by the reciprocal of the divisor. Reduce.

$\dfrac{x}{x+2} \cdot \dfrac{x+2}{2} = \dfrac{x}{2}$ — Multiply remaining factors.

Simplify a complex rational expression: 1. Rewrite the complex rational expression as a division then convert to an equivalent multiplication of rational expressions. **2.** Multiply the rational expressions.

Simplify.

$\dfrac{\dfrac{x-3}{5x}}{\dfrac{4x-12}{15x^2}}$ — Write as division.

$\dfrac{x-3}{5x} \div \dfrac{4x-12}{15x^2} =$ — Rewrite as multiplication, factor, and reduce.

$\dfrac{x-3}{5x} \cdot \dfrac{15x^2}{4(x-3)} =$ — Multiply remaining factors.

$\dfrac{3x}{4}$

---

2 Use multiplication and conjugates to rationalize a numerator or denominator of a fraction that has a binomial with an irrational term (pp. 587–588).

1. Determine which part of the fraction (numerator or denominator) is to be cleared of a binomial with an irrational term. **2.** Multiply by 1 in the form of $\frac{n}{n}$, where n is the conjugate of the binomial with the irrational term. **3.** Simplify the resulting expression.

Rationalize the denominator in the expression $\dfrac{5x}{3+2\sqrt{5}}$

$\dfrac{5x}{3+2\sqrt{5}} \cdot \dfrac{3-2\sqrt{5}}{3-2\sqrt{5}} =$ — Multiply by 1 in the form of $\dfrac{3-2\sqrt{5}}{3-2\sqrt{5}}$.

$\dfrac{5x(3-2\sqrt{5})}{3^2 - (2\sqrt{5})^2} =$ — Distribute and simplify.

$\dfrac{15x - 10x\sqrt{5}}{9-4(5)} = \dfrac{15x - 10x\sqrt{5}}{9-20}$

$\dfrac{15x - 10x\sqrt{5}}{-11}$ or $-\dfrac{15x - 10x\sqrt{5}}{11}$

LEARNING OUTCOMES	KEY CONCEPTS AND EXAMPLES

Section 14-3

1 Add and subtract rational expressions (pp. 590–592).

Add (or subtract) only rational expressions with the same denominator.

Add or subtract rational expressions with unlike denominators: 1. Find the *least common denominator (LCD)*. **2.** Change *each* fraction to an equivalent fraction with the least common denominator. **3.** Add or subtract the numerators. **4.** Keep the same (common) denominator. **5.** Reduce (or simplify) if possible.

Perform the following operations:

(a) $\dfrac{x}{x+4} + \dfrac{2x}{x+4} = \dfrac{3x}{x+4}$ Add numerators. Keep like denominator.

(b) $\dfrac{3}{x+2} - \dfrac{2}{x+1}$ LCD $= (x+2)(x+1)$. Change each fraction to an equivalent fraction with the LCD.

$\dfrac{3}{x+2} = \dfrac{3(x+1)}{(x+2)(x+1)}$ First fraction

$= \dfrac{3x+3}{(x+2)(x+1)}$

$\dfrac{2}{x+1} = \dfrac{2(x+2)}{(x+2)(x+1)}$ Second fraction

$= \dfrac{2x+4}{(x+2)(x+1)}$

$\dfrac{3x+3}{(x+2)(x+1)} - \dfrac{2x+4}{(x+2)(x+1)} = \dfrac{(3x+3)-(2x+4)}{(x+2)(x+1)}$ Subtract numerators. Keep common denominator.

$\dfrac{3x+3-2x-4}{(x+2)(x+1)} = \dfrac{x-1}{(x+2)(x+1)}$

2 Use addition and subtraction of rational expressions to simplify complex fractions (pp. 592–593).

1. Use addition or subtraction to combine terms in the numerator or denominator or both.
2. Rewrite the complex expression as a multiplication of the reciprocal of the denominator.
3. Proceed using the procedure for multiplying rational expressions.

Simplify $\dfrac{\dfrac{3}{x}+5}{\dfrac{2}{3x}-\dfrac{1}{x}}$.   Change to equivalent fractions in numerator with LCD of x.

Change to equivalent fractions in denominator with LCD of $3x$.

$\dfrac{\dfrac{3}{x}+\dfrac{5}{1}\left(\dfrac{x}{x}\right)}{\dfrac{2}{3x}-\dfrac{1}{x}\left(\dfrac{3}{3}\right)}$

$\dfrac{\dfrac{3}{x}+\dfrac{5x}{x}}{\dfrac{2}{3x}-\dfrac{3}{3x}}$ Add fractions in the numerator.

Subtract fractions in the denominator.

$\dfrac{\dfrac{3+5x}{x}}{\dfrac{-1}{3x}}$ Rewrite as multiplication of reciprocal.

LEARNING OUTCOMES	KEY CONCEPTS AND EXAMPLES

$$\frac{3 + 5x}{\cancel{x}} \cdot \frac{3\cancel{x}}{-1} = \qquad \text{Reduce and multiply.}$$

$$\frac{(3 + 5x)3}{-1} = \qquad \text{Distribute.}$$

$$\frac{9 + 15x}{-1} = -(9 + 15x) \quad \text{or} \quad -9 - 15x$$

Section 14–4

1 Exclude certain values as solutions of rational equations (pp. 594–595).

Find excluded values of rational equations: 1. See each denominator containing a variable equal to zero and solve for the variable. **2.** Each equation in Step 1 produces an excluded value; however, some values may be repeats.

Find the excluded values for the equation:

$$4 = \frac{2}{3 + y} \qquad \text{Set denominator equal to zero.}$$

$$3 + y = 0$$

$$y = -3 \qquad \text{Excluded value}$$

2 Solve rational equations with variable denominators (pp. 595–597).

Solve rational equations: 1. Determine the excluded values. **2.** Clear the equation of all denominators by multiplying the entire equation by the least common multiple (LCM) of the denominators. **3.** Complete the solution using previously learned strategies. **4.** Eliminate any excluded values as solutions. **5.** Check the solutions.

Solve $\dfrac{2}{x - 3} - \dfrac{1}{x + 3} = \dfrac{1}{x^2 - 9}$.

Excluded values are 3 and -3.

LCM is $x^2 - 9$ or $(x - 3)(x + 3)$.

$$(x - 3)(x + 3)\frac{2}{x - 3} - (x - 3)(x + 3)\frac{1}{x + 3} = (x - 3)(x + 3)\frac{1}{(x - 3)(x + 3)}$$

$$2(x + 3) - 1(x - 3) = 1$$

$$2x + 6 - x + 3 = 1$$

$$x + 9 = 1$$

$$x = -8$$

Check:

$$\frac{2}{-8 - 3} - \frac{1}{-8 + 3} = \frac{1}{(-8)^2 - 9}$$

$$\frac{2}{-11} - \frac{1}{-5} = \frac{1}{64 - 9}$$

$$-\frac{2}{11} + \frac{1}{5} = \frac{1}{55}$$

$$-\frac{10}{55} + \frac{11}{55} = \frac{1}{55}$$

$$\frac{1}{55} = \frac{1}{55} \qquad \text{The solution } x = -8 \text{ checks.}$$

LEARNING OUTCOMES	KEY CONCEPTS AND EXAMPLES

3 Solve rational inequalities in one variable (pp. 597–599).

Solve a rational inequality like $\dfrac{x + a}{x + b} < 0$ **or** $\dfrac{x + a}{x + b} > 0$: **1.** Find critical values by setting both the numerator and denominator equal to zero. **2.** Solve each equation from Step 1 and use the solutions (critical values) to divide the number line into three regions. **3.** Evaluate the numerator and denominator for a test point in *each* region. Examine the signs of the results. **4.** Regions that have like signs for both the numerator and denominator are in the solution set of $\dfrac{x + a}{x + b} > 0$ for $x \neq -b$ ($x = -b$ is an excluded value). **5.** Regions that have unlike signs for the numerator and denominator are in the solution set of $\dfrac{x + a}{x + b} < 0$ for $x \neq -b$ ($x = -b$ is an excluded value).

The process for solving rational inequalities for $\leq$ or $\geq$ relationships is similar to the process for solving $<$ or $>$ relationships.

Solve the rational inequality $\dfrac{x + 3}{x - 2} \leq 0$.

$x = 2$ is an excluded value.   Set numerator and denominator equal to zero.

$x + 3 = 0$ $x - 2 = 0$ Solve each equation.

$x = -3$ $x = 2$ Critical values

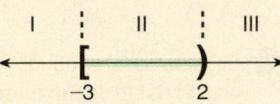

The critical value 2 is not included in the solution set because it is an excluded value.

Region I has like signs $\left(\dfrac{-}{-}\right)$.

Region II has unlike signs $\left(\dfrac{+}{-}\right)$.   Unlike signs indicate < 0.

Region III has like signs $\left(\dfrac{+}{+}\right)$.

The solution set is Region II.

$-3 \leq x < 2; [-3, 2)$

FIGURE 14–5

---

14 CHAPTER REVIEW EXERCISES

Section 14–1   **MyLab Math** **For additional practice go to your study plan in MyLab Math.**

Simplify the fractions.

1. $\dfrac{18}{24}$

2. $\dfrac{24}{42}$

3. $\dfrac{5a^2b^3c}{10a^3bc^2}$

4. $\dfrac{3x^2y}{9x^3y^3}$

5. $\dfrac{4xy(x - 3)}{8xy(x + 3)}$

6. $\dfrac{(x + 7)(3 - x)}{(x - 3)(x - 7)}$

7. $\dfrac{(x - 4)(x + 2)}{(x + 2)(4 - x)}$

8. $\dfrac{x + 1}{3x + 3}$

9. $\dfrac{m^2 - n^2}{m^2 + n^2}$

10. $\dfrac{3x^2 + 8x - 3}{2x^2 + 5x - 3}$

11. $\dfrac{x}{x + xy}$

12. $\dfrac{2x}{4x + 6}$

13. $\dfrac{5x + 15}{x + 3}$

14. $\dfrac{x^3 + 2x^2 - 3x}{3x}$

15. $\dfrac{y^2 + 2y + 1}{y + 1}$

16. $\dfrac{y - 1}{y^2 - 2y + 1}$

17. $\dfrac{2x - 6}{x^2 + 3x - 18}$

18. $\dfrac{x^2 - 4x + 4}{x^2 - 2x}$

19. $\dfrac{3x - 9}{x - 3}$

20. $\dfrac{4x - 12}{x - 3}$

Section 14–2

Multiply or divide the fractions. Reduce to simplest terms.

21. $\dfrac{3x^2}{2y} \cdot \dfrac{5x}{6y}$

22. $\dfrac{x^2 - y^2}{6} \cdot \dfrac{18}{x - y}$

23. $\dfrac{9}{x + b} \cdot \dfrac{5x + 5b}{3}$

24. $\dfrac{81 - x^2}{16 - f^2} \cdot \dfrac{4 - f}{9 + x}$

25. $\dfrac{4y^2 - 4y + 1}{6y - 6} \cdot \dfrac{24}{2y - 1}$

26. $\dfrac{x - 3}{x + 5} \cdot \dfrac{2x^2 + 10x}{2x - 6}$

27. $\dfrac{5 - x}{x - 5} \cdot \dfrac{x - 1}{1 - x}$

28. $\dfrac{3 - x}{x - 2} \cdot \dfrac{2x - 4}{x - 3}$

29. $\dfrac{x^2 + 6x + 9}{x^2 - 4} \cdot \dfrac{x - 2}{x + 3}$

30. $\dfrac{x^2}{x^2 - 9} \cdot \dfrac{x^2 - 5x + 6}{x^2 - 2x}$

31. $\dfrac{2a + b}{8} \div \dfrac{2a + b}{2}$

32. $\dfrac{17a^2}{21y^2} \div \dfrac{34a^2}{68}$

33. $\dfrac{y^2 - 2y + 1}{y} \div \dfrac{1}{y - 1}$

34. $\dfrac{x^2 y^2}{3x^2 - 3y^2} \div \dfrac{8xy}{x - y}$

35. $\dfrac{y^2 + 6y + 9}{y^2 + 4y + 4} \div \dfrac{y + 3}{y + 2}$

36. $\dfrac{2x + 2y}{3} \div \dfrac{x^2 - y^2}{y - x}$

37. $\dfrac{3x^2 + 6x}{x} \div \dfrac{2x + 4}{x^2}$

38. $\dfrac{x^2 - 7x}{x^2 - 3x - 28} \div \dfrac{1}{-x - 4}$

39. $\dfrac{y^2 - 16}{y + 3} \div \dfrac{y - 4}{y^2 - 9}$

40. $\dfrac{12x + 24}{36x - 36} \div \dfrac{6x + 12}{8x - 8}$

Simplify.

41. $\dfrac{\frac{5}{x - 3}}{4}$

42. $\dfrac{6ab}{\frac{3a}{4}}$

43. $\dfrac{\frac{x^2 - 4x}{6x}}{\frac{x - 4}{8x^2}}$

44. $\dfrac{\frac{2}{7}}{\frac{3}{4}}$

Rationalize the denominator of each fraction.

45. $\dfrac{12}{6 - \sqrt{5}}$

46. $\dfrac{8}{3 - \sqrt{2}}$

47. $\dfrac{7 + \sqrt{3}}{7 - \sqrt{3}}$

48. $\dfrac{4 - \sqrt{7}}{3 + \sqrt{5}}$

49. $\dfrac{5 + \sqrt{2}}{7 - 3\sqrt{5}}$

50. $\dfrac{4 - \sqrt{7}}{8 + 2\sqrt{11}}$

Rationalize the numerator of each fraction.

51. $\dfrac{8 + 2\sqrt{3}}{5}$

52. $\dfrac{7 - \sqrt{5}}{11}$

53. $\dfrac{4 - \sqrt{13}}{12}$

54. $\dfrac{1 - 3\sqrt{7}}{8}$

55. $\dfrac{5 - \sqrt{7}}{16}$

56. $\dfrac{2 + \sqrt{15}}{15}$

Section 14–3

Add or subtract the fractions. Reduce to simplest terms.

57. $\dfrac{2}{9} + \dfrac{4}{9}$

58. $\dfrac{2}{3} + \dfrac{5}{12}$

59. $\dfrac{3x}{7} + \dfrac{2x}{14}$

60. $\dfrac{5x}{3} - \dfrac{2x}{4}$

61. $\dfrac{3x}{4} + \dfrac{5x}{6}$

62. $\dfrac{3}{4} + \dfrac{7}{x}$

63. $\dfrac{5}{x} - \dfrac{7}{3}$

64. $\dfrac{3}{x} + \dfrac{5}{7}$

65. $\dfrac{3}{4x} + \dfrac{2}{x} + \dfrac{3}{6x}$

66. $\dfrac{6}{2x + 3} + \dfrac{2}{2x - 3}$

67. $\dfrac{7}{x - 3} + \dfrac{3}{x + 2}$

68. $\dfrac{4}{3x + 2} - \dfrac{3}{x - 2}$

69. $\dfrac{8}{x + 3} - \dfrac{2}{x - 4}$

70. $\dfrac{3}{x - 2} + \dfrac{5}{2 - x}$

71. $\dfrac{x}{x - 5} - \dfrac{3}{5 - x}$

Simplify.

72. $\dfrac{2 + \dfrac{3}{x}}{\dfrac{2}{x}}$

73. $\dfrac{\dfrac{5}{x} - \dfrac{3}{4x}}{\dfrac{1}{3x} + \dfrac{2}{x}}$

74. $\dfrac{5 - \dfrac{x - 2}{x}}{\dfrac{x - 4}{2x} - 2}$

75. $\dfrac{\dfrac{3x}{6} - \dfrac{5}{x}}{\dfrac{x}{3} + \dfrac{4}{2x}}$

Section 14-4

Determine the value or values that must be excluded as possible solutions of the following.

76. $\dfrac{3}{x} - \dfrac{4}{5} = 2$

77. $\dfrac{4}{x} = \dfrac{3}{x - 2}$

78. $\dfrac{3}{x - 5} - \dfrac{4}{x + 5} = \dfrac{1}{25}$

79. $\dfrac{5}{2x - 1} - \dfrac{6}{x} = \dfrac{4}{3x}$

Solve the equations. Check for extraneous roots.

80. $\dfrac{3}{x} + \dfrac{2}{3} = 1$

81. $\dfrac{4}{x} = \dfrac{1}{x + 5}$

82. $\dfrac{3}{x - 4} + \dfrac{1}{x + 4} = \dfrac{1}{x^2 - 16}$

83. $-\dfrac{4x}{x + 1} = 3 - \dfrac{4}{x + 1}$

Use rational equations to solve the following problems. Check for extraneous or inappropriate roots.

84. **INDTEC** Fugita can complete the assembly of 50 widgets in 6 days. Ohn can complete the same job in 8 days. How many days will it take to complete the project if Fugita and Ohn work together? (*Hint:* Rate × Time = Amount of work.)

85. Several students chipped in a total of $120 to buy a small refrigerator to use in the dorm. If the group was increased to four students, the cost to the original group dropped $10 per person. How many students were in the original group?

86. Pipe 1 fills a tank in 4 min and pipe 2 fills the same tank in 3 min. What part of the tank will be filled if both pipes are open for 1 min? (*Hint:* Rate × Time = Amount of work.)

87. One machine can do a job in 5 h alone. How many hours would it take a second machine to complete the job alone if both machines together can do the job in 3 h?

88. $\dfrac{x + 1}{x - 3} > 0$

89. $\dfrac{x}{x + 8} > 0$

90. $\dfrac{x - 8}{x + 1} < 0$

91. $\dfrac{x - 7}{x + 1} < 0$

14   TEAM PROBLEM-SOLVING EXERCISES

1. Applications that involve several facts and relationships may vary in the information that is known and unknown. For example, an application of two pipes working together to fill a tank may be interested in filling the tank for only a certain period of time. Solve the following problem by adapting your strategy to the given information.

   Pipe 1 fills a tank in 4 min and pipe 2 fills the same tank in 3 min. What portion of the tank will be filled if both pipes are open for 1 min?

2. Use your strategy for two quantities working together and your knowledge of setting up an equation to model the conditions of an application to solve the following.

One machine can do a job in 5 h alone. How many hours would it take a second machine to complete the job alone if both machines together can do the job in 3 h?

14   CONCEPTS ANALYSIS

1. How are the properties $\frac{n}{n} = 1$ and $1 \times n = n$ applied in the following example: $\frac{4}{6} = \frac{2}{3}$?

2. Why is the following problem incorrect?

$$\frac{5}{10} = \frac{2 + 3}{2 + 8} = \frac{3}{8}$$

3. Write a brief comment for each step of the following example.

$$\frac{x^2 + 3x + 2}{x^2 - 9} \div \frac{2x^2 + 3x - 2}{2x^2 - 7x + 3} =$$

$$\frac{x^2 + 3x + 2}{x^2 - 9} \cdot \frac{2x^2 - 7x + 3}{2x^2 + 3x - 2} =$$

$$\frac{(x + 1)(x + 2)}{(x + 3)(x - 3)} \cdot \frac{(2x - 1)(x - 3)}{(2x - 1)(x + 2)} = \frac{x + 1}{x + 3}$$

Find the mistake or mistakes in each problem and briefly explain each mistake. Then, rework the problem correctly.

4. $\dfrac{x}{x + 3} = \dfrac{\cancel{x}}{\cancel{x} + 3} = \dfrac{1}{3}$

5. $\dfrac{x^2 - 4}{x^2 + 4x + 4} \div x + 2 =$

$$\frac{(x + 2)(x - 2)}{(x + 2)(x + 2)} \cdot \frac{x + 2}{1} = x - 2$$

6. $\dfrac{5}{x + 2} + \dfrac{3}{x - 3} = \dfrac{8}{(x + 2)(x - 3)}$

7. $\dfrac{3}{x - 1} - \dfrac{5}{x + 1} =$

$$\frac{3(x + 1)}{(x - 1)(x + 1)} - \frac{5(x - 1)}{(x + 1)(x - 1)} =$$

$$\frac{3x + 3 - 5x - 5}{(x + 1)(x - 1)} =$$

$$\frac{-2x - 2}{(x + 1)(x - 1)} =$$

$$\frac{-2(x + 1)}{(x + 1)(x - 1)} = \frac{-2}{x - 1}$$

8.

$$\frac{14}{3x} = \frac{12}{7x}$$

$$\frac{\overset{2}{\cancel{14}}}{\underset{1}{\cancel{3x}}} = \frac{\overset{4}{\cancel{12}}}{\underset{1}{\cancel{7x}}}$$

$$8 = x^2$$

$$\pm\sqrt{8} = x$$

$$\pm 2\sqrt{2} = x$$

9. How is a division of rational expressions related to a multiplication of rational expressions?

10. In your own words, write a rule for multiplying rational expressions.

14 | PRACTICE TEST

Perform the indicated operations.

1. $\dfrac{x-3}{2x-6}$

2. $\dfrac{x^2-16}{x-4}$

3. $\dfrac{6x^2-11x+4}{2x^2+5x-3}$

4. $\dfrac{x^2-6x+8}{2-x}$

5. $\dfrac{(x-2)(x-4)}{(4-x)(x+2)}$

6. $\dfrac{y^2+x^2}{y^2-x^2}$

7. $\dfrac{6xy}{ab}\cdot\dfrac{a^2b}{2xy^2}$

8. $\dfrac{x^2-y^2}{2x+y}\cdot\dfrac{4x^2+2xy}{y-x}$

9. $\dfrac{x-2y}{x^3-3x^2y}\div\dfrac{x^2-4y^2}{x-3y}$

10. $\dfrac{2a^2-ab-b^2}{6x^2+x-1}\div\dfrac{a^2-b^2}{8x+4}$

11. $\dfrac{2x^2+3x+1}{x}\div\dfrac{x+1}{1}$

12. $\dfrac{4y^2}{2x}\cdot\dfrac{x}{8x}$

13. $\dfrac{1}{x+2}-\dfrac{1}{x-3}$

14. $\dfrac{2}{x+2}+\dfrac{3}{x-1}$

15. $\dfrac{3}{x}+\dfrac{1}{4}$

16. $\dfrac{2}{3y}-\dfrac{7}{y}$

17. $\dfrac{5}{3x-2}+\dfrac{7}{2-3x}$

18. $\dfrac{2x}{3}-\dfrac{5x}{2}$

19. $\dfrac{x-2y}{x^2-4y^2}$

20. $\dfrac{3}{2y}+\dfrac{2}{y}+\dfrac{1}{5y}$

21. $\dfrac{2x}{1-\dfrac{3}{x}}$

22. $\dfrac{\dfrac{1}{a}+\dfrac{1}{b}}{ab}$

Determine the excluded values of x.

23. $\dfrac{5}{x}=\dfrac{2}{x+3}$

24. $\dfrac{3}{x-4}+\dfrac{5}{x+4}=6$

Solve the rational equations. Check for extraneous roots.

25. $\dfrac{3x}{x-2}+4=\dfrac{3}{x-2}$

26. $\dfrac{x}{x+3}=5$

27. $\dfrac{5}{x-2}=\dfrac{-4}{x+1}$

Use rational equations to solve the problems. Check for extraneous roots.

28. Henry can assemble four computers in 3 h. Lester can assemble four computers in 4 h. How long will it take to assemble four computers if both work together? (*Hint:* Rate × Time = Amount of work.)

29. A medical group purchased land in Colorado for $100,000. When the size of the group doubled, the cost dropped $10,000 per person in the original group. How many persons were in the original group?

30. Cedric Partee drove 300 mi in one day while on a vacation to the mountains, but on the return trip by the same route he drove 10 mi per hour less and the return trip took 1 h longer. Find Partee's speed to and from the mountains. (*Hint:* Time = Distance ÷ Rate.)

31. $\dfrac{x-2}{x+5}<0$

32. $\dfrac{x-3}{x}>0$

15

Quadratic and Other Nonlinear Equations and Inequalities

SIR ISAAC NEWTON

Science History Images/Alamy Stock Photo

In Great Company

Isaac Newton Was An Idiot Too (April 1987)

Have you ever looked at a problem solution and asked yourself, "OK, where did *that* number come from?" Of course you have. Anyone who has studied math for more than an hour has said this. And of course you don't raise your hand in class and ask about it, because everyone else probably knows where the number came from, right? So, you just shake your head, turn in the assignment, and move on.

You are not alone.

In 1687, Isaac Newton wrote an incredibly important mathematical work called the *Principia Mathematica*. It laid the foundation for what we now call the *calculus*, helped establish the mathematical foundations for the laws of motion and gravity, and is often hailed as the most important mathematical work in the history of humankind. Because the work is so important, it has served as the basis for mathematical studies for over three centuries.

In 1987, exactly 300 years after the *Principia* was first published, Robert Garisto was a senior physics student taking a history of science class that studied the *Principia*. One of the professor's assignments included an analysis of Proposition Eight of Book Three. In that section, Newton had calculated an angle to be 10.5 s in width, but for some unknown reason, actually used 11 s in his equation. Garisto found the error: "When I found the discrepancy, my initial reaction was 'Wow!'" He assumed, however, that everyone knew it was there, so he identified it, and turned in his paper along with the others. His professor gave him an A+, but didn't notice what Garisto had discovered.

This was a little ironic. The professor had given the assignment precisely because he could never get Newton's equations in that section to work. It wasn't until Nobel Laureate Subrahmanyan Chandrasekhar gave a public lecture on the *Principia* in April 1987 that Garisto and his professor realized what Garisto had done. Garisto wrote up the discovery for publication and won the Excellence in Science prize that year, beating out 18 Nobel laureates.

For 300 years, mathematicians and physicists could not get that section to work, but for 300 years, the finest scientific and mathematical minds in history didn't discover Newton's error. Why? Because every last one of them was afraid he would look stupid if he admitted he couldn't work the math. It took some kid who didn't know he wasn't supposed to find an error to . . . find the error.

15–1   Solving Quadratic Equations by the Square-Root Method

LEARNING OUTCOMES

1   Write quadratic equations in standard form.

2   Identify the coefficients of the quadratic, linear, and constant terms of a quadratic equation.

3   Solve pure quadratic equations ($ax^2 + c = 0$) by the square-root method.

LC LEARNING CATALYTICS

1. Write $3x - 7 + 4x^2$ in descending order of x.

STOP AND CHECK

Arrange the quadratic equations in standard form.

1. $x^2 - 3 + 2x = 0$
2. $3x + 5 = 5x^2$

Answers:
1. $x^2 + 2x - 3 = 0$
2. $5x^2 - 3x - 5 = 0$

1 Write Quadratic Equations in Standard Form. There are several methods for solving quadratic equations. Each method has strengths and weaknesses depending on the characteristics of the equation being solved. First, let's examine the basic characteristics of quadratic equations, starting with quadratic equations having only one variable.

A **quadratic equation** is an equation in which at least one variable term is raised to the second power and no variable term has a power more than 2 or less than 0.

The **standard form of a quadratic equation** is $ax^2 + bx + c = 0$, where a, b, and c are real numbers and $a > 0$.

EXAMPLE 1

Arrange the quadratic equations in standard form.

(a) $3x^2 - 3 + 5x = 0$ **(b)** $7x - 2x^2 - 8 = 0$

(c) $12 = 7x^2 - 4x$ **(d)** $7x^2 = 3$

(a) $3x^2 - 3 + 5x = 0$ Arrange the terms in descending powers of x.
  $\mathbf{3x^2 + 5x - 3 = 0}$ Standard form

Quadratic equation: an equation in which at least one variable term is raised to the second power and no variable term has a power more than 2 or less than 0

(b) $7x - 2x^2 - 8 = 0$ Arrange the terms in descending powers of x.

$-2x^2 + 7x - 8 = 0$ Multiply both sides by -1 to make the coefficient of the x^2 term positive.

$-1(-2x^2 + 7x - 8) = -1(0)$

$\mathbf{2x^2 - 7x + 8 = 0}$ Standard form

Standard form of a quadratic equation: the standard form of a quadratic equation is $ax^2 + bx + c = 0$, where a, b, and c are real numbers and $a > 0$

(c) $12 = 7x^2 - 4x$ Rearrange the terms so that one side of the equation equals zero.

$0 = 7x^2 - 4x - 12$ Interchange sides of equation.

$\mathbf{7x^2 - 4x - 12 = 0}$

(d) $7x^2 = 3$ Rearrange so that one side equals zero.

$\mathbf{7x^2 - 3 = 0}$ There is no first-degree variable term.

See Exercises 1–12.

A quadratic equation is a second-degree equation, which means that it can have as many as two solutions. As we examine the process for solving quadratic equations, we shall see quadratic equations that have two, one, or no solutions.

2 **Identify the Coefficients of the Quadratic, Linear, and Constant Terms of a Quadratic Equation.** Some methods for solving quadratic equations can be used for any type of quadratic equation but are time consuming. Other methods are quicker but apply only to certain types of quadratic equations. In choosing an appropriate method, it is helpful to be able to recognize similar and different characteristics of quadratic equations.

The standard form of quadratic equations, $ax^2 + bx + c = 0$, has three types of terms.

1. ax^2 is a *quadratic term;* that is, the degree of the term is 2. In standard form, this term is the *leading term* and a is the *leading coefficient*.
2. bx is a *linear term;* that is, the degree of the term is 1 and the coefficient is b.
3. c is a *number term* or *constant term;* that is, the degree of the term is 0. The coefficient is c ($c = cx^0$).

A quadratic equation that has all three types of terms is sometimes referred to as a **complete quadratic equation.**

$$ax^2 + bx + c = 0 \qquad \text{Complete quadratic equation}$$

Complete quadratic equation: a quadratic equation that has a quadratic term, a linear term, and a constant term; symbolically, in standard form $ax^2 + bx + c = 0$

A quadratic equation that has a quadratic term (ax^2) and a linear term (bx) but no constant term (c) is sometimes referred to as an **incomplete quadratic equation.**

$$ax^2 + bx = 0 \qquad \text{Incomplete quadratic equation}$$

Incomplete quadratic equation: a quadratic equation that has a quadratic term and a linear term but no constant term; symbolically, $ax^2 + bx = 0$

A quadratic equation that has a quadratic term (ax^2) and a constant term (c) but no linear term (bx) is sometimes referred to as a **pure quadratic equation.**

$$ax^2 + c = 0 \qquad \text{Pure quadratic equation}$$

Pure quadratic equation: a quadratic equation that has a quadratic term and a constant term but no linear term; symbolically, $ax^2 + c = 0$

It is helpful in planning our strategy for solving a quadratic equation to be able to identify the types of terms in a quadratic equation and to identify the coefficients a, b, and c.

Write each equation in standard form and identify a, b, and c.

1. $3x^2 - 2x = 7$
2. $7x = 3x^2$
3. $5 - 2x = 2x^2 - x$
4. $16 = x^2$

Answers:

1. $3x^2 - 2x - 7 = 0$; $a = 3$; $b = -2$; $c = -7$
2. $3x^2 - 7x = 0$; $a = 3$; $b = -7$; $c = 0$
3. $2x^2 + x - 5 = 0$; $a = 2$; $b = 1$; $c = -5$
4. $x^2 - 16 = 0$; $a = 1$; $b = 0$; $c = -16$

EXAMPLE 2

Write each equation in standard form and identify a, b, and c.

(a) $3x^2 + 5x - 2 = 0$ **(b)** $5x^2 = 2x - 3$

(c) $3x = 5x^2$ **(d)** $-6x^2 + 2x = 0$

(e) $5x^2 - 4 = 0$ **(f)** $x^2 = 9$

(g) $x^2 + 3x = 5x$ **(h)** $7 - x^2 = 3 + x$

(a) $3x^2 + 5x - 2 = 0$ In standard form

$\quad a = 3, \quad b = 5, \quad c = -2$

(b) $\qquad\qquad 5x^2 = 2x - 3$ Write in standard form.

$\quad 5x^2 - 2x + 3 = 0$

$\quad a = 5, \quad b = -2, \quad c = 3$

(c) $\qquad\qquad 3x = 5x^2$ Write in standard form.

$\qquad\qquad 0 = 5x^2 - 3x$ Interchange sides of equation.

$\quad 5x^2 - 3x = 0$

$\quad a = 5, \quad b = -3, \quad c = 0$ No constant term means $c = 0$.

(d) $\qquad -6x^2 + 2x = 0$ Multiply by -1.

$\quad -1(-6x^2 + 2x) = (-1)0$

$\qquad\qquad 6x^2 - 2x = 0$ Standard form

$\quad a = 6, \quad b = -2, \quad c = 0$ No constant term

(e) $5x^2 - 4 = 0$ Standard form

$\quad a = 5, \quad b = 0, \quad c = -4$ No linear term means $b = 0$.

(f) $\qquad x^2 = 9$ Write in standard form.

$\quad x^2 - 9 = 0$

$\quad a = 1, \quad b = 0, \quad c = -9$ No linear term

(g) $\qquad x^2 + 3x = 5x$ Write in standard form.

$\quad x^2 + 3x - 5x = 0$ Combine like terms.

$\qquad\quad x^2 - 2x = 0$

$\quad a = 1, \quad b = -2, \quad c = 0$ No constant term

(h) $\qquad\qquad 7 - x^2 = 3 + x$ Rearrange.

$\quad -x^2 - x + 7 - 3 = 0$ Combine like terms.

$\qquad -x^2 - x + 4 = 0$ Multiply by -1.

$\quad -1(-x^2 - x + 4) = (-1)0$

$\qquad\qquad x^2 + x - 4 = 0$ Standard form

$\quad a = 1, \quad b = 1, \quad c = -4$

See Exercises 13–31.

Square-root property of equality: equality is maintained if the square root is taken of both sides of the equation

3 **Solve Pure Quadratic Equations ($ax^2 + c = 0$) by the Square-Root Method.**

To solve pure quadratic equations, solve for the squared variable and apply the **square-root property of equality.** That is, take the square root of both sides of the equation.

> **To solve a pure quadratic equation ($ax^2 + c = 0$):**
>
> 1. Rearrange the equation, if necessary, so that the quadratic term is on one side and the constant is on the other side of the equation.
>
> 2. Combine like terms, if appropriate.
>
> 3. Rewrite the equation so that the quadratic term has a coefficient of $+1$.
>
> 4. Apply the square-root property of equality by taking the square root of both sides.
>
> 5. Check each solution in the original equation.

STOP AND CHECK

1. Solve $7x^2 = 28$.

Answer:

1. $x = \pm 2$

EXAMPLE 3

Solve $3y^2 = 27$.

$$3y^2 = 27 \qquad \text{Divide both sides by the coefficient of the quadratic term.}$$
$$\sqrt{y^2} = \pm\sqrt{9} \qquad \text{Apply the square-root property of equality.}$$
$$y = \pm 3 \qquad \text{Solutions are } y = 3 \text{ and } y = -3.$$

Check $y = \boxed{3}$ Check $y = \boxed{-3}$
$$3y^2 = 27 \qquad\qquad\qquad 3y^2 = 27$$
$$3(\boxed{3})^2 = 27 \qquad\qquad 3(\boxed{-3})^2 = 27$$
$$3(9) = 27 \qquad\qquad\quad 3(9) = 27$$
$$27 = 27 \qquad\qquad\quad 27 = 27 \qquad \text{Solutions check.} \qquad \textbf{See Exercises 32–37.}$$

STOP AND CHECK

1. Solve $25x^2 - 9 = 0$.

Answer:

1. $x = \pm\dfrac{3}{5}$ or ± 0.6

EXAMPLE 4

Solve $4x^2 - 9 = 0$.

$$4x^2 - 9 = 0 \qquad \text{Sort terms.}$$
$$4x^2 = 9 \qquad \text{Divide by 4.}$$
$$x^2 = \frac{9}{4} \qquad \text{Apply the square-root property of equality.}$$
$$x = \pm\frac{3}{2} \qquad \text{Divide if a decimal answer is desired.}$$
$$x = \pm 1.5 \qquad \text{Decimal solutions are } x = 1.5 \text{ and } x = -1.5.$$

Check $x = \boxed{1.5}$ Check $x = \boxed{-1.5}$
$$4x^2 - 9 = 0 \qquad\qquad\qquad 4x^2 - 9 = 0$$
$$4(\boxed{1.5})^2 - 9 = 0 \qquad\qquad 4(\boxed{-1.5})^2 - 9 = 0$$
$$4(2.25) - 9 = 0 \qquad\qquad 4(2.25) - 9 = 0$$
$$9 - 9 = 0 \qquad\qquad\quad 9 - 9 = 0$$
$$0 = 0 \qquad\qquad\qquad 0 = 0 \qquad \text{Solutions check.}$$

In some applications, the fractional answer may be more convenient. In other applications, the decimal answer may be more convenient. **See Exercises 38–47.**

Did You Know? In a pure quadratic equation, the value of b is zero: $ax^2 + 0x + c = 0$.

The characteristics of the roots of a pure quadratic equation are:

▶ The equation has real number solutions only if c is negative when the equation is in standard form. Remember, standard form requires that a be positive.

▶   The two solutions have the same absolute value and are opposites.

▶   There will be either two solutions or no real solution unless $c = 0$.

Double root: two roots of an equation are the same value

▶   When $c = 0$, the solution is $x = 0$ and it is **double root.**

▶   A double root occurs when the two roots of the equation are the same value.

15–1 EXERCISES   MyLab Math For additional practice go to your study plan in MyLab Math.

1 Arrange the quadratic equations in standard form. *See Example 1.*

1. $5 - 4x + 7x^2 = 0$

2. $8x^2 - 3 = 6x$

3. $7x^2 = 5$

4. $x^2 = 6x - 8$

5. $x^2 - 3x = 6x - 8$

6. $8 - x^2 + 6x = 2x$

7. $3x^2 + 5 - 6x = 0$

8. $5 = x^2 - 6x$

9. $x^2 - 16 = 0$

10. $8x^2 - 7x = 8$

11. $8x^2 + 8x - 2 = 8$

12. $0.3x^2 - 0.4x = 3$

2 Write each equation in standard form and identify a, b, and c. *See Example 2.*

13. $5x = x^2$

14. $7x - 3x^2 = 5$

15. $7x^2 - 4x = 0$

16. $8 + 3x^2 = 5x$

17. $5x - 6 = x^2$

18. $11x^2 = 8x$

19. $x = x^2$

20. $9x^2 - 7x = 12$

21. $5 = x^2$

22. $-x^2 - 6x + 3 = 0$

23. $0.2x - 5x^2 = 1.4$

24. $\dfrac{2}{3}x^2 - \dfrac{5}{6}x = \dfrac{1}{2}$

25. $1.3x^2 - 8 = 0$

26. $\sqrt{3}x^2 + \sqrt{5}x - 2 = 0$

27. $5 = -3x - 15x^2$

Write equations in standard form.

28. The coefficient of the quadratic term is 8, the coefficient of the linear term is -2, the constant term is -3.

29. The constant is 0, the coefficient of the linear term is 3, and the coefficient of the quadratic term is 1.

30. $a = 5, b = 2, c = -7$

31. $a = 2.5, c = -0.8$

3 Solve the equations. Round to thousandths when necessary. *See Example 3.*

32. $x^2 = 9$

33. $28x^2 = 112$

34. $9x^2 = 64$

35. $0.04x^2 = 0.81$

36. $2x^2 = 10$

37. $0.09y^2 = 0.81$

See Example 4.

38. $x^2 - 49 = 0$

39. $16x^2 - 49 = 0$

40. $5x^2 - 8 = 12$

41. $7x^2 - 2 = 19$

42. $12x^2 - 27 = 0$

43. $3x^2 + 4 = 7$

44. **AG/H** The label on a new swimming pool cover shows it is a square and has an area of 529 ft². What is the length of each side of the pool cover?

45. **AG/H** A farmer has a field designed in the shape of a circle for efficiency in irrigation. The area of the field is known to be 21,000 yd². What length of irrigation line is required to provide water to the entire field? (*Note:* The line moves from the center of the field in a circle around the field. The area of a circle is $A = \pi r^2$.)

46. Describe the process for solving a pure quadratic equation.

47. What is the relationship between the roots of a pure quadratic equation?

<table>
<tr><td>**15–2**</td><td># Solving Quadratic Equations by Factoring</td></tr>
</table>

LEARNING OUTCOMES

1 Solve incomplete quadratic equations ($ax^2 + bx = 0$) by factoring.

2 Solve complete quadratic equations ($ax^2 + bx + c = 0$) by factoring.

LC LEARNING CATALYTICS

1. Factor $10x^2 - 15x$ for x.
2. Factor $3x^2 + 7x - 6$ for x.

zero-product property: If $ab = 0$, then $a = 0$ or $b = 0$ where a and b are real numbers

1 **Solve Incomplete Quadratic Equations ($ax^2 + bx = 0$) by Factoring.** An incomplete quadratic equation in standard form always has a variable common factor on one side of the equation. We can factor the left side of the equation $x^2 + 2x = 0$ to $x(x + 2) = 0$. We have two factors that have a product of 0. The **zero-product property** can be used to solve the equation.

> **Zero-product property**
>
> If $ab = 0$, then $a = 0$ or $b = 0$ where a and b are real numbers.

> **To solve an incomplete quadratic equation ($ax^2 + bx = 0$) by factoring:**
>
> 1. Use the addition axiom to write the equation in standard form.
>
> 2. Factor out all common factors and set each factor containing a variable equal to zero using the zero-product property.
>
> 3. Solve for the variable in each equation formed in Step 2.
>
> 4. Check each solution in the original equation.

STOP AND CHECK

Solve for x.
1. $3x^2 = 6x$
2. $8x^2 + 2x = 0$

Answers:
1. $x = 0; x = 2$
2. $x = 0; x = -\dfrac{1}{4}$

EXAMPLE 1

Solve **(a)** $2x^2 = 5x$ for x and **(b)** $4x^2 - 8x = 0$ for x.

(a)

$$2x^2 = 5x \qquad \text{Write in standard form.}$$
$$2x^2 - 5x = 0 \qquad \text{Factor common factor.}$$
$$x(2x - 5) = 0 \qquad \text{Set each variable factor equal to zero.}$$
$$x = 0 \quad 2x - 5 = 0 \qquad \text{Solve each equation.}$$
$$x = \boxed{0} \qquad 2x = 5$$
$$x = \frac{5}{2}$$

$$\text{Check } x = \boxed{0} \qquad\qquad \text{Check } x = \frac{5}{2}$$
$$2x^2 = 5x \qquad\qquad 2x^2 = 5x$$
$$2(\boxed{0})^2 = 5(\boxed{0}) \qquad\qquad 2\left(\frac{5}{2}\right)^2 = 5\left(\frac{5}{2}\right)$$
$$2(0) = 5(0) \qquad\qquad \overset{1}{2}\left(\frac{25}{\underset{2}{\cancel{4}}}\right) = \frac{25}{2}$$
$$0 = 0 \qquad\qquad \frac{25}{2} = \frac{25}{2} \qquad \text{Solutions check.}$$

(b)

$$4x^2 - 8x = 0 \qquad \text{Standard form. Factor common factors.}$$
$$4x(x - 2) = 0 \qquad \text{Set each variable factor equal to zero.}$$
$$4x = 0 \quad x - 2 = 0 \qquad \text{Solve each equation.}$$
$$x = \frac{0}{4} \qquad x = \boxed{2}$$
$$x = \boxed{0}$$

$$\text{Check } x = \boxed{0}$$
$$4(\boxed{0})^2 - 8(\boxed{0}) = 0$$
$$4(0) - 8(0) = 0$$
$$0 - 0 = 0$$
$$0 = 0$$

$$\text{Check } x = \boxed{2}$$
$$4(\boxed{2})^2 - 8(\boxed{2}) = 0$$
$$4(4) - 8(2) = 0$$
$$16 - 16 = 0$$
$$0 = 0 \qquad \text{Solutions check.}$$

See Exercises 1–6.

Did You Know? In an incomplete quadratic equation, the value of $c = 0$: $ax^2 + bx + 0 = 0$.

$$x(ax + b) = 0$$
$$x = 0 \qquad ax + b = 0$$
$$ax = -b$$
$$\frac{ax}{a} = \frac{-b}{a}$$
$$x = -\frac{b}{a}$$

The characteristics of the roots of an incomplete quadratic equation are:

▶ One root is always zero.

▶ The other root is $-\dfrac{b}{a}$.

---

2   Solve Complete Quadratic Equations ($ax^2 + bx + c = 0$) by Factoring. Complete quadratic equations have all three types of terms. If the expression on the left will factor into two binomials, we can apply the zero-product property. If the trinomial will not factor, we must use another method for solving the quadratic equation.

To solve a complete quadratic equation ($ax^2 + bx + c = 0$) by factoring:

1. Arrange the equation in standard form.

2. Factor the trinomial into the product of two binomials, and set each binomial factor equal to zero using the zero-product property.

3. Solve for the variable in each equation formed in Step 2.

4. Check if desired.

STOP AND CHECK

Solve for x.

1. $3x^2 + 7x - 6 = 0$

2. $2x^2 + 5 = -7x$

Answers:

1. $x = \dfrac{2}{3}; x = -3$

2. $x = -\dfrac{5}{2}; x = -1$

EXAMPLE 2

(a) Solve $x^2 + 6x + 5 = 0$ for x. **(b)** Solve $6x^2 + 4 = 11x$ for x.

(a)

$$x^2 + 6x + 5 = 0 \qquad \text{Standard form. Factor as the product of two binomials.}$$
$$(x + 5)(x + 1) = 0 \qquad \text{Set each factor equal to 0.}$$
$$x + 5 = 0 \quad x + 1 = 0 \qquad \text{Solve for } x \text{ in each equation.}$$
$$x = \boxed{-5} \quad x = \boxed{-1}$$

$$\text{Check } x = -5 \qquad\qquad \text{Check } x = -1$$

$$x^2 + 6x + 5 = 0 \qquad\qquad x^2 + 6x + 5 = 0$$

$$(-5)^2 + 6(-5) + 5 = 0 \qquad (-1)^2 + 6(-1) + 5 = 0$$

$$25 + (-30) + 5 = 0 \qquad\qquad 1 + (-6) + 5 = 0$$

$$-5 + 5 = 0 \qquad\qquad\qquad -5 + 5 = 0$$

$$0 = 0 \qquad\qquad\qquad\qquad 0 = 0 \qquad \text{Solutions check.}$$

(b)

$6x^2 + 4 = 11x$	Write in standard form.
$6x^2 - 11x + 4 = 0$	Factor trinomial by grouping. Write $-11x$ as $-3x - 8x$.
$6x^2 - 3x - 8x + 4 = 0$	Group.
$(6x^2 - 3x) + (-8x + 4) = 0$	Factor common factors.
$3x(2x - 1) - 4(2x - 1) = 0$	Factor the common grouping.
$(2x - 1)(3x - 4) = 0$	Set each factor equal to 0.
$2x - 1 = 0 \qquad 3x - 4 = 0$	Solve each equation.

$$2x = 1 \qquad\qquad 3x = 4$$

$$x = \frac{1}{2} \qquad\qquad x = \frac{4}{3}$$

$$\text{Check } x = \frac{1}{2} \qquad\qquad \text{Check } x = \frac{4}{3}$$

$$6x^2 + 4 = 11x \qquad\qquad 6x^2 + 4 = 11x$$

$$6\left(\frac{1}{2}\right)^2 + 4 = 11\left(\frac{1}{2}\right) \qquad 6\left(\frac{4}{3}\right)^2 + 4 = 11\left(\frac{4}{3}\right)$$

$$\overset{3}{\cancel{6}}\left(\frac{1}{\underset{2}{\cancel{4}}}\right) + 4 = \frac{11}{2} \qquad\quad \overset{2}{\cancel{6}}\left(\frac{16}{\underset{3}{\cancel{9}}}\right) + 4 = \frac{44}{3}$$

$$\frac{3}{2} + \frac{8}{2} = \frac{11}{2} \qquad \left(4 = \frac{8}{2}\right) \qquad \frac{32}{3} + \frac{12}{3} = \frac{44}{3} \qquad \left(4 = \frac{12}{3}\right)$$

$$\frac{11}{2} = \frac{11}{2} \qquad\qquad\qquad \frac{44}{3} = \frac{44}{3} \qquad \text{Solutions check.}$$

See Exercises 17–42.

Did You Know? Because all three types of terms are included in complete quadratic equations, the roots do *not* have the same characteristics as the roots of pure or incomplete quadratic equations.

The characteristics of a complete quadratic equation are:

▶ Zero is not a root.

▶ The two roots are not opposites.

▶ There can be just one distinct root. If an equation has two equal roots, the one distinct root is called a double root.

15–2 EXERCISES

MyLab Math For additional practice go to your study plan in MyLab Math.

1 Solve by factoring. *See Example 1.*

1. $x^2 - 3x = 0$

2. $x^2 - 6x = 0$

3. $5x^2 - 10x = 0$

4. $2x^2 + x = 0$

5. $8x^2 - 4x = 0$

6. $5x^2 - 15x = 0$

7. $3x^2 - 7x = 0$

8. $y^2 + 4y = 0$

9. $3x^2 + 2x = 0$

10. $9x^2 = 12x$

11. $6x^2 = 18x$

12. $9x^2 = 6x$

13. The square of a number is 8 times the number. Find the number.

14. A square rug has an area 15 times the length of one of the sides. What is the length of a side?

15. How does an incomplete quadratic equation differ from a pure quadratic equation?

16. Will one of the two roots of an incomplete quadratic equation always be zero? Justify your answer.

2 Solve the equations by factoring. *See Example 2.*

17. $x^2 + 5x + 6 = 0$

18. $x^2 - 6x + 9 = 0$

19. $x^2 - 5x - 14 = 0$

20. $x^2 + 3x - 18 = 0$

21. $x^2 + 7x + 12 = 0$

22. $y^2 - 8y = -15$

23. $a^2 - 13a = 14$

24. $b^2 - 9b = -18$

25. $2x^2 - 7x + 3 = 0$

26. $3x^2 + 13x + 4 = 0$

27. $10x^2 - x - 3 = 0$

28. $6x^2 + 11x + 3 = 0$

29. $2x^2 + 13x + 15 = 0$

30. $3x^2 - 10x + 8 = 0$

31. $6x^2 + 17x - 3 = 0$

32. $3x^2 + 14x + 8 = 0$

33. $2x^2 - 13x + 15 = 0$

34. $6x^2 - 7x + 2 = 0$

35. $8x^2 + 3 = 10x$

36. $5x^2 + 3x = 2$

37. $6x^2 = x + 15$

38. $9x^2 + 18x = -5$

39. $6x^2 + 3 = 11x$

40. $5x^2 + 13x = 6$

41. **CON**   A rectangular hallway is 6 ft longer than its width. The area is 55 ft^2. What are the length and width of the hallway?

42. **AUTO**   A rectangular metal plate covering a spare tire well that has an opening of 378 in^2 is broken and must be reconstructed. The width is 3 in. less than the length. What dimensions should the metalsmith use when making the replacement part?

15–3   Solving Quadratic Equations by Completing the Square or Using the Formula

LEARNING OUTCOMES

1 Solve quadratic equations by completing the square.

2 Solve quadratic equations using the quadratic formula.

LC LEARNING CATALYTICS

1. Expand $(2x - 3)^2$.

Completing the square: a procedure for solving a quadratic equation that is used most often when an equation cannot be solved by factoring

1 **Solve Quadratic Equations by Completing the Square.** Not all complete quadratic equations can be solved by factoring. One procedure for solving these equations is called **completing the square.**

This method incorporates manipulations that result in a perfect-square trinomial that factors into the square of a binomial. Look again at the relationship between a perfect-square trinomial and the square of a binomial.

$$a^2 + 2ab + b^2 = (a + b)^2$$

To solve a quadratic equation by completing the square:

1. Write the equation in the form $ax^2 + bx = -c$. That is, isolate the terms with variable factors.

2. Divide the equation by a so that the coefficient of x^2 is 1.

$$x^2 + \frac{b}{a}x = -\frac{c}{a}$$

3. Form a perfect-square trinomial on the left by adding $\left(\dfrac{b}{2a}\right)^2$ or $\dfrac{b^2}{4a^2}$ to both sides of the equation.

$$x^2 + \frac{b}{a}x + \frac{b^2}{4a^2} = -\frac{c}{a} + \frac{b^2}{4a^2}$$

4. Factor the perfect-square trinomial into the square of a binomial.

$$\left(x + \frac{b}{2a}\right)^2 = -\frac{c}{a} + \frac{b^2}{4a^2}$$

5. Take the square root of both sides of the equation.

$$x + \frac{b}{2a} = \pm\sqrt{-\frac{c}{a} + \frac{b^2}{4a^2}}$$

6. Rearrange terms under the radical.

$$x + \frac{b}{2a} = \pm\sqrt{\frac{b^2}{4a^2} - \frac{c}{a}}$$

7. Make equivalent fractions using the common denominator for terms under the radical.

$$x + \frac{b}{2a} = \pm\sqrt{\frac{b^2}{4a^2} - \frac{4ac}{4a^2}}$$

8. Write terms under the radical as one fraction.

$$x + \frac{b}{2a} = \pm\sqrt{\frac{b^2 - 4ac}{4a^2}}$$

9. Solve for x.

$$x = \frac{-b}{2a} \pm \frac{\sqrt{b^2 - 4ac}}{2a}$$

STOP AND CHECK

Solve by completing the square.

1. $x^2 + 3x - 1 = 0$
2. $2x^2 - 9x + 3 = 0$

Answers:

1. $x = -\dfrac{3}{2} + \dfrac{\sqrt{13}}{2}$ or $x = -\dfrac{3}{2} - \dfrac{\sqrt{13}}{2}$;
$x \approx 0.30$ or $x \approx -3.30$ rounded to the nearest hundredth
2. $x = \dfrac{9}{4} + \dfrac{\sqrt{57}}{4}$ or $x = \dfrac{9}{4} - \dfrac{\sqrt{57}}{4}$;
$x \approx 4.14$ or $x \approx 0.36$ rounded to the nearest hundredth

---

TIP Exact Root Versus Approximate Root Since the decimal equivalent of an irrational number will never terminate or repeat, in real-world settings it is often desirable to express the root of an equation as an approximate root. As a general rule, we will round approximate roots to the nearest hundredth unless indicated otherwise. Keeping the exact root is also desirable to be used with additional calculations, especially when a more accurate approximation is needed. The end user can then decide what level of accuracy is needed.

EXAMPLE 1

Solve **(a)** $x^2 - 5x - 2 = 0$ and **(b)** $3x^2 - 3x - 7 = 0$ by completing the square.

(a) $\qquad x^2 - 5x - 2 = 0$ — The coefficient of x^2 is 1. Isolate the terms with variable factors.

$$x^2 - 5x = 2$$

Add $\left(\dfrac{b}{2a}\right)^2$ to both sides $(a = 1, b = -5)$. $\left(\dfrac{-5}{2(1)}\right)^2$

$$x^2 - 5x + \left(\frac{-5}{2}\right)^2 = 2 + \left(\frac{-5}{2}\right)^2 \qquad \text{Simplify.}$$

$$x^2 - 5x + \frac{25}{4} = 2 + \frac{25}{4} \qquad \text{Combine like terms.}$$

$$x^2 - 5x + \frac{25}{4} = \frac{8}{4} + \frac{25}{4}$$

$$x^2 - 5x + \frac{25}{4} = \frac{33}{4}$$

Write the perfect-square trinomial as the square of a binomial. $\left(x + \dfrac{b}{2a}\right)^2$

$$\left(x - \frac{5}{2}\right)^2 = \frac{33}{4} \qquad \text{Take the square root of both sides.}$$

$$x - \frac{5}{2} = \pm\sqrt{\frac{33}{4}} \qquad \text{Simplify and solve for } x.$$

$$x = \frac{5}{2} \pm \frac{\sqrt{33}}{2} \qquad \text{Identify each root.}$$

$x = \dfrac{5}{2} + \dfrac{\sqrt{33}}{2}$ or $x = \dfrac{5}{2} - \dfrac{\sqrt{33}}{2}$ — Exact roots

$x \approx 2.5 + 2.872281323$ $\qquad$ $x \approx 2.5 - 2.872281323$

$x \approx 5.37$ or $x \approx -0.37$ — Approximate roots

(b) $3x^2 - 3x - 7 = 0$ Make the leading coefficient 1 by dividing the equation by 3.

$$\frac{3x^2}{3} - \frac{3x}{3} - \frac{7}{3} = 0$$

$$x^2 - x - \frac{7}{3} = 0$$ Isolate the terms with variable factors.

$$x^2 - x = \frac{7}{3}$$ Add $\left(\frac{-1}{2(1)}\right)^2$ to both sides.

$$x^2 - x + \left(\frac{-1}{2}\right)^2 = \frac{7}{3} + \left(\frac{-1}{2}\right)^2$$ Simplify.

$$x^2 - x + \frac{1}{4} = \frac{7}{3} + \frac{1}{4}$$ $\frac{7}{3} + \frac{1}{4} = \frac{28}{12} + \frac{3}{12} = \frac{31}{12}$

$$x^2 - x + \frac{1}{4} = \frac{31}{12}$$ Write the perfect-square trinomial as the square of the binomial.

$$\left(x - \frac{1}{2}\right)^2 = \frac{31}{12}$$ Take the square root of both sides.

$$x - \frac{1}{2} = \pm\sqrt{\frac{31}{12}}$$ Simplify and solve for x.

$\sqrt{\frac{31}{12}} = \frac{\sqrt{31}}{2\sqrt{3}} \cdot \frac{\sqrt{3}}{\sqrt{3}} = \frac{\sqrt{93}}{6}$

$$x = \frac{1}{2} \pm \frac{\sqrt{93}}{6}$$ Identify each root.

$x = \frac{1}{2} + \frac{\sqrt{93}}{6}$ or $x = \frac{1}{2} - \frac{\sqrt{93}}{6}$ Exact solutions

$x \approx 0.5 + 1.607275127$ $x \approx 0.5 - 1.607275127$

$\boldsymbol{x \approx 2.11}$ or $\boldsymbol{x \approx -1.11}$ Approximate solutions

See Exercises 1–18.

TIP   **Completing-the-Square Method Works for All Types of Quadratic Equations**   The completing-the-square method works for all types of quadratic equations. However, factoring is more efficient for incomplete quadratic equations and complete quadratic equations that factor.

Solve the incomplete quadratic equation $x^2 - 7x = 0$.

By Factoring

$$x^2 - 7x = 0$$

$$x(x - 7) = 0$$

$$x = 0 \quad x - 7 = 0$$

$$x = 7$$

By Completing the Square

$$x^2 - 7x + \left(\frac{-7}{2}\right)^2 = 0 + \left(\frac{-7}{2}\right)^2$$

$$x^2 - 7x + \frac{49}{4} = \frac{49}{4}$$

$$\left(x - \frac{7}{2}\right)^2 = \frac{49}{4}$$

$$x - \frac{7}{2} = \pm\frac{7}{2}$$

$$x = \frac{7}{2} \pm \frac{7}{2}$$

$x = \frac{7}{2} + \frac{7}{2}$ $x = \frac{7}{2} - \frac{7}{2}$

$x = \frac{14}{2}$ $x = 0$

$x = 7$

Solve the complete quadratic equation $x^2 + 5x + 6 = 0$.

By Factoring

$$x^2 + 5x + 6 = 0$$

$$(x + 2)(x + 3) = 0$$

$$x + 2 = 0 \quad x + 3 = 0$$

$$x = -2 \qquad x = -3$$

By Completing the Square

$$x^2 + 5x + 6 = 0$$

$$x^2 + 5x = -6$$

$$x^2 + 5x + \left(\frac{5}{2}\right)^2 = -6 + \left(\frac{5}{2}\right)^2$$

$$x^2 + 5x + \frac{25}{4} = -6 + \frac{25}{4}$$

$$\left(x + \frac{5}{2}\right)^2 = \frac{-24}{4} + \frac{25}{4}$$

$$\left(x + \frac{5}{2}\right)^2 = \frac{1}{4}$$

$$x + \frac{5}{2} = \pm\sqrt{\frac{1}{4}}$$

$$x = -\frac{5}{2} \pm \frac{1}{2}$$

$$x = \frac{-5}{2} + \frac{1}{2} \qquad x = \frac{-5}{2} - \frac{1}{2}$$

$$x = -\frac{4}{2} \qquad\qquad x = -\frac{6}{2}$$

$$x = -2 \qquad\qquad x = -3$$

Solve the pure quadratic equation $5x^2 = 45$.

A pure quadratic equation is solved by dividing by the coefficient of x^2 and then applying the square-root principle.

$$5x^2 = 45$$

$$\frac{5x^2}{5} = \frac{45}{5}$$

$$x^2 = 9$$

$$x = \pm 3$$

Quadratic formula: a formula for finding the solutions of a quadratic equation

2 Solve Quadratic Equations Using the Quadratic Formula. Another method for solving quadratic equations is to use the **quadratic formula.** This formula results from solving the standard quadratic equation, $ax^2 + bx + c = 0$, by completing the square.

Quadratic formula:

$$x = \frac{-b \pm \sqrt{b^2 - 4ac}}{2a}$$

where a, b, and c are real-number coefficients of a quadratic equation in the form $ax^2 + bx + c = 0$.

> **TIP**   **Quadratic Formula Versus Completing the Square**   Look again at the completing-the-square method for solving quadratic equations on pages 618–619. The symbolic representation of the method shows how the quadratic formula is derived from this method. On Step 9, the equation is solved for x and the fractions can be combined to match the form used in the quadratic formula.
>   Both methods can be used to solve any type of quadratic equation.

To solve quadratic equations using the quadratic formula:

1. Write the equation in standard form.

2. Identify a, b, and c.

3. Substitute numbers for a, b, and c in the quadratic formula.

4. Use the order of operations to simplify the expression under the radical.

5. Simplify the radical.

6. Factor any common factors in the numerator.

7. Simplify by reducing.

8. Write as two distinct solutions.

9. Write as exact solution or approximate solution as desired.

STOP AND CHECK

1. Use the quadratic formula to solve $x^2 + 5x - 6 = 0$ for x.

Answer:

1. $x = -6; x = 1$

EXAMPLE 2

Use the quadratic formula to solve $x^2 + 5x + 6 = 0$ for x.

$a = 1, \quad b = 5, \quad c = 6$ Identify a, b, and c.

$x = \dfrac{-b \pm \sqrt{b^2 - 4ac}}{2a}$ Quadratic formula. Substitute for a, b, and c.

$x = \dfrac{-5 \pm \sqrt{5^2 - 4(1)(6)}}{2 \cdot 1}$ Evaluate power and multiply in radicand.

$x = \dfrac{-5 \pm \sqrt{25 - 24}}{2}$ Combine terms in radicand.

$x = \dfrac{-5 \pm \sqrt{1}}{2}$ Evaluate radical.

$x = \dfrac{-5 \pm 1}{2}$

At this point, we separate the solutions into two parts, one using the $+1$ and the other using the -1.

$x = \dfrac{-5 + 1}{2}$ $\qquad\qquad$ $x = \dfrac{-5 - 1}{2}$

$x = \dfrac{-4}{2}$ $\qquad\qquad$ $x = \dfrac{-6}{2}$

$x = -2$ $\qquad\qquad$ $x = -3$

$\qquad$ Check $x = -2$ $\qquad\qquad\qquad$ Check $x = -3$

$\qquad x^2 + 5x + 6 = 0$ $\qquad\qquad\qquad x^2 + 5x + 6 = 0$

$\quad (-2)^2 + 5(-2) + 6 = 0$ $\qquad\qquad (-3)^2 + 5(-3) + 6 = 0$

$\qquad\quad 4 - 10 + 6 = 0$ $\qquad\qquad\qquad 9 - 15 + 6 = 0$

$\qquad\qquad\quad -6 + 6 = 0$ $\qquad\qquad\qquad\quad -6 + 6 = 0$

$\qquad\qquad\qquad\quad 0 = 0$ $\qquad\qquad\qquad\qquad\quad 0 = 0$ Solutions check.

See Exercises 19–24.

> **TIP** **Common Cause for Error** When you use the quadratic formula to solve problems, begin by writing the formula to help you remember it. When you write the formula, be sure to extend the fraction bar beneath the *entire* numerator. This omission is a common cause for errors.

STOP AND CHECK

1. Use the quadratic formula to solve the second equation in the Stop and Check for Example 1 (*see Outcome 1 in this section*), $2x^2 - 9x + 3 = 0$, for x.

Answer:

1. $x = \dfrac{9 \pm \sqrt{57}}{4}$;

$x \approx 4.14$ or $x \approx 0.36$

EXAMPLE 3

Use the quadratic formula to solve the equation in Example 1b (*see Outcome 1 in this section*), $3x^2 - 3x - 7 = 0$, for x. Round answers to the nearest hundredth.

$a = 3, \quad b = -3, \quad c = -7$ Identify a, b, and c.

$x = \dfrac{-b \pm \sqrt{b^2 - 4ac}}{2a}$ Quadratic formula. Substitute.

                       $-(b) = -(-3) = +3$

$x = \dfrac{3 \pm \sqrt{(-3)^2 - 4(3)(-7)}}{2 \cdot 3}$ Perform calculations in radicand.

$x = \dfrac{3 \pm \sqrt{9 + 84}}{6}$ Combine terms in the radicand.

$x = \dfrac{3 \pm \sqrt{93}}{6}$ Exact solutions

Find the approximate solutions:

$x \approx \dfrac{3 \pm 9.643650761}{6}$ Evaluate the radical. Separate the two solutions.

$x \approx \dfrac{3 + 9.643650761}{6}$ $x \approx \dfrac{3 - 9.643650761}{6}$

$x \approx \dfrac{12.643650761}{6}$ $x \approx \dfrac{-6.643650761}{6}$

$x \approx \mathbf{2.11}$ Rounded $x \approx \mathbf{-1.11}$ Approximate solutions

Notice that we round to hundredths *after* making the final calculation.

        Check $x \approx 2.11$ Check $x \approx -1.11$

$3x^2 - 3x - 7 \approx 0$ $3x^2 - 3x - 7 \approx 0$

$3(2.11)^2 - 3(2.11) - 7 \approx 0$ $3(-1.11)^2 - 3(-1.11) - 7 \approx 0$

$3(4.4521) - 6.33 - 7 \approx 0$ $3(1.2321) + 3.33 - 7 \approx 0$

$13.3563 - 6.33 - 7 \approx 0$ $3.6963 + 3.33 - 7 \approx 0$

            $0.0263 \approx 0$ $0.0263 \approx 0$ Solutions check.

See Exercises 25–30.

> **TIP** **Rounding Discrepancies and Checking with a Calculator** The symbol $\approx$ means "is approximately equal to." Another symbol for approximations is $\doteq$. If we had checked *without* rounding in the previous example, the checks would have been $0 = 0$.
>
> It is best to use the full calculator value to check. The full calculator value improves the accuracy of the check. Calculate the first root in the previous example.
>
> $3 \boxed{+} \boxed{\sqrt{}} \; 93 \boxed{=} \boxed{\div} 6 \boxed{=} \Rightarrow 2.107275127$

Check:

If your calculator allows you to insert the answer of your last calculation, use this function in checking. Otherwise, you can store an answer in memory and recall as appropriate.

$$3 \boxed{\text{ANS}} \boxed{x^2} \boxed{-} 3 \boxed{\text{ANS}} \boxed{-} 7 \boxed{=} \Rightarrow 0$$

Calculate the second root. $3 \boxed{-} \boxed{\sqrt{}} 93 \boxed{=} \boxed{\div} 6 \boxed{=} \Rightarrow -1.107275127$

Check: $3 \boxed{\text{ANS}} \boxed{x^2} \boxed{-} 3 \boxed{\text{ANS}} \boxed{-} 7 \boxed{=} \Rightarrow 0$

Even though quadratic equations have two roots, a root may be disregarded in an applied problem because it is not appropriate within the context of the problem.

Nuwatphoto/Shutterstock

EXAMPLE 4

Find the length and width of a rectangular table if the length is 8 in. more than the width and the area is 260 in^2.

Unknown facts

Length and width of rectangle

Known facts

Area $= 260$ in^2, length $= 8$ in. more than width

Relationships

Let $x =$ number of inches in the width
$x + 8 =$ number of inches in the length
Area $=$ length times width, or $A = lw$

Estimation

The square root of 260 is between 16 and 17 ($16^2 = 256$, $17^2 = 289$); the width should be less than 16 and the length more than 16.

Calculations

$260 = (x + 8)(x)$	Substitute into the area formula and distribute.
$260 = x^2 + 8x$	Write in standard form.
$0 = x^2 + 8x - 260 \quad$ or $\quad x^2 + 8x - 260 = 0$	
$a = 1, \quad b = 8, \quad c = -260$	
$x = \dfrac{-b \pm \sqrt{b^2 - 4ac}}{2a}$	Quadratic formula. Substitute.
$x = \dfrac{-8 \pm \sqrt{(8)^2 - 4(1)(-260)}}{2(1)}$	Multiply in the radicand.
$x = \dfrac{-8 \pm \sqrt{64 + 1,040}}{2}$	Combine terms in radicand.
$x = \dfrac{-8 \pm \sqrt{1,104}}{2}$	Simplify radical.
$x = \dfrac{-8 \pm 4\sqrt{69}}{2}$	Factor and reduce.
$x = -4 \pm 2\sqrt{69}$	Exact solutions

Find the approximate solutions:

$$x \approx -4 \pm 2\sqrt{69}$$ Evaluate the radical.

$$x \approx -4 \pm 2(8.306623863)$$ Perform the multiplication.

$$x \approx -4 + 16.61324773 \qquad x \approx -4 - 16.61324773$$ Separate into two solutions.

$$x \approx 12.61324773 \qquad\qquad x \approx -20.61324773$$

$$x \approx 12.6 \ \text{width} \qquad\qquad\quad x \approx -20.6$$ Disregard negative solution.

$$x + 8 \approx 20.6 \ \text{length}$$

Interpretation

Measurements are positive, so disregard the negative solution.
The width is approximately 12.6 in. and the length is approximately 20.6 in.

See Exercises 31–36.

15-3 EXERCISES   MyLab Math   For additional practice go to your study plan in MyLab Math.

1 Solve by completing the square. Give the exact solutions. Simplify radicals and reduce. *See Example 1.*

1. $x^2 + 2x - 48 = 0$ **2.** $x^2 - 8x - 9 = 0$ **3.** $x^2 - 10x + 9 = 0$

4. $x^2 - 10x + 24 = 0$ **5.** $x^2 + 2x = 24$ **6.** $x^2 + 16x = -28$

7. $x^2 - 3x - 5 = 0$ **8.** $x^2 - 5x - 2 = 0$ **9.** $x^2 - 5x - 3 = 0$

10. $x^2 - 3x - 1 = 0$ **11.** $2x^2 - 4x - 3 = 0$ **12.** $2x^2 - 2x - 1 = 0$

13. $3x^2 - 6x + 2 = 0$ **14.** $3x^2 + 12x + 8 = 0$ **15.** $4x^2 - 12 = 8x$

16. $7x^2 = 3x + 7$ **17.** $5x^2 - 11x + 2 = 0$ **18.** $3x^2 - x = 7$

2 Solve the quadratic equations by using the quadratic formula. *See Example 2.*

19. $3x^2 - 7x - 6 = 0$ **20.** $x^2 + x - 12 = 0$ **21.** $5x^2 - 6x = 11$

22. $x^2 - 6x + 9 = 0$ **23.** $x^2 - x - 6 = 0$ **24.** $8x^2 - 2x = 3$

Solve the quadratic equations by using the quadratic formula. Round each approximate answer to the nearest hundredth. *See Example 3.*

25. $x^2 = -9x + 20$ **26.** $3x^2 + 6x + 1 = 0$ **27.** $\dfrac{2}{x} + \dfrac{5}{2} = x$

28. $\dfrac{2}{x} + 3x = 8$ **29.** $2x^2 + 3x + 3 = 0$ **30.** $x^2 - x + 2 = 0$

Solve the applied problems. *See Example 4.*

31. **CAD/ARC** A rectangular tabletop is 3 cm longer than it is wide. Find the length and width to the nearest hundredth if the area is 47.5 cm². Round to hundredths. (Area = length × width, or $A = lw$.)

32. **INDTEC** A rectangular instrument case has an area of 40 in². If the length is 6 in. more than the width, find the dimensions (length and width) of the instrument case.

33. **CON**   A bricklayer plans to build an arch with a span (s) of 8 m and a radius (r) of 4 m. How high (h) is the arch? (Use the formula $h^2 - 2hr + \dfrac{s^2}{4} = 0$.)

34. **CON** A rectangular patio slab is 180 ft^2. If the length is 1.5 times the width, find the width and length of the slab to the nearest whole number.

35. **AG/H**   A farmer normally plants a rectangular field 80 ft by 120 ft. This year the government requires the planting area to be decreased by 20%, and the farmer chooses to decrease the width and length of the field by an equal amount. Draw both the original field and the reduced-size field. If x is the amount by which the length and width are decreased, find the length and width of the new field. Round to the nearest whole number.

36. **HLTH/N**   The recommended dosage of a certain type of medicine is determined by the patient's weight. The formula to determine the dosage is given by $D = 0.1w^2 + 5w$, where D is the dosage in milligrams (mg) and w is the patient's body weight in kilograms. Find the weight of a patient for whom the recommended dosage is 1,800 mg. Round to the nearest tenth.

15–4   Graphing Quadratic Functions

LEARNING OUTCOMES

1 Graph quadratic functions using the table-of-solutions method.

2 Graph quadratic functions by examining properties.

3 Solve a quadratic equation from a graph of a corresponding quadratic function.

4 Graph quadratic functions using a graphing calculator.

5 Determine the nature of the roots of a quadratic equation by examining the discriminant.

LC **LEARNING CATALYTICS**

1. Evaluate $y = x^2 - 4x + 3$ for $x = -3$.

---

1   **Graph Quadratic Functions Using the Table-of-Solutions Method.** The graphs of linear equations are *straight* lines. We will examine the graphs of some equations that have a degree higher than 1, which are not linear equations.

The graph of an equation of a degree higher than 1 is a *curved* line. The curved line can be a parabola, hyperbola, circle, ellipse, or irregular curved line. We do not define these terms at this time, but Fig. 15–1 illustrates them. The equation for a *parabola* that opens up or down is distinguished from other quadratic equations. The y variable has degree 1. The x variable must have one term with degree 2 and no term with a degree higher than 2. The quadratic equation in two variables for a parabola can be written as the function $y = ax^2 + bx + c$ or $f(x) = ax^2 + bx + c$.

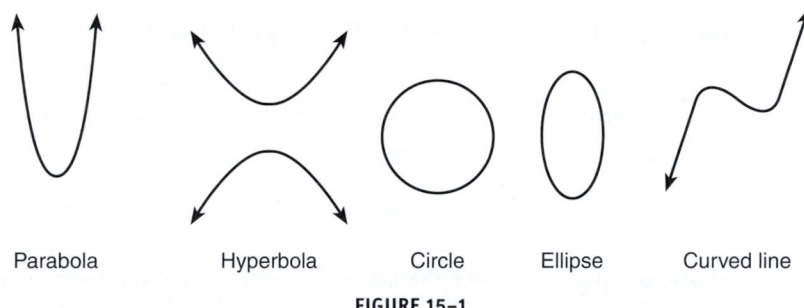

Parabola   Hyperbola   Circle   Ellipse   Curved line

FIGURE 15–1

One method of graphing quadratic functions is to form a table of solutions and plot the points. This method requires *more* points than we customarily use in graphing linear equations.

To graph quadratic functions using the table-of-solutions method:

1. Prepare a table of solutions of at least five ordered pairs.

2. Plot the points on a rectangular coordinate system.

3. Connect the points with a smooth, continuous curve.

1. Prepare a table of solutions and graph the function $y = x^2 - 4x + 3$ using integers between -1 and 5 for x, inclusive. Determine the domain and range.

Answer:

1.

x	y
-1	8
0	3
1	0
2	-1
3	0
4	3
5	8

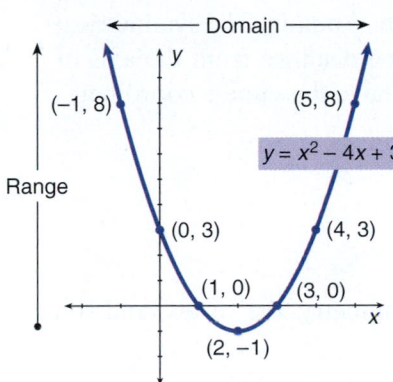

The domain is the set of all real numbers. The range is the set of real numbers greater than or equal to -1.

EXAMPLE 1

Prepare a table of solutions and graph the function $y = x^2 + x - 6$ using integers between -3 and 3 for x, inclusive. Determine the domain and range.

$y = x^2 + x - 6$

x	y
-3	0
-2	-4
-1	-6
0	-6
1	-4
2	0
3	6

At $x = -3$:
$$y = (-3)^2 + (-3) - 6$$
$$y = 9 - 3 - 6$$
$$\boldsymbol{y = 0}$$

At $x = -1$:
$$y = (-1)^2 + (-1) - 6$$
$$y = 1 - 1 - 6$$
$$\boldsymbol{y = -6}$$

At $x = 1$:
$$y = 1^2 + 1 - 6$$
$$y = 1 + 1 - 6$$
$$\boldsymbol{y = -4}$$

At $x = 3$:
$$y = 3^2 + 3 - 6$$
$$y = 9 + 3 - 6$$
$$\boldsymbol{y = 6}$$

At $x = -2$:
$$y = (-2)^2 + (-2) - 6$$
$$y = 4 - 2 - 6$$
$$\boldsymbol{y = -4}$$

At $x = 0$:
$$y = 0^2 + 0 - 6$$
$$y = 0 + 0 - 6$$
$$\boldsymbol{y = -6}$$

At $x = 2$:
$$y = 2^2 + 2 - 6$$
$$y = 4 + 2 - 6$$
$$\boldsymbol{y = 0}$$

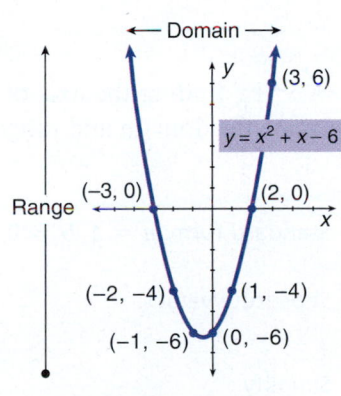

FIGURE 15-2

Plot the points indicated in the table of solutions and connect the points with a smooth, continuous curve. The curve is symmetrical (Fig. 15-2).

The lowest point is halfway between $x = -1$ and $x = 0$ or at $x = -\dfrac{1}{2}$. The corresponding y-value is $-6\dfrac{1}{4}$. $\left(-\dfrac{1}{2}\right)^2 + \left(-\dfrac{1}{2}\right) - 6 = \dfrac{1}{4} - \dfrac{1}{2} - 6 = \dfrac{1}{4} - \dfrac{2}{4} - 6 = -6\dfrac{1}{4}$

The domain is the set of all real numbers. The range is the set of real numbers greater than or equal to the lowest point of the graph $\left(-6\frac{1}{4}\right)$.   See Exercises 1–4.

Axis of symmetry of a parabola: the fold line of a parabola when the parabola is folded in half and the two halves match. Symbolically,
$$x = -\frac{b}{2a}$$

Vertex of a parabola: the point of the graph of a parabola that crosses the axis of symmetry

2 Graph Quadratic Functions by Examining Properties. As with linear equations, graphing quadratic equations that form a parabola by the table-of-solutions method can be time-consuming. The graph can be drawn and other important characteristics determined by examining some key properties of the parabola.

A parabola is *symmetrical;* that is, it can be folded in half and the two halves will match. The fold line is called the **axis of symmetry of a parabola.** For a parabola in

the form $y = ax^2 + bx + c$, the equation of the axis of symmetry is $x = -\dfrac{b}{2a}$. The coefficient a does *not* have to be positive.

The point of the graph that crosses the axis of symmetry is the **vertex of a parabola**.

Thus, the x-coordinate of the vertex of the parabola is $-\dfrac{b}{2a}$.

To graph a quadratic function in the form $y = ax^2 + bx + c$:

1. Find the axis of symmetry: $x = -\dfrac{b}{2a}$.

2. Find the vertex: x-coordinate of vertex $= -\dfrac{b}{2a}$. To find the y-coordinate of the vertex, substitute the value for the x-coordinate into the original function and solve for y.

3. Find one or two additional points that are to the right of the axis of symmetry.

4. Apply the property of symmetry to find additional points. The symmetrical point will have an x-coordinate that is the same distance from the axis of symmetry on the opposite side. The two points have the same y-coordinate.

5. Connect the plotted points with a smooth, continuous curve.

EXAMPLE 2

Graph the function $y = x^2$ by finding the axis of symmetry, the vertex, and some additional points. Determine the domain and range.

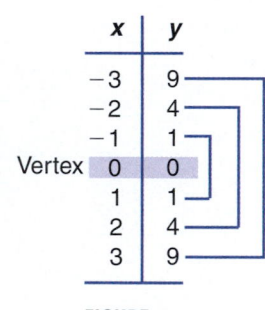

x	y
-3	9
-2	4
-1	1
Vertex 0	0
1	1
2	4
3	9

FIGURE 15–3

$$y = x^2$$
$$y = x^2 + 0x + 0 \qquad \text{Standard form: } a = 1, b = 0, c = 0$$
$$x = -\dfrac{b}{2a} \qquad \text{Substitute values.}$$
$$x = -\dfrac{0}{2(1)} \qquad \text{Simplify.}$$
$$x = 0 \qquad \text{Axis of symmetry and } x\text{-coordinate of vertex}$$

Axis of symmetry: $x = 0$

$$y = 0^2 \qquad \text{Substitute 0 for } x \text{ in } y = x^2.$$
$$y = 0 \qquad y\text{-coordinate of vertex}$$

Vertex: (0, 0)

Find y for $x = 1$: $y = 1^2$ Find y for $x = 2$: $y = 2^2$ Find y for $x = 3$: $y = 3^2$

               $y = 1$ $y = 4$ $y = 9$

Apply the principle of symmetry to complete the table (Fig. 15–3). Plot the points and connect them with a smooth, continuous curve (Fig. 15–4).

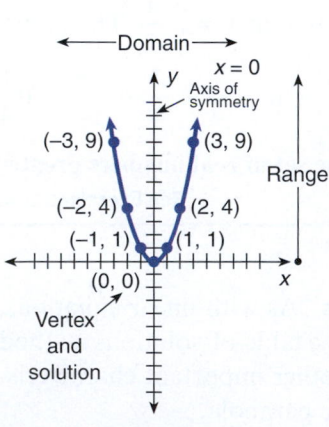

FIGURE 15–4

The domain is the set of all real numbers. The range is the set of real numbers greater than or equal to 0. See Exercises 5–9.

STOP AND CHECK

1. Graph the function $y = x^2 - 4x + 3$ by using the axis of symmetry, the vertex, and the x-intercepts. Determine the domain and the range.

Answer:

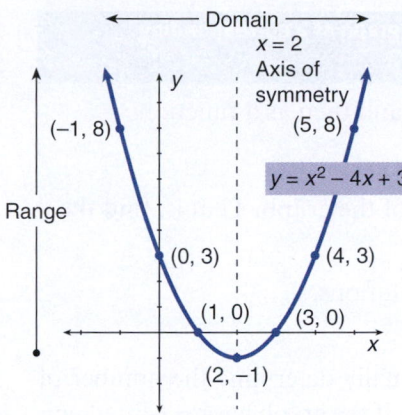

The domain is the set of all real numbers. The range is the set of real numbers greater than or equal to -1.

EXAMPLE 3

Graph the function $y = -x^2 + 3$ by using the axis of symmetry, the vertex, and the x-intercepts. Determine the domain and range.

Axis of symmetry: $\quad x = -\dfrac{b}{2a}$ $\qquad$ $y = -x^2 + 0x + 3$. Substitute values.
$a = -1, b = 0$

$\qquad\qquad\qquad\quad x = -\dfrac{0}{2(-1)}$ $\qquad$ Simplify.

$\qquad\qquad\qquad\quad \mathbf{x = 0}$ $\qquad\qquad$ Also, x-coordinate of vertex

Vertex: $(0, y)$ $\qquad\qquad\qquad\qquad\qquad$ Find y-coordinate of vertex.

$\qquad\qquad\qquad\quad y = -0^2 + 0 + 3$ $\qquad$ Substitute $x = 0$ and solve for y.

$\qquad\qquad\qquad\quad y = 3$ $\qquad\qquad\qquad$ y-coordinate of vertex

$\qquad\qquad \mathbf{(0, 3)}$

x-intercepts: $\qquad\quad 0 = -x^2 + 3$ $\qquad$ Substitute $y = 0$ and solve for x.

$\qquad\qquad\qquad\quad x^2 = 3$ $\qquad\qquad\quad$ Apply the square-root property.

$\qquad\qquad\qquad\quad x = \pm\sqrt{3}$ $\qquad\qquad$ Exact value of x-intercept

$\qquad\qquad\qquad\quad x \approx \pm 1.7$ $\qquad\qquad$ Approximate value of x-intercept

$\mathbf{(1.7, 0); (-1.7, 0)}$ $\qquad\qquad\qquad$ Point notation of x-intercepts

Plot the points and connect them with a smooth, continuous curve. Two additional points are plotted to give a more complete view of the parabola (Fig. 15–5).

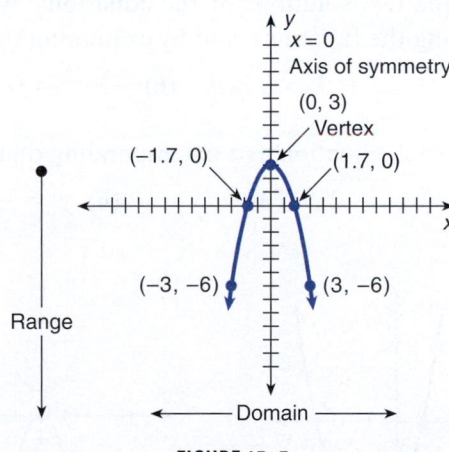

When $x = 3$: $y = -(3)^2 + 3$
$\qquad\qquad\quad y = -9 + 3$
$\qquad\qquad\quad y = -6$
$(3, -6)$ is on the graph.
Then, $(-3, -6)$ is on the graph by the property of symmetry.

FIGURE 15–5

The domain is the set of all real numbers. The range is the set of real numbers less than or equal to 3.

See Exercises 10–15.

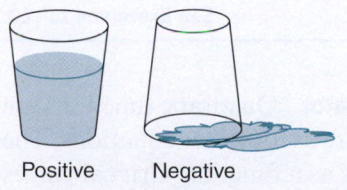

Positive $\qquad$ Negative

FIGURE 15–6

TIP **Tips for Sketching Curves** When the coefficient of the quadratic term is negative as in the preceding example, the graph of the parabola opens downward. When the coefficient of the quadratic term is positive, the graph of the parabola opens upward. Think of the parabola as a cup or glass: positive holds water; negative spills water (Fig. 15–6).

Zero of a function: the x-coordinate of an x-intercept of a function

3   Solve a Quadratic Equation from a Graph of a Corresponding Quadratic Function.
A quadratic equation in one variable can be written as a function by writing the equation in standard form ($ax^2 + bx + c = 0$) and then writing it as a function [$y = ax^2 + bx + c$ or $f(x) = ax^2 + bx + c$]. As with linear equations, the real-number solutions of a quadratic equation are at the x-intercepts of the graph of the equation written as a function. The x-coordinate of an x-intercept is also called a **zero of a function**.

> **To find the real solutions of a quadratic equation from the graph of a corresponding quadratic function:**
>
> 1. Write the quadratic equation in standard form and then as a function.
> 2. Graph the function.
> 3. Determine the x-coordinate of all x-intercepts of the graph. That is, find the zeros of the function.
> 4. If there are no x-intercepts, there are no real solutions.

When we graph a quadratic function, we can visually determine the number of real solutions of the corresponding quadratic equation. If the graph has no x-intercept, the quadratic equation has *no real solution*. When the graph has exactly one x-intercept, the equation has a *double root* (one distinct root or solution). If the graph has two x-intercepts, the corresponding quadratic equation has *two real solutions*.

STOP AND CHECK

Find the real solutions of the equations by writing the equations as functions, by graphing the functions, and by examining the graphs.

1. $x^2 + x - 2 = 0$
2. $x^2 - 4x = -4$
3. $x^2 - x - 1 = 0$

Answers:
1. $x = -2; x = 1$
2. $x = 2$
3. No real solutions

EXAMPLE 4

Find the real solutions of the equations by writing the equations as functions, by graphing the functions, and by examining the graphs.

(a) $x^2 - 2x - 3 = 0$ (b) $-3x^2 = 0$ (c) $x^2 + 5 = 0$

Write each equation as a corresponding quadratic function.

(a) $y = x^2 - 2x - 3$ (b) $y = -3x^2$ (c) $y = x^2 + 5$

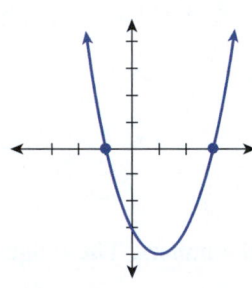

FIGURE 15–7

Solutions: $x = -1$
             $x = 3$

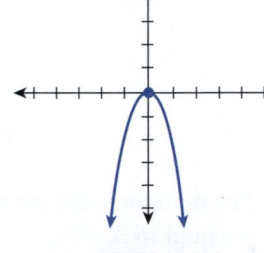

FIGURE 15–8

Solution: $x = 0$

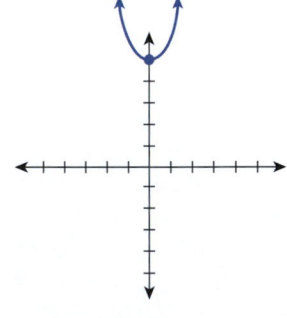

FIGURE 15–9

No real solutions

See Exercises 16–20.

4   Graph Quadratic Functions Using a Graphing Calculator. Quadratic functions are graphed on graphing calculators using the same procedure as for linear equations. The equation must first be written in standard form and then as a function. Critical values on the graph such as the vertex, x-intercepts or solutions of the corresponding quadratic equation, and the y-intercepts can be determined using calculator functions.

> **To graph quadratic functions using a graphing calculator:**
>
> 1. Clear the graphing screen.
> 2. Enter the equation that has been solved for y and press the graph function.

TIP **To Find the Vertex of a Quadratic Function from the Calculator Display of the Graph**

1. Graph the function on the calculator.
2. Adjust the view of the graph by using the Window or Zoom feature so that the vertex is visible. If the graph turns upward, the vertex is a low point or **minimum of a parabola.** If the graph turns downward, the vertex is a high point or **maximum of a parabola** (Fig. 15–10).

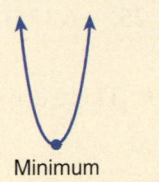 

Maximum

Minimum

FIGURE 15–10

3. Select the appropriate minimum or maximum option from the CALC menu.
4. Move the cursor and press ENTER to a mark point to the left of the vertex (left bound?), to the right of the vertex (right bound?), and near the vertex (guess?).
5. Read the coordinates of the vertex at the bottom of the display screen.

TIP **To Find the Real Solutions of a Quadratic Equation from the Calculator Display of the Graph**

1. Write the equation in standard form and then as a function.

2. Graph the function on the calculator.

3. Adjust the view of the graph by using the Window or Zoom feature so that one or two x-intercepts are visible or so that it can be determined that there are no x-intercepts.

4. If there are no x-intercepts, there are no real solutions.

5. If there are one or two intercepts, select the zero option from the CALC menu.

6. Move the cursor and press ENTER to mark points to the left (left bound?), right (right bound?), and near each x-intercept (guess?). Each x-intercept is one real solution.

7. Each solution is represented by the x-coordinate of each x-intercept at the bottom of the display screen.

EXAMPLE 5

Graph the equation $2x^2 - 5x - 3 = 0$ by writing it as a function and by using a graphing calculator. Find the vertex and the zeros of the function from the graph (Fig. 15–11).

Enter the equation into the calculator.

$\boxed{Y=}$ 2 $\boxed{X, \theta, T, n}$ $\boxed{x^2}$ $\boxed{-}$ 5 $\boxed{X, \theta, T, n}$ $\boxed{-}$ 3 $\boxed{ZOOM}$ $\boxed{6\text{:ZStandard}}$

Minimum of a parabola: the vertex or low point of a parabola that turns upward

Maximum of a parabola: the vertex or high point of a parabola that turns downward

STOP AND CHECK

1. Graph the equation $3x^2 + 8x + 4 = 0$ by writing it as a function and by using a graphing calculator. Find the vertex and the zeros of the function from the graph.

Answer:

1. Vertex: $(-1.33, -1,33)$; zeros or roots: $x = -2$ and $x = -0.67$

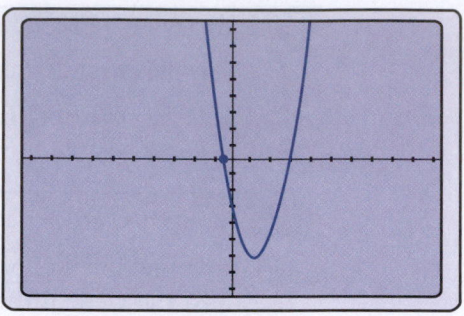

FIGURE 15–11

Find the vertex: The vertex is a minimum.	CALC 3:minimum
Vertex = (1.25, −6.125)	Mark left, right, and guess. $x \approx 1.25$ (rounded), $y \approx -6.125$
Find the zeros: There are two x-intercepts.	CALC 2:zero
	Mark left, right, and guess for leftmost solution.
x-intercept 1: $x = -0.5$	$x = -0.5, y = 0$
	CALC 2:zero
	Mark left, right, and guess for rightmost solution.
x-intercept 2: $x = 3$	$x = 3, y = 0$ **See Exercises 21–24.**

5 Determine the Nature of the Roots of a Quadratic Equation by Examining the Discriminant. We have made various observations about the roots of the different types of quadratic equations from both algebraic and graphical methods. Now, we make additional observations about the roots of quadratic equations.

In general, follow these suggestions for solving quadratic equations.

1. Write the equation in standard form: $ax^2 + bx + c = 0$.
2. To solve the equation algebraically:
   (a) Identify the type of quadratic equation.
   (b) Solve pure quadratic equations by the square-root method.
   (c) Solve incomplete quadratic equations by finding common factors.
   (d) Solve complete quadratic equations by factoring into two binomials, if possible.
   (e) If factoring is not possible or is difficult, use the completing-the-square method or the quadratic formula to solve complete quadratic equations.
3. To solve the equation graphically:
   (a) Write the equation as a function.
   (b) Graph the function.
   (c) Find the x-coordinates of the x-intercepts (zeros) of the function.

The nature of the roots of a quadratic equation can be determined before solving the equation. All types of quadratic equations can be solved using the completing-the-square method or the quadratic formula. Only certain types of quadratic equations can be solved by factoring or applying the square-root property. The radicand of the radical portion of the quadratic formula, $b^2 - 4ac$, is called the **discriminant.** The general characteristics of the roots of a quadratic equation can be determined by examining the discriminant.

Discriminant: the radicand of the radical portion of the quadratic formula, $b^2 - 4ac$

> **Properties of the discriminant, $b^2 - 4ac$:**
>
> 1. If $b^2 - 4ac \geq 0$, the equation has real-number roots.
>   (a) If $b^2 - 4ac$ is a perfect square, there are *two rational roots.*
>   (b) If $b^2 - 4ac = 0$, there are two equal rational roots, sometimes called a *double root.*
>   (c) If $b^2 - 4ac$ is not a perfect square, there are two irrational roots.
> 2. If $b^2 - 4ac < 0$, the equation has no real-number roots. The roots are *imaginary* or *complex.*

STOP AND CHECK

Examine the discriminant of each equation and determine the nature of the roots. Then solve the equation.

1. $3x^2 + 20x - 7 = 0$
2. $7x^2 - 8x + 3 = 0$
3. $2x^2 - 5x + 1 = 0$
4. $4x^2 + 12x + 9 = 0$

Answers:

1. Two rational roots, $\dfrac{1}{3}$ and -7
2. Two complex roots, $\dfrac{4 \pm i\sqrt{5}}{7}$
3. Two irrational roots, $\dfrac{5 \pm \sqrt{17}}{4}$
4. Double root, $-\dfrac{3}{2}$

EXAMPLE 6

Examine the discriminant of each equation and determine the nature of the roots. Then solve the equation.

(a) $5x^2 + 3x - 1 = 0$ **(b)** $3x^2 + 5x = 2$ **(c)** $4x^2 + 2x + 3 = 0$

(a) $5x^2 + 3x - 1 = 0$	$a = 5, b = 3, c = -1$.
$b^2 - 4ac = 3^2 - 4(5)(-1)$	Examine the discriminant.
$= 9 + 20$	
$= \boxed{29}$	There are two irrational roots. Use the quadratic formula.
$x = \dfrac{-b \pm \sqrt{b^2 - 4ac}}{2a}$	Substitute 29 for the discriminant.
$x = \dfrac{-3 \pm \sqrt{29}}{2(5)}$	Identify each root.
$x = \dfrac{-3 + \sqrt{29}}{10}$ or $x = \dfrac{-3 - \sqrt{29}}{10}$	Exact irrational roots
(b) $\qquad 3x^2 + 5x = 2$	
$3x^2 + 5x - 2 = 0$	Standard form: $a = 3, b = 5, c = -2$
$b^2 - 4ac = 5^2 - 4(3)(-2)$	Examine the discriminant.
$= 25 + 24$	
$= 49$	The discriminant is a perfect square. **There are two rational roots** and the trinomial will factor.
$(3x - 1)(x + 2) = 0$	Factor. Set each factor equal to zero.
$3x - 1 = 0 \qquad x + 2 = 0$	Solve each equation.
$3x = 1$	
$x = \dfrac{1}{3} \qquad\qquad x = -2$	Exact rational roots
(c) $4x^2 + 2x + 3 = 0$	$a = 4, b = 2, c = 3$.
$b^2 - 4ac = 2^2 - 4(4)(3)$	Examine the discriminant.
$= 4 - 48$	
$= \boxed{-44}$	The discriminant is negative. There are no real roots. **The roots are imaginary or complex.**

$$x = \frac{-2 \pm \sqrt{-44}}{2 \cdot 4}$$

Substitute -44 for the discriminant in the quadratic formula. Simplify the radical.

$$x = \frac{-2 \pm 2i\sqrt{11}}{8}$$

Factor a common factor in the numerator.

$$x = \frac{2(-1 \pm i\sqrt{11})}{8}$$

Reduce.

$$x = \frac{-1 \pm i\sqrt{11}}{4}$$

The roots are complex.

See Exercises 25–30.

15–4 EXERCISES   MyLab Math   For additional practice go to your study plan in MyLab Math.

1 Use a table of solutions to graph the quadratic equations. Determine the domain and range. *See Example 1.*

1. $y = x^2$

2. $y = 3x^2$

3. $y = \frac{1}{3}x^2$

2 *See Example 2.*

4. $y = x^2 - 4$

5. $y = -4x^2$

6. $y = -\frac{1}{4}x^2$

7. $y = x^2 + 4$

8. $y = x^2 - 6x + 9$

9. $y = -x^2 + 6x - 9$

Graph the equations using the vertex, x-intercepts, and one other point on the graph. Determine the domain and range. *See Example 3.*

10. $y = x^2 - 4x + 4$

11. $y = x^2 + 4x - 2$

12. $y = 2x^2 + 10x + 8$

13. $y = -3x^2 - 6x + 9$

14. $y = -3x^2 - 6x - 6$

15. $y = \frac{1}{2}x^2 - 6x + 3$

3 Find all real solutions of the equations by writing and graphing the equations of the corresponding quadratic functions. *See Example 4.*

16. $x^2 - 5x + 6 = 0$

17. $4x^2 = 0$

18. $3x^2 = -5$

19. **BUS**   The revenue R for a leather wallet is based on the price P of the item and is given by the formula $R = -4p^2 + 36p$. What prices generate \$0 revenue?

20. **TELE**   The distance y that an object falls in a vacuum because of gravity in t seconds after it is released is given by the formula $d = 4.9t^2$. How much time is elapsed when $d = 0$?

4 Graph using a graphing calculator. Adjust the window if necessary to show the vertex of the parabola. *See Example 5.*

21. $y = 3x^2 + 5x - 2$

22. $y = (2x - 3)(x - 1)$

23. $y = 2x^2 - 9x - 5$

24. $y = -2x^2 + 9x + 5$

5 Examine the discriminant of each equation and determine the nature of the roots. Solve each equation. Round to hundredths if necessary. *See Example 6.*

25. $3x^2 + x - 2 = 0$

26. $x^2 - 3x = -1$

27. $2x^2 + x = 2$

28. $3x^2 - 2x + 1 = 0$

29. $x^2 - 3x - 7 = 0$

30. $3x^2 + 5x - 6 = 0$

31. Describe the discriminant of a quadratic equation that has real roots.

32. Describe the discriminant of a quadratic equation that has rational and unequal roots.

15–5 Solving Higher-Degree Equations by Factoring

LEARNING OUTCOMES

1 Identify the degree of an equation.

2 Solve higher-degree equations by factoring.

3 Graph higher-degree equations.

4 Distinguish between a function and a relation using the vertical line test.

5 Find the domain and range of a relation from a graph.

LC LEARNING CATALYTICS
State the degree of each term.
1. $5x^2$
2. $3x$
3. 6

Cubic equation: an equation that has at least one variable term raised to the third power and no variable terms having a power higher than 3

Third-degree equation: another name for a cubic equation

Degree of an equation in one variable: the highest degree of any variable term that appears in the equation

STOP AND CHECK
State the type of equation according to the degree of each of the equations.
1. $x^3 + 3x = 0$
2. $3x + 7 = 0$
3. $x^4 = 16$

Answers:
1. Cubic or third-degree equation
2. Linear or first-degree equation
3. Fourth-degree equation

Some equations contain terms that have a higher degree than 2.

1 Identify the Degree of an Equation. In Section 15–1, we defined a *quadratic* equation as an equation that has at least one variable term raised to the second power and no variable terms having a power higher than 2. Similarly, a **cubic equation** is an equation that has at least one variable term raised to the third power and no variable terms having a power higher than 3. A quadratic equation can also be referred to as a *second-degree equation,* and a cubic equation as a **third-degree equation.** The **degree of an equation in one variable** is the highest power of any variable term that appears in the equation.

When we solved linear or first-degree equations, we obtained at most one solution for the equation. Quadratic or second-degree equations have at most two solutions. Cubic or third-degree equations have at most three solutions, and fourth-degree equations have at most four solutions.

EXAMPLE 1

State the type of equation according to the degree of each of the equations.

(a) $x^2 + 3x + 4 = 0$ (b) $x^3 = 27$ (c) $x^4 + 3x^3 + 2x^2 + x + 4 = 0$
(d) $x^8 = 256$ (e) $3x + 7 = 5x - 3$

(a) $x^2 + 3x + 4 = 0$
Quadratic or second-degree equation

(b) $x^3 = 27$
Cubic or third-degree equation

(c) $x^4 + 3x^3 + 2x^2 + x + 4 = 0$
Fourth-degree equation

(d) $x^8 = 256$
Eighth-degree equation

(e) $3x + 7 = 5x - 3$
Linear or first-degree equation   See Exercises 1–6.

2 Solve Higher-Degree Equations by Factoring. Higher-degree equations that are written in factored form or can be written in factored form are presented here. We solve these equations using the *zero-product property;* that is, if the product of the factors equals 0, then any factor containing a variable may be equal to 0. (See Section 15–2.)

To solve a higher-degree equation by factoring:

1. Write the equation in standard form.

2. Factor the polynomial side of the equation.

3. Set each factor containing a variable equal to zero.

4. Solve each equation from Step 3.

EXAMPLE 2

Solve the equations for the exact roots using the zero-product property.

(a) $x(x + 4)(x - 3) = 0$ **(b)** $(x + 5)(x - 7)(2x - 1) = 0$

(c) $2x^3 - 14x^2 + 20x = 0$ **(d)** $3x^3 - 15x = 0$

(a) $x\ (x + 4)\ (x - 3) = 0$ Already factored. Set each factor equal to zero.

$x + 4 = 0$ $x - 3 = 0$ Solve each equation.

$x = 0$ $x = -4$ $x = 3$

(b) $(x + 5)\ (x - 7)\ (2x - 1) = 0$ Already factored. Set each factor equal to zero.

$x + 5 = 0$ $x - 7 = 0$ $2x - 1 = 0$ Solve each equation.

$x = -5$ $x = 7$ $2x = 1$

$\dfrac{2x}{2} = \dfrac{1}{2}$

$x = \dfrac{1}{2}$

(c) $2x^3 - 14x^2 + 20x = 0$ Factor the common factors.

$2x(x^2 - 7x + 10) = 0$ Factor the trinomial.

$2x(x - 5)\ (x - 2) = 0$ Set each factor equal to zero.

$2x = 0$ $x - 5 = 0$ $x - 2 = 0$ Solve each equation.

$\dfrac{2x}{2} = \dfrac{0}{2}$ $x = 5$ $x = 2$

$x = 0$

(d) $3x^3 - 15x = 0$ Factor the common factors.

$3x(x^2 - 5) = 0$

$3x(x^2 - 5) = 0$ Set each factor equal to zero.

$3x = 0$ $x^2 - 5 = 0$

$\dfrac{3x}{3} = \dfrac{0}{3}$ $x^2 = 5$ Solve each equation.

$x = 0$ $x = \pm\sqrt{5}$ The exact roots are $x = 0, x = \sqrt{5}$, or $x = -\sqrt{5}$.

See Exercises 7–18.

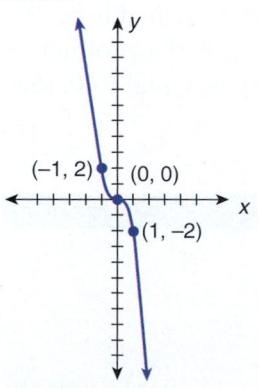
3 **Graph Higher-Degree Equations.** Any equation can be graphed by writing it as a function and using the table-of-solutions method. It is important to include enough points to get a complete view of the graph.

EXAMPLE 3

Prepare a table of solutions and graph the function $f(x) = x^3$ using integral values between -2 and 2 for x, inclusive. Determine the domain and range.

$f(-2) = (-2)^3$ $f(-1) = (-1)^3$ $f(0) = 0^3$
$f(-2) = -8$ $f(-1) = -1$ $f(0) = 0$

$f(1) = 1^3$ $f(2) = 2^3$
$f(1) = 1$ $f(2) = 8$

x	$f(x)$
−2	−8
−1	−1
0	0
1	1
2	8

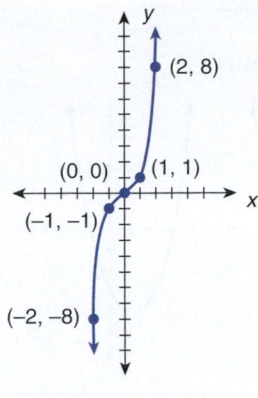

FIGURE 15–12

Plot the points indicated in the table of solutions and connect the points with a smooth, continuous curve (Fig. 15–12).

The domain and range are the set of all real numbers. See Exercises 19–24.

EXAMPLE 4

Use a calculator to graph $y = 2x^3 - 1$. Determine the range for a domain from -3 to 3 inclusive.

Enter the equation into the calculator (Fig. 15–13).

| Y= | CLEAR | 2 | X, T, θ, n | ∧ | 3 | → | − | 1 | ZOOM | 6:ZStandard |

Examine a table for the domain from -3 to 3 inclusive (Fig. 15–14).

| TBLSET | (−) | 3 | ENTER | 1 | − | TABLE |

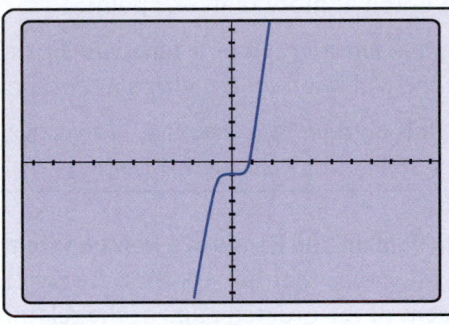

FIGURE 15–13

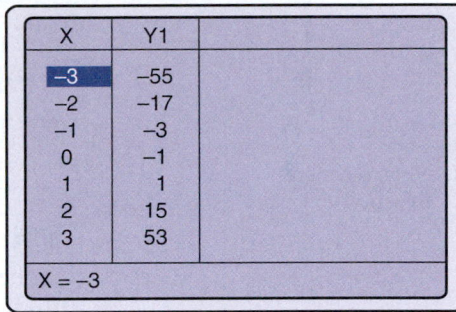

FIGURE 15–14

The range for a domain from -3 to 3 inclusive is -55 to 53 inclusive.

See Exercises 19–24.

Relation: an equation in two variables. All functions are also relations, but not all relations are functions

4 Distinguish Between a Function and a Relation Using the Vertical Line Test.
Equations in two variables describe a **relation** between the two variables. This connection between the two variables defines a set of ordered pairs. The equations in two variables that we have examined thus far have been functions. A *function* is a relation that has exactly one value of the dependent variable (*y*-value) for every value of the independent variable (*x*-value). In other words, all functions are also relations, but not all relations are functions. A relation that is not a function will have at least one case where a value of the independent variable (*x*-value) corresponds to *more than one* value of the dependent variable (*y*-value).

| To distinguish between a function and a relation using the vertical line test: |

1. Graph the relation on a coordinate system.
2. Apply the vertical line test.
    (a) If a vertical line can be drawn so that it intersects a graph at more than one point, then the graph is the graph of a relation that is not a function.
    (b) If no vertical line can be drawn so that it intersects a graph at more than one point, then the graph is the graph of a function.

STOP AND CHECK

Determine if the graph is the graph of a function.

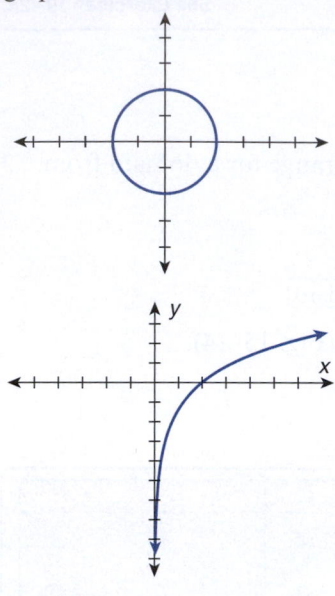

Answers:
1. Not a function
2. Function

EXAMPLE 5

Determine which graphs in Fig. 15–15 are graphs of functions.

(a) 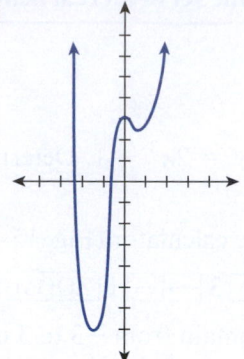   (b) 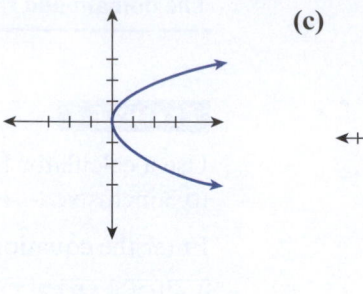   (c)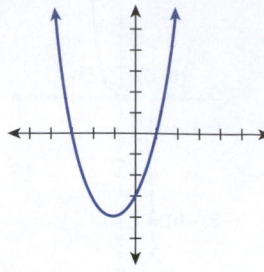

FIGURE 15–15

(a) **The graph is a graph of a function.** No vertical line can be drawn so that it intersects the graph at more than one point.

(b) **The graph is not a graph of a function.** For all values of x greater than zero, a vertical line will intersect the graph at two points.

(c) **The graph is a graph of a function.** No vertical line can be drawn so that it intersects the graph at more than one point.   **See Exercises 25–30.**

5   Find the Domain and Range of a Relation from a Graph. The *domain* of a relation is the set of all values that are appropriate for the independent variable and is the first component of the ordered pairs of the relation. The *range* of a relation is the set of all resulting values of the dependent variable that correspond to the independent variable and is the second component of the ordered pairs of the relation.

TIP   Relation Versus Function

▶ For every value of the domain of a relation there is at least one corresponding value of the range.

▶ For every value of the domain of a function there is exactly one corresponding value of the range.

$x = y^2$

x	y
4	−2
1	−1
0	0
1	1
4	2

Domain	Range
0	−2
*1	−1
	0
*4	1
	2

*x-values with two different y-values
$x = y^2$ **is a relation, but not a function.**

$y = x^2$

x	y
−2	4
−1	1
0	0
1	1
2	4

Domain	Range
−2	0
−1	1
0	
1	4
2	

Each x-value only has one y-value.
Two different x-values can have the same y-value.
$y = x^2$ **is a function.**

To find the domain and range of a relation from a graph:

1. Graph the relation on a coordinate system.

2. The domain is the set of all the values on the graph for the independent variable (*x*-values).

3. The range is the set of all the values on the graph for the dependent variable (*y*-values).

EXAMPLE 6

Determine the domain and range of each relation (Fig. 15–16).

(a)

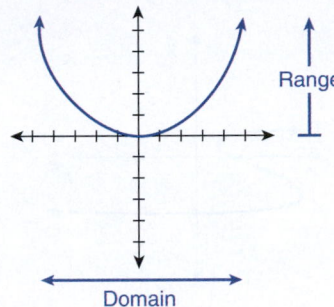

(b)

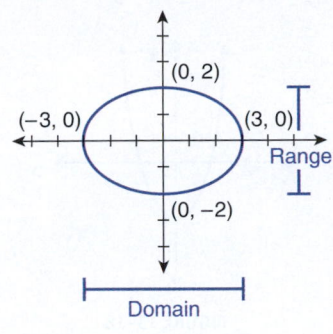

(c)

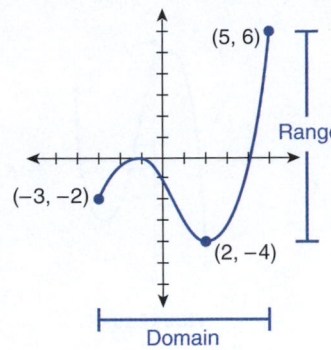

FIGURE 15–16

(a) **Domain:** the set of all real numbers

Range: the set of real numbers greater than or equal to zero

(b) **Domain:** the set of real numbers from −3 to 3 inclusive

Range: the set of real numbers from −2 to 2 inclusive

(c) **Domain:** the set of real numbers from −3 to 5 inclusive

Range: the set of real numbers from −4 to 6 inclusive

See Exercises 31–36.

15–5 EXERCISES

MyLab Math For additional practice go to your study plan in MyLab Math.

1 State the degree of the equations. *See Example 1.*

1. $x^2 + 2x - 3 = 0$

2. $3x + 2x = 5$

3. $x^4 = 42$

4. $2x - 7 - 4x = 3$

5. $3x^3 + 2x^4 + 3 = x^2$

6. $x^7 = 128$

2 Find the roots of the equations. Factor if necessary. *See Example 2.*

7. $x(x - 2)(x + 3) = 0$

8. $2x(2x - 1)(x + 3) = 0$

9. $3x(2x - 5)(3x - 2) = 0$

10. $x^3 - 7x^2 + 10x = 0$

11. $3x^3 - 3x^2 = 18x$

12. $4x^3 + 10x^2 + 4x = 0$

13. $2x^3 - 18x = 0$

14. $12x^3 = 3x$

15. $16x^3 = 9x$

16. 32 times a number is subtracted from twice the cube of the number and the result is zero. How many numbers meet these conditions? What are they?

17. Find all the numbers that satisfy the following conditions: A number cubed is increased by 5 times the number and the result is zero.

18. **BUS** A shipping container is a large cardboard box. The volume is found by multiplying the length times the width times the height. The length of the box is 3 ft more than the width, and the height is 7 ft more than the width. The volume is 421 ft³. Write an equation to find the length, width, and height. Can this problem be solved by the methods found in Section 15–5? If so, solve it. If not, explain why it can't be solved.

3 Graph the equations using the table-of-solutions method. Determine the domain and range. Check by using a graphing calculator. *See Examples 3 and 4.*

19. $y = 2x^3$

20. $y = \frac{1}{2}x^3$

21. $y = x^3 - x^2 - 4x + 4$

22. $y = x^3 + 2$

23. $y = -x^3 + 4$

24. $y = x^3 - 3x$

4 Use the vertical line test to determine which are graphs of functions. *See Example 5.*

25. $y = x^3 - x^2 - 8x + 8$

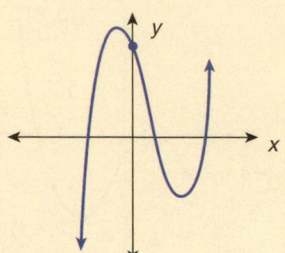

FIGURE 15–17

26. $y = 3x^2 - 2x - 5$

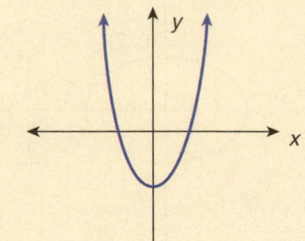

FIGURE 15–18

27. $x = y^3 - y^2 - 3y - 3$

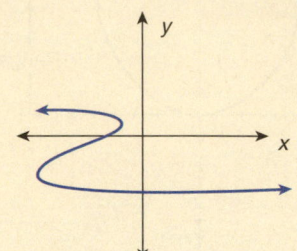

FIGURE 15–19

28. $x = y^2 - y - 5$

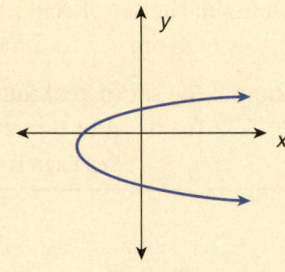

FIGURE 15–20

29. $y = x^4 - 3x^3 + x^2$

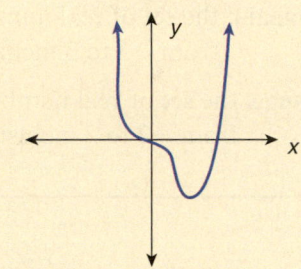

FIGURE 15–21

30. $x = 2y + 3$

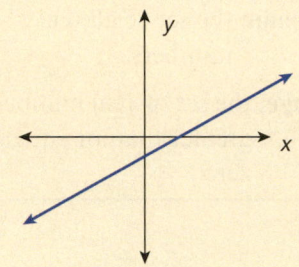

FIGURE 15–22

5 Find the domain and range of the functions or relations from the graph. *See Example 6.*

31.

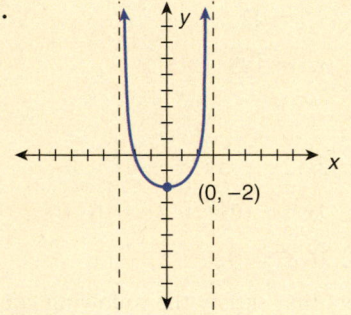

FIGURE 15–23

32.

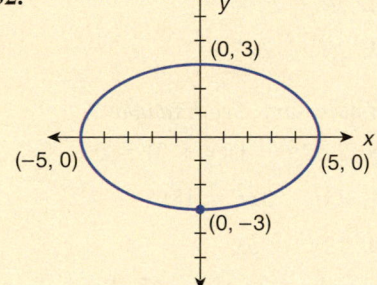

FIGURE 15–24

33.

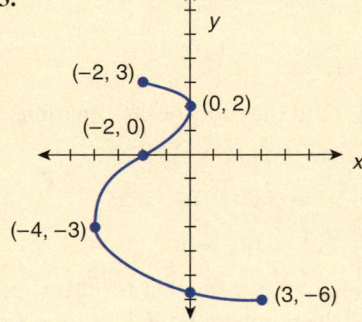

FIGURE 15–25

34.

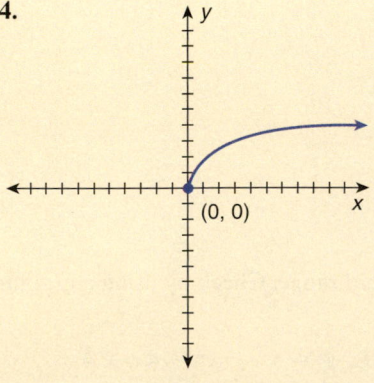

FIGURE 15–26

35.

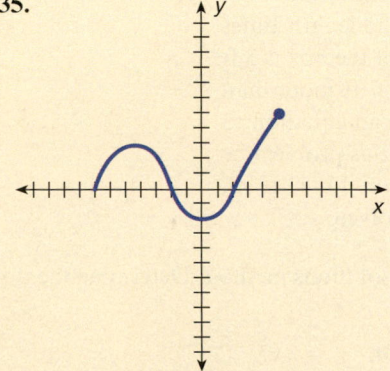

FIGURE 15–27

36.

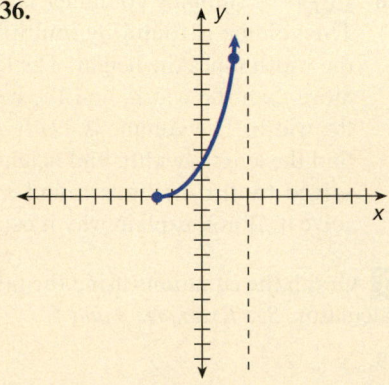

FIGURE 15–28

15–6 Solving Quadratic Inequalities

LEARNING OUTCOMES

1 Solve quadratic inequalities in one variable.

2 Graph quadratic inequalities in two variables.

LC LEARNING CATALYTICS
1. Factor $8x^2 - 2x - 3$.

Critical values of an inequality: values that are solutions to a similar equation. The critical values become boundaries for the solution set of values of an inequality

Boundaries: another name for the critical values of an inequality

Solving *quadratic inequalities* is similar to solving quadratic equations. The solution to these inequalities, like the solution to linear inequalities, is a *set* of numbers with specific boundaries.

1 **Solve Quadratic Inequalities in One Variable.** Quadratic inequalities can be solved by all of the same methods we used to solve quadratic equations. You first treat the inequality as an equation to find the **critical values of an inequality** or **boundaries.**

To solve quadratic inequalities:

1. Determine the critical values by solving an equation in which an equal sign is substituted for the inequality sign.

2. Plot the critical values on a number line to form the three regions.

3. Test *each* region of values by selecting any point within the region, substituting that value into the original inequality, solving the inequality, and deciding if the resulting inequality is a *true* statement.

4. The solution set for the quadratic inequality is the region or regions that produce a *true* statement in Step 3.

5. The critical values or boundary points are included in the solution set if the inequality is "inclusive" ($\leq$ or $\geq$). The critical values or boundary points are not included in the solution set if the inequality is "exclusive" ($<$ or $>$).

STOP AND CHECK

1. Solve the inequality $2x^2 - 3x - 20 < 0$. Write the solution set in both symbolic and interval notation.

Answer:

1. $-\dfrac{5}{2} < x < 4; \left(-\dfrac{5}{2}, 4\right)$

EXAMPLE 1

Solve the inequality $x^2 + 5x + 6 \leq 0$. Write the solution set in both symbolic and interval notation.

$x^2 + 5x + 6 = 0$	Write a similar equation and solve. Write in factored form.
$(x + 3)(x + 2) = 0$	Set each factor equal to 0.
$x + 3 = 0 \quad x + 2 = 0$	Determine the critical values by solving each equation.
$x = -3 \quad\quad x = -2$	Critical values, or boundary points

Plot the critical values and label the corresponding regions (Fig. 15–29).

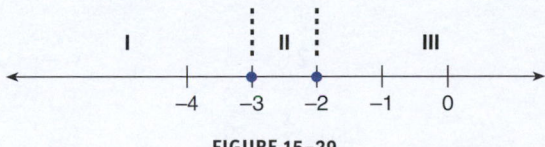

FIGURE 15–29

Region I:	$x < -3$	To the left of the smaller critical value, -3
Region II:	$-3 < x < -2$	Between the critical values, -3 and -2
Region III:	$x > -2$	To the right of the larger critical value, -2

Test one point in each region.

Region I: $x < -3$	Region II: $-3 < x < -2$

Region I test point: $x = -4$

$$(x + 3)(x + 2) \le 0$$
$$(-4 + 3)(-4 + 2) \le 0$$
$$(-1)(-2) \le 0$$
$$2 \le 0$$

The inequality is **false**, so Region I is *not* in the solution set.

Region II test point: $x = -2.5$

$$(-2.5 + 3)(-2.5 + 2) \le 0$$
$$(0.5)(-0.5) \le 0$$
$$-0.25 \le 0$$

The inequality is **true**, so Region II and the boundaries *are* in the solution set.

Region III: $x > -2$

Region III test point: $x = -1$

$$(-1 + 3)(-1 + 2) \le 0$$
$$(2)(1) \le 0$$
$$2 \le 0$$

The inequality is **false**, so Region III is *not* in the solution set.

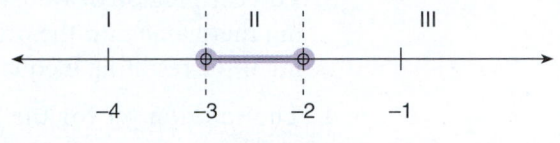

FIGURE 15–30

The solution set is $-3 \le x \le -2$ or $[-3, -2]$ (Fig. 15–30).   **See Exercises 1–7.**

TIP **Selecting Test Points** Boundary points should not be used as test points. When possible, select integers as test points, and if zero is in a region, it makes an excellent test point.

STOP AND CHECK

1. Solve the inequality $5x^2 + 16x + 3 \ge 0$. Write the solution set in both symbolic and interval notation.

Answer:

1. $x \le -3$ or $x \ge \dfrac{1}{5}$;

$(-\infty, -3] \cup \left[-\dfrac{1}{5}, \infty\right)$

EXAMPLE 2

Solve the inequality $2x^2 + x - 6 > 0$. Write the solution set in both symbolic and interval notation.

$2x^2 + x - 6 = 0$	Write as an equation and solve.
$(2x - 3)(x + 2) = 0$	Write in factored form. Set each factor equal to 0.
$2x - 3 = 0 \qquad x + 2 = 0$	Determine the critical values by solving each equation.
$2x = 3 \qquad\qquad x = -2$	
$x = \dfrac{3}{2}$	

Plot the critical values and label the corresponding regions (see Fig. 15–31).

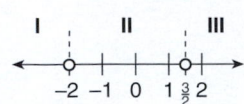

FIGURE 15–31

Region I: $x < -2$ Region II: $-2 < x < \frac{3}{2}$ Region III: $x > \frac{3}{2}$

Test each region.

Region I: $x < -2$

Region II: $-2 < x < \dfrac{3}{2}$

Region I test point: $x = -3$

$$(2x - 3)(x + 2) > 0$$
$$[2(-3) - 3](-3 + 2) > 0$$
$$(-6 - 3)(-1) > 0$$
$$(-9)(-1) > 0$$
$$9 > 0$$

The inequality is **true**, so Region I *is* in the solution set.

Region II test point: $x = 0$

$$(2x - 3)(x + 2) > 0$$
$$[2(0) - 3](0 + 2) > 0$$
$$(0 - 3)(2) > 0$$
$$(-3)(2) > 0$$
$$-6 > 0$$

The inequality is **false**, so Region II is *not* in the solution set.

Region III: $x > \dfrac{3}{2}$

Region III test point: $x = 2$

$$(2x - 3)(x + 2) > 0$$
$$[2(2) - 3][2 + 2] > 0$$
$$(4 - 3)(4) > 0$$
$$(1)(4) > 0$$
$$4 > 0$$

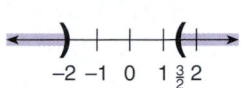

The inequality is **true**, so Region III *is* in the solution set.

FIGURE 15–32

The solution set is $x < -2$ or $x > \frac{3}{2}$. In interval notation the solution is $(-\infty, -2) \cup (\frac{3}{2}, \infty)$. See Fig. 15–32.

See Exercises 8–12.

After examining several problems, you may notice a pattern for determining the solution set by inspection once the critical values are known. This pattern or generalization also can be applied to quadratic inequalities that do not factor.

TIP **Finding Solution Sets by Inspection for Quadratic Inequalities** Patterns and generalizations may allow shortcuts, but they are appropriate only under special conditions. For quadratic inequalities, the inequality **must** be written in standard form ($ax^2 + bx + c < 0$ or $ax^2 + bx + c > 0$), where $a > 0$ (a is positive). The critical values are represented symbolically as s_1 and s_2, where $s_1 < s_2$.

Figures 15–33 and 15–34 illustrate the patterns for these conditions.

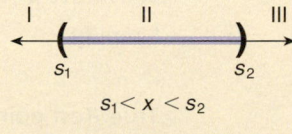

FIGURE 15–33

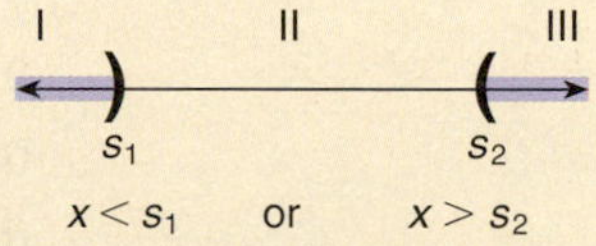

FIGURE 15–34

If you solve a quadratic equation by *any* method (square-root method, factoring, completing the square, or using the quadratic formula), you can determine the solution set of the associated quadratic inequality by inspection after you have found the critical values. The solution can be checked by testing one point in the projected solution set. Similar generalizations can be made for $ax^2 + bx + c \leq 0$ and $ax^2 + bx + c \geq 0$.

2 **Graph Quadratic Inequalities in Two Variables.** The solution set for a quadratic inequality in two variables can be represented by a shaded region of a graph on a rectangular coordinate system.

> **To graph a quadratic inequality in two variables:**
>
> 1. Write the inequality in function notation or solve for y.
>
> 2. Graph the boundary by substituting an equal sign for the inequality sign and solving the equation.
>
> 3. Select a test point that is *not* on the boundary.
>
> 4. Substitute the values from the test point into the original inequality and evaluate. The test point will be either *inside* or *outside* the boundary.
>
> 5. If the evaluated inequality makes a *true* statement, the solution set is the portion of the graph (inside or outside) that includes the test point.
>
> 6. If the evaluated inequality makes a *false* statement, the solution set is the portion of the graph (inside or outside) that does *not* include the test point.
>
> 7. The boundary *is not* included if the inequality is $<$ or $>$. The boundary *is* included if the inequality is $\le$ or $\ge$.

STOP AND CHECK

1. Graph the inequality
   $y \le 2x^2 - 5x - 3$.

Answer:

1.

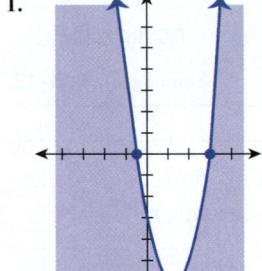

EXAMPLE 3

Graph the inequality $y \ge 2x^2 - 5x - 3$.

Graph the equation $y = 2x^2 - 5x - 3$ $\boxed{y=}$ 2 $\boxed{\text{X, T, }\theta, n}$ $\boxed{\wedge}$ 2 $\boxed{-}$ 5 $\boxed{\text{X, T, }\theta, n}$ $\boxed{-}$ 3
(Fig. 15–35).

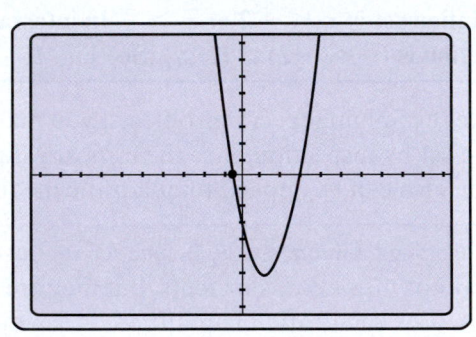

FIGURE 15–35

$y \ge 2x^2 - 5x - 3$
$\leftarrow y = 2x^2 - 5x - 3$
$(-0.5, 0)$ $(3, 0)$
$(1.25, -6.25)$

FIGURE 15–36

Test point: $(0, 0)$ Select a test point that is *not* on the boundary.

$y \ge 2x^2 - 5x - 3$ Substitute $x = 0$ and $y = 0$.

$0 \ge 2(0)^2 - 5(0) - 3$ Simplify.

$0 \ge 0 - 0 - 3$

$0 \ge -3$ True. Test point is included in the solution set.

Sketch the graph and shade inside the graph. The boundary is included (Fig. 15–36).   **See Exercises 13–18.**

15–6 EXERCISES   **MyLab Math** For additional practice go to your study plan in MyLab Math.

1 Graph the solution of the quadratic inequalities on a number line and write the solution in symbolic and interval notation. *See Example 1.*

1. $(x - 2)(x + 3) < 0$ **2.** $(2x - 5)(3x - 2) < 0$ **3.** $(a + 4)(a - 6) < 0$

4. $x^2 - 7x + 10 \leq 0$ **5.** $x^2 - 4x + 3 \leq 0$ **6.** $4x^2 - 21x + 5 \leq 0$

7. $2x^2 + 9x \leq 5$

See Example 2.

8. $(2x - 1)(x + 3) > 0$ **9.** $(y + 6)(y - 2) \geq 0$ **10.** $x^2 - x - 6 \geq 0$

11. $6x^2 + x > 1$ **12.** $6x^2 + 2 \geq 7x$

2 Graph using a graphing calculator. Adjust the window if necessary to show the vertex of the parabola. *See Example 3.*

13. $y < 3x^2 + 5x - 2$ **14.** $y > (2x - 3)(x - 1)$

15. $y \leq 2x^2 - 9x - 5$ **16.** $y \geq -2x^2 + 9x + 5$

17. The revenue range for a leather wallet is based on the price of the item given by the formula $R \leq -4x^2 + 52x$, where R is the revenue and x is the price. Graph the revenue function and determine the greatest possible revenue.

18. The formula $d \geq 4.9t^2$ gives the distance d an object falls in t seconds. Graph the solution set to show the possible range of solutions.

15–7   Solving Equations and Inequalities Containing One Absolute-Value Expression

LEARNING OUTCOMES

1 Solve equations containing one absolute-value expression.

2 Graph equations containing one absolute-value expression.

3 Solve absolute-value inequalities using the "less than" relationship.

4 Solve absolute-value inequalities using the "greater than" relationship.

LC LEARNING CATALYTICS

1. Evaluate $|4 - 15|$.

1 **Solve Equations Containing One Absolute-Value Expression.** The equation $|x| = 5$ has two roots, $+5$ and -5. To solve an equation that contains an absolute-value term, we must examine each of two cases. One case is when the expression within the absolute-value symbol is positive. The other case is when the expression within the absolute-value symbol is negative. Recall that the symbolic definition of the absolute value of a number is $|a| = a$ for $a \geq 0$, and $|a| = -a$ for $a < 0$.

To solve an equation containing one absolute-value expression:

1. Isolate the absolute-value expression on one side of the equation.

2. Separate the equation into two cases. One case considers the expression within the absolute-value symbol to be positive. The other case considers it to be negative.

3. Solve each case to obtain the *two* roots of the equation.

4. $|x| = b$ has no solution if b is negative ($b < 0$).

5. $|x| = b$ has one root when $b = 0$.

STOP AND CHECK

Find the roots for the equations.

1. $|x + 3| = 5$

2. $|x - 2| - 4 = 2$

Answers:

1. 2 and -8 **2.** 8 and -4

EXAMPLE 1

Find the roots for the equations   **(a)** $|x - 4| = 7$ and   **(b)** $|y + 3| - 5 = 6$.

(a)   *Case 1: $x - 4 = 7$*   *Case 2: $x + 4 = -7$*

                  $x = 7 + 4$ $x - 4 = -7$

                  $x = 11$ $x = -7 + 4$

                                                    $x = -3$

The roots of the equation are 11 and −3.

If $x = 11$,	If $x = -3$,	Check each solution.
$\lvert x - 4 \rvert = 7$	$\lvert x - 4 \rvert = 7$	
$\lvert 11 - 4 \rvert = 7$	$\lvert -3 - 4 \rvert = 7$	
$\lvert 7 \rvert = 7$	$\lvert -7 \rvert = 7$	
$7 = 7$	$7 = 7$	Each root checks.

(b) $\lvert y + 3 \rvert - 5 = 6$ Isolate the absolute-value expression.

$\lvert y + 3 \rvert = 6 + 5$ Combine like terms.

$\lvert y + 3 \rvert = 11$ Separate into two cases.

Case 1: $y + 3 = 11$	Case 2: $y + 3 = -11$
$y = 11 - 3$	$y = -11 - 3$
$y = 8$	$y = -14$

The roots of the equation are 8 and −14. Each root checks.   **See Exercises 1–18.**

TIP **The Importance of Isolating the Absolute-Value Expression** The recommended procedure for finding the root for the case when the expression inside the absolute-value grouping is negative (Case 2) is to isolate the absolute-value expression first. Let's look at an alternative correct procedure and an incorrect procedure.

Find the root of $\lvert y + 3 \rvert - 5 = 6$ when $(y + 3)$ is negative (Case 2).

Alternative Correct Procedure Without Isolating

$\lvert y + 3 \rvert - 5 = 6$

Case 2: $-(y + 3) - 5 = 6$

$-y - 3 - 5 = 6$

$-y - 8 = 6$

$-y = 6 + 8$

$-y = 14$

$y = -14$

Check: $y = -14$

$\lvert y + 3 \rvert - 5 = 6$

$\lvert -14 + 3 \rvert - 5 = 6$

$\lvert -11 \rvert - 5 = 6$

$11 - 5 = 6$

$6 = 6$

Incorrect Procedure Without Isolating

$\lvert y + 3 \rvert - 5 = 6$

Case 2: $y + 3 - 5 = -6$

$y - 2 = -6$

$y = -6 + 2$

$y = -4$

Check: $y = -4$

$\lvert y + 3 \rvert - 5 = 6$

$\lvert -4 + 3 \rvert - 5 = 6$

$\lvert -1 \rvert - 5 = 6$

$1 - 5 = 6$

$-4 = 6$ Incorrect.

If we choose *not* to isolate the absolute-value expression first, then we *must* take the opposite of the absolute-value grouping rather than the opposite of the value on the other side of the equation.

STOP AND CHECK

1. Solve the equation $\lvert 3x - 2 \rvert + 12 = 9$.

Answer:

1. No solution

EXAMPLE 2

Solve the equation $\lvert 2x - 5 \rvert + 14 = 7$.

$\lvert 2x - 5 \rvert + 14 = 7$ Isolate the absolute-value expression.

$\lvert 2x - 5 \rvert = 7 - 14$ Combine like terms.

$\lvert 2x - 5 \rvert = -7$ Cannot continue because the absolute value cannot be negative.

$\lvert 2x - 5 \rvert + 14 = 7$ has no solution.   **See Exercises 19–21.**

2 Graph Equations Containing One Absolute-Value Expression. An equation containing one absolute value term can be written as a function and graphed using the table-of-solutions method.

Look at the equation in Example 1(a). We can rearrange the equation so that one side equals zero and then write in function notation.

$$|x - 4| = 7$$
$$|x - 4| - 7 = 0$$
$$f(x) = |x - 4| - 7$$

> **To graph an equation containing one absolute-value expression.**
>
> 1. Rearrange the equations so that one side equals zero.
>
> 2. Write the equation from Step 1 in function notation.
>
> 3. Determine the value of x inside the absolute-value symbols that will make the value of the absolute-value expression zero.
>
> 4. Use the value of x found in Step 3 and other values of x that are less than and greater than the value found in Step 3 and prepare a table of solutions with at least five solutions.
>
> 5. Draw the graph using the table of solutions.

STOP AND CHECK

1. Graph the function $f(x) = |x + 3| - 2$ using the table-of-solutions method.

Answer:

1.

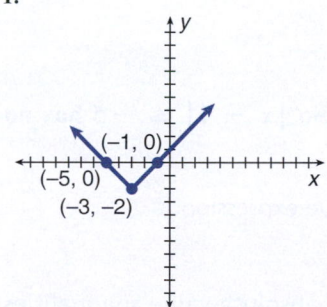

EXAMPLE 3

Graph the function $f(x) = |x - 4| - 7$ using the table-of-solutions method.

Values of x that are greater than 4 will give a positive result for the expression inside the absolute-value symbol. A value less than 4 will give a negative result for the expression inside the absolute-value symbol. To have a representative graph for the equation, find ordered pairs for values of x that are both less than and greater than 4.

$$f(-3) = |-3 - 4| - 7 = 7 - 7 = 0$$
$$f(0) = |0 - 4| - 7 = 4 - 7 = -3$$
$$f(2) = |2 - 4| - 7 = 2 - 7 = -5$$
$$f(4) = |4 - 4| - 7 = 0 - 7 = -7$$
$$f(6) = |6 - 4| - 7 = 2 - 7 = -5$$
$$f(8) = |8 - 4| - 7 = 4 - 7 = -3$$
$$f(11) = |11 - 4| - 7 = 7 - 7 = 0$$

x	f(x)
−3	0
0	−3
2	−5
4	−7
6	−5
8	−3
11	0

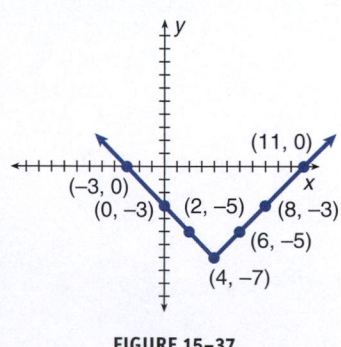

FIGURE 15-37

Note that graph in Fig. 15-37 crosses the x-axis at $(-3, 0)$ and $(11, 0)$. These points represent the solutions of the equation, $x = -3$ and $x = 11$. **See Exercises 22-25.**

3 Solve Absolute-Value Inequalities Using the "Less Than" Relationship. An inequality that contains an absolute-value expression is interpreted by the sense of the inequality. If the inequality is a "less than" or "less than or equal to" relationship, the following property is used.

> **"Less than" relationships for absolute-value inequalities:**
>
> If $|x| < b$ and $b > 0$, then $-b < x < b$.
>
> or
>
> If $|x| \leq b$ and $b > 0$, then $-b \leq x \leq b$.

On a number line (Fig. 15–38), the solution set is a continuous set of values. The values in the solution set of $|x| < b$ are between b and $-b$ when b is positive. If $b = 0$, $|x| = 0$. Since zero is unsigned, we do not create an interval from -0 to 0. The solution is a single value, 0.

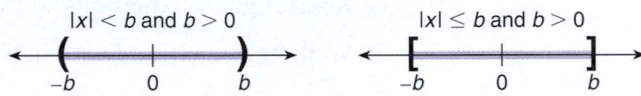

FIGURE 15–38

STOP AND CHECK

Find the solution set for each inequality.

1. $|x + 2| < -3$
2. $|5x| + 3 \leq 18$

Answer:
1. No solution
2. $-3 \leq x \leq 3$ or $[-3, 3]$

EXAMPLE 4

Find the solution set for each inequality.

(a) $|x| < 3$ **(b)** $|x - 1| \leq -5$ **(c)** $|3x| + 2 \leq 14$

(a) $|x| < 3$ Apply the property of absolute-value inequalities having $<$ relationship.

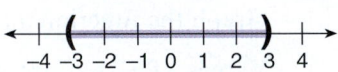

FIGURE 15–39

$-3 < x < 3$ **or** $(-3, 3)$ See Fig. 15–39.

(b) $|x - 1| \leq -5$

An absolute-value expression cannot be negative. So $|x - 1| \leq -5$ **has no solution.**

(c) $|3x| + 2 \leq 14$ Isolate the absolute-value expression.

$\qquad |3x| \leq 14 - 2$ Combine like terms.

$\qquad |3x| \leq 12$ Apply the property of absolute-value inequalities having $\leq$ relationship.

$-12 \leq 3x \leq 12$ Separate into two inequalities.

$-12 \leq 3x$ and $3x \leq 12$ Divide.

$-4 \leq x$ and $x \leq 4$ Find where the inequalities overlap.

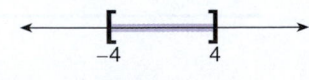

FIGURE 15–40

$-4 \leq x \leq 4$ or $[-4, 4]$ **is the solution.** See Fig. 15–40.   **See Exercises 26–40.**

4 **Solve Absolute-Value Inequalities Using the "Greater Than" Relationship.** The solution set for an absolute-value inequality with a "greater than" or "greater than or equal to" relationship is represented by the extreme values on the number line.

> **"Greater than" relationships for absolute-value inequalities:**
>
> If $|x| > b$ and $b > 0$, then $x < -b$ or $x > b$.
>
> or
>
> If $|x| \geq b$ and $b > 0$, then $x \leq -b$ or $x \geq b$.

The solution set for an absolute-value inequality with a $>$ or $\geq$ relationship is represented on the number lines in Figs. 15–41 and 15–42.

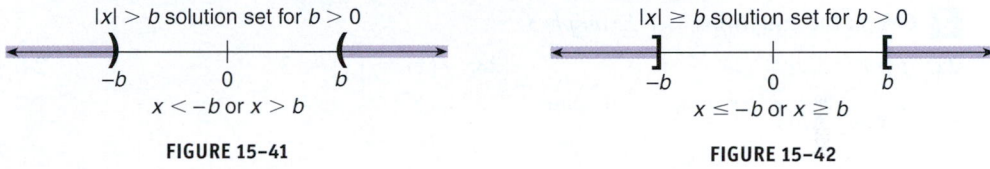

$|x| > b$ solution set for $b > 0$

$x < -b$ or $x > b$

FIGURE 15-41

$|x| \geq b$ solution set for $b > 0$

$x \leq -b$ or $x \geq b$

FIGURE 15-42

STOP AND CHECK
Find the solution set for each inequality. Write the solution set symbolically and in interval notation.
1. $|x| + 3 > 7$
2. $|x + 2| > 3$

Answer:
1. $x < -4$ or $x > 4$; $(-\infty, -4) \cup (4, \infty)$
2. $x \leq -5$ or $x \geq 1$; $(-\infty, -5] \cup [1, \infty)$

EXAMPLE 5

Find the solution set for each inequality. Write the solution set symbolically and in interval notation.

(a) $|x| > 5$ **(b)** $|x| + 1 \geq 4$

(a) $|x| > 5$ Apply the property of absolute-value inequalities having $>$ relationship.

$x < -5$ or $x > 5$

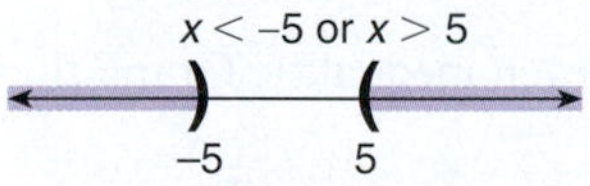

FIGURE 15-43

$x < -5$ or $x > 5$; $(-\infty, -5) \cup (5, \infty)$. See Fig. 15–43.

(b) $|x| + 1 \geq 4$ Isolate the absolute-value expression.

$|x| \geq 4 - 1$ Combine like terms.

$|x| \geq 3$ Apply the property of inequalities having $\geq$ relationship.

$x \leq -3$ or $x \geq 3$

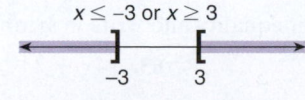

$x \leq -3$ or $x \geq 3$

FIGURE 15-44

$x \leq -3$ or $x \geq 3$; $(-\infty, -3] \cup [3, \infty)$. See Fig. 15–44.   **See Exercises 41–70.**

15–7 EXERCISES MyLab Math For additional practice go to your study plan in MyLab Math.

1 Solve the equations containing absolute values. *See Example 1.*

1. $|x| = 8$ **2.** $|x| = 15$ **3.** $|x - 3| = 5$

4. $|x - 7| = 2$ **5.** $|2x - 3| = 9$ **6.** $|3x - 7| = 2$

7. $|x| - 4 = 7$ **8.** $-5 + |x - 4| = -2$ **9.** $|2x - 1| - 3 = 4$

10. $|3x - 5| + 12 = 18$ **11.** $|6 - 3x| + 2 = 8$ **12.** $|5 - 2x| = 15$

13. $|3x - 8| + 2 = 12$ **14.** $|7x - 9| - 1 = 8$ **15.** $|5x - 3| - 7 = 8$

16. $|5x - 1| - 8 = 7$ **17.** $|2x - 3| - 8 = -5$ **18.** $|3x - 2| + 2 = 2$

See Example 2.

19. $|x - 2| + 5 = 3$ **20.** $|3x - 2| + 1 = -6$ **21.** $|4x - 3| + 11 = 6$

2 Graph the equations. *See Example 3.*

22. $f(x) = |x|$ **23.** $f(x) = |x| - 3$ **24.** $f(x) = |x - 3|$

25. $f(x) = |3x|$

3 Find the solution set for each inequality. Graph the solution set on a number line and write it in symbolic and interval notation. *See Example 4.*

26. $|x| < 5$ **27.** $|x| < 1$ **28.** $|x| < 2$

29. $|x| \leq 7$ **30.** $|x + 3| < 2$ **31.** $|x + 4| < 3$

32. $|x - 3| < 4$ **33.** $|x - 2| < 3$ **34.** $|x - 5| \leq 6$

35. $|x - 1| \leq 7$ **36.** $|3x - 5| < 7$ **37.** $|2x - 4| < 3$

38. $|5x + 2| \leq 3$ **39.** $|4x + 7| \leq 5$ **40.** $|3x - 1| + 2 < 5$

4 Find the solution set for each inequality. Graph the solution set on a number line and write it in symbolic and interval notation. *See Example 5.*

41. $|x| > 2$ **42.** $|x| > 3$ **43.** $|x| \geq 5$

44. $|x| \geq 6$ **45.** $|x - 5| > 2$ **46.** $|x - 4| > 6$

47. $|x + 3| \geq 5$ **48.** $|x + 4| \geq 6$ **49.** $|2x - 5| \geq 0$

50. $|3x - 2| \geq 4$ **51.** $|5x + 8| \geq 2$ **52.** $|4x - 3| \geq 9$

53. $|3x - 1| + 3 \geq 5$ **54.** $|2x - 3| + 1 \geq 5$ **55.** $|4x - 2| - 3 \geq 7$

56. $|3x - 2| + 4 \leq 5$ **57.** $|2x - 5| - 3 \leq 2$ **58.** $|5x - 3| - 6 \geq 1$

59. $|4x - 6| + 4 \geq 5$ **60.** $|3x - 8| - 2 \leq -6$ **61.** $|5x + 7| + 3 \leq -6$

Find the solution set for each inequality and write in symbolic notation. *See Example 5.*

62. $|x - 3| \geq 3$ **63.** $|4x + 1| > 1$ **64.** $|x - 4| \leq 3$

65. $|2x - 1| < 0$ **66.** $|x| < 7$ **67.** $|x + 8| < 3$

68. $|x| > 12$ **69.** $|x + 8| > 2$ **70.** $|2x - 7| < -5$

15 | CHAPTER REVIEW OF KEY CONCEPTS

LEARNING OUTCOMES	KEY CONCEPTS AND EXAMPLES

Section 15–1

1 Write quadratic equations in standard form (pp. 610–611).

Write the equation with all terms on the left side of the equation and zero on the right and arrange with the terms in descending powers. The leading coefficient should be positive.

Write the equation $5x + 3 = 4x^2$ in standard form.

$$5x + 3 = 4x^2 \qquad \text{Rearrange all terms on the left and arrange in descending order.}$$
$$-4x^2 + 5x + 3 = 0 \qquad \text{Make leading coefficient positive.}$$
$$-1(-4x^2 + 5x + 3) = -1(0)$$
$$4x^2 - 5x - 3 = 0$$

2 Identify the coefficients of the quadratic, linear, and constant terms of a quadratic equation (pp. 611–612).

The coefficient of the quadratic term is the coefficient of the squared variable, the coefficient of the linear term is the coefficient of the first-power variable, and the constant term is number term.

Identify the coefficient of the quadratic and linear terms and identify the constant term in the following:

$$5x = 4x^2 - 3 \qquad \text{Write the equation in standard form.}$$
$$4x^2 - 5x - 3 = 0 \qquad a = 4, b = -5, c = -3$$

The coefficient of the quadratic term is 4, the coefficient of the linear term is -5, and the constant term is -3.

3 Solve pure quadratic equations $(ax^2 + c = 0)$ by the square-root method (pp. 612–614).

1. Rearrange the equation, if necessary, so that the quadratic term is on one side and the constant is on the other side of the equation.   **2.** Combine like terms, if appropriate.   **3.** Rewrite the equation so that the quadratic term has a coefficient of +1.   **4.** Apply the square-root property of equality by taking the square root of both sides.   **5.** Check each solution in the original equation.

Solve $2x^2 - 72 = 0$.

$$2x^2 - 72 = 0 \qquad \text{Sort.}$$
$$2x^2 = 72 \qquad \text{Divide by 2.}$$
$$x^2 = 36 \qquad \text{Take the square root of both sides.}$$
$$x = \pm 6$$

Section 15–2

1 Solve incomplete quadratic equations $(ax^2 + bx = 0)$ by factoring (pp. 615–616).

1. Use the addition axiom to write the equation in standard form.   **2.** Factor out all common factors and set each factor containing a variable equal to zero using the zero-product property.   **3.** Solve for the variable in each equation formed in Step 2.   **4.** Check each solution in the original equation.

LEARNING OUTCOMES	KEY CONCEPTS AND EXAMPLES

Solve $5x^2 - 15x = 0$.

$5x^2 - 15x = 0$ Factor common factors.

$5x(x - 3) = 0$ Set each factor equal to 0.

$5x = 0$ $x - 3 = 0$ Solve each equation.

$x = 0$ $x = 3$

2 Solve complete quadratic equations $(ax^2 + bx + c = 0)$ by factoring (pp. 616–617).

1. Arrange the equation in standard form.   **2.** Factor the trinomial into the product of two binomials, and set each binomial factor equal to zero using the zero-product property. **3.** Solve for the variable in each equation formed in Step 2.   **4.** Check if desired.

Solve $2x^2 - 5x - 3 = 0$.

$2x^2 - 5x - 3 = 0$ Factor the trinomial.

$(x - 3)(2x + 1) = 0$ Set each factor equal to 0.

$x - 3 = 0$ $2x + 1 = 0$ Solve each equation.

$x = 3$ $2x = -1$

$x = -\dfrac{1}{2}$

Section 15-3

1 Solve quadratic equations by completing the square (pp. 618–621).

1. Write the equation in the form $ax^2 + bx = -c$. That is, isolate the terms with variable factors.   **2.** Divide the equation by a so that the coefficient of x^2 is 1. $x^2 + \dfrac{b}{a}x = -\dfrac{c}{a}$
3. Form a perfect-square trinomial on the left by adding $\left(\dfrac{b}{2a}\right)^2$ or $\dfrac{b^2}{4a^2}$ to both sides of the equation.   **4.** Factor the perfect-square trinomial into the square of a binomial.   **5.** Take the square root of both sides of the equation.   **6.** Rearrange terms under the radical.
7. Make equivalent fractions using the common denominator for terms under the radical.
8. Write terms under the radical as one fraction.   **9.** Solve for x.

Solve $2x^2 - x + 4 = 0$ by completing the square.

$2x^2 - x + 4 = 0$ Isolate the variable terms.

$2x^2 - x = -4$ Divide equation by 2.

$x^2 - \dfrac{x}{2} = \dfrac{-4}{2}$ Simplify and add $\left(\dfrac{-1}{2}\right)^2$ to both sides.

$x^2 - \dfrac{x}{2} + \left(\dfrac{-1}{4}\right)^2 = -2 + \left(\dfrac{-1}{4}\right)^2$ Simplify.

$x^2 - \dfrac{x}{2} + \dfrac{1}{16} = -2 + \dfrac{1}{16}$ Write the left side of the equation as the square of a binomial.

$\left(x - \dfrac{1}{4}\right)^2 = \dfrac{-32}{16} + \dfrac{1}{16}$ Add like terms.

$\left(x - \dfrac{1}{4}\right)^2 = \dfrac{-31}{16}$ Take the square root of both sides.

$x - \dfrac{1}{4} = \pm\sqrt{\dfrac{-31}{16}}$ Solve for x.

$x = \dfrac{1}{4} \pm \dfrac{i\sqrt{31}}{4}$ Exact roots

LEARNING OUTCOMES	KEY CONCEPTS AND EXAMPLES

2 Solve quadratic equations using the quadratic formula (pp. 621–625).

In the quadratic formula, $x = \dfrac{-b \pm \sqrt{b^2 - 4ac}}{2a}$, a is the coefficient of the quadratic term, b is the coefficient of the linear term, and c is the constant when the equation is written in standard form. The variable is x.

1. Write the equation in standard form.   **2.** Identify a, b, and c.   **3.** Substitute numbers for a, b, and c in the quadratic formula.   **4.** Use the order of operations to simplify the expression under the radical.   **5.** Simplify the radical.   **6.** Factor any common factors in the numerator.   **7.** Simplify by reducing.   **8.** Write as two distinct solutions.   **9.** Write as exact solution or approximate solution as desired.

Solve using the quadratic formula:

$5x^2 + 7x - 6 = 0$ Identify a, b, and c.

$a = 5, \qquad b = 7, \qquad c = -6$

$x = \dfrac{-b \pm \sqrt{b^2 - 4ac}}{2a}$ Substitute values.

$x = \dfrac{-7 \pm \sqrt{7^2 - 4(5)(-6)}}{2(5)}$ Simplify the radicand.

$x = \dfrac{-7 \pm \sqrt{49 + 120}}{10}$ Add in the radicand.

$x = \dfrac{-7 \pm \sqrt{169}}{10}$ Evaluate the square root.

$x = \dfrac{-7 \pm 13}{10}$ Separate into two cases.

$x = \dfrac{-7 + 13}{10} \qquad x = \dfrac{-7 - 13}{10}$ Solve each equation.

$x = \dfrac{6}{10} \qquad\qquad x = -\dfrac{20}{10}$

$x = \dfrac{3}{5} \qquad\qquad x = -2$ Exact roots

Solving applied problems requires knowledge of other mathematical formulas, for example, $A = lw$ (area of a rectangle).

Find the length and width of a rectangular parking lot if the length is to be 12 m longer than the width and the area is to be 6,205 m^2.

Let x = number of meters in the width
$x + 12$ = number of meters in the length

Estimate: If the parking lot were a square, one side would be $s = \sqrt{A}$ or $s = 78.8$ m. In a rectangle where the length and width are different, the length is more than 78.8 m and the width is less than 78.8 m.

$$lw = A$$
$$x(x + 12) = 6{,}205$$
$$x^2 + 12x = 6{,}205$$
$$x^2 + 12x - 6{,}205 = 0$$

LEARNING OUTCOMES

KEY CONCEPTS AND EXAMPLES

Use a calculator with the formula.

$$x = \frac{-12 \pm \sqrt{12^2 - 4(1)(-6{,}205)}}{2(1)}$$ Substitute values for a, b, and c.

$$x = \frac{-12 \pm \sqrt{144 + 24{,}820}}{2}$$ Simplify radicand.

$$x = \frac{-12 \pm \sqrt{24{,}964}}{2}$$ Evaluate the square root.

$$x = \frac{-12 \pm 158}{2}$$ Separate into two cases.

$$x = \frac{-12 + 158}{2} \qquad x = \frac{-12 - 158}{2}$$ Solve each case.

$$x = \frac{146}{2} = 73 \qquad x = \frac{-170}{2} = -85$$ Disregard the negative root for a width measure.

$$x = 73 \text{ width}$$

$$x + 12 = 85 \text{ length}$$

Section 15–4

1 Graph quadratic functions using the table-of-solutions method (pp. 626–627).

1. Prepare a table of solutions of approximately five ordered pairs.   **2.** Plot the points on a rectangular coordinate system.   **3.** Connect the points with a smooth, continuous curve.

Graph $y = x^2 - 6x + 8$ using a table of solutions. Start with three values of x, -2, 0, and 2.

At $x = -2$:   At $x = 0$:   At $x = 2$:

$y = (-2)^2 - 6(-2) + 8$ $y = 0^2 - 6(0) + 8$ $y = 2^2 - 6(2) + 8$

$y = 4 + 12 + 8$ $y = 0 - 0 + 8$ $y = 4 - 12 + 8$

$y = 24$ $y = 8$ $y = 0$

Try two more, $x = 4$ and $x = 6$.

At $x = 4$:   At $x = 6$:

$y = 4^2 - 6(4) + 8$ $y = 6^2 - 6(6) + 8$

$y = 16 - 24 + 8$ $y = 36 - 36 + 8$

$y = 0$ $y = 8$

Finally, see what happens between $x = 2$ and $x = 4$ (Fig. 15–45).

At $x = 3$:

$y = 3^2 - 6(3) + 8$

$y = 9 - 18 + 8$

$y = -1$

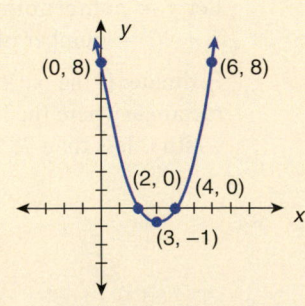

x	y
-2	24
0	8
2	0
3	-1
4	0
6	8

FIGURE 15–45

LEARNING OUTCOMES	KEY CONCEPTS AND EXAMPLES

2 Graph quadratic functions by examining properties (pp. 627–629).

Graph a quadratic function in the form $y = ax^2 + bx + c$: **1.** Find the axis of symmetry: $x = -\dfrac{b}{2a}$. **2.** Find the vertex: x-coordinate of vertex $= -\dfrac{b}{2a}$. To find the y-coordinate of the vertex, substitute the value for the x-coordinate into the original function and solve for y. **3.** Find one or two additional points that are to the right of the axis of symmetry. **4.** Apply the property of symmetry to find additional points. The symmetrical point will have an x-coordinate that is the same distance from the axis of symmetry on the opposite side. The two points have the same y-coordinate. **5.** Connect the plotted points with a smooth, continuous curve.

Graph $y = x^2 - 6x + 8$ by examining properties (Fig. 15–46).

$x = -\dfrac{b}{2a}$ Substitute $a = 1$ and $b = -6$.

$x = -\dfrac{-6}{2(1)}$ Simplify.

$x = \dfrac{6}{2}$

$x = 3$ Axis of symmetry: $x = 3$

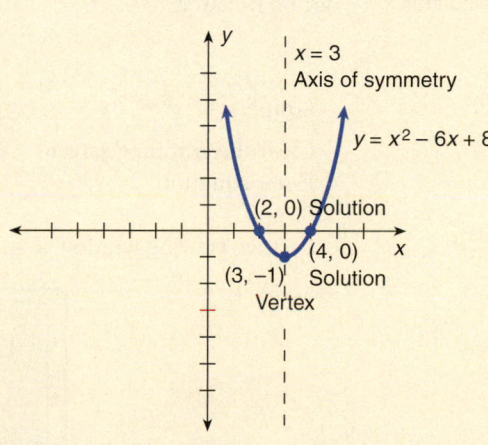

FIGURE 15–46

To find the vertex:

$y = (3)^2 - 6(3) + 8$ Substitute $x = 3$.

$y = 9 - 18 + 8$ Simplify.

$y = -1$ Vertex: $(3, -1)$

To find the x-intercepts:

$0 = x^2 - 6x + 8$ Substitute $y = 0$.

$0 = (x - 2)(x - 4)$ Factor. Solve for x.

$x - 2 = 0$ $x - 4 = 0$

$x = 2$ $x = 4$ x-intercepts: $(2, 0)$ and $(4, 0)$

3 Solve a quadratic equation from a graph of a corresponding quadratic function (p. 630).

1. Write the quadratic equation in standard form and then as a function. **2.** Graph the function. **3.** Determine the x-coordinate of all x-intercepts of the graph. That is, find the zeros of the function. **4.** If there are no x-intercepts, there are no real solutions.

LEARNING OUTCOMES	KEY CONCEPTS AND EXAMPLES

Find the solutions for the equation $x^2 - 6x + 8 = 0$ from the graph (Fig. 15–47).

The corresponding quadratic function is $y = x^2 - 6x + 8$. The x-intercepts are $(2, 0)$ and $(4, 0)$. The solutions of $x^2 - 6x + 8 = 0$ are $x = 2$ and $x = 4$.

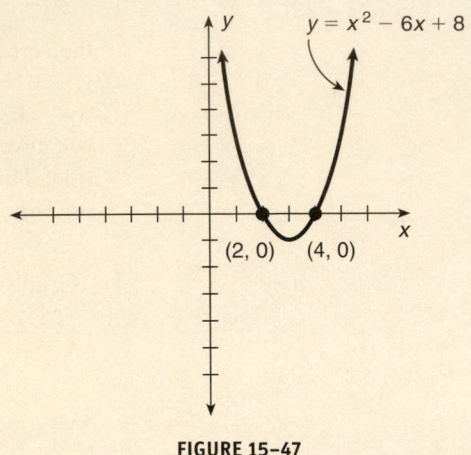

FIGURE 15–47

4 Graph quadratic functions using a graphing calculator (pp. 630–632).

1. Clear the graphing screen. **2.** Enter the equation that has been solved for y and press the graph function.

Graph $y = x^2 - 6x + 8$.

Clear the graphing screen.
Enter equation.
Graph.
Change viewing window if appropriate.

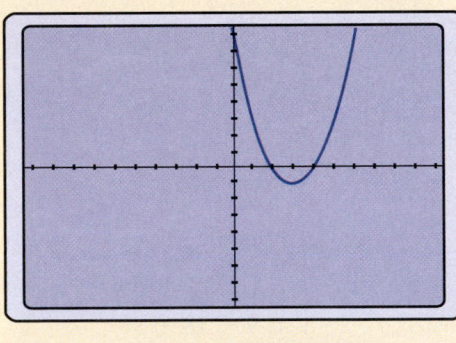

FIGURE 15–48

5 Determine the nature of the roots of a quadratic equation by examining the discriminant (pp. 632–634).

The radicand of the quadratic formula, $b^2 - 4ac$, is the discriminant of the quadratic equation.

Properties of the discriminant, $b^2 - 4ac$:

1. If $b^2 - 4ac \geq 0$, the equation has real-number roots.
   (a) If $b^2 - 4ac$ is a perfect square, there are *two rational roots*.
   (b) If $b^2 - 4ac = 0$, there are two equal rational roots, sometimes called a *double root*.
   (c) If $b^2 - 4ac$ is not a perfect square, there are two irrational roots.

2. If $b^2 - 4ac < 0$, the equation has no real-number roots. The roots are *imaginary* or *complex*.

LEARNING OUTCOMES

KEY CONCEPTS AND EXAMPLES

Use the discriminant to determine the characteristics of the roots of the equation $3x^2 - 5x + 7 = 0$.

$a = 3, \quad b = -5, \quad c = 7$ Identify a, b, and c.

$(-5)^2 - 4(3)(7) = 25 - 84$ Substitute values into $b^2 - 4ac$ and simplify.

$= -59$

Since -59 is less than zero, the   Interpret result.
roots are not real; they are complex.

Section 15–5

1   Identify the degree of an equation (p. 635).

The degree of an equation in one variable is the highest power of any term that appears in the equation.

State the degree: $x^3 + 4x = 0$ is a cubic or third-degree equation.

$x^4 = 81$ is a fourth-degree equation.

2   Solve higher-degree equations by factoring (pp. 635–636).

The higher-degree equations discussed in this section have a common variable factor and can be solved by factoring.

1. Write the equation in standard form.   **2.** Factor the polynomial side of the equation.
3. Set each factor containing a variable equal to zero.   **4.** Solve each equation from Step 3.

Solve $x^3 + 2x^2 - 3x = 0$.

$x^3 + 2x^2 - 3x = 0$ Factor common factor.

$x(x^2 + 2x - 3) = 0$ Factor trinomial.

$x(x + 3)(x - 1) = 0$ Set each factor equal to 0.

$x = 0 \quad\quad x + 3 = 0 \quad\quad x - 1 = 0$ Solve each equation.

$x = -3 \quad\quad\quad x = 1$

3   Graph higher-degree equations (pp. 636–637).

Prepare a table of solutions with enough points to get a complete view of the graph or use a graphing calculator.

Prepare a table of solutions and graph the function $f(x) = x^3 + 2$ (Fig. 15–49).

x	$f(x)$
-2	-6
-1	1
0	2
1	3
2	10

$f(-2) = (-2)^3 + 2 = -8 + 2 = -6$

$f(-1) = (-1)^3 + 2 = -1 + 2 = 1$

$f(0) = 0^3 + 2 = 2$

$f(1) = 1^3 + 2 = 1 + 2 = 3$

$f(2) = 2^3 + 2 = 8 + 2 = 10$

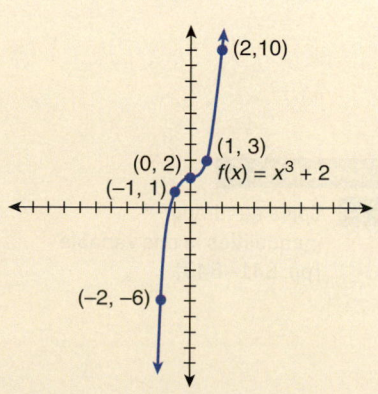

FIGURE 15–49

LEARNING OUTCOMES	KEY CONCEPTS AND EXAMPLES

4 Distinguish between a function and a relation using the vertical line test (pp. 637–638).

1. Graph the relation on a coordinate system.
2. Apply the vertical line test.
   (a) If a vertical line can be drawn so that it intersects a graph at more than one point, then the graph is the graph of a relation that is not a function.
   (b) If no vertical line can be drawn so that it intersects a graph at more than one point, then the graph is the graph of a function.

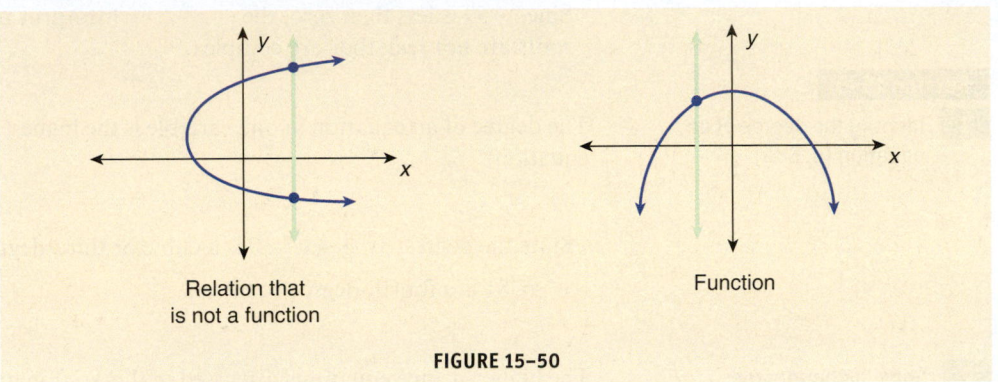

FIGURE 15–50

5 Find the domain and range of a relation from a graph (pp. 638–639).

1. Graph the relation on a coordinate system.   2. The domain is the set of all the values on the graph for the independent variable (*x*-values).   3. The range is the set of all the values on the graph for the dependent variable (*y*-values).

Determine the domain and range (Fig. 15–51).

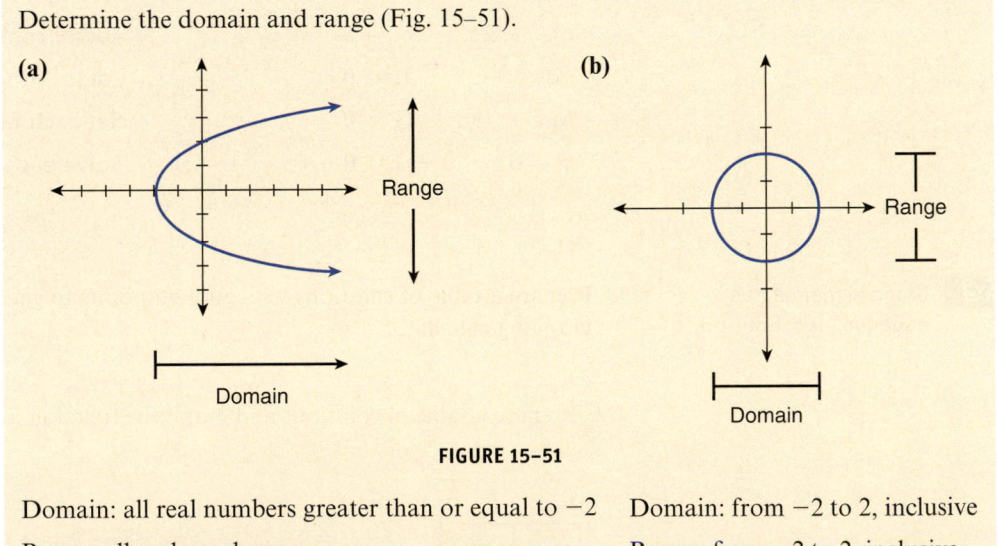

FIGURE 15–51

Domain: all real numbers greater than or equal to -2 Domain: from -2 to 2, inclusive

Range: all real numbers   Range: from -2 to 2, inclusive

Section 15–6

1 Solve quadratic inequalities in one variable (pp. 641–643).

1. Determine the critical values by solving an equation in which an equal sign is substituted for the inequality sign.   2. Plot the critical values on a number line to form the three regions.   3. Test *each* region of values by selecting any point within the region, substituting that value into the original inequality, solving the inequality, and deciding if the resulting inequality is a *true* statement.   4. The solution set for the quadratic inequality is the region or regions that produce a *true* statement in Step 3.   5. The critical values or boundary points are included in the solution set if the inequality is "inclusive" ($\leq$ or $\geq$). The critical values or boundary points are not included in the solution set if the inequality is "exclusive" ($<$ or $>$).

LEARNING OUTCOMES   **KEY CONCEPTS AND EXAMPLES**

Solve $x^2 + 4x - 21 < 0$.

$x^2 + 4x - 21 = 0$	Write as an equation in factored form.
$(x + 7)(x - 3) = 0$	Set each factor equal to zero.
$x + 7 = 0 \qquad x - 3 = 0$	Solve each equation.
$x = -7 \qquad\quad x = 3$	Critical values.

Test Region I, $x < -7$. Let $x = -8$.   Test Region II, $-7 < x < 3$. Let $x = 0$.

$(-8)^2 + 4(-8) - 21 < 0$ $0^2 + 4(0) - 21 < 0$

$\qquad 64 - 32 - 21 < 0$ $\qquad 0 + 0 - 21 < 0$

$\qquad\qquad\quad 11 < 0.$ *False.* $\qquad\qquad\quad -21 < 0.$ *True.*

Test Region III, $x > 3$. Let $x = 4$.

$4^2 + 4(4) - 21 < 0?$

$\quad 16 + 16 - 21 < 0?$

$\qquad\qquad 11 < 0.$ *False.*

FIGURE 15–52

The solution set is Region II.

$-7 < x < 3$ or $(-7, 3)$.

2 Graph quadratic inequalities in two variables (pp. 643–644).

1. Write the inequality in function notation or solved for y.   **2.** Graph the boundary by substituting an equal sign for the inequality sign and solving the equation.   **3.** Select a test point that is *not* on the boundary.   **4.** Substitute the values from the test point into the original inequality and evaluate. The test point will be either *inside* or *outside* the boundary.   **5.** If the evaluated inequality makes a *true* statement, the solution set is the portion of the graph (inside or outside) that includes the test point.   **6.** If the evaluated inequality makes a *false* statement, the solution set is the portion of the graph (inside or outside) that does *not* include the test point.   **7.** The boundary *is not* included if the inequality is $<$ or $>$. The boundary *is* included if the inequality is $\le$ or $\ge$.

Graph $y \le x^2 - 6x + 8$.

Graph the boundary $y = x^2 - 6x + 8$.

Clear the graphing screen.
Enter equation.
Graph.
Change viewing window if appropriate.

Select a test point such as $(0, 0)$.

$y \le x^2 - 6x + 8$

$0 \le 0^2 - 6(0) + 8$

$0 \le 8$ True.

Shade the portion of the graph that includes the test point. The boundary is included in the solution set because the inequality is $\le$ (Fig. 15–53).

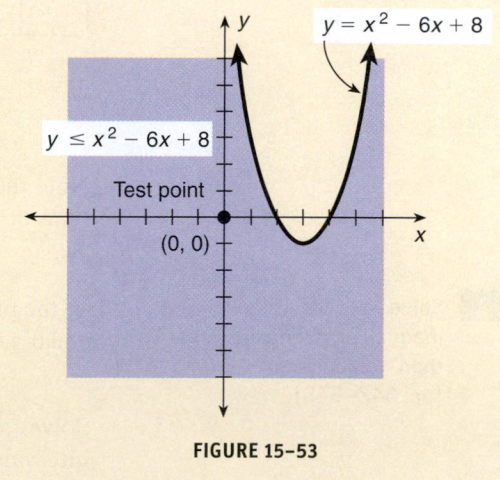

FIGURE 15–53

Section 15–7

1 Solve equations containing one absolute-value expression (pp. 645–646).

1. Isolate the absolute-value expression on one side of the equation.   **2.** Separate the equation into two cases. One case considers the expression within the absolute-value symbol to be positive. The other case considers it to be negative.   **3.** Solve each case to obtain the *two* roots of the equation.   **4.** $|x| = b$ has no solution if b is negative ($b < 0$).   **5.** $|x| = b$ has one root when $b = 0$.

LEARNING OUTCOMES	KEY CONCEPTS AND EXAMPLES

Solve $|x - 3| + 1 = 5$.

$|x - 3| + 1 = 5$ Isolate the absolute-value expression.

$|x - 3| = 4$

Case 1: $x - 3 = 4$ Case 2: $x - 3 = -4$ or $-(x - 3) = 4$

$x = 7$ $x = -4 + 3$ $-x + 3 = 4$

$x = -1$ $-x = 1$

$x = -1$

2 Graph equations containing one absolute-value expression (p. 647).

1. Rearrange the equations so that one side equals zero. **2.** Write the equation from Step 1 in function notation. **3.** Determine the value of x inside the absolute-value symbols that will make the value of the absolute-value expression zero. **4.** Use the value of x found in Step 3 and other values of x that are less than and greater than the value found in Step 3 and prepare a table of solutions with at least five solutions. **5.** Draw the graph using the table of solutions.

Graph the function $f(x) = |x| - 4$ using the table-of-solutions method.

Values of x that are greater than 0 will give a positive result for the expression inside the absolute-value symbols. A value less than 0 will give a negative result for the expression inside the absolute-value symbols. To have a representative graph for the equation, find ordered pairs for values of x that are both less than and greater than 0.

x	$f(x)$
-4	0
-2	-2
0	-4
2	-2
4	0

$f(-4) = |-4| - 4 = 4 - 4 = 0$

$f(-2) = |-2| - 4 = 2 - 4 = -2$

$f(0) = |0| - 4 = 0 - 4 = -4$

$f(2) = |2| - 4 = 2 - 4 = -2$

$f(4) = |4| - 4 = 4 - 4 = 0$

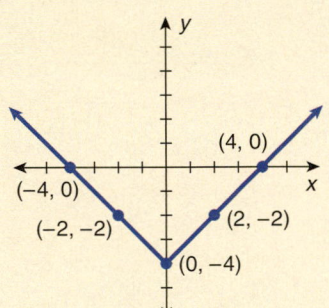

FIGURE 15–54

Note that graph crosses the x-axis at $(-4, 0)$ and $(4, 0)$. These points represent the solutions of the equations, $x = -4$ and $x = 4$.

3 Solve absolute-value inequalities using the "less than" relationship (pp. 647–648).

Use the property if $|x| < b$ and $b > 0$, then $-b < x < b$.
Similarly, if $|x| \le b$ and $b > 0$, then $-b \le x \le b$.

Solve $|x - 3| < 4$. Show the solution set graphically, in symbolic notation, and in interval notation.

$-4 < x - 3 < 4$ Separate into two simple inequalities connected by *and*.

$-4 < x - 3$ and $x - 3 < 4$ Solve each inequality.

$-1 < x$ $x < 7$

$-1 < x < 7; (-1, 7)$

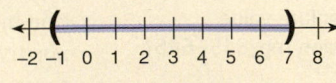

FIGURE 15–55

LEARNING OUTCOMES	KEY CONCEPTS AND EXAMPLES

4 Solve absolute-value inequalities using the "greater than" relationship (p. 649).

Use the property if $|x| > b$ and $b > 0$, then $x < -b$ or $x > b$.
Similarly, if $|x| \geq b$ and $b > 0$, then $x \leq -b$ or $x \geq b$.

Solve $|x + 7| > 1$. Show the solution set graphically, in symbolic notation, and in interval notation.

$x + 7 < -1$ or $x + 7 > 1$ Separate into two simple inequalities connected by *or*.

$\qquad x < -8 \qquad\qquad x > -6$

$(-\infty, -8) \cup (-6, \infty)$

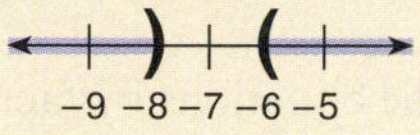

$$\xleftarrow{\quad}\!)\!\mid\!(\!\xrightarrow{\quad}$$
$$\;-9\;-8\;-7\;-6\;-5$$

FIGURE 15–56

---

15 CHAPTER REVIEW EXERCISES

Section 15–1   MyLab Math **For additional practice go to your study plan in MyLab Math.**

Identify the quadratic equations as pure, incomplete, or complete.

1. $x^2 = 49$

2. $x^2 - 5x = 0$

3. $5x^2 - 45 = 0$

4. $3x^2 + 2x - 1 = 0$

5. $8x^2 + 6x = 0$

6. $5x^2 + 2x + 1 = 0$

7. $x^2 - 32 = 0$

8. $x^2 + x = 0$

9. $3x^2 + 6x + 1 = 0$

Write the equations in standard form.

10. $x^2 - 7x = 8$

11. $2x^2 - 5 = 8x$

12. $7x = 2x^2 - 3$

13. $5 + x^2 - 7x = 0$

14. $x - 4 = 3x^2$

15. $3x = 1 - 4x^2$

Solve the equations by the square-root method. Round to thousandths when necessary.

16. $x^2 = 121$

17. $x^2 = 100$

18. $x^2 - 64 = 0$

19. $4x^2 = 9$

20. $64x^2 - 49 = 0$

21. $0.36y^2 = 1.09$

22. $0.16x^2 = 0.64$

23. $5x^2 = 40$

24. $2x^2 - 5 = 3$

25. $6x^2 + 4 = 34$

26. $5x^2 - 6 = 19$

27. $3x^2 = 12$

28. $3x^2 - 4 = 8$

29. $2x^2 = 34$

30. $5x^2 - 9 = 30$

31. $3y^2 - 36 = -8$

32. $2x^2 + 3 = 51$

33. $\frac{1}{2}x^2 = 8$

34. $\frac{2}{3}x^2 = 24$

35. $\frac{1}{4}x^2 - 1 = 15$

36. $\frac{2}{5}x^2 + 2 = 8$

37. A circle has an area of 845 cm^2. What is the radius of the circle? Use the formula $A = \pi r^2$ and the calculator key for π. Round to the nearest tenth centimeter.

38. The area of an oil painting that is a square is known to be 9,072 cm^2. What are the inside dimensions of its picture frame?

Section 15–2

Solve the equations by factoring.

39. $x^2 - 5x = 0$

40. $4x^2 = 8x$

41. $6x^2 - 12x = 0$

42. $3x^2 + x = 0$

43. $10x^2 + 5x = 0$

44. $3y^2 = 12y$

45. $y^2 - 7y = 0$

46. $x^2 = 16x$

47. $12x^2 + 8x = 0$

48. $8x^2 - 12x = 0$

49. $x^2 + 3x = 0$

50. $4x^2 - 28x = 0$

51. $5x^2 = 45x$

52. $7x^2 = 28x$

53. $y^2 + 8y = 0$

54. $z^2 - 6z = 0$

55. $3m^2 - 5m = 0$

56. $4n^2 - 3n = 0$

57. $2x^2 = x$

58. $5y^2 = y$

59. 3 times the square of a number is the same as 12 times the number. What is the number?

60. Describe the steps for solving an incomplete quadratic equation. Include a clear description of the nature of the roots.

Solve the equations by factoring.

61. $x^2 - 4x + 3 = 0$

62. $x^2 + 7x + 12 = 0$

63. $x^2 + 3x = 10$

64. $x^2 - 7x + 12 = 0$

65. $x^2 + 7x = -6$

66. $x^2 + 3 = -4x$

67. $x^2 - 6x + 8 = 0$

68. $6y + 7 = y^2$

69. $6y^2 - 5y - 6 = 0$

70. $5y^2 + 23y = 10$

71. $10y^2 - 21y - 10 = 0$

72. $6x^2 - 16x + 8 = 0$

73. $4x^2 + 7x + 3 = 0$

74. $3x^2 = -7x + 6$

75. $12y^2 - 5y - 3 = 0$

76. $x^2 - 3x = 18$

77. $x^2 + 19x = 42$

78. $3x^2 + x - 2 = 0$

79. $3y^2 + y - 2 = 0$

80. $2x^2 - 4x - 6 = 0$

81. $2x^2 - 10x + 12 = 0$

82. $y^2 + 18y + 45 = 0$

83. $x^2 - 3x - 18 = 0$

84. $3x^2 - 9x - 30 = 0$

85. $2y^2 + 22y + 60 = 0$

86. $x^2 + 13x + 12 = 0$

87. $x^2 + 7x - 18 = 0$

88. **AG/H**   An office building in the shape of a rectangle is known to have 47,500 ft² of space on the ground floor. The tenant wants to landscape the two longer sides of the building. The tenant also knows that the building is about 60 ft longer than it is wide. How many feet of land along the building need to be landscaped?

89. **CAD/ARC**   Jerri Amour is an architect who is designing a hospital. She knows that a kidney dialysis machine needs a space that is 7 ft longer than it is wide. The total area needed is 228 ft² of space. Find the length and width of the space needed for the machine.

Section 15–3

Solve each equation by completing the square. Give the answer as an exact solution.

90. $x^2 + 2x - 3 = 0$

91. $x^2 - 4x + 4 = 0$

92. $x^2 - 6x + 8 = 0$

93. $x^2 - 8x + 12 = 0$

94. $x^2 - 6x + 7 = 0$

95. $x^2 - 8x + 14 = 0$

96. $x^2 - 6x + 4 = 0$

97. $x^2 - 6x + 12 = 0$

98. $x^2 - 2x + 6 = 0$

99. $x^2 - 5x + 4 = 0$

100. $x^2 - 3x - 18 = 0$

101. $x^2 - 3x = 7$

Indicate the values for a, b, and c in the quadratic equations.

102. $5x^2 + x + 6 = 0$

103. $x^2 - 2x = 8$

104. $x^2 - 7x + 12 = 0$

105. $x^2 + 3x = 4$

106. $3x^2 = 2x + 7$

107. $x^2 - 3x = -2$

Solve the quadratic equations by using the quadratic formula.

108. $x^2 - 9x + 20 = 0$

109. $x^2 - 8x - 9 = 0$

110. $x^2 - 5x = -6$

111. $x^2 + 2x = 8$

112. $x^2 - x - 12 = 0$

113. $2x^2 - 3x - 2 = 0$

Solve the quadratic equations by using the quadratic formula. Round each final answer to the nearest hundredth.

114. $3x^2 + 6x + 2 = 0$

115. $2x^2 - 3x - 1 = 0$

116. $5x^2 + 4x - 8 = 0$

117. $3x^2 + 5x + 1 = 0$

118. **CON**   A bricklayer plans to build an arch with a span (s) of 10 m and a radius (r) of 5 m. How high (h) is the arch?

(Use the formula $h^2 - 2hr + \dfrac{s^2}{4} = 0$.)

119. **CON**   A rectangular kitchen contains 240 ft². If the length is 2 times the width, find the length and width of the room to the nearest whole number. (Area = length × width, or $A = lw$.)

120. **CON**   What are the dimensions of a rectangular tool storage room if the area is 45.5 m² and the room is 0.5 m longer than it is wide?

121. **INDTEC**   Find the length and width of a rectangular piece of fiberglass if its length is 3 times the width and the area is 591 in². Round to the nearest inch.

Section 15–4

Graph the quadratic functions. Determine the domain and range.

122. $y = x^2 - 1$

123. $y = -x^2 - 1$

124. $y = x^2 + 3x - 10$

125. $y = x^2 - 6x + 8$

126. $y = x^2 - 2x + 1$

127. $y = -x^2 + 2x - 1$

Graph the quadratic functions by using the vertex, x-intercepts, and one other point.

128. $y = x^2 - 4x + 3$

129. $y = x^2 - 2x - 8$

130. $y = x^2 + 12x + 35$

131. $y = -x^2 + 8x - 12$

Find all real solutions of the equations by writing and graphing the equations of the corresponding quadratic functions.

132. $x^2 + x - 12 = 0$

133. $x^2 - 4x - 12 = 0$

134. $x^2 - 5x = 14$

135. $x^2 + 8x = -16$

Use a calculator to graph the following. Adjust the window to display the vertex.

136. $y = x^2 - 2x + 1$

137. $y = -2x^2$

138. $y = \frac{1}{2}x^2 - 3$

Use the discriminant of each equation to determine the nature of the roots of the equations.

139. $x^2 + 8x + 16 = 0$

140. $2x^2 - 3x - 5 = 0$

141. $5x^2 - 100 = 0$

142. $3x^2 - 2x + 4 = 0$

143. $2x = 5x^2 - 3$

Section 15–5

State the degree of each equation.

144. $2x + 5x = 15$

145. $3x - 2x^3 + 8 = 0$

146. $16 = x^4$

147. $6 - 3x - 3 = 2x + 4$

148. $y^6 = 729$

149. $5y^8 + 2y^3 - 6 = y^2$

Find the roots of the following equations. Factor if necessary. Round to thousandths if appropriate.

150. $x(x + 2)(x - 3) = 0$

151. $2x(3x - 2)(x - 2) = 0$

152. $3x(2x + 1)(x + 4) = 0$

153. $2x^3 + 10x^2 + 12x = 0$

154. $x^3 = 2x^2$

155. $2x^3 + 9x^2 = 5x$

156. $6x^3 + 3x^2 - 18x = 0$

157. $3x^3 - 6x^2 = 0$

158. $3x^3 - x^2 - 2x = 0$

159. $x^3 + 6x^2 + 8x = 0$

160. $x^3 - 8x^2 + 15x = 0$

161. $x^3 - x^2 - 20x = 0$

162. $x^3 + x^2 - 20x = 0$

163. $y^3 - 6y^2 + 7y = 0$

164. $y^3 + 2y^2 + 5y = 0$

165. $x^3 - 3x^2 - 4x = 0$

166. $y^3 + 7y^2 + 12y = 0$

167. $2y^3 + 6y^2 + 4y = 0$

Graph the equations using the table-of-solutions method. Determine the domain and range.

168. $y = x^3 - x^2 - 8x + 1$

169. $y = 5x^3$

Use the vertical line test to determine which are graphs of functions.

170. $x = 6 - 3y$

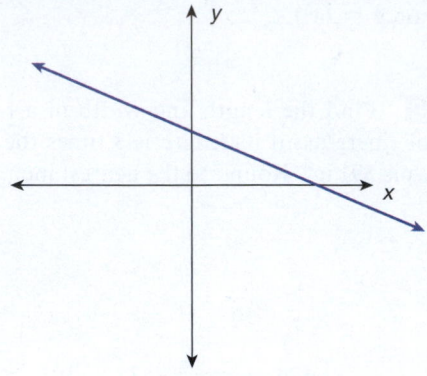

FIGURE 15–57

171. $x = -y^3 + 2y^2 + 2y + 3$

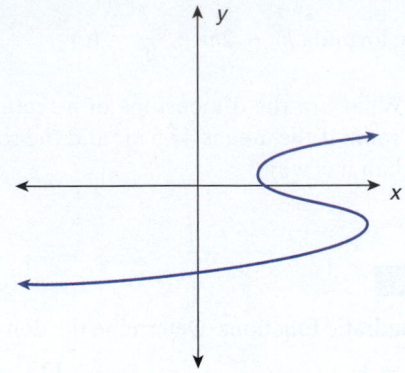

FIGURE 15–58

172. $y = x^4 - 3x^3 + x$

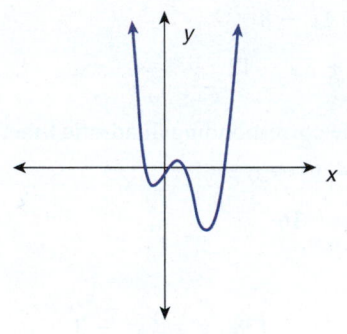

FIGURE 15–59

173. $\dfrac{x^2}{25} + \dfrac{y^2}{4} = 1$

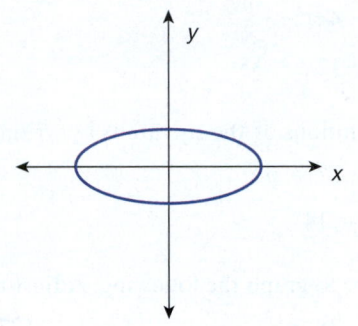

FIGURE 15–60

Find the domain and range of the relations from the graph.

174.

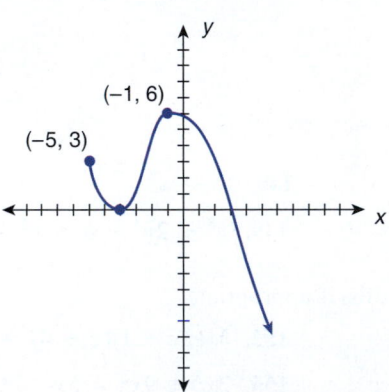

FIGURE 15–61

175.

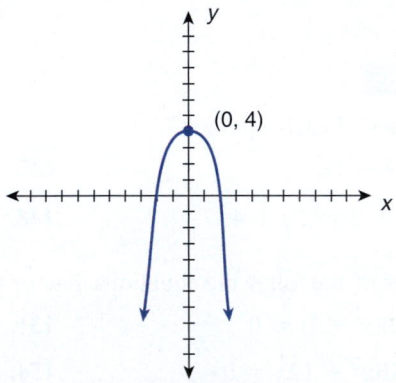

FIGURE 15–62

176.

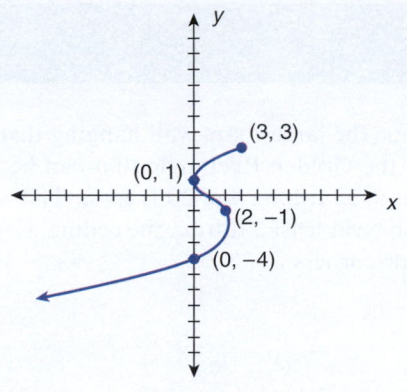

FIGURE 15–63

177.

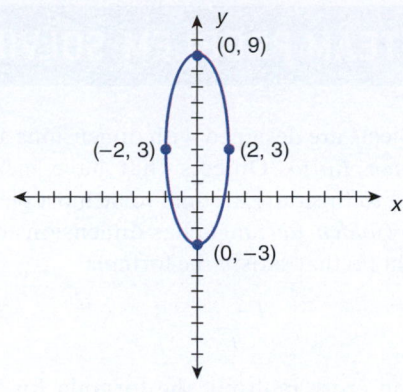

FIGURE 15–64

Section 15–6

Solve each inequality and graph the solution on a number line and in symbolic and interval notation.

178. $(3x + 2)(x - 2) < 0$

179. $(3x + 1)(2x - 3) < 0$

180. $(5x - 6)(x + 1) \geq 0$

181. $(x + 1)(x - 2) \leq 0$

182. $x^2 + x - 12 < 0$

183. $2x^2 \leq 5x + 3$

184. $2x^2 - 3x \geq -1$

185. $2x^2 + 7x - 15 < 0$

186. $x^2 - 2x - 8 \geq 0$

Use a calculator to graph the following. Adjust the window to display the vertex.

187. $y \geq -2x^2$

188. $y < x^2 - 2x + 1$

189. $y \geq x^2 - 6x + 9$

190. $y \leq \frac{1}{2}x^2 - 3$

Section 15–7

Solve each equation containing an absolute value.

191. $|x| = 12$

192. $|x - 9| = 2$

193. $|x + 3| = 7$

194. $|x + 4| = 11$

195. $|x - 8| = 12$

196. $|x + 7| = 3$

197. $|4x - 7| = 17$

198. $|2x + 3| = 5$

199. $|7x + 8| = 15$

200. $|4x + 1| = 9$

201. $|7x - 4| = 17$

202. $|6x - 2| = 3$

203. $|3x - 9| = 2$

204. $|x| + 8 = 10$

205. $|x| + 12 = 19$

206. $|2x| - 1 = 9$

207. $|x| - 9 = 7$

208. $|x - 4| - 10 = 6$

209. $-5 + |x - 3| = 2$

210. $|4x - 2| + 1 = 5$

211. $|4x - 3| - 12 = -7$

212. $|3x + 4| - 5 = 17$

Graph the functions.

213. $f(x) = |2x|$

214. $f(x) = |x| - 1$

Graph the solution set for each inequality and write in symbolic and interval notation.

215. $|x - 3| < 4$

216. $|x + 7| > 3$

217. $|x - 4| - 3 < 5$

218. $|x - 1| < 9$

219. $|x - 3| < -4$

15 | TEAM PROBLEM-SOLVING EXERCISES

1. Many objects are designed with dimensions according to the *Golden Ratio*. Objects that have measurements according to this ratio are said to be most pleasing to the eye. The *Golden Rectangle* has dimensions of length (*l*) and height (*h*) that satisfy the formula

$$\frac{l + h}{l} = \frac{l}{h}$$

When you cross multiply the formula for the Golden Rectangle, a quadratic equation results.
   (a) An artist wants a canvas proportioned according to the Golden Rectangle. If the length is to be 36 in., find the height to the nearest inch.
   (b) What should the length of a canvas be if the height is 20 in.?

2. Determine the largest-size wall hanging that has dimensions of the Golden Rectangle that can be placed on a wall that is 32 ft long and 20 ft high. The wall hanging must also be at least 2 ft from the ceiling, floor, and each of the side corners.

3. The tip for finding the solution set for a quadratic inequality by inspection (p. 643) is appropriate only when *a,* the coefficient of the leading coefficient, is positive.
   (a) How would the tip need to be modified if the leading coefficient were negative?
   (b) Illustrate your modified tip with a specific example for the "less than" relationship.
   (c) Illustrate your modified tip with a specific example for the "greater than" relationship.

15 | CONCEPTS ANALYSIS

1. State the zero-product property and explain how it applies to solving quadratic equations.

2. Under what conditions would the square-root method be used to solve an equation?

3. Use the equation $y = ax^2 + x + 1$ to graph equations when $a = 1, 2, 3, 4, 5$. Describe the changes in the graph caused by changes in a.

4. Under what conditions will a quadratic equation have no real solutions?

5. Under what conditions will a quadratic equation have irrational roots?

6. Under what conditions will a quadratic equation have rational roots?

Find a mistake in each of the following. Correct and briefly explain the mistake.

7. $2x^2 - 2x - 12 = 0$
   $2(x^2 - x - 6) = 0$
   $2(x + 2)(x - 3) = 0$
   $2 = 0 \quad x + 2 = 0 \quad x - 3 = 0$
   $\qquad\qquad x = -2 \qquad x = 3$
   The roots are 0, −2, and 3.

8. $-2x^3 + 5x^2 + 2x = 0$
   $\dfrac{x(2x^2 + 5x + 2)}{x} = \dfrac{0}{x}$
   $2x^2 + 5x + 2 = 0$
   $(2x + 1)(x + 2) = 0$
   $2x + 1 = 0 \qquad x + 2 = 0$
   $2x = -1 \qquad\qquad x = -2$
   $x = -\dfrac{1}{2}$
   The roots are $-\frac{1}{2}$ and −2.

9. Under what conditions would the quadratic formula be used to solve a quadratic equation?

10. What is the maximum number of roots the equation $x^4 = 16$ *could* have? How many real roots does the equation have?

11. How is the graph of a quadratic function different from the graph of a linear function?

12. What do *axis of symmetry* and *vertex* refer to on the graph of a quadratic function that represents a parabola?

13. In the inequality $x^2 + 6x + 5 < 0$, what roles do the numbers -5 and -1 play? Explain the solution for the inequality in words.

14. If p_1 and p_2 are real numbers that solve the equation $ax^2 + bx + c = 0$ and $p_1 < p_2$, when will the solution of $ax^2 + bx + c < 0$ be $p_1 < x < p_2$? When will the solution be $x < p_1$ or $x > p_2$? Give an example to illustrate each answer.

---

15 PRACTICE TEST

Identify the quadratic equations as pure, incomplete, or complete.

1. $3x^2 = 42$

2. $7x^2 - 3x + 2 = 0$

3. $5x^2 = 7x$

4. $4x^2 - 1 = 0$

Solve the quadratic equations using the square-root method. Give the exact roots.

5. $x^2 = 81$

6. $81x^2 - 64 = 0$

Solve the equations by factoring.

7. $3x^2 - 6x = 0$

8. $2x^2 + 3x + 1 = 0$

9. $x^2 - 5x + 6 = 0$

10. $3x^2 - 2x - 1 = 0$

11. $2x^2 + 12 = 11x$

12. $x^2 - 3x - 4 = 0$

Solve the equations using the quadratic formula. Round to the nearest hundredth when necessary.

13. $2x^2 + 3x - 5 = 0$

14. $3x^2 - 5x + 4 = 0$

15. $x^2 - 3x - 5 = 0$

16. $4x^2 - 2x = 3$

Describe the roots of the equations without solving.

17. $x^2 - 8x + 12 = 0$

18. $3x^2 - 2x + 3 = 0$

Graph the quadratic equations by examining properties. Show the axis of symmetry, vertex, and x-intercepts.

19. $y = x^2 + 2x + 1$

20. $y = x^2 + 4x + 4$

21. **ELEC** Find the diameter (d) in mils to the nearest hundredth of a copper wire conductor whose resistance (R) is 1.314 Ω and whose length (L) is 3,642.5 ft. (Formula: $R = \dfrac{KL}{d^2}$, where K is 10.4 for copper wire.)

22. Find the radius (r) of a circle whose area (A) is 35.15 cm^2. Round the answer to the nearest hundredth centimeter. (Formula: $A = \pi r^2$.)

23. **ELEC** What is the current (I) in amps if the resistance (R) of the circuit is 52.29 Ω and 205 watts (W) are used? Round to the nearest hundredth. $\left(\text{Formula: } R = \dfrac{W}{I^2} \right)$

24. **AG/H** A square parcel of land has an area of 156.25 m^2. What is the length of a side? (Use the formula $A = s^2$, where A is the area and s is the length of a side.)

25. **ELEC** In the formula $E = 0.5\,mv^2$, solve for v if $E = 180$ and $m = 10$ and $v > 0$.

Solve each equation.

26. $x(2x - 5)(x - 3) = 0$

27. $6x^3 + 21x^2 = 45x$

28. $2x^3 - x^2 - 6x = 0$

29. $6x^3 - 18x^2 = 0$

Use the vertical line test to identify the graphs as functions or relations. Determine the domain and range.

30. $y = x^5 + 2x^4 - x^3$

31. $x = y^3 - 2y^2 - y$

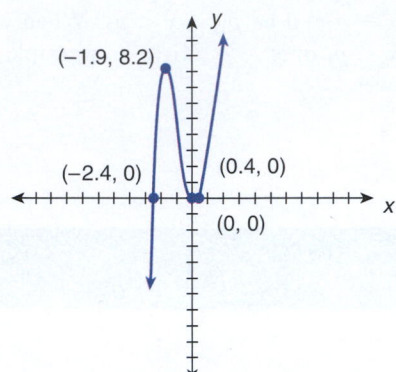

FIGURE 15–65

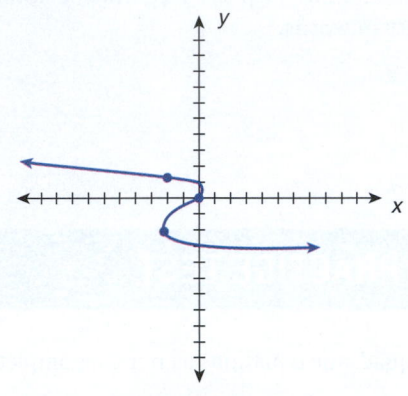

FIGURE 15–66

32. $(x + 4)(x - 2) < 0$

33. $(2x + 3)(x - 1) > 0$

34. $2y^2 + y < 15$

Solve.

35. $|x| = 15$

36. $|x + 8| = 7$

37. $|x| + 8 = 10$

Solve and graph the solution set on a number line and write the solution set in interval notation.

38. $|4x - 7| < 17$

39. $|x + 8| > 10$

40. $|x + 1| - 3 < 2$

Graph the quadratic inequalities.

41. $y \leq x^2 - 6x + 8$

42. $y > 2x^2$

43. $y < -\frac{1}{2}x^2$

11–15   CUMULATIVE PRACTICE TEST

Perform the indicated operations.

1. $5x^7 \cdot 8x^3$

2. $\dfrac{x^3 y^{-3}}{x^4 y^{-5}}$

3. $(x^4)^8$

Arrange in descending order.

4. $15 - 32x^3 + 6x^4 - 8x^2 + 40x$

Simplify.

5. $5x^4 - 3x^2 + 7x^4 + 8x^2$

Multiply.

6. $-8x^2(3x^2 - 4x + 7)$

Divide and simplify.

7. $\dfrac{10x^5 - 20x^6 + 15x^4}{5x^3}$

Factor by removing the greatest common factor.

8. $12x^3 - 18x^2 y$

9. $24x^3 - 16x^2 - 8x$

Find the product.

10. $(3x - 2)(x + 1)$

11. $(4x - 5)(3x - 2)$

Find the quotient.

12. $x + 3 \,)\overline{\, x^2 - 4x - 21 \,}$

13. $x - 5 \,)\overline{\, x^2 - 8x + 15 \,}$

Factor completely.

14. $49y^2 - 36$

15. $x^2 - 16x + 64$

16. $x^2 - 8x + 15$

17. $x^2 + x - 42$

18. $3x^2 - 17x + 10$

19. $6x^2 + 17x + 12$

20. $x^3 - 5x^2 - x$

Solve by the square-root method. Round to thousandths if necessary.

21. $3x^2 - 7 = 20$

22. $4x^2 - 9 = 39$

Solve by factoring.

23. $x^2 - 8x = 0$

24. $x^2 - x - 12 = 0$

25. $2x^2 - 7x + 3 = 0$

26. $4x^2 + 5x = 6$

Solve using the quadratic formula. Round to the nearest hundredth.

27. $4x^2 - 3x - 2 = 0$

28. $x^2 + 6x = 2$

Solve.

29. $x^3 - 3x^2 - 10x = 0$

30. $2x^3 - 5x^2 + 3x = 0$

31. $y \geq x^2 + x - 6$

32. $y < -x^2 + x + 6$

33. $|x + 3| < 2$

Exponential and Logarithmic Equations

Rawpixel.com/Shutterstock

In Great Company

How Nobel Prize Winners Crashed the World's Banks

The economists Fischer Black, Robert Merton, and Myron Scholes developed the Black-Scholes formula in 1973 to help stock traders. In stock trading, there are some contracts that take time to complete—30, 60, or 90 days, for instance. It used to be the case that if you bought such a contract, you couldn't sell it until the contract had completed. Why not? Well, no one could tell how much it was worth until it was close to completion. But what if you wanted to sell it earlier? How would you know how much it was worth?

The Black-Scholes formula, as it came to be known, was intended to help stock traders price a financial contract that had not yet completed. As one wag put it, "It was like buying or selling a bet on a horse, halfway through the race." The work was so innovative and complex that Scholes shared the 1997 Nobel Prize in

Economics with Robert Merton for his help in inventing it (Black had died by then, so he wasn't eligible).

This formula is the basis for the enormous derivatives market. What's a derivative? Well, it's an investment on an investment, a bet on a bet. If used sensibly, it gives traders much more flexibility in holding a market position.

If it was only used sensibly.

It wasn't. Not only was it used in situations that didn't call for it, stock traders began hiring mathematicians to generate variations of the Black-Scholes formula.

By 2007, the world economic community was trading derivatives valued at one quadrillion dollars a year. One quadrillion is 10 times more than the entire value of the whole world's production for the last 100 years. The whole world together does not have anywhere close to that much money—but everyone was expecting payment when the bets came in. And 75 to 90% of the time, the traders lost their bets.

When they lost their bets, they had to liquidate their holdings to pay the losses. They didn't have enough. They went bankrupt. The people to whom they owed money didn't get the money that was owed them. They, in turn, didn't have enough to pay their debts, and so on down the line.

Because people used the Black-Scholes formula when it should not have been used, and then used it to create derivative formulas that were even more problematic than Black-Scholes, the world's banks had made bets that could not be covered. As a result, the banks lost hundreds of billions of dollars when the real estate bubble burst.

16–1 Exponential Expressions, Equations, and Formulas

LEARNING OUTCOMES

1 Evaluate formulas with at least one exponential term.

2 Evaluate formulas that contain a power of the natural exponential, e.

3 Solve exponential equations in the form $b^x = b^y$, where $b > 0$ and $b \neq 1$.

4 Graph an exponential function.

LC LEARNING CATALYTICS

1. Evaluate $10,000\left(\dfrac{0.03}{(1 + 0.03)^5 - 1}\right)$ to the nearest cent.

Many scientific, technical, and business phenomena have the property of exponential growth; that is, the growth rate does not remain constant as certain physical properties increase. Instead, the growth rate increases exponentially. For example, notice the difference between $2x$ and 2^x when x increases. If we write the expressions in function notation, we have two different functions of x.

$$f(x) = 2x$$
$$g(x) = 2^x$$

Examine the values in Table 16–1 and the graphical representation of these two functions (Fig. 16–1). 2^x is said to increase *exponentially*. $g(x) = 2^x$ is an *exponential function*. On the other hand, $2x$ increases at a constant rate. $f(x) = 2x$ is a *linear function* since $2x$ is a linear polynomial.

Table 16-1 Values of $f(x)$ and $g(x)$

x	$f(x) = 2x$	$g(x) = 2^x$
1	$f(1) = 2(1) = 2$	$g(1) = 2^1 = 2$
2	$f(2) = 2(2) = 4$	$g(2) = 2^2 = 4$
3	$f(3) = 2(3) = 6$	$g(3) = 2^3 = 8$
4	$f(4) = 2(4) = 8$	$g(4) = 2^4 = 16$
5	$f(5) = 2(5) = 10$	$g(5) = 2^5 = 32$
6	$f(6) = 2(6) = 12$	$g(6) = 2^6 = 64$

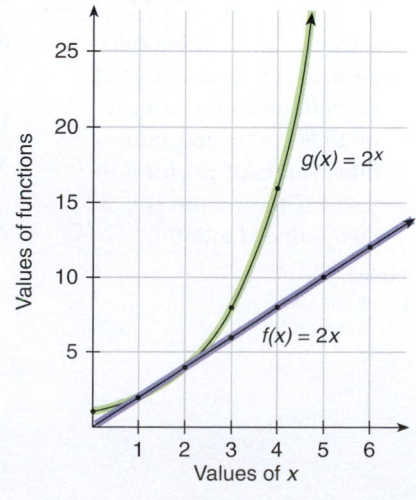

FIGURE 16–1

Exponential expression: an expression that contains at least one term that has a variable exponent

Variable exponent: an exponent that has at least one variable factor

Exponential equation or formula: an equation or formula that contains at least one term with a variable exponent

STOP AND CHECK

Evaluate using a scientific or graphing calculator.

1. 8^5
2. 3^{-3}
3. $9^{1.5}$

Answers:

1. 32,768   2. 0.037037037
3. 27

Compound interest: the interest for an investment or loan that is calculated at the end of each period and then added to the principal for the next period

Compound amount: the accumulated amount that is the combined principal and interest accumulated over a period of time

1 **Evaluate Formulas with at Least One Exponential Term.** Many formulas have terms that contain exponents. An **exponential expression** is an expression that contains at least one term that has a variable exponent. A **variable exponent** is an exponent that has at least one variable factor. Exponential expressions can be evaluated on a calculator by using the general power key $\boxed{\wedge}$ or other power keys as appropriate. An **exponential equation or formula** contains at least one term that has a variable exponent.

EXAMPLE 1

Evaluate using a scientific or graphing calculator.

(a) 5^6 (b) 4^{-3} (c) $8^{2.5}$

(a) $5^6 = 15,625$ $5\boxed{\wedge}6\boxed{\text{ENTER}} \Rightarrow 15625$

(b) $4^{-3} = 0.015625$ $4\boxed{\wedge}\boxed{(-)}3\boxed{\text{ENTER}} \Rightarrow .015625$

(c) $8^{2.5} = 181.019336$ $8\boxed{\wedge}2\boxed{.}5\boxed{\text{ENTER}} \Rightarrow 181.019336$ **See Exercises 1–5.**

A commonly used formula that contains a variable exponent is the formula for calculating the compound amount for compound interest. **Compound interest** for an investment or loan is the interest calculated at the end of each period and then added to the principal for the next period. The **compound amount** or the accumulated amount is the combined principal and interest accumulated over a period of time.

To find the compound amount (accumulated amount):

$$A = P\left(1 + \frac{r}{n}\right)^{nt}$$

where $A =$ accumulated amount
$P =$ original principal
$t =$ time in years
$r =$ rate per year expressed as a decimal equivalent
$n =$ compounding periods per year

STOP AND CHECK

1. Using the formula $A = P\left(1 + \frac{r}{n}\right)^{nt}$ and a scientific or graphing calculator, find the accumulated amount on an investment of $5,000, invested at an interest rate of 3% for 5 years, if the interest is compounded monthly.

Answer:

1. $5,808.08

EXAMPLE 2

Using the formula $A = P\left(1 + \frac{r}{n}\right)^{nt}$ and a scientific or graphing calculator, find the accumulated amount on an investment of $1,500, invested at an interest rate of 9% for 3 years, if the interest is compounded quarterly.

Estimation

We expect to have more than $1,500.

$A = P\left(1 + \dfrac{r}{n}\right)^{nt}$ $P = \$1,500; r = 9\%$ or $0.09; n =$ quarterly or 4 times a year; $t = 3$ years.

$A = 1,500\left(1 + \dfrac{0.09}{4}\right)^{(4)(3)}$ Simplify exponent and division term in grouping.

$A = 1,500(1 + 0.0225)^{12}$ Combine terms in grouping.

$A = 1,500(1.0225)^{12}$ $1.0225\boxed{\wedge}12\boxed{=} \Rightarrow 1.30604999$

$A = 1,500(1.30604999)$ Multiply.

$A = 1,959.07$ Rounded

Interpretation

The accumulated amount of the $1,500 investment after 3 years is $1,959.07 to the nearest cent. See Exercises 6–19.

TIP **Your Mind Is Often Quicker (and More Accurate) Than Your Fingers** Should we do every step on the calculator? The following sequence of keystrokes for a scientific calculator is one option for performing all the calculations in the preceding example in a continuous sequence.

$$1500 \;\boxed{\times}\;\boxed{(}\;1\;\boxed{+}\;.09\;\boxed{\div}\;4\;\boxed{)}\;\boxed{\wedge}\;\boxed{(}\;4\;\boxed{\times}\;3\;\boxed{)}\;\boxed{=}\; \Rightarrow 1959.074985$$

Whenever possible, it is advisable to do some calculations mentally. This can greatly decrease the complexity of the calculator sequence. Note, 4(3) = 12.

$$1500 \;\boxed{\times}\;\boxed{(}\;1\;\boxed{+}\;.09\;\boxed{\div}\;4\;\boxed{)}\;\boxed{\wedge}\;12\;\boxed{=}\; \Rightarrow 1959.074985$$

Try each sequence on your calculator.

Future value: another term for accumulated amount, compound amount, or maturity value; generally associated with investments

Maturity value: another term for accumulated amount, compound amount, or future value; generally associated with loans or annuities

Compounding periods: the number of times compounding occurs each year; number can be implied with words such as *annually* (1 time), *semiannually* (2 times), *quarterly* (4 times), and *monthly* (12 times)

Interest rate per period: the annual interest rate (*r*) divided by the compounding periods per year (*n*)

The accumulated amount is also the **future value**, or **maturity value**. Many business calculators or computer software programs have a future value function (FV).

TIP **Interpreting a Formula** In using formulas it is important to understand what each letter of the formula represents. The compound interest formula can also be given in terms of the number of **compounding periods** and the **interest rate per period**.

$$\text{Total compounding periods } (N) = \frac{\text{Compounding periods per year } (n)}{\text{Number of years } (t)} \text{ times}$$

$$N = nt$$

$$\text{Interest rate per period } (R) = \frac{\text{Interest rate per year } (r)}{\text{Compounding periods per year } (n)}$$

$$R = \frac{r}{n}$$

The compound amount or future value formula can also be written as

$$A = P(1 + R)^N \quad \text{or} \quad FV = P(1 + R)^N$$

where A or FV = accumulated amount or future value
P = original principal
R = interest rate per compounding period
N = total number of compounding periods

To find the compound interest:

1. Find the accumulated amount or future value using the formula

$$A = P(1 + R)^N$$

where A or FV = accumulated amount or future value
P = original principal
R = interest rate per compounding period
N = total number of compounding periods

2. Subtract the original principal from the accumulated amount.

$$I = A - P$$

Combining both formulas:

$$I = P(1 + R)^N - P$$

or

$$I = P((1 + R)^N - 1)$$

EXAMPLE 3

BUS/PFIN   Find the compound interest on an investment of $10,000 at 2% annual interest compounded semiannually for 3 years.

$$P = 10,000; \; R = \frac{0.02}{2} = 0.01 \text{ per period}; \; N = 2 \times 3 = 6 \text{ periods}$$

$I = P(1 + R)^N - P$	Substitute values.
$I = 10,000(1 + 0.01)^6 - 10,000$	Combine terms inside grouping.
$I = 10,000(1.01)^6 - 10,000$	Raise to the power.
$I = 10,000(1.061520151) - 10,000$	Multiply.
$I = 10,615.20 - 10,000$	Subtract.
$I = 615.20$	

The interest is $615.20.   **See Exercises 10–19.**

Present value: the *lump sum* amount that should be invested now at a given interest rate for a specific period of time to *yield* a specific accumulated amount in the future

The **present value** of an investment is the *lump sum* amount that should be invested now at a given interest rate for a specific period of time to *yield* a specific accumulated amount in the future.

To find the present value:

$$PV = \frac{FV}{(1 + R)^N}$$

where PV = present value
FV = future value
R = interest rate per compounding period
N = total number of compounding periods

Aumm Graphixphoto/Shutterstock

EXAMPLE 4

BUS   The 7th Inning Sports Shop needs $20,000 in 10 years to replace engraving equipment. Find the amount the firm must invest at the present if it receives 5% interest compounded annually.

$$FV = 20,000; \; R = \frac{0.05}{1} = 0.05; \; N = 1(10) = 10$$

$PV = \dfrac{20,000}{(1 + 0.05)^{10}}$	Combine terms inside grouping.
$PV = \dfrac{20,000}{(1.05)^{10}}$	Raise denominator to the power.
$PV = \dfrac{20,000}{1.628894627}$	Divide.
$PV = \$12,278.27$	Rounded

$12,278.27 should be invested now.   **See Exercises 20–27.**

In advertising or in stating the terms of an investment or loan it is common to equate the compound interest rate to a comparable simple interest rate. This rate is referred to as the **effective rate**, *annual percentage rate (APR)*, and *annual percentage yield (APY)*.

Effective rate: the simple interest rate equivalent to a compound interest rate; also referred to as annual percentage rate (APR) and annual percentage yield (APY)

To find the effective rate:

$$E = \left(1 + \frac{r}{n}\right)^n - 1$$

where E = effective rate
r = interest rate per year
n = number of compounding periods per year

STOP AND CHECK

1. Find the effective interest rate for a loan at 12% compounded monthly.

Answer:
1. 12.68%

EXAMPLE 5

Find the effective interest rate for a loan at 10% compounded semiannually.

$E = \left(1 + \dfrac{r}{n}\right)^n - 1$ Substitute values. $r = 0.1; n = 2$

$E = \left(1 + \dfrac{0.1}{2}\right)^2 - 1$ Simplify inside grouping.

$E = (1 + 0.05)^2 - 1$

$E = (1.05)^2 - 1$ Raise to the power.

$E = 1.1025 - 1$ Subtract.

$E = 0.1025$

Effective interest rate is 10.25%.   **See Exercises 28–31.**

An **annuity** is a fund that accumulates compound interest as periodic payments are added to the principal. An **ordinary annuity** has periodic payments that are made at the end of each payment period.

Annuity: a fund that accumulates compound interest as periodic payments are added to the principal

Ordinary annuity: an annuity that has periodic payments made at the end of each payment period

To find the future value of an ordinary annuity:

Apply the formula

$$FV = P\left[\frac{(1 + R)^N - 1}{R}\right]$$

where FV = future value of an ordinary annuity
P = amount of the periodic payment
R = interest rate per period
N = total number of periods

STOP AND CHECK

1. Find the future value of an ordinary annuity of $500 for 10 years at a 4% annual interest rate.

Answer:
1. $6,003.05

EXAMPLE 6

Find the future value of an ordinary annuity of $6,000 payment made semiannually for 5 years at 6% annual interest compounded semiannually.

$$FV = P\left[\frac{(1 + R)^N - 1}{R}\right], \quad P = \$6,000, \quad R = \frac{0.06}{2} = 0.03, \quad N = 2(5) = 10$$

$$FV = 6,000\left[\frac{(1 + 0.03)^{10} - 1}{0.03}\right]$$ Simplify innermost grouping.

$$FV = 6,000\left[\frac{(1.03)^{10} - 1}{0.03}\right]$$ Raise to the power.

$$FV = 6,000\left[\frac{1.343916379 - 1}{0.03}\right]$$ Subtract in numerator.

$$FV = 6,000\left[\frac{0.343916379}{0.03}\right]$$ Divide in grouping.

$$FV = 6,000(11.46387931)$$ Multiply.

$$FV = \$68,783.28$$ **See Exercises 32–41.**

When you have a specific future goal or target amount that you want to accumulate, a **sinking fund payment** is the amount you would invest in periodic payments to reach this goal. To determine the sinking fund payment, the *known values* are the future goal or amount, the amount of time, and the expected or guaranteed interest rate.

Sinking fund payment: the amount to be invested in periodic payments to reach a specific future value

> **To find the sinking fund payment to produce a specified future value:**
>
> Apply the formula
>
> $$P = FV\left[\frac{R}{(1 + R)^N - 1}\right]$$
>
> where P = sinking fund payment
> FV = future value or goal
> R = interest rate per period
> N = total number of periods

STOP AND CHECK

1. Find the amount of a sinking fund payment to be made semiannually that will generate a future value of $25,000 in 10 years at a 5% annual interest rate.

Answer:
1. $978.68

EXAMPLE 7

A municipality has established a sinking fund to retire a bond issue of $500,000, which is due in 10 years. The account pays 8% quarterly interest. Find the amount of the quarterly sinking fund payment.

$$P = FV\left[\frac{R}{(1 + R)^N - 1}\right]$$ Substitute known values.

$$FV = \$500,000, \quad R = \frac{0.08}{4} = 0.02, \quad N = 4(10) = 40$$

$$P = 500,000\left[\frac{0.02}{(1 + 0.02)^{40} - 1}\right]$$ Simplify grouping in denominator.

$$P = 500,000\left[\frac{0.02}{(1.02)^{40} - 1}\right]$$

$$P = 500,000\left[\frac{0.02}{2.208039664 - 1}\right]$$

$$P = 500,000\left[\frac{0.02}{1.208039664}\right]$$ Divide.

$$P = 500,000[0.0165557478]$$ Multiply.

$$P = \$8,277.87$$ **See Exercises 42–43.**

Finding the monthly payment to repay a loan is similar to the process for finding the sinking fund payment. The repayment of the loan in equal installments that are applied to the principal and interest over a specified amount of time is called the **amortization of a loan**.

Amortization of a loan: the repayment of a loan in equal installments that are applied to the principal and interest over a specified amount of time

To find the monthly payment for an amortized loan:

Apply the formula

$$M = P\left[\frac{R}{1 - (1 + R)^{-N}}\right]$$

where M = monthly payment
P = principal or initial amount of the loan
R = interest rate per month
N = total number of months

STOP AND CHECK

1. Find the monthly payment on a 30-year home mortgage of $225,000 at 3%.

Answer:
1. $948.61

EXAMPLE 8

PFIN Find the monthly payment on a 25-year home mortgage of $135,900 at 4%.

$$P = \$135{,}900, \qquad R = \frac{0.04}{12} = 0.0033333333, \qquad N = 12(25) = 300$$

$$M = P\left[\frac{R}{1 - (1 + R)^{-N}}\right] \qquad \text{Substitute values.}$$

$$M = 135{,}900\left[\frac{0.0033333333}{1 - (1 + 0.0033333333)^{-300}}\right] \qquad \text{Simplify denominator.}$$

$$M = 135{,}900\left[\frac{0.0033333333}{1 - (1.0033333333)^{-300}}\right]$$

$$M = 135{,}900\left[\frac{0.0033333333}{1 - 0.3684917271}\right]$$

$$M = 135{,}900\left[\frac{0.0033333333}{0.6315082729}\right] \qquad \text{Divide in grouping and multiply.}$$

$$M = \$717.33 \qquad\qquad\qquad \textbf{See Exercises 44–45.}$$

2 **Evaluate Formulas That Contain a Power of the Natural Exponential, e.** In many applications involving circles, the irrational number π (approximately equal to 3.141592654) is used. Another irrational number, e, arises in the discussion of many physical phenomena. Many formulas contain a power of the natural exponential, e.

Exponential change is an interesting phenomenon. Let's look at the value of the expression $\left(1 + \dfrac{1}{n}\right)^{n}$ as n gets larger and larger. See Table 16–2.

The value of the expression changes very little as the value of n gets larger. We can say that the value approaches a given number. We call the number e, the **natural exponential**. The natural exponential, e, like π, is an irrational number and will never terminate nor repeat as more decimal places are examined.

Natural exponential, e: an irrational number that is the limit that the value of the expression $(1 + 1/n)^n$ approaches as n gets larger and larger without bound

The natural exponential, e, like π, is a constant. That is, the value is always the same; it does not vary. The natural exponential, e, is the limit that the value of the expression $\left(1 + \dfrac{1}{n}\right)^{n}$ approaches as n gets larger and larger without bound. The value of e to nine decimal places is 2.718281828.

Table 16-2   Values of $\left(1 + \frac{1}{n}\right)^n$

n	$\left(1 + \frac{1}{n}\right)^n$	Result
1	$\left(1 + \frac{1}{1}\right)^1$	2
2	$\left(1 + \frac{1}{2}\right)^2$	2.25
3	$\left(1 + \frac{1}{3}\right)^3$	2.37037037
10	$\left(1 + \frac{1}{10}\right)^{10}$	2.59374246
100	$\left(1 + \frac{1}{100}\right)^{100}$	2.704813829
1,000	$\left(1 + \frac{1}{1,000}\right)^{1,000}$	2.716923932
10,000	$\left(1 + \frac{1}{10,000}\right)^{10,000}$	2.718145927
100,000	$\left(1 + \frac{1}{100,000}\right)^{100,000}$	2.718268237
1,000,000	$\left(1 + \frac{1}{1,000,000}\right)^{1,000,000}$	2.718280469

To evaluate formulas containing the natural exponential, *e,* we can use a calculator or computer.

> **TIP** **The Natural Exponential, *e*, on Your Calculator** The natural exponential key is generally labeled $\boxed{e^x}$. On most calculators, the exponent is entered after pressing the $\boxed{e^x}$ key. To find $e^{2.3}$, enter the following sequence:
>
> $$\boxed{e^x} \; 2.3 \; \boxed{=} \Rightarrow 9.974182455$$
>
> Always test your calculator. A good test for the $\boxed{e^x}$ key is to find e^0.
>
> $e^0 = 1$ Also, $e^1 = 2.718281828$

STOP AND CHECK

Evaluate using a calculator.
1. $e^{3.1}$
2. $e^{-4.3}$

Answers:
1. 22.19795128   **2.** 0.013568559

EXAMPLE 9

Evaluate using a calculator.

(a) $e^{2.7}$ **(b)** $e^{-3.2}$

(a) $e^{2.7} = $ **14.87973172** $\boxed{e^x}\,2\,\boxed{\cdot}\,7\,\boxed{\text{ENTER}} \Rightarrow 14.87973172$

(b) $e^{-3.2} = $ **0.040762204** $\boxed{e^x}\,\boxed{(-)}\,3\,\boxed{\cdot}\,2\,\boxed{\text{ENTER}} \Rightarrow .040762204$

See Exercises 46–50.

STOP AND CHECK

1. Find the atmospheric pressure at 500 m above sea level.

Answer:
1. 712.17 m

EXAMPLE 10

The formula for the atmospheric pressure (in millimeters of mercury) is $P = 760e^{-0.00013h}$, where h is the height above sea level in meters. Find the atmospheric pressure at 100 m above sea level ($h = 100$).

Estimation

Developing your number sense with powers of e comes after much examination of the power of e. For now you will minimize errors by making the calculations in the calculator two times and preferably two different ways. A negative exponent means you will multiply by a number less than 1 and the product will be smaller than the original value.

$P = 760e^{-0.00013h}$ $h = 100$

$P = 760e^{-0.00013(100)}$ Simplify exponent. Multiply -0.00013 times 100 mentally.

$P = 760e^{-0.013}$ $e^{-0.013} = 0.987084135$ (leave in calculator)

$P = 760(0.987084135)$ Multiply by 760.

One continuous calculator sequence: $760 \boxed{\times} \boxed{e^x} \boxed{(-)}.013 \boxed{=}$

$P = 750.1839426$

Interpretation

The atmospheric pressure at 100 m above sea level is 750.18 mm.   **See Exercises 51–53.**

Continuous compounding: the compounding period interpreted as daily, every minute, every second, and so on

As the number of compounding periods per year increases, the effect of compounding levels off or reaches a limit. Therefore, the natural exponential e can be substituted into the compound interest formula to accomplish **continuous compounding**. Continuous compounding can be interpreted as daily compounding, compounding every minute, compounding every second, and so on.

> **To find the accumulated amount (future value) for continuous compounding:**
>
> 1. Determine the principal (P), rate per year (r), and the number of years (t).
>
> 2. Evaluate the formula
>
> $$A = Pe^{rt}$$
>
> where A = accumulated amount or future value
> P = principal
> r = rate per year
> t = time in years

STOP AND CHECK

1. Find the compound amount of $15,000 invested at an annual rate of 3% compounded continuously for 10 years.

Answer:

1. $20,247.88

EXAMPLE 11

Find the compound amount of $5,000 invested at an annual rate of 4% compounded continuously for 5 years.

$P = 5,000, r = 0.04$ (from 4%), $t = 5$

$A = Pe^{rt}$ Substitute values.

$A = 5,000e^{(0.04)(5)}$ Simplify exponents.

$A = 5,000e^{0.2}$ Raise to the power.

$A = 5,000(1.221402758)$ Multiply.

$A = 6,107.01$

The accumulated amount is $6,107.01.   **See Exercises 54–55.**

---

3 **Solve Exponential Equations in the Form $b^x = b^y$, where $b > 0$ and $b \neq 1$.**
Some applications of exponents may require finding the value of an unknown exponent. If we can write an equation in a specific form, $b^x = b^y$, where $b > 0$ and $b \neq 1$, we will be able to solve the equation using many of our previously learned skills for solving linear equations. Otherwise, we will need to learn some new strategies involving logarithms, which are introduced in the next section.

To illustrate a property of exponential equations, we look at the equation $2^x = 32$. In the equation, the value of x is the power of 2 that gives a result of 32. We can rewrite 32 as 2^5. Thus, $2^x = 2^5$. When an equation can be written in this form, a special property applies.

> **To solve an exponential equation in the form of $b^x = b^y$:**
>
> 1. Rewrite the equation in the form $b^x = b^y$.
>
> 2. Apply the following property and solve for x. If $b^x = b^y$, and $b > 0$ and $b \neq 1$, then $x = y$ where x and y are any real numbers.
>
> Note: This property applies only when bases are *like bases*, the bases are positive, and the bases are not equal to 1.

> **TIP**   **Using Conditional Properties (if . . . then)**   Many mathematical properties are phrased using an "if . . . then" format. That means, before the property can be used, the conditions or restrictions must be examined. It also means that the property is not appropriate under other conditions.
>   What are the conditions of the previous property?
>
> ▶ Bases must be like.
>
> ▶ Bases must be positive.
>
> ▶ Bases cannot equal 1.
>
>   To illustrate why the property does not work unless these conditions are met, look at this equation, $1^5 = 1^8$. This is a true statement because $1^5 = 1$ and $1^8 = 1$. But, applying the property "if $b^x = b^y$, then $x = y$," does $5 = 8$? No!

STOP AND CHECK

Solve the equations.

1. $5^x = 125$
2. $4^{2x-1} = 64$

Answers:
1. $x = 3$ 2. $x = 2$

EXAMPLE 12

Solve the equation: **(a)** $2^x = 32$ **(b)** $3^{x+1} = 27$

(a) $2^x = 32$ Rewrite 32 as a power of 2.
$\qquad 2^x = 2^5$ $b = 2$, so $b > 0$, $b \neq 1$
$\quad$ If $2^x = 2^5$, then $x = 5$.   Apply property.
$\quad$ Check to see if the solution is appropriate.   Does $2^5 = 32$? Yes, so 5 is the correct solution.

(b) $3^{x+1} = 27$ Rewrite 27 as a power of 3.
$\qquad 3^{x+1} = 3^3$ $b = 3$, so $b > 0$, $b \neq 1$
$\quad$ If $3^{x+1} = 3^3$, then $x + 1 = 3$.   Apply property and solve for x.
$\qquad x + 1 = 3$ Solve for x.
$\qquad\quad x = 3 - 1$
$\qquad\quad x = 2$
$\quad$ Check to see if the solution, 2, is correct.   Does $3^{2+1} = 27$?
$$3^{2+1} = 27$$
Thus, the solution, $x = 2$, is correct.   See Exercises 56–71.

It will not always be possible to rewrite an exponential equation as an equation with like bases. In such cases, other methods for solving the exponential equation are used.

4   **Graph an Exponential Function.**   A table of solutions can be used to graph an exponential equation. As with other nonlinear functions, several points should be included in the table.

1. Graph the function $f(x) = 3^x$ using the table-of-solutions method.

Answer:

1.

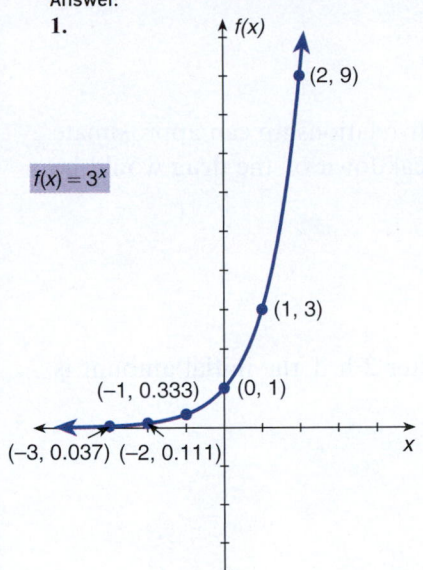

x	f(x)
−3	0.037
−2	0.111
−1	0.333
0	1
1	3
2	9

EXAMPLE 13

Prepare a table of solutions and graph the function $f(x) = 2^x$ using integral values of x between -3 and 3, inclusive. Determine the domain and range.

$$f(-3) = 2^{-3} \qquad\qquad f(-2) = 2^{-2} \qquad\qquad f(-1) = 2^{-1}$$

$$f(-3) = \frac{1}{2^3} \qquad\qquad f(-2) = \frac{1}{2^2} \qquad\qquad f(-1) = \frac{1}{2^1}$$

$$f(-3) = \frac{1}{8} \text{ or } \mathbf{0.125} \qquad f(-2) = \frac{1}{4} \text{ or } \mathbf{0.25} \qquad f(-1) = \frac{1}{2} \text{ or } \mathbf{0.5}$$

$$f(0) = 2^0 \qquad\qquad f(1) = 2^1 \qquad\qquad f(2) = 2^2 \qquad\qquad f(3) = 2^3$$

$$f(0) = 1 \qquad\qquad f(1) = 2 \qquad\qquad f(2) = 4 \qquad\qquad f(3) = 8$$

Plot the points indicated in the table of solutions and connect the points with a smooth, continuous curve (Fig. 16–2).

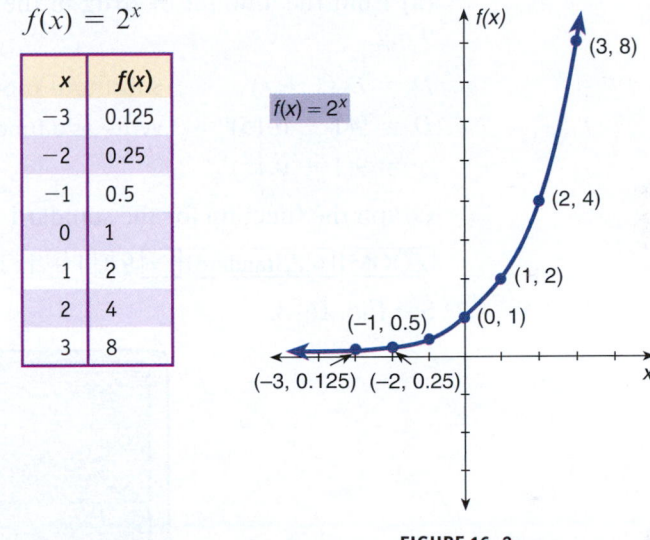

$f(x) = 2^x$

x	f(x)
−3	0.125
−2	0.25
−1	0.5
0	1
1	2
2	4
3	8

FIGURE 16–2

The domain is all real numbers. The range is real numbers greater than 0.

See Exercises 72–80.

The GRAPH and TABLE features of a graphing calculator can be used to evaluate an exponential function.

EXAMPLE 14

HLTH/N The exponential equation for the breakdown of a particular drug in the human body is $D = D_0(1 + r)^t$, where D is the amount of drug in the bloodstream after a specified amount of time, D_0 is the initial amount of drug in the bloodstream, r is the decimal equivalent of the rate of change, and t is the amount of lapsed time. A certain drug breaks down at the rate of 15% per hour for the first 24 h. **(a)** Find the amount of drug in the bloodstream after 2 h if the initial amount is 9 mg. **(b)** After approximately how many hours will the amount of drug in the bloodstream be less than 1 mg?

Known facts

The breakdown of a drug represents a negative rate of change.
Rate of change $= -15\%$
Initial amount of drug in the bloodstream $= 9$ mg

Unknown facts

Amount of drug in the bloodstream after 2 h
How long it will take for the drug level to be less than 1 mg

Relationships

$D = D_0(1 + r)^t$

Estimation

Even though the rate of change is not linear, a linear relationship can approximate
the breakdown of the drug. After two hours the breakdown of the drug would be
approximately 30%.
30% of 9 mg is 2.7 mg.
$D \approx 9 - 2.7 \approx 6.3 \text{ mg}$

Calculation

(a) Find the amount of drug in the bloodstream after 2 h if the initial amount is
9 mg.

$D = D_0(1 + r)^t$ Substitute known values.
$D = 9(1 - 0.15)^t$ Write as a function of x.
$y = 9(1 - 0.15)^x$

Graph the function for the standard window.

| ZOOM | | 6:ZStandard | | $Y=$ | 9 | (| 1 | – | . | 15 |) | ^ | X, T, θ, n | | GRAPH |

See Fig. 16–3.

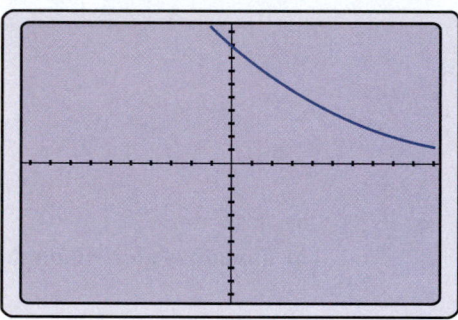

FIGURE 16–3

Evaluate the functions for $x = 2$.

| CALC | | 1:value | | ENTER | 2 | ENTER | See Fig. 16–4.

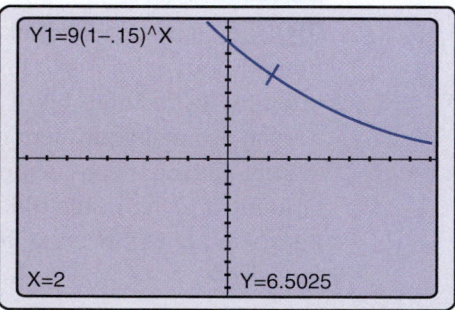

FIGURE 16–4

Interpretation

The drug level in the bloodstream after 2 h is 6.5025 mg.

Calculation

(b) After approximately how many hours will the drug level in the bloodstream be less than 1 mg?

TBLSET , TblStart = 0 , △Tbl = 1 TABLE

View TABLE . Scroll using arrow keys until Y_1 is less than 1.

See Fig. 16–5.

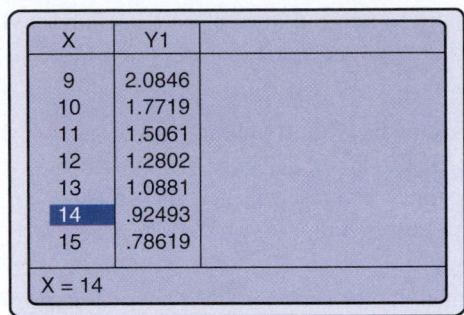

X	Y1	
9	2.0846	
10	1.7719	
11	1.5061	
12	1.2802	
13	1.0881	
14	.92493	
15	.78619	

X = 14

FIGURE 16–5

Interpretation

The drug level in the bloodstream will be less than 1 mg between 13 h and 14 h after taking 9 mg of the drug.

See Exercises 81–86.

16–1 EXERCISES   MyLab Math For additional practice go to your study plan in MyLab Math.

1 Evaluate using a scientific or graphing calculator. *See Example 1.*

1. 4^3 **2.** 3^{-5} **3.** 5^{10} **4.** 8^{-3} **5.** $9^{2.5}$

Using the compound amount formula, $A = P\left(1 + \dfrac{r}{n}\right)^{nt}$, find the accumulated amount. *See Example 2.*

6. **BUS** Principal = $1,500, rate = 10%, compounded annually, time = 5 years.

7. **BUS** Principal = $1,750, rate = 8%, compounded quarterly, time = 2 years.

8. **INDTEC** The number of grams of a chemical that will dissolve in a solution is given by the formula $C = 100e^{0.02t}$, where t = temperature in degrees Celsius. Evaluate when
(a) $t = 10$ **(b)** $t = 20$
(c) $t = 25$ **(d)** $t = 30$

9. **BUS** The compound amount when an investment is compounded continually (every instant) is expressed by the formula $A = Pe^{rt}$, where A = compounded amount, P = principal, t = time in years, and r = rate per year. Find the compound amount when
(a) Principal = $1,000, interest = 9%, for 2 years
(b) Principal = $1,500, interest = 10%, for 6 months

Use the compound interest formula for Exercises 10–19. *See Examples 2 and 3.*

10. **PFIN/BUS** First State Bank loaned Doug Morgan $2,000 for 4 years compounded annually at 8%. How much interest was Doug required to pay on the loan?

11. **PFIN/BUS** A loan of $8,000 for 2 acres of woodland is compounded quarterly at an annual rate of 12% for 5 years. Find the compound amount and the compound interest.

12. **PFIN/BUS** Compute the compound amount and the interest on a loan of $10,500 compounded annually for 4 years at 10%.

13. **PFIN/BUS** Find the future value of an investment of $10,500 if it is invested for 4 years and compounded quarterly at an annual rate of 8%.

14. **PFIN/BUS** You have $8,000 that you plan to invest in a compound-interest-bearing instrument. Your investment agent advises you that you can invest $8,000 at 8% compounded quarterly for 3 years or you can invest $8,000 at $8\frac{1}{4}$% compounded annually for 3 years. Which investment should you choose to receive the most interest?

15. **PFIN/BUS** Find the future value of $50,000 at 6%, compounded semiannually for 10 years.

16. **PFIN/BUS** Find the compound interest on $2,500 at $6\frac{3}{4}$%, compounded daily by Leader Financial Bank for 20 days.

17. **PFIN/BUS** How much compound interest is earned on a deposit of $1,500 at 6.25%, compounded daily for 30 days?

18. **PFIN/BUS** Ezell Allen has found a short-term investment opportunity. He can invest $8,000 at 8.5% interest for 15 days. How much interest will he earn on this investment if the interest is compounded daily?

19. **PFIN/BUS** What is the compound interest on $8,000 invested at 8% for 180 days if it is compounded daily?

Use the present value formula for Exercises 20–27. *See Example 4.*

20. **PFIN/BUS** Compute the amount of money to be set aside today to ensure a future value of $2,500 in 1 year if the interest rate is 11% annually, compounded annually.

21. **PFIN/BUS** How much should Latonia Shegog set aside now to buy equipment that costs $8,500 in 1 year? The current interest rate is 7.5% annually, compounded annually.

22. **PFIN/BUS** Ronnie Cox has just inherited $27,000. How much of this money should he set aside today to have $21,000 to pay cash for a Ventura Van, which he plans to purchase in 1 year? He can invest at 7.9% annually, compounded annually.

23. **PFIN/BUS** Shirley Riddle received a $10,000 gift from her mother and plans a minor renovation to her home and an investment for 1 year, at which time she plans to take a trip projected to cost $6,999. The current interest rate is 8.3% annually, compounded annually. How much should be set aside today for her trip?

24. **PFIN/BUS** Joe Brozovich needs $2,000 in 3 years to make the down payment on a new car. How much must he invest today if he receives 8% interest annually, compounded annually?

25. **PFIN/BUS** Calculate the amount of money that must be invested now at 6% annually, compounded quarterly, to obtain $1,500 in 3 years.

26. **PFIN/BUS** Kristen Bieda plans to open a business in 4 years when she retires. How much must she invest today to have $10,000 when she retires if the bank pays 10% annually, compounded quarterly?

27. **PFIN/BUS** Charlie Bryant has a child who will be college age in 5 years. How much must he set aside today to have $20,000 for college tuition in 5 years if he gets 8% annually, compounded annually?

Find the effective rate for Exercises 28–31. *See Example 5.*

28. **PFIN/BUS** Find the effective interest rate for the investment described in Exercise 19. Use the formula for the effective rate.

29. **PFIN/BUS** What is the effective interest rate for a loan of $5,000 at 10%, compounded semiannually for 3 years? Use the effective rate formula.

30. **PFIN/BUS** Betty Veteto has a loan of $8,500, compounded quarterly for 4 years at 6%. What is the effective interest rate for the loan? Use the formula.

31. **PFIN/BUS** What is the effective interest rate for a loan of $20,000 for 3 years if the interest is compounded quarterly at a rate of 12%?

Use the future value formula for Exercises 32–41. *See Example 6.*

32. **PFIN/BUS** Find the future value of an ordinary annuity of $3,000 annually for 2 years at 9% annual interest. Find the total interest earned.

33. **PFIN/BUS** Len and Sharron Smith are saving money for their daughter Heather to attend college. They set aside an ordinary annuity of $4,000 annually for 2 years at 7% annual interest. How much will Heather have for college when she graduates from high school in 2 years? Find the total interest earned.

34. **PFIN/BUS** Jake Drewrey plans to pay an ordinary annuity of $5,000 annually for 3 years so he can take a year's sabbatical to study for a master's degree in business. The annual rate of interest is 8%. How much will Jake have at the end of 3 years?

35. **BUS** Joe Freeman is planning to establish a small business to provide consulting services in computer networking. He is committed to an ordinary annuity of $3,000 annually at 8.5% annual interest. How much will Joe have to establish the business after 3 years?

36. **PFIN/BUS** Find the future value of an ordinary annuity of $6,500 semiannually for 7 years at 10% annual interest, compounded semiannually. How much was invested?

37. **PFIN/BUS** Jimmie Van Alphen pays an ordinary annuity of $2,500 quarterly at 8% annual interest, compounded quarterly, to establish supplemental income for retirement. How much will Jimmie have available at the end of 5 years?

38. **PFIN/BUS** Rosa Kavanaugh established an ordinary annuity of $1,000 annually at 7% annual interest. What is the future value of the annuity after 15 years?

39. **PFIN/BUS** You invest in an ordinary annuity of $500 annually at 8% annual interest. Find the future value of the annuity at the end of 10 years.

40. **PFIN/BUS** You invest in an ordinary annuity of $1,000 annually at 8% annual interest. What is the future value of the annuity at the end of 5 years?

41. **PFIN/BUS** Make a chart comparing your results for Exercises 39 and 40. Use these headings: Years, Total Investment, Total Interest. What general conclusion might you draw about effective investment strategy?

See Example 7.

42. Rogers, AR, has established a sinking fund to retire a municipal bond issue of $1,500,000, which is due in 20 years. The account pays 2% annual interest. Find the amount of the annual sinking fund payment.

43. Bowling Green, KY, has established a sinking fund to retire a municipal bond issue of $5,000,000, which is due in 25 years. The account pays 2.8% annual interest. Find the semiannual amount of the sinking fund payment.

See Example 8.

44. **PFIN/BUS** Find the monthly payment on a home mortgage of $128,600 if the home is financed for 25 years at 7% interest.

45. **PFIN/BUS** Find the monthly payment on a home mortgage of $128,600 if the home is financed for 15 years at 7% interest.

2 Evaluate using a scientific or graphing calculator. Round to hundredths. *See Example 9.*

46. e^2 **47.** e^{-3} **48.** $e^{0.21}$ **49.** $e^{-3.5}$ **50.** e^{10}

A formula for electric current is $I = 1.50e^{-200t}$, where t is the time in seconds. Calculate the current for each time. Express answers in scientific notation. *See Example 10.*

51. **ELEC** 1 s   **52.** **ELEC** 1.1 s   **53.** **ELEC** 0.5 s

See Example 11.

54. **PFIN/BUS** Micheal Temal invested $20,000 at an annual rate of 1.08% compounded continuously for 5 years. How much will his investment be worth at the end of 5 years?

55. **PFIN/BUS** Nikita Phillips invested $15,000 at an annual rate of 2.25% compounded continuously for 10 years. How much will her investment be worth at the end of 7 years?

3 Solve for x. *See Example 12.*

56. $3^x = 3^7$ **57.** $5^x = 5^{-3}$ **58.** $3^{x+4} = 3^6$ **59.** $2^{x-3} = 2^7$

60. $6^x = 6^7$ **61.** $3^x = 27$ **62.** $2^x = 64$ **63.** $3^x = \frac{1}{81}$

64. $2^x = \frac{1}{64}$ **65.** $4^{3x} = 128$ **66.** $3^{2x} = 243$ **67.** $4^{3-x} = \frac{1}{16}$

68. $2^{4-x} = \frac{1}{16}$ **69.** $\left(\frac{1}{3}\right)^x = 27$ **70.** $\frac{8}{27} = \left(\frac{2}{3}\right)^{5x-12}$ **71.** $\left(\frac{1}{16}\right)^{2x} = (64)^{6x-10}$

4 Graph the functions using the table-of-solutions method. *See Example 13.*

72. $f(x) = 3^x$

73. $f(x) = -3^x$

74. $f(x) = 2^x - 3$

75. $f(x) = -2^x - 3$

76. $f(x) = 2(2)^x$

77. $f(x) = 4(2)^x$

78. $f(x) = 6(2)^x$

79. $f(x) = -2(2)^x$

80. $f(x) = -4(2)^x$

Use a graphing calculator for Exercises 81–86. *See Example 14.*

81. **HLTH/N** Insulin-delivery systems are designed to release insulin slowly. Even though the rate varies among individuals, the typical pattern of insulin decrease is modeled by the equation $y = 10(0.95)^x$, where x is the number of minutes after insulin enters the bloodstream and y is the number of insulin units in the blood. (a) How many insulin units remain in the bloodstream after 18 min? (b) After approximately how many minutes will the bloodstream contain less than 7.5 units of insulin?

82. **HLTH/N** Using the equation $y = 10(0.95)^x$, how many minutes will it take for the insulin to be reduced to half the original dosage (see Exercise 81)? This length of time is called the *half-life* of the drug.

83. **HLTH/N** Penicillin, the most famous antibiotic, discovered in 1929, remains in the bloodstream on average according to the formula $y = I(0.6)^x$, where I is the initial dosage, x is the number of hours after the initial injection, and y is the number of milligrams remaining in the bloodstream. Make a table that shows the amount of penicillin in the bloodstream up to 5 h later for a 250-mg injection that was administered at 3:00 P.M. How many milligrams remained in the bloodstream at 6:00 P.M.?

84. **HLTH/N** What is the half-life of penicillin? See Exercise 82 for *half-life*.

85. **PFIN** If you invest $1,000 upon graduating from college and each year the accumulated amount earns on average 5% per year, how much will you have after 1 year? After 2 years? After 10 years? After 30 years? After 50 years? An investment grows according to the exponential equation $y = P(1.05)^x$, where P is the initial amount invested, x is the number of years of the investment, and y is the amount at the end of the time period.

86. Ultrasound is mechanical vibration or sound waves with a typical frequency of between 1.0 and 3.0 MHz (1 MHz = 1 million cycles per second). The intensity of a 3-MHz therapeutic ultrasound can be approximated by the equation $y = 100(0.5)^{0.5x}$, where x represents the depth of penetration in centimeters and y represents the ultrasound intensity at the given depth with 100% intensity at skin surface. What is the ultrasound intensity 6 cm below the skin surface? Use a calculator table to approximate the half-value depth for therapeutic ultrasound with a 3-MHz frequency.

16–2 Logarithmic Expressions, Equations, and Formulas

LEARNING OUTCOMES

1 Write exponential equations as equivalent logarithmic equations.

2 Write logarithmic equations as equivalent exponential equations.

Logarithms were first introduced as a relatively fast way of carrying out lengthy calculations. A **logarithm** is an alternate method for relating a base, an exponent, and the resulting power. The word *power* has more than one meaning so we will use the terminology **result of exponentiation** in our discussion of logarithms. The calculator and computer have diminished the importance of logarithms as a computational device; however, the importance of logarithms in advanced mathematics, electronics, and theoretical work is more evident than ever. Many formulas use logarithms to express the relationships of physical properties. Logarithms are also used to solve many exponential equations.

3 Evaluate common and natural logarithmic expressions using a calculator.

4 Evaluate logarithms with a base other than 10 or *e*.

5 Evaluate formulas containing at least one logarithmic term.

6 Graph a logarithmic function.

7 Simplify logarithmic expressions using the properties of logarithms.

LC LEARNING CATALYTICS
Evaluate.
1. 3^{-2} **2.** 10^5 **3.** 10^{-3}

Logarithm: an alternate method for relating a base, an exponent, and the resulting power

Result of exponentiation: an alternate expression for a power

Logarithmic expression: an expression that contains at least one term with a logarithm

1 **Write Exponential Equations as Equivalent Logarithmic Equations.** A **logarithmic expression** is an expression that contains at least one term with a logarithm. The three basic components of a power are the base, exponent, and result of exponentiation or power.

$$\text{exponent}$$
$$3^4 = 81 \leftarrow \text{result of exponentiation (power)}$$
$$\text{base}$$

When finding a *power* or the result of exponentiation, you are given the base and exponent. When finding a *root,* you know the base (radicand) and the index of the root (exponent). A third type of calculation involves finding the exponent when the base and the result of exponentiation are given. The exponent in this process is called the *logarithm.* In the logarithmic form $\log_b x = y$, *b* is the *base, y* is the *expo-nent* or *logarithm,* and *x* is the *result of exponentiation.* This equation is read, "The log to the base *b* of *x* is *y*." Logarithm is written in abbreviated form as "log." In the exponential form $x = b^y$, *b* is also the base, *y* is the exponent, and *x* is the result of the exponentiation.

Algebraic expressions written in logarithmic or exponential form have the same three components: base, exponent, result of exponentiation. Let's examine the mapping from one form to the other.

exponential form   **logarithmic form**

$$x = b^y \quad \overbrace{\text{exponent (logarithm)}} \quad \log_b x = y$$
$$\text{base}$$
$$\text{result of exponentiation}$$
$$\text{(argument)}$$

Argument: the result of exponentiation or the dependent variable of an expression in the exponential form

To write an exponential equation as an equivalent logarithmic equation:

If $x = b^y$, then $\log_b x = y$, provided that $b > 0$ and $b \neq 1$.

1. The exponent in the exponential form is the dependent variable in the logarithmic form.

2. The base in the exponential form is the base in the logarithmic form.

3. The dependent variable in the exponential form is the result of exponentiation in the logarithmic form. This result is sometimes referred to as the **argument**.

STOP AND CHECK
Write in logarithmic form.
1. $4^2 = 16$
2. $5^3 = 125$
3. $3^{-3} = \dfrac{1}{27}$

Answers:
1. $\log_4 16 = 2$ **2.** $\log_5 125 = 3$
3. $\log_3 \dfrac{1}{27} = -3$

EXAMPLE 1

Write in logarithmic form.

(a) $2^4 = 16$ **(b)** $3^2 = 9$ **(c)** $2^{-2} = \dfrac{1}{4}$

(a) $2^4 = 16$ converts to $\log_2 16 = 4$ Base = 2, exponent = 4, result of exponentiation = 16.

(b) $3^2 = 9$ converts to $\log_3 9 = 2$ Base = 3, exponent = 2, result of exponentiation = 9.

(c) $2^{-2} = \dfrac{1}{4}$ converts to $\log_2 \dfrac{1}{4} = -2$ Base = 2, exponent = -2, result of exponentiation = $\frac{1}{4}$.

See Exercises 1–6.

2 **Write Logarithmic Equations as Equivalent Exponential Equations.** The inverse of the process in the preceding learning outcome changes a logarithmic expression to an exponential expression.

To write a logarithmic equation as an equivalent exponential equation:

$\log_b x = y$ converts to $x = b^y$, provided that $b > 0$ and $b \neq 1$.

1. The dependent variable in the logarithmic form is the exponent in the exponential form.

2. The base in the logarithmic form is the base in the exponential form.

3. The result of the exponentiation in the logarithmic form is the dependent variable in the exponential form.

STOP AND CHECK
Write in exponential form.

1. $\log_2 \dfrac{1}{8} = -3$

2. $\log_6 36 = 2$

3. $\log_2 64 = 6$

Answers:

1. $2^{-3} = \dfrac{1}{8}$ 2. $6^2 = 36$

3. $2^6 = 64$

EXAMPLE 2

Write in exponential form.

(a) $\log_2 32 = 5$ **(b)** $\log_3 81 = 4$ **(c)** $\log_5 \dfrac{1}{25} = -2$ **(d)** $\log_{10} 0.001 = -3$

(a) $\log_2 32 = 5$ converts to $\mathbf{2^5 = 32}$ Base = 2, exponent = 5, result of exponentiation = 32.

(b) $\log_3 81 = 4$ converts to $\mathbf{3^4 = 81}$ Base = 3, exponent = 4, result of exponentiation = 81.

(c) $\log_5 \dfrac{1}{25} = -2$ converts to $\mathbf{5^{-2} = \dfrac{1}{25}}$ Base = 5, exponent = -2, result of exponentiation = $\frac{1}{25}$.

(d) $\log_{10} 0.001 = -3$ converts to $\mathbf{10^{-3} = 0.001}$ Base = 10, exponent = -3, result of exponentiation = 0.001.

See Exercises 7–12.

---

3 **Evaluate Common and Natural Logarithmic Expressions Using a Calculator.**
When the base of a logarithm is 10, the logarithm is referred to as a **common logarithm.** If the base is omitted in a logarithmic expression, the base is assumed to be 10. Thus, $\log_{10} 1,000 = 3$ is normally written as $\log 1,000 = 3$. On a calculator, expressions containing common logarithms can be evaluated using the $\boxed{\log}$ key.

Common logarithm: a logarithm with a base of 10; abbreviated as log

A logarithm with a base of e is a **natural logarithm** and is abbreviated as ln. Calculators normally have a $\boxed{\ln}$ key. Thus, $\log_e 1 = 0$ is normally written as $\ln 1 = 0$. Expressions containing natural logarithms can be evaluated using the $\boxed{\ln}$ key.

Natural logarithm: a logarithm with a base of e; abbreviated as ln

---

TIP **Finding Logarithms Using Your Calculator** Calculators generally have two logarithm keys, $\boxed{\log}$ for *common logarithms* and $\boxed{\ln}$ for *natural logarithms.*
Evaluate log 2 and ln 2.

Most Calculators:

To find the common log of 2, press the $\boxed{\log}$ key, followed by the result of exponentiation, 2.

$$\boxed{\log}\, 2 \,\boxed{=} \Rightarrow 0.3010299957$$

We can interpret this as $10^{0.3010299957} = 2$.

To find the natural log of 2, press the $\boxed{\ln}$ key, followed by the result of exponentiation, 2.

$$\boxed{\ln}\ 2\ \boxed{=}\ \Rightarrow\ 0.6931471806$$

That is, $e^{0.6931471806} = 2$.

Experiment with your calculator to determine the necessary keystrokes. A good test is to verify that $\log 1 = 0$ and $\ln 1 = 0$.

EXAMPLE 3

Evaluate using a calculator or exponential equations.

(a) $\log 10{,}000$ (b) $\log 0.000001$ (c) $\log 9$ (d) $\log 5.4$

(a) $\log 10{,}000 = 4$ $\boxed{\log}\ 10000\ \boxed{=}$. Check: $10^4 = 10{,}000$

(b) $\log 0.000001 = -6$ Enter $\boxed{\log}$. $000001\ \boxed{=}$. Check: $10^{-6} = 0.000001$

(c) $\log 9 = 0.9542425094$ or **0.9542 rounded to the nearest ten-thousandth**

(d) $\log 5.4 = 0.7323937598$ or **0.7324 rounded to the nearest ten-thousandth**

See Exercises 13–24.

EXAMPLE 4

Evaluate using a calculator. Express the answers to the nearest ten-thousandth.

(a) $\ln 5$ (b) $\ln 4.5$ (c) $\ln 948$

(a) $\ln 5 = $ **1.6094** (b) $\ln 4.5 = $ **1.5041** (c) $\ln 948 = $ **6.8544**

See Exercises 25–29.

4 Evaluate Logarithms with a Base Other Than 10 or e. The natural logarithms have e as their base. To distinguish natural logarithms from logarithms with a base other than e, we use the abbreviation "ln" to indicate logarithms that have base e. Most calculators have a key for the natural logarithm noted by $\boxed{\text{LN}}$ on the key.

Calculators generally do not have a key for logarithms with a base different from 10 or e. However, many logarithms with bases other than 10 or e can be evaluated by writing them in exponential form and using the property: If $x^a = x^b$, then $a = b$.

EXAMPLE 5

Evaluate using the property: If $x^a = x^b$, then $a = b$.

(a) $\log_4 256 = x$ (b) $\log_3 \dfrac{1}{27} = x$

(a) $\log_4 256 = x$ Write 256 as a power of 4. We have $256 = 4^4$.
$\qquad 4^x = 256$
$\qquad 4^x = 4^4$ Like bases.
$\qquad \boxed{x = 4}$ Check: $4^4 = 256$

(b) $\log_3 \dfrac{1}{27} = x$ Write $\frac{1}{27}$ as a power of 3. That is, $27 = 3^3$; $\frac{1}{27} = 3^{-3}$.

$\qquad 3^x = \dfrac{1}{27}$

$\qquad 3^x = 3^{-3}$

$\qquad \boxed{x = -3}$ Check: $3^{-3} = \frac{1}{27}$

See Exercises 30–35.

There is another way to find a logarithm with a base other than 10 or e when we cannot convert to an equivalent exponential equation. A conversion formula is used to find the logarithm for a base other than 10 or e using a calculator.

To evaluate a logarithm with a base b other than 10 or e:

1. Use the property $\log_b a = \dfrac{\log a}{\log b}$

2. A similar property can be used with natural logarithms.

$$\log_b a = \frac{\ln a}{\ln b}$$

3. To check, evaluate b^x where b is the base of the logarithm and x is the logarithm. The result should equal the argument.

STOP AND CHECK

Evaluate using a calculator.
1. $\log_6 216$
2. $\log_8 4{,}096$

Answers:
1. 3   2. 4

EXAMPLE 6

Evaluate using a calculator: **(a)** $\log_7 343$ **(b)** $\log_4 256$ **(c)** $\log_3 81$ **(d)** $\log_5 28$.

(a) $\log_7 343 = \dfrac{\log 343}{\log 7} = 3$ $\boxed{\log}\,343\,\boxed{\div}\,\boxed{\log}\,7\,\boxed{=} \Rightarrow 3$; Check: $7^3 = 343$

(b) $\log_4 256 = \dfrac{\log 256}{\log 4} = 4$ $\boxed{\log}\,256\,\boxed{\div}\,\boxed{\log}\,4\,\boxed{=} \Rightarrow 4$

(c) $\log_3 81 = \dfrac{\log 81}{\log 3} = 4$ $\boxed{\log}\,81\,\boxed{\div}\,\boxed{\log}\,3\,\boxed{=} \Rightarrow 4$

See Exercises 36–41.

---

5 **Evaluate Formulas Containing at Least One Logarithmic Term.** Many formulas use common and natural logarithms.

Decibel (dB): a unit for measuring the loudness of sound (dB)

Threshold sound: a very faint sound indicated as an intensity of I_0

EXAMPLE 7

AVIA   The loudness of sound is measured by a unit called a **decibel** (dB). A very faint sound, called the **threshold sound**, is assigned an intensity of I_0. Other sounds have an intensity of I, which is a specified number times the threshold sound ($I = nI_0$).

Then the decibel rate is given by the formula $dB = 10 \log \frac{I}{I_0}$. Find the decibel rating to the nearest decibel for sounds having the following intensities (I).

(a) A whisper, $110 I_0$

(b) A speaking voice, $230 I_0$

(c) A jet plane at takeoff, $109{,}000{,}000{,}000{,}000 I_0$

Aleksandar Mijatovic/123RF

Alexey Y. Petrov/Shutterstock

(a) A whisper, $I = 110 I_0$ $n = 110$

$$dB = 10 \log \frac{110 I_0}{I_0} \qquad \frac{I_0}{I_0} = 1$$

$$dB = 10 \log 110 \qquad \log 110 = 2.041392685$$

$$\mathbf{dB = 20 \text{ (rounded)}}$$

(b) A speaking voice, $I = 230 I_0$ $n = 230$

$$dB = 10 \log \frac{230 I_0}{I_0} \qquad \frac{I_0}{I_0} = 1$$

$$dB = 10 \log 230 \qquad \log 230 = 2.361727836$$

$$\mathbf{dB = 24 \text{ (rounded)}}$$

(c) A jet plane at takeoff, $I = 109{,}000{,}000{,}000{,}000I_0 = (1.09 \times 10^{14})I_0$

$$dB = 10 \log \frac{(1.09 \times 10^{14})I_0}{I_0} \qquad n = 1.09 \times 10^{14}$$

$$dB = 10 \log (1.09 \times 10^4) \qquad \frac{I_0}{I_0} = 1$$

$$dB = 10(14.0374265)$$

dB = 140 (rounded)   See Exercises 42–43.

To find the time for an investment to reach a specified amount if compounded continuously:

$$t = \frac{(\ln A - \ln P)}{r}$$

$t = $ time (in years) of investment
$A = $ accumulated amount
$P = $ initial invested amount
$r = $ annual rate of interest

STOP AND CHECK

1. Find the length of time for $25,000 to reach $35,000 when invested at 4% compounded continuously.

Answer:

1. 8.4 years

EXAMPLE 8

BUS   Find the length of time for $32,750 to reach $47,650.97 when invested at 5% compounded continuously.

$$A = \$47{,}650.97, \qquad P = \$32{,}750, \qquad r = 0.05$$

$$t = \frac{\ln A - \ln P}{r} \qquad\qquad\qquad \text{Substitute known values.}$$

$$t = \frac{\ln 47{,}650.97 - \ln 32{,}750}{0.05} \qquad\qquad \text{Evaluate numerator.}$$

$$t = \frac{10.77165827 - 10.39665824}{0.05} \qquad \text{Subtract in numerator.}$$

$$t = \frac{0.3750000246}{0.05} \qquad\qquad\qquad \text{Divide.}$$

$$t = 7.5 \text{ years} \qquad\qquad\qquad\qquad \text{Rounded} \qquad\qquad \text{See Exercises 44–45.}$$

EXAMPLE 9

PFIN   How long does it take an investment of $10,000 to double if it is invested at 5% annual interest compounded continuously?

$$A = \$20{,}000, \qquad P = 10{,}000, \qquad r = 0.05$$

$$t = \frac{\ln A - \ln P}{r} \qquad\qquad\qquad \text{Substitute known values.}$$

$$t = \frac{\ln 20{,}000 - \ln 10{,}000}{0.05} \qquad\qquad \text{Evaluate numerator.}$$

$$t = \frac{9.903487553 - 9.210340372}{0.05} \qquad \text{Subtract in numerator.}$$

$$t = \frac{0.6931471806}{0.05} \qquad\qquad\qquad \text{Divide.}$$

$$t = 13.9 \text{ years} \qquad\qquad\qquad\qquad \text{Rounded} \qquad\qquad \text{See Exercises 46–47.}$$

6 Graph a Logarithmic Function. Logarithmic functions can be graphed using a table of solutions or a graphing calculator.

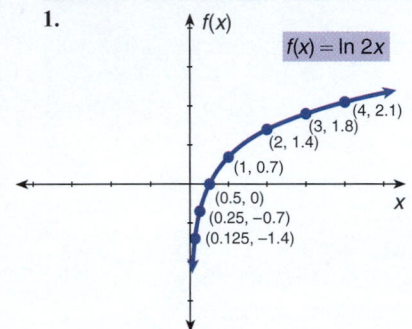
EXAMPLE 10

Prepare a table of solutions and graph the function $f(x) = \ln 4x$ using the values of x: 0.125, 0.25, 0.5, 1, 2, 3, 4. Determine the domain and range.

Using a calculator:

$f(0.125) = \ln 4(0.125)$ $f(0.25) = \ln 4(0.25)$ $f(0.5) = \ln 4(0.5)$
$f(0.125) = \ln 0.5$ $f(0.25) = \ln 1$ $f(0.5) = \ln 2$
$f(0.125) = -0.6931471806$ $f(0.25) = 0$ $f(0.5) = 0.6931471806$
$f(1) = \ln 4(1)$ $f(2) = \ln 4(2)$ $f(3) = \ln 4(3)$
$f(1) = \ln 4$ $f(2) = \ln 8$ $f(3) = \ln 12$
$f(1) = 1.386294361$ $f(2) = 2.079441542$ $f(3) = 2.48490665$
                                 $f(4) = \ln 4(4)$
                                 $f(4) = \ln 16$
                                 $f(4) = 2.772588722$

Plot the points indicated in the table of solutions and connect the points with a smooth, continuous curve (Fig. 16–6).

$f(x) = \ln 4x$

x	$f(x)$
0.125	−0.7
0.25	0
0.5	0.7
1	1.4
2	2.1
3	2.5
4	2.8

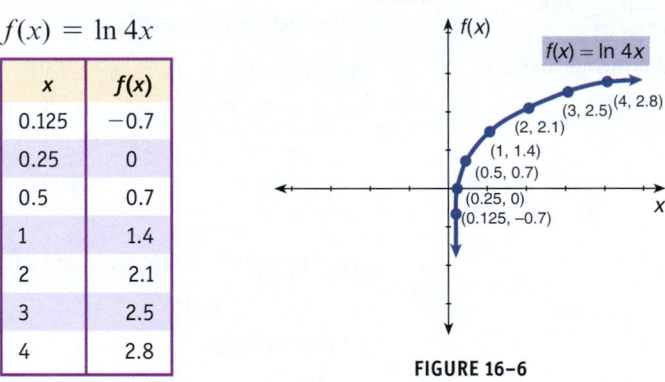

FIGURE 16–6

The domain is the set of real numbers greater than 0. The range is the set of all real numbers.

See Exercises 48–53.

EXAMPLE 11

A scale for measuring the intensity of an earthquake is the **Richter scale.** Richter scale rating $= \log \dfrac{I}{I_0}$, where I_0 is the measure of the intensity of a very small (faint) earthquake and I is the measure of an earthquake times I_0.

Find the Richter scale ratings for intensity values of powers of 10 from 10^3 to 10^8. Describe the relationship between the Richter scale rating and the intensity of an earthquake.

Richter scale rating $= \log \dfrac{I}{I_0}$

Values of I include a factor of I_0, which will reduce with the denominator. Let x represent the quotient of $\dfrac{I}{I_0}$ and write the Richter scale as a function of x.

$y = \log x$

Enter function in a graphing calculator.

Set the Window for x-values 0–100,000,000 with a scale of 1,000 and for y-values 0–10 with a scale of 1.

Find the corresponding y-values for powers of 10 from 10^3 to 10^8. Use the $\boxed{\text{CALC}}$ and the $\boxed{\text{1:VALUE}}$ features for each value of x.

x	y
1,000	3
10,000	4
100,000	5
1,000,000	6
10,000,000	7
100,000,000	8

The Richter scale rating for the intensity of an earthquake is equal to the exponent of the intensity expressed as a power of 10.   See Exercises 54–58.

7 **Simplify Logarithmic Expressions Using the Properties of Logarithms.** Many applications of logarithms and exponential expressions require the understanding and use of the properties of logarithms. The laws of exponents are similar to the properties of logarithms because of the relationship between logarithms and exponents. Similar laws are appropriate for natural logarithms.

Properties of logarithms:

Equality of logarithms	If $\log_b x = \log_b y$, then $x = y$.	This is similar to the exponential property, if $b^x = b^y$, then $x = y$.
Logarithm of product	$\log_b mn = \log_b m + \log_b n$	The log of a product is the sum of the logs of the factors.
Logarithm of quotient	$\log_b \dfrac{m}{n} = \log_b m - \log_b n$	The log of a quotient is the difference of the logs of the numerator and denominator.
Logarithm of power	$\log_b m^n = n \log_b m$	The log of a quantity raised to an exponent is the exponent times the log of the quantity.
Logarithm of quantity with same base	$\log_b b = 1$	The log of a quantity with the same base as the quantity equals 1.

These laws can be illustrated with examples using a calculator.

EXAMPLE 12

Show that the statements are true by using a calculator.

(a) $\log 6 = \log 2 + \log 3$ **(b)** $\log 2 = \log 6 - \log 3$

(c) $\log 2^3 = 3 \log 2$ **(d)** $\log 20 = 1 + \log 2$

(a) $\log 6 = \log 2(3) = \log 2 + \log 3$

$$\boxed{\log}\, 6\, \boxed{=} \Rightarrow 0.7781512504, \qquad\qquad \boxed{\log}\, 2\, \boxed{+}\, \boxed{\log}\, 3\, \boxed{=} \Rightarrow 0.7781512504$$

(b) $\log 2 = \log \dfrac{6}{3} = \log 6 - \log 3$

$$\boxed{\log}\, 2\, \boxed{=} \Rightarrow 0.3010299957, \qquad\qquad \boxed{\log}\, 6\, \boxed{-}\, \boxed{\log}\, 3\, \boxed{=} \Rightarrow 0.3010299957$$

(c) $\log 2^3 = 3 \log 2$

$$\boxed{\log}\, \boxed{(}\, 2\, \boxed{\wedge}\, 3\, \boxed{)}\, \boxed{=} \Rightarrow 0.903089987, \qquad 3\, \boxed{\log}\, 2\, \boxed{=} \Rightarrow 0.903089987$$

(d) $\log 20 = \log 10(2) = \log 10 + \log 2 = 1 + \log 2$

$$\boxed{\log}\, 20\, \boxed{=} \Rightarrow 1.301029996, \qquad 1 + \boxed{\log}\, 2\, \boxed{=} \Rightarrow 1.301029996$$

See Exercises 59–62.

Supply and demand: a concept in business and economics that examines the effects of the availability of a product (supply) and the desire for that product (demand)

An important concept in business and economics is **supply and demand**. The availability of a product (supply) and the desire for that product (demand) influence the price of the product. One formula relating the demand (D) of a product with the price (P) of the product is

$$\log_a D = \log_a c - k \log_a P,$$

where a, c, and k are positive constants.

STOP AND CHECK

1. Solve the equation
$\ln y = \ln x + a \ln z$ for y, assuming all variables are positive.

Answer:
1. $y = xz^a$

EXAMPLE 13

BUS   Solve the equation $\log_a D = \log_a c - k \log_a P$ for D and explain how increasing or decreasing the price affects the demand.

$\log_a D = \log_a c - k \log_a P$

$\log_a D = \log_a c - \log_a P^k$ Apply the logarithm of power property.

$\log_a D = \log_a \dfrac{c}{P^k}$ Apply the logarithm of quotient property.

$D = \dfrac{c}{P^k}$ Apply the equality of logarithms property.

If the price (P) increases, the denominator becomes larger, which means the value of the fraction is smaller or the demand (D) decreases. If the price (P) decreases, the denominator becomes smaller, which means the value of the fraction is larger or the demand (D) increases.

See Exercises 63–64.

16–2 EXERCISES

MyLab Math   For additional practice go to your study plan in MyLab Math.

1   Write as logarithmic equations. *See Example 1.*

1. $3^2 = 9$

2. $2^5 = 32$

3. $9^{\frac{1}{2}} = 3$

4. $16^{\frac{1}{4}} = 2$

5. $4^{-2} = \dfrac{1}{16}$

6. $3^{-4} = \dfrac{1}{81}$

2   Write in exponential form. *See Example 2.*

7. $\log_3 81 = 4$

8. $\log_{12} 144 = 2$

9. $\log_2 \dfrac{1}{8} = -3$

10. $\log_5 \dfrac{1}{25} = -2$

11. $\log_{25} \dfrac{1}{5} = -0.5$

12. $\log_4 \dfrac{1}{2} = -0.5$

3 Evaluate using a calculator or exponential equation. *See Example 3.*

13. log 0.001

14. log 100,000

15. log 10,000,000

16. log 0.00001

17. log 1

18. log 0.000001

Evaluate with a calculator. Express answers to the nearest ten-thousandth. *See Example 3.*

19. log 3

20. log 6

21. log 2.4

22. log 4.2

23. log 150

24. log 0.0012

Evaluate using a calculator. Express answers to the nearest ten-thousandth. *See Example 4.*

25. ln 4

26. ln 2.5

27. ln 0.15

28. ln 275

29. ln 100

4 Solve for x by using an equivalent exponential expression. *See Example 5.*

30. $\log_3 x = -4$

31. $\log_6 36 = x$

32. $\log_7 \dfrac{1}{49} = x$

33. $\log_5 x = 4$

34. $\log_4 \dfrac{1}{256} = x$

35. $\log_4 64 = x$

Evaluate with a calculator. *See Example 6.*

36. $\log_5 125$

37. $\log_3 729$

38. $\log_{\frac{1}{2}} 0.03125$

39. $\log_7 49$

40. $\log_8 56$

41. $\log_2 128$

5 Find the decibel rating to the nearest decibel for the sounds having the given intensities. *See Example 7.*

42. A busy street, $9,000,000 I_0$

43. Loud music, $879,000,000,000 I_0$

See Example 8.

44. How long does it take an investment of $20,000 to double if it is invested at 5% annual interest, compounded continuously? Round to tenths.

45. How long does it take for $48,000 to reach $52,000 when invested at 4%, compounded continuously? Round to tenths.

See Example 9.

46. **PFIN/BUS** How long does it take an investment of $15,000 to double if it is invested at 1% annual interested compounded continuously?

47. **PFIN/BUS** How long does it take an investment of $15,000 to double if it is invested at 3% annual interest compounded continuously?

6 Graph the functions using the table-of-solutions method. Determine the domain and range. *See Example 10.*

48. $y = \ln 2x$

49. $y = -\ln 2x$

50. $y = \ln 5x$

51. $y = -\log 3x$

52. $y = -\log 6x$

53. $y = \log(-3x)$

The intensity of an earthquake is measured on the *Richter scale* by the formula

$$\text{Richter scale rating} = \log \frac{I}{I_0}$$

where I_0 is the measure of the intensity of a very small (faint) earthquake. Find the Richter scale rating of earthquakes having the following intensities: *See Example 11.*

54. $1,000,000,000 I_0$

55. $1,000 I_0$

56. $100,000 I_0$

57. $100,000,000 I_0$

58. Graph the equation, Richter scale rating $= \log\left(\dfrac{I}{I_0}\right)$, where I represents the intensity of an earthquake and I_0 represents the intensity of a very small earthquake. Use the TABLE function to find the Richter scale rating of an earthquake that has an intensity of $19{,}500{,}000I_0$.

7 Show that the statements are true by using a calculator. *See Example 12.*

59. $\log 10 = \log 2 + \log 5$

60. $\log 6 = \log 3 - \log 2$

61. $\log 5^2 = 2 \log 5$

62. $\log 30 = 1 + \log 3$

See Example 13.

63. Solve the equation $\ln m = a \ln n - \ln p$ for m, assuming all variables are positive.

64. Solve the equation $\ln a = r \ln b - 3 \ln c$ for a, assuming all variables are positive.

---

16 | CHAPTER REVIEW OF KEY CONCEPTS

LEARNING OUTCOMES	KEY CONCEPTS AND EXAMPLES

Section 16–1

1 Evaluate formulas with at least one exponential term (pp. 672–677).

A scientific or graphing calculator can be used along with the order of operations to evaluate formulas with at least one exponential term.

Use the formula for compound interest to find the compound amount for a loan of $5,000 for 3 years at an annual interest rate of 6% if the principal is compounded semiannually.

$A = P\left(1 + \dfrac{r}{n}\right)^{nt}$ $P = \$5{,}000; r = 0.06; n = 2; t = 3$

$A = 5{,}000\left(1 + \dfrac{0.06}{2}\right)^{2(3)}$ Perform operations inside grouping. Perform operation in exponent.

$A = 5{,}000(1.03)^6$ Raise to power.

$A = 5{,}000(1.194052297)$ Multiply.

$A = \$5{,}970.26$ Compound amount (rounded)

2 Evaluate formulas that contain a power of the natural exponential, e (pp. 677–679).

Use a calculator to evaluate formulas containing the natural exponential, e.

Use the formula $P = 760\,e^{-0.00013h}$ for atmospheric pressure to find P at $h = 50$ m above sea level.

$P = 760e^{-0.00013(50)}$

$P = 760e^{-0.0065}$

$P = 760(0.9935210793)$

$P = 755.08$ Rounded

LEARNING OUTCOMES	KEY CONCEPTS AND EXAMPLES

3 Solve exponential equations in the form $b^x = b^y$, where $b > 0$ and $b \neq 1$ (pp. 679–680).

1. Rewrite the equation in the form $b^x = b^y$. **2.** Apply the following property and solve for x. If $b^x = b^y$, and $b > 0$ and $b \neq 1$, then $x = y$ where x and y are any real numbers.

Note: This property applies only when bases are *like bases*, the bases are positive, and the bases are not equal to 1.

Solve the equation $4^{3x+1} = 8$. Rewrite the bases: $4 = 2^2$ and $8 = 2^3$.

$2^{2(3x+1)} = 2^3$	The bases are equal, so exponents are equal.
$2(3x + 1) = 3$	Distribute.
$6x + 2 = 3$	Sort terms and combine like terms.
$6x = 1$	Divide.
$x = \dfrac{1}{6}$	

4 Graph an exponential function (pp. 680–683).

Use a table of values or a graphing calculator to graph exponential functions.

The exponential equation for the breakdown of a particular drug in the human body is $D = D_0(1 + r)^t$, where D is the amount of drug in the bloodstream after a specified amount of time, D_0 is the initial amount of drug in the bloodstream, r is the rate of change as a decimal, and t is the amount of lapsed time. A certain drug breaks down at the rate of 25% per hour for the first 24 h. Find the amount of drug in the bloodstream after 6 h if the initial amount is 50 mg.

$D = 50(1 - 0.25)^t$	Write as a function of x. Simplify grouping.
$y = 50(0.75)^x$	Graph.

$\boxed{\text{y=}}\ 50\ \boxed{(}\ .75\ \boxed{)}\ \boxed{\wedge}\ \boxed{\text{X, T, }\theta, n}\ \boxed{\text{TBLSET}}\ \text{TblStart=}\ \boxed{0}\ \Delta\text{Tbl=}\ \boxed{1}\ \boxed{\text{TABLE}}$

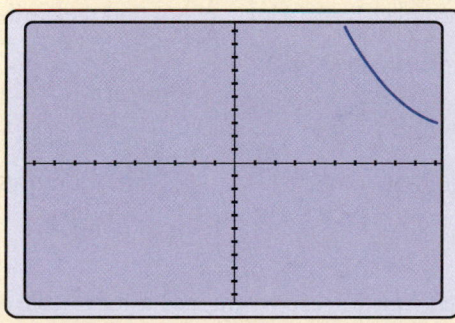
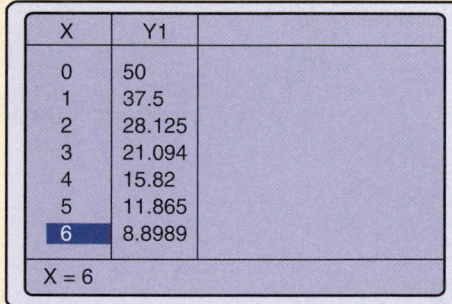

X	Y1	
0	50	
1	37.5	
2	28.125	
3	21.094	
4	15.82	
5	11.865	
6	8.8989	

X = 6

Approximately 8.9 mg of drug remains in the bloodstream after 6 h.

Section 16-2

1 Write exponential equations as equivalent logarithmic equations (p. 687).

If $x = b^y$, then $\log_b x = y$, provided that $b > 0$ and $b \neq 1$. **1.** The exponent in the exponential form is the dependent variable in the logarithmic form. **2.** The base in the exponential form is the base in the logarithmic form. **3.** The dependent variable in the exponential form is the result of exponentiation in the logarithmic form. This result is sometimes referred to as the **argument.**

Write $16 = 2^4$ in logarithmic form. $\log_2 16 = 4$

LEARNING OUTCOMES	KEY CONCEPTS AND EXAMPLES
2 Write logarithmic equations as equivalent exponential equations (p. 688).	$\log_b x = y$ converts to $x = b^y$, provided that $b > 0$ and $b \neq 1$. **1.** The dependent variable in the logarithmic form is the exponent in the exponential form. **2.** The base in the logarithmic form is the base in the exponential form. **3.** The result of the exponentiation in the logarithmic form is the dependent variable in the exponential form.

Write $\log_5 125 = 3$ in exponential form. $5^3 = 125$

| **3** Evaluate common and natural logarithmic expressions using a calculator (pp. 688–689). | Use the $\boxed{\log}$ key to find common logarithms and the $\boxed{\ln}$ key to find natural logarithms. Some calculators require the number to be entered before the $\boxed{\log}$ or $\boxed{\ln}$ key. Most calculators require the $\boxed{\log}$ or $\boxed{\ln}$ key to be entered, followed by the number. If a parenthesis is opened, entering a closing parenthesis after entering the argument of the logarithm may be required. |

Use a calculator to find log 25. Scientific calculator steps:

$$\boxed{\log}\ 25\ \boxed{=} \qquad \text{or} \qquad 25\ \boxed{\log} \Rightarrow 1.397940009$$

| **4** Evaluate logarithms with a base other than 10 or e (pp. 689–690). | **1.** Use the property $\log_b a = \dfrac{\log a}{\log b}$ **2.** A similar property can be used with natural logarithms. $\log_b a = \dfrac{\ln a}{\ln b}$ **3.** To check, evaluate b^x where b is the base of the logarithm and x is the logarithm. The result should equal the argument. |

Evaluate $\log_4 64$:

$$\log_4 64 = \frac{\log 64}{\log 4} = 3 \qquad \boxed{\log}\ 64\ \boxed{)}\ \boxed{\div}\ \boxed{\log}\ 4\ \boxed{=} \Rightarrow 3$$

| **5** Evaluate formulas containing at least one logarithmic term (pp. 690–691). | Use a calculator and follow the order of operations. |

Use the formula $\text{dB} = 10 \log\left(\dfrac{I}{I_0}\right)$ to find the decibel rating for a sound that is 350 times the threshold sound ($350I_0$).

$$\text{dB} = 10 \log\left(\frac{350I_0}{I_0}\right)$$

$\text{dB} = 10 \log 350$ 10 $\boxed{\times}$ $\boxed{\log}$ 350 $\boxed{=}$

$\text{dB} = 25$ Rounded

| **6** Graph a logarithmic function (pp. 691–693). | A logarithmic function can be graphed using a table of values or a graphing calculator. |

Graph the function, $y = \log(0.4x)$ and find the value of y when $x = 150$.

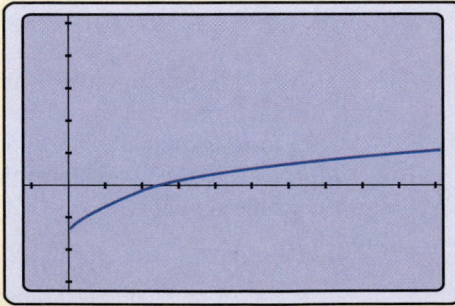

LEARNING OUTCOMES	KEY CONCEPTS AND EXAMPLES

The preceding figure shows the graph for x between 0 and 10.

To find the value as 150:

TBLSET , TblStart = 145, Δ Tbl = 1, ENTER

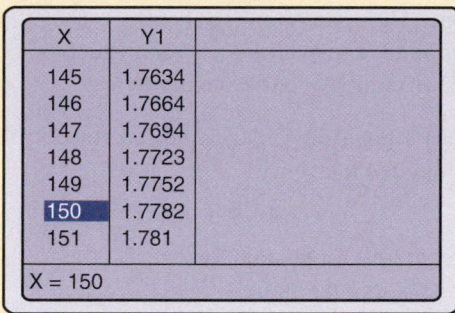

X	Y1
145	1.7634
146	1.7664
147	1.7694
148	1.7723
149	1.7752
150	1.7782
151	1.781

X = 150

When $x = 150, y = 1.7782$.

7 Simplify logarithmic expressions using the properties of logarithms (pp. 693–694).

Properties of logarithms:

If $\log_b x = \log_b y$, then $x = y$.

$\log_b mn = \log_b m + \log_b n$

$\log_b \dfrac{m}{n} = \log_b m - \log_b n$

$\log_b m^n = n \log_b m$

$\log_b b = 1$

Write $\log (3)(8)$ in another way. Use a calculator to verify that the equation is true.

$\log (3)(8) = \log 3 + \log 8$

log (3 $\times$ 8) = $\Rightarrow$ 1.380211242

log 3) + log 8 = $\Rightarrow$ 1.380211242

16   CHAPTER REVIEW EXERCISES

Section 16–1   **MyLab Math**   For additional practice go to your study plan in MyLab Math.

Solve for x.

1. $5^x = 5^8$

2. $2^x = 2^6$

3. $3^x = 3^{-2}$

4. $2^{x+3} = 2^7$

5. $4^{x-2} = 4^2$

6. $5^{2x-1} = 5^2$

7. $6^{3x+2} = 6^{-3}$

8. $2^x = 16$

9. $3^x = 81$

10. $3^x = \frac{1}{9}$

11. $2^x = \frac{1}{32}$

12. $4^{2x} = 64$

13. $5^{3x} = 125$

14. $3^{4-x} = \frac{1}{27}$

15. $6^{2-x} = \frac{1}{36}$

Evaluate.

16. e^3

17. e^{-4}

18. $e^{-0.12}$

19. e^{-10}

20. **PFIN/BUS** Calculate the compound interest on a loan of $200 at 6%, compounded annually for 4 years.

21. **PFIN/BUS** Calculate the compound interest on a loan of $1,600 for 3 years at 13% if the interest is compounded annually.

22. **PFIN/BUS** Calculate the compound interest on a loan of $6,150 at $11\frac{1}{2}$% annual interest, compounded annually for 3 years.

23. **PFIN/BUS** Maria Sanchez invested $2,000 for 2 years at 12% annual interest, compounded semiannually. Calculate the interest she earned on her investment.

24. **PFIN/BUS** EZ Loan Company loaned $500 at 8% annual interest, compounded quarterly for 1 year. Calculate the amount the loan company will earn in interest.

25. **PFIN/BUS** Find the future value on an investment of $3,000 made by Ling Lee for 5 years at 12% annual interest, compounded semiannually.

26. **PFIN/BUS** Find the interest on $2,500 invested for 3 years at 3.75%, compounded quarterly.

27. **PFIN/BUS** Find the interest on a certificate of deposit (CD) of $10,000 for 5 years at 4%, compounded semiannually.

28. **PFIN/BUS** Find the interest on $10,000 for 2 years compounded monthly at a 6% annual interest rate.

29. **PFIN/BUS** Find the future value on an investment of $8,000, compounded quarterly for 7 years at 8%.

30. **PFIN/BUS** Find the compound interest on $10,000 for 2 years if the interest is compounded continuously at 6%.

31. **PFIN/BUS** Find the compound interest on $5,000 for 2 years if the interest is compounded continuously at 12%.

32. **PFIN/BUS** Find the effective interest rate for the loan described in Exercise 28.

33. **PFIN/BUS** Find the effective interest rate for the loan described in Exercise 29.

34. **PFIN/BUS** Find the accumulated amount on an investment of $8,000 invested for 5 years compounded quarterly at 5%.

35. **PFIN/BUS** Find the amount of interest on $1,500 invested for 4 years at a 3% annual rate compounded monthly.

36. **PFIN/BUS** Tommye Adams wishes to have $8,000 1 year from now to make a down payment on a lake house. How much should she invest at 9.5% annual interest to have her payment in 1 year?

37. **PFIN/BUS** Billy Hill wishes to have $4,000 in 4 years to tour Europe. How much must he invest today at 8% annual interest, compounded quarterly, to have the $4,000 in 4 years?

38. **BUS** Kristin Ammons was offered $15,000 cash or $19,500 to be paid after 2 years for a resort cabin. If money can be invested in today's market for 3% annual interest, compounded quarterly, which offer should Kristin accept?

39. **BUS** An art dealer offered a collector $8,000 for a painting. The collector could sell the painting to an individual for $11,000 to be paid in 18 months. Currently investments yield 12% annual interest, compounded monthly. Which is the better deal for the collector?

40. **BUS** If you were offered $700 today or $800 in 2 years, which would you accept if money can be invested at 12% annual interest, compounded monthly?

41. **PFIN** How much should a family invest now at 10% compounded annually to have a $7,000 house down payment in 4 years?

42. **PFIN** Greg and Brienne Jackson had a baby in 2013. At the end of that year they began putting away $900 per year at 10% compounded annual interest for a college fund. When their child is 18 years old in 2031, the cost for 4 years of college at a public college is estimated to be about $20,000 per year.
    (a) How much money will be in the account when the child is 18 years old?
    (b) Will the Jacksons have enough saved to send their child to college for 4 years?

43. **PFIN** Skip Quinn plans to deposit $2,000 at the end of every 6 months for the next 5 years to save for a boat. If the interest rate is 12% annually, compounded semiannually, how much money will Skip have in his boat fund after 5 years?

44. **BUS** A business deposits $4,500 at the end of each quarter in an account that earns 8% annual interest, compounded quarterly. What is the value of the annuity in 5 years?

45. **BUS** How much must be set aside at the end of each six months in a sinking fund by the Fabulous Toy Company to replace a $155,000 piece of equipment at the end of 8 years if the account pays 8% annual interest, compounded semiannually?

46. **BUS**   Tasty Food Manufacturers, Inc., has a bond issue of $1,400,000 due in 30 years. If it wants to establish a sinking fund to meet this obligation, how much must be set aside at the end of each year if the annual interest rate is 6%?

47. **BUS**   Lausanne Private School System needs to set aside funds for a new computer system. What monthly sinking fund payment would be required to amount to $45,000, the approximate cost of the computer system, in $1\frac{1}{2}$ years at 12% annual interest, compounded monthly?

48. **BUS**   Zachary Alexander owns a limousine that will need to be replaced in 4 years at a cost of $65,000. How much must he put aside each year in a sinking fund at 8% annual interest to be able to afford the new limousine?

49. **BUS**   Braddy's Department Store has a fleet of delivery trucks that needs to be replaced at a cost of $75,000. How much must it set aside every 3 months for 3 years in a sinking fund at 8% annual interest, compounded quarterly, to have enough money to replace the trucks?

50. **BUS**   Danny Lawrence Properties, Inc., has a bond issue that matures in 25 years for $1 million. How much must the company set aside each year in a sinking fund at 12% annual interest to meet this future obligation?

51. **PFIN**   Brenda Pearson wants to save $25,000 for a new car in 6 years. How much must be put aside in equal payments each year in an account earning 8% annual interest for Brenda to be able to purchase the car?

52. **PFIN**   How much money needs to be set aside today at 10% annual interest, compounded semiannually, to have the same amount as a semiannual ordinary annuity of $500 for 5 years at 10% annual interest, compounded semiannually?

53. **PFIN**   Find the monthly payment on a mortgage of $238,000 if it is financed for 20 years at 7.5% annual interest.

54. **PFIN**   Pam Murphrey purchased a property with a mortgage of $528,260 at 6.8% annual interest. Find the monthly payment if the mortgage is amortized for 30 years.

A formula for electric current is $I = 1.50e^{-200t}$, where t is time in seconds. Calculate the current for each time. Express the answers in scientific notation.

55. **ELEC**   0.07 s

56. **ELEC**   0.2 s

57. **ELEC**   0.4 s

Graph the equations using the table-of-solutions method or a calculator.

58. **INDTEC**   The number of grams of a chemical that will dissolve in a solution is given by the formula $C = 100e^{0.05t}$, where t = temperature in degrees Celsius. Evaluate when
(a) $t = 10$ (b) $t = 20$
(c) $t = 45$ (d) $t = 50$

59. $y = 5^x$

60. $y = -5^x$

61. $y = 5^{(x+4)}$

62. $y = 5^{(x-4)}$

63. $y = 5^{-x}$

64. $y = -5^{-x}$

65. **HLTH/N**   In the equation $D = D_0(1 + r)^t$, D is the amount of drug in the bloodstream after a specified amount of time. D_0 is the initial amount of the drug in the bloodstream, r is the decimal equivalent of the rate of change, and t is the amount of lapsed time. A drug breaks down at the rate of 4% per hour for the first 24 h. Find the amount of drug after 5 h if the initial amount of drug in the bloodstream is 20 mg. (Round to tenths.)

66. **HLTH/N**   Use the exponential model in Exercise 65 to find the approximate number of hours it will take to have less than 60 mg of a drug in the bloodstream if the drug breaks down at the rate of 25% per hour when the initial amount is 250 mg.

Section 16-2

Rewrite the following as logarithmic equations.

67. $2^3 = 8$

68. $5^2 = 25$

69. $3^4 = 81$

70. $81^{1/2} = 9$

71. $27^{1/3} = 3$

72. $5^{-3} = \frac{1}{125}$

73. $4^{-3} = \frac{1}{64}$

74. $8^{-1/3} = \frac{1}{2}$

75. $9^{-1/2} = \frac{1}{3}$

76. $121^{1/2} = 11$

77. $12^{-2} = \frac{1}{144}$

78. $49^{-1/2} = \frac{1}{7}$

Rewrite the following as exponential equations. Verify if the equation is true.

79. $\log_{11} 121 = 2$

80. $\log_3 81 = 4$

81. $\log_{15} 1 = 0$

82. $\log_{25} 5 = \frac{1}{2}$

83. $\log_7 7 = 1$

84. $\log_3 3 = 1$

85. $\log_4 \frac{1}{16} = -2$

86. $\log_2 \frac{1}{16} = -4$

87. $\log_9 \frac{1}{3} = -0.5$

88. $\log_{16} \frac{1}{4} = -0.5$

89. $\log_{10} 1,000 = 3$

90. $\log_{10} 100 = 2$

Evaluate the following with a calculator. Express the answers to the nearest ten-thousandth.

91. $\log 5$

92. $\log 3.8$

93. $\log 180$

94. $\log 0.0015$

95. $\log 0.4$

96. $\ln 12$

97. $\ln 270$

98. $\ln 0.134$

99. $\ln 0.8$

100. $\ln 80$

101. $\log_5 30$

102. $\log_7 120$

Solve for x by using an equivalent exponential expression.

103. $\log_4 16 = x$

104. $\log_7 49 = x$

105. $\log_7 x = 3$

106. $\log_5 x = -2$

107. $\log_6 \frac{1}{36} = x$

108. $\log_4 \frac{1}{64} = x$

109. $\log_2 x = 5$

110. $\log_x 125 = 3$

111. The intensity of an earthquake is measured on the Richter scale by the formula

$$\text{Richter scale rating} = \log \frac{I}{I_0}$$

where I_0 is the measure of the intensity of a very small (faint) earthquake. Find the Richter scale rating of the earthquakes having the following intensities:
(a) $100I_0$ **(b)** $10,000I_0$ **(c)** $150,000,000I_0$

Solve for x by using an equivalent logarithmic expression and the facts that $\log_2 3 = 1.585$ and $\log_2 7 = 2.807$.

112. $\log_2 21$

113. $\log_2 9$

114. $\log_2 6$

115. **PFIN/BUS**   An investment of $100,000 is invested at 4.1%, compounded continuously. How long will it take for the investment to accumulate to $150,000?

116. **PFIN/BUS**   Quenesha McGee has $10,000 invested at 5.9%, compounded continuously. She needs to know how long it will take for the investment to accumulate to $18,000.

Graph the equations using a table of solutions or calculator.

117. $y = \ln 4x$

118. $y = -\ln 4x$

119. $y = \log 8x$

120. $y = \log (x - 8)$

121. **HLTH/N**   Use the equation $\text{pH} = 6.1 + \log \left(\dfrac{50}{c} \right)$ to find the value of c that produces a neutral pH (7) to the nearest tenth.

---

16   TEAM PROBLEM-SOLVING EXERCISES

1. The Environmental Protection Agency (EPA) monitors atmosphere and soil contamination by dangerous chemicals. When possible, the chemical contamination is decomposed using microorganisms that change the chemicals so they are no longer harmful. A particular microorganism can reduce the contamination level to about 65% of the existing level every 30 days. A soil test for a contaminated site shows 72,000,000 units per cubic meter of soil.

(a) Write a formula for determining the contamination level after x 30-day periods: the initial contamination times the percent reduction (65% or 0.65) raised to the xth power.

(b) What is the level of contamination after 60 days? After 150 days?

(c) A "safe level" is 60,000 units of contamination per cubic meter of soil. Estimate how long it would take for the soil to reach this safe level. Discuss your method of arriving at the estimate.

2. Complete the spreadsheet showing the Accumulated Amount and Compound Interest for each entry (Fig. 16–7). The formula written for the spreadsheet is: $= P * (1 + r) \wedge t$, where P is principal, r is rate per period, and t is number of periods.

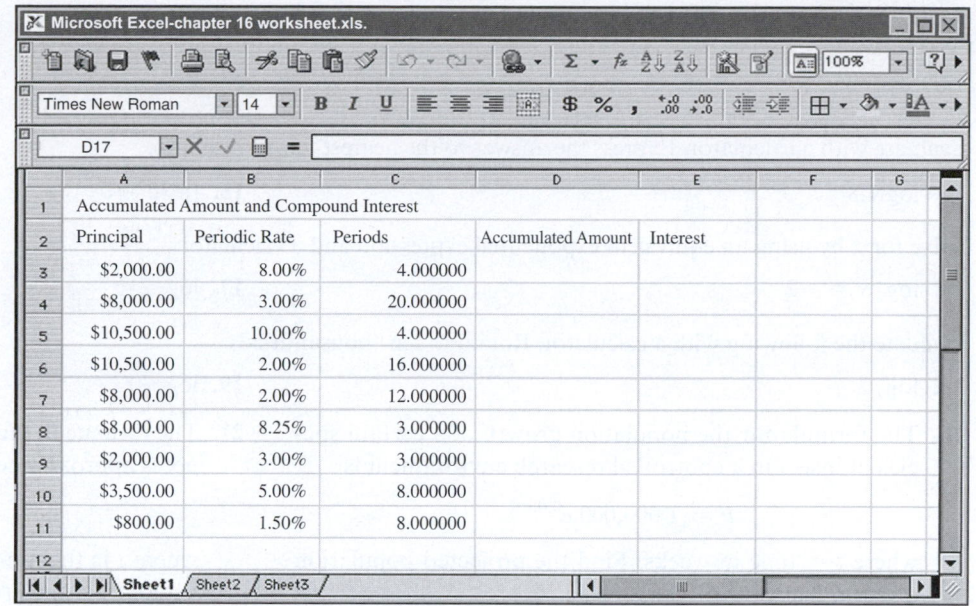

FIGURE 16–7

16 | CONCEPTS ANALYSIS

1. Give an example of an application that has a linear relationship and an application that has an exponential relationship.

2. Briefly describe how a linear relationship is different from an exponential relationship.

3. Describe how to recognize an equation as an exponential equation.

4. Explain why an equation of the form $b^x = b^y$ is only appropriate when $b > 0$ and $b \neq 1$.

5. Show symbolically the relationship between an exponential equation and a logarithmic equation. Give a numerical example illustrating this relationship.

6. Explain the difference between a common logarithm and a natural logarithm.

Find the mistakes in the examples. Explain each mistake and correct it.

7. $2^{x-3} = 4^2$
$x - 3 = 2$
$x = 2 + 3$
$x = 5$

8. Write $\log_5 125 = 3$ as an equivalent exponential equation.
$3^5 = 125$

16 | PRACTICE TEST

Evaluate using a calculator.

1. 1.2^{45}

2. 5^{-8}

3. $15^{\frac{3}{2}}$

4. $e^{-0.25}$

5. 12^5

Solve for x.

6. $4^x = 4^{-3}$

7. $2^{x-4} = 2^5$

8. $3^x = \dfrac{1}{9}$

9. $2^{2x-1} = 8$

Write as logarithmic equations.

10. $2^8 = 256$

11. $4^{-\frac{1}{2}} = \dfrac{1}{2}$

Write as exponential equations.

12. $\log_5 625 = 4$

13. $\log_3 \dfrac{1}{27} = -3$

Evaluate with a calculator. Express the answer to the nearest ten-thousandth.

14. $\log 4.8$

15. $\ln 32$

Solve for x by using an equivalent exponential expression and a calculator.

16. $\log_4 x = -2$

17. $\log_6 216 = x$

Evaluate the following with a calculator. Round to ten-thousandths.

18. $\log_7 2$

19. $\log_8 21$

20. The formula for the population growth of a certain species of insect in a controlled research environment is

$$P = 1{,}000{,}000 \, e^{0.05t}$$

where t = time in weeks. Find the projected population after **(a)** 2 weeks and **(b)** 3 weeks.

21. The revenue in thousands of dollars from sales of a product is approximated by the formula

$$S = 125 + 83 \log (5t + 1)$$

where t is the number of years after a product was marketed. Find the projected revenue from sales of the product after 3 years.

22. The height in meters of the male members of a certain group is approximated by the formula

$$h = 0.4 + \log t$$

where t represents age in years for the first 20 years of life ($1 \leq t \leq 20$). Find the projected height of a 10-year-old male.

23. Find the accumulated amount on an investment of $5,000 at a rate of 5.8% per year, compounded annually for 2 years. Use the formula

$$A = P\left(1 + \frac{r}{n}\right)^{nt}$$

24. What is the future value of an ordinary annuity of $300 every 3 months for 4 years at 8% annual interest, compounded quarterly?

25. Find the future value of an ordinary annuity of $9,000 per year for 2 years at 15% annual interest.

26. What is the future value of an ordinary annuity of $985 every 6 months for 8 years at 8% annual interest, compounded semiannually?

27. What is the sinking fund payment required at the end of each year to accumulate to $125,000 in 16 years at 4% annual interest?

28. Find the compound interest on a loan of $3,000 for 1 year at 12% annual interest if the interest is compounded quarterly.

29. Find the effective interest rate for the loan described in Exercise 28.

30. Find the compound interest on an investment of $2,000 invested at 5.75% for 28 years compounded quarterly.

31. If you were offered $600 today or $680 in 1 year, which would you accept if money can be invested at 12% annual interest, compounded monthly?

32. Mabel Langston needs $12,000 in 10 years for her daughter's college education. How much must be invested today at 8% annual interest, compounded semiannually, to have the needed funds?

33. Which of the two options yields the greater return on your investment of $2,000?

Option 1: 8% annual interest compounded quarterly for 4 years

Option 2: $8\frac{1}{4}$% annual interest compounded annually for 4 years

34. Harvey Barton plans to buy a house in 4 years. He will make an $8,000 down payment on the property. How much should he invest today at 6% annual interest, compounded quarterly, to have the required amount in 4 years?

35. If you invest $1,000 today at 4% annual interest compounded continuously, how much will you have after 20 years?

36. If you invest $2,000 today at 8% annual interest, compounded quarterly, how much will you have after 3 years?

37. How much money should Bryan Trailer Sales set aside today to have $15,000 in 1 year to purchase a forklift if the interest rate is 10.4%, compounded annually?

38. Ginger Canoy has purchased a home for $122,000. She plans to finance $100,000 for 15 years at $5\frac{1}{2}$% interest. Calculate the monthly payment and the total interest.

17

Geometry

Trond Runar Solevaag/123RF

In Great Company

Understanding Geometry (August 23, 1991)

The Sleipner A platform produces oil and gas in the North Sea. It is supported on the seabed at a water depth of 82 m. The concrete structure is made up of individual cells, four of which elongate into legs that support the structure. But the structure that is there today is not the one that was originally meant to be there.

By August 23, 1991, the first version of the Sleipner A platform had been floated into place in Gandsfjorden outside Stavanger, Norway. Once in position, the operators began taking on sea water ballast to lower the structure in the water and stabilize it. It was supposed to sink 80 m at a rate of 1 m every 20 min. Unfortunately, when the platform had reached about 65 m depth, the operators heard rumbling and the noise of water cascading into the ballast tanks. They flipped on the de-ballasting pumps, but the water inflow was too great. The whole structure sank.

To recognize what happened next, you have to understand just how big this thing was. The top deck was roughly 200 by 500 ft, and weighed 57,000 tons.

It was built to accommodate between 160 and 200 crew members, who were to operate the rig itself and the drilling equipment on the rig. While the people were not on board, the equipment was and it weighed about 40,000 tons.

The fjord is 220 m (690 ft) deep. The buoyancy chambers weren't designed to sink to the depth they eventually reached as the structure sank—they imploded. When what was left of the structure finally hit the ocean floor, it caused a seismic event registering 3.0 on the Richter scale, and left nothing but a pile of debris at a depth of 220 m. The failure involved a total economic loss of about $700 million.

What happened? Well, the engineers misunderstood the geometry. As a result, they underestimated the weight the water ballast volume would put on the supporting structure, and not just by a little bit, either. They were off by a whopping 47%.

When engineers used the *correct* numbers to calculate the additional water ballast pressure, the model predicted the rig would fail at a depth of 62 m. This second estimate was much better, as the rig actually did fail at 65 m. Of course, they figured this out about $700 million too late.

17–1   Lines and Angles

LC LEARNING CATALYTICS

1. Find the perimeter of a rectangle that is 15 ft by 12 ft.
2. Find the area of a rectangle that is 15 ft by 12 ft.

Geometry: the mathematical science that involves the study and measurement of shapes according to their size, volume, and position

Point: a location or position that has no size or dimension

Line: a line extends indefinitely in both directions and contains an infinite number of points; it has length but no width

Plane: a flat, smooth surface that extends indefinitely in all directions. It contains an infinite number of points and lines

Geometry is one of the oldest and most useful of the mathematical sciences. **Geometry** involves the study and measurement of shapes according to their size, volume, and position. A knowledge of geometry is necessary in many careers. Notice how the meanings of common words with which you already are familiar change when they are precisely defined for geometry.

1 Use Various Notations to Represent Points, Lines, Line Segments, Rays, Planes, and Angles. Geometry is the study of size, shape, position, and other properties of the objects around us. The basic terms used in geometry are *point, line,* and *plane.* Generally, these terms are not defined. Instead, they are only described. Once described, they are used in definitions of other terms and concepts.

A **point** is a location or position that has no size or dimension. A dot is used to represent a point, and a capital letter may be used to label the point (Fig. 17–1).

A **line** extends indefinitely in both directions and contains an infinite number of points. It has length but no width. In our discussions, the word *line* always refers to a straight line unless otherwise specified. In Fig. 17–1, the line can be identified by naming any two points on the line (such as *A* and *B*).

A **plane** is a flat, smooth surface that extends indefinitely in all directions. A plane contains an infinite number of points and lines (Fig. 17–1).

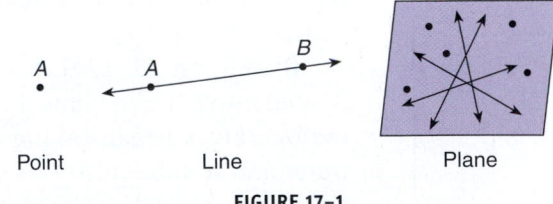

Point   Line   Plane

FIGURE 17–1

Since a line extends indefinitely in both directions, most geometric applications deal with parts of lines. A part of a line is called a **line segment** or **segment.** A line segment starts and stops at distinct points that we call *end points.* A *line segment,* or *segment,* consists of all points on the line between and including the two *end points* (Fig. 17–2).

Line segment or segment: a part of a line; it starts and stops at distinct points called end points

Line AB

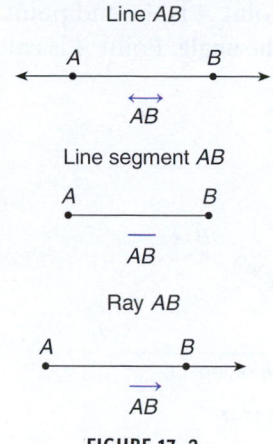

FIGURE 17-2

Ray: a point on a line and all points of the line on one side of the point

Interior points of a ray: all points on the ray except the end point

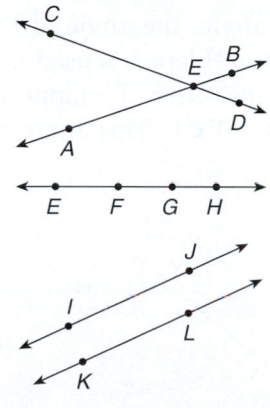

FIGURE 17-4

STOP AND CHECK

Use Fig. 17–5 to answer the following:

1. Name a ray with an end point of M and passing through N.
2. Do $\overrightarrow{PM}$ and $\overrightarrow{PQ}$ represent the same ray?
3. Name a line segment with end points at L and R two different ways.

Answers:

1. $\overrightarrow{MN}$ 2. Yes   3. $\overline{LR}$ and $\overline{RL}$

The notation for a line that extends through points A and B is $\overleftrightarrow{AB}$ (read "line AB"). The notation for the line segment including points A and B and all the points between is $\overline{AB}$ (read "line segment AB").

Another term used in connection with parts of a line is *ray.* Before we give the definition of a ray, consider the beam of light from a flashlight. The beam is like a ray. It seems to continue indefinitely in only one direction. A **ray** consists of a point on a line and all points of the line on one side of the point (Fig. 17–2).

The point from which the ray originates is called the *end point,* and all other points on the ray are called **interior points of the ray.** A ray is named by its end point and any interior point of the ray. In Fig. 17–2, we use the notation $\overrightarrow{AB}$ to denote the ray whose end point is A and that passes through B.

To illustrate appropriate notations for lines, segments, and rays, consider a line with several points designated on the line (Fig. 17–3).

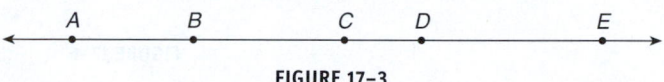

FIGURE 17-3

Any two points can be used to name the line in Fig. 17–3. For example, $\overleftrightarrow{AB}$, $\overleftrightarrow{AC}$, $\overleftrightarrow{CE}$, $\overleftrightarrow{BD}$, and $\overleftrightarrow{BC}$ are some of the possible ways to name the line. However, a segment is named *only* by its end points. Thus, in Fig. 17–3, $\overline{AB}$ is not the same segment as $\overline{AC}$, but $\overleftrightarrow{AB}$ and $\overleftrightarrow{AC}$ represent the same line.

In Fig. 17–3, $\overrightarrow{BC}$ and $\overrightarrow{BD}$ represent the same ray, but $\overline{BC}$ and $\overline{BD}$ do not represent the same segment.

A line extends indefinitely in either direction. If two lines are drawn in the same plane, one of three situations may occur:

1. The two lines *intersect* in *one and only one point.* In Fig. 17–4, $\overleftrightarrow{AB}$ and $\overleftrightarrow{CD}$ intersect at point E.
2. The two lines *coincide;* that is, one line fits exactly on the other. In Fig. 17–4, $\overleftrightarrow{EF}$ and $\overleftrightarrow{GH}$ coincide.
3. The two lines never intersect. In Fig. 17–4, $\overleftrightarrow{IJ}$ and $\overleftrightarrow{KL}$ are the same distance from each other along their entire lengths and so never touch.

The relationship described in the third situation has a special name, *parallel lines.* The symbol $\|$ is used for parallel lines, $\overleftrightarrow{IJ} \parallel \overleftrightarrow{KL}$.

EXAMPLE 1

Use Figure 17–5 to answer the following:

(a) Name a ray with an end point of Q and passing through P.

(b) Do $\overleftrightarrow{LM}$ and $\overleftrightarrow{MN}$ represent the same line?

(c) Do $\overline{PQ}$ and $\overline{QP}$ represent the same line segment?

(d) Name two lines that are parallel.

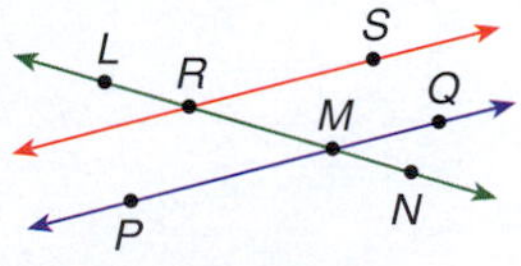

FIGURE 17-5

(a) $\overrightarrow{QP}$ starts at Q and continues indefinitely passing through the point P. It is different from the ray $\overrightarrow{PQ}$.

(b) Yes, both notations show two points on the same line.

(c) Yes, both segments have the same end points. It does not matter which end point is given first.

(d) $\overleftrightarrow{PQ}$ and $\overleftrightarrow{RS}$ are parallel. Line $\overleftrightarrow{PQ}$ could be named using other letter pairs such as $\overleftrightarrow{PM}$ and $\overleftrightarrow{MQ}$.

See Exercises 1–12.

When two lines intersect in a point, four *angles* are formed, as shown in Fig. 17–6. An **angle** is a geometric figure formed by two rays that intersect in a point, and the point of intersection is the end point of each ray.

In Fig. 17–7, rays $\overrightarrow{AB}$ and $\overrightarrow{AC}$ intersect at point A. Point A is the end point of $\overrightarrow{AB}$ and $\overrightarrow{AC}$. $\overrightarrow{AB}$ and $\overrightarrow{AC}$ are called the **sides** or **legs of the angle**. Point A is called the **vertex of the angle.**

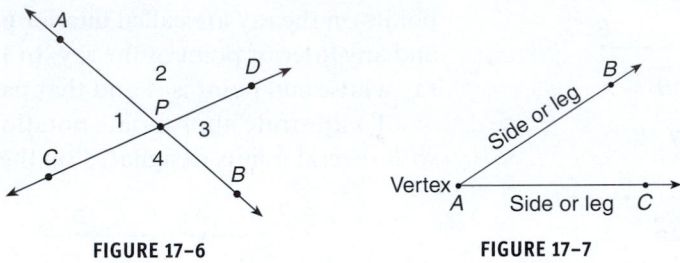

FIGURE 17–6   FIGURE 17–7

Angles can be named in several ways. An angle can be named by using a number or lowercase letter, by using the capital letter that names the vertex point, or by using three capital letters. If three capital letters are used, two of the letters name interior points of each of the two rays, and the middle letter names the vertex point of the angle. Using the symbol $\angle$ for angle, the angle in Fig. 17–8 can be named $\angle 1$, $\angle KLM$, $\angle MLK$, or $\angle L$. One capital letter is used only when it is perfectly clear which angle is designated by the letter. To name the angle in Fig. 17–9 with three letters, we write $\angle XZY$ or $\angle YZX$. This angle can also be named $\angle a$ or $\angle Z$.

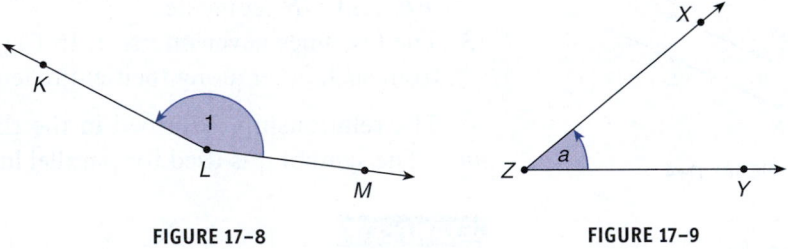

FIGURE 17–8   FIGURE 17–9

Fig. 17–10 illustrates how the intersection of two rays actually forms two angles. In this text, we refer to the smaller of the two angles formed by two rays unless the other angle is specifically indicated. Arcs (curved lines) and arrows are often used to clarify which angle we are considering.

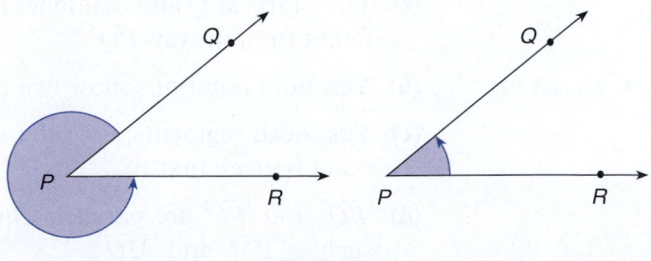

FIGURE 17–10

EXAMPLE 2

Use Fig. 17–11 to answer the following:

(a) Name the angle with one capital letter.

(b) Name the angle with three capital letters.

(c) Name the angle with a number.

(a) $\angle X$ **(b)** $\angle YXZ$ or $\angle ZXY$ **(c)** $\angle 1$

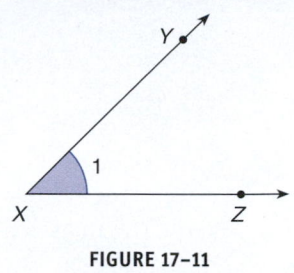

FIGURE 17–11

See Exercises 13–15.

2 Classify Angles According to Size. The measure of an angle is determined by the amount of opening between the two sides of the angle. The length of the sides does not affect the angle measure. Two units are commonly used to measure angles, *degrees* and *radians*. In this section, only degrees are used to measure angles. Radians are discussed in Section 17–3.

Consider the hands of a clock as the sides of an angle. When the two hands both point to the same number, the measure of the angle formed is 0 degrees (0°). An angle of 0 degrees is used in trigonometry but is seldom used in geometric applications. During 1 h, the minute hand makes one complete revolution. Ignoring the movement of the hour hand, this revolution of the minute hand contains 360 degrees (360°). Fig. 17–12 shows a revolution or rotation of 360°. Note that A is kept as a fixed point and B rotates around point A. This rotation can be either clockwise or counterclockwise.

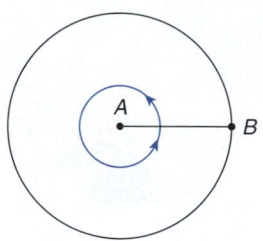

FIGURE 17–12

Counterclockwise: a circular rotation in the opposite direction as the normal rotation of the hands of a clock. This corresponds to the positive rotation of an angle

Clockwise: a circular rotation in the same direction as the normal rotation of the hands of a clock. This corresponds to the negative rotation of an angle

The customary rotation around a point progresses *opposite* to the normal rotation of the hands of a clock. The direction is called **counterclockwise** (Fig. 17–13). **Clockwise** is a circular rotation in the same direction as the normal rotation of the hands of a clock. This corresponds to the negative rotation of an angle.

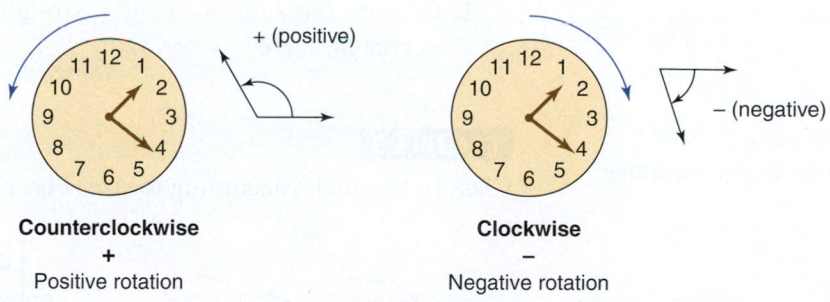

FIGURE 17–13

Degree: a unit for measuring angles. A degree represents 1/360 of a complete rotation about the vertex

Right angle: an angle of 90° that is one-fourth of a complete rotation

A **degree** is a unit for measuring angles. It represents $\frac{1}{360}$ of a complete rotation about the vertex. Suppose, in Fig. 17–14, that $\overrightarrow{AC}$ rotates from $\overrightarrow{AB}$ through one-fourth of a circle. Then $\overrightarrow{AC}$ and $\overrightarrow{AB}$ form a 90° angle ($\frac{1}{4}$ of $360 = 90$). This angle is a **right angle.** The symbol for a right angle is ⌐.

If two lines intersect so that right angles (90° angles) are formed, the lines are *perpendicular* to each other. (See Section 9–4.) The symbol for "perpendicular" is ⊥. However, the right-angle symbol also implies the lines forming the angle are perpendicular.

If a string is suspended at one end and weighted at the other (Fig. 17–15), the line it forms is a *vertical line.* A line that is perpendicular to the vertical line is a *horizontal line.* In Fig. 17–16, $\overleftrightarrow{AB}$ is a vertical line, and $\overleftrightarrow{AB}$ and $\overleftrightarrow{CD}$ form right angles. Thus, $\overleftrightarrow{AB} \perp \overleftrightarrow{CD}$, and $\overleftrightarrow{CD}$ is a horizontal line.

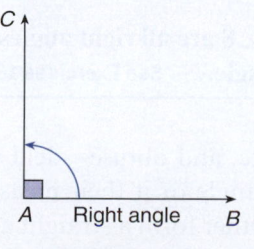

FIGURE 17–14

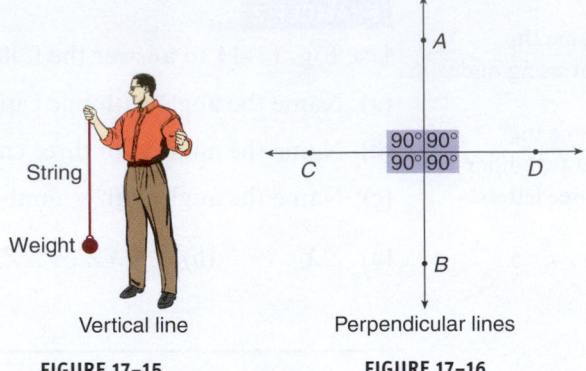

String

Weight

Vertical line

FIGURE 17–15

Perpendicular lines

FIGURE 17–16

Straight angle: a straight angle (180°) represents one-half of a circle or one-half of a complete rotation

Acute angle: an angle that is less than 90° but more than 0°

Obtuse angle: an angle that is more than 90° but less than 180°

Now look at Fig. 17–17. When $\overrightarrow{ML}$ rotates one-half a circle from $\overrightarrow{ML}$, an angle of 180° is formed ($\frac{1}{2}$ of 360 = 180). This angle is a **straight angle.**

An angle that is less than 90° but more than 0° is an **acute angle.** An angle that is more than 90° but less than 180° is an **obtuse angle** (see Fig. 17–18).

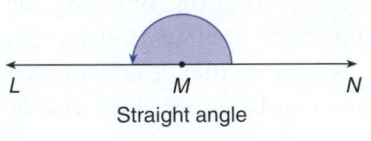

Straight angle

FIGURE 17–17

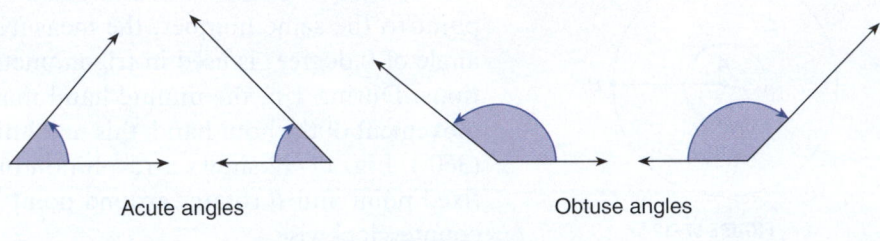

Acute angles

Obtuse angles

FIGURE 17–18

> **To classify angles according to size:**
>
> 1. Examine the angle to determine the number of degrees it measures.
> 2. Classify the angle as a right, straight, acute, or obtuse angle based on its degree measure.

EXAMPLE 3

Classify the angles according to size (Fig. 17–19).

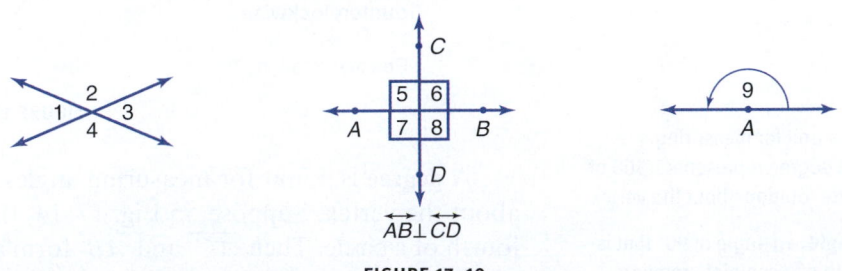

$AB \perp CD$

FIGURE 17–19

∠**1 is acute.**   ∠**3 is acute.**   ∠**5,** ∠**6,** ∠**7, and** ∠**8 are all right angles.**
∠**2 is obtuse.**   ∠**4 is obtuse.**   ∠**9 is a straight angle.**   **See Exercises 16–25.**

Complementary angles: two angles that together form a right angle or their measures total 90°

Supplementary angles: two angles that together form a straight angle or their measures total 180°

The angle classifications used so far—right, straight, acute, and obtuse—deal with one angle at a time. If two angles together form a right angle or if their measures total 90°, they are **complementary angles.** If two angles together form a straight angle or if their measures total 180°, they are **supplementary angles** (Fig. 17–20).

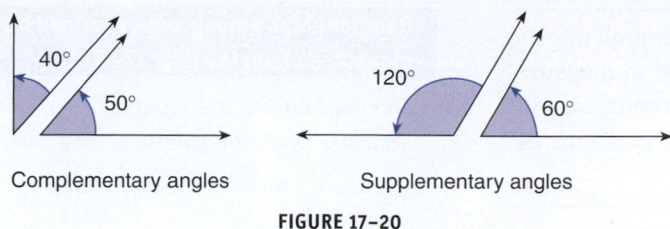

Complementary angles   Supplementary angles

FIGURE 17–20

> **To find the complement or supplement of an angle:**
>
> 1. To find the complement of an angle, subtract its measure from 90°.
>
> 2. To find the supplement of an angle, subtract its measure from 180°.

STOP AND CHECK

1. Find the complement of an angle that measures 48°.
2. Find the supplement of an angle that measures 48°.

Answers:

1. 42°   2. 132°

EXAMPLE 4

Find the complement and supplement of an angle that measures 57°.

Complement

$90° - 57° = 33°$

The complement of 57° is 33°.

Supplement

$180° - 57° = 123°$

The supplement of 57° is 123°.

See Exercises 26–32.

When the word *equal* is used to describe the relationship between two angles, it implies the measures of the angles are equal. Another word often used in geometry is *congruent*. When geometric figures are **congruent,** one figure can be placed on top of the other, and the two figures will match perfectly. If two angles are congruent, the measures of the angles are equal. Also, if the measures of two angles are equal, the angles are congruent. The symbol for congruence is $\cong$. For instance, in Fig. 17–21, the two angles have equal measures, both 45°. Because they have the same measures, they are congruent; that is, $\angle BAC \cong \angle FED$.

Congruent: when geometric figures are congruent, they have the same shape and size; if one figure is placed on top of the other, the two figures will match perfectly

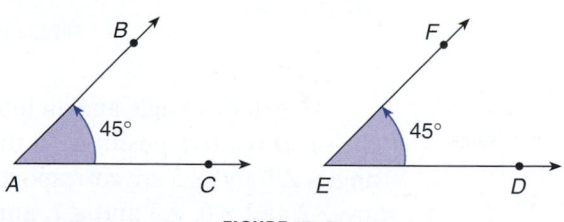

FIGURE 17–21

Traditionally, we indicate the measures of angles by placing the letter m before the angle symbol. Thus, $m\angle BAC = m\angle FED$ tells us that the measures of angles BAC and FED are equal. We interpret $\angle BAC = \angle FED$ and $m\angle BAC = m\angle FED$ as giving us the same information.

Vertical angles: two pairs of angles formed by two intersecting lines. The angles share a common vertex but not a common side

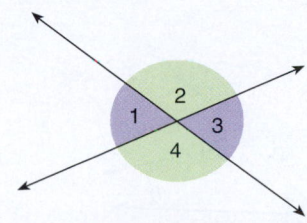

FIGURE 17–22

Adjacent angles: two angles that share a common side

3 **Determine the Measure of an Angle Using Relationships Among Intersecting Lines.** Intersecting lines generate several angle relationships. When two lines intersect, two pairs of **vertical angles** are formed. In Fig. 17–22, angles 1 and 3 are vertical angles. Also, angles 2 and 4 are vertical angles. Angles 1 and 2 share a common side and are **adjacent angles.** Other adjacent angles are 2 and 3, 3 and 4, and 4 and 1.

Relationships among angles formed by intersecting lines can be used to find missing measures of angles (Fig. 17–22).

Vertical angles are equal in measure and congruent.	$\angle 1 = \angle 3$	$\angle 2 = \angle 4$
Adjacent angles are supplementary.	$\angle 1 + \angle 2 = 180°$	$\angle 2 + \angle 3 = 180°$
	$\angle 3 + \angle 4 = 180°$	$\angle 1 + \angle 4 = 180°$

Transversal: a line that intersects two or more lines in different points

A **transversal** is a line that intersects two or more lines in different points. When a transversal intersects (cuts) two lines, eight angles are formed (Fig. 17–23).

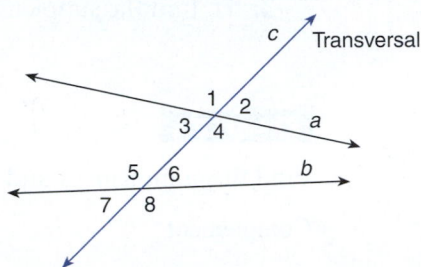

FIGURE 17–23

Interior angles: four angles between two lines cut by a transversal

Exterior angles: four angles outside the two lines cut by a transversal

The four blue-shaded angles *between* lines *a* and *b* ($\angle 3$, $\angle 4$, $\angle 5$, and $\angle 6$) are called **interior angles** (Fig. 17–24). The four green-shaded angles *outside* lines *a* and *b* ($\angle 1$, $\angle 2$, $\angle 7$, and $\angle 8$) are called **exterior angles** (Fig. 17–24).

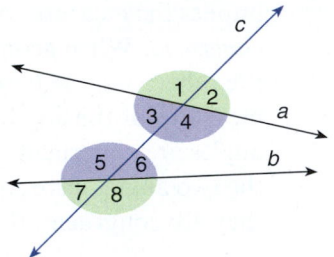

FIGURE 17–24 Interior and exterior angles.

Corresponding angles: an exterior angle and an interior angle on same side of the transversal and in the same relative position on the intersecting lines

Alternate angles: angles on opposite sides of the transversal

Alternate exterior angles: exterior angles that are on opposite sides of the transversal but are not adjacent angles

Alternate interior angles: interior angles that are on opposite sides of the transversal but are not adjacent angles

An exterior angle and an interior angle on the same side of the transversal and in the same relative position on the intersecting lines are **corresponding angles.** Shaded angles $\angle 1$ and $\angle 5$ are corresponding angles (Fig. 17–25). Other corresponding angles are $\angle 2$ and $\angle 6$, $\angle 3$ and $\angle 7$, and $\angle 4$ and $\angle 8$. **Alternate angles** are on opposite sides of the transversal. In Fig. 17–26, $\angle 1$ and $\angle 8$ are **alternate exterior angles** and $\angle 3$ and $\angle 6$ are **alternate interior angles.** There are other pairs of alternate angles.

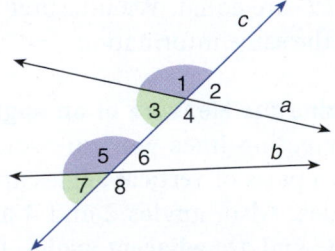

FIGURE 17–25 Corresponding angles.

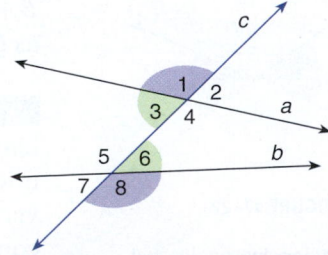

FIGURE 17–26 Alternate angles.

Relationships among angles on parallel lines cut by a transversal (Fig. 17–27) can be used to find missing measures of angles.

Corresponding angles are equal.

$\angle 1 = \angle 5$ $\angle 2 = \angle 6$

$\angle 3 = \angle 7$ $\angle 4 = \angle 8$

Adjacent angles are supplementary.

$\angle 1 + \angle 2 = 180°$ $\angle 2 + \angle 4 = 180°$

$\angle 3 + \angle 4 = 180°$ $\angle 1 + \angle 3 = 180°$

Alternate angles are equal.

$\angle 1 = \angle 8$ $\angle 2 = \angle 7$ (alternate exterior angles)

$\angle 3 = \angle 6$ $\angle 4 = \angle 5$ (alternate interior angles)

Interior angles on the same side of the transversal are supplementary.

$\angle 3 + \angle 5 = 180°$ $\angle 4 + \angle 6 = 180°$

Exterior angles on the same side of the transversal are supplementary.

$\angle 1 + \angle 7 = 180°$ $\angle 2 + \angle 8 = 180°$

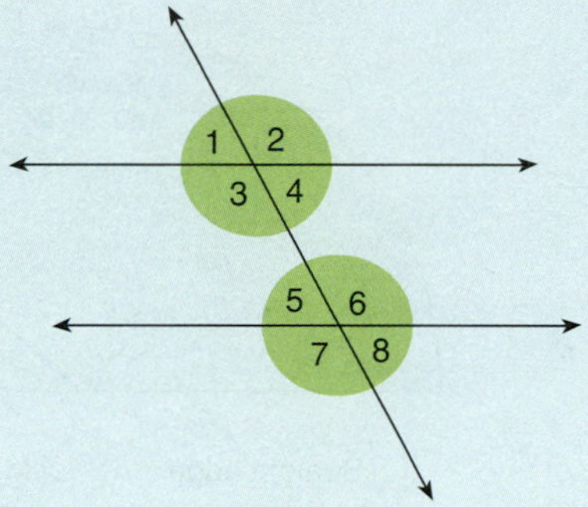

FIGURE 17–27

STOP AND CHECK

1. In Fig. 17–27, if $\angle 5$ is 38°, find the measures of the other angles.
2. In Fig. 17–27, if $\angle 3$ is $x°$, write an expression representing the other angles.

Answers:

1. $\angle 1 = 38°$; $\angle 2 = 142°$; $\angle 3 = 142°$; $\angle 4 = 38°$; $\angle 6 = 142°$; $\angle 7 = 142°$; $\angle 8 = 38°$
2. $\angle 1 = (180 - x)°$; $\angle 2 = x°$; $\angle 4 = (180 - x)°$; $\angle 5 = (180 - x)°$; $\angle 6 = x°$; $\angle 7 = x°$; $\angle 8 = (180 - x)°$

EXAMPLE 5

(a) In Fig. 17–27, $\angle 2$ is 125°. Find the measures of the other angles.

$\angle 1 = 180° - \angle 2$	$\angle 1$ and $\angle 2$ are adjacent angles
$\angle 1 = 180° - 125°$	and are supplementary.
$\angle 1 = 55°$	
$\angle 3 = \angle 2$	Vertical angles are equal.
$\angle 3 = 125°$	
$\angle 4 = \angle 1$	Vertical angles are equal.
$\angle 4 = 55°$	
$\angle 5 = \angle 1$	Corresponding angles are equal.
$\angle 5 = 55°$	
$\angle 6 = \angle 2$	Corresponding angles are equal.
$\angle 6 = 125°$	
$\angle 7 = \angle 3$	Corresponding angles are equal.
$\angle 7 = 125°$	
$\angle 8 = \angle 4$	Corresponding angles are equal.
$\angle 8 = 55°$	

(b) In Fig. 17–27, if $\angle 1$ is $x°$, write an expression representing the other angles.

$\angle 2 = (180 - x)°$	Adjacent angle to $\angle 1$
$\angle 3 = (180 - x)°$	Adjacent angle to $\angle 1$ or vertical angel to $\angle 2$
$\angle 4 = x°$	Vertical angle to $\angle 1$
$\angle 5 = x°$	Corresponding angle to $\angle 1$

$$\angle 6 = (180 - x)°$$ Corresponding angle to $\angle 2$
$$\angle 7 = (180 - x)°$$ Exterior angle on same side of transversal as $\angle 1$
$$\angle 8 = x°$$ Alternate exterior angle to $\angle 1$

Other justifications could be made to come to the same conclusions.

See Exercises 33–40.

---

4 | **Convert Angle Measures Between Decimal Degrees and Degrees, Minutes, and Seconds.** A device used to measure angles is called a **protractor.** The most common protractor is a semicircle with two scales from 0° to 180°. An **index mark on a protractor,** the mark to align with the vertex and one side of the angle, is in the middle of the straight edge of the protractor (Fig. 17–28).

Protractor: a device used to measure angles

Index mark on a protractor: the mark to align with the vertex and one side of the angle; it is in the middle of the straight edge of the protractor

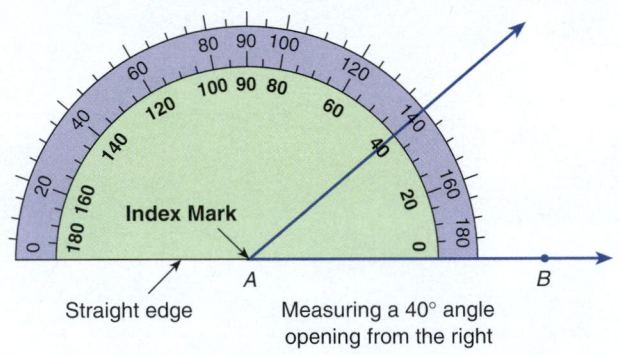

Measuring a 40° angle opening from the right

FIGURE 17–28

To measure an angle that opens from the right, read the degree measure from the lower scale. Notice in Fig. 17–28 the lower scale starts with 0 at the right. The index mark of the protractor is aligned with the vertex of the angle, and the straight edge lies along the lower side of the angle.

To measure an angle that opens from the left, read the degree measure from the upper scale. In Fig. 17–29, the upper scale starts with 0 at the left. The index mark of the protractor is aligned with the vertex of the angle, and the straight edge lies along the lower side of the angle.

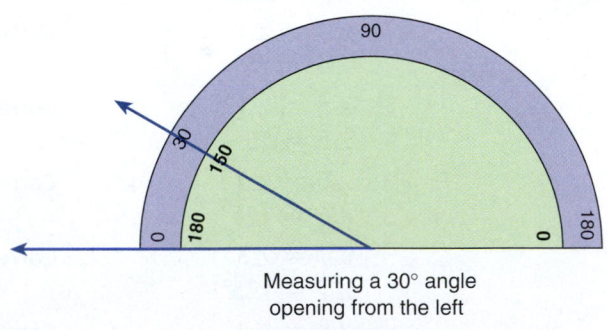

Measuring a 30° angle opening from the left

FIGURE 17–29

Minute of a degree: 1/60 of a degree; 60 minutes = 1 degree

Degrees are divided into 60 equal parts. Each part is called a **minute.**

$$1 \text{ degree } (1°) = 60 \text{ minutes } (60')$$

The symbol ' is used for minutes. Similarly, 1 minute is divided into 60 equal parts called **seconds.**

Second of a minute: 1/60 of a minute; 60 seconds = 1 minute; 1/3,600 of a degree

$$1 \text{ minute } (1') = 60 \text{ seconds } (60'')$$

The symbol ″ is used for seconds. Thus,

$$1° = 60' = 3,600″$$

With the increased popularity of the calculator, it is sometimes desirable to change minutes or seconds to decimal equivalents. Minutes or seconds can be changed to a fractional part of a degree. Then the fraction is changed to its decimal equivalent by dividing the numerator by the denominator. Remember, $1' = \frac{1}{60}$ of a degree, and $1″ = \frac{1}{3,600}$ of a degree.

Scientific and graphing calculators have a key or menu choice that automatically converts between degrees, minutes, and seconds and decimal degrees. A key may be labeled $\boxed{°\,'\,″}$ or $\boxed{\text{DMS}}$, or a menu choice on the angle menu may be DMS. This function key is also called a **sexagesimal function key** because of its relationship with the number 60. The same key is also used for hours, minutes, and seconds.

Sexagesimal function key: a calculator function key for converting from decimal-degree format to a degrees, minutes, and seconds format and vice versa

---

To convert from degrees, minutes, and seconds to a decimal-degree equivalent:

1. Find the sum of the degrees, the minutes divided by 60, and the seconds divided by 3,600.

$$\text{degrees} + \frac{\text{minutes}}{60} + \frac{\text{seconds}}{3,600}$$

2. Round to the desired decimal place.

STOP AND CHECK

1. Change 65°20′15″ to a decimal-degree equivalent. Round to the nearest thousandth.

Answer:
1. 65.338°

EXAMPLE 6

Change 115°40′15″ to a decimal-degree equivalent. Round to the nearest thousandth.

$115 + 40/60 + 15/3,600 = 115.6708333$ Perform the series of calculations.
$\qquad\qquad\qquad 115°40′15″ = 115.671°$ Round.   **See Exercises 41–45.**

To convert from a decimal-degree format to degrees, minutes, and seconds:

1. The whole-number part of the decimal-degree format will be the degrees.

2. Multiply the decimal part of the degree in decimal-degree format by 60 and the whole-number part of the product will be the minutes.

3. Multiply the decimal part of the product from Step 2 by 60 to obtain the seconds.

4. Round the seconds to the desired decimal place.

STOP AND CHECK

1. Change 54.482° to degrees, minutes, and seconds to the nearest second.

Answer:
1. 54°28′55″

EXAMPLE 7

Change 25.375° to degrees, minutes, and seconds.

$0.375(60) = 22.5$ Multiply decimal portion of 25.375 by 60.
$\quad 0.5(60) = 30$ Multiply decimal portion of 22.5 by 60.

25°22′30″

Using a calculator with a sexagesimal function key:

$25.375\ \boxed{2^{nd}}\ \boxed{\text{ANGLE}}\ \boxed{4:\blacktriangleright\text{DMS}}\ \boxed{\text{ENTER}}$ Using a TI-83 or TI-84.
$25.375° = \mathbf{25°22′30″}$ **See Exercises 46–50.**

TIP **To Change the Format of Angle Measures Using a Calculator:** From degrees in decimal notation to degrees, minutes, and seconds (calculator steps will vary with different calculators):

1. Set the calculator to degree mode.

2. Enter the angle measure in decimal notation.

3. Access the sexagesimal function key or menu choice on the calculator (most often labeled $\boxed{°\,'\,''}$ or $\boxed{\text{DMS}}$).

4. Press $\boxed{\text{ENTER}}$ or $\boxed{=}$.

From degrees, minutes, and seconds to decimal notation (calculator steps will vary with different calculators):

Option 1

1. Enter the series of calculations—degrees + minutes ÷ 60 + seconds ÷ 3,600.

Option 2

1. Enter the degrees followed by the degree symbol ($\boxed{\text{2}^{\text{nd}}}$, $\boxed{\text{ANGLE}}$, $\boxed{\text{1:°}}$).

2. Enter the minutes followed by the minute symbol ($\boxed{\text{2}^{\text{nd}}}$, $\boxed{\text{ANGLE}}$, $\boxed{\text{2:}'}$).

3. Enter the seconds followed by the second symbol ($\boxed{\text{ALPHA}}$, $\boxed{''}$).

4. Press $\boxed{\text{ENTER}}$ or $\boxed{=}$.

17–1 EXERCISES   MyLab Math For additional practice go to your study plan in MyLab Math.

1 Use Fig. 17–30 for Exercises 1–3. *See Example 1.*

1. Name the line in three different ways.

2. Name in two different ways the ray with end point P and with interior points Q and R.

3. Name the segment with end points Q and R.

Use Fig. 17–31 for Exercises 4–10.

4. Does $\overleftrightarrow{XY}$ represent the same line as $\overleftrightarrow{YZ}$?

5. Does $\overrightarrow{XY}$ represent the same ray as $\overrightarrow{YZ}$?

6. Does $\overline{XY}$ represent the same segment as $\overline{YZ}$?

7. Is $\overrightarrow{WX}$ the same as $\overleftarrow{WY}$?

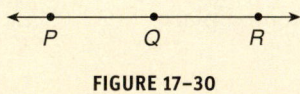

FIGURE 17–30

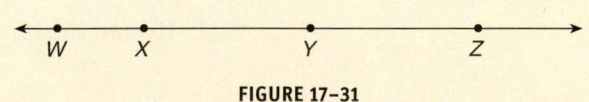

FIGURE 17–31

Use Fig. 17–32 for Exercises 8–12. *See Example 1.*

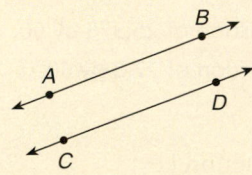
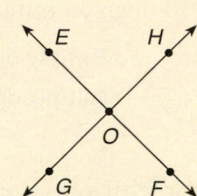
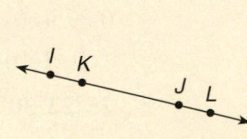

FIGURE 17–32

8. $\overleftrightarrow{AB}$ and $\overleftrightarrow{CD}$ are _____ lines.

9. $\overleftrightarrow{EF}$ and $\overleftrightarrow{GH}$ _____ at point O.

10. $\overleftrightarrow{IJ}$ and $\overleftrightarrow{KL}$ _____.

11. $\overleftrightarrow{AB}$ and $\overleftrightarrow{CD}$ will never _____.

12. Name two lines that intersect in exactly one point.

Use Fig. 17–33 for Exercises 13–15. *See Example 2.*

13. Name the angle in two different ways using three capital letters.

14. Name the angle using one capital letter.

15. Name the angle using a number.

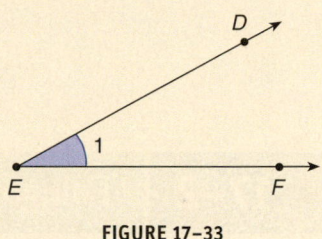

FIGURE 17–33

2   Fill in the blanks regarding angle rotation. *See Example 3.*

16. One complete rotation is _____°.

17. One-half a complete rotation is _____°.

18. One-fourth a complete rotation is _____°.

19. One-eighth a complete rotation is _____°.

Classify the angle measures using the terms *right, straight, acute,* and *obtuse. See Example 3.*

20. 38°   **21.** 95°   **22.** 90°   **23.** 153°   **24.** 10°   **25.** 180°

Tell whether the angle pairs are complementary, supplementary, or neither. *See Example 4.*

26. 42°, 80°   **27.** 17°, 73°   **28.** 38°, 142°   **29.** 52°, 48°   **30.** 60°, 30°   **31.** 110°, 70°   **32.** 42°, 138°

3   Use Fig. 17–34 for Exercises 33–35. *See Example 5.*

33. Name two pairs of equal angles.

34. What are angles a and b called?

35. What is the sum of $\angle c$ and $\angle d$?

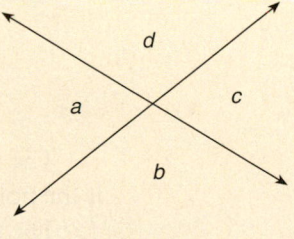

FIGURE 17–34

In Fig. 17–35, $l \parallel m$ and n is a transversal. *See Example 5.*

36. Name two pairs of alternate interior angles.

37. What is the sum of $\angle a$ and $\angle g$?

38. What name is given to $\angle d$ and $\angle h$?

39. If $\angle a = 150°$ in Fig. 17–35, find the measures of the other angles.

40. If $\angle d = y°$ in Fig. 17–35, write an expression to represent the measures of $\angle e$, $\angle f$, $\angle g$, and $\angle h$.

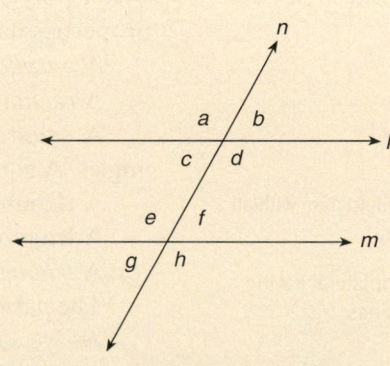

FIGURE 17–35

4 Change to decimal-degree equivalents. Round to the nearest ten-thousandth if necessary. *See Example 6.*

41. 47′   **42.** 36″   **43.** 5′14″   **44.** 10′15″   **45.** 10°18′15″

Change to equivalent minutes and seconds. Round to the nearest second when necessary. *See Example 7.*

46. 0.35°   **47.** 0.20°   **48.** 0.12°   **49.** 0.213°   **50.** 0.3149°

17–2   Polygons

LEARNING OUTCOMES

1 Find the perimeter of a polygon using the appropriate formula.

2 Find the area of a polygon using the appropriate formula.

LC LEARNING CATALYTICS

1. Solve the formula $A = \frac{1}{2}bh$ for h.

1 Find the Perimeter of a Polygon Using the Appropriate Formula. In Chapter 1, Section 3, Outcome 2, we introduced three polygons: the parallelogram, rectangle, and square. To review, a *polygon* is a plane or flat, closed figure described by straight-line segments and angles. Polygons have different numbers of sides and different properties. Some common polygons are the parallelogram, rectangle, square, rhombus, trapezoid, and triangle (Fig. 17–36).

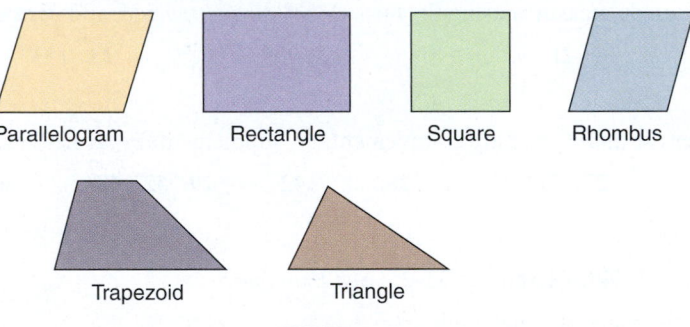

Parallelogram   Rectangle   Square   Rhombus

Trapezoid   Triangle

FIGURE 17–36

The *base* of any polygon is the horizontal side or a side that would be horizontal if the polygon's orientation were modified.

The *adjacent side* of any polygon is the side that has an end point in common with the base.

Quadrilateral: a polygon that has four sides

A polygon that has four sides is a **quadrilateral.** Quadrilaterals with specific properties have other names.

A *parallelogram* is a quadrilateral with opposite sides that are parallel.

A *rectangle* is a parallelogram with angles that are all right angles.

A *square* is a parallelogram with sides all of equal length and with all right angles. A square can also be described as a rectangle with all sides of equal length.

Rhombus: a parallelogram with all sides of equal length

A **rhombus** is a parallelogram with all sides of equal length.

A **trapezoid** is a quadrilateral having only two parallel sides.

Trapezoid: a quadrilateral having only two parallel sides

A *triangle* is a polygon that has three sides.

The *perimeter* is the total length of the sides of a plane figure.

As we saw in Chapter 1, formulas provide a shortcut or guide for the calculations required to find the perimeter of a polygon. We will review the perimeter formulas for a parallelogram, a rectangle, and a square and introduce new formulas for the perimeter of a rhombus, a trapezoid, and a triangle.

Perimeter

Parallelogram	$P_{\text{parallelogram}} = 2b + 2s$ or $P_{\text{parallelogram}} = 2(b + s)$		b is the base s is the adjacent side
Rectangle	$P_{\text{rectangle}} = 2l + 2w$ or $P_{\text{rectangle}} = 2(l + w)$		l is length w is width
Square	$P_{\text{square}} = 4s$		s is length of each side
Rhombus	$P_{\text{rhombus}} = 4s$		s is length of each side
Trapezoid	$P_{\text{trapezoid}} = b_1 + b_2 + a + c$		b_1 and b_2 are bases, a and c are the adjacent sides
Triangle	$P_{\text{triangle}} = a + b + c$		a, b, and c are sides

To find the perimeter of a polygon:

1. Select the appropriate formula.
2. Substitute the known values into the formula.
3. Evaluate the formula (perform the indicated operations).

A *trapezoid* is a four-sided polygon having only two parallel sides. Unlike a parallelogram, the trapezoid's four sides may all be of unequal size, as illustrated in Fig. 17–37.

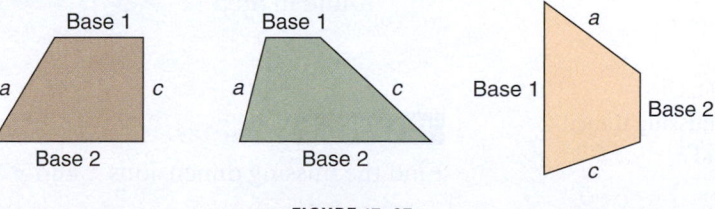

FIGURE 17–37

Bases of a trapezoid: the two parallel sides of a trapezoid

The *parallel* sides are called **bases of a trapezoid** (b_1 and b_2) with the subscripts 1 and 2 used to distinguish them. The nonparallel sides are designated side a and side c.

STOP AND CHECK

1. Find the perimeter of a trapezoid with bases of 12 in. and 15 in. and adjacent sides of 4 in. and 5 in.

Answer:

1. 36 in.

EXAMPLE 1

AUTO The rear window of a pickup truck is a trapezoid whose shorter base is 40 in., whose longer base is 48 in., and whose nonparallel sides are each 16 in. How many inches of rubber gasket are needed to surround the window?

$P_{\text{trapezoid}} = b_1 + b_2 + a + c$ Select appropriate formula.

Estimation

By rounding, the perimeter is $40 + 50 + 20 + 20$ or 130 in.

$P_{\text{trapezoid}} = 40 + 48 + 16 + 16$ Substitute in formula.

$P_{\text{trapezoid}} = 120$ in.   Add.

Interpretation

The rubber gasket to surround the rear window must be 120 in. long.   **See Exercises 1–6.**

A *triangle* is a three-sided polygon. The perimeter of a triangle is the sum of the lengths of the three sides. The triangle, like the trapezoid, may have unequal lengths for all three sides.

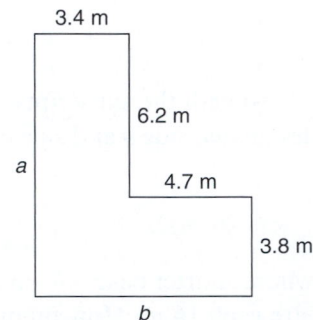

18 ft   18 ft

32 ft

FIGURE 17–38

EXAMPLE 2

CON   The triangular gables of an apartment unit under construction are outlined in contrasting trim. If each gable is 32 ft wide and 18 ft on each side, how many feet of contrasting wood trim are needed for each gable? Disregard overlap at the corners (Fig. 17–38).

$P_{\text{triangle}} = a + b + c$ Select the appropriate formula and substitute in formula.

$P_{\text{triangle}} = 18 + 32 + 18$ Evaluate.

$P_{\text{triangle}} = 68$ ft

Each gable will require 68 linear feet of trim.   **See Exercises 7–28.**

Composite figure: a figure made up of two or more geometric figures

Shapes called **composite figures** are figures made up of two or more geometric figures.

> **To find a missing dimension of a composite shape:**
>
> 1. Determine how the missing dimension is related to known dimensions.
>
> 2. Make a calculation of known dimensions according to the relationship found in Step 1.

STOP AND CHECK

1. Find the missing dimensions a and b.

3.4 m

6.2 m

a

4.7 m

3.8 m

b

EXAMPLE 3

Find the missing dimensions x and y on the slab foundation (Fig. 17–39).

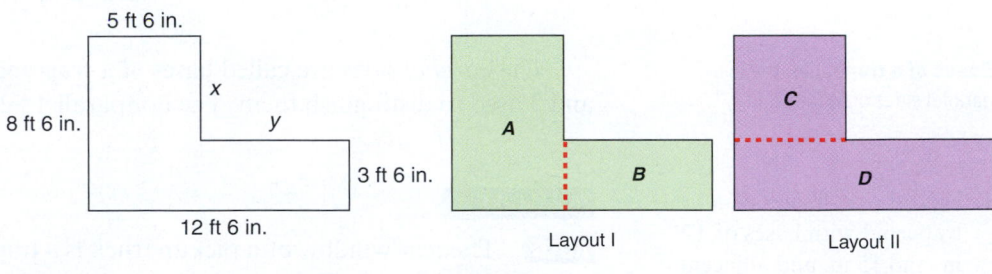

5 ft 6 in.

x

8 ft 6 in. y

3 ft 6 in.

12 ft 6 in.

A

B

Layout I

C

D

Layout II

FIGURE 17–39

Answer:

1. $a = 10$ m   **2.** $b = 8.1$ m

Separate the figure into parts.

In layout I (Fig. 17–39) the side of B opposite its 3′6″ side is also 3′6″ because opposite sides of a rectangle are equal. The side of A opposite its 8′6″ side is, for the same reason, 8′6″. Dimension x must therefore be the difference between 8′6″ and 3′6″.

$$x = 8'6'' - 3'6''$$
$$x = 5'$$

If we think of layout II as two horizontal rectangles, we can find dimension y. The side opposite the 5′6″ side of C must be 5′6″. The side of D opposite the 12′6″ side must also be 12′6″. Dimension y must therefore be the difference between 12′6″ and 5′6″.

$$y = 12'6'' - 5'6''$$
$$y = 7'$$

The missing dimensions are $x = 5'$ and $y = 7'$. See Exercise 29.

The *perimeter* of a composite figure is the sum of the lengths of the sides of a figure. The number and the length of the sides vary from one composite figure to the next, so no specific formula covers the entire variety of composite shapes that exist. However, we can use a general formula.

Perimeter of a composite figure:

$$P = a + b + c + \cdots$$

where $a, b, c, \ldots$ are lengths of all the sides.

STOP AND CHECK

1. Find the perimeter of the composite figure in the Stop and Check for Example 3 on p. 720.

Answer:
1. 36.2 m

EXAMPLE 4

Find the number of feet of 4-in. stock needed for the base plates of a room that has the layout shown in Fig. 17–40. Make no allowances for openings when calculating the linear footage of the base plates.

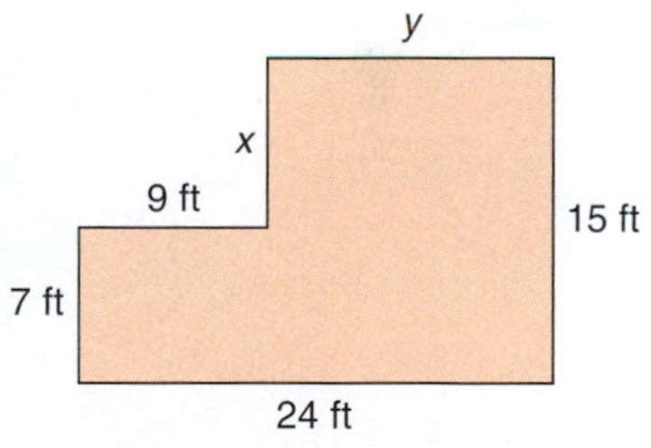

FIGURE 17–40

1. Find the missing dimensions.

$x = 15\text{ ft} - 7\text{ ft}$ $\quad$ $y = 24\text{ ft} - 9\text{ ft}$

$x = 8\text{ ft}$ $\quad$ $y = 15\text{ ft}$

2. Apply the general formula for the perimeter of a polygon.

$$P = a + b + c + \cdots$$
$$P = 7\text{ ft} + 24\text{ ft} + 15\text{ ft} + 15\text{ ft} + 8\text{ ft} + 9\text{ ft}$$
$$P = 78\text{ ft}$$

3. Count the number of sides on the layout to make sure that each is substituted into the formula.

The room needs 78 ft of 4-in. stock for the base plates. See Exercise 30.

2 Find the Area of a Polygon Using the Appropriate Formula. In Chapter 1, Section 3, Outcome 2, the concept of the area of a polygon was introduced and formulas were given for the area of a parallelogram, a rectangle, and a square. To review, the area of a polygon is the amount of surface of a plane figure. Area is expressed in square units. We will add formulas for the area of the rhombus, trapezoid, and triangle to our list of area formulas.

Area

Parallelogram	$A_{\text{parallelogram}} = bh$	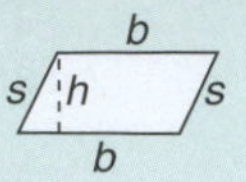	b is base h is height
Rectangle	$A_{\text{rectangle}} = lw$	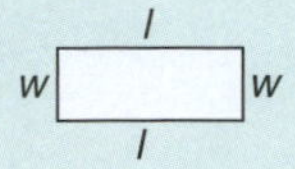	l is length w is width
Square	$A_{\text{square}} = s^2$	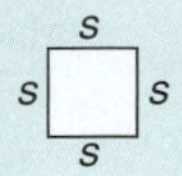	s is length of a side
Rhombus	$A_{\text{rhombus}} = bh$	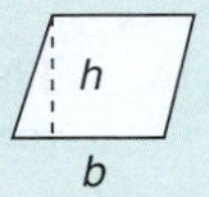	b is base h is height
Trapezoid	$A_{\text{trapezoid}} = \dfrac{1}{2}h(b_1 + b_2)$	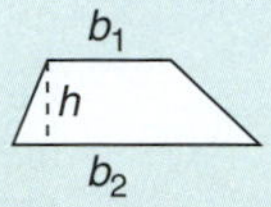	b_1 and b_2 are bases h is height
Triangle	$A_{\text{triangle}} = \dfrac{1}{2}bh$	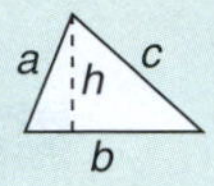	b is base h is height
	$A_{\text{triangle}} = \sqrt{s(s-a)(s-b)(s-c)}$ (Heron's formula)		$s = \dfrac{1}{2}(a+b+c)$ a, b, and c are sides

To find the area of a polygon:

1. Visualize the polygon.

2. Select the appropriate formula.

3. Substitute the known values into the formula.

4. Evaluate the formula (perform the indicated operations).

1. Find the area of a trapezoid with bases of 22 ft and 37 ft and a height of 12 ft.

Answer:
1. 354 ft^2

EXAMPLE 5

Find the area of the trapezoid in Fig. 17–41.

$A_{\text{trapezoid}} = \dfrac{1}{2}h(b_1 + b_2)$ Select the appropriate formula and substitute the known values.

$A_{\text{trapezoid}} = \dfrac{1}{2}(4)(9 + 13)$ Add inside grouping.

$A_{\text{trapezoid}} = \dfrac{1}{2}(4)(22)$ Multiply.

$A_{\textbf{trapezoid}} = \textbf{44 cm}^2$ cm × cm = cm^2

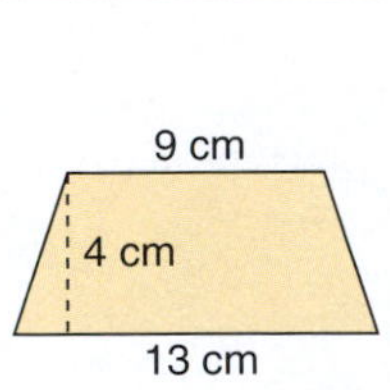

9 cm

4 cm

13 cm

FIGURE 17–41

See Exercises 31–48.

1. Find the area of a triangle with a base of 32 cm and a height of 25 cm.

Answer:
1. 400 cm^2

EXAMPLE 6

Find the area of the triangle in Fig. 17–42.

$A_{\text{triangle}} = \dfrac{1}{2}bh$ Select the appropriate formula and substitute the known values.

$A_{\text{triangle}} = \dfrac{1}{2}(8)(6)$ Multiply.

$A_{\textbf{triangle}} = \textbf{24 in}^2$ in. × in. = in^2

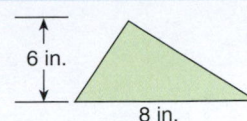

FIGURE 17–42

See Exercises 49–50.

In some triangles, the height is measured along an imaginary line outside the triangle from the base to the highest point of the triangle (the *apex*) (Fig. 17–43).

EXAMPLE 7

The base of ΔDEF in Fig. 17–43 is 15 cm and the height is 21 cm. Find the area.

$A_{\text{triangle}} = \dfrac{1}{2}bh$ Substitute into the formula.

$A_{\text{triangle}} = \dfrac{1}{2}(15)(21)$ Multiply.

$A_{\text{triangle}} = 157.5 \text{ cm}^2$

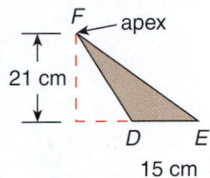

The area of ΔDEF is 157.5 cm^2.

FIGURE 17–43

See Exercises 51–52.

When the three sides of a triangle are known and the height is not known, the area of a triangle can be found using *Heron's formula* (p. 722).

1. Find the area of a triangle to the nearest hundredth by using Heron's formula if the sides are 15 in., 21 in., and 28 in.

Answer:
1. 154.71 in^2

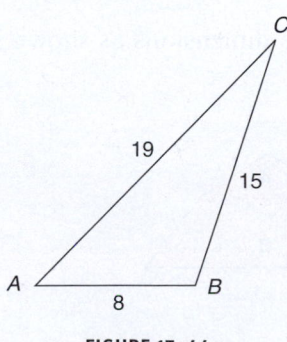

FIGURE 17–44

EXAMPLE 8

Find the area of the triangle in Fig. 17–44 to the nearest hundredth by using Heron's formula.

$s = \dfrac{1}{2}(a + b + c)$ Find s. Substitute length of sides.

$s = \dfrac{1}{2}(19 + 15 + 8)$ Add, then multiply.

$s = 21$

Area $= \sqrt{s(s - a)(s - b)(s - c)}$ Heron's formula

Area $= \sqrt{21(21 - 19)(21 - 15)(21 - 8)}$ Substitute for s, a, b, and c, and perform each subtraction.

Area $= \sqrt{21(2)(6)(13)}$ Multiply.

Area $= \sqrt{3{,}276}$ Take the square root using a calculator.

Area $= 57.23635209$ Principal square root

Area $= $ 57.24 square units   Round to the nearest hundredth.

See Exercises 53–54.

As with the perimeter, the *area* of a composite figure is found by finding the sum of the areas of the parts of the figure.

> **Area of a composite figure:**
>
> $$A = A_1 + A_2 + A_3 + \cdots$$
>
> where A_1, A_2, A_3, ... are the areas of all the parts of the composite figure.

EXAMPLE 9

Find the number of square yards of carpeting required for the room in Fig. 17–45.

1. Divide the composite shape into two polygons with areas we can compute (Fig. 17–46). In this case, A is a rectangle and B is a square.

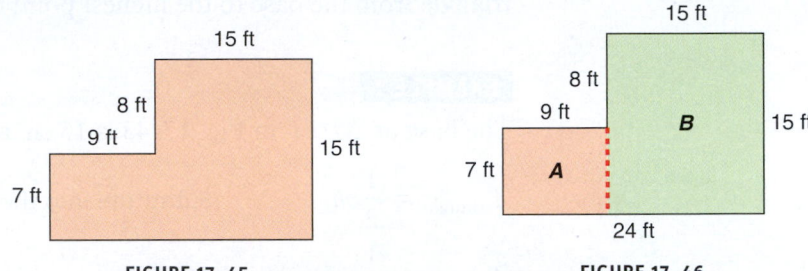

FIGURE 17–45   FIGURE 17–46

2. Find the areas of the smaller polygons and add them.

Rectangle A	Square B
$A_1 = lw$	$A_2 = s^2$
$A_1 = 9 \times 7$	$A_2 = 15^2$
$A_1 = 63 \text{ ft}^2$	$A_2 = 225 \text{ ft}^2$

$$A_1 + A_2 = \text{total area}$$
$$63 + 225 = 288 \text{ ft}^2$$

3. Convert square feet to square yards using a unit ratio.

$$\frac{\overset{32}{\cancel{288}} \, \cancel{\text{ft}^2}}{1} \times \frac{1 \text{ yd}^2}{\underset{1}{\cancel{9}} \, \cancel{\text{ft}^2}} = 32 \text{ yd}^2$$

The room requires 32 yd^2 of carpeting.   **See Exercises 55–56.**

EXAMPLE 10

Find the area of a gable end of a gambrel roof that has dimensions as shown in Fig. 17–47.

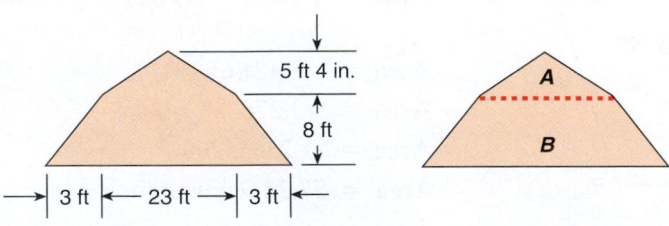

FIGURE 17–47

Marjorie Tietjen/123RF

1. Divide the gable end of the gambrel roof into polygons with areas that can be calculated.

2. Calculate the areas of the triangle and the trapezoid. Find the sum of the areas.

Triangle A	**Trapezoid** B
Base $= 23'$	Base$_1 = 3' + 23' + 3' = 29'$
Height $= 5'4'' = 5\dfrac{1'}{3}$	Base$_2 = 23'$
$A_{\text{triangle}} = \dfrac{1}{2}\,bh$	Height $= 8'$
$A_{\text{triangle}} = \dfrac{1}{2}\,(23)\left(5\dfrac{1}{3}\right)$	$A_{\text{trapezoid}} = \dfrac{1}{2}\,h(b_1 + b_2)$
$A_{\text{triangle}} = \dfrac{1}{\overset{}{\underset{1}{2}}}\,(23)\left(\dfrac{\overset{8}{16}}{3}\right)$	$A_{\text{trapezoid}} = \dfrac{1}{2}\,(8)(29 + 23)$
$A_{\text{triangle}} = \dfrac{184}{3} = 61\dfrac{1}{3}\ \text{ft}^2$	$A_{\text{trapezoid}} = 4(52)$
$A_{\text{triangle}} + A_{\text{trapezoid}} = A_{\text{gable}}$	$A_{\text{trapezoid}} = 208\ \text{ft}^2$

$$61\dfrac{1}{3}\ \text{ft}^2 + 208\ \text{ft}^2 = 269\dfrac{1}{3}\ \text{ft}^2$$

The area of the gable end of the gambrel roof is $269\frac{1}{3}$ ft^2.   **See Exercise 57.**

Sometimes when we figure area, we find it more convenient to calculate an overall area and subtract a smaller area.

STOP AND CHECK

1. Find the area of the composite figure.

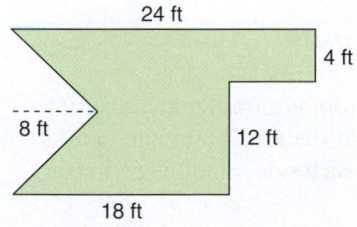

Answer:
1. 248 ft^2

EXAMPLE 11

Find the area of the flat metal piece shown in Fig. 17–48.

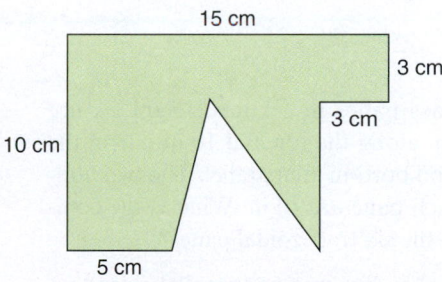

FIGURE 17–48

1. Divide the figure into polygons for which we can calculate areas (Fig. 17–49). Other strategies could be used.

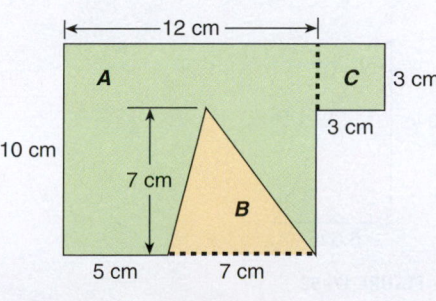

FIGURE 17–49

2. Find the missing dimensions.

3. Find the area of the square C (A_C), rectangle A (A_A), and triangle B (A_B). The area of the piece of metal is $A_C + A_A - A_B$.

Square C	**Rectangle A**	**Triangle B**
$A_C = s^2$	$A_A = lw$	$A_B = \dfrac{1}{2}bh$
$A_C = 3^2$	$A_A = 12(10)$	$A_B = \dfrac{1}{2}(7)(7)$
$A_C = 9 \text{ cm}^2$	$A_A = 120 \text{ cm}^2$	$A_B = \dfrac{1}{2}(49)$
		$A_B = 24.5 \text{ cm}^2$

Area of piece $= A_C + A_A - A_B = 9 + 120 - 24.5 = 104.5 \text{ cm}^2$

The area of the flat piece of metal is 104.5 cm².

See Exercise 58.

17-2 EXERCISES   MyLab Math For additional practice go to your study plan in MyLab Math.

1 Find the perimeter of Figs. 17–50 and 17–51. *See Example 1.*

1.

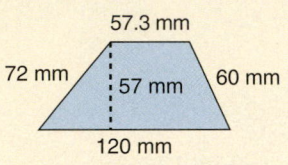

57.3 mm
72 mm   57 mm   60 mm
120 mm

FIGURE 17–50

2.

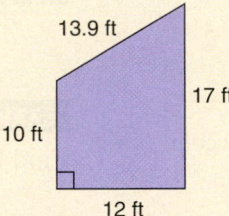

13.9 ft
17 ft
10 ft
12 ft

FIGURE 17–51

3. CON   The six glass panes in a kitchen light fixture each measure $4\frac{1}{2}$ in. along the top and 10 in. along the bottom. The top and bottom are parallel. The two non-parallel sides of each pane are 10 in. What is the combined perimeter of the six trapezoidal panes?

4. CON   A section of a hip roof is a trapezoid measuring 38 ft at the bottom, 14 ft at the top, 10 ft high, and 11 and 12 ft, respectively, on each side. Find the perimeter of this section of the roof.

5. A lot in an urban area has two nonparallel sides that are each 64 ft. The parallel sides are 120 ft and 154 ft. Find the perimeter of the trapezoidal property.

6. A swimming pool is fashioned in a trapezoidal design. The parallel sides are $18\frac{1}{2}$ ft and 31 ft. The other sides are $24\frac{1}{2}$ ft and 24 ft. What is the perimeter of the pool?

Find the perimeter of the triangles in Figs. 17–52 and 17–53. *See Example 2.*

7.

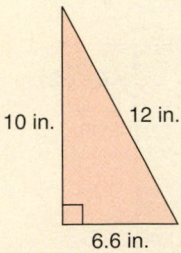

10 in.   12 in.
6.6 in.

FIGURE 17–52

8.

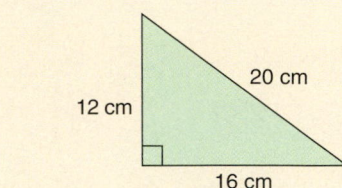

20 cm
12 cm
16 cm

FIGURE 17–53

9. **CON** Selena Henson is planning a patio that will adjoin the sides of her L-shaped home. One side of the patio is 24 ft and the other is 18 ft. The shape of the patio is triangular. Draw a representation of the patio and find the perimeter if the length of the third side is 30 ft.

10. Find the perimeter of a triangle that has sides measuring 2 ft 6 in., 1 yd 8 in., and 4 ft 6 in.

11. Find the perimeter of the parallelogram in Fig. 17–54.

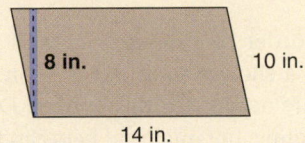

8 in.   10 in.

14 in.

FIGURE 17–54

12. Find the perimeter of the parallelogram in Fig. 17–55.

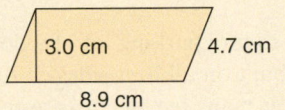

3.0 cm   4.7 cm

8.9 cm

FIGURE 17–55

Solve the problems involving perimeter.

13. An illuminated sign in the main entrance of a hospital is a parallelogram with a base of 56 in. and an adjacent side of 42 in. How many inches of aluminum molding are needed to frame the sign?

14. A customized van has a window cut in each side in the shape of a parallelogram with a base of 30 in. and an adjacent side of 14 in. How many inches of trim are needed to surround the two windows?

15. **CON** A contemporary building has a window in the shape of a parallelogram with a base of 80 in. and an adjacent side of 40 in. How many inches of trim are needed to surround the window?

16. **CON** A table for a mathematics lab has a top in the shape of a parallelogram with a base of 40 in. and an adjacent side of 24 in. How many inches of edge trim are needed to surround the tabletop?

17. Find the perimeter of the rectangle in Fig. 17–56.

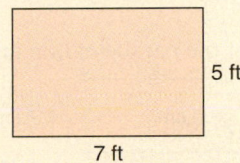

5 ft

7 ft

FIGURE 17–56

18. Find the perimeter of the rectangle in Fig. 17–57.

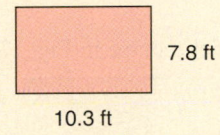

7.8 ft

10.3 ft

FIGURE 17–57

19. **CON** A rectangular parking lot is 200 ft by 145 ft. Find the perimeter of the parking lot.

20. **CON** A room is 18 ft by 15 ft. How many feet of chair rail are needed for the room? Disregard openings.

21. **CON** How many feet of quarter-round molding are needed to finish around the baseboard after sheet vinyl flooring is installed if the room is 14 ft by 20 ft and there are three 3.5-ft-wide doorways?

22. **CON** The swimming pool in Fig. 17–58 measures 36 ft by 20 ft. How much fencing is needed, including material for a gate, if the fence is to be built 9 ft from each side of the pool?

Fence

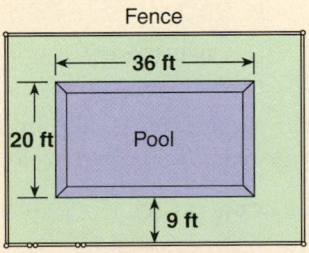

36 ft

20 ft   Pool

9 ft

FIGURE 17–58

23. **CON** A Formica tabletop measures 40 in. by 62 in. How many feet of edge trim are needed (12 in. = 1 ft)?

24. **CON** A countertop requires rolled edging to be installed on all four sides. How much rolled edging material is needed if the countertop measures 25 in. by 40 in.?

25. Find the perimeter of Fig. 17–59.

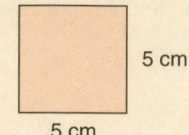

5 cm

5 cm

FIGURE 17–59

26. Find the perimeter of Fig. 17–60.

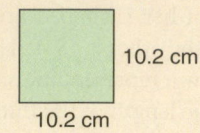

10.2 cm

10.2 cm

FIGURE 17–60

27. **CON**   The square parking lot of a doctor's office is to have curbs built on all four sides. If the lot is 160 ft on each side, how many feet of curb are needed? Allow 12 ft for a driveway into the parking lot.

See Example 3.

29. Find the perimeter of Fig. 17–61.

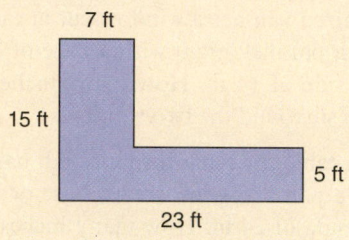

7 ft

15 ft

5 ft

23 ft

FIGURE 17–61

28. **CON**   A border of 4-in. × 4-in. wall tiles surrounds the floor of a shower stall that is 52 in. × 52 in. How many tiles are needed for this border? Disregard spaces for grout (connecting material between the tiles).

See Example 4.

30. Find the perimeter of Fig. 17–62.

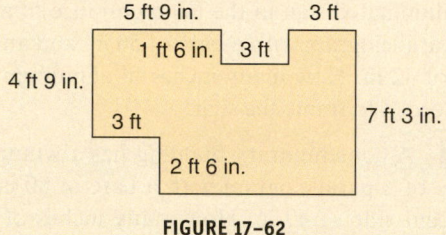

5 ft 9 in.   3 ft

1 ft 6 in.   3 ft

4 ft 9 in.

3 ft   7 ft 3 in.

2 ft 6 in.

FIGURE 17–62

2   *See Example 5.*

31. Find the area of the shape in Fig. 17–63.

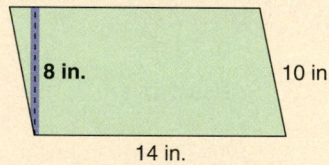

8 in.   10 in.

14 in.

FIGURE 17–63

33. Find the area of the shape in Fig. 17–65.

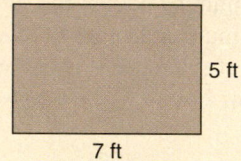

5 ft

7 ft

FIGURE 17–65

35. A rectangular parking lot is 340 ft by 125 ft. Find the number of square feet in the parking lot.

37. Find the area of Fig. 17–67.

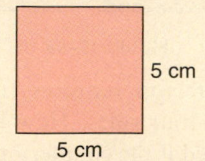

5 cm

5 cm

FIGURE 17–67

32. Find the area of the parallelogram in Fig. 17–64.

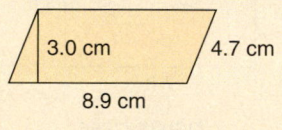

3.0 cm   4.7 cm

8.9 cm

FIGURE 17–64

34. Find the area of the shape in Fig. 17–66.

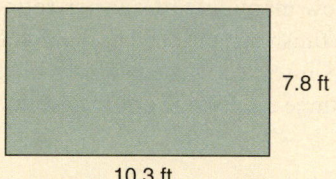

7.8 ft

10.3 ft

FIGURE 17–66

36. A room is 15 ft by 12 ft. How many square feet of flooring are needed for the room?

38. Find the area of the square in Fig. 17–68.

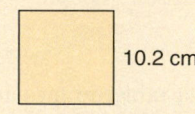

10.2 cm

FIGURE 17–68

39. **CON** Madison Duke is wallpapering a laundry room 8 ft by 8 ft by 8 ft high. How many square feet of paper will she need if there are 63 ft² of openings in the room?

40. **AG/H** Making no allowances for bases, the pitcher's mound, or the home plate area, how many square yards of artificial turf are needed to resurface an infield at an indoor baseball stadium? The infield is 90 ft on each side (9 ft² = 1 yd²).

41. **AG/H** Ted Davis is a farmer who wants to apply fertilizer to a 40-acre field with dimensions $\frac{1}{4}$ mi × $\frac{1}{4}$ mi. Find the area in square miles.

42. **CON** A 36-in. × 36-in. ceramic tile shower stall is being installed. How many 4-in. × 4-in. tiles are needed to cover the floor? Disregard the drain opening and grout spaces.

43. **CON** Tiles that are 6 in. × 6 in. cover the floor of a shower. How many tiles are needed for the floor if the shower measures 4.5 ft by 6 ft?

44. **INDTR** A 20-in. × 20-in. central heating and air conditioning return air vent is being installed in a wall. Find the area of the wall opening.

45. Find the area of Fig. 17-69.

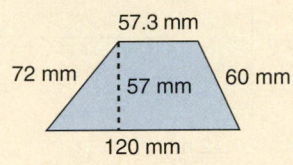

FIGURE 17-69

46. Find the area of Fig. 17-70.

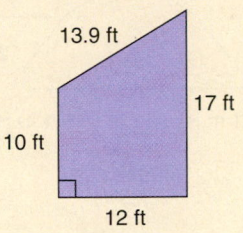

FIGURE 17-70

47. **CON** The six glass panes in a kitchen light fixture each measure $4\frac{1}{2}$ in. along the top and 10 in. along the bottom. The top and bottom are parallel. The height of each pane is 8 in. What is the combined area of the six trapezoidal panes?

48. **CON** A section of a hip roof is a trapezoid measuring 38 ft at the bottom, 14 ft at the top, and 10 ft high. Find the area of this section of the roof in square feet.

See Example 6.

49. Find the area of the triangle in Fig. 17-71.

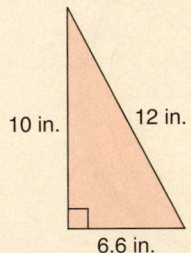

FIGURE 17-71

50. Find the area of the triangle in Fig. 17-72.

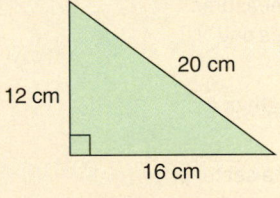

FIGURE 17-72

See Example 7.

51. Find the area of Fig. 17-73 for a height of 9.3 m and a base of 4.8 m.

52. Find the area of Fig. 17-73 for a height of 4 ft 3 in. and a base of 3 ft.

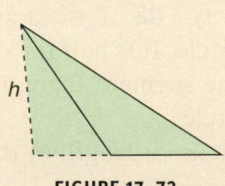

FIGURE 17-73

See Example 8.

53. **CON** Carlee McAnally is planning a patio. The triangular-shaped patio has sides 24 ft, 18 ft, and 22 ft. Find the number of square feet of surface area to be covered with concrete. Round to the nearest square foot.

54. **CON** If aluminum siding costs $6.75 a square yard installed, how much does it cost to put the siding on the two triangular gable ends of a roof under construction? Each gable has a span (base) of 30 ft 6 in. and a rise (height) of 7 ft 6 in. Any portion of a square yard is rounded to the next highest square yard for each gable.

See Example 9.

55. Find the area of Fig. 17–61 in Exercise 29.

56. Find the area in square feet of Fig. 17–62 in Exercise 30.

See Example 10.

57. Find the area of Fig. 17–74.

See Example 11.

58. Find the area of Fig. 17–75.

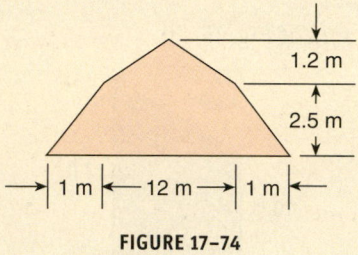

FIGURE 17–74

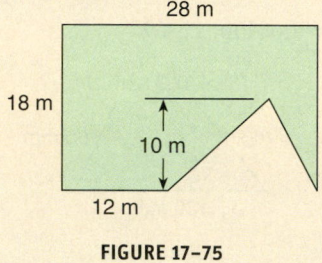

FIGURE 17–75

17–3 | Circles and Radians

A **circle** is a closed curved line with points that lie in a plane and are the same distance from the *center* of the figure (Fig. 17–76).

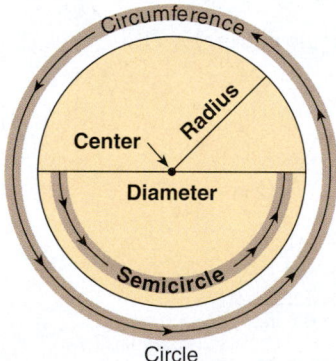

FIGURE 17–76

The **center of a circle** is the point that is the same distance from every point on the circle.

The **radius** (plural: *radii,* pronounced "ray · dē · ī") is a straight-line segment from the center of a circle to a point on the circle. It is half the diameter.

The **diameter** of a circle is a straight-line segment from a point on the circle through the center to another point on the circle.

The **circumference** of a circle is the perimeter or length of the closed curved line that forms the circle.

A **semicircle** is half a circle and is created by drawing a diameter.

Radius: a straight-line segment from the center of a circle to a point on the circle. It is half the diameter; plural: *radii*

Diameter: a straight-line segment from a point on the circle through the center to another point on the circle

Circumference: the perimeter or length of the closed curved line that forms the circle

Semicircle: half a circle; it is created by drawing a diameter

1 **Find the Circumference or Area of a Circle Using the Appropriate Formula.** The circle is a geometric form with a special relationship between its circumference and its diameter. If we divide the circumference of any circle by its diameter, the quotient is always the same number.

$$\pi = \frac{\text{circumference}}{\text{diameter}} = \frac{C}{d}$$

This number is a nonrepeating, nonterminating decimal approximately equal to 3.1415927 to seven decimal places. The Greek letter π or pi (pronounced "pie") represents this value. Convenient approximations often used in calculations involving π are $3\frac{1}{7}$ and 3.14. Many calculators have a π key.

The formulas for the circumference and area of a circle are:

Circumference (C)	Area (A)		
$C = \pi d$	$A = \pi r^2$		d is diameter $(d = 2r)$
$C = 2\pi r$			r is radius $(r = \frac{1}{2}d)$

To find the circumference or area of a circle:

1. Select the appropriate formula.

2. Substitute values for r or d as appropriate.

3. Evaluate the formula. Use the calculator value for π.

STOP AND CHECK

1. Find the circumference (to the nearest tenth of a cm) of a circle that has a diameter of 24 cm.

Answer:
1. 75.4 cm

EXAMPLE 1

Find the circumference (to the nearest tenth of a meter) of a circle that has a diameter of 1.3 m.

$C = \pi d$	Select the circumference formula with diameter d.
$C = \pi(1.3)$	Use the π key on calculator.
$C = 4.08407045$ m	Evaluate.

The circumference is 4.1 m (rounded).   Circumference is a linear measure.

See Exercises 1–3.

> **TIP** **Calculator Values of π** Calculations involving π are always approximations. Scientific or graphing calculators include a π function or menu option where π to seven or more decimal places is computed by pressing a single key. Other calculators require the use of more than one key to activate the $\boxed{\pi}$ key. However, for computations by hand or a calculator without the $\boxed{\pi}$ function, 3.14 is sometimes adequate. **We use the calculator value 3.141592654 for π in all examples and exercises unless stated otherwise.**

> **TIP** **Fractions Versus Decimals** When you use your calculator, you may find it convenient to convert mixed U.S. customary linear measurements to their decimal equivalents. For instance, if a diameter is 7 ft 6 in., convert it to 7.5 ft (from $7\frac{6}{12}$ ft, in which $\frac{6}{12} = 0.5$).

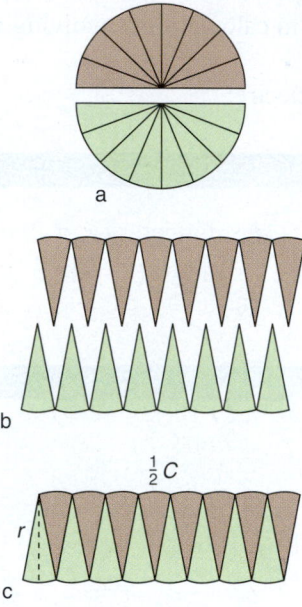

a

b

$\frac{1}{2}C$

r

c

FIGURE 17–77

EXAMPLE 2

Find to the nearest hundredth the circumference of a circle with a radius of 1 ft 9 in.

$C = 2\pi r$ Substitute for r: $1\frac{9}{12}$ ft = 1.75 ft. Use calculator value for π.

$C = 2\pi(1.75)$

$C = 10.99557429$

$C = \mathbf{11.00}$ **ft (rounded)**   Circumference is a linear measure.   **See Exercises 4–6.**

The area of a circle, like the circumference, is obtained from relationships within the circle. If we divide a circle into two semicircles and then subdivide each semicircle into pie-shaped pieces, we get something like Fig. 17–77a. If we then spread the upper and lower pie-shaped pieces, we get Fig. 17–77b. Now if we push the upper and lower pieces together, the result approximates the parallelogram in Fig. 17–77c, whose base is $\frac{1}{2}$ the circumference and whose height is the radius. Thus, the area of the circle is approximately the area of a parallelogram, that is, base times height. Since the base of the parallelogram is one-half the circumference and the height is the radius, the area of a circle equals one-half the circumference times the radius.

$A = \dfrac{1}{2}C \times r$ The formula for circumference is $C = 2\pi r$, so we substitute $2\pi r$ for C.

$A = \dfrac{1}{2}(2\pi r)(r)$ Multiply. Reduce where possible.

$A = \dfrac{1}{\underset{1}{2}}(\overset{1}{2}\pi r)(r)$ Area is a square measure: $r \cdot r = r^2$

$A = \pi r^2$

EXAMPLE 3

Find the area of a circle whose radius is 8.5 m. Round to tenths.

$A = \pi r^2$ Select appropriate formula and substitute for r.

$A = \pi(8.5)^2$ Square the radius.

$A = \pi(72.25)$ Multiply by π. Use the π key on your calculator.

$A = 226.9800692$ m²   Area is a square measure.

The area of the circle is 227.0 m².   **See Exercises 7–10.**

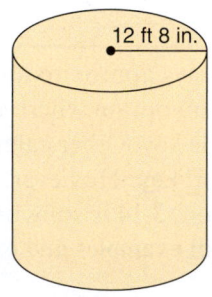

12 ft 8 in.

FIGURE 17–78

EXAMPLE 4

Find the area to the nearest hundredth of the top of a circular tank with a diameter of 12 ft 8 in. (Fig. 17–78).

$A = \pi r^2$

$A = (\pi)\left(\dfrac{12 + \frac{8}{12}}{2}\right)^2$ Follow the order of operations. 8 in. $= \frac{8}{12}$ ft. The diameter, $12 + \frac{8}{12}$, divided by 2 is the radius.

$A = 126.012772$ ft²   Calculator result

$A = 126.01$ ft² (rounded)   Area is a square measure.

The area of the top of the tank is 126.01 ft².   **See Exercises 11–14.**

Ring: the area between two circles (one inside the other) with the same center

EXAMPLE 5

A 15-in.-diameter wheel has a 3-in. hole in the center. Find the area of a side of the wheel to the nearest tenth (Fig. 17–79).

We are asked to find the area of the colored portion of the wheel in Fig. 17–79. To do so, we find A_{outside}, the area of the larger circle (diameter 15 in.), and *subtract* the area of A_{inside}, the smaller circle (diameter 3 in.). The colored portion is called a **ring.**

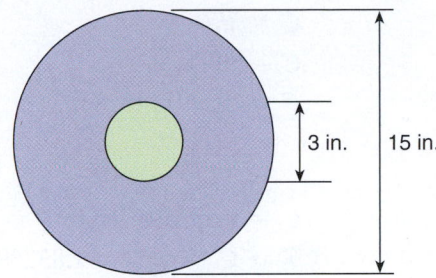

3 in. 15 in.

FIGURE 17–79

$$A_{\text{outside}} = \pi r^2 \quad \left(r = \frac{15}{2} = 7.5\right) \qquad\qquad A_{\text{inside}} = \pi r^2 \quad \left(r = \frac{3}{2} = 1.5\right)$$
$$A_{\text{outside}} = \pi(7.5)^2 \qquad\qquad\qquad\qquad A_{\text{inside}} = \pi(1.5)^2$$
$$A_{\text{outside}} = \pi(56.25) \qquad\qquad\qquad\qquad A_{\text{inside}} = \pi(2.25)$$
$$A_{\text{outside}} = 176.7145868 \text{ in}^2 \qquad\qquad A_{\text{inside}} = 7.068583471 \text{ in}^2$$

area of wheel (ring) $= A_{\text{outside}} - A_{\text{inside}}$

$A_{\text{wheel}} = 176.7145868 - 7.068583471 = 169.6460033 \qquad$ or $\qquad 169.6 \text{ in}^2$

The area of the wheel (ring) is 169.6 in².

See Exercises 15–19.

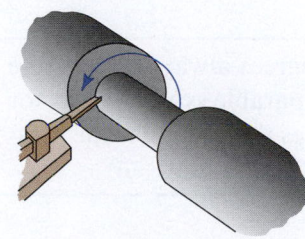

25 cm   25 cm

90 cm

25 cm   25 cm

90 cm

FIGURE 17–80

Semicircular-sided figure: a composite figure that consists of a semicircle at each end and a rectangle in the middle

EXAMPLE 6

INDTR   A bandsaw has two 25-cm wheels spaced 90 cm between centers (Fig. 17–80). Find the length of the saw blade.

    This layout is a composite figure that consists of a semicircle at each end and a rectangle in the middle. It is called a **semicircular-sided figure.** The two semicircles equal one whole circle, so we need to find the circumference of one circle (wheel) and add it to the lengths of the two sides of the rectangle.

$C = \pi d$ total length of blade $= C + 2l$

$C = \pi(25)$ total length of blade $= 78.53981634 + 2(90)$

$C = 78.53981634$ cm   total length of blade $= 258.5$ cm (rounded)

The bandsaw blade is 258.5 cm in length.

See Exercise 20–21.

Specific applications for a particular industry or career often use the formulas for area, perimeter, or circumference.

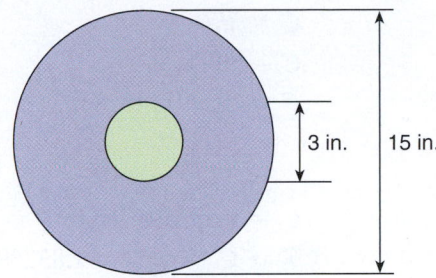

FIGURE 17–81

Cutting speed: the speed of a tool that passes over the work that is spinning in a circular motion

EXAMPLE 7

INDTEC   Find the cutting speed of a lathe if a piece of work that has a 7-in. diameter turns on a lathe at 75 revolutions per minute (rpm) (Fig. 17–81).

    The **cutting speed** is the speed of a tool that passes over the work, such as the speed of a lathe or a sander as it sands (passes over) a piece of wood. If the cutting speed is too fast or too slow, safety and quality are impaired. The formula for cutting speed is CS = C (in feet) × rpm.

$$CS = \text{cutting speed}$$
$$C = \text{circumference or one revolution (in feet)}$$
$$\text{rpm} = \text{revolutions per minute (r/min)}$$

Cutting speed is measured in *feet per minute* (ft/min)

$C = \pi d$	Find the circumference.
$C = \pi(7)$	
$C = 21.99114858$ in.	Convert 21.99114858 in. into feet using a unit ratio.

$$\frac{21.99114858 \text{ in.}}{1}\left(\frac{1 \text{ ft}}{12 \text{ in.}}\right) = 1.832595715 \text{ ft}$$

$C = 1.832595715$ ft	Circumference in feet or feet per revolution
CS = C × rpm = 1.832595715 ft/r (75 r/min) = 137 ft/min	Rounded

The cutting speed of the lathe is approximately 137 ft/min.   **See Exercises 23–24.**

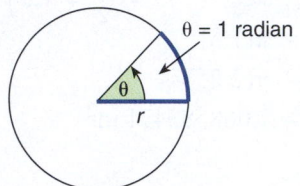

FIGURE 17–82

Radian, rad: the measure of a central angle of a circle whose intercepted arc is equal in length to the radius of the circle. The abbreviation for radian is *rad*

Central angle: an angle with its vertex at the center of the circle; its legs or sides are radii of the circle

2 **Convert Angle Measures Between Degrees and Radians.** In Section 17–1, we learned that angles can be measured in units called *degrees.* Angles can also be measured in *radians.* A **radian** is the measure of a central angle of a circle whose intercepted arc is equal in length to the radius of the circle (Fig. 17–82). The abbreviation for radian is **rad.** A **central angle** of a circle is an angle with its vertex at the center of the circle. The legs or sides of a central angle are radii of the circle.

The circumference of a circle is related to the radius by the formula $C = 2\pi r$. Thus, the ratio of the circumference to the radius of any circle is $\frac{C}{r} = 2\pi$; that is, the radius could be measured off 2π times (about 6.28 times) along the circumference. A complete rotation is 2π radians (see Fig. 17–83).

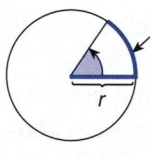 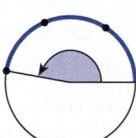 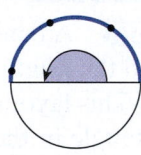 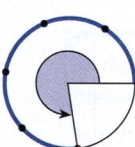 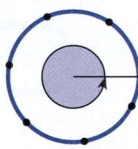

| 1 radian | 3 radians | 3.14 or π radians | 5 radians | 6.28 or 2π radians |

FIGURE 17–83

How are radians and degrees related? A central angle measuring 1 radian makes an arc length equal to the radius.

TIP **Degree and Radian Notation** Angles measured in degrees always require the word *degree* or the degree symbol to be written. No comparable symbol exists for the radian. The abbreviation *rad* or no unit at all indicates that radian is the measuring unit. Some calculators use a raised *r* to indicate a radian unit.

A complete rotation is 360° or 2π rad. To convert from one unit of angle measure to another, we multiply by a *unit ratio* that relates degrees and radians. Because 360° = 2π rad, we can simplify the relationship to 180° = π rad $\left(\dfrac{360°}{2} = \dfrac{2\pi}{2} \text{ rad}\right)$.

To convert degrees to radians:

1. Multiply degrees by the unit ratio $\dfrac{\pi \text{ rad}}{180°}$.

2. Write the product in simplest form.

Using a calculator (steps may vary):

1. Set angle MODE to radian.

2. Enter the degree measure followed by the degree symbol. Use $\boxed{° ' ''}$, $\boxed{\text{DMS}}$, or the angle menu, option $\boxed{1:°}$.

3. Enter the minute measure followed by the minute symbol. Use $\boxed{° ' ''}$, $\boxed{\text{DMS}}$, or the angle menu, option $\boxed{2:'}$.

4. Enter the second measure followed by the second symbol. Use $\boxed{° ' ''}$, $\boxed{\text{DMS}}$, or $\boxed{\text{ALPHA}}\ \boxed{''}$.

5. Press $\boxed{\text{ENTER}}$.

STOP AND CHECK

1. Convert an angle measuring 78° to radians to the nearest hundredth.

Answer:
1. 1.36 rad

EXAMPLE 8

Convert the angle measures to radians. Use the calculator value for π to change to a decimal equivalent rounded to the nearest hundredth.

(a) 20°   **(b)** 175°

(a) $20°\left(\dfrac{\pi \text{ rad}}{180°}\right) = \dfrac{20(\pi)}{180} = \textbf{0.35 rad}$ Multiply by $\dfrac{\pi \text{ rad}}{180°}$.

(b) $175° = 3.054326191 = \textbf{3.05 rad}$ Set calculator in radian mode.
175 $\boxed{\text{2}^{\text{nd}}}$ $\boxed{\text{ANGLE}}$ $\boxed{\text{1:°}}$ $\boxed{\text{ENTER}}$ $\boxed{\text{ENTER}}$

See Exercises 25–30.

When converting from radians to degrees, we multiply by a unit ratio so that the radians cancel and are replaced by degrees.

To convert radians to degrees:

1. Multiply radians by the unit ratio $\dfrac{180°}{\pi \text{ rad}}$.

2. Write the product in simplest form.

Using a calculator (steps may vary):

1. Set angle MODE to degree.

2. Enter radian amount and radian symbol (angle menu, option $\boxed{3:^{\text{r}}}$).

3. If degrees, minutes, and seconds are desired, use DMS function (angle menu, option $\boxed{4:\blacktriangleright\text{DMS}}$).

STOP AND CHECK

1. Convert an angle measuring 1.5 rad to a degree measure to the nearest ten-thousandth of a degree.

2. Convert an angle measuring $\dfrac{\pi}{6}$ rad to a degree measure.

Answers:
1. 85.9437°   2. 30°

EXAMPLE 9

Convert to degrees. Round to the nearest ten-thousandth of a degree.

(a) 2 rad   **(b)** $\dfrac{\pi}{2}$ rad

(a) $2 \text{ rad}\left(\dfrac{180°}{\pi \text{ rad}}\right) = \dfrac{360°}{\pi} = \mathbf{114.5916°}$ Multiply by $\dfrac{180°}{\pi \text{ rad}}$. Use calculator value of π and round.

(b) $\dfrac{\pi}{2} \text{ rad} = 90°$ In degree mode:
(π ÷ 2) ANGLE 3:ʳ ENTER.

See Exercises 31–33.

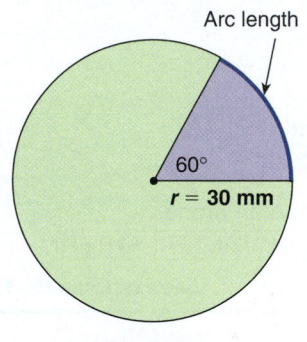

Arc length

60°

$r = 30$ mm

FIGURE 17–84

Sector of a circle: the portion of the area of a circle cut off by two radii

Arc: the portion of the circumference cut off by a central angle

Theta θ: Greek letter normally used to denote an angle measure

Arc length of a sector: the portion of the circumference of a circle intercepted by the sides of the sector

EXAMPLE 10

Convert to degrees, minutes, and seconds. Round to the nearest second.

(a) 1 rad   **(b)** 2.5 rad

(a) 1 rad $= 57°17′44.806″$ In degree mode:
Press 1 ANGLE 3:ʳ ANGLE 4:▶DMS ENTER ENTER.

1 rad $= \mathbf{57°17′45″}$ Round.

(b) 2.5 rad $= 143°14′22.016″$ In degree mode:
Press 2.5 ANGLE 3:ʳ ANGLE 4:▶DMS ENTER ENTER.

2.5 rad $= \mathbf{143°14′22″}$ Round.

Thus, 2.5 rad $= 143°14′22″$.

See Exercises 34–36.

3 **Find the Arc Length of a Sector.** We often work with figures that are less than a whole circle. For example, earlier we worked with the semicircle in composite figures. A **sector of a circle** is the portion of the area of a circle cut off by two radii. The two radii form a *central angle*. The portion of the circumference cut off by a central angle is called an **arc**. The sector is formed by a central angle of the circle and the arc connecting the sides of the angle (radii). See Fig. 17–84. We use the Greek letter **theta** (θ) to represent angle measures.

The **arc length of a sector** is the portion of the circumference intercepted by the sides of the sector (see Fig. 17–84).

To find the arc length of a sector:

Using degrees:

1. Substitute known values into the formula.

$$s = \frac{\theta}{360}(2\pi r) \quad \text{or} \quad s = \frac{\theta}{360}(\pi d)$$

where θ is a central angle measured in degrees and r is the radius of the circle.

2. Evaluate.

Using radians:

1. Substitute the given radian measure of the central angle and the radius in the formula

$$s = \theta r$$

where θ is the central angle measured in radians and r is the radius of the circle.

2. Simplify the expression.

EXAMPLE 11

Find the arc length of the sector formed by a 60° central angle if the radius of the circle is 30 mm (Fig. 17–86).

$s = \dfrac{\theta}{360}(2\pi r)$ Select the formula for degrees. Substitute values.

$s = \dfrac{60}{360}(2)(\pi)(30)$ Evaluate.

$s = 31.41592654$ Round.

$s = \mathbf{31.42\ mm}$ **See Exercises 37–40.**

TIP **Exact Solutions Involving π** There are times when exact solutions are desirable. An exact solution involving π leaves the factor of π in the solution. In the previous example, the exact solution is found by simplifying the expression

$$\frac{60}{360}(2)(\pi)(30) = \frac{60}{360}(2)(30)\pi = 10\pi$$

EXAMPLE 12

Find the exact arc length and the arc length to the nearest hundredth that is intercepted on the circumference of the circle in Fig. 17–85 by a central angle of $\dfrac{\pi}{3}$ radians (rad) if the radius of the circle is 10 cm.

$s = \theta r$ Select the formula for radians. Substitute $\frac{\pi}{3}$ for θ and 10 cm for r.

$s = \dfrac{\pi}{3}(10\ \text{cm})$ Evaluate.

$s = \dfrac{10}{3}\pi\ (\text{cm})$ Exact arc length

$s = \dfrac{\pi(10\ \text{cm})}{3}$ Use calculator value for π.

$s = 10.47197551\ \text{cm}$ Round to the nearest hundredth.

$s = 10.47\ \text{cm}$

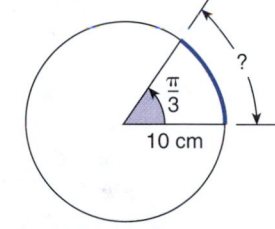

FIGURE 17–85

Thus, the exact arc length of the intercepted arc is $\dfrac{10}{3}\pi$ cm and the approximate arc length is 10.47 cm.

See Exercises 41–42.

TIP **Analyzing Arc Length Dimensions** In the preceding example, $\left(\dfrac{\pi}{3}\ \text{rad}\right)(10\ \text{cm}) = 10.47\ \text{cm}$, what happened to the radians? In the definition of a radian, we relate the measure of an arc connecting the end points of a central angle to the measure of the radius of the circle. Therefore, the arc length will have the same measuring unit as the radius.

To find the central angle or radius of a sector given the arc length:

1. Substitute given values in the formula $s = \theta r$, where s is the arc length, θ is the central angle measured in radians, and r is the radius of the circle.

2. Solve for the missing value.

EXAMPLE 13

Find the radian measure of an angle at the center of a circle of radius 5 m. The angle intercepts an arc length of 12.5 m (see Fig. 17–86).

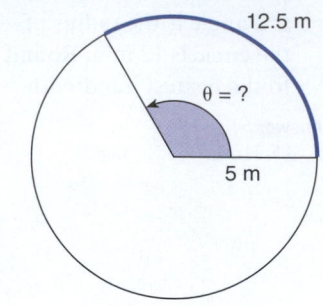

12.5 m

$\theta = ?$

5 m

$s = \theta r$ Formula for arc length. Solve for θ.

$\theta = \dfrac{s}{r}$ Substitute 12.5 m for s and 5 m for r. Arc length and radius measuring units are compatible.

$\theta = \dfrac{12.5 \text{ m}}{5 \text{ m}}$

$\theta = 2.5$ rad

The angle is 2.5 rad.

FIGURE 17–86

See Exercises 43–44.

EXAMPLE 14

Find the radius of an arc if the length of the arc is 8.22 cm and the intercepted central angle is 3 rad (see Fig. 17–87).

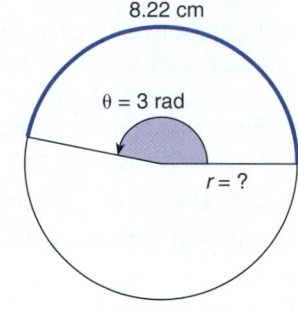

8.22 cm

$\theta = 3$ rad

$r = ?$

$s = \theta r$ Formula for arc length. Solve for r.

$r = \dfrac{s}{\theta}$ Substitute 8.22 cm for s and 3 rad for θ.

$r = \dfrac{8.22 \text{ cm}}{3}$ Simplify.

$r = 2.74$ cm   Same measuring unit as arc length

The radius of the arc is 2.74 cm.

FIGURE 17–87

See Exercises 45–46.

4   Find the Area of a Sector or Segment.

To find the area of a sector:

Using degrees:

1. Calculate the portion of the circle included in the sector. θ is a central angle measured in degrees.

$$\frac{\theta}{360} = \text{fractional part of circle}$$

2. Find the fractional part of the area of the circle.

$$A_{\text{sector}} = \frac{\theta}{360} \pi r^2 \text{ where } \pi r^2 = \text{area of circle}$$

Using radians:

1. Substitute known values into the formula $A_{\text{sector}} = \dfrac{1}{2}\theta r^2$, where θ is the central angle measured in radians and r is the radius of the circle.

2. Solve for the missing value.

STOP AND CHECK

1. Find the area of a sector with a central angle of 80° in a circle with a radius of 7 cm. Round to hundredths.

Answer:
1. 34.21 cm^2

EXAMPLE 15

Find the area of a sector with a central angle of 45° in a circle with a radius of 10 in. Round to hundredths (see Fig. 17–88).

$A_{sector} = \frac{\theta}{360}\pi r^2$ — Substitute for θ and *r*.

$A_{sector} = \frac{45}{360}(\pi)(10)^2$ — Perform the indicated operations.

$A_{sector} = 0.125(\pi)(100)$

$A_{sector} = 12.5\pi \text{ in}^2$ — Exact solution

$A_{sector} = 39.26990817 \text{ in}^2$ — Round.

The area of the sector is 39.27 in^2.

45°
FIGURE 17–88

See Exercises 47–50.

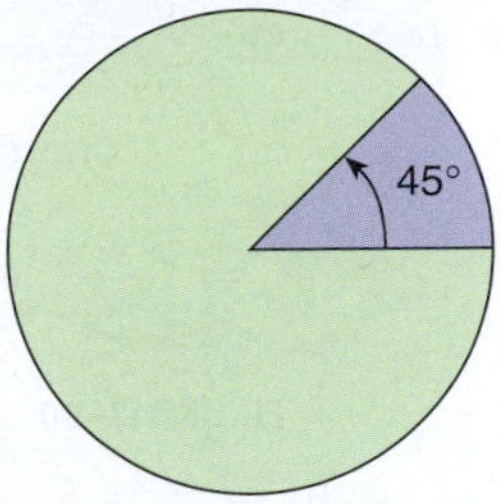

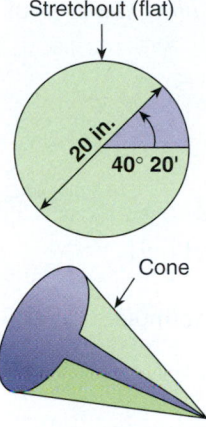

Stretchout (flat)
20 in.
40° 20'
Cone
FIGURE 17–89

EXAMPLE 16

INDTR A cone is made from sheet metal. To form a cone, a sector with a central angle of 40°20′ is cut from a metal circle whose diameter is 20 in. Find the area of the stretchout (portion of the circle) that is used to form the cone. Round to the nearest hundredth (Fig. 17–89).

area used for cone = area of circle − area of sector

Circle	Sector	
$A_{circle} = \pi r^2$	$A_{sector} = \frac{\theta}{360}\pi r^2$	Substitute known values. Convert 40°20′ to 40.333333333°.
$A_{circle} = \pi(10)^2$	$A_{sector} = \frac{40.3\overline{3}}{360}(314.1592654)$	Substitute A_{circle} for πr^2.
$A_{circle} = \pi(100)$	$A_{sector} = 0.112037037(314.1592654)$	
$A_{circle} = 314.1592654 \text{ in}^2$	$A_{sector} = 35.19747324 \text{ in}^2$	

area used for cone = $A_{circle} - A_{sector}$

$A_{cone} = 314.1592654 - 35.19747324$

$A_{cone} = 278.9617922 \text{ in}^2$ — Round.

The area of the metal sector used to form the cone is 278.96 in^2.

See Exercises 51–52.

STOP AND CHECK

1. Find the area to the nearest tenth of a sector that has a central angle of 2.3 rad and has a radius of 6.6 cm.

Answer:
1. 50.1 cm^2

EXAMPLE 17

Find the area to the nearest tenth of a sector that has a central angle of 1.6 rad and has a radius of 7.2 in.

$A_{sector} = \frac{1}{2}\theta r^2$ — Substitute θ = 1.6 and *r* = 7.2.

$A_{sector} = \frac{1}{2}(1.6)(7.2 \text{ in.})^2$ — Simplify.

$A_{sector} = 41.5 \text{ in.}^2$ — Area of sector

The area of the sector is 41.5 in^2.

See Exercises 53–58.

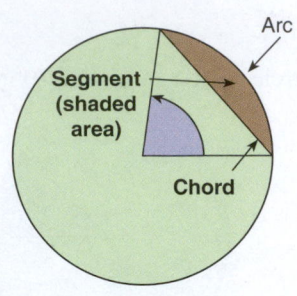

FIGURE 17–90

Chord: a line segment joining two points on the circumference of a circle

Segment of a circle: the portion of the area of a circle bounded by a chord and an arc

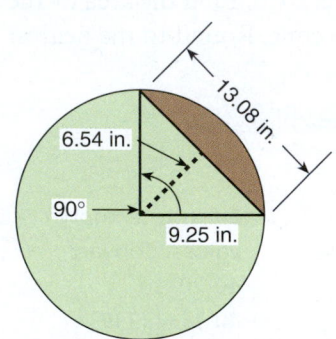

FIGURE 17–91

STOP AND CHECK

1. Find the area to the nearest hundredth of a segment if the chord has a length of 9.3 cm, a central angle of 90°, a radius of 6 cm, and the height of the triangle portion of the corresponding sector is 4.24 cm.

Answer:
1. 8.56 cm²

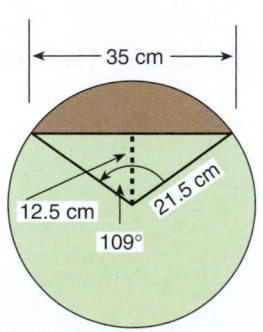

FIGURE 17–92

If a line segment (called a *chord*) joins the end points of the radii that form a sector, the sector is divided into two figures, a triangle and a *segment* (Fig. 17–90). A **chord** is a line segment joining two points on the circumference of a circle. The portion of the circumference cut off by a chord is an *arc*. A **segment of a circle** is the portion of the area of a circle bounded by a chord and an arc.

Because the chord divides the sector into a triangle and a segment, we can calculate the area of the segment by subtracting the area of the triangle from the area of the sector.

To find the area of a segment:

1. Substitute known values into the formula.

Using degrees:
$$A_{segment} = \frac{\theta}{360}\pi r^2 - \frac{1}{2}bh$$

Using radians:
$$A_{segment} = \frac{1}{2}\theta r^2 - \frac{1}{2}bh$$

where $\frac{\theta}{360}\pi r^2$ or $\frac{1}{2}\theta r^2$ is the area of the sector and $\frac{1}{2}bh$ is the area of the triangle.

2. Evaluate.

EXAMPLE 18

Find the area to the nearest hundredth of the segment in Fig. 17–91.

$$A_{segment} = \frac{\theta}{360}\pi r^2 - \frac{1}{2}bh \qquad \text{Substitute in formula.}$$
$$A_{segment} = \frac{90}{360}(\pi)(9.25)^2 - \frac{1}{2}(13.08)(6.54) \qquad \text{Evaluate.}$$
$$A_{segment} = 67.20063036 - 42.7716$$
$$A_{segment} = 24.42903036 \text{ in}^2 \qquad \text{Round.}$$

The area of the segment is 24.43 in².   **See Exercises 59–62.**

EXAMPLE 19

INDTR   A segment in Fig. 17–92 is removed so that a template for a cam is made from the rest of the circle. What is the area of the template? Give the answer to the nearest hundredth.

area of template = area of circle − area of segment

Circle

$$A_{circle} = \pi r^2$$
$$A_{circle} = \pi(21.5)^2$$
$$A_{circle} = 1,452.201204 \text{ cm}^2$$

Segment

$$A_{segment} = \frac{\theta}{360}\pi r^2 - \frac{1}{2}bh$$
$$A_{segment} = \frac{109}{360}(\pi)(21.5)^2 - \frac{1}{2}(35)(12.5)$$
$$A_{segment} = 439.6942435 - 218.75$$
$$A_{segment} = 220.9442535 \text{ cm}^2$$

$$A_{template} = A_{circle} - A_{segment}$$
$$A_{template} = 1,452.201204 - 220.9442535$$
$$A_{template} = 1,231.256951 \text{ cm}^2 \qquad \text{Round.}$$

The area of the template is 1,231.26 cm².   **See Exercises 63–64.**

17–3 EXERCISES   MyLab Math For additional practice go to your study plan in MyLab Math.

1 Find the circumference of circles with the following dimensions. Round to tenths. *See Example 1.*

1. Diameter = 8 cm

2. Diameter = 15 m

See Example 2.

3. Diameter = 5.5 m

4. Radius = 3 in.

5. Radius = 1.5 ft

6. Radius = $8\frac{1}{2}$ ft

Find the area of circles with the following dimensions. Round to tenths. *See Example 3.*

See Example 4

7. Radius = 3 in.

8. Radius = 16 yd

9. Radius = 1.5 ft

10. Radius = $8\frac{1}{2}$ ft

11. Diameter = 8 cm

12. Diameter = 15 m

13. Diameter = 5.5 m

14. Diameter = $5\frac{1}{4}$ in.

Find the color area of Figs. 17–93 through 17–96 to the nearest tenth. *See Example 5.*

15.

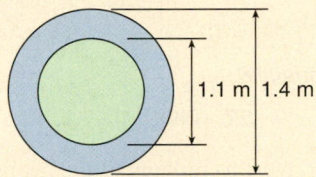

1.1 m   1.4 m

FIGURE 17–93

16.

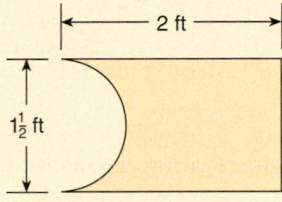

2 ft

$1\frac{1}{2}$ ft

FIGURE 17–94

17.

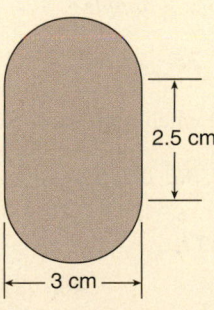

2.5 cm

3 cm

FIGURE 17–95

18.

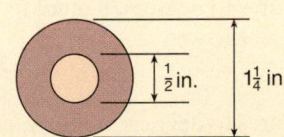

$\frac{1}{2}$ in. $1\frac{1}{4}$ in.

FIGURE 17–96

See Example 6.

19. A swimming pool is in the form of a semicircular-sided figure. Its width is 20 ft and the parallel portions of the sides are each 20 ft (Fig. 17–97). What is the area of a 5-ft-wide walk surrounding the pool? Round to tenths.

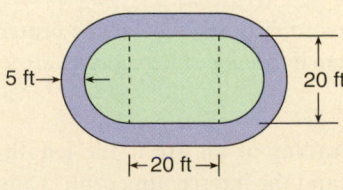

5 ft   20 ft

|← 20 ft →|

FIGURE 17–97

20. **INDTR**   A belt connecting two 9-in.-diameter drums on a conveyor system needs replacing. How many inches must the new belt be if the centers of the drums are 10 ft apart (Fig. 17–98)? Round to tenths.

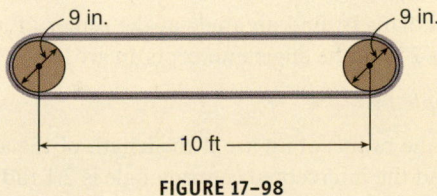

9 in.   9 in.

10 ft

FIGURE 17–98

21. INDTR A 2-in.-inside-diameter pipe and a 4-in.-inside-diameter pipe empty into a third pipe whose inside diameter is 5 in. (Fig. 17–99). Is the third pipe large enough for the combined flow? (Justify your answer.)

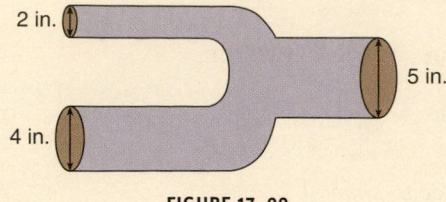

FIGURE 17–99

22. INDTR A large pipe whose interior cross-sectional area is 20 in² empties into two smaller pipes that each have an interior diameter of 4 in. (Fig. 17–100). Are the smaller pipes together large enough to carry off the flow from the larger pipe? (Justify your answer.)

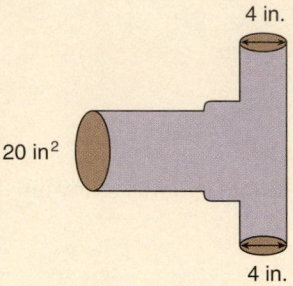

FIGURE 17–100

See Example 7.

23. INDTR Cutting speed, when applied to a grinding wheel, is called *surface speed*. What is the surface speed in ft/min of a 9-in.-diameter grinding wheel revolving at 1,200 rpm? (Surface speed = circumference in feet × rpm.)

24. INDTR A 12-in.-diameter polishing wheel revolves at 500 rpm. What is the surface speed? (See Exercise 23 for formula.)

2 Convert the measures to radians rounded to the nearest hundredth. *See Example 8.*

25. 45°   **26.** 56°   **27.** 140°

Convert the measures to degrees rounded to the nearest ten-thousandth. Then convert to radians to the nearest hundredth.

28. 21°45′   **29.** 177°33′   **30.** 44°54′12″

Convert the measures to degrees. Round to the nearest ten-thousandth of a degree. *See Example 9.*

31. $\dfrac{\pi}{4}$ rad   **32.** $\dfrac{\pi}{6}$ rad   **33.** 2.5 rad

Convert the measures to degrees, minutes, and seconds. Round to the nearest second when necessary. *See Example 10.*

34. 0.5 rad   **35.** $\dfrac{\pi}{8}$ rad   **36.** 0.75 rad

3 Find the arc length of the sectors of a circle. Round to hundredths. *See Example 11.*

37. ∠ = 54°   **38.** ∠ = 150°   **39.** ∠ = 120°30′   **40.** ∠ = 25°16′
   $r = 30$ mm $r = 5$ in. $r = 1.52$ ft $r = 16$ cm

Solve the following problems. Round answers to hundredths if necessary.

See Example 12.

41. Find the arc length intercepted on the circumference of a circle by a central angle of 2.15 rad if the radius of the circle is 3 in.

42. Find the arc length intercepted on the circumference of a circle by a central angle of 4 rad if the radius of the circle is 3.5 cm.

See Example 13.

43. Use radians to find an angle at the center of a circle of radius 2 in. if the angle intercepts an arc length of 8.5 in.

44. Use radians to find an angle at the center of a circle of radius 4.3 cm if the angle intercepts an arc length of 15 cm.

See Example 14.

45. Find the radius of an arc if the length of the arc is 14.7 cm and the intercepted central angle is 2.1 rad.

46. Find the radius of an arc if the length of the arc is 12.375 in. and the intercepting central angle is 2.75 rad.

4 Find the area of the sector of a circle using Fig. 17–101. Round to hundredths. *See Example 15.*

47. $\angle = 54°$
$r = 16$ cm

48. $\angle = 25°16'$
$r = 30$ mm

49. $\angle = 120°30'$
$r = 1.52$ ft

50. $\angle = 65°$
$r = 5$ in.

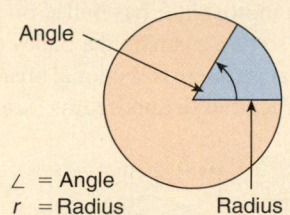

$\angle$ = Angle
r = Radius   Radius

FIGURE 17–101

See Example 16.

51. **CON** The library of a contemporary elementary school is circular (Fig. 17–102). The floor plan includes sectors reserved for science materials, literary materials, reference materials, and so on. Find the area of the reference section excluding its storage area.

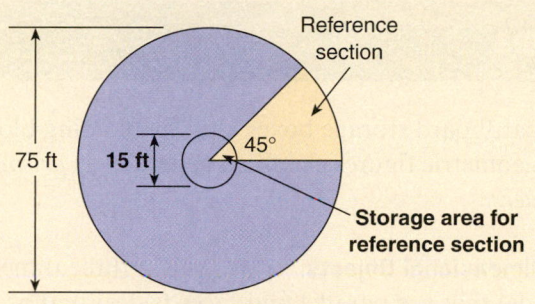

FIGURE 17–102

52. **CON** A mason lays a tile mosaic featuring a four-sector design (color portion of Fig. 17–103). What is the area of the design to the nearest hundredth?

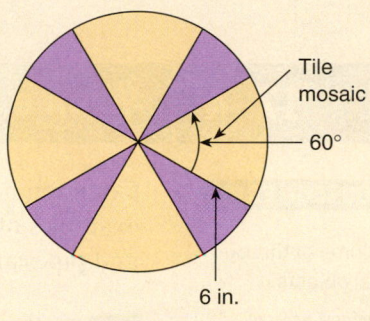

FIGURE 17–103

See Example 17.

53. Find the area of a sector whose central angle is 2.14 rad and whose radius is 4 in.

54. Find the area of a sector that has a central angle of 6 rad and a radius of 1.2 cm.

55. Find the radius of a sector if the area of the sector is 7.5 cm² and the central angle is 3 rad.

56. How many radians does the central angle of a sector measure if its area is 1.7 in² and its radius is 2 in.?

57. Find the area of a sector whose central angle is 1.83 rad and whose radius is 7.2 cm. Round to hundredths.

58. Find the number of radians of a central angle of a sector whose area is 5.6 cm² and whose radius is 4 cm.

Find the area of the segments of a circle using Fig. 17–104. Round to hundredths. *See Example 18.*

59. $\angle = 60°$
$r = 13.3$ cm
$h = 11.52$ cm
$b = 13.3$ cm

60. $\angle = 110°$
$r = 10$ in.
$h = 5.74$ in.
$b = 16.38$ in.

61. $\angle = 105°$
$r = 11.25$ in.
$h = 6.85$ in.
$b = 17.85$ in.

62. $\angle = 60°$
$r = 24$ cm
$h = 20.8$ cm
$b = 24$ cm

$\angle$ = Angle
r = Radius
h = Height
b = Base

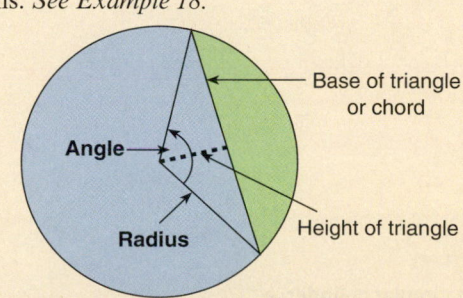

FIGURE 17–104

See Example 19.

63. **INDTEC**   A motor shaft has milled on it a flat for a set-screw to rest so that it can hold a pulley on the shaft (Fig. 17–105). What is the cross-sectional area of the shaft after being milled? Round to hundredths. *See Example 19.*

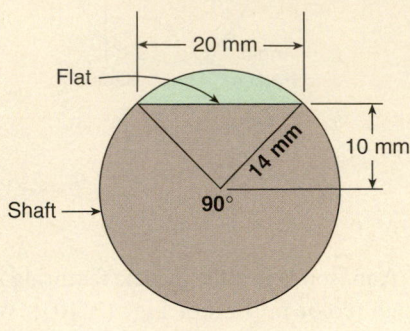

FIGURE 17–105

64. **CON**   A contractor pours a concrete patio in the shape of a circle except where the patio touches the exterior wall of the house (Fig. 17–106). What is the area of the patio in square feet? Round to hundredths.

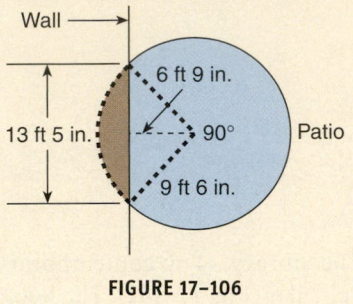

FIGURE 17–106

17–4   Volume and Surface Area

Common household items like cardboard storage boxes and toy building blocks are examples of three-dimensional geometric figures classified generally as *prisms*. Cans and pipes are examples of *cylinders*.

1   Find the Volume of Three-Dimensional Objects.   A *prism* is a three-dimensional figure with polygonal bases (ends) that are parallel and faces (sides) that are parallelograms, rectangles, or squares (Fig. 17–107). In a *right rectangular prism*, the faces are perpendicular to the bases.

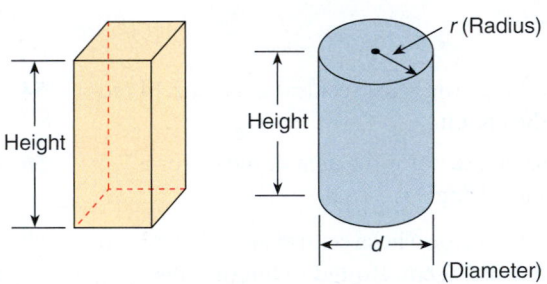

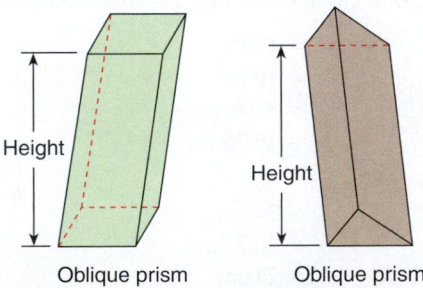

FIGURE 17–107

Right circular cylinder: a three-dimensional figure with a curved surface and two circular bases such that the height is perpendicular to the bases

A **right circular cylinder** is a three-dimensional figure with a curved surface and two circular bases such that the height is perpendicular to the bases.

Oblique prism: a prism for which the faces are *not* perpendicular to the bases

Oblique cylinder: a cylinder for which the faces are *not* perpendicular to the bases

Height of a prism or cylinder: the shortest distance between the two bases

In an **oblique prism** or an **oblique cylinder** the faces are *not* perpendicular to the bases.

The **height of a prism or cylinder** with two bases is the shortest distance between the two bases.

In right circular cylinders and in right rectangular prisms, the height is the same as the length of a side or face. However, in oblique prisms and in oblique cylinders the height is the perpendicular distance between the bases and is shorter than the length of a side or face.

As we saw in Chapter 1, Section 3, Outcome 2, the *volume* of an object, such as a container, is used to estimate how many containers can be loaded into a given-size storage area or shipped in a container of certain dimensions.

The *volume* of a three-dimensional geometric figure is the amount of space it occupies, measured in terms of three dimensions (length, width, and height). Volume is expressed as a *cubic measure*.

A general formula can be used for the volume of *any* right prism or right cylinder. If the base of the prism is a triangle, trapezoid, or other polygon, we use the appropriate formula for the area of the base.

> **Formula for the volume of right prism or right cylinder:**
>
> $$V = Bh$$
>
> where B is the area of the base and h is the height of the prism or cylinder.

STOP AND CHECK

1. Find the volume to the nearest tenth of a triangular prism if the height is 5 ft and the bases are triangles 1 ft on a side and 0.87 ft in height. Round to the nearest tenth.

Answer:
1. 2.2 ft³

EXAMPLE 1

Find the volume of the triangular prism in Fig. 17–108 if the height is 15 cm and the bases are triangles 3 cm on a side and 2.6 cm in height.

$V_{\text{prism}} = Bh$ Substitute the formula for the area of the triangular base for B.

$V_{\text{prism}} = \left(\dfrac{1}{2}bh_1\right)h_2$ $h_1 = 2.6$ cm (height of prism base)
    $h_2 = 15$ cm (height of prism)

$V_{\text{prism}} = \left[\dfrac{1}{2}(3)(2.6)\right]15$ Substitute values and evaluate.

$V_{\text{prism}} = 58.5$ cm³

The volume of the prism is 58.5 cm³.

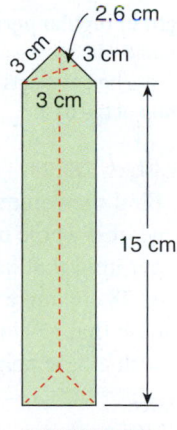

FIGURE 17–108

See Exercises 1–3.

STOP AND CHECK

1. Find the volume of a cylinder with a diameter of 0.5 m and a height of 3.2 m. Round to the nearest tenth.

Answer:
1. 0.6 m³

EXAMPLE 2

What is the cubic-inch displacement (space occupied) of a cylinder (Fig. 17–109) whose diameter is 5 in. and whose height is 4 in.? Round to the nearest tenth.

$V = Bh$ Substitute the formula for the area of the circular base for B.
$V = \pi r^2 h$ Substitute values; $r = \frac{1}{2}$ diameter, or 2.5.
$V = \pi(2.5)^2(4)$ Evaluate.
$V = 25\pi$ in³   Exact solution
$V = 78.5$ in³   Round.

The displacement of the cylinder is 78.5 in³.

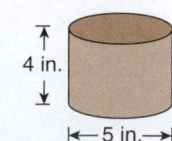

FIGURE 17–109

See Exercises 4–6.

Right cone: a three-dimensional figure whose base is a circle and whose side surface tapers to a point, called the apex, and whose height is the perpendicular line segment between the base and the apex

Apex of a cone or pyramid: the vertex formed by the sides that is the tip of the height

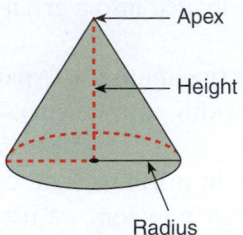

FIGURE 17–110

Pyramid: a three-dimensional geometric shape that has a polygon for a base and lateral faces that are triangles with a common vertex (apex)

Lateral faces of a pyramid: the triangular sides of a pyramid with a common vertex (apex)

Right or regular pyramid: a pyramid with a base of equal sides and the height meets the base at the center of the base

STOP AND CHECK

1. Find the volume to the nearest whole number of a pyramid that has a height of 28 cm and a triangular base that is 9 in. on a side with a base height of 7.8 in.

Answer:
1. 246 in³

Frustum of a cone: a part of a cone between the base and a plane passing through the cone parallel to the base

A **right cone** is a three-dimensional figure whose base is a circle and whose side surface tapers to a point, called the **apex.** The apex is the vertex formed by the sides that is the tip of the height, and the height is the perpendicular line segment between the base and apex (Fig. 17–110). One example of a *cone* is the funnel or the circular rain cap placed on top of stove vent pipes extending through the roof of some homes.

A **pyramid** is a three-dimensional geometric shape that has a polygon for a *base* and **lateral faces** that are triangles with a common vertex (apex) (Fig. 17–111). The *height or altitude* of the pyramid is the perpendicular distance from the vertex to the base. If the base of the pyramid is a polygon that has all sides equal, such as a square, the height meets the base at the center of the base and is a **right or regular pyramid.** Fig. 17–113 shows right pyramids with 3-, 4-, and 5-sided bases.

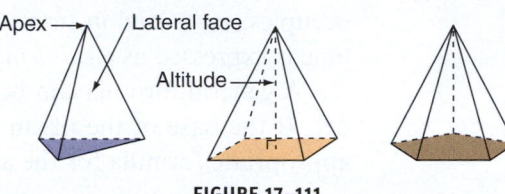

FIGURE 17–111

Volume of any cone or pyramid:

$$V = \frac{1}{3}Bh$$

where V is the volume, B is the area of the base, and h is the height of the cone or pyramid.

EXAMPLE 3

Find the volume of a pyramid that has a height of 28 cm and a square base that is 15 cm on a side (Fig. 17–112).

$V = \frac{1}{3}Bh$ Find the area of the base and substitute values for B and h. $B = s^2 = 15^2 = 225$ cm²

$V = \frac{1}{3}(225)(28)$ Multiply.

$V = \mathbf{2,100}$ cm³   Volume is cubic units.

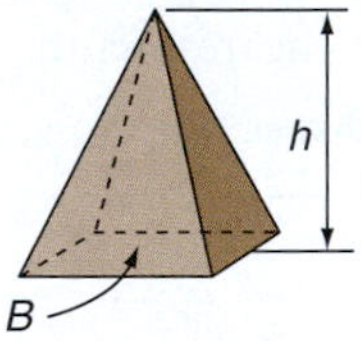

FIGURE 17–112

See Exercises 7–2.

A **frustum of a cone** is a part of a cone between the base and a plane passing through the cone parallel to the base (Fig. 17–113).

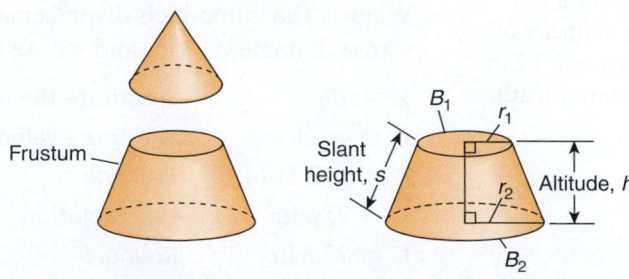

FIGURE 17–113

Frustum of a pyramid: a part of a pyramid between the base and a plane passing through the pyramid parallel to the base

Similarly, a **frustum of a pyramid** is a part of a pyramid between the base and a plane passing through the pyramid parallel to the base (Fig. 17–114).

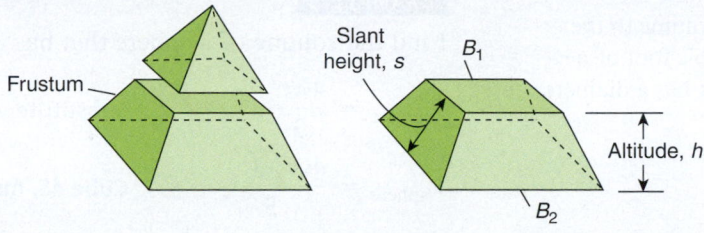

FIGURE 17-114

Volume of the frustum of a cone or pyramid:

$$V_{\text{frustum}} = \frac{1}{3} h (B_1 + B_2 + \sqrt{B_1 B_2})$$

where V is the volume, B_1 is the area of the base, B_2 is the area of the top, and h is the height of the frustum.

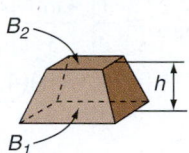

FIGURE 17-115

STOP AND CHECK

1. Find the volume to the nearest cubic centimeter of a frustum of a cone with a bottom base with a diameter of 16 cm and a top base with a diameter of 10 cm and a height of 5 cm.

Answer:
1. 675 cm³

EXAMPLE 4

Find the volume of the frustum of a pyramid with a square base that is 4 in. on each side, a square top that is 2 in. on each side, and a height of 3 in. (Fig. 17–115).

$V_{\text{frustum}} = \frac{1}{3} h (B_1 + B_2 + \sqrt{B_1 B_2})$ Substitute the values for B_1, B_2, and h.
$B_1 = 4^2 = 16 \text{ in}^2$; $B_2 = 2^2 = 4 \text{ in}^2$

$V_{\text{frustum}} = \frac{1}{3}(3)(16 + 4 + \sqrt{16 \cdot 4})$ Simplify the radicand.

$V_{\text{frustum}} = \frac{1}{3}(3)(16 + 4 + \sqrt{64})$ Take the square root.

$V_{\text{frustum}} = \frac{1}{3}(3)(16 + 4 + 8)$ Simplify the grouping.

$V_{\text{frustum}} = \frac{1}{3}(3)(28)$ Multiply.

$\mathbf{V_{\text{frustum}} = 28 \text{ in}^3}$ Volume is cubic units.   **See Exercises 13–15.**

Soccer balls, golf balls, tennis balls, baseballs, and ball bearings are *spheres*. Spheres are also used as tanks to store gas and water because spheres hold the greatest volume for a specified amount of surface area. A **sphere** is a three-dimensional figure formed by a curved surface with points that are all equidistant from a point inside called the center (Fig. 17–116). A **great circle** divides the sphere in half at its greatest diameter and is formed by a plane passing through the center of the sphere.

A sphere does not have bases like prisms and cylinders. Because of the relationship of the sphere to the circle, the formula for the volume of a sphere includes elements of the formula for the area of a circle.

Sphere: a three-dimensional figure formed by a curved surface with points that are all equidistant from a point inside called the center

Great circle: a great circle divides a sphere in half at its greatest diameter and is formed by a plane passing through the center of the sphere

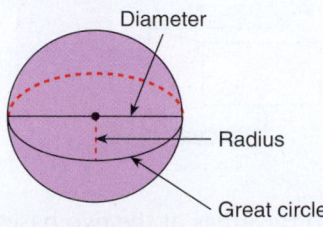

FIGURE 17-116 Sphere.

Volume of a sphere:

$$V_{\text{sphere}} = \frac{4\pi r^3}{3}$$

where r = radius.

Note that the radius is *cubed,* or raised to the power of 3, indicating volume.

EXAMPLE 5

Find the volume of a sphere that has a diameter of 90 cm.

$$V_{\text{sphere}} = \frac{4\pi r^3}{3}$$ Substitute value $r = 1/2\ d = 1/2(90) = 45$ cm.

$$V_{\text{sphere}} = \frac{4(\pi)(45)^3}{3}$$ Cube 45, multiply, and divide.

$$V_{\text{sphere}} = 381{,}704 \text{ cm}^3$$ Round.   **See Exercises 16–19.**

EXAMPLE 6

INDTR   Find the weight of the solid cast-iron object shown in Fig. 17–117 if cast iron weighs 0.26 lb per cubic inch. Round to the nearest whole pound.

The solution requires finding the volume of the cone that forms the top of the object, the volume of the cylinder that forms the middle portion of the object, and the volume of the **hemisphere** (half sphere) that forms the bottom of the object.

Hemisphere: half of a sphere

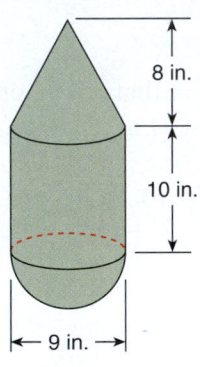

8 in.

10 in.

← 9 in. →

FIGURE 17–117

$$V_{\text{cone}} = \frac{\pi r^2 h}{3} \qquad V_{\text{cylinder}} = \pi r^2 h \qquad V_{\text{hemisphere}} = \frac{1}{2}\left[\frac{4\pi r^3}{3}\right]$$

$$V_{\text{cone}} = \frac{(\pi)(4.5)^2(8)}{3} \qquad V_{\text{cylinder}} = (\pi)(4.5)^2(10) \qquad V_{\text{hemisphere}} = \frac{1}{2}\left[\frac{4(\pi)(4.5)^3}{3}\right]$$

$$V_{\text{cone}} = 169.6460033 \text{ in}^3 \quad V_{\text{cylinder}} = 636.1725124 \text{ in}^3 \quad V_{\text{hemisphere}} = 190.8517537 \text{ in}^3$$

$$\text{total volume} = V_{\text{cone}} + V_{\text{cylinder}} + V_{\text{hemisphere}}$$
$$\text{total volume} = 996.6702694 \text{ in}^3$$

Convert to pounds:

$$\frac{996.6702694 \text{ in}^3}{1}\left(\frac{0.26 \text{ lb}}{1 \text{ in}^3}\right) = 259 \text{ lb} \qquad \text{Rounded}$$

To the nearest whole pound, the cast-iron object weighs 259 lb.   **See Exercises 20–21.**

2 **Find the Surface Area of Three-Dimensional Objects.** The surface area of a three-dimensional figure can refer to just the area of the *sides* of the figure. Or surface area can refer to the overall area, including the bases along with the sides.

The **lateral surface area (LSA)** of a three-dimensional figure is the area of its sides only.

The **total surface area (TSA)** of a three-dimensional figure is the area of the sides plus the area of its base or bases.

Lateral surface area (LSA): the lateral surface area of a three-dimensional figure is the area of its sides only (bases excluded)

Total surface area (TSA): the total surface area of a three-dimensional figure is the area of the sides plus the area of its base or bases

Lateral surface area of a right prism or a right circular cylinder:

$$\text{LSA} = ph$$

where p is the perimeter of the base and h is the height of the three-dimensional figure (Fig. 17–118).

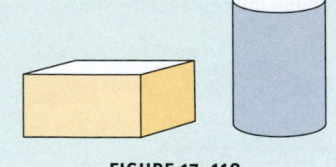

FIGURE 17–118

To get the total surface area (TSA) (Fig. 17–119), add the areas of the two bases to the lateral surface area.

> **Total surface area of a right prism or a right circular cylinder:**
>
> $$\text{TSA} = ph + 2B$$
>
> where p is the perimeter of the base, h is the height of the three-dimensional figure, and B is the area of the base (Fig. 17–119).

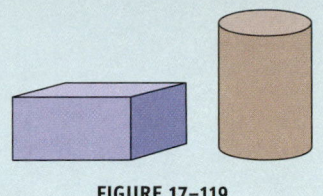

FIGURE 17–119

STOP AND CHECK

1. Find the total surface area to the nearest tenth of a square foot of a cube that is 4 ft 5 in. on a side.

Answer:
1. 117.0 ft^2

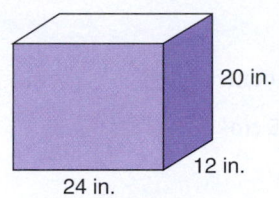

20 in.

12 in.

24 in.

FIGURE 17–120

EXAMPLE 7

INDTR Find the lateral surface area of a rectangular shipping carton measuring 24 in. in length, 12 in. in width, and 20 in. in height (Fig. 17–120).

$\text{LSA} = p\,h$	Substitute the formula for the perimeter of a rectangle, $p = 2l + 2w$.
$\text{LSA} = (2l + 2w)h$	Substitute numerical values.
$\text{LSA} = [2(24) + 2(12)]20$	Perform calculations inside grouping.
$\text{LSA} = [72]20$	Multiply.
$\text{LSA} = 1{,}440 \text{ in}^2$	Area requires square units.

The lateral surface area of the carton is 1,440 in².   **See Exercise 22.**

EXAMPLE 8

Find the total surface area of the triangular prism shown in Fig. 17–121.

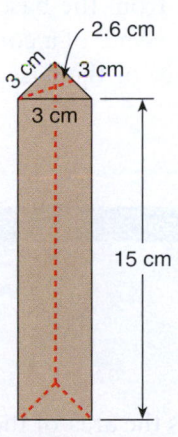

2.6 cm

3 cm · 3 cm

3 cm

15 cm

FIGURE 17–121

$\text{TSA} = ph + 2B$	Perimeter of triangular base is $a + b + c$.
	Area of triangular base B is $\frac{1}{2}bh_1$, where h_1 is the height of the triangular base.
$\text{TSA} = (a + b + c)h + 2\left(\frac{1}{2}bh_1\right)$	Substitute values.
$\text{TSA} = (3 + 3 + 3)(15) + 2\left(\frac{1}{2}\right)(3)(2.6)$	Add inside grouping.
$\text{TSA} = 9(15) + (2)\left(\frac{1}{2}\right)(3)(2.6)$	Perform multiplications.
$\text{TSA} = 135 + 7.8$	Add.
$\text{TSA} = 142.8 \text{ cm}^2$	Area requires square units.

The total surface area of the triangular prism is 142.8 cm².   **See Exercise 23.**

STOP AND CHECK

1. Find the lateral surface area of a rectangular prism with a length of 45 cm, a width of 28 cm, and a height of 24 cm.

Answer:
1. $3{,}504 \text{ cm}^2$

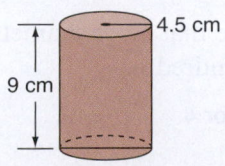

4.5 cm

9 cm

FIGURE 17–122

EXAMPLE 9

INDTR How many square centimeters of sheet metal are required to manufacture a can that has a radius of 4.5 cm and height of 9 cm? Assume no waste or overlap and round to the nearest square centimeter (Fig. 17–122).

$\text{TSA} = p\,h + 2\,B$	Total surface area is needed. Substitute formulas $p = 2\pi r$, $B = \pi r^2$.
$\text{TSA} = 2\pi r\,h + 2\pi r^2$	Substitute values.
$\text{TSA} = 2(\pi)(4.5)(9) + 2(\pi)(4.5)^2$	Square 4.5 and perform multiplications.

TSA = 254.4690049 + 127.2345025   Add.

TSA = 382 cm^2 Round. Area requires square units.

The can requires 382 cm^2 of sheet metal.   **See Exercises 24–26.**

A sphere does not have bases like prisms and cylinders. The surface area of a sphere includes *all* the surface so there is only one formula. Because of the relationship of the sphere to the circle, the formula includes elements of the formula for the area of a circle. The total surface area of the sphere is 4 times the area of a circle with the same radius.

Total surface area of a sphere:
$$TSA = 4\pi r^2$$
where r = radius.

STOP AND CHECK

1. Find the surface area to the nearest square foot of a sphere that has a radius of 15 ft.

Answer:
1. 2,827 ft^2

EXAMPLE 10

Find the surface area of a sphere that has a diameter of 90 cm.

TSA = $4\pi r^2$ Substitute values. $\frac{1}{2}d = r$, so r = 45 cm.

TSA = $4(\pi)(45)^2$ Square 45 and multiply.

TSA = 25,447 cm^2   Rounded   **See Exercises 27–29.**

Slant height: the distance along the side of a cone from the base to the apex

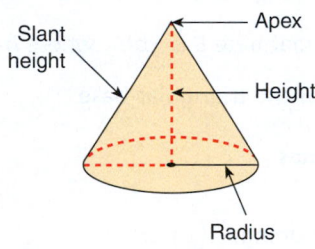

FIGURE 17–123 Circular cone.

The **slant height** of a cone is the distance along the side from the base to the apex. Fig. 17–123 shows the *perpendicular height* and the *slant height* of a cone.

The lateral surface area of a cone equals the circumference of the base times $\frac{1}{2}$ the slant height, or LSA = $2\pi r\left(\dfrac{1}{2}s\right)$, which simplifies as πrs.

Lateral surface area of a cone:
$$LSA = \pi rs$$
where r is the radius and s is the slant height.

The total surface area, then, is the lateral surface area plus the area of the base.

Total surface area of a cone:
$$TSA = \pi rs + \pi r^2$$
where r is the radius of the circular base and s is the slant height. The area of the base is πr^2.

STOP AND CHECK

1. Find the lateral surface area and total surface area of a cone that has a diameter of 18 in., height of 12 in., and slant height of 15 in. Round to hundredths.

Answer:
1. LSA = 424.12 in^2;
TSA = 678.58 in^2

EXAMPLE 11

Find the lateral surface area and total surface area of a cone that has a diameter of 8 cm, height of 6 cm, and slant height of 7 cm. Round to hundredths.

LSA = πrs Substitute values: $r = \frac{1}{2}d$, or 4.

LSA = $(\pi)(4)(7)$ Multiply.

LSA = 87.96 cm^2   Rounded from 87.9645943

$$TSA = \pi rs + \pi r^2$$ Substitute values. Use full calculator value for πrs.

$$TSA = 87.9645943 + (\pi)(4)^2$$ Perform operations using the proper order of operations.

$$TSA = 138.23 \text{ cm}^2$$ Round.   **See Exercises 30–32.**

17–4 EXERCISES

MyLab Math For additional practice go to your study plan in MyLab Math.

1 Find the volume in Fig. 17–124. Round to the nearest hundredth if necessary. *See Example 1.*

1.

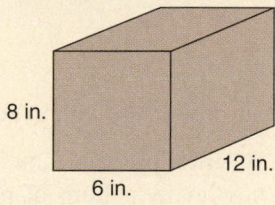

8 in.

12 in.

6 in.

FIGURE 17–124

2. A right pentagonal (five-sided) prism is 10 cm high. If the area of each pentagonal base is 32 cm², what is the volume of the prism?

3. What is the volume of a triangular prism that has a height of 8 in., a triangular base that measures 4 in. on each side, and a height of 3.46 in.? Round to hundredths.

Find the volume in Fig. 17–125. Round to the nearest hundredth if necessary. *See Example 2.*

4.

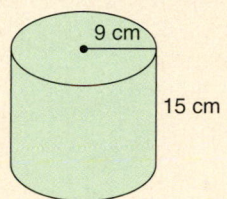

9 cm

15 cm

FIGURE 17–125

5. How many cubic inches are in an aluminum can with a $2\frac{1}{2}$ in. diameter and $4\frac{3}{4}$ in. height? Round to tenths.

6. What is the volume of a cylindrical oil storage tank that has a 40-ft diameter and 15-ft height? Round to the nearest whole number.

Solve. Round to tenths. *See Example 3.*

7. Find the volume of the cone in Fig. 17–126.

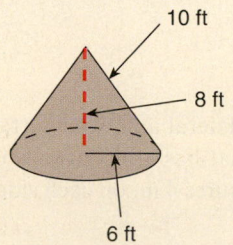

10 ft

8 ft

6 ft

FIGURE 17–126

8. **CON**   How many cubic feet are in a conical pile of sand that is 30 ft in diameter and 20 ft high?

9. **BUS**   A cone-shaped storage container holds a photographic chemical. If the container is 80 cm wide and 30 cm high, how many liters of the chemical does it hold if 1 L=1,000 cm³?

10. Find the volume of a pyramid that has a square base of 30 cm on a side and a height of 42 cm.

11. Find the volume of a pyramid that has a square base of 48 m on a side and a height of 100 m.

12. Find the volume of a pyramid that has an equilateral triangular base with altitude 10.39 m and a side of 12 m. The height is 20 m.

See Example 4.

13. A frustum of a pyramid has a square base that is 18 in. on each side, a square top that is 10 in. on each side, and a height of 13 in. Find the volume of the frustum.

14. A frustum of a pyramid has a triangular base that has an area of 32 cm² and a triangular top that has a surface area of 28 cm². The height of the frustum is 81 cm. Find the volume of the frustum.

15. **CON**   Cap blocks for a fence are molded in the shape of a frustum of a pyramid that has a square base and top. The base of the frustum is 30 in. on each side and the top is 24 in. on each side. The cap block is 5 in. thick (height of frustum). What is the volume of the frustum?

See Example 5.

16. Find the volume of a sphere with a radius of 5 cm.

17. Find the volume of a sphere with a radius of 6 in.

18. If 1 ft³ = 7.48 gal, how many gallons can a spherical water tank hold if its diameter is 45 ft?

19. **INDTR**   If a spherical propane tank is filled to 90% of its capacity, how many gallons of propane does the tank hold if its diameter is 4 ft? (1 ft³ = 7.48 gal.)

See Example 6.

20. A cylindrical storage tower for liquids with a conical top and hemispheric bottom (see Fig. 17–127) will be filled with a herbicide for preemergent weed control in the cylindrical and hemispheric portions of the tower. How many cubic feet to the nearest tenth of a cubic foot of herbicide will it hold?

21. If the herbicide in Exercise 20 can be purchased in bulk for $68.95 per gallon (1 ft³ ≈ 7.48 gal), how much will it cost to fill the storage tower in Fig. 17–129?

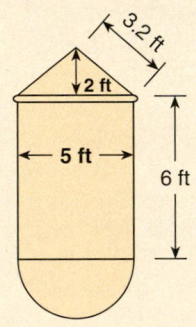

FIGURE 17–127

2 Solve. Round to tenths unless otherwise indicated.

See Example 7.

22. A right pentagonal prism is 10 cm high. If the area of each pentagonal base is 32 cm² and the perimeter is 20 cm, what are the lateral and total surface areas of the prism?

See Example 8.

23. What are the lateral and total surface areas of a triangular prism that has a height of 8 in. and a triangular base that measures 4 in. on each side with an altitude of 3.46 in.?

See Example 9.

24. **INDTEC**   How many square inches are in the lateral surface area and total surface area of an aluminum can with a $2\frac{1}{2}$ in. diameter and $4\frac{3}{4}$ in. height?

25. What are the lateral surface area and total surface area of a cylindrical oil storage tank that has a 40-ft diameter and 15-ft height?

26. **CON**   A cylindrical water well is 1,200 ft deep and 6 in. across. Find the lateral surface area of the well. Round to the nearest square foot. (*Hint:* Convert measures to a common unit.)

See Example 10.

27. Find the surface area of a sphere with a radius of 5 cm.

28. **INDTR**   How many square feet of steel are needed to manufacture a spherical water tank with a diameter of 45 ft?

29. **INDTR**   A spherical propane tank has a diameter of 4 ft. How many square feet of surface area need to be painted?

See Example 11.

30. Find the lateral surface area and total surface area of the cone that has a radius of 6 ft, slant height of 10 ft, and height of 8 ft.

32. Find the total surface area of a conical tank that has a radius of 15 ft and a slant height of 20 ft.

31. **INDTR**   How many square centimeters of sheet metal are needed to form a conical rain cap 25 cm in diameter if the slant height is 15 cm?

33. **INDTR**   A cylindrical water tower with a conical top and hemispheric bottom (see Fig. 17–128) needs to be painted. If the cost is $2.19 per square foot, how much does it cost (to the nearest dollar) to paint the tank?

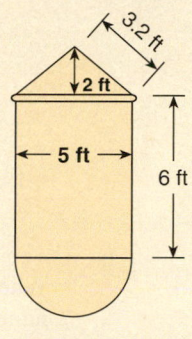

FIGURE 17–128

---

17   CHAPTER REVIEW OF KEY CONCEPTS

LEARNING OUTCOMES	KEY CONCEPTS AND EXAMPLES

Section 17–1

1   Use various notations to represent points, lines, line segments, rays, planes, and angles (pp. 706–709).

A dot represents a point and a capital letter names the point. A line extends in both directions and is named by any two points on the line (with double arrow above the letters, such as $\overleftrightarrow{AB}$). A line segment has a beginning point and an ending point, which are named by letters (with a bar above the letters, such as $\overline{CD}$). Line segments are sometimes named by letters only, such as CD. A ray extends from a point on a line and includes all points on one side of the point and is named by the end point and any other point on the line forming the ray (with an arrow above the letters, such as $\overrightarrow{AC}$). A plane contains an infinite number of points and lines on a flat surface.

Use proper notation for the following:

(a) Line *GH*   **(b)** Line segment *OP*   **(c)** Ray *ST*

(a) $\overleftrightarrow{GH}$ **(b)** $\overline{OP}$ **(c)** $\overrightarrow{ST}$

LEARNING OUTCOMES   **KEY CONCEPTS AND EXAMPLES**

Lines that intersect meet at only one point. Lines that coincide fit exactly on top of one another. Lines that are parallel are the same distance from one another along their entire lengths and never intersect.

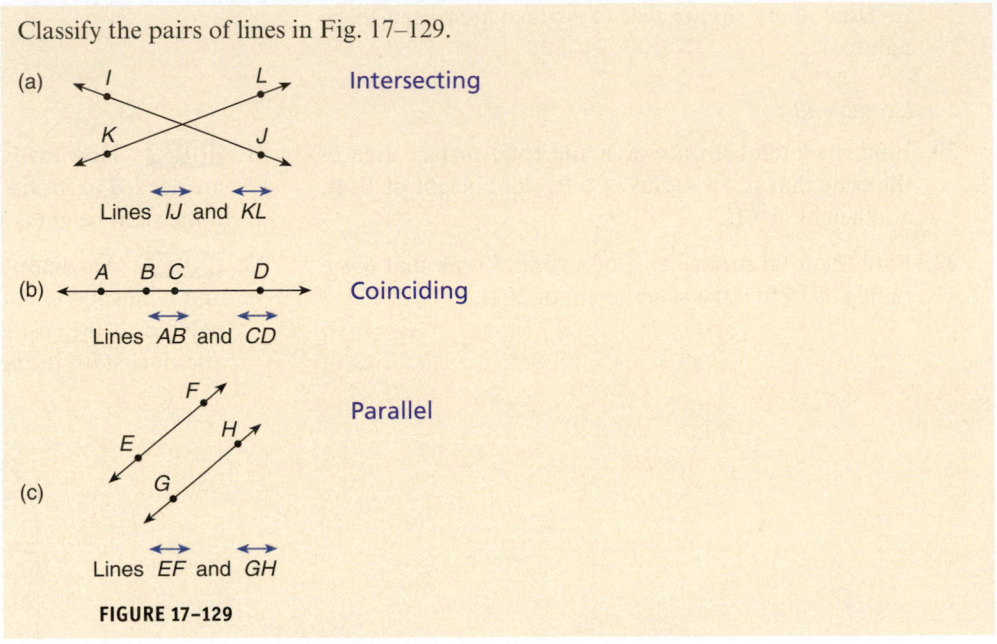

Classify the pairs of lines in Fig. 17–129.

(a) Lines *IJ* and *KL*   Intersecting

(b) Lines *AB* and *CD*   Coinciding

(c) Lines *EF* and *GH*   Parallel

FIGURE 17–129

When two rays intersect in a point (end point of rays), an angle is formed. Angles may be named with three capital letters (end point in middle and one point on each ray), such as $\angle ABC$. They may be named by only the middle letter (vertex of the angle), such as $\angle B$. They may be assigned a number or lowercase letter placed within the vertex, such as $\angle 2$ or $\angle d$.

Name the angle in Fig. 17–130 three ways.

FIGURE 17–130

$\angle DEF$ or $\angle FED$, $\angle E$, $\angle a$

2 Classify angles according to size (pp. 709–711).

A right angle contains 90°, or one-fourth a rotation. A straight angle contains 180°, or one-half a rotation. An acute angle contains less than 90° but more than 0°. An obtuse angle contains more than 90° but less than 180°. The sum of the measures of complementary angles is 90°. The sum of the measures of supplementary angles is 180°.

1. Examine the angle to determine the number of degrees it measures. **2.** Classify the angle as a right, straight, acute, or obtuse angle based on its degree measure.

Identify the following angles:

(a) 30°	(b) 100°	(c) 90°	(d) 180°
(a) acute	(b) obtuse	(c) right	(d) straight

LEARNING OUTCOMES	KEY CONCEPTS AND EXAMPLES

1. To find the complement of an angle, subtract its measure from 90°.   **2.** To find the supplement of an angle, subtract its measure from 180°.

> Find the complement and supplement of an angle that measures 43°.
>
> Complement   Supplement
>
> $90° - 43° = 47°$ $180° - 43° = 137°$

3 Determine the measure of an angle using relationships among intersecting lines (pp. 711–714).

Relationships among angles formed by intersecting lines: Vertical angles are equal. Adjacent angles are supplementary.

When two parallel lines are cut by a transversal: Corresponding angles are equal. Adjacent angles are supplementary. Interior angles on the same side of the transversal are supplementary. Alternate angles are equal. Exterior angles on the same side of the transversal are supplementary.

> In Fig. 17–131 identify at least one pair of angles that are corresponding, alternate interior, and exterior on the same side of the transversal. Then state the relationship between each pair of angles.
>
>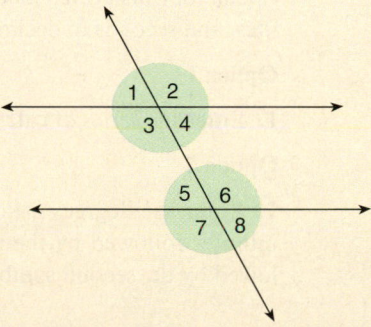
>
> **FIGURE 17–131**
>
> Corresponding angles: $\angle 1$ and $\angle 5$ Corresponding angles are equal.
>
> Alternate interior angles: $\angle 3$ and $\angle 6$ Alternate interior angles are equal.
>
> Exterior angles on the
> same side of the   Exterior angles on the same side of
> transversal: $\angle 1$ and $\angle 7$ the transversal are supplementary.

4 Convert angle measures between decimal degrees and degrees, minutes, and seconds (pp. 714–716).

Convert from degrees, minutes, and seconds to a decimal-degree equivalent: **1.** Find the sum of the degrees, the minutes divided by 60, and the seconds divided by 3,600.

$$\text{degrees} + \frac{\text{minutes}}{60} + \frac{\text{seconds}}{3,600}$$

2. Round to the desired decimal place.

> Change 45°15′45″ to a decimal-degree equivalent. Round to the nearest hundredth.
>
> $45 + 15/60 + 45/3,600 = 45.2625$
>
> $45°15′45″ = 45.26°$

LEARNING OUTCOMES	KEY CONCEPTS AND EXAMPLES

Convert from a decimal-degree format to degrees, minutes, and seconds: **1.** The whole-number part of the decimal-degree format will be the degrees. **2.** Multiply the decimal part of the degree in decimal-degree format by 60 and the whole-number part of the product will be the minutes. **3.** Multiply the decimal part of the product from Step 2 by 60 to obtain the seconds. **4.** Round the seconds to the desired decimal place.

> Change $110.625°$ to degrees, minutes, and seconds.
>
> $0.625(60) = 37.5'$ $0.5(60) = 30''$
>
> **$110°37'30''$**
>
> Using a calculator with a sexagesimal function key:
>
> 110.625 $\boxed{\text{2}^{\text{nd}}}$ $\boxed{\text{ANGLE}}$ $\boxed{\text{4:▶DMS}}$ $\boxed{\text{ENTER}}$
>
> $110.625° = 110°37'30''$

Change the format of angle measures using a calculator:

Convert from degrees in decimal notation to degrees, minutes, and seconds (calculator steps will vary with different calculators): 1. Set the calculator to degree mode. **2.** Enter the angle measure in decimal notation. **3.** Access the sexagesimal function key or menu choice on the calculator (most often labeled $\boxed{° \,' \,''}$ or $\boxed{\text{DMS}}$). **4.** Press $\boxed{\text{ENTER}}$ or $\boxed{=}$. From degrees, minutes, and seconds to decimal notation (calculator steps will vary with different calculators):

Option 1

1. Enter the series of calculations—degrees + minutes ÷ 60 + seconds ÷ 3,600.

Option 2

1. Enter the degrees followed by the degree symbol ($\boxed{\text{2}^{\text{nd}}}$, $\boxed{\text{ANGLE}}$, $\boxed{\text{1:°}}$). **2.** Enter the minutes followed by the minute symbol ($\boxed{\text{2}^{\text{nd}}}$, $\boxed{\text{ANGLE}}$, $\boxed{\text{2:'}}$). **3.** Enter the seconds followed by the second symbol ($\boxed{\text{ALPHA}}$, $\boxed{''}$). **4.** Press $\boxed{\text{ENTER}}$ or $\boxed{=}$.

Section 17–2

1 Find the perimeter of a polygon using the appropriate formula (pp. 718–721).

1. Select the appropriate formula. **2.** Substitute the known values into the formula. **3.** Evaluate the formula (perform the indicated operations).

Formulas for perimeter:

$P_{\text{parallelogram}} = 2(b + s)$: b is the base, or the length of a side that is or can be rotated to a horizontal position; and s is the length of a side that joins with the base.
$P_{\text{rectangle}} = 2(l + w)$: l is length of the long side; w is width of the short side.
$P_{\text{square}} = 4s$: s is the length of a side.
$P_{\text{rhombus}} = 4s$, where s is the length of a side.
$P_{\text{trapezoid}} = b_1 + b_2 + a + c$, where b_1 and b_2 are the lengths of the bases or parallel sides and a and c are the lengths of the sides adjacent to the bases.
$P_{\text{triangle}} = a + b + c$, where a, b, and c are the lengths of the three sides.

> What is the perimeter (P) of a triangular roof vent that has sides a, b, and c of 6 ft, 4 ft, and 4 ft, respectively?
>
> $P_{\text{triangle}} = a + b + c$ Select the appropriate formula and substitute values.
>
> $P_{\text{triangle}} = 6 + 4 + 4$ Add.
>
> $P_{\text{triangle}} = 14$ ft

LEARNING OUTCOMES	KEY CONCEPTS AND EXAMPLES

2 Find the area of a polygon using the appropriate formula (pp. 721–726).

1. Visualize the polygon.   **2.** Select the appropriate formula.   **3.** Substitute the known values into the formula.   **4.** Evaluate the formula (perform the indicated operations).

Formulas for area:

$A_{rectangle} = lw$, where l is the length and w the width.

$A_{square} = s^2$, where s is the length of a side.

$A_{parallelogram} = bh$: b is the base; h is the height, or length of a perpendicular distance between the bases.

$A_{rhombus} = bh$, where b is the base and h is the height.

$A_{trapezoid} = \frac{1}{2}h(b_1 + b_2)$, where b_1 and b_2 are the lengths of the bases or two parallel sides and h is the height between the parallel bases.

$A_{triangle} = \frac{1}{2}(bh)$, where b is the base and h is the height.

Heron's formula for the area of a triangle: $A = \sqrt{s(s-a)(s-b)(s-c)}$, where $s = \frac{1}{2}(a + b + c)$ and a, b, and c are sides.

Find the area of a trapezoid that has bases of 11 in. and 18 in. and height of 12 in.

$A_{trapezoid} = \frac{1}{2}h(b_1 + b_2)$ Select the appropriate formula and substitute values.

$A_{trapezoid} = \frac{1}{2}(12)(11 + 18)$ Perform operation in grouping.

$A_{trapezoid} = \frac{1}{2}(12)(29)$ Multiply (in. × in. = in^2).

$A_{trapezoid} = 174$ in^2

Section 17–3

1 Find the circumference or area of a circle using the appropriate formula (pp. 731–734).

1. Select the appropriate formula.   **2.** Substitute values for r or d as appropriate.
3. Evaluate the formula. Use the calculator value for π.

Formula for circumference: $C = \pi d$, or $C = 2\pi r$; d is the diameter or distance across the center of a circle; r is the radius, or half the diameter; π is approximated on a calculator as 3.141592654.

Formula for area: $A_{circle} = \pi r^2$; r is the radius.

Find the area of a circle whose diameter is 3 m.

First, find the radius: $r = \dfrac{d}{2}$; 3 m ÷ 2 = 1.5 m.

$$A_{circle} = \pi r^2$$

$$A_{circle} = \pi(1.5 \text{ m})^2$$

$$A_{circle} = 7.07 \text{ m}^2 \text{ (rounded)}$$

---

2 Convert angle measures between degrees and radians (pp. 734–736).

Convert degrees to radians: 1. Multiply degrees by the unit ratio $\dfrac{\pi \text{ rad}}{180°}$.   **2.** Write the product in simplest form.

Using a calculator (steps may vary):

1. Set angle MODE to radian.   **2.** Enter the degree measure followed by the degree symbol. Use $\boxed{° ' ''}$, $\boxed{\text{DMS}}$, or the angle menu, option $\boxed{1:°}$.   **3.** Enter the minute measure followed by the minute symbol. Use $\boxed{° ' ''}$, $\boxed{\text{DMS}}$, or the angle menu, option $\boxed{2:'}$.   **4.** Enter the second measure followed by the second symbol. Use $\boxed{° ' ''}$, $\boxed{\text{DMS}}$, or $\boxed{\text{ALPHA}}\boxed{''}$.   **5.** Press $\boxed{\text{ENTER}}$.

LEARNING OUTCOMES	KEY CONCEPTS AND EXAMPLES

Change 125°30′30″ to radians to the nearest hundredth.

125 [2nd] [ANGLE] [1:°] 30 [2nd] [ANGLE] [2:′] 30 [ALPHA] [″] [2nd] [ANGLE], [1:°] [ENTER]
= 2.190533655 = 2.19 (rounded)

Convert radians to degrees: 1. Multiply radians by the unit ratio $\dfrac{180°}{\pi \text{ rad}}$. **2.** Write the product in simplest form.

Using a calculator (steps may vary):

1. Set angle MODE to degree. **2.** Enter radian amount and radian symbol (angle menu, option [3:ʳ]). **3.** If degrees, minutes, and seconds are desired, use DMS function (angle menu, option [4:▶DMS]).

Change 1.245 rad to a decimal degree to the nearest thousandth.

1.245 [[2nd] [ANGLE] menu, option 3] [ENTER] = 71.33324549 = 71.333° (rounded)

3 Find the arc length of a sector (pp. 736–738).

Arc length of a sector is the portion of the circumference formed by the sides of a sector.

Using degrees:

1. Substitute known values into the formula.

$$s = \frac{\theta}{360}(2\pi r) \quad \text{or} \quad s = \frac{\theta}{360}(\pi d)$$

where θ is a central angle measured in degrees and r is the radius of the circle.

2. Evaluate.

Using radians:

1. Substitute the given radian measure of the central angle and the radius in the formula

$$s = \theta r$$

where θ is the central angle measured in radians and r is the radius of the circle.

2. Simplify the expression.

Find the arc length of a sector formed by a 50° central angle if the radius is 20 cm (Fig. 17–132).

$s = \dfrac{\theta}{360}(2\pi r)$ Substitute $\theta = 50°$ and $r = 20$ cm.

$s = \dfrac{50}{360}(2)(\pi)(20)$ Evaluate.

$s = 17.45329252$ or   17.45 cm (rounded)

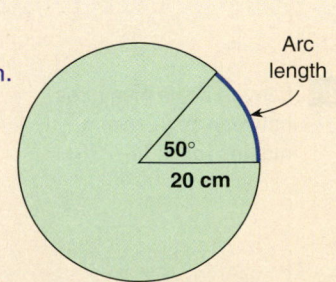

FIGURE 17–132

LEARNING OUTCOMES	KEY CONCEPTS AND EXAMPLES

4 Find the area of a sector or segment (pp. 738–740).

A sector is a portion of a circle cut off by two radii (r).
Using degrees:

1. Calculate the portion of the circle included in the sector. θ is a central angle measured in degrees.

$$\frac{\theta}{360} = \text{fractional part of circle}$$

2. Find the fractional part of the area of the circle.

$$A_{\text{sector}} = \frac{\theta}{360}\,\pi r^2 \text{ where } \pi r^2 = \text{area of circle}$$

Using radians:

1. Substitute known values into the formula $A_{\text{sector}} = \frac{1}{2}\theta r^2$, where θ is the central angle measured in radians and r is the radius of the circle.   2. Solve for the missing value.

Find the area of a sector that has a central angle of 0.87 rad and a radius of 20 cm (Fig. 17–133).

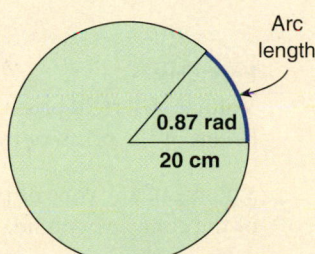

Arc length

0.87 rad

20 cm

FIGURE 17–133

$\text{Area} = \dfrac{1}{2}\theta r^2$ Formula for area of a sector of a circle. Substitute $\theta = 0.87$ rad and $r = 20$ cm.

$\text{Area} = \dfrac{1}{2}(0.87)(20)^2$ Evaluate. (cm × cm = cm²)

$\text{Area} = 174 \text{ cm}^2$ Round.

A chord is a line segment joining two points on a circle. An arc is the portion of the circumference cut off by a chord. A segment is the portion of a circle bounded by a chord and an arc.

1. Substitute known values into the formula.

Using degrees:

$$A_{\text{segment}} = \frac{\theta}{360}\,\pi r^2 - \frac{1}{2}bh$$

Using radians:

$$A_{\text{segment}} = \frac{1}{2}\theta r^2 - \frac{1}{2}bh$$

where $\dfrac{\theta}{360}\,\pi r^2$ or $\dfrac{1}{2}\theta r^2$ is the area of the sector and $\dfrac{1}{2}bh$ is the area of the triangle.

2. Evaluate.

| LEARNING OUTCOMES | KEY CONCEPTS AND EXAMPLES |

Find the area of the segment formed by a 12-cm chord if the height of the triangle part of the sector is 6 cm and the radius is 8.5 cm. The central angle is 90° (Fig. 17–134).

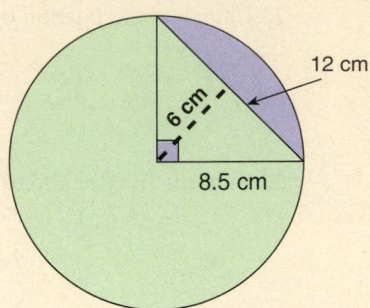

FIGURE 17–134

$$A_{\text{segment}} = \frac{\theta}{360}\,\pi r^2 - \frac{1}{2}bh \qquad \text{Substitute } \theta = 90°, r = 8.5, b = 12, h = 6.$$

$$A_{\text{segment}} = \frac{90}{360}\,(\pi)(8.5)^2 - \frac{1}{2}(12)(6) \qquad \text{Evaluate.}$$

$$A_{\text{segment}} = 0.25(\pi)(72.25) - 0.5(72) \qquad \text{Multiply and subtract.}$$

$$A_{\text{segment}} = 20.74501731 \quad \text{or} \quad 20.75\ \text{cm}^2 \qquad \text{Round.}$$

Section 17–4

1 Find the volume of three-dimensional objects (pp. 744–748).

Formulas:

Volume of a right prism or right cylinder: $V = Bh$, where B is the area of the base and h is the height of the prism or cylinder.

Volume of any cone or pyramid: $V = \frac{1}{3}Bh$, where B is the area of the base and h is the height of the cone or pyramid.

Volume of a frustum of a cone or pyramid: $V = \frac{1}{3}h(B_1 + B_2 + \sqrt{B_1 B_2})$, where h is the height of the frustum, B_1 is the area of the base, and B_2 is the area of the top.

Volume of a sphere: $V = \frac{4\pi r^3}{3}$, where r is the radius.

Find the volume of a cylinder that has a diameter of 20 mm and a height of 80 mm.

$V = Bh$	$B = \pi r^2.$
$V = \pi r^2 h$	$r = \frac{1}{2}d = 10\ \text{mm}.$
$V = \pi(10)^2(80)$	
$V = \pi(100)(80)$	
$V = 25{,}132.74123\ \text{mm}^3 \approx 25{,}133\ \text{mm}^3$	Round.

2 Find the surface area of three-dimensional objects (pp. 748–751).

Formulas:

Lateral surface area (area of sides) of a right prism or a right circular cylinder: $\text{LSA} = ph$, where p is the perimeter of the base and h is the height.

Total surface area (area of sides plus bases) of a right prism or a right circular cylinder: $\text{TSA} = ph + 2B$, where p is the perimeter of the base, h is the height, and B is the area of a base.

Total surface area of a sphere: $\text{TSA} = 4\pi r^2$, where r is the radius.

LEARNING OUTCOMES	KEY CONCEPTS AND EXAMPLES

Lateral surface area of a cone: LSA $= \pi rs$, where r is the radius and s is the slant height.
Total surface area of a cone: TSA $= \pi rs + \pi r^2$, where r is the radius of the circular base and s is the slant height. The area of the base is πr^2.

> A conical pile of gravel has a diameter of 30 ft and a slant height of 40 ft. What is the lateral surface area?
>
> LSA $= \pi rs$ $r = \dfrac{1}{2}d = \dfrac{1}{2}(30) = 15.$
>
> LSA $= \pi(15)(40)$
>
> LSA $= 1{,}884.96 \text{ ft}^2$ Rounded from 1,884.955592

17   CHAPTER REVIEW EXERCISES

Section 17–1   MyLab Math   **For additional practice go to your study plan in MyLab Math.**

Use Fig. 17–135 for Exercises 1–3.

1. Which lines are parallel?

2. Which lines coincide?

3. Which lines intersect?

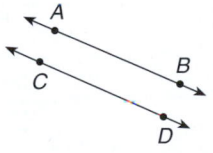 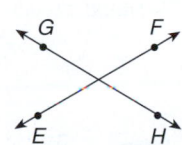 

FIGURE 17–135

Use Fig. 17–136 for Exercises 4–5.

4. Name $\angle a$ using three capital letters.

5. Name $\angle a$ using one capital letter.

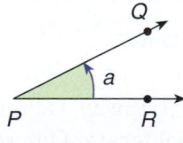

FIGURE 17–136

Classify the angle measures using the terms *right, straight, acute,* and *obtuse.*

6. 50°   **7.** 90°   **8.** 120°   **9.** 180°

State whether the angle pairs are complementary, supplementary, or neither.

10. 98°, 62°   **11.** 135°, 45°   **12.** 45°, 35°   **13.** 21°, 79°   **14.** 90°, 90°

Use Fig. 17–137 to answer Exercises 15–16.

15. Name two pairs of supplementary angles.

16. What are angles *a* and *c* called?

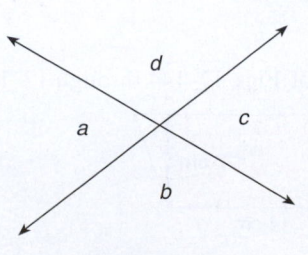

FIGURE 17–137

In Fig. 17–138, $l \parallel m$ and n is a transversal.

17. Name two pairs of alternate exterior angles.

18. If $\angle a = 147°$ what is the measure of $\angle e$?

19. If $\angle d = 135°$ what is the measure of $\angle f$?

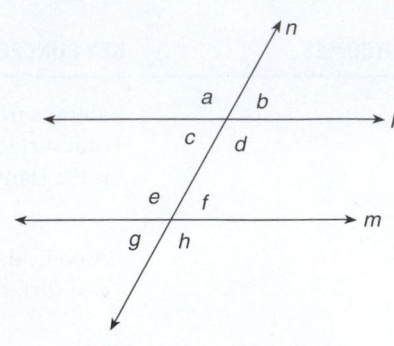

FIGURE 17–138

Change to decimal-degree equivalents. Express the decimals to the nearest ten-thousandth.

20. $32°14'15''$ **21.** $29'$ **22.** $47''$ **23.** $7'34''$

Change to equivalent degrees, minutes, and seconds. Round to the nearest second when necessary.

24. $20.6°$ **25.** $0.75°$ **26.** $0.46°$ **27.** $0.2176°$

Section 17–2

Find the perimeter of Figs. 17–139 through 17–143.

28.

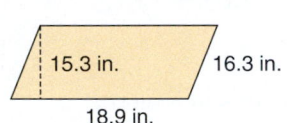

FIGURE 17–139

29.

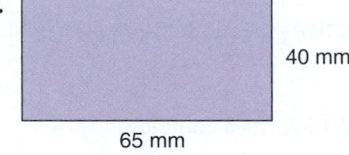

FIGURE 17–140

30.

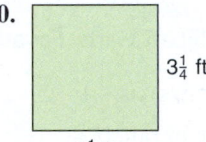

FIGURE 17–141

31.

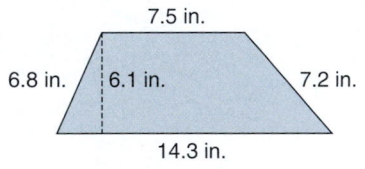

FIGURE 17–142

32.

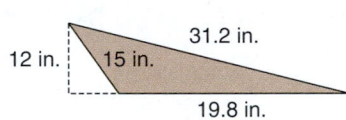

FIGURE 17–143

33. The Tennessee Highway Department has signs in the form of a parallelogram. One set of parallel sides each measures 15 ft and one set of parallel sides each measures 18 ft. Find the perimeter of the sign.

34. Antique tiles were often made in the form of a square that is 6 in. on each side. What is the perimeter of a tile?

35. A rectangular tablecloth measures 84 in. by 60 in. What length of lace is required to trim the edges of the cloth?

36. A classroom table in the form of a trapezoid has parallel sides that measure 26 in. and 48 in. The two nonparallel sides both measure 21.1 in. Find the length of trim needed to encase the edges of the table.

Find the area of Figs. 17–144 through 17–148.

37.

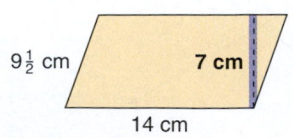

FIGURE 17–144

38.

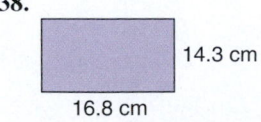

FIGURE 17–145

39.

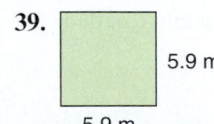

FIGURE 17–146

40.
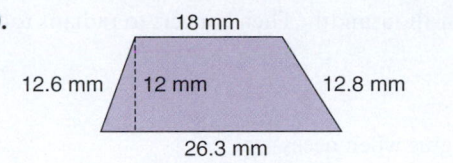
FIGURE 17–147

41.
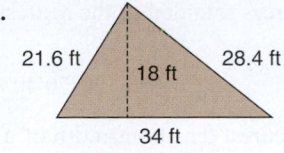
FIGURE 17–148

42. Find the area of a scalene triangle with sides of 15 in., 31.2 in., and 19.8 in. long.

43. **CON** If a parking lot for a new hospital in the shape of a parallelogram measures 300 ft by 175 ft and has a height of 120 ft, how many square feet need to be paved?

44. **CON** A hall wall with no windows or doors measures 30 ft long by 9 ft high. Find the number of square feet to be covered if paneling is installed on the two walls.

45. **CON** A den 21 ft by $18\frac{1}{2}$ ft is to be carpeted. How many square yards of carpeting are needed?

46. A square area of land is 10,000 m². If a baseball field must be at least 99.1 m along each foul line to the park fence, is the square adequate for regulation baseball?

47. **CON** Debbie Murphy is building a contemporary home with four front windows, each in the form of a parallelogram. If each window has a 5-ft base and a height of 2 ft, how many square feet of the 25 ft × 11 ft wall will require stain?

48. **CON** If the stain used on the wall in Exercise 47 is applied at a cost of $2.75 per square yard, find the cost to the nearest dollar of staining the front wall.

Section 17–3

Find the circumference (perimeter) and area of Figs. 17–149 through 17–152. Round to hundredths when necessary.

49.
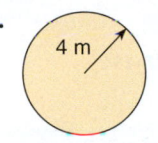
FIGURE 17–149

50.
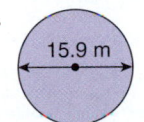
FIGURE 17–150

51.
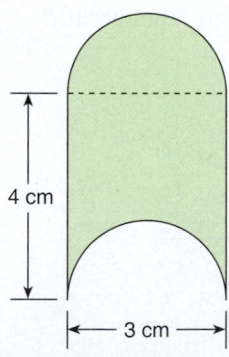
FIGURE 17–151

52.
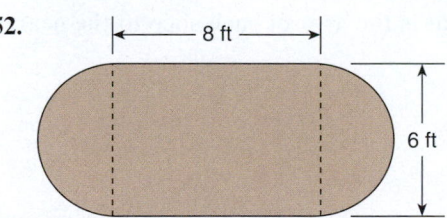
FIGURE 17–152

53. **INDTR** A $\frac{1}{4}$-in. electric drill with variable speed control turns as slowly as 25 rpm. If an abrasive disk with a 5-in. diameter is attached to the drill driveshaft, what is the disk's slowest cutting speed in ft/min? Round to the nearest whole number. (Cutting speed = circumference measured in feet × rpm.)

54. **CON** What is the cross-sectional area of the opening in a round flue tile whose inside diameter is 8 in.? Round any part of an inch to the next tenth of an inch.

Convert the degree measures to radians rounded to the nearest hundredth.

55. 60°

56. 212°

Convert the degree, minute, and second measures to degrees rounded to the nearest ten-thousandth. Then convert to radians to the nearest hundredth.

57. 99°45′

58. 120°20′40″

Convert the radian measures to degrees. Round to the nearest ten-thousandth of a degree when necessary.

59. $\dfrac{5\pi}{6}$ rad

60. 2.4 rad

Find the area of the sectors of a circle using Fig. 17–153. Round to hundredths.

61. ∠ = 45°9′
    $r = 2.58$ cm

62. ∠ = 2.88 rad
    $r = 15$ in.

63. ∠ = 40°
    $r = 2\frac{1}{2}$ ft

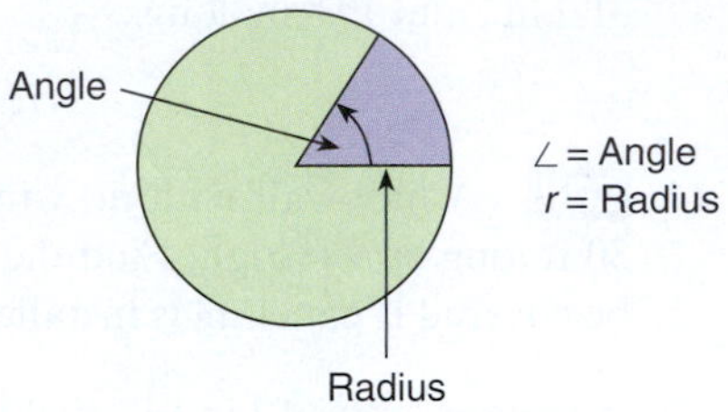

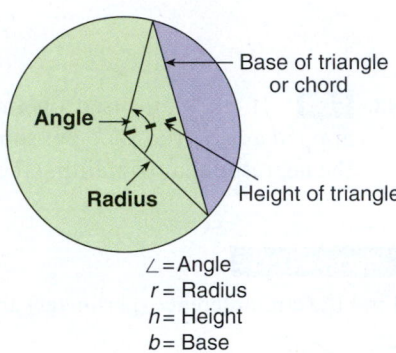

FIGURE 17–153

Find the arc length of the sectors of a circle using Fig. 17–153. Round to hundredths.

64. ∠ = 0.7 rad
    $r = 1.45$ ft

65. ∠ = 180°
    $r = 10$ in.

Find the area of the segments of a circle using Fig. 17–154. Round to hundredths.

66. ∠ = 55°
    $r = 10″$
    $h = 8.9″$
    $b = 9.2″$

67. ∠ = 30°
    $r = 24$ cm
    $h = 23.2$ cm
    $b = 12.4$ cm

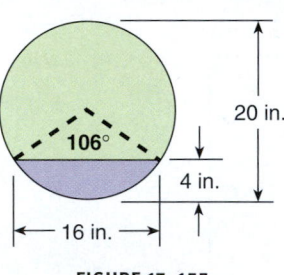

FIGURE 17–154

Solve.

68. **INDTR**   A machine cuts a 12-in.-diameter frozen pizza into slices with sides that form 72° angles at the center of the pizza. What is the area of each slice to the nearest square inch?

69. **CON**   A drain pipe with a 20-in. diameter has 4 in. of water in it (Fig. 17–155). What is the cross-sectional area of the water in the pipe to the nearest hundredth?

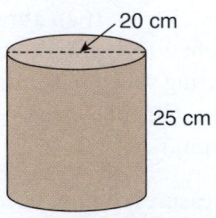

FIGURE 17–155

Section 17–4

70. Find the volume of the prism in Fig. 17–156.

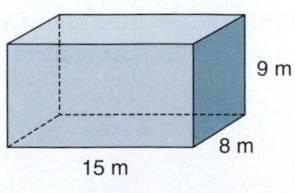

FIGURE 17–156

71. Find the volume of the cylinder in Fig. 17–157.

FIGURE 17–157

72. **CON**   If concrete weighs 160 lb per cubic foot, what is the weight of a concrete circular slab 4 in. thick and 15 ft across? Round to the nearest pound.

73. **CON**   How many cubic yards of topsoil are needed to cover an 85-ft by 65-ft area for landscaping if the topsoil is 6 in. deep? Round to the nearest whole number. $(27 \text{ ft}^3 = 1 \text{ yd}^3)$

74. **CON**   An interstate highway is repaired in one section 48 ft across, 25 ft long, and 8 in. deep. If concrete costs $25.50 per cubic yard, what is the cost of the concrete needed to repair the highway rounded to the nearest dollar? $(27 \text{ ft}^3 = 1 \text{ yd}^3)$

75. Find the total surface area of the triangular prism in Fig. 17–158.

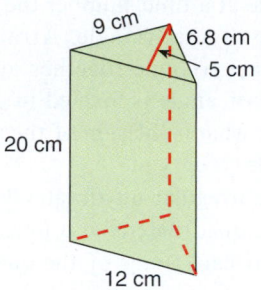

FIGURE 17–158

76. Find the lateral surface area of the cylinder in Fig. 17–159. Round to the nearest cm^2.

77. Find the total surface area of the cylinder in Fig. 17–159. Round to the nearest cm^2.

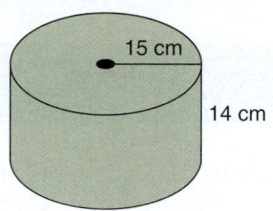

FIGURE 17–159

78. Find the total surface area of a cylinder that has a radius of 10 cm and a height of 30 cm. Round to the nearest cm^2.

79. **CON**   A pipeline to carry oil between two towns 5 mi apart has an outside diameter of 18 in. If 1 mi = 5,280 ft, find the lateral surface area to the nearest ft^2.

80. How many barrels of oil will a cylindrical tank hold if its height is $65\frac{1}{2}$ ft and its radius is 20 ft? Round to the nearest whole barrel. (31.5 gal = 1 barrel and 1 ft^3 = 7.48 gal.)

81. Find the volume of a pyramid that has a square base of 12 m on a side and a height of 36 m.

82. Find the volume of a frustum of a pyramid that has a square base of 15 yd on a side and a top of 12 yd on a side. The height is 12 yd.

Solve. Round the final answer to the nearest tenth unless otherwise specified.

83. Find the total surface area of a sphere with a radius of 9 m.

84. Find the volume of a sphere that has a diameter of 30 cm.

85. Find the lateral surface area of a cone with a radius of 6 cm and a slant height of 9 cm.

86. Find the total surface area of a cone that has a radius of 4 m and a slant height of 8 m.

87. **CON**   The entire exterior surface of a conical tank with a slant height of 12 ft and a diameter of 18 ft is being painted. If the paint covers at a rate of 350 ft^2 per gallon, how many gallons of paint are needed for the job? Round any fraction of a gallon to the next whole gallon.

88. **AG/H**   A hopper deposits grain in a cone-shaped pile with a diameter of 9′6″ and a height of 8′3″. To the nearest cubic foot, how much grain is deposited?

17 | TEAM PROBLEM-SOLVING EXERCISES

1. Use an inductive (experimental) process to determine the sum of the angles of irregular triangles and quadrilaterals.
   (a) Draw three scalene triangles and cut them out. Taking one triangle at a time, number the angles and tear off each angle of the triangle. Arrange the vertices of each original triangle together at a common point. What type of angle is formed in each case? Without measuring, what is the sum of the angles of each original triangle?
   (b) Draw three irregular quadrilaterals and cut them out. Taking one quadrilateral at a time, number the angles and tear off each angle of the quadrilateral. Arrange the vertices of the original quadrilateral together at a common point. What type of angle is formed? Without measuring, what is the sum of the angles of each original quadrilateral?

2. Use an inductive (experimental) process to determine the sum of the angles of irregular polygons.
   (a) Draw an irregular pentagon. From one vertex draw lines connecting this vertex with each nonadjacent vertex. What types of polygons are formed? Use your knowledge of the sum of the angles of a triangle to determine the sum of the angles of the pentagon.
   (b) Draw an irregular hexagon. From one vertex draw lines connecting this vertex with each nonadjacent vertex. Use your knowledge of the sum of the angles of a triangle to determine the sum of the angles of the hexagon.
   (c) Use your findings from Exercise 1 and parts (a) and (b) of Exercise 2 to state the pattern that is developing. Test your proposed generalization (formula) with at least two other irregular polygons like a *septagon* (seven-sided polygon), *octagon* (eight-sided polygon), *nonagon* (nine-sided polygon), and so on.

17 | CONCEPTS ANALYSIS

1. Explain the process for finding the perimeter of a composite figure.

2. Explain the process for finding the area of a composite figure.

3. Draw a composite figure for which the area can be found by using either addition or subtraction and explain each approach.

4. Write in words the formula for finding the degrees in each angle of a regular polygon.

5. Describe a shortcut for finding the perimeter of an *L*-shaped figure. Illustrate your procedure with a problem.

6. Two pieces of property have the same area (acreage) and equal desirability for development. Both require expensive fencing. One piece is a square 600 ft on each side. The other piece is a rectangle 400 ft by 900 ft. Which piece of property requires the lesser amount of fencing and is thus more desirable?

7. Discuss the differences in a sector and a segment.

8. Why are two formulas needed to find the area of a sector? Create an example using each formula and solve your examples.

9. If the height of a cone is doubled, what effect does this have on the volume of the cone? Create several examples of cones with the same area of base but different heights to validate your answer.

10. If the height of a right circular cylinder is doubled, what effect does this have on the surface area of the cylinder? Create several examples of right circular cylinders that have the same area of base but different heights to validate your answer.

17 | PRACTICE TEST

1. Change 0.3125° to minutes and seconds.

2. Change 15′32″ to a decimal degree to the nearest ten-thousandth.

Convert the degree measures to radians rounded to the nearest hundredth.

3. 35°

4. 122°

Convert the radian measures to degrees. Round to the nearest ten-thousandth of a degree when necessary.

5. $\dfrac{5\pi}{8}$ rad

6. 3.1 rad

Find the perimeter and area of Figs. 17–160 through 17–163. Round to the nearest tenth when necessary.

7.

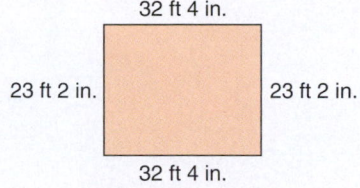

32 ft 4 in.

23 ft 2 in.   23 ft 2 in.

32 ft 4 in.

FIGURE 17–160

8.

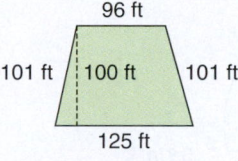

96 ft

101 ft   100 ft   101 ft

125 ft

FIGURE 17–161

9.

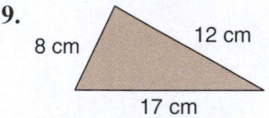

8 cm   12 cm

17 cm

FIGURE 17–162

10.

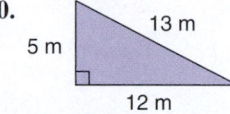

13 m

5 m

12 m

FIGURE 17–163

Find the circumference and area (Figs. 17–164 and 17–165).

11.

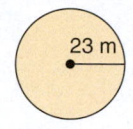

23 m

FIGURE 17–164

12.

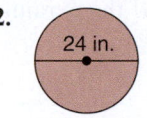

24 in.

FIGURE 17–165

Use the relationships of arc length s, area A, central angle (in radians) θ, and radius r to solve the problems relating to sectors. Round to hundredths if necessary.

13. Find s if $\theta = 0.5$ and $r = 2$ in.

14. Find θ if $s = 5.3$ m and $r = 7$ m.

15. Find r if $\theta = 1.7$ and $s = 2.9$ m.

16. Find A if $\theta = 0.6$ and $r = 7.3$ cm.

17. How many square feet of floor space are there in the plan shown in Fig. 17–166?

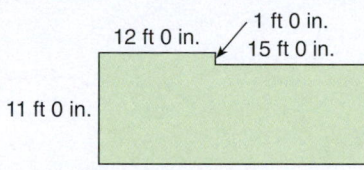

12 ft 0 in.   1 ft 0 in.
                15 ft 0 in.

11 ft 0 in.

FIGURE 17–166

Solve and when necessary, round to hundredths.

18. Find the area of the colored portion of the tiled walk that surrounds a rectangular swimming pool (Fig. 17–167).

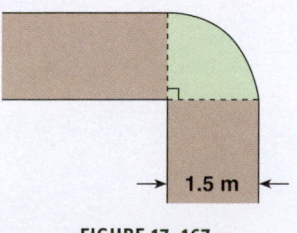

1.5 m

FIGURE 17–167

19. Find the area of the composite figure in Fig. 17–168.

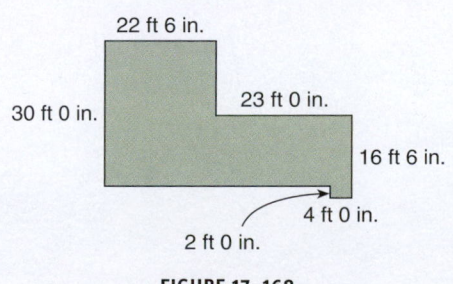

22 ft 6 in.

30 ft 0 in.   23 ft 0 in.

16 ft 6 in.

2 ft 0 in.   4 ft 0 in.

FIGURE 17–168

20. A section of a hip roof is a trapezoid measuring 35 ft at the bottom, 15 ft at the top, and 10 ft high. Find the area of this section of the roof in square feet.

21. Find the area of a sector of a circle that has a radius of 14 cm if the sector has an angle of 42°. Round to hundredths.

22. A segment is removed from a flat metal circle so that the piece rests on a horizontal base (Fig. 17–169). Find the area of the segment that is removed.

23. A pentagonal prism (five sides) measures 1 in. on each side of its base and has a height of 10 in. What is its lateral surface area?

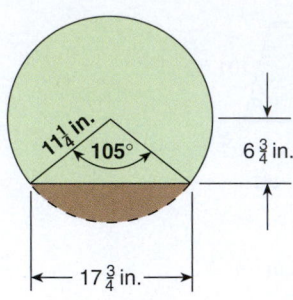

FIGURE 17–169

24. Find the total surface area of the pentagonal prism in Exercise 23 if the area of the base is 2.17 in².

25. A spherical tank 12 ft in diameter can hold how many gallons of fluid if 1 ft³ = 7.48 gal? Answer to the nearest whole gallon.

26. The base of a brass pyramid is an equilateral triangle with sides of 3 in. and altitude of 2.6 in. If the pyramid's height is 8 in., what is the volume of the pyramid?

18

Triangles

Masisyan/Shutterstock

In Great Company

A Snowy Evening (1978)

In 1970, Hartford, Connecticut, decided to build a new civic center. Vincent Kling designed the architecture and the engineering firm of Fraoli, Blum, and Yesselman (FB&Y) built the solution. To save on construction costs, the engineers proposed a new "space frame" design for the roof. Space frame design was not a new technique. The variations were introduced by FB&Y.

Not to worry. Even though this was a new design, the engineering firm had tested it extensively using state-of-the-art computer analysis to verify the safety of the design.

Construction began in 1971. The entire roof was assembled on the ground so it could be lifted into place, thus saving time and reducing risk to the men building it. After assembly, the men noticed portions of the roof had odd deflections. The engineers told them not to worry about it.

The roof was slowly lifted into place and secured. The deflections got worse. In fact, they became twice as bad as the computer analysis had predicted. The engineers weren't worried.

As a subcontractor tried to fit steel fascia panels to the supports, he found he couldn't. The deflections were getting worse. He was told to get the panels attached or pay fines for being late, as per the contract. The subcontractor didn't want to damage the exterior panels, so he cut into the roof supports to make the panels fit, a procedure called "coping."

The roof was completed January 16, 1973. Some visitors to events at the center noticed the odd downward deflection in the center of the roof and thought it unsafe. The engineers insisted everything was fine.

On the evening of January 17, 1978, with the civic center packed for a major hockey game, the largest snowstorm in a half-decade hit Hartford, Connecticut. Shout after shout came from the fans as the heavy snow began to accumulate on the roof. Finally, as more and more snow swirled heavily down, the game concluded; the fans went home. And at 4:15 A.M. on January 18, the roof buckled under the tons of accumulated snow. It cracked. It collapsed.

By then, the arena was empty. No one was injured. The engineers had been very, very lucky.

Later investigation showed the roof had been slowly failing since the day it was installed. Between the weight of the snow and the failed design, the east and west faces of the roof's top layer were overloaded by a whopping 852%. The parties settled out of court.

18-1   Special Triangle Relationships

LEARNING OUTCOMES

1. Classify triangles by the relationship of the sides or angles.

2. Determine if two triangles are congruent using inductive and deductive reasoning.

3. Solve problems that involve similar triangles.

LC LEARNING CATALYTICS

1. Evaluate $\dfrac{x}{14} = \dfrac{5}{27}$ to the nearest tenth.

2. Evaluate $\dfrac{x+1}{6} = \dfrac{3}{2}$.

Equilateral triangle: a triangle with three equal sides and three equal angles

Isosceles triangle: a triangle with *exactly* two equal sides. The angles opposite the equal sides are also equal

Scalene triangle: a triangle with *all* three sides unequal

We examined various relationships among lines and angles in the previous chapter. In this chapter, we study one of the most useful mathematical figures for any technician who uses geometry—the *triangle.* The relationships among the sides and angles in the triangle enable us to obtain much information that is implied but not always expressed in certain applications.

Triangles can be classified according to the relationship of their sides. The symbol $\triangle$ is used to indicate a triangle.

1   Classify Triangles by the Relationship of the Sides or Angles. Three relationships are possible among the three sides of a triangle and result in special names for each type of triangle.

An **equilateral triangle** is a triangle with three equal sides. The three angles of an equilateral triangle are also equal. Each angle measures 60° (Fig. 18–1, left).

An **isosceles triangle** is a triangle with *exactly* two equal sides. The angles opposite these equal sides are also equal (Fig. 18–1, middle).

A **scalene triangle** is a triangle with *all* three sides unequal (Fig. 18–1, right).

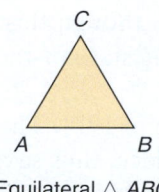

Equilateral $\triangle$ ABC
$AB = BC = AC$
$\angle A = \angle B = \angle C$

Isosceles $\triangle ABC$
$AC = BC$
$\angle A = \angle B$

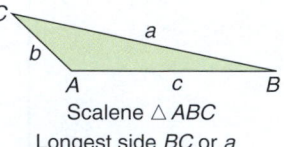

Scalene $\triangle ABC$
Longest side BC or a
Middle side AB or c
Shortest side AC or b
Largest angle $\angle A$
Middle angle $\angle C$
Smallest angle $\angle B$

FIGURE 18–1

In a scalene triangle, where no sides are equal, we can state an important relationship between the sides and their opposite angles (Fig. 18–1).

> **To determine the longest and shortest sides of a triangle:**
>
> If the three sides of a triangle are unequal, the *largest* angle is opposite the *longest* side and the *smallest* angle is opposite the *shortest* side.

EXAMPLE 1

Identify the longest and shortest sides of the triangle in Fig. 18–2.

The longest side is *AB* or *c* (87° is the largest angle). **The shortest side is *AC* or *b*** (43° is the smallest angle).   **See Exercises 1–7.**

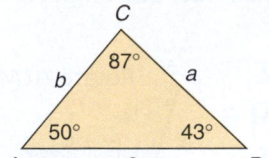

FIGURE 18-2

STOP AND CHECK

1. If the angle measures in Fig. 18–2 are changed to $\angle A = 48°$, $\angle B = 85°$, and $\angle C = 47°$, identify the longest and shortest sides.

Answer:

1. Longest side, *AC* or *b*; shortest side, *AB* or *c*

2 Determine if Two Triangles Are Congruent Using Inductive and Deductive Reasoning. Triangles that have the same size and shape are *congruent* triangles. These triangles fit exactly on top of each other.

In Fig. 18–3, $\triangle ABC$ fits exactly over $\triangle RST$. They are congruent triangles. The symbol ≅ means congruent. That is, $\triangle ABC \cong \triangle RST$. Each angle in $\triangle ABC$ has an angle in $\triangle RST$ that is its equal. We say that these pairs of angles *correspond*. In Fig. 18–3, $\angle A$ corresponds to $\angle R$. Also, $\angle C$ corresponds to $\angle T$, and $\angle B$ corresponds to $\angle S$. The equal sides also correspond. $\overline{AB}$ corresponds to $\overline{RS}$, $\overline{CB}$ corresponds to $\overline{TS}$, and $\overline{AC}$ corresponds to $\overline{RT}$.

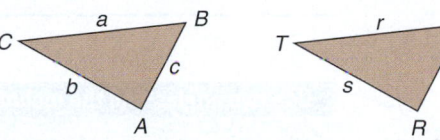

FIGURE 18-3

Congruent triangles: triangles in which corresponding sides and angles are equal; they have the same size and shape

Congruent triangles are triangles in which the corresponding sides and angles are equal.

> **Use three sides to determine congruent triangles:**
>
> If the three sides of one triangle are equal to the corresponding three sides of another triangle, the triangles are congruent (side-side-side or SSS).

STOP AND CHECK

1. Reposition $\triangle GHI$ to align with $\triangle ABC$ in Fig. 18–4.

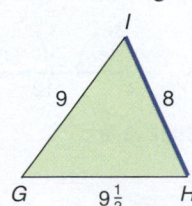

Answer:
1.

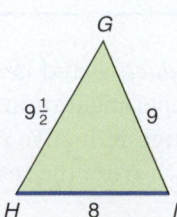

EXAMPLE 2

Use the side-side-side property to list the corresponding, equal sides of the two triangles in Fig. 18–4.

$\triangle ABC \cong \triangle FDE$ Rotate $\triangle DEF$.

$\overline{AB}$ corresponds to $\overline{FD}$
$\overline{BC}$ corresponds to $\overline{DE}$
$\overline{AC}$ corresponds to $\overline{FE}$

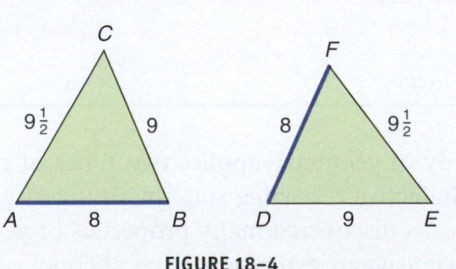

FIGURE 18-4

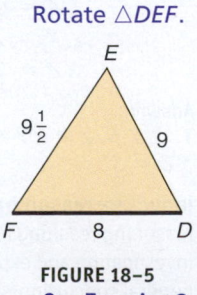

FIGURE 18-5
See Exercise 8.

Use two sides and the included angle to determine congruent triangles:

If two sides and the *included* angle of one triangle are equal to two sides and the included angle of another triangle, the triangles are congruent (side-angle-side or SAS).

STOP AND CHECK

1. Reposition $\triangle PQR$ to align with $\triangle ABC$ in Fig. 18–6.

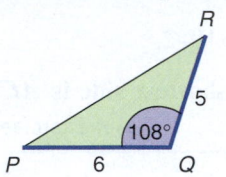

Answer:
1.

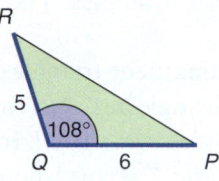

EXAMPLE 3

Use the side-angle-side property to list the corresponding, equal sides and angles of the two triangles in Fig. 18–6.

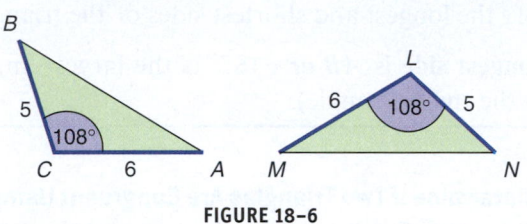

$$\overline{BC} = \overline{NL}$$
$$\overline{CA} = \overline{LM}$$
$$\angle C = \angle L$$

Rotate $\triangle LMN$.

FIGURE 18–6

FIGURE 18–7

The triangles are congruent because two sides and the included angle of one triangle are equal to two sides and the included angle of the other (SAS). Then, the other corresponding sides and angles are equal.

$\overline{AB} = \overline{MN}$	The sides are equal because they are opposite equal angles.
$\angle A = \angle M$	The angles are equal because they are opposite equal sides.
$\angle B = \angle N$	The angles are equal because they are opposite equal sides.

See Exercise 9.

STOP AND CHECK

1. Write the corresponding parts not given for the given congruent triangles.

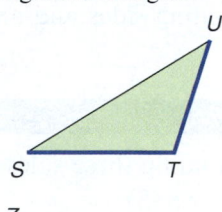

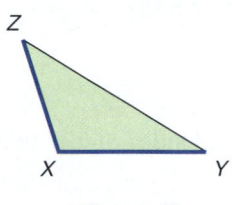

$$\angle Z = \angle U$$
$$\angle X = \angle T$$
$$XZ = TU$$

Answer:
1. $\angle Y = \angle S; XY = TS;$
$YZ = SU$

Use two angles and the included side to determine congruent triangles:

If two angles and the common side of one triangle are equal to two angles and the common side of another triangle, the triangles are congruent (angle-side-angle or ASA).

EXAMPLE 4

Use the angle-side-angle property to list the corresponding, equal angles and sides of the two triangles in Fig. 18–8.

$$\angle X = \angle A$$
$$\angle Y = \angle B$$
$$\overline{XY} = \overline{AB}$$

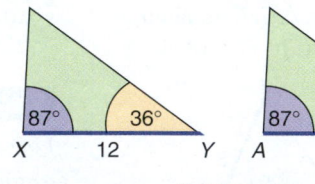

FIGURE 18–8

The triangles are congruent because of the ASA relationship. Then,

$$\angle Z = \angle C$$
$$\overline{XZ} = \overline{AC}$$
$$\overline{YZ} = \overline{BC}$$

See Exercise 10.

Inductive reasoning: the reasoning resulting from investigation and experimentation; general conclusions are drawn from the results of specific cases

The study of geometry applies two types of reasoning, *inductive* and *deductive* reasoning. **Inductive reasoning** starts with investigation and experimentation. Early mathematicians discovered many properties of geometry through inductive reasoning. After extensive investigation, general conclusions are drawn from the results of specific cases.

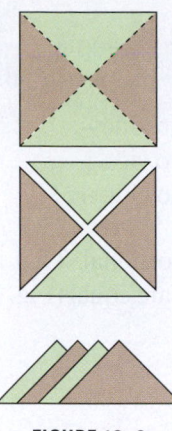

FIGURE 18–9

Deductive reasoning: the reasoning resulting from applying accepted properties; additional properties are concluded from these original properties

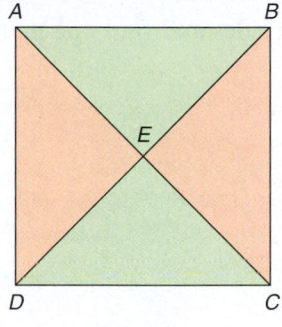

A   B

E

D   C

FIGURE 18–10

Let's use inductive reasoning to examine the results of cutting a square into four parts by cutting along the two diagonals (Fig. 18–9). Compare the resulting four triangles. Are they congruent triangles? Make a square of a different size and repeat the exercise. Continued repetitions lead you to conclude that the diagonals of a square divide the square into four congruent triangles.

Deductive reasoning starts with accepted properties. Additional properties are concluded from these original properties. We accept the congruent triangle properties and an additional property that the diagonals of a square bisect (cut in half) the angles of the square. We can use deductive reasoning to show that the two diagonals of a square form four congruent triangles.

A two-column format can be used to show the flow of the deductive-reasoning process.

$ABCD$ is a square and the diagonals AC and BD are equals (Fig. 18–10). The diagonals also bisect each other. Are $\triangle AED$, $\triangle AEB$, $\triangle BEC$, and $\triangle DEC$ congruent?

Accepted Properties:

$AB = BC = CD = AD$	The sides of a square are equal.
$AE = BE = CE = DE$	The diagonals of a square are equal and bisect each other.

Deductions: **Reason:**

$\triangle AED \cong \triangle AEB$	SSS; $AB = AD$, $AE = AE$, $DE = BE$
$\triangle BEC \cong \triangle DEC$	SSS; $CD = BC$, $CE = CE$, $DE = BE$
$\triangle AED \cong \triangle BEC$	SSS; $AD = BC$, $DE = BE$, $AE = CE$

All other pairings of triangles can be shown to be congruent using a similar deductive-reasoning process.

Conclusion:

Yes, $\triangle AED$, $\triangle AEB$, $\triangle BEC$, and $\triangle DEC$ are congruent.

In a systematic or axiomatic study of geometric concepts, all properties are developed or proved from a limited number of basic properties. In our study of geometric concepts, we apply both inductive and deductive reasoning in our arguments for solving problems. We do not introduce all the concepts necessary to make formal proofs of all geometric properties.

3 **Solve Problems That Involve Similar Triangles.** As we saw in the preceding learning outcome, *congruent triangles* have the same size and shape. **Similar triangles** have the same shape but not the same size (Fig. 18–11).

The corresponding angles in similar triangles are equal. The corresponding sides of similar triangles are directly proportional.

The symbol for showing similarity is ~ and is read "is similar to."

Similar triangles: two triangles are similar if they have the same shape but not the same size. Each angle of one triangle is equal to its corresponding angle in the other triangle. Each side of one triangle is directly proportional to its corresponding side in the other triangle

Properties of similar triangles:

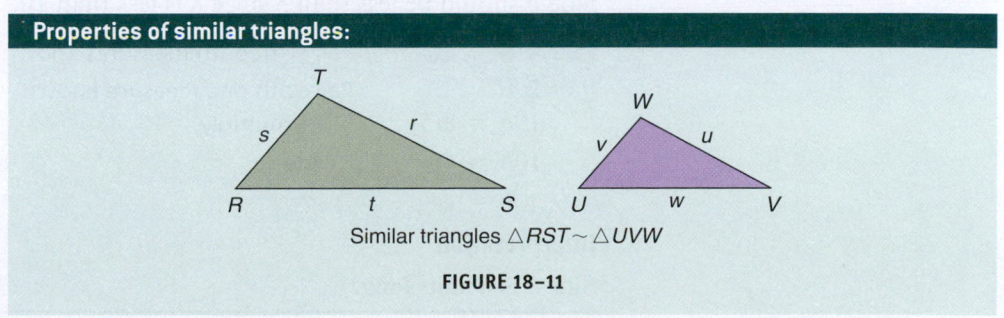

Similar triangles $\triangle RST \sim \triangle UVW$

FIGURE 18–11

Corresponding angles of similar triangles are equal in size.	Corresponding sides of similar triangles are directly proportional.
$\angle R = \angle U$ $\angle S = \angle V$ $\angle T = \angle W$	Side r corresponds to side u. Side s corresponds to side v. Side t corresponds to side w.

The corresponding sides of similar triangles are directly proportional. That is, each pair of corresponding sides form a ratio, or fraction, and the ratios are equal (Fig. 18–12).

$$\frac{a}{m} = \frac{b}{n} = \frac{c}{q}$$

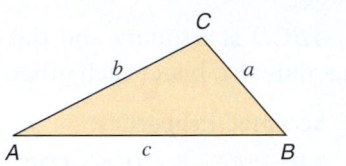 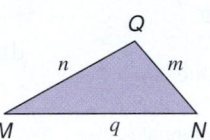

FIGURE 18–12

To find a missing side of a similar triangle:

1. Examine the given dimensions of the corresponding sides and find one pair for which both measures are known.

2. Pair the unknown side with its corresponding known side.

3. Set up a direct proportion with the two pairs.

4. Solve the proportion.

EXAMPLE 5

Find the unknown side in Fig. 18–13 if $\triangle ABC \sim \triangle DEF$.

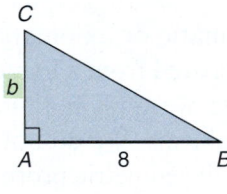 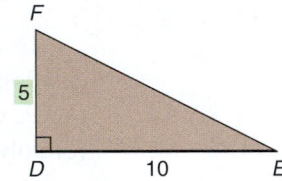

FIGURE 18–13

Write ratios of corresponding sides in a proportion. Use single lowercase letters to identify the sides. Side a is opposite $\angle A$, side b is opposite $\angle B$, and so on.

Pair 1: 8 from $\triangle ABC$ corresponds and is proportional to 10 from $\triangle DEF$.

Pair 2: b from $\triangle ABC$ corresponds and is proportional to 5 from $\triangle DEF$.

Estimation

Side b should be less than 5 since 8 is less than 10.

$$\frac{\text{Pair 1}}{\text{Pair 2}} \frac{8}{b} = \frac{10}{5} \qquad \text{Pair with both measures known} \\ \qquad\qquad\qquad \text{Pair with one measure known}$$

$$10b = 8(5) \qquad \text{Cross multiply.}$$
$$10b = 40 \qquad \text{Divide.}$$
$$b = 4$$

Interpretation

Side b is 4 units long.

See Exercises 11–14.

EXAMPLE 6

AG/H   A arborist must know the height of a tree to determine which way to fell it so that it does not endanger lives, traffic, or property. A 6-ft pole casts a 4-ft shadow when the tree casts a 20-ft shadow (Fig. 18–14). What is the height of the tree?

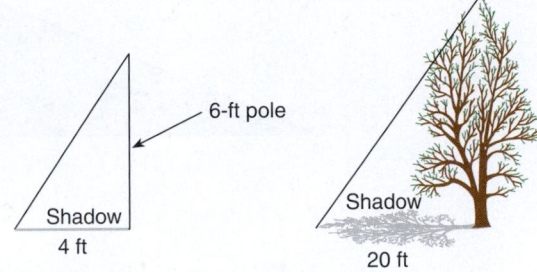

FIGURE 18–14

The triangles formed are similar. Let x = the height of the tree.

Pair 1: 6-ft pole casts a 4-ft shadow.   Both measures known

Pair 2: x-ft tree casts a 20-ft shadow.   One measure known

Estimation

The tree is more than 20 ft tall.   Since the pole is taller than its shadow, the tree will be taller than its shadow of 20 ft.

$$\frac{6 \text{ (height of pole)}}{x \text{ (height of tree)}} = \frac{4 \text{ (shadow of pole)}}{20 \text{ (shadow of tree)}}$$ Pair 1 / Pair 2

$$6(20) = 4x$$ Cross multiply.

$$120 = 4x$$ Divide.

$$30 = x$$

Interpretation

The tree is 30 ft tall.   **See Exercise 15.**

EXAMPLE 7

CAD/ARC   A building lies between points A and B, so the distance between these points cannot be measured directly by a surveyor. Find the distance using the similar triangles shown in Fig. 18–15.

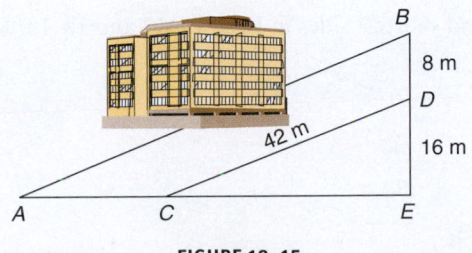

FIGURE 18–15

$\triangle ABE \sim \triangle CDE$. Note that CD must be made parallel to AB for the triangles to be similar. Parallel means the lines are the same distance apart from end to end.

Visualize the triangles as separate triangles (Fig. 18–16). Using the lowercase letter *e* for the missing *AB* measure is confusing because in △*CDE*, side *CD* can also be thought of as side *e*.

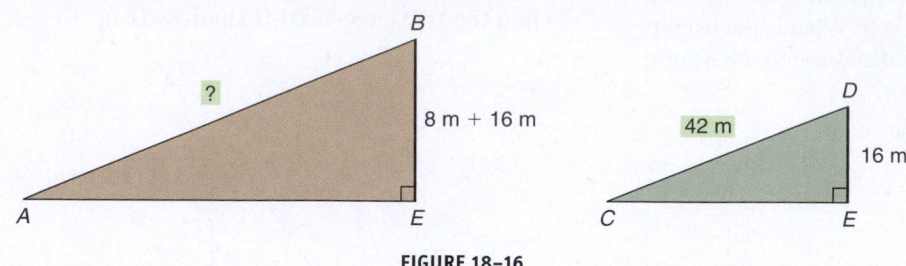

FIGURE 18–16

Pair 1: *BE* from △*ABE* corresponds to *DE* from △*CDE*.
Pair 2: *AB* from △*ABE* corresponds to *CD* from △*CDE*.

Estimation

AB is more than 42 m.

$$\frac{BE}{AB} = \frac{DE}{CD}$$ Substitute values. *CD* = 42; *DE* = 16
 BE = *BD* + *DE* = 8 + 16 = 24

$$\frac{24}{AB} = \frac{16}{42}$$ Cross multiply.

$16AB = (24)42$ *AB* is interpreted as a single variable.

$16AB = 1{,}008$ Divide.

$AB = 63$ m

Interpretation

The distance from *A* to *B* is 63 m.   **See Exercise 16.**

18–1 EXERCISES   MyLab Math For additional practice go to your study plan in MyLab Math.

1 Fill in the blanks.

1. A triangle with no equal sides is called a(n) _____ triangle.

2. A triangle with three equal sides is called a(n) _____ triangle.

3. A triangle with only two equal sides is called a(n) _____ triangle.

Identify the longest and shortest sides in Figs. 18–17 and 18–18. *See Example 1.*

4.

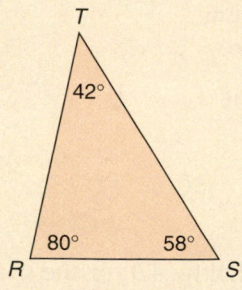

FIGURE 18–17

5.

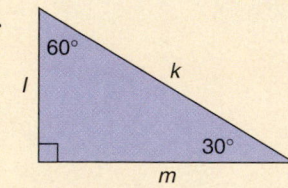

FIGURE 18–18

List the angles in order of size from largest to smallest in Figs. 18–19 and 18–20.

6.

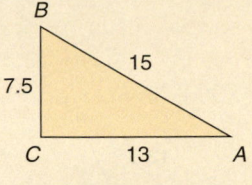

FIGURE 18–19

7.

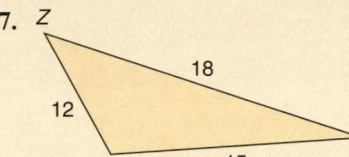

FIGURE 18–20

2 Write the corresponding parts not given for the congruent triangles in Figs. 18–21 to 18–23.

8. *See Example 2.*

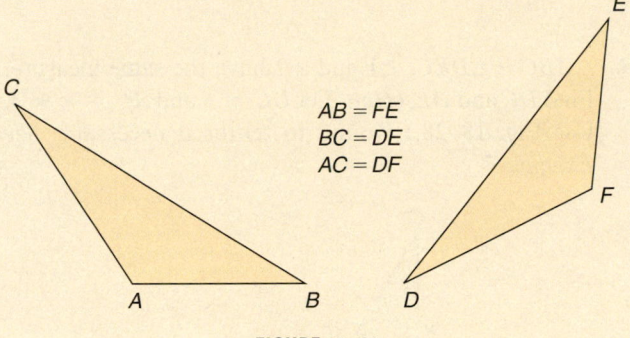

$AB = FE$
$BC = DE$
$AC = DF$

FIGURE 18–21

9. *See Example 3.*

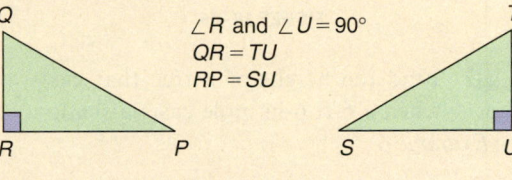

$\angle R$ and $\angle U = 90°$
$QR = TU$
$RP = SU$

FIGURE 18–22

10. *See Example 4.*

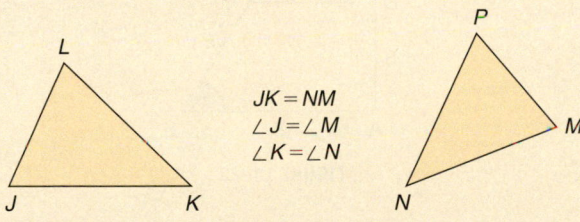

$JK = NM$
$\angle J = \angle M$
$\angle K = \angle N$

FIGURE 18–23

3 Find the indicated parts for the similar triangles in Exercises 11–14 (Figs. 18–24 to 18–27). *See Example 5.*

11. $AB = 12$ cm
$BC = 26$ cm
$DE = 20.8$ cm
Find EF.

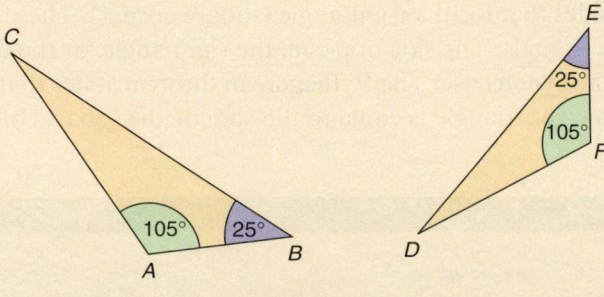

FIGURE 18–24

12. $\angle R$ and $\angle U = 90°$
$QR = 24$ in.
$RP = 28$ in.
$US = 22.4$ in.
Find TU.

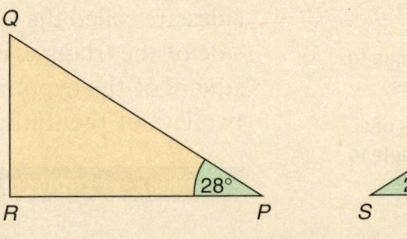

FIGURE 18–25

13. $JK = 8$ ft
$MN = 6$ ft
$JL = 5$ ft
$\angle J = \angle M = 50°$
$\angle K = \angle N = 18°$
Find PM.

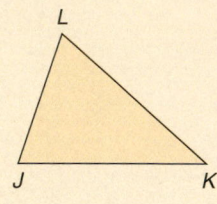

FIGURE 18-26

14. $\triangle ABC \sim \triangle EDF$. $\angle A = \angle E$, $\angle C = \angle F$. Find a and d.
Use Fig. 18-27.

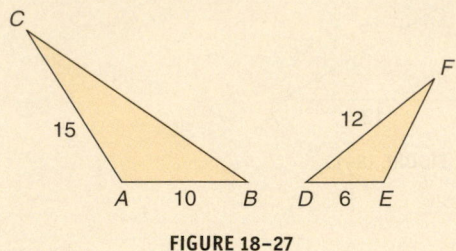

FIGURE 18-27

15. **AG/H** Find the height of a tree that casts a 30-ft shadow when a 6-ft 6-in. pole casts a shadow of 3 ft. *See Example 6.*

16. $\triangle ABC \sim \triangle DEC$. $\angle 1$ and $\angle 2$ have the same measure. Find DC and DE. (*Hint:* Let $DC = x$ and $AC = x + 3$. Use Fig. 18-28.) Round to tenths if necessary. *See Example 7.*

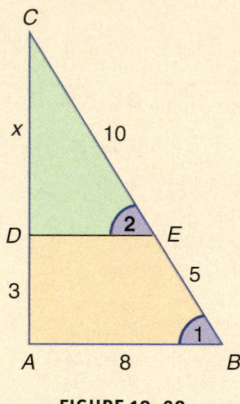

FIGURE 18-28

18-2 Pythagorean Theorem

LEARNING OUTCOMES

1 Use the Pythagorean theorem to find the unknown side of a right triangle.

2 Use the properties of a 45°, 45°, 90° triangle to find unknown parts.

3 Use the properties of a 30°, 60°, 90° triangle to find unknown parts.

LC LEARNING CATALYTICS
1. Evaluate $\sqrt{3^2 + 4^2}$.
2. Evaluate to the nearest tenth $\sqrt{12^2 - 4^2}$.

1 Use the Pythagorean Theorem to Find the Unknown Side of a Right Triangle. One of the most famous and useful theorems in mathematics is the Pythagorean theorem. It is named for the Greek mathematician, Pythagoras. The theorem applies to a right triangle.

A **right triangle** has two sides that form a right angle (square corner). These two sides are called the **legs** of the triangle. The side opposite the right angle, or the third side of the triangle, is called the **hypotenuse:** The **Pythagorean theorem** states that the square of the hypotenuse of a right triangle is equal to the sum of the squares of the two legs of the triangle.

Formula for Pythagorean theorem:

$$c^2 = a^2 + b^2$$

where c is the hypotenuse of a right triangle, and a and b are legs (Fig. 18-29).

Right triangle: a triangle that has two sides that form a right (90°) angle.

Leg: one of the two sides of a right triangle that is not the hypotenuse. Or, one of the two sides that form the right angle

Hypotenuse: the side of a right triangle that is opposite the right angle

Pythagorean theorem: the theorem that states that the square of the hypotenuse of a right triangle is equal to the sum of the squares of the two legs of the triangle

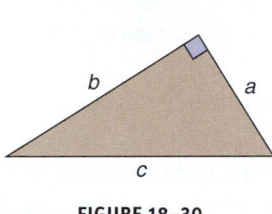

FIGURE 18-29

STOP AND CHECK

Use the Pythagorean theorem and Fig. 18–30 to find:
1. a when $b = 24$ cm and $c = 25$ cm
2. c when $a = 12$ and $b = 16$ mm

Answer:
1. $a = 7$ cm 2. $c = 20$ mm

PITAGORA.

Offscreen/Shutterstock

STOP AND CHECK
1. In Fig. 18–31, if $AB = 18$ in. and $BC = 12$ in., find AC to the nearest tenth of an inch.

Answer:
1. 21.6 in.

To find a leg or the hypotenuse of a right triangle:

1. Identify the two known sides of the triangle and the unknown side, and state the theorem symbolically.
2. Substitute known values in the Pythagorean theorem. $c^2 = a^2 + b^2$
3. Solve for the unknown value.
4. The solution is the principal square root.

EXAMPLE 1

Use the Pythagorean theorem to find b when $a = 8$ mm and $c = 17$ mm (Fig. 18–30).

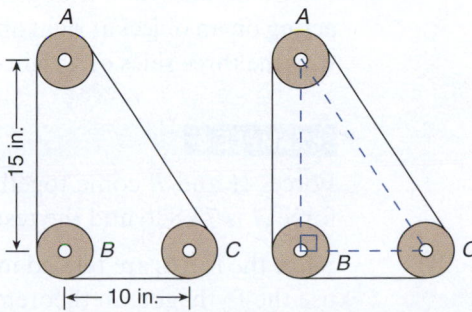

FIGURE 18-30

$c^2 = a^2 + b^2$ State the theorem symbolically and substitute known values.

$17^2 = 8^2 + b^2$ Evaluate powers.

$289 = 64 + b^2$ Use the addition axiom to isolate b^2.

$289 - 64 = b^2$ Combine like terms.

$225 = b^2$ Take the square root of both sides.

$\sqrt{225} = b$ Solution is the principal square root.

$\mathbf{15\ mm = b}$ **See Exercises 1–5.**

Many times we are given problems that contain "hidden" triangles. In these cases, we need to visualize the triangle or triangles in the problems. Drawing one or more of the sides of the "hidden" triangle helps solve the problem.

EXAMPLE 2

INDTR   Find the center-to-center distance between pulleys A and C (Fig. 18–31).

FIGURE 18-31

Connect the center points of the three pulleys to form a right triangle. The hypotenuse is the distance between the centers of pulleys A and C. Use the Pythagorean theorem and substitute the given values for the two known sides.

$$(AC)^2 = (AB)^2 + (BC)^2 \qquad \text{State theorem symbolically and substitute values.}$$
$$(AC)^2 = 15^2 + 10^2 \qquad \text{Square both constants.}$$
$$(AC)^2 = 225 + 100 \qquad \text{Combine like terms.}$$
$$(AC)^2 = 325 \qquad \text{Take the square root of both sides.}$$
$$AC = \sqrt{325} \qquad \text{Find the approximate principal square root.}$$
$$AC = 18.02775638$$

The distance from pulley A to pulley C is 18.0 in. (to the nearest tenth).

See Exercise 6.

EXAMPLE 3

INDTR The head of a bolt is a square 0.5 in. on each side (distance across flats). What is the distance from corner to corner (distance across corners)? (See Fig. 18–32.)

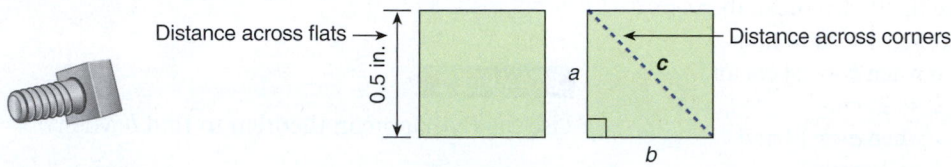

FIGURE 18–32

If a diagonal line is drawn from one corner to the opposite corner, the sides of the bolt head form the legs of a triangle. This diagonal forms the hypotenuse of the right triangle. Because the legs of the triangle are known to be 0.5 in. each, we can substitute in the Pythagorean theorem to find the length of the diagonal line, the hypotenuse, which is the distance across corners.

$$c^2 = a^2 + b^2 \qquad \text{State theorem symbolically.}$$
$$c^2 = 0.5^2 + 0.5^2 \qquad \text{Substitute values and evaluate.}$$
$$c^2 = 0.25 + 0.25 \qquad \text{Combine like terms.}$$
$$c^2 = 0.5 \qquad \text{Take the square root of both sides.}$$
$$c = \sqrt{0.5} \qquad \text{Evaluate.}$$
$$c = 0.7071067812$$

The distance across corners is 0.71 in. (to the nearest hundredth).   **See Exercises 7–12.**

Sometimes right triangles are used to represent certain relationships, such as forces acting on an object at right angles or electrical and electronic phenomena related in the way the three sides of a right triangle are related. Let's look at an example.

EXAMPLE 4

Forces A and B come together at a right angle to produce force C (Fig. 18–33). If force A is 74.8 lb and the resulting force C is 91.5 lb, what is force B?

Since the forces are related in the way the sides of a right triangle are related, we may use the Pythagorean theorem to find the unknown force B, a leg of the triangle.

$$A^2 + B^2 = C^2 \qquad \qquad \text{State theorem symbolically and substitute values.}$$
$$74.8^2 + B^2 = 91.5^2 \qquad \text{Square both constants.}$$
$$5{,}595.04 + B^2 = 8{,}372.25 \qquad \text{Apply the addition axiom to isolate } B^2.$$

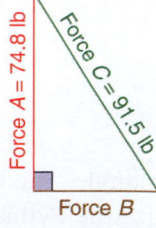

FIGURE 18–33

$$B^2 = 8,372.25 - 5,595.04 \qquad \text{Combine constants.}$$
$$B^2 = 2,777.21 \qquad \text{Take the square root of both sides.}$$
$$B = \sqrt{2,777.21} \qquad \text{Find the approximate principal square root.}$$
$$B = 52.7 \qquad \text{Round to the nearest tenth.}$$

Force B is 52.7 lb. See Exercise 13.

2 **Use the Properties of a 45°, 45°, 90° Triangle to Find Unknown Parts.** An isosceles triangle is a triangle with two equal sides. An **isosceles right triangle** is a right triangle that has equal legs. The angles opposite the equal legs are **base angles of an isoceles triangle.** The two base angles are also equal because they are the angles opposite the equal sides (Fig. 18–34).

The sum of the angles of a triangle is 180°. The sum of the two base angles of a right triangle is $180° - 90°$ or 90°. If both angles are equal, as in the isosceles right triangle, then each angle is $\frac{1}{2}(90°)$ or 45°. These base angles are complementary.

A **45°, 45°, 90° triangle** is an isosceles right triangle. The two base angles are each 45°. This triangle is frequently used in applications of the Pythagorean theorem.

All isosceles right triangles (45°, 45°, 90°) have the same relationship between each leg of the triangle and the hypotenuse (Fig. 18–35).

Applying the Pythagorean theorem to find the hypotenuse when $x = $ length of each leg:

$$\text{hypotenuse}^2 = x^2 + x^2$$
$$\text{hypotenuse} = \sqrt{x^2 + x^2}$$
$$\text{hypotenuse} = \sqrt{2x^2}$$
$$\text{hypotenuse} = x\sqrt{2}$$

Isosceles right triangle: a right triangle that has two equal legs; the base angles are also equal and are 45° each

Base angles of an isosceles right triangle: the angles opposite the equal legs; these base angles are also equal

45°, 45°, 90° triangle: an isosceles right triangle

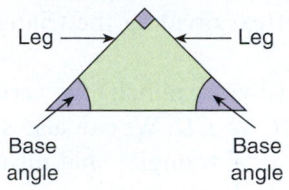

Leg ——→ ←—— Leg

Base angle   Base angle

FIGURE 18–34

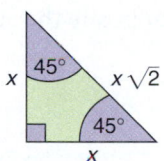

x 45° $x\sqrt{2}$

45°

x

FIGURE 18–35

To find the hypotenuse or leg of a 45°, 45°, 90° triangle:
1. Identify the known and unknown parts of the triangle and substitute in the formula:
$$\text{Hypotenuse} = \text{Leg}\sqrt{2} \text{ or } x\sqrt{2}, \text{ where } x = \text{length of leg}$$
2. Solve for the unknown value.
3. Write as an exact solution or approximate solution, as desired.

STOP AND CHECK

1. Find the hypotenuse to the nearest thousandth of an isosceles right triangle that has equal sides of 5 ft.

Answer:
1. 7.071 ft

EXAMPLE 5

Find the hypotenuse to the nearest thousandth of an isosceles right triangle that has equal sides of 2 m (Fig. 18–36).

$$\text{hypotenuse} = \text{leg}\sqrt{2} \qquad \text{Substitute into appropriate formula.}$$
$$\text{hypotenuse} = 2\sqrt{2} \text{ m} \qquad \text{Exact solution}$$
$$\text{hypotenuse} = 2.828 \text{ m} \qquad \text{Approximate solution}$$

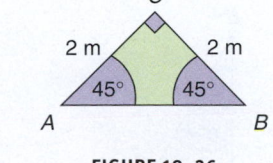

C

2 m   2 m

45°   45°

A B

FIGURE 18–36

See Exercises 14–18.

B

45°   5 cm

45°

C A

FIGURE 18–37

EXAMPLE 6

Find AC and BC if $AB = 5$ cm (Fig. 18–37).

$$\text{Hypotenuse} = \text{Leg}\sqrt{2} \qquad \text{Substitute into appropriate formula.}$$
$$5 = AC\sqrt{2} \qquad \text{Solve for } AC, \text{ a leg. Divide both sides by } \sqrt{2}.$$

1. Find the length of each leg to the nearest thousandth of a 45°, 45°, 90° triangle if the hypotenuse is 15 m.

Answer:

1. 10.607 m

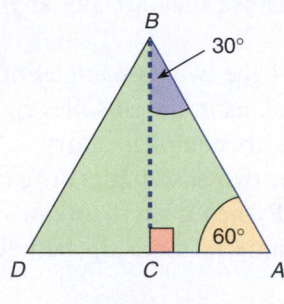

FIGURE 18-38

Altitude of an equilateral triangle: a line drawn from the midpoint of the base to the opposite vertex, dividing the triangle into two congruent 30°, 60°, 90° triangles

Height of a 30°, 60°, 90° triangle: the side opposite the 60° angle

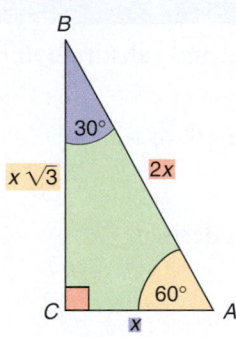

FIGURE 18-39

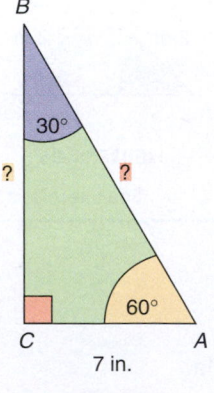

FIGURE 18-40

$$\frac{5}{\sqrt{2}} = \frac{AC\sqrt{2}}{\sqrt{2}}$$

$$\frac{5}{\sqrt{2}} = AC$$

Rationalize the denominator. Multiply numerator and denominator by 1 in the form of $\frac{\sqrt{2}}{\sqrt{2}}$.

$$\frac{5}{\sqrt{2}} \cdot \frac{\sqrt{2}}{\sqrt{2}} = AC$$

$$\frac{5\sqrt{2}}{2} \text{ cm} = AC \qquad \text{Exact solution}$$

$$3.536 \text{ cm} = AC \qquad \text{Approximate solution}$$

Since $AC = BC$, then $BC = 3.536$ cm. See Exercises 19–20.

3 **Use the Properties of a 30°, 60°, 90° Triangle to Find Unknown Parts.** Another special case of the Pythagorean theorem is the 30°, 60°, 90° *triangle,* which arises often in applications. If we draw the altitude of an *equilateral* triangle, we form two 30°, 60°, 90° triangles (Fig. 18–38). The **altitude of an equilateral triangle** is a line drawn from the midpoint of the base to the opposite vertex, dividing the triangle into two congruent right triangles.

Because the altitude of an equilateral triangle bisects (divides in half) the vertex angle and the base, two 30° angles are formed at B, and $AC = CD$. We can also say that AC is $\frac{1}{2}AD$ or one-half any side of the original equilateral triangle. That means $AC = \frac{1}{2}AB$, or the side opposite the 30° angle is one-half the hypotenuse.

Apply the Pythagorean theorem to find the **height of a 30°, 60°, 90° triangle,** or the side opposite the 60° angle. When x is the length of the side opposite the 30° angle and $2x$ is the length of the hypotenuse (Fig. 18–39):

$$(2x)^2 = x^2 + \text{height}^2$$
$$4x^2 - x^2 = \text{height}^2$$
$$3x^2 = \text{height}^2$$
$$\text{height} = \sqrt{3x^2}$$
$$\text{height} = x\sqrt{3}$$

To find the unknown sides of a 30°, 60°, 90° triangle if one side is known:

1. Identify the known side of the triangle.

2. If the side opposite the 30° angle is known, substitute in the formulas,

$$\text{hypotenuse} = 2x$$
$$\text{height} = x\sqrt{3}$$

where x is the measure of the side opposite the 30° angle.

3. If either the hypotenuse or height is known, use the appropriate formula and solve for x. Use x and the remaining formula to find the remaining side.

EXAMPLE 7

Find AB and BC if $AC = 7$ in. (Fig. 18–40).

$$\text{Hypotenuse} = 2x \qquad \text{Substitute known values.}$$
$$AB = 2(7 \text{ in.}) = \textbf{14 in.}$$

STOP AND CHECK

1. Find both legs of a 30°, 60°, 90° triangle if the hypotenuse is 18 cm.

Answer:

1. Side opposite 30° angle $= 9$ cm; height $= 15.6$ cm to the nearest tenth

STOP AND CHECK

1. Find the hypotenuse and the side opposite the 30° angle to the nearest tenth of a 30°, 60°, 90° triangle if the height is 21 ft.

Answer:

1. Side opposite 30° angle $= 12.1$ ft; hypotenuse $= 24.2$ ft to the nearest tenth

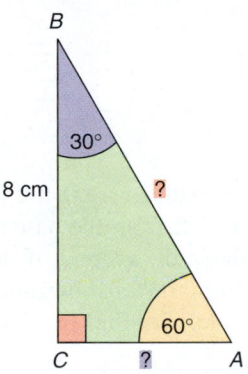

FIGURE 18-41

Height $= x\sqrt{3}$	Substitute known values.
$BC = 7$ in.$(\sqrt{3})$	
$BC = 7\sqrt{3}$ in.	Exact solution
$BC = 12.1$ in.	Approximate solution to the nearest tenth

See Exercises 21–28.

EXAMPLE 8

Find AC and AB when $BC = 8$ cm (Fig. 18–41).

Find AC (side opposite the 30° angle):

$$\text{height} = x\sqrt{3} \qquad x = AC; BC = \text{height} = 8$$
$$8 = x\sqrt{3} \qquad \text{Solve for } x. \text{ Divide both sides by } \sqrt{3}.$$
$$\frac{8}{\sqrt{3}} = \frac{x\sqrt{3}}{\sqrt{3}}$$
$$\frac{8}{\sqrt{3}} = x \qquad \text{Rationalize the denominator.}$$
$$x = \frac{8}{\sqrt{3}} \cdot \frac{\sqrt{3}}{\sqrt{3}}$$
$$x = \frac{8\sqrt{3}}{3} \text{ cm}$$
$$AC = \frac{8\sqrt{3}}{3} \text{ cm (exact)} \qquad \text{or} \qquad \textbf{4.6 cm (approximate)}$$

Find AB (hypotenuse):

$$\text{hypotenuse} = 2x \qquad \text{Hypotenuse} = AB; x = \frac{8\sqrt{3}}{3}$$
$$AB = 2\left(\frac{8\sqrt{3}}{3}\right) \qquad \text{Substitute for } x.$$
$$\mathbf{AB = \frac{16\sqrt{3}}{3}} \text{ cm (exact)} \qquad \text{or} \qquad \textbf{9.2 cm (approximate)}$$

See Exercises 21–28.

TIP **Special Triangle Relationships** Visualize the relationships among the sides of these special triangles.

30°, 60°, 90° triangle (Fig. 18–42):   45°, 45°, 90° triangle (Fig. 18–43):

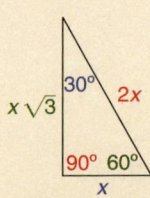

FIGURE 18-42

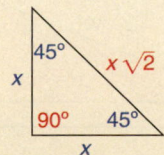

FIGURE 18-43

18–2 EXERCISES   MyLab Math For additional practice go to your study plan in MyLab Math.

---

1 Find the unknown side of the right triangle. Round the final answers to the nearest thousandth. Use Fig. 18–44 for Exercises 1–2. *See Example 1.*

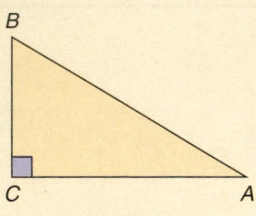

FIGURE 18–44

1. $AC = 24$ mm
   $BC = ?$
   $AB = 26$ mm

2. $AC = 15$ yd
   $BC = 8$ yd
   $AB = ?$

Use Fig. 18–45 for Exercises 3–5. Find the unknown dimensions. Round to the nearest thousandth where appropriate. *See Example 1.*

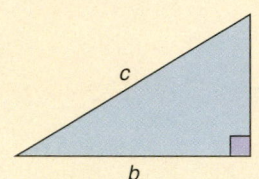

FIGURE 18–45

3. $a = 5$ cm
   $b = 4$ cm
   $c = ?$

4. $a = ?$
   $b = 9$ m
   $c = 11$ m

5. $a = 9$ ft
   $b = ?$
   $c = 15$ ft

Solve the problems. Round final answers to the nearest thousandth if necessary.

See Example 2.

6. **INDTR** Find the center-to-center distance between holes A and C in a sheet-metal plate if the distance between the centers of A and B is 16.5 cm and the distance between the centers of B and C is 36.2 cm (Fig. 18–46).

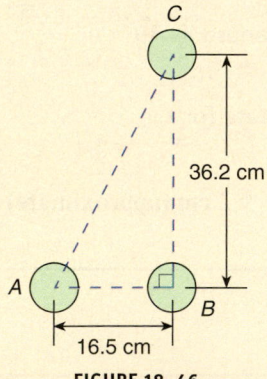

FIGURE 18–46

See Example 3.

7. **CON** A light pole will be braced with a wire that is to be tied to a stake in the ground 18 ft from the base of the pole, which extends 26 ft above the ground. If the wire is attached to the pole 2 ft from the top, how much wire must be used to brace the pole? (See Fig. 18–47.)

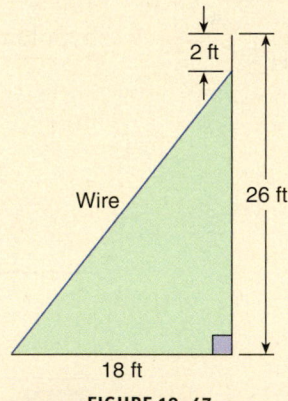

FIGURE 18–47

8. **CON** Find the length of a rafter that has a 10-in. overhang if the rise of the roof is 10 ft and the joists are 48 ft long (Fig. 18–48).

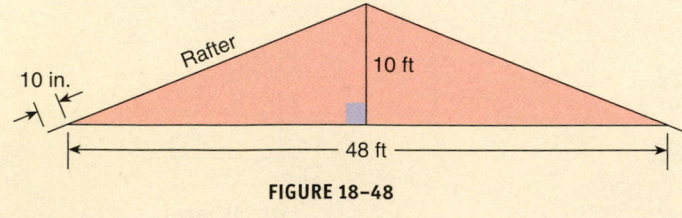

FIGURE 18–48

9. **CON** A stair stringer is 8 ft high and extends 10 ft from the wall (Fig. 18–49). How long will the stair stringer be?

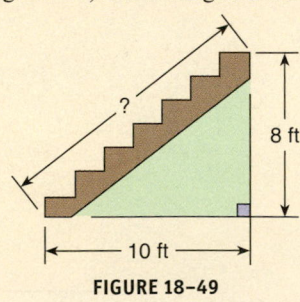

FIGURE 18–49

10. **INDTR** A machinist wishes to strengthen an L bracket that is 5 cm by 12 cm by welding a brace (hypotenuse of the triangle) to each end of the bracket. How much metal rod is needed for the brace?

See Example 4.

12. **ELEC** A rigid length of electrical conduit must be shaped as shown in Fig. 18–50 to clear an obstruction. What total length of the conduit is needed? (*Hint:* Don't forget to include *AB* and *CD* in the total length.)

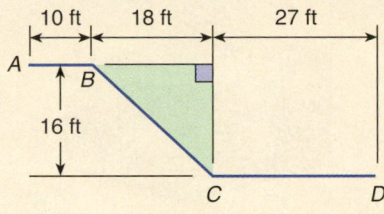

FIGURE 18-50

11. **CON** To make a rectangular table more stable, a diagonal brace is attached to the underside of the table surface. If the table is 27 dm by 36 dm, how long is the brace?

13. Find the length of the side of the largest square nut that can be milled from a piece of round stock whose diameter is 15 mm (Fig. 18–51).

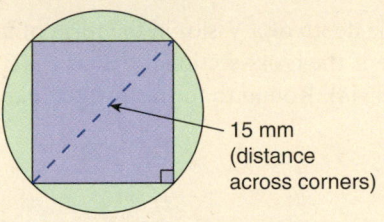

15 mm (distance across corners)

FIGURE 18-51

2 Use Fig. 18–52 to solve. Round the final answers to the nearest thousandth if necessary. *See Example 5.*

14. *AC* = 12 cm; find *BC* and *AB*.

15. *AB* = 10 m; find *AC* and *BC*.

16. *BC* = $7\sqrt{2}$ m; find *AC* and *AB*.

17. *AB* = $8\sqrt{2}$ m; find *AC* and *BC*.

18. *AB* = $12\sqrt{3}$ mm; find *AC* and *BC*.

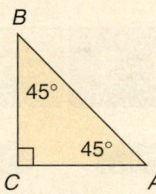

FIGURE 18-52

See Example 6.

19. A rafter 19.5 ft long makes a 45° angle with a joist. If the rafter has an 18-in. overhang, find the length of the joist to the nearest inch (Fig. 18–53).

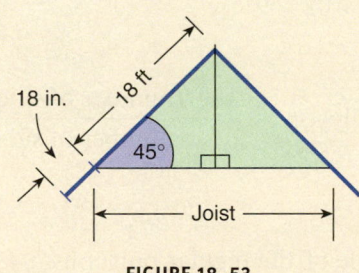

FIGURE 18-53

20. An elevated tank is connected to a pipe. The connecting pipe is 53.5 ft long and forms a 45° angle where it connects to the tank (Fig. 18–54). At what horizontal distance from the tank should the pipe stop to make the connection?

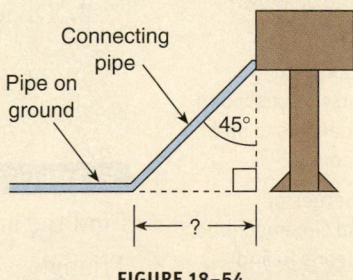

FIGURE 18-54

3 Use Fig. 18–55 to solve. Round the final answers to the nearest thousandth if necessary. *See Examples 7–8.*

21. *AC* = 6 cm; find *AB* and *BC*.

22. *AB* = 18 mm; find *AC* and *BC*.

23. *BC* = 8 in.; find *AC* and *AB*.

24. *BC* = $7\sqrt{2}$ cm; find *AC* and *AB*.

25. *AC* = 2 ft 9 in.; find *AB* and *BC* to the nearest inch.

FIGURE 18-55

26. Find the length of the conduit *ABCD* (Fig. 18–56) if *AB* = 18 ft, *CK* = 6 ft, *CD* = 5 ft, and ∠*KCB* = 60°. Round to the nearest foot.

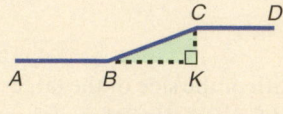

FIGURE 18–56

27. A rafter makes a 30° angle with the horizontal. If the rise is 9 ft, find the rafter length and the run to the nearest inch (Fig. 18–57).

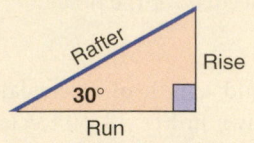

FIGURE 18–57

28. Find the depth of a V-slot in the form of an equilateral triangle if the cross-sectional opening is 5.2 cm across (Fig. 18–58). Round to the nearest tenth.

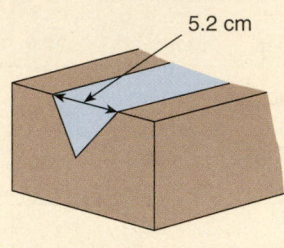

FIGURE 18–58

18–3   Inscribed and Circumscribed Regular Polygons and Circles

LEARNING OUTCOMES

1 Find the number of degrees in each angle of a regular polygon.

2 Find the area of a regular polygon.

3 Use the properties of inscribed and circumscribed squares to find unknown amounts.

4 Use the properties of inscribed and circumscribed equilateral triangles to find unknown amounts.

5 Use the properties of inscribed and circumscribed regular hexagons to find unknown amounts.

LC LEARNING CATALYTICS
1. Rearrange the formula to solve for *r*: $A = \pi r^2$. Do not rationalize the denominator.

Regular polygon: a polygon with equal sides and equal angles

STOP AND CHECK
1. Find the number of degrees in each angle of a regular polygon having 10 sides.

Answer:
1. 144°

1 Find the Number of Degrees in Each Angle of a Regular Polygon. Polygons with equal sides and angles, such as squares, are called *regular polygons*. Several other regular polygons with definite shapes and specific names are treated as composite figures when calculating their areas. Let's examine regular polygons more closely.

A **regular polygon** is a polygon with equal sides and equal angles.

To find the number of degrees in each angle of a regular polygon:

1. Multiply the number of sides less 2 by 180°.

2. Divide by the number of sides.

$$\text{degrees per angle of a regular polygon} = \frac{180°(\text{number of sides} - 2)}{\text{number of sides}}$$

EXAMPLE 1

Find the number of degrees in each angle of the regular polygons.

Triangle (3 sides)   Pentagon (5 sides)   Octagon (8 sides)
Square   (4 sides)   Hexagon   (6 sides)

 Equilateral Triangle: $\dfrac{180°(3-2)}{3} = \dfrac{180°(1)}{3} = 60°$ 3 sides

Square: $\dfrac{180°(4-2)}{4} = \dfrac{180°(2)}{4} = \dfrac{360°}{4} = 90°$ 4 sides

Regular Pentagon: $\dfrac{180°(5-2)}{5} = \dfrac{180°(3)}{5} = \dfrac{540°}{5} = 108°$ 5 sides

Regular Hexagon: $\dfrac{180°(6-2)}{6} = \dfrac{180°(4)}{6} = \dfrac{720°}{6} = 120°$ 6 sides

Regular Octagon: $\dfrac{180°(8-2)}{8} = \dfrac{180°(6)}{8} = \dfrac{1{,}080°}{8} = 135°$ 8 sides

See Exercises 1–6.

2 Find the Area of a Regular Polygon. A regular polygon can be divided into as many congruent triangles as there are sides of the polygon.

> **To form congruent triangles in a regular polygon:**
>
> 1. Draw line segments from the center of the regular polygon to each vertex.
>
> 2. Congruent triangles are formed.

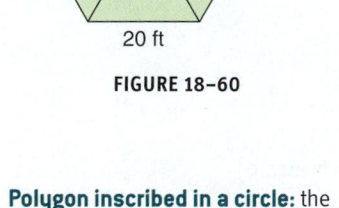

FIGURE 18-59

Fig. 18–59 shows how this property applies to regular polygons: the equilateral triangle, square, regular pentagon, and regular hexagon. To accept this property using inductive reasoning, construct various regular polygons, draw lines from each vertex to the center, and compare the resulting triangles.

> **To find the area of a regular polygon:**
>
> 1. Find the area of one of the congruent triangles formed by connecting each vertex with the center of the polygon.
>
> 2. Multiply the result from Step 1 by the number of congruent triangles.

STOP AND CHECK

1. Find the area of a regular hexagon with each side measuring 12 in. and the perpendicular distance from one side to the center is 10.4 in.

Answer:
1. 374.4 in^2

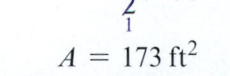

FIGURE 18-60

EXAMPLE 2

Find the floor area of a recreational building at a park if it forms a regular hexagon and each side is 20 ft long. The perpendicular distance from one side to the center of the building is 17.3 ft (Fig. 18–60).

Divide the regular hexagon into congruent triangles by connecting the center of the hexagon with each vertex of the hexagon.

$$A = \frac{1}{2}bh$$ Find the area of one triangle. Substitute known values. Use the given distance from the side of the hexagon to the center for the height of the triangle.

$$A = \frac{1}{2}(\overset{10}{\underset{1}{20}})(17.3)$$ Simplify.

$$A = 173 \text{ ft}^2$$

$$173(6) = 1,038 \text{ ft}^2$$ Multiply the area of the one triangle by 6 because the hexagon has been divided into six congruent triangles.

The floor area of the recreational building is 1,038 ft².   **See Exercises 7–12.**

3 Use the Properties of Inscribed and Circumscribed Squares to Find Unknown Amounts. Previously, we studied polygons and their areas and perimeters, and the area and circumference of a circle. In this section, we examine regular polygons *inscribed* in a circle and *circumscribed* about a circle.

A **polygon is inscribed in a circle** when it is inside the circle and all its *vertices* (points where sides of each angle meet) are on the circle. The **circle is circumscribed about the polygon** (Fig. 18–61, left).

Polygon inscribed in a circle: the polygon is inside the circle and all its *vertices* (points where sides of each angle meet) are on the circle. The circle is circumscribed about the polygon

Circle circumscribed about a polygon: the circle is outside the polygon and each vertex of the polygon is on the circumference of the circle. The polygon is inscribed in the circle

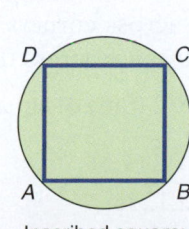 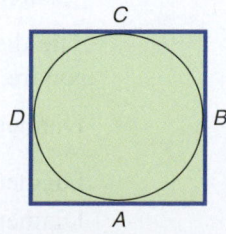

Inscribed square:
Vertices lie on circle
at points *A*, *B*, *C*, and *D*.

Circumscribed square:
Sides of square are tangent
at points *A*, *B*, *C*, and *D*.

FIGURE 18-61

Polygon circumscribed about a circle: the polygon is outside the circle and each side of the polygon is tangent to the circumference of the circle. The circle is inscribed in the polygon

Tangent to: when an intersecting line on the outside of a circle intersects in exactly one point the circumference

Circle inscribed in a polygon: the circle is inside the polygon and each side of the polygon is tangent to the circumference of the circle. The polygon is circumscribed about the circle

A **polygon is circumscribed about a circle** when it is outside the circle and all its sides are **tangent** to (intersecting in exactly one point) the circumference of the circle. The **circle is inscribed in the polygon** (Fig. 18–61, right).

In this section, we examine the three most common polygons inscribed in or circumscribed about a circle: the triangle, the square, and the hexagon.

A square has four equal sides and four equal angles, each 90°. If a diagonal is drawn, the result is two congruent right triangles that have angles of 45°, 45°, 90°. A second diagonal divides the square into four 45°, 45°, 90° congruent right triangles, as shown in Fig. 18–62. Diagonals *AC* and *BD* are equal in length and bisect each other. Point *O,* where the diagonals intersect, is the center of the square and the center of any inscribed or circumscribed circle (Fig. 18–63).

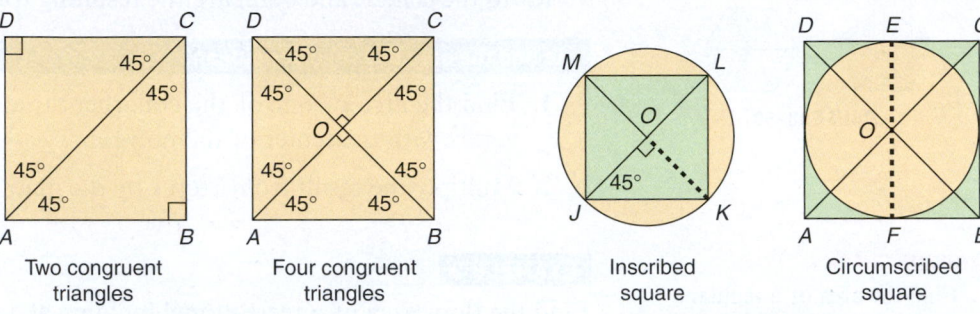

| Two congruent triangles | Four congruent triangles | Inscribed square | Circumscribed square |

FIGURE 18–62   **FIGURE 18–63**

Square–circle relationships

Square Inscribed in Circle

Radius of circle $= \dfrac{1}{2}$ Diagonal of square

Diameter of circle $= \dfrac{\text{Diagonal of square}}{\text{(across corners)}}$

$\dfrac{\text{Side of square}}{\text{(across flats)}} = \dfrac{\text{Diameter of circle}}{\sqrt{2}}$

Diameter of circle $= \sqrt{2}$ (Side of square)

Square Circumscribed About a Circle

Side of square (across flats) $=$ Diameter of circle

Radius of circle $= \dfrac{1}{2}$ Side of square

Diagonal of square (across corners) $= \sqrt{2}$ (Diameter of circle)

$\qquad\qquad\qquad\qquad\qquad\qquad\quad = 2\sqrt{2}$ (Radius of circle)

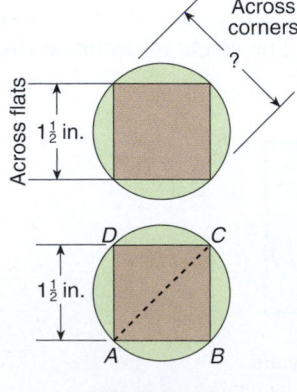

FIGURE 18–64

EXAMPLE 3

To mill a square bolt head $1\frac{1}{2}$ in. across flats, round stock of what diameter is needed? Answer to the nearest hundredth (see Fig. 18–64).

The distance across corners is the diameter of the circle and the diagonal of the square. The distance across flats is the side of a square inscribed in a circle.

Diameter $= \sqrt{2}$ (Side of square)   Substitute 1.5 for the side of the square.

$1\frac{1}{2}$ in. $= 1.5$ in

Diameter $= \sqrt{2}$ (1.5)

Diameter $= 2.12$ in.   Rounded

The diameter of the round stock is 2.12 in.   **See Exercises 13–21.**

STOP AND CHECK

1. Find the diagonal of an inscribed square if the area of the circle is 18 cm². Round to the nearest hundredth.
2. Find the side of an inscribed square if the area of the circle is 18 cm².

Answers:
1. 4.79 cm   2. 3.39 cm

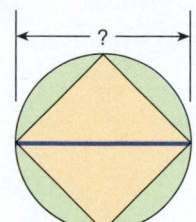

FIGURE 18–65

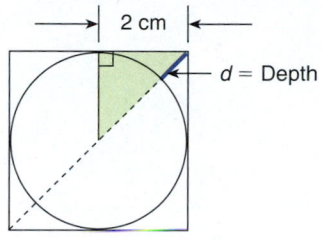

FIGURE 18–66

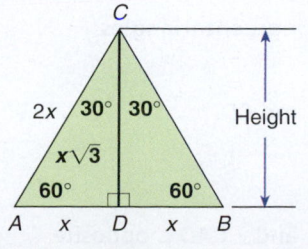

FIGURE 18–67

EXAMPLE 4

Find the diagonal of an inscribed square if the area of the circle is 22 in². Round to hundredths (Fig. 18–65).

$$A = \pi r^2$$ Formula for the area of circle

Substitute $A = 22$.

$$22 = \pi r^2$$ Solve for r.

$$\frac{22}{\pi} = r^2$$

$$\sqrt{\frac{22}{\pi}} = r$$

$$2.646283714 = r$$ Radius of the circle

Diagonal $=$ Diameter $= 2 \times$ Radius

$$= 2(2.646283714) = 5.29 \text{ in.}$$ Rounded

The diagonal of the inscribed square is 5.29 in.   **See Exercises 13–21.**

EXAMPLE 5

What is the maximum depth of cut required to mill a circle on the end of a square piece of stock with a side measuring 4 cm? Answer to hundredths (Fig. 18–66).

The side of the square is 4 cm, so the diameter of the circle is 4 cm and the radius is 2 cm. If the radius is drawn perpendicular to the top side, a right triangle is formed as indicated (shading).

Let $d =$ depth of milling, which equals the hypotenuse of the shaded triangle minus the radius of the circle.

$$c^2 = a^2 + b^2$$ Find the hypotenuse using the Pythagorean theorem. Substitute $a = 2$ and $b = 2$.

$$c^2 = 2^2 + 2^2$$ Solve for c.

$$c^2 = 4 + 4$$

$$c^2 = 8$$

$$c = \sqrt{8}$$

$$c = 2.828427125 \text{ cm}$$ Hypotenuse

Depth of milling $=$ Hypotenuse $-$ Radius

$$d = c - r$$ Substitute $c = 2.828427125$ and $r = 2$.

$$d = 2.828427125 - 2$$

$$d = 0.83 \text{ cm}$$ Rounded

The maximum depth of milling is 0.83 cm.   **See Exercises 22–24.**

4 Use the Properties of Inscribed and Circumscribed Equilateral Triangles to Find Unknown Amounts. An equilateral triangle can be inscribed in a circle or circumscribed about a circle.

The height (or altitude) of the equilateral triangle has a special relationship with the radius of the circle. The height of an equilateral triangle divides the triangle into two 30°, 60°, 90° triangles (Fig. 18–67). By relating the radius of a circle to the height, we can then determine the length of the side of the equilateral triangle. Examine the relationships in Fig. 18–68.

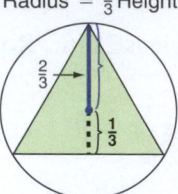

Radius = $\frac{2}{3}$ Height

$\frac{2}{3}$
$\frac{1}{3}$

Inscribed triangle

Radius = $\frac{1}{3}$ Height

$\frac{2}{3}$
$\frac{1}{3}$

Circumscribed triangle

FIGURE 18–68

Triangle–circle relationships

Equilateral Triangle Inscribed in a Circle

Radius of circle $= \dfrac{2}{3}$ (Height of triangle)

Height of triangle $= \dfrac{3}{2}$ (Radius of circle)

Side of triangle $= \dfrac{2(\text{Height of triangle})}{\sqrt{3}}$

Side of triangle $= \dfrac{3(\text{Radius of circle})}{\sqrt{3}}$

Radius of circle $= \dfrac{\sqrt{3}(\text{Side of triangle})}{3}$

Equilateral Triangle Circumscribed About a Circle

Radius of circle $= \dfrac{1}{3}$ (Height of triangle)

Height of triangle $= 3$(Radius of circle)

Side of triangle $= \dfrac{2(\text{Height of triangle})}{\sqrt{3}}$

Side of triangle $= \dfrac{6(\text{Radius of circle})}{\sqrt{3}}$

Radius of circle $= \dfrac{\sqrt{3}(\text{Side of triangle})}{6}$

STOP AND CHECK

1. Find the height to the nearest tenth of an equilateral triangle inscribed in a circle if the radius of the circle is 5 in.
2. Find the length of a side to the nearest tenth of an equilateral triangle inscribed in a circle if the radius of the circle is 5 in.
3. Find the radius to the nearest tenth of a circle if an equilateral triangle that is 10 cm on each side is circumscribed about the circle.

Answers:
1. 7.5 in.   2. 8.7 in.   3. 2.9 cm

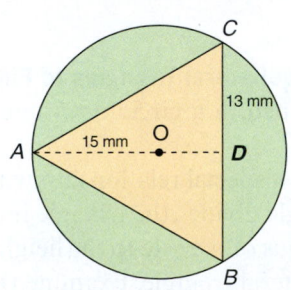

FIGURE 18–69

EXAMPLE 6

Find the dimensions for the inscribed equilateral triangle in Fig. 18–69, where $CD = 13$ mm and $AO = 15$ mm.

(a) Height of $\triangle ABC$ (b) BD (c) BC (d) $\angle CAD$
(e) $\angle ACD$ (f) Area of $\triangle ABC$

(a) Height of $\triangle ABC = \dfrac{3}{2}$ (radius of circle)

$AD = \dfrac{3}{2}(AO)$ Substitute $AO = 15$ mm.

$AD = \dfrac{3}{2}(15)$ Simplify.

$AD = \dfrac{45}{2}$

$AD = 22.5$ mm

(b) $BD = CD$ Corresponding sides of congruent triangles
$BD = 13$ mm   Substitute $CD = 13$ mm.

(c) $BC = BD + CD$ Substitute 13 mm for BD and CD.
$BC = 13$ mm $+ 13$ mm   Simplify.
$BC = 26$ mm

(d) $\angle CAD = 30°$ $\triangle CAD$ is a 30°, 60°, 90° $\triangle$, and $\angle CAD$ is opposite shortest side.

(e) $\angle ACD = 60°$ | $\angle ACD$ is opposite height or other leg of $\triangle$.

(f) $A = \dfrac{1}{2}bh$ | Formula for area of triangle. Substitute for base and height.

$A = \dfrac{1}{2}(BC)(AD)$ | $BC = 26, AD = 22.5$

$A = \dfrac{1}{2}(26)(22.5)$ | Simplify.

$A = 292.5 \text{ mm}^2$ | Area of $\triangle ABC$ **See Exercises 25–34.**

EXAMPLE 7

One end of a shaft 35 mm in diameter is milled as shown in Fig. 18–70. Find the radius of the smaller circle.

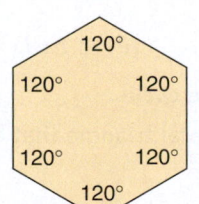

FIGURE 18–70

$BC = \text{Radius} = \dfrac{1}{2}\text{Diameter of large circle}$

$BC = \dfrac{1}{2}(35)$ | Substitute 35 for diameter.

$BC = 17.5 \text{ mm}$ | Radius of large circle

$AB = \dfrac{1}{2}BC$ | From $AB = \dfrac{1}{3}AC$ and $BC = \dfrac{2}{3}AC$

$AB = \dfrac{1}{2}(17.5)$

$AB = 8.75 \text{ mm}$ | Radius of small circle

The radius of the smaller circle is 8.75 mm.   **See Exercises 35–36.**

---

5 **Use the Properties of Inscribed and Circumscribed Regular Hexagons to Find Unknown Amounts.** A regular hexagon is a figure with six equal sides and six equal angles of 120° each. Three diagonals joining pairs of opposite vertices divide the hexagon into six congruent equilateral triangles (all angles 60°) (see Fig. 18–71).

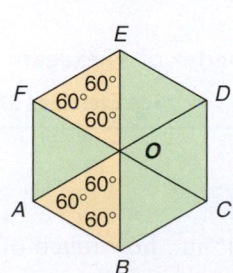

FIGURE 18–71

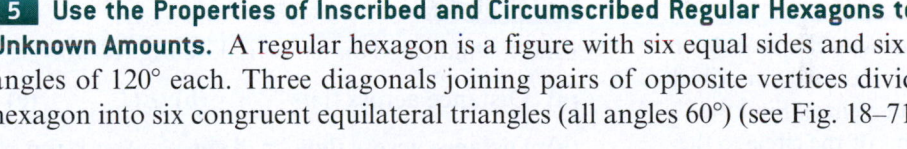

Inscribed hexagon   Circumscribed hexagon   Across flats   Across corners

FIGURE 18–72

Hexagon–circle relationships (Fig. 18–72)

Hexagon Inscribed in Circle

Diameter of circle = diagonal of hexagon or across corners

Across corners = (side of hexagon)(2)

Radius of circle = side of hexagon

Across flats = (side of hexagon)($\sqrt{3}$)

Hexagon Circumscribed About a Circle

Across flats = diameter of circle

Across corners = 2 (side of hexagon)

Radius of circle = height of one triangle formed by the three diagonals

$$= \frac{\text{side of hexagon } (\sqrt{3})}{2}$$

STOP AND CHECK

1. A hexagon that is 2 ft on each side is inscribed in a circle. Find the radius of the circle.
2. Find the distance across flats to the nearest tenth of a hexagon inscribed in a circle with a radius of 4 in.

Answers:

1. 2 ft   2. 6.9 in.

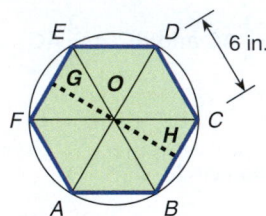

FIGURE 18–73

STOP AND CHECK

1. A hexagon that is 7 cm on each side is circumscribed about a circle. Find the radius of the circle to the nearest tenth.

Answer:

1. 6.1 cm

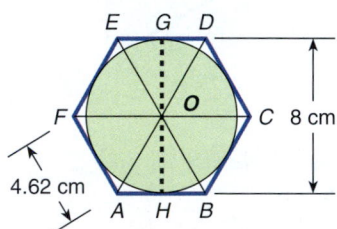

FIGURE 18–74

STOP AND CHECK

1. Find the area to the nearest tenth between a circle and a regular hexagon inscribed in the circle if the radius of the circle 3.5 m.

Answer:

1. 6.7 m²

EXAMPLE 8

Find the indicated dimensions. Use Fig. 18–73.

(a) ∠EOD   (b) ∠COH   (c) ∠BHO   (d) Radius
(e) Distance across corners   (f) FO

(a) ∠EOD = 60°   This is an angle of an equilateral triangle.

(b) ∠COH = 30°   Height OH bisects the 60° angle to form a 30°, 60°, 90° triangle.

(c) ∠BHO = 90°   The height OH forms a right angle with the base of △BOC.

(d) **Radius = 6 in.**   The radius is a side of an equilateral triangle or a side of the hexagon.

(e) **Distance across corners = 12 in.**   The distance across corners is the diameter.

(f) **FO = 6 in.**   FO is a radius or a side of an equilateral triangle.   **See Exercises 37–46.**

EXAMPLE 9

Find the indicated dimensions. Use Fig. 18–74.

(a) Distance across flats   (b) EO   (c) Diagonal   (d) ∠FED

(a) **Distance across flats = 8 cm**   This is the diameter of the circle.

(b) **EO = 4.62 cm**   EO is a side of an equilateral triangle that equals the side of the hexagon.

(c) **Diagonal = 9.24 cm**   The diagonal is twice the side of an equilateral triangle.

(d) ∠FED = 120°   This is the angle at the vertex of a hexagon.

                                       See Exercises 37–46.

EXAMPLE 10

If a regular hexagon is cut from a circle with a radius of 6.45 in., how much of the circle is not used? Round to hundredths (Fig. 18–75).

To solve, we find the area of the circle and the area of the hexagon and subtract to find the difference (portion not used).

$A_{\text{circle}} = \pi r^2$ Substitute r = 6.45 in.

$A_{\text{circle}} = \pi (6.45)^2$ Simplify.

$A_{\text{circle}} = \pi (41.6025)$

$A_{\text{circle}} = 130.6981084 \text{ in}^2$

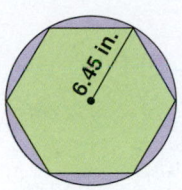

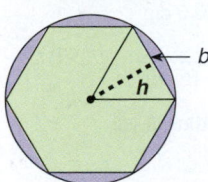

FIGURE 18–75

$$A_{\text{hexagon}} = 6\left(\frac{1}{2}bh\right)$$

A_{hexagon} = area of six triangles. The height of an equilateral triangle forms a 30°, 60°, 90° triangle,

$$h = \frac{b}{2}\sqrt{3} = 5.585863854 \text{ in.}$$

$$A_{\text{hexagon}} = 6\left[\frac{1}{2}(6.45)(5.585863854)\right]$$

Substitute $b = 6.45$ and $h = 5.585863854$.

$$A_{\text{hexagon}} = 108.0864656 \text{ in}^2$$

Waste $= A_{\text{circle}} - A_{\text{hexagon}}$ Difference in areas

Waste $= 130.6981084 - 108.0864656$

Waste $= 22.61 \text{ in}^2$ Rounded

There are 22.61 in² of waste from the circle.   **See Exercises 37–46.**

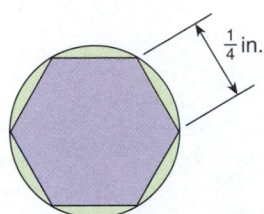

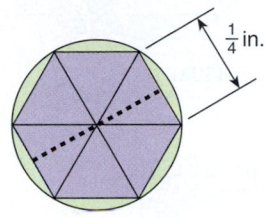

FIGURE 18–76

EXAMPLE 11

Use Fig. 18–76 to answer the questions. Round answers to hundredths.

(a) What is the smallest-diameter round stock from which a hex-bolt head $\frac{1}{4}$ in. on a side can be milled?

(b) What is the distance across the corners of the hex-bolt head?

(c) What is the distance across the flats?

(a) Diameter $= 2 \times$ Side $=$ Diagonal   Side = radius. $2\left(\frac{1}{4}\right) = \frac{1}{2}$

Smallest-diameter round stock $= \frac{1}{2}$ in.

(b) Distance across corners $=$ Diagonal

Distance across corners $= \frac{1}{2}$ in.

(c) Distance across flats $=$ (side of hexagon) $(\sqrt{3})$ Side of hexagon $= \frac{1}{4}$ in.

       $= 0.25(\sqrt{3})$ $= 0.25$ in.

       $= 0.43$ To nearest hundredth

The distance across the flats is 0.43 in.   **See Exercises 47–48.**

18-3 EXERCISES   **MyLab Math** For additional practice go to your study plan in MyLab Math.

1 Give the specific name of each regular polygon in Figs. 18–77 through 18–82 and the number of degrees in each angle. *See Example 1.*

1.

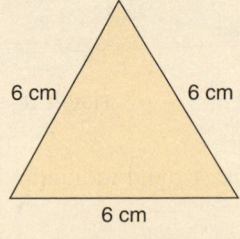

FIGURE 18–77

2.

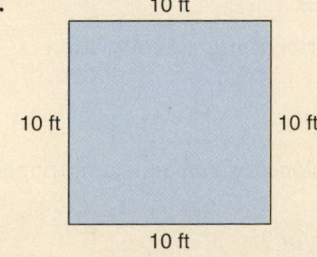

FIGURE 18–78

3.

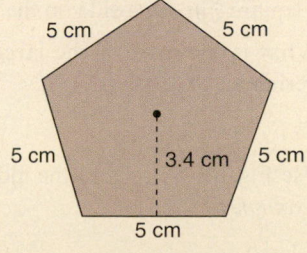

FIGURE 18–79

4.

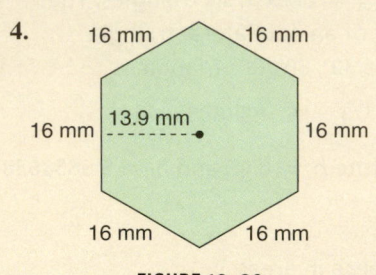

FIGURE 18–80

5.

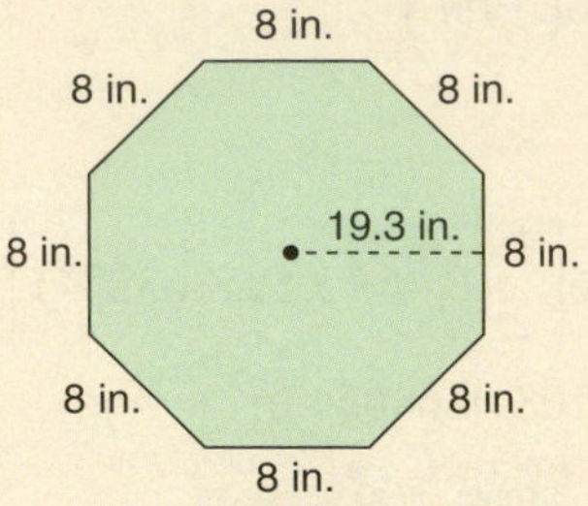

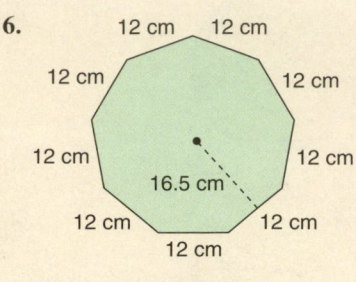

FIGURE 18–81

6.

FIGURE 18–82

2 *See Example 2.*

7. Find the area of Fig. 18–77.

8. Find the area of Fig. 18–78.

9. Find the area of Fig. 18–79.

10. Find the area of Fig. 18–80.

11. Find the area of Fig. 18–81.

12. Find the area of Fig. 18–82.

3 Find the measures of the inscribed square in Fig. 18–83. *See Examples 3 and 4.*

13. ∠*DOC*

14. ∠*BCO*

15. *BO*

16. ∠*BOE*

17. *CE*

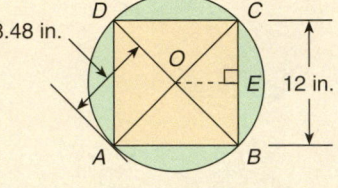

FIGURE 18–83

Find the measures of the circumscribed square in Fig. 18–84. *See Examples 3 and 4.*

18. *DO*

19. Radius of circle

20. ∠*AOB*

21. *BE*

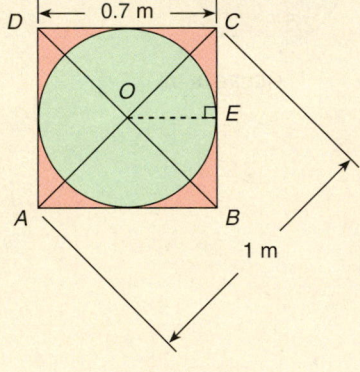

FIGURE 18–84

Solve the problems involving squares and circles. Round the answers to hundredths.

22. To what depth must a 2-in. shaft (Fig. 18–85) be milled on an end to form a square 1.414 in. on a side? *See Example 5.*

23. What is the smallest-diameter round stock (Fig. 18–86) needed to mill a square $\frac{3}{4}$ in. on a side on the end of the stock?

24. What is the area of the largest circle inscribed in a square with a perimeter of 36 dkm?

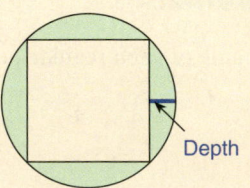

FIGURE 18–85

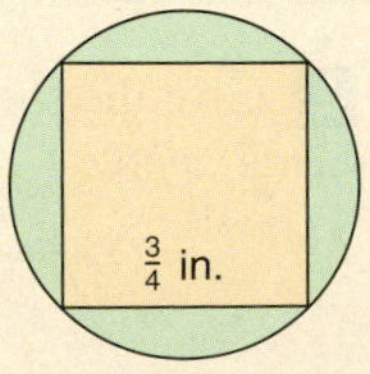

FIGURE 18–86

4 Use Fig. 18–87 to find the indicated dimensions for an equilateral triangle inscribed in a circle. Round to hundredths. *See Example 6.*

25. ∠*CAB*

26. ∠*BCD*

27. ∠*ADC*

28. *CB*

29. *AC*

30. Diameter

31. Area of △ABC   **32.** AD   **33.** Radius

34. Circumference

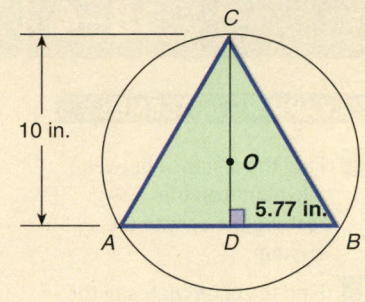

FIGURE 18-87

Solve the problems involving inscribed or circumscribed equilateral triangles. Round to hundredths. *See Example 7.*

35. The area of an equilateral triangle with a base of 10 cm is 43.5 cm². The triangle is circumscribed about a circle. What is the radius of the inscribed circle?

36. One end of a circular steel rod is milled into an equilateral triangle whose height is $\frac{3}{4}$ in. What is the diameter of the rod if the smallest diameter possible was used for the job?

5 Find the dimensions for Fig. 18–88 if *GO* = 17.32 mm and *FO* = 20 mm. *See Examples 8–10.*

37. Distance across flats   **38.** ∠CDE

39. ∠EFO   **40.** Distance across corners

41. ∠FGO   **42.** GA

43. AB

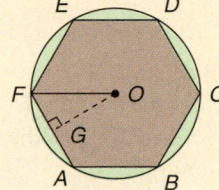

FIGURE 18-88

Find the dimensions for Fig. 18–89 if *FO* = 3 in. *See Examples 8–10.*

44. ∠ODC

45. ∠EOG

46. OC

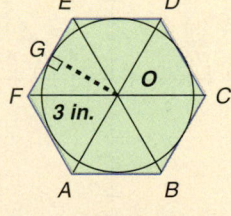

FIGURE 18-89

See Example 11.

47. What is the distance across the flats of a hexagon cut from a circular blank with a circumference of 145 mm (Fig. 18–90)? Round to hundredths.

48. The end of a hexagonal rod is milled into the largest circle possible (Fig. 18–91). What is the distance across the corners of the hexagon if the area of the circle is 4.75 in²? Round to hundredths.

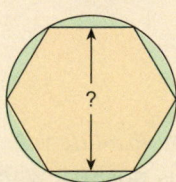

FIGURE 18-90

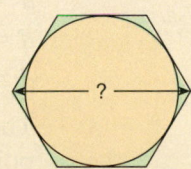

FIGURE 18-91

18–4   Distance and Midpoints

LC LEARNING CATALYTICS

1. Evaluate $\sqrt{(5-3)^2 + (2-7)^2}$ to the nearest hundredth.

STOP AND CHECK

1. Find the distance from $(3, 7)$ to $(5, 2)$ to the nearest hundredth.

Answer:

1. 5.39

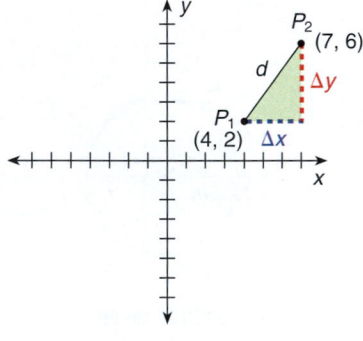

FIGURE 18–92

1 Find the Distance Between Two Points on the Rectangular Coordinate System. In Chapter 4 we found the distance between two points on a line. To expand this concept to the distance between two points on the rectangular coordinate system, we will apply the Pythagorean theorem.

$$c^2 = a^2 + b^2 \quad \text{or} \quad c = \sqrt{a^2 + b^2}$$

The distance between two points on the rectangular coordinate system is the shortest distance between the points.

To find the distance between two points on the rectangular coordinate system:

1. Use the formula

$$d = \sqrt{(\triangle x)^2 + (\triangle y)^2} \quad \text{or} \quad \sqrt{(x_2 - x_1)^2 + (y_2 - y_1)^2}$$

where P_1 and P_2 are two points with coordinates (x_1, y_1) and (x_2, y_2), respectively.

2. Substitute values for x_1, x_2, y_1, and y_2.

3. Evaluate to find d.

EXAMPLE 1

Find the distance from $(4, 2)$ to $(7, 6)$.

These points are shown in Fig. 18–92. Point $(4, 2)$ is labeled P_1, and point $(7, 6)$ is labeled P_2.

$d = \sqrt{(x_2 - x_1)^2 + (y_2 - y_1)^2}$ Substitute: $x_1 = 4, x_2 = 7, y_1 = 2$, and $y_2 = 6$

$d = \sqrt{(7 - 4)^2 + (6 - 2)^2}$ Combine in each grouping.

$d = \sqrt{3^2 + 4^2}$ Square each term in the radicand.

$d = \sqrt{9 + 16}$ Add terms in the radicand.

$d = \sqrt{25}$ Find the principal square root.

$d = 5$

The distance from $(4, 2)$ to $(7, 6)$ is 5 units.   **See Exercises 1–10.**

TIP Point Selection Does Not Matter The distance formula can be used to calculate the distance between two points, no matter which point we designate as P_1 and which point we designate as P_2. Let's rework the preceding example, letting $P_1 = (7, 6)$ and $P_2 = (4, 2)$. Thus, $x_1 = 7, x_2 = 4, y_1 = 6$, and $y_2 = 2$.

$d = \sqrt{(4 - 7)^2 + (2 - 6)^2}$ Substitute.

$d = \sqrt{(-3)^2 + (-4)^2}$ $(-3)^2 = +9; (-4)^2 = +16$.

$d = \sqrt{9 + 16}$

$d = \sqrt{25}$

$d = 5$

The change in vertical movement has the same *absolute value,* regardless of which point is used first. The result after squaring is positive whether the difference is positive or negative. For similar reasons, the change in horizontal movement squared is the same, regardless of which point is used first.

2 Find the Coordinates of the Midpoint of a Line Segment if Given the Coordinates of the End Points. In Chapter 4 we found the midpoint between two points on a line. We will expand that concept to find the midpoint between two points on the rectangular coordinate system.

To find the coordinates of the midpoint of a line segment if given the coordinates of the end points:

1. Average the respective coordinates of the end points of the segment using the formula

$$\text{midpoint} = \left(\frac{x_1 + x_2}{2}, \frac{y_1 + y_2}{2}\right)$$

where P_1 and P_2 are end points of the segment, $P_1 = (x_1, y_1)$ and $P_2 = (x_2, y_2)$.

2. Substitute values for x_1, x_2, y_1, and y_2.
3. Evaluate to find the coordinates of the midpoint.

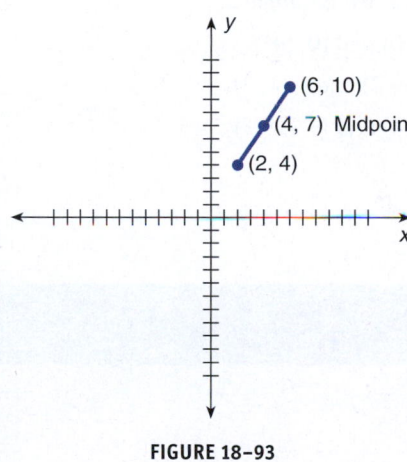

STOP AND CHECK

1. Find the midpoint between the points (3, 7) and (5, 2).

Answer:
1. (4, 4.5)

EXAMPLE 2

Find the midpoint of each segment with the given end points.

(a) (2, 4) and (6, 10) (Fig. 18–93) **(b)** (3, −5) and origin (Fig. 18–94)

(a) (2, 4) and (6, 10)

$$\text{midpoint} = \left(\frac{x_1 + x_2}{2}, \frac{y_1 + y_2}{2}\right) \quad \text{Substitute: } x_1 = 2, x_2 = 6, y_1 = 4, y_2 = 10$$

$$\text{midpoint} = \left(\frac{2 + 6}{2}, \frac{4 + 10}{2}\right) \quad \text{Simplify.}$$

$$\text{midpoint} = \left(\frac{8}{2}, \frac{14}{2}\right) \quad \text{Reduce.}$$

$$\textbf{midpoint} = \mathbf{(4, 7)}$$

(b) (3, −5) and origin.

$$\text{midpoint} = \left(\frac{x_1 + x_2}{2}, \frac{y_1 + y_2}{2}\right) \quad \text{Substitute: } x_1 = 3, x_2 = 0, y_1 = -5, y_2 = 0$$

$$\text{midpoint} = \left(\frac{3 + 0}{2}, \frac{-5 + 0}{2}\right)$$

$$\textbf{midpoint} = \left(\frac{3}{2}, -\frac{5}{2}\right) \text{ or } \mathbf{(1.5, -2.5)}$$

See Exercises 11–20.

FIGURE 18–93

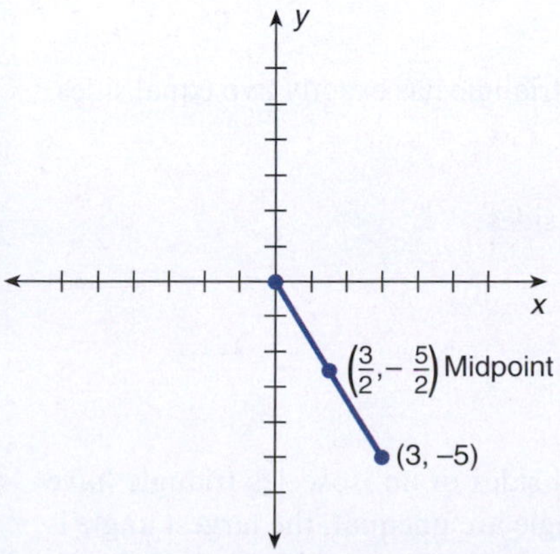

FIGURE 18–94

TIP Midpoint Between the Origin and Another Point To find the coordinates of the midpoint of a segment from the origin to any point, take one-half of each coordinate of the point that is not at the origin. Apply this tip to part (b) of the preceding example.

TIP Visualize and Estimate Whenever Practical When formulas are developed to find desired information, we often rely totally on our calculation skills and ability to manipulate formulas. We can strengthen our understanding of the concepts and avoid mistakes by visualizing the problem and by estimating.

In some of the examples in this section, we provided a visualization of the problem. Now, let's examine some estimates. To estimate the distance between points (4, 2) and (7, 6), we find the vertical and horizontal changes. The vertical change is 4, from (6 − 2), and the horizontal change is 3, from (7 − 4). Applying properties of the Pythagorean theorem, the distance between the points (hypotenuse) is more than the longest leg (4) and less than the sum of the legs, 3 + 4 = 7. So, we estimate the distance to be between 4 and 7 units.

To estimate the coordinates of the midpoint between these points, find the range of values. The x-coordinate of the midpoint is halfway between 4 and 7. The y-coordinate of the midpoint is halfway between 2 and 6.

18-4 EXERCISES

MyLab Math For additional practice go to your study plan in MyLab Math.

1 Graph the line segment determined by the given end points and find the distance between each of the pairs of points. Express the answer to the nearest thousandth when necessary. *See Example 1.*

1. (7, 10) and (1, 2)

2. (7, −7) and (2, 5)

3. (5, 0) and (0, 5)

4. (−2, −2) and (3, −4)

5. (5, 7) and (0, −3)

6. (8, −2) and (−4, 3)

7. (−4, 6) and (2, −2)

8. (3, 5) and (0, 1)

9. (5, 4) and (−7, −5)

10. (7, 2) and (−2, −3)

2 Find the coordinates of the midpoints of the segments determined by the given end points. *See Example 2.*

11. (7, 10) and (1, 2)

12. (7, −7) and (2, 5)

13. (5, 0) and (0, 5)

14. (−2, −2) and (3, −4)

15. (5, 7) and (0, −3)

16. (8, −2) and (−4, 3)

17. (−4, 6) and (2, −2)

18. (3, 5) and (0, 1)

19. (5, 4) and (−7, −5)

20. (7, 2) and (22, 23)

18   CHAPTER REVIEW OF KEY CONCEPTS

LEARNING OUTCOMES	KEY CONCEPTS AND EXAMPLES

Section 18–1

1 Classify triangles by the relationship of the sides or angles (pp. 770–771).

An equilateral triangle has three equal sides. An isosceles triangle has exactly two equal sides. A scalene triangle has three unequal sides.

Identify the following triangles by the measures of their sides:

(a) 6 cm, 6 cm, 6 cm   Equilateral

(b) 12 in., 10 in., 12 in.   Isosceles

(c) 10 cm, 12 cm, 15 cm   Scalene

An equilateral triangle has three equal angles. The equal sides of an isosceles triangle have opposite angles that are equal. If the three sides of a triangle are unequal, the largest angle is opposite the longest side, and the smallest angle is opposite the shortest side. In a right triangle, the hypotenuse is always the longest side.

LEARNING OUTCOMES	KEY CONCEPTS AND EXAMPLES

Identify the largest and the smallest angles in triangles with the following sides: 12 in., 10 in., 9 in.

The largest angle is opposite the 12-in. side; the smallest angle is opposite the 9-in. side.

2 Determine if two triangles are congruent using inductive and deductive reasoning (pp. 771–773).

Triangles are congruent (one can fit exactly over the other) (a) if the three sides of one triangle are equal to the corresponding three sides of the other triangle (SSS), (b) if two sides and the included angle of one triangle are equal to two sides and the included angle of the other triangle (SAS), or (c) if two angles and the common side of one triangle are equal to two angles and the common side of the other triangle (ASA).

Determine whether the pairs of triangles in Fig. 18–95 are congruent.

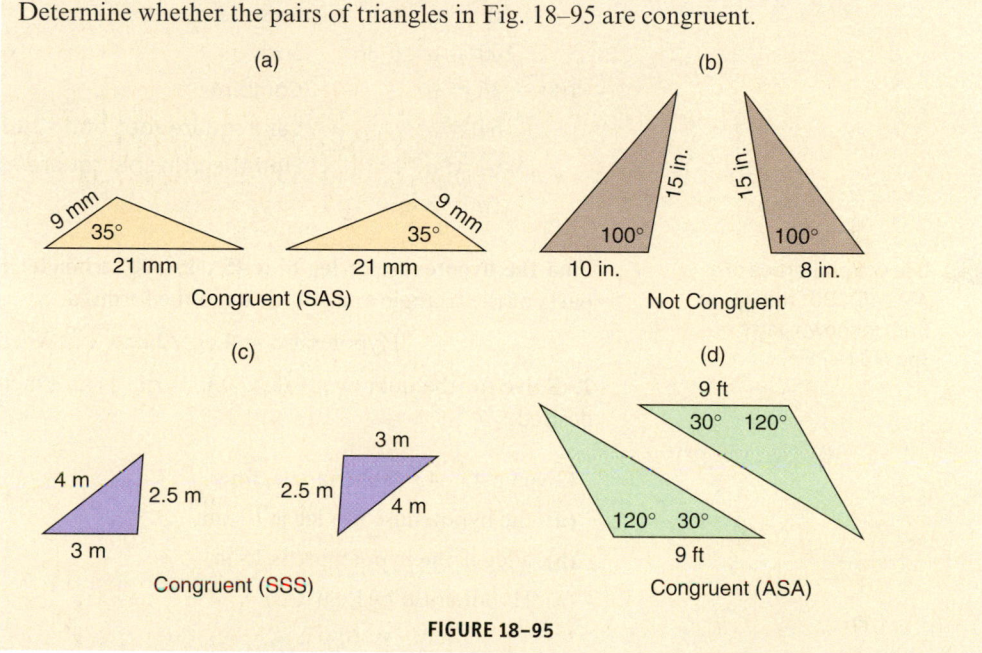

FIGURE 18–95

3 Solve problems that involve similar triangles (pp. 773–776).

Each angle of one triangle is equal to its corresponding angle in the other similar triangle. Each side of one triangle is directly proportional to its corresponding side in the other similar triangle.

Find a missing side of a similar triangle: 1. Examine the given dimensions of the corresponding sides and find one pair for which both measures are known. **2.** Pair the unknown side with its corresponding known side. **3.** Set up a direct proportion with the two pairs. **4.** Solve the proportion.

Find the height of a building that makes a shadow of 25 m when a meter stick makes a shadow of 1.5 m.

Direct proportion: As shadow increases, height increases.

$$\frac{x\text{-m building}}{1\text{-m stick}} = \frac{25\text{-m building shadow}}{1.5\text{-m stick shadow}}$$ Pair 1. Estimation: meter stick.
    Pair 2. Stick < shadow, so building < 25 m.

$$\frac{x}{1} = \frac{25}{1.5}$$ Cross multiply.

$$1.5x = 25$$ Divide.

$$x = \frac{25}{1.5}$$

$$x = 16.67 \text{ m}$$

LEARNING OUTCOMES   **KEY CONCEPTS AND EXAMPLES**

1 Use the Pythagorean theorem to find the unknown side of a right triangle (pp. 778–781).

The square of the hypotenuse of a right triangle equals the sum of the squares of the other two legs. $c^2 = a^2 + b^2$

Find a leg or the hypotenuse of a right triangle: 1. Identify the two known sides of the triangle and the unknown side, and state the theorem symbolically.   **2.** Substitute known values in the Pythagorean theorem. $c^2 = a^2 + b^2$ **3.** Solve for the unknown value.   **4.** The solution is the principal square root.

The hypotenuse (c) of a right triangle is 10 in. If side b is 6 in., find side a.

$$c^2 = a^2 + b^2 \quad \text{Substitute values.}$$
$$10^2 = a^2 + 6^2 \quad \text{Raise to powers.}$$
$$100 = a^2 + 36 \quad \text{Sort.}$$
$$100 - 36 = a^2 \quad \text{Combine.}$$
$$64 = a^2 \quad \text{Take square root of both sides.}$$
$$\sqrt{64} = a \quad \text{Find the principal square root.}$$
$$a = 8 \text{ in.}$$

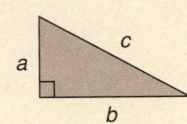

FIGURE 18-96

2 Use the properties of a 45°, 45°, 90° triangle to find unknown parts (pp. 781–782).

Find the hypotenuse or leg of a 45°, 45°, 90° triangle: 1. Identify the known and unknown parts of the triangle and substitute in the formula:

$$\text{Hypotenuse} = \text{Leg}\sqrt{2} \text{ or } x\sqrt{2}, \text{ where } x = \text{length of leg}$$

2. Solve for the unknown value.   **3.** Write as an exact solution or approximate solution, as desired.

Given a 45°, 45°, 90° triangle, find

(a) the hypotenuse if a leg is 10 cm.

(b) a leg if the hypotenuse is 15 in.

(a) Hypotenuse $= \text{Leg}(\sqrt{2})$
$$= 10\sqrt{2}$$
$$= 14.14213562 \text{ or } 14.1 \text{ cm} \qquad \textbf{Rounded}$$

(b) Leg $= \dfrac{\text{Hypotenuse} (\sqrt{2})}{2}$
$$= \dfrac{15\sqrt{2}}{2}$$
$$= 10.60660172 \text{ or } 10.6 \text{ in.} \qquad \textbf{Rounded}$$

3 Use the properties of a 30°, 60°, 90° triangle to find unknown parts (pp. 782–783).

Find the unknown sides of a 30°, 60°, 90° triangle if one side is known: 1. Identify the known side of the triangle.   **2.** If the side opposite the 30° angle is known, substitute in the formulas,

$$\text{hypotenuse} = 2x$$
$$\text{height} = x\sqrt{3}$$

where x is the measure of the side opposite the 30° angle.   **3.** If either the hypotenuse or height is known, use the appropriate formula and solve for x. Use x and the remaining formula to find the remaining side.

Given a 30°, 60°, 90° triangle, find

(a) the hypotenuse if the side opposite the 30° angle $= 4.5$ cm.

(b) the side opposite the 30° angle if the hypotenuse is 20 in.

(c) the side opposite the 60° angle if the side opposite the 30° angle is 35 mm.

LEARNING OUTCOMES   **KEY CONCEPTS AND EXAMPLES**

(a) Hypotenuse = 2 (Side opposite 30°)
              = 2(4.5)
              = 9 cm

(b) Side opposite 30° = $\dfrac{\text{Hypotenuse}}{2}$

                    = $\dfrac{20}{2}$

                    = 10 in.

(c) Side opposite 60° = (Side opposite 30°)$(\sqrt{3})$
                    = 35$(\sqrt{3})$
                    = 60.62177826 or 60.6 mm   **Rounded**

Section 18–3

1 Find the number of degrees in each angle of a regular polygon (p. 786).

1. Multiply the number of sides less 2 by 180°.   **2.** Divide by the number of sides.

$$\text{degrees per angle of a regular polygon} = \frac{180°(\text{number of sides} - 2)}{\text{number of sides}}$$

Find the number of degrees in each angle of a pentagon.

A pentagon has five sides.

$$\text{degrees per angle} = \frac{180°(5 - 2)}{5} = \frac{180°(3)}{5} = \frac{540°}{5} = 108°$$

2 Find the area of a regular polygon (p. 787).

A regular polygon has all sides equal. Regular polygons with 3, 4, 5, 6, and 8 sides are equilateral triangles, squares, pentagons, hexagons, and octagons, respectively. Lines drawn from the center of a regular polygon to each vertex will form as many congruent triangles as the polygon has sides.

Form congruent triangles in a regular polygon: 1. Draw line segments from the center of the regular polygon to each vertex.   **2.** Congruent triangles are formed.

Find the area of a regular polygon: 1. Find the area of one of the congruent triangles formed by connecting each vertex with the center of the polygon.   **2.** Multiply the result from Step 1 by the number of congruent triangles.

Find the area of the regular hexagon in Fig. 18–97.

Each congruent triangle formed is an equilateral triangle (Fig. 18–98). Six congruent central angles make each central angle 60°: $\frac{360°}{6}$ = 60°. Each base angle of the triangles equals one-half the angle at each vertex of the hexagon: $\frac{1}{2}$(120°) = 60°. The congruent triangles are 60°, 60°, 60° triangles, or equilateral.

Find the area of one congruent triangle of the hexagon.

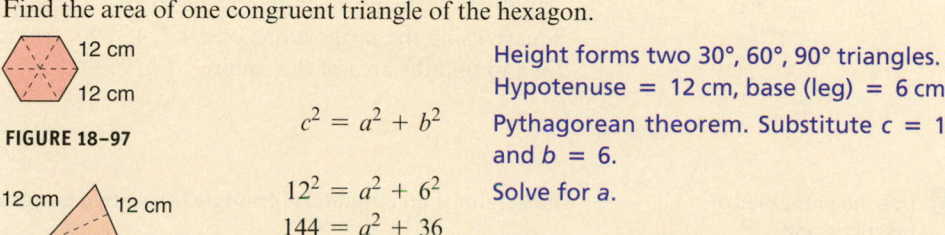

FIGURE 18–97

FIGURE 18–98

$c^2 = a^2 + b^2$ Height forms two 30°, 60°, 90° triangles.
                     Hypotenuse = 12 cm, base (leg) = 6 cm.
$12^2 = a^2 + 6^2$ Pythagorean theorem. Substitute c = 12 and b = 6.
$144 = a^2 + 36$ Solve for a.
$144 - 36 = a^2$
$108 = a^2$ Take the square root of both sides.
$10.4 \text{ cm} = a$ Height of triangle. Rounded

| LEARNING OUTCOMES | KEY CONCEPTS AND EXAMPLES |

$$A_{\text{triangle}} = \frac{1}{2}bh$$ Substitute $b = 12$ and $h = 10.4$.

$$A_{\text{triangle}} = \frac{1}{2}(12)(10.4)$$ Evaluate.

$$A_{\text{triangle}} = 62.4 \text{ cm}^2$$ Area of one congruent triangle

$$A_{\text{hexagon}} = 6(A_{\text{triangle}})$$ Multiply by the number of congruent

$$A_{\text{hexagon}} = 6(62.4)$$ triangles. Substitute the area of one

$$A_{\text{hexagon}} = 374.4 \text{ cm}^2$$ triangle.

3 Use the properties of inscribed and circumscribed squares to find unknown amounts (pp. 787–789).

The diagonal of a square inscribed in or circumscribed about a circle forms congruent $45°, 45°, 90°$ triangles.

The diagonal of a square inscribed in a circle equals the diameter of the circle.

The side of a square circumscribed about a circle equals the diameter of the circle.

The radius of a circle when a square is inscribed in the circle equals the height of a $45°, 45°, 90°$ triangle formed by one diagonal.

The radius of a circle when a square is circumscribed about the circle equals the height of a $45°, 45°, 90°$ triangle formed by two diagonals or it equals one-half the length of a side of the square.

The end of a 20-mm diameter round rod is milled as shown in Fig. 18–99. Find the area of the square.

The radius of the circle equals the height of a right triangle formed by one diagonal. The diameter is the base. Find the area of the triangle and double it.

$$A_{\text{triangle}} = \frac{1}{2}bh$$

$$A_{\text{triangle}} = \frac{1}{2}(20)(10)$$

$$A_{\text{triangle}} = 100 \text{ mm}^2 \quad \text{Area of one triangle}$$

Area of square: $100 \times 2 = 200 \text{ mm}^2$

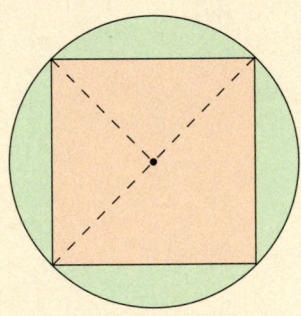

FIGURE 18–99

An alternative approach would be to find the side of the square using the proportions of a $45°, 45°, 90°$ triangle and then to find the area of the square.

4 Use the properties of inscribed and circumscribed equilateral triangles to find unknown amounts (pp. 789–791).

The height of an equilateral triangle forms two congruent $30°, 60°, 90°$ triangles.

The radius of a circle circumscribed about an equilateral triangle is two-thirds the height of the triangle or twice the distance from the center of the circle to the base of the triangle.

The radius of a circle inscribed in an equilateral triangle is one-third the height of the triangle or one-half the distance from the vertex of the triangle to the center of the circle.

LEARNING OUTCOMES	KEY CONCEPTS AND EXAMPLES

One end of a shaft 30 mm in diameter is milled as shown in Fig. 18–100.

Find the diameter of the smaller circle.

BC is the radius (half the diameter) of the large circle.

$$BC = \frac{1}{2}(30)$$

$BC = 15\,\text{mm}$ **Radius of large circle**

$$AB = \frac{1}{2}BC$$ **Radius of small circle**

$$AB = \frac{1}{2}(15)$$

$AB = 7.5\,\text{mm}$

The diameter of the small circle $= 2(7.5) = 15\,\text{mm}$.

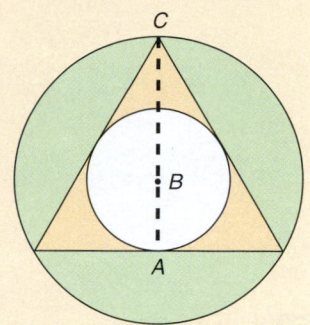

FIGURE 18–100

5 Use the properties of inscribed and circumscribed regular hexagons to find unknown amounts (pp. 791–793).

The three diagonals that join opposite vertices of a regular hexagon form six congruent equilateral triangles.

The height of each of the equilateral triangles forms two congruent 30°, 60°, 90° triangles. The diagonal of a regular hexagon inscribed in a circle is the diameter of the circle (distance across corners).

The radius of the circle when the regular hexagon is inscribed in the circle is a side of an equilateral triangle formed by the three diagonals and is one-half the distance across corners.

The radius of the circle when a regular hexagon is circumscribed about the circle is the height of one of the six equilateral triangles formed by the three diagonals.

The diameter of the circle when a regular hexagon is circumscribed about the circle is the distance across flats or twice the height of one of the six equilateral triangles formed by the three diagonals.

Find the following dimensions for Fig. 18–101 if the diameter is 10 cm.

(a) $\angle AOF$

(b) Radius

(c) Distance across corners

(a) $\angle AOF = 60°$ (angle of equilateral triangle)

(b) 5 cm (radius of a circle is one-half the diameter)

(c) 10 cm (diameter of a circle is the distance across corners of the hexagon)

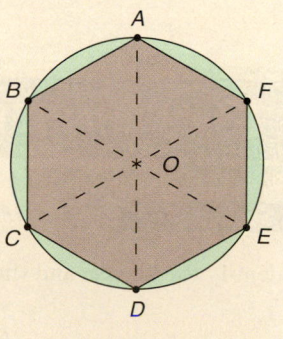

FIGURE 18–101

Section 18–4

1 Find the distance between two points on the rectangular coordinate system (p. 796).

1. Use the formula

$$d = \sqrt{(\triangle x)^2 + (\triangle y)^2} \quad \text{or} \quad \sqrt{(x_2 - x_1)^2 + (y_2 - y_1)^2}$$

where P_1 and P_2 are two points with coordinates (x_1, y_1) and (x_2, y_2), respectively.   **2.** Substitute values for $x_1, x_2, y_1,$ and y_2.   **3.** Evaluate to find d.

LEARNING OUTCOMES	KEY CONCEPTS AND EXAMPLES

Find the distance between the points $(3, 2)$ and $(-1, 5)$.

$d = \sqrt{(x_2 - x_1)^2 + (y_2 - y_1)^2}$ Substitute values.

$d = \sqrt{(-1 - 3)^2 + (5 - 2)^2}$ Simplify groupings in radicand.

$d = \sqrt{(-4)^2 + 3^2}$ Square terms in radicand.

$d = \sqrt{16 + 9}$ Combine terms in radicand.

$d = \sqrt{25}$ Find principal square root.

$d = 5$

2   Find the coordinates of the midpoint of a line segment if given the coordinates of the end points (pp. 797–798).

1. Average the respective coordinates of the end points of the segment using the formula

$$\text{midpoint} = \left(\frac{x_1 + x_2}{2}, \frac{y_1 + y_2}{2} \right)$$

where P_1 and P_2 are end points of the segment, $P_1 = (x_1, y_1)$ and $P_2 = (x_2, y_2)$.   **2.** Substitute values for x_1, x_2, y_1, and y_2.   **3.** Evaluate to find the coordinates of the midpoint.

Find the midpoint of the segment joining the points $(3, 2)$ and $(-1, 5)$.

$\left(\dfrac{x_1 + x_2}{2}, \dfrac{y_1 + y_2}{2} \right)$ Substitute values.

$\left(\dfrac{3 + (-1)}{2}, \dfrac{2 + 5}{2} \right)$ Simplify numerators.

$\left(\dfrac{2}{2}, \dfrac{7}{2} \right)$ Reduce or simplify.

$\left(1, \dfrac{7}{2} \right)$ or $(1, 3.5)$

18   CHAPTER REVIEW EXERCISES

Section 18–1   MyLab Math   For additional practice go to your study plan in MyLab Math.

Identify the longest and shortest sides in Figs. 18–102 and 18–103.

1.

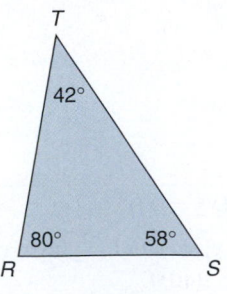

FIGURE 18–102

2.

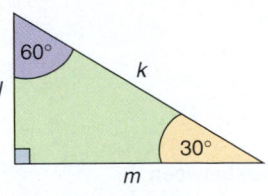

FIGURE 18–103

List the angles in order of size from largest to smallest in Figs. 18–104 and 18–105.

3.

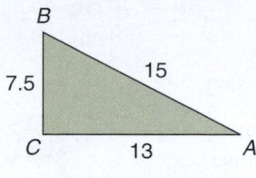

FIGURE 18–104

4.

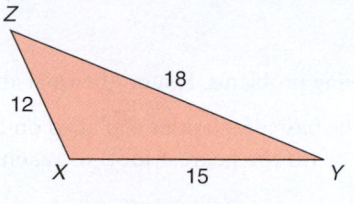

FIGURE 18–105

Write the corresponding parts that are not given for the congruent triangles in Figs. 18–106 through 18–108.

5.

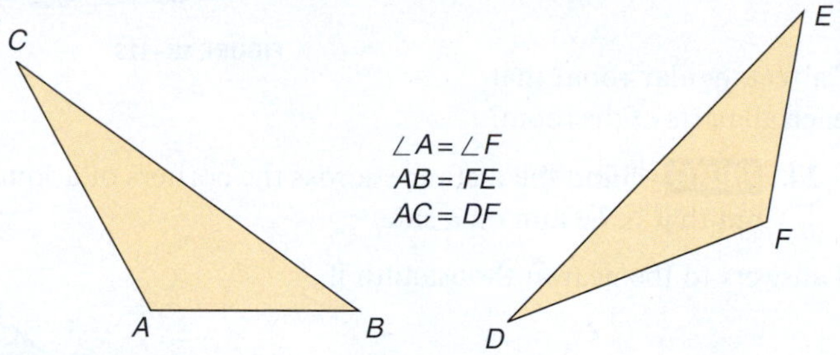

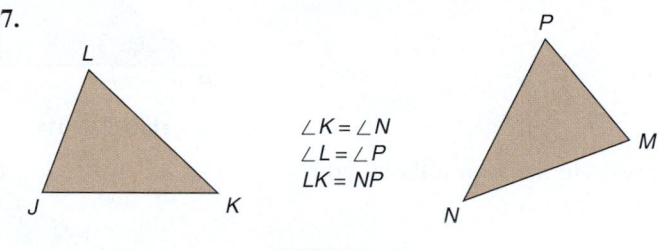

$\angle A = \angle F$
$AB = FE$
$AC = DF$

FIGURE 18–106

6.

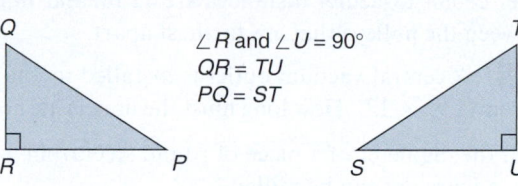

$\angle R$ and $\angle U = 90°$
$QR = TU$
$PQ = ST$

FIGURE 18–107

7.

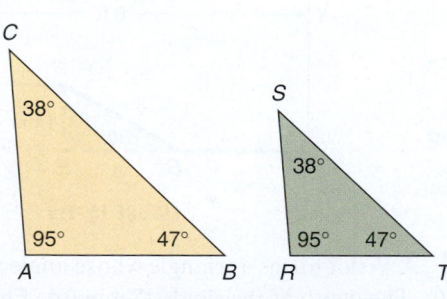

$\angle K = \angle N$
$\angle L = \angle P$
$LK = NP$

FIGURE 18–108

8. $\triangle LMN \sim \triangle XYZ$, $\angle L = \angle X$, $\angle M = \angle Y$. Find m and x (see Fig. 18–109).

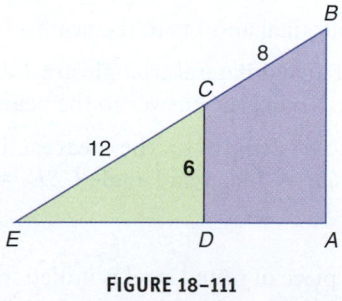

FIGURE 18–109

9. Write the proportions for the sides of the similar triangles in Fig. 18–110.

FIGURE 18–110

10. Find AB if $\triangle DEC \sim \triangle AEB$ (see Fig. 18–111).

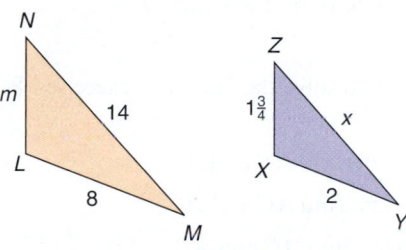

FIGURE 18–111

Section 18–2

Use Fig. 18–112 to solve the following exercises. Round the final answers to the nearest thousandth if necessary.

11. $a = 9$ in.
$b = 12$ in.
$c = ?$

12. $a = 8$ cm
$b = 15$ cm
$c = ?$

13. $a = 7$ ft
$b = ?$
$c = 10$ ft

14. $a = 8$ mm
$b = ?$
$c = 17$ mm

15. $a = ?$
$b = 15$ yd
$c = 17$ yd

16. $a = ?$
$b = 12$ km
$c = 15$ km

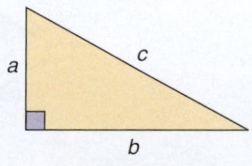

FIGURE 18–112

17. $a = 11$ mi
 $b = 17$ mi
 $c = ?$

18. $a = 10$ in.
 $b = 24$ in.
 $c = ?$

19. $a = ?$
 $b = 40$ cm
 $c = 50$ cm

Solve the following problems. Round the final answers to the nearest thousandth if necessary.

20. **CON**   If the base of a ladder is placed on the ground 4 ft from a house, how tall must the ladder be (to the nearest foot) to reach the chimney top that extends $18\frac{1}{2}$ ft above the ground?

21. **AUTO**   In an automobile, three pulleys are connected by one belt. The center-to-center distance between the pulleys farthest apart cannot be measured conveniently. The other center-to-center distances are 12 in. and 18 in. (Fig. 18–113). Find the distance between the pulleys that are farthest apart.

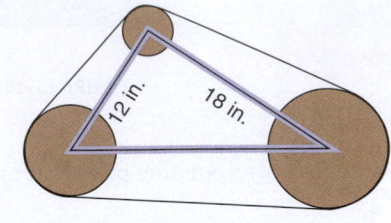

FIGURE 18–113

22. **CON**   A central vacuum outlet is installed in one corner of a rectangular room that measures $9' \times 12'$. How long must the nonelastic hose be to reach all parts of the room?

23. Find the diameter of a piece of round steel from which a 3-in. square nut can be milled.

24. **INDTR**   Find the distance across the corners of a square nut that is 7.9 mm on a side.

Use Fig. 18–114 to solve the following exercises. Round your final answers to the nearest thousandth if necessary.

25. $RT = 15$ cm; find RS and ST.

26. $ST = 7.2$ ft; find RS and RT.

27. $RT = 9\sqrt{2}$ hm; find RS and ST.

28. $RT = 8\sqrt{5}$ dkm; find RS and ST.

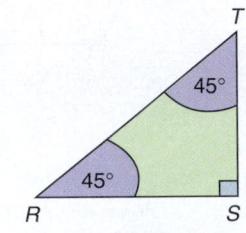

FIGURE 18–114

Use Fig. 18–115 to solve the following exercises. Round your final answers to the nearest thousandth if necessary.

29. $AC = 12$ dm; find AB and BC.

30. $AB = 15$ km; find AC and BC.

31. $BC = 10$ in.; find AC and AB.

32. $BC = 8\sqrt{2}$ hm; find AC and AB.

33. $AC = 40$ ft 7 in.; find AB and BC to the nearest inch.

Solve. Round your final answers to the nearest thousandth if necessary.

34. The sides of an equilateral triangle are 4 dm in length. Find the altitude of the triangle. Round the answer to the nearest thousandth of a decimeter.

35. Find the total length to the nearest inch of the conduit $ABCD$ if $XY = 8$ ft, $CE = 14$ in., and angle $CBE = 30°$ (Fig. 18–116).

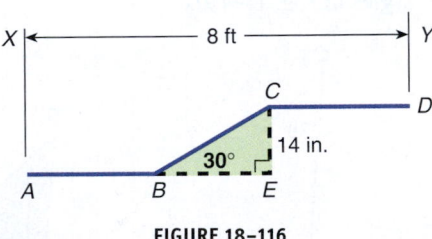

FIGURE 18–116

36. **INDTR**   A piece of round steel is milled to a point at one end. The angle formed by the point is the angle of taper. If the steel has a 16-mm diameter and a 60° angle of taper, find the length c of the taper (Fig. 18–117).

37. **INDTR**   A V-slot forms a triangle whose angle at the vertex is 60°. The depth of the slot is 17 mm (see Fig. 18–118). Find the width of the V-slot.

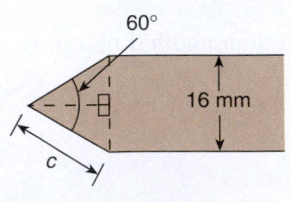

FIGURE 18–117

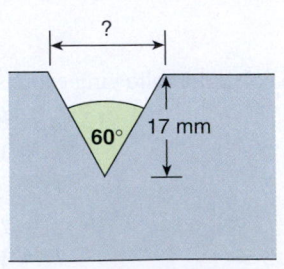

FIGURE 18–118

38. A manufacturer recommends attaching a guy wire to its 30-ft antenna at a 45° angle. If the antenna is installed on a flat surface, how long must the guy wire be (to the nearest foot) if it is attached to the antenna 4 ft from the top?

Section 18–3

Find the area of Figs. 18–119 and 18–120.

39.

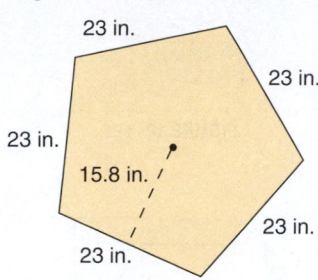

FIGURE 18–119

40.

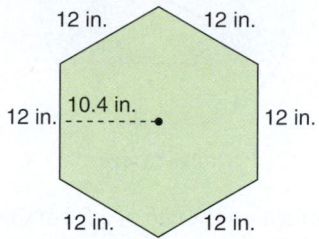

FIGURE 18–120

41. Identify the figure in Exercise 39 by giving its specific name and the number of degrees in each angle.

42. Identify the regular polygon in Exercise 40 by giving its specific name and the number of degrees in each angle.

For each exercise, supply the missing dimensions. For an inscribed triangle, see Fig. 18–121.

43. If $DO = 7$, $OB =$ _____

44. If $FB = 9$, $FO =$ _____

45. If $AB = 10$, $AE =$ _____

46. $\angle ABF =$ _____

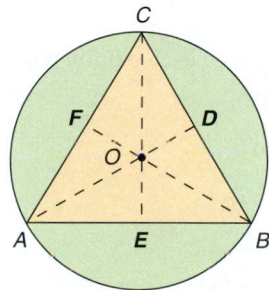

FIGURE 18–121

For a circumscribed square, see Fig. 18–122.

47. $\angle GJO =$ _____

48. If $GO = 7$, $HO =$ _____

49. If $KO = 10$, $IJ =$ _____

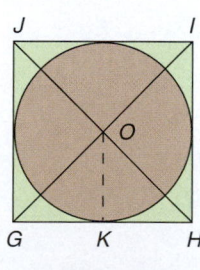

FIGURE 18–122

For an inscribed hexagon, see Fig. 18–123.

50. If $QO = 15$, $MN =$ _____

51. $\angle MOP =$ _____

52. $\angle MNO =$ _____

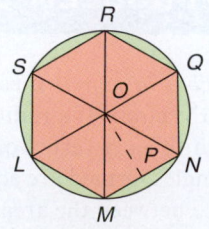

FIGURE 18–123

Solve the problems involving polygons and circles. Round the answers to hundredths.

53. A shaft with a 20-mm diameter is milled at one end, as illustrated in Fig. 18–124. What is the radius of the inscribed circle?

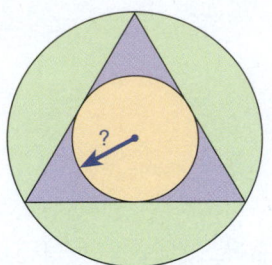

FIGURE 18–124

54. A hex nut 0.87 in. on a side is milled from the smallest-diameter round stock possible (Fig. 18–125). What is the diameter of the stock?

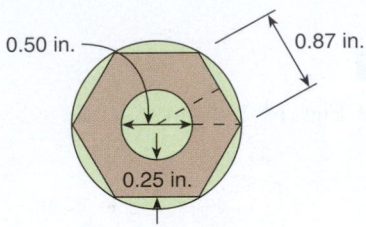

FIGURE 18–125

55. A square is milled on the end of a 5-cm shaft (Fig. 18–126). If the square is the largest that can be milled on the 5-cm shaft, what is the length of a side to hundredths?

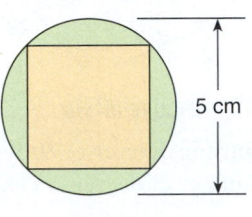

FIGURE 18–126

Section 18–4

Find the distance between each pair of points. Express the answer to the nearest thousandth when necessary.

56. (8, 3) and (3, −4)

57. (3, 6) and (−1, 4)

58. (4, 0) and (0, −1)

59. (3, −3) and (0, 7)

60. (2, 7) and (−3, −2)

61. (0, 0) and (−3, 5)

62. (−5, −5) and (4, 0)

63. (5, 2) and (−3, −3)

64. (1, −5) and (−2, −5)

Calculate the coordinates for the midpoint of each segment whose end points are given.

65. (3, 6) and (−1, 4)

66. (4, 0) and (0, −1)

67. (3, −3) and (0, 7)

68. (2, 7) and (−3, −2)

69. (0, 0) and (−3, 5)

70. (−5, −5) and (4, 0)

71. (5, 2) and (−3, −3)

72. (1, −5) and (−2, −5)

73. (−5, −4) and (2, −2)

74. (8, 3) and (3, −4)

18 TEAM PROBLEM-SOLVING EXERCISES

1. A *polygon is inscribed in a circle* if each vertex lies on the circumference of the circle. If a radius from the center of the circle is drawn to each vertex of the inscribed polygon, the central angles formed are equal.

The difference between the area of the circle and the area of the inscribed polygon can be found by calculating the areas of the circle and the polygon and subtracting.

Another approach is to find the sum of the areas of the segments of the circle. There will be the same number of segments as sides of the polygon.

For a circle with a radius of 5 cm:

(a) Find the area of an inscribed equilateral triangle and the area of the circle. Then, find the difference between the areas of the circle and triangle.

(b) Find the area of an inscribed square and find the difference between the areas of the circle and square.

(c) Find the area of the inscribed hexagon (six equal sides) in Fig. 18–127 and the area of the circle. Then, find the difference between the areas of the circle and hexagon.

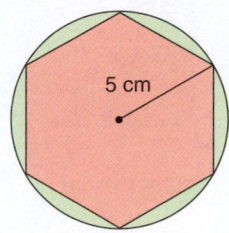

FIGURE 18–127

2. Using the information found in Exercise 1:
(a) Estimate the area of the regular pentagon (five equal sides) in Fig. 18–128 by giving two values the area is between. Present a convincing argument for your estimate.
(b) Estimate the area of an inscribed regular octagon (eight equal sides) that has a radius of 5 cm and give an argument for your estimate.

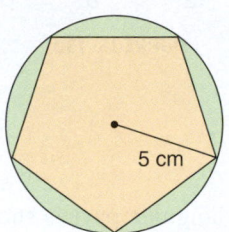

FIGURE 18–128

18   CONCEPTS ANALYSIS

1. Can the measures 8 cm, 12 cm, and 25 cm represent the sides of a triangle? Illustrate and explain your answer.

2. The Pythagorean theorem, $a^2 + b^2 = c^2$, applies to right triangles where a and b are legs and c is the hypotenuse. For what type of triangle is the statement $a^2 + b^2 > c^2$ true? Illustrate your answer.

3. For what type of triangle is the statement $a^2 + b^2 < c^2$ true? Illustrate your answer.

4. A triangle has 6, 3, and $3\sqrt{3}$ as the measures of its sides. Sketch the triangle, place the measures on the appropriate sides, and show the measures of each of the three angles of the triangle.

5. Explain how the Pythagorean theorem is used to find the distance between two points on a coordinate grid.

6. Use the Internet to research Pythagorean triples and give 5 sets of Pythagorean triples that are natural numbers. Discuss the results of your research.

7. Find three sets of Pythagorean triples that are "primitive"—that is, they are not multiples of some other Pythagorean triple.

8. The formula for the radius of the inscribed circle in a triangle whose sides are length a, b, and c is:

$$r = \frac{\sqrt{(a + b + c)(a + b - c)(a - b + c)(-a + b + c)}}{2(a + b + c)}$$

What is the radius of the inscribed circle of a triangle that has sides 3, 4, and 5?

9. What is the radius of the inscribed circle of a triangle that has sides 5, 12, and 13?

10. Use the theorem to show that $p^2 - q^2$, $2pq$, and $p^2 + q^2$ is a Pythagorean triple if p and q are positive integers and q is less than p.

18   PRACTICE TEST

1. Identify the largest and the smallest angles in Fig. 18–129.

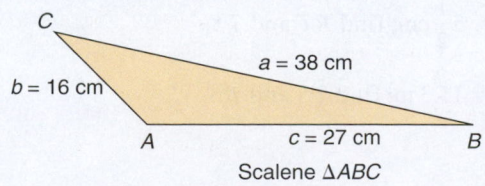

FIGURE 18–129

2. Find *HI* if △*ABC* ~ △*GHI* (Fig. 18–130).

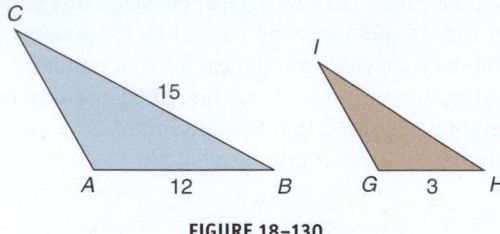

FIGURE 18–130

3. Find *DB* if △*ABE* ~ △*CDE*, *CD* = 9, *AB* = 12, and *DE* = 15 (Fig. 18–131).

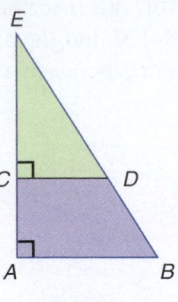

FIGURE 18–131

4. A model of a triangular part is shown in Fig. 18–132. Find side *x* if the part is to be similar to the model.

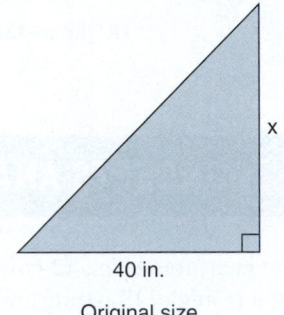

FIGURE 18–132

Use Figure 18–133 and the Pythagorean theorem to find the missing value. Round to the nearest whole number.

5. $a = 20, b = 48$, find c.

6. $b = 10, c = 18$, find a.

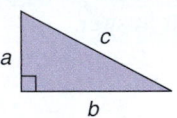

FIGURE 18–133

7. Three pulleys are designed so their centers form a right triangle when connecting lines are drawn. The distances between the centers forming the sides of the triangle are 8 in. and 15 in., respectively. What is the distance between the centers of the pulleys that form the hypotenuse?

Use Fig. 18–134 for Exercises 8–10. Round to the nearest hundredth if necessary.

8. $AC = 18$ ft; find AB and BC.

9. $AB = 62$ cm; find AC and BC.

10. $BC = 13.7$ mm; find AC and AB.

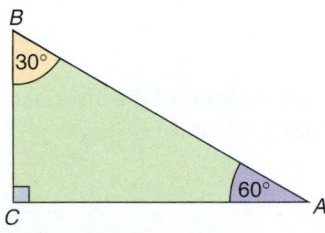

FIGURE 18–134

Use Fig. 18–135 for Exercises 11–12. Round to the nearest hundredth if necessary.

11. $RS = 5\frac{1}{2}$ cm; find RT and TS.

12. $RT = 15.3$ m; find TS and RS.

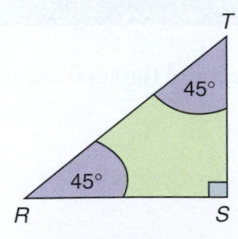

FIGURE 18–135

13. A conduit *ABCD* is made so that ∠*CBE* is 45° (Fig. 18–136). If *BE* = 4 cm and *AK* = 12 cm, find the length of the conduit to the nearest thousandth centimeter.

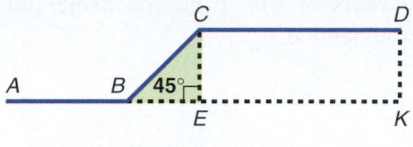

FIGURE 18–136

14. The rafters of a house make a 30° angle with the joists (Fig. 18–137). If the rafters have an 18-in. overhang and the center of the roof is 10 ft above the joists, how long must the rafters be cut?

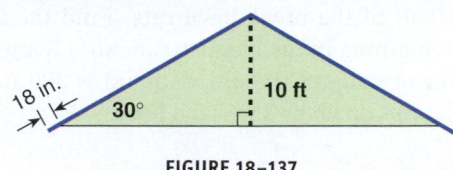

FIGURE 18–137

15. Find the circumference of a circle inscribed in an equilateral triangle if the height of the triangle is 12 in. (Fig. 18–138).

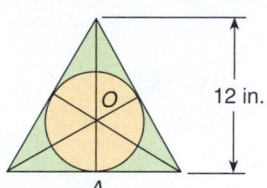

FIGURE 18–138

16. Ken Bennett milled a square metal rod so that a circle is formed at the end (see Fig. 18–139). If the square cross-section is 1.8 in. across the corners, what is the diameter of the circle?

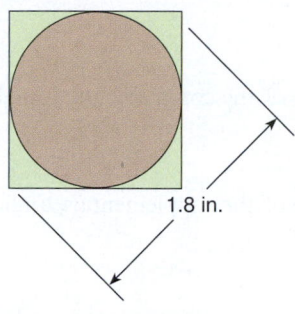

FIGURE 18–139

17. Laura Deskin milled a circle on the cross-sectional end of a hexagonal rod (Fig. 18–140). If the distance across the flats of the cross-section is $\frac{1}{2}$ in., what is the circumference of the circle?

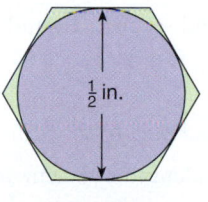

FIGURE 18–140

Find the distance between each pair of points and find the coordinates of the midpoint of the line segment made by the pairs of points. Round to the tenths.

18. (5, 2) and (−7, −3)

19. (−2, 1) and (3, 3)

20. (3, 5) and (6, 9)

16–18   CUMULATIVE PRACTICE TEST

Write as logarithmic equations.

1. $5^3 = 125$

2. $36^{\left(-\frac{1}{2}\right)} = \frac{1}{6}$

Write as exponential equations.

3. $\log_6 36 = 2$

4. $\log_2 \frac{1}{32} = -5$

5. Use the exponential model $D = D_0(1 + r)^t$, where D is the amount of drug in the bloodstream after a specified amount of time and D_0 is the initial amount of drug in the bloodstream. The elapsed time is t, and r is the decimal equivalent of the breakdown rate. Find the amount of drug remaining in the bloodstream after 8 h if the initial amount of drug in the bloodstream is 300 mg and the drug breaks down by 12% each hour.

6. The height in meters of the female members of a population group is approximated by the formula $h = 0.3 + \log t$, where t represents age in years for the first 20 years of life. Find the projected height of a 12-year-old female.

7. If you invest $12,000 today at 4.5% annual interest compounded continuously, how much will you have after 15 years?

8. Find the perimeter and area of a trapezoid that has a height of 12 cm if the lower base is 20 cm and the upper base is 10 cm. The nonparallel sides are 13 cm each.

Classify the angle measures using the terms *right, straight, acute,* and *obtuse*.

9. 28°

10. 150°

11. 180°

Find the measure of the complementary angle.

12. 46°

13. 28°

Find the measure of the supplementary angle.

14. 52°

15. 143°

Change to decimal-degree equivalent rounded to the nearest ten-thousandth.

16. 42′15″

17. 30°12′20″

Change to minutes and seconds. Round to the nearest second when necessary.

18. 0.86°

19. 0.352°

Find the perimeter and area to the nearest whole unit.

20. A rectangle with length of 30 cm and width of 18 cm.

21. A triangle with sides that measure 42 m, 36 m, and 30 m.

22. Find the circumference and area of a circle that has a radius of 40 cm.

23. Find the circumference and area of a circle that has a diameter of 150 cm.

Convert to radians. Round to the nearest ten-thousandth radian.

24. 45°

25. 15°22′

Convert to degrees. Round to the nearest ten-thousandth of a degree.

26. $\dfrac{3\pi}{4}$ *rad*

27. 1.8 rad

28. Find the area of a sector of a circle that has a radius of 12 cm if the angle of the sector is 60°.

29. Find the volume to the nearest centimeter of a cylinder that has a circular base with a radius of 12 cm if the cylinder has a height of 40 cm.

30. Find the lateral surface area of a cone with a radius of 72 cm and a slant height of 100 cm.

31. Use the Pythagorean theorem to find the hypotenuse to the nearest inch of a right triangle that has legs measuring 12 in. and 15 in.

Use Fig. 18–141 for Exercises 32–34. Round to tenths.

32. $AC = 28$ m; find AB and BC.

33. $AB = 32.5$ m; find AC and BC.

34. $BC = 17.32$ ft; find AC and AB.

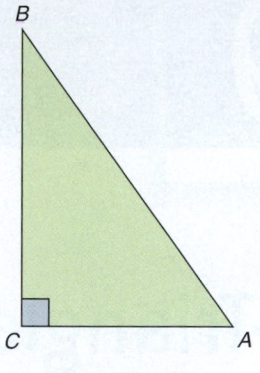

FIGURE 18–141

35. A square picture is 38 in. on each side. Find the length of the diagonal. Round to the nearest whole number.

36. The diagonal of a square is 48 in. Find the length of each side. Round to the nearest whole number.

37. Find the distance between two points described by (3, 14) and (9, 8).

38. Find the midpoint of the line segment described by (−4, 5) and (5, 14). Round to tenths.

Right-Triangle Trigonometry

Johnbraid/Shutterstock

In Great Company

Math Humiliates Us All—Why It's Called "The Bleeding Edge" (1953)

The British Overseas Airway Corporation (BOAC) was quite proud of itself. In 1949, it flew the initial prototype of what became the first modern commercial jet airliner, the de Havilland Comet. By May 1952, BOAC had the world's first commercial jet route, between London and Johannesburg, South Africa. The Comet's pressurized cabin and jet engines guaranteed an incredibly smooth ride at 35,000 feet and cut hours off the flight time.

The Comet was cutting edge in nearly every aspect of its design, so some minor problems were expected. For instance, pilots discovered the jet sometimes had difficulty in becoming airborne at takeoff. In March 1953, five crew members and nine passengers became the first fatalities aboard a commercial jetliner during a failed takeoff in Pakistan.

But something else was wrong. On May 2, 1953, almost exactly one year after it had entered commercial service, a Comet took off from Calcutta, India, and flew into a severe storm. It crashed six minutes later, killing all aboard. Witnesses said the wings had torn off the plane. The crash was attributed to the storm.

Seven months later, on January 10, 1954, a Comet flying out of Rome broke up 20 minutes into the flight and crashed near the island of Elba. There were no witnesses. Four months later, on April 4, 1954, while the British were still recovering wreckage from the Elba flight, another Comet disintegrated in flight, going down near Naples.

Something was wrong. No one knew what.

British engineers suspected some kind of pressurization failure, so they built an enormous water tank capable of submerging an entire Comet at once. Then they dunked the plane repeatedly—hundreds of times—to simulate the pressurization cycles the plane went through each time it flew.

The test plane disintegrated.
They had found the problem.
It was the windows.
They were square.

Each of the four corners in each of the jet's square windows was subjected to pressures many times higher than anyone expected. The metal at the corners became fatigued, cracked, then tore itself apart at high altitudes. After roughly 1,000 pressurization cycles, every fuselage would fail.

The solutions: (1) use stronger, thicker metal to encase the windows, and (2) make sure window corners were rounded, not squared, to reduce the pressure.

Engineers from Boeing and McDonnell Douglas, the other two major airline companies, later admitted that their designs originally had the same flaw. If the Comet hadn't beaten them to market, their planes would have disintegrated in mid-flight just like the Comet.

19–1 Trigonometric Functions

LC LEARNING CATALYTICS

1. Express $\frac{23}{29.3}$ as a decimal to the nearest ten-thousandth.

1 **Find the Sine, Cosine, and Tangent of Angles of Right Triangles, Given the Measures of at Least Two Sides.** In geometry, we studied the basic properties of similar triangles and right triangles that allowed us to find missing measures of the sides of the triangle. Using trigonometry, we can determine the measure of either acute angle of a right triangle if we know the measure of at least two sides of the right triangle.

Fig. 19–1 shows a right triangle, *ABC*, with the sides of the triangle labeled according to their relationship to angle *A*. The *hypotenuse* is the side opposite the right angle of the triangle and the hypotenuse forms one side of angle *A*. The other side that forms angle *A* is the adjacent side of angle *A*. The third side of the triangle is the opposite side of angle *A*.

In Fig. 19–2, the sides of the right triangle *ABC* are labeled according to their relationship to angle *B*. The hypotenuse forms one side of angle *B*, the other side

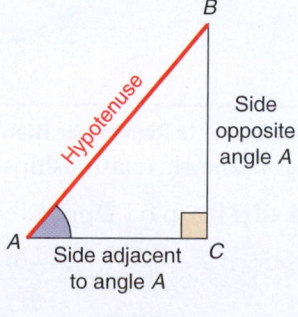

FIGURE 19–1

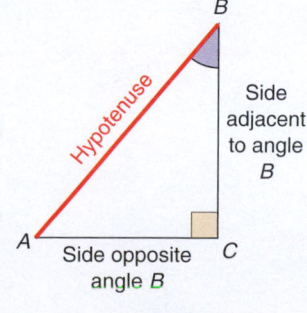

FIGURE 19–2

that forms angle *B* is the adjacent side of angle *B,* and the third side is the opposite side of angle *B.* In Chapter 18, we used the term *hypotenuse,* but the terms *adjacent side* and *opposite side* are also very important in understanding trigonometric functions. The **adjacent side** of an acute angle of a right triangle is the side that forms the angle with the hypotenuse. The **opposite side** of an acute angle of a right triangle is the side that does not form the given angle.

The three most commonly used trigonometric functions are the **sine, cosine,** and **tangent of an angle.** The sine, cosine, and tangent of angle *A* in Fig. 19–3 are defined as ratios of the sides of the right triangle.

For convenience, the sine, cosine, and tangent functions are abbreviated as **sin, cos,** and **tan,** respectively. We will use letters to designate the sides of the **standard right triangle,** as shown in Fig. 19–3. In the standard right triangle, side *a* is opposite angle *A,* side *b* is opposite angle *B,* and side *c* (hypotenuse) is opposite angle *C* (right angle).

Adjacent side: the adjacent side of a given acute angle of a right triangle is the side that forms the angle with the hypotenuse

Opposite side: the opposite side of a given acute angle of a right triangle is the side that does not form the given angle

Sine of an angle (sin): the ratio of the side opposite the angle divided by the hypotenuse

Cosine of an angle (cos): the ratio of the side adjacent to the angle divided by the hypotenuse

Tangent of an angle (tan): the ratio of the side opposite the angle divided by the adjacent side

Standard right triangle: a standardized labeling of a right triangle with side *a* opposite angle *A,* side *b* opposite angle *B,* and side *c* (hypotenuse) opposite angle *C* (right angle)

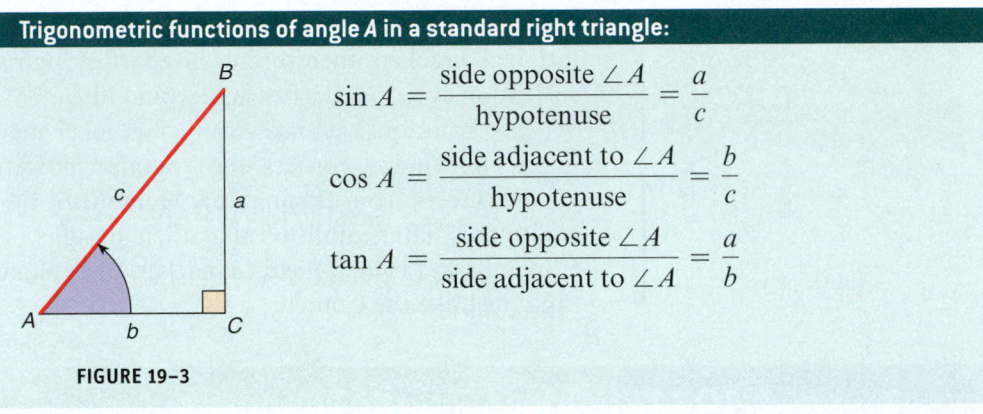

Trigonometric functions of angle *A* in a standard right triangle:

$$\sin A = \frac{\text{side opposite } \angle A}{\text{hypotenuse}} = \frac{a}{c}$$

$$\cos A = \frac{\text{side adjacent to } \angle A}{\text{hypotenuse}} = \frac{b}{c}$$

$$\tan A = \frac{\text{side opposite } \angle A}{\text{side adjacent to } \angle A} = \frac{a}{b}$$

FIGURE 19–3

Similarly, we can identify the sine, cosine, and tangent relationships for the other acute angle, angle *B* (Fig. 19–4).

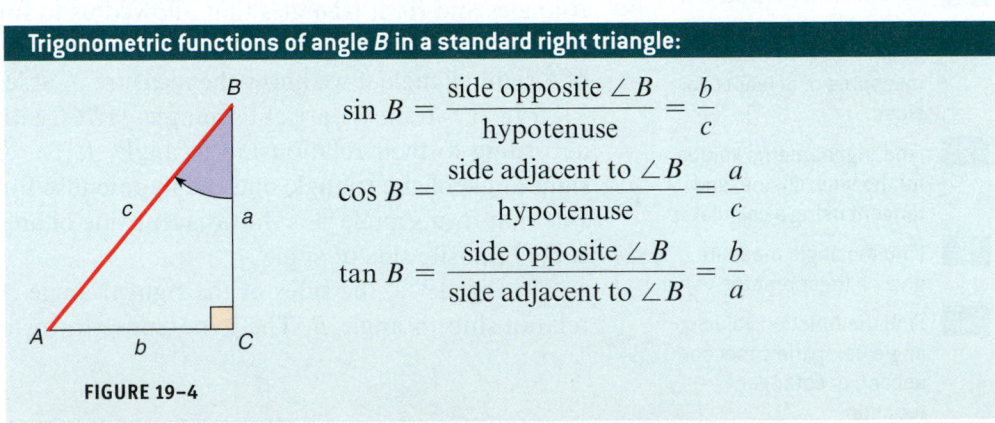

Trigonometric functions of angle *B* in a standard right triangle:

$$\sin B = \frac{\text{side opposite } \angle B}{\text{hypotenuse}} = \frac{b}{c}$$

$$\cos B = \frac{\text{side adjacent to } \angle B}{\text{hypotenuse}} = \frac{a}{c}$$

$$\tan B = \frac{\text{side opposite } \angle B}{\text{side adjacent to } \angle B} = \frac{b}{a}$$

FIGURE 19–4

TIP   Words Are Easier to Remember Than Letters Use these common tips for remembering the trigonometric relationships of sin, cos, and tan.

▶ "**O**scar **h**ad **a**h**eap o**f **a**pples."

$$\sin = \frac{\text{\textbf{o}pposite}}{\text{\textbf{h}ypotenuse}} = \frac{\text{\textbf{O}scar}}{\text{\textbf{h}ad}}$$

$$\cos = \frac{\text{a djacent}}{\text{h ypotenuse}} = \frac{\text{a}}{\text{h eap}}$$

$$\tan = \frac{\text{o pposite}}{\text{a djacent}} = \frac{\text{o f}}{\text{a pples}}$$

▶ "Chief Soh-Cah-Toa" or "Some old hogs come around here tasting our apples."

$$\text{s in} = \frac{\text{o pposite}}{\text{hypotenuse}} \qquad \text{c os} = \frac{\text{a djacent}}{\text{hypotenuse}} \qquad \text{t an} = \frac{\text{o pposite}}{\text{a djacent}}$$

To write the trigonometric ratios for a right triangle given the measures of at least two sides:

1. Identify the acute angle that is being used.

2. Identify the hypotenuse, opposite side, and adjacent side in relation to the acute angle selected in Step 1.

3. Write the appropriate trigonometric ratio.

4. Simplify the ratio or convert to a decimal equivalent.

STOP AND CHECK

Use a standard right triangle (Fig. 19–3) when $a = 9$ in., $b = 2$ ft, and $c = 2$ ft 2 in. Find the trigonometric values as fractions in lowest terms.

1. $\sin A$ 2. $\cos A$
3. $\tan A$ 4. $\tan B$

Answers:

1. $\dfrac{9}{26}$ 2. $\dfrac{12}{13}$ 3. $\dfrac{3}{8}$ 4. $\dfrac{8}{3}$

EXAMPLE 1

Find the sine, cosine, and tangent of angles A and B in Fig. 19–5. Leave the answers as fractions in lowest terms.

$$\sin A = \frac{\text{opposite}}{\text{hypotenuse}} = \frac{7 \text{ in.}}{2 \text{ ft 1 in.}} = \frac{7 \text{ in.}}{25 \text{ in.}} = \frac{7}{25}$$

$$\cos A = \frac{\text{adjacent}}{\text{hypotenuse}} = \frac{2 \text{ ft}}{2 \text{ ft 1 in.}} = \frac{24 \text{ in.}}{25 \text{ in.}} = \frac{24}{25}$$

$$\tan A = \frac{\text{opposite}}{\text{adjacent}} = \frac{7 \text{ in.}}{2 \text{ ft}} = \frac{7 \text{ in.}}{24 \text{ in.}} = \frac{7}{24}$$

$$\sin B = \frac{\text{opposite}}{\text{hypotenuse}} = \frac{2 \text{ ft}}{2 \text{ ft 1 in.}} = \frac{24 \text{ in.}}{25 \text{ in.}} = \frac{24}{25}$$

$$\cos B = \frac{\text{adjacent}}{\text{hypotenuse}} = \frac{7 \text{ in.}}{2 \text{ ft 1 in.}} = \frac{7 \text{ in.}}{25 \text{ in.}} = \frac{7}{25}$$

$$\tan B = \frac{\text{opposite}}{\text{adjacent}} = \frac{2 \text{ ft}}{7 \text{ in.}} = \frac{24 \text{ in.}}{7 \text{ in.}} = \frac{24}{7}$$

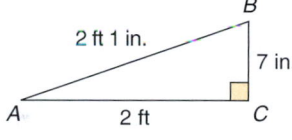

FIGURE 19–5

See Exercises 1–18.

STOP AND CHECK

Use a standard right triangle (Fig. 19–3) when $a = 8$ mm, $b = 2.1$ cm, and $c = 22.5$ mm. Find the trigonometric values as decimals rounded to the nearest ten-thousandth.

1. $\sin A$ 2. $\cos A$
3. $\tan A$ 4. $\tan B$

Answers:

1. 0.3556   2. 0.9333   3. 0.3810
4. 2.625

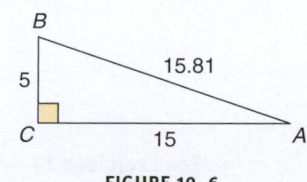

FIGURE 19–6

EXAMPLE 2

Find the sine, cosine, and tangent of angles A and B in Fig. 19–6. Express answers as decimals. Round to the nearest ten-thousandth.

$$\sin A = \frac{\text{opposite}}{\text{hypotenuse}} = \frac{5}{15.81} = \mathbf{0.3163} \qquad \sin B = \frac{\text{opposite}}{\text{hypotenuse}} = \frac{15}{15.81} = \mathbf{0.9488}$$

$$\cos A = \frac{\text{adjacent}}{\text{hypotenuse}} = \frac{15}{15.81} = \mathbf{0.9488} \qquad \cos B = \frac{\text{adjacent}}{\text{hypotenuse}} = \frac{5}{15.81} = \mathbf{0.3163}$$

$$\tan A = \frac{\text{opposite}}{\text{adjacent}} = \frac{5}{15} = \mathbf{0.3333} \qquad \tan B = \frac{\text{opposite}}{\text{adjacent}} = \frac{15}{5} = \mathbf{3}$$

See Exercises 19–24.

Did You Know? **Trigonometric Ratios Have No Unit of Measure.** When both terms of a ratio are expressed in the same unit of measure, that is, *like* units, then the ratio itself is unitless because the common units cancel, for example, $\dfrac{3 \text{ in.}}{5 \text{ in.}} = \dfrac{3}{5}$. Thus, *trigonometric ratios are numerical values with no unit of measure.*

2 **Find Trigonometric Values for the Sine, Cosine, and Tangent Using a Calculator.**
In studying similar triangles, we found that corresponding angles of similar triangles are equal and that corresponding sides are directly proportional. These properties of similar triangles lead us to a very important property of trigonometric functions.

Values of trigonometric functions:

Every angle of a specified measure has a specific set of values for its trigonometric functions.

We recommend the use of a calculator to find trigonometric values. To find trigonometric values on your calculator, the keys $\boxed{\text{SIN}}$, $\boxed{\text{COS}}$, and $\boxed{\text{TAN}}$ are used. The angle measure can be entered in degrees or radians, depending on the selected mode.

To find trigonometric values using the calculator:

1. Set your calculator to the desired mode of angle measure. Calculators generally have at least two angle modes: degrees and radians.

2. Press the appropriate function key (sin, cos, tan) and then enter the angle measure.

3. Display the result by pressing the $\boxed{=}$ or $\boxed{\text{ENTER}}$ key.

STOP AND CHECK

Using a calculator, find the trigonometric values.

1. sin 72°
2. cos 40°15′
3. $\sin \dfrac{\pi}{6}$
4. tan 0.78

Answers:
1. 0.9510565163
2. 0.7632324698
3. 0.5
4. 0.9892615369

EXAMPLE 3

Using a calculator, find the trigonometric values.

(a) sin 27°   **(b)** sin 20°30′   **(c)** cos 1.34   **(d)** $\tan \dfrac{\pi}{4}$

(a) Be sure your calculator is in degree mode.

$\boxed{\text{SIN}}\ 27\ \boxed{=} \Rightarrow$ **0.4539904997**   Keystrokes may vary.

(b) Be sure your calculator is in degree mode.

$\boxed{\text{SIN}}\ 20.5\ \boxed{=} \Rightarrow$ **0.3502073813**   20°30′ = 20.5°

With some calculators, angle measures can be entered using degrees, minutes, and seconds *or* using the decimal equivalent.

(c) Reset your calculator to radian mode.

$\boxed{\text{COS}}\ 1.34\ \boxed{=} \Rightarrow$ **0.2287528078**

(d) Be sure your calculator is in radian mode.

$\boxed{\text{TAN}}\ \boxed{(}\ \boxed{\pi}\ \boxed{\div}\ 4\ \boxed{)}\ \boxed{=} \Rightarrow$ **1**

See Exercises 25–44.

> **TIP** **Automatic Parentheses in Calculator Functions** Some calculators automatically insert an open parenthesis after certain functions. If a close parenthesis is not entered before the calculation is executed, the close parenthesis is automatically interpreted at the end of the entry. Test your calculator with and without the close parenthesis to verify the interpretation made by your calculator.
>
> In degree mode:
>
> Entry: SIN 27 ENTER   Entry: SIN 27) ENTER
> Display: SIN (27   Display: SIN (27)
>   .4539904997   .4539904997

3 **Find the Angle Measure, Given a Trigonometric Value.** Finding one of the acute angle measures of a right triangle when given the trigonometric value is an *inverse operation* to finding the trigonometric value when given the angle measure. The notation most commonly used is $\sin^{-1}$, $\cos^{-1}$, or $\tan^{-1}$.

> **To find angle measures using a calculator:**
>
> 1. Set your calculator to the desired mode of angle measure (degrees or radians).
>
> 2. Select the appropriate inverse trigonometric function key or menu option ($\boxed{\text{SIN}^{-1}}$, $\boxed{\text{COS}^{-1}}$, or $\boxed{\text{TAN}^{-1}}$). Enter the trigonometric value and $\boxed{=}$ or $\boxed{\text{ENTER}}$.
>
> Some calculators use an inverse key INV, shift key SHIFT, or second $\boxed{2^{nd}}$ key to access the inverse keys. Test your calculator with a known value. For example, in the previous example we found $\sin 27° = 0.4539904997$. And, $\sin^{-1} 0.4539904997$ is 27°.

EXAMPLE 4

Find the angle in degrees given the trigonometric values in parts (a) and (b). θ represents the unknown angle measure. Round to the nearest tenth of a degree. For part (c), find the radians to the nearest thousandth.

(a) $\sin \theta = 0.6561$ (b) $\cos \theta = 0.4226$ (c) $\tan \theta = 2.825$

(a) Be sure your calculator is in degree mode.

$\boxed{\text{SIN}^{-1}}$ .6561 $\boxed{=}$ ⇒ 41.0031105 ≈ **41.0°**   Round.

(b) Be sure your calculator is in degree mode.

$\boxed{\text{COS}^{-1}}$ .4226 $\boxed{=}$ ⇒ 65.00115448 ≈ **65.0°**   Round.

(c) Be sure your calculator is in radian mode.

$\boxed{\text{TAN}^{-1}}$ 2.825 $\boxed{=}$ ⇒ 1.230578215 ≈ **1.231 rad**   Round.   **See Exercises 45–62.**

4 **Find the function value or angle using the cosecant, secant, or cotangent function.** There are three other trigonometric functions. These functions are not used as often as the sine, cosine, and tangent functions; but, because of the relationships between these new functions and the previously learned functions, these new functions are useful to know.

Each of the three basic trigonometric functions (sine, cosine, and tangent) has a related trigonometric function for which the defining ratio is the reciprocal of that of the basic function.

The ratio for the **cosecant (csc)** function is the reciprocal of the ratio for the sine function; the ratio for the **secant (sec)** function is the reciprocal of the ratio for the

Cotangent (cot): a trigonometric function that is the ratio of the adjacent side to the opposite side. The ratios of the tangent and cotangent functions are reciprocals

cosine function; and the ratio of the **cotangent (cot)** function is the reciprocal of the ratio for the tangent function. As you might imagine, if you examine the relationship of the sides of the triangle for each of these related functions, the function will be the reciprocal of the corresponding basic function.

<table>
<tr><td colspan="2" align="center">**Basic Trigonometric Functions**</td><td colspan="2" align="center">**Related Trigonometric Functions**</td></tr>
<tr>
<td align="center">sine $A =$</td>
<td>$\dfrac{\text{opposite side}}{\text{hypotenuse}} = \dfrac{a}{c}$</td>
<td align="center">cosecant $A =$</td>
<td>$\dfrac{\text{hypotenuse}}{\text{opposite side}} = \dfrac{c}{a}$</td>
</tr>
<tr>
<td align="center">cosine $A =$</td>
<td>$\dfrac{\text{adjacent side}}{\text{hypotenuse}} = \dfrac{b}{c}$</td>
<td align="center">secant $A =$</td>
<td>$\dfrac{\text{hypotenuse}}{\text{adjacent side}} = \dfrac{c}{b}$</td>
</tr>
<tr>
<td align="center">tangent $A =$</td>
<td>$\dfrac{\text{opposite side}}{\text{adjacent side}} = \dfrac{a}{b}$</td>
<td align="center">cotangent $A =$</td>
<td>$\dfrac{\text{adjacent side}}{\text{opposite side}} = \dfrac{b}{a}$</td>
</tr>
</table>

STOP AND CHECK

Find the following trigonometric values when $a = 4$, $b = 12$, and $c = 12.65$ (Fig. 19–3). Round to the nearest ten-thousandth.

1. csc A
2. sec A
3. cot B

Answers:
1. 3.1625   2. 1.0542   3. 0.3333

EXAMPLE 5

Find the following trigonometric values when $a = 5$, $b = 15$, and $c = 15.81$ (Fig. 19–3). Round to the nearest ten-thousandth.

(a) csc A **(b)** sec A **(c)** cot A **(d)** csc B **(e)** sec B **(f)** cot B

(a) $\csc A = \dfrac{\text{hypotenuse}}{\text{opposite side}} = \dfrac{c}{a} = \dfrac{15.81}{5} = 3.162$

(b) $\sec A = \dfrac{\text{hypotenuse}}{\text{adjacent side}} = \dfrac{c}{b} = \dfrac{15.81}{15} = 1.054$

(c) $\cot A = \dfrac{\text{adjacent side}}{\text{opposite side}} = \dfrac{b}{a} = \dfrac{15}{5} = 3$

(d) $\csc B = \dfrac{\text{hypotenuse}}{\text{opposite side}} = \dfrac{c}{b} = \dfrac{15.81}{15} = 1.054$

(e) $\sec B = \dfrac{\text{hypotenuse}}{\text{adjacent side}} = \dfrac{c}{a} = \dfrac{15.81}{5} = 3.162$

(f) $\cot B = \dfrac{\text{adjacent side}}{\text{opposite side}} = \dfrac{a}{b} = \dfrac{5}{15} = 0.3333$

See Exercises 63–68.

TIP   Memory Tip for Pairing Related Functions   Each pair of related functions has one of the functions starting with the prefix "co" **but** never both.

Sine pairs with	Cosine pairs with	Tangent pairs with
Cosecant	Secant	Cotangent

As with the basic trigonometric functions, you can determine the value of the acute angles of a right triangle using the related functions. Using the calculator to find the angle from the value of the cosecant, secant, or cotangent requires that we apply our knowledge of reciprocals in general. Since most calculators have only the SIN, COS, and TAN trigonometric functions and the SIN⁻¹, COS⁻¹, and TAN⁻¹ inverse trigonometric functions, we must take the reciprocal of a cosecant, secant, or cotangent function *before* using the inverse operation for the corresponding related function.

> **To find the angle measure given the value of the cosecant, secant, or cotangent:**
>
> 1. Find the reciprocal of the given value of the cosecant, secant, or cotangent.
>
>   (a) 1 ⊡ (given value) ENTER OR (given value) x^{-1} ENTER.
>
>   (b) Select the appropriate inverse trigonometric function key for the given reciprocal. Be sure your calculator is in the appropriate angle mode (degrees or radians).
>
>   For CSC use SIN⁻¹.
>   For SEC use COS⁻¹.
>   For COT use TAN⁻¹.
>
> 2. Enter the value of the reciprocal found in Step 1 and press ANS ENTER.

STOP AND CHECK

Find the angle in degrees or radians as indicated when given the trigonometric value. θ represents the unknown angle measure. Round degree measures to the nearest tenth of a degree. Round radian measures to the nearest thousandth.

1. csc θ = 2.289, θ in degrees
2. sec θ = 3.045, θ in degrees
3. cot θ = 0.582, θ in radians

Answers:
1. 25.9° 2. 70.8° 3. 1.044 rad

EXAMPLE 6

Find the angle in degrees or radians as indicated when given the trigonometric value. θ represents the unknown angle measure. Round degree measures to the nearest tenth of a degree. Round radian measures to the nearest thousandth.

(a) csc θ = 1.439, θ in degrees   (b) sec θ = 2.793, θ in degrees
(c) cot θ = 0.412, θ in radians

(a) Be sure your calculator is in degree mode.
   1 ⊡ 1.439 ENTER SIN⁻¹ ANS ENTER ≈ **44.0°**   Round.

(b) Be sure your calculator is in degree mode.
   2.793 x^{-1} ENTER COS⁻¹ ANS ENTER ≈ **69.0°**   Round.

(c) Be sure your calculator is in radian mode.
   .412 x^{-1} ENTER TAN⁻¹ ANS ENTER ≈ **1.180 radians**   Round.

See Exercises 69–71.

TIP **Calculator Keys Are Sometimes Confusing** On a calculator the inverse trigonometric function keys are most often labeled with an exponent of −1 but they have nothing to do with a reciprocal. Instead these functions are the equivalent of the functions also called *arcsin*, *arccos*, and *arctan*. The arcsin is interpreted as the "angle whose sine is."

   SIN⁻¹ is the same as ARCSIN.
   COS⁻¹ is the same as ARCCOS.
   TAN⁻¹ is the same as ARCTAN.

On the other hand, the key x^{-1} is actually a reciprocal key.

19–1 EXERCISES   MyLab Math For additional practice go to your study plan in MyLab Math.

1 Use Fig. 19–7 to find the indicated trigonometric ratios.

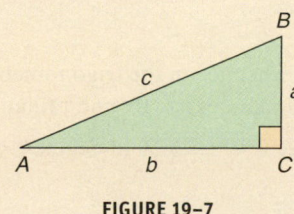

FIGURE 19–7

See Example 1.

Use $a = 5, b = 12$, and $c = 13$ in Exercises 1–6 and express the ratios as fractions in lowest terms.

1. sin A	**2.** cos A	**3.** tan A	**4.** sin B	**5.** cos B	**6.** tan B

Use $a = 9, b = 12$, and $c = 15$ in Exercises 7–12 and express the ratios as fractions in lowest terms.

7. sin A	**8.** cos A	**9.** tan A	**10.** sin B	**11.** cos B	**12.** tan B

Use $a = 16, b = 30$, and $c = 34$ in Exercises 13–18 and express the ratios as fractions in lowest terms.

13. sin A	**14.** cos A	**15.** tan A	**16.** sin B	**17.** cos B	**18.** tan B

See Example 2.

Use $a = 9, b = 14$, and $c = 16.64$ in Exercises 19–24 and express the ratios as decimals to the nearest ten-thousandth.

19. sin A	**20.** cos A	**21.** tan A	**22.** sin B	**23.** cos B	**24.** tan B

2 Use your calculator to find the trigonometric values. Express your answers in ten-thousandths. *See Example 3.*

25. sin 21°	**26.** cos 3.5°	**27.** tan 47°	**28.** cos 52.5°
29. sin 0.5498	**30.** cos 21°30′	**31.** cos 1.1519	**32.** cos 0.3665
33. sin 53°30′	**34.** tan 42.5°	**35.** tan 47.7°	**36.** sin 62°10′
37. cos 12°40′	**38.** cos 1.0530	**39.** tan 73°14′	**40.** sin 1.2363
41. cos 46.8°	**42.** cos 0.3549	**43.** cos 1.1636	**44.** tan 12.4°

3 Find the angles of the trigonometric values in degrees. θ represents the unknown angle measure. Express each answer to the nearest tenth of a degree. *See Example 4, parts (a) and (b).*

45. sin θ = 0.3420	**46.** cos θ = 0.9239	**47.** tan θ = 2.356
48. cos θ = 0.4617	**49.** cos θ = 0.540	**50.** sin θ = 0.5712
51. tan θ = 1.265	**52.** cos θ = 0.137	**53.** sin θ = 0.6298

Find the angles of the trigonometric values in radians. Express each answer to the nearest ten-thousandth. *See Example 4, part c.*

54. tan θ = 0.8098	**55.** cos θ = 0.6947	**56.** cos θ = 0.3907
57. cos θ = 0.968	**58.** sin θ = 0.9959	**59.** tan θ = 0.3160
60. tan θ = 2.430	**61.** cos θ = 0.9610	**62.** cos θ = 0.4210

4 Find the following trigonometric values when $a = 3, b = 4$, and $c = 5$ (Fig. 19–3). Round to the nearest ten-thousandth.

See Example 5.

63. csc A	**64.** sec A	**65.** cot A
66. csc B	**67.** sec B	**68.** cot B

See Example 6.

Find the angle in degrees or radians as indicated when given the trigonometric values. θ represents the unknown angle measure. Round degree measures to the nearest tenth of a degree. Round radian measures to the nearest thousandth.

69. csc θ = 1.269, θ in degrees	**70.** sec θ = 1.236, θ in degrees	**71.** cot θ = 0.800, θ in radians

19-2 Solving Right Triangles Using the Sine, Cosine, and Tangent Functions

LC LEARNING CATALYTICS

1. Rearrange the equation for b:

$$\tan 48.5° = \frac{b}{21}.$$

We often use the sine, cosine, and tangent functions to find unknown parts of a right triangle. To do this, we manipulate formulas and use other algebraic principles depending on what information is given and what information needs to be found.

1 Find the Unknown Parts of a Right Triangle Using the Sine Function. Using the relationship $\sin \theta = \dfrac{\text{opp}}{\text{hyp}}$, we can find parts of right triangles when we know any two parts that involve the sine function: one acute angle, the side opposite the known acute angle, and the hypotenuse.

Use the sine function to find unknown parts of a right triangle:

1. Two of these three parts of a right triangle must be known:

    (a) One acute angle

    (b) The side opposite the known acute angle

    (c) Hypotenuse

2. Substitute the two known values in the ratio $\sin \theta = \dfrac{\text{opp}}{\text{hyp}}$.

3. Solve for the unknown part.

TIP Does It Matter Which Acute Angle Is Known? No. The acute angles of a right triangle are complementary. Therefore, if we know either acute angle (A), we can find the other one ($90° - A$ or $\frac{\pi}{2} - A$). Thus, we can use the sine function if we know *either* acute angle and any side.

STOP AND CHECK

1. In a standard right triangle, find angle B if $b = 9$ and $c = 25$. Round to the nearest tenth of a degree.

Answer:

1. $B = 21.1°$

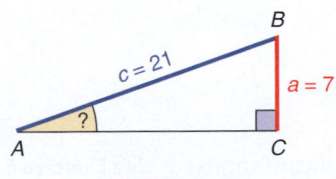

FIGURE 19-8

EXAMPLE 1

Find angle A to the nearest tenth of a degree if $a = 7$ and $c = 21$ (see Fig. 19–8).

The two known values are the side opposite angle A and the hypotenuse, so we use the sine function for the acute angle A.

$\sin A = \dfrac{\text{opp}}{\text{hyp}}$ Substitute the known values. opp = 7, hyp = 21

$\sin A = \dfrac{7}{21}$ Convert ratio to a decimal equivalent.

$\sin A = 0.3333333333$ Find $\boxed{\text{SIN}^{-1}}$ of 0.3333333333.

$A = \mathbf{19.47122063°}$ or $\mathbf{19.5°}$ Round to the nearest tenth of a degree.

See Exercises 1–3.

TIP How Do I Round My Answers? Rounding practices are generally dictated by the context of the problem or industry standards; however, **for consistency, we round all lengths of sides to four significant digits and all angle values to the nearest 0.1° or the nearest thousandth radian throughout Chapters 19 and 20 unless otherwise indicated.**

STOP AND CHECK

1. In a standard right triangle, find side b if angle $B = 85.2°$ and $c = 19$ m. Round to four significant digits.

Answer:
1. $b = 18.93$ m

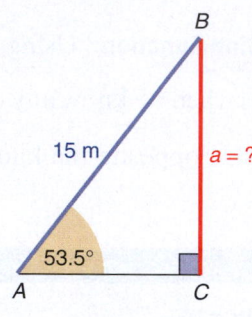

FIGURE 19–9

EXAMPLE 2

Find side a in triangle ABC (see Fig. 19–9).

We are given $\angle A$ and the hypotenuse and are asked to find the measure of the side opposite $\angle A$, so we use the sine function.

$$\sin A = \frac{\text{opp}}{\text{hyp}}$$
Substitute known values.
$\angle A = 53.5°$, hyp $= 15$ m

$$\sin 53.5° = \frac{a}{15}$$
$\sin 53.5° = 0.8038568606$

$$0.8038568606 = \frac{a}{15}$$
Solve for a.

$$15(0.8038568606) = a$$

$$a = \mathbf{12.05785291} \quad \text{or} \quad \mathbf{12.06\ m}$$
Round to four significant digits.
See Exercises 4–5.

TIP **Rearrange the Formula Before You Calculate** In the preceding example and those to follow, we can visualize a continuous sequence of calculator steps if we rearrange the formula for the missing part *before* we make any calculations.

$$\sin 53.5° = \frac{a}{15} \qquad \text{Rearrange for } a.$$

$$15(\sin 53.5°) = a \qquad \text{Evaluate.}$$

Then a continuous series of calculations is made. In degree mode,

$$15 \boxed{\times} \boxed{\text{SIN}} 53.5 \boxed{=} \Rightarrow 12.05785291$$

STOP AND CHECK

1. In a standard right triangle, find side c if angle $B = 1.34$ rad and $b = 32$ m. Round to four significant digits.

Answer:
1. $c = 32.87$ m

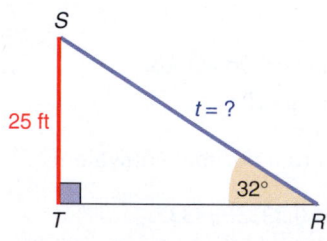

FIGURE 19–10

EXAMPLE 3

Find the hypotenuse in triangle RST (see Fig. 19–10).

We are given an acute angle and the side opposite the angle and are asked to find the hypotenuse, so we use the sine function.

$$\sin R = \frac{\text{opp}}{\text{hyp}}$$
Substitute known values. $\angle R = 32°$, opp $= 25$ ft

$$\sin 32° = \frac{25}{t}$$
Rearrange for t.

$$t(\sin 32°) = 25$$

$$t = \frac{25}{\sin 32°}$$
Evaluate.

$$t = 47.17699787$$

$$t = \mathbf{47.18\ ft}$$
Round to four significant digits. See Exercise 6.

TIP **Choose Given Values over Calculated Values Whenever Possible** It is best to use given values rather than calculated values when finding missing parts of a triangle. Because the rounded value for one missing part is sometimes used to find other missing parts, final answers may vary slightly due to rounding discrepancies. For instance, the sum of the angles of a triangle may be as little as 179° or as much as 181°. The length of a side may be slightly different in the last significant digit. If the full calculator value of a side or angle is used to find other missing parts, the rounding discrepancy is reduced.

Solve a triangle: to find the measures of all sides and all angles of the triangle

To **solve a triangle** means to find the measures of all sides and all angles. A right triangle can be solved if we know one side and any other part besides the right angle.

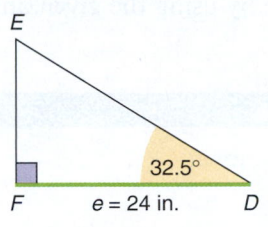

FIGURE 19–11

EXAMPLE 4

Solve triangle *DEF* (see Fig. 19–11). One side and one other part besides the right angle are known.

Because we have one acute angle and a side that is not opposite the known angle, we must find the other angle of the triangle.

$$\angle E = 90° - 32.5° = \mathbf{57.5°} \qquad \text{Find the complement of } \angle D.$$

To find the hypotenuse f:

$$\sin E = \frac{\text{opp}}{\text{hyp}} \qquad \text{Use the sine function and substitute } E = 57.5° \text{ and opp} = 24.$$

$$\sin 57.5° = \frac{24}{f} \qquad \text{Solve for } f.$$

$$f(\sin 57.5°) = 24$$

$$f = \frac{24}{\sin 57.5°} \qquad \text{Evaluate.}$$

$$f = 28.45653714$$

$$\mathbf{f = 28.46 \text{ in.}} \qquad \text{Round to four significant digits.}$$

To find side d:

$$\sin D = \frac{\text{opp}}{\text{hyp}} \qquad \text{Substitute } D = 32.5° \text{ and hyp} = 28.45653714. \text{ Use full calculator value for } f \text{ to get the most accurate result.}$$

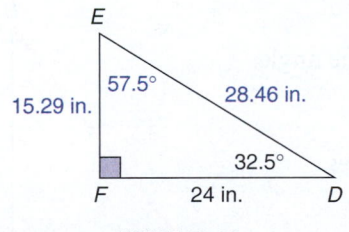

FIGURE 19–12

$$\sin 32.5° = \frac{d}{28.45653714} \qquad \text{Solve for } d.$$

$$28.45653714(\sin 32.5°) = d \qquad \text{Evaluate.}$$

$$15.28968626 = d$$

$$\mathbf{15.29 \text{ in.} = d} \qquad \text{Round to four significant digits.}$$

The solved triangle is shown in Fig. 19–12.   **See Exercises 7–10.**

TIP **Check Computations Using the Pythagorean Theorem** In this and other problems involving right triangles, you can check your computations using the Pythagorean theorem, $(\text{hyp})^2 = (\text{leg})^2 + (\text{leg})^2$. Let's check the solution to the preceding example.

Using Values to Four Significant Digits	**Using Full Calculator Values**
$(28.46)^2 = (15.29)^2 + (24)^2$	$(28.45653714)^2 = (15.28968626)^2 + (24)^2$
$809.9716 = 233.7841 + 576$	$809.774506 = 233.7745059 + 576$
$809.9716 = 809.7841$	$809.774506 = 809.7745059$

Rounding discrepancies are minimized when more significant digits are used.

> **TIP**   **See the BIG Picture Then Focus on the Little Parts**   In solving a right triangle, we are generally given the values of three parts and are asked to find the values of the three missing parts. Look at the big picture first.
>
> ▶   Identify the three given parts, one of which is the right angle.
>
> ▶   Identify the missing parts.
>
> ▶   Plan a strategy to find each missing part.
>
> ▶   Focus on one part at a time.
>
> ▶   Check by using the Pythagorean theorem and the property that the three angles of a triangle add to 180°.

2   **Find the Unknown Parts of a Right Triangle Using the Cosine Function.**   In some of the previous examples, when we found a side not opposite the given angle, we had to find the other angle first by subtracting the given acute angle from 90°. If we use the cosine function, however, we can find the desired side by using the given angle, rather than its complement.

> **Use the cosine function to find unknown parts of a right triangle:**
>
> 1. Two of these three parts of a right triangle must be known:
>
>   **(a)** One acute angle
>
>   **(b)** The side adjacent to the known acute angle
>
>   **(c)** Hypotenuse
>
> 2. Substitute two known values in the ratio $\cos \theta = \dfrac{\text{adj}}{\text{hyp}}$.
>
> 3. Solve for the unknown part.

EXAMPLE 5

Find angle A of Fig. 19–13.

We are given the hypotenuse and the side adjacent to the angle, so we use the cosine function to find the desired angle.

$$\cos A = \frac{\text{adj}}{\text{hyp}}$$ Substitute known values.

$$\cos A = \frac{1.9}{3.6}$$ Use the inverse cosine function. Evaluate.

$$\cos^{-1}\left(\frac{1.9}{3.6}\right) = A$$

$$A = 58.14456918$$

$$A = 58.1°$$ Round to nearest 0.1°.   **See Exercise 11.**

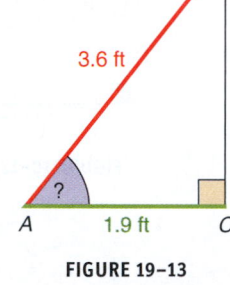

FIGURE 19–13

EXAMPLE 6

Find side b of Fig. 19–14 on page 827.

We can use either the sine or the cosine function because we are given the hypotenuse and an angle. However, we do not have to find the complement of the given angle if we use the cosine function.

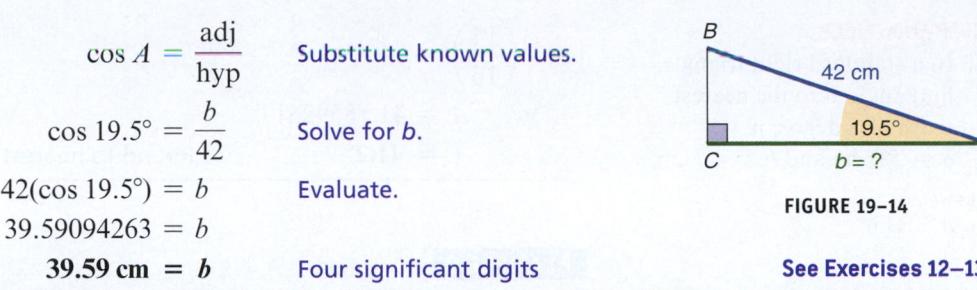

$$\cos A = \frac{adj}{hyp} \qquad \text{Substitute known values.}$$

$$\cos 19.5° = \frac{b}{42} \qquad \text{Solve for } b.$$

$$42(\cos 19.5°) = b \qquad \text{Evaluate.}$$

$$39.59094263 = b$$

$$\mathbf{39.59 \text{ cm}} = \boldsymbol{b} \qquad \text{Four significant digits} \qquad \text{See Exercises 12–13.}$$

FIGURE 19–14

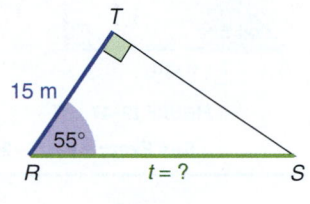
EXAMPLE 7

Find side t of Fig. 19–15.

The cosine function is the most efficient function to use because we are finding the hypotenuse and given an acute angle and its adjacent side.

$$\cos R = \frac{adj}{hyp} \qquad \text{Substitute known values.}$$

$$\cos 55° = \frac{15}{t} \qquad \text{Solve for } t.$$

$$t(\cos 55°) = 15$$

$$t = \frac{15}{\cos 55°} \qquad \text{Evaluate.}$$

$$t = 26.15170193$$

$$\boldsymbol{t = 26.15 \text{ m}} \qquad \text{Four significant digits} \qquad \text{See Exercises 14–16.}$$

3 **Find the Unknown Parts of a Right Triangle Using the Tangent Function.** If we know only the length of the two legs of a triangle, we should use the tangent function.

> **Use the tangent function to find unknown parts of a right triangle:**
>
> 1. Two of these three parts of a right triangle must be known:
>
>   **(a)** One acute angle
>
>   **(b)** The side opposite the known acute angle
>
>   **(c)** The side adjacent to the known acute angle
>
> 2. Substitute two known values in the ratio $\tan \theta = \dfrac{opp}{adj}$.
>
> 3. Solve for the unknown part.

EXAMPLE 8

Find angle A of Fig. 19–16.

We are looking for an angle and we are given the side opposite and the side adjacent to the angle, so we use the tangent function.

$$\tan A = \frac{opp}{adj} \qquad \text{Substitute known values.}$$

$$\tan A = \frac{14}{16} \qquad \text{Use the inverse tangent function. Evaluate.}$$

FIGURE 19–16

1. In a standard right triangle, find angle B to the nearest tenth of a degree if $a = 5.3$ cm and $b = 4.7$ cm.

Answer:

1. $B = 41.6°$

$$\tan^{-1}\left(\frac{14}{16}\right) = A$$

$$A = 41.18592517°$$

$$A = \mathbf{41.2°} \qquad \text{Round to nearest } 0.1°. \qquad \textbf{See Exercises 21–22.}$$

---

1. In a standard right triangle, find side a if angle $A = 36.1°$ and $b = 28$ ft. Round to four significant digits.

Answer:

1. $a = 20.42$ ft

EXAMPLE 9

Find side a of Fig. 19–17.

We are given an acute angle and the side adjacent to the acute angle. We are looking for the opposite side, so use the tangent function.

$$\tan A = \frac{\text{opp}}{\text{adj}} \qquad \text{Substitute known values.}$$

$$\tan 44.5° = \frac{a}{6} \qquad \text{Solve for } a.$$

$$6(\tan 44.5°) = a \qquad \text{Evaluate.}$$

$$5.896183579 = a$$

$$\mathbf{5.896 \text{ cm}} = \mathbf{a} \qquad \text{Four significant digits}$$

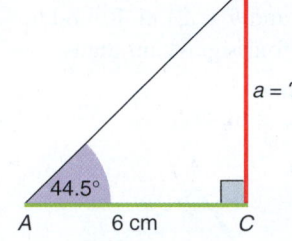

FIGURE 19–17

See Exercises 23–25.

---

1. In a standard right triangle, find side b if angle $A = 0.93$ rad and $a = 24.3$ m. Round to four significant digits.

Answer:

1. $b = 18.12$ m

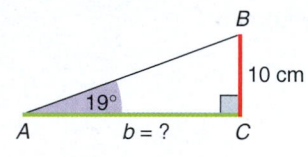

FIGURE 19–18

EXAMPLE 10

Find side b of Fig. 19–18.

We know an acute angle and its opposite side. We are looking for the side adjacent to the acute angle.

$$\tan A = \frac{\text{opp}}{\text{adj}} \qquad \text{Substitute known values.}$$

$$\tan 19° = \frac{10}{b} \qquad \text{Solve for } b.$$

$$b(\tan 19°) = 10$$

$$b = \frac{10}{\tan 19°} \qquad \text{Evaluate.}$$

$$b = 29.04210878$$

$$\mathbf{b = 29.04 \text{ cm}} \qquad \text{Four significant digits} \qquad \textbf{See Exercise 26.}$$

---

4 Select the Most Direct Method for Solving Right Triangles. Our first task in solving any problem involving right triangles is to *select the most convenient and efficient function.*

> **To select the most direct method for solving a right triangle:**
>
> **1.** Where possible, choose the function that uses given parts rather than unknown parts that must be calculated.
>
> **2.** Where possible, choose the function that gives the desired part directly, that is, without having to find other parts first.

STOP AND CHECK

1. Find side c to four significant digits and angle B to the nearest tenth of a degree.

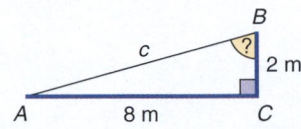

Answer:

1. $c = 8.246$ m; $\angle B = 76.0°$

EXAMPLE 11

In $\triangle ABC$ of Fig. 19–19, **(a)** find side c to four significant digits and **(b)** angle A to the nearest tenth of a degree.

(a) The most direct way of finding side c is to use the *Pythagorean theorem*. The two legs of a right triangle are given.

$c^2 = a^2 + b^2$	Substitute known values.
$c^2 = 6^2 + 9^2$	Solve for c.
$c^2 = 36 + 81$	
$c^2 = 117$	Take square root of both sides.
$c = 10.81665383$	
$c = \mathbf{10.82}$	Four significant digits

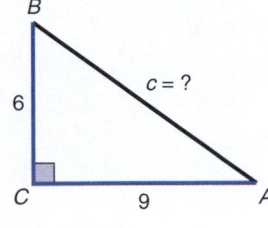

FIGURE 19–19

(b) Use the tangent function to find angle A using given values.

$\tan \theta = \dfrac{\text{opp}}{\text{adj}}$	Substitute known values.
$\tan A = \dfrac{6}{9}$	Reduce.
$\tan A = \dfrac{2}{3}$	Use the inverse tangent function.
$\tan^{-1}\left(\dfrac{2}{3}\right) = B$	Evaluate.
$B = 33.69006753$	
$B = 33.7°$	

See Exercises 31–35.

5 **Solve Applied Problems Using Right-Triangle Trigonometry.** In solving technical problems, it's a good idea to draw diagrams or pictures to visualize the various relationships.

To solve a problem using trigonometric functions:

1. First determine the given parts and the unknown part or parts; then identify the trigonometric function that relates the given and unknown parts.

2. Use this function and the given information to find the missing information.

EXAMPLE 12

AVIA A jet takes off at a 30° angle (see Fig. 19–20). If the runway (from takeoff) is 875 ft long, find the altitude of the airplane as it flies over the end of the runway.

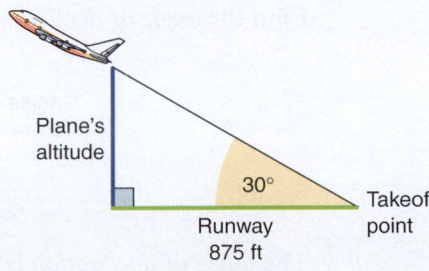

FIGURE 19–20

Known facts

One acute angle $= 30°$; adjacent side $= 875$ ft

Unknown fact

Plane's altitude at end of runway or opposite side

Relationship

$$\tan \theta = \frac{\text{opp}}{\text{adj}}$$

Estimation

Since this is a $30°$-$60°$-$90°$ triangle, the altitude is the side opposite the $30°$ angle and the smallest side of the triangle.

Calculations

$\tan \theta = \dfrac{\text{opp}}{\text{adj}}$	Substitute known values.
$\tan 30° = \dfrac{a}{875}$	Solve for a.
$875(\tan 30°) = a$	Evaluate.
$505.1814855 = a$	

Interpretation

505.2 ft $=$ plane's altitude   Four significant digits   **See Exercises 36–37.**

Angle of elevation: an angle that moves in a positive direction from the point of sight or horizontal line. Also described as looking up

Angle of depression or declination: an angle that moves in a negative direction from the point of sight or horizontal line. Also described as a location below our sight level

Many right-triangle applications use the terminology **angle of elevation** and **angle of depression or declination**. See Fig. 19–21. The angle of elevation is generally used when we are looking *up* at an object. We use the angle of depression to describe the location of an objective *below* our eye level. *Both* angles are formed by a line of sight and a horizontal line from the point of sight.

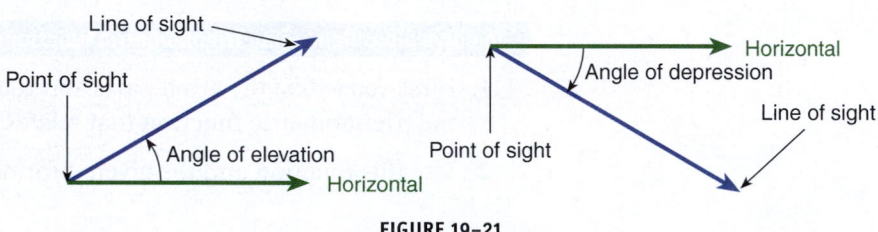

FIGURE 19–21

EXAMPLE 13

CON A stretch of roadway drops 30 ft for every 300 ft of road (see Fig. 19–22). Find the *angle of declination* of the road.

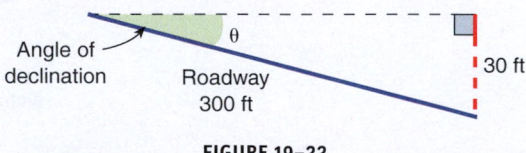

FIGURE 19–22

The *angle of declination* is the angle of depression. The opposite side and the hypotenuse are given.

$$\sin \theta = \frac{opp}{hyp} \qquad \text{Substitute known values.}$$

$$\sin \theta = \frac{30}{300} \qquad \text{Reduce.}$$

$$\sin \theta = 0.1 \qquad \text{Use the inverse sine function.}$$

$$\sin^{-1} 0.1 = \theta$$

$$\theta = 5.739170477°$$

The angle of declination of the road is 5.7° rounded to the nearest 0.1°.

See Exercises 38–39.

EXAMPLE 14

CON A surveyor locates two points on a steel column so that it can be set plumb (perpendicular to the horizon). If the angle of elevation is 15° and the surveyor's transit is 175 ft from the column (see Fig. 19–23), find the distance from the transit (point A) to the upper point (point B) on the column. (A **transit** is a surveying instrument used for measuring angles.)

An acute angle and the adjacent side are given. To find the distance from the transit to point B on the column (hypotenuse), we use the *cosine function*.

Dmitry Kalinovsky/123RF

Transit: a surveying instrument for measuring angles

$$\cos \theta = \frac{adj}{hyp} \qquad \text{Substitute known values.}$$

$$\cos 15° = \frac{175}{hyp} \qquad \text{Solve for hypotenuse.}$$

$$hyp(\cos 15°) = 175$$

$$hyp = \frac{175}{\cos 15°} \qquad \text{Evaluate.}$$

$$hyp = 181.1733316$$

$$hyp = 181.2 \text{ ft} \qquad \text{Four significant digits}$$

FIGURE 19–23

Point B is 181.2 ft from the transit.

See Exercises 40–41.

EXAMPLE 15

CON Find the angle a rafter makes with a joist of a house if the rise is 12 ft and the span is 30 ft (see Fig. 19–24). Also, find the length of the rafter.

The span is twice the distance from the outside end to the center point of the joist. Therefore, to solve the right triangle for the desired angle, we draw the triangle shown in Fig. 19–25. The horizontal leg of the right triangle formed is one-half the length of the joist, or 15 ft. Because the legs of a right triangle are given, the tangent function is used.

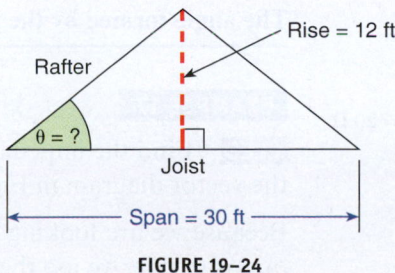

FIGURE 19–24

Find θ.

$$\tan \theta = \frac{\text{opp}}{\text{adj}}$$ Substitute known values.

$$\tan \theta = \frac{12}{15}$$ Use the inverse tangent function.

$$\tan^{-1}\left(\frac{12}{15}\right) = \theta$$

$$\theta = 38.65980825$$ Evaluate.

$$\theta = 38.7°$$ Round to nearest 0.1°.

Find the length of the rafter (hypotenuse) (Fig. 19–25).

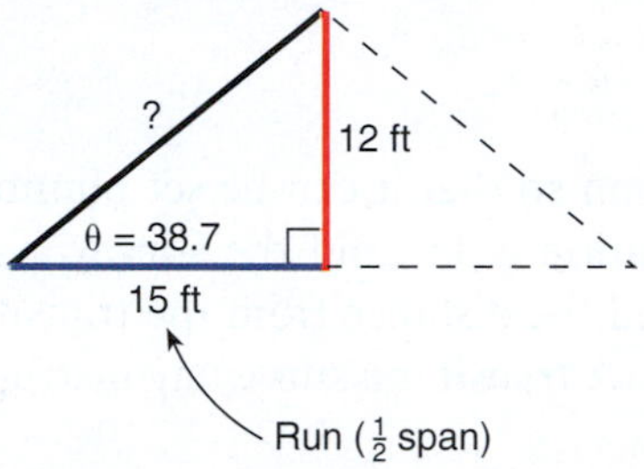

We use the *Pythagorean theorem* to find the length of the rafter directly.

$$(\text{hyp})^2 = (\text{leg})^2 + (\text{leg})^2$$ Substitute known values.

$$(\text{hyp})^2 = 12^2 + 15^2$$ Evaluate.

$$(\text{hyp})^2 = 144 + 225$$

$$(\text{hyp})^2 = 369$$ Take square root of both sides.

$$\text{hyp} = 19.20937271$$

$$\text{hyp} = 19.21 \text{ ft}$$ Four significant digits

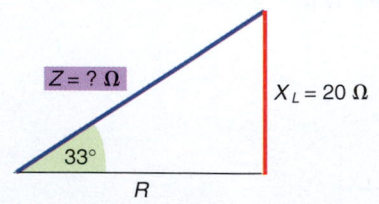

? 12 ft **θ = 38.7** **15 ft** Run ($\frac{1}{2}$ span)

FIGURE 19–25

The angle the rafter makes with the joist is 38.7° and the length of the rafter is 19.21 ft.

See Exercises 42–43.

EXAMPLE 16

INDTEC   Find the angle formed by the rod in the mechanical assembly shown in Fig. 19–26.

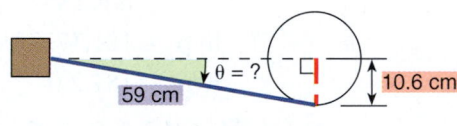

59 cm   **θ = ?**   **10.6 cm**

FIGURE 19–26

Given the hypotenuse and the side opposite the desired angle, we use the sine function.

$$\sin \theta = \frac{\text{opp}}{\text{hyp}}$$ Substitute known values.

$$\sin \theta = \frac{10.6}{59}$$ Use the inverse sine function. Evaluate.

$$\sin^{-1}\left(\frac{10.6}{59}\right) = \theta$$

$$\theta = 10.35001563$$ Round.

$$\theta = 10.4°$$

The angle formed by the rod is 10.4°.

See Exercises 44–45.

EXAMPLE 17

ELEC   Find the impedance Z of a circuit with 20 Ω of reactance X_L represented by the vector diagram in Fig. 19–27.

Because we are looking for the hypotenuse Z and are given an acute angle and the opposite side, we use the sine function.

Z = ? Ω   **X_L = 20 Ω**   **33°**   **R**

FIGURE 19–27

$$\sin \theta = \frac{\text{opp}}{\text{hyp}}$$ Substitute known values.

$$\sin 33° = \frac{20}{Z}$$ Solve for Z.

$$Z(\sin 33°) = 20$$

$$Z = \frac{20}{\sin 33°}$$ Evaluate.

$$Z = 36.72156918$$

$$Z = 36.72 \ \Omega$$ Four significant digits

The impedance Z is 36.72 Ω.   See Exercises 46–47.

19–2 EXERCISES   MyLab Math For additional practice go to your study plan in MyLab Math.

1 Use the sine function to find the indicated parts of the triangle *LMN* in Fig. 19–28. Round lengths of sides to four significant digits and angles to the nearest 0.1°. *See Example 1.*

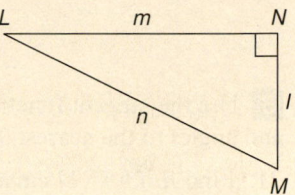

1. Find M if $n = 15$ m and $m = 7$ m.

2. Find M if $m = 13$ cm and $n = 19$ cm.

3. Find M if $n = 3.7$ in. and $l = 2.4$ in.

FIGURE 19–28

See Example 2.

4. Find l if $n = 13$ in. and $L = 32°$.   5. Find m if $l = 15$ m and $L = 28°$.

See Example 3.

6. Find n if $l = 12$ ft and $M = 42°$.

Solve triangle *STU* in Fig. 19–29 for the given values using the sine function. Check the measures of the sides by using the Pythagorean theorem. Round as above. *See Example 4.*

7. Solve if $t = 18$ yd and $s = 14$ yd.

8. Solve if $U = 45°$ and $u = 4.7$ m.

9. Solve if $S = 34.5°$ and $t = 8.5$ mm.

10. Solve if $S = 16°$ and $s = 14$ m.

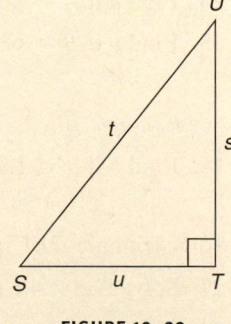

FIGURE 19–29

2 Use the cosine function to find the indicated parts of triangle *KLM* in Fig. 19–30. Round sides to four significant digits and angles to the nearest 0.1°. *See Example 5.*

11. Find M if $k = 13$ m and $l = 16$ m.

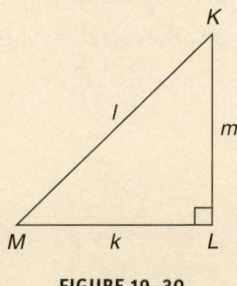

FIGURE 19–30

See Example 6.

12. Find k if $l = 11$ cm and $M = 24°$.

13. Find k if $K = 72°$ and $l = 16.7$ mm.

See Example 7.

14. Find l if $m = 15$ dm and $M = 25°$.

15. Find l if $M = 31°$ and $k = 27$ ft.

16. Find l if $K = 67°$ and $k = 13$ yd.

Use the sine *or* cosine function to solve triangle QRS in Fig. 19–31. Round as above. *See Examples 5–7.*

17. Solve if $s = 23$ ft and $q = 16$ ft.

18. Solve if $s = 17$ cm and $R = 46°$.

19. Solve if $q = 14$ dkm and $Q = 73.5°$.

20. Solve if $R = 59.5°$ and $q = 8$ m.

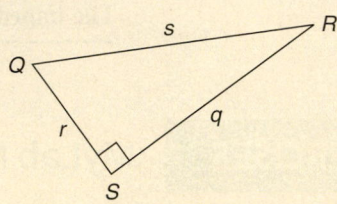

FIGURE 19–31

3 Use the tangent function to find the indicated parts of triangle ABC in Fig. 19–32. Round sides to four significant digits and angles to the nearest 0.1°. *See Example 8.*

21. Find A if $b = 11$ cm and $a = 6$ cm.

22. Find A if $b = 10.8$ m and $a = 4.7$ m.

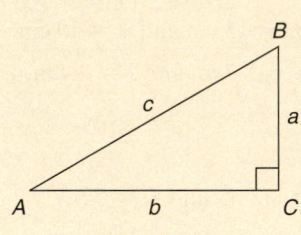

FIGURE 19–32

See Example 9.

23. Find a if $A = 40.5°$ and $b = 7$ ft.

24. Find a if $A = 43°$ and $b = 0.05$ cm.

25. Find a if $B = 68°$ and $b = 0.03$ m.

See Example 10.

26. Find b if $a = 1.9$ m and $A = 25°$.

Solve triangle DEF in Fig. 19–33. Round sides to four significant digits and angles to the nearest 0.1°. *See Examples 8–10.*

27. Solve if $e = 4.6$ m and $d = 3.2$ m.

28. Solve if $D = 42°$ and $e = 7$ ft.

29. Solve if $E = 73.5°$ and $e = 20.13$ in.

30. Solve if $d = 11$ ft and $e = 8$ ft.

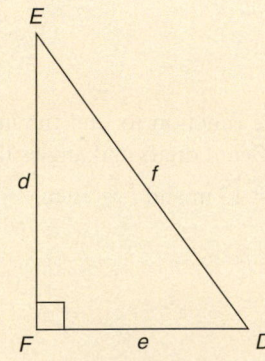

FIGURE 19–33

4 Find the indicated part of the right triangles in Figs. 19–34 through 19–38, by the most direct method. *See Example 11.*

31.

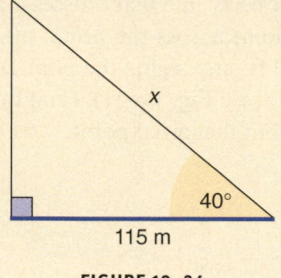

FIGURE 19–34

32.

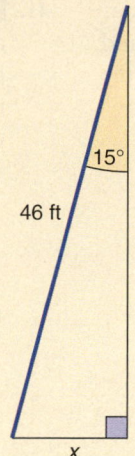

FIGURE 19–35

33.

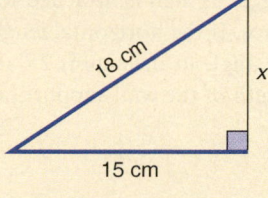

FIGURE 19–36

34.

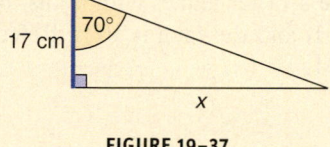

FIGURE 19–37

35.

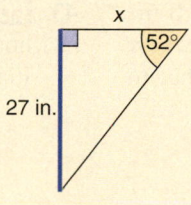

FIGURE 19–38

5 Use the trigonometric functions to solve the problems. Round side lengths to four significant digits and angles to 0.1° unless otherwise indicated. *See Example 12.*

36. **AVIA**   A plane takes off from Reagan National Airport at an angle of elevation of 20°. If the plane travels 4,000 ft/min, how far above the ground will the plane be after 2 min (Fig. 19–39)?

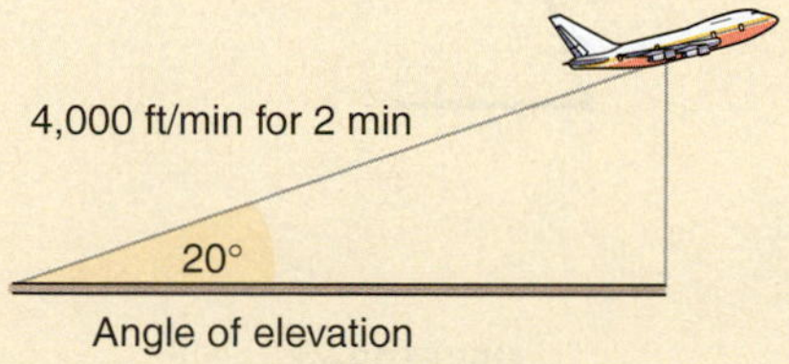

4,000 ft/min for 2 min

20°

Angle of elevation

FIGURE 19–39

37. **AVIA**   How far above the ground will a plane be if its angle of elevation is 22° and it travels 10,000 ft (in the air) from takeoff?

See Example 13.

38. **AVIA**   At what angle must a jet descend if it is 900 ft above the beginning of the runway and must touch down 1,500 ft from the runway's beginning?

39. **CON**   A roadway rises 4 ft for every 15 ft along the road. What is the angle of inclination of the roadway?

See Example 14.

40. CON A sign is attached to a building by a triangular brace. If the horizontal length of the brace is 48 in. and the angle at the sign is 25° (see Fig. 19–40), what is the length of the wall support piece?

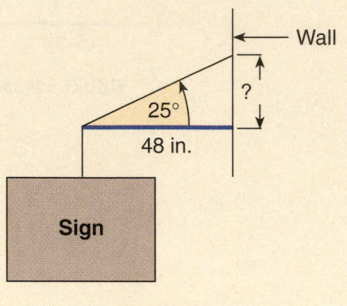

FIGURE 19–40

41. CON To measure a property line that crosses a pond, a surveyor sights to a point across the pond, makes a right angle, measures 50 ft, and sights the point across the pond with a 47° angle (see Fig. 19–41). Find the distance across the pond from the initial point.

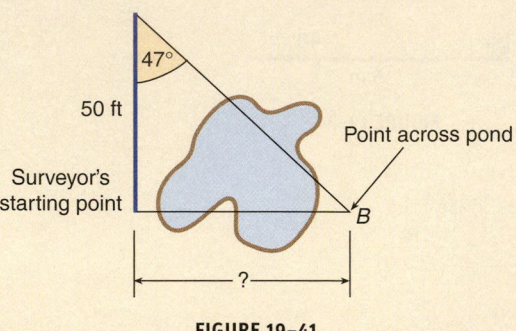

FIGURE 19–41

See Example 15.

42. CON A 50-ft wire is used to brace a utility pole. If the wire is attached 4 ft from the top of the 35-ft pole, how far from the base of the pole will the wire be attached to the ground?

43. CON Find the angle a rafter makes with a joist of a house if the rise is 18 ft and the span is 50 ft. Refer to Fig. 19–24 on page 831.

See Example 16.

44. ELEC The vector diagram of the ac circuit in Fig. 19–42 shows the relationship between the impedance Z, the resistance R, the reactance X, and the phase angle θ. Find the reactance X. All units are in ohms. The impedance is 20 Ω and the phase angle is 30°.

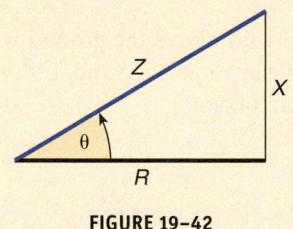

FIGURE 19–42

45. INDTR A piston assembly at the midpoint of its stroke forms a right triangle (see Fig. 19–43). Find the length of R.

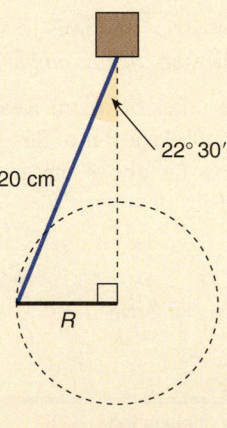

FIGURE 19–43

See Example 17.

46. ELEC An ac circuit has a resistance of 30 Ω and an impedance of 38 Ω. What is the phase angle (Fig. 19–42)?

48. ELEC Using Fig. 19–42, find resistance R, when the impedance is 20 Ω and the phase angle is 30°.

47. ELEC Find the impedance and reactance of an ac circuit that has a resistance of 26 Ω and a phase angle of 25° (Fig. 19–42). Round to the nearest tenth.

49. AG/H A shadow cast by a tree is 32 ft long when the angle of inclination of the sun is 36°. How tall is the tree?

50. **CON** A 20-ft ladder is set against a building at a 70° angle to the ground in Fig. 19–44. How high up the wall does the ladder touch? Round to whole feet.

51. **CON** A cell tower is 100 m high. A guy wire that makes a 60° angle with the ground must be fastened on the tower 90 m above the ground. What length guy wire is needed? Round to whole feet.

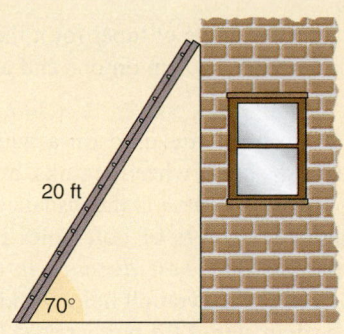

20 ft

70°

FIGURE 19–44

52. **INDTR** An L brace is installed as shown in Fig. 19–45. Find the length of *x* and angle *A* in degrees. Round the length to the nearest tenth inch and the angle to the nearest tenth degree.

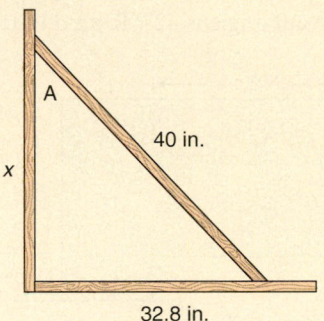

A

40 in.

x

32.8 in.

FIGURE 19–45

53. **INDTEC** Three holes must be drilled in a plastic plate, as shown in Fig. 19–46. Find distance *a*. Round to the nearest tenth.

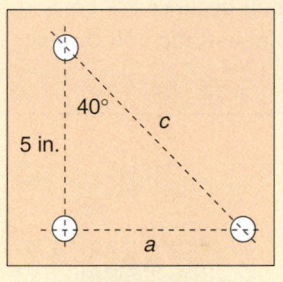

40°

c

5 in.

a

FIGURE 19–46

Use Fig. 19–47 for Exercises 54–56.

54. **CON** A sewage drain pipe has a 60° offset angle and a run of 24 ft. Find the length of the offset and the diagonal. Round to the nearest foot.

55. **ELEC** An electrical conduit has an offset of 46 in. and an offset angle of 56°. Find the length of the run and of the diagonal. Round to the nearest inch.

56. **CON** A 2-in. diameter pipe has a 15-in. offset and a 40° bend (angle of offset). Find the run and the diagonal. Round to the nearest inch.

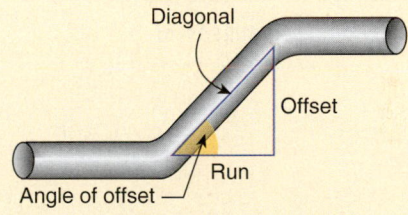

Diagonal

Offset

Angle of offset

Run

FIGURE 19–47

57. **INDTEC** A V-slot gauge is constructed to enable machinists to check the diameter of bearings. Find the radius of the bearing (*AC*) and the distance (*BC*) from the vertex (*B*) to the point where the bearing touches the gauge (*C*) in Fig. 19–48 if *AB* = 2.8 in. and angle *ABC* is 45°. Round to tenths.

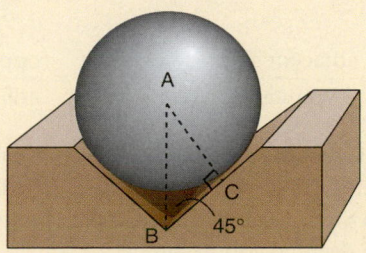

A

C

B 45°

FIGURE 19–48

58. **INDTEC** Find the angle of taper for the tapered steel rod shown in Fig. 19–49.

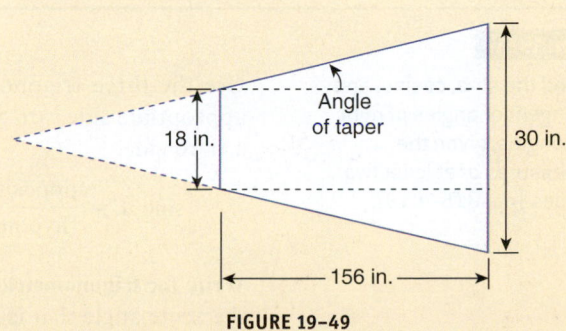

18 in.

Angle of taper

30 in.

156 in.

FIGURE 19–49

59. **INDTEC** Find the angle of taper for a flagpole that is 5 m tall and measures 20 cm on one end and 10 cm on the other end.

60. **INDTEC** Bolt circles are used on a wide variety of structures such as hubs, wheels, cranks, and metal and plastic covers. A *bolt circle* is the visualized circle that runs through the centers of bolt holes in a piece of metal or plastic. The *bolt distance* is the distance between the centers of two bolt holes. Find the bolt distance *E* if the diameter of the bolt circle is 12 in. and the angle formed by the bolt circle diameter and bolt distance is 67.5° (Fig. 19–50). Round to tenths.

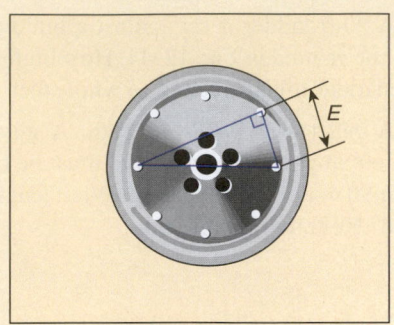

FIGURE 19–50

61. **INDTEC** Find the diameter of the bolt circle in Fig. 19–51 if the bolt distance is 8 cm and the angle formed by the diameter and the bolt distance is 72°. Round to tenths.

62. **CON** What is the depth of a dovetail in Fig. 19–52 if the larger opening is 5.2 cm, the smaller opening is 3.6 cm, and the cut angle is 42°? Round to tenths.

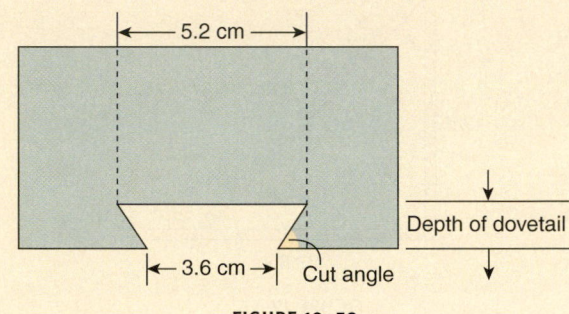

FIGURE 19–52

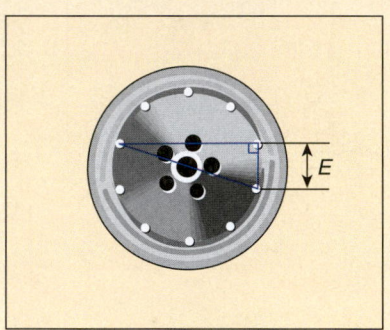

FIGURE 19–51

19   CHAPTER REVIEW OF KEY CONCEPTS

LEARNING OUTCOMES	KEY CONCEPTS AND EXAMPLES

Section 19–1

1 Find the sine, cosine, and tangent of angles of right triangles, given the measures of at least two sides (pp. 815–818).

Use the three trigonometric functions to calculate the value of the function when the appropriate sides are given. The Pythagorean theorem may be needed to find the length of a third side.

$$\text{sine } A = \frac{\text{opposite side}}{\text{hypotenuse}} \qquad \text{cosine } A = \frac{\text{adjacent side}}{\text{hypotenuse}} \qquad \text{tangent } A = \frac{\text{opposite side}}{\text{adjacent side}}$$

Write the trigonometric ratios for a right triangle given the measures of two sides: **1.** Identify the acute angle that is being used. **2.** Identify the hypotenuse, opposite side, and adjacent side in relation to the acute angle selected in Step 1. **3.** Write the appropriate trigonometric ratio. **4.** Simplify the ratio or convert to a decimal equivalent.

LEARNING OUTCOMES   **KEY CONCEPTS AND EXAMPLES**

Find the sine, cosine, and tangent of angle A in Fig. 19–53 and round to four significant digits.

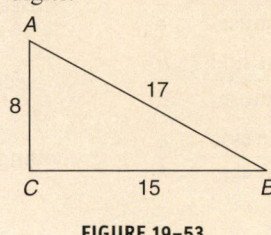

$$\sin A = \frac{15}{17} = 0.8824$$

$$\cos A = \frac{8}{17} = 0.4706$$

$$\tan A = \frac{15}{8} = 1.875$$

FIGURE 19–53

2 Find trigonometric values for the sine, cosine, and tangent using a calculator (pp. 818–819).

Use the ⎡SIN⎤, ⎡COS⎤, and ⎡TAN⎤ keys to find the trigonometric value of a specified angle. Be sure your calculator is set to the appropriate mode: degree or radian.

1. Set your calculator to the desired mode of angle measure. Calculators generally have at least two angle modes: degrees and radians.   **2.** Press the appropriate function key (sin, cos, tan) and then enter the angle measure.   **3.** Display the result by pressing the ⎡=⎤ or ⎡ENTER⎤ key.

Use your calculator to find the following: sin 35°, cos 18°, tan 55°, sin 1.3, cos 0.87, tan 1.1.

Most calculators:
Be sure your calculator is in degree mode:

⎡SIN⎤ 35 ⎡=⎤ ⇒ 0.5735764364

⎡COS⎤ 18 ⎡=⎤ ⇒ 0.9510565163

⎡TAN⎤ 55 ⎡=⎤ ⇒ 1.428148007

Change to radian mode:

⎡SIN⎤ 1.3 ⎡=⎤ ⇒ 0.9635581854

⎡COS⎤ 0.87 ⎡=⎤ ⇒ 0.6448265472

⎡TAN⎤ 1.1 ⎡=⎤ ⇒ 1.964759657

3 Find the angle measure, given a trigonometric value (p. 819).

1. Set your calculator to the desired mode of angle measure (degrees or radians).   **2.** Select the appropriate inverse trigonometric function key or menu option (⎡SIN⁻¹⎤, ⎡COS⁻¹⎤, or ⎡TAN⁻¹⎤). Enter the trigonometric value and ⎡=⎤ or ⎡ENTER⎤.

Some calculators use an inverse key ⎡INV⎤, shift key ⎡SHIFT⎤, or second ⎡2ⁿᵈ⎤ key to access the inverse keys. Test your calculator with a known value. For example, in the previous example we found sin 27° = 0.4539904997. And, $\sin^{-1} 0.4539904997$ is 27°.

Find the value of x in degrees: $\cos x = 0.906307787$ or $x = \cos^{-1}(0.906307787)$.

⎡COS⁻¹⎤ 0.906307787 ⎡=⎤   **Be sure calculator is in degree mode.**

$$x = 25°$$

4 Find the function value or angle using the cosecant, secant, or cotangent function (pp. 819–821).

Examine the relationships between the basic trigonometric functions and the related trigonometric functions.

LEARNING OUTCOMES	KEY CONCEPTS AND EXAMPLES

Basic Trigonometric Functions

$$\text{sine } A = \frac{\text{opposite side}}{\text{hypotenuse}} = \frac{a}{c}$$

$$\text{cosine } A = \frac{\text{adjacent side}}{\text{hypotenuse}} = \frac{b}{c}$$

$$\text{tangent } A = \frac{\text{opposite side}}{\text{adjacent side}} = \frac{a}{b}$$

Related Trigonometric Functions

$$\text{cosecant } A = \frac{\text{hypotenuse}}{\text{opposite side}} = \frac{c}{a}$$

$$\text{secant } A = \frac{\text{hypotenuse}}{\text{adjacent side}} = \frac{c}{b}$$

$$\text{cotangent } A = \frac{\text{adjacent side}}{\text{opposite side}} = \frac{b}{a}$$

Use a standard right triangle when $a = 6$ m, $b = 7$ m, and $c = 9.2$ m. Find the trigonometric value of csc A, sec A, and cot A rounded to four significant digits.

$$\csc A = \frac{9.2 \text{ m}}{6 \text{ m}} = 1.533 \qquad \sec A = \frac{9.2 \text{ m}}{7 \text{ m}} = 1.314 \qquad \cot A = \frac{7 \text{ m}}{6 \text{ m}} = 1.167$$

Find the angle measure given the value of the cosecant, secant, or cotangent: 1. Find the reciprocal of the given value of the cosecant, secant, or cotangent. **(a)** 1 $\div$ (given value) $\boxed{\text{ENTER}}$ OR (given value) $\boxed{x^{-1}}$ $\boxed{\text{ENTER}}$. **(b)** Select the appropriate inverse trigonometric function key for the given reciprocal. Be sure your calculator is in the appropriate angle mode (degrees or radians). For $\boxed{\text{CSC}}$ use $\boxed{\text{SIN}^{-1}}$. For $\boxed{\text{SEC}}$ use $\boxed{\text{COS}^{-1}}$. For $\boxed{\text{COT}}$ use $\boxed{\text{TAN}^{-1}}$. **2.** Enter the value of the reciprocal found in Step 1 and press $\boxed{\text{ANS}}$ $\boxed{\text{ENTER}}$.

Using a standard right triangle, find the angle in degrees or radians as indicated when given the trigonometric values. θ represents the unknown angle measure. Round degree measures to the nearest tenth of a degree. Round radian measures to the nearest thousandth.

$$\csc \theta = 1.531, \theta \text{ in degrees} \qquad \sec \theta = 1.432, \theta \text{ in degrees} \qquad \cot \theta = 0.615, \theta \text{ in radians}$$

$$\sin \theta = \frac{1}{1.531}; \text{ in degree mode, } \theta = \sin^{-1}\left(\frac{1}{1.531}\right) = 40.8°$$

$$\cos \theta = \frac{1}{1.432}; \text{ in degree mode, } \theta = \cos^{-1}\left(\frac{1}{1.432}\right) = 45.7°$$

$$\tan \theta = \frac{1}{0.615}; \text{ in radian mode, } \theta = \tan^{-1}\left(\frac{1}{0.615}\right) = 1.019 \text{ rad}$$

Section 19–2

1 Find the unknown parts of a right triangle using the sine function (pp. 823–826).

1. Two of these three parts of a right triangle must be known: **(a)** One acute angle **(b)** The side opposite the known acute angle **(c)** Hypotenuse **2.** Substitute the two known values in the ratio $\sin \theta = \dfrac{\text{opp}}{\text{hyp}}$. **3.** Solve for the unknown part.

Use the sine function to find side x (Fig. 19–54).

$$\sin \theta = \frac{\text{opp}}{\text{hyp}} \qquad \text{Substitute known values.}$$

$$\sin 25° = \frac{x}{12} \qquad \text{Solve for } x.$$

$$x = 12 \sin 25° \qquad \text{Evaluate.}$$

$$x = 5.071419141$$

$$x = 5.071 \text{ in.} \qquad \text{Four significant digits}$$

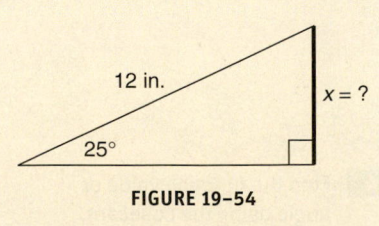

FIGURE 19–54

LEARNING OUTCOMES	**KEY CONCEPTS AND EXAMPLES**

2 Find the unknown parts of a right triangle using the cosine function (pp. 826–827).

1. Two of these three parts of a right triangle must be known: **(a)** One acute angle **(b)** The side adjacent to the known acute angle **(c)** Hypotenuse   **2.** Substitute two known values in in the ratio $\cos \theta = \dfrac{adj}{hyp}$.   **3.** Solve for the unknown part.

Use the cosine function to find side x (Fig. 19–55).

$$\cos \theta = \frac{adj}{hyp} \qquad \text{Substitute known values.}$$

$$\cos 37° = \frac{26}{x} \qquad \text{Solve for } x.$$

$$x = \frac{26}{\cos 37°} \qquad \text{Evaluate.}$$

$$x = 32.55552711$$

$$x = 32.56 \text{ m}$$

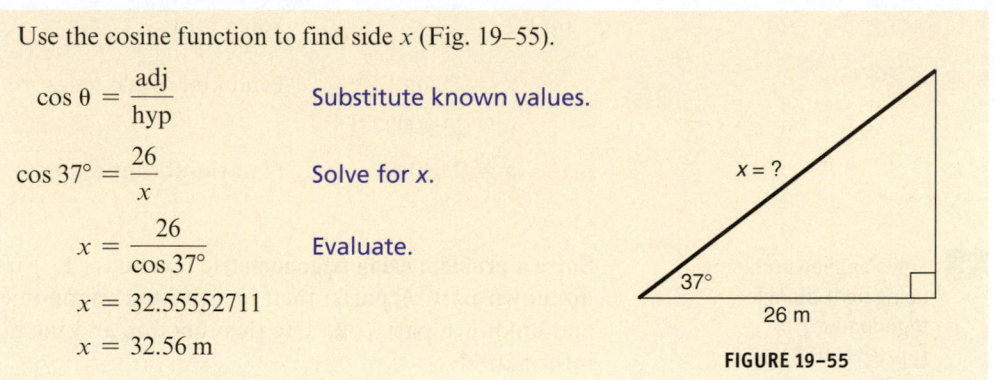

FIGURE 19–55

3 Find the unknown parts of a right triangle using the tangent function (pp. 827–828).

1. Two of these three parts of a right triangle must be known: **(a)** An acute angle **(b)** The side opposite the known acute angle **(c)** The side adjacent to the known acute angle   **2.** Substitute two known values in the ratio $\tan \theta = \dfrac{opp}{adj}$.   **3.** Solve for the unknown part.

Use the tangent function to find θ in degrees (Fig. 19–56).

$$\tan \theta = \frac{opp}{adj} \qquad \text{Substitute known values.}$$

$$\tan \theta = \frac{8}{11} \qquad \text{Use inverse tangent function.}$$

$$\theta = \tan^{-1}\left(\frac{8}{11}\right) \qquad \text{Evaluate.}$$

$$\theta = 36.02737339$$

$$\theta = 36.0° \qquad \text{Round.}$$

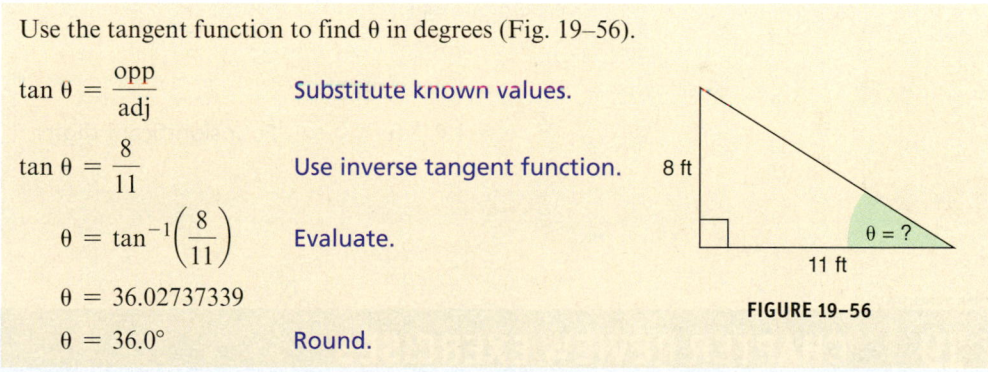

FIGURE 19–56

4 Select the most direct method for solving right triangles (pp. 828–829).

1. Where possible, choose the function that uses given parts rather than unknown parts that must be calculated.   **2.** Where possible, choose the function that gives the desired part directly, that is, without having to find other parts first.

Find the parts indicated in Fig. 19–57 on page 842.

$$\cos \theta = \frac{adj}{hyp} \qquad \begin{array}{l} \theta \text{ and hypotenuse are known.} \\ x = \text{ adjacent side.} \end{array}$$

$$\cos 1.2 = \frac{x}{25} \qquad \text{Solve for } x.$$

$$x = 25(\cos 1.2) \qquad \text{Evaluate. Angle measure is in radians.}$$

LEARNING OUTCOMES   **KEY CONCEPTS AND EXAMPLES**

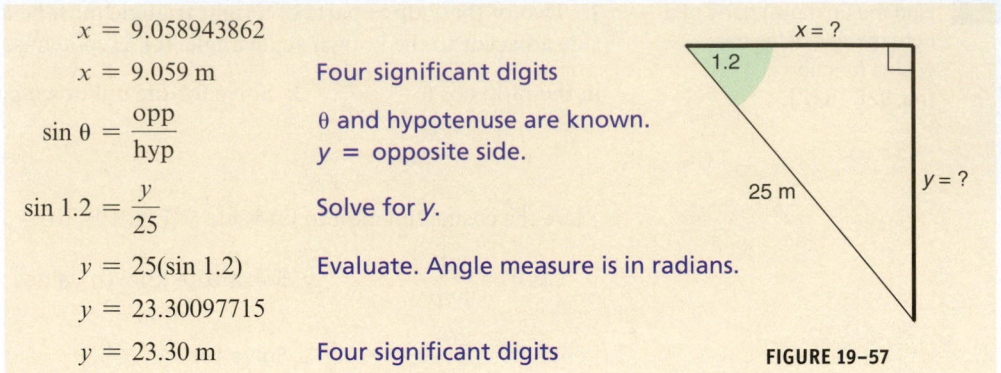

$$x = 9.058943862$$

$$x = 9.059 \text{ m} \qquad \text{Four significant digits}$$

$$\sin \theta = \frac{\text{opp}}{\text{hyp}} \qquad \theta \text{ and hypotenuse are known.}$$
$$\qquad\qquad\qquad y = \text{opposite side.}$$

$$\sin 1.2 = \frac{y}{25} \qquad \text{Solve for } y.$$

$$y = 25(\sin 1.2) \qquad \text{Evaluate. Angle measure is in radians.}$$

$$y = 23.30097715$$

$$y = 23.30 \text{ m} \qquad \text{Four significant digits}$$

FIGURE 19–57

5 Solve applied problems using right-triangle trigonometry (pp. 829–833).

Solve a problem using trigonometric functions: 1. First determine the given parts and the unknown part or parts; then identify the trigonometric function that relates the given and unknown parts. **2.** Use this function and the given information to find the missing information.

A jet takes off at a 25° angle. Find the distance traveled by the plane from the takeoff point to point directly above the end of the runway if the runway is 950 ft long (Fig. 19–58).

$$\cos \theta = \frac{\text{adj}}{\text{hyp}} \qquad \theta \text{ and adjacent side are known. Find the hypotenuse.}$$

$$\cos 25° = \frac{950}{c} \qquad \text{Solve for } c.$$

$$c = \frac{950}{\cos 25°} \qquad \text{Evaluate.}$$

$$c = 1,048.209023$$

$$c = 1,048 \text{ ft} \qquad \text{Four significant digits}$$

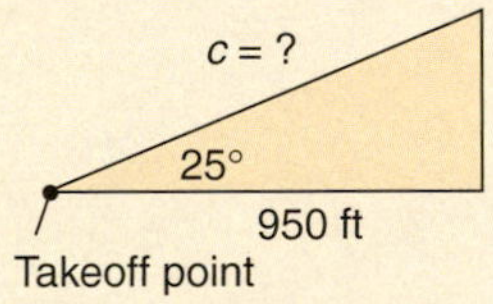

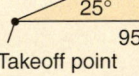

FIGURE 19–58

19 | CHAPTER REVIEW EXERCISES

Section 19–1   MyLab Math   For additional practice go to your study plan in MyLab Math.

Find the trigonometric ratios using Fig. 19–59.

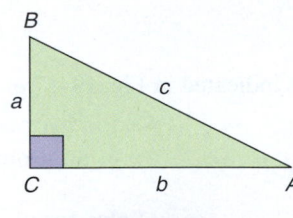

FIGURE 19–59

Use $a = 15$, $b = 20$, $c = 25$ in Exercises 1–6 and express the ratios as fractions in lowest terms.

1. $\sin A$ **2.** $\cos A$ **3.** $\tan A$ **4.** $\tan B$ **5.** $\sin B$ **6.** $\cos B$

Use $a = 2$ ft, $b = 10$ in., $c = 2$ ft 2 in. in Exercises 7–10 and express the ratios as fractions in lowest terms.

7. $\sin A$ **8.** $\tan A$ **9.** $\tan B$ **10.** $\cos A$

Use $a = 7$, $b = 10.5$, $c = 12.62$ in Exercises 11–16 and express the ratios in decimals to the nearest ten-thousandth.

11. $\cos B$ **12.** $\cos A$ **13.** $\tan A$ **14.** $\sin B$ **15.** $\tan B$ **16.** $\sin A$

Using a standard right triangle, find the following trigonometric values when $a = 5$, $b = 12$, and $c = 13$. Round to the nearest ten-thousandth.

17. $\csc A$ **18.** $\sec A$ **19.** $\cot B$

Using a standard right triangle, find the angle in degrees or radians as indicated when given the trigonometric values. θ represents the unknown angle measure. Round degree measures to the nearest tenth of a degree. Round radian measures to the nearest thousandth.

20. $\csc \theta = 1.269$, θ in degrees   **21.** $\sec \theta = 1.236$, θ in degrees   **22.** $\cot \theta = 0.800$, θ in radians

Section 19–2

Use your calculator to find the trigonometric values. Round to the nearest ten-thousandth.

23. $\cos 32°50'$ **24.** $\cos 42.5°$ **25.** $\sin 0.4712$ **26.** $\tan 1.0210$

27. $\tan 47°$ **28.** $\tan 15.6°$ **29.** $\sin 0.8610$ **30.** $\tan 25°40'$

Find an angle (in degrees) having the trigonometric values. θ represents the unknown angle measure. Express answers to the nearest tenth of a degree.

31. $\sin \theta = 0.5446$ **32.** $\tan \theta = 0.8720$ **33.** $\cos \theta = 0.6088$ **34.** $\cos \theta = 0.8897$

Find an angle (in radians) having the trigonometric values. Round to the nearest ten-thousandth.

35. $\tan \theta = 2.723$ **36.** $\sin \theta = 0.9205$ **37.** $\tan \theta = 0.3440$ **38.** $\cos \theta = 0.9450$

Section 19–3

Find the indicated parts of the triangles in Figs. 19–60 through 19–65. Round sides to four significant digits and round angles to the nearest 0.1°.

39. Find A.   **40.** Find r.   **41.** Find h.

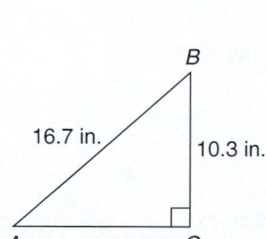

FIGURE 19–60

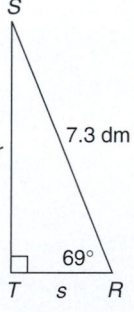

FIGURE 19–61

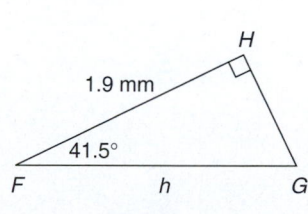

FIGURE 19–62

42. Find c.   **43.** Find y.   **44.** Find B.

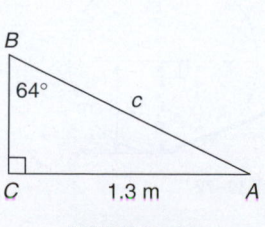

FIGURE 19–63

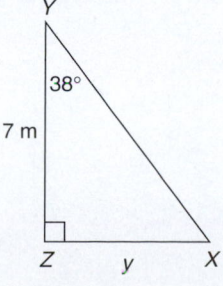

FIGURE 19–64

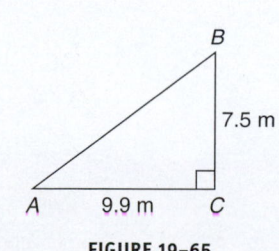

FIGURE 19–65

Solve the triangles in Figs. 19–66 through 19–67. Round sides to four significant digits and round angles to the nearest 0.1°.

45.

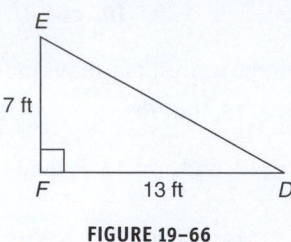

FIGURE 19–66

46.

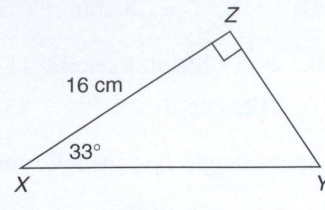

FIGURE 19–67

Find the indicated parts of the triangles in Figs. 19–68 through 19–70. Round sides to four significant digits and angle measures to the nearest 0.1°.

47. Find A.

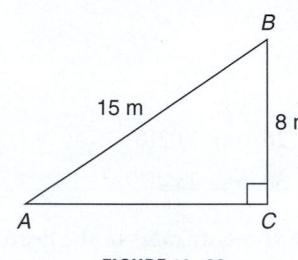

FIGURE 19–68

48. Find A.

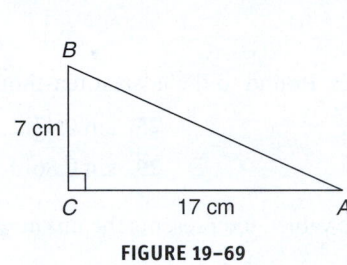

FIGURE 19–69

49. Find a.

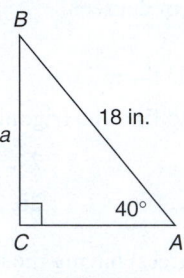

FIGURE 19–70

50. **CON**   A railway inclines 14°. How many feet of track must be laid if the hill is 15 ft high?

51. **CON**   A corner shelf is cut so that the sides placed on the wall are 37 in. and 42 in. What are the measures of the acute angles? Round to tenths.

52. **CON**   From a point 5 ft above the ground and 20 ft from the base of a building, a surveyor uses a transit to sight an angle of 38° to the top of the building. Find the height of the building.

53. **CON**   A surveyor makes the measures indicated in Fig. 19–71. Solve the triangle. All parts of the triangle should be included in a report. Round sides to four significant digits.

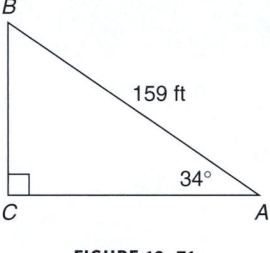

FIGURE 19–71

54. **CON**   What length of rafter is needed for a roof if the rafters form an angle of 35.5° with a joist and the rise is 8 ft? Round to hundredths.

55. **INDTEC**   Find the angle formed by the rod and the horizontal in the mechanical assembly shown in Fig. 19–72. Round to tenths.

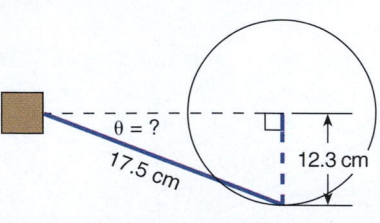

FIGURE 19–72

56. **CON** A utility pole is 40 ft above ground level. A wire must be attached to the pole 3 ft from the top to give it support. If the wire forms a 20° angle with the ground, how long is the wire? Disregard the length needed for attaching the wire to the pole or ground. Round to hundredths.

57. **ELEC** Solve for reactance X_L and resistance R in Fig. 19–73. All units are in ohms. Round to hundredths.

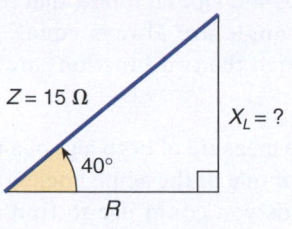

$Z = 15\ \Omega$

$X_L = ?$

40°

R

FIGURE 19–73

---

19 | TEAM PROBLEM-SOLVING EXERCISES

1. Draw triangle ABC so that angle C is 90° and angle A is 70°. Make a point E on side AB that is 10 cm from A. Make a point D on BC so that DE is parallel to CA. If CA is 24 cm, find the length of BD.

2. You are designing a stairway for a home and need to determine how many steps the stairway will have. If the stairway is 8 ft high and the floor space allotted is 14 ft, determine how many steps are needed and determine the optimum measure (rise and run) for each step.

3. Pairs of trigonometric functions have special relationships.
 (a) Is there an angle between 0° and 360° for which the sine and cosine of an angle are equal? Justify your answer by making a table of values for the sine and cosine of angles between 0° and 360° with 15° increments and by graphing sin x and cos x for these values.

 (b) Investigate the sine and tangent functions with a table of values and graph for values between 0° and 360° with 15° increments.

---

19 | CONCEPTS ANALYSIS

1. Explain how the sine function is a ratio.

2. Use your calculator to find the value of the sine of several angles in each of the four quadrants and make a general statement about the greatest and least values the sine function can have.

3. Draw three right triangles. The acute angles are 30° and 60° for triangle 1; 25° and 65° for triangle 2; and 80° and 10° for triangle 3. For each triangle, find the sine of the first angle and the cosine of the second angle. Compare the sine and cosine values for each triangle. What generalization can you make from these three triangles?

4. Draw three right triangles of your choice and verify the generalization you made in Exercise 3.

5. Use your calculator to find the sine of the angles given below and make a table. List the given angle in column 1 and the sine of the angle in column 2. What is the angle measure at which the sine of the angle begins to repeat?

 30°, 60°, 90°, 120°, 150°, 180°, 210°,

 240°, 270°, 300°, 330°, 360°, 390°, 420°

6. Find the sine of several angles larger than those given in Exercise 5 to determine where, if anywhere, the sine begins to repeat again.

 Make a general statement about the sine function as it repeats. Compare your general statement with the graph of the sine function.

7. Use a calculator to find sin 45° and cos 45°. Using your knowledge of geometry, explain the relationship between sin 45° and cos 45°. Does it follow that the sine and cosine of the same angle are always equal? Are there other angles for which the two functions are equal? If so, discuss them.

8. If you are given an angle and an adjacent side of a right triangle, explain how you would find the hypotenuse of the triangle.

9. You know the measure of both legs of a right triangle and the measure of one of the acute angles. Describe two different methods you could use to find the length of the hypotenuse.

10. What is the fewest number of parts of a triangle that must be known to find the measures of all the other parts? What are the parts required?

11. Explain the differences and similarities between angle of inclination and angle of declination.

12. Using a calculator, compare the sine of an acute angle with the sine of its complement for several angles. Compare the sine of an acute angle with the cosine of its complement for several angles. Generalize your findings.

19 PRACTICE TEST

Write the trigonometric ratios as fractions in lowest terms for the following trigonometric functions using triangle ABC in Fig. 19–74. $a = 10, b = 24, c = 26$

1. sin A

2. tan B

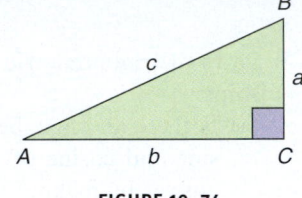

FIGURE 19–74

Express the trigonometric values in decimals to the nearest ten-thousandth. Refer to Fig. 19–74 and use $a = 5, b = 11.5, c = 12.54$.

3. cos A

4. sin B

Use your calculator to find the trigonometric values. Round to the nearest ten-thousandth.

5. sin 53°

6. sin 61°10′

Find an angle (in degrees) having the indicated trigonometric values; θ represents the unknown angle measure. Express your answers to the nearest tenth of a degree.

7. sin θ = 0.2756

8. tan θ = 1.280

Find an angle (in radians) having the indicated trigonometric values. Round to the nearest ten-thousandth.

9. sin θ = 0.7660

10. cos θ = 0.8387

Round angles to the nearest tenth of a degree and the unknown sides to the four significant digits.

11. Solve triangle ABC in Fig. 19–75.

12. Solve triangle DEF in Fig. 19–76.

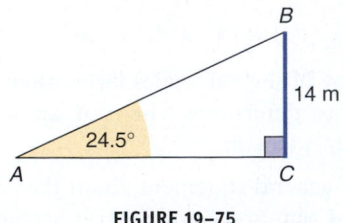

FIGURE 19–75

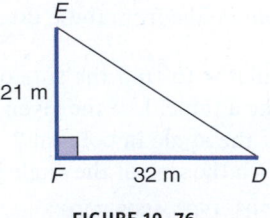

FIGURE 19–76

13. A stair has a rise of 4 in. for every 5 in. of run. What is the angle of inclination of the stair? Round to tenths.

14. A surveyor uses indirect measurement to find the width of a river at a certain point. The surveyor marks off 50 ft along the riverbank at right angles with the river and then sights an angle of 43° to point A across the river (Fig. 19–77). Find the width of the river. Round to hundredths.

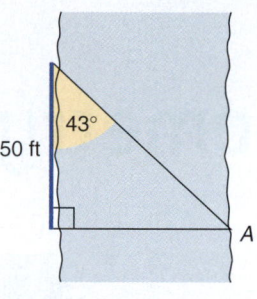

FIGURE 19–77

15. Steel girders are reinforced by placing steel supports between two runners so that right triangles are formed (see Fig. 19–78). If the runners are 24 in. apart and a 30° angle is desired between the support and a runner, find the length of the support to be placed at a 30° angle.

16. In the circuit represented by the diagram of Fig. 19–79, find the total current I_t. All units are in amps. Round to hundredths.

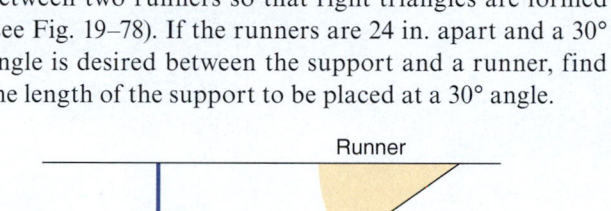

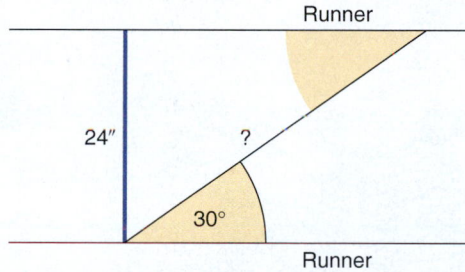

FIGURE 19–78

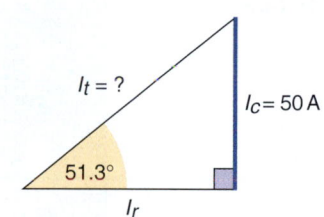

FIGURE 19–79

17. Find the current in the resistance branch I_r of the circuit in Fig. 19–79 of Exercise 16. Round to four significant digits.

18. The minimum clearances for the installation of a metal chimney pipe are shown in Fig. 19–80. How far from the ridge should the hole be cut for the pipe to pass through the roof? Round to hundredths.

19. Refer to Fig. 19–80. What angle is formed by the chimney pipe and the roof where the pipe passes through the roof? Round to the nearest tenth of a degree.

20. Elbows are used to form bends in rigid pipe and are measured in degrees. What is the angle of bend of the elbow in the installation shown in Fig. 19–81 to the nearest whole degree?

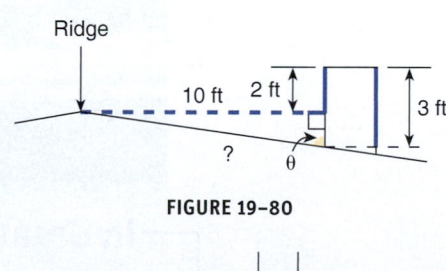

FIGURE 19–80

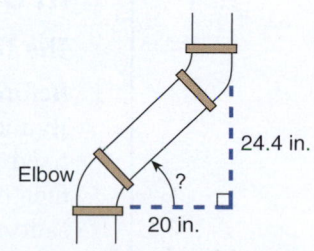

FIGURE 19–81

Trigonometry with Any Angle

Richard Semik/123RF

In Great Company

The Torrey Canyon Triangulation (1967)

Before GPS was invented, sailors determined positions at sea by triangulation. By measuring the angle between a star and the horizon, or the angle between fixed points on land, the ship's position would be determined. Of course, crewmen had to be relatively experienced in doing the necessary calculations. Not all sailors were.

It was 6 A.M. on March 18, 1967. Pastrengo Rugiati was captain of one of the world's first supertankers. Three hundred meters long, it carried 119,328 metric tons of Kuwaiti crude oil. This experienced sea captain was headed for Wales. To get there, he had to pass the Scilly Islands, near the southwest coast of England.

The captain had a new helmsman and new third officer. Rugiati wanted the ship steered to the right of the islands. His supertanker would maneuver through the 6-mi-wide channel. He would carefully avoid a collection of rocky reefs known as the Seven Stones lying just to the right of the isles.

Unfortunately, as the ship entered the channel, Rugiati was forced to make course corrections to avoid fishing boats. This brought the ship to the right side of the channel.

His third officer took a fix on the Scillies' lighthouses and plotted the ship's position. As Rugiati watched, he suddenly realized the new man had marked an impossible position, at least a mile off where the captain thought they were.

"Stop using the Scillies for bearings," he shouted. "Use the lightship!" They needed two fixed landmarks. He knew the Seven Stones were close, but now he wasn't sure where they were relative to the ship. Frightened, the third officer confused the visual fixes and had to retake them. The ship continued to plow forward.

The third officer finally got the calculation correct and marked the ship's position. It was less than a mile from the Seven Stones. The captain ordered hard port. It was too late.

The Pollard Rock reef tore the bottom out of the ship as it ran aground. Over the next few days, every drop of the crude oil would spill, producing the worst environmental disaster ever seen up to that time. One hundred twenty miles of Cornish coastline was contaminated; 15,000 seabirds were killed. The oil slick reached as far as Brittany, France. As of December 2010, nearly 45 years later, the last remnants of the oil were still being cleaned up. That's the bad news. On the bright side, the disaster sparked the creation of international maritime regulations on pollution.

20–1 Vectors

LEARNING OUTCOMES

1 Find the magnitude of a vector in standard position, given the coordinates of the end point.

2 Find the direction of a vector in standard position, given the coordinates of the end point.

3 Convert vectors in standard position between rectangular coordinate notation and polar coordinate notation.

4 Add vectors and multiply a vector by a scalar.

LC LEARNING CATALYTICS
1. Add $4.1 + 2.7i$ and $2.7 + 3.7i$.

Scalar: a quantity described by magnitude only

Vector: a quantity described by magnitude (or length) and direction (angle). It represents a shift or movement from one point to another

Although right triangles are perhaps the most frequently used triangles, we often work with other kinds of triangles. In this chapter, we apply the laws of sines and cosines to oblique triangles and use *vectors* for applications with angles.

1 Find the Magnitude of a Vector in Standard Position, Given the Coordinates of the End Point. Quantities that we have discussed so far in this text have been described by specifying their size or magnitude. Quantities such as area, volume, length, and temperature, which are characterized by magnitude only, are called **scalars**. Other quantities such as electrical current, force, velocity, and acceleration are called *vectors*. A quantity described by magnitude (or length) and direction is a **vector**. A vector represents a shift or movement from one point to another.

When we consider the speed of a plane to be $500 \frac{\text{mi}}{\text{h}}$, we are considering a scalar quantity. However, when we consider the speed of a plane *traveling northeast from a given location,* we are concerned with both the distance traveled and the direction. This is a vector quantity. We have seen such quantities in the vector diagrams used to solve electronic problems by means of right triangles in Chapter 19 and elsewhere.

Vectors are represented by straight arrows. The length of the arrow represents the **magnitude of the vector** (see Fig. 20–1). The symbol for the magnitude of a vector v is $\|v\|$. The curved arrow shows the counterclockwise *direction* of the vector (see Fig. 20–2), with the end of the arrow being the beginning point and the arrowhead being the end point.

Magnitude of a vector: the length of a vector represented by an arrow. The symbol for the magnitude of a vector v is $\|v\|$

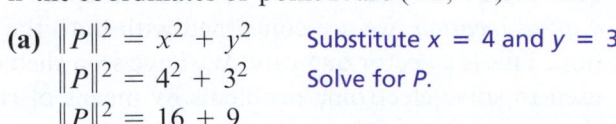

FIGURE 20–1

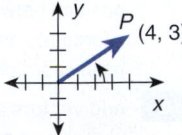

FIGURE 20–2

Standard position of a vector: the beginning point of a vector is at the origin of a rectangular coordinate system, and the direction of the vector is the angle of counter-clockwise rotation from the positive x-axis (horizontal axis)

Direction of a vector: the angle of counterclockwise rotation from the horizontal axis

In relating trigonometric functions to vector quantities, we will place the vectors in standard position. The **standard position of a vector** places the beginning point of the vector at the origin of a rectangular coordinate system. The **direction of the vector** is the counterclockwise angle measured from the positive x-axis (horizontal axis).

In Fig. 20–2, vector P is a first-quadrant vector that has a direction of θ_1 and an end point at the point P. The magnitude of a vector can be determined if the vector is in standard position and the x- and y-coordinates of the end point are known. Vector R is a third-quadrant vector that has a direction of θ_2 and an end point at point R. Its magnitude may be determined using the x- and y-coordinates of the end point.

Magnitude of a vector in standard position:

The *magnitude* of a vector in standard position is the length of the hypotenuse of the right triangle formed by the vector, the x-axis, and the vertical line from the end point of the vector to the x-axis.

Symbolically, for a vector v with an end point at (x_1, y_1),

$$\|v\| = \sqrt{x_1^2 + y_1^2}$$

To find the magnitude of a vector in standard position:

1. Substitute into the Pythagorean theorem the coordinates of the end points of the vector (x_1, y_1) as the legs of a right triangle.

2. Solve for the hypotenuse (magnitude).

STOP AND CHECK

1. Find the magnitude of vector Q to the nearest tenth if the coordinates of Q are (2, 7).
2. Find the magnitude of vector S to the nearest tenth if the coordinates of S are $(-3, 6)$.

Answers:
1. 7.3 2. 6.7

EXAMPLE 1

Find the magnitude to the nearest tenth of **(a)** vector P in Fig. 20–3 if the coordinates of P are (4, 3) and **(b)** vector R in Fig. 20–4 if the coordinates of point R are $(-2, -5)$.

(a) $\|P\|^2 = x^2 + y^2$ Substitute $x = 4$ and $y = 3$.
  $\|P\|^2 = 4^2 + 3^2$ Solve for P.
  $\|P\|^2 = 16 + 9$
  $\|P\|^2 = 25$
  $\|P\| = \pm\sqrt{25}$
  $\|P\| = \pm 5$ We use only $+5$ because length or magnitude is positive.

The magnitude of vector P is 5 units.

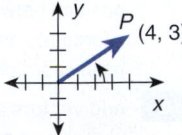

FIGURE 20–3

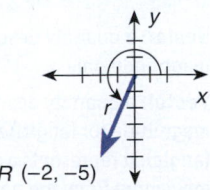

FIGURE 20–4

(b) The magnitude of a vector is always positive; however, the x- and y-coordinates can be negative.

$\|R\|^2 = x^2 + y^2$ Substitute $x = -2$ and $y = -5$.

$\|R\|^2 = (-2)^2 + (-5)^2$ Solve for R.

$\|R\|^2 = 4 + 25$

$\|R\|^2 = 29$

$\|R\| = \sqrt{29}$ Principal square root

$\|R\| = 5.4$ Rounded

The magnitude of vector R is 5.4. See Exercises 1–5.

2 Find the Direction of a Vector in Standard Position, Given the Coordinates of the End Point. If the vector in standard position falls in Quadrant I, we can use right-triangle trigonometry to find the direction of the vector. For a vector in standard position with an end point in Quadrant I, the tangent function can be used to find the direction or angle of the vector.

To find the direction of a quadrant I vector in standard position:

1. Identify the coordinates of the end point of the vector.
2. Substitute the coordinates of the end point into the tangent function: $\tan\theta = \frac{\text{opp}}{\text{adj}}$ or $\tan\theta = \frac{y}{x}$.
3. Solve for θ.

STOP AND CHECK

1. Find the direction of vector Q to the nearest tenth if the coordinates of Q are (2, 7).
2. Find the direction of vector S to the nearest tenth if the coordinates of S are (3, 6).

Answers:
1. 74.1° **2.** 63.4°

EXAMPLE 2

Find the direction to the nearest tenth of a degree of vector P in Fig. 20–5 if the coordinates of P are (4, 3).

$\tan\theta = \frac{y}{x}$ Substitute values for end point of vector P: $x = 4$ and $y = 3$.

$\tan\theta = \frac{3}{4}$ Use the inverse tangent function and the decimal equivalent of $\frac{3}{4}$.

$\tan^{-1} 0.75 = \theta$ Evaluate.

$\theta = 36.9°$ Rounded

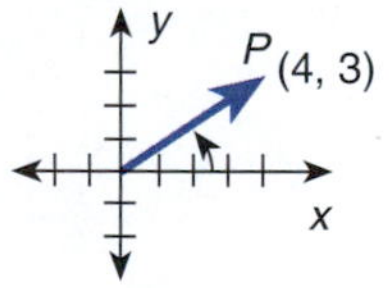

FIGURE 20–5

The direction of vector P is 36.9°. See Exercises 6–10.

Polar coordinate system: a coordinate system relating the fixed point called the pole or origin and a fixed line called the polar axis

Pole or origin: the fixed reference point in the polar coordinate system

Polar axis: a fixed line in the polar coordinate system

Polar notation: an ordered pair in the polar coordinate system written as (r, θ), where r is the magnitude and θ is the direction of the vector

3 Convert Vectors in Standard Position Between Rectangular Coordinate Notation and Polar Coordinate Notation. Another coordinate system that often is used with vectors is the polar coordinate system. In the **polar coordinate system**, points are in relation to a fixed point called the **pole, or origin**, and a fixed line called the **polar axis** (Fig. 20–6).

A vector in standard position has the beginning point at the pole and its end point r units from the pole. The magnitude of the vector is r. The angle measure θ from the polar axis to the vector is the direction of the vector. A point on a polar coordinate system is written in **polar notation** as (r, θ). The end point of the vector OP in Fig. 20–7 is (3, 45°).

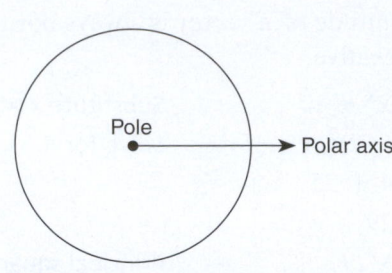

FIGURE 20-6 Polar coordinate system.

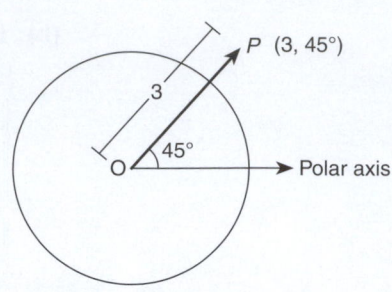

FIGURE 20-7 Vector *OP* at (3, 45°).

The polar coordinate system can be related to the rectangular (Cartesian) coordinate system by using the Pythagorean theorem and trigonometric functions. For now we will examine only Quadrant I vectors.

In polar notation, *r* is the hypotenuse of a triangle from the end point of the vector to the polar axis, and θ is the angle formed by the polar axis and the vector. The value of *r* is found by using the Pythagorean theorem with the *x*- and *y*-coordinates of the end point of the vector in rectangular notation. The value of θ is found by solving the tangent relationship, $\tan\ \theta = \dfrac{y}{x}$, for θ. In rectangular notation, the value of *x* can be found from polar notation, where *r* is the hypotenuse, by solving the cosine relationship, $\cos\ \theta = \dfrac{x}{r}$, for *x*. The value of *y* can be found from polar notation, where *r* is the hypotenuse, by solving the sine relationship, $\sin\ \theta = \dfrac{y}{r}$, for *y*.

Relationships between the rectangular coordinate system and the polar coordinate system:

From Rectangular to Polar:	From Polar to Rectangular:
$r = \sqrt{x^2 + y^2}$	$x = r\cos\ \theta$
$\theta = \tan^{-1}\left(\dfrac{y}{x}\right)$	$y = r\sin\ \theta$

To convert the end point of a vector in rectangular notation to polar notation:

1. Find *r* using the relationship $r = \sqrt{x^2 + y^2}$.

2. Find θ using the relationship $\theta = \tan^{-1}\left(\dfrac{y}{r}\right)$.

STOP AND CHECK

1. Find the polar coordinates to the nearest tenth for a vector with the end point of (1, 7).

Answer:
1. (7.1, 81.9°)

EXAMPLE 3

Given the end point of a vector in standard position, find the polar coordinates.

(a) (3, 4)   **(b)** (5, 8)

(a) To find *r*:

$r = \sqrt{x^2 + y^2}$ Substitute known values: *x* = 3, *y* = 4
$r = \sqrt{3^2 + 4^2}$ Evaluate.
$r = \sqrt{9 + 16}$
$r = \sqrt{25}$
$r = 5$ Principal square root

To find θ:

$$\theta = \tan^{-1}\frac{y}{x}$$ Substitute known values: $x = 3$, $y = 4$

$$\theta = \tan^{-1}\frac{4}{3}$$ $\boxed{\text{TAN}^{-1}}$ 4 $\boxed{\div}$ 3 $\boxed{\text{ENTER}}$

$$\theta = 53.13010235°$$ Round to the nearest tenth of a degree.

$$\theta = 53.1°$$

The polar coordinates are (5, 53.1°).

(b) To find r:

$$r = \sqrt{x^2 + y^2}$$ Substitute known values: $x = 5$, $y = 8$
$$r = \sqrt{5^2 + 8^2}$$ Evaluate.
$$r = \sqrt{25 + 64}$$
$$r = \sqrt{89}$$
$$r = 9.433981132$$ Principal square root
$$r = 9.4$$ Round to the nearest tenth of a degree.

To find θ:

$$\theta = \tan^{-1}\frac{y}{x}$$ Substitute known values: $x = 5$, $y = 8$

$$\theta = \tan^{-1}\frac{8}{5}$$ $\boxed{\text{TAN}^{-1}}$ 8 $\boxed{\div}$ 5 $\boxed{\text{ENTER}}$

$$\theta = 57.99461679°$$ Round to the nearest tenth of a degree.

$$\theta = 58.0°$$

The polar coordinates are (9.4, 58.0°).   See Exercises 11–15.

To convert the end point of a vector in polar notation to rectangular notation:

1. Find x using the relationship $x = r\cos\theta$.

2. Find y using the relationship $y = r\sin\theta$.

EXAMPLE 4

Given the polar coordinates, find the rectangular coordinates.

(a) $(2, 30°)$ **(b)** $\left(7, \frac{\pi}{2}\right)$

(a) To find x:

$$x = r\cos\theta$$ Substitute known values: $r = 2$, $\theta = 30°$
$$x = 2\cos 30°$$ In degree mode: 2 $\boxed{\text{COS}}$ 30 $\boxed{\text{ENTER}}$
$$x = 1.732050808$$ x-coordinate

To find y:

$$y = r\sin\theta$$ Substitute known values: $r = 2$, $\theta = 30°$
$$y = 2\sin 30°$$ In degree mode: 2 $\boxed{\text{SIN}}$ 30 $\boxed{\text{ENTER}}$
$$y = 1$$ y-coordinate
$$(1.7, 1)$$ Rectangular notation rounded

(b) To find x:

$x = r \cos \theta$ — Substitute known values: $r = 7, \theta = \dfrac{\pi}{2}$

$x = 7 \cos \dfrac{\pi}{2}$ — In radian mode: 7 [COS] [(] π [÷] 2 [)] [ENTER]

$x = 0$ — x-coordinate

To find y:

$y = r \sin \theta$ — Substitute known values: $r = 7, \theta = \dfrac{\pi}{2}$

$y = 7 \sin \dfrac{\pi}{2}$ — In radian mode: 7 [SIN] [(] π [÷] 2 [)] [ENTER]

$y = 7$ — y-coordinate

(0, 7) — Rectangular notation — See Exercises 16–20.

Some calculators have keys or menu choices to convert between rectangular and polar notation. On the TI-84, the appropriate menu is accessed by the [ANGLE] key.

TIP **Polar and Rectangular Conversions on the TI-84** Set [MODE] to degrees or radians, as desired.
Access the [ANGLE] menu:

Menu Choices	Feature	Syntax
5:R▶Pr(	rectangular to polar r-coordinate	x [,] y [)] [ENTER]
6:R▶Pθ(	rectangular to polar θ-coordinate	x [,] y [)] [ENTER]
7:P▶Rx(	polar to rectangular x-coordinate	r [,] θ [)] [ENTER]
8:P▶Ry(	polar to rectangular y-coordinate	r [,] θ [)] [ENTER]

STOP AND CHECK
Use polar and rectangular functions on a calculator to make the conversions. Round to four significant digits.
1. (2, 9) to polar in degrees
2. $\left(3, \dfrac{\pi}{6}\right)$ to rectangular

Answers:
1. (9,220, 77.47°)
2. (2.598, 1.5)

EXAMPLE 5

Use a calculator to make the following conversions. Round to four significant digits.

(a) (5, 12) to polar in degrees **(b)** $\left(6, \dfrac{\pi}{3}\right)$ to rectangular

(a) (5, 12) to polar in degrees

Find the r coordinate:

Set to degree mode and enter x- and y-coordinates.

[ANGLE] [5:R▶Pr(] 5 [,] 12 [)] [ENTER] $\Rightarrow$ 13

Find the θ coordinate:

[ANGLE] [6:R▶Pθ(] 5 [,] 12 [)] [ENTER] $\Rightarrow$ 67.38013505

(13, 67.38°)   Polar notation

(b) $\left(6, \dfrac{\pi}{3}\right)$ to rectangular

Find the x-coordinate:

Set to radian mode and enter r- and θ-coordinates.

[ANGLE] [7:P▶Rx(] 6 [,] π ÷ 3 [)] [ENTER] $\Rightarrow$ 3

Find the y-coordinate:

[ANGLE] [8:P▶Ry(] 6 [,] π ÷ 3 [)] [ENTER] $\Rightarrow$ 5.196152423

(3, 5.196)   Rectangular notation rounded   See Exercises 11–20.

4 Add Vectors and Multiply a Vector by a Scalar. Two vectors with the same direction can be added by aligning the beginning point of one vector with the end point of the other. The vector represented by the sum is called the **resultant vector,** or **resultant.** The resultant has the same direction as the vectors being added and a magnitude that is the sum of the two magnitudes.

Resultant vector or resultant: the vector represented by the sum of vectors. If two vectors are being added that have the same direction, the resultant vector has the same direction as the vectors being added and a magnitude that is the sum of the two magnitudes

> **To add vectors of the same direction:**
>
> 1. Add the magnitudes of the vectors to get the magnitude of the resultant vector.
>
> 2. The resultant vector will have the same direction as the original vectors.

STOP AND CHECK

1. Add two vectors with a direction of 47° if the magnitudes of the vectors are 2.8 and 5.2, respectively.

Answer:

1. The resultant vector has a magnitude of 8.0 and a direction of 47°.

EXAMPLE 6

Add two vectors with a direction of 35° if the magnitudes of the vectors are 4 and 5, respectively (see Fig. 20–8).

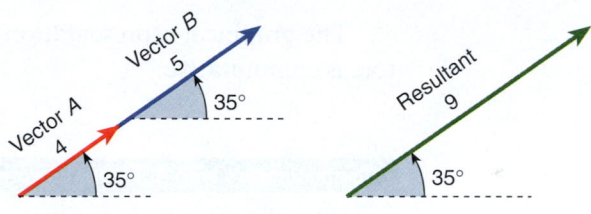

FIGURE 20-8

The resultant vector has a magnitude of 9 and a direction of 35°. **See Exercises 21–22.**

Two vectors have opposite directions if the directions of the vectors differ by 180°. Two vectors with opposite directions can be added by aligning the beginning point of the vector that has the smaller magnitude with the end point of the vector that has the larger magnitude. The resultant has the same direction as the vector that has the larger magnitude. The magnitude of the resultant is the difference of the two magnitudes.

> **To add vectors of opposite direction:**
>
> 1. Subtract the magnitude of the smaller vector from the magnitude of the larger vector to get the magnitude of the resultant vector.
>
> 2. The resultant vector has the direction of the vector that has the larger magnitude.

STOP AND CHECK

1. Add a vector with a direction of 70° and a magnitude of 9 to a vector with a direction of 250° and a magnitude of 12.

Answer:

1. The resultant has a magnitude of 3 and a direction of 250°.

EXAMPLE 7

Add a vector with a direction of 60° and a magnitude of 6 to a vector with a direction of 240° and a magnitude of 4 (see Fig. 20–9).

Subtract the magnitudes: $6 - 4 = 2$.

The resultant has a magnitude of 2 and a direction of 60°.

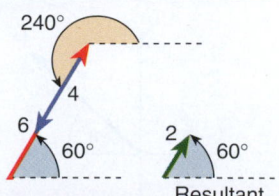

FIGURE 20-9

See Exercises 23–24.

Adding any two vectors that have different directions is accomplished by performing the shifts of each vector in succession.

Examine the graphical representation of the sum of a 45° vector with a magnitude of 5 (vector *A*) and a 60° vector with a magnitude of 6 (vector *B*) (see Fig. 20–10).

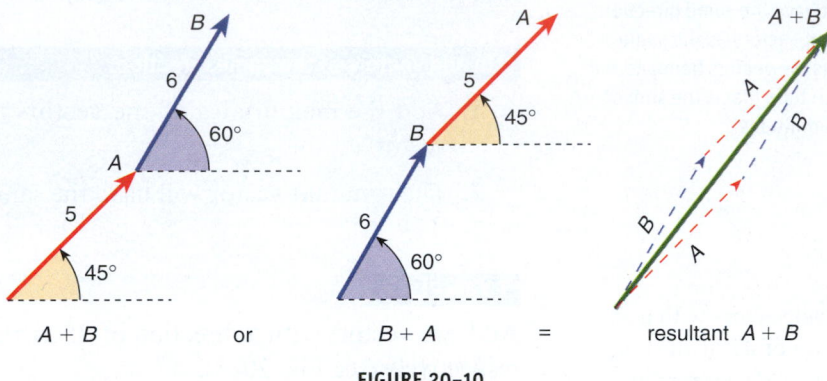

A + B   or   B + A   =   resultant A + B

FIGURE 20–10

The graphical representation in Fig. 20–10 illustrates that the addition of vectors is commutative.

To add vectors with different directions:

1. Determine the *x*- and *y*-coordinates of the end point of each vector in standard position.

2. Add the *x*-coordinates of the end points for the *x*-coordinate of the end point of the resultant vector.

3. Add the *y*-coordinates of the end points for the *y*-coordinate of the end point of the resultant vector.

4. Find the magnitude using the Pythagorean theorem and the end point of the resultant vector.

5. Find the direction using the inverse tangent function and the end point of the resultant vector.

Complex notation of a vector: the *x*-coordinate of a vector in standard position is represented as the real part *a*. The *y*-coordinate of a vector in standard position is represented as the imaginary part *bi*

Complex numbers in the form $a + bi$ are sometimes used to represent vectors. For a vector to be written in **complex notation,** the *x*-coordinate of a vector in standard position is represented as the real component *a*. The *y*-coordinate of a vector in standard position is represented as the imaginary component *bi*.

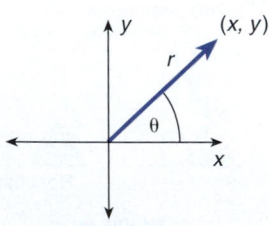

FIGURE 20–11

To write a vector in complex notation:

1. Find the *x*-coordinate of the end point of the vector (Fig. 20–11). Use the relationship $x = r \cos \theta$.

2. Find the *y*-coordinate of the end point of the vector. Use the relationship $y = r \sin \theta$.

3. Write the vector in complex notation as $x + yi$.

Vector C is a 30° vector with a magnitude of 9 m and vector D is a 50° vector with a magnitude of 5 m.

1. Write vector C in complex notation. Round each component to four significant digits.
2. Write vector D in complex notation. Round each component to four significant digits.
3. Write resultant $C + D$ in complex notation.

Answers:
1. $7.794 + 4.5i$
2. $3.214 + 3.830i$
3. $11.01 + 8.330i$

EXAMPLE 8

Write the vectors A, B, and $A + B$ from Fig. 20–10 in complex notation.

Vector A:	**x-coordinate**	**y-coordinate**	
magnitude = 5	$x = r \cos \theta$	$y = r \sin \theta$	Solve for x or y.
direction = 45°	$x = 5(\cos 45°)$	$y = 5(\sin 45°)$	Evaluate.
	$x = 3.535533906$	$y = 3.535533906$	

Vector A in complex form: 3.535533906 + 3.535533906i

Vector B:	**x-coordinate**	**y-coordinate**	
magnitude = 6	$x = r \cos \theta$	$y = r \sin \theta$	Solve for x or y.
direction = 60°	$x = 6(\cos 60°)$	$y = 6(\sin 60°)$	Evaluate.
	$x = 3$	$y = 5.196152423$	

Vector B in complex form: 3 + 5.196152423i

$$\text{Resultant } A + B = (3.535533906 + 3.535533906i) \qquad \text{Add } x\text{-values.}$$
$$+ (3 + 5.196152423i) \qquad \text{Add } y\text{-values.}$$

Resultant $A + B$ = 6.535533906 + 8.731686329i   or   **6.536 + 8.732i**

See Exercises 25–26.

---

1. Find the magnitude to four significant digits and direction to the nearest tenth of a degree of the resultant $C + D$ from the complex notation in the preceding Stop and Check problem.

Answer:
1. magnitude = 13.81; direction = 37.1°

EXAMPLE 9

Find the magnitude and direction of the resultant $A + B$ from the complex notation in the preceding example.

The resultant in complex form is $6.535533906 + 8.731686329i$; thus, the x-coordinate of the end point is 6.535533906 and the y-coordinate is 8.731686329.

$\text{magnitude} = \sqrt{x^2 + y^2}$	Substitute for x and y.
$\text{magnitude} = \sqrt{(6.535533906)^2 + (8.731686329)^2}$	Evaluate.
$\text{magnitude} = \sqrt{118.9555496}$	
magnitude = 10.91	Round to 4 significant digits.
$\tan \theta = \dfrac{y}{x}$	Substitute for x and y.
$\tan \theta = \dfrac{8.731686329}{6.535533906}$	Evaluate.
$\tan \theta = 1.336032596$	Use inverse tangent function.
$\tan^{-1}(1.336032596) = \theta$	Evaluate.
$\tan \theta = 53.18570658$	
direction: $\theta = 53.2°$	Round to nearest tenth of a degree. **See Exercises 27–28.**

Did You Know? Vectors written in complex form are often used in electronics. The imaginary part is referred to as the *j-factor*.

Resultant vector $A + B$ from the example on the preceding page would be written as $6.536 + j8.732$. The customary notation is for j to be followed by the coefficient.

When a vector is multiplied times a scalar, the resulting vector has the same direction as the original vector. The magnitude of the original vector is multiplied by the scalar.

> **To multiply a vector by a scalar:**
>
> 1. Multiply the magnitude of the original vector by the scalar to give the magnitude of the resultant vector.
>
> 2. The direction of the resultant will be the same as the direction of the original vector.

STOP AND CHECK

1. Find the resultant vector if a vector of magnitude 5.3 and a direction of $\frac{\pi}{3}$ is multiplied by a scalar of magnitude 2.

Answer:

1. magnitude = 10.6;

direction = $\frac{\pi}{3}$

EXAMPLE 10

Find the resultant vector if a vector of magnitude 2.8 and a direction of $\frac{\pi}{4}$ is multiplied by a scalar of magnitude 3.

Find the magnitude of the resultant vector:

$3(2.8) = 8.4$ Multiply the magnitude 2.8 by the scalar 3.

Find the direction of the resultant vector:

$\theta = \dfrac{\pi}{4}$ The resultant vector has the same direction as the original vector.

The resultant vector has a magnitude of 8.4 and a direction of $\dfrac{\pi}{4}$.   See Exercises 29–32.

20-1 EXERCISES   MyLab Math For additional practice go to your study plan in MyLab Math.

1 Find the magnitude of the vectors in standard position with end points at the indicated points. Round to the nearest thousandth if necessary. *See Example 1.*

  1. $(5, 12)$ **2.** $(-12, 9)$ **3.** $(2, -7)$ **4.** $(-8, -3)$ **5.** $(1.5, 2.3)$

2 Find the direction of the vectors in standard position with end points at the indicated points. Round to the nearest hundredth. *See Example 2.*

  6. $(5, 12)$ **7.** $(6, 8)$ **8.** $(8, 3)$ **9.** $(2, 5)$ **10.** $(1, 4)$

3 Given the end point of a vector in standard position, find the polar coordinates. Round lengths and degrees to tenths. *See Examples 3 and 5(a).*

11. $(4, 3)$ **12.** $(4, 4)$ **13.** $(7, 2)$ **14.** $(5, 12)$ **15.** $(8, 15)$

Given the polar coordinates, find the rectangular coordinates. Round to tenths. *See Examples 4 and 5(b).*

16. $(3, 45°)$ **17.** $(5, 80°)$ **18.** $\left(3, \dfrac{\pi}{2}\right)$ **19.** $\left(3, \dfrac{\pi}{6}\right)$ **20.** $\left(4, \dfrac{\pi}{4}\right)$

4 *See Example 6.*

21. Find the resultant vector of two vectors that have a direction of 42° and magnitudes of 7 and 12, respectively.

22. Two vectors have a direction of 72°. Find the sum of the vectors if their magnitudes are 1 and 7, respectively.

See Example 7.

23. Find the sum of two vectors if one has a direction of 45° and a magnitude of 7 and the other has a direction of 225° and a magnitude of 8.

24. Find the sum of two vectors if one has a direction of 75° and a magnitude of 15 and the other has a direction of 255° and a magnitude of 9.

See Example 8.

25. Write in complex notation the vectors and the sum of the two vectors in Exercise 23.

26. Write in complex notation the vectors and the sum of the two vectors in Exercise 24.

Find the magnitude and direction of the resultant of the two given vectors in complex notation. *See Example 9.*

27. $5 + 3i$ and $7 + 2i$

28. $1 + 2i$ and $5 + 2i$

Find the resultant vectors. *See Example 10.*

29. Find the resultant vector if a vector of magnitude 2.3 and a direction of 75° that is multiplied by the scalar of magnitude 2.

30. Find the resultant vector if a vector of magnitude 3.8 and a direction of 60° that is multiplied by the scalar of magnitude 2.5.

31. Find the resultant vector if a vector of magnitude 3.7 and a direction of $\frac{\pi}{6}$ that is multiplied by the scalar of magnitude 3.

32. Find the resultant vector if a vector of magnitude 5.1 and a direction of $\frac{\pi}{2}$ that is multiplied by the scalar of magnitude 3.1.

20–2 | Trigonometric Functions for Any Angle

LEARNING OUTCOMES

1 Find related acute angles for angles or vectors in Quadrants II, III, and IV.

2 Determine the signs of trigonometric functions of angles of more than 90°.

3 Find the trigonometric functions of angles of more than 90° using a calculator.

LC LEARNING CATALYSTS

1. Identify the quadrant of point $(2, -4)$.
2. Identify the quadrant of point $(-3, 5)$.

Reference triangle: a right triangle formed by drawing a vertical line from the end point of a vector to the x-axis

Related angle: the acute angle formed by the x-axis and the vector

1 **Find Related Acute Angles for Angles or Vectors in Quadrants II, III, and IV.** The direction of any vector in standard position can be determined, if the coordinates of the end point are known, by applying our knowledge of trigonometric functions. A right triangle that we will refer to as a **reference triangle** can be formed by drawing a vertical line from the end point of the vector to the x-axis. See the example of a reference triangle in Fig. 20–12. The angle θ is called the *related angle*. The **related angle** is the acute angle formed by the x-axis and the vector.

In Quadrant I, the related angle is the same as the direction of the vector. Therefore, the direction of the vector in Quadrant I is always less than 90°. For vectors in Quadrants II, III, and IV, see Figs. 20–12, 20–13, and 20–14.

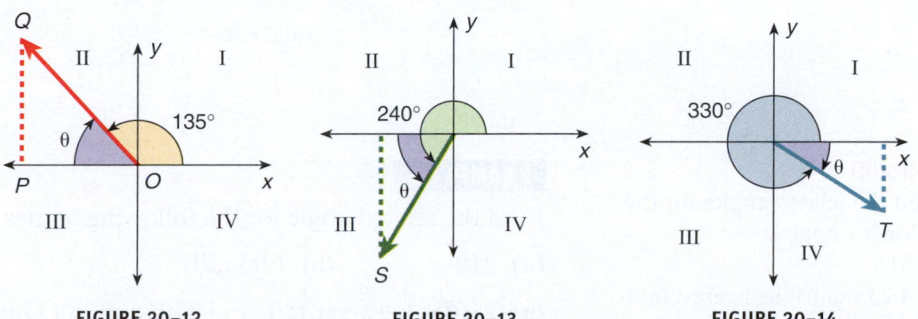

FIGURE 20–12　　　**FIGURE 20–13**　　　**FIGURE 20–14**

In Fig. 20–12, $\overline{PQ}$, the x-axis, and vector Q form a right triangle. The direction of the vector is 135°; therefore, the related angle is 45° (180° − 135°). 135° is a second-quadrant angle. Second-quadrant angles are more than 90° and less than 180°, or more than $\frac{\pi}{2}$ rad (1.57) and less than π rad (3.14). The related angle for any second-quadrant vector can be found by subtracting the direction of the vector from 180° (or π radians).

Third-quadrant angles are more than 180° and less than 270°, or more than π radians (3.14) and less than $\frac{3\pi}{2}$ radians (4.71). In Fig. 20–13, vector S is a third-quadrant vector. The related angle θ is 60° (240° − 180°). The related angle for any third-quadrant vector is found by subtracting 180° (or π radians) from the direction of the vector.

Vector T in Fig. 20–14 is a fourth-quadrant vector. Fourth-quadrant vectors or angles are more than 270° and less than 360°, or more than $\frac{3\pi}{2}$ rad (4.71) and less than 2π rad (6.28). The related angle θ is 30° (360° − 330°). The related angle for any fourth-quadrant vector is found by subtracting the direction of the vector from 360° (or 2π radians).

To find a related acute angle for an angle (θ) in any quadrant:

1. Draw the angle and form the third side of a triangle by drawing a line perpendicular to the *x*-axis from the vector or ray forming the angle.

2. The related angle is the angle formed by the ray and the *x*-axis portion of the triangle. (See Fig. 20–15.)

Quadrant II angle ($90° < \theta_2 < 180°$ or $\frac{\pi}{2}$ rad $< \theta_2 < \pi$ rad): $180° - \theta_2$ or $\pi - \theta_2$

Quadrant III angle ($180° < \theta_3 < 270°$ or π rad $< \theta_3 < \frac{3\pi}{2}$ rad): $\theta_3 - 180°$ or $\theta_3 - \pi$

Quadrant IV angle ($270° < \theta_4 < 360°$ or $\frac{3\pi}{2}$ rad $< \theta_4 < 2\pi$ rad): $360° - \theta_4$ or $2\pi - \theta_4$

Related angles are always angles less than 90° or $\frac{\pi}{2}$ rad (1.57 rad).

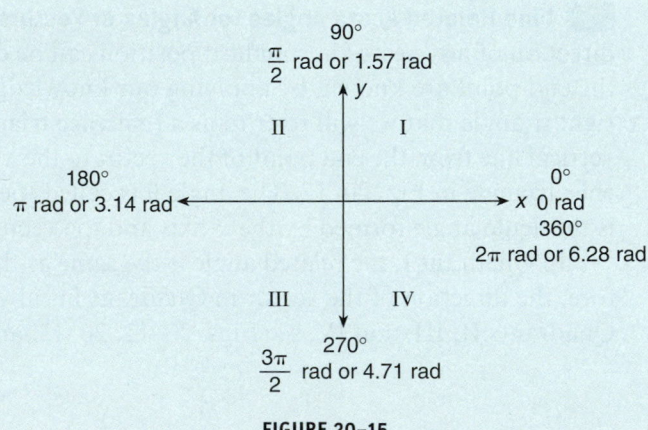

FIGURE 20–15

STOP AND CHECK

Find the related angles for the following angles.

1. 315°

2. 4.23 rad to the nearest hundredth

Answers:

1. 45°   2. 1.09 rad

EXAMPLE 1

Find the related angle for the following angles.

(a) 210°   **(b)** 1.93 rad

(a) 210° is between 180° and 270° and is a Quadrant III angle.

$\theta_3 - 180° =$

$210° - 180° = 30°$

The related angle is 30°.

(b) 1.93 rad is between $\frac{\pi}{2}$ (1.57) and π (3.14) rad and is a Quadrant II angle.

$$\pi - 1.93 = 1.211592654 \text{ rad} \qquad \text{Use calculator value for } \pi.$$

The related angle is 1.21 rad to the nearest hundredth. See Exercises 1–10.

2 **Determine the Signs of Trigonometric Functions of Angles of More Than 90°.** To determine the appropriate sign of trigonometric functions of angles more than 90°, we examine the trigonometric functions in each quadrant. The sign of the function indicates the direction or quadrant of the vector.

To understand the signs of the trigonometric functions in each quadrant, we fit together information we already know:

> r, the hypotenuse is always positive.
> Two negatives in division result in a positive quotient.
> One negative and one positive result in a negative quotient.

Use these three tools along with the fractional relationship for sine, cosine, and tangent to determine the sign of the sine, cosine, and tangent functions for angles in each of the four quadrants (Fig. 20–16).

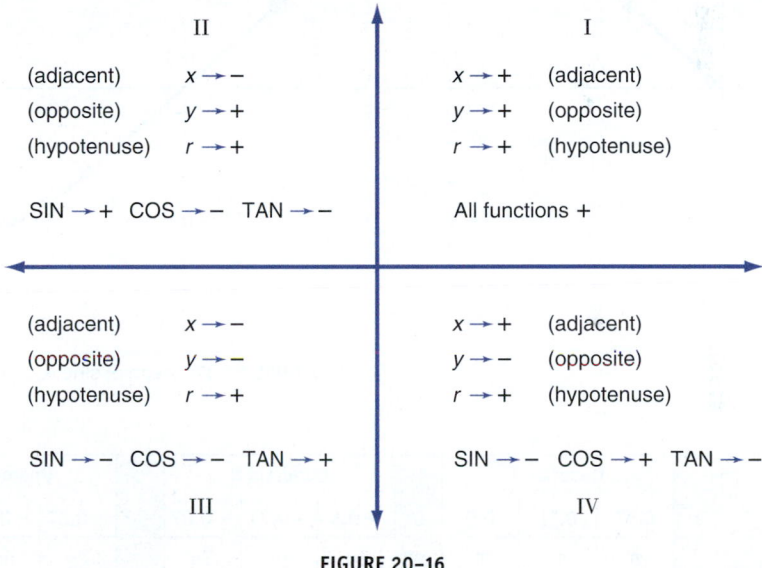

FIGURE 20–16

TIP **Signs of Trigonometric Functions** To help remember the signs of the trigonometric functions in the various quadrants, you may find the following reminder helpful:

ALL-SIN-TAN-COS

The reminder gives the positive functions in Quadrants I to IV, respectively. Also, the functions cosecant, secant, and cotangent have the same signs as their respective reciprocal functions.

TIP **Using Signs of Trigonometric Functions in Graphical Representation** Knowing the signs of the trigonometric functions in the various quadrants is helpful in checking calculations and making estimations. As you examine the graphs of the sine, cosine, and tangent functions, pay particular attention when the graph is above the x-axis (positive) or below the x-axis (negative).

The graphical representation of trigonometric functions can help you visualize the sign patterns of the various functions and draws attention to other properties of trigonometric functions. To graph these functions, we make a table of values from 0° to 360°, or from 0 rad to 2π rad. See Figs. 20–17 through 20–19.

		Quadrant I				Quadrant II				Quadrant III				Quadrant IV			
sin θ	0	0.5	0.71	0.87	1	0.87	0.71	0.5	0	−0.5	−0.71	−0.87	−1	−0.87	−0.71	−0.5	0
θ	0	$\frac{\pi}{6}$	$\frac{\pi}{4}$	$\frac{\pi}{3}$	$\frac{\pi}{2}$	$\frac{2\pi}{3}$	$\frac{3\pi}{4}$	$\frac{5\pi}{6}$	π	$\frac{7\pi}{6}$	$\frac{5\pi}{4}$	$\frac{4\pi}{3}$	$\frac{3\pi}{2}$	$\frac{5\pi}{3}$	$\frac{7\pi}{4}$	$\frac{11\pi}{6}$	2π
θ°	0	30	45	60	90	120	135	150	180	210	225	240	270	300	315	330	360

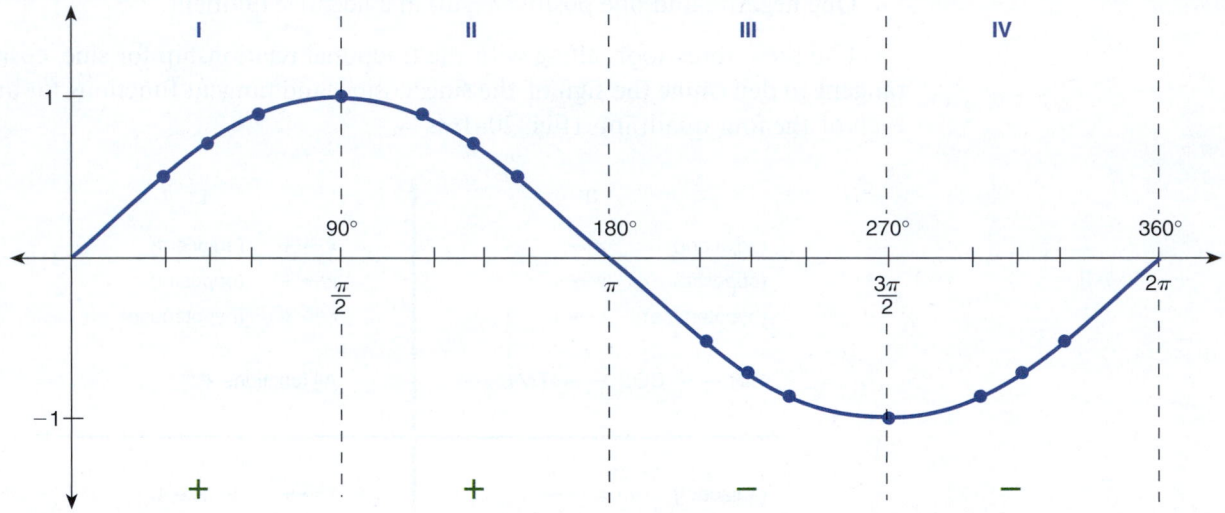

FIGURE 20–17 Graph of sine θ.

		Quadrant I				Quadrant II				Quadrant III				Quadrant IV			
cos θ	1	0.87	0.71	0.5	0	−0.5	−0.71	−0.87	−1	−0.87	−0.71	−0.5	0	0.5	0.71	0.87	1
θ	0	$\frac{\pi}{6}$	$\frac{\pi}{4}$	$\frac{\pi}{3}$	$\frac{\pi}{2}$	$\frac{2\pi}{3}$	$\frac{3\pi}{4}$	$\frac{5\pi}{6}$	π	$\frac{7\pi}{6}$	$\frac{5\pi}{4}$	$\frac{4\pi}{3}$	$\frac{3\pi}{2}$	$\frac{5\pi}{3}$	$\frac{7\pi}{4}$	$\frac{11\pi}{6}$	2π
θ°	0	30	45	60	90	120	135	150	180	210	225	240	270	300	315	330	360

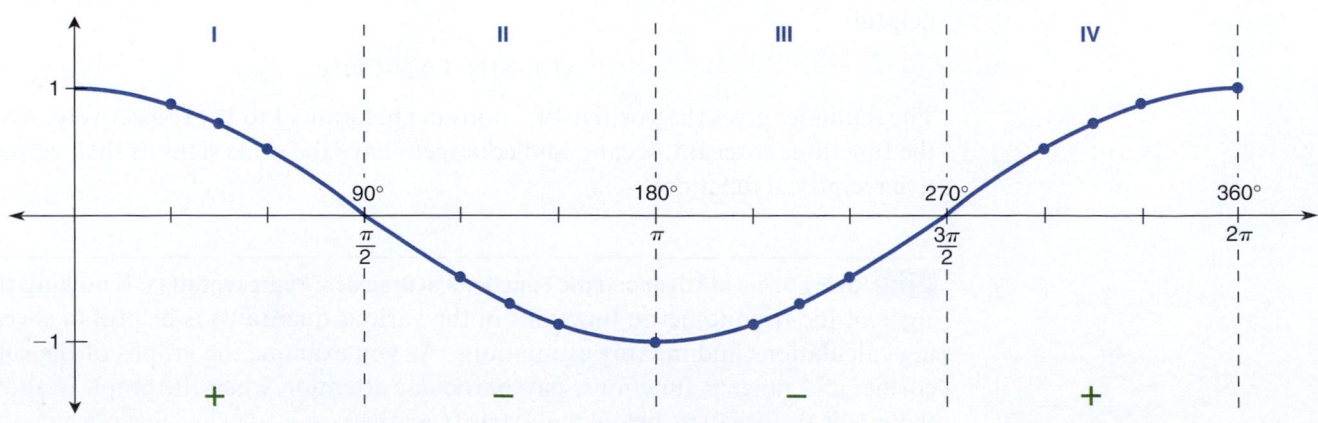

FIGURE 20–18 Graph of cosine θ.

		Quadrant I				Quadrant II				Quadrant III				Quadrant IV			
tan θ	0	0.58	1	1.7	∞	−1.7	−1	−0.58	0	0.58	1	1.7	∞	−1.7	−1	−0.58	0
θ	0	$\frac{\pi}{6}$	$\frac{\pi}{4}$	$\frac{\pi}{3}$	$\frac{\pi}{2}$	$\frac{2\pi}{3}$	$\frac{3\pi}{4}$	$\frac{5\pi}{6}$	π	$\frac{7\pi}{6}$	$\frac{5\pi}{4}$	$\frac{4\pi}{3}$	$\frac{3\pi}{2}$	$\frac{5\pi}{3}$	$\frac{7\pi}{4}$	$\frac{11\pi}{6}$	2π
θ°	0	30	45	60	90	120	135	150	180	210	225	240	270	300	315	330	360

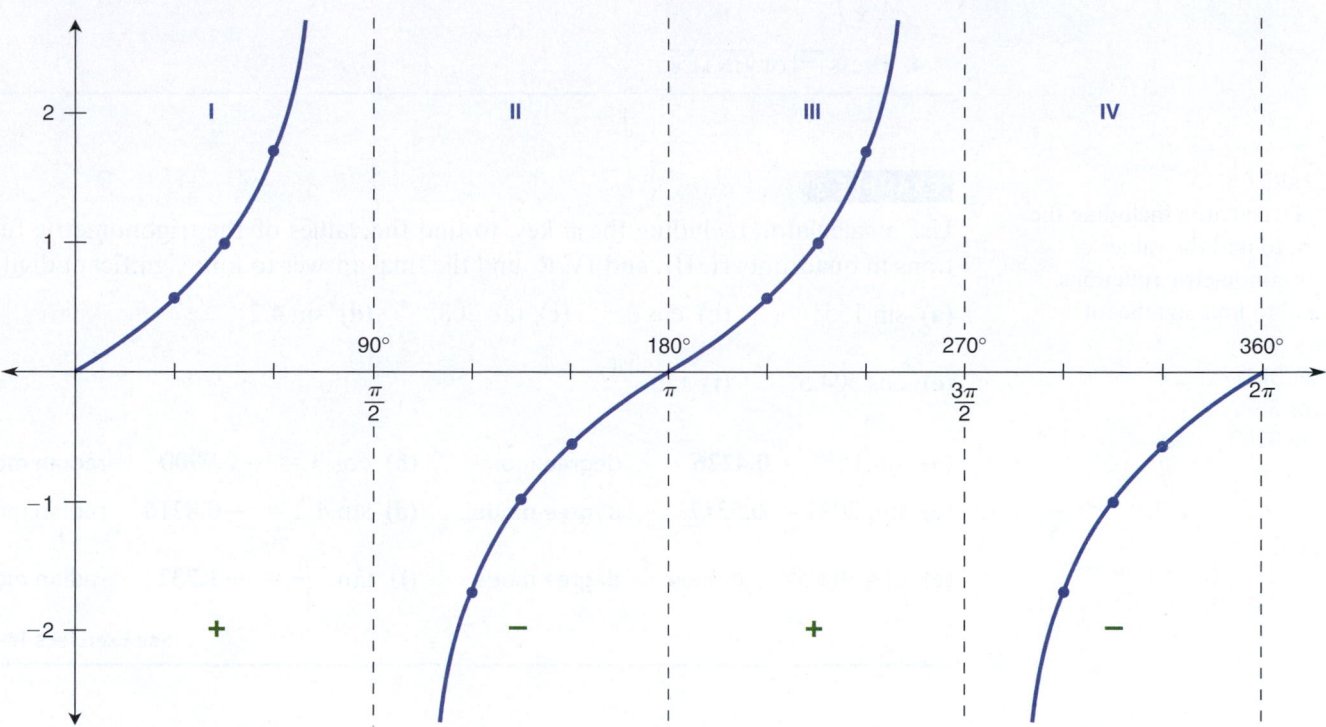

FIGURE 20–19 Graph of tangent θ.

STOP AND CHECK

Give the signs of the sine, cosine, and tangent functions of a vector in standard position with the given end point.

1. $(5, 4)$
2. $(2, -8)$
3. $(-6, 1)$

Answers:
1. positive sine, positive cosine, positive tangent
2. negative sine, positive cosine, negative tangent
3. positive sine, negative cosine, negative tangent

EXAMPLE 2

Give the signs of the sine, cosine, and tangent functions of a vector in standard position that has an end point of $(-2, -4)$.

$(-2, -4)$ is in Quadrant III.
The sine is negative in Quadrant III.
The cosine is negative in Quadrant III.
The tangent is positive in Quadrant III.

See Exercises 11–15.

3 **Find the Trigonometric Functions of Angles of More Than 90° Using a Calculator.**

TIP **General Tips for Using the Calculator** To find the value of the trigonometric function of an angle that is more than 90° using a calculator:

1. Set the calculator to degree or radian mode as desired.

2. Enter the trigonometric function.

3. Enter the angle measure in degrees or radians.

4. Press = or ENTER.

To find the related angle, given the end point of a vector:

1. Set the calculator to degree or radian mode as desired.

2. Press the inverse tangent $\boxed{\text{TAN}^{-1}}$ or $\boxed{\text{ARCTAN}}$ key.

3. Enter the coordinates of the end point as appropriate in the tangent ratio $\left(\dfrac{y}{x}\right)$.

4. Press $\boxed{=}$ or $\boxed{\text{ENTER}}$.

STOP AND CHECK

Use a calculator, including the π key, to find the values of the trigonometric functions. Round to four significant digits.

1. $\sin 212°$
2. $\cos 3.9$
3. $\tan 320°$
4. $\sin \dfrac{7\pi}{4}$

Answers:
1. -0.5299 **2.** -0.7259
3. -0.8391 **4.** -0.7071

EXAMPLE 3

Use a calculator, including the π key, to find the values of the trigonometric functions in quadrants II, III, and IV. Round the final answer to four significant digits.

(a) $\sin 155°$ **(b)** $\cos 3$ **(c)** $\tan 208°$ **(d)** $\sin 4.2$

(e) $\cos 304.5°$ **(f)** $\tan \dfrac{5\pi}{3}$

(a) $\sin 155° = \mathbf{0.4226}$ degree mode   **(b)** $\cos 3 = \mathbf{-0.9900}$ radian mode

(c) $\tan 208° = \mathbf{0.5317}$ degree mode   **(d)** $\sin 4.2 = \mathbf{-0.8716}$ radian mode

(e) $\cos 304.5° = \mathbf{0.5664}$ degree mode   **(f)** $\tan \dfrac{5\pi}{3} = \mathbf{-1.732}$ radian mode

See Exercises 16–23.

STOP AND CHECK

1. Find the direction in degrees of a vector in standard position if the coordinates of the end point are $(-5, -7)$. Round to the nearest tenth of a degree.

Answer:
1. $234.5°$

EXAMPLE 4

Find the direction in degrees of a vector in standard position if the coordinates of its end point are $(6, -8)$ (Fig. 20–20).

$$\tan \theta = \frac{y}{x}$$ Substitute values of the end points.

$$\tan \theta = \frac{-8}{6}$$ Quadrant IV

$$\tan \theta = -1.333333333$$ Use inverse tangent function.

$$\tan^{-1}(-1.333333333) = \theta$$ Evaluate.

$$\theta = -53.1°$$ To the nearest tenth of a degree

$$\text{related angle} = 53.1°$$ Fig. 20–21. Positive direction from vector to x-axis

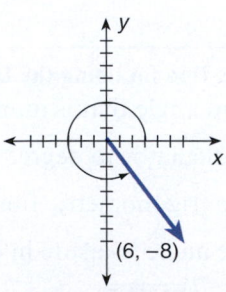

FIGURE 20–20

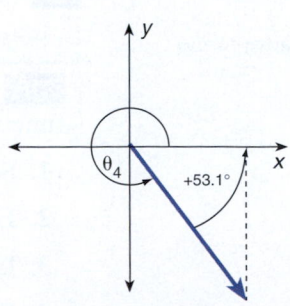

FIGURE 20–21

The angle θ is in Quadrant IV. Find θ_4 as an equivalent Quadrant IV angle with a positive direction.

$$360° - \theta_4 = 53.1° \qquad \text{Solve for } \theta_4, \text{ direction of vector.}$$
$$360° - 53.1° = \theta_4$$
$$306.9° = \theta_4$$

Thus, the direction of the vector is 306.9°. See Exercises 24–25.

20-2 EXERCISES   MyLab Math For additional practice go to your study plan in MyLab Math.

1 Find the related angle for the angles. Use the calculator value for π for radian measures and round to hundredths. *See Example 1.*

1. 120°	**2.** 195°	**3.** 290°	**4.** 345°	**5.** 148°
6. 250°	**7.** 212°	**8.** 118°	**9.** 2.18 rad	**10.** 5.84 rad

2 Give the signs of the sine, cosine, and tangent functions of vectors with the indicated end points. *See Example 2.*

11. (3, 5)	**12.** (−2, 6)	**13.** (−4, −2)	**14.** (5, −3)	**15.** (5, 0)

3 Using a calculator and the π key, evaluate the trigonometric functions. Round to four significant digits. *See Example 3.*

16. sin 210°	**17.** tan 140°	**18.** cos 2.5	**19.** cos 4	**20.** sin 300°
21. tan 6	**22.** cos 100°	**23.** $\sin \dfrac{5\pi}{6}$		

See Example 4.

24. Find the direction in degrees of a vector in standard position if the coordinates of its end point are (−3, 2). Round to the nearest 0.1°.

25. Find the direction in radians of a vector in standard position if the coordinates of its end point are (−2, −1). Round to the nearest hundredth.

20-3 | Period and Phase Shift

LEARNING OUTCOMES

1 Graph a sine or cosine function using a calculator.

2 Find the period of a sine or cosine function.

3 Find the phase shift of a sine or cosine function.

LC LEARNING CATALYTICS

Evaluate.

1. sin 0°   **2.** cos 0°
3. sin 90°   **4.** cos 90°

Period of a function: one complete cycle of the graph of a trigonometric function. A cycle is completed when it begins to repeat

1 Graph a Sine or Cosine Function Using a Calculator. The graphs of sine and cosine functions continue to repeat in both directions as the graphs are extended to values less than 0° and more than 360°. One complete cycle of the graph of a trigonometric function is called the **period of the function.** The period of the sine and cosine functions is 360° or 2π rad.

To graph two or more periods or repetitions of a sine or cosine function using a calculator:

1. Set MODE to degrees or radians as desired.

2. From the ZOOM function select the option ZTrig or manually set the window as desired.

3. Enter the function using the Y = function.

4. View the graph using GRAPH function.

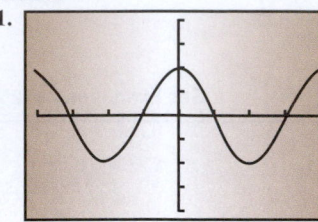
EXAMPLE 1

Use a calculator to graph the functions $y = \sin x$ and $y = \cos x$. See Figs. 20–22 and 20–23. Use the ZTrig function or manually set the window from $-360°$ to $360°$ or -2π to 2π.

|Y=| |CLEAR| |sin| |x,T,θ,n| |ZOOM| 7 |Y=| |CLEAR| |cos| |x,T,θ,n| |ZOOM| 7

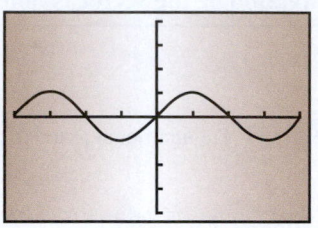   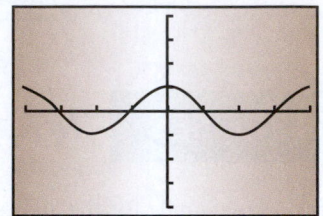

FIGURE 20–22 $y = \sin x$ **FIGURE 20–23** $y = \cos x$

See Exercises 1–19.

Waves: the repetitions of a graph

Wavelength: the *distance* along the *x*-axis of one repetition of a wave. Represented by the Greek letter lambda

Lambda, λ: the Greek letter used to represent wavelength

Frequency of a wavelength: the number of waves that pass through a given point in one second. The reciprocal of the *period* of a function

We refer to the repetitions of the graph as **waves**. The **wavelength** is the *distance* along the *x*-axis of one repetition of a wave. Wavelength is represented by the Greek letter **lambda**, λ. If the *x*-axis is a *time* variable, the time it takes for one complete wave to pass through a given point is the *period, T*. The number of waves that pass through a given point in one second is called **frequency of a wavelength,** *f*. Frequency and period are reciprocals; that is,

$$f = \frac{1}{T} \quad \text{or} \quad T = \frac{1}{f}$$

Frequency is measured in hertz (Hz). That is, 1 Hz = 1 wave or cycle per second. Common measuring units for frequency are kilohertz (1 kHz = 10^3 Hz), megahertz (1 MHz = 10^6 Hz), and gigahertz (1 GHz = 10^9 Hz).

EXAMPLE 2

Find the frequency of a wave that passes through a given point once each 0.00008 s.

$$f = \frac{1}{T}$$

$$f = \frac{1}{0.00008} \qquad \text{The period is 0.00008 s per cycle.}$$

$$f = 12{,}500 \text{ Hz or } 12.5 \text{ kHz}$$

12,500 waves pass through the point each second.   See Exercises 20–23.

Wave velocity: the rate at which a position on a wave changes or moves

Wave velocity, *v*, is the rate at which a position on the wave changes or moves. Wave velocity is related to frequency or period and wavelength by the formulas

$$v = \lambda f \quad \text{and} \quad v = \frac{\lambda}{T}$$

where *v* is the wave velocity, λ is the wavelength, *f* is the frequency, and *T* is the period.

SingjaiStock/Shutterstock

EXAMPLE 3

ELEC A National Public Radio (NPR) station has an FM band (frequency) of 91.1 MHz. The speed of a radio wave is the same as the speed of light, which is 300,000 km/s, or 300,000,000 m/s. Find the station's wavelength in meters.

$$v = \lambda f$$ Solve for **λ**. Divide both sides by *f*.

$$\lambda = \frac{v}{f}$$ Substitute 300,000,000 m/s for *v* and 91.1 MHz for *f*.

$$\lambda = \frac{300,000,000 \text{ m/s}}{91.1 \text{ MHz}}$$ 1 MHz = 10^6 Hz, or 10^6 waves per second

$$\lambda = \frac{300,000,000 \text{ m/s}}{91,100,000 \text{ waves/s}}$$

$$\lambda = 3.293084523 \text{ m/wave}$$ Round to hundredths.

The length of each wave is 3.29 m. See Exercises 24–27.

One of the most common applications of waves is in alternating current. For example, in a generator, the current *i* changes according to the equation

$$i = I \sin x$$

where *I* is the maximum current and *x* is the angle through which the coil rotates in a magnetic field to generate an electric current. The voltage *v* changes according to the equation

$$v = V \sin x$$

where the maximum voltage is *V* and the angle through which the coil rotates is *x*. Both these equations are examples of an equation in the form

$$y = A \sin x \qquad \text{for } A > 0$$

Amplitude: the maximum value that a graph will reach on the *y*-axis. In a function in the form $y = A \sin x$ or $y = A \cos x$, the amplitude is *A*

In this equation, *A* represents the **amplitude**, or maximum value that the graph will reach on the *y*-axis. The minimum value is represented by $-A$.

To find the amplitude of a graph using a calculator:

1. Set the window of the graph to be what you expect to be greater than the maximum value of the graph.

2. Graph the function.

3. Use the CALC function and select *maximum* (option 4).

4. Set the left and right bounds and guess and press ENTER.

5. The amplitude will be the *y*-value.

EXAMPLE 4

ELEC The maximum voltage *V* in a simple generator is 30 V. As the coil rotates, the voltage changes according to the equation $v = 30 \sin x$. Use your calculator to graph the voltage changes for one complete revolution. Set the minimum Y-value to be less than the amplitude and the maximum Y-value to be greater than the amplitude to show the graph in the calculator window. Determine the voltage at 210°.

Set MODE to degree measure for angles.

Set Window: Xmin = 0   Settings for one period
            Xmax = 360
            Xscl = 30
            Ymin = −40   Set Ymin to be less than amplitude of −30.
            Ymax = 40   Set Ymax to be greater than amplitude of 30.
            Yscl = 1
            Xres = 1

Enter function: Y = 30 sin (X)   Use Y = function.
View graph.   Use GRAPH function.
Find voltage at 210° (Fig. 20–24).   Use CALC function and VALUE option.

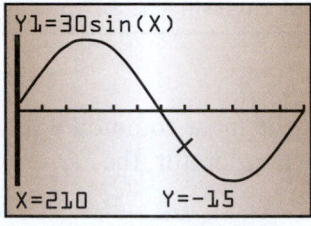

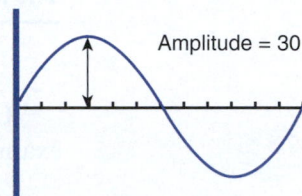

Amplitude = 30

FIGURE 20–24   **FIGURE 20–25**

The voltage at 210° is − 15. Fig. 20–25 shows the amplitude.   **See Exercises 28–35.**

---

2 **Find the Period of a Sine or Cosine Function.** In the equations $y = \sin x$ and $y = \cos x$, the coefficient of the variable x is 1, which indicates that the period of the function is 360°, or 2π rad. A change in the coefficient of x will result in a change of the period. In the equations $y = A \sin Bx$ and $y = A \cos Bx$, the period P is found by the formula $P = \dfrac{360°}{B}$.

> **To find the period of a function in the form $y = A \sin Bx$ or $y = A \cos Bx$:**
>
> 1. Identify B (the coefficient of x).
>
> 2. Divide 360° or 2π radians by the coefficient of x.
>
> $$P = \frac{360°}{B} \text{ for degrees} \quad \text{or} \quad P = \frac{2\pi}{B} \text{ for radians}$$

EXAMPLE 5

Find the amplitude and period in degrees of the function $y = 3 \cos 2x$. Graph the equation to show two periods.

Amplitude = 3 $A = 3$; coefficient of cos 2x

Period $= \dfrac{360°}{2} = 180°$ $B = 2$; coefficient of x

The amplitude is 3 and the period is 180°.

Set calculator to degree mode.

Set Window: $X_{min} = 0$ Set X minimum and maximum to show two periods;

$X_{max} = 360$ Multiply degrees in period by 2. $180° \times 2 = 360°$.

$X_{scl} = 30$ Set X_{scl} (tick marks) to 30° each.

$Y_{min} = -4$ Set Y_{min} to be < -3.

$Y_{max} = 4$ Set Y_{max} to be > 3.

$Y_{scl} = 1$ Set Y_{scl} (tick marks) each 1 unit.

Enter equation: $y = 3 \cos 2x$
Graph (Fig. 20–26).

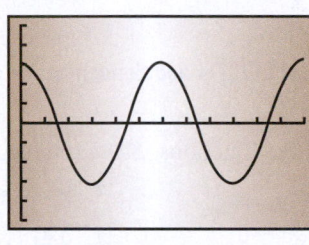

FIGURE 20–26

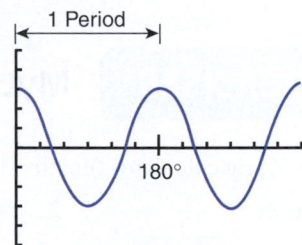

FIGURE 20–27

Fig. 20–27 shows the period.   **See Exercises 36–41.**

3   **Find the Phase Shift of a Sine or Cosine Function.** All graphs in the form $y = A \sin Bx$ pass through the origin, and all graphs in the form $y = A \cos Bx$ pass through the point $(0, A)$. If the respective graphs do not pass through the appropriate point, the function is *out of phase*. The **phase shift** is the distance between two corresponding points on the appropriate reference graph of $y = A \sin Bx$ or $y = A \cos Bx$ and the out-of-phase function. The equations for out-of-phase functions are

$$y = A \sin (Bx + C) \quad \text{or} \quad y = A \cos (Bx + C)$$

Phase shift: the distance between two corresponding points on an appropriate reference graph of $y = A \sin Bx$ or $y = A \cos Bx$ and the out-of-phase function. Equations for out-of-phase functions are $y = A \sin (Bx + C)$ or $y = A \cos (Bx + C)$. The phase shift is the value C/B

where $C \neq 0$.

> **To find the phase shift for the function $y = A \sin (Bx + C)$ or $y = A \cos (Bx + C)$:**
>
> 1. Find the value $\dfrac{C}{B}$.
>
> 2. If $\dfrac{C}{B}$ is positive, shift the graph $\dfrac{C}{B}$ units to the *left*.
>
> 3. If $\dfrac{C}{B}$ is negative, shift the graph $\dfrac{C}{B}$ units to the *right*.

EXAMPLE 6

Graph $y = 5 \sin 2x$ and $y = 5 \sin (2x + 90°)$ in the same rectangular coordinate system.

Amplitude $= 5$ Coefficient of $\sin (2x + 90°)$

Period $= \dfrac{360°}{2} = 180°$ $\dfrac{360°}{B}$; B is the coefficient of x

Phase shift $= \dfrac{90°}{2} = 45°$ $\dfrac{C}{B}$ is positive.

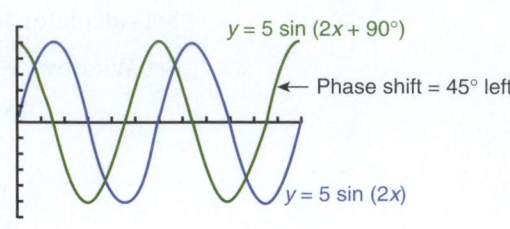

FIGURE 20–28   **FIGURE 20–29**

Graph of $y = 5 \sin(2x + 90°)$ shifts 45° to the left from the graph $y = 5 \sin 2x$
(Figs. 20–28 and 20–29).   **See Exercises 42–47.**

20–3 EXERCISES   MyLab Math For additional practice go to your study plan in MyLab Math.

1 Use the ZTrig calculator function to graph the functions. *See Example 1.*

1. $y = \sin 2x$ **2.** $y = \sin 3x$ **3.** $y = \sin 4x$ **4.** $y = \sin 8x$

5. $y = \cos 2x$ **6.** $y = \cos 4x$ **7.** $y = \cos 5x$ **8.** $y = \cos 6x$

9. Compare the graphs for Exercises 1–4.   **10.** Compare the graphs for Exercises 5–8.

Graph using a graphing calculator with *x*-values between −90° and 360° in multiples of 30° (*x*-scale) and *y*-values between
−8 and 8 and a *y*-scale of 1 unit.

11. $y = 2 \sin x$ **12.** $y = 4 \sin x$ **13.** $y = 6 \sin x$

14. $y = 3 \cos x$ **15.** $y = 4 \cos x$ **16.** $y = 5 \cos x$

17. Compare the graphs for Exercises 11–13.   **18.** Compare the graphs for Exercises 14–16.

19. Use the CALC function and VALUE option to find the
value of *y* in the function $y = 10 \sin x$ when $x = 100°$.

See Example 2.

20. **ELEC** Find the frequency of a wave that passes through   **21.** **ELEC** Find the frequency of a wave that has a period
a given point once each 0.045 s. Round to thousandths.   of 0.006 s.

22. **ELEC** What is the period of a wave if 25×10^4 waves   **23.** **ELEC** What is the period of a wave that has a fre-
pass through a given point each second?   quency of 99.8 MHz? Round to thousandths.

See Example 3.

24. **ELEC** WHBQ AM broadcasts at 1,390 KHz. What is   **25.** **ELEC** Cellular phone waves travel at the same speed
the wavelength of its radio waves? Radio waves travel   of light, 300,000 km/s. What is the frequency of a cel-
at the speed of light, 300,000 km/s.   lular phone wave if the phone operates at a wavelength
                                                                   of 300 mm?

26. **ELEC** What is the velocity of a wave that has a period   **27.** **ELEC** What is the velocity of a wave that has a wave-
of 0.00004 s and a wavelength of 0.82 m?   length of 1.2 m and a frequency of 88.9 MHz?

Use a graphing calculator to show the changing voltage *v* for a simple generator that has the given maximum voltage *V*;
$v = V \sin x$. *See Example 4.*

28. **ELEC** $V = 42$ V   **29.** **ELEC** $V = 32$ V

30. **ELEC** Use the CALC function and VALUE option   **31.** **ELEC** Use the CALC function and VALUE option
to estimate to the nearest tenth the changing voltage *v*   to estimate to the nearest tenth the changing voltage *v*
when $x = 45°$ and $x = 60°$ if the maximum voltage of   when $x = 120°$ and $x = 300°$ if the maximum voltage
the generator is $V = 42$ V.   of the generator is $V = 32$ V.

Use the equation $i = I \sin x$ to graph the changing current in a simple generator that has the given maximum current.

32. **ELEC** $I = 3.5\,\text{A}$

33. **ELEC** $I = 8.4\,\text{A}$

34. **ELEC** Use the CALC function and VALUE option to estimate to the nearest tenth the changing current i when $x = 225°$ and $x = 315°$ if the maximum voltage of the generator is $I = 3.5\,\text{A}$.

35. **ELEC** Use the CALC function and VALUE option to estimate to the nearest tenth the changing current i when $x = 115°$ and $x = 270°$ if the maximum voltage of the generator is $I = 8.4\,\text{A}$.

2 Find the amplitude and period in degrees for the functions. Graph each function to show two periods. *See Example 5.*

36. $y = 4 \sin 6x$

37. $y = 30 \sin 10x$

38. $y = 3 \cos 5x$

39. $y = 4 \cos 8x$

40. $y = 2.5 \cos \dfrac{3}{4} x$

41. $y = 1.9 \sin \dfrac{8}{5} x$

3 Find the amplitude, period, and phase shift for the functions. Graph each function to show two periods. *See Example 6.*

42. $y = \sin (x + 45°)$

43. $y = \cos (x - 180°)$

44. $y = 2 \sin (x + 120°)$

45. $y = 6 \sin (3x - 240°)$

46. $y = 10 \cos (4x + 270°)$

47. $y = 8 \cos (6x - 45°)$

20–4 Law of Sines

LEARNING OUTCOMES

1 Find the unknown parts of an oblique triangle, given two angles and a side.

2 Find the unknown parts of an oblique triangle, given two sides and an angle opposite one of them.

LC **LEARNING CATALYSTS**

1. Evaluate $\dfrac{9 \sin 50°}{\sin 80°}$ to four significant digits.

Oblique triangle: any triangle that does not contain a right angle

Law of sines: the ratios of the sides of a triangle to the sines of the angles opposite these respective sides are equal

An **oblique triangle** is a triangle that does not contain a right angle. Because these triangles do not have right angles, we cannot use the Pythagorean theorem or trigonometric functions directly as we did previously to find sides and angles of right triangles. However, two formulas based on the trigonometric functions of right triangles can be used to solve oblique triangles. In this section, we examine one of these formulas, the *law of sines*.

1 **Find the Unknown Parts of an Oblique Triangle, Given Two Angles and a Side.**

Law of sines:

The **law of sines** states that the ratios of the sides of a triangle to the sines of the angles opposite these respective sides are equal (see Fig. 20–30).

$$\frac{a}{\sin A} = \frac{b}{\sin B} = \frac{c}{\sin C}$$

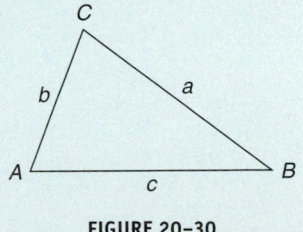

FIGURE 20–30

TIP **Conditions for Using the Law of Sines** The law of sines is used to solve triangles when either of these conditions exists:

1. Two angles and a side are known.

2. Two sides and the angle opposite one of the sides are known.

When using the law of sines you must have at least one complete ratio—an angle and side opposite. However, if we know two angles of a triangle, we can always find the third angle. Then, we can form a complete ratio.

> **To find the unknown parts of an oblique triangle, given two angles and a side:**
>
> 1. Find the third angle of the triangle by adding the two known angles and subtracting the sum from 180° or 2π radians.
>
> 2. Set up a ratio with the given side and sine of the opposite angle.
>
> 3. Set up a second ratio with an unknown side and the sine of the opposite angle.
>
> 4. Form a proportion using the ratios in Steps 2 and 3 and solve the proportion for the unknown side.
>
> 5. Repeat Steps 2–4 if the value of the third side is needed.

In the examples that follow, all digits of the calculated value that show in the calculator display will be given. Rounding should be done after the last calculation has been made.

EXAMPLE 1

Solve triangle ABC if $A = 50°$, $B = 75°$, and $b = 12$ ft.

Sketch the triangle and label the parts, as shown in Fig. 20–31.

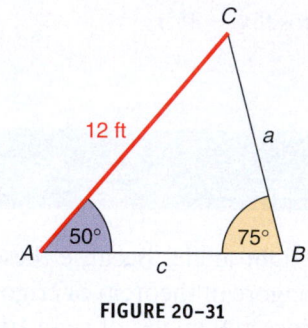

FIGURE 20–31

Find $\angle C$.

$$A + B + C = 180° \qquad \text{Substitute } A = 50° \text{ and } B = 75°.$$
$$50° + 75° + C = 180° \qquad \text{Solve for } C.\ 50° + 75° = 125°$$
$$C = 180° - 125°$$
$$\boldsymbol{C = 55°}$$

Find a.

$$\frac{a}{\sin A} = \frac{b}{\sin B} \qquad \text{Side } b \text{ and } \angle B \text{ form the ratio with both values known. Substitute } b = 12, A = 50°, B = 75°.$$

$$\frac{a}{\sin 50°} = \frac{12}{\sin 75°} \qquad \text{Cross multiply.}$$

$$a \sin 75° = 12 \sin 50° \qquad \text{Solve for } a.$$

$$a = \frac{12 \sin 50°}{\sin 75°} \qquad \text{Perform calculations.}$$

$$a = 9.516810781$$

$$\boldsymbol{a = 9.517 \text{ ft}} \qquad \text{Four significant digits}$$

Find c.

$$\frac{b}{\sin B} = \frac{c}{\sin C} \qquad \text{Substitute } b = 12, B = 75°, \text{ and } C = 55°.$$

$$\frac{12}{\sin 75°} = \frac{c}{\sin 55°} \qquad \text{Cross multiply.}$$

$$12 \sin 55° = c \sin 75° \qquad \text{Solve for } c.$$

$$\frac{12 \sin 55°}{\sin 75°} = c \qquad \text{Perform calculations.}$$

$$c = 10.1765832$$

$$\boldsymbol{c = 10.18 \text{ ft}} \qquad \text{Four significant digits}$$

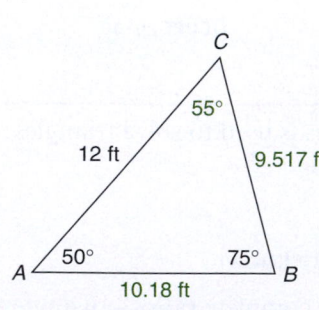

FIGURE 20–32

As a quick check for our work, the longest side should be opposite the largest angle, and the shortest side should be opposite the smallest angle (see Fig. 20–32).

See Exercises 1–13.

STOP AND CHECK

1. Solve triangle ABC if
$A = 42°$, $B = 63°$, and
$b = 11$ in. Round sides to
four significant digits.

Answer:

1.
$C = 75°$; $a = 8.261$ in.; $c = 11.92$ in.

EXAMPLE 2

Solve triangle ABC if $A = 35°$, $a = 7$ cm, and $B = 40°$.

Sketch the triangle and label its parts (see Fig. 20–33).

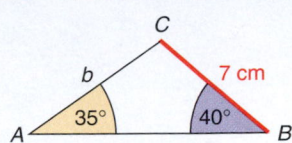

FIGURE 20-33

Find $\angle C$.

$A + B + C = 180°$	Substitute $A = 35°$ and $B = 40°$.
$35° + 40° + C = 180°$	Solve for C.
$C = 180° - 35° - 40°$	
$\mathbf{C = 105°}$	

Find b.

$\dfrac{a}{\sin A} = \dfrac{b}{\sin B}$	Side a and $\angle A$ form the ratio with both values known. Substitute $a = 7$, $A = 35°$, and $B = 40°$.
$\dfrac{7}{\sin 35°} = \dfrac{b}{\sin 40°}$	Cross multiply.
$7 \sin 40° = b \sin 35°$	Solve for b.
$\dfrac{7 \sin 40°}{\sin 35°} = b$	Evaluate.
$b = 7.844661989$	
$\mathbf{b = 7.845 \ cm}$	Four significant digits

Find c.

$\dfrac{a}{\sin A} = \dfrac{c}{\sin C}$	Substitute $a = 7$, $A = 35°$, and $C = 105°$.
$\dfrac{7}{\sin 35°} = \dfrac{c}{\sin 105°}$	Cross multiply.
$7 \sin 105° = c \sin 35°$	Solve for c.
$\dfrac{7 \sin 105°}{\sin 35°} = c$	Evaluate.
$c = 11.78828201$	
$\mathbf{c = 11.79 \ cm}$	Four significant digits

As a quick check for our work, the longest side should be opposite the largest angle, and the shortest side should be opposite the smallest angle (see Fig. 20–34).

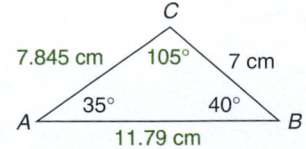

FIGURE 20-34

See Exercises 1–13.

2 **Find the Unknown Parts of an Oblique Triangle, Given Two Sides and an Angle Opposite One of Them.** When two sides of a triangle and an angle opposite one of the sides are known, we do not always have a single triangle. If the given sides are a and b and the given angle is B, three possibilities may exist (see Fig. 20–35).

▶ If $b < a$ and $\angle A \neq 90°$, we have two possible solutions (case 1).

▶ If $b < a$ and $\angle A = 90°$, we have one solution (case 2).

▶ If $b \geq a$ and $\angle A \neq 90°$, we have one solution (case 3).

Ambiguous case: when two sides and an angle opposite one of them are given, there could be two different solutions of the triangle; that is, the given information can be interpreted in more than one way

Note that side b above could be any side of the triangle that is opposite the given angle. Side a is then the other given side. Because of the lack of clarity when two sides and an angle opposite one of them are given, the situation is called the **ambiguous case.** (*Ambiguous* means that the given information can be interpreted in more than one way.)

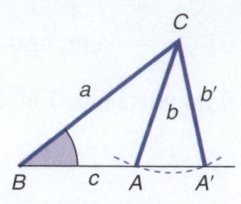
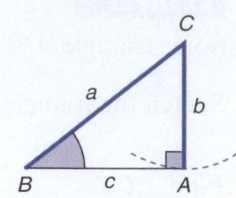
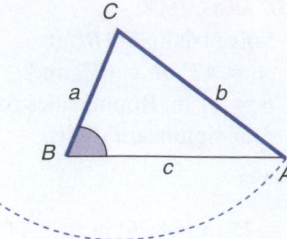

Case 1 ($b < a$)

Two solutions for A, C, and c
since b can meet side c in
either of two points, A or A'.

Case 2 ($A = 90°$)

One solution since b meets
side c in exactly one point.

Case 3 ($b \geq a$)

One solution since b meets
side c in only one point.

FIGURE 20–35

To find the unknown parts of an oblique triangle, given two sides and an angle opposite one of them:

1. Use the law of sines to find the measure of the angle opposite the other given side.

2. Determine if there are two angle values that are possible for the solution.

3. Find the third angle by subtracting the sum of the two known angles from 180°.

4. Find the third side using the law of sines.

5. If two angles were found in Step 2, find the second possible solution by repeating Steps 3 and 4.

EXAMPLE 3

Solve triangle ABC if $A = 25°$, $a = 8$, and $b = 11.426$ (see Fig. 20–36).

Find $\angle B$.

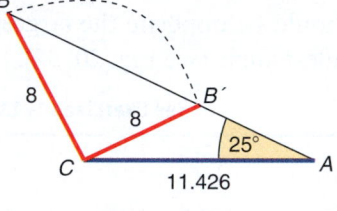

FIGURE 20–36

$$\frac{a}{\sin A} = \frac{b}{\sin B}$$ Substitute $A = 25°$, $a = 8$, and $b = 11.426$, $a < b$. There will be two solutions.

$$\frac{8}{\sin 25°} = \frac{11.426}{\sin B}$$ Cross multiply.

$$8 \sin B = 11.426 \sin 25°$$ Solve for B.

$$\sin B = \frac{11.426 \sin 25°}{8}$$ Evaluate.

$$\sin B = 0.6036045323$$

$$B = \sin^{-1}(0.6036045323)$$ Use inverse sine function.

$$B = 37.1284918°$$ or $$B' = 180° - 37.1284918°$$

$$B' = 142.8715082°$$

$$\mathbf{B' = 142.9°}$$

$$\mathbf{B = 37.1°}$$ Round to the nearest tenth of a degree.

Possible solution 1:

Find $\angle C$.

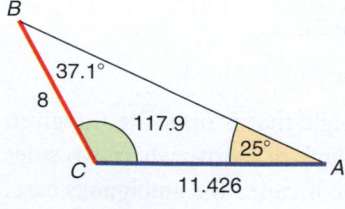

FIGURE 20–37

$$A + B + C = 180°$$ Substitute $A = 25°$ and $B = 37.1284918°$.

$$25° + 37.1284918° + C = 180°$$ Solve for C.

$$C = 180° - (25° + 37.1284918°)$$

$$\mathbf{C = 117.8715082°}$$

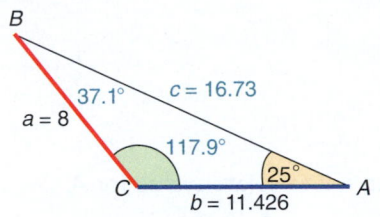

FIGURE 20–38

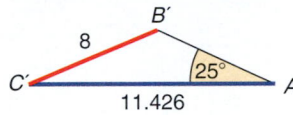

FIGURE 20–39a

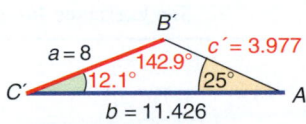

FIGURE 20–39b

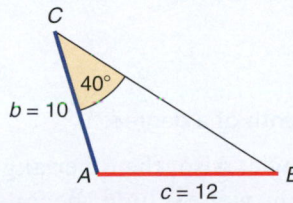

FIGURE 20–40

To find the third side of this right triangle, we can use the Pythagorean theorem or the law of sines.

Find c.

$$\frac{a}{\sin A} = \frac{c}{\sin C}$$

Substitute $a = 8$, $A = 25°$, and $C = 117.8715082°$.

$$\frac{8}{\sin 25°} = \frac{c}{\sin 117.8715082°}$$

Cross multiply.

$$8 \sin 117.8715082° = c \sin 25°$$

Solve for c.

$$\frac{8 \sin 117.8715082°}{\sin 25°} = c$$

Evaluate.

$$c = 16.73374373$$

$$c = \mathbf{16.73}$$

Four significant digits

The solved triangle for this possibility is shown in Fig. 20–38.

Possible solution 2:

Find $\angle C'$.

$$C' = 180° - (25° + 142.8715082°) = 12.1284918°$$

$$C' = \mathbf{12.1°}$$

See Fig. 20–39a.

Find c'.

$$\frac{a}{\sin A} = \frac{c'}{\sin C'}$$

Substitute $a = 8$, $A = 25°$, and $C' = 12.1284918°$.

$$\frac{8}{\sin 25°} = \frac{c'}{\sin 12.1284918°}$$

Cross multiply.

$$c' \sin 25° = 8 \sin 12.1284918°$$

Solve for C'.

$$c' = \frac{8 \sin 12.1284918°}{\sin 25°}$$

$$c' = 3.977201817$$

$$c' = \mathbf{3.977}$$

Rounded

The solved triangle for this possibility is shown in Fig. 20–39.   **See Exercises 14–21.**

EXAMPLE 4

Solve triangle ABC if $b = 10$, $c = 12$, and $C = 40°$ (Fig. 20–40).

Find B.

Because c, the side opposite the given angle C, is *longer* than the other given side, b, we expect *one* solution.

$$\frac{c}{\sin C} = \frac{b}{\sin B}$$

Substitute $c = 12$, $C = 40°$, and $b = 10$.

$$\frac{12}{\sin 40°} = \frac{10}{\sin B}$$

Cross multiply.

$$12 \sin B = 10 \sin 40°$$

Solve for B.

$$\sin B = \frac{10 \sin 40°}{12}$$

Evaluate.

$$\sin B = 0.5356563414 \qquad \text{Use inverse sine function.}$$
$$B = \sin^{-1}(0.5356563414)$$
$$B = 32.38843382°$$
$$\boldsymbol{B = 32.4°} \qquad \text{Round to the nearest tenth of a degree.}$$

Both 32.38843382° and 147.6115662° have a sine of approximately 0.5356563414; however, we cannot have an angle of 147.6115662° in this triangle because the triangle already has a 40° angle. 147.6115662° + 40° = 187.6115662°. The *three* angles of a triangle total only 180°. Therefore, we have only *one* solution.

Find *A*.

$$A = 180° - 40° - 32.38843382° \qquad \text{Use full calculator value for } B.$$
$$A = 107.6115662°$$
$$\boldsymbol{A = 107.6°}$$

Find *a*.

$$\frac{a}{\sin A} = \frac{c}{\sin C} \qquad \text{Substitute full calculator value for } A.$$
$$\frac{a}{\sin 107.6115662°} = \frac{12}{\sin 40°} \qquad \text{Cross multiply.}$$
$$a \sin 40° = 12 \sin 107.6115662° \qquad \text{Solve for } a.$$
$$a = \frac{12 \sin 107.6115662°}{\sin 40°} \qquad \text{Evaluate.}$$
$$a = 17.79367732$$
$$\boldsymbol{a = 17.79} \qquad \text{Four significant digits}$$

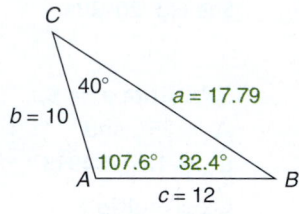

FIGURE 20–41

The solved triangle is shown in Fig. 20–41.   **See Exercises 14–21.**

EXAMPLE 5

CON   A technician checking a surveyor's report is given the information shown in Fig. 20–42. Calculate the missing information.

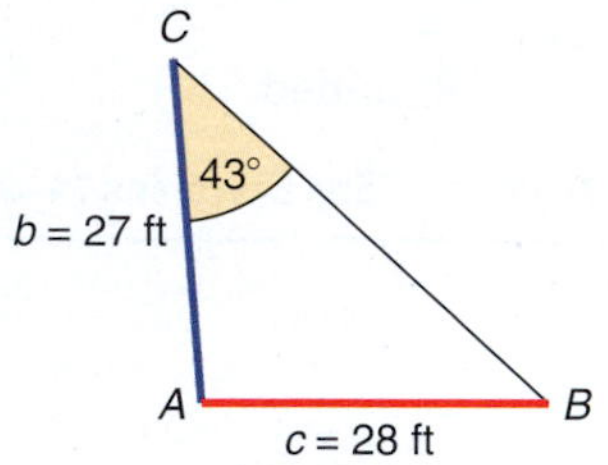

FIGURE 20–42

Find *B*.

$$\frac{c}{\sin C} = \frac{b}{\sin B} \qquad \text{Substitute given values.}$$
$$\frac{28}{\sin 43°} = \frac{27}{\sin B} \qquad \text{Cross multiply.}$$
$$28 \sin B = 27 \sin 43° \qquad \text{Solve for } \sin B.$$
$$\sin B = \frac{27 \sin 43°}{28} \qquad \text{Evaluate.}$$
$$\sin B = 0.6576412758 \qquad \text{Use inverse sine function.}$$
$$B = \sin^{-1}(0.6576412758)$$
$$B = 41.12023021°$$
$$\boldsymbol{B = 41.1°} \qquad \text{Round to the nearest tenth of a degree.}$$

There is only one case here because we are given the triangle. Also, the other angle whose sine is 0.6576412758 is 138.8797698°, which we exclude because 138.8797698° + 43° = 181.8797698°.

Dinis Tolipov/123RF

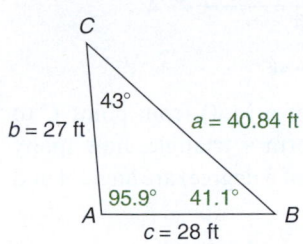

FIGURE 20–43

Find angle A.

$A = 180° - (43° + 41.12023021°)$

$A = 95.87976979°$

$A = 95.9°$ Round to the nearest tenth of a degree.

Find a.

$$\frac{a}{\sin A} = \frac{c}{\sin C}$$ Substitute.

$$\frac{a}{\sin 95.87976979°} = \frac{28}{\sin 43°}$$ Cross multiply.

$a \sin 43° = 28 \sin 95.87976979°$ Solve for a.

$$a = \frac{28 \sin 95.87976979°}{\sin 43°}$$ Evaluate.

$a = 40.83982458$

$a = 40.84$ ft   Four significant digits

The completed survey should have the measures shown in Fig. 20–43.

See Exercise 22.

20–4 EXERCISES   MyLab Math   For additional practice go to your study plan in MyLab Math.

1 In Exercises 1–6, solve each of the oblique triangles using the law of sines. Round the final answer for sides to four significant digits. *See Examples 1 and 2.*

1. $a = 46$ m, $A = 65°$, $B = 52°$

2. $b = 7.2$ mi, $B = 58°$, $C = 72°$

3. $a = 65$ cm, $B = 60°$, $A = 87°$

4. $c = 3.2$ ft, $A = 120°$, $C = 30°$

5. $b = 12$ km, $A = 95°$, $B = 35°$

6. $b = 142$ dkm, $A = 135°$, $B = 25°$

7. Find b if $a = 14$ in., $B = 72°$, and $C = 608$.

8. Find a if $c = 148$ m, $A = 100°$, and $B = 30°$.

9. Find C if $c = 2.8$ m, $A = 42°$, and $B = 62°$.

10. Find c if $a = 5\frac{3}{8}$ ft, $B = 61.7°$, and $C = 72.5°$.

11. Find b if $a = 7{,}896$ ft, $B = 42.8°$, and $C = 51.9°$.

12. Find a if $c = 14.32$ km, $A = 38.2°$, and $B = 42.6°$ (Fig. 20–44).

13. **AVIA** A satellite sends a signal to cellular towers A and B in two cities and makes an angle of elevation of 88° with tower A and an angle of 89° with tower B. The satellite is 18,000 mi from tower A. What is the approximate distance between the two cities?

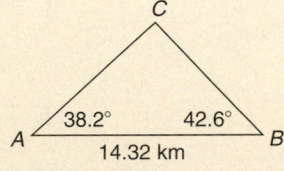

FIGURE 20–44

2 Use the law of sines to solve the triangles. If a triangle has two possibilities, find both solutions. Round sides to four significant digits and angles to the nearest 0.1°. *See Examples 3 and 4.*

14. $a = 42$ in., $b = 24$ in., $A = 40°$

15. $b = 15$ yd, $c = 3$ yd, $B = 70°$

16. $a = 18$ ft, $c = 9$ ft, $C = 20°$

17. $a = 8$ m, $b = 4$ m, $A = 30°$

18. $a = 65.2$ cm, $b = 42.5$ cm, $A = 30°$

19. $b = 22.9$ m, $c = 45.6$ m, $C = 120°$

20. **AG/H**   Find the missing angle and sides of the plot of land described by Figure 20–45.

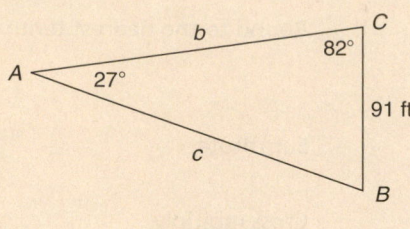

FIGURE 20-45

21. **CON**   Find the distance from *A* to *B* on the surveyed plot shown in Fig. 20–46.

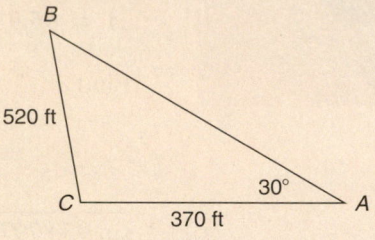

FIGURE 20-46

22. **CON**   A surveyor marks points *A* and *B* 35 ft apart. He then moves to point *C*, which is 20 ft from point *A*. If he determines from his transit that from point *C*, lines to points *A* and *B* form a 64° angle, what is the distance

(to the nearest hundredth of a foot) from point *C* to point *B*? If *A*, *B*, and *C* form a triangle, how many degrees (to the nearest tenth of a degree) are angle *A* and angle *B*? *See Example 5.*

20–5   Law of Cosines

LEARNING OUTCOMES

1   Find the unknown parts of an oblique triangle, given three sides of the triangle.

2   Find the unknown parts of an oblique triangle, given two sides and the included angle of the triangle.

LC LEARNING CATALYTICS

1. Evaluate $\cos^{-1} 0.63$ and round to four significant digits.

Law of cosines: the square of any side of a triangle equals the sum of the squares of the other sides minus twice the product of the other two sides and the cosine of the angle opposite the first side

1   Find the Unknown Parts of an Oblique Triangle, Given Three Sides of the Triangle.   In some cases, our given information does not allow us to use the law of sines. For example, if we know only all three sides of a triangle, we cannot use the law of sines to find the angles. In a case such as this, however, we can use the **law of cosines**. This law is based on the trigonometric functions just as the law of sines is.

Law of cosines:

The square of any side of a triangle equals the sum of the squares of the other sides minus twice the product of the other two sides and the cosine of the angle opposite the first side. For triangle *ABC*:

$$a^2 = b^2 + c^2 - 2\,bc \cos A$$
$$b^2 = a^2 + c^2 - 2\,ac \cos B$$
$$c^2 = a^2 + b^2 - 2\,ab \cos C$$

To use the law of cosines efficiently, we are given

1. Three sides of a triangle

    or

2. Two sides and the included angle of a triangle.

We can sometimes use *either* the law of sines or the law of cosines to solve certain triangles. However, whenever possible, the law of sines is generally preferred because it involves fewer calculations.

To find the unknown parts of an oblique triangle, given three sides of the triangle:

1. Substitute the given values of the three sides into any version of the law of cosines.

2. Solve for the unknown angle in the version of the law of cosines selected in Step 1.

3. Repeat Steps 1 and 2 using a different version of the law of cosines and solving for the different angle. Or, use the angle measure found in Step 2 and find a second angle using the law of sines.

4. Find the third angle of the triangle by subtracting from 180° the sum of the two angles found in Steps 2 and 3.

STOP AND CHECK

1. Solve triangle ABC if $a = 11$ ft, $b = 12$ ft, and $c = 14$ ft. Round angles to the nearest 0.1°.

Answer:

1. $A = 49.3°$; $B = 55.8°$; $C = 74.9°$

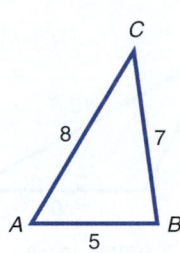

FIGURE 20–47

EXAMPLE 1

Find the angles in triangle ABC (see Fig. 20–47).

We may find any one of the angles first. If we choose to find angle A first, we must use the formula that contains cos A: $a^2 = b^2 + c^2 - 2bc \cos A$. Rearrange the formula to solve for cos A.

Find A.

$$a^2 - b^2 - c^2 = -2bc \cos A \qquad \text{Multiply each term on both sides by } -1 \text{ to reduce the number of negative signs.}$$

$$-a^2 + b^2 + c^2 = 2bc \cos A \qquad \text{Solve for cos } A.$$

$$\frac{-a^2 + b^2 + c^2}{2bc} = \cos A \qquad \text{Substitute } a = 7, b = 8, \text{ and } c = 5.$$

$$\frac{-(7)^2 + 8^2 + 5^2}{2(8)(5)} = \cos A \qquad \text{Solve for } A.$$

$$\frac{-49 + 64 + 25}{80} = \cos A$$

$$\frac{40}{80} = \cos A$$

$$0.5 = \cos A$$

$$A = \cos^{-1} 0.5 \qquad \text{Use inverse cosine function.}$$

$$\mathbf{60° = A}$$

Find B.

To find angle B, we use the law of cosines and only given values. The law of sines could be used but would involve using a calculated value.

$$b^2 = a^2 + c^2 - 2ac \cos B \qquad \text{Solve for cos } B.$$

$$2ac \cos B = a^2 + c^2 - b^2$$

$$\cos B = \frac{a^2 + c^2 - b^2}{2ac} \qquad \text{Substitute } a = 7, b = 8, \text{ and } c = 5.$$

$$\cos B = \frac{7^2 + 5^2 - (8)^2}{2(7)(5)} \qquad \text{Solve for } B.$$

$$\cos B = \frac{49 + 25 - 64}{70}$$

$$\cos B = \frac{10}{70}$$

$$\cos B = 0.1428571429 \qquad \text{Use inverse cosine function.}$$

$$B = \cos^{-1} 0.1428571429$$

$$B = 81.7867893°$$

$$\mathbf{B = 81.8°} \qquad \text{Round to the nearest tenth of a degree.}$$

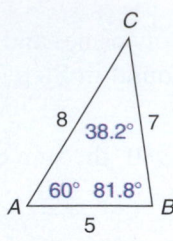

FIGURE 20–48

Find C.

The law of cosines or the law of sines can be used to find the third angle. However, the quickest way to find this angle is to subtract the sum of A and B from $180°$.

$$C = 180° - (A + B)$$
$$C = 180° - (60° + 81.7867893°)$$
$$C = 180° - 141.7867893°$$
$$C = 38.2132107°$$
$$\mathbf{C = 38.2°} \qquad \text{Nearest } 0.1°$$

The solved triangle is shown in Fig. 20–48. **See Exercises 1–6.**

STOP AND CHECK

1. Solve triangle ABC if $a = 15$ in., $b = 14$ in., and $c = 24$ in. Round angle to the nearest $0.1°$.

Answer:
1. $A = 35.5°$; $B = 32.8°$; $C = 111.7°$

EXAMPLE 2

Find the angles in triangle ABC (see Fig. 20–49).

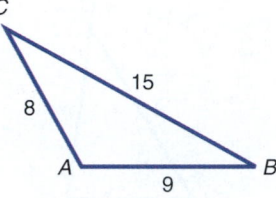

FIGURE 20–49

Find A.

$$a^2 = b^2 + c^2 - 2bc \cos A \qquad \text{Solve for } \cos A.$$

$$\cos A = \frac{b^2 + c^2 - a^2}{2bc} \qquad \text{Substitute.}$$

$$\cos A = \frac{8^2 + 9^2 - 15^2}{2(8)(9)} \qquad \text{Evaluate.}$$

$$\cos A = \frac{64 + 81 - 225}{144}$$

$$\cos A = \frac{-80}{144}$$

$$\cos A = -0.5555555556 \qquad \text{Use inverse cosine function.}$$

$$A = \cos^{-1}(-0.5555555556)$$

$$A = 123.7489886°$$

$$\mathbf{A = 123.7°} \qquad \text{Round to the nearest tenth of a degree.}$$

Recall from Section 20–2 that the cosine is *negative* in the second and third quadrants. Because A is either acute ($< 90°$) or obtuse ($> 90°$ but $< 180°$) and because its cosine is negative (see Fig. 20–50), it must be in the second quadrant. If a calculator is used and -0.5555555556 is entered, $123.7489886°$ will appear in the display. We can use the law of sines to find the second angle.

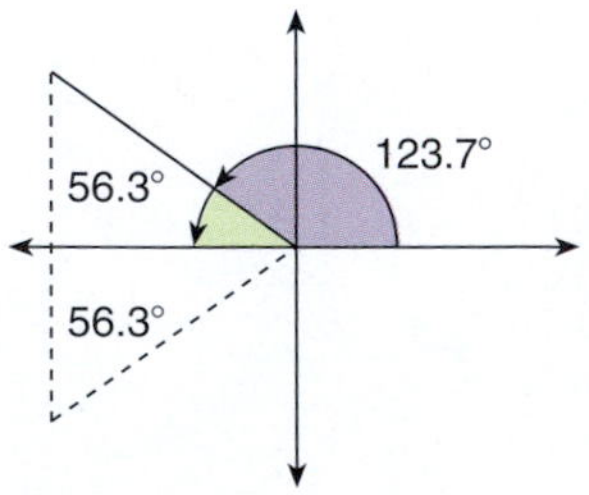

Find B.

$$\frac{a}{\sin A} = \frac{b}{\sin B} \qquad \text{Substitute.}$$

$$\frac{15}{\sin 123.7489886°} = \frac{8}{\sin B} \qquad \text{Solve for } B.$$

$$15 \sin B = 8 \sin 123.7489886°$$

$$\sin B = \frac{8 \sin 123.7489886°}{15}$$

$$\sin B = 0.4434556903 \qquad \text{Use inverse sine function.}$$

$$B = \sin^{-1}(0.4434556903)$$

$$B = 26.32457654°$$

$$\mathbf{B = 26.3°} \qquad \text{Round to the nearest tenth of a degree.}$$

FIGURE 20–50

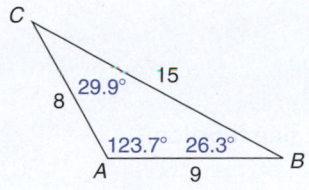

FIGURE 20-51

Find C.

$C = 180° - A - B$	Substitute.
$C = 180° - (123.7489886° + 26.32457654°)$	Evaluate.
$C = 29.92643486°$	
$C = \mathbf{29.9°}$	

The solved triangle is given in Fig. 20–51.   See Exercises 1–6.

TIP Rounding discrepancies like in Example 2 may result in the sum of the rounded angle measures of a triangle seeming to be less than 180°.

$29.9° + 123.7° + 26.3°$
$= 179.7°$

The full calculator values result in a more accurate sum.

$29.92643486°$
$+ 123.74898886°$
$+ 26.32457654° = 180°$

2 Find the Unknown Parts of an Oblique Triangle, Given Two Sides and the Included Angle of the Triangle. The law of cosines is needed to solve oblique triangles if two sides and the included angle are given.

To find the unknown parts of an oblique triangle, given two sides and the included angle of the triangle:

1. Find the side opposite the given angle using the law of cosines.

2. Use the law of sines to find one of the two unknown angles.

3. Find the third angle by subtracting from 180° the sum of the two known angles.

STOP AND CHECK

1. Solve triangle ABC if $C = 78°$, $a = 8$ ft, and $b = 10$ ft. Round sides to four significant digits and angles to the nearest 0.1°.

Answer:
1. $A = 43.2°$; $B = 58.8°$; $c = 11.43$ ft

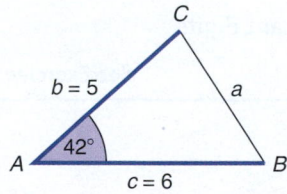

FIGURE 20-52

EXAMPLE 3

Solve triangle ABC in Fig. 20–52.

Find a.

$a^2 = b^2 + c^2 - 2bc \cos A$	Substitute into the appropriate version of the law of cosines.
$a^2 = 5^2 + 6^2 - 2(5)(6) \cos 42°$	Evaluate.
$a^2 = 25 + 36 - 60 \cos 42°$	$\cos 42° = 0.7431448255.$
$a^2 = 25 + 36 - 44.58868953$	Simplify.
$a^2 = 16.41131047$	Solve for a.
$a = 4.051087566$	
$a = \mathbf{4.051}$	Four significant digits

Find B.

$\dfrac{a}{\sin A} = \dfrac{b}{\sin B}$	Substitute $a = 4.051087566$, $b = 5$, and $A = 42°$.
$\dfrac{4.051087566}{\sin 42°} = \dfrac{5}{\sin B}$	Solve for B.
$4.051087566 \sin B = 5 \sin 42°$	
$\sin B = \dfrac{5 \sin 42°}{4.051087566}$	Evaluate.
$\sin B = 0.8258653947$	Use inverse sine function.
$B = \sin^{-1}(0.8258653947)$	
$B = 55.67632739°$	
$B = \mathbf{55.7°}$	Round to the nearest tenth of a degree.

Find C.

$\dfrac{a}{\sin A} = \dfrac{c}{\sin C}$	Substitute $a = 4.051087566$, $c = 6$, and $A = 42°$.
$\dfrac{4.051087566}{\sin 42°} = \dfrac{6}{\sin C}$	Solve for C.

$$4.051087566 \sin C = 6 \sin 42°$$

$$\sin C = \frac{6 \sin 42°}{4.051087566} \qquad \text{Evaluate.}$$

$$\sin C = 0.9910384737 \qquad \text{Use inverse sine function.}$$

$$C = \sin^{-1}(0.9910384737)$$

$$C = 82.32367267°$$

$$\mathbf{C = 82.3°} \qquad \text{Round to the nearest tenth of a degree.}$$

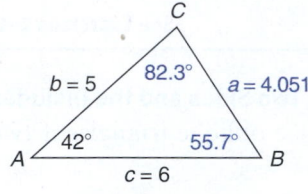

FIGURE 20–53

The solved triangle is given in Fig. 20–53.

To check, verify that the sum of the angles adds to 180°, the longest side is opposite the largest angle, and the shortest side is opposite the smallest angle.

See Exercises 7–10.

EXAMPLE 4

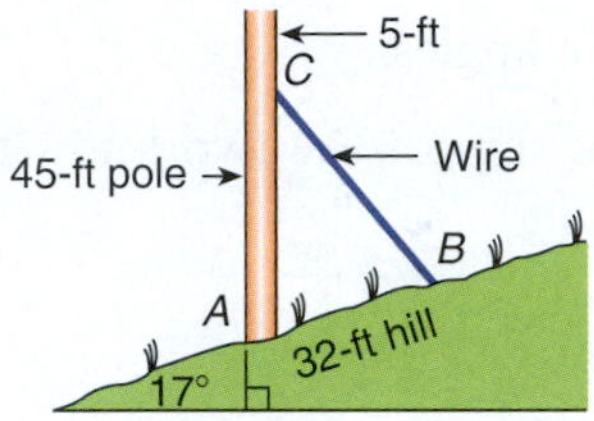

CON A vertical 45-ft pole is placed on a hill that is inclined 17° to the horizontal (see Fig. 20–54). How long a wire is needed if the wire is placed 5 ft from the top of the pole and attached to the ground at a point 32 ft uphill from the base of the pole?

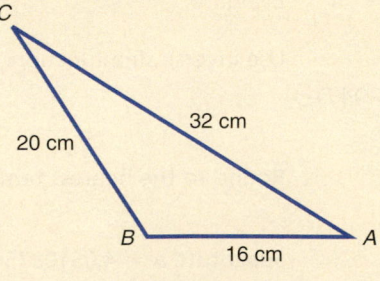

FIGURE 20–54

Point A is where the pole enters the ground. Point B is where the wire is attached to the ground. Point C is where the wire is attached to the pole (5 ft from the top of the pole). Side b is the length of the pole from the wire to the ground ($45 - 5 = 40$ ft). The length of the wire is side a in the triangle. Because AC makes a 90° angle with the horizontal, we know that angle A is $90° - 17°$, or 73°. Using the law of cosines, we have

$$a^2 = b^2 + c^2 - 2bc \cos A \qquad \text{Substitute } b = 40, c = 32, \text{ and } A = 73°.$$

$$a^2 = 40^2 + 32^2 - 2(40)(32) \cos 73° \qquad \text{Evaluate.}$$

$$a^2 = 1{,}600 + 1{,}024 - 2{,}560(0.2923717047)$$

$$a^2 = 1{,}600 + 1{,}024 - 748.4715641$$

$$a^2 = 1{,}875.528436$$

$$a = 43.30737161$$

$$a = 43.31 \text{ ft} \qquad \text{Four significant digits}$$

The length of the wire is 43.31 ft. **See Exercise 11.**

20–5 EXERCISES

MyLab Math For additional practice go to your study plan in MyLab Math.

1 Solve the triangles in Figs. 20–55 and 20–56. Round angles to the nearest 0.1°. *See Examples 1 and 2.*

1.

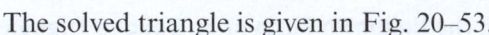

FIGURE 20–55

2.

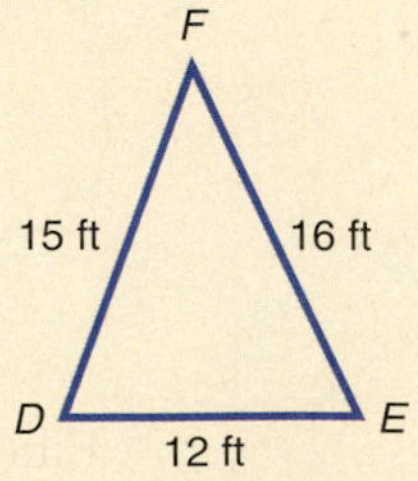

FIGURE 20–56

Find the three angles of each triangle. Round angle measures to the nearest 0.1°. *See Examples 1 and 2.*

3. $a = 26$ cm, $b = 28$ cm, $c = 32$ cm

4. $a = 12.5$ m, $b = 16.1$ m, $c = 14.6$ m

5. $a = 8.6$ ft, $b = 9.8$ ft, $c = 7.2$ ft

6. $a = 128.5$ in., $b = 120.4$ in., $c = 114.3$ in.

2 Solve the triangles in Figs. 20–57 through 20–60. Round sides to four significant digits. Round angles to the nearest 0.1°.
See Example 3.

7.

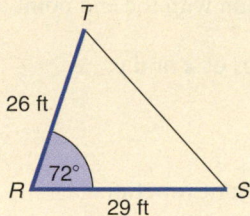

FIGURE 20–57

8.

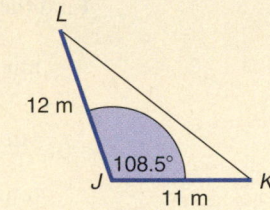

FIGURE 20–58

9.

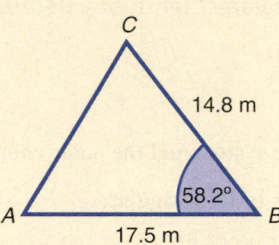

FIGURE 20–59

10.

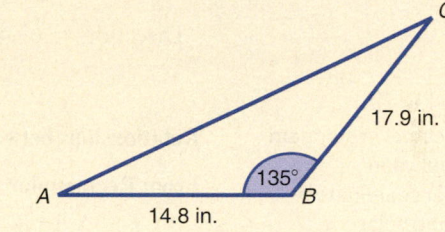

FIGURE 20–60

11. `CON` A hill is inclined 20° to the horizontal. A pole stands vertically on the side of the hill with 35 ft above the ground. How much wire will it take to reach from a point 2 ft from the top of the pole to a point on the ground 27 ft downhill from the base of the pole? *See Example 4.*

12. `CON` A triangular tabletop is to be 8.4 ft by 6.7 ft by 9.3 ft. What angles must be cut?

---

20 │ CHAPTER REVIEW OF KEY CONCEPTS

LEARNING OUTCOMES   **KEY CONCEPTS AND EXAMPLES**

Section 20–1

1 Find the magnitude of a vector in standard position, given the coordinates of the end point (pp. 849–851).

1. Substitute into the Pythagorean theorem the coordinates of the end points of the vector (x_1, y_1) as the legs of a right triangle. **2.** Solve for the hypotenuse (magnitude).

Find the magnitude of a vector in standard position with the end point at (4, 2) (Fig. 20–61).

$r = \sqrt{x^2 + y^2}$ Substitute values of *x* and *y*.

$r = \sqrt{4^2 + 2^2}$ Evaluate.

$r = \sqrt{16 + 4}$

$r = \sqrt{20}$

Magnitude = 4.472   Round to four significant digits.

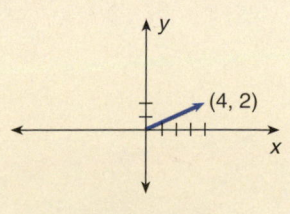

FIGURE 20–61

LEARNING OUTCOMES	KEY CONCEPTS AND EXAMPLES

2 Find the direction of a vector in standard position, given the coordinates of the end point (p. 851).

1. Identify the coordinates of the end point of the vector. **2.** Substitute the coordinates of the end point into the tangent function: $\tan\theta = \dfrac{\text{opp}}{\text{adj}}$ or $\tan\theta = \dfrac{y}{x}$. **3.** Solve for θ.

Find the direction of a vector in standard position with the end point at $(2, 4)$.

$$\tan\theta = \frac{y}{x} \qquad \text{Substitute values of } x \text{ and } y.$$

$$\tan\theta = \frac{4}{2} \qquad \text{Simplify.}$$

$$\tan\theta = 2 \qquad \text{Use inverse tangent function.}$$

$$\tan^{-1} 2 = 63.43494882°$$

$$\theta = 63.4° \qquad \text{Round to the nearest tenth of a degree.}$$

Direction $= 63.4°$

3 Convert vectors in standard position between rectangular coordinate notation and polar coordinate notation (pp. 851–854).

Relationships between the rectangular coordinate system and the polar coordinate system:

From Rectangular to Polar:

$$r = \sqrt{x^2 + y^2}$$

$$\theta = \tan^{-1}\frac{y}{x}$$

From Polar to Rectangular:

$$x = r\cos\theta$$

$$y = r\sin\theta$$

Convert the end point of a vector in rectangular notation to polar notation: 1. Find r using the relationship $r = \sqrt{x^2 + y^2}$. **2.** Find θ using the relationship $\theta = \tan^{-1}\dfrac{y}{r}$.

The end point of a vector in standard position is at $(4, 6)$. Find the polar coordinates in degree mode.

To find r:

$$r = \sqrt{x^2 + y^2} \qquad \text{Substitute known values.}$$

$$r = \sqrt{4^2 + 6^2} \qquad \text{Evaluate.}$$

$$r = \sqrt{16 + 36}$$

$$r = \sqrt{52}$$

$$r = 7.2 \qquad \text{Principal square root rounded.}$$

To find θ:

$$\theta = \tan^{-1}\frac{y}{x} \qquad \text{Substitute known values.}$$

$$\theta = \tan^{-1}\frac{6}{4}$$

$$\theta = \tan^{-1} 1.5 \qquad \text{In degree mode. } \boxed{\text{TAN}^{-1}}\ 1.5\ \boxed{\text{ENTER}}$$

$$\theta = 56.30993247° \qquad \text{Round to the nearest tenth of a degree.}$$

$$\theta = 56.3°$$

The polar coordinates are $(7.2, 56.3°)$.

LEARNING OUTCOMES	KEY CONCEPTS AND EXAMPLES

Convert the end point of a vector in polar notation to rectangular notation: **1.** Find x using the relationship $x = r \cos \theta$.   **2.** Find y using the relationship $y = r \sin \theta$.

Given the polar coordinates $(4, 75°)$, find the rectangular coordinates.

$x = r \cos \theta$	$y = r \sin \theta$
$x = 4 \cos 75°$	$y = 4 \sin 75°$
$x = 1.03527618$	$y = 3.863703305$
$x = 1.035$	$y = 3.864$

$(1.035, 3.864)$ Rectangular notation

4 Add vectors and multiply a vector by a scalar (pp. 854–858).

Add vectors of the same direction: **1.** Add the magnitudes of the vectors to get the magnitude of the resultant vector.   **2.** The resultant vector will have the same direction as the original vectors.

Add vectors of opposite direction: **1.** Subtract the magnitude of the smaller vector from the magnitude of the larger vector to get the magnitude of the resultant vector.   **2.** The resultant vector has the direction of the vector that has the larger magnitude.

Add vectors with different directions: **1.** Determine the x- and y-coordinates of the end point of each vector in standard position.   **2.** Add the x-coordinates of the end points for the x-coordinate of the end point of the resultant vector.   **3.** Add the y-coordinates of the end points for the y-coordinate of the end point of the resultant vector.   **4.** Find the magnitude using the Pythagorean theorem and the end point of the resultant vector.   **5.** Find the direction using the inverse tangent function and the end point of the resultant vector.

Write a vector in complex notation: **1.** Find the x-coordinate of the end point of the vector. Use the relationship $x = r \cos \theta$.   **2.** Find the y-coordinate of the end point of the vector. Use the relationship $y = r \sin \theta$.   **3.** Write the vector in complex notation as $x + yi$.

Add vectors A and B. See Fig. 20–62.

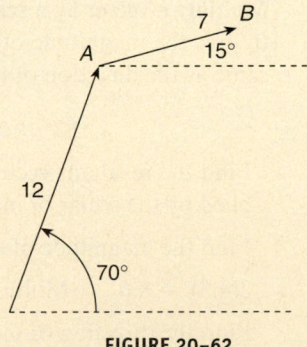

FIGURE 20–62

vector A:

x-coordinate: $\cos \theta = \dfrac{x}{r}$ $r =$ magnitude. Substitute $\theta = 70°$ and $r = 12$.

$\cos 70° = \dfrac{x}{12}$ Solve for x.

$12(\cos 70°) = x$

$4.10424172 = x$

$x = 4.104$ Round.

LEARNING OUTCOMES	KEY CONCEPTS AND EXAMPLES

y-coordinate: $\sin\theta = \dfrac{y}{r}$ r = magnitude. Substitute $\theta = 70°$ and $r = 12$.

$$\sin 70° = \dfrac{y}{12}$$ Solve for y.

$$12(\sin 70°) = y$$

$$11.27631145 = y$$

$$y = 11.28$$ Round.

vector $A = 4.104 + 11.28i$

vector B:

x-coordinate: $\cos\theta = \dfrac{x}{r}$ Substitute $\theta = 15°$ and $r = 7$.

$$\cos 15° = \dfrac{x}{7}$$ Solve for x.

$$7(\cos 15°) = x$$

$$6.761480784 = x$$

$$x = 6.761$$ Round.

y-coordinate: $\sin\theta = \dfrac{y}{r}$ Substitute $\theta = 15°$ and $r = 7$.

$$\sin 15° = \dfrac{y}{7}$$ Solve for y.

$$7(\sin 15°) = y$$

$$1.811733316 = y$$

$$y = 1.812$$ Round.

vector $B = 6.761 + 1.812j$

vector $A + B = (4.104 + 11.28j) + (6.761 + 1.812j)$ Add like vector

$$= 10.87 + 13.09j$$ components.

Multiply a vector by a scalar: 1. Multiply the magnitude of the original vector by the scalar to give the magnitude of the resultant vector.   **2.** The direction of the resultant will be the same as the direction of the original vector.

Find the resultant vector of a vector of magnitude 4.3 and a direction of $\dfrac{\pi}{3}$ that is multiplied by the scalar of magnitude 2.

Find the magnitude of the resultant vector:

$2(4.3) = 8.6$ Multiply the magnitude 4.3 by the scalar 2.

Find the direction of the resultant vector:

$\theta = \dfrac{\pi}{3}$ The resultant vector has the same magnitude as the original vector.

The resultant vector has a magnitude of 8.6 and a direction of $\dfrac{\pi}{3}$.

Section 20-2

1 Find related acute angles for angles or vectors in Quadrants II, III, and IV (pp. 859–861).

Find a related acute angle for an angle (θ) in any quadrant: 1. Draw the angle and form the third side of a triangle by drawing a line perpendicular to the x-axis from the vector or ray forming the angle.   **2.** The related angle is the angle formed by the ray and the x-axis portion of the triangle.

LEARNING OUTCOMES	KEY CONCEPTS AND EXAMPLES

Quadrant II angle (Fig. 20–63):
$\theta = 180° - \theta_2$ or $\theta = \pi - \theta_2$

Quadrant III angle (Fig. 20–64):
$\theta = \theta_3 - 180°$ or $\theta = \theta_3 - \pi$

Quadrant IV angle (Fig. 20–65):
$\theta = 360° - \theta_4$ or $\theta = 2\pi - \theta_4$

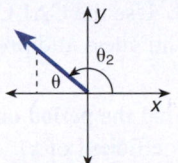 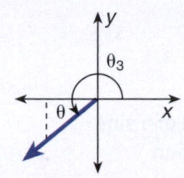 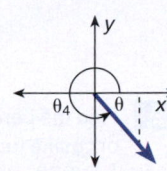

FIGURE 20–63   **FIGURE 20–64**   **FIGURE 20–65**

> Find the related angle for an angle of 156°.
>
> The angle is in Quadrant II, thus, $180° - 156° = 24°$.

2 Determine the signs of trigonometric functions of angles of more than 90° (pp. 861–863).

The memory device ALL-SIN-TAN-COS can be used to remember the signs of the trigonometric functions in the various quadrants. The signs of *all* trigonometric functions are positive in the first quadrant. The sign of the SIN function is positive in the second quadrant (COS and TAN are negative). The sign of the TAN function is positive in the third quadrant. The sign of the COS function is positive in the fourth quadrant.

> What is the sign of cos 145°?
>
> 145° is a Quadrant II angle and the cosine function is negative in Quadrant II. Thus, the sign of cos 145° is negative.

3 Find the trigonometric functions of angles of more than 90° using a calculator (pp. 863–865).

Most calculators give the trigonometric value of an angle regardless of the quadrant in which it is found.

> Use a calculator to find the value of the following trigonometric functions. Round to four significant digits.
>
> $$\sin 125° = 0.8192$$
> $$\cos 215° = -0.8192$$
> $$\tan 335° = -0.4663$$

Section 20–3

1 Graph a sine or cosine function using a calculator (pp. 865–868).

1. Set MODE to degrees or radians as desired.   **2.** From the ZOOM function select the option ZTrig or manually set the window as desired.   **3.** Enter the function using the $Y =$ function.   **4.** View the graph using GRAPH function.

> Graph $y = 2 \cos x$.
> Set ZOOM to option ZTrig.
> Enter $y = 2 \cos x$.
> Press GRAPH (Fig. 20–66).
>
>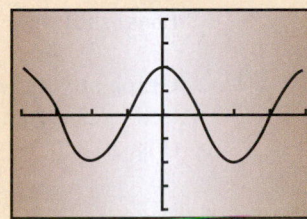
>
> **FIGURE 20–66**

Find the amplitude of a graph using a calculator: 1. Set the window of the graph to be what you expect to be greater than the maximum value of the graph.   **2.** Graph the function.

LEARNING OUTCOMES	KEY CONCEPTS AND EXAMPLES

3. Use the CALC function and select *maximum* (option 4).   **4.** Set the left and right bounds and guess and press ENTER.   **5.** The amplitude will be the *y*-value.

2 Find the period of a sine or cosine function (pp. 868–869).

Find the period of a function in the form $y = A \sin Bx$ or $y = A \cos Bx$: 1. Identify B (the coefficient of x).   **2.** Divide by 360° or 2π radians by the coefficient of x.

$$P = \frac{360°}{B} \text{ for degrees} \qquad \text{or} \qquad P = \frac{2\pi}{B} \text{ for radians}$$

Find the amplitude and period in degrees of the function $y = 2 \cos 3x$. Graph the equation from $-90°$ to $270°$.

Amplitude = 2   *A* = 2 (coefficient of cos 3x)

$$\text{Period} = \frac{360°}{3} = 120° \quad \textit{B} = 3 \text{ (coefficient of } x\text{)}$$

Set calculation to Degree mode.

Set Window; $X_{min} = -90$
$X_{max} = 270$
$X_{scl} = 30$
$Y_{min} = -3$
$Y_{max} = 3$
$Y_{scl} = 1$

Enter equation: $y = 2 \cos 3x$

Graph (Fig. 20–67).

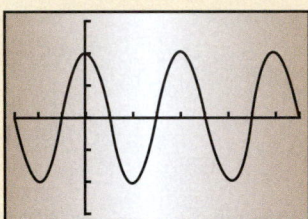

FIGURE 20–67

3 Find the phase shift of a sine or cosine function (pp. 869–870).

Find the phase shift for the function $y = A \sin(Bx + C)$ or $y = A \cos(Bx + C)$:
1. Find the value $\frac{C}{B}$.   **2.** If $\frac{C}{B}$ is positive, shift the graph $\frac{C}{B}$ units to the *left*.   **3.** If $\frac{C}{B}$ is negative, shift the graph $\frac{C}{B}$ units to the *right*.

Graph $y = 3 \sin(3x - 180°)$

Amplitude = 3   Coefficient of sine

$$\text{Period} = \frac{360°}{3} = 120° \qquad \frac{360°}{B}; \textit{B} \text{ is the coefficient of } x.$$

$$\text{Phase shift} = \frac{-180°}{3} = -60° \qquad \frac{C}{B} \text{ is negative.}$$

Graph of $y = 3 \sin(3x - 180°)$ shifts 60° to the right from the graph $y = 3 \sin 3x$ (Fig. 20–68).

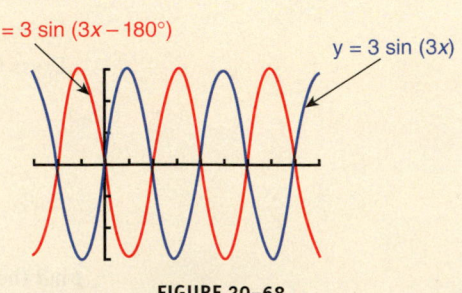

y = 3 sin (3x – 180°)

y = 3 sin (3x)

FIGURE 20–68

| **LEARNING OUTCOMES** | **KEY CONCEPTS AND EXAMPLES** |

Section 20–4

1 Find the unknown parts of an oblique triangle, given two angles and a side (pp. 871–873).

Use the law of sines to find the missing part of an oblique triangle when two angles and a side are given.

$$\frac{a}{\sin A} = \frac{b}{\sin B} = \frac{c}{\sin C}$$

1. Find the third angle of the triangle by adding the two known angles and subtracting the sum from 180° or 2π radians. **2.** Set up a ratio with the given side and sine of the opposite angle. **3.** Set up a second ratio with an unknown side and the sine of the opposite angle. **4.** Form a proportion using the ratios in Steps 2 and 3 and solve the proportion for the unknown side. **5.** Repeat Steps 2–4 if the value of the third side is needed.

Find side c in the triangle of Fig. 20–69.

$$\frac{a}{\sin A} = \frac{c}{\sin C}$$ Substitute $a = 13$, $A = 22°$, and $C = 97°$.

$$\frac{13}{\sin 22°} = \frac{c}{\sin 97°}$$ Solve for c.

$$c \sin 22° = 13 \sin 97°$$

$$c = \frac{13 \sin 97°}{\sin 22°}$$ Evaluate.

$$c = 34.44440167 \text{ cm}$$

$$AB = c = 34.44 \text{ cm}$$ Round to four significant digits.

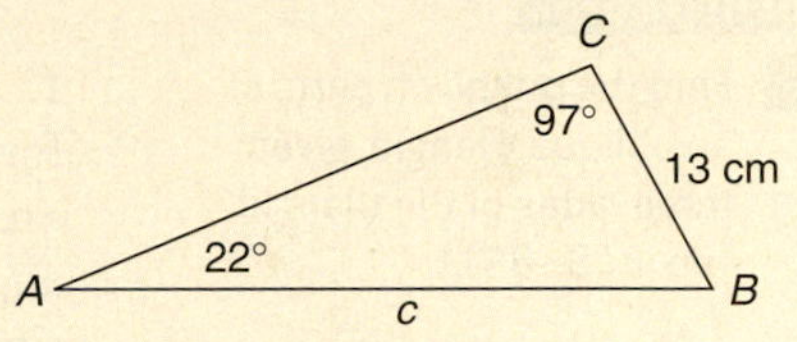

FIGURE 20–69

2 Find the unknown parts of an oblique triangle, given two sides and an angle opposite one of them (pp. 873–877).

1. Use the law of sines to find the measure of the angle opposite the other given side. **2.** Determine if there are two angle values that are possible for the solution. **3.** Find the third angle by subtracting the sum of the two known angles from 180°. **4.** Find the third side using the law of sines. **5.** If two angles were found in Step 2, find the second possible solution by repeating Steps 3 and 4.

Find the measure of angle A in the triangle of Fig. 20–70.

$$\frac{a}{\sin A} = \frac{b}{\sin B}$$ Substitute $a = 18$, $b = 16$, and $B = 32°$.

$$\frac{18}{\sin A} = \frac{16}{\sin 32°}$$ Solve for A.

$$16 \sin A = 18 \sin 32°$$

$$\sin A = \frac{18 \sin 32°}{16}$$ Evaluate.

$$\sin A = 0.5961591723$$ Use inverse sine function.

$$A = \sin^{-1}(0.5961591723)$$

$$A = 36.59531105°$$

$$A = 36.6°$$ Round to the nearest tenth of a degree.

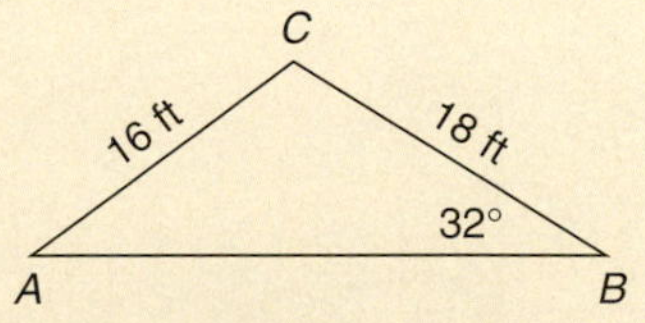

FIGURE 20–70

Given $\triangle ABC$ and b is a known side and a is a known side opposite a known angle, if $b < a$ and $\angle A \neq 90°$, we have two possible solutions. If $b < a$ and $\angle A = 90°$, we have one solution. If $b \geq a$, we have one solution.

LEARNING OUTCOMES

KEY CONCEPTS AND EXAMPLES

Determine how many solutions exist for the triangle in Fig. 20–71 based on the given facts. Explain your response.

$B = 87°$ and $A \neq 90°$

$b = 15$ cm (known side opposite known angle)

$a = 8$ cm (known side opposite unknown angle)

Because $b > a$, there is one possible solution.

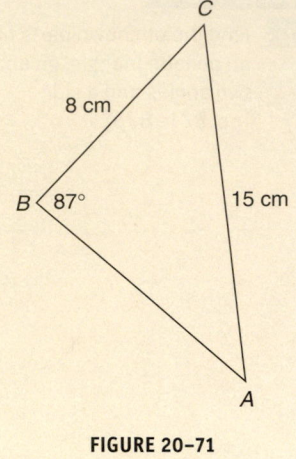

FIGURE 20–71

Section 20–5

1 Find the unknown parts of an oblique triangle, given three sides of the triangle (pp. 878–881).

1. Substitute the given values of the three sides into any version of the law of cosines. **2.** Solve for the unknown angle in the version of the law of cosines selected in Step 1. **3.** Repeat Steps 1 and 2 using a different version of the law of cosines and solving for the different angle. Or, use the angle measure found in Step 2 and find a second angle using the law of sines. **4.** Find the third angle of the triangle by subtracting from 180° the sum of the two angles found in Steps 2 and 3.

Find the measure of angle A in the triangle of Fig. 20–72.

$a = 20$ cm

$b = 9$ cm

$c = 15$ cm

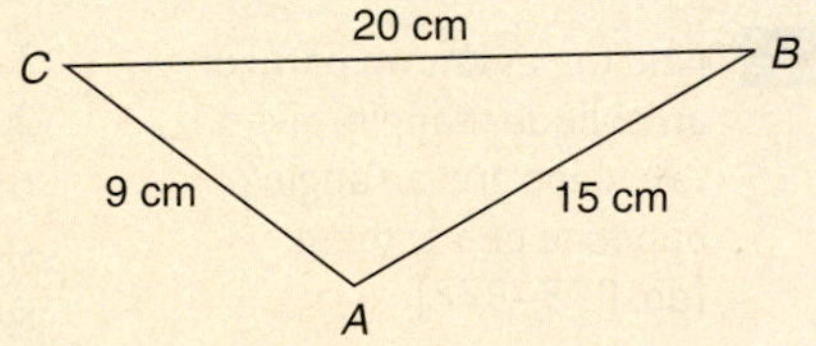

$a^2 = b^2 + c^2 - 2bc \cos A$

$\cos A = \dfrac{-a^2 + b^2 + c^2}{2bc}$ Substitute known values.

$\cos A = \dfrac{-20^2 + 9^2 + 15^2}{2(9)(15)}$ Evaluate.

$\cos A = \dfrac{-400 + 81 + 225}{270}$

$\cos A = -0.3481481481$ Use inverse cosine function.

$A = \cos^{-1}(-0.3481481481)$

$A = 110.3740893°$

$A = 110.4°$ Round to the nearest tenth of a degree.

FIGURE 20–72

2 Find the unknown parts of an oblique triangle, given two sides and the included angle of the triangle (pp. 881–882).

1. Find the side opposite the given angle using the law of cosines. **2.** Use the law of sines to find one of the two unknown angles. **3.** Find the third angle by subtracting from 180° the sum of the two known angles.

LEARNING OUTCOMES   **KEY CONCEPTS AND EXAMPLES**

Find the side opposite the given angle of the triangle in Fig. 20–73.

$a^2 = b^2 + c^2 - 2bc \cos A$

$a = \sqrt{b^2 + c^2 - 2bc \cos A}$ Substitute $A = 70°$, $b = 17$, and $c = 23$.

$a = \sqrt{17^2 + 23^2 - 2(17)(23) \cos 70°}$

$a = \sqrt{289 + 529 - 267.4597521}$

$a = \sqrt{550.5402479}$

$a = 23.4635941$

$a = 23.46$ cm   Round.

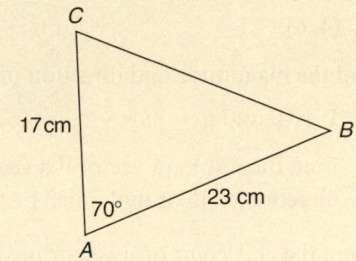

FIGURE 20–73

For most the accurate results, use as given values instead of calculated values, when possible.

Find the measures of all the angles and sides of the triangle in Fig. 20–74.

$\dfrac{a}{\sin A} = \dfrac{c}{\sin C}$ Substitute $a = 6.8$, $A = 48°$, and $c = 7.5$.

$\dfrac{6.8}{\sin 48°} = \dfrac{7.5}{\sin C}$ Solve for C.

$\sin C = \dfrac{7.5 \sin 48°}{6.8}$

$\sin C = 0.8196450281$ Use inverse sine
function.

$C = \sin^{-1}(0.8196450281)$

$C = 55.04927548°$

$C = 55.0°$ Round to the nearest tenth of a degree.

$\angle B = 180° - (\angle A + \angle C)$

$\angle B = 180° - 103.04927548°$

$\angle B = 76.95072452°$

$\angle B = 77.0°$ Round to the nearest tenth of a degree.

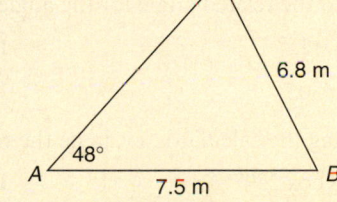

FIGURE 20–74

Use the law of sines or law of cosines to find b or AC.

$\dfrac{a}{\sin A} = \dfrac{b}{\sin B}$ Using the law of sines. Substitute
$a = 6.8$, $B = 76.95072452°$, and $A = 48°$.

$\dfrac{6.8}{\sin 48°} = \dfrac{b}{\sin 76.95072452}$ Solve for b.

$b = \dfrac{6.8 \sin 76.95072452°}{\sin 48°}$

$b = 8.914007364$

$b = 8.914$ m

20   CHAPTER REVIEW EXERCISES

Section 20–1   MyLab Math   **For additional practice go to your study plan in MyLab Math.**

Find the direction and magnitude of vectors in standard position with the indicated end points. Round lengths to four significant digits and angles to the nearest tenth of a degree.

1. $(4, 6)$ **2.** $(2, 8)$

Find the magnitude and direction of the resultant of the two given vectors in complex notation. Round as above.

3. $1 + 4j$ and $6 + 8j$ **4.** $3 + 5j$ and $2 + 3j$

5. Find the resultant vector if a vector of magnitude 2.4 and a direction of $45°$ is multiplied by the scalar of magnitude 3.5.

Given the end point of a vector in standard position, find the polar coordinates. Give the angle measure in radians and round the length and angle to tenths.

6. $(6, 8)$ **7.** $(3, 5.2)$

Given the polar coordinates, find the rectangular coordinates. Round to tenths.

8. $(4, 50°)$ **9.** $\left(11, \dfrac{3\pi}{8}\right)$

Section 20–2

Find the related angle for the angles. Express radian measures to the nearest ten-thousandth.

10. $115°$ **11.** 3.04 rad   **12.** 4.75 rad   **13.** $221°$

14. $305°$ **15.** 5.4 rad   **16.** $138.5°$ **17.** $212°15'10''$

Using the calculator, evaluate the trigonometric functions. Round to four significant digits.

18. $\cos 250°$ **19.** $\sin 2.1$ **20.** $\tan 175°$ **21.** $\sin 340°$

22. $\tan 4.5$ **23.** $\cos 290°$ **24.** $\sin\dfrac{3\pi}{4}$ **25.** $\tan\dfrac{5\pi}{4}$

26. **ELEC**   Find the magnitude and direction of a vector of an electrical current in standard position if the coordinates of the end point of the vector are $(2, -3)$. Round the magnitude (in amps) to the nearest hundredth. Express the direction in degrees to the nearest $0.1°$.

27. Find the direction and magnitude of a vector in standard position if the coordinates of the end point of the vector are $(-2, 2)$. Express its direction in radians and round its direction and its magnitude to the nearest hundredth.

Section 20–3

28. **ELEC**   What is the frequency of a wave that passes through a given point once each 0.025 s?

29. **ELEC**   What is the period of a wave that has a frequency of $15,000$ Hz?

30. **ELEC**   WMC AM has a frequency of 790 KHz. Find the station's wavelength in meters.

31. Find the amplitude and period of the function $y = 5 \cos 4x$. Graph the function to show two periods.

32. Find the amplitude, period, and phase shift and graph the function $y = \frac{1}{2} \sin (8x - 180°)$.

33. **ELEC**   The maximum voltage V in a generator is 50 V. The voltage charges according to the equation $v = 50 \sin x$. Use a calculator to find the voltage at $120°$.

Section 20–4

Solve the triangles. If a triangle has two possible solutions, find both solutions. Round sides to tenths and angles to the nearest $0.1°$.

34. $A = 60°, B = 40°, b = 20$ cm   **35.** $B = 120°, C = 20°, a = 8$ m

36. $A = 60°, B = 60°, a = 10$ ft   **37.** $a = 5$ dkm, $c = 7$ dkm, $C = 45°$

38. $b = 10\,\text{m}$, $c = 8\,\text{m}$, $B = 52°$

39. $a = 9.2\,\text{cm}$, $b = 6.8\,\text{cm}$, $B = 28°$

40. A surveyor needs the measure of JK in Fig. 20–75. Find JK to the nearest foot.

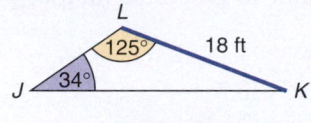

FIGURE 20–75

41. In Fig. 20–76, find RS.

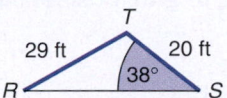

FIGURE 20–76

Section 20–5

Solve the triangles in Figs. 20–77 through 20–82. Round sides to four significant digits and angles to the nearest 0.1°.

42.

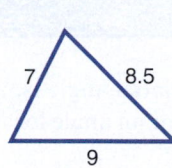

FIGURE 20–77

43.

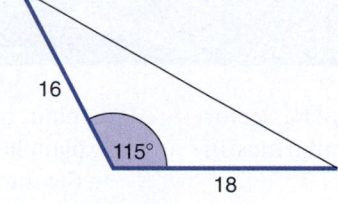

FIGURE 20–78

44.

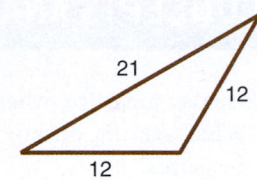

FIGURE 20–79

45.

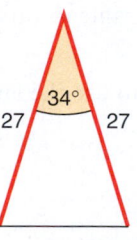

FIGURE 20–80

46.

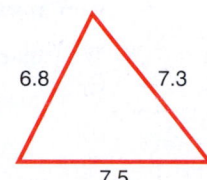

FIGURE 20–81

47.

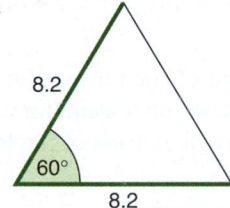

FIGURE 20–82

Solve the exercises. Round sides to four significant digits and angles to the nearest 0.1°.

48. **CON**   A triangular lot has sides 180 ft long, 160 ft long, and 123.5 ft long. Find the angles of the lot.

49. **CON**   A vertical 50-ft pole stands on top of a hill inclined 18° to the horizontal. What length of wire is needed to reach from a point 6 ft from the pole's top to a point 75 ft downhill from the base of the pole?

50. **AVIA**   A ship sails from a harbor 35 nautical miles east, and then 42 nautical miles in a direction 32° south of east (Fig. 20–83). How far is the ship from the harbor?

51. **CON**   A hill with a 35° grade (inclined to the horizontal) is cut down for a roadbed to a 10° grade. If the distance from the base to the top of the original hill is 800 ft (Fig. 20–84), how many vertical feet will be removed from the top of the hill, and what is the distance from the bottom to the top of the hill for the roadbed?

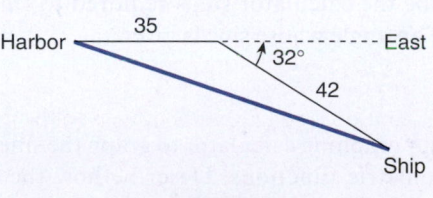

FIGURE 20–83

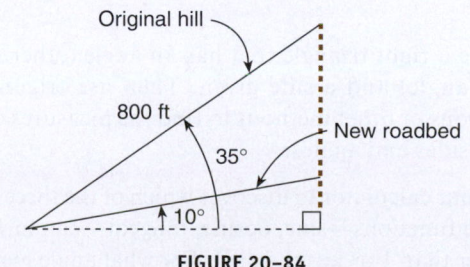

FIGURE 20–84

20   TEAM PROBLEM-SOLVING EXERCISES

1. Use a calculator to graph the following in radian angle mode:
   (a) $y = \sin x$
   (b) $y = \sin(x + 1)$
   (c) $y = \sin(x + 2)$
   (d) $y = \sin(x - 1)$
   (e) $y = \sin(x - 2)$
   (f) Discuss the similarities and differences of the graphs.

2. Use a calculator to graph the following in radian angle mode:
   (a) $y = \sin x$
   (b) $y = \sin(x) + 1$
   (c) $y = \sin(x) + 2$
   (d) $y = \sin(x) - 1$
   (e) $y = \sin(x) - 2$
   (f) Discuss the similarities and differences of the graphs.
   (g) Discuss how these graphs are different from the graphs in Exercise 1.

20   CONCEPTS ANALYSIS

1. Speed is a scalar measure, whereas velocity is a vector measure. What are the differences and similarities in these two measures?

2. Explain in your own words what a related angle is, and explain how to find the related angle for an angle located in the third quadrant.

3. Locate a point in the second quadrant by a set of coordinates. Find the magnitude and direction of the vector determined by the point, and find the related angle for the vector.

4. Give the sign of the sine function in each of the four quadrants. Use your calculator to find the value of the sine of an angle in each of the four quadrants to validate your answer.

5. Give the sign of the tangent function in each of the four quadrants. Use your calculator to find the value of the tangent of an angle in each of the four quadrants to validate your answer.

6. What parts of a triangle must be given to use the law of sines?

7. Draw a triangle and assign values to three parts of the triangle, then use the law of sines and other methods to find the remaining angles and sides.

8. (a) Write the law of sines in words.
   (b) Write the law of cosines in words.

9. How would you know whether to select the law of sines or the law of cosines to find missing angles or sides of a triangle?

10. Use a graphing calculator to graph the following functions on the same grid: $\sin x$; $2 \sin x$; $3 \sin x$; $\frac{1}{2} \sin x$; $\frac{1}{3} \sin x$. Describe how the coefficient of the sine function impacts the graph of the function. This coefficient is called the amplitude. Look up the definition of the word *amplitude*. Does the dictionary definition make sense in view of your description? Set your graphing window for x as $0°$ to $360°$ and for y as -3 to $+3$.

11. Graph $y = \sin x$ and $y = \sin 2x$ on the same graph. Compare these graphs. How do they differ and how are they alike?

12. Graph $y = \sin x$ and $y = \sin\left(\frac{1}{2}x\right)$ on the same graph. Compare these two graphs to the graphs for Exercise 11. What effect does the coefficient of x have on the graph of the sine function? Graph other sine functions in which the coefficient of x is different to verify your description.

13. Devise a right triangle that has an angle (other than the right angle) and a side given. Then use trigonometric functions or other methods to find the measures of all the other sides and angles.

14. Describe the calculator steps required to find the measure of an angle whose sine is given.

15. Use your calculator to discover which of the three trigonometric functions—sine, cosine, tangent—can have values greater than 1 or less than -1. For what angle measures is the value of this function greater than 1 or less than -1?

16. Use your graphing calculator to graph the sine and cosine trigonometric functions. Describe how the graphs are similar and how they are different.

20 | PRACTICE TEST

Find the values. Round to four significant digits.

1. sin 125°

2. tan 140°

3. cos 160°

4. Find the length of the vector that has end point coordinates of (8, 15).

5. Find the angle of the vector whose end point coordinates are (−8, 8).

Use the law of sines or the law of cosines to find the side or angle indicated in Figs. 20–85 through 20–92. Round sides to four significant digits and angles to the nearest 0.1°.

6.

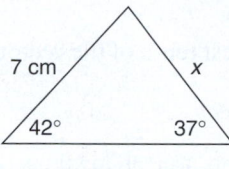

FIGURE 20–85

7.

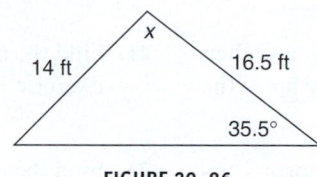

FIGURE 20–86

8.

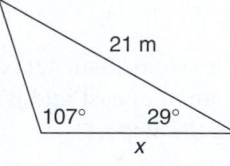

FIGURE 20–87

9.

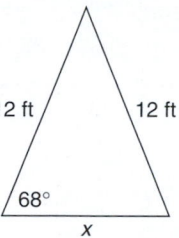

FIGURE 20–88

10.

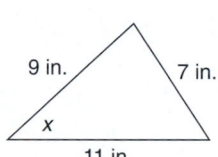

FIGURE 20–89

11.

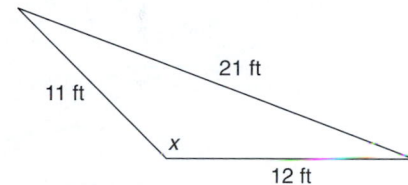

FIGURE 20–90

12.

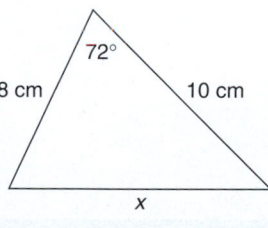

FIGURE 20–91

13.

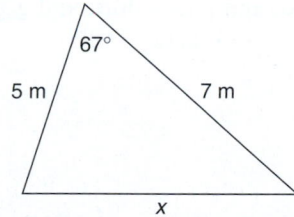

FIGURE 20–92

14. Find the *AC* of triangle *ABC* in Fig. 20–93. Round to four significant digits.

15. Find ∠*R* of triangle *RST* in Fig. 20–94. Round to the nearest 0.1°.

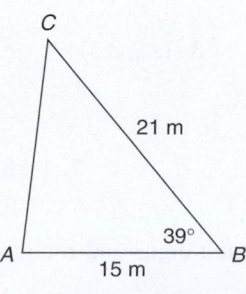

FIGURE 20–93

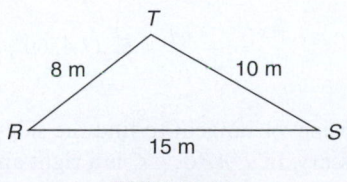

FIGURE 20–94

16. Find the direction in degrees (to the nearest 0.1°) of the vector I_t (total current) in Fig. 20–95.

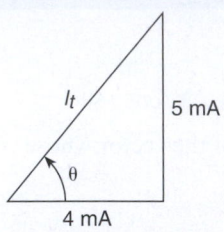

FIGURE 20–95

17. A surveyor measures two sides of a triangular lot and the angle formed by these two sides. What is the length of the third side if the two measured sides are 76 ft and 110 ft and the angle formed by these two sides is 107°?

18. The sides of a triangle cast measure 31 mm, 42 mm, and 27 mm. Find the measure of the angle opposite the 31 mm side.

19. Find the third side of a triangular flower bed if two sides measure 12 ft and 15 ft and the included angle measures 48°.

20. A plane flies from an airport due east for 32 mi, and then turns 15° north of east and travels 72 mi. How far is the plane from the airport?

21. Find the magnitude to the nearest tenth of the vector I_t in Exercise 16.

22. A rod 11 cm long joins a crank 20 cm long to form a triangle with a third, imaginary line (Fig. 20–96). If the crank and the rod form an angle of 150°, find the angle formed by the rod and the imaginary line.

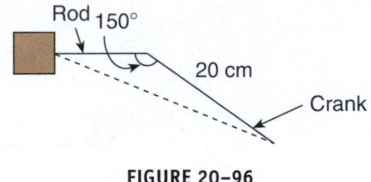

FIGURE 20–96

23. Find the magnitude of vector Z in Fig. 20–97 if vectors R and X_L form a right angle. All units are in ohms.

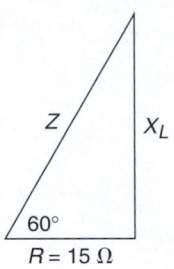

FIGURE 20–97

24. Find the amplitude, period, and phase shift, and graph two periods of the function $y = 4 \sin (x - 90°)$.

19–20 CUMULATIVE PRACTICE TEST

Given the end point of a vector in standard position, find the polar coordinates. Round lengths and degrees to tenths when necessary.

1. (9, 12)

2. (10, 24)

3. (2.7, 3.5)

Given the polar coordinates, find the rectangular coordinates. Round to tenths.

4. (10, 35°)

5. (7.4, 60°)

6. $\left(4, \dfrac{\pi}{3}\right)$

7. $\left(4.8, \dfrac{\pi}{8}\right)$

Use the sine, cosine, or tangent to find the sides and angles. Round sides to four significant digits and round angles to the nearest 0.1° when necessary. In $\angle ABC$, $\angle C$ is a right angle.

8. $BC = 4.6$ cm; $\angle A = 40°$. Find AB.

9. $AC = 23.5$ cm; $\angle A = 60°$. Find AB.

10. $AC = 32.1$ cm; $BC = 28.9$ cm. Find $\angle A$.

11. $AB = 15.8$ cm; $AC = 12.5$ cm. Find $\angle B$.

Find the magnitude and direction of the vector in standard position with the indicated end point. Round the magnitude to the nearest thousandth and the direction to the nearest tenth of a degree.

12. (3, 4)

13. (5, 2)

14. Find the resultant vector of the vectors that have a direction of 30° and magnitudes of 8 and 10, respectively.

Find the amplitude, period, and phase shift for the functions.

15. $y = \cos(x + 90°)$

16. $y = 5\sin(x - 180°)$

17. $y = 3\sin(2x - 240°)$

18. A triangle has sides that measure 8.3 cm, 8.9 cm, and 9.2 cm. Find the measure of each of the three angles. Round to the nearest 0.1°.

19. In $\triangle ABC$, $\angle C$ measures 38°. $AC = 34$ cm and $BC = 36$ cm. Find AB to the nearest centimeter.

20. In $\triangle ABC$, $\angle C$ measures 55°. $AC = 20.5$ cm and $BC = 26.2$ cm. Find AB to the nearest centimeter.

Selected Answers to Student Exercise Material

Answers to all Section Exercises and answers to odd-numbered exercises in the Chapter Review Exercises and Practice Tests are included here. Solutions for the Chapter Review Exercises and Practice Tests are available separately in the Student Solutions Manual (odd-numbered) or the Instructor's Resource Manual (even-numbered).

1   REVIEW OF BASIC CONCEPTS

Section 1–1   Exercises, p. 24

1   **1.** $6 < 8; 8 > 6$ **2.** $32 < 42; 42 > 32$ **3.** $148 < 196; 196 > 148$
4. $2,517 < 2,802; 2,802 > 2,517$ **5.** $7,809 < 8,902; 8,902 > 7,809$
6. $42,999 < 44,000; 44,000 > 42,999$
7. $\$183,500 < \$198,900; \$198,900 > \$183,500$
8. 583 flats $<$ 786 flats; 786 flats $>$ 583 flats
9. 758 rooms $<$ 893 rooms; 893 rooms $>$ 758 rooms
10. 2,807 e-mails $<$ 5,982 e-mails; 5,982 e-mails $>$ 2,807 e-mails

2   **11.** 0.5   **12.** 0.23   **13.** 0.07   **14.** 6.83   **15.** 0.079   **16.** 0.468

3   **17.** 3.72   **18.** 7.08   **19.** 0.3   **20.** 0.837 in. > 0.81 in.
21. the micrometer reading   **22.** yes   **23.** 0.04 in.   **24.** No. 10 wire
25. 72.3 kg   **26.** 1.87, 1.9, 1.92   **27.** 72.07, 72.1, 73
28. 0.026 kW > 0.003 kW; 800-W toaster   **29.** $0.394 < 0.621; 0.621$

4   **30.** 20,000,000,000,000 or 20 trillion   **31.** 300,000,000 or
300 million   **32.** 500   **33.** 430,000   **34.** 83,000,000,000
35. 300,000,000   **36.** \$550   **37.** \$3,295.86   **38.** 0.784 in.   **39.** 2.8 A
40. \$3   **41.** Drivers A, B, and D

5   **42.** 97,614   **43.** 7,007   **44.** 30,133   **45.** 15.7   **46.** 34.18
47. 129.97   **48.** 3.077 in.   **49.** 15.503 A   **50.** \$181.25   **51.** 391.4 ft
52. 100.9°   **53.** \$28,334.01   **54.** 402,000; 402,199   **55.** 84,000;
84,213   **56.** 180; 183.405   **57.** 60; 56.365   **58.** 80; 81.401
59. \$3,440; \$3,443.60   **60.** \$3,400; \$3,365   **61.** 159 lb   **62.** Yes,
total capacity available is 115 gal.   **63.** Yes, 479 pages are needed.
64. 350 ft   **65.** 4   **66.** 14   **67.** 1,020   **68.** 115   **69.** 53,036
70. 22 bags   **71.** 341 boxes   **72.** 291.82   **73.** 7.5   **74.** 310.8
75. 4.4   **76.** 12.08 in.; 12.10 in.   **77.** 4.189 in.; 4.201 in.   **78.** 200;
198.924   **79.** 20; 25.49   **80.** 18,497.55   **81.** 92,410.53   **82.** 5°F
83. 15.1 lb   **84.** 59.83 cm   **85.** 186 bricks   **86.** 45 m   **87.** 378
88. 1,260   **89.** 0   **90.** 0   **91.** 216,954   **92.** 102,612   **93.** 4,096
washers   **94.** 84 automobiles   **95.** 700 tickets   **96.** \$288   **97.** 56.55
98. 3.2445   **99.** 0.05805   **100.** 0.08672   **101.** 170.12283
102. 0.38381871   **103.** 4.9386274   **104.** 30.66   **105.** 5.25 in.
106. 624 ft   **107.** \$60; \$64.20   **108.** \$297.14   **109.** 630,000,000
110. 2,286,000,000   **111.** 239,200,000   **112.** 43   **113.** 47   **114.** 3
115. 3   **116.** 1.28   **117.** 15.21   **118.** 30 ft; 32 ft   **119.** \$1,245 per
tank   **120.** 6 in.   **121.** 10.9   **122.** 0.19   **123.** 25   **124.** 6 lb
125. 23 rolls   **126.** 600 revolutions   **127.** 0.7 ft   **128.** 0.16   **129.** 9
130. 80; 69 R 6   **131.** 30; 28 R 1   **132.** \$575   **133.** \$1.98   **134.** 85
135. 454 lb   **136.** \$765.32   **137.** 1.69 in.   **138.** 3.5 A
139. 473 crates

Section 1–2   Exercises, p. 33

1   **1.** 4; 3   **2.** 9; 4   **3.** 2.7; 9   **4.** 15; 2   **5.** 1,000   **6.** 16
7. 11.56   **8.** 15   **9.** 8   **10.** 81   **11.** 64   **12.** 324   **13.** 1.96   **14.** 169
15. Use base as a factor 5 times.   **16.** Use base as a factor 3 times.

2   **17.** 64   **18.** 144   **19.** 18.49   **20.** 79.21   **21.** 5   **22.** 7   **23.** 9
24. 6   **25.** 14   **26.** 21   **27.** 23   **28.** 32   **29.** 10^6 **30.** 10^{12}
31. 10^9 **32.** 10^{10} **33.** 100,000,000,000,000   **34.** 1,000,000,000
35. 1,000,000,000,000   **36.** 10,000

3   **37.** 1,000   **38.** 1,200,000   **39.** 20,000   **40.** 10,200   **41.** 25
42. 21   **43.** 3   **44.** 250   **45.** Shift the decimal to the *right* in the
number being multiplied by the power of 10 as indicated by the
exponent.   **46.** Shift the decimal to the *left* in the number being
divided by the power of 10 as indicated by the exponent.   **47.** 225
48. 343   **49.** 78,125   **50.** 20,736   **51.** 18   **52.** 28   **53.** 33
54. 14   **55.** 6   **56.** 8   **57.** 14   **58.** 33

Section 1–3   Exercises, p. 43

1   **1.** 26   **2.** 18   **3.** 9   **4.** 12.5   **5.** 24   **6.** 32   **7.** 18   **8.** 9
9. 10   **10.** 9   **11.** 30   **12.** 15   **13.** 156   **14.** 109   **15.** 23
16. 145   **17.** 7   **18.** 15   **19.** 160   **20.** 1,458   **21.** 80   **22.** 10
23. 24   **24.** 15.6   **25.** 25.04   **26.** 9   **27.** 5   **28.** 70   **29.** 13
30. 11   **31.** 29   **32.** 5   **33.** 39   **34.** 20   **35.** 30   **36.** 20.52
37. 483   **38.** 443   **39.** 79.16

2   **40.** 38 in.   **41.** 21.2 cm   **42.** 40 ft   **43.** 61 m   **44.** 45.8 ft
45. 218 m   **46.** 12 cm   **47.** 35.6 cm   **48.** 156 in.   **49.** 124 in.   **50.** 160 in.
51. 108 in.   **52.** 930 ft   **53.** 54 ft   **54.** 156 ft   **55.** 59 ft   **56.** 590 ft
57. 48 tiles   **58.** 6 ft^2 **59.** 116.84 ft^2 **60.** 9 cm^2 **61.** 79.21 cm^2
62. 72 in^2 **63.** 13.8 cm^2 **64.** 9,175 ft^2 **65.** 300 bundles
66. 103.823 ft^3 **67.** 2,097.152 cm^3 **68.** 6,426 cm^3 **69.** 16.704 m^3

3   **70.** 1,080 boxes of cards   **71.** 29 boxes

Chapter 1 Review Exercises, p. 53

1. $142 < 187; 187 > 142$ **3.** (a) 0.3   (b) 0.15   (c) 0.04   **5.** 4.79
7. 0.02; 0.021; 0.0216   **9.** $\dfrac{7}{8}$ **11.** (a) 500   (b) 50,000   (c) 41.4
(d) 6.90   (e) 23.4610   **13.** \$83   **15.** (a) 28   (b) 25   **17.** 34.9 kW
19. 10   **21.** (a) 4.61   (b) 3.127   (c) 204.899   (d) 12,140
23. \$100; \$87.25   **25.** 64.8   **27.** 0.43 in.   **29.** 8.930 in.; 8.940 in.   **31.** 84
33. 13,725   **35.** 394,254,080   **37.** \$1,407   **39.** \$43,920   **41.** \$14,800,
\$13,140   **43.** 1,140,000 ft^2, 1,204,010.28 ft^2 **45.** 0.12096 in.
47. 0.1 cm   **49.** \$40; \$50.14   **51.** 348,000   **53.** 24,075,000
55. 2,008.4   **57.** 3   **59.** 60.8   **61.** 664.0   **63.** 23 envelopes, 11 left
over   **65.** 48.79 ft   **67.** \$53   **69.** (a) 7, 3, 343   (b) 2.3, 4, 27.9841
(c) 8, 4, 4,096   **71.** (a) 1   (b) 15,625   (c) 31.36   (d) 441
73. (a) 10^1 (b) 10^3 (c) 10^4 (d) 10^5 **75.** (a) 7   (b) 0.04056
(c) 0.605   (d) 2.3079   **77.** (a) 6.16   (b) 3.81   **79.** 75   **81.** 54
83. 15   **85.** 15   **87.** 13   **89.** 13.6   **91.** 113.608   **93.** 3   **95.** 18
97. 61   **99.** 57 cm   **101.** 210 mm   **103.** 28.8 m   **105.** 66 ft
107. 288 in.   **109.** 2,450 mm^2 **111.** 51.84 m^2 **113.** 33 yd^2
115. \$14.25   **117.** 162 cm^2 **119.** 33,000 ft^2 **121.** 517.28 cm^3
123. \$38,339.49   **125.** \$53.82

Chapter 1 Practice Test, p. 57

1. 5.09   **3.** 48.3   **5.** 1,007   **7.** \$9,271,314   **9.** 134   **11.** 106
13. \$310, \$310   **15.** \$10, \$14   **17.** 42,730   **19.** 11.6   **21.** 83
23. 0.6   **25.** \$17,500   **27.** $P = 91$ ft; $A = 514.5$ ft^2
29. $P = 12.2$ in.; $A = 6.08$ in^2

2   REVIEW OF FRACTIONS

Section 2–1   Exercises, p. 68

1 1. $\dfrac{9}{248}$ **2.** $\dfrac{7}{15}$ **3.** $5 = 5 \times 1, 10 = 5 \times 2, 15 = 5 \times 3,$ $20 = 5 \times 4, 25 = 5 \times 5, 30 = 5 \times 6$ **4.** $6 = 6 \times 1, 12 = 6 \times 2,$ $18 = 6 \times 3, 24 = 6 \times 4, 30 = 6 \times 5, 36 = 6 \times 6$ **5.** $8 = 8 \times 1,$ $16 = 8 \times 2, 24 = 8 \times 3, 32 = 8 \times 4, 40 = 8 \times 5, 48 = 8 \times 6$
6. $9 = 9 \times 1, 18 = 9 \times 2, 27 = 9 \times 3, 36 = 9 \times 4, 45 = 9 \times 5,$ $54 = 9 \times 6$ **7.** $10 = 10 \times 1, 20 = 10 \times 2, 30 = 10 \times 3,$ $40 = 10 \times 4, 50 = 10 \times 5, 60 = 10 \times 6$ **8.** $30 = 30 \times 1,$ $60 = 30 \times 2, 90 = 30 \times 3, 120 = 30 \times 4, 150 = 30 \times 5,$ $180 = 30 \times 6$ **9.** $6 \times 1 = 6, 6 \times 2 = 12, 6 \times 3 = 18,$ $6 \times 4 = 24, 6 \times 5 = 30$; answers may vary. **10.** $12 \times 1 = 12,$ $12 \times 2 = 24, 12 \times 3 = 36, 12 \times 4 = 48, 12 \times 5 = 60$; answers may vary. **11.** $13 \times 1 = 13, 13 \times 2 = 26, 13 \times 3 = 39, 13 \times 4 = 52,$ $13 \times 5 = 65, 13 \times 6 = 78$; answers may vary. **12.** $3 \times 1 = 3,$ $3 \times 2 = 6, 3 \times 3 = 9, 3 \times 4 = 12, 3 \times 5 = 15$; answers may vary.
13. $50 \times 1 = 50, 50 \times 2 = 100, 50 \times 3 = 150, 50 \times 4 = 200,$ $50 \times 5 = 250$; answers may vary. **14.** $4 \times 1 = 4, 4 \times 2 = 8,$ $4 \times 3 = 12, 4 \times 4 = 16, 4 \times 5 = 20$; answers may vary.
15. No; not divisible by 6 **16.** Yes; ends in zero **17.** No, last two digits not divisible by 4 **18.** Yes; sum of digits divisible by 3 **19.** Yes; no remainder **20.** Yes; the three digits divisible by 8 **21.** Yes; sum of digits divisible by 3 **22.** Yes; divisible by both 2 and 3 **23.** Yes; last digit even number

2 24. $1 \cdot 4, 2 \cdot 2$ **25.** $1 \cdot 8, 2 \cdot 4$ **26.** $1 \cdot 12, 2 \cdot 6, 3 \cdot 4$
27. $1 \cdot 15, 3 \cdot 5$ **28.** $1 \cdot 16, 2 \cdot 8, 4 \cdot 4$ **29.** $1 \cdot 20, 2 \cdot 10, 4 \cdot 5$
30. $1 \cdot 24, 2 \cdot 12, 3 \cdot 8, 4 \cdot 6$ **31.** $1 \cdot 30, 2 \cdot 15, 3 \cdot 10, 5 \cdot 6$
32. $1 \cdot 36, 2 \cdot 18, 3 \cdot 12, 4 \cdot 9, 6 \cdot 6$ **33.** $1 \cdot 38, 2 \cdot 19$ **34.** 1, 2, 4, 5, 10, 20, 40 **35.** 1, 2, 23, 46 **36.** 1, 2, 4, 13, 26, 52 **37.** 1, 2, 4, 8, 16, 32, 64 **38.** 1, 2, 3, 4, 6, 8, 9, 12, 18, 24, 36, 72 **39.** 1, 3, 9, 27, 81 **40.** 1, 5, 17, 85 **41.** 1, 2, 4, 23, 46, 92 **42.** 1, 2, 3, 4, 6, 8, 12, 16, 24, 32, 48, 96 **43.** 1, 2, 7, 14, 49, 98

3 44. Prime **45.** Composite **46.** Composite **47.** Prime **48.** Composite **49.** Composite **50.** Composite **51.** Composite **52.** Composite **53.** Prime **54.** Composite **55.** Prime **56.** $2 \cdot 2 \cdot 3$ **57.** $2 \cdot 3 \cdot 3$ **58.** $2 \cdot 2 \cdot 5$ **59.** $2 \cdot 2 \cdot 2 \cdot 3$ **60.** $5 \cdot 5$ **61.** $3 \cdot 3 \cdot 3$ **62.** 29 **63.** $2 \cdot 3 \cdot 5$ **64.** $5 \cdot 7$ **65.** $2 \cdot 2 \cdot 2 \cdot 5$ **66.** 47 **67.** $7 \cdot 7$ **68.** $2 \cdot 5 \cdot 5$ **69.** $2 \cdot 2 \cdot 13$ **70.** $5 \cdot 13$ **71.** $3 \cdot 5 \cdot 5$ **72.** $2 \cdot 2 \cdot 5 \cdot 5$ **73.** $3 \cdot 5 \cdot 7$ **74.** $2 \cdot 2 \cdot 3 \cdot 3 \cdot 3$ **75.** $5 \cdot 23$ **76.** $11 \cdot 11$ **77.** $2 \cdot 2 \cdot 2 \cdot 2 \cdot 3$ **78.** $2 \cdot 2 \cdot 3 \cdot 13$ **79.** 157 **80.** $2^3 \cdot 3^2$ **81.** $2^4 \cdot 7$ **82.** $2^2 \cdot 31$ **83.** $2^2 \cdot 41$ **84.** $2^3 \cdot 71$ **85.** $2^2 \cdot 3^2 \cdot 5^2$

4 86. 6 **87.** 30 **88.** 56 **89.** 12 **90.** 72 **91.** 60 **92.** 24 **93.** 18 **94.** 24 **95.** 700 **96.** 27 **97.** 16 **98.** 90 **99.** 120 **100.** 180 **101.** 30 **102.** 96 **103.** 72 **104.** 60 **105.** 300 **106.** 66 **107.** 312 **108.** 14 **109.** 144 **110.** 1 **111.** 1 **112.** 1 **113.** 6 **114.** 5 **115.** 12 **116.** 9 **117.** 5 **118.** 2 **119.** 6 **120.** 5 **121.** 3 **122.** 2 **123.** 2 **124.** 2

Section 2–2   Exercises, p. 80

1 1. $\dfrac{8}{10}, \dfrac{12}{15}, \dfrac{16}{20}, \dfrac{20}{25}, \dfrac{24}{30}$ **2.** $\dfrac{14}{20}, \dfrac{21}{30}, \dfrac{28}{40}, \dfrac{35}{50}, \dfrac{42}{60}$ **3.** $\dfrac{3}{4} = \dfrac{18}{24}$
4. $\dfrac{6}{16}$ **5.** $\dfrac{12}{21}$ **6.** $\dfrac{36}{44}$ **7.** $\dfrac{5}{15}$ **8.** $\dfrac{20}{24}$ **9.** $\dfrac{21}{24}$ **10.** $\dfrac{12}{30}$ **11.** $\dfrac{1}{2}$
12. $\dfrac{3}{5}$ **13.** $\dfrac{3}{4}$ **14.** $\dfrac{5}{16}$ **15.** $\dfrac{1}{2}$ **16.** $\dfrac{7}{8}$ **17.** $\dfrac{5}{16}$ **18.** $\dfrac{1}{5}$ **19.** $\dfrac{1}{5}$
20. $\dfrac{6}{25}$ **21.** $\dfrac{5}{8}$ **22.** $\dfrac{1}{4}$ **23.** $\dfrac{3}{4}$ **24.** False **25.** True **26.** $4\dfrac{9}{16}$ in.

27. $4\dfrac{1}{16}$ in. **28.** $3\dfrac{13}{16}$ in. **29.** $3\dfrac{3}{8}$ in. **30.** $2\dfrac{1}{4}$ in. **31.** 2 in.
32. $1\dfrac{3}{4}$ in. **33.** $1\dfrac{3}{16}$ in. **34.** $\dfrac{3}{4}$ in. **35.** $\dfrac{3}{8}$ in. or $\dfrac{7}{16}$ in.

2 36. $2\dfrac{2}{5}$ **37.** $1\dfrac{3}{7}$ **38.** 1 **39.** $4\dfrac{4}{7}$ **40.** 4 **41.** $2\dfrac{1}{7}$ **42.** $2\dfrac{5}{9}$
43. $9\dfrac{2}{5}$ **44.** $9\dfrac{5}{9}$ **45.** $1\dfrac{17}{21}$ **46.** $3\dfrac{4}{5}$ **47.** 16 **48.** $7\dfrac{1}{5}$ **49.** $9\dfrac{1}{2}$
50. 9

3 51. $\dfrac{15}{3}$ **52.** $\dfrac{18}{2}$ **53.** $\dfrac{56}{8}$ **54.** $\dfrac{32}{4}$ **55.** $\dfrac{48}{16}$ **56.** $\dfrac{7}{3}$ **57.** $\dfrac{25}{8}$
58. $\dfrac{15}{8}$ **59.** $\dfrac{77}{12}$ **60.** $\dfrac{77}{8}$ **61.** $\dfrac{31}{8}$ **62.** $\dfrac{89}{12}$ **63.** $\dfrac{103}{16}$ **64.** $\dfrac{257}{32}$
65. $\dfrac{69}{64}$ **66.** $\dfrac{73}{10}$ **67.** $\dfrac{26}{3}$ **68.** $\dfrac{100}{3}$ **69.** $\dfrac{200}{3}$ **70.** $\dfrac{25}{2}$

4 71. $\dfrac{1}{2}$ **72.** $\dfrac{1}{10}$ **73.** $\dfrac{1}{5}$ **74.** $\dfrac{7}{10}$ **75.** $\dfrac{1}{4}$ **76.** $\dfrac{1}{40}$ **77.** $3\dfrac{9}{10}$
78. $4\dfrac{4}{5}$ **79.** $\dfrac{189}{500}$ **80.** $\dfrac{7}{8}$ **81.** $\dfrac{3}{8}$ **82.** $\dfrac{5}{8}$ **83.** $\dfrac{3}{4}$ in. **84.** $\dfrac{3}{16}$ ft
85. $2\dfrac{3}{8}$ in. **86.** $\dfrac{83}{100}$ lb **87.** $\dfrac{5}{16}$ in. **88.** $3\dfrac{1}{8}$ in. **89.** 0.6 **90.** 0.3
91. 0.875 **92.** 0.375 **93.** 0.45 **94.** 0.98 **95.** 0.21 **96.** 3.875
97. 1.4375 **98.** 4.5625 **99.** 0.17 **100.** 0.44 **101.** 0.83 **102.** 0.58
103. 0.67 **104.** 0.27 **105.** 0.78 **106.** 0.38 **107.** 1.43 **108.** 3.45
109. 2.38 **110.** 5.57 **111.** 2.046875 in. **112.** \$0.045 **113.** 0.125 in.
114. $47\dfrac{3}{5}$ m

5 115. 72 **116.** 30 **117.** 50 **118.** 48 **119.** 24 **120.** $\dfrac{2}{3}$
121. $\dfrac{7}{16}$ **122.** $\dfrac{8}{9}$ **123.** $\dfrac{11}{16}$ **124.** $\dfrac{15}{32}$ **125.** $\dfrac{7}{12}$ **126.** $\dfrac{4}{5}$ **127.** $\dfrac{9}{10}$
128. $\dfrac{4}{15}$ **129.** $\dfrac{1}{2}$ **130.** Yes, $\dfrac{3}{8}$ is greater **131.** No, $\dfrac{15}{64}$ is greater
132. Yes **133.** Yes, $\dfrac{7}{16}$ is greater **134.** No **135.** $\dfrac{5}{8}$ in. end
136. Yes **137.** No **138.** Too small **139.** No **140.** 0.37 **141.** $\dfrac{4}{5}$
142. $\dfrac{5}{12}$ **143.** $\dfrac{1}{2}$ **144.** 0.34

Section 2–3   Exercises, p. 87

1 1. $\dfrac{3}{8}$ **2.** $1\dfrac{3}{8}$ **3.** $\dfrac{5}{8}$ **4.** $\dfrac{25}{32}$ **5.** $\dfrac{9}{16}$ **6.** $1\dfrac{7}{16}$ **7.** $\dfrac{11}{64}$ **8.** $1\dfrac{19}{40}$
9. $1\dfrac{23}{36}$ **10.** $1\dfrac{1}{12}$ **11.** $1\dfrac{2}{5}$ **12.** $\dfrac{19}{21}$ **13.** $\dfrac{15}{16}$ in. **14.** $1\dfrac{27}{32}$ in.
15. $1\dfrac{15}{16}$ in. **16.** $1\dfrac{7}{8}$ in. **17.** $1\dfrac{1}{4}$ in. **18.** $6\dfrac{4}{5}$ **19.** $4\dfrac{1}{8}$ **20.** $16\dfrac{13}{16}$
21. $1\dfrac{11}{18}$ **22.** $4\dfrac{13}{16}$ **23.** $5\dfrac{17}{32}$ **24.** $4\dfrac{11}{16}$ **25.** $13\dfrac{1}{2}$ **26.** $1{,}001\dfrac{31}{36}$
27. $370\dfrac{37}{120}$ **28.** $1{,}115\dfrac{2}{45}$ **29.** $125\dfrac{21}{160}$ **30.** $8\dfrac{13}{16}$ in. **31.** $3\dfrac{15}{16}$ in.
32. $11\dfrac{5}{8}$ gal **33.** $24\dfrac{5}{32}$ in. **34.** $22\dfrac{7}{8}$ in. **35.** $5\dfrac{7}{8}$ c

2 36. $\dfrac{3}{16}$ **37.** $\dfrac{3}{16}$ **38.** $\dfrac{1}{8}$ **39.** $\dfrac{9}{64}$ **40.** $17\dfrac{3}{4}$ **41.** $4\dfrac{15}{16}$
42. $5\dfrac{21}{32}$ **43.** $1\dfrac{5}{7}$ **44.** $8\dfrac{13}{16}$ in. **45.** $3\dfrac{1}{10}$ lb **46.** $10\dfrac{1}{8}$ in.
47. $34\dfrac{3}{4}$ ft **48.** $3\dfrac{19}{24}$ **49.** $18\dfrac{17}{48}$ **50.** $350\dfrac{79}{175}$ **51.** $759\dfrac{19}{56}$
52. Yes; $60\dfrac{5}{16}$ in. < 60.4 in. **53.** $\dfrac{3}{16}$ in. **54.** $\dfrac{29}{32}$ in.

Section 2–4 Exercises, p. 99

1 1. $\frac{3}{32}$ 2. $\frac{7}{32}$ 3. $\frac{7}{16}$ 4. $\frac{7}{12}$ 5. $\frac{1}{3}$ 6. $\frac{5}{32}$ 7. $21\frac{7}{8}$ 8. 75
9. $4\frac{1}{8}$ 10. $36\frac{1}{10}$ 11. $1\frac{21}{40}$ 12. $2\frac{1}{6}$ 13. $29\frac{3}{4}$ in. 14. $1\frac{5}{8}$ in.
15. $18\frac{3}{4}$ L 16. 90 in. 17. 264 kg copper, 84 kg tin, 36 kg zinc
18. 148 in.

2 19. $\frac{4}{25}$ 20. $\frac{1}{16}$ 21. $\frac{27}{125}$ 22. $\frac{1}{27}$ 23. $\frac{49}{81}$

3 24. $\frac{8}{5}$ 25. $\frac{5}{11}$ 26. $\frac{1}{8}$ 27. $\frac{10}{9}$ 28. $\frac{5}{9}$ 29. $\frac{6}{7}$ 30. $\frac{9}{10}$
31. $\frac{11}{12}$ 32. 2 33. $4\frac{2}{3}$ 34. 4 rotations 35. 6 turns. 36. $13\frac{1}{3}$
37. $12\frac{1}{2}$ 38. $\frac{5}{8}$ 39. 4 40. $2\frac{1}{2}$ 41. 12 shovelfuls 42. 20 ft × 15 ft
43. 7 strips 44. 4 whole pieces 45. 23 straws, $3\frac{3}{4}$ in. left
46. $23\frac{11}{12}$ in. 47. $1\frac{1}{2}$ ft 48. $3\frac{1}{5}$ 49. $8\frac{1}{3}$ 50. $\frac{5}{6}$

4 51. $17\frac{7}{40}$ 52. $82\frac{26}{245}$ 53. $32\frac{44}{45}$ 54. $5\frac{8}{9}$ 55. $5\frac{4}{9}$ 56. 3
57. $\frac{3}{4}$ 58. $4\frac{1}{6}$ 59. $3\frac{37}{108}$ 60. $\frac{9}{14}$

Chapter 2 Review Exercises, p. 108

1. $\frac{7}{10}$ 3. $11 = 1(11), 22 = 2(11), 33 = 3(11), 44 = 4(11), 55 = 5(11)$
5. 21, 42, 63, 84, 105; answers may vary. 7. 7, 14, 21, 28, 35; answers may vary. 9. 8, 16, 24, 32, 40; answers may vary.
11. Yes; sum of digits divisible by 3 13. No, divisible by 2 but not by 3 15. Yes; ends in zero 17. No; sum of digits not divisible by 9
19. $1 \times 48, 2 \times 24, 3 \times 16, 4 \times 12, 6 \times 8$; 1, 2, 3, 4, 6, 8,12, 16, 24, 48
21. $1 \times 51, 3 \times 17$; 1, 3, 17, 51 23. $1 \times 74, 2 \times 37$; 1, 2, 37, 74
25. Composite 27. Composite 29. $2 \cdot 3 \cdot 7$ 31. $2 \cdot 7 \cdot 7$ or $2 \cdot 7^2$
33. 360 35. 180 37. 2 39. 6 41. $\frac{4}{6}, \frac{6}{9}, \frac{8}{12}, \frac{10}{15}, \frac{12}{18}$; answers may vary. 43. $\frac{25}{60}$ 45. $\frac{10}{15}$ 47. $\frac{24}{32}$ 49. $\frac{11}{55}$ 51. $\frac{16}{20}$ 53. $\frac{1}{2}$ 55. $\frac{1}{8}$
57. $\frac{1}{4}$ 59. $\frac{17}{32}$ 61. $\frac{3}{8}$ 63. True 65. $5\frac{1}{4}$ in. 67. $4\frac{7}{16}$ in.
69. $3\frac{15}{16}$ in. 71. $3\frac{9}{16}$ in. 73. $2\frac{3}{4}$ in. 75. $1\frac{1}{2}$ in. 77. $2\frac{1}{8}$ in.
79. $3\frac{3}{5}$ 81. $4\frac{7}{8}$ 83. $5\frac{3}{8}$ 85. $87\frac{1}{2}$ 87. $1\frac{1}{2}$ 89. $\frac{8}{1}$ 91. $\frac{57}{8}$
93. $\frac{147}{16}$ 95. $\frac{23}{5}$ 97. $\frac{12}{1}$ 99. $\frac{16}{3}$ 101. $\frac{20}{10}$ 103. $\frac{33}{3}$ 105. $\frac{7}{10}$
107. $\frac{19}{20}$ 109. $\frac{109}{125}$ 111. $\frac{1}{50}$ 113. 0.2 115. 0.625 117. 0.818
119. 16 121. 60 123. $\frac{3}{8}$ 125. $\frac{3}{8}$ 127. $\frac{3}{16}$ 129. $\frac{27}{32}$ 131. $\frac{9}{19}$
133. Larger 135. No 137. Smaller 139. $\frac{2}{5}$ 141. $\frac{3}{4}$ 143. $\frac{3}{11}$
145. $\frac{21}{64}$ 147. $1\frac{13}{30}$ 149. $\frac{7}{8}$ in. 151. $10\frac{3}{8}$ 153. $8\frac{19}{32}$ 155. $10\frac{9}{32}$
157. $18\frac{1}{16}$ mi 159. $12\frac{19}{32}$ in. 161. $15\frac{25}{32}$ in. 163. $\frac{1}{3}$ 165. $1\frac{5}{8}$
167. $5\frac{31}{32}$ 169. $7\frac{11}{16}$ 171. $34\frac{3}{4}$ 173. $\frac{7}{16}$ in. 175. $1\frac{29}{64}$ in.

177. $\frac{7}{24}$ 179. $\frac{7}{24}$ 181. $\frac{1}{2}$ 183. $7\frac{7}{8}$ 185. $1\frac{1}{5}$ 187. 2
189. $100\frac{1}{2}$ in. 191. $2\frac{3}{4}$ cups 193. 150 cm 195. $\frac{9}{16}$ 197. $\frac{16}{81}$
199. $\frac{1}{8}$ 201. $\frac{81}{100}$ 203. $\frac{1}{4}$ 205. $\frac{10}{7}$ or $1\frac{3}{7}$ 207. $1\frac{1}{6}$ 209. $9\frac{1}{3}$
211. 24 213. 2 215. $\frac{3}{5}$ 217. $2\frac{55}{64}$ in. 219. $\frac{3}{16}$ yd 221. 36 lengths
223. $\frac{1}{18}$ 225. $5\frac{1}{3}$ 227. $\frac{1}{4}$ 229. $\frac{1}{8}$ 231. $\frac{11}{18}$ 233. $\frac{6}{19}$

Chapter 2 Practice Test, p. 113

1. $\frac{3}{4}$ 3. 3 5. $\frac{34}{7}$ 7. $2^5 \cdot 3$ 9. $2 \cdot 31$ 11. 540 13. 6 15. $1\frac{1}{4}$ in.
17. 10.5 19. $\frac{7}{10}$ 21. $7\frac{17}{30}$ 23. $3\frac{8}{9}$ 25. $3\frac{1}{2}$ 27. $13\frac{1}{2}$ 29. $\frac{2}{7}$
31. $3\frac{5}{6}$ c 33. 8 yd

3 PERCENTS

Section 3–1 Exercises, p. 119

1 1. 20% 2. 14% 3. 0.7% 4. 1.25% 5. 500% 6. 800%
7. 305% 8. 720% 9. 1,510% 10. 3,625% 11. 27.3%
12. 75.2% 13. 62.5% 14. 77.8% 15. 0.7% 16. 0.3%
17. 133.3% 18. 350% 19. 430% 20. 220% 21. 40%
22. 70% 23. 60% 24. 90%

2 25. 0.36 26. 0.45 27. 0.2 28. 0.75 29. 0.0625 30. 0.625
31. 0.6667 32. 0.006 33. 0.002 34. 0.0005 35. 0.0833
36. 0.1875 37. 8 38. 4 39. 2.5 40. 4.25 41. 1.76 42. 3.8
43. 1.375 44. 3.875 45. 1.153 46. 2.125 47. 1.0625 48. 0.548
49. 0.257 under 18 years old; 0.004 Native Hawaiian or other Pacific
Islander 50. $\frac{9}{20}$ 51. $\frac{9}{25}$ 52. $\frac{3}{4}$ 53. $\frac{1}{5}$ 54. $\frac{5}{8}$ 55. $\frac{1}{16}$ 56. $\frac{3}{500}$
57. $\frac{2}{3}$ 58. $\frac{1}{2,000}$ 59. $\frac{1}{500}$ 60. $\frac{3}{16}$ 61. $\frac{1}{12}$ 62. $4\frac{1}{4}$ 63. $2\frac{1}{2}$
64. $3\frac{4}{5}$ 65. $1\frac{19}{25}$ 66. $3\frac{7}{8}$ 67. $1\frac{3}{8}$ 68. $1\frac{2}{3}$ 69. $3\frac{1}{6}$ 70. $\frac{3}{10}$
women ownership; $\frac{3}{20}$ Hispanic ownership 71. $\frac{1}{200}$ American
Indian or Alaska Native

Section 3–2 Exercises, p. 134

1 1. $R = 40\%$; $B = 18$; P missing 2. $R = 66\%$; $B = \$35.99$;
P missing 3. $P = \$12.21$; $R = 27\%$; B missing 4. $R = 83\%$; $B = 12$;
P missing 5. R missing; $B = 10$; $P = 2$ 6. $P = 2$; $R = 20\%$;
B missing 7. $P = 3$; R missing; $B = 4$ 8. R missing;
$B = 25$; $P = 5$ 9. $P = 6$; $B = 15$; R missing 10. P missing;
$R = 20\%$; $B = 15$ 11. $R = 35\%$; B missing; $P = 70$
12. R missing; $B = \$45$; $P = \$3.15$

2 13. 5.4 14. 252 15. 1.8 16. 26 17. 50% 18. $333\frac{1}{3}\%$
19. 87.5% 20. 150% 21. 64 22. 70,000 23. $51.66 saved
24. $1.04 per hour raise 25. 83 gallons 26. $44,167 bonus
27. $75,000 28. 55 shirts 29. $6,373.91 30. 890 shirts
31. 92% correct 32. 14.43 % vacant seats 33. 6% 34. 8.75%
35. 300% 36. 225% 37. $2,800 38. 342 39. $4.67 40. $8.60

3 **41.** 75 **42.** 63 **43.** 206 **44.** 115.92 **45.** 0.675 **46.** 0.94 **47.** 154.1 **48.** 231 **49.** 924 **50.** 345 **51.** 72 **52.** 50 **53.** 344 **54.** 46 **55.** 360 **56.** 275 **57.** 75 **58.** 250 **59.** 18.4 **60.** 261

61. 25% **62.** 30% **63.** $33\frac{1}{3}$% **64.** 32.9% **65.** 15.75% **66.** 16%

67. 0.8% **68.** $0.66\frac{2}{3}$% or $\frac{2}{3}$% **69.** 500% **70.** 111.25% **71.** 3.75

72. 0.625 **73.** 9.375 **74.** 0.021 lb **75.** 1.0625 lb **76.** $575

77. 100,880 welds **78.** 0.4 **79.** 0.4 **80.** $37\frac{1}{2}$% **81.** $66\frac{2}{3}$%

82. 3% **83.** 20% **84.** 11% **85.** 5% **86.** 350 **87.** 14.25 **88.** 220 **89.** 200 hp **90.** 200 **91.** 1,200 lb **92.** 9,625 swabs

4 **93.** $4.55, $80.38 **94.** $829.06 **95.** $23.30 **96.** 25% **97.** 23% **98.** $41.93 **99.** $11,832 **100.** $17.52 **101.** $3,866.00 **102.** $944.50 **103.** $434.84 **104.** $3,205.75 **105.** $9.75 **106.** $150 **107.** $1,460 **108.** $171.50 **109.** 1.75% **110.** 2.5%

111. 6.8% **112.** 30.1% **113.** 111 lb **114.** $8\frac{3}{4}$ lb **115.** $32.91

116. 21,600 lb **117.** 170 lb **118.** $47.94 **119.** $4,285.71 **120.** $1,364.16

Section 3-3 Exercises, p. 144

1 **1.** $3,600 **2.** $7.99 **3.** 108 **4.** 109.2 **5.** 5,496; 23,696 **6.** 66,383; 273,183 **7.** 10.2 **8.** 33.75

2 **9.** 2,393 board feet **10.** 1,815 board feet **11.** $16,802.50 **12.** 25,908 bricks **13.** $53,350.20 **14.** $2,041.46 **15.** 27 in. **16.** 1,366.4 yd³ **17.** $3,961.82 **18.** $58.80 **19.** 1,470 programs

20. 1,545 programs **21.** 60% **22.** 82% **23.** 13.7% **24.** $66\frac{2}{3}$%

25. 99.91% **26.** 99% **27.** Estimate: $40(0.7) = $28 **28.** Estimate: $52 + $50(0.1) = $57

3 **29.** 7% **30.** 8% **31.** 20% **32.** 5.0% **33.** 17% **34.** 20% **35.** 12% **36.** 10 yd³ **37.** 350 hp **38.** 25 lb **39.** 40 in. **40.** 150 hp

Chapter 3 Review Exercises, p. 150

1. 70% **3.** 83% **5.** 12,500% **7.** 72% **9.** 23% **11.** $83\frac{1}{3}$%

13. 320% **15.** 2.272 **17.** 3.4 **19.** 0.83 **21.** 0.625 **23.** $\frac{18}{25}$

25. $\frac{1}{8}$ **27.** $\frac{1}{150}$ **29.** $2\frac{3}{4}$ **31.** $1\frac{1}{8}$ **33.** $R = 5\%$; $B = 180$;
P missing **35.** $R = 45\%$; B missing; $P = 36 **37.** $P = 6$;
R missing; $B = 25$ **39.** $R = 18\%$; $B = 150$; P missing **41.** 24
43. 0.4375 **45.** 60% **47.** 250% **49.** $16 **51.** $7 **53.** 135.29

55. 83 **57.** 152 **59.** 84 **61.** 500% **63.** 37.5% or $37\frac{1}{2}$%

65. 1.14% **67.** $266 **69.** 4% **71.** 200 students **73.** $204.35 **75.** $439.84 **77.** $365.66 **79.** $1,707.60 **81.** 1.75% **83.** $3,935.63 **85.** $0.89 **87.** 8.5% **89.** $369.69 **91.** $355 **93.** 0.14 **95.** 0.07 m **97.** 7% **99.** 14% **101.** 142.1 kg **103.** 62 cm, 63 cm **105.** 87.1%

107. 78.5% **109.** $67\frac{3}{5}$% **111.** 8% **113.** $3,200 **115.** $1.69

117. 12.5% **119.** 57.5 lb **121.** $33\frac{1}{3}$% **123.** 4.8% **125.** 289 hp

127. 52% **129.** 26.6% **131.** 28.6%

Chapter 3 Practice Test, p. 155

1. 80% **3.** 0.003 **5.** $R = 40\%$; $B = 10$; P missing **7.** $P = 9$;
R missing; $B = 27$ **9.** $R = 12\%$; $B = 50$; P missing **11.** 9 **13.** 305
15. 115 **17.** 67.10 **19.** 12% **21.** $525 **23.** $2,500 **25.** 21.14%
27. 13.17% **29.** 5% **31.** $104.87

Chapters 1–3 Cumulative Practice Test, p. 155

1. 14 **3.** $P = 128$cm **5.** 42.82 **7.** Factored form = $(2)(2)(3)(5)(7)$;
Exponential notation = $2^2(3)(5)(7)$ **9.** 81.76 **11.** 454.11938

13. $19\frac{1}{8}$ **15.** 2 **17.** $1\frac{4}{17}$ **19.** 12.25 **21.** 1.4 **23.** 60

25. $10\frac{5}{12}$ ft or 10 ft 5 in. **27.** 1,775 in. **29.** $R = 83\%$ **31.** 7.14%

33. 86% **35.** 44.4% protein

4   MEASUREMENT

Section 4-1   Exercises, p. 169

1 **1.** $\frac{2\ pt}{1\ qt}$; $\frac{1\ qt}{2\ pt}$ **2.** $\frac{5,280\ ft}{1\ mi}$; $\frac{1\ mi}{5,280\ ft}$ **3.** $\frac{12\ in.}{1\ ft}$; $\frac{1\ ft}{12\ in.}$

4. $\frac{3\ ft}{1\ yd}$; $\frac{1\ yd}{3\ ft}$ **5.** 48 in. **6.** 21 ft **7.** 4,400 yd **8.** $9\frac{1}{3}$ yd

9. $2\frac{3}{10}$ lb or 2.3 lb **10.** 732.8 oz **11.** 20 qt **12.** 13 pt

13. 48 c **14.** $2\frac{1}{4}$ qt

2 **15.** feet to yards = $\frac{1}{3}$ or 0.3333333; yards to feet = 3

16. quarts to gallons = $\frac{1}{4}$ or 0.25; gallons to quarts = 4

17. 28.75 pt **18.** 92 pt **19.** 36.25 lb **20.** 153 in. **21.** 7,920 ft

22. 28 qt **23.** 3 ft 8 in. **24.** 2 mi 1,095 ft **25.** 3 lb $3\frac{1}{2}$ oz

26. 2 gal 1 qt **27.** 2 ft 10 in. **28.** 6 lb 9 oz **29.** 4 gal 2 qt 16 oz or 4 gal 2 qt 1 pt **30.** 6 qt 20 oz or 6 qt 1 pt 4 oz

3 **31.** 2 lb 5 oz or 37 oz **32.** 4 ft 7 in. or 55 in. **33.** 15 lb 11 oz

34. 18 ft 3 in. **35.** 8 qt $\frac{1}{2}$ pt **36.** 14 gal 1 qt **37.** 9 yd 1 ft **38.** 7 yd
1 ft 5 in. **39.** 31 in. **40.** 11 ft **41.** 5 lb 15 oz **42.** 12 lb **43.** 6 in.
44. 1 pt **45.** 2 ft **46.** 4 lb 14 oz **47.** 1 lb 6 oz **48.** 7 lb 2 oz

49. 2 in. **50.** 3 gal 3 qt 1 $\frac{1}{2}$ pt or 3 gal 3 qt 1 pt 1 c **51.** 4 ft 2 in.

52. 2 ft 3 in. **53.** 45 s **54.** 39 s **55.** 69 lb 7 oz **56.** 16 lb 14 oz

4 **57.** 60 mi **58.** 108 gal **59.** 57 lb 8 oz **60.** 58 ft **61.** 36 lb
62. 7 qt 1 pt or 1 gal 3 qt 1 pt **63.** 35 in² **64.** 108 ft² **65.** 180 yd²
66. 108 mi² **67.** 378 tiles **68.** 46 gal 18 oz **69.** 7 qt 1 pt or 1 gal
3 qt 1 pt **70.** 6 lb 4 oz **71.** 10 yd 1 ft 3 in. **72.** 1 day 15 h
73. 1 yd 1 ft 7 in. **74.** 7 ft 6 in. **75.** 3 gal 2 qt **76.** 9 pieces
77. 5 ft **78.** 3 gal 1 qt 5 oz **79.** 24 lb 3 oz **80.** 3 **81.** 18 **82.** 3
83. 8 pieces **84.** 9 boxes **85.** 24 cans **86.** 9 tickets

5 **87.** $\frac{3}{4}\frac{lb}{min}$ **88.** $15,840\frac{ft}{h}$ **89.** $2,304\frac{oz}{min}$ **90.** $2\frac{qt}{s}$

91. $3\frac{qt}{min}$ **92.** $53.3\frac{lb}{min}$

Section 4–2   Exercises, p. 182

1   1. (a) 1,000 m   (b) 10 L   (c) $\frac{1}{10}$ of a gram   (d) $\frac{1}{1,000}$ of a meter
(e) 100 g   (f) $\frac{1}{100}$ of a liter   2. b   3. b   4. c   5. b   6. a   7. c
8. b   9. c   10. a   11. b   12. c   13. b   14. a   15. b   16. b

2   17. 40   18. 70   19. 580   20. 80   21. 2.5   22. 210   23. 85
24. 142   25. 153 mL   26. 460 m   27. 75 dkg   28. 160 mm
29. 400   30. 8,000   31. 58,000   32. 800   33. 250   34. 2,100
35. 102,500   36. 8,330   37. 2,000,000   38. 70   39. 236 L
40. 467 cm   41. 38,000 dg   42. 13,000 cm   43. 2.8   44. 23.8
45. 10.1   46. 6   47. 2.9   48. 19.25   49. 1.7   50. 438.9 dm
51. 4.7 g   52. 0.225 dL   53. 2.743   54. 0.385   55. 0.15   56. 0.08
57. 2,964.84   58. 0.2983   59. 0.0003   60. 0.004   61. 0.002857
62. 15.285   63. 0.0297 hm   64. 0.00003 L

3   65. 11 m   66. 12 hL   67. 6 cg   68. 2.4 dm or 24 cm
69. 5.9 cL or 59 mL   70. cannot add   71. 10.1 kL or 101 hL
72. 0.55 g or 55 cg   73. cannot subtract   74. 7.002 km or 7,002 m
75. 1,000 mL   76. 1.47 kL or 147 dkL   77. 516 m   78. 40.8 m
79. 150.96 dm   80. 969.5 m   81. 4,680 mm   82. 16 g   83. 13 m
84. 9 cL   85. 163 g   86. 0.4 m or 4 dm   87. 5   88. 30
89. 16 prescriptions   90. 80 containers   91. 5.74 dL or 57.4 cL
92. 2.3 dkm or 23 m   93. 165.7 hm or 16.57 km   94. 9.1 kL or 91 hL
95. 1.25 cL   96. 70 mm   97. 7.5 dm   98. 50 mL
99. 21,250 containers   100. 19 vials

Section 4–3   Exercises, p. 188

1   1. 1.5 days   2. 9.67 min   3. 150 min   4. 318 s   5. 210 min
6. 3.03 min   7. 432 min   8. 1.47 min

2   9. 4,320 lb/h   10. 2,400 gal/h   11. 10,950 lb/yr
12. 1,680 vehicles/h   13. 0.0167 mi/s   14. 2.0833 gal/min
15. 60 lb/min   16. 1.6 gal/s   17. 50,000 species   18. 215,000 acres

3   19. 35°C   20. 0°C   21. 45°C   22. 5°C   23. 15°C   24. 10°C   25. 65°C
26. 50°C   27. 80°C   28. 120°C   29. 32.2° to 33.9°C   30. 35° to 37°C
31. 158°F   32. 59°F   33. 113°F   34. 122°F   35. 68°F   36. 419°F
37. 590°F   38. 770°F   39. 365°F   40. 32°F   41. 90°F   42. 68°F

4   43. 0.045 s   44. 0.000000805 s   45. 500,000,000 Hz   46. 0.42 H
47. 1,400,000 W   48. 20 images

Section 4–4   Exercises, p. 191

1   1. 354.33 in.   2. 131.232 yd   3. 26.0988 mi   4. 6.3402 qt
5. 9.463 L   6. 59.5242 lb   7. 22.68 kg   8. 17.78 cm   9. 5.4864 m
10. 1.3188 oz   11. 21.7649 L   12. 23.9498 ft   13. 4.99999 in.
14. 0.4724 in.   15. 146.029 mi   16. 310.7 mi   17. 711.2 mm
18. 91.44 cm   19. 1.2192 m   20. 1,609.344 m   21. 91.44 m
22. 321.86 km   23. 52.835 qt   24. 94.63 L   25. 0.8472 oz
26. 44.092 lb   27. 68.04 kg   28. 20.412 kg   29. 56.6572 g
30. 14.1643 g   31. 6.6138 lb   32. 30.48 m   33. 27.216 kg
34. 241.395 km   35. 32.808 yd   36. 236.132 mi   37. 7.4568 mi
38. 11.3556 L   39. 328.08 ft   40. 22   41. 18

Section 4–5   Exercises, p. 198

1   1. 3   2. 4   3. 2   4. 4   5. 4   6. 2   7. 3   8. 5   9. 5   10. 3

2   11. $\frac{1}{32}$ in.   12. 0.05 mm   13. $\frac{1}{8}$ in.   14. $\frac{1}{2}$ oz or 0.5 oz
15. $\frac{1}{2}$ L or 0.5 L   16. $\frac{1}{16}$ in.   17. $\frac{1}{64}$ in.   18. $\frac{1}{16}$ oz   19. 0.05 mi
20. 0.05 cm   21. 0.005 dg   22. 0.05 cg   23. 0.05 cL   24. 0.05 km
25. 0.05 km   26. 0.05 km

3   27. absolute error = 0.02 cm   28. absolute error = 0.2 mm
       relative error = 0.00038   relative error = 0.0038
       percent error = 0.038%   percent error = 0.38%
29. absolute error = 1.5 cm   30. absolute error = 0.29 in.
       relative error = 0.0308   relative error = 0.0204
       percent error = 3.08%   percent error = 2.04%
31. absolute error = 0.2 L   32. absolute error = 0.4 in.
       relative error = 0.0040   relative error = 0.0267
       percent error = 0.40%   percent error = 2.67%

4   33. 215.4 m   34. 64.4 g   35. 600 cm or 6.0 m   36. 5.34 kg or
5,340 g   37. 900,000 m³   38. 17,000,000 m³   39. 43   40. 70

5   41. 115 mm or 11.5 cm   42. 102 mm or 10.2 cm   43. 96 mm
or 9.6 cm   44. 85 mm or 8.5 cm   45. 57 mm or 5.7 cm   46. 50 mm
or 5 cm   47. 44 mm or 4.4 cm   48. 30 mm, 30.5 mm, or 31 mm
(3.0 cm, 3.05 cm, or 3.1 cm)   49. 19 mm or 1.9 cm   50. 10 mm or 1.0 cm
51. 8.3 cm   52. 5.1 cm   53. 2.8 cm   54. 7.0 cm

6   55. 3 in.   56. $2\frac{3}{4}$ in.   57. 7.5 cm   58. 13.6 cm   59. $11\frac{1}{4}$ in.
60. $10\frac{7}{8}$ in.   61. 4.4 cm   62. 4.9 cm   63. 7.9 cm   64. 4.1 cm
65. $11\frac{3}{8}$ in.   66. $5\frac{7}{8}$ in.

Chapter 4 Review Exercises, p. 207

1. $\frac{1\text{ ft}}{12\text{ in.}}$; $\frac{12\text{ in.}}{1\text{ ft}}$ 3. $\frac{2,000\text{ lb}}{1\text{ T}}$; $\frac{1\text{ T}}{2,000\text{ lb}}$ 5. 80 oz   7. $42\frac{1}{2}$ lb
9. 6,600 ft   11. 2 ft 7 in.   13. 13 lb $1\frac{1}{2}$ oz   15. 8 gal 2 qt   17. 14 oz
19. 17 in. or 1 ft 5 in.   21. 63 in²   23. 10 yd 1 ft 3 in.   25. 4 lb 8 oz
27. $3\frac{1}{2}$ 29. 300 $\frac{\text{mi}}{\text{h}}$ 31. $1\frac{1}{2}$ qt or 1 qt 1 pt   33. kilo-   35. milli-
37. centi-   39. 10 times   41. $\frac{1}{1,000}$ of   43. 1,000 times   45. a
47. a   49. c   51. b   53. 6.71 dkm   55. 2,300 mm   57. 12,300 mm
59. 230,000 mm   61. 413.27 km   63. 3.945 hg   65. 30.00974 kg
67. Cannot add unlike measures   69. 748 cg or 7.48 g   71. 61.47 cg
73. 15   75. 8.5 hL   77. 18.9 m   79. 245 mL or 24.5 cL   81. 6 m
83. 100 servings   85. 3 days   87. 2 h 38 min or 2.63 h   89. 4 days
91. 3 yr 3 mo or 3.25 yr   93. 59°F   95. 203°F   97. 104°F
99. 185°C   101. 100.2°F   103. a trillion bytes   105. 235.124 yd
107. 15.851 qt   109. 70.547 lb   111. 22.86 cm   113. 156.392 qt
115. 60.96 m   117. 281.628 km   119. 21   121. 6   123. 4
125. $\frac{1}{4}$ in.   127. $\frac{1}{32}$ ft   129. 0.05 cm   131. 0.05 cm   133. 1%
135. 4 significant digits   137. 0.05 cg   139. 46.5 m or 4,650 cm
141. 117 mm or 118 mm   143. 99 mm   145. 60 mm   147. 45 mm
149. 20 mm   151. $5\frac{1}{4}$ in.   153. $4\frac{27}{32}$ in.   155. 2.7 cm   157. $20\frac{1}{4}$ in.
159. 5.6 cm   161. 7.4 cm

Chapter 4 Practice Test, p. 211

1. 23 ft   **3.** 2 yd 1 ft 10 in.   **5.** 0.298 km   **7.** 9.48 L or 94.8 dL
9. 120.6975 km   **11.** 8.4536 pt   **13.** 9°C   **15.** 2,520 min   **17.** 235 h
19. 2   **21.** relative error $= 0.0012$; percent error $= 0.12\%$
23. 99 mm   **25.** 747 m/s   **27.** 4.85 cm

5	SIGNED NUMBERS AND POWERS OF 10

Section 5-1   Exercises, p. 219

1   **1.** $<$ **2.** $<$ **3.** $>$ **4.** $<$ **5.** $<$ **6.** $>$ **7.** $<$ **8.** $>$
9. 7   **10.** 17   **11.** 8   **12.** 7   **13.** -42 **14.** 17   **15.** 78   **16.** -57

2   **17.** 23   **18.** -16 **19.** -13 **20.** -29 **21.** -55 **22.** -124
23. 99   **24.** -59 **25.** -48 **26.** -161 **27.** 29   **28.** -20
29. -53 **30.** 232   **31.** $-1,005$

3   **32.** 2   **33.** 4   **34.** 6   **35.** -6 **36.** -2 **37.** -2 **38.** 7
39. -40 **40.** 2   **41.** -4 **42.** -10 **43.** 2   **44.** 4   **45.** 15   **46.** -8
47. -33 **48.** 9   **49.** -11 **50.** 13   **51.** 9   **52.** 3   **53.** 0   **54.** -3
55. -12 **56.** -14 **57.** 13   **58.** -12 **59.** 24   **60.** -98 **61.** 2
62. 10   **63.** 0   **64.** 0   **65.** 0   **66.** 0   **67.** -3 **68.** -7 **69.** \$66
70. \$36   **71.** $-295°C$ **72.** $-23°F$

Section 5-2   Exercises, p. 224

1   **1.** -12 **2.** 6   **3.** -6 **4.** 4   **5.** -25 **6.** -3 **7.** 8   **8.** -3
9. -9 **10.** 13   **11.** -1 **12.** -8 **13.** 62   **14.** -11 **15.** -58
16. -57 **17.** -15 **18.** -105 **19.** 18   **20.** -17 **21.** -167
22. 83   **23.** 1,413   **24.** -103 **25.** 15   **26.** -8 **27.** -12 **28.** 8
29. 7   **30.** 10   **31.** 56   **32.** -92 **33.** 14   **34.** -36 **35.** 0   **36.** 0

2   **37.** 4   **38.** 7   **39.** 6   **40.** -4 **41.** 1   **42.** 14   **43.** -3
44. -9 **45.** -5 **46.** 13   **47.** 2   **48.** -1 **49.** 1   **50.** -20
51. 26   **52.** -99 **53.** -45 **54.** -32 **55.** -60 **56.** -80
57. -35 **58.** 50   **59.** -21 **60.** 186   **61.** \$115,054   **62.** 107°F
63. 107°F   **64.** Subtracting zero from a number results in the same
number with the same sign. Subtracting a number from zero results in
the opposite of the number.   **65.** $+\$146.9$ million
66. \$63.6 million   **67.** $-\$96.8$ million   **68.** \$89.1 million

Section 5-3   Exercises, p. 229

1   **1.** 40   **2.** 12   **3.** 35   **4.** 21   **5.** 24   **6.** 6   **7.** -15 **8.** -10
9. -32 **10.** -12 **11.** -56 **12.** -24 **13.** 336   **14.** -60 **15.** 36
16. 0   **17.** 90   **18.** 0   **19.** -42 **20.** 54   **21.** -210 **22.** 2,268
23. $-20,160$ **24.** 840   **25.** -240 **26.** 0   **27.** 0   **28.** 0   **29.** 0
30. 0   **31.** 0   **32.** 0   **33.** 0   **34.** 0   **35.** 0   **36.** -30 **37.** 168
38. 42   **39.** -8 **40.** 0   **41.** -16 **42.** $4(-\$28) = -\$112$
43. $(7)(-\$40) = -\$280$ **44.** The multiplicative inverse of a number
is the number that, when multiplied by the original number, results in
1, the multiplicative identity.   **45.** Answers will vary,
$5(-3) = -3(5) = -15.$

2   **46.** 9   **47.** -8 **48.** 25   **49.** 0   **50.** -8 **51.** -512 **52.** 625
53. 81   **54.** 2,401   **55.** 1,764   **56.** 5   **57.** -28 **58.** 12   **59.** -8
60. -10 **61.** -8 **62.** -27 **63.** -121 **64.** 121   **65.** -125
66. 17,576,000

3   **67.** 5   **68.** -3 **69.** -4 **70.** -4 **71.** -4 **72.** 8   **73.** 5
74. -6 **75.** 8   **76.** -37 **77.** $-\$1,800; -\$300$ per month

78. $-6°C$ per hour   **79.** undefined   **80.** undefined   **81.** 0   **82.** 0
83. 0   **84.** undefined   **85.** undefined   **86.** undefined   **87.** 0
88. undefined   **89.** multiplication and division   **90.** $-10°F$ per hour
91. \$3,472.50 per month   **92.** The numerator must be zero and the
denominator must be any number except zero.

Section 5-4   Exercises, p. 237

1   **1.** $\dfrac{-5}{8}, \dfrac{5}{-8}, \dfrac{-5}{-8}$ **2.** $\dfrac{-3}{-4}, \dfrac{-3}{4}, \dfrac{3}{-4}$ **3.** $\dfrac{2}{-5}, \dfrac{-2}{5}, \dfrac{2}{5}$
4. $\dfrac{7}{8}, \dfrac{-7}{8}, \dfrac{7}{-8}$ **5.** $\dfrac{-7}{8}, \dfrac{7}{-8}, \dfrac{-7}{-8}$

2   **6.** $-\dfrac{1}{4}$ **7.** $-1\dfrac{1}{10}$ or $-\dfrac{11}{10}$ **8.** $-\dfrac{1}{16}$ **9.** $\dfrac{3}{16}$ **10.** $-\dfrac{2}{9}$
11. $-10\dfrac{5}{8}$ or $-\dfrac{85}{8}$ **12.** $-2\dfrac{13}{24}$ or $-\dfrac{61}{24}$ **13.** $-3\dfrac{7}{8}$ or $-\dfrac{31}{8}$
14. $1\dfrac{1}{10}$ or $\dfrac{11}{10}$ **15.** $-1\dfrac{3}{8}$ or $-\dfrac{11}{8}$ **16.** $-\dfrac{1}{16}$ **17.** $2\dfrac{17}{24}$ or $\dfrac{65}{24}$
18. $-\dfrac{1}{2}$ **19.** $\dfrac{6}{11}$ **20.** $\dfrac{10}{27}$ **21.** 30   **22.** $-\dfrac{1}{32}$ **23.** -2 **24.** $-\dfrac{1}{10}$
25. $-\dfrac{1}{7}$ **26.** $-\dfrac{25}{32}$ **27.** -1 **28.** -81 **29.** 81   **30.** 64   **31.** -16
32. 32   **33.** -26.30 **34.** -1.11 **35.** -91.44 **36.** -110.72
37. -59.04 **38.** 340.71   **39.** -27.73 **40.** 0.41   **41.** 4.6   **42.** 0.03
43. -2.3 **44.** -5.5

3   **45.** -9 **46.** 20   **47.** 24   **48.** -1 **49.** -13 **50.** 21   **51.** -13
52. 13   **53.** 15   **54.** 8   **55.** 24   **56.** -5 **57.** -7 **58.** -3 **59.** -28
60. -12 **61.** $-1,260$ **62.** 27   **63.** -14 **64.** 28   **65.** $-8\dfrac{3}{4}$ or $-\dfrac{35}{4}$
66. -29 **67.** -132 **68.** $-\dfrac{1}{5}$ **69.** $-5\dfrac{7}{8}$ or $-\dfrac{47}{8}$ **70.** -7 **71.** $-3\dfrac{3}{5}$
72. -10.38 **73.** $-\dfrac{127}{245}$ **74.** $\dfrac{37}{144}$ **75.** -3.992 **76.** -11.88
77. -1 **78.** -4 **79.** 18   **80.** 102   **81.** 46   **82.** 2   **83.** 8

Section 5-5   Exercises, p. 242

1   **1.** 45,300   **2.** 27   **3.** 5,820   **4.** 0.897   **5.** 5.23   **6.** 0.00806
7. 37   **8.** 1,820   **9.** 0.56   **10.** 1.42   **11.** 780,000   **12.** 62
13. 0.573   **14.** 0.00293   **15.** 0.0457   **16.** 857.9   **17.** 4,370
18. 8,370   **19.** 0.00046   **20.** 0.61   **21.** 0.72   **22.** 42   **23.** 10^{12}
24. 10   **25.** 10^2 **26.** 10^4 **27.** 1   **28.** 10^{-2} or $\dfrac{1}{10^2}$ **29.** 10^{-12} or $\dfrac{1}{10^{12}}$
30. 10^{-6} or $\dfrac{1}{10^6}$ **31.** 10^{10} **32.** 10^{-7} or $\dfrac{1}{10^7}$ **33.** 10^{-3} or $\dfrac{1}{10^3}$
34. 10   **35.** 10^3 **36.** 10^2 **37.** 10^{-3} or $\dfrac{1}{10^3}$ **38.** 1   **39.** 10^{-5} or $\dfrac{1}{10^5}$
40. 10^{-1} or $\dfrac{1}{10}$

2   **41.** 10^6 **42.** 10^{12} **43.** 10^{-4} or $\dfrac{1}{10^4}$ **44.** 10^{10} **45.** 10^{-8} or $\dfrac{1}{10^8}$

Section 5-6   Exercises, p. 251

1   **1.** Yes. The absolute value of the first factor is greater than 1
and less than 10. The second factor is a power of ten.   **2.** No. The
first factor is greater than 10.   **3.** 430   **4.** 0.0065   **5.** 2.2   **6.** 83,000
7. 0.0058   **8.** 80,000   **9.** 6.732   **10.** 0.00589   **11.** 78.3   **12.** 1,590
13. 397,000   **14.** 0.0004723   **15.** 0.00000991   **16.** 1,030,000

2   **17.** 3.92×10^2 **18.** 2×10^{-2} **19.** 7.03×10^0 **20.** 4.2×10^4
21. 8.1×10^{-2} **22.** 2.1×10^{-3} **23.** 2.392×10^1 **24.** 1.01×10^{-1}

25. 1.002×10^0 **26.** 7.21×10^2 **27.** 4.2×10^5 **28.** 3.26×10^4
29. 2.13×10^1 **30.** 6.2×10^{-6} **31.** 5.6×10^1 **32.** 1.97×10^{-6}
33. 7.45×10^1 **34.** 1.8×10^4 **35.** 7.01×10^1 **36.** 7.25×10^{-1}

3 **37.** 2.144×10^7 **38.** 5.6×10^3 **39.** 2.36×10^{-2}
40. 4.73×10^{-3} **41.** 7×10^2 **42.** 6.5×10^{-5} **43.** 7×10^2
44. 8×10^{-3} **45.** 2.45×10^1 **46.** 3.2285×10^{13} mi
47. 4.2×10^2 Å

4 **48.** 5.7×10^8 **49.** 2×10^{-5} **50.** 3.72×10^{-8} **51.** 5.0×10^8
52. 3.6×10^{-19} **53.** 4×10^8 **54.** 5×10^{10} **55.** 7×10^{-1}
56. 2.7×10^9 **57.** 5.9×10^7 **58.** 2.9×10^8 **59.** 1.9×10^{17}
60. 2×10^0 **61.** 2×10^4 **62.** 1×10^9 **63.** 2.9×10^8
64. 2.3×10^{-9} **65.** 3.8×10^6 **66.** 5.6×10^3 **67.** 78×10^0
68. 52×10^3 **69.** 80×10^6 **70.** 5.83×10^6 **71.** 1.7365×10^9
72. 41.98×10^6 **73.** 780×10^{-3} **74.** 330×10^{-3} **75.** 1.1×10^{-6}
76. 8×10^{-9} **77.** 983.2×10^{-6} **78.** 71.9×10^{-6}
79. 12.0307×10^{-3} **80.** 675×10^{-6} **81.** 428 kΩ **82.** 5.7 MV
83. 3.52 GW **84.** 79 MHz **85.** 81 μs **86.** 97.3 mÅ **87.** 5.41 ns
88. 890 ms **89.** 580 Ω **90.** 770 ps **91.** 2.98 μs **92.** 7.81 kW
93. 42.3 MV **94.** 1.572 GW **95.** 5.096 MΩ **96.** 8 nÅ
97. 182 kHz **98.** 1.6 kΩ **99.** 5.2 MV **100.** 97 MW

Chapter 5 Review Exercises, p. 259

1. > **3.** 5 **5.** 7 **7.** 12 **9.** 2 **11.** −87 **13.** −26 **15.** −6
17. gain of 8 yd; +8 yd **19.** $569 **21.** −13 **23.** −14 **25.** 14
27. −15 **29.** 70°F **31.** −5.4°F **33.** 36°C **35.** −394,000
37. Japan **39.** −14 **41.** 0 **43.** −168 **45.** 343 **47.** −16
49. 25 **51.** −10° **53.** higher, 20, $-4 \times 5 = -20$ points
55. 4 **57.** 4 **59.** 17 **61.** undefined **63.** −3
65. $\dfrac{-2}{3}, \dfrac{2}{-3}, -\dfrac{-2}{-3}$ **67.** $1\dfrac{5}{7}$ or $\dfrac{12}{7}$ **69.** −62 **71.** $-1\dfrac{7}{24}$ or $-\dfrac{31}{24}$
73. $-11\dfrac{13}{16}$ or $-\dfrac{189}{16}$ **75.** 56 **77.** 1 **79.** 73 **81.** 23.76 **83.** 46
85. $-\dfrac{4}{5}$ **87.** 1.181 **89.** $-\dfrac{7}{18}$ **91.** 1,000,000,000,000 **93.** $\dfrac{1}{1,000}$
or 0.001 **95.** 8,730 **97.** yes **99.** 375,000 **101.** 0.0000387
103. 5.2×10^4 **105.** 1.7×10^{-4} **107.** 8×10^{-9} **109.** 4.73×10^4
111. 1.7×10^0 **113.** 3.4×10^{10} (rounded to tenths)
115. 5.7×10^{-4} **117.** 3.25×10^8 **119.** 920×10^{-9}
121. 8.4×10^6 **123.** 41×10^0 **125.** 17×10^6 **127.** 3.084×10^9
129. 1.8×10^{-6} **131.** 7×10^{-3} **133.** 350×10^{-12}
135. 490×10^{-6} s or 490 μs **137.** 588×10^{-3} Å or 588 mÅ
139. 246.7 V **141.** 42 W **143.** 5.729 mW **145.** 4.8 THz

Chapter 5 Practice Test, p. 262

1. $-8 < 0$ **3.** $-5 > -10$ **5.** −8 **7.** 4 **9.** −48 **11.** 2.1 **13.** 0
15. undefined **17.** −2 **19.** 4 **21.** $\dfrac{1}{6}$ **23.** 10^6 **25.** 42,000
27. 0.059 **29.** 2.1×10^{-2} **31.** 7.83×10^{-3} **33.** 3.5×10^2
35. 1,500,000 ohms **37.** 175°F **39.** 4.7×10^{-3} **41.** 23 ys

6 STATISTICS

Section 6-1 Exercises, p. 267

1 **1.** 10% **2.** 36.8% **3.** 54.3% **4.** 20.8% **5.** 33.3% **6.** 75%
7. 230,000,000 bushels **8.** 1,150,000,000 bushels **9.** 34,776,000,000 lb
10. 47 bushels/acre **11.** Routine discharges **12.** 17.5%

2 **13.** Debt retirement **14.** Misc. expenses and general government
15. Social projects and education costs **16.** 84,000,000 barrels

17. Approximately 4% increase **18.** Approximately 27,000,000 barrels
19. Approximately 30% increase **20.** 2005–2010 **21.** 2010
22. 67.96 million barrels; **23.** 31.0% **24.** 3.5%

3 **25.** 5 A **26.** 50 V **27.** 35 V **28.** 25 Ω **29.** 2006
30. Motor gasoline **31.** 1996–2001 **32.** 2006–2011 **33.** 16.1%
34. Cattle **35.** Hogs and pigs **36.** 1,067,279 metric tons
37. Approximately 4.3% increase **38.** 2007–2011 **39.** Dramatically
increasing **40.** Fluctuating with slight overall increase **41.** About
216% increase **42.** 1990

Section 6-2 Exercises, p. 277

1 **1.** 16 **2.** 17 **3.** 66 **4.** 73.75 **5.** 33.7 **6.** 66.6 **7.** 42.33°F
8. 12.67°C **9.** $34.80 **10.** $39.20 **11.** 14.67 in. **12.** 8 in. **13.** 20
14. 76 **15.** 17.3 runs **16.** 98 **17.** $24.6 \dfrac{\text{mi}}{\text{gal}}$ **18.** $10.2 \dfrac{\text{mi}}{\text{gal}}$
19. 737 thousand tons or 737,000 tons **20.** 1,300 million tons or
1,300,000,000 tons **21.** 3.43 **22.** 100

2 **23.** 44 **24.** 43 **25.** $30 **26.** 26.5 **27.** 15.5 **28.** $66
29. $8.25 **30.** $8.85 **31.** 745 thousand **32.** 1,238 million **33.** 2
34. 5 **35.** No mode **36.** No mode **37.** $67 **38.** $32 **39.** 4 h
40. $1.97 **41.** No mode **42.** No mode

3 **43.** 10 **44.** 2 **45.** $\dfrac{1}{3}$ **46.** $\dfrac{1}{7}$ **47.** 12% **48.** 28%
49. 35–37 and 38–40 **50.** 20–22 and 23–25 **51.** 7 **52.** 16

	Midpoint	Tally	Class Frequency
53.	15	||	2
54.	12	||||	4
55.	9	⊣⊦⊦||	7
56.	6	⊣⊦⊦ ⊣⊦⊦ ⊣⊦⊦ ⊣⊦⊦	20

	Midpoint	Tally	Class Frequency
57.	93	||	2
58.	88	⊣⊦⊦	5
59.	83	⊣⊦⊦ ||	7
60.	78	⊣⊦⊦ ⊣⊦⊦	10
61.	73	|||	3
62.	68	⊣⊦⊦		6
63.	63	||	2
64.	58	⊣⊦⊦	5

	Class Interval	Midpoint	Tally	Class Frequency
65.	15–19	17	⊣⊦⊦ ||	7
	10–14	12	⊣⊦⊦ |||	8
	5–9	7	⊣⊦⊦ ||||	9
	0–4	2	⊣⊦⊦		6

	Class Interval	Midpoint	Tally	Class Frequency
66.	51–60	55.5	||	2
	41–50	45.5	||||	4
	31–40	35.5	||	2
	21–30	25.5	⊣⊦⊦ ||	7
	11–20	15.5	⊣⊦⊦	5

	Class Interval	Midpoint	Tally	Class Frequency
67.	801–900	850.5	||	2
	701–800	750.5	⊣⊦⊦	5
	601–700	650.5	|||	3

	Class Interval	Midpoint	Tally	Class Frequency
68.	1,751–2,000	1,875.5	||	1
	1,501–1,750	1,625.5	⊮	5
	1,251–1,500	1,375.5	||||	4
	1,001–1,250	1,125.5	⊮ ||	7
	751–1,000	875.5	|||	3

69. 25 students **70.** $8 **71.** 76 **72.** 10 h

Section 6–3 Exercises, p. 290

1 **1.** 24 **2.** 23 **3.** 13 **4.** 18 **5.** $25 **6.** $33 **7.** 48°F
8. 117,400 T **9.** 136,100 T **10.** 199,300 T **11.** 159,000 T
12. 54,900 T **13.** 990 T **14.** 42,825 tons **15.** 99,650 tons
16. 79,500 tons **17.** 19,200 tons

2

18.

Class Interval	Class Frequency	Relative Frequency	Cumulative Frequency	Cumulative Relative Frequency
30–39	4	$\frac{4}{32} = \frac{1}{8} = 0.125$	4	0.125
40–49	3	$\frac{3}{32} = 0.09375$	7	0.21875
50–59	2	$\frac{2}{32} = \frac{1}{16} = 0.0625$	9	0.28125
60–69	8	$\frac{8}{32} = \frac{1}{4} = 0.25$	17	0.53125
70–79	5	$\frac{5}{32} = 0.15625$	22	0.6875
80–89	6	$\frac{6}{32} = \frac{3}{16} = 0.1875$	28	0.875
90–99	4	$\frac{4}{32} = \frac{1}{8} = 0.125$	32	1

19.

Class	Class Frequency	Relative Frequency	Cumulative Frequency	Cumulative Relative Frequency
100–119	45	$\frac{45}{360} = \frac{1}{8} = 0.125$	45	0.125
120–139	54	$\frac{54}{360} = \frac{3}{20} = 0.15$	99	0.275
140–159	72	$\frac{72}{360} = \frac{1}{5} = 0.2$	171	0.475
160–179	63	$\frac{63}{360} = \frac{7}{40} = 0.175$	234	0.65
180–199	81	$\frac{81}{360} = \frac{9}{40} = 0.225$	315	0.875
200–219	45	$\frac{45}{360} = \frac{1}{8} = 0.125$	360	1

20. Median = 6.0; lower-quartile boundary = 5.8; upper-quartile boundary = 6.35 **21.** Median = 209.5; lower-quartile boundary = 196.5; upper-quartile boundary = 231 **22.** Inner-quartile range = 0.55 **23.** Inner-quartile range = 34.5 **24.** (a) 5.8 (b) 7.1
25. (a) 203 (b) 267

3 **26.** 3.16 **27.** 8.29 **28.** 11.06°F **29.** $17.36 **30.** 4 **31.** 8
32. 51,444 T **33.** 16,685 T **34.** 321 T **35.** 15.87% **36.** 2.28%
37. 2.28% **38.** Approximately 2 pediatricians **39.** Approximately
32 patients **40.** Approximately 11 patients

Section 6–4 Exercises, p. 298

1 **1.**

Keaton Brienne Renee	
Keaton Renee Brienne	
Brienne Keaton Renee	
Brienne Renee Keaton	
Renee Keaton Brienne	
Renee Brienne Keaton	
$3 \cdot 2 \cdot 1 = 6$ ways	

2.

ABCD	BACD
ABDC	BADC
ACBD	BCAD
ACDB	BCDA
ADBC	BDAC
ADCB	BDCA
CABD	DABC
CADB	DACB
CBAD	DBAC
CBDA	DBCA
CDAB	DCAB
CDBA	DCBA
$4 \cdot 3 \cdot 2 \cdot 1 = 24$ ways	

3. $4 \cdot 3 \cdot 2 \cdot 1 = 24$ ways **4.** $5 \cdot 4 \cdot 3 \cdot 2 \cdot 1 = 120$ ways
5. $5 \cdot 4 \cdot 3 \cdot 2 \cdot 1 = 120$ ways **6.** $5 \cdot 4 \cdot 3 = 60$ ways **7.** 20 words

2 **8.** $\frac{1}{6}$ **9.** $\frac{1}{2}$ **10.** $\frac{2}{5}$ **11.** $\frac{1}{3}$ **12.** $\frac{1}{4}$ **13.** $\frac{11}{48}$ **14.** $\frac{4}{143}$
15. $\frac{1}{24}, \frac{1}{23}$ **16.** 1 **17.** 0

3 **18.** 6 to 18 or 1 to 3 **19.** 18 to 6 or 3 to 1

Chapter 6 Review Exercises, p. 306

1. 12.9% **3.** 17.4% **5.** 7,684,500 barrels **7.** 27.3%
9. 2013, 2015, 2016 **11.** 2014, 2017, 2018 **13.** 12% **15.** 2003
17. 7-10-2018 @ 4 P.M. **19.** 2002 **21.** 2008–2009 **23.** 12.6 cars
25. 13.8 $\frac{\text{mi}}{\text{gal}}$ **27.** 3.67 **29.** 83 **31.** $10.03 **33.** $1.85

	Midpoint	Tally	Class Frequency
35.	60.5	⊮⊮	10
37.	40.5	⊮⊮||	12
39.	20.5	⊮||	7

41. 36–45 **43.** 15 **45.** $\frac{1}{2}$ **47.** $\frac{10}{54} = 18.5\%$

49.

Miles per Gallon	Midpoint	Tally	Frequency
20–24	22	⊮||	7
25–29	27	⊮|	6
30–34	32	||||	4

51. 26.1 **53.** Range: 59 **55.** 68.5

57.

Class Interval	Class Frequency	Relative Frequency	Cumulative Frequency	Cumulative Relative Frequency
66–75	5	$\frac{5}{54} = 0.09259$	5	0.09259
56–65	10	$\frac{10}{54} = \frac{5}{27} = 0.18519$	15	0.27778
46–55	10	$\frac{10}{54} = \frac{5}{27} = 0.18519$	25	0.46297
36–45	12	$\frac{12}{54} = \frac{2}{9} = 0.22222$	37	0.68519
26–35	10	$\frac{10}{54} = \frac{5}{27} = 0.18519$	47	0.87038
16–25	7	$\frac{7}{54} = 0.12963$	54	1.00001*

*Due to rounding.

59. lower-quartile boundary $= 23$; middle-quartile boundary (median) $= 26$; upper-quartile boundary $= 29.5$ **61.** 6.5 **63.** The 14th score is 30, that is, 80% of the scores are at or below 30. **65.** 21.47   **67.** 68.26%   **69.** Approximately 841 men   **71.** $\dfrac{10}{13}; \dfrac{3}{4}$

73. 16 combinations   **75.** 5,040 with no repeats; 10,000 with repeats **77.** $\dfrac{1}{2}$ **79.** 0   **81.** $\dfrac{28}{188} = \dfrac{7}{47}$ or 7 to 47

Chapter 6 Practice Test, p. 311

1. Bar   **3.** Line   **5.** 2°C   **7.** $65   **9.** 25%   **11.** $\dfrac{15}{200} = \dfrac{3}{40}$

13. English dept., electronics dept.   **15.** Approximately 104%   **17.** $\dfrac{1}{2}$

19. Range: 25
Mean: 77.9
Median: 78
Mode: 81

21.

Class Interval	Class Frequency	Relative Frequency	Cumulative Frequency	Cumulative Relative Frequency
26–30	3	$\dfrac{3}{20} = 0.15$	3	0.15
21–25	4	$\dfrac{4}{20} = \dfrac{1}{5} = 0.2$	7	0.35
16–20	7	$\dfrac{7}{20} = 0.35$	14	0.7
11–15	3	$\dfrac{3}{20} = 0.15$	17	0.85
6–10	2	$\dfrac{2}{20} = \dfrac{1}{10} = 0.1$	19	0.96
1–5	1	$\dfrac{1}{20} = 0.05$	20	1

23. 15.5   **25.** The 17th score is 90. Therefore, 80% of scores are at or below 90.   **27.** 8.49   **29.** 6   **31.** $\dfrac{3}{5}$ **33.** 3 to 1   **35.** 34.13%

Chapters 4–6 Cumulative Practice Test, p. 312

1. 1.43 m   **3.** 3.0002 kg   **5.** $C = 30°$ **7.** $\dfrac{1}{8}$ cm **9.** 3 significant digits   **11.** 30.21 cm rounds to 30.2 cm   **13.** -14 **15.** -6 **17.** $-1\dfrac{1}{4}$ **19.** 0.053   **21.** 3.5×10^{-3} **23.** 145.7×10^{-6} **25.** 2×10^6 **27.** 64,000 cwt   **29.** Mississippi and Missouri **31.** China   **33.** Mean, 489; median, 477; mode, 456 **35.** $p = \dfrac{4}{15}$ **37.** 22   **39.** $\dfrac{40}{22} = \dfrac{20}{11}$ or 20 to 11

---

7   **LINEAR EQUATIONS AND INEQUALITIES**

Section 7–1   **Exercises, p. 322**

1   **1.** $-15 = -15$ **2.** $11 = 11$ **3.** $2 = 2$ **4.** $-7 = -7$ **5.** $-35 = -35$ **6.** $-34 = -34$ **7.** $n = 11$ **8.** $m = 6$ **9.** $y = -4$ **10.** $x = 6$ **11.** $p = 9$ **12.** $b = -9$ **13.** $\boxed{7} + \boxed{c}$

14. $\boxed{4a} - \boxed{7}$ **15.** $\boxed{3x} - \boxed{2(x+3)}$ **16.** $\boxed{\dfrac{a}{3}}$

17. $\boxed{7xy} + \boxed{3x} - \boxed{4} + \boxed{2(x+y)}$ **18.** $\boxed{14x} + \boxed{3}$ **19.** $\boxed{\dfrac{7}{(a+5)}}$

20. $\boxed{\dfrac{4x}{7}} + \boxed{5}$ **21.** $\boxed{11x} - \boxed{5y} + \boxed{15xy}$ (Answers may vary.)

22. 5   **23.** -4 **24.** $\dfrac{1}{5}$ **25.** $\dfrac{2}{7}$ **26.** 6   **27.** $-\dfrac{4}{5}$ **28.** 7   **29.** -1 **30.** $-15c$ (Answers may vary.)

2   **31.** Four more than a number is 7. (Answers may vary.) **32.** Five less than a number is 2. (Answers may vary.) **33.** Three times a number is 15. (Answers may vary.) **34.** One more than 3 times a number is 7. (Answers may vary.)

3   **35.** $x + 5 = 12$ **36.** $\dfrac{x}{6} = 9$ **37.** $4(x - 3) = 12$ **38.** $4x - 3 = 12$ **39.** $12 + 7 + x = 17$ **40.** $2x + 7 = 21$ **41.** $x + 15 = 48$ **42.** $x + 15 \text{ mL} = 45 \text{ mL}$ **43.** $x \cdot 5 = 45$ **44.** $\dfrac{18 - 3}{5} = x$ or $5x = 18 - 3$ **45.** $3(x + 12) = 45$ **46.** $3x + 12 = 45$

4   **47.** $-3a$ **48.** $-8x - 2y$ **49.** $7x - 2y$ **50.** $-3a + 6$ **51.** 10 **52.** $3a + 6b + 8c + 1$ **53.** $10x - 20$ **54.** $6x + 13$ **55.** $-4x + 5$ **56.** $-12a + 3$ **57.** $-5x + 5y - 10z$ **58.** $-2 - 2a + 3b$ **59.** $-13x + 25y - 15$ **60.** $-6x - 20y + 20$

Section 7–2   **Exercises, p. 334**

1   **1.** $x = 8$ **2.** $x = 15$ **3.** $x = -3$ **4.** $x = -8$ **5.** $x = 4$ **6.** $x = -11$ **7.** $x = -26$ **8.** $x = 3$ **9.** $x = 5$ **10.** $x = -7$ **11.** $x = 16$ **12.** $x = 9$ **13.** $x = -2$ **14.** $x = 4$ **15.** $x = 6$ **16.** $x = 9$

2   **17.** $x = 8$ **18.** $x = 9$ **19.** $x = 5$ **20.** $b = \dfrac{28}{3}$ **21.** $a = -9$

22. $b = -2$ **23.** $c = 7$ **24.** $x = -9$ **25.** $n = -\dfrac{5}{2}$ **26.** $n = 30$

27. $x = 64$ **28.** $x = 45$ **29.** $x = \dfrac{10}{3}$ **30.** $y = -\dfrac{20}{3}$ **31.** $x = 12$

32. $n = 56$ **33.** $x = -7$ **34.** $y = 2$ **35.** $x = 15$ **36.** $y = 8$

37. $\quad -\dfrac{1}{2}x = -6$ **38.** $21 = \dfrac{3}{8}n$

$\quad \left(-\dfrac{1}{2}\right)(12) = -6$ $21 = \left(\dfrac{3}{8}\right)\left(\dfrac{56}{1}\right)$

$\quad\quad\quad -6 = -6$ $21 = 21$

3   **39.** $x = 6$ **40.** $m = 7$ **41.** $a = -3$ **42.** $m = -\dfrac{1}{2}$

43. $y = 8$ **44.** $x = 0$ **45.** $y = -7$ **46.** $y = -4$ **47.** $x = -2$

4   **48.** $x = -6$ **49.** $b = -1$ **50.** $x = 8$ **51.** $t = 6$ **52.** $x = 3$

53. $y = 5$ **54.** $x = -\dfrac{7}{2}$ **55.** $a = -2$ **56.** $x = -8$

57. No solution   **58.** All real numbers   **59.** $t = 7$ **60.** $x = -1$

61. $y = 11$ **62.** $y = -2$

5   **63.** $x = 7$ **64.** $y = 2$ **65.** $x = -5$ **66.** $x = -2$ **67.** $x = 5$ **68.** $x = -6$ **69.** $x = 1$ **70.** $x = 8$ **71.** $x = 1$ **72.** $x = 0$ **73.** No solution   **74.** All real numbers   **75.** No solution

76. No solution   **77.** $x = 2$ **78.** $x = 5$ **79.** $x = \dfrac{1}{2}$ **80.** $x = -15$

81. $x = -9$ **82.** $x = -1$ **83.** $x = \dfrac{16}{5}$ **84.** $x = 5$ **85.** $x = 1$

86. $x = -2$ **87.** $x = 27$ **88.** $P = \dfrac{5}{6}$ **89.** $x = 14$ **90.** $x = -\dfrac{4}{15}$

91. $m = \dfrac{49}{18}$ **92.** $s = \dfrac{8}{5}$ **93.** $m = \dfrac{8}{3}$ **94.** $T = \dfrac{68}{9}$ **95.** $x = 2$

96. $x = 2$ **97.** $R = 0.8$ **98.** $x = 6.03$ **99.** $x = 0.08$ **100.** $R = 0.34$
101. $x + 4 = 12; x = 8$ **102.** $2x - 4 = 6; 5$
103. $x + x + (x - 3) = 27$; two parts weigh 10 lb each; third part weighs 7 lb **104.** $24 + x = 60; 36$ gal **105.** $4.03 - 3.97 = x$ or $4.03 - x = 3.97; 0.06$ kg **106.** $x + 3x = 400$; 100 ft of solid pipe; 300 ft of perforated pipe **107.** $x + 2x + 70 = 235$; Spanish text: \$55; calculator: \$110 **108.** $C = \$14(4)(150) = \$8,400$

109. $x + 2x = 325$;   tank 1: $216\frac{2}{3}$ gal tank 2: $108\frac{1}{3}$ gal

110. $A = 10w; 60 = 10w; w = 6$ ft

Section 7-3   Exercises, p. 346

1 **1.** 28 **2.** -11 **3.** $\frac{18}{7}$ **4.** 0 **5.** $-\frac{27}{5}$ **6.** $-\frac{5}{27}$ **7.** $\frac{8}{5}$ **8.** 48

9. -350 **10.** 27 **11.** 0 **12.** 576 **13.** 28 **14.** $\frac{1}{3}; Q \neq 0$ **15.** $\frac{1}{9}; P \neq 0$

16. No solution **17.** No solution **18.** $\frac{25}{8}; p \neq 4$ **19.** $4; B \neq 6$

20. -68 **21.** 13 **22.** $\frac{11}{15}$ **23.** $\frac{27}{35}$ **24.** $\frac{108}{5}$ **25.** $\frac{217}{24}$ **26.** 8

27. $-5; R \neq 0$ **28.** $\frac{3}{10}$ **29.** 70 **30.** 0 **31.** $\frac{32}{63}$ **32.** $\frac{32}{5}$ **33.** $-\frac{36}{11}$

34. $-\frac{1}{108}; x \neq 0$ **35.** 21 **36.** $\frac{4}{25}$ **37.** $\frac{2}{21}$ **38.** $\frac{9}{7}$ **39.** 8.57 Ω
40. 5 Ω **41.** 7.38 Ω **42.** 3.6 Ω

2 **43.** 80 vases **44.** $3\frac{3}{4}$ h **45.** $3\frac{1}{13}$ h **46.** $3\frac{3}{7}$ days

47. $\frac{3}{5}$ h or 36 min **48.** 3 min **49.** 15 min **50.** $1\frac{1}{3}$ h

51. Pipe 2 empties at a faster rate than pipe 1 fills; therefore, the tank will never be filled. **52.** $2\frac{1}{3}$ min

3 **53.** 2 **54.** 0.8 **55.** 6.03 **56.** 16 **57.** 4.7 **58.** 0.08 **59.** 0.035
60. 0.34 **61.** -3.3 **62.** 38.8 **63.** 3.3 **64.** -0.2 **65.** 0.6
66. 1.128 **67.** 16 **68.** \$136 **69.** \$264. **70.** \$3,325 **71.** 1.2%
72. 12.25% **73.** \$3,325 **74.** \$11.30 **75.** 1.2% **76.** 12.25%
77. 8.3 Ω **78.** 156.75 V

Section 7-4   Exercises, p. 352

1 **1.** F **2.** F **3.** T **4.** T **5.** $\{6, 7, 8, 9, 10, 11\}$ **6.** $\{4, 6\}$
7. $\{15, 20, 25, 30, 35\}$ **8.** $\{-4, -3, -2, -1\}$
9. $\{x \mid x < 10 \text{ and } x \text{ is odd natural number}\}$
10. $\{x \mid -7 \leq x \leq -1 \text{ and } x \text{ is odd integer}\}$

2 **11.** $(5, 9)$ **12.** $(-7, -3)$

13. $[-5, -3]$ **14.** $[8, 12]$

15. $(-\infty, 7)$ **16.** $(-\infty, -2)$

17. $[2, \infty)$ **18.** $(-\infty, 3]$

19. $(5, \infty)$ **20.** $[-3, \infty)$

21. $(-\infty, -2]$

22. $(-6, \infty)$ **23.** $(843{,}000, 1{,}000{,}000)$

24. $[4, 18]$

25. $0 \leq n < 12$ $[0, 12)$

Section 7-5   Exercises, p. 356

1 **1.** $y > 8; (8, \infty)$ **2.** $x < 1; (-\infty, 1)$

3. $x \leq -3; (-\infty, 3]$ **4.** $x < 6; (-\infty, 6)$

5. $b > -1; (-1, \infty)$ **6.** $t < 6; (-\infty, 6)$

7. $y < 5; (-\infty, 5)$ **8.** $t \leq 7; (-\infty, 7]$

9. $y > 11; (11, \infty)$ **10.** $a \leq -2; (-\infty, -2]$

11. $x \geq 1; [1, \infty)$ **12.** $x \geq 3; [3, \infty)$

13. $x < 1; (-\infty, 1)$ **14.** $x > \frac{1}{3}; \left(\frac{1}{3}, \infty\right)$

15. $a \leq -3; (-\infty, -3]$ **16.** $x < \frac{1}{2}; \left(-\infty, \frac{1}{2}\right)$

17. $a \geq -3; [-3, \infty)$ **18.** $y \leq 8; (-\infty, 8]$

19. $2x + 196 \geq 52,800$
$x \geq 26,302$

20. $x > \$3.60 \cdot 2$
$x > \$7.20$

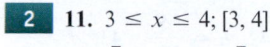

0 2 4 6 8 10

Section 7–6 Exercises, p. 363

1 **1.** T; all elements in B also in U   **2.** F; 2 is not in A
3. T; ϕ is a subset of all sets.   **4.** F; No element of A is in B
5. { } or ϕ **6.** {2, 0, −2}   **7.** {−4, −2, 0, 1, 2, 3, 4, 5}
8. {−5, −3, −1, 1, 2, 3, 4, 5}   **9.** {−5, −4, −3, −2, −1, 1, 2, 3, 4, 5}
10. {−5, −4, −3, −1, 0}

2 **11.** $3 \leq x \leq 4$; [3, 4]

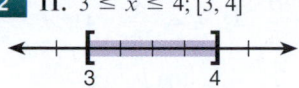
3   4

12. $5 \leq x \leq 7$; [5, 7]

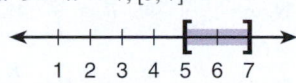
1 2 3 4 5 6 7

13. $-1 \leq x \leq 3$; [−1, 3]

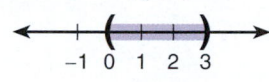
−2 −1 0 1 2 3 4

14. $-1 \leq x \leq 2$; [−1, 2]

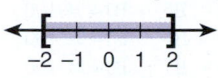
−2 −1 0 1 2 3

15. $0 < x < 3$; (0, 3)

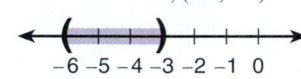
−1 0 1 2 3

16. $-2 \leq x \leq 2$; [−2, 2]

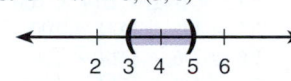
−2 −1 0 1 2

17. $-6 < x < -3$; (−6, −3)

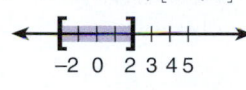
−6 −5 −4 −3 −2 −1 0

18. $3 < x < 5$; (3, 5)

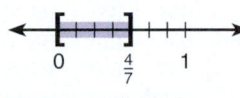
2 3 4 5 6

19. $-2 \leq x \leq 2$; [−2, 2]

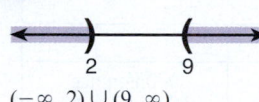
−2 0 2 3 4 5

20. $0 \leq x \leq \frac{4}{7}$; $\left[0, \frac{4}{7}\right]$

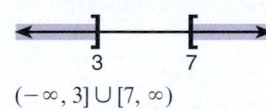
0   $\frac{4}{7}$ 1

21. No solution; ϕ

22. $3 < x < 7$; (3, 7)

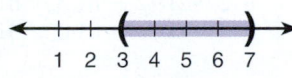
1 2 3 4 5 6 7

3 **23.** $x < 2$ or $x > 9$

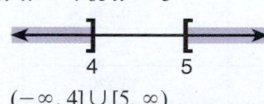
2   9
$(-\infty, 2) \cup (9, \infty)$

24. $x \leq 3$ or $x \geq 7$

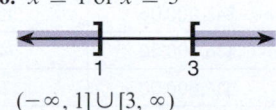
3   7
$(-\infty, 3] \cup [7, \infty)$

25. $x \leq 4$ or $x \geq 5$

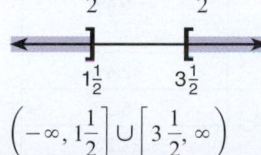
4   5
$(-\infty, 4] \cup [5, \infty)$

26. $x \leq 1$ or $x \geq 3$

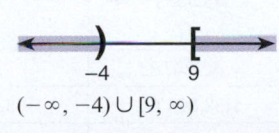
1   3
$(-\infty, 1] \cup [3, \infty)$

27. $x \leq 1\frac{1}{2}$ or $x \geq 3\frac{1}{2}$

$1\frac{1}{2}$ $3\frac{1}{2}$
$\left(-\infty, 1\frac{1}{2}\right] \cup \left[3\frac{1}{2}, \infty\right)$

28. $x < -4$ or $x \geq 9$

−4   9
$(-\infty, -4) \cup [9, \infty)$

29. $5.22 \leq x \leq 5.32$;

30. $14.7 \leq x \leq 15.7$

Chapter 7 Review Exercises, p. 371

1. $12 = 12$ **3.** $8 = 8$ **5.** $x = 14$ **7.** $y = -8$

9. $\boxed{15x} - \boxed{\frac{3a}{7}} + \boxed{\frac{(x-7)}{5}}$ **11.** A number increased by 5 equals 2.
Answers will vary.   **13.** The quotient of a number and 8 equals 7.
Answers will vary.   **15.** $2x + 7 = 11$ **17.** $2(x + 8) = 40$ **19.** $5a$
21. $7y - 12$ **23.** $-3a - 11$ **25.** $-2x + 21$ or $21 - 2x$
27. $11 - 2x + 4y$ **29.** $30a - 21b + 12$ **31.** $x = 13$ **33.** $x = -2$
35. $x = 19$ **37.** $x = 3$ **39.** $x = -1$ **41.** $a = 5$ **43.** $x = 5$
45. $x = 7$ **47.** $x = 4$ **49.** $x = -3$ **51.** $x = 7$ **53.** $b = -\frac{15}{2}$
55. $x = 15$ **57.** $y = 7$ **59.** $x = 64$ **61.** $x = -49$ **63.** $x = 84$
65. $b = -2$ **67.** $x = 7$ **69.** $x = 5$ **71.** $x = -4$ **73.** $x = 2$
75. no real solution   **77.** $y = -3$ **79.** $x = \frac{20}{13}$ **81.** $y = -12$
83. $y = 7$ **85.** $x = 9$ **87.** $x = \frac{7}{20}$ **89.** $c = -\frac{9}{32}$ **91.** $R = 0.61$
93. $y = -1$ **95.** $x = 1$ **97.** $x = 2$ **99.** $x = -9$ **101.** $x = 3$
103. $x = 2$ **105.** $x = -4$ **107.** $x = 0$ **109.** $x = 3$ **111.** $x = 3$
113. $x = -1$ **115.** $x = -\frac{1}{8}$ **117.** $x - 6 = 8$; $x = 14$
119. $5(x + 6) = x + 42$; $x = 3$ **121.** $x + (x - 3) = 51$; 27 h, 24 h
123. $720 = 2(2w + w)$; $w = 120$ ft; $l = 2w = 240$ ft   **125.** $\frac{1}{2}$
127. $\frac{5}{6}$ **129.** $-\frac{4}{15}$ **131.** $\frac{49}{18}$ **133.** $\frac{8}{3}$ **135.** 6; P $\neq$ 0   **137.** $2\frac{1}{10}$ h
139. 25 fixtures   **141.** 1.33 Ω   **143.** 2   **145.** −11.8   **147.** 17 Ω
149. $5 \in W$

151. $(-7, \infty)$

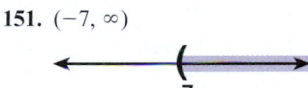
−7

153. $[-4, 2)$

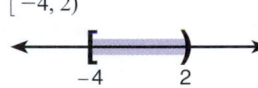
−4   2

155. $(-2, \infty)$

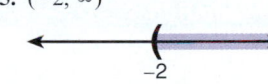
−2

157. $m < 7$; $(-\infty, 7)$

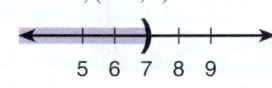
5 6 7 8 9

159. $x > 0$; $(0, \infty)$

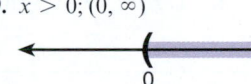
0

161. $x \leq 3$; $(-\infty, 3]$

2 3 4 5

163. $x > -9$; $(-9, \infty)$

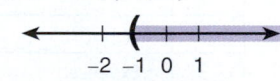
−10 −9 −8 −7

165. $x > -1$; $(-1, \infty)$

−2 −1 0 1

167. $x > -2$; $(-2, \infty)$

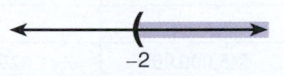
−2

169. $x \geq -3$; $[-3, \infty)$

−3

171. $y \geq -1$; $[-1, \infty)$

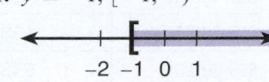
−2 −1 0 1

173. $x \leq 6$; $(-\infty, 6]$

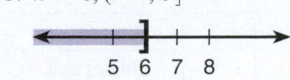
5 6 7 8

175. $D = P + \$5.60$; $2D + 12P \leq \$59.80$; therefore,
$P \leq \$3.47$; $D \leq \$9.07$ **177.** { } or ϕ **179.** {−1}   **181.** {5}

183. $2 < x < 4$; (2, 4)

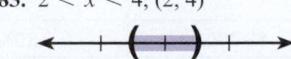
1 2 3 4 5

185. no solution; ϕ

187. $2 < x < 6$; (2, 6)

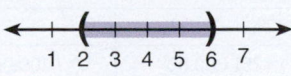
1 2 3 4 5 6 7

189. $\frac{1}{2} < x < \frac{9}{2}; \left(\frac{1}{2}, \frac{9}{2}\right)$

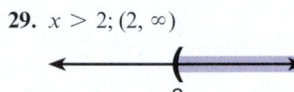

191. $-3 < x \le 2; (-3, 2\,]$

193. $-2 \le x \le -\frac{3}{4}; \left[-2, -\frac{3}{4}\right]$

195. $x < 2$ or $x > 8; (-\infty, 2) \cup (8, \infty)$

197. $x < -9$ or $x > 8; (-\infty, -9) \cup (8, \infty)$

199. $10 < x < 19$

Chapter 7 Practice Test, p. 376

1. 14   **3.** 10   **5.** 3   **7.** 2   **9.** $-\frac{22}{7}$ **11.** $\frac{7}{25}$ **13.** $\frac{10}{21}$ **15.** 6.17

17. 254.24   **19.** -0.15 **21.** $2(x - 9) = 30; x = 24$ **23.** \$1,255.50

25. $x \ge -12; [-12, \infty)$

27. $x > 3; (3, \infty)$

29. $x > 2; (2, \infty)$

31. $x \le -6; (-\infty, -6\,]$

33. $x > -2; (-2, \infty)$

35. $-8 < x < 4; (-8, 4)$

37. No solution   **39.** $A \cup B = \{1, 2, 3, 4, 5, 6, 7, 8, 9\}$

41. $A \cap B' = \{9\}$

8 | FORMULAS, PROPORTION, AND VARIATION

Section 8–1 | Exercises, p. 385

1 1. 280 mi   **2.** \$120   **3.** 2,826 in^2 **4.** 117.6 m^2 **5.** 138 cm

6. 61.6%   **7.** 153.9 in.^{2} **8.** 4.8 Ω   **9.** 66 in.   **10.** \$310

11. 14.25%   **12.** 3 years   **13.** \$2,600   **14.** 90 lb   **15.** 8%

16. \$3,378.38   **17.** 13.6   **18.** 17%   **19.** 1,600 rpm   **20.** 71.71 in.

21. \$48.75   **22.** \$10.50   **23.** 4 mi   **24.** 6.25 km^2 **25.** 1.5 A

26. 6 ft^3 **27.** 3 A   **28.** 2 cylinders   **29.** 9,326.6 cm^2

30. 35 mi/h   **31.** 17 Ω

2 32. $S = I - I_n$ **33.** $D = P - S$ **34.** $b = y - mx$

35. $C = S - M$ **36.** $h = \dfrac{V}{\pi r^2}$ **37.** $r = \dfrac{S}{2\pi h}$ **38.** $l = \dfrac{A}{w}$

39. $r = \dfrac{C}{2\pi}$ **40.** $R = \dfrac{E}{I}$ **41.** $R = \dfrac{D}{T}$ **42.** $h = \dfrac{S}{2\pi r}$ **43.** $d = \dfrac{C}{\pi}$

44. $b = \dfrac{P - 2s}{2}$ **45.** $C = \dfrac{R + B}{A}$ **46.** $R = \dfrac{100\,P}{B}$

47. $I = A - P$ **48.** $T = \dfrac{I}{PR}$

49. $B14 = B5 + B6 + B7 + B8 + B9 + B10 + B11 + B12$
$D14 = D5 + D6 + D7 + D8 + D9 + D10 + D11 + D12$
$C5 = B5 \div B14 \times 100$
$C6 = B6 \div B14 \times 100$
$C7 = B7 \div B14 \times 100$
$\vdots$
$C12 = B12 \div B14 \times 100$
$E5 = D5 \div D14 \times 100$
$E6 = D6 \div D14 \times 100$
$E7 = D7 \div D14 \times 100$
$\vdots$
$E12 = D12 \div D14 \times 100$
$F5 = (D5 - B5) \div B5 \times 100$
$F6 = (D6 - B6) \div B6 \times 100$
$F7 = (D7 - B7) \div B7 \times 100$
$\vdots$
$F12 = (D12 - B12) \div B12 \times 100$
$F14 = (D14 - B14) \div B14 \times 100$

50.

	A	B	C	D	E	F
1	The 7th Inning: Budget Operating Expenses and Actual Expenses					
2						
3	Expense	Budget Amount	Percent of Total Budget	Actual Expenses	% of Actual Total Expense	% Difference from Budget
4						
5	Salaries	\$45,000.00	17.93%	\$42,000.00	16.57%	−6.7%
6	Rent	\$37,000.00	14.74%	\$36,000.00	14.20%	−2.7%
7	Depreciation	\$12,000.00	4.78%	\$14,000.00	5.52%	16.7%
8	Utilities and Phone	\$13,000.00	5.18%	\$10,862.56	4.28%	−16.4%
9	Taxes and Insurance	\$15,000.00	5.98%	\$13,583.29	5.36%	−9.4%
10	Advertising	\$2,000.00	0.80%	\$2,847.83	1.12%	42.4%
11	Purchases	\$125,000.00	49.80%	\$132,894.64	52.41%	6.3%
12	Other	\$2,000.00	0.80%	\$1,356.35	0.53%	−32.2%
13						
14	Total	\$251,000.00	100.01%	\$253,544.67	99.99%	

Section 8–2 Exercises, p. 389

2 **1.** $x = 3$ **2.** $x = 2$ **3.** $x = 29$ **4.** $x = \dfrac{5}{8}$ or 0.625

5. $x = \dfrac{34}{5}$ or 6.8 **6.** $x = \dfrac{17}{14}$ or 1.214 **7.** $x = 3$ **8.** $x = 21$

9. $x = 4$ **10.** $x = -\dfrac{4}{5}$ or -0.8 **11.** $x = 1$ **12.** $x = \dfrac{9}{5}$ or 1.8

13. $x = 30$ **14.** $x = 2.698$ **15.** $x = 836.9$ **16.** $x = -2$
17. $x = 2.835$ **18.** $x = 1.174$ **19.** $x = 2.769$ **20.** $x = 3.25$
21. $x = 214.7$ **22.** $x = 1.438$ **23.** $x = 29.18$ **24.** $x = 10^2$ or 100
25. $x = 10^{12}$ **26.** $x = 1.875$ **27.** $x = 1.25$ **28.** $x = 2.556$
29. $x = 10$ in. **30.** $x = 22.5$ ft **31.** $x = 9\,\text{k}\Omega$ **32.** $x = 257.1$ W
33. $x = 0.26$ V **34.** $x = 0.7317$ mA **35.** $x = 1.346$ **36.** $x = 1$

37. $x = \dfrac{5}{6}$ or 0.8333 **38.** $x = \dfrac{5}{24}$ or 0.2083 **39.** $x = \dfrac{10}{7}$ or $1\dfrac{3}{7}$ or

1.429 **40.** $x = \dfrac{28}{5}$ or $5\dfrac{3}{5}$ or 5.6 **41.** $x = 8$ h **42.** $x = 0.76$ h

43. $x = \$98$ **44.** $x = \$95.19$ **45.** $x = \$0.21$ **46.** $x = 0.375$ gal
47. $x = 44.87$ mg **48.** $x = 5.833$ mg **49.** $x = 0.25$ gal
50. $x = 132$ gal **51.** $x = 555.6$ mi **52.** $x = 20$ mg
53. $x = 8.333$ mg **54.** $x = 8.036$ mg

Section 8–3 Exercises, p. 395

1 **1.** $5.90 **2.** $74.13 **3.** 48 engines **4.** 15 headpieces
5. 1,425 mi **6.** 3,360 mi **7.** 550 lb **8.** 1.7 gal **9.** 12 teeth
10. 32 teeth **11.** 4 cm **12.** 384 cm **13.** 7 ft **14.** 6 ft **15.** 10 ft
16. 14 ft **17.** 355 mg **18.** 277.5 mg **19.** 10 mg **20.** 20 mL

2 **21.** $0.0019 per gram **22.** $0.0022 per gram **23.** The larger
can has lower cost per gram. **24.** 320 bottles per hour **25.** 250 mL
26. 0.4 mL **27.** 15 oz **28.** 12.5 V **29.** 1 mL **30.** 15 ft

31. $11\dfrac{2}{3}$ ft or 11 ft 8 in. **32.** 30 in. **33.** 40 in. **34.** 0.02325 Ω

35. 0.576 Ω **36.** About 45 mi **37.** About 23 mi **38.** About 32 mi
39. 12.5 in. **40.** 30 in. **41.** 7.5 in.

42.
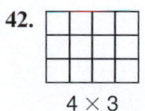
4 × 3

43.
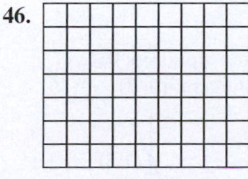
10 × 10

44.
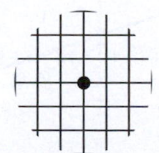
Diameter = 6 squares

45.
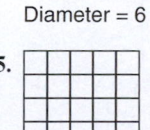
5 × 4

46.

9 × 7

3 **47.** 1,080 **48.** 750 **49.** 324 **50.** $2,400 **51.** $1,960
52. $P = 108$ W

Section 8–4 Exercises, p. 404

1 **1.** 10 h **2.** 200 in^3 **3.** 5 machines **4.** 18 painters
5. 4 helpers (5 workers total) **6.** 4 h **7.** 900 rpm **8.** 50 rpm
9. 20 in. **10.** 120 rpm **11.** 5 in. **12.** 9 in.

2 **13.** 360 **14.** 960 **15.** 240 rpm **16.** 225 rpm **17.** 700 cm
18. 4 cm **19.** 36 cm **20.** 600 rpm **21.** 400 rpm **22.** 140 rpm
23. 300 teeth **24.** 35 teeth **25.** 37.5 rpm **26.** 21 teeth
27. 10 rpm **28.** 38 rpm **29.** 180 teeth **30.** 108 teeth **31.** 9

32. 15 **33.** 6 **34.** 1,296 **35.** $A = \dfrac{kB^2}{C}; k = 25$ **36.** 5,000

37. 504 lb **38.** 1,372 lb

Chapter 8 Review Exercises, p. 410

1. 2 years **3.** 2.75 A **5.** 7 A **7.** $w = \dfrac{V}{lh}$ **9.** $r = s + d$

11. $t = \dfrac{v_0 - v}{32}$ **13.** $P = S + D$ **15.** $\dfrac{7}{6}$ **17.** $\dfrac{1}{2}$ **19.** $\dfrac{49}{12}$ **21.** $-\dfrac{21}{2}$

23. $\dfrac{3}{14}$ **25.** $-\dfrac{21}{23}$ **27.** 4,500 women **29.** $4\dfrac{1}{5}$ ft **31.** 129.7 gal

33. 5.0 h **35.** 1,351 ft **37.** $b = 240$ **39.** 700 in^2 **41.** 23 machines

43. $2\dfrac{1}{2}$ h **45.** 1,500 rpm **47.** 3 days **49.** 60 rpm **51.** $p = 100$

Chapter 8 Practice Test, p. 413

1. $I = 2.2$ A **3.** $L = \dfrac{AR}{P}$ **5.** $s = \dfrac{d}{\pi r^2 n}$ **7.** $\dfrac{15}{2}$ **9.** 9,600 **11.** $\dfrac{32}{5}$

13. 200 rpm **15.** 168.75 rpm **17.** $x = 88.74$ in^2 **19.** 54.7 L
21. $x = 8.8$ A **23.** 107 lb **25.** $450

9 LINEAR EQUATIONS, FUNCTIONS, AND INEQUALITIES IN TWO VARIABLES

Section 9–1 Exercises, p. 425

1 **1.** R: horizontal, 4; vertical, 2 **2.** S: horizontal, -4; vertical, 3
3. T: horizontal, 5; vertical, -3 **4.** U: horizontal, -2; vertical, -5
5. V: horizontal, 7; vertical, 0 **6.** W: horizontal, 0; vertical, 5
7. X: horizontal, 0; vertical, 0 **8.** Y: horizontal, -3; vertical, 0
9. $(4, 2)$ **10.** $(-4, 3)$ **11.** $(5, -3)$ **12.** $(-2, -5)$ **13.** $(7, 0)$
14. $(0, 5)$ **15.** $(0, 0)$ **16.** $(-3, 0)$
17. $A = (4, 2)$ $B = (-3, 2)$ $C = (-2, -1)$ $D = (3, -2)$
18-23.
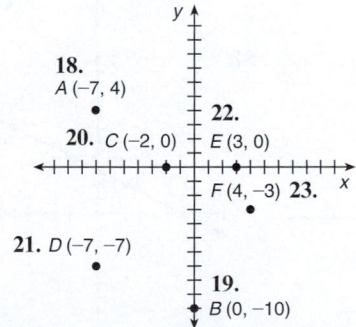

18. $A\,(-7, 4)$
20. $C\,(-2, 0)$ $E\,(3, 0)$
22.
21. $D\,(-7, -7)$
$F\,(4, -3)$ **23.**
19. $B\,(0, -10)$

24. y-value $= 0$ **25.** x-value $= 0$ **26.** $(-x, +y)$ **27.** $(+x, -y)$

2 **28.** $f(x) = 5x - 7$ **29.** $f(x) = -3x + 2$ **30.** $f(x) = 4x - 5$

31. $f(x) = -\dfrac{7}{4}x - \dfrac{3}{4}$ **32.** $f(x) = -3x + 12$ **33.** $f(x) = 3x - 4$

34. $f(x) = -\dfrac{3}{2}x + \dfrac{9}{2}$ **35.** $f(x) = \dfrac{1}{7}x - 2$ **36.** $f(x) = \dfrac{5}{3}x - \dfrac{7}{3}$

3 Values chosen for table of values may vary. At least three
points are required.

37. x	y
0	-7
1	-2
2	3

38. x	y
-1	5
0	2
1	-1

39. x	y
0	-5
1	-1
2	3

40. x	y
-1	1
0	$\dfrac{3}{4}$
1	$\dfrac{5}{2}$

41.

x	y
1	9
2	6
3	3

42.

x	y
−1	−7
0	−4
1	−1

43.

x	y
0	−7
1	−4
2	−1

44.

x	y
−1	5
0	3
1	1
2	−1

45.

x	y
−1	9
0	5
1	1
2	−3
3	−7

46.

x	y
−6	0
−2	2
0	3
2	4
6	6

47.

x	y
−6	3
−3	1
0	−1
3	−3
6	−5

48.

x	y
−1	−3
0	2
1	7

49.

x	f(x)
−2	3
0	2
2	1
4	0

50.

x	f(x)
−6	0
−3	1
0	2
3	3
6	4

57.

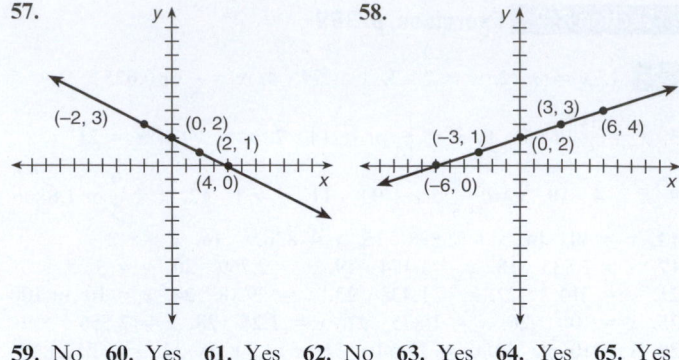

58.

59. No **60.** Yes **61.** Yes **62.** No **63.** Yes **64.** Yes **65.** Yes
66. No **67.** Yes **68.** No **69.** Yes **70.** Yes **71.** (8, 5)
72. (2, 5) **73.** (−6, 1) **74.** (21, 3) **75.** (5, 15) **76.** (2, 0)
77. (1, 3); (1, −2); (1, 0); (1, 1) **78.** (−1, 4); (3, 4); (0, 4); (2, 4)
79. $12 **80.** $2 **81.** $34 **82.** $13,800

Section 9–2   **Exercises, p. 438**

1 **1.** *x*-intercept, (3, 0)
y-intercept, (0, −15)

2. *x*-intercept, $\left(\frac{9}{2}, 0\right)$
y-intercept, (0, 3)

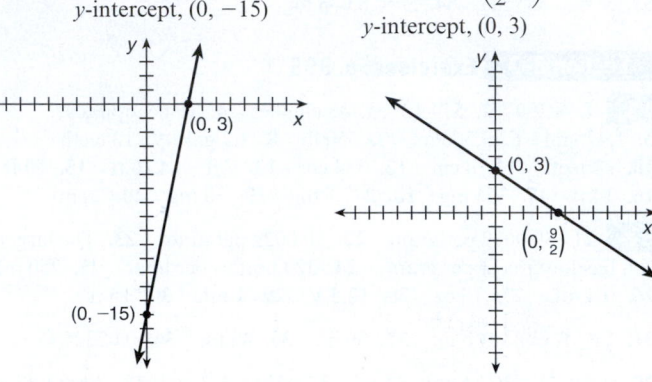

3. *x*-intercept, (5, 0)
y-intercept, (0, 5)

4. *x*-intercept, (5, 0)
y-intercept, $\left(0, \frac{5}{3}\right)$

4

51.

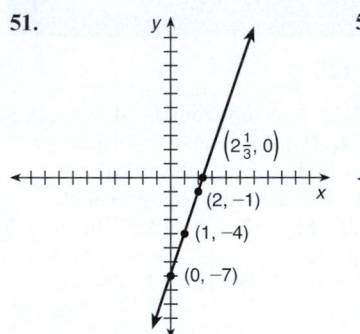

52.

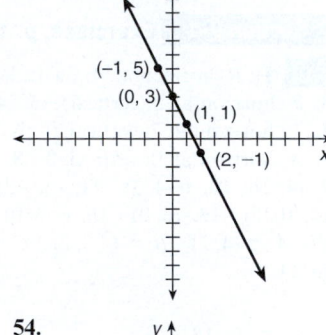

53.

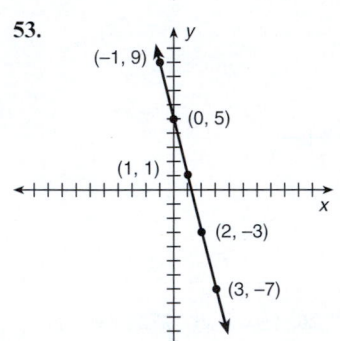

54.

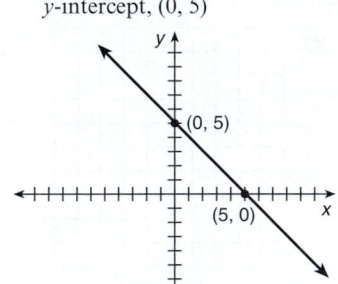

55.

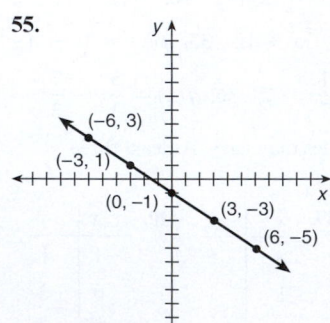

56.

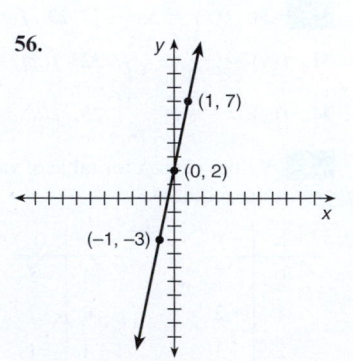

5. *x*-intercept, (−4, 0)
y-intercept, (0, 8)

6. *x*-intercept, $\left(\frac{1}{3}, 0\right)$
y-intercept, (0, −1)

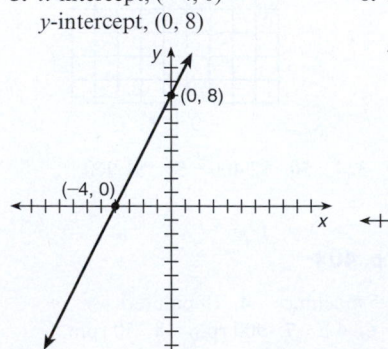

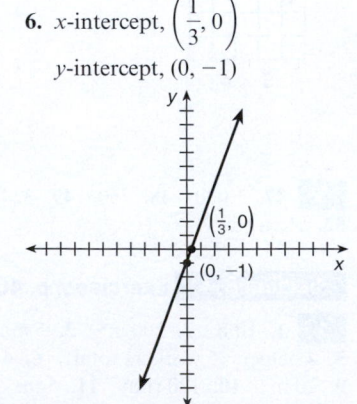

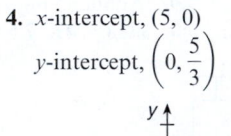

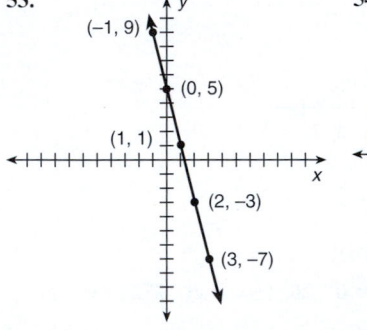

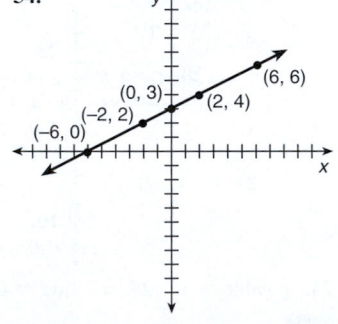

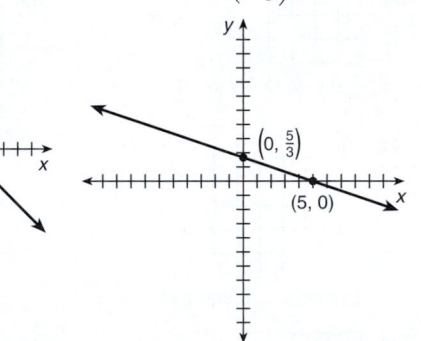

7. x-intercept, $\left(\dfrac{2}{5}, 0\right)$

y-intercept, $(0, -2)$

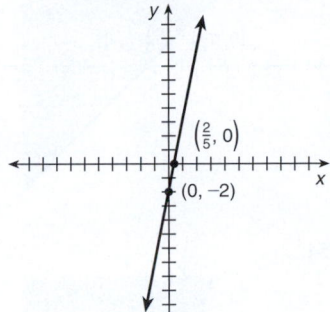

8. x-intercept, $(3, 0)$

y-intercept, $\left(0, \dfrac{3}{2}\right)$

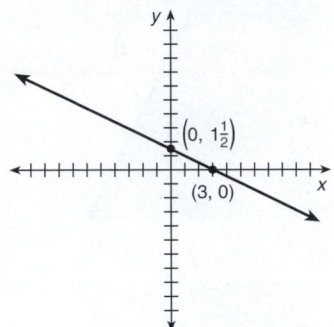

23. slope $= -\dfrac{3}{5}$; y-intercept, $(0, 0)$ **24.** slope $= \dfrac{1}{2}$;

y-intercept, $\left(0, -1\dfrac{1}{2}\right)$

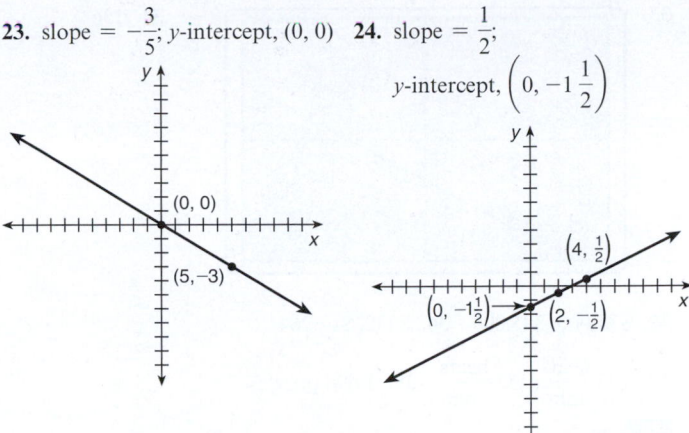

9.

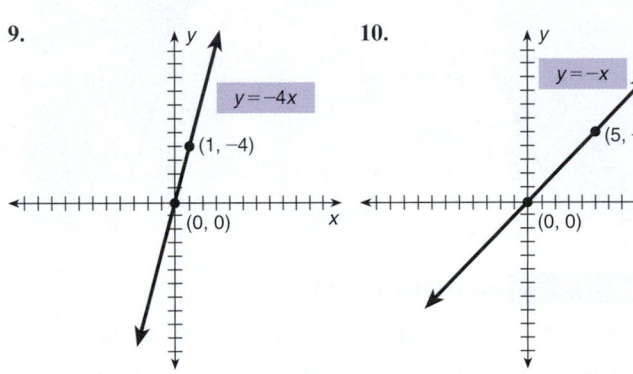

10.

25. slope $= -\dfrac{2}{1}$; y-intercept, $(0, 1)$ **26.** slope $= \dfrac{3}{4}$; y-intercept, $(0, 2)$

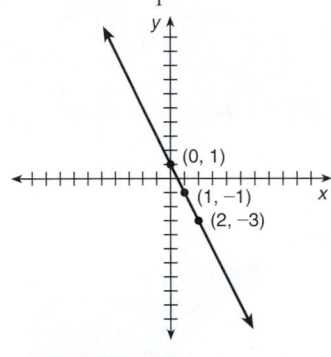

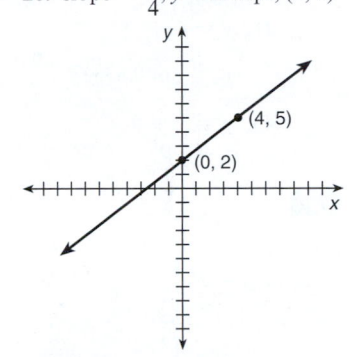

2 **11.** slope $= 4$; y-intercept $= 3$ **12.** slope $= -5$; y-intercept $= 6$

13. slope $= -\dfrac{7}{8}$; y-intercept $= -3$ **14.** slope $= 0$; y-intercept $= 3$

15. $y = 2x - 5$; slope $= 2$; y-intercept $= -5$

16. $y = \dfrac{5}{2}$; slope $= 0$; y-intercept $= \dfrac{5}{2}$ **17.** $y = -2x + 6$;

slope $= -2$; y-intercept $= 6$ **18.** $y = 2x + 5$; slope $= 2$;

y-intercept $= 5$ **19.** $x = 4$; slope not defined; no y-intercept

20. $x = 9$; slope not defined; no y-intercept

21. slope $= \dfrac{2}{1}$; y-intercept, $(0, -3)$ **22.** slope $= -\dfrac{1}{2}$; y-intercept, $(0, -2)$

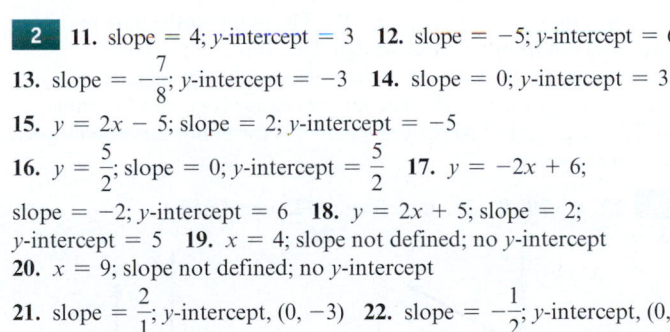

3

27.

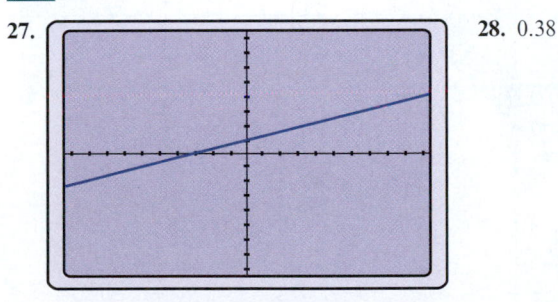

28. 0.38

29.

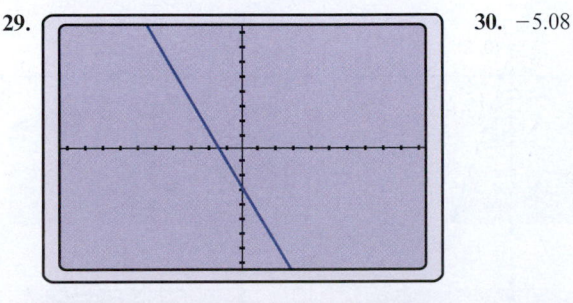

30. -5.08

31.

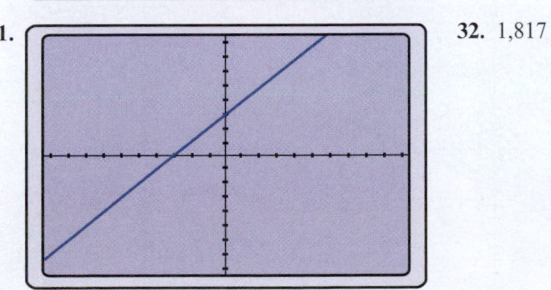

32. $1{,}817$

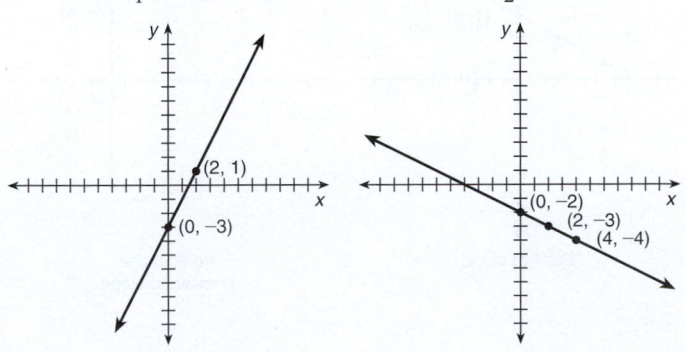

33.

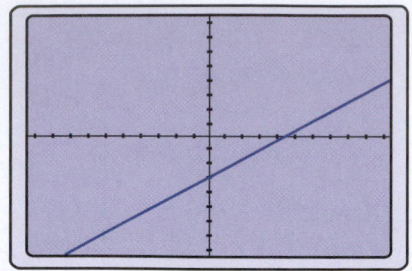

34. 136

35. $35,000; $95,000   **36.** $332; $17,264

37. $195 \dfrac{\text{beats}}{\text{min}}$; $155 \dfrac{\text{beats}}{\text{min}}$ **38.** 1,096 puzzles

39. $y = 3x - 5$
$x = \dfrac{5}{3}$ or 1.66667

40. $y = 5x - 6$
$x = \dfrac{6}{5}$ or 1.2

41. $y = 6x - 5$
$x = \dfrac{5}{6}$ or 0.83333

42. $y = x - 1$
$x = 1$

43. $y = x + 9$
$x = -9$

44. $y = 2x + 16$
$x = -8$

5

45.

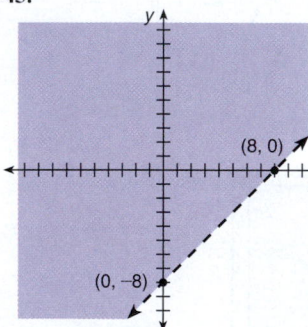

46.

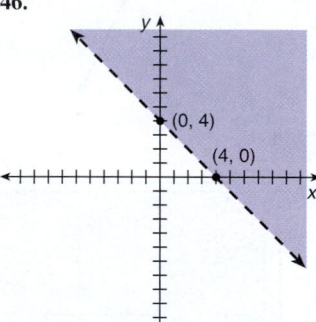

47.

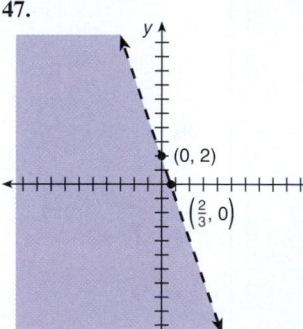

48.

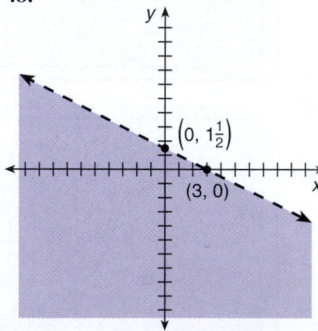

49.

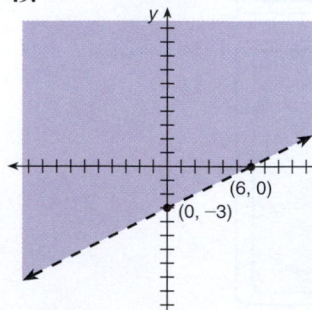

50.

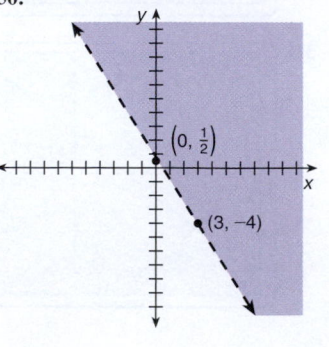

51.

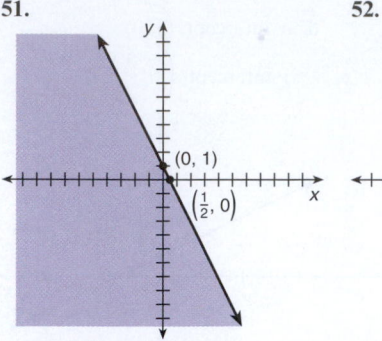

52.

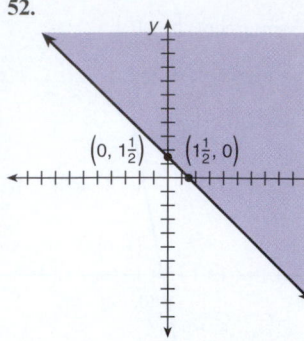

53.

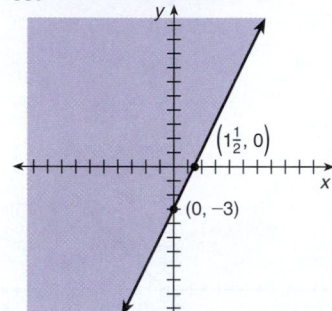

54.

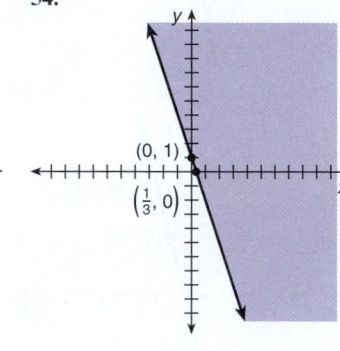

Section 9–3   **Exercises, p. 444**

1   **1.** $\dfrac{1}{2}$ **2.** $-\dfrac{5}{3}$ **3.** 3   **4.** 1   **5.** $\dfrac{3}{4}$ **6.** $-\dfrac{5}{4}$ **7.** -1 **8.** -3

9. $1,400/year   **10.** $10/backpack   **11.** 2,000 ft/min   **12.** $144.40 per year   **13.** $197.40 per year   **14.** $466.00 per year   **15.** $33.00 per year   **16.** $-$12.50 per year   **17.** The graph of the data in this table does not form a perfect straight line. If the slope (rate of change) remains the same between every two data points, the graph will be a straight line.   **18.** The rate of change is greater for public 4-year colleges ($172.50 per year) than it is for public 2-year colleges ($42.60 per year).

2   **19.** 0   **20.** 0   **21.** undefined   **22.** undefined

23.

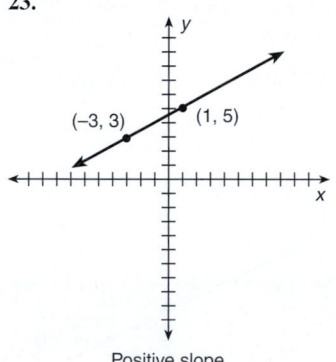

Positive slope

24.

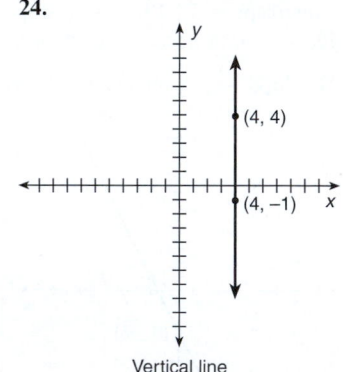

Vertical line
Undefined slope

25.

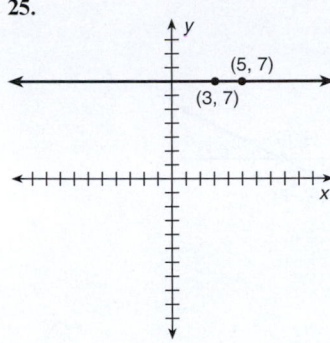

Horizontal line
Zero slope

26.

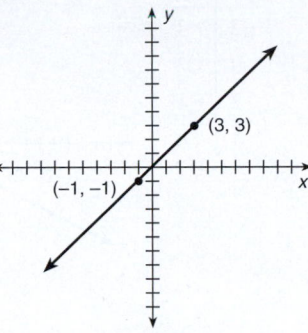

Positive slope

33.

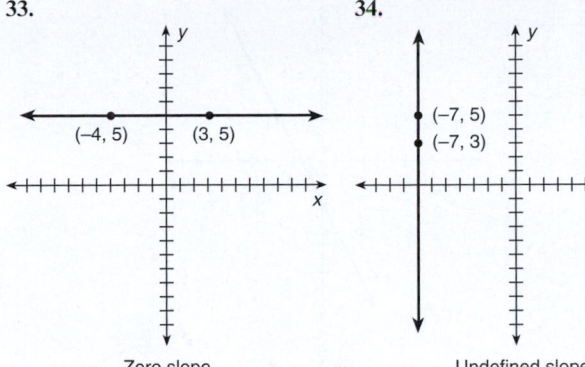

Zero slope

34.

Undefined slope

27.

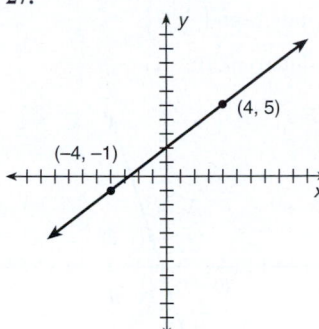

Positive slope

28.

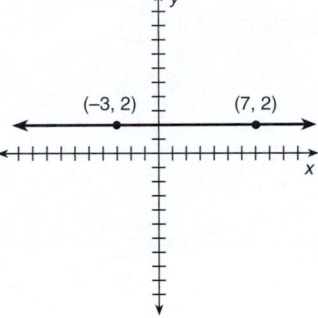

Zero slope

29.

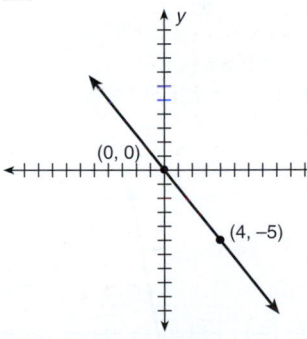

Negative slope

30.

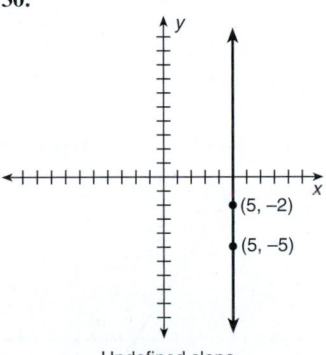

Undefined slope

31.

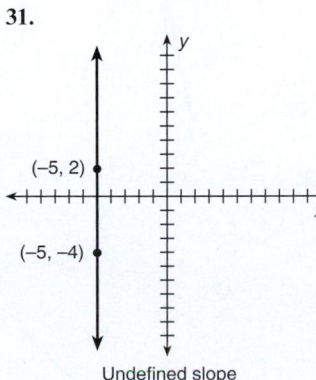

Undefined slope

32.

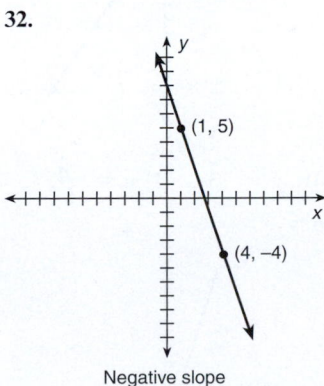

Negative slope

Section 9–4 Exercises, p. 452

1 **1.** $y = \frac{2}{3}x + \frac{25}{3}$; $2x - 3y = -25$ **2.** $y = -\frac{1}{2}x + 3$; $x + 2y = 3$ **3.** $y = 2x + 1$; $2x - y = -3$ **4.** $y = x - 1$; $x - y = 1$

2 **5.** $y = -\frac{5}{3}x + \frac{38}{3}$ **6.** $y = 2x$ **7.** $y = -3$ **8.** $y = 0$ **9.** $x = -1$ **10.** $x = -4$ **11.** $y = 5x + 1{,}935$ **12.** $y = 22x + 50{,}000$

3 **13.** $y = \frac{1}{4}x + 7$ **14.** $y = -8x - 4$ **15.** $y = -2x + 3$ **16.** $y = \frac{3}{5}x - 2$ **17.** $y = x$ **18.** $y = 5x - \frac{1}{5}$ **19.** $y = 2x - 2$ **20.** $y = -\frac{3}{4}x$ **21.** $y = 0.2x + 3$; \$12 **22.** $y = 8x + 12{,}000$; \$76,000 **23.** $y = 2x + 3$ **24.** \$23 **25.** $y = -0.9x + 40$; 31 mi/h **26.** $y = -3x + 1$ **27.** $y = \frac{2}{3}x - 2$ **28.** $y = 4$

4 **29.** $x + y = -1$ **30.** $2x + y = 9$ **31.** $x - 3y = 4$ **32.** $3x - y = -7$ **33.** $x + 3y = 7$ **34.** $3x - y = -3$ **35.** $2x + 3y = 5$ **36.** $3x + 2y = 6$ **37.** $2x - 5y = 11$ **38.** $6x - 8y = 3$ **39.** $y = 12x + 20{,}000$ **40.** $y = 0.5x + 45$

5 **41.** $x - y = -1$ **42.** $x - 2y = -13$ **43.** $3x + y = 12$ **44.** $x + 3y = -9$ **45.** $3x - y = 11$ **46.** $2x - y = -7$ **47.** $3x - 2y = 11$ **48.** $x - 4y = 0$ **49.** $2x - 2y = -3$ **50.** $x + 5y = 4$

Chapter 9 Review Exercises, p. 462

1.–6. See figure below.

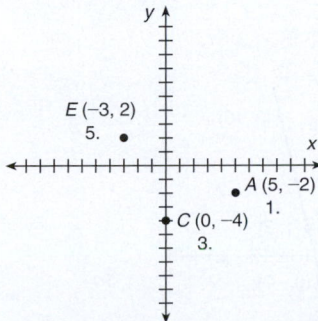

7. $(0, 0)$ **9.** $A(3, 0)$; $B(2, 2)$; $C(2, -5)$; $D(-4, -1)$; $E(-3, 1)$ Since plotted solutions will vary, check graphs by comparing x- and y-intercepts.

11.

x	y
-1	-5
1	-1
3	3

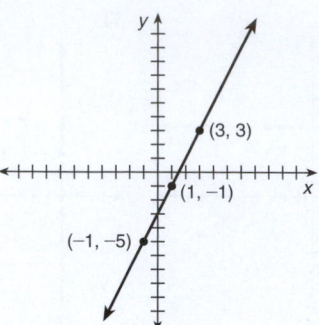

13.

x	y
-1	-3
0	0
1	3

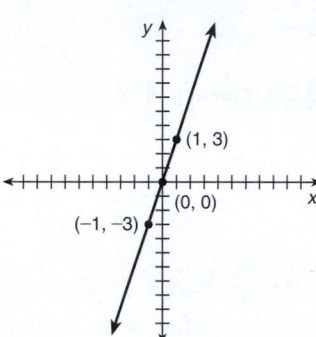

15.

x	y
-1	3
0	0
1	-3

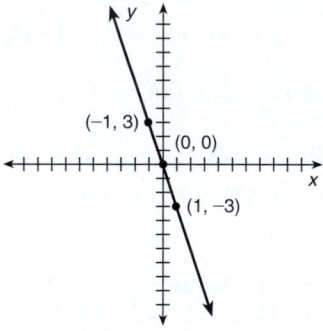

17.

x	y
-1	-1
0	1
1	3

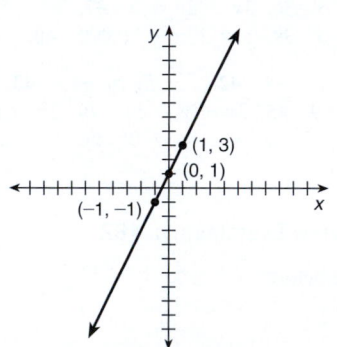

19.

x	y
-2	-10
-1	-6
0	-2
1	2
3	10

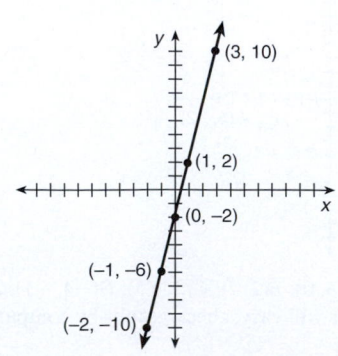

21.

x	y
-2	-3
0	-2
2	-1
4	0

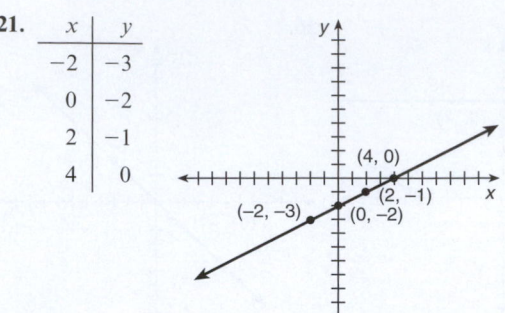

23. No **25.** Yes **27.** Yes **29.** $y = -7$ **31.** $y = 1$

33. $x = -4y - 1$
x-intercept, $(-1, 0)$;
y-intercept, $\left(0, -\frac{1}{4}\right)$

35. $3x - y = 1$
x-intercept, $\left(\frac{1}{3}, 0\right)$;
y-intercept, $(0, -1)$

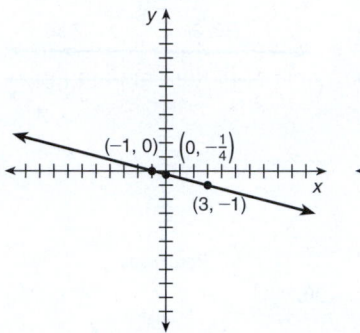

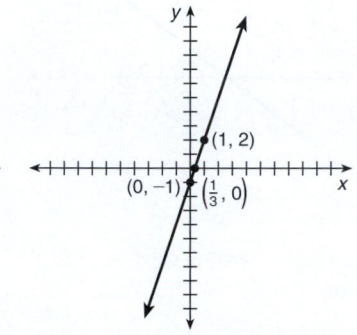

37. $\frac{1}{2}x + \frac{1}{3}y = 1$
x-intercept $(2, 0)$; y-intercept $(0, 3)$

39. $y = 5x - 2; m = \frac{5}{1}; b = -2$

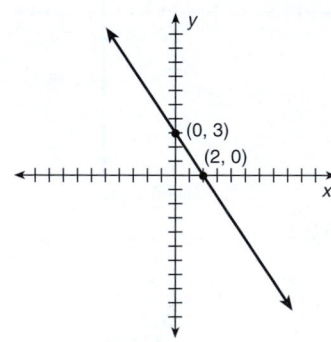

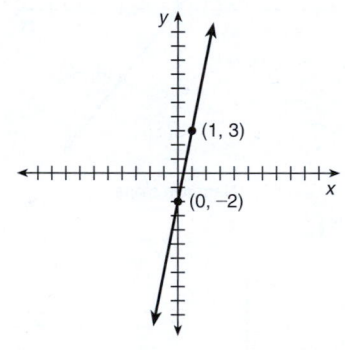

41. $y = -3x - 1; m = \frac{-3}{1}$;
$b = -1$

43. $y = x - 4; m = \frac{1}{1}; b = -4$

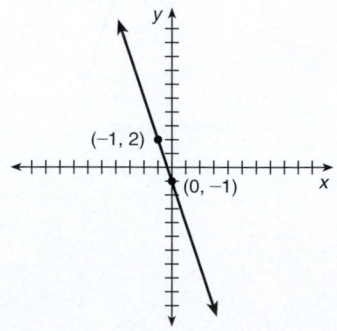

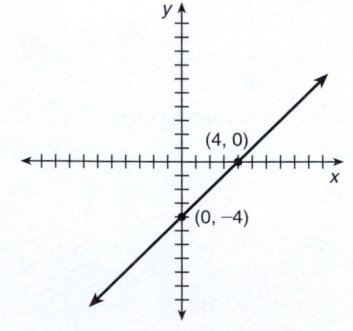

45. $y = \frac{1}{2}x + \frac{1}{2}; m = \frac{1}{2}; b = \frac{1}{2}$ **47.** $y = 2x - 2; m = 2; b = -2$

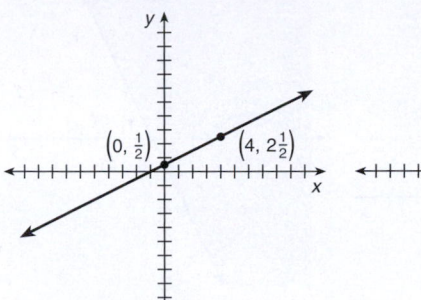

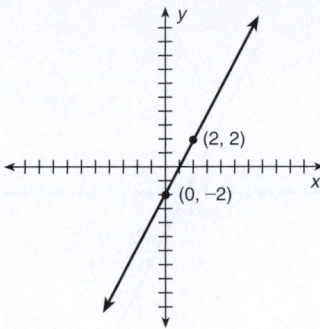

49.   **51.**

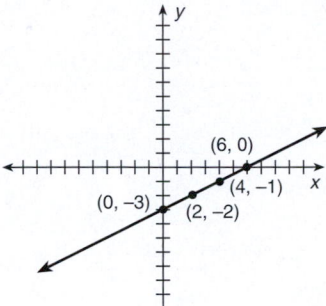

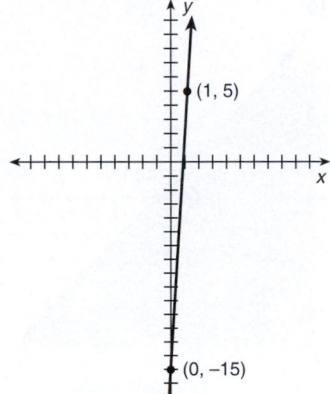

53. \$58,000

55. $y = x - 6; x = 6$ **57.** $y = 3x + 6; x = -2$

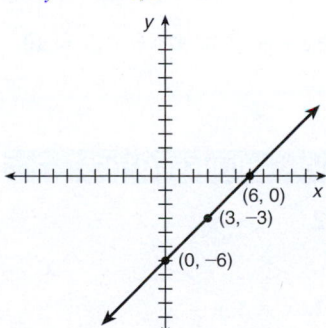

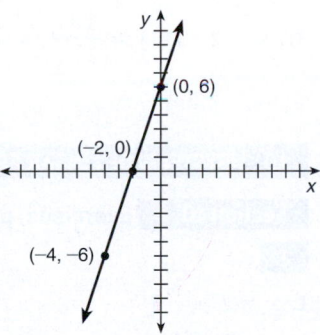

59. $y = 2x - 2; x = 1$

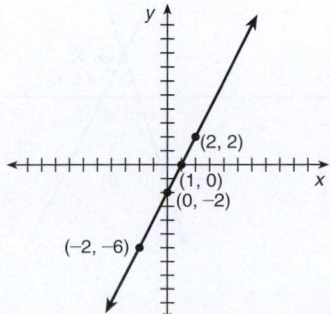

61. $m = 3; b = \frac{1}{4}$ **63.** $m = -5; b = 4$ **65.** Undefined slope; no

y-intercept   **67.** $m = \frac{1}{8}; b = -5$ **69.** $y = -2x + 8; m = -2; b = 8$

71. $y = \frac{3}{2}x - 3; m = \frac{3}{2}; b = -3$ **73.** $y = \frac{3}{5}x - 4; m = \frac{3}{5};$

$b = -4$ **75.** $y = \frac{5}{3}; m = 0; b = \frac{5}{3}$

77.   **79.**

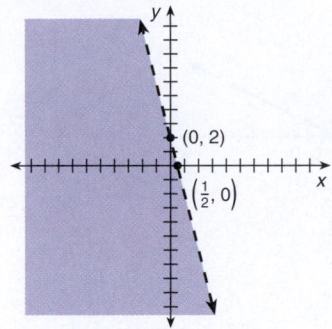

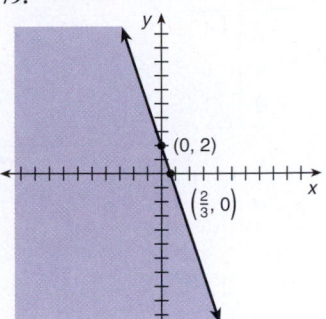

81.   **83.**

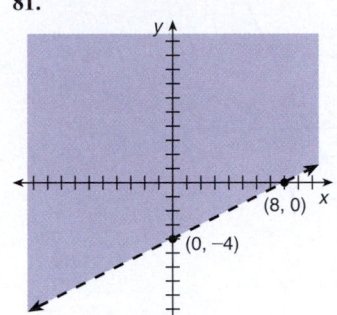

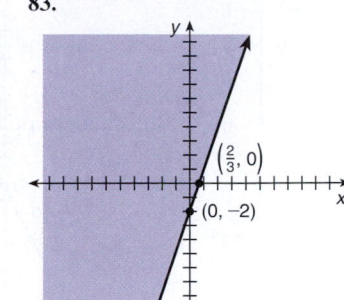

85.

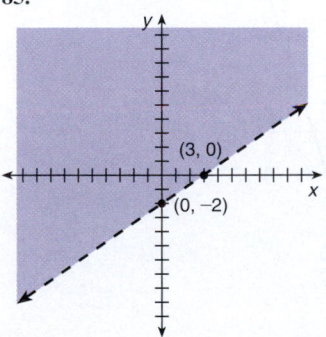

87. 2   **89.** $\frac{5}{8}$ **91.** $-\frac{4}{3}$ **93.** Undefined   **95.** $-\frac{4}{7}$ **97.** $\frac{7}{2}$ **99.** 0

101. Undefined   **103.** $(-1, 2) (3, 2)$; answers will vary.   **105.** \$56

107. \$37   **109.** The table of values is not a perfect linear function.

111. $y = \frac{1}{3}x + 4$ **113.** $y = \frac{3}{4}x - 3$ **115.** $4x - y = 5$

117. $y = -\frac{1}{11}x + \frac{17}{11}$ **119.** $y = \frac{7}{4}x - \frac{5}{4}$ **121.** $y = \frac{9}{5}x + \frac{3}{5}$

123. $y = x - 3$ **125.** $y = x - 1$ **127.** $y = -2$

129. $S = 10x + 3,000$ **131.** $y = 3x - 2$ **133.** $y = 2x - 2$

135. $x + y = 7$ **137.** $x - 2y = 8$ **139.** $x - 3y = 20$

141. $x + 3y = -10$ **143.** $3x - 4y = -7$ **145.** $x - y = -4$

147. $2x - y = -4$ **149.** $x - 5y = -11$ **151.** $2x + 10y = 31$

153. $x + 4y = 2$

Chapter 9 Practice Test, p. 466

Since plotted solutions will vary, check graphs by comparing *x*- and *y*-intercepts.

1.

x	y
−2	−1
0	0
2	1

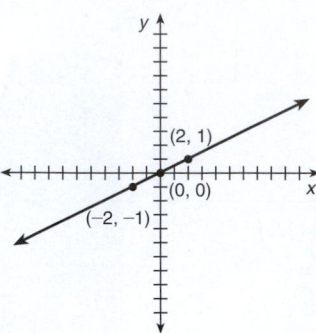

3.

x	y
−1	−6
0	−4
1	−2
2	0

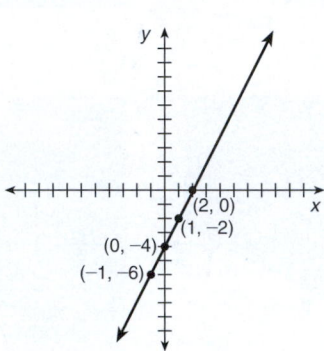

5. $y = 3x - 3; x = 1$

x	y
−1	−6
0	−3
1	0

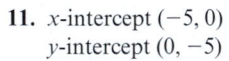

7. (2, 5)   **9.** \$23,200

11. *x*-intercept (−5, 0)
    y-intercept (0, −5)

13. *x*-intercept (8, 0)
    y-intercept (0, 4)

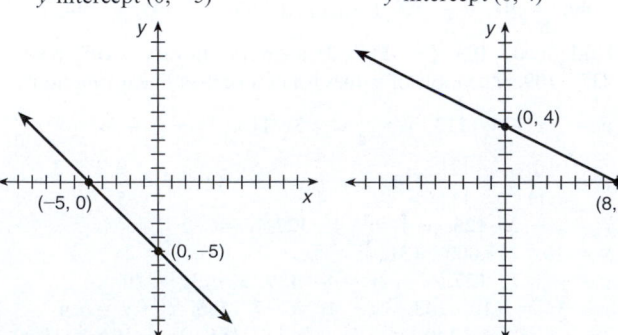

15. *y*-intercept = (0, −3)
    slope = $\dfrac{-2}{1}$

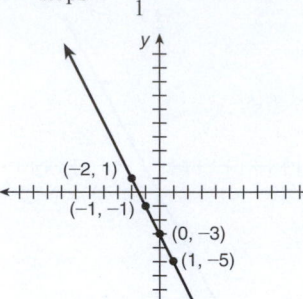

17.

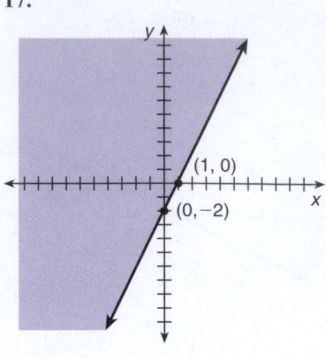

19.

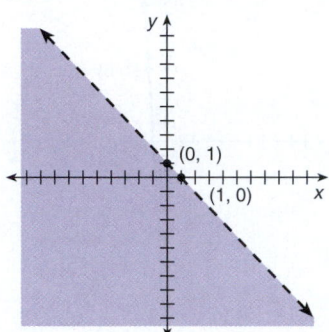

21. $-\dfrac{2}{3}$ **23.** $y = 2x + 34; m = 2; b = 34$

25. $y = \dfrac{1}{4}x; m = \dfrac{1}{4}; b = 0$ **27.** $y = \dfrac{2}{3}x - 7$ **29.** $y = \dfrac{2}{3}x + \dfrac{7}{3}$

31. $y = 2$ **33.** $y = \dfrac{3}{2}x + 3$ **35.** $2x + y = 5$ **37.** $x - 2y = 10$

10 SYSTEMS OF LINEAR EQUATIONS AND INEQUALITIES

Section 10–1 Exercises, p. 472

1

1. $x = 7, y = 5$

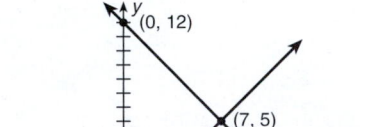

2. $x = 3, y = 3$

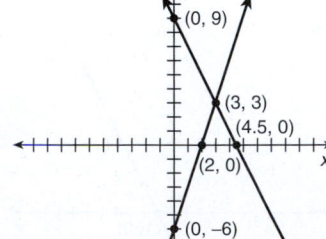

3. No solution; no intersection

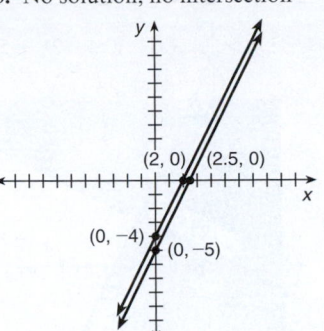

(2, 0) (2.5, 0)
(0, −4) (0, −5)

4. Many solutions; lines coincide

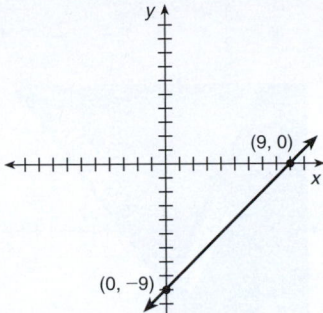

(9, 0)
(0, −9)

11. $x = 1, y = -7$

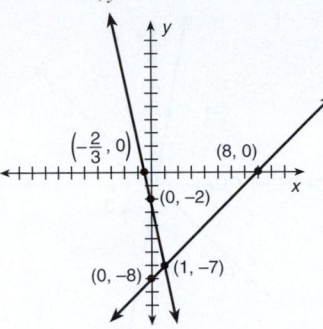

$\left(-\frac{2}{3}, 0\right)$ (8, 0)
(0, −2)
(0, −8) (1, −7)

12. $x = 0, y = 3$

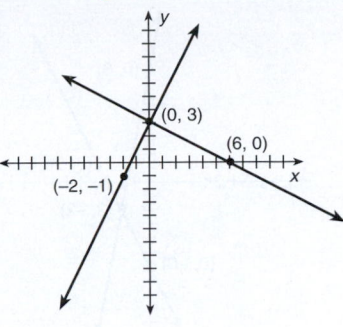

(0, 3) (6, 0)
(−2, −1)

5. $x = 4, y = -2$

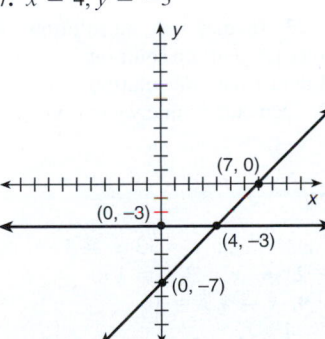

(0, 4)
(0, 2) $\left(2\frac{2}{3}, 0\right)$
(2, 0) (4, −2)

6. $x = 3, y = 1$

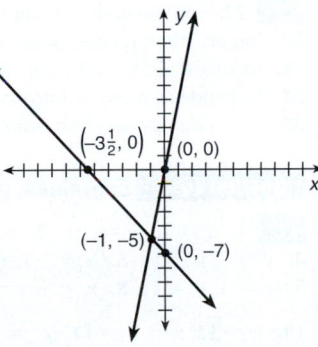

(0, 7)
(3, 1)
(0, −2)

13. $x = 1, y = -5$

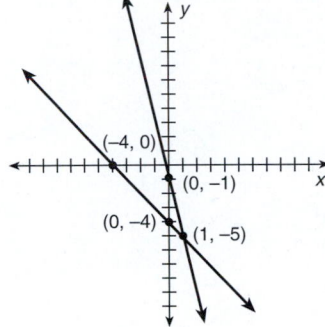

(−4, 0)
(0, −1)
(0, −4) (1, −5)

14. $x = 2, y = 1$

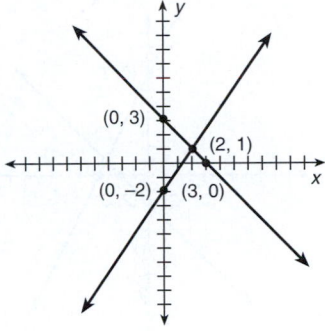

(0, 3)
(2, 1)
(0, −2) (3, 0)

7. $x = 4, y = -3$

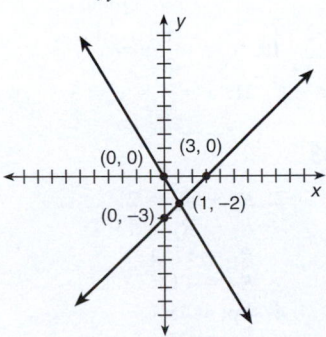

(7, 0)
(0, −3)
(4, −3)
(0, −7)

8. $x = -1, y = -5$

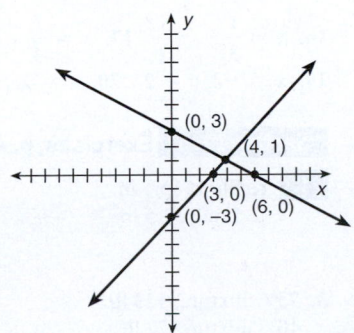

$\left(-3\frac{1}{2}, 0\right)$ (0, 0)
(−1, −5) (0, −7)

15. $x = 7, y = 9$

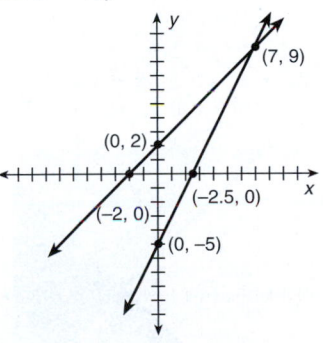

(7, 9)
(0, 2)
(−2.5, 0)
(−2, 0) (0, −5)

16. $x = -2, y = 4$

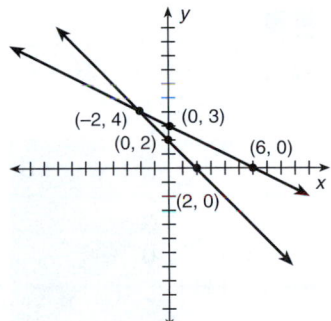

(−2, 4) (0, 3)
(0, 2) (6, 0)
(2, 0)

17. $x = 4, y = 1$

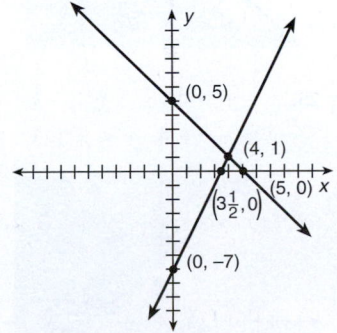

(0, 5)
(4, 1)
$\left(3\frac{1}{2}, 0\right)$ (5, 0)
(0, −7)

18. $x = 3, y = -4$

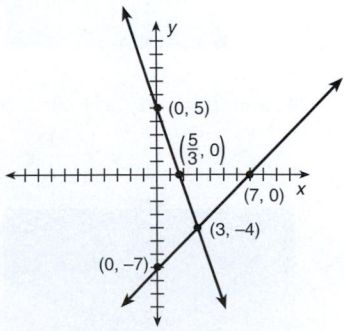

(0, 5)
$\left(\frac{5}{3}, 0\right)$
(7, 0)
(3, −4)
(0, −7)

9. $x = 1, y = -2$

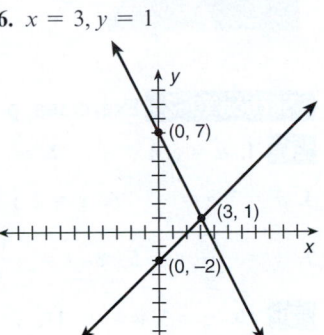

(0, 0) (3, 0)
(0, −3) (1, −2)

10. $x = 4, y = 1$

(0, 3)
(4, 1)
(3, 0) (6, 0)
(0, −3)

19. $x = 2, y = -2$

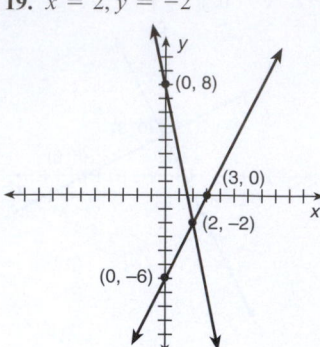

20. $x = 0, y = 0$

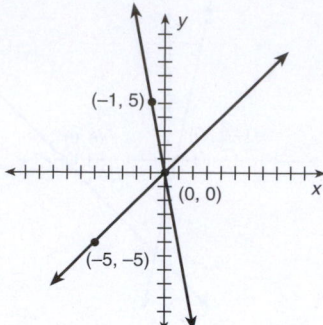

21. $x = 3, y = 1$

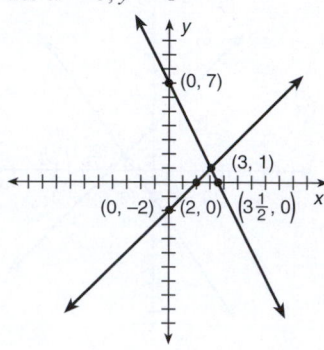

2

22. $x + y > 5$ $y < 2x + 1$
$y > -x + 5$

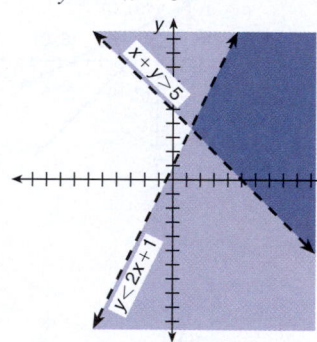

23. $x + y \geq 3$ $x - y \geq 2$
$y \leq -x + 3$ $y \leq x - 2$

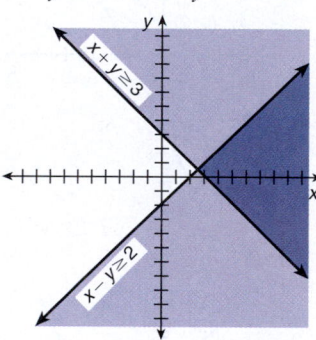

24. $x - 2y \leq -1$ $x + 2y \geq 3$
$y \geq \dfrac{1}{2}x + \dfrac{1}{2}$ $y \geq -\dfrac{1}{2}x + \dfrac{3}{2}$

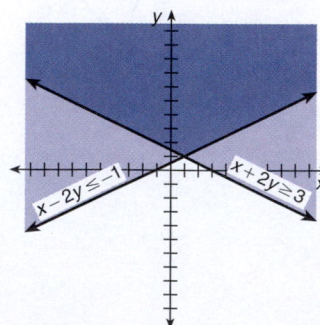

25. $x + y > 4$ $x - y > -3$
$y > -x + 4$ $y < x + 3$

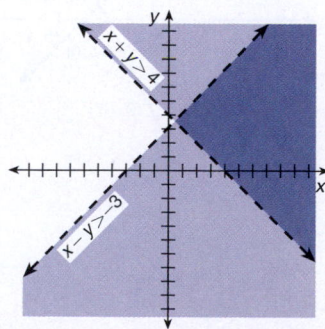

26. $3x - 2y < 8$ $2x + y \leq -4$
$y > \dfrac{3}{2}x - 4$ $y \leq -2x - 4$

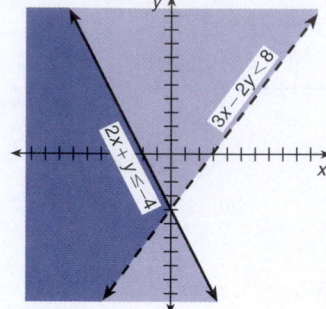

27. $x - 2y < -6$ $2x + y \leq 5$
$y > \dfrac{1}{2}x + 3$ $y \leq -2x + 5$

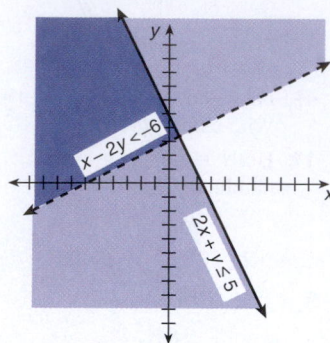

Section 10–2 Exercises, p. 477

1 **1.** $a = 5, b = -1$ **2.** $m = 4, n = -1$ **3.** $x = 1, y = -1$
4. $a = 3, b = -3$ **5.** $x = 2, y = \dfrac{3}{2}$ **6.** $x = 4, y = -\dfrac{3}{5}$
7. $x = -\dfrac{13}{5}, y = 5$ **8.** $x = -\dfrac{11}{4}, y = 5$ **9.** $x = -4, y = 3$

2 **10.** $x = 3, y = 0$ **11.** $x = 1, y = 3$ **12.** $a = 2, b = \dfrac{8}{3}$
13. $x = 1, y = 5$ **14.** $a = 6, y = 1$ **15.** $x = -1, y = 2$
16. $x = 2, y = -2$ **17.** $x = -4, y = 1$ **18.** $x = 5, y = -2$
19. $x = 0, y = 3$ **20.** $x = -1, y = 3$ **21.** $x = 13, y = 3$

3 **22.** Inconsistent; no solution **23.** Inconsistent; no solution
24. Inconsistent; no solution **25.** Inconsistent; no solution
26. Inconsistent; no solution **27.** Inconsistent; no solution
28. Dependent; many solutions **29.** Dependent; many solutions
30. Dependent; many solutions

Section 10–3 Exercises, p. 479

1 **1.** $a = 20, b = 10$ **2.** $r = 5, c = 7$ **3.** $x = 13, y = 5$
4. $x = 7, y = 5$ **5.** $a = -7, x = -2$ **6.** $x = 3, y = 1$
7. $x = 1, y = 1$ **8.** $x = 1, y = 6$ **9.** $x = 4, y = 10$
10. $x = 13, y = 11$ **11.** $p = \dfrac{1}{2}, k = \dfrac{1}{3}$ **12.** $x = 8, y = 1$
13. $x = 3, y = 2$ **14.** $x = 3, y = 3$ **15.** $x = 1, y = \dfrac{3}{2}$
16. $x = \dfrac{1}{3}, y = \dfrac{1}{2}$ **17.** $x = \dfrac{1}{4}, y = \dfrac{2}{3}$ **18.** $x = -\dfrac{1}{2}, y = \dfrac{1}{2}$
19. $x = -2, y = 2$ **20.** $x = 2, y = -1$ **21.** $x = -8, y = 3$

Section 10–4 Exercises, p. 483

1 **1.** Short 15.5 in.
Long 32.5 in.

2. $R_1 + R_2 = 21$
$R_1 - R_2 = 13$
$R_1 = 17 \, \Omega$
$R_2 = 4 \, \Omega$

3. 75% mixture, 123 lb
10% mixture, 77 lb
Amounts rounded to nearest pound.

4. 4 pt at 75%
4 pt at 25%

5. Scientific \$8
Graphing \$72

6. Reserved \$20
General \$15

7. \$18,000 at 4%
\$17,000 at 5%

8. \$4,000 at 10%
\$6,000 at 15%

9. $3,000 at 7%
    $5,000 at 9%

10. Seven 8-cylinder jobs
    Three 4-cylinder jobs

11. Resistor $0.25
    Capacitor $0.30

12. Rate of plane = 130 mi/h
    Rate of wind = 10 mi/h

13. $21 per shirt
    $16 per hat

14. Rate of current = 1.67 mi/h
    Rate of motorboat = 11.67 mi/h

15. Dark roast, $4.10
    With chicory, $3.75

16. Potassium = 120 lb
    Nitrogen = 360 lb

17. Holly = 40 shrubs,
    nandina = 20 shrubs

18. Water = 2 ft^3, sand = 10 ft^3

51. $(12, 2)$ **53.** $(1, -2)$ **55.** electrician = $75
                                        apprentice = $35

57. shellac = $2.50   **59.** 119°, 56°   **61.** $2,000 at 5%
    thinner = $3.50 $3,000 at 6%

63. Colombian = $5   **65.** Ohio = $1.95
    blended = $4   Alaska = $2.10

67. name brand = $320   **69.** telephone = $15,000
    generic = $175   showroom = $25,000

Chapter 10 Review Exercises, p. 487

1.

3.

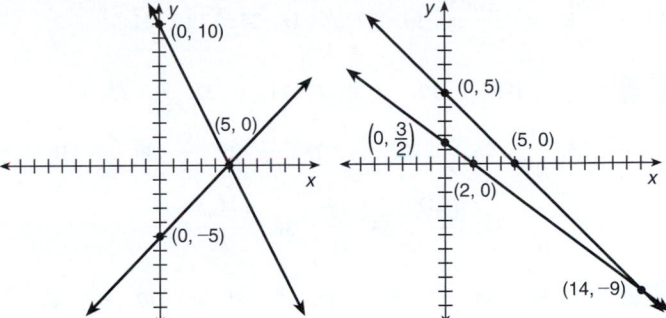

Solution: $x = 2, y = 9$

Solution: $x = -1, y = -4$

5.

7.

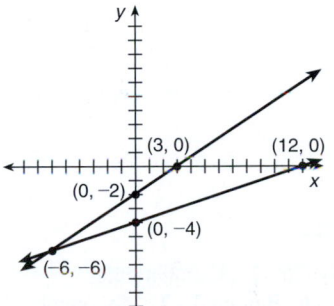

Solution: $x = 5; y = 3$

Dependent; many solutions, lines coincide

9.

11.

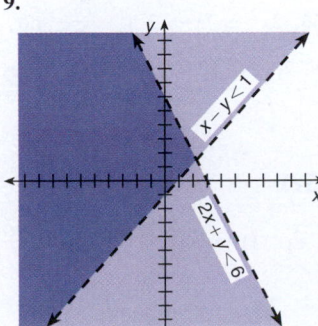
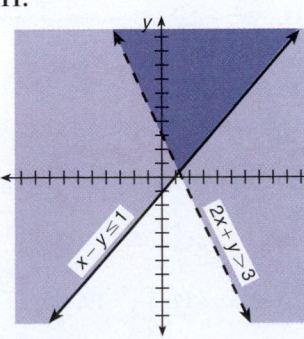

13. $(3, 0)$ **15.** $(1, -1)$ **17.** $(-2, 4)$ **19.** $(2, -2)$ **21.** $(6, 3)$

23. Dependent; many solutions   **25.** $(5, 1)$ **27.** $\left(2, \dfrac{8}{3}\right)$ **29.** $(2, 0)$

31. $(3, 2)$ **33.** $(2, 1.5)$ **35.** $(4, 4)$ **37.** $(7, 5)$ **39.** $(4, -6)$

41. $(3, -4)$ **43.** $\left(\dfrac{4}{5}, \dfrac{2}{5}\right)$ **45.** $(28, 44)$ **47.** $\left(-3, \dfrac{5}{2}\right)$ **49.** $(4, -6)$

Chapter 10 Practice Test, p. 490

1. $(5, 0)$

3. $(14, -9)$

5. $(-6, -6)$

7. $(8, 4)$ **9.** $(0, 0)$ **11.** $(1, 2)$ **13.** $(2, 2)$ **15.** $(-3, 15)$

17. $(-9, -11)$ **19.** 20 A, 15 A

21. $L = 51$ in.
    $W = 34$ in.

23. $20,000 at 3.5%
    $5,000 at 4%

25. 0.000215 F
    0.000055 F

Chapters 7–10 Cumulative Practice Test, p. 491

1. $-6x + 19$ **3.** $x = 1$ **5.** $x = \dfrac{31}{39}$ **7.** $Z = \sqrt{65}\ \Omega$ or 8.1 Ω

9. $x = \dfrac{21}{10}$ **11.** 160 teeth

13.

15.

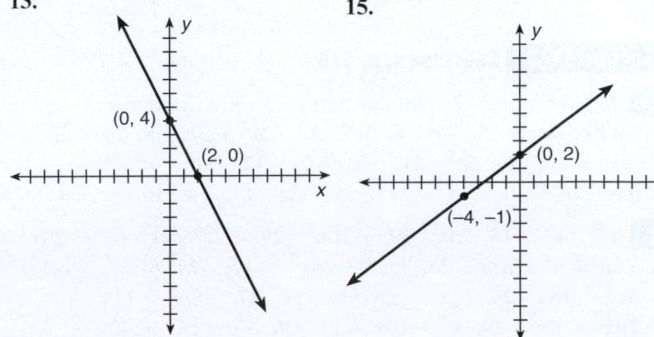

17. slope $= \dfrac{3}{2}$; y-intercept $= -5$ **19.** $y = -\dfrac{2}{3}x - \dfrac{2}{3}$

21. $y = 2x - 4$ **23.** $y = 4x + 3$ **25.** $x - y = 1$ **27.** $(1, 3)$

29. larger resistance $= 20\ \Omega$; smaller resistance $= 12\ \Omega$

11   POWERS AND POLYNOMIALS

Section 11–1   Exercises, p. 499

1   **1.** x^7 **2.** m^4 **3.** a^2 **4.** x^{10} **5.** y^6 **6.** a^2b **7.** $12a^5b^7$

8. $10x^{11}$ **9.** $-10x^3y^3z^4$ **10.** $24x^3y^4$ **11.** $\dfrac{2}{7}a^3b^5$ **12.** $\dfrac{3}{8}a^4b^3$

13. $-\dfrac{4.2n^{10}}{m^2}$ or $-\dfrac{21n^{10}}{5m^2}$ **14.** $-72a^6$ **15.** $24x^7$ **16.** $24ab^4$

2   **17.** y^5 **18.** x^4 **19.** $\dfrac{1}{a}$ **20.** b **21.** 1 **22.** $\dfrac{1}{x^2}$ **23.** y^4

24. n^7 **25.** $\dfrac{1}{x^{10}}$ **26.** $\dfrac{2}{n}$ **27.** $\dfrac{3x^7}{2}$ **28.** $\dfrac{1}{x^5}$ **29.** $\dfrac{x}{y^6}$ **30.** $\dfrac{a^3}{b^2}$ **31.** xy^2

32. $\dfrac{a}{b}$ **33.** $\dfrac{3a}{b^5}$ **34.** $\dfrac{3rs^{12}}{2}$ **35.** $\dfrac{2y^2}{3x^4}$ **36.** $\dfrac{3m^6n^3}{2}$

3   **37.** $4{,}096$ **38.** x^8 **39.** 1 **40.** x^{40} **41.** a^{21} **42.** $-\dfrac{1}{8}$ **43.** $\dfrac{4}{49}$

44. $\dfrac{a^4}{b^4}$ **45.** $\dfrac{x^6}{y^3}$ **46.** $8m^6n^3$ **47.** x^6y^{12} **48.** $4a^2$ **49.** x^6y^3

50. $-27a^3b^6$ **51.** $-16{,}807x^{10}$ **52.** $16x^4y^{16}$ **53.** $-\dfrac{7}{x^{17}}$ **54.** $-x^4$

55. x^4 **56.** $-6x^8$ **57.** $-x^{13}y$ **58.** $-3x^3$ **59.** $-\dfrac{x^2}{y^7}$ **60.** $-27a^6b^3c^3$

61. $-\dfrac{3}{5}$ **62.** $\dfrac{4y^{13}}{13}$ **63.** d^9 **64.** $\dfrac{5}{x^2z}$

Section 11–2   Exercises, p. 504

1   **1.** Not a polynomial   **2.** Polynomial   **3.** Polynomial
4. Not a polynomial   **5.** Binomial   **6.** Binomial   **7.** Monomial
8. Monomial   **9.** Monomial   **10.** Monomial   **11.** Trinomial
12. Trinomial   **13.** Monomial   **14.** Binomial   **15.** Binomial
16. Binomial

2   **17.** 1 **18.** 2 **19.** $2, 1, 0$ **20.** $3, 1, 0$ **21.** $1, 0$ **22.** $2, 0$
23. 0 **24.** 0 **25.** $1, 0$ **26.** $2, 0$ **27.** $5, 2$ **28.** $5, 8$ **29.** 2 **30.** 3
31. 6 **32.** 5 **33.** 2 **34.** 1 **35.** 6 **36.** 6

3   **37.** $-3x^2 + 5x$; 2; $-3x^2$; -3 **38.** $-x^3 + 7$; 3; $-x^3$; -1
39. $9x^2 + 4x - 8$; 2; $9x^2$; 9 **40.** $5x^2 - 3x + 8$; 2; $5x^2$; 5
41. $7x^3 + 8x^2 - x - 12$; 3; $7x^3$; 7 **42.** $-15x^4 + 12x + 7$; 4; $-15x^4$; -15
43. $8x^6 - 7x^3 - 7x$; 6; $8x^6$; 8 **44.** $-14x^8 + x + 15$; 8; $-14x^8$; -14
45. $8x^4 + 15x^3 + 12x$; 4; $8x^4$; 8 **46.** $-7x^4 + 5x^3 + 3x - 8$; 4; $-7x^4$; -7
47. $-5x^4 + 2x^2 + 3x - 5$; 4; $-5x^4$; -5
48. $-8x^4 - x^3 + 7x - 3$; 4; $-8x^4$; -8

Section 11–3   Exercises, p. 515

1   **1.** Like terms   **2.** Not like terms   **3.** Not like terms
4. Not like terms   **5.** $7a^2$ **6.** $3x^3$ **7.** $-2a^2 + 3b^2$ **8.** $2a - 2b$
9. $-5x^2 - 2x$ **10.** $5a^2$ **11.** $-x^2 - 2y$ **12.** $2m^2 + n^2$
13. $9a + 2b + 10c$ **14.** $-2x + 3y + 2z$ **15.** -4 **16.** $3x + 1$

2   **17.** $14x^3$ **18.** $2m^3$ **19.** $-21m^2$ **20.** $-2y^6$ **21.** $8x^3 - 28x^2$
22. $-6a^3b^2 + 15a^2b^3$ **23.** $35x^4 - 21x^3 - 49x$ **24.** $6x^3y^2 - 10x^2y^3$
25. $3x^2 - 18x$ **26.** $12x^3 - 28x^2 + 32x$ **27.** $-8x^2 + 12x$
28. $10x^2 + 4x^3$ **29.** $x^2 + 10x + 21$ **30.** $x^2 + 13x + 40$

31. $2x^2 + 3x - 2$ **32.** $3x^2 - 8x - 35$ **33.** $x^3 + 7x^2 + 9x - 5$
34. $x^3 + 4x^2 - 19x + 14$ **35.** $x^3 - 12x^2 + 37x - 14$
36. $x^3 - 11x^2 + 25x - 3$ **37.** $2x^3 - 7x^2 + 7x - 5$
38. $3x^3 + x^2 - 11x + 6$ **39.** $6x^3 + 11x^2 - 31x + 14$
40. $20x^3 - 22x^2 - 9x + 9$ **41.** $x^3 - 8$ **42.** $8x^3 - 27$ **43.** $x^3 + 27$
44. $27x^3 + 8$ **45.** $x^3 + 125$ **46.** $1{,}331x^3 - 27$
47. $3x^4 - 2x^3 - 7x^2 + 8x - 2$ **48.** $8x^4 - 10x^3 + 7x^2 - 8x + 3$

3   **49.** $a^2 + 11a + 24$ **50.** $x^2 + x - 20$ **51.** $y^2 - 11y + 18$
52. $y^2 - 10y + 21$ **53.** $2a^2 + 11a + 12$ **54.** $3a^2 - 2a - 5$
55. $3a^2 - 8ab + 4b^2$ **56.** $5x^2 - 26xy + 5y^2$ **57.** $6x^2 - 17x + 12$
58. $2a^2 - 7ab + 5b^2$ **59.** $21 - 52m + 7m^2$ **60.** $40 - 21x + 2x^2$
61. $x^2 + 11x + 28$ **62.** $y^2 - 12y + 35$ **63.** $m^2 - 4m - 21$
64. $3bx + 18b - 2x - 12$ **65.** $12r^2 - 7r - 10$ **66.** $35 - 22x + 3x^2$
67. $4 - 14m + 6m^2$ **68.** $6 + 13x + 6x^2$ **69.** $2x^2 + x - 15$
70. $20x^2 - 13xy - 21y^2$ **71.** $14a^2 + 19ab - 3b^2$
72. $30a^2 - 13ab - 10b^2$ **73.** $27x^2 + 30xy - 8y^2$
74. $20x^2 - 47xy + 24y^2$ **75.** $21m^2 + 29mn - 10n^2$

4   **76.** $a^2 - 9$ **77.** $4x^2 - 9$ **78.** $a^2 - y^2$ **79.** $16r^2 - 25$
80. $25x^2 - 4$ **81.** $49 - m^2$ **82.** $9y^2 - 25$ **83.** $64y^2 - 9$
84. $9a^2 - 121b^2$ **85.** $25y^2 - 9$ **86.** $x^2 - 49$ **87.** $x^2 - 121$
88. $4 - 12x + 9x^2$ **89.** $9x^2 + 24x + 16$ **90.** $Q^2 + 2QL + L^2$
91. $a^4 + 2a^2 + 1$ **92.** $4d^2 - 20d + 25$ **93.** $9a^2 + 12ax + 4x^2$
94. $9x^2 - 42x + 49$ **95.** $36 + 12Q + Q^2$ **96.** $y^2 + 10xy + 25x^2$
97. $16 - 24j + 9j^2$ **98.** $9m^2 - 12mp + 4p^2$ **99.** $m^4 + 2m^2p^2 + p^4$
100. $4a^2 - 28ac + 49c^2$ **101.** $81 - 234a + 169a^2$
102. $4x^2 - 20xy + 25y^2$ **103.** $x^3 + p^3$ **104.** $Q^3 + L^3$ **105.** $27 + a^3$
106. $8x^3 + 125p^3$ **107.** $27m^3 + 8$ **108.** $216 - p^3$ **109.** $125y^3 - p^3$
110. $x^3 - 8y^3$ **111.** $Q^3 - 216$ **112.** $27t^3 - 8$ **113.** $8x^3 + 27y^3$
114. $27x^3 - 64y^3$

5   **115.** $2x^2$ **116.** $-\dfrac{a}{2}$ **117.** $\dfrac{1}{2x^2}$ **118.** $-\dfrac{3x^3}{4}$ **119.** $3x - 2$

120. $4x^3 - 2x - 1$ **121.** $7x - \dfrac{1}{x}$ **122.** $4x^2 + 3x$ **123.** $\dfrac{x^2}{3} + \dfrac{4}{9}$

124. $5ab^2 - 3b - \dfrac{7}{ab}$ **125.** $\dfrac{5ab^2}{2} - \dfrac{b}{2}$ **126.** $\dfrac{3x}{y} - \dfrac{2y}{x} - 1$

127. $-4 - 4x^2$ **128.** $24x^9y^6 + 4$ **129.** $5x - 14x^4$

130. $\dfrac{x}{5y} - \dfrac{y}{3x} - 2x^2$ **131.** $x + 4$, yes   **132.** $x + 6$, yes

133. $x^2 - 3x + 2$, yes   **134.** $x^2 + x - 5$, yes

135. $3x^2 - 2x + 3 - \dfrac{1}{x + 3}$, no   **136.** $2x^2 + 7x + 2 + \dfrac{18}{x - 5}$, no

137. $x^2 + x - 6$, yes   **138.** $x^2 - x - 2$, yes   **139.** $x^2 + 3x + 9$, yes
140. $x^2 - 5x + 25$, yes

Chapter 11 Review Exercises, p. 521

1. x^{10} **3.** $21x^9$ **5.** x^3 **7.** $7x^3$ **9.** $\dfrac{x}{y^3}$ **11.** x^{12} **13.** x^{15}

15. $-27x^6$ **17.** $\dfrac{1}{y^2}$ **19.** $40x^2$ **21.** 4 **23.** 5

25. $3x^3 + x^2 + 5x - 8$; 3; $3x^3$; 3

27. $2x^4 + 5x^2 - 12x - 32$; 4; $2x^4$; 2

29. $-2x^4 - 2x^3 + 3x$ **31.** $-3x^2 + 5y^2$ **33.** $21x^6$

35. $2x^3 + 6x^2 - 10x$ **37.** $-2x^4 + 14x^3 - 30x$ **39.** $m^2 - 4m - 21$

41. $12r^2 - 7r - 10$ **43.** $4 - 14m + 6m^2$ **45.** $2x^2 + x - 15$

47. $14a^2 + 19ab - 3b^2$ **49.** $27x^2 + 30xy - 8y^2$

51. $21m^2 + 29mn - 10n^2$ **53.** $5x^3 - 12x^2 - 4x + 3$ **55.** $36x^2 - 25$
57. $49y^2 - 121$ **59.** $64a^2 - 25b^2$ **61.** 4 **63.** $9 - x^2$
65. $x^2 + 18x + 81$ **67.** $x^2 - 6x + 9$ **69.** $16x^2 - 120x + 225$
71. $64 + 112m + 49m^2$ **73.** $16x^2 - 88x + 121$ **75.** $g^3 - h^3$
77. $8H^3 - 27T^3$ **79.** $216 - x^3$ **81.** $z^3 + 8t^3$ **83.** $343T^3 + 8$
85. $-\dfrac{2x^3}{3}$ **87.** $-\dfrac{14}{5y^2}$ **89.** $2x^2 - 4x + 7$ **91.** $\dfrac{x^3}{2} - 6x^6$
93. $-4x - 3x^2 + 5x^3$ **95.** $x - 2$ **97.** $3x - 2$
99. $x^2 + x - 5$ **101.** $3x + 1$

Chapter 11 Practice Test, p. 523

1. x^5 **3.** $\dfrac{16}{49}$ **5.** $\dfrac{1}{x^{10}}$ **7.** $\dfrac{x^4}{y^2}$ **9.** $12a^3 - 8a^2 + 20a$ **11.** $x^2 - 5x$
13. $56x^7$ **15.** $4xy - 3x + 6$ **17.** 1 **19.** $-3x^4 - 2x^3 + 6$
21. $m^2 - 49$ **23.** $a^2 + 6a + 9$ **25.** $2x^2 - 11x + 15$ **27.** $x^3 - 8$
29. $125a^3 - 27$ **31.** $2x + 7$

12 ROOTS AND RADICALS

Section 12-1 Exercises, p. 530

1 **1.** $\sqrt{36}$; $36^{1/2}$ **2.** $\sqrt[3]{27}$; $27^{1/3}$ **3.** $\sqrt[4]{81}$; $81^{1/4}$ **4.** $\sqrt[5]{32}$; $32^{1/5}$
5. $\sqrt{81}$; $81^{1/2}$ **6.** $\sqrt[3]{1,331}$; $1,331^{1/3}$

2 **7.** 6 and 7 **8.** 8 and 9 **9.** 3 and 4 **10.** 10 and 11 **11.** 2 and 3
12. 3 and 4 **13.** 3 and 4 **14.** 4 and 5 **15.–22.** See number line and
answers that follow. Numbers given for 16.–22. are approximate square
roots. **15.** 5 **16.** 4.1 **17.** 4.5 **18.** 3.2 **19.** 4.2 **20.** 5.7 **21.** 2.6
22. 6.3

3 **23.** $x^{3/2}$; $\sqrt{x^3}$ **24.** $x^{2/3}$; $\sqrt[3]{x^2}$ **25.** $x^{4/3}$; $(\sqrt[3]{x})^4$ **26.** $x^{2/3}$; $(\sqrt[3]{x})^2$
27. $x^{3/4}$; $(\sqrt[4]{x})^3$ **28.** $x^{5/3}$; $(\sqrt[3]{x})^5$ **29.** $\sqrt[5]{x^3}$ **30.** $\sqrt[5]{x^2}$ **31.** $\sqrt[4]{x^3}$
32. $\sqrt[8]{x^5}$ **33.** $\sqrt[5]{y}$ **34.** $\sqrt[8]{y^7}$ **35.** $\sqrt[8]{y^3}$ **36.** $\sqrt[5]{y^4}$ **37.** $x^{2/3}$ or $x^{0.6}$
38. $x^{1/2}$ or $x^{0.5}$ **39.** $x^{2/5}$ or $x^{0.4}$ **40.** $x^{3/5}$ or $x^{0.6}$ **41.** $x^{5/6}$ or $x^{0.8\overline{3}}$
42. $x^{1/2}$ or $x^{0.5}$ **43.** $x^{2/3}$ or $x^{0.\overline{6}}$ **44.** $x^{4/7}$ or $\approx x^{0.57}$

Section 12-2 Exercises, p. 535

1 **1.** x^5 **2.** x^3 **3.** $\pm x^8$ **4.** $\dfrac{x^7}{y^{12}}$ **5.** $a^3 b^5$ **6.** $-ab^2c^6$
7. $\dfrac{xy^2}{z^5}$ **8.** $\dfrac{a^2}{b^5c^6}$

2 **9.** $y^{1/3}$ **10.** $(5x)^{1/3}$ or $5^{1/3}x^{1/3}$ **11.** $(xy)^{5/2}$ or $x^{5/2}y^{5/2}$ **12.** $r^{7/2}$
13. $3x^2$ **14.** $2r^3$ **15.** $16x^4y^8$ **16.** $32x^{5/4}y^{25/4}$ **17.** $x^{1/4}$
18. $(7x)^{1/5}$ or $7^{1/5}x^{1/5}$ **19.** $(xy)^{4/3}$ or $x^{4/3}y^{4/3}$ **20.** $3b^5$ **21.** $4x^{2/3}y^{14/3}$
22. x^4 **23.** y^3 **24.** $64ab^6$ **25.** $8x^2y^{5/2}$ **26.** $\dfrac{1}{x^{1/3}}$ **27.** $x^{3/5}$
28. $\dfrac{1}{y^{1/6}}$ **29.** $b^{1/6}$ **30.** $4x^{7/2}$ **31.** $5x^{7/3}$ **32.** 2 **33.** 192 **34.** $a^{5.2}$
35. $49a^8$ **36.** x^2 **37.** $3x^2$ **38.** $6x^6$

3 **39.** $2\sqrt{6}$ **40.** $7\sqrt{2}$ **41.** $4\sqrt{3}$ **42.** $x^4\sqrt{x}$ **43.** $y^7\sqrt{y}$
44. $2x\sqrt{3x}$ **45.** $2a^2\sqrt{14a}$ **46.** $6ax^2\sqrt{2a}$ **47.** $2x^2yz^3\sqrt{11xz}$
48. $\sqrt{8x^3} = 2x\sqrt{2x}$; Answers will vary.

Section 12-3 Exercises, p. 543

1 **1.** $12\sqrt{3}$ **2.** $-4\sqrt{5}$ **3.** $2\sqrt{7}$ **4.** $9\sqrt{3} - 8\sqrt{5}$ **5.** $-3\sqrt{11} + \sqrt{6}$
6. $43\sqrt{2} + 6\sqrt{3}$ **7.** $12\sqrt{3}$ **8.** $13\sqrt{5}$ **9.** $11\sqrt{7}$ **10.** $-3\sqrt{6}$
11. $8\sqrt{2} - 6\sqrt{7}$ **12.** $6\sqrt{3}$ **13.** $27\sqrt{5}$ **14.** $47\sqrt{2}$ **15.** $6\sqrt{10}$
16. $-\sqrt{3}$ **17.** $17\sqrt{7}$ **18.** $2\sqrt{2}$

2 **19.** $\sqrt{42}$ **20.** 6 **21.** $15\sqrt{10}$ **22.** 240 **23.** $20x\sqrt{15x}$
24. $28x^5\sqrt{6x}$ **25.** $7\sqrt{2} + 35$ **26.** $48 - 4\sqrt{3}$ **27.** $\sqrt{10} - 3\sqrt{5}$
28. $4\sqrt{7} + \sqrt{21}$ **29.** $\sqrt{15} + \sqrt{6}$ **30.** $\sqrt{22} + \sqrt{33} - 5\sqrt{2} - 5\sqrt{3}$
31. $2\sqrt{6} + 6\sqrt{3} - \sqrt{2} - 3$ **32.** $\sqrt{10} - 3\sqrt{5} - 4\sqrt{2} + 12$
33. $7 + 3\sqrt{35} + \sqrt{14} + 3\sqrt{10}$ **34.** -1 **35.** 5 **36.** $37 - 20\sqrt{3}$
37. $76 + 24\sqrt{2}$ **38.** $76 - 42\sqrt{3}$ **39.** $x\sqrt{3x}$ **40.** $\dfrac{2\sqrt{2}}{\sqrt{15y}}$ or $\dfrac{2\sqrt{30y}}{15y}$
41. $\dfrac{2\sqrt{2}}{x\sqrt{7}}$ or $\dfrac{2\sqrt{14}}{7x}$ **42.** $\dfrac{2}{3y\sqrt{2y}}$ or $\dfrac{\sqrt{2y}}{3y^2}$ **43.** $\dfrac{4x\sqrt{x}}{3}$ **44.** $x\sqrt{7}$
45. $4\sqrt{3}$ **46.** $\dfrac{x}{\sqrt{6}}$ or $\dfrac{x\sqrt{6}}{6}$ **47.** $\dfrac{y}{\sqrt{7}}$ or $\dfrac{y\sqrt{7}}{7}$ **48.** $2x\sqrt{2x}$
49. $3x\sqrt{6x}$ **50.** $6xy^2\sqrt{3x}$ **51.** 3 **52.** $\dfrac{\sqrt{10}}{2}$ **53.** $2\sqrt{2}$ **54.** $\dfrac{10\sqrt{3}}{3}$
55. $\dfrac{8}{5\sqrt{3}}$ or $\dfrac{8\sqrt{3}}{15}$ **56.** $\dfrac{15x^3\sqrt{x}}{4}$ **57.** $\dfrac{5\sqrt{3}}{3}$ **58.** $\dfrac{6\sqrt{5}}{5}$ **59.** $\dfrac{\sqrt{2}}{4}$
60. $\dfrac{\sqrt{55}}{11}$ **61.** $\dfrac{\sqrt{6}}{3}$ **62.** $\dfrac{5\sqrt{3x}}{3}$ **63.** $\dfrac{2x\sqrt{7}}{7}$ **64.** $\dfrac{2x^3\sqrt{3x}}{3}$
65. $2x^4\sqrt{3x}$ **66.** $\dfrac{3\sqrt{2x}}{5x}$

Section 12-4 Exercises, p. 548

1 **1.** $5i$ **2.** $6i$ **3.** $8xi$ **4.** $4y^2i\sqrt{2y}$

2 **5.** i **6.** i **7.** 1 **8.** -1 **9.** 1 **10.** i **11.** 1 **12.** $-i$

3 **13.** $15 + 0i$ **14.** $17 + 0i$ **15.** $0 + 7i$ **16.** $0 + 9i$ **17.** $0 + 33i$
18. $0 + 7i$ **19.** $-4 + 0i$ **20.** $5 + 2i$ **21.** $8 + 4i\sqrt{2}$ **22.** $7 - i\sqrt{3}$

4 **23.** $11 + 5i$ **24.** $(2\sqrt{3} + 2\sqrt{2}) - 8i\sqrt{3}$ **25.** $4 + i$
26. $12 + 20i$ **27.** $1 + 9i$ **28.** $12 - 8i$

5 **29.** $-9 + 7i$ **30.** $-16 - 3i$ **31.** 65 **32.** 26 **33.** -2 **34.** 17
35. -74 **36.** -7 **37.** $13 - 13i$ **38.** $8 - 6i$ **39.** $45 - 28i$
40. $7 + 24i$

Chapter 12 Review Exercises, p. 552

1. $\sqrt{49}$; $49^{1/2}$ **3.** $\sqrt[4]{16}$; $16^{1/4}$ **5.** $\sqrt{121}$; $121^{1/2}$ **7.** 6 and 7
9. 11 and 12 **11.** 3 and 4 **13.–15.** See graph and answers that
follow. **13.** approximately 3.9 **15.** approximately 2.2
17. $x^{7/2}$; $\sqrt{x^7}$ **19.** $x^{2/3}$; $(\sqrt[3]{x})^2$ **21.** $x^{1/2}$

23. $x^{4/5}$ **25.** $x^{4/3}y^{4/3}$ **27.** $7^{1/2}$ **29.** $\sqrt[5]{y^3}$ **31.** y^6 **33.** $-b^9$
35. $x^{1/3}$ **37.** $(4y)^{1/5}$ or $2^{2/5}y^{1/5}$ **39.** $2b^4$ **41.** a^2 **43.** y
45. $27x^{3/4}y^6$ **47.** $64a^3x^{3/2}$ **49.** $x^{1/2}$ **51.** $a^{7/6}$ **53.** $\dfrac{1}{x^{1/8}}$ **55.** $a^{8/3}$
57. $2a^{7/2}$; approximately 22.627 **59.** $\dfrac{3}{2a^{22/5}}$; approximately 0.071
61. $a^{6.3}$; approximately 78.793 **63.** $2a^3b^4$; 16 **65.** x **67.** $3P\sqrt{P}$
69. $3a\sqrt{2b}$ **71.** $4x^2y\sqrt{2x}$ **73.** $5x^5y^4\sqrt{3y}$ **75.** $7y^4\sqrt{3x}$ **77.** $-2\sqrt{3}$
79. $-\sqrt{7}$ **81.** $11\sqrt{6}$ **83.** $-28\sqrt{3}$ **85.** $-5\sqrt{2}$ **87.** $-17\sqrt{2}$

89. $24\sqrt{3}$ **91.** $40\sqrt{21}$ **93.** $-160\sqrt{6}$ **95.** $6 - 5\sqrt{3}$ **97.** $3\sqrt{2} - 3\sqrt{5}$
99. -59 **101.** $\dfrac{3}{4}$ **103.** $\dfrac{3\sqrt{3}}{4\sqrt{2}}$ or $\dfrac{3\sqrt{6}}{8}$ **105.** $\dfrac{\sqrt{3}}{2\sqrt{5}}$ or $\dfrac{\sqrt{15}}{10}$ **107.** $\sqrt{3}$
109. $\dfrac{9}{16}$ **111.** $\dfrac{6x^4}{9y^5}$ **113.** $\dfrac{\sqrt{21}}{6}$ **115.** $\dfrac{\sqrt{6}}{4}$ **117.** $\dfrac{5\sqrt{2}}{4}$ **119.** $\dfrac{\sqrt{10}}{3}$
121. $10i$ **123.** $\pm 2y^3 i\sqrt{6y}$ **125.** -1 **127.** i **129.** $0 + 15i$
131. $0 - 12i$ **133.** $7 - 4i$ **135.** $11 + i$ **137.** -25

Chapter 12 Practice Test, p. 555

1. $6\sqrt{14}$ **3.** $6\sqrt{3}$ **5.** $\dfrac{4\sqrt{6}}{3}$ **7.** $2\sqrt{6}$ **9.** $\dfrac{7\sqrt{15x}}{2x}$ **11.** $3x^5$
13. $x^2 y^3 z^6$ **15.** $5x^{\frac{1}{6}}y^2$ **17.** $2x$ **19.** $-i$ **21.** $-3 + 5i$
23. $-2\sqrt{3} + 8\sqrt{5}$ **25.** $\sqrt{35} - 4\sqrt{7}$ **27.** 41

13   FACTORING

Section 13–1   Exercises, p. 560

1 **1.** $7(a + b)$ **2.** $12(x + y)$ **3.** $m(m + 2)$ **4.** $y^2(5y + 8)$
5. $3x(2x + 1)$ **6.** $6y^3(2 + 3y)$ **7.** $6x^4(2x - 1)$ **8.** $5x(1 - 3y)$
9. $5(ab + 2a + 4b)$ **10.** $2(2ax + 3x + 5a)$ **11.** Prime
12. $3(4a^2 - 5b^2 + 2a)$ **13.** $3x(x^2 - 3x - 2)$
14. $2ab(4a + 7b^2 + 14a^2 b^2)$ **15.** $3m^2(1 - 2m + 4m^2)$
16. $6xy(2x - 3y^2 + 4xy)$ **17.** $3a^2 bc(5 + 6abc^2 - 7a^2 c^4)$
18. $5x^2 y^2 z(4y - 7x - 8)$ **19.** $4x^2 y^2(2x^2 - 3y^2 - 1)$ **20.** Prime
or $5y + 3z$ **21.** Prime or $18a - 7b$ **22.** $(x + 3)(5x + 8y)$
23. $(2x - 1)(3x + 5)$ **24.** $(3y - 5)(4y + 7)$ **25.** $(a - b)(7a + 2b)$
26. $(2m - 3n)(5m - 7n)$ **27.** $(y - 2)(y - 3)$ **28.** $(2x - 7)(3x - 8)$
29. $(9y - 2)(7y - 5)$ **30.** $-(5x - 2)$ **31.** $-(12x - 7)$
32. $-(x^2 - 3x + 8)$ **33.** $-(2x^2 + 7x + 11)$ **34.** $-2(x^2 - 3x + 4)$
35. $-3(x^2 + 3x - 5)$ **36.** $-7(x^2 + 3x - 2)$ **37.** $-6(2x^2 - 3x - 1)$

Section 13–2   Exercises, p. 564

1 **1.** Yes   **2.** No, not a binomial   **3.** Yes   **4.** Yes   **5.** No, not
difference   **6.** Yes   **7.** $(y + 7)(y - 7)$ **8.** $(4x + 1)(4x - 1)$
9. $(3a + 10)(3a - 10)$ **10.** $(2m + 9n)(2m - 9n)$
11. $(3x + 8y)(3x - 8y)$ **12.** $(5x + 8)(5x - 8)$ **13.** $(10 + 7x)(10 - 7x)$
14. $(2x + 7y)(2x - 7y)$ **15.** $(11m + 7n)(11m - 7n)$
16. $(9x + 13)(9x - 13)$ **17.** $(2a + 3)(2a - 3)$ **18.** $(5r + 4)(5r - 4)$
19. $(6x + 7y)(6x - 7y)$ **20.** $(7 + 12x)(7 - 12x)$
21. $(4x + 9y)(4x - 9y)$

2 **22.** No, not perfect square. The middle term is not twice the
product of the square roots of the first and last terms.   **23.** Yes
24. No, not perfect square. The last term is negative.   **25.** No, not
perfect square. The first term is negative.   **26.** Yes   **27.** Yes
28. $(x + 3)^2$ **29.** $(x + 7)^2$ **30.** $(x - 6)^2$ **31.** $(x - 8)^2$
32. $(2a + 1)^2$ **33.** $(5x - 1)^2$ **34.** $(3m - 8)^2$ **35.** $(2x - 9)^2$
36. $(x - 6y)^2$ **37.** $(2a - 5b)^2$ **38.** $(y - 5)^2$ **39.** $(3x + 10y)^2$
40. $-(x + 6)^2$ **41.** $-(3x - 1)^2$ **42.** $-(x + 4)^2$

3 **43.** Difference of two cubes   **44.** Not a sum or difference of
cubes   **45.** Sum of two cubes   **46.** Not a sum or difference of cubes
47. Difference of two cubes   **48.** Sum of two cubes
49. $(m - 2)(m^2 + 2m + 4)$ **50.** $(y - 5)(y^2 + 5y + 25)$
51. $(Q + 3)(Q^2 - 3Q + 9)$ **52.** $(c + 1)(c^2 - c + 1)$
53. $(5d - 2p)(25d^2 + 10dp + 4p^2)$ **54.** $(a + 4)(a^2 - 4a + 16)$
55. $(6a - b)(36a^2 + 6ab + b^2)$ **56.** $(x + Q)(x^2 - xQ + Q^2)$
57. $(2p - 5)(4p^2 + 10p + 25)$ **58.** $(3 - 2y)(9 + 6y + 4y^2)$
59. $-(a + 2)(a^2 - 2a + 4)$ **60.** $-(x + 3)(x^2 - 3x + 9)$

Section 13–3   Exercises, p. 572

1 **1.** $(x + 6)(x + 1)$ **2.** $(x - 6)(x - 1)$ **3.** $(x - 2)(x - 3)$
4. $(x + 3)(x + 2)$ **5.** $(x - 7)(x - 4)$ **6.** $(x + 6)(x + 2)$
7. $(x - 6)(x - 2)$ **8.** $(x + 12)(x + 1)$ **9.** $(x - 12)(x - 1)$
10. $(x + 4)(x + 3)$ **11.** $(x - 4)(x - 3)$ **12.** $(x - 3)(x - 1)$
13. $(x + 7)(x + 1)$ **14.** $(x + 5)(x + 2)$ **15.** $(x - 3)(x + 2)$
16. $(x + 3)(x - 2)$ **17.** $(x - 6)(x + 1)$ **18.** $(x + 6)(x - 1)$
19. $(x - 4)(x + 3)$ **20.** $(x + 4)(x - 3)$ **21.** $(x + 6)(x - 2)$
22. $(x - 6)(x + 2)$ **23.** $(x - 12)(x + 1)$ **24.** $(x + 12)(x - 1)$
25. $(y - 5)(y + 2)$ **26.** $(y - 5)(y + 4)$ **27.** $(b + 3)(b - 1)$
28. $(b - 7)(b + 2)$ **29.** $(4 - x)(3 - x)$ or $(x - 4)(x - 3)$
30. $-(5 + x)(6 - x)$ or $(x - 6)(x + 5)$ **31.** $(x + 9)(x + 2)$
32. $(x - 6)(x - 3)$ **33.** $(x - 9)(x + 2)$ **34.** $(x + 18)(x - 1)$
35. $(x + 5)(x + 4)$ **36.** $-(x - 10)(x - 2)$ **37.** $-(x - 8)(x - 2)$
38. $-(x - 16)(x - 1)$ **39.** $(x - 14)(x + 1)$ **40.** $(x - 7)(x + 2)$

2 **41.** $(x + y)(x + 4)$ **42.** $(3x + 2)(2x - y)$ **43.** $(3x + 5)(m - 2n)$
44. $(6x - 7)(5y - 6)$ **45.** $(x - 2)(x + 8)$ **46.** $(3x - 1)(2x - 7)$
47. $(x - 4)(x + 1)$ **48.** $(2x - 1)(4x + 3)$ **49.** $(x - 5)(x + 4)$
50. $(x - 2)(3x + 5)$ **51.** $(x + 2)(4x - 3)$ **52.** $(x - 2)(4x + 3)$
53. $(x + 2)(4x + 3)$ **54.** $(x - 2)(4x - 3)$ **55.** $(2x + 1)(4x - 3)$

3 **56.** $(3x + 1)(x + 2)$ **57.** $(3x + 2)(x + 4)$ **58.** $(3x + 2)(2x + 3)$
59. $(4x + 3)(2x + 1)$ **60.** $(4x + 3)(2x + 5)$ **61.** $(4x + 1)(3x + 2)$
62. $(2a + 19)(a + 1)$ **63.** $(2x - 5)(x - 2)$ **64.** $(3x - 5)(2x - 1)$
65. $(2x - 1)(4x - 3)$ **66.** $(3x - 4)(2x - 3)$ **67.** $(3x - 1)(8x - 1)$
68. $(6x - 5)(x - 1)$ **69.** $(4x - 1)(3x - 2)$ **70.** $(3x + 2)(4x - 1)$
71. $(8x - 1)(3x + 1)$ **72.** $(6x - 5y)(x + 2y)$ **73.** $(7x + 8)(x - 1)$
74. $(6x - 5)(2x + 3)$ **75.** $(4x + 3)(2x - 1)$ **76.** $(5x + 3)(2x - 1)$
77. $(2x - 7)(x + 1)$ **78.** $(3x - 5)(5x + 1)$ **79.** $(3x - 2)(4x + 1)$
80. $(5x - 4y)(4x + 3y)$ **81.** $(6x - 5)(3x + 2)$ **82.** $(3a + 2b)(2a - 7b)$

4 **83.** $4(x - 1)$ **84.** $(x + 8)(x - 7)$ **85.** $(2x + 3)(x - 1)$
86. $(x + 3)(x - 3)$ **87.** $4(x + 2)(x - 2)$ **88.** $(m + 5)(m - 3)$
89. $2(a + 2)(a + 1)$ **90.** $(b + 3)^2$ **91.** $(4m - 1)^2$ **92.** $(x - 7)(x - 1)$
93. $(2m + 1)(m + 2)$ **94.** $(2m + 1)(m - 3)$ **95.** $(2a - 5)(a + 1)$
96. $(3x - 2)(x + 4)$ **97.** $(3x + 5)(2x - 3)$ **98.** $(4x - 1)(2x + 3)$
99. $-2(x - 2)(x - 1)$ **100.** $(x^2 + 4)(x + 2)(x - 2)$
101. $(x^3 + 9)(x^3 - 9)$ **102.** $(x^2 - 3)(x^4 + 3x^2 + 9)$
103. $3(x^2 + 4)(x + 2)(x - 2)$ **104.** $6(x - 2)(x^2 + 2x + 4)$
105. $3(x - 4)(x + 2)$ **106.** $(4a^3 + b)(16a^6 - 4a^3 b + b^2)$

Chapter 13 Review Exercises, p. 576

1. $5(x + y)$ **3.** $4(3m^2 - 2n^2)$ **5.** $2a(a^2 - 7a - 1)$ **7.** $5x(3x^2 - x - 4)$
9. $6a^2(3a - 2)$ **11.** $-2x(3x + 5)$ **13.** $5(\sqrt{3} + 3\sqrt{7})$
15. $2(\sqrt{3} - 5\sqrt{7})$ **17.** Not difference   **19.** Difference
21. Difference   **23.** Not perfect-square trinomial   **25.** Perfect-
square trinomial   **27.** Perfect-square trinomial   **29.** Difference of
two cubes   **31.** Difference of two cubes   **33.** Sum of two cubes
35. $(5y + 2)(5y - 2)$ **37.** *NSP*, this is a sum of two squares, not a
difference.   **39.** $(a + 1)^2$ **41.** $(4c - 3b)^2$ **43.** $(n - 13)^2$
45. $(6a + 7b)^2$ **47.** $(7 - x)^2$ or $(x - 7)^2$ **49.** *NSP*, this is a sum of
two squares, not a difference.   **51.** *NSP*, the middle term needs a y
factor.   **53.** $(7 + 9y)(7 - 9y)$ **55.** $(3x + 10y)(3x - 10y)$
57. $(3x - y)^2$ **59.** $(3xy + z)(3xy - z)$ **61.** $(x + 2)^2$
63. $\left(\dfrac{2}{5}x + \dfrac{1}{4}y\right)\left(\dfrac{2}{5}x - \dfrac{1}{4}y\right)$ **65.** $(T - 2)(T^2 + 2T + 4)$
67. $(d + 9)(d^2 - 9d + 81)$ **69.** $(3k + 4)(9k^2 - 12k + 16)$
71. $(x + 3)(x + 8)$ **73.** $(x + 3)(x + 10)$ **75.** $(x - 1)(x - 8)$

77. $(x + 2)(x - 13)$ **79.** $(x + 8)(x - 3)$ **81.** $(6x + 1)(x + 4)$
83. $(5x - 4)(x - 6)$ **85.** $(3x + 7)(2x - 5)$ **87.** $(7x + 8)(x - 3)$
89. $(3a + 10)(3a - 10)$ **91.** $(2x + 1)(x - 2)$ **93.** $(a + 9)(a - 9)$
95. $(y - 7)^2$ **97.** $(b + 3)(b + 5)$ **99.** $(13 + m)(13 - m)$
101. $(x + 4)(x - 8)$ **103.** $(x + 20)(x - 1)$ **105.** $2(x + 2)(x - 4)$
107. $2x(x + 1)(x - 6)$

Chapter 13 Practice Test, p. 578

1. $x(7x + 8)$ **3.** $7ab(a - 2)$ **5.** $(3x + 5)(3x - 5)$ **7.** $(x - 9)^2$
9. $(3r + 4s)(9r^2 - 12rs + 16s^2)$ **11.** $(3x + 2)(2x - 3)$ **13.** $(a + 8b)^2$
15. $(b + 2)(b - 5)$ **17.** $(3m - 2)(m - 1)$ **19.** $3(x + 2)(x - 2)$
21. $5(x + 2)(x - 2)$ **23.** $3(x + 2)^2$

14 RATIONAL EXPRESSIONS, EQUATIONS, AND INEQUALITIES

Section 14-1 Exercises, p. 584

1 **1.** $\dfrac{4}{9}$ **2.** $\dfrac{3}{8}$ **3.** $2ab^2$ **4.** $\dfrac{3a}{2b^3}$ **5.** $\dfrac{x}{x + 2}$ **6.** $\dfrac{3(3x + 2)}{x(2x - 1)}$ or
$\dfrac{9x + 6}{2x^2 - x}$ **7.** $\dfrac{(x + 2)(x - 5)}{(x + 5)(x - 2)}$ or $\dfrac{x^2 - 3x - 10}{x^2 + 3x - 10}$ **8.** $\dfrac{1}{2}$ **9.** $\dfrac{1}{a - b}$
10. $\dfrac{x - 2}{x + 2}$ **11.** $\dfrac{2(x - 2)}{3}$ or $\dfrac{2x - 4}{3}$ **12.** -1 **13.** $-(x + 2)$ or
$-x - 2$ **14.** $-\dfrac{1}{2}$ **15.** -3 **16.** $-\dfrac{1}{2}$ **17.** -5 **18.** $-\dfrac{2}{x - 2}$
19. $-\dfrac{x - 5}{x + 1}$ **20.** -1 **21.** $\dfrac{x + 3}{x + 2}$ **22.** $\dfrac{x - 3}{x + 3}$ **23.** $\dfrac{x + 1}{x + 4}$
24. $\dfrac{2x - 1}{3x - 2}$

Section 14-2 Exercises, p. 589

1 **1.** $\dfrac{1}{2}$ **2.** $\dfrac{1}{10}$ **3.** $\dfrac{10x^3}{9y^2}$ **4.** $\dfrac{7}{15}$ **5.** $\dfrac{1}{12a}$ **6.** $\dfrac{3}{2}$ **7.** $\dfrac{2}{3}$ **8.** $\dfrac{3y^3}{2x}$
9. $\dfrac{3ab}{10}$ **10.** $\dfrac{3}{x - 2}$ **11.** $\dfrac{4}{x - 2}$ **12.** $3(a - b)$ or $3a - 3b$ **13.** 4
14. $\dfrac{a - 7}{b + 5}$ **15.** $\dfrac{3(x + 1)}{x - 1}$ or $\dfrac{3x + 3}{x - 1}$ **16.** $\dfrac{1}{2}$ **17.** -2 **18.** 1
19. $\dfrac{x}{x - 2}$ **20.** $\dfrac{50(a - b)}{ab}$ or $\dfrac{50a - 50b}{ab}$ **21.** 1 **22.** $\dfrac{25}{27}$
23. $\dfrac{21}{20}$ **24.** $\dfrac{x - 5}{12x}$ **25.** $\dfrac{2}{3x(x - 2)}$ or $\dfrac{2}{3x^2 - 6x}$ **26.** $\dfrac{x + 3}{2}$
27. $\dfrac{x - 3}{3}$ **28.** $x + y$ **29.** $-\dfrac{x - 1}{x + 1}$ or $\dfrac{-x + 1}{x + 1}$ **30.** $3(x - 6)$ or
$3x - 18$ **31.** $\dfrac{3x^2}{4}$

2 **32.** $-3(2 + \sqrt{5})$ or $-6 - 3\sqrt{5}$ **33.** $5 + 2\sqrt{3}$
34. $\dfrac{22 - 9\sqrt{2}}{14}$ or $-\dfrac{9\sqrt{2} - 22}{14}$ **35.** $\dfrac{14 - \sqrt{7}}{9}$
36. $\dfrac{14 + 4\sqrt{3} + 21\sqrt{5} + 6\sqrt{15}}{37}$ **37.** $13 + 5\sqrt{6}$ **38.** $-\dfrac{1}{14 + 7\sqrt{5}}$
39. $\dfrac{14}{75 + 15\sqrt{11}}$ **40.** $\dfrac{1}{8 - 2\sqrt{6}}$ **41.** $\dfrac{1}{3 - \sqrt{2}}$ **42.** $-\dfrac{2}{28 - 21\sqrt{2}}$
or $\dfrac{2}{21\sqrt{2} - 28}$ **43.** $\dfrac{13}{40 - 16\sqrt{3}}$

Section 14-3 Exercises, p. 593

1 **1.** $\dfrac{5}{7}$ **2.** $\dfrac{13}{16}$ **3.** $\dfrac{3x}{2}$ **4.** $\dfrac{23x}{24}$ **5.** $\dfrac{4 + 3x}{2x}$ **6.** $\dfrac{55}{6x}$

7. $\dfrac{9x - 1}{(x + 1)(x - 1)}$ or $\dfrac{9x - 1}{x^2 - 1}$ **8.** $\dfrac{10x + 6}{(x + 3)(x - 1)}$ or $\dfrac{10x + 6}{x^2 + 2x - 3}$
9. $\dfrac{x - 7}{(2x + 1)(x - 1)}$ or $\dfrac{x - 7}{2x^2 - x - 1}$ **10.** $\dfrac{x + 1}{(x - 6)(x - 5)}$ or
$\dfrac{x + 1}{x^2 - 11x + 30}$

2 **11.** $\dfrac{5(2x + 1)}{3x}$ or $\dfrac{10x + 5}{3x}$ **12.** $\dfrac{8(3x + 2)}{5x}$ or $\dfrac{24x + 16}{5x}$
13. $-\dfrac{28}{9}$ **14.** $-\dfrac{1}{7}$ **15.** $\dfrac{11}{3x^2}$ **16.** $\dfrac{10 + 9x}{3 + 10x}$
17. $\dfrac{-2(x - 2)}{x + 2}$ or $\dfrac{4 - 2x}{x + 2}$ **18.** 9

Section 14-4 Exercises, p. 599

1 **1.** 0 **2.** 0 **3.** $5, 0$ **4.** $0, -9$ **5.** $-8, 8$ **6.** $2, -2$ **7.** $\dfrac{3}{4}, 0$
8. $0, -3$

2 **9.** 35 **10.** 5 **11.** 1 **12.** $\dfrac{1}{4}$ **13.** $x = -17$ **14.** $x = \dfrac{65}{11}$
15. $1\dfrac{5}{7}$ days **16.** 8 investors **17.** 25 mph going **18.** 3 min
19. 50 lb dark-roast, 75 lb medium-roast, $2 per pound **20.** 6 ohms

3 **21.** $-7 < x < 3; (-7, 3)$ **22.** $x < -8$ or $x > 3$;
$(-\infty, -8) \cup (3, \infty)$

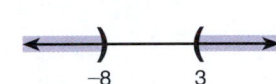

23. $x < \dfrac{1}{3}$ or $x > 3$; **24.** $-7 < x \le 7; (-7, 7]$
$\left(-\infty, \dfrac{1}{3}\right) \cup (3, \infty)$

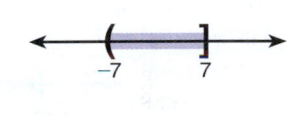

25. $0 < x < 1; (0, 1)$ **26.** $x \le 2$ or $x > 3$;
$(-\infty, 2] \cup (3, \infty)$

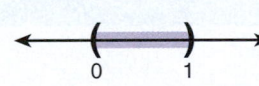

27. $2 < x \le 3; (2, 3]$ **28.** $x \le 0$ or $x > 4$;
$(-\infty, 0] \cup (4, \infty)$

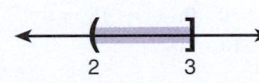

29. $x < -1$ or $x \ge 3$;
$(-\infty, -1) \cup [3, \infty)$

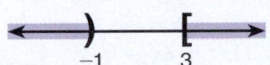

Chapter 14 Review Exercises, p. 604

1. $\dfrac{3}{4}$ **3.** $\dfrac{b^2}{2ac}$ **5.** $\dfrac{x - 3}{2(x + 3)}$ or $\dfrac{x - 3}{2x + 6}$ **7.** -1 **9.** $\dfrac{m^2 - n^2}{m^2 + n^2}$ or
$\dfrac{(m + n)(m - n)}{m^2 + n^2}$ **11.** $\dfrac{1}{1 + y}$ **13.** 5 **15.** $y + 1$ **17.** $\dfrac{2}{x + 6}$

19. 3 **21.** $\dfrac{5x^3}{4y^2}$ **23.** 15 **25.** $\dfrac{4(2y-1)}{y-1}$ or $\dfrac{8y-4}{y-1}$ **27.** 1

29. $\dfrac{x+3}{x+2}$ **31.** $\dfrac{1}{4}$ **33.** $\dfrac{(y-1)^3}{y}$ **35.** $\dfrac{y+3}{y+2}$ **37.** $\dfrac{3x^2}{2}$

39. $(y+4)(y-3)$ **41.** $\dfrac{5}{4(x-3)}$ or $\dfrac{5}{4x-12}$ **43.** $\dfrac{4x^2}{3}$

45. $\dfrac{72+12\sqrt{5}}{31}$ **47.** $\dfrac{26+7\sqrt{3}}{23}$ **49.** $\dfrac{35+15\sqrt{5}+7\sqrt{2}+3\sqrt{10}}{4}$

51. $\dfrac{26}{20-5\sqrt{3}}$ **53.** $\dfrac{1}{16+4\sqrt{13}}$ **55.** $\dfrac{9}{40+8\sqrt{7}}$ **57.** $\dfrac{2}{3}$ **59.** $\dfrac{4x}{7}$

61. $\dfrac{19x}{12}$ **63.** $\dfrac{15-7x}{3x}$ **65.** $\dfrac{13}{4x}$ **67.** $\dfrac{10x+5}{(x-3)(x+2)}$

69. $\dfrac{6x-38}{(x+3)(x-4)}$ or $\dfrac{2(3x-19)}{(x+3)(x-4)}$ **71.** $\dfrac{x+3}{x-5}$ **73.** $\dfrac{51}{28}$

75. $\dfrac{3x^2-30}{2x^2+12}$ **77.** 0, 2 **79.** $\dfrac{1}{2}$, 0 **81.** $-\dfrac{20}{3}$ **83.** $\dfrac{1}{7}$

85. 3 students **87.** $7\dfrac{1}{2}\,\text{h}$

89. $x<-8 \cup x>0$; **91.** $-1<x<7$; $(-1,7)$
$(-\infty,-8)\cup(0,\infty)$

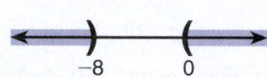
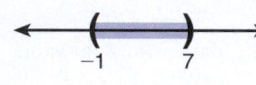

Chapter 14 Practice Test, p. 608

1. $\dfrac{1}{2}$ **3.** $\dfrac{3x-4}{x+3}$ **5.** $-\dfrac{x-2}{x+2}$ or $\dfrac{2-x}{x+2}$ **7.** $\dfrac{3a}{y}$ **9.** $\dfrac{1}{x^2(x+2y)}$

11. $\dfrac{2x+1}{x}$ **13.** $-\dfrac{5}{(x+2)(x-1)}$ **15.** $\dfrac{12+x}{4x}$ **17.** $-\dfrac{2}{3x-2}$ or

$\dfrac{2}{2-3x}$ **19.** $\dfrac{1}{x+2y}$ **21.** $\dfrac{2x^2}{x-3}$ **23.** 0, -3 **25.** $\dfrac{11}{7}$ **27.** $\dfrac{1}{3}$

29. 5 persons **31.** $-5<x<2$; $(-5,2)$

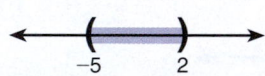

15 **QUADRATIC AND OTHER NONLINEAR EQUATIONS AND INEQUALITIES**

Section 15–1 Exercises, p. 614

1 **1.** $7x^2-4x+5=0$ **2.** $8x^2-6x-3=0$ **3.** $7x^2-5=0$
4. $x^2-6x+8=0$ **5.** $x^2-9x+8=0$ **6.** $x^2-4x-8=0$
7. $3x^2-6x+5=0$ **8.** $x^2-6x-5=0$ **9.** $x^2-16=0$
10. $8x^2-7x-8=0$ **11.** $8x^2+8x-10=0$
12. $0.3x^2-0.4x-3=0$

2 **13.** $x^2-5x=0$ **14.** $3x^2-7x+5=0$
$a=1; b=-5; c=0$ $a=3; b=-7; c=5$
15. $7x^2-4x=0$ **16.** $3x^2-5x+8=0$
$a=7; b=-4; c=0$ $a=3, b=-5, c=8$
17. $x^2-5x+6=0$ **18.** $11x^2-8x=0$
$a=1, b=-5, c=6$ $a=11, b=-8, c=0$
19. $x^2-x=0$ **20.** $9x^2-7x-12=0$
$a=1, b=-1, c=0$ $a=9, b=-7, c=-12$
21. $x^2-5=0$ **22.** $x^2+6x-3=0$
$a=1, b=0, c=-5$ $a=1, b=6, c=-3$
23. $5x^2-0.2x+1.4=0$ **24.** $\dfrac{2}{3}x^2-\dfrac{5}{6}x-\dfrac{1}{2}=0$
$a=5, b=-0.2, c=1.4$ $a=\dfrac{2}{3}, b=-\dfrac{5}{6}, c=-\dfrac{1}{2}$

25. $1.3x^2-8=0$ **26.** $\sqrt{3}x^2+\sqrt{5}x-2=0$
$a=1.3, b=0, c=-8$ $a=\sqrt{3}, b=\sqrt{5}, c=-2$
27. $15x^2+3x+5=0$ **28.** $8x^2-2x-3=0$
$a=15, b=3, c=5$ **29.** $x^2+3x=0$
30. $5x^2+2x-7=0$ **31.** $2.5x^2-0.8=0$

3 **32.** $x=\pm3$ **33.** $x=\pm2$ **34.** $x=\pm\dfrac{8}{3}$ **35.** $x=\pm\dfrac{9}{2}$
36. $x=\pm\sqrt{5}$ or ±2.236 **37.** $y=\pm3$ **38.** $x=\pm7$
39. $x=\pm\dfrac{7}{4}$ **40.** $x=\pm2$ **41.** $x=\pm1.732$ **42.** $x=\pm\dfrac{3}{2}$
43. $x=\pm1$ **44.** 23 ft **45.** 81.759 yd **46.** Isolate the quadratic term; then take the square root of both sides of the equation and simplify when necessary. **47.** opposites

Section 15–2 Exercises, p. 617

1 **1.** $x=3$ or 0 **2.** $x=0$ or 6 **3.** $x=0$ or 2
4. $x=0$ or $-\dfrac{1}{2}$ **5.** $x=0$ or $\dfrac{1}{2}$ **6.** $x=0$ or 3 **7.** $x=0$ or $\dfrac{7}{3}$
8. $y=0$ or -4 **9.** $x=0$ or $-\dfrac{2}{3}$ **10.** $x=0$ or $\dfrac{4}{3}$ **11.** $x=0$ or 3
12. $x=0$ or $\dfrac{2}{3}$ **13.** 8 or 0 **14.** 15 units **15.** An incomplete quadratic equation is missing the constant while a pure quadratic equation is missing the linear term. **16.** Yes, the common factor of x will be set equal to zero.

2 **17.** -3 or -2 **18.** 3 (double root) **19.** 7 or -2 **20.** 3 or -6
21. -4 or -3 **22.** 5 or 3 **23.** 14 or -1 **24.** 3 or 6 **25.** $\dfrac{1}{2}$ or 3
26. $-\dfrac{1}{3}$ or -4 **27.** $\dfrac{3}{5}$ or $-\dfrac{1}{2}$ **28.** $-\dfrac{1}{3}$ or $-\dfrac{3}{2}$ **29.** $-\dfrac{3}{2}$ or -5
30. $\dfrac{4}{3}$ or 2 **31.** $\dfrac{1}{6}$ or -3 **32.** -4 or $-\dfrac{2}{3}$ **33.** 5 or $\dfrac{3}{2}$ **34.** $\dfrac{1}{2}$ or $\dfrac{2}{3}$
35. $\dfrac{3}{4}$ or $\dfrac{1}{2}$ **36.** $\dfrac{2}{5}$ or -1 **37.** $\dfrac{5}{3}$ or $-\dfrac{3}{2}$ **38.** $-\dfrac{5}{3}$ or $-\dfrac{1}{3}$ **39.** $\dfrac{1}{3}$ or $\dfrac{3}{2}$
40. -3 or $\dfrac{2}{5}$ **41.** $w=5$ ft **42.** $l=21$ in.
$l=11$ ft $w=18$ in.

Section 15–3 Exercises, p. 625

1 **1.** $-8, 6$ **2.** $-1, 9$ **3.** $1, 9$ **4.** $4, 6$ **5.** $-6, 4$ **6.** $-14, -2$
7. $\dfrac{3\pm\sqrt{29}}{2}$ **8.** $\dfrac{5\pm\sqrt{33}}{2}$ **9.** $\dfrac{5\pm\sqrt{37}}{2}$ **10.** $\dfrac{3\pm\sqrt{13}}{2}$ **11.** $\dfrac{2\pm\sqrt{10}}{2}$
12. $\dfrac{1\pm\sqrt{3}}{2}$ **13.** $\dfrac{3\pm\sqrt{3}}{3}$ **14.** $\dfrac{-6\pm2\sqrt{3}}{3}$ **15.** $-1, 3$ **16.** $\dfrac{3\pm\sqrt{205}}{14}$
17. $2, \dfrac{1}{5}$ **18.** $\dfrac{1\pm\sqrt{85}}{6}$

2 **19.** $3, -\dfrac{2}{3}$ **20.** $3, -4$ **21.** $\dfrac{11}{5}, -1$ **22.** 3 (double root)

23. $3, -2$ **24.** $\dfrac{3}{4}, -\dfrac{1}{2}$ **25.** $1.84, -10.84$ **26.** $-0.18, -1.82$
27. $3.14, -0.64$ **28.** $2.39, 0.28$ **29.** $-0.75\pm0.97i$ or no real solution **30.** $0.5\pm1.32i$ or no real solution
31. width $=5.55$ cm, length $=8.55$ cm
32. width $=4$ in., length $=10$ in. **33.** 4 m
34. width $=11$ ft, length $=16$ ft **35.** length $=110$ ft, width $=70$ ft (nearest ft); see figure below. A solution that is larger than the original field must be disregarded. **36.** 111.5 kg

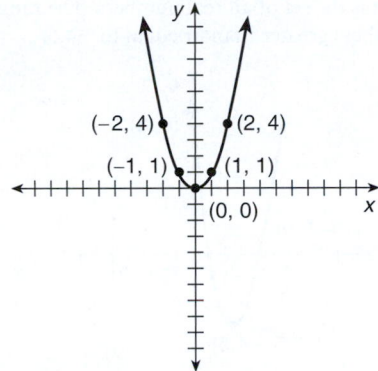

Section 15–4 **Exercises, p. 634**

1 **1.** The domain is the set of all real numbers. The range is the set of all real numbers greater than or equal to zero.

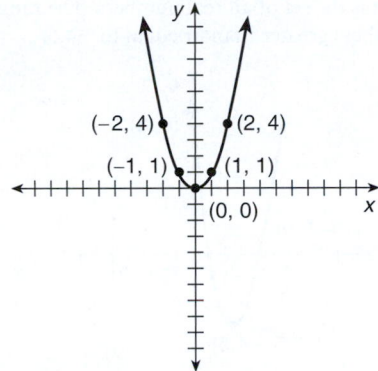

2. The domain is the set of all real numbers. The range is the set of all real numbers greater than or equal to zero.

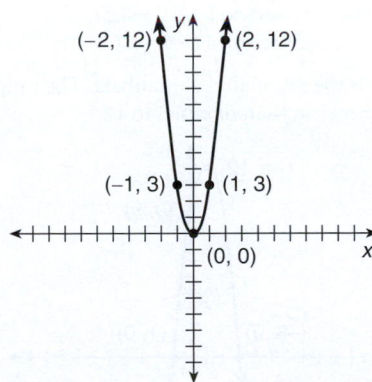

3. The domain is the set of all real numbers. The range is the set of all real numbers greater than or equal to zero.

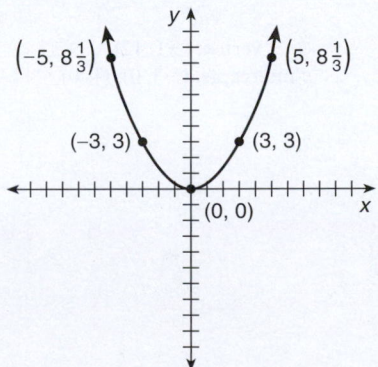

4. The domain is the set of all real numbers. The range is the set of all real numbers greater than or equal to -4.

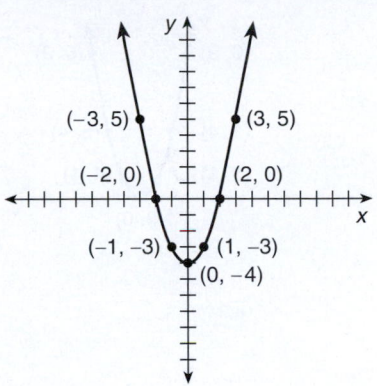

2 **5.** The domain is the set of all real numbers. The range is the set of all real numbers less than or equal to zero.

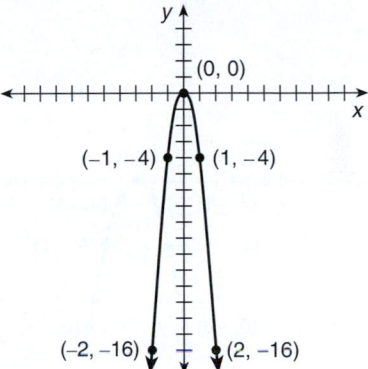

6. The domain is the set of all real numbers. The range is the set of all real numbers less than or equal to zero.

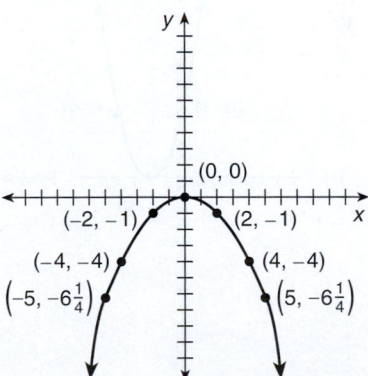

7. The domain is the set of all real numbers. The range is the set of all real numbers greater than or equal to 4.

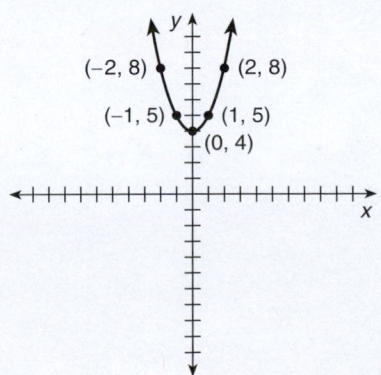

8. The domain is the set of all real numbers. The range is the set of all real numbers greater than or equal to zero.

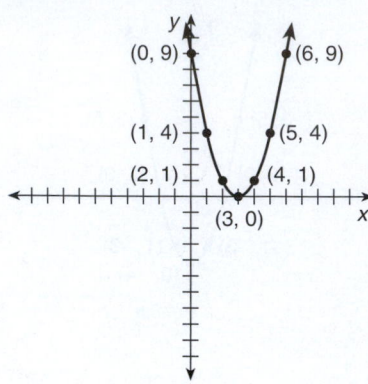

9. The domain is the set of all real numbers. The range is the set of all real numbers less than or equal to zero.

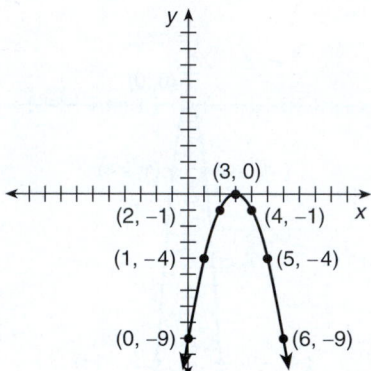

10. The domain is the set of all real numbers. The range is the set of all real numbers greater than or equal to zero.

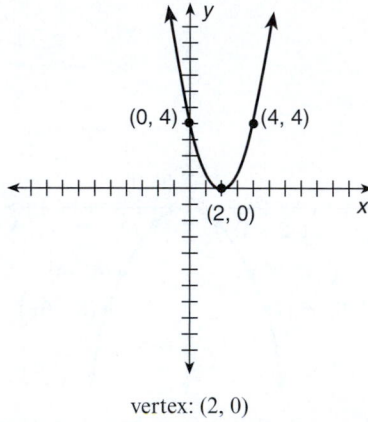

vertex: (2, 0)
x-intercept: (2, 0)

11. The domain is the set of all real numbers. The range is the set of all real numbers greater than or equal to −6.

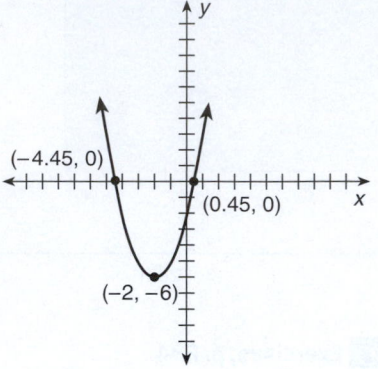

vertex: (−2, −6)
x-intercepts: (−4.45, 0); (0.45, 0)

12. The domain is the set of all real numbers. The range is the set of all real numbers greater than or equal to −4.5.

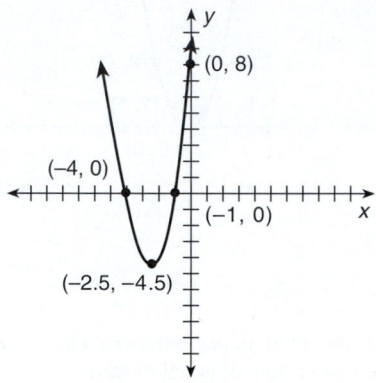

vertex: (−2.5, −4.5)
x-intercepts: (−4, 0); (−1, 0)

13. The domain is the set of all real numbers. The range is the set of all real numbers less than or equal to 12.

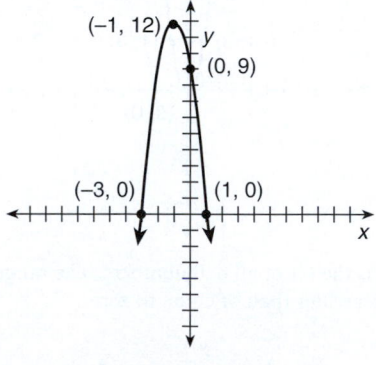

vertex: (−1, 12)
x-intercepts: (−3, 0); (1, 0)

14. The domain is the set of all real numbers. The range is the set of all real numbers less than or equal to −3.

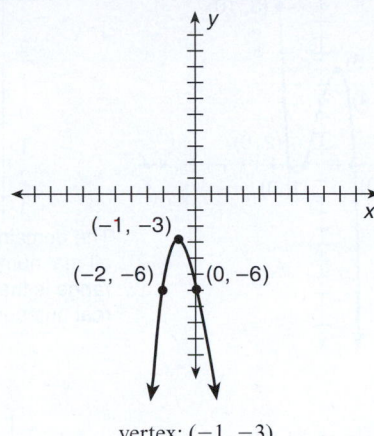

vertex: (−1, −3)
no *x*-intercepts

15. The domain is the set of all real numbers. The range is the set of all real numbers greater than or equal to −15.

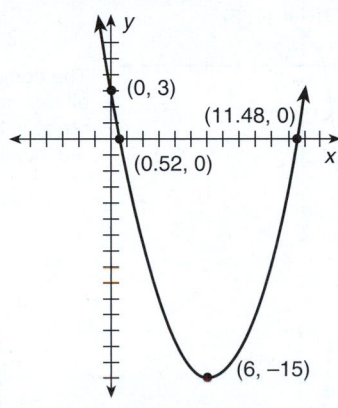

vertex: (6, −15)
x-intercepts: (0.52, 0); (11.48, 0)

3 **16.** $x = 2, x = 3$ **17.** $x = 0$ **18.** No real solution

19. $0, $9   **20.** $t = 0$ s

4 **21.** $y = 3x^2 + 5x - 2$

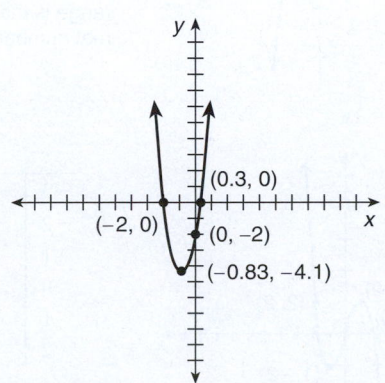

22. $y = (2x - 3)(x - 1)$

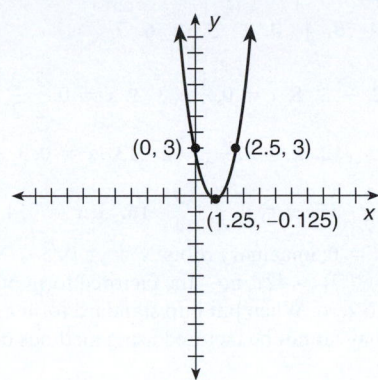

23. $y = 2x^2 - 9x - 5$

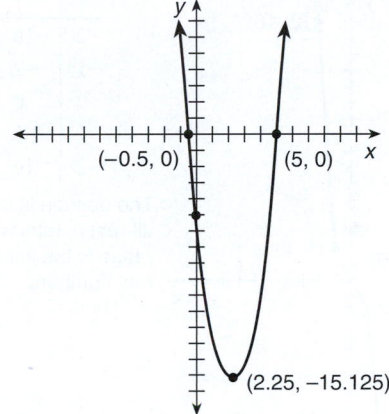

24. $y = -2x^2 + 9x + 5$

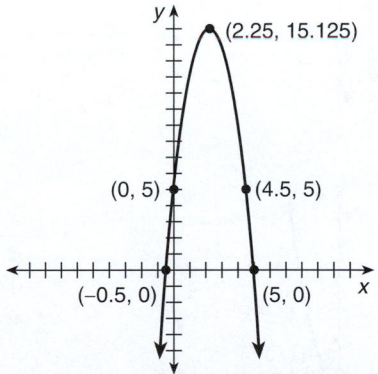

5 **25.** Real, rational, $x = \dfrac{2}{3}, -1$

26. Real, irrational, $x = \dfrac{3 \pm \sqrt{5}}{2}$ or 2.62, 0.38

27. Real, irrational, $x = \dfrac{-1 \pm \sqrt{17}}{4}$ or 0.78, −1.28

28. No real solutions or $x = \dfrac{1 \pm i\sqrt{2}}{3}$ or $0.33 \pm 0.47i$

29. Real, irrational, $x = \dfrac{3 \pm \sqrt{37}}{2}$ or 4.54, −1.54

30. Real, irrational, $x = \dfrac{-5 \pm \sqrt{97}}{6}$ or 0.81 or −2.47

31. The discriminant must be greater than or equal to zero.
32. The discriminant must be greater than zero and a perfect square.

Section 15–5 Exercises, p. 639

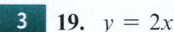

1. 1. 2 2. 1 3. 4 4. 1 5. 4 6. 7

2. 7. $x = 0, 2, -3$ 8. $x = 0, \dfrac{1}{2}, -3$ 9. $x = 0, \dfrac{5}{2}, \dfrac{2}{3}$ 10. $x = 0, 2, 5$

11. $x = 0, 3, -2$ 12. $x = 0, -\dfrac{1}{2}, -2$ 13. $x = 0, 3, -3$

14. $x = 0, \dfrac{1}{2}, -\dfrac{1}{2}$ 15. $x = 0, \dfrac{3}{4}, -\dfrac{3}{4}$ 16. $3; x = 0, 4, -4$

17. Real root: $x = 0$; imaginary roots: $x = \pm i\sqrt{5}$

18. $w(w + 3)(w + 7) = 421$; no—the factored form of the equation does not equal to zero. When put into standard form a cubic equation is formed that cannot be factored using methods developed in this text.

3. 19. $y = 2x^3$

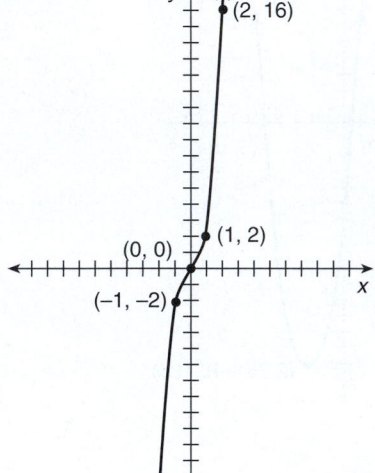

x	y
-2	-16
-1	-2
0	0
1	2
2	16

The domain is the set of all real numbers. The range is the set of all real numbers.

20. $y = \dfrac{1}{2}x^3$

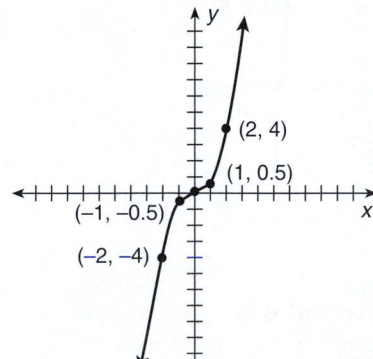

x	y
-2	-4
-1	-0.5
0	0
1	0.5
2	4

The domain is the set of all real numbers. The range is the set of all real numbers.

21. $y = x^3 - x^2 - 4x + 4$

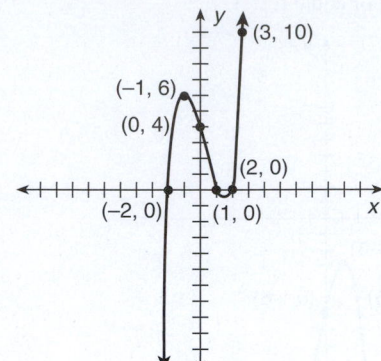

x	y
-3	-20
-2	0
-1	6
0	4
1	0
2	0
3	10

The domain is the set of all real numbers. The range is the set of all real numbers.

22. $y = x^3 + 2$

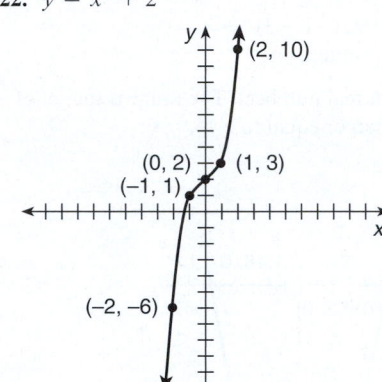

x	y
-2	-6
-1	1
0	2
1	3
2	10

The domain is the set of all real numbers. The range is the set of all real numbers.

23. $y = -x^3 + 4$

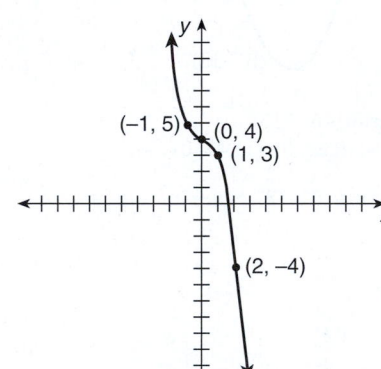

x	y
-2	12
-1	5
0	4
1	3
2	-4
3	-23

The domain is the set of all real numbers. The range is the set of all real numbers.

24. $y = -x^3 - 3x$

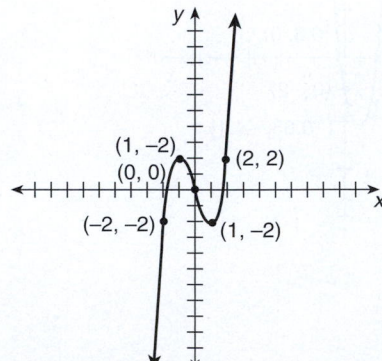

x	y
-3	-18
-2	-2
-1	2
0	0
1	-2
2	2

The domain is the set of all real numbers. The range is the set of all real numbers.

4 **25.** The graph is the graph of a function. **26.** The graph is the graph of a function. **27.** The graph is the graph of a relation but not a function. **28.** The graph is the graph of a relation but not of a function. **29.** The graph is the graph of a function. **30.** The graph is the graph of a function.

5 **31.** Domain—all real numbers between -3 and 3; range—all real numbers greater than or equal to -2 **32.** Domain—the real numbers between and including -5 and 5; range—the real numbers between and including -3 and 3 **33.** Domain—the real numbers between and including -4 and 3; range—the real numbers between and including -6 and 3 **34.** Domain—the real numbers greater than or equal to 0; range—the real numbers greater than or equal to 0 **35.** Domain—the real numbers greater than or equal to -7 and less than or equal to 5; range—the real numbers greater than or equal to -2 and less than or equal to 5 **36.** Domain—the real numbers greater than or equal to -3 and less than 3; range—the real numbers greater than or equal to 0

Section 15–6 Exercises, p. 644

1 **1.** $-3 < x < 2; (-3, 2)$

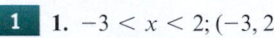

2. $\frac{2}{3} < x < \frac{5}{2}; \left(\frac{2}{3}, \frac{5}{2}\right)$

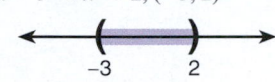

3. $-4 < a < 6; (-4, 6)$

4. $2 \le x \le 5; [2, 5]$

5. $1 \le x \le 3; [1, 3]$

6. $\frac{1}{4} \le x \le 5; \left[\frac{1}{4}, 5\right]$

7. $-5 \le x \le \frac{1}{2}; \left[-5, \frac{1}{2}\right]$

8. $x < -3$ or $x > \frac{1}{2}; (-\infty, -3) \cup \left(\frac{1}{2}, \infty\right)$

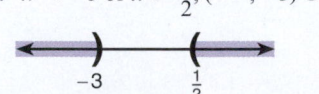

9. $y \le -6$ or $y \ge 2; (-\infty, -6] \cup [2, \infty)$

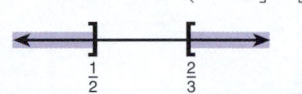

10. $x \le -2$ or $x \ge 3; (-\infty, -2] \cup [3, \infty)$

11. $x < -\frac{1}{2}$ or $x > \frac{1}{3}; \left(-\infty, -\frac{1}{2}\right) \cup \left(\frac{1}{3}, \infty\right)$

12. $x \le \frac{1}{2}$ or $x \ge \frac{2}{3}; \left(-\infty, \frac{1}{2}\right] \cup \left[\frac{2}{3}, \infty\right)$

2 **13.**

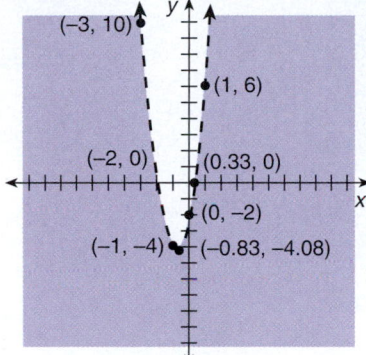

14.

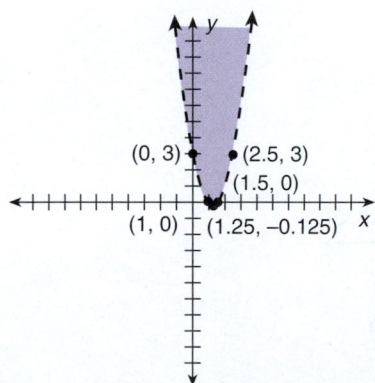

15.

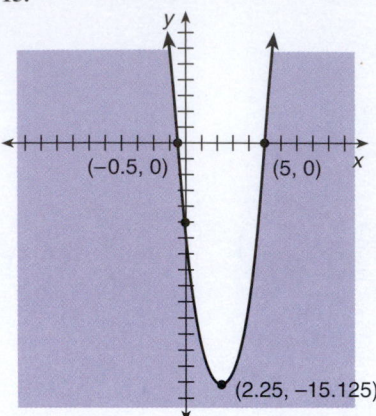

16.

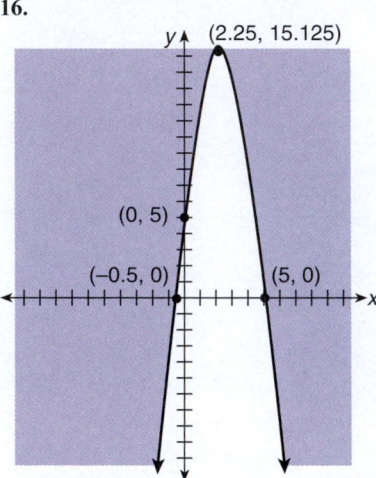

17.

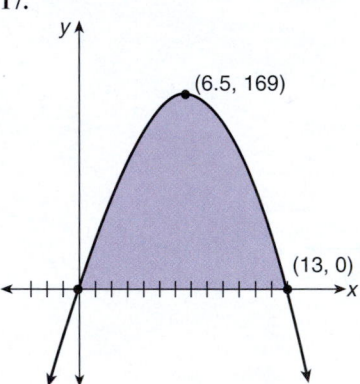

18.

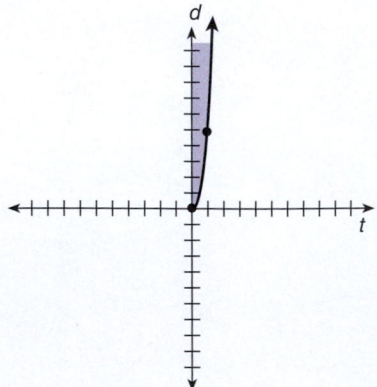

Section 15–7 Exercises, p. 650

1 **1.** ± 8 **2.** ± 15 **3.** $8, -2$ **4.** $9, 5$ **5.** $6, -3$ **6.** $3, \dfrac{5}{3}$

7. ± 11 **8.** $7, 1$ **9.** $4, -3$ **10.** $-\dfrac{1}{3}, +3\dfrac{2}{3}$ **11.** $0, 4$ **12.** $-5, 10$

13. $-\dfrac{2}{3}, 6$ **14.** $0, \dfrac{18}{7}$ **15.** $-\dfrac{12}{5}, \dfrac{18}{5}$ **16.** $-\dfrac{14}{5}, \dfrac{16}{5}$ **17.** $0, 3$ **18.** $\dfrac{2}{3}$

19. No solution   **20.** No solution   **21.** No solution

2 **22.**

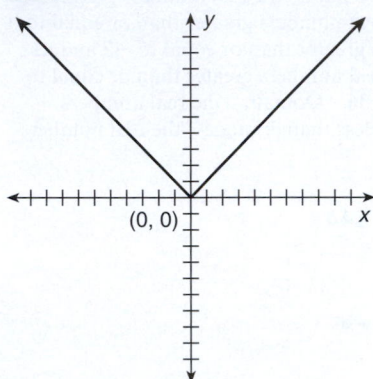

23.

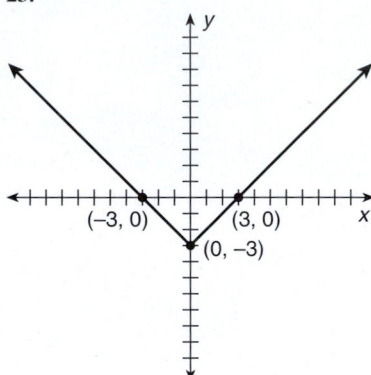

24.

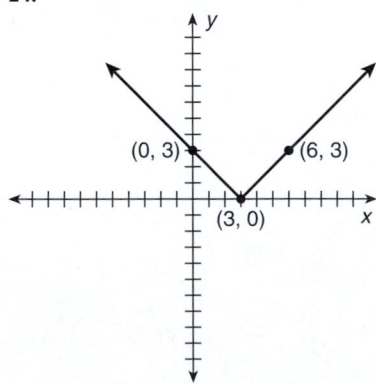

25.

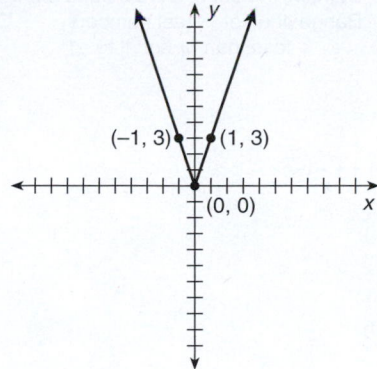

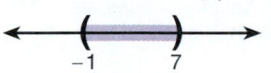

3 **26.** $-5 < x < 5; (-5, 5)$ **27.** $-1 < x < 1; (-1, 1)$

28. $-2 < x < 2; (-2, 2)$ **29.** $-7 \le x \le 7; [-7, 7]$

30. $-5 < x < -1; (-5, -1)$ **31.** $-7 < x < -1; (-7, -1)$

32. $- > 1 < x < 7; (-1, 7)$ **33.** $-1 < x < 5; (-1, 5)$

34. $-1 \le x \le 11; [-1, 11]$ **35.** $-6 \le x \le 8; [-6, 8]$

36. $-\frac{2}{3} < x < 4; \left(-\frac{2}{3}, 4\right)$ **37.** $\frac{1}{2} < x < \frac{7}{2}; \left(\frac{1}{2}, \frac{7}{2}\right)$

38. $-1 \le x \le \frac{1}{5}; \left[-1, \frac{1}{5}\right]$ **39.** $-3 \le x \le -\frac{1}{2}; \left[-3, -\frac{1}{2}\right]$

40. $-\frac{2}{3} < x < \frac{4}{3}; \left(-\frac{2}{3}, \frac{4}{3}\right)$

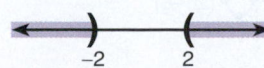

4 **41.** $x < -2$ or $x > 2; (-\infty, -2) \cup (2, \infty)$

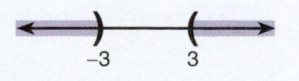

42. $x < -3$ or $x > 3; (-\infty, -3) \cup (3, \infty)$

43. $x \le -5$ or $x \ge 5; (-\infty, -5] \cup [5, \infty)$

44. $x \le -6$ or $x \ge 6; (-\infty, -6] \cup [6, \infty)$

45. $x < 3$ or $x > 7; (-\infty, 3) \cup (7, \infty)$

46. $x < -2$ or $x > 10; (-\infty, -2) \cup (10, \infty)$

47. $x \le -8$ or $x \ge 2; (-\infty, -8] \cup [2, \infty)$

48. $x \le -10$ or $x \ge 2; (-\infty, -10] \cup [2, \infty]$

49. $x \le \frac{5}{2}$ or $x \ge \frac{5}{2}$ or all real numbers; $(-\infty, \infty)$

50. $x \le -\frac{2}{3}$ or $x \ge 2; \left(-\infty, -\frac{2}{3}\right] \cup [2, \infty)$

51. $x \le -2$ or $x \ge -\frac{6}{5}; \left(-\infty, -2\right] \cup \left[-\frac{6}{5}, \infty\right)$

52. $x \le -\frac{3}{2}$ or $x \ge 3; \left(-\infty, -\frac{3}{2}\right] \cup [3, \infty)$

53. $x \le -\frac{1}{3}$ or $x \ge 1; \left(-\infty, -\frac{1}{3}\right] \cup [1, \infty)$

54. $x \le -\frac{1}{2}$ or $x \ge \frac{7}{2}; \left(-\infty, -\frac{1}{2}\right] \cup \left[\frac{7}{2}, \infty\right)$

55. $x \le -2$ or $x \ge 3; (-\infty, -2] \cup [3, \infty)$

56. $\frac{1}{3} \le x \le 1; \left[\frac{1}{3}, 1\right]$

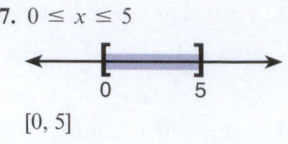

57. $0 \le x \le 5$

[0, 5]

58. $x \le -\frac{4}{5}$ or $x \ge 2$

$\left(-\infty, -\frac{4}{5}\right] \cup \left[2, \infty\right)$

59. $x \le 1\frac{1}{4}$ or $x \ge 1\frac{3}{4}$

$\left(-\infty, 1\frac{1}{4}\right] \cup \left[1\frac{3}{4}, \infty\right)$

60. No solution; { } or φ   **61.** No solution; { } or φ

62. $x \le 0$ or $x \ge 6$ **63.** $x < -\frac{1}{2}$ or $x > 0$ **64.** $1 \le x \le 7$

65. No solution   **66.** $-7 < x < 7$ **67.** $-11 < x < -5$

68. $x < -12$ or $x > 12$ **69.** $x < -10$ or $x > -6$ **70.** No solution

Chapter 15 Review Exercises, p. 661

1. Pure   **3.** Pure   **5.** Incomplete   **7.** Pure   **9.** Complete

11. $2x^2 - 8x - 5 = 0$ **13.** $x^2 - 7x + 5 = 0$

15. $4x^2 + 3x - 1 = 0$ **17.** $x = \pm10$ **19.** $x = \pm\frac{3}{2}$ or ±1.5

21. $y = \pm1.740$ **23.** $x = \pm2.828$ **25.** $x = \pm2.236$

27. $x = \pm2$ **29.** $x = \pm4.123$ **31.** $y = \pm3.055$ **33.** $x = \pm4$

35. $x = \pm8$ **37.** 16.4 cm   **39.** 0 or 5   **41.** 0 or 2   **43.** 0 or $-\frac{1}{2}$

45. 0 or 7   **47.** 0 or $-\frac{2}{3}$ **49.** 0 or -3 **51.** 0 or 9   **53.** 0 or -8

55. 0 or $\frac{5}{3}$ **57.** 0 or $\frac{1}{2}$ **59.** 0 or 4   **61.** 3 or 1   **63.** -5 or 2

65. -6 or -1 **67.** 2 or 4   **69.** $-\frac{2}{3}$ or $\frac{3}{2}$ **71.** $-\frac{2}{5}$ or $\frac{5}{2}$

73. $-\frac{3}{4}$ or -1 **75.** $\frac{3}{4}$ or $-\frac{1}{3}$ **77.** -21 or 2   **79.** $\frac{2}{3}$ or -1

81. 3 or 2   **83.** 6 or -3 **85.** -6 or -5 **87.** -9 or 2

89. width $= 12$ ft
length $= 19$ ft

91. 2 (double root)   **93.** 2 or 6   **95.** $4 - \sqrt{2}$ or $4 + \sqrt{2}$

97. $3 - i\sqrt{3}$ or $3 + i\sqrt{3}$ or no real solutions   **99.** 1 or 4

101. $\frac{3 - \sqrt{37}}{2}$ or $\frac{3 + \sqrt{37}}{2}$

103. $a = 1$ **105.** $a = 1$ **107.** $a = 1$
$b = -2$ $b = 3$ $b = -3$
$c = -8$ $c = -4$ $c = 2$

109. 9 or -1 **111.** 2 or -4 **113.** 2 or $-\frac{1}{2}$ **115.** 1.78 or -0.28

117. -0.23 or -1.43 **119.** $w = 11$ ft, $l = 22$ ft

121. $w = 14$ in., $l = 42$ in.

123.

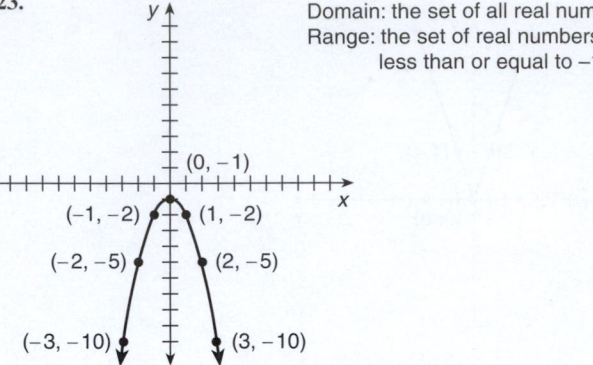

Domain: the set of all real numbers
Range: the set of real numbers
less than or equal to -1

(0, −1)   (−1, −2)   (1, −2)   (−2, −5)   (2, −5)   (−3, −10)   (3, −10)

125.

Domain: the set of all real numbers
Range: the set of real numbers
greater than -1

(0, 8)   (6, 8)   (2, 0)   (4, 0)   (3, −1)

127.

Domain: the set of all real numbers
Range: the set of real numbers
less than or equal to 0

(1, 0)   (0, −1)   (2, −1)   (−1, −4)   (3, −4)   (−2, −9)   (4, −9)

129. vertex: $(1, -9)$; axis of symmetry: $x = 1$

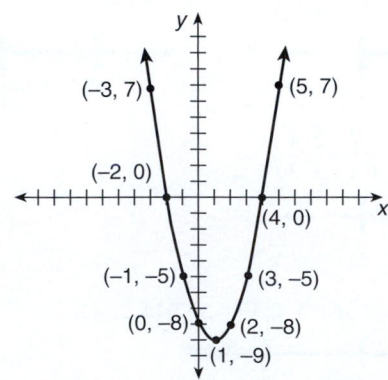

(−3, 7)   (5, 7)   (−2, 0)   (4, 0)   (−1, −5)   (3, −5)   (0, −8)   (2, −8)   (1, −9)

131. vertex: (4, 4); axis of symmetry: $x = 4$

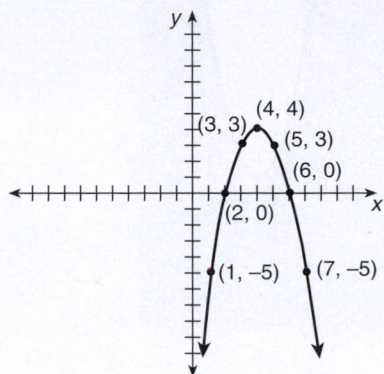

133. $x = -2$ or 6 **135.** $x = -4$ (double root)
137. $y = -2x^2$

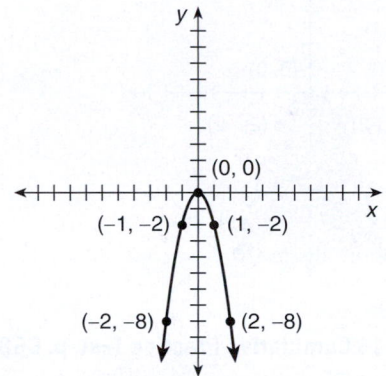

139. Real, rational, double root **141.** Real, irrational, 2 unequal roots
143. Real, rational, 2 unequal roots **145.** 3 **147.** 1 **149.** 8

151. $0, 2, \dfrac{2}{3}$ **153.** $0, -2, -3$ **155.** $0, -5, \dfrac{1}{2}$ **157.** $0, 2$

159. $0, -2, -4$ **161.** $0, 5, -4$ **163.** $0, 3 \pm \sqrt{2}$ or 4.414, 1.586

165. $0, -1, 4$ **167.** $-2, -1, 0$

169. $y = 5x^3$

x	y
-2	-40
-1	-5
0	0
1	5
2	40
3	135

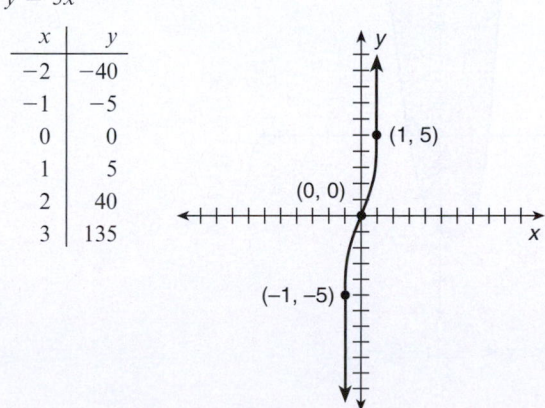

The domain is the set of all real numbers. The range is the set of all real numbers.

171. The graph is the graph of a relation but not a function since at least one vertical line can be drawn that intersects the graph in more than one point.

173. The graph is the graph of a relation but not a function. A vertical line intersects the graph in two points everywhere except at $(-5, 0)$ and $(5, 0)$.

175. Domain—the set of all real numbers
Range—the set of all real numbers less than or equal to 4

177. Domain—the set of all real numbers from -2 to 2 inclusive
Range—all real numbers from -3 to 9 inclusive

179. $-\dfrac{1}{3} < x < \dfrac{3}{2}; (\dfrac{1}{3}, \dfrac{3}{2})$

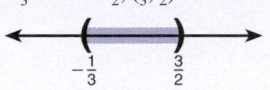

181. $-1 \le x \le 2; [-1, 2]$

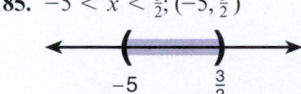

183. $-\dfrac{1}{2} \le x \le 3; [-\dfrac{1}{2}, 3]$

185. $-5 < x < \dfrac{3}{2}; (-5, \dfrac{3}{2})$

187.

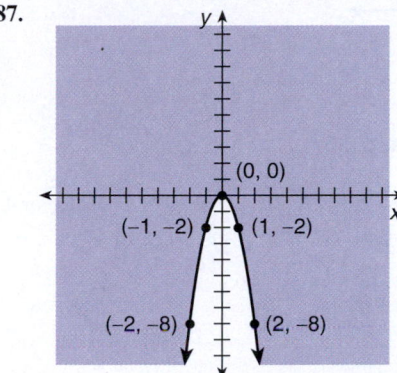

189.

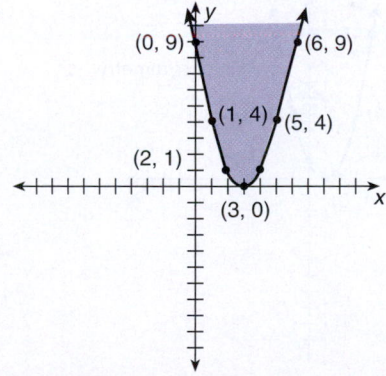

191. ± 12 **193.** $4, -10$ **195.** $20, -4$ **197.** $6, -\dfrac{5}{2}$ **199.** $1, -\dfrac{23}{7}$

201. $3, -\dfrac{13}{7}$ **203.** $\dfrac{11}{3}, \dfrac{7}{3}$ **205.** ± 7 **207.** ± 16 **209.** $10, -4$

211. $2, -\dfrac{1}{2}$

213.

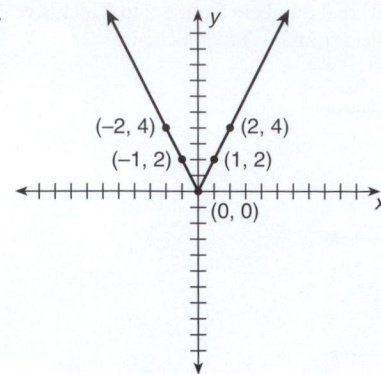

215. $-1 < x < 7; (-1, 7)$

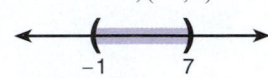

217. $-4 < x < 12; (-4, 12)$

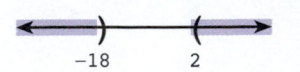

219. No solution

Chapter 15 Practice Test, p. 667

1. Pure   **3.** Incomplete   **5.** ± 9 **7.** 0 or 2   **9.** 3 or 2   **11.** $\frac{3}{2}$ or 4

13. 1 or $-\frac{5}{2}$ **15.** $\frac{3 + \sqrt{29}}{2}$ or 4.19, $\frac{3 - \sqrt{29}}{2}$ or -1.19

17. Real, rational, 2 unequal roots

19. $y = x^2 + 2x + 1$

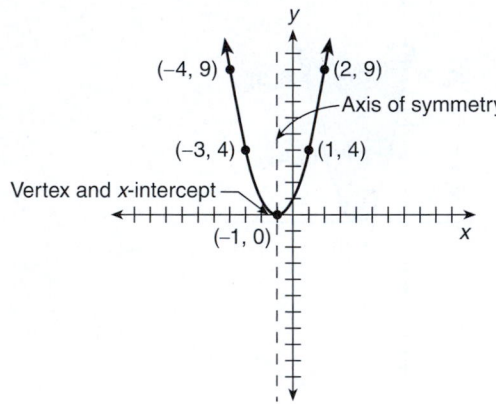

21. 169.79 mils   **23.** 1.98 amps   **25.** 6   **27.** 0 or $\frac{3}{2}$ or -5 **29.** 0 or 3

31. The graph is the graph of a relation but not a function. The domain is the set of all real numbers. The range is the set of all real numbers.

33. $x < -\frac{3}{2} \cup x > 1; \left(-\infty, -\frac{3}{2}\right) \cup (1, \infty)$

35. ± 15 **37.** ± 2

39. $x < -18 \cup x > 2; (-\infty, -18) \cup (2, \infty)$

41.

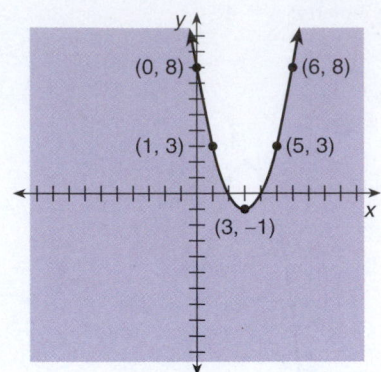

43.

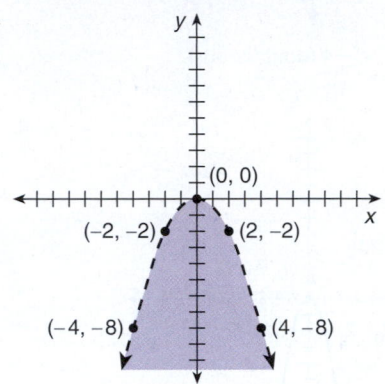

Chapters 11–15 Cumulative Practice Test, p. 668

1. $40x^{10}$ **3.** x^{32} **5.** $12x^4 + 5x^2$ **7.** $2x^2 - 4x^3 + 3x$

9. $8x(3x^2 - 2x - 1)$ **11.** $12x^2 - 23x + 10$ **13.** $x - 3$

15. $(x - 8)^2$ **17.** $(x + 7)(x - 6)$ **19.** $(2x + 3)(3x + 4)$

21. $x = \pm 3$ **23.** $x = 0; x = 8$ **25.** $x = \frac{1}{2}; x = 3$

27. $x \approx 1.18; x \approx -0.43$

29. $x = -2; x = 5; x = 0$

31.

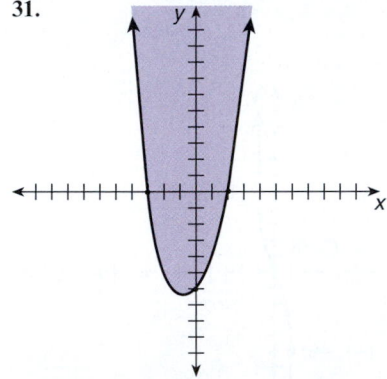

33. $-5 < x < -2; (-5, -2)$

16   **EXPONENTIAL AND LOGARITHMIC EQUATIONS**

Section 16–1   **Exercises, p. 683**

1   **1.** 64   **2.** 0.0041   **3.** 9,765,625   **4.** 0.002   **5.** 243
6. $2,415.77   **7.** $2,050.40   **8.** (a) 122.14 (b) 149.18 (c) 164.87
(d) 182.21   **9.** (a) $1,197.22 (b) $1,576.91   **10.** $720.98
11. $14,448.89; $6,448.89   **12.** $15,373.05; $4,873.05
13. $14,414.25   **14.** $8\frac{1}{4}$% annually is the better deal.   **15.** $90,305.56

16. $9.26 **17.** $7.72 **18.** $27.99 **19.** $321.89 **20.** $2,252.25
21. $7,906.98 **22.** $19,462.47 **23.** $6,462.60 **24.** $1,587.66
25. $1,254.58 **26.** $6,736.25 **27.** $13,611.66 **28.** 8.24%
29. 10.25% **30.** 6.14% **31.** 12.55% **32.** $6,270; $270 **33.** $8,280;
$280 **34.** $16,232 **35.** $9,786.68 **36.** $127,391.11; $91,000
37. $60,743.42 **38.** $25,129.02 **39.** $7,243.28 **40.** $5,866.60

41.

Years	Total Investment	Total Interest
Ten-year	$5,000	$2,243.28
Five-year	$5,000	$866.60

The 10-year investment earned more interest even though half as much money was invested per year. At the same period interest rate, investing for twice as long gives a better yield on your investment than investing the same amount for half as long. Thus, the earlier you start saving, the better.

42. $61,735.08 **43.** $69,721.11 **44.** $908.92 **45.** $1,155.89

2 **46.** 7.39 **47.** 0.05 **48.** 1.23 **49.** 0.03 **50.** 22,026.47
51. 2.08×10^{-87} **52.** 4.28×10^{-96} **53.** 5.58×10^{-44}
54. $21,883.49 **55.** $17,558.71

3 **56.** $x = 7$ **57.** $x = -3$ **58.** $x = 2$ **59.** $x = 10$ **60.** $x = 7$
61. $x = 3$ **62.** $x = 6$ **63.** $x = -4$ **64.** $x = -6$ **65.** $x = \dfrac{7}{6}$
66. $x = \dfrac{5}{2}$ **67.** $x = 5$ **68.** $x = 8$ **69.** $x = -3$ **70.** $x = 3$
71. $x = \dfrac{15}{11}$

4 **72.**

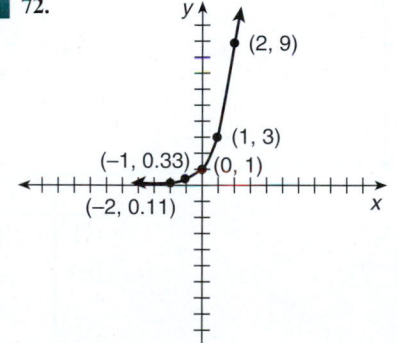

x	y
−2	0.11
−1	0.33
0	1
1	3
2	9

73.

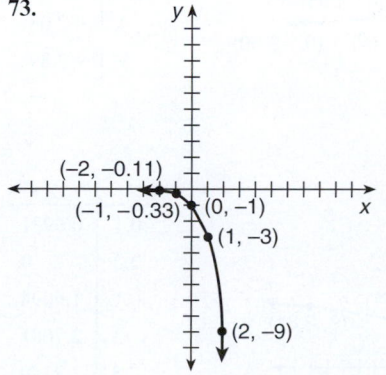

x	y
−2	−0.11
−1	−0.33
0	−1
1	−3
2	−9

74.

x	y
−2	−2.75
−1	−2.5
0	−2
1	−1
2	1
3	5

75.

x	y
−2	−3.25
−1	−3.5
0	−4
1	−5
2	−7
3	−11

76.

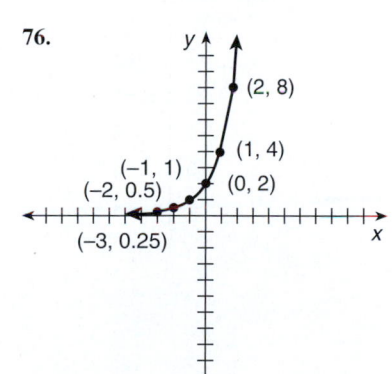

x	y
−3	0.25
−2	0.5
−1	1
0	2
1	4
2	8

77.

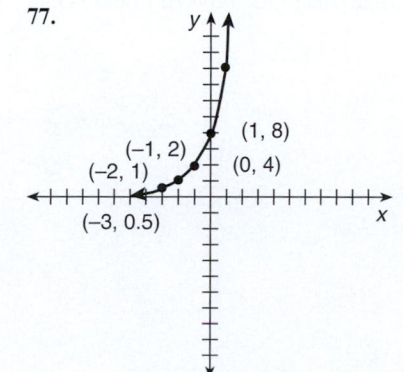

x	y
−3	0.5
−2	1
−1	2
0	4
1	8

78.

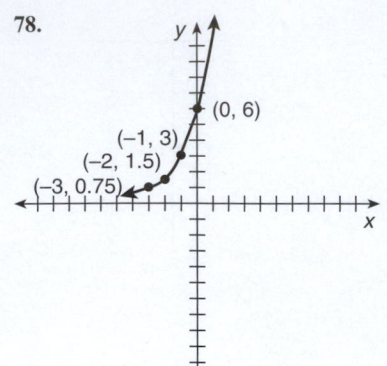

x	y
−3	0.75
−2	1.5
−1	3
0	6
1	12

79.

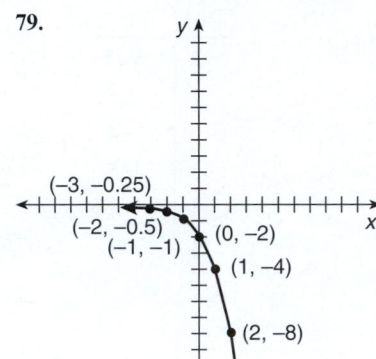

x	y
−3	−0.25
−2	−0.5
−1	−1
0	−2
1	−4
2	−8

80.

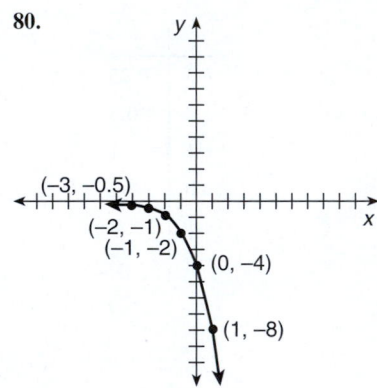

x	y
−3	−0.5
−2	−1
−1	−2
0	−4
1	−8

81. (a) 3.9721 units (b) between 5 and 6 min   **82.** between 13 and 14 min

83.

x	y
0	250
1	150
2	90
3	54
4	32.4
5	19.44
6	11.664

$x = 0$

54 mg remained in the bloodstream at 6:00 P.M.

84. between 1 and 2 h   **85.** after 1 year: $1,050; after 2 years: $1,102.50; after 10 years: $1,628.89; after 30 years: $4,321.94; after 50 years: $11,467.40   **86.** 12.5; 2 cm

---

Section 16–2 **Exercises, p. 694**

1 **1.** $\log_3 9 = 2$ **2.** $\log_2 32 = 5$ **3.** $\log_9 3 = \dfrac{1}{2}$ **4.** $\log_{16} 2 = \dfrac{1}{4}$

5. $\log_4 \dfrac{1}{16} = -2$ **6.** $\log_3 \dfrac{1}{81} = -4$

2 **7.** $3^4 = 81$ **8.** $12^2 = 144$ **9.** $2^{-3} = \dfrac{1}{8}$ **10.** $5^{-2} = \dfrac{1}{25}$

11. $25^{-0.5} = \dfrac{1}{5}$ **12.** $4^{-0.5} = \dfrac{1}{2}$

3 **13.** −3   **14.** 5   **15.** 7   **16.** −5   **17.** 0   **18.** −6   **19.** 0.4771
20. 0.7782   **21.** 0.3802   **22.** 0.6232   **23.** 2.1761   **24.** −2.9208
25. 1.3863   **26.** 0.9163   **27.** −1.8971   **28.** 5.6168   **29.** 4.6052

4 **30.** $x = \dfrac{1}{81}$ **31.** $x = 2$ **32.** $x = -2$ **33.** $x = 625$
34. $x = -4$ **35.** $x = 3$ **36.** 3   **37.** 6   **38.** 5   **39.** 2   **40.** 1.9358
41. 7

5 **42.** 70 dB   **43.** 119 dB   **44.** 13.9 years   **45.** 2.0 years
46. 69.3 years   **47.** 23.1 years

6 **48.**

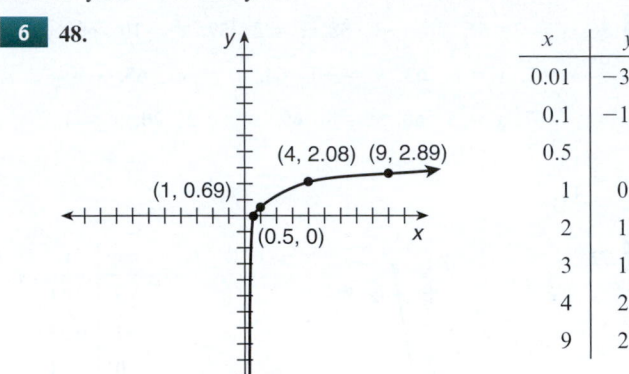

x	y
0.01	−3.91
0.1	−1.61
0.5	0
1	0.69
2	1.39
3	1.79
4	2.08
9	2.89

49.

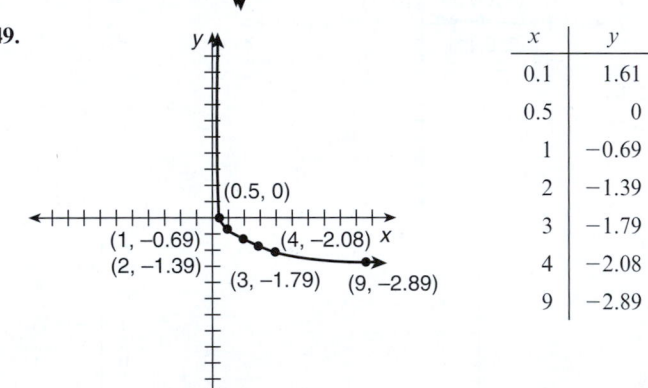

x	y
0.1	1.61
0.5	0
1	−0.69
2	−1.39
3	−1.79
4	−2.08
9	−2.89

50.

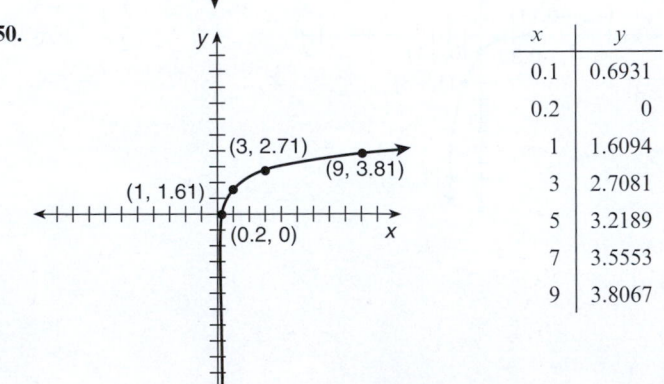

x	y
0.1	0.6931
0.2	0
1	1.6094
3	2.7081
5	3.2189
7	3.5553
9	3.8067

51.

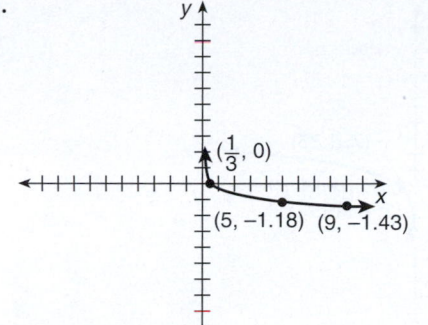

x	y
0.1	0.52288
1/3	0
1	−0.47712
3	−0.9542
5	−1.1761
7	−1.3222
9	−1.4314

52.

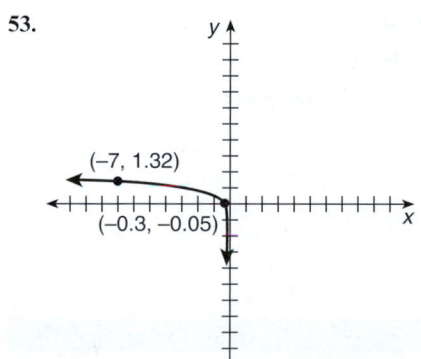

x	y
0.1	0.22185
1/6	0
1	−0.7782
3	−1.255
6	−1.556
9	−1.737

53.

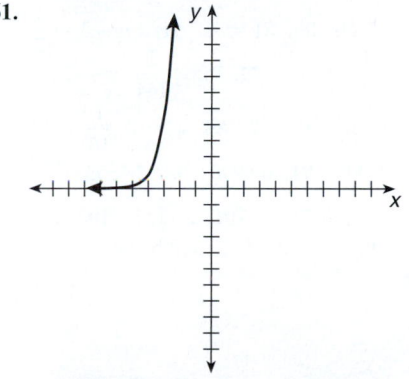

x	y
−7	1.3222
−5	1.1761
−3	0.95424
−1	0.47712
−0.4	0.0792
−0.3	−0.0458
−0.1	−0.5229

54. 9　**55.** 3　**56.** 5　**57.** 8　**58.** 7.3

7　**59.** 1 = 0.3010299957 + 0.698970043
60. 0.7781512504 = 0.4771212547 + 0.3010299957
61. 1.397940009 = 2(0.6989700043)

62. 1.477121255 = 1 + 0.477121255　**63.** $m = \dfrac{n^a}{p}$　**64.** $a = \dfrac{b^r}{c^3}$

Chapter 16　Review Exercises, p. 699

1. $x = 8$　**3.** $x = -2$　**5.** $x = 4$　**7.** $x = -\dfrac{5}{3}$　**9.** $x = 4$

11. $x = -5$　**13.** $x = 1$　**15.** $x = 4$　**17.** 0.0183　**19.** 0.0000454
21. \$708.64　**23.** \$524.95　**25.** \$5,372.54　**27.** \$2,189.94
29. \$13,928.19　**31.** \$1,356.25　**33.** 8.24%　**35.** \$190.99
37. \$2,913.78　**39.** \$11,000 in 18 months is better.　**41.** \$4,781.09
43. \$26,361.59　**45.** \$7,102.10　**47.** \$2,294.19　**49.** \$5,591.97
51. \$3,407.88　**53.** \$1,917.31　**55.** 1.25×10^{-6}　**57.** 2.71×10^{-35}

59.

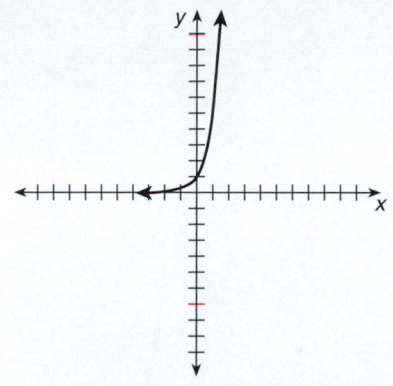

X	Y₁
−3	0.008
−2	0.04
−1	0.2
0	1
1	5
2	25
3	125

X = −3

61.

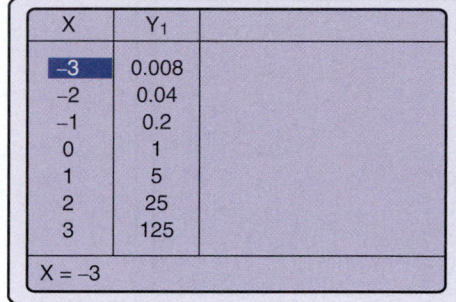

X	Y₁
−7	0.008
−6	0.04
−5	0.2
−4	1
−3	5
−2	25
−1	125

X = −7

63.

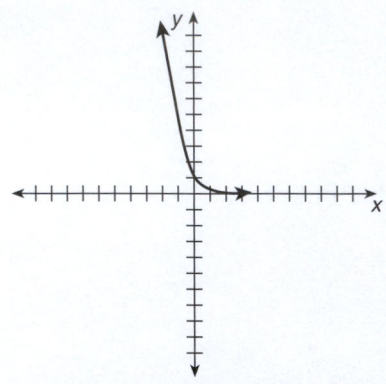

X	Y₁	
-3	125	
-2	25	
-1	5	
0	1	
1	0.2	
2	0.04	
3	0.008	
X = -3		

65. 16.3 mg **67.** $\log_2 8 = 3$ **69.** $\log_3 81 = 4$ **71.** $\log_{27} 3 = \frac{1}{3}$

73. $\log_4 \frac{1}{64} = -3$ **75.** $\log_9 \frac{1}{3} = -\frac{1}{2}$ **77.** $\log_{12} \frac{1}{144} = -2$

79. $11^2 = 121$ **81.** $15^0 = 1$ **83.** $7^1 = 7$ **85.** $4^{-2} = \frac{1}{16}$

87. $9^{-0.5} = \frac{1}{3}$ **89.** $10^3 = 1,000$ **91.** 0.6990 **93.** 2.2553

95. −0.3979 **97.** 5.5984 **99.** −0.2231 **101.** 2.1133 **103.** 2

105. 343 **107.** $x = -2$ **109.** $x = 32$ **111.** a. 2 b. 4 c. 8.2

113. 3.17 **115.** 9.89 years

117.

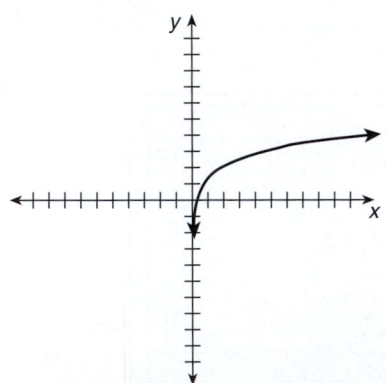

X	Y₁	
1	1.3863	
2	2.0794	
3	2.4849	
4	2.7726	
5	2.9957	
6	3.1781	
7	3.3322	
X = 1		

119.

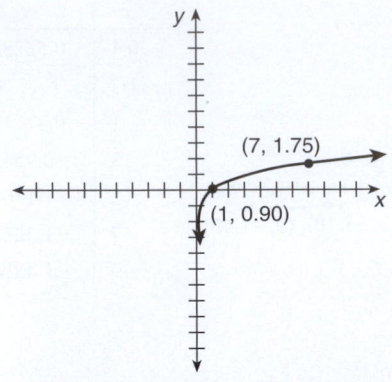

(7, 1.75)
(1, 0.90)

X	Y₁	
1	.90309	
2	1.2041	
3	1.3802	
4	1.5051	
5	1.6021	
6	1.6812	
7	1.7482	
X = 1		

121. $c = 6.3$

Chapter 16 Practice Test, p. 703

1. 3,657.26 **3.** 58.09 **5.** 248,832 **7.** $x = 9$ **9.** $x = 2$

11. $\log_4 \frac{1}{2} = -\frac{1}{2}$ **13.** $3^{-3} = \frac{1}{27}$ **15.** 3.4657 **17.** $x = 3$

19. 1.4641 **21.** $S = \$224.94$ thousands or $\$224,942$ **23.** $5,596.82

25. $19,350 **27.** $5,727.50 **29.** 12.55% **31.** $680 in 1 year is better.

33. Option 2 yields the greater return by $0.69. **35.** $2,225.54

37. $13,586.96

---

17   GEOMETRY

Section 17–1 Exercises, p. 716

1 **1.** $\overleftrightarrow{PQ}, \overleftrightarrow{QR}$ & $\overleftrightarrow{PR}$ or $\overleftrightarrow{QP}, \overleftrightarrow{RQ}$, and $\overleftrightarrow{RP}$ **2.** $\overrightarrow{PQ}, \overrightarrow{PR}$

3. $\overline{QR}$ or $\overline{RQ}$ **4.** Yes **5.** No, endpoints are different. **6.** No

7. Yes **8.** Parallel **9.** Intersect **10.** Coincide **11.** Intersect

12. $\overleftrightarrow{EF}$ and $\overleftrightarrow{GH}$ **13.** $\angle DEF, \angle FED$ **14.** $\angle E$ **15.** $\angle 1$

2 **16.** 360° **17.** 180° **18.** 90° **19.** 45° **20.** Acute **21.** Obtuse

22. Right **23.** Obtuse **24.** Acute **25.** Straight **26.** Neither

27. Complementary **28.** Supplementary **29.** Neither

30. Complementary **31.** Supplementary **32.** Supplementary

3 **33.** $\angle a$ and $\angle c$; $\angle b$ and $\angle d$ **34.** Supplementary angles

35. 180° **36.** $\angle c$ and $\angle f$; $\angle d$ and $\angle e$ **37.** 180° **38.** Corresponding

angles **39.** $\angle b = 30°$; $\angle c = 30°$; $\angle d = 150°$; $\angle e = 150°$;

$\angle f = 30°$; $\angle g = 30°$; $\angle h = 150°$ **40.** $\angle e = y°$; $\angle f = (180° - y°)$;

$\angle g = (180° - y°)$; $\angle h = y°$

4 **41.** 0.7833° **42.** 0.01° **43.** 0.0872° **44.** 0.1708° **45.** 10.3042°

46. 21′ **47.** 12′ **48.** 7′12″ **49.** 12′47″ **50.** 18′54″

Section 17–2 Exercises, p. 726

1   **1.** 309.3 mm   **2.** 52.9 ft   **3.** 207 in.   **4.** 75 ft   **5.** 402 ft
6. 98 ft   **7.** 28.6 in.   **8.** 48 cm   **9.** 72 ft   **10.** 10 ft 8 in. or 3 yd 1 ft
8 in.   **11.** 48 in.   **12.** 27.2 cm   **13.** 196 in.   **14.** 176 in.   **15.** 240 in.
16. 128 in.   **17.** 24 ft   **18.** 36.2 ft   **19.** 690 ft   **20.** 66 ft
21. 57.5 ft   **22.** 184 ft   **23.** 17 ft   **24.** 130 in.   **25.** 20 cm
26. 40.8 cm   **27.** 628 ft   **28.** 52 tiles   **29.** 76 ft   **30.** 44 ft

2   **31.** 112 in^2 **32.** 26.7 cm^2 **33.** 35 ft^2 **34.** 80.34 ft^2 **35.** 42,500 ft^2
36. 180 ft^2 **37.** 25 cm^2 **38.** 104.04 cm^2 **39.** 193 ft^2 **40.** 900 yd^2
41. $\frac{1}{16}$ mi^2 **42.** 81 tiles   **43.** 108 tiles   **44.** 400 in^2
45. 5,053.05 mm^2 **46.** 162 ft^2 **47.** 348 in^2 **48.** 260 ft^2 **49.** 33 in^2
50. 96 cm^2 **51.** 22.32 m^2 **51.** 6.375 ft^2 **53.** 189 ft^2 **54.** $175.50
55. 185 ft^2 **56.** 73.1875 ft^2 **57.** 39.7 m^2 **58.** 424 m^2

Section 17–3 Exercises, p. 741

1   **1.** 25.1 cm   **2.** 47.1 m   **3.** 17.3 m   **4.** 18.8 in.   **5.** 9.4 ft
6. 53.4 ft   **7.** 28.3 in^2 **8.** 804.2 yd^2 **9.** 7.1 ft^2 **10.** 227.0 ft^2
11. 50.3 cm^2 **12.** 176.7 m^2 **13.** 23.8 m^2 **14.** 21.6 in^2 **15.** 0.6 m^2
16. 2.1 ft^2 **17.** 14.6 cm^2 **18.** 1.0 in^2 **19.** 592.7 ft^2 **20.** 268.3 in.
21. Yes, the cross-sectional area of the third pipe is larger than the
combined area of the other two pipes.   **22.** Yes, the combined cross-
sectional area of the two pipes is 25.1 in^2, which is greater than 20 in^2—
the area of the large pipe.   **23.** 2,827 ft/min   **24.** 1,571 ft/min

2   **25.** 0.79 rad   **26.** 0.98 rad   **27.** 2.44 rad   **28.** 21.75°; 0.38 rad
29. 177.55°; 3.10 rad   **30.** 44.9033°; 0.78 rad   **31.** 45°   **32.** 30°
33. 143.2394°   **34.** 28°38′52″   **35.** 22°30′   **36.** 42°58′19″

3   **37.** 28.27 mm   **38.** 13.09 in.   **39.** 3.20 ft   **40.** 7.06 cm
41. 6.45 in.   **42.** 14 cm   **43.** 4.25 rad   **44.** 3.49 rad   **45.** 7 cm
46. 4.5 in.

4   **47.** 120.64 cm^2 **48.** 198.44 mm^2 **49.** 2.43 ft^2 **50.** 14.18 in^2
51. 530.14 ft^2 **52.** 75.40 in^2 **53.** 17.12 in^2 **54.** 4.32 cm^2
55. 2.24 cm   **56.** 0.85 rad   **57.** 47.43 cm^2 **58.** 0.7 rad
59. 16.01 cm^2 **60.** 48.98 in^2 **61.** 54.83 in^2 **62.** 51.99 cm^2
63. 561.81 mm^2 **64.** 257.93 ft^2

Section 17–4 Exercises, p. 751

1   **1.** 576 in^3 **2.** 320 cm^3 **3.** 55.36 in^3 **4.** 3,817.04 cm^3
5. 23.3 in^3 **6.** 18,850 ft^3 **7.** $V = 301.6$ ft^3 **8.** 4,712.4 ft^3 **9.** 50.3 L
10. 12,600 cm^3 **11.** 76,800 m^3 **12.** 415.6 m^3 **13.** 2,617.3 in^3
14. 2,428.2 cm^3 **15.** 3,660 in^3 **16.** 523.6 cm^3 **17.** 904.8 in^3
18. 356,892.8 gal   **19.** 225.6 gal   **20.** 150.5 ft^3 **21.** $77,619.77

2   **22.** $LSA = 37.3$ in^2; $TSA = 47.1$ in^2
23. $LSA = 96$ in^2; $TSA = 109.8$ in^2
24. $LSA = 200$ cm^2; $TSA = 264$ cm^2
25. $LSA = 1,885.0$ ft^2; $TSA = 4,398.2$ ft^2 **26.** 1,885 ft^2
27. 314.2 cm^2 **28.** 6,361.7 ft^2 **29.** 50.3 ft^2
30. $LSA = 188.5$ ft^2; $TSA = 301.6$ ft^2
31. 589.0 cm^2 **32.** 1,649.3 ft^2 **33.** $347

Chapter 17 Review Exercises, p. 761

1. $\overleftrightarrow{AB}$ & $\overleftrightarrow{CD}$ **3.** $\overleftrightarrow{EF}$ & $\overleftrightarrow{GH}$ **5.** $\angle P$ **7.** Right   **9.** Straight
11. Supplementary   **13.** Neither   **15.** $\angle a$ and $\angle b$; $\angle b$ and $\angle c$; $\angle c$
and $\angle d$; $\angle d$ and $\angle a$ **17.** $\angle a$ and $\angle h$; $\angle b$ and $\angle g$ **19.** 45°
21. 0.4833°   **23.** 0.1261°   **25.** 45′   **27.** 13′3″   **29.** 210 mm
31. 35.8 in.   **33.** 66 ft   **35.** 288 in.   **37.** 98 ft^2 **39.** 34.81 m^2
41. 306 ft^2 **43.** 36,000 ft^2 **45.** 44 yd^2 **47.** 235 ft^2

49. $A = 50.27$ m^2; $c = 25.13$ m^2 **51.** $A = 12$ cm^2; $p = 17.42$ cm
53. 33 ft/min   **55.** 1.05 rad   **57.** 99.75°; 1.74 rad   **59.** 150°
61. 2.62 cm^2 **63.** 2.18 ft^2 **65.** 31.42 in.   **67.** 6.96 cm^2
69. 44.50 in^2 **71.** 7,854 cm^3 **73.** 102 yd^3 **75.** 616 cm^2 **77.** 2,733 cm^2
79. 124,407 ft^2 **81.** 1,728 m^3 **83.** 1,017.9 m^2 **85.** 169.6 cm^2
87. 2 gal

Chapter 17 Practice Test, p. 766

1. 18′45″   **3.** 0.61 rad   **5.** 112.5°   **7.** $P = 111$ ft; $A = 749.1$ ft^2
9. $P = 37$ cm; $A = 43.5$ cm^2 **11.** $C = 144.5$ m; $A = 1,661.9$ m^2
13. 1 in.   **15.** 1.71 m   **17.** 282 ft^2 **19.** 1,016.5 ft^2
21. $A = 71.84$ cm^2 **23.** 50 in^2 **25.** 6,768 gal

18 TRIANGLES

Section 18–1 Exercises, p. 776

1   **1.** Scalene   **2.** Equilateral   **3.** Isosceles   **4.** Longest TS,
shortest RS **5.** Longest k, shortest l **6.** $\angle C, \angle B, \angle A$
7. $\angle X, \angle Z, \angle Y$

2   **8.** $\angle A = \angle F, \angle C = \angle D, \angle B = \angle E$
9. $\angle P = \angle S, \angle Q = \angle T, QP = TS$
10. $\angle L = \angle P, PM = LJ, LK = PN$

3   **11.** $EF = 9.6$ cm   **12.** $TU = 19.2$ in.   **13.** $PM = 3.75$ ft
14. $a = 20$; $d = 9$ **15.** 65 ft   **16.** $DC = 6, DE = 5.3$

Section 18–2 Exercises, p. 784

1   **1.** $BC = 10$ mm   **2.** $AB = 17$ yd   **3.** $c = 6.403$ cm
4. $a = 6.325$ m   **5.** $b = 12$ ft   **6.** $AC = 39.783$ cm   **7.** 30 ft
8. 26 ft 10 in.   **9.** 12.806 ft   **10.** 13 cm   **11.** 45 dm   **12.** 61.083 ft
13. 10.607 mm

2   **14.** $BC = 12$ cm, $AB = 16.971$ cm
15. $AC = 7.071$ m, $BC = 7.071$ m   **16.** $AC = 9.899$ m, $AB = 14.0$ m
17. $AC = 8$ m, $BC = 8$ m   **18.** $AC = 14.697$ mm, $BC = 14.697$ mm
19. 25′5″   **20.** 37.830 ft

3   **21.** $AB = 12$ cm, $BC = 10.392$ cm
22. $AC = 9.0$ mm, $BC = 15.588$ mm
23. $AC = 4.619$ in., $AB = 9.238$ in.
24. $AC = 5.715$ cm, $AB = 11.431$ cm
25. $AB = 5$ ft 6 in., $BC = 4$ ft 9 in.   **26.** 35 ft
27. Rafter $= 18$ ft 0 in., run $= 15$ ft 7 in.   **28.** 4.5 cm

Section 18–3 Exercises, p. 793

1   **1.** Equilateral triangle; $\angle 60°$ **2.** Square; $\angle 90°$
3. Regular pentagon; $\angle 108°$ **4.** Regular hexagon; $\angle 120°$
5. Regular octagon; 135°   **6.** Regular nonagon; 140°

2   **7.** 15.6 cm^2 **8.** 100 ft^2 **9.** 42.5 cm^2 **10.** 667.2 mm^2
11. 617.6 in^2 **12.** 891 cm^2

3   **13.** 90°   **14.** 45°   **15.** 8.49 in.   **16.** 45°   **17.** 6 in.   **18.** 0.5 m
19. 0.35 m   **20.** 90°   **21.** 0.35 m   **22.** 0.29 in.   **23.** 1.06 in.
24. 63.62 dkm^2

4   **25.** 60°   **26.** 30°   **27.** 90°   **28.** 11.54 in.   **29.** 11.54 in.
30. 13.33 in.   **31.** 57.7 in^2 **32.** 5.77 in.   **33.** 6.67 in.   **34.** 41.89 in.
35. 2.9 cm   **36.** 1 in.

5 **37.** 34.64 mm   **38.** 120°   **39.** 60°   **40.** 40 mm   **41.** 90°
42. 10 mm   **43.** 20 mm   **44.** 60°   **45.** 30°   **46.** 3 in.
47. 39.97 mm   **48.** 2.84 in.

Section 18–4 Exercises, p. 798

1 **1.** 10   **2.** 13

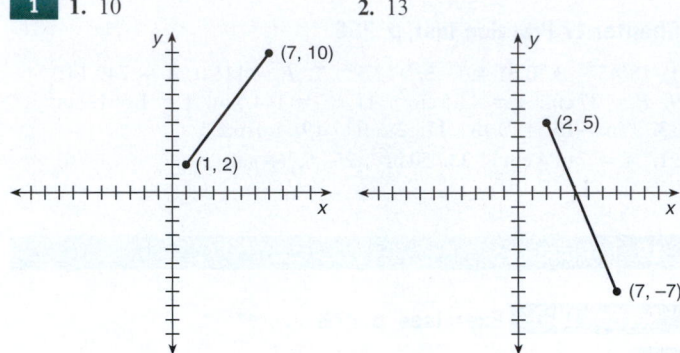

3. 7.071   **4.** 5.3857

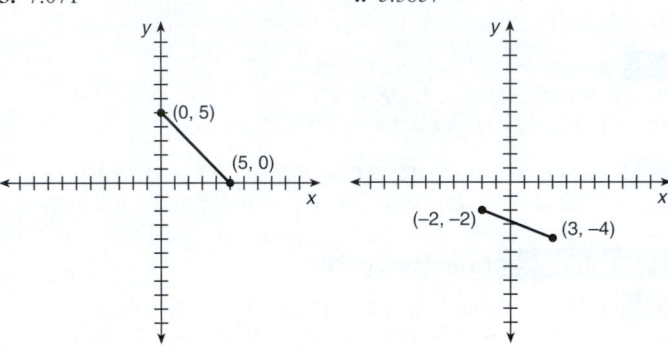

5. 11.180   **6.** 13

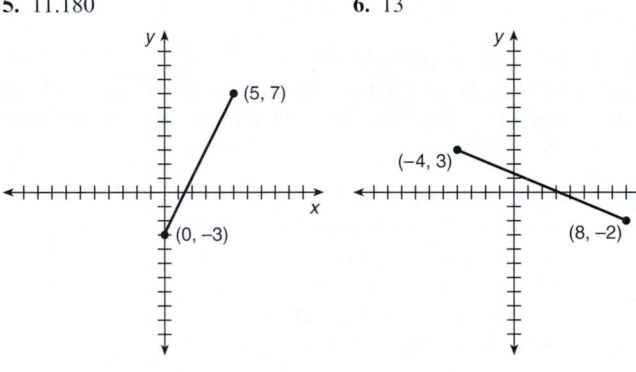

7. 10   **8.** 5

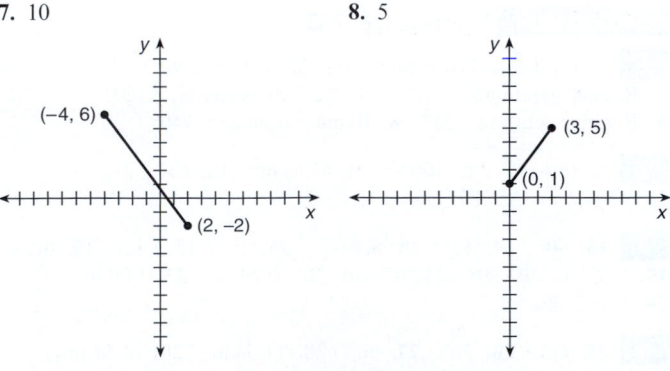

9. 15   **10.** 10.296

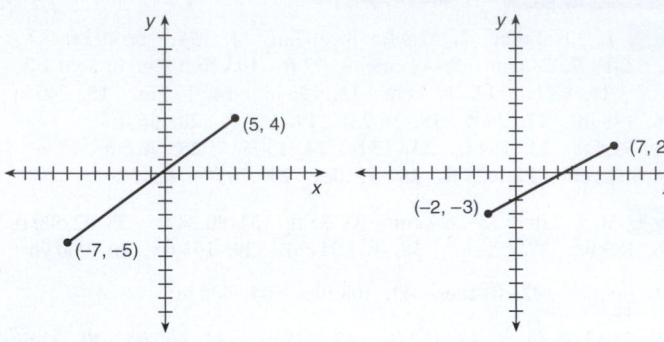

2 **11.** (4, 6)   **12.** $\left(\dfrac{9}{2}, -1\right)$ **13.** $\left(\dfrac{5}{2}, \dfrac{5}{2}\right)$ **14.** $\left(\dfrac{1}{2}, -3\right)$

15. $\left(\dfrac{5}{2}, 2\right)$ **16.** $\left(2, \dfrac{1}{2}\right)$ **17.** (−1, 2)   **18.** $\left(\dfrac{3}{2}, 3\right)$ **19.** $\left(-1, -\dfrac{1}{2}\right)$

20. $\left(\dfrac{5}{2}, -\dfrac{1}{2}\right)$

Chapter 18 Review Exercises, p. 804

1. Longest TS, shortest RS **3.** $\angle C, \angle B, \angle A$
5. $\angle C = \angle D, \angle B = \angle E, CB = DE$
7. $\angle M = \angle J, JL = PM, JK = MN$
9. $\dfrac{AB}{RT} = \dfrac{BC}{ST} = \dfrac{AC}{RS}$ **11.** 15 in.   **13.** 7.141 ft   **15.** 8 yd
17. 20.248 mi   **19.** 30 cm   **21.** 21.633 in.   **23.** 4.243 in.
25. $RS = 10.607$ cm, $ST = 10.607$ cm   **27.** $RS = 9$ hm, $ST = 9$ hm
29. $AB = 13.856$ dm, $BC = 6.928$ dm
31. $AC = 17.321$ in., $AB = 20.0$ in.
33. $AB = 46$ ft 10 in., $BC = 23$ ft 5 in.   **35.** 8 ft 4 in.   **37.** 19.630 mm
39. 908.5 in^2 **41.** Regular pentagon, 108°   **43.** 14   **45.** $AE = 5$
47. $\angle GJO = 45°$ **49.** $IJ = 20$ **51.** $\angle MOP = 30°$ **53.** 5 mm
55. 3.54 cm   **57.** 4.472   **59.** 10.440   **61.** 5.831   **63.** 9.434   **65.** (1, 5)

67. $\left(1\dfrac{1}{2}, 2\right)$ **69.** $\left(-1\dfrac{1}{2}, 2\dfrac{1}{2}\right)$ **71.** $\left(1, -\dfrac{1}{2}\right)$ **73.** $\left(-1\dfrac{1}{2}, -3\right)$

Chapter 18 Practice Test, p. 809

1. Largest $\angle A$, smallest $\angle B$ **3.** $DB = 5$ **5.** $c = 52$ **7.** 17 in.
9. $AC = 31$ cm; $BC = 53.69$ cm   **11.** $AB = 7.78$ cm; $BC = 5.5$ cm
13. 13.657 cm   **15.** 25.13 in.   **17.** 1.57 in.   **19.** distance = 5.4;
midpoint = $\left(\dfrac{1}{2}, 2\right)$

Chapters 16–18 Cumulative Practice Test, p. 811

1. $\log_5 125 = 3$ **3.** $6^2 = 36$ **5.** 107.9 mg   **7.** $23,568.40
9. acute   **11.** straight   **13.** 62°   **15.** 37°   **17.** 30.2056°   **19.** 21′7″
21. $P = 108$ m; $A = 529$ m^2 **23.** $C = 471$ cm; $A = 17{,}671$ cm^2
25. 0.2682 rad   **27.** 103.1324°   **29.** 18,096 cm^3 **31.** 19 in.
33. $AC = 16.25$ m; $BC = 28.1$ m   **35.** 54 in.   **37.** 8.5

19   RIGHT-TRIANGLE TRIGONOMETRY

Section 19–1 Exercises, p. 821

1 **1.** $\dfrac{5}{13}$ **2.** $\dfrac{12}{13}$ **3.** $\dfrac{5}{12}$ **4.** $\dfrac{12}{13}$ **5.** $\dfrac{5}{13}$ **6.** $\dfrac{12}{5}$ **7.** $\dfrac{9}{15} = \dfrac{3}{5}$

8. $\dfrac{12}{15} = \dfrac{4}{5}$ **9.** $\dfrac{9}{12} = \dfrac{3}{4}$ **10.** $\dfrac{12}{15} = \dfrac{4}{5}$ **11.** $\dfrac{9}{15} = \dfrac{3}{5}$ **12.** $\dfrac{12}{9} = \dfrac{4}{3}$

13. $\frac{16}{34} = \frac{8}{17}$ **14.** $\frac{30}{34} = \frac{15}{17}$ **15.** $\frac{16}{30} = \frac{8}{15}$ **16.** $\frac{30}{34} = \frac{15}{17}$

17. $\frac{16}{34} = \frac{8}{17}$ **18.** $\frac{30}{16} = \frac{15}{8}$ **19.** $\frac{9}{16.64} = 0.5409$

20. $\frac{14}{16.64} = 0.8413$ **21.** $\frac{9}{14} = 0.6429$ **22.** $\frac{14}{16.64} = 0.8413$

23. $\frac{9}{16.64} = 0.5409$ **24.** $\frac{14}{9} = 1.5556$

2 **25.** 0.3584 **26.** 0.9981 **27.** 1.0724 **28.** 0.6088 **29.** 0.5225
30. 0.9304 **31.** 0.4068 **32.** 0.9336 **33.** 0.8039 **34.** 0.9163
35. 1.0990 **36.** 0.8843 **37.** 0.9757 **38.** 0.4950 **39.** 3.3191
40. 0.9446 **41.** 0.6845 **42.** 0.9377 **43.** 0.3960 **44.** 0.2199

3 **45.** 20.0° **46.** 22.5° **47.** 67.0° **48.** 62.5° **49.** 57.3°
50. 34.8° **51.** 51.7° **52.** 82.1° **53.** 39.0° **54.** 0.6807 rad
55. 0.8028 rad **56.** 1.1694 rad **57.** 0.2537 rad **58.** 1.4802 rad
59. 0.3061 rad **60.** 1.1804 rad **61.** 0.2802 rad **62.** 1.1362 rad

4 **63.** 1.667 **64.** 1.25 **65.** 1.333 **66.** 1.25 **67.** 1.667
68. 0.75 **69.** 52.0° **70.** 36.0° **71.** 0.896 rad

Section 19–2 Exercises, p. 833

1 **1.** 27.8° **2.** 43.2° **3.** 49.6° **4.** 6.889 in. **5.** 28.21 m **6.** 16.15 ft
7. $S = 51.1°$ **8.** $S = 45°$
$\quad U = 38.9°$ $\qquad t = 6.647$ m
$\quad u = 11.31$ yd $\qquad s = 4.700$ m
$\quad 18^2 \approx 14^2 + (11.31)^2$ $\qquad (6.647)^2 = (4.7)^2 + (4.7)^2$
$\quad 324 \approx 323.9161$ $\qquad 44.18 = 44.18$
9. $U = 55.5°$ **10.** $U = 74°$
$\quad s = 4.814$ mm $\qquad t = 50.79$ m
$\quad u = 7.005$ mm $\qquad u = 48.82$ m
$\quad (8.5)^2 \approx (4.814)^2 + (7.005)^2$ $\qquad (50.79)^2 \approx (48.82)^2 + (14)^2$
$\quad 72.25 \approx 23.17 + 49.07$ $\qquad 2{,}579.62 \approx 2{,}383.39 + 196$
$\quad 72.25 \approx 72.24$ $\qquad 2{,}579.62 \approx 2{,}579.39$

2 **11.** 35.7° **12.** 10.05 cm **13.** 15.88 mm **14.** 35.49 dm
15. 31.50 ft **16.** 14.12 yd
17. $R = 45.9°$ **18.** $Q = 44°$
$\quad Q = 44.1°$ $\qquad r = 12.23$ cm
$\quad r = 16.52$ ft $\qquad q = 11.81$ cm
19. $R = 16.5°$ **20.** $Q = 30.5°$
$\quad r = 4.147$ dkm $\qquad r = 13.58$ m
$\quad s = 14.60$ dkm $\qquad s = 15.76$ m

3 **21.** 28.6° **22.** 23.5° **23.** 5.979 ft **24.** 0.04663 cm
25. 0.01212 m **26.** 4.075 m
27. $D = 34.8°$ **28.** $E = 48°$
$\quad E = 55.2°$ $\qquad d = 6.303$ ft
$\quad f = 5.604$ m $\qquad f = 9.419$ ft
29. $D = 16.5°$ **30.** $D = 54.0°$
$\quad d = 5.963$ in. $\qquad E = 36.0°$
$\quad f = 20.99$ in. $\qquad f = 13.60$ ft

4 **31.** 150.1 m **32.** 11.91 ft **33.** 9.950 cm **34.** 46.71 cm
35. 21.09 in.

5 **36.** 2,736 ft **37.** 3,746 ft **38.** 31.0° **39.** 15.5° **40.** 22.38 in.
41. 53.62 ft **42.** 39.23 ft **43.** 35.8° **44.** 10 Ω **45.** 7.654 cm
46. 37.9° **47.** impedance = 28.7 Ω; reactance = 12.1 Ω
48. 17.32 Ω **49.** 23.25 ft **50.** 19 ft **51.** 104 m
52. $x = 22.9$ in.; 55.1° **53.** 4.2in. **54.** offset = 42 ft; diagonal = 48 ft
55. run = 31 in.; diagonal = 55 in. **56.** run = 18 in.;
diagonal = 23 in. **57.** $AC = 2.0$ in.; $BC = 2.0$ in. **58.** 2.2° **59.** 5.7°
60. 4.6 in. **61.** 25.9 cm **62.** 0.7 cm

Chapter 19 Review Exercises, p. 842

1. $\frac{15}{25} = \frac{3}{5}$ **3.** $\frac{15}{20} = \frac{3}{4}$ **5.** $\frac{20}{25} = \frac{4}{5}$ **7.** $\frac{2 \text{ ft}}{2 \text{ ft } 2 \text{ in.}} = \frac{24 \text{ in.}}{26 \text{ in.}} = \frac{12}{13}$

9. $\frac{10 \text{ in.}}{2 \text{ ft}} = \frac{10 \text{ in.}}{24 \text{ in.}} = \frac{5}{12}$ **11.** $\frac{7}{12.62} = 0.5547$ **13.** $\frac{7}{10.5} = 0.6667$

15. $\frac{10.5}{7} = 1.5$ **17.** 2.6 **19.** 0.417 **21.** 79.0° **23.** 0.8403

25. 0.4540 **27.** 1.0724 **29.** 0.7585 **31.** 33.0° **33.** 52.5°
35. 1.2188 rad **37.** 0.3313 rad **39.** $A = 38.1°$ **41.** $h = 2.537$ mm
43. $y = 5.469$ m **45.** $D = 28.3°, E = 61.7°, f = 14.76$ ft **47.** 32.2°
49. 11.57 in. **51.** 48.6° and 41.4° **53.** $B = 56°, a = 88.91$ ft,
$b = 131.8$ ft **55.** 44.7° **57.** $X_L = 9.64 \ \Omega; R = 11.49 \ \Omega$

Chapter 19 Practice Test, p. 846

1. $\frac{10}{26} = \frac{5}{13}$ **3.** $\frac{11.5}{12.54} = 0.9171$ **5.** 0.7986 **7.** 16.0°
9. 0.8726 rad **11.** $B = 65.5°, b = 30.72$ m, $c = 33.76$ m
13. 38.7° **15.** 48 in. **17.** 40.06 A **19.** 84.3°

20 **TRIGONOMETRY WITH ANY ANGLE**

Section 20–1 Exercises, p. 858

1 **1.** 13 **2.** 15 **3.** 7.280 **4.** 8.544 **5.** 2.746

2 **6.** 67.38° **7.** 53.13° **8.** 20.56° **9.** 68.20° **10.** 75.96°

3 **11.** (5, 36.9°) **12.** (5.7, 45°) **13.** (7.3, 15.9°) **14.** (13, 67.4°)
15. (17, 61.9°) **16.** (2.1, 2.1) **17.** (0.9, 4.9) **18.** (0, 3)
19. (2.6, 1.5) **20.** (2.8, 2.8)

4 **21.** 19; 42° **22.** 8; 72° **23.** 1; 225° **24.** 6; 75°
25. $4950 + 4.950i; -5.657 - 5.657i; -0.707 - 0.707i;$
26. $3.882 + 14.49i; -2.329 - 8.693i; 1.553 + 5.797i;$ **27.** $12 + 5j;$
$13; 22.62°$ **28.** $6 + 4j; 7.211; 33.69°$ **29.** 4.6; 75° **30.** 9.5; 60°
31. $11.1; \frac{\pi}{6}$ **32.** $15.81; \frac{\pi}{2}$

Section 20–2 Exercises, p. 865

1 **1.** 60° **2.** 15° **3.** 70° **4.** 15° **5.** 32° **6.** 70° **7.** 32°
8. 62° **9.** 0.96 rad **10.** 0.44 rad

2 **11.** all positive **12.** sin positive; others negative
13. tan positive; others negative **14.** cos positive; others negative
15. cos positive; tan, sin zero

3 **16.** -0.5000 **17.** -0.8391 **18.** -0.8011 **19.** -0.6536
20. -0.8660 **21.** -0.2910 **22.** -0.1736 **23.** 0.5000 **24.** 146.3°
25. 3.61 rad

Section 20–3 Exercises, p. 870

1 **1.** $y = \sin 2x$ **2.** $y = \sin 3x$

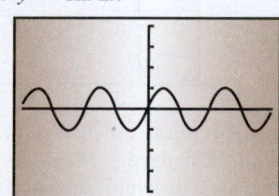

3. $y = \sin 4x$

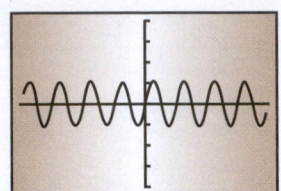

4. $y = \sin 8x$

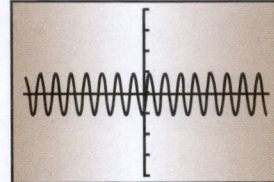

5. $y = \cos 2x$

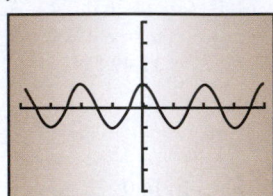

6. $y = \cos 4x$

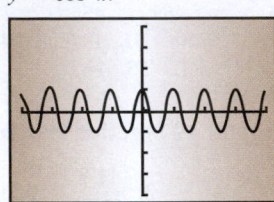

7. $y = \cos 5x$

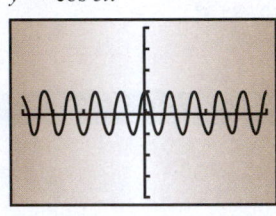

8. $y = \cos 6x$

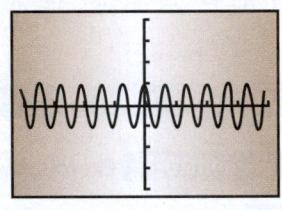

9. Answers will vary. The amplitude remains the same, but the period decreases as the coefficient of x increases. All graphs cross the y-axis at the origin.   **10.** Answers will vary. Each graph has the same maximum and minimum, but the number of repeats of the graph increases as the coefficient of x increases. All graphs cross the y-axis at 1.

11. $y = 2 \sin x$

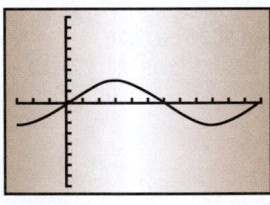

12. $y = 4 \sin x$

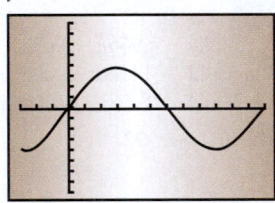

13. $y = 6 \sin x$

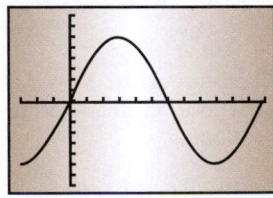

14. $y = 3 \cos x$

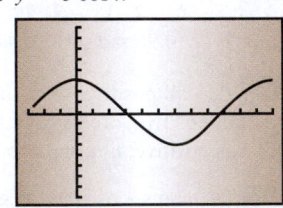

15. $y = 4 \cos x$

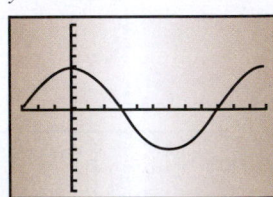

16. $y = 5 \cos x$

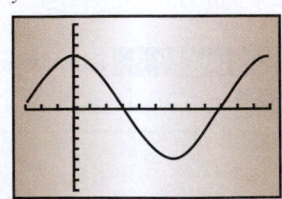

17. Answers will vary. As the coefficient of $\sin x$ increases, the amplitude increases but the period remains unchanged.   **18.** Answers will vary. As the coefficient of $\cos x$ increases, the amplitude increases but the period remains unchanged.   **19.** 9.8480775   **20.** 22.222 Hz

21. 166.667 Hz   **22.** 4×10^{-6} s   **23.** 1.002×10^{-8} s   **24.** 215.8 m

25. 1×10^9 Hz   **26.** $v = 20{,}500$ m/s   **27.** $v = 106{,}680{,}000$ m/s or 1.0668×10^8 m/s

28. $v = 42 \sin x$; $y = \pm 50$

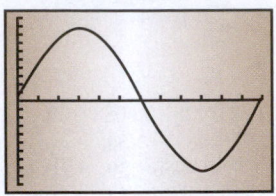

29. $v = 32 \sin x$; $y = \pm 50$

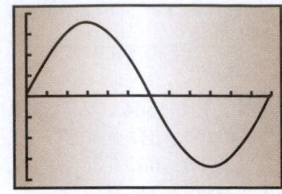

30. $x = 45°$; $v = 29.7$ volts
$x = 60°$; $v = 36.4$ volts

31. $x = 120°$; $v = 27.7$ volts
$x = 300°$; $v = -27.7$ volts

32. $i = 3.5 \sin x$; $y = \pm 4$

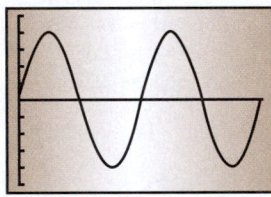

33. $i = 8.4 \sin x$; $y = \pm 10$

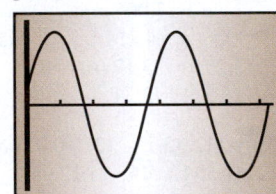

34. $x = 225°$; $i = -2.5$ amps
$x = 315°$; $i = -2.5$ amps

35. $x = 115°$; $i = 7.6$ amps
$x = 270°$; $i = -8.4$ amps

36. amplitude $= 4$;
period $= 60°$

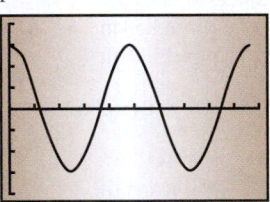

37. amplitude $= 30$;
period $= 36°$

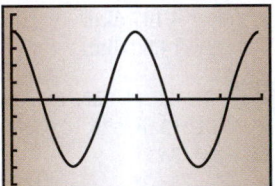

38. amplitude $= 3$;
period $= 72°$

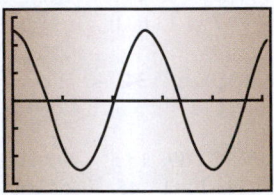

39. amplitude $= 4$;
period $= 45°$

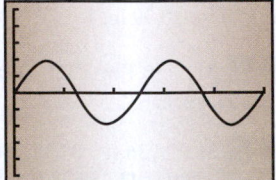

40. amplitude $= 2.5$;
period $= 480°$

41. amplitude $= 1.9$;
period $= 225°$

3 **42.** amplitude = 1;
period = 360°;
phase shift = 45° left

43. amplitude = 1;
period = 360°;
phase shift = 180° right

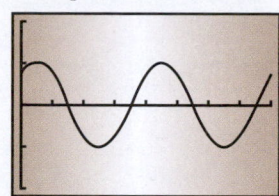

44. amplitude = 2;
period = 360°;
phase shift = 120° left

45. amplitude = 6;
period = 120°;
phase shift = 80° right

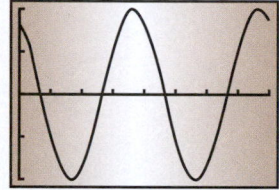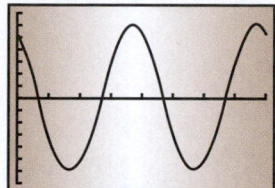

46. amplitude = 10;
period = 90°;
phase shift = 67.5° left

47. amplitude = 8;
period = 60°;
phase shift = 7.5° right

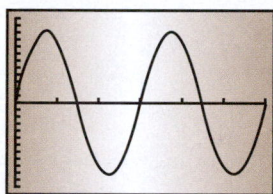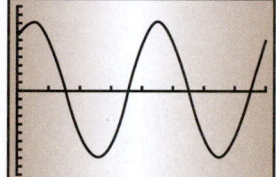

Section 20–4 Exercises, p. 877

Rounding discrepancies will be minimized if full calculator values are used in all calculations.

1 **1.** $C = 63°$; $b = 40.00$ m; $c = 45.22$ m **2.** $A = 50°$;
$a = 6.504$ mi; $c = 8.075$ mi **3.** $C = 33°$; $b = 56.37$ cm;
$c = 35.45$ cm **4.** $B = 30°$; $a = 5.543$ ft; $b = 3.200$ ft **5.** $C = 50°$;
$a = 20.84$ km; $c = 16.03$ km **6.** $C = 20°$; $c = 114.9$ dkm;
$a = 237.6$ dkm **7.** $b = 17.92$ in. **8.** $a = 190.3$ m **9.** $C = 76°$
10. $c = 7.150$ ft **11.** $b = 5,383$ ft **12.** 8.971 km **13.** 942 mi

2 **14.** $B = 21.5°$; $C = 118.5°$; $c = 57.45$ in. **15.** $C = 10.8°$;
$A = 99.2°$; $a = 15.76$ yd **16.** $A = 43.2°$; $B = 116.8°$;
$b = 23.48$ ft or $A = 136.8°$; $B = 23.2°$; $b = 10.35$ ft **17.** $B = 14.5°$;
$C = 135.5°$; $c = 11.21$ m **18.** $B = 19.0°$; $C = 131.0°$; $c = 98.45$ cm
19. $B = 25.8°$; $A = 34.2°$; $a = 29.61$ m **20.** $B = 71°$; $b = 189.5$ ft;
$c = 198.5$ ft **21.** 805.9 ft **22.** $\angle B = 38.80$ ft, $\angle A = 85.1°$;
$\angle B = 30.9°$

Section 20–5 Exercises, p. 882

1 **1.** $A = 30.8°$
$B = 125.1°$
$C = 24.1°$

2. $D = 71.7°$
$E = 62.9°$
$F = 45.4°$

3. $A = 50.8°$; $B = 56.6°$; $C = 72.6°$ **4.** $A = 47.7°$; $B = 72.4°$;
$C = 59.8°$ **5.** $A = 58.4°$; $B = 76.1°$; $C = 45.5°$ **6.** $A = 66.3°$;
$B = 59.1°$; $C = 54.6°$

2 **7.** $r = 32.42$ ft
$S = 49.7°$
$T = 58.3°$

8. $j = 18.68$ m
$K = 37.5°$
$L = 34.0°$

9. $b = 15.88$ m; $A = 52.4°$; $C = 69.4°$ **10.** $A = 24.7°$; $C = 20.3°$;
$b = 30.23$ m **11.** 49.27 ft **12.** 60.8°; 44.1°; 75.1°

Chapter 20 Review Exercises, p. 892

1. 7.211; 56.3° **3.** 13.89; 59.7° **5.** 8.4; 45° **7.** (6.0, 1.0)
9. (4.2, 10.2) **11.** 0.1016 rad **13.** 41° **15.** 0.8832 rad
17. 32°15′10″ **19.** 0.8632 **21.** −0.3420 **23.** 0.3420 **25.** 1.000
27. 2.83; 2.36 rad **29.** 0.000067 s
31. $y = 5 \cos 4x$; amplitude = 5; period = 90°

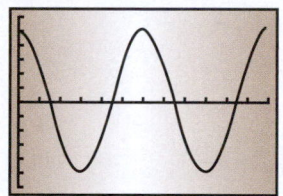

33. 43.3 V **35.** $A = 40°$; $b = 10.8$ m; $c = 4.3$ m **37.** $A = 30.3°$;
$B = 104.7°$; $b = 9.6$ dkm **39.** 1st solution: $A = 39.4°$; $C = 112.6°$;
$c = 13.4$ cm; 2nd solution: $A = 140.6°$; $C = 11.4°$; $c = 2.9$ cm
41. 42.0 ft
43.

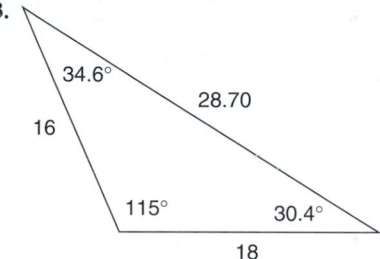

45. 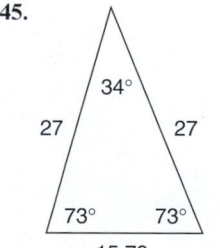 **47.**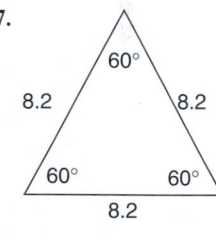

49. 97.98 ft **51.** 343.3 ft; 665.4 ft

Chapter 20 Practice Test, p. 895

1. 0.8192 **3.** −0.9397 **5.** 135° **7.** 101.3° or 7.7° **9.** 8.991 ft
11. 131.8° **13.** 6.830 m **15.** 38.0° **17.** 150.9 ft **19.** 11.32 ft
21. 6.4 mA **23.** 30 Ω

Chapters 19–20 Cumulative Practice Test, p. 896

1. (15, 53.1°) **3.** (4.4, 52.4°) **5.** (3.7, 6.4) **7.** (4.4, 1.8)
9. $AB = 47$ cm **11.** $\angle B = 52.3°$ **13.** magnitude = 5.385;
direction = 21.8° **15.** amplitude = 1; period = 360°;
phase shift = 90° left **17.** amplitude = 3; period = 180°;
phase shift = 120° right **19.** $AB = 23$ cm

Glossary and Index

List of Career Applications

Page references to Examples, Section Exercises, Chapter Exercises, and Practice Tests are indicated by EX, SE, CE, and PT respectively.

Auto/Diesel Technology AUTO

Cad/Drafting/Architecture/Surveying CAD/ARC

Computer Technologies COMP

Electronics Technology ELEC

Helping Professionals/Education-Criminal Justice/Fire Fighting `HELPP`

Hospitality/Culinary/Food Technology `HOSP`

Industrial Technology/Manufacturing/Machine Technology/Engineering Technology `INDTEC`

Industrial Trades/Welding/Machine Tool/Industrial Maintenance INDTR

Geometric Formulas

Perimeter and Area of Two-Dimensional Figures

Square

Perimeter $= 4s$, where $s =$ length of a side of the square
Area $= s^2$, where $s =$ length of the side of the square

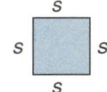

Rectangle

Perimeter $= 2l + 2w$, where $l =$ length of the rectangle and $w =$ width of the rectangle
Area $= lw$, where $l =$ length of the rectangle and $w =$ width of the rectangle

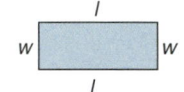

Parallelogram

Perimeter $= 2b + 2s$, where $b =$ the base and $s =$ length of the adjacent side
Area $= bh$, where $b =$ length of the base and $h =$ height of the parallelogram

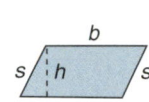

Triangle

Perimeter $= a + b + c$, where a, b, and c are the lengths of the sides of the triangle
Area $= \frac{1}{2}bh$, where $b =$ base of triangle and $h =$ height of triangle
Area of a triangle if the *height* is not known (Heron's formula):

Area $= \sqrt{s(s - a)(s - b)(s - c)}$, where $s = \frac{1}{2}(a + b + c)$ and $a =$ side 1, $b =$ side 2, and $c =$ side 3 of the triangle

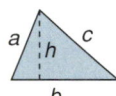

Trapezoid

Perimeter $= a + b + c + d$, where a, b, c, and d are the lengths of the sides

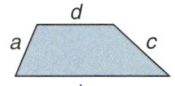

Area $= \frac{1}{2}h(b_1 + b_2)$, where $h =$ height of trapezoid, $b_1 =$ length of short base, and $b_2 =$ length of long base

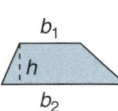

Circle

Circumference $= 2\pi r$ or πd, where $r =$ radius and $d =$ diameter
Area $= \pi r^2$, where $r =$ radius

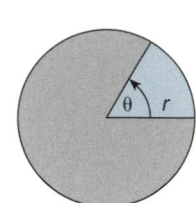

Sector

Area of a sector $= \frac{\theta}{360°}(\pi r^2)$, where $\theta =$ angle (measured in degrees) that forms the sector and $r =$ radius of circle
Area of a sector $= \frac{\theta r^2}{2}$, where $\theta =$ angle (measured in radians) that forms the sector and $r =$ radius of circle

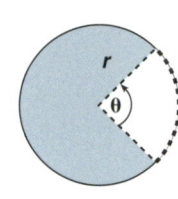

Arc Length

Length of an arc $= \frac{\theta}{360°}(2\pi r)$, where $\theta =$ angle (measured in degrees) that forms the arc and $r =$ radius of circle
Length of an arc $= r\theta$, where $\theta =$ angle (measured in radians) that forms the sector and $r =$ radius of circle

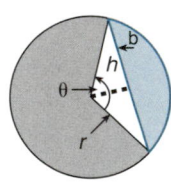

Segment

Area of a segment $= \frac{\theta}{360°}(\pi r^2) - \frac{1}{2}bh$, $\theta =$ angle (measured in degrees), $r =$ radius of the circle, $b =$ base of triangle formed by two radii and the chord and $h =$ height of the triangle.

Area and Volume of Three-Dimensional Objects

Cube

Lateral Surface Area $= 4s^2$, where $s =$ length of one side of cube
Total Surface Area $= 6s^2$, where $s =$ length of one side of cube
Volume $= s^3$, where $s =$ length of one side of cube

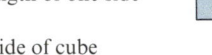

Right Rectangular Prism or Right Circular Cylinder

Lateral Surface Area (LSA) $= ph$ where $p =$ perimeter of the base and $h =$ height of the three-dimensional figure

Total Surface Area (TSA) $= ph + 2B$, where $p =$ perimeter of the base and $h =$ height of the three-dimensional figure and $B =$ area of the base.

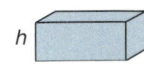